ENCYCLOPAEDIA OF THE UNIVERSE

ENCYCLOPAEDIA OF THE UNIVERSE

A POPULAR WORK ON THE MARVELS OF

THE HEAVENS :: THE EARTH :: PLANT LIFE :: ANIMAL LIFE :: THE MIGHTY DEEP

WITH 1000 FINE ILLUSTRATIONS

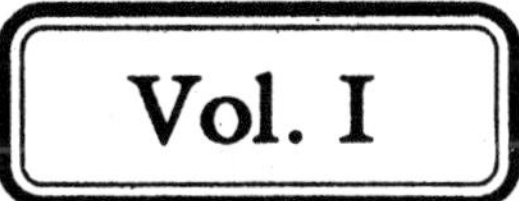

Editor

WALTER HUTCHINSON

B.A., F.R.G.S., F.R.A.I., Etc.

WITH AN INTRODUCTION BY

LORD AVEBURY, P.C.

AND WITH CONTRIBUTIONS BY LEADING SPECIALISTS, INCLUDING :

Sir Harry Johnston, G.C.M.G., K.C.B.
Richard Lydekker, F.R.S.
Frank Finn, F.Z.S.
Edward Step, F.L.S.
E.W. Maunder, F.R.A.S.
W.P. Pycraft, A.L.S., F.Z.S.
Russell F. Gwinnell, F.C.S.
Frank T. Bullen, F.R.G.S.
Tickner Edwardes
J.J. Ward, F.E.S.
C. Flammarion
Lionel E. Adams, B.A.
Harold Bastin, F.E.S.
S. Leonard Bastin
K. G. Blair, F.E.S.
Sir Herbert Maxwell, F.R.S.
E. B. Poulton, F.R.S.
G. F. Chambers, F.R.A.S.
H. St. J. K. Donisthorpe, F.E.S.
Richard Kerr, F.G.S.
W.F. Kirby, F.L.S.
Hugh Main, F.E.S.
Ernest Marriage
E.A. Martin, F.G.S.
T. E. R. Phillips, F.R.A.S.
R. I. Pocock, F.R.S.
Dr. E. J. Spitta, F.R.M.S.
E. W. Swanton
J. Sinel
W. Mark Webb, F.L.S.

DISCOVERY PUBLISHING HOUSE PVT. LTD.

NEW DELHI-110 002

Published by:
Namit Wasan
DISCOVERY PUBLISHING HOUSE PVT. LTD.
4383/4B, Ansari Road, Darya Ganj
New Delhi-110 002 (India)
Phone : +91-11-23279245; 23253475; 43596065
E-mail : discoverybooksindia@gmail.com
discoverypublishinghouse@gmail.com
namitwasan9@gmail.com
web : www.discoverypublishinggroup.com

Edition: 2020

ISBN: 978-81-8356-873-9 (Set)

Encyclopaedia of the Universe

Printed at:
Infinity Imaging Systems
Delhi

CONTENTS OF VOLUME I.

Some curious deep-sea fishes which have very highly-developed organs of touch.

INTRODUCTION

By Lord AVEBURY, P.C., F.R.S., D.C.L., LL.D., For. Sec. Roy. Acad.

O wunderschön ist Gottes Erde
Und schön auf ihr ein Mensch zu sein.
(Beautiful is God's Earth, and beautiful it is to be a man thereon.)

WE live in a wonderful and beautiful world, full of interest; a world which it is most important to understand; and dangerous, if not fatal, to misunderstand. It is a fairy-land of exquisite beauty; our very existence is a miracle in itself, and yet few of us enjoy as we might, and none as yet appreciate fully, the beauties and wonders which surround us. The greatest traveller cannot hope even in a long life to visit more than a very small part of our earth, and even of that which is under our very eyes how little we see!

Photo by] [Gimpel.

A MARVELLOUS SPIDER.

This remarkable photograph shows a spider holding up the silken cradle containing its eggs, and revolving it in the Sun's rays.

"Paradise indeed might," as Luther said, "apply to the whole world." What more is there we could ask for ourselves? "Every sort of beauty," says Mr. Greg,* "has been lavished on our allotted home; beauties to enrapture every sense, beauties to satisfy every taste; forms the noblest and the loveliest, colours the most gorgeous and the most delicate, odours the sweetest and subtlest, harmonies the most soothing and the most stirring; the sunny glories of the day; the pale Elysian grace of moonlight; the lake, the mountain, the primeval forest, and the boundless ocean. 'Silent pinnacles

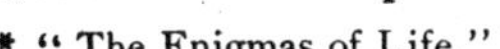

* "The Enigmas of Life."

of aged snow' in one hemisphere, the marvels of tropical luxuriance in another; the serenity of sunsets; the sublimity of storms; everything is bestowed in boundless profusion on the scene of our existence; we can conceive or desire nothing more exquisite or perfect than what is round us every hour; and our perceptions are so framed as to be consciously alive to all. The provision made for our sensuous enjoyment is in overflowing abundance; so is that for the other elements of our complex nature. Who that has revelled in the opening ecstasies of a young imagination, or the rich marvels of the world of Thought, does not confess that the Intelligence has been dowered at least with as profuse a beneficence as the Senses? Who that has truly tasted and fathomed human Love in its dawning and crowning joys has not thanked God for a felicity which indeed 'passeth understanding.' If we had set our fancy to picture a Creator occupied solely in devising delight for children whom He loved, we could not conceive one single element of bliss which is not here."

[*By Prof. Lowell.*

MARS.

The lines on the map represent the numerous canals which appear to irrigate the otherwise barren land of Mars.

A friend of mine, on a scientific mission some years ago, to his great surprise met an old French Abbé in one of the most remote parts of the Rocky Mountains, and could not help showing his surprise. The Abbé observed this, and in the course of conversation explained his presence in that distant region.

"You were," he said, "surprised to find me here. The fact is, that some months ago I was very ill. My physicians gave me up; one morning I seemed to faint, and thought that I was already in the arms of the *Bon Dieu*. I fancied one of the angels came and asked me, 'Well, Monsieur l'Abbé, how did you like the beautiful world you have just left?' And then it occurred to me that I, who had been all my life preaching about Heaven, had seen almost nothing of the world in which I was living. I determined, therefore, if it pleased Providence to spare me, to see something of this world; and so here I am."

Few of us are free, however much we might wish it, to follow the example of the worthy Abbé. But although it may not be possible for us to reach the Rocky Mountains, there are other countries nearer home which most of us might find time to visit.

There is no single substance in Nature the properties of which are fully known to us. There is no animal or plant which would not well repay, I do not say merely the attention of an hour, but even the devotion of a lifetime. I often grieve to think how much happiness our fellow countrymen lose from their ignorance of science. Man, we know, is born to sorrow and suffering, but he is not born to be dull, and no one with any knowledge of science could ever be. If any one is

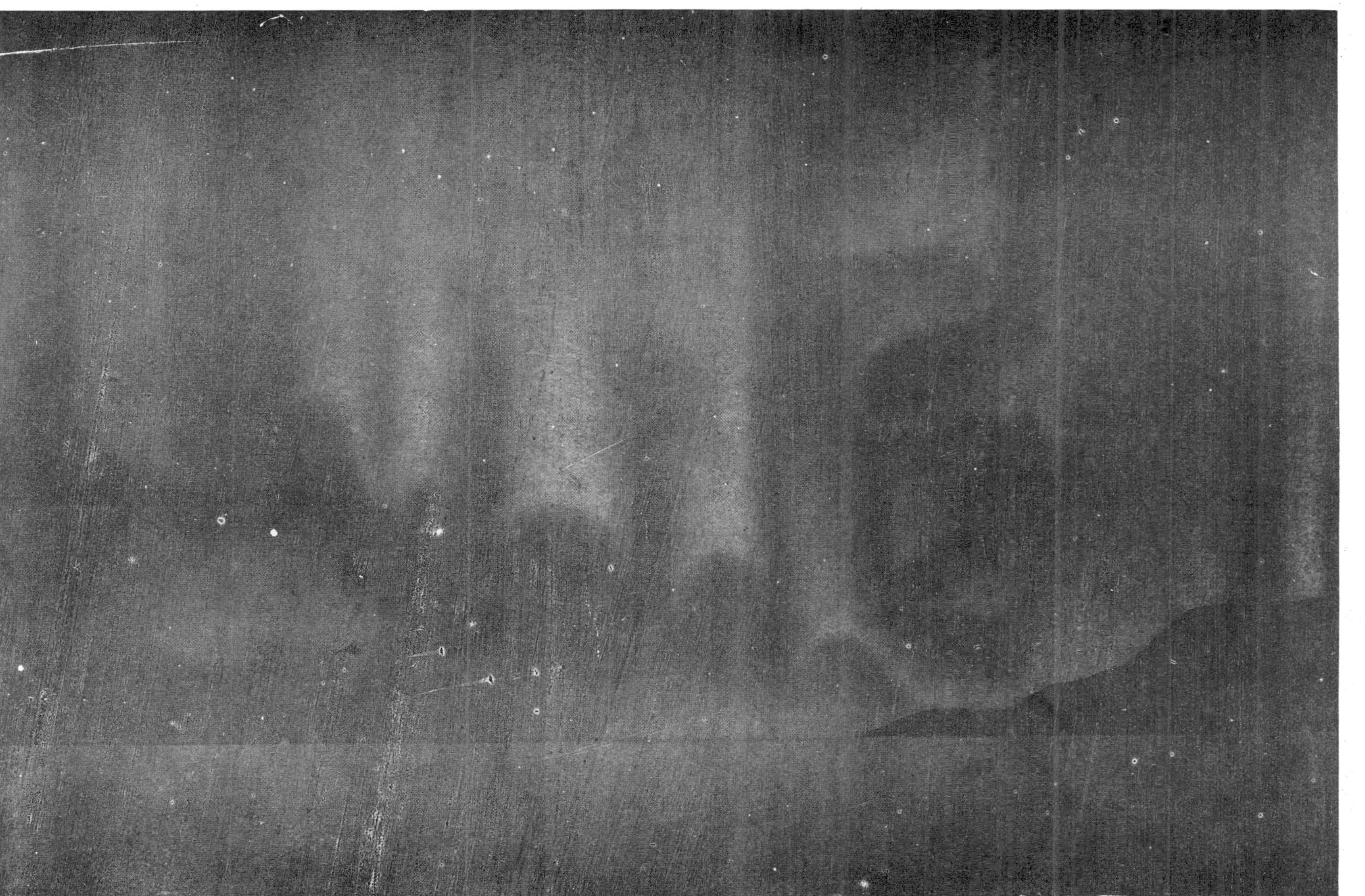

[By permission of the Royal Society.

THE AURORAL CURTAIN.

One of the most magnificent sights in the Polar Regions is the Auroral Curtain, which is of a pale straw-yellow colour. As new folds are formed in one direction the curtain vanishes away in another. The varied rays of light which compose it—at first brilliant and then imperceptibly fading away—render it a difficult matter to describe.

ever dull it is his own fault. Every wood, every field, every garden, every stream, every pond, is full of interest to those who have eyes to see.

Let me take a single subject :

Even about our own senses how little we know! No doubt it is always difficult to place on paper what we do not know. It may illustrate the writer's, rather than general, ignorance. Moreover, even if true at the moment it was written, the answer may be discovered before the ink is dry. I speak, therefore, with all hesitation.

Nevertheless, I believe it may be said that the sensibility of our skin appears to be mainly due to a plexus of fine nerve-fibres, which end in free terminations between the cells of the skin (*rete mucosum*). There are also in some parts of the skin two sets of minute corpuscles, which are called after their discoverers, the first Vaterian, or more commonly Pacinian, corpuscles; the second, Meissner's or Wagner's corpuscles.

The Pacinian corpuscles consist of a capsule formed of several layers, one enveloping the other. The undulating nerve-fibres, after several windings, enter the capsule, which, indeed, seems to be nothing more than a much-thickened end of the outer nerve-coat. These corpuscles measure from 1·1 to 4·5 mm. They occur principally on the hands and feet, and in the flexures of the joints, but occasionally also elsewhere. Meissner's or Wagner's corpuscles are cone-like or egg-shaped bodies, in each of which a nerve terminates, after several convolutions. They are especially numerous at the tips of the fingers, where there may be as many as a hundred in a square line. They occupy the papillæ (which, however, do not always contain one), which give the surface of the hand its peculiar striped appearance. They also occur, though less numerously, elsewhere, as on the feet, breast and lips.

Nevertheless, these observations, interesting as they are, give us in reality but little information as to the sense of touch.

THE WILLIAMETTE METEOR.

This enormous mass of meteoric metal, which fell at Williamette, Oregon, U.S.A., weighs over fifteen and a half tons. It is ten feet in length and six and a half feet high. The force with which such a mass strikes the earth is terrific; and one can imagine the fearful havoc it would cause should a meteor fall upon one of our famous buildings.

Let us pass on to the sense of taste. Loven and Schwalbe detected, independently and almost simultaneously, in the epithelium of the papillæ of the tongue, many small bud-like groups of cells, which are probably connected with the ultimate fibres of the glosso-pharyngeal nerves. These have been supposed to be the special seats of the sense of taste, and thence termed "taste-buds"; they are in a man shaped like a flask; in some other animals they are more slender.

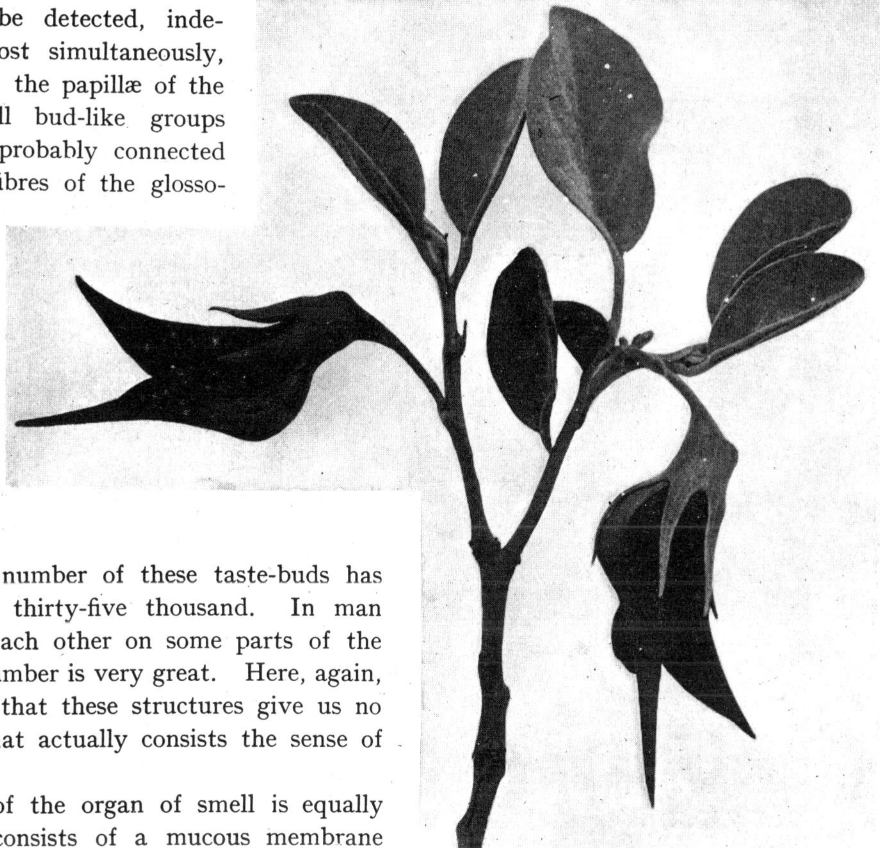

THE BIRD-PEA OF WESTERN AUSTRALIA.

This flower bears a striking resemblance to a small bird. The colour is yellowish-green with purple streaks which give an idea of feathers.

In the cow the number of these taste-buds has been estimated at thirty-five thousand. In man they almost touch each other on some parts of the tongue, and their number is very great. Here, again, it will be admitted that these structures give us no help to realize in what actually consists the sense of taste.

Our knowledge of the organ of smell is equally unsatisfactory. It consists of a mucous membrane containing (1) cylindrical epithelial cells, with a broad, flat termination at the free end; and (2) of rod-like filaments which, some little distance below the surface, swell out into a nut-shaped expansion, and then contract again into a fine thread, which is probably continuous with the fibrils of the olfactory nerve.

At first sight the sense of hearing may seem to be better understood. The waves of sound in the air play on the drum of the ear. Across the drum stretches a chain of little bones, the last of which is shaped somewhat like a stirrup, the flat plate of which rests against the *fenestra ovalis* of the drum. The sounds are thus probably intensified. Behind the *fenestra ovalis* is the "labyrinth," which is filled with fluid, and on which the final filaments of the auditory nerve are distributed. The labyrinth consists mainly of two parts—the cochlea and the semi-circular canals, which are three in number and stand at right angles to one another. It is considered that these act as balancing organs, and it is found that if they are injured or diseased the body cannot be properly balanced. Dean Swift was sometimes unable to walk straight, and was, therefore, as it would now appear, quite

A single flower detached and an eye added to make the resemblance more pronounced

unjustly suspected of secret drinking. It is probable that one or more of his semi-circular canals was out of order.

The cochlea contains a wonderful and complicated organ, discovered by Count Corti. This appears to be, in fact, a microscopic musical instrument, composed of some four thousand complex arches, increasing regularly in length and diminishing in height from the base to the summit of the cochlea. The waves of sound have been supposed to play on this organ, almost like the fingers of a performer on the keys of a musical instrument.

Photo by] [*A. E. Smith.*

The claws of a spider, as seen under a microscope.

We thus obtain a glimpse, though but a glimpse, of the manner in which the organ of Corti may possibly act, but how these vibrations are detached and transmitted to the brain, how they are then transmuted into sounds, we have really no conception.

Lastly, we come to the eye: a marvellous and most complex organ. Externally is the cornea, then comes the vitreous humor, followed by the lens, the aqueous humor, and, lastly, the retina. The lens throws an image on the retina, which is very complicated, and though no thicker than a sheet of thin paper, consists of no fewer than nine separate layers, the innermost being the rods and cones, which are the immediate recipients of the undulations of light. The number of rods and cones in the human eye is enormous. At a moderate computation the cones may be estimated at over three millions, and the rods at thirty millions.

It will be observed that the nerve does not, as one might naturally have expected, enter the eye and then spread itself out at the back of the retina; but, on the contrary, pierces the retina and spreads itself out on the front, so that the cones and rods look inwards, and not outwards—towards the back of the eye, and not at the object itself. In fact, we do not look outwards at the actual object, but we see the object as reflected from the base of our own eye. Knowing then, as we do, so little about our own senses, it is natural that those of other animals should be still more of a mystery. Moreover, why should we assume that they have only our five senses?

Many years ago I showed that the ultra-violet rays which are invisible to us, are on the contrary visible to some of the lower animals. Moreover, looking at the question from the other side, we find in animals complex organs of sense, richly supplied with nerves, but the functions of which we are as yet unable to explain. There may be fifty other senses as different from ours as sound is from sight; and even within the boundaries of our own senses there may be endless sounds which we cannot hear, and colours of which we have no conception.

But although what we know is such an infinitesimal fraction of what we have still to learn, on the other hand, the inventions and discoveries of the last century have been really marvellous. To name only a few—railways and steamers, telegraphs and photography, gas, petroleum, and electric

A SNAIL AS LARGE AS A MAN'S FIST.

The African Agate Snail is the largest land snail known to exist. The illustration is life-size, but specimens occur eight inches long. The snail's egg shown to the left of the picture is an inch long, and has a chalky shell like that of a bird's egg

light, spectrum analysis and the Röntgen rays, the telephone and the phonograph, the liquefaction of air and even of hydrogen, the far-reaching discoveries of Darwin, the foundation of geology, the discovery of anæsthetics, of helium and radium, the successful campaign against malaria, constitute a galaxy of marvellous discoveries to which no other century can afford a parallel.

It is sometimes said that science is prosaic, but geologists have shown us more wonderful things in the depths of the earth than Homer or Virgil ever imagined, and the modern views of the origin of volcanoes have revealed to us much more magnificent conceptions than the mere workshop of Vulcan. There are, it seems to me, strong reasons for looking forward to the future with hope. If, indeed, the world was fairly well known to us—if our knowledge bore any considerable

THE SIBERIAN MAMMOTH.

The Mammoth, or Woolly Elephant, that formerly ranged over Europe, has been preserved for many thousands of years in the ice of Siberia. During last century many examples were discovered with the flesh intact and the wool still attached to the hide. This is no fossil, but an actual preservation of the animal in Nature's cold storage.

proportion to what we have still to learn—the case would be very different. But what we know is an infinitesimal fraction of what we do not know. There is no single substance in Nature the uses and the properties of which are as yet completely known to us. There is no animal or plant the whole life-history of which we have yet unravelled. We are surrounded by forces and influences of which we understand nothing, and which we are as yet but dimly commencing to perceive.

The benefits which Science confers on Man are innumerable and cannot be exaggerated. It is difficult, indeed, to say which branch of Science has done most for us.

Lastly, Science in the troubles and sorrows of life will do much to soothe, comfort and console. If we contemplate the immeasurable lapse of time indicated by geology; the almost infinitely small, and quite infinitely complex and beautiful structures rendered visible by the microscope, or the depths of space revealed by the telescope, we cannot but be carried out of ourselves.

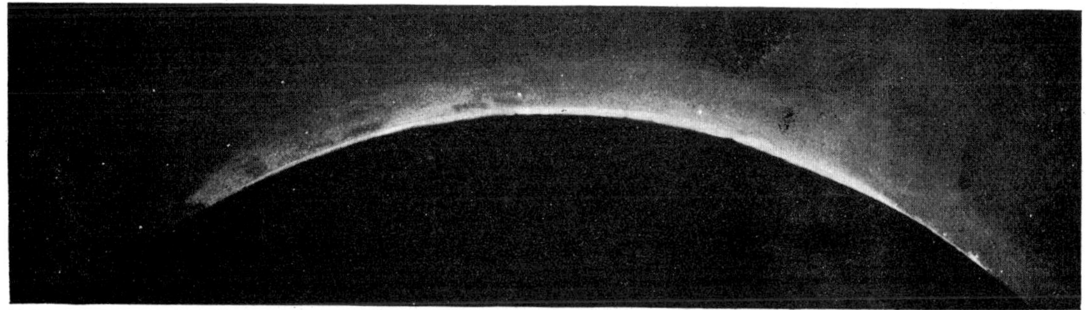

A GROUP OF THE SUN'S PROMINENCES.

At ordinary times these prominences are not visible on account of the superior glare of the sun, but during a total eclipse the temporary disappearance of the sun itself enables the flames surrounding it to be clearly seen.

Encyclopaedia of the Universe

FLAMES OF THE SUN

BY E. WALTER MAUNDER, F.R.A.S.

"THERE sinks the nebulous star we call the sun," said the Princess Ida; but she gave the descriptive epithet a little doubtfully, and qualified it with the proviso, "If that hypothesis of theirs be sound."

Nothing less like a "nebulous star" could be well conceived than the sun, as it appears either in the telescope or to the unassisted eye. Its outline is seen against the sky, hard and clear-cut, a perfect circle without irregularities; the moon itself is not more sharply defined and is far less regular. Yet "nebulous star" well describes the sun, not only in respect of its long past history—the sense in which the Princess used the term—but in its present condition. Only, it is necessary for us to be specially favoured by circumstances, or to develop some special means of seeing, before the fact becomes obvious.

The special circumstances are provided when, in a total eclipse, the dark body of the moon intervenes so as to completely hide from us the solar disc. Then it becomes clear that what we ordinarily see is only part of the sun; beyond it stretch out wide-spread appendages, irregular and ill-defined. The sun is literally a "nebulous star."

Of these appendages the one that attracts chief attention is the "corona," a pearl-coloured light surrounding the sun on every side, and drawn out here and there into long streamers frequently leaf-like in shape. But on the background of this corona there often appear bright spots of crimson light, glowing like rubies.

These are the "red flames," as they used to be called when observed in eclipses of sixty or seventy years ago—the "prominences," as they are termed to-day. A fine group of these prominences on the background of the lowest and brightest stratum of the corona is shown on the accompanying photograph, part of one taken by Professor E. E. Barnard during the eclipse of May 28th, 1900.

Eruptive prominences of the sun as seen in the spectroscope.

As the " flames " were only seen during eclipses, when the moon was exactly before the sun, it was disputed for some time to which of these bodies they belonged—an important question for decision; for, as the sun is four hundred times more distant than the moon, the prominences, if solar, must be on four hundred times the scale that they would be if lunar. The eclipses of 1842 and 1851 proved that they were really parts of the sun, but not solid parts; for one prominence was bent over like a boomerang, and another floated free like a cloud. It was evident, therefore, that they were something of the nature of vapour, gas, or flame.

In the eclipse of 1868 the spectroscope was brought to bear upon these objects, and revealed that they consisted of intensely heated gases—" flames," therefore, in no mere metaphorical sense—hydrogen being the first gas to be recognized in them.

But the spectroscope enabled the prominences to be seen without an eclipse. The hindrance to their being seen in ordinary daylight lies in the overpowering glare from the sky in the immediate neighbourhood of the sun; but with a spectroscope of high dispersion this glare can be greatly weakened without much reducing the brightness of the prominence, as it is seen in one or other of the four bright lines of the spectrum of hydrogen. When the slit of the spectroscope can be

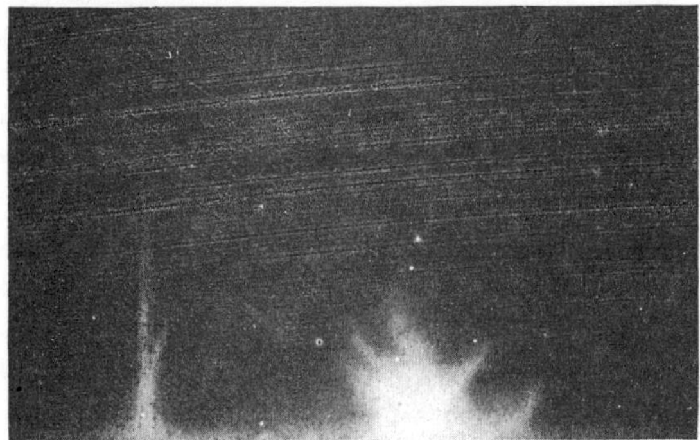
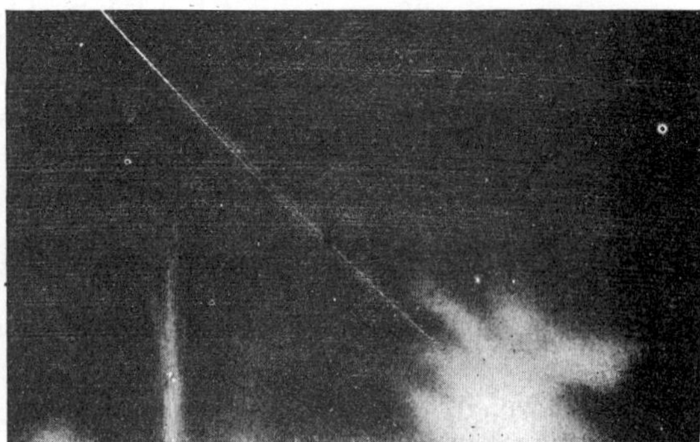

These are two views of the same solar eruption, the second one being taken twenty-five minutes after the first.

opened comparatively wide, the observer looks out through it, as it were, to see the flames which appear on the spectrum somewhat as shown in the diagram.

A quarter of a century later an ingenious arrangement was devised by several astronomers independently, amongst them by Professor Hale, then Director of the Kenwood Observatory, U.S.A., by which one of the bright prominence lines could be isolated from the rest of the spectrum, and a photograph in its light could be taken of the entire surface of the sun, together with its immediate surroundings. The wonderful achievements of this new instrument, the " Spectroheliograph," will form the subject of a later paper. Two of Professor Hale's earliest prominence photographs are, however, reproduced here.

When the spectroscopic method of observing the " flames " was established in 1868, it was seen, on examining the circumference of the sun, that it is everywhere surrounded by a layer of bright hydrogen, some five thousand miles in depth, the upper surface of which is notched like a saw, and hence is sometimes called the " sierra," but more generally the " chromosphere." It suggests the turbulent surface of a sea of flames, from whose jagged flickering points there rise irregularly jets and streamers of the same material in the most fantastic forms and of every variety of size. Sometimes they resemble groves of trees, or quiet hedgerows; sometimes the delicate lace-work of cirrus, or the rounded masses of cumulus clouds; sometimes they are jets of fire, rushing forth under tremendous pressure, or torn and driven as by some terrible explosion. Thus they readily fall into two great classes—the quiet and eruptive prominences.

The quiet prominences are chiefly composed of hydrogen, and change their shape slowly. They

Photographed at the] [*Kenwood Observatory, U.S.A.*

A QUIESCENT PROMINENCE ON THE SUN.

These Prominences float like red clouds high above the solar surface and change their shapes slowly.

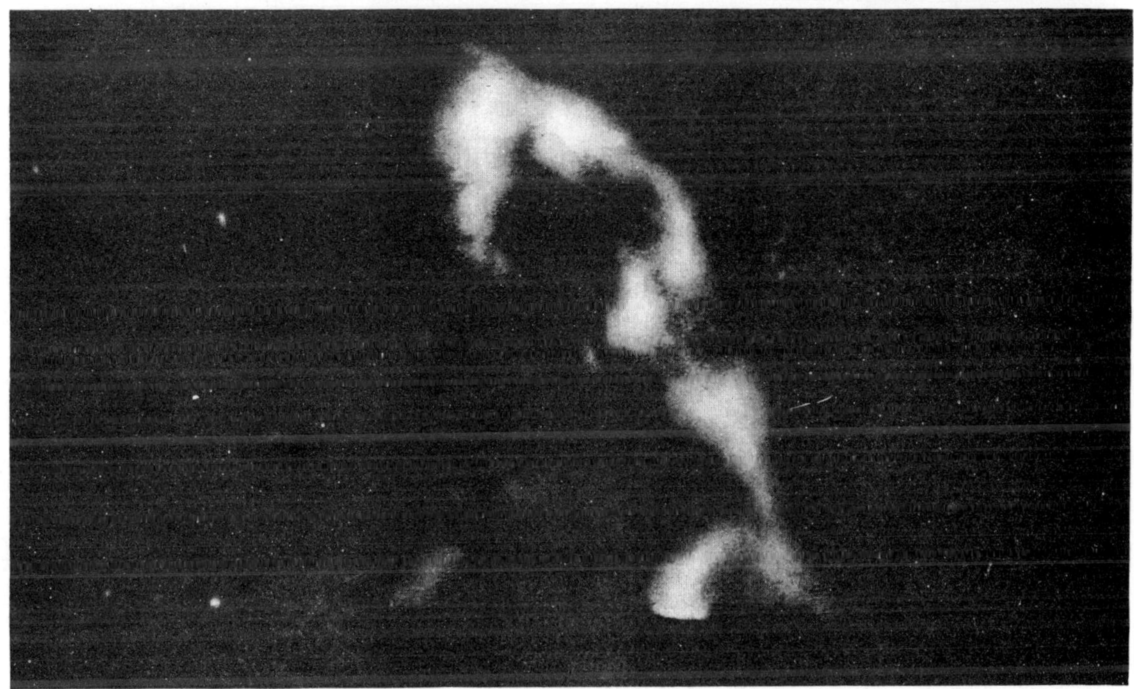

Photographed at the] [*Kenwood Observatory, U.S.A.*

AN ERUPTIVE PROMINENCE ON THE SUN.

As indicated by their name, Eruptive Prominences are subject to changes and movements of the most violent nature. The one shown above rose to a height of over seventy thousand miles.

rise like smoke-wreaths in summer, or float like clouds high above the solar surface, but are often connected with it by narrow stems or pillars. The eruptive prominences often show the lines of many metals in their spectra—of iron, sodium, magnesium, titanium, etc., and are subject to changes and movements of the most violent nature. Thus a group, sketched by M. Fényi on September 30th, 1895, went through numerous changes in a very short time. The greatest height attained was more than three hundred thousand miles—about forty times the diameter of the earth, or considerably more than its distance from the moon. During four minutes of the time that it was being watched it increased in height nearly one hundred thousand miles, an upward motion of some two hundred and seventy-eight miles *per second.* After this violent development it faded and disappeared almost as quickly.

The solar " flames " are not isolated features of the sun ; they form an intimate part of the solar organism, having a close connection with the spots of the general surface, lying below them, and with the corona rising above them. From time to time they give indications of a connection with the general structure of the sun more intimate still. Thus on June 26th, 1885, M. Trouvelot saw two huge prominences, each more than three hundred and fifty thousand miles in height, rising from the sun. Flames of such dimensions are exceedingly rare ; it therefore makes it more significant that they rose exactly opposite to each other, from the ends of the same diameter. In cases like these we may be sure that the origin of such tremendous energy does not lie in the mere shell or surface of the sun, but far deeper.

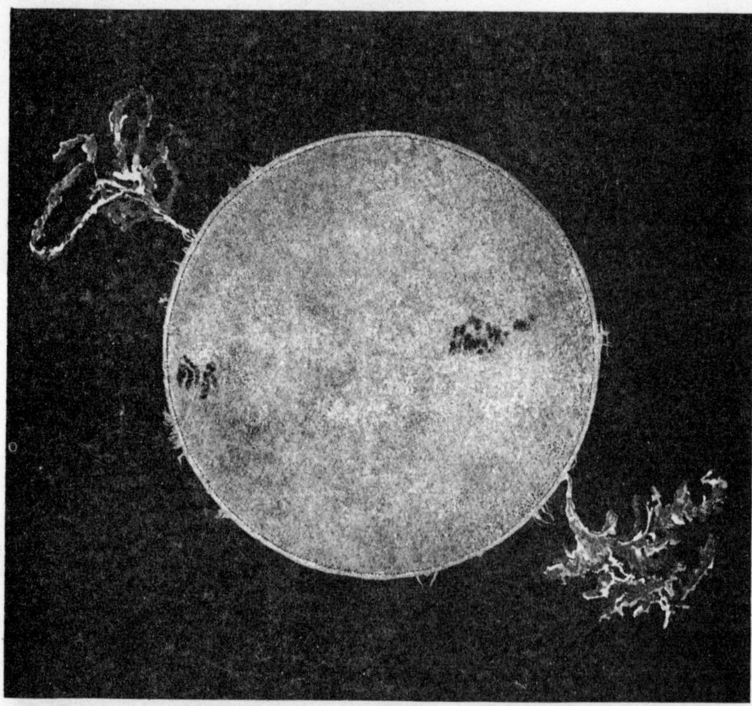

ENORMOUS ERUPTIONS ON THE SUN.

These great prominences consist of flame rising sometimes to a height of over three hundred thousand miles, which is about forty times the diameter of the earth.

THE X-RAY SPIDER

BY R. I. POCOCK, F.R.S.

In the late summer and early autumn our bushes and hedgerows are decorated with the wonderful orbicular webs, consisting of radiating and concentric threads, which are spun by a large fat-bodied spider, known from the white pattern on its back as the Diadem or Cross-Spider. Spiders related to this species are very abundant in the tropics, where they attain far larger dimensions than our British examples. One of the finest, both in colour and size, is the kind known as *Argiope*, all the species of which have lustrous, silvery-white backs and a mottled underside. They spin the same kind of web as the Diadem spider, but many of them decorate the centre with broad white ribbons of closely-woven silk, the arrangement of which differs according to the species. In the web of an Australian species these ribbons form an X-like pattern, whence the trivial name of X-Ray applied to this spider. In this instance the ribbons are placed so as to correspond exactly with the position of the legs in the spider's habitual attitude of rest in the middle of her snare, two legs being extended along each ribbon; and since the spider always hangs on the side of the web turned towards the bush, it results that the ribbons conceal the greater part of her limbs from any insect that approaches the web. Thus, concealment of the occupant of the snare both from flies it is designed to catch and from marauding wasps, which prey largely upon spiders, seems to be one of the chief purposes of these ribbons. Probably, also, they serve to strengthen the middle of the web, enabling it the better to support the weight of the spider.

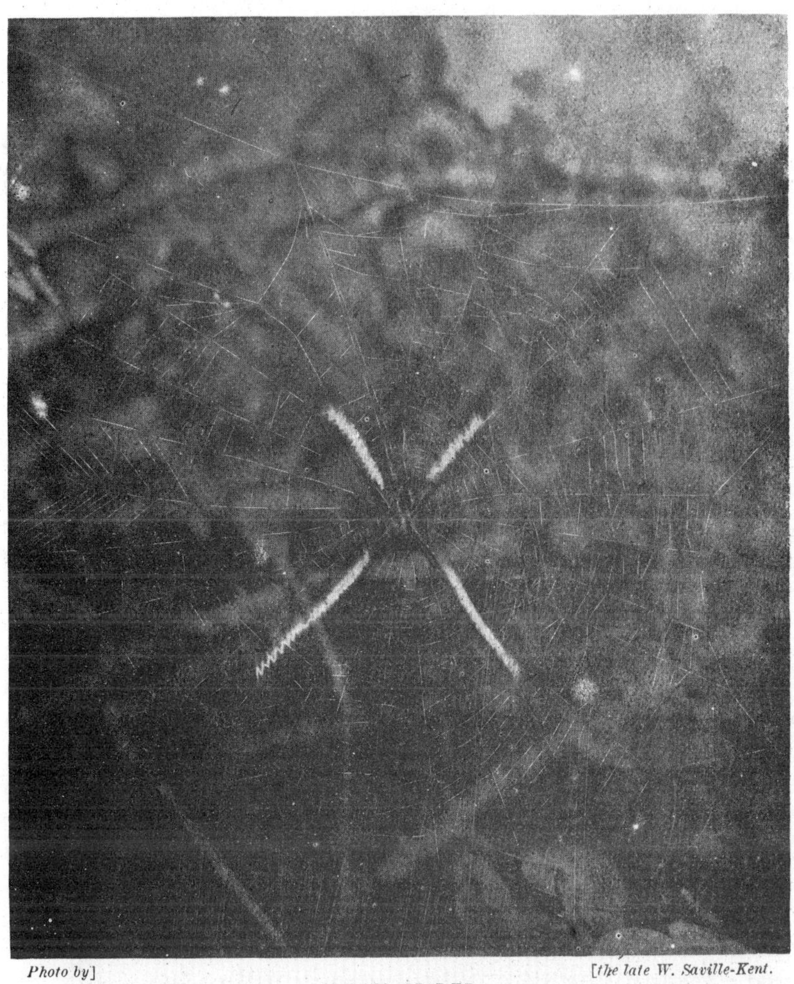

Photo by] [*the late W. Saville-Kent.*

X-RAY SPIDER.

This spider builds a web much like that of the common garden spider, but strengthens the middle—where she sits with four zigzag ribbons, and extending her limbs to each of these completes the X.

SELF-LUMINANT FISHES OF THE DEEP

BY FRANK T. BULLEN, F.R.G.S.

ONE of the many vexed questions concerning life in the abysmal ocean depths which science has set at rest, by the ardent work of her votaries at sea during the last thirty years, is that of the supply of light in those vast profundities. It was well known that the depth to which, under the most favourable circumstances—say, in mid-sea at tropical high noon—the sun's light could penetrate was under a hundred fathoms, a mere surface layer where depths run into four or five thousand fathoms. It was known, too, that in varying degrees of intensity, the conditions of which is not understood, the surface of the sea is luminous, so much so, in fact, that sometimes on luminous night its effulgence pales the stars, makes the heavens appear coal black, and causes luminous that breaks to appear as a shower or living light.

In the midst of this splendour the track of every fish, however small, may be seen very much a large size of the creature, whose passage is its cause, would warrant; until the passing of

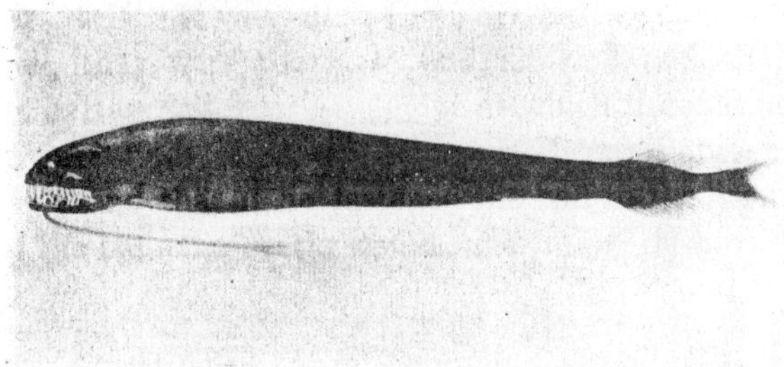

Photograph of a self-luminant fish.

a whale will leave a belt of glowing greenish luminosity spreading far behind the animal to a hundred feet in length. The origin of this light has been accounted for by the presence of countless myriads of *Pyrosoma* and *Noctiluca*, which, though microscopically small, are capable, like the glow-worm and the fire-fly, of giving forth a light, which, in proportion to their size, is amazingly brilliant.

And, loosely, perhaps, because of its resemblance to the pale glow of phosphorus in the dark and the appearance of dried or stale fish, when seen in a dark place, which lambent flame will come off upon the fingers like some forms of electricity, but without the slightest sensation to the experimenter ; it has become the practice to call this glow phosphorescence. Therefore, without committing ourselves to any statement as to the origin or causes of fluctuation of this sea-light, we will use the trivial name where necessary.

It is not even now known how far down the illumination of the sea by the minute creatures named extends, but comparatively recent discoveries go to show that some of the denizens of those vast depths carry about their own electric light installation, so to speak, being able to render themselves luminous or the reverse at will. To what extent this faculty prevails is unknown, because our knowledge of life in those awful depths is very limited ; but it may be noted that the majority of the abyssal fish have no means of lighting themselves up, as it were. This incapacity is not confined to the more hideous or horrible-looking among them, but some quite innocent-looking creatures are left in the dark, owing to the absence of any system of lighting on their bodies.

It must also be borne in mind that the glow which is noticeable in many surface fish, pervading the whole body of the creature, has nothing in common with the extraordinary system of animal lighting which we are about to consider, which is without any exaggeration more akin to an electrical installation than aught else. Let us take, for instance, the very queer little fish with the euphonious title of *Collettia rafinesquei*, which was caught not far from the coast of the United States in a depth of nearly eighteen thousand feet. It has enormous eyes in proportion to its body, and a shape not unlike a mullet, but on its underside it is provided with a number of tiny lamps, little globular points arranged in a certain sequence of longitudinal lines from gills to tail with short spurs running up the sides.

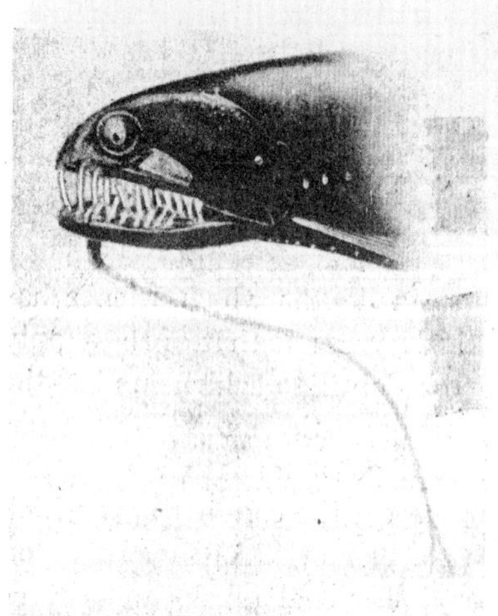

The head of a fierce-looking fish; the photograph shows clearly four of the phosphorescent spots which line the body on each side.

It is evident from the arrangement of the light vesicles on this fish that it does not use its power for hunting purposes. Beyond that, being, as we are, completely in the dark as to its habits, we can say little eccept that it swims serenely above shedding a beneficent glow beneath itsself to assist others for their avocations. It also affords a proof of the whole of the habitat of all deep sea fishes, since other examples of creatures have been found in the Mediterranean.

Come we now to a stately creature with a now so frequent name, *Aethroprora effulgens*. In shape of this fish is nothing out of the common, but in complete affinity with his poetical name, he bears upon his snout a signal light, which is indeed a "flashing light ahead," and it shows sundry of his approach. Down the median of this fish either side runs a row of incandescent phosphorescent spot in which

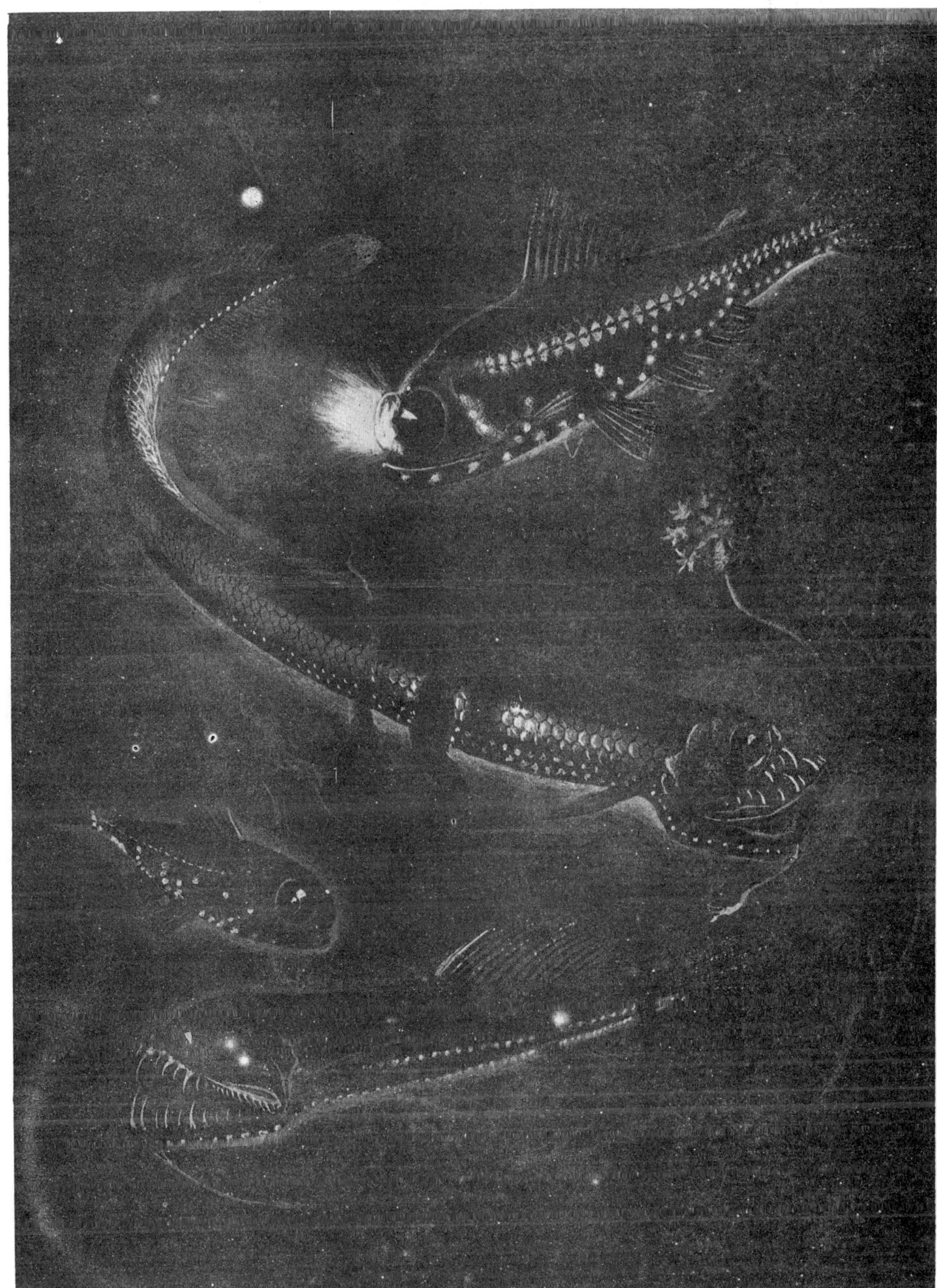

[By Arthur Twidle.

SELF-LUMINANT FISHES.

The fish shown nearest the top is a splendid light-bearer, and has a name signifying a flashing light on the bow; the next one, a fearsome-looking snake-like creature, inhabits the lowest depths of the ocean; the third is a comparatively insignificant little fish with only a feeble illumination; while the fourth is a black ferocious monster which, in order to assist it during its attacks on other inhabitants of the deep, produces only a subdued light.

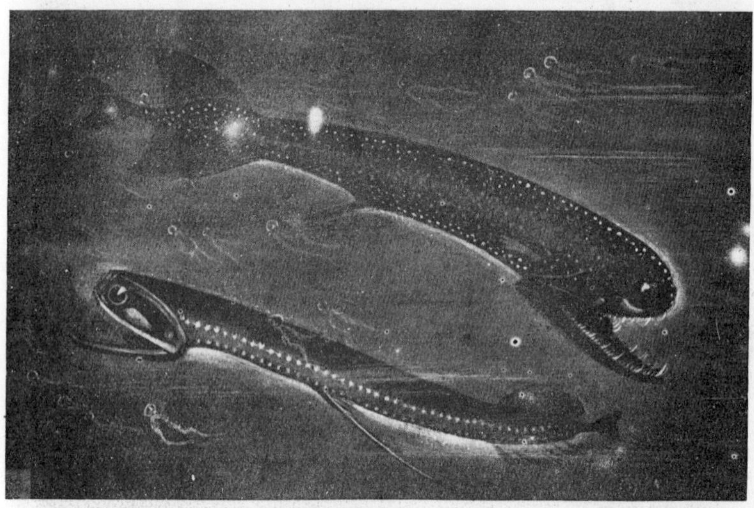

[*By Arthur Twidle.*

Two forms, apparently of allied species, with a truly awful appearance. They are sluggish and badly illuminated, and have enormous mouths.

he carries a few scattered lamps, arranged as in *Collettia.* But his other illuminating factors are so powerful that the small array of electric light globules make very little difference. This gorgeously-lighted creature was captured by the United States steamer *Albatross* within the tropics at a depth of ten thousand feet ; but, as usual with the deep sea inhabitants, nothing can be said of its habits.

Our next subject is a truly ferocious looking individual with a mouth capable of enormous gape; the lower jaw opening until at right angles with the body, while both jaws are armed with formidable teeth. It is an elegant fish, obviously able to travel at tremendous speed, although its formidable name, *Cyclothone elongata* has no allusion to the dreaded meteor of the tropics, but merely signifies, after the manner of naturalists, that it has rounded opercular flaps to its head like a veil. This creature has been taken by the *Challenger* south of New Guinea in four thousand eight hundred feet of water and again in the Banda sea at two thousand feet, while many specimens have been secured by the United States ships in the North Atlantic in depths varying from five thousand feet to thirteen thousand feet. Its electric installation is severely simple, consisting of but two lines of lamps on each side of its belly as far aft as the middle, and thence one line of lights to the tail. But its lamps are larger and more brilliant than ordinary.

Appalling is the only epithet that one can apply suitably to the appearance of *Astronesthes niger*, a black creature of prey possessing a double line of luminous points on its under side. It is the first of these deep-sea creatures noticed here that has a barbel, or tentacle-like appendage, from the lower jaw, which is so characteristic of the denizens of the lower depths. Very little is known about the creature, for only half a dozen specimens altogether have been taken in various parts of the ocean. and even these have been divided into three species. One of these was taken from the stomach of a halibut caught at a depth of eighteen hundred feet, so that in spite of its awe-inspiring appearance and its devilish teeth it is

[*By Arthur Twidle.*

The fish at the top of the picture is a harmless, eel-like one, and almost transparent; the centre one is swift and ferocious; and the lower one a modest, snake-like form, only illuminated about the head.

obviously preyed upon by other and more useful fish.

The *Stomias*, of which we give one illustration, *Stomias boa*, are obviously feeble swimmers, and their illuminating powers may very probably be for the purpose of luring their prey within reach of their relatively immense jaws.

[*By Arthur Twidle.*

In certain localities this fish is very common, and has the power of living, apparently with equal ease, near the surface or at a depth of four to five thousand feet.

The teeth are as in *Astronesthes*, large and curved inwards, capable of being bent sufficiently to allow of the ingress of prey, but rigidly refusing to yield in the opposite direction so as to allow of its escape. These fish are black and almost serpent-like in shape, and have been found in the North Atlantic from the shores of Greenland to the Equator at depths varying from one to ten thousand feet.

A curious form, the hook of whose lower jaw suggests a salmon gone wrong, has also been found in the North Atlantic, but not at great depths—at least, not greater than two thousand five hundred feet, which, for abysmal fish, is shallow. It is fairly well lighted, and, like the two last-described fish, has a dependent barbel, but it is apparently a lumpy, innocent creature, with a tail so insignificant that it is doomed to inactivity and probably lives upon tiny forms which immolate themselves in its stomach unconsciously. To this idea the insignificance of its teeth lends the appearance of good reason.

Eustomias obscurus is the first of our subjects which permits the notice of a strange sequence of the luminous vesicles upon the body. It will be seen from the illustration that there is a certain regular irregularity, if I may so term it, in the disposition of these tiny globes, and it has been noticed that this arrangement is peculiar to a species. The grouping of the lights, in other words, marks the family to which the fish belongs. 'Twere to consider too curiously to suppose that the exhibition of a row of lights specifically arranged upon the body is intended for a kind of standing code of signals to friends in those gloomy profundities, yet if it were true it would not be one whit more marvellous than many other of the facts we have been considering.

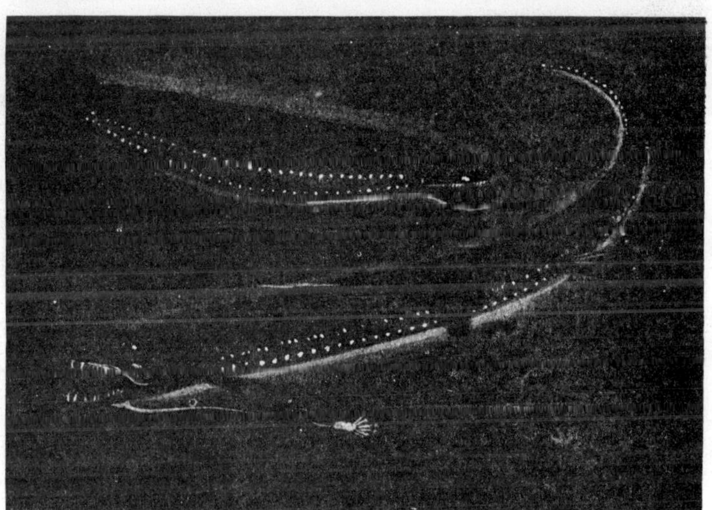

[*By Arthur Twidle.*

While the smaller fish is an inoffensive sluggish creature and feeds upon very lowly organisms, which it collects without effort, the one below is swift and fierce and catches its victims in a distinctly opposite manner.

This curious denizen of the depths has an enormously long barbel, tufted at the end, which is obviously used as an antenna, for the movements of the fish must be of the very slowest and feeblest, its tail fin being but a rudiment. It remains only to add that it is not at all predaceous or ferocious, for its teeth are hardly visible to the naked eye, it is velvety black in colour, and the single specimen upon which the foregoing information is based was captured at a depth of about eight thousand feet near the Azores.

These few specimens out of an enormous number will serve to give an idea of the extraordinary system of self-illumination existing among the inhabitants of those profound depths. But it must be understood that we have only touched the fringe of the subject, feeble and tentative as is all the knowledge extant. Moreover, wonderful and beautiful as are the light systems, in strong distinction to what in the majority of cases is the hideousness of the creatures bearing them, it is impossible to think of them without the mind dwelling for a time upon the invisible gliding forms that, unlit,

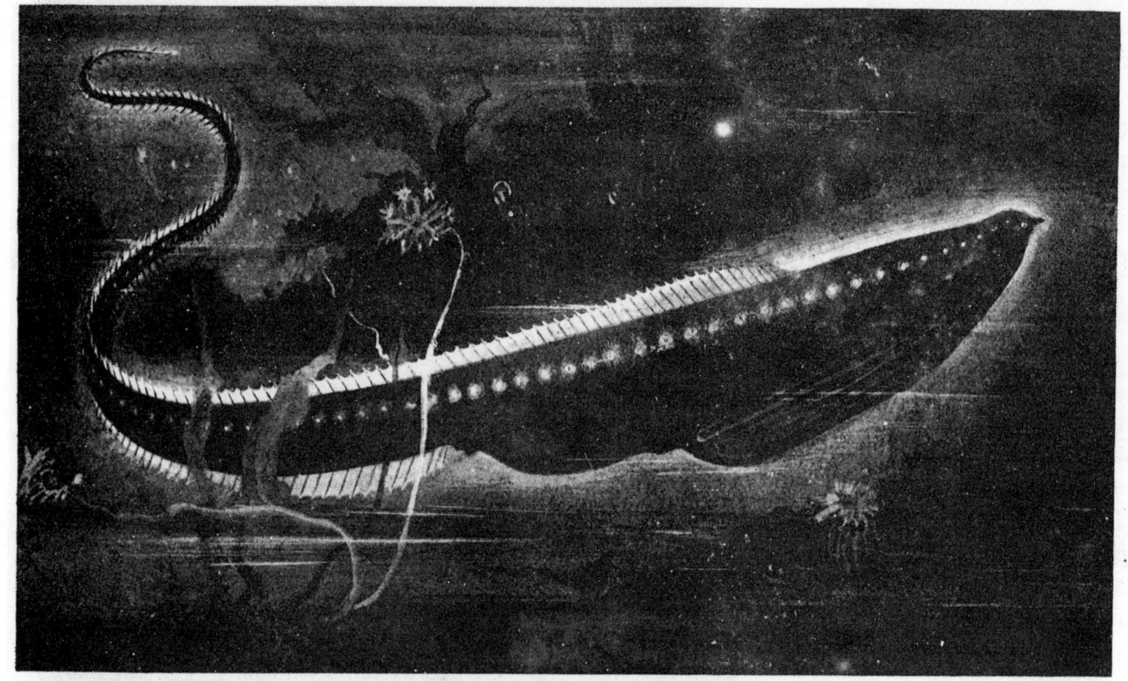

[*By Arthur Twidle.*

This fish is one of the most paradoxical of all the deep-sea denizens. It is sluggish, blind, and brilliantly lit, in fact a complex collection of marine puzzles.

prowl those mighty depths unseen. A few, just one or two, here and there have been brought to the light of day and given names in accordance with their terror-striking appearance, but it must be at once evident, to even the most superficial reader, how few and feeble are those specimens captured to those monstrous forms which no trawl, however strongly constructed, could imprison.

For this reason all the specimens we have are small, and we are, moreover, entirely uncertain whether they have been caught on the bottom or not. We can only say that at such and such a depth a specimen was taken without knowing whether it was entrapped while the trawl was going down or coming up. It is not possible then for any man, however great his knowledge of oceanology, to say with confidence what living forms there are or are not at the bottom of the deep sea. It is quite unlikely that any of the fearsome monsters which may there have their abode, either dwelling in the thick darkness or surrounded by glowing incandescent globes furnished with light by their own vitality, will ever voluntarily rise to the surface. But it is quite certain that what we do know

[*By Arthur Twidle.*

SELF-LUMINANT FISHES.

The top fish, although in appearance more suited to shallow water, is generally found in immense depths; the centre form is a brilliant specimen of sluggish movement, with two antennae depending from the middle of its body for the purpose of feeling the bottom; the lower one is the same fish as the one depicted on the previous page, the huge mouth in this case being open.

of the marvels of the mysterious ocean depths has rendered us entirely unsceptical as to the wonders it may yet conceal from us, wonders which do not, and will not, in the least conflict with sober judgment and scientific knowledge. After all, the ocean is the birthplace of all life, and it is only fitting, therefore, that it should contain within its mighty bosom stranger forms of life than any of which man has yet dreamed.

SOME PLANTS THAT FEED ON INSECTS

BY JOHN J. WARD, F.E.S.

PLANTS, by means of their leaves, can absorb light and the carbonic acid gas of the atmosphere. In the chemical laboratory of each leaf the energy of this intercepted light is expended in the production of organic materials ; the carbon is changed into sugar, starch, and similar substances which sustain the living plant during its growth and development.

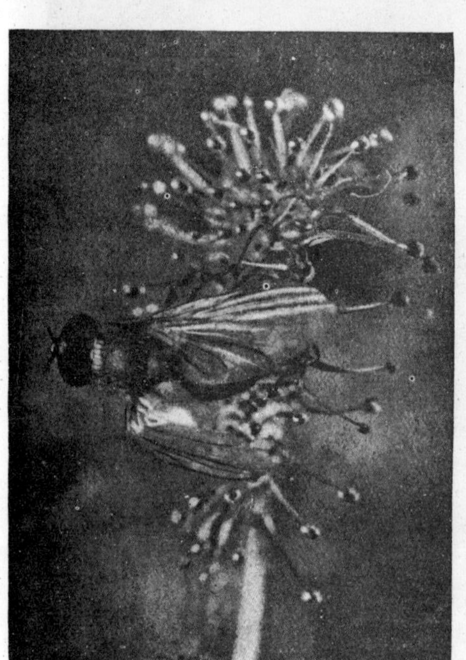

Photo by] [*S. L. Bastin.*

A common fly captured by the Sundew.

In this way the plant becomes a builder of living, or energy-yielding, substances from the dead, or inorganic, carbonic acid gas and water. Just how this wonderful change comes about is not clearly understood, but the green-coloured granules which give the green hue to the leaf are an essential factor in the working of the process.

A plant, therefore, given light, air and water, can manufacture energy-yielding material, or food stuff. In this it differs essentially from an animal. The animal cannot manufacture its own food ; every particle of its structure and every movement that it makes, is at the expense of living material afforded by the leaf or stem or fruit of a plant. To put the matter briefly, the plant is a builder of life-sustaining substances which support either plant or animal life—and these substances the animal can only utilize and destroy.

Although the plant possesses the power to manufacture its own food substances, yet sometimes its situation is such that, although it has light, air and water, the water lacks a sufficient amount of nitrates and mineral substances, which are very essential in effecting the necessary changes of its structure. A plant so placed is comparable to a man trying to live on bread alone—the material is not sufficient to carry on the natural functions.

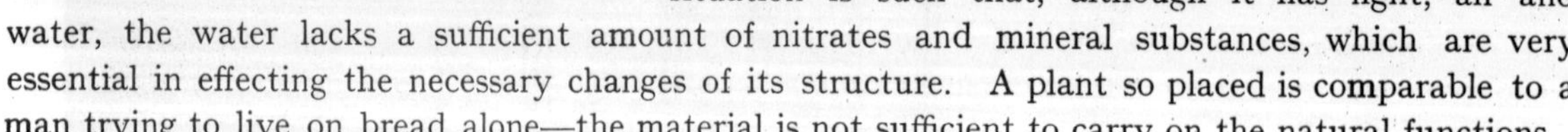

Such deficiencies have been compensated by the ability of the plant to capture and eat small flying and creeping animals. They are thus enabled to add to their tissues the necessary substances which the soil lacks.

The non-botanist can hardly imagine a plant setting itself to catch a fly, yet a visit to almost any bog will reveal thousands of plants so engaged. Bog-land is the natural home of these plant insect-eaters, and they are especially abundant on peat bogs. These peat mosses are so saturated with moisture that access of oxygen is prevented, and various antiseptic acids are formed, so that minute animal life cannot exist amongst them ; and the soil becomes altogether untenable for ordinary plants, especially as nitrogen and essential mineral matters are almost entirely absent.

Photo by] [*E. J. Wallis.*

THE HUNTSMAN'S HORN PITCHER-PLANT.

The hollow leaves of this plant are usually found half-filled with water and decaying insects, the latter being digested and used as food by the plant. It inhabits the bogs of parts of North America.

In spite of all these difficulties, the little Sundew (a distant relative of the common London Pride) lives and flourishes amazingly among the sphagnum mosses on the peat bogs. How it learnt to thrive so well where most other plants would perish, is not difficult to understand when we remember that its remote ancestors were almost certainly saxifrages—a family of plants noted for possessing sticky glandular hairs about their stems and leaves. If then the parent form of the Sundew happened to be placed on peaty ground, and, as often happens with hairy plants, small flies became entangled in the sticky hairs of its leaves and perished there, their decaying bodies would render good service to such plants by supplying their tissues with those materials which the soil lacked. In this manner the plant would benefit, and so those individuals which produced the most sticky and the most absorbent hairs about their leaves, would thrive best, and by a process of natural selection the hairs would develop into complex and sensitive glands capable of movement and of secreting a digestive fluid.

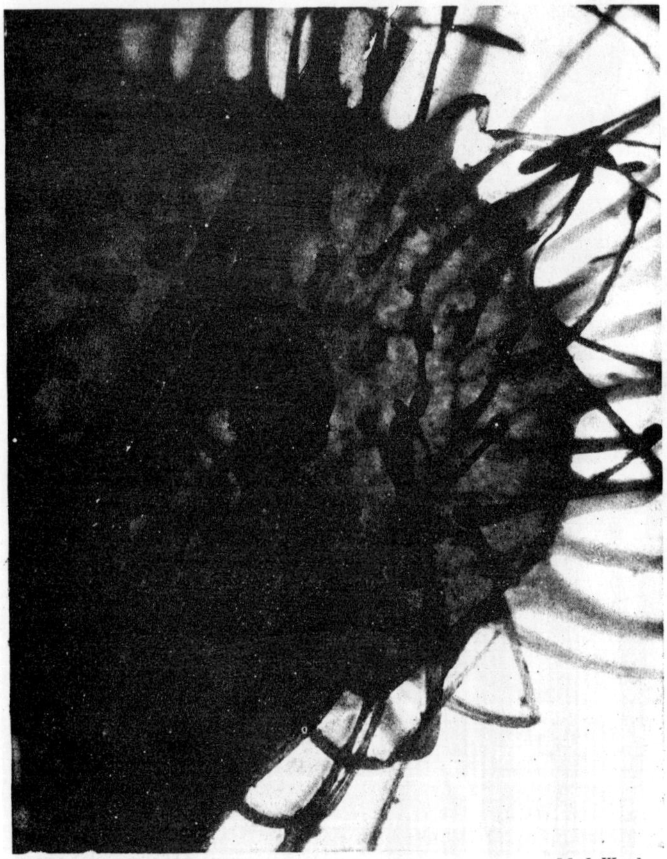

Photo by] [J. J. Ward.

A SUNDEW LEAF AND ITS CAPTIVE ANT.

A microscopical photograph showing how the tentacles approach, and, by their sticky knobs, hold any insects that alight on the leaf.

An examination of one of the leaves with a pocket lens will show how it contrives to obtain sufficient nourishment. Its green surface is studded over with numerous movable glandular hairs of a reddish colour. The swollen gland at the head of each hair glistens in the sunlight as if fresh with moisture from the morning dew — hence its poetical name Sundew. The whole structure of the newly-opened leaf makes a charming object when seen through a lens. But the great Charles Darwin discovered some extraordinary details about these tentacles of the Sundew which leave no doubt of their blood-sucking propensities.

He has conclusively shown that these glandular structures are of a most complex character, and that when the long external tentacles were touched, or when placed in contact with organic and inorganic substances, they became inflected. If the object adhering is a fly, the one or two tentacles engaged heave it into the centre of the leaf. The moment the object so lifted comes in contact with the central area of the leaf, the glands there affect the external tentacles so that they also slowly bend inwards. At the same time the blade of the leaf will often become hollowed, and so form a kind of cup in which the unfortunate caterpillar, fly, or other prey, is held while the glands pour upon it their viscid fluid. It is obvious, therefore, that the more the quarry struggles the less chance it has of escaping, for, the more the glands are excited, the more fluid do they pour out.

A more curious discovery made by Darwin was that, when the tentacles were bent, this secretion was changed into an acid. Now this was important, for the gastric juice of animals contains both an acid and a ferment, and so Darwin was led to inquire if the leaves of the Sundew really ate and digested the prey they captured ? After a long series of experiments he was able to affirm that the secretion of the tentacles acted upon by acid became a ferment and produced the same change in a variety of food substances as does the gastric juice of an animal, and also that the digested matter was absorbed by the glands. This was a striking discovery, for until then a true digestion and the power to absorb the digested material were unknown in plant life.

Photo by] [Edward Step, F.L.S.

A DRAGON-FLY CAPTURED BY A SUNDEW.

The Sundew appears too feeble a plant to hold so powerful an insect as a Dragon-fly; but this photograph, taken on one of the West Surrey bogs, shows that it is at least an occasional occurrence.

There are two other species of Sundews found in the British Isles besides the common round-leaved form here described, the leaves of these species being more oblong and standing almost erect. Also, in plant houses, several larger forms may frequently be seen under cultivation. An Australian example is shown amongst the illustrations. One North American species has thread-like leaves a foot or more in length which grow thickly entangled about the ground. Growing with the Sundew amongst the mountain boglands, we may not infrequently find the Butterwort. This plant, which is shown in one of the accompanying photographs, is a somewhat distant relative of the snapdragon, and is in nowise connected with the Sundew, although we find it carrying on the same uncanny practice of ensnaring and eating insect prey. Here we have one of those instances which frequently occur in Nature, where, as it were, a satisfactory solution of a difficult problem having been discovered, it has then been adopted by entirely different creatures to attain a similar end. It is interesting to note, however, that, so far as we know, both the Sundew and the Butterwort had their origin in families of plants whose existing species are characterized by hairs and glands on the leaves and stems. Darwin has experimentally proved that members of both the saxifrage and the primrose families possess glands with the power of absorbing liquid food. There is,

Photo by] S. L. Dustin.

THE VICTIMS OF A PITCHER-PLANT.

The interior of a pitcher of the American plant called Trumpets reveals quite a number of dead captives at the bottom.

therefore, good reason for assuming that the complex glands of these carnivorous plants have evolved from more simple absorbent hairs.

The leaves of the Butterwort are also arranged in rosette fashion, and their upper surfaces are covered with tiny glands of two kinds, both of which are so small as to be scarcely visible to the unaided eye. In the section of the leaf shown in the illustration these glands will be seen. The larger ones are elevated on pedicels, while the smaller ones are placed very close to the surface of the leaf. Both kinds of glands secrete a viscid fluid, which, like that of the Sundew, becomes acid in the presence of material containing nitrogen. The Butterwort glands are not, however, movable like those of the Sundew. The captured insect is brought into contact with a large number of the glands by the incurving of the margins of the leaf. In this manner it is slowly pushed towards the centre of the leaf, and this causes a considerable secretion of the digestive fluid, which runs towards the apex of the leaf and often carries the insect with it. If the photograph of the Butterwort is observed, the margins of the leaves will be seen to be incurved and turned downwards at their tips.

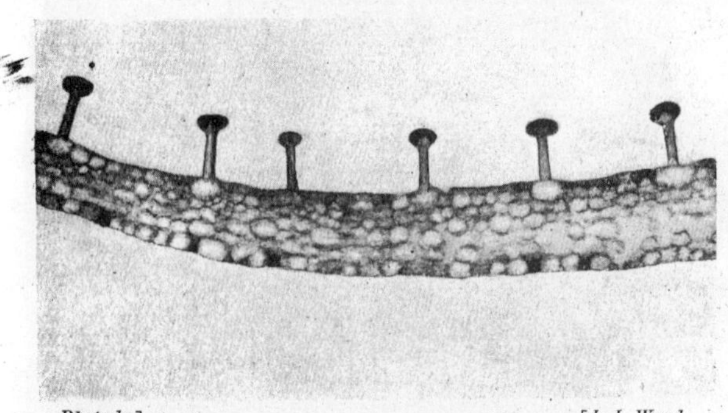

Photo by] [*J. J. Ward.*

THE GLANDS OF A BUTTERWORT'S LEAF.

A slice through the thickness of a leaf, very highly magnified under the microscope, in order to show how the leaf is studded with sensitive glands.

Darwin also made some experiments with this plant and one of the most interesting things which he was able to prove was that it is not only carnivorous, but is also a vegetarian feeder; for he found that it would digest bits of leaves, pollen from flowers, seeds, etc.

On the marshy soils of other countries still more striking examples of insect-eating plants may be found. The Venus' Fly-trap, of tropical North America, is one of the most remarkable plants in the world, for, as its name implies, it captures its prey by means of a trap, having no viscid secretion like the Sundew or the Butterwort; indeed, considering that it is a plant, its movements are most uncanny.

The curious leaves stand more or less erect and are fringed along their margins with conspicuous teeth, as the photograph on page 17 shows. It will also be observed that the lobes of the leaves are held like a half-opened book. Presently a beetle alights upon the footstalk of one of them; although in alighting and closing its wings it may vibrate the leaf considerably, yet nothing happens. It travels up and down the leaf stalk and right on the back of the leaf itself, even walking with difficulty amongst the teeth at the margins, and still nothing happens. Then suddenly the beetle disappears from view and we find that the two lobes of the leaf have closed, and that their teeth have interlocked, the beetle being enclosed between the lobes.

Photo by] [*J. J. Ward.*

THE BUTTERWORT.

An insect-catching plant, common on mountain bogs. Its leaves exude a sticky substance, and then curl their edges over to capture alighting insects.

What really happened was that the beetle, in moving over the surface of the leaf, touched one of three extremely sensitive filaments which are arranged on each lobe—the spring to the trap, as it were. You may handle the leaf quite carelessly and it will not close, but you may close it with a hair, providing the hair touches one of the sensitive filaments. On this page one of the leaves is shown closed and holding a pin, with which I touched one of these sensitive filaments. If, however, the capture is not large enough, the insect is allowed to crawl away unhurt.

Finally, there are those insectivorous plants which set water traps in which to drown their prey. The American *Sarracenias*, or Side-saddle plants, hold up to the sunlight their coloured vase-like leaves, about the mouths of which are glands which secrete honey. Unwary insects, smelling the honey, are attracted towards the leaves, and to further insure that there shall be no turning back when once they approach, the leaves are beautifully streaked and shaded with reds and purples, which contrast with the green shades and make an attractive display of colour. In this way the thirsty and unsuspecting insects are lured to the edges of the deadly leaves, and are then led further and further down the tube by still more honey-secreting glands. It is, however, a fatal journey, for detentive hairs are arranged to prevent them climbing back again; and, in some species, the mouth of the tube is shaded from view by the lid, and curious translucent spots appear in the walls of the tube, against which the trapped insects are continually striking to effect their escape, until completely exhausted they fall into the pool at the base of the vase-like leaf.

The *Darlingtonia*, a native of California, behaves in a very similar manner; while the Pitcher-plants work on almost exactly the same principle, producing beautifully-coloured leaf-structures of pitcher-like form. These secrete honey around their rims and beneath their lids, and from glands within an acid fluid, in

Photo by] [*J. J. Ward.*

THE VENUS' FLY-TRAP OPEN.

The leaves of this plant are constructed on the principle of the rat-gin; the fringing hairs interlock to prevent the escape of any captive.

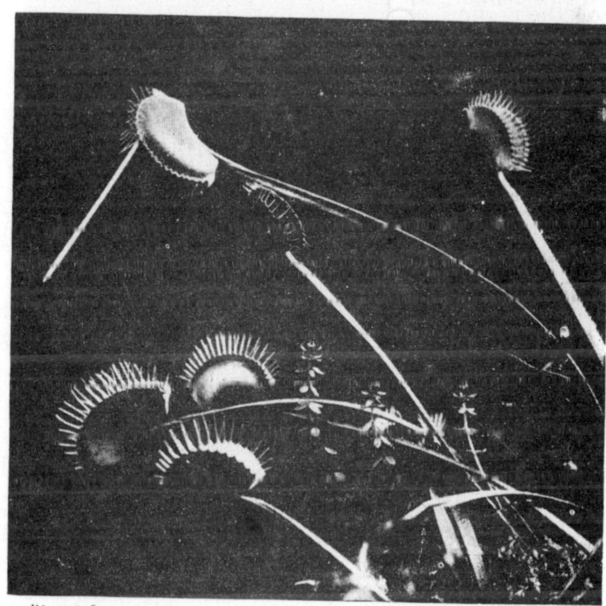

Photo by] [*J. J. Ward.*

THE VENUS' FLY-TRAP CLUTCHING A PIN.

This photograph shows the retentive power of the leaf, which has been induced to snap at and hold a pin, as seen in the upper left-hand corner.

Photo by] [S. L. Bastin.

THE CAPE SUNDEW.

This plant, a native of South Africa, has long, narrow leaf blades, which are densely clothed with tentacle-like hairs for holding insects.

which the numerous insects captured, and even small birds, are drowned and decomposed.

A TREE-CLIMBING CRAB

BY EDWARD STEP, F.L.S.

THE most marvellous departure from orthodox crustacean habits is that of the Robber Crab, or Coco-nut Crab of Keeling Island, and neighbouring coral islands. This crab, which is often a foot in length, is clearly a hermit crab, that has given up the habit of lodging in a cast-off sea-shell, and, taking to a terrestrial life, has now to protect its soft hinder parts by inhabiting holes in the ground. It has so modified its gill-chamber that while the upper part serves for breathing air, the lower portion is still available for aquatic respiration when it pays an occasional visit to the sea. At different periods it has been variously asserted that the Robber Crab does and does not climb coco-nut trees; but recent observations leave no doubt that it does, though not, as formerly asserted, for the purpose of gathering the nuts. These it finds in plenty on the ground, and sets to work to strip off the thick coat of fibres in which each is packed. Some of this fibre it uses as lining for its burrow. It begins at the end that will expose the three eye-like depressions in the true shell, and hammers at one of these with a big pincer-claw until it has made a hole; then it introduces one of the smaller pincers and extracts the flesh of the nut piecemeal. Upon this fare it thrives, and lays up such store of fat, that it is much esteemed by the natives, who melt it down and often get as much as a quart of oil from one large crab. The eggs are laid in the sea, where the young pass through their larval stages.

Mr. H. O. Forbes says that the Robber Crab has such muscular power that with one blow of its large pincer-claw it can break either a coco-nut or a man's arm. Its burrow is larger than that made by the rabbit. It needs such a retreat, for apart from man's appreciation of it as food, on those islands where there are wild pigs the latter root out and devour the Robber Crab.

It appears that its appreciation of coco-nut is an acquired taste of the Robber Crab's, for the coco-nut tree is not a true native of these islands, but is believed to have been introduced from Mexico by the Polynesian mariners, who anticipated the discovery of America by Columbus. As the Robber Crab has never been found away from the islands of the Indo-Pacific, it is certain that the crab did not come with the tree. As an aboriginal native, it must have had another kind of food before the introduction of the fruit. Possibly some light may have been thrown upon its past habits by Mr. Arthur Adams' statement that in the Meïa-co-shimah group of islands the Robber Crab frequents cemeteries and feeds upon what is chiefly found in cemeteries! A crab-advocate could plead that this gruesome practice is merely a reprisal, for, as already indicated, the Malays consume many of the crabs, digging them out of their burrows and appropriating their fibre beds for use as oakum. Darwin, who had opportunities for observing the habits of this crab during that

[*By Theo. Carreras.*

THE ROBBER CRAB.

This great crab lives on land and climbs coco-nut palms. It uses the outer husk to line its burrows, and feeds on the flesh of the nut, hammering in the eye-holes with its big claws, and extracting the fruit by means of its small pincers.

Photo by] [*W. Saville Kent, F.Z.S.*

THE ROBBER CRAB.

The crab that climbs trees is here seen engaged in pulling off the outer fibrous coating or oakum with which the coconut is covered. The crab to the left is carrying away a portion to its burrow.

memorable voyage of the *Beagle*, gives an example of its strength. He says: "Captain Moresby confined one in a strong tin box . . . the lid being secured with wire; but the crab turned down the edges and escaped. In turning down the edges, it actually punched many small holes quite through the tin."

THE MAKING OF A BEE-CITY

BY TICKNER EDWARDES.

Author of "The Lore of the Honey-Bee," etc.

To one knowing nothing about hive-life, the first sight of a swarm of bees hanging from the branch of a tree in the soft May sunlight brings a feeling little short of amazement. We have always thought of the hive-bee as a creature hovering from flower to flower, a winged atom making music in the summer sunshine—useful as a honey-giver, but otherwise wholly inconsiderable in the scheme of mundane affairs. But here suddenly she shows herself in a new and almost incredible aspect. In a moment, as it were, her whole nature appears to have undergone a revolution. From the apple-bough in the cottage-garden depends a brown, torpedo-shaped mass, bending the branch almost double with its weight. It droops inert, and it is difficult to believe that it is composed of twenty thousand living creatures, collectively engaged in what is the most important enterprise of their existence.

With the issue of the swarm, not only does a new bee-colony come into existence, but every bee in the cluster begins an entirely new life. Only yesterday each was absorbed in the affairs of the old home, passing busily to and fro between the hive and the clover-fields through every moment of daylight; intent only on its prosperity, and ready to defend it, if need be, at the cost of her very life. But now all this is past. When the signal for the departure of the swarm rang through the crowded hive the old home was completely forgotten. It is not merely that each bee commences an entirely new existence ; for her the old life is completely wiped out of memory.

Hanging there torpid and almost silent in the May sunshine, while all other bees in the garden are obsessed by a perfect fury of work, it looks as though, for once, the busy bee has discarded her principles. But this is not really so. Swarming bees cluster in this way with a two-fold object: Their first and principal care is to find a suitable place, sheltered and dry, where the new home can be fashioned. To this end, as soon as the party is free of the mother-hive, it makes this temporary halt while certain bees go forth to look for a site wherein to establish the permanent quarters of the new colony. And then, ordinarily, the cluster remains on the branch only so long as is necessary to get the party together, rising on the wing again after a few minutes and sailing straight away to the predetermined spot.

The second reason for the close clustering of the swarm is the necessity for preserving a high temperature among those of the worker-bees who are presently to build the waxen walls of the new home. The wax, which is secreted from certain glands in the bee's body, will generate only when the wax-makers lie together for many hours quiescent and in great heat. This condition is secured by close clustering while the scouts are abroad on their prospecting journey. To the uninformed observer, the swarm on the apple-bough seems idle enough, but it is in reality busy already at one indispensable part of the work to come.

Whether the bees are now left to their natural instinct to fly away to some hollow in a tree-trunk, there to set up their new house, or whether fate, in the guise of the old-fashioned village bee-keeper, intervenes to provide them with a home, the task before them is much the same. There is the queen in their midst, who must be furnished with brood-comb in which to lay her eggs ; there are the store-cells to make, where the honey and pollen are to be warehoused for the sustenance of the colony during the coming winter and early spring. And here, at the threshold of affairs, the student of bee-life is brought face to face with a problem little less than bewildering in its complexity. Once settled in its new quarters, the great army of worker-bees divides itself into two main gangs. The one clusters together in the roof of the hive, and at once begins the work of wax-making and comb-building ; the other sets off in its thousands towards the fields with the first light of morning, to bring in provisions for the hive. From the outset it is plain that the bees work together on some well-considered, orderly system. Each bee has her allotted part to play, and plays it without hesitation or demur.

In the making of a bee-city, however, though we see ingenious works carried through on every hand for the attainment of intelligent objects, we fail altogether to discover any central directing mind. The old idea that the queen--the single large bee to be found in every hive—was herself the controlling spirit of the colony must now be laid aside. All recent investigation goes to prove that

Photo by Tickner Edwardes.] [*By permission of Methuen & Co. Ltd.*

THE ARRANGEMENT OF THE COMBS.

An upturned skep-hive, showing natural arrangement of combs. On being hived, a swarm of bees commences at once to build the comb. The bees get together in the upper part of the hive, and there remain in a dense cluster while their wax is forming. The wax is a secretion from the bee's own body.

far from being a ruler, a dominating personality in the hive, she is really the least intelligent and the most subordinate member of the whole community. It is the worker-bees which retain among themselves all power of inception and direction of the public polity. It is the worker-bees who, when the old hive became too populous, decided that a greater part of its inhabitants should leave and set up housekeeping for themselves; they select those who should join the emigrating party and those who should remain; they give the signal for the departure of the swarm and command the old queen to accompany it, leaving her ancient realm to the charge of a younger and more vigorous successor.

The truth is that in hive-life, though wise laws are made, ingenious devices originated and complex works carried through triumphantly in the face of unforeseen difficulties, there is no central guiding intelligence at all. In some mysterious way, the whole worker-bee population of the hive seems to unite in the solution of the common difficulties.

Photo by] [*Tickner Edwardes.*

THE FANNING ARMY.

A bee-hive is ventilated by gangs of worker-bees, who station themselves at the entrance in open order, with their heads towards the hive, vibrating their wings and thereby causing a circulation of air through the hive. These fanning bees are regularly relieved by others, the fanning being incessant during hot weather.

As we watch the swarm at work day after day, week after week, transforming the empty straw skep into a convenient and well-appointed bee-dwelling, the feeling of wonder at this fact only deepens. When filled, the straw hive is seen to contain nine or ten combs side by side, suspended from the roof. Between the surfaces of the combs is left a space which will just allow the bees covering each comb to pass back to back. The comb-ends are attached to the sides of the skep half-way down: but for the rest of the way, and at the bottom, they hang perfectly free. It will also be seen that the whole interior of the hive has been plastered with a thin, almost transparent substance, like resin, hermetically sealing every crevice in the hive-walls.

At this point, probably, the observer will be struck with what, according to human notions, is a manifest fault in the bees' system of house construction. Having to accommodate twenty or thirty thousand insects in a single building, they appear to have left out of account all provision for ventilation. How can one small aperture, which is all that is reserved, suffice not only for the

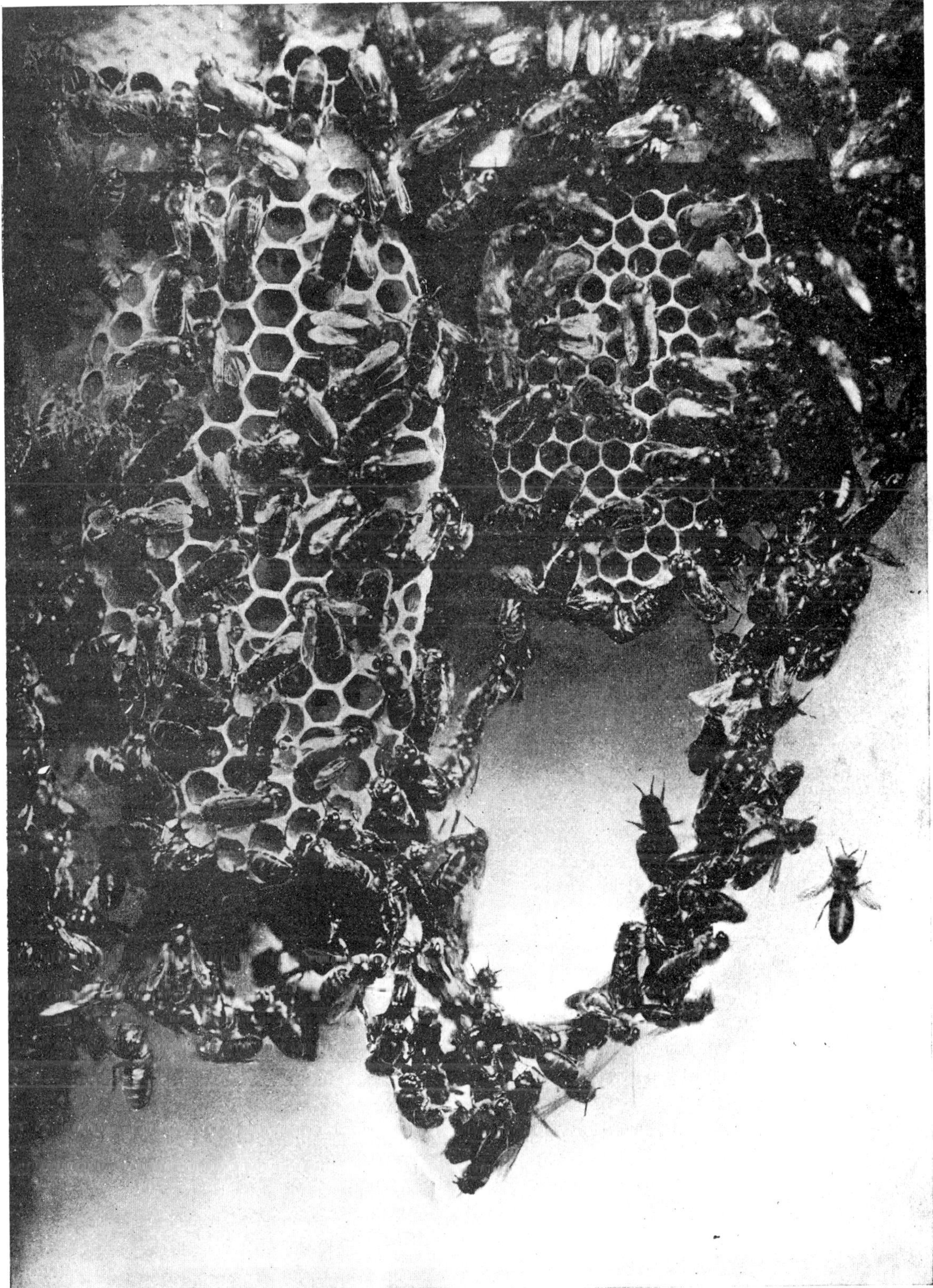

Photo by Tickner Edwardes.] [*By permission of Methuen & Co. Ltd.*

A LIVING CHAIN.

Bees forming themselves into a cluster for wax-generating. This remarkable photograph shows the drone and worker combs being constructed side by side by the same party of builders. At nearly all times bees carry through these complicated works in total darkness, comb-building being generally undertaken at night.

Photo by] [*Tickner Edwardes.*

DEPOSITING THE EGGS IN THE BROOD-COMB.

A brood-comb, showing the two sizes of cell made by bees. The large cell is used for raising the drones, or male bees, and the smaller cell for the workers. The queen, in laying the eggs, which can be seen in the illustration, attaches them to the base of the cells.

passage of the bees, but also for the ingress of the fresh air and escape of the foul? The impulse of the inexperienced bee-keeper here is to remedy what he considers the bees' want of intelligence and foresight by piercing a row of ventilation-holes in the hive-roof. If he does so, he will find that within an hour or two the bees will have blocked up the holes again.

But the fact is we do not at first realize the whole ingenuity of this branch of hive-policy, because we have not taken into account all the conditions. In a normal hive, beside the vast population of adult bees, there are innumerable bee-grubs — perhaps ten or twelve thousand at a time — maturing in the brood-cells; there are also the store-combs filled with brewing honey, each little vat incessantly giving forth a heavy steam. Not only must fresh air be continually provided for the bees, young and old, but there must be a fairly brisk air-current through the honey-combs to carry off this vapour, or it would soon condense and the hive become a quagmire. Yet this is not all. Both young brood and maturing honey require a high and equable temperature for their proper development, so that the air must be warmed before it reaches the nurseries and brewing-quarters of the hive.

Throughout the winter, when the conditions governing bee-life are fundamentally changed, there is no need for mechanical ventilation of the hive, because the natural motive power of the warm air given off by the clustering bees is sufficient to create the gentle air-circulation then alone needful. But at the first return of warm weather a new form of activity is inaugurated. Gangs of bees are stationed at the hive-entrance, both inside and out, whose duty it is to fan the air in different directions according to the position taken up. The combined result of this fanning system is to draw in air at one side and expel it at the other, after it has made the whole circuit of the interior of the hive. And now we see the wisdom of allowing only one aperture. Any method of through ventilation must be partly automatic, and, therefore, not entirely under the control of the workers.

Although the work of these fanning parties never ceases while the necessity for it continues, the bees forming the squads are constantly being changed. Each bee fans for about a minute, after which another worker comes up behind her, whereupon the first bee ceases work and marches off. All this time the foraging bees are hurrying in and out of the hive, passing in both directions through the open ranks of the fanners.

In the making of a bee-city the worker-bees do not have their tasks allotted to them indiscriminately. The age of each worker seems largely to determine the branch of activity in which she engages. In the composition of the swarm we find a queen, or mother-bee, whose single duty it is to keep all brood-cells furnished with eggs; a few score of drones, or male bees, at present nearly useless appendages to the company, and from ten to thirty thousand workers, on whom the whole future progress and well-being of the colony will devolve. This great army of workers seems to consist in about equal proportions of young bees who have never previously flown abroad, and those experienced in outdoor work who will be upwards of a fortnight old.

The wax-making bees of the swarm unite themselves into a dense cluster in the upper regions of the hive and there remain while their wax is forming. The temperature of the cluster varies from ninety-five degrees to ninety-eight degrees, which is considerably higher than that of the rest of the hive. Each bee hangs motionless in the throng until her eight abdominal wax-pockets have filled themselves, and then she moves up to the edge of the growing comb, labouring there so long

Photo by] [*Tickner Edwardes.*

AN UPWARD-BUILT COMB.

The almost invariable practice of the honey-bee is to suspend her combs from the roof of the hive, but in rare instances the rule may be reversed. The photograph shows an unique example of this. The triangular piece of comb at base of picture was built upwards, story by story, exactly as a human dwelling is built.

as her supply of building material lasts, and returning to the cluster for another period of quiescence while her wax-pockets are replenishing.

The entire conception of a honey-comb is perhaps the most perfect example in existence of the triumph of mind over matter, whether we look for it among the marvels of the natural universe or the works of man.

THE COMMON OCTOPUS

BY FRANK T. BULLEN, F.R.G.S.

Beauty of form and radiance of colour are undoubtedly the possession of a vast number of the inhabitants of the ocean, and in many cases the beautiful are also the useful in every sense. But in the opposite direction it is doubtful whether any of the inhabitants of earth, even among the reptiles, can begin to compare with the majority of the deep-sea folk in the matter of absolute hideousness of detail both of form and colour. And this is the case with all the immense varieties of the mollusca, one of the chief of which is the subject of the present brief sketch.

The Octopus is, as his name indicates, an eight-armed creature, which differentiates him from all the mighty army of squid or cuttle-fish, all of whom possess, in addition to the eight equally lengthy arms owned by the Octopus, two additional tentacles of double the length of the others and of a different shape. They spring from the crown of the head like the rest, but for nearly the whole of their length have a smooth stem. At a short distance from the remote end from the head they are armed with a few sucking discs, but they do not appear to be used for any other purpose than that of antennæ or feelers.

These organs are not possessed by the Octopus, which is indeed different from the squid in other important details. The shape of the body of the Octopus is nearly globular and is vastly dispro-

Photo by] [*J. Sinel.*

THE OCTOPUS.

The Octopus is here shown at rest on the sea-bottom, waiting for a victim or digesting his last meal. If suddenly alarmed he can conceal himself by ejecting a little ink, producing a dense black cloud, under cover of which he escapes.

Photo by] [*F. Martin Duncan*

OCTOPUS AND CRAB.

A shore-crab prowling about the sea-bottom in search of food comes into the vicinity of the resting Octopus, who quickly spreads out its long arms, and the suckers attach themselves to the crab's shell. The victim is drawn within the web, and the strong beak quickly pierces his shell.

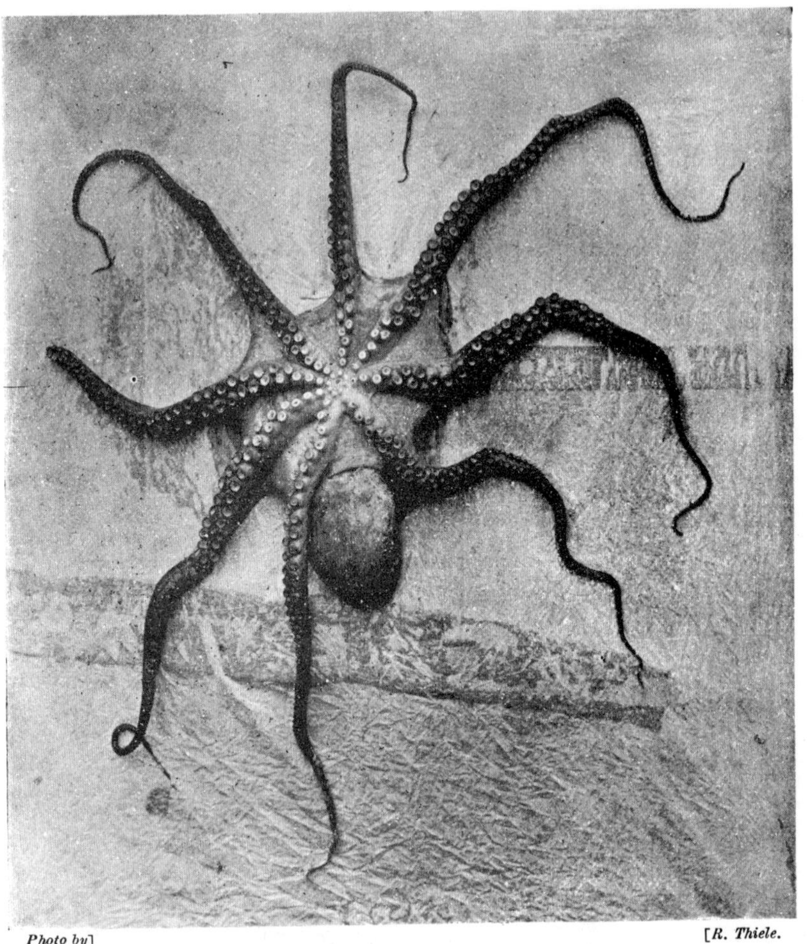

Photo by] [*R. Thiele.*

AN OCTOPUS ON ITS BACK.

The eight "arms" of this powerful creature have all their bases connected by a web. The parrot-like beak is in the centre.

portioned to the size, length and mass of the tentacles, while the squid has a cylindrical body, with a pair of fin-like appendages at its point which are used for steering purposes when this creature propels itself rapidly through the water *backwards*, as we should think, by ejecting a jet of water from a pipe in its neck.

Now the Octopus has no such organs, for its mode of life does not call for them. It is an inhabitant of shallow waters, lurking among the rocks near the surface by day, and at night warping itself along by means of its tentacles, which rove and quest into every nook and cranny in search of anything eatable which they may grip and present to the gaping mouth with its parrot-like beak in the centre of that deadly group of arms. Nothing in animated nature approaches in appearance these never-resting purveyors. Varying in size, of course, according to that of the animal they serve, they spring from the top of the head in a circle disposed around the mouth, and they are about one quarter of the thickness of the head itself. From the root they taper off gradually to the ends, which are fine as a whip-lash. On their inner sides they are furnished with a number of sucking discs set closely together, each of which acts like the receiver of an air-pump, clinging to any object it touches with an amazing tenacity.

Having once touched an object desired by the creature, another set of forces comes into play to aid the adhesive power of the discs. The muscular power of these arms is enormous when the size of the creature is considered, as many an incautious bather knows to his cost. It costs me a shudder even now when I remember how, when wading in a shallow bay in the South of New Zealand, with the object of catching flounders, Maori fashion, by holding them down with my bare feet, then stooping and securing them, I suddenly felt a pricking all over my bare legs (the water was just above my knees). Stooping to ascertain the cause, I had a sensation of nausea when I found the arms of an octopus wound around my legs and creeping upward. I started to rip them off my flesh, detaching the skin as I did so; but as fast as I did, others attached themselves, and I had only two hands to the creature's eight.

The cold sweat broke out upon me; I drew my sheath knife, and, stooping, groped for the body of the animal. It was not much larger than my fist, but the malignity of its eyes belonged to the Pit. I sawed all the tentacles off (they were tough, too), then stripped them from my legs, waded ashore, and vomited. A paltry adventure, since the thing could not have done me any harm in that shallow water ; but if it had been a large one ?

Now in size, as far as our fairly full knowledge of the Octopus carries us, this creature is fortunately very limited, the largest known specimens having bodies not more than a foot in diameter when distended, and tentacles six feet in length. Such an animal as this is sufficiently terrible, especially when to its enormous muscular power is added a curiously malignant intelligence. Without encroaching in the least upon the realms of fiction, it is impossible to avoid the conviction, upon looking into the eyes of an Octopus, that it *knows*, that its ferocious character is buttressed by an intelligence such as is possessed by no other animal except man.

The eye of the Octopus is entirely black and quite circular. It has no lids, not even a nictitating membrane such as birds possess, and these strange characteristics, added to the slimy horror of its convolutions, are quite sufficient to account for the many amazing perversions of fact connected with it that are to be found in the pages of ancient writers upon the subject, splendid inventions based upon real facts and feebly copied by moderns.

One exception must be made to this latter sweeping statement. Victor Hugo, with that splendid disregard of facts which makes him such good reading, has given many thousands of his admirers a totally false idea of the magnitude of the *Pieuvre* of the Channel Islands in his "Toilers of the Sea." These horrible creatures are big enough in all conscience, and a body of a foot in diameter and tentacles of from five to six feet in length would serve to destroy the strongest of the sons of men if it seized him in water up to his waist, but those *pieuvres* of Victor Hugo's would pull down an elephant. The size I have indicated is about the limit, and is not likely to be exceeded, since our knowledge of this creature, owing to its habit of living near the shore, is as I have said, fairly full.

Photo by] [*R. Thiele.*

THE OCTOPUS VIEWED FROM ABOVE.

This view gives a good idea of the bag-like body and shows how the eyes protrude from the body.

Another brilliant Frenchman, Jules Verne, has, in his wonderful

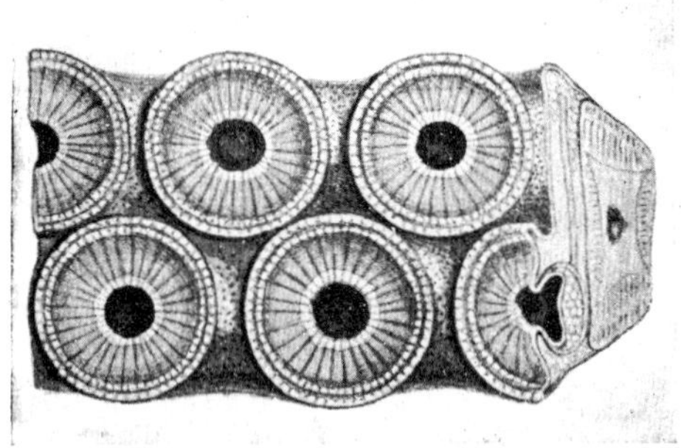

SUCKERS OF THE OCTOPUS.
The arm is drawn in section at the right of the figure to show how the suckers are worked.

romance of anticipation, 'Twenty Thousand Leagues Under the Sea," also given a magnificent description of a fight between the voyagers and a number of "Poulps," as he calls them, but although it is difficult to believe that he means anything but Octopods, he may be alluding to gigantic squid, which really do attain to enormous proportions.

Lying in the centre of all those writhing arms, a point to which all their energies are directed, all they capture brought, is a pair of horny mandibles exactly like the beak of a parrot, but very much more powerful. This apparatus is capable of crushing an oyster or a whelk, to say nothing of the carapace of a crab or a lobster, and it is most probable that, in spite of its omnivorous character, the Octopus finds its chief sustenance in these comparatively helpless and sluggish crustacea. This horny beak is common to all the varieties of shell-less cephalopods and is found intact in that amazing substance, ambergris, which is secreted by the sperm whale.

The parrot-like beak of the Octopus.

Perhaps the most remarkable thing about all these molluscs, Octopods and Decapods alike, is the way in which all the fish of the sea love to devour them. In fact, they may truly be said to form the base of the great pyramid of the higher life of the sea; for while they do undoubtedly prey upon all that lives in the ocean, there is no fish or mammal that does not love their gelatinous musky flesh. They are oviparous, and their rate of production is immense, as it need be, considering the demand made upon them as a food supply for the other sea-creatures. But so terrible are they when they have grown to any size, that few of their neighbours care to attack them then. Occasionally a monster conger-eel, that python of the sea, gets enwrapped in deadly conflict with an Octopus, and usually comes off victor, his slimy body offering but a poor holding surface for the clutching discs, and his razor-edged teeth cutting through the gristly tissues with ease.

Another characteristic, which the Octopus shares with the squid, is the power to secrete sepia in such quantities as to enable the creature at will to conceal itself by darkening all the water adjacent, creating a murkiness through which its powerful eyes are able to discern either enemy or prey. It is the possession of this well-known substance which makes the creature commercially so valuable, for its indelibility and unfading quality is such that even the sepia from fossil Octopodia is as good as that from the recent mollusc. The creature's habit, too, of varying its tint like the chameleon, either for disguise or in irritation, is not the least marvellous feature of this extraordinary denizen of the deep sea.

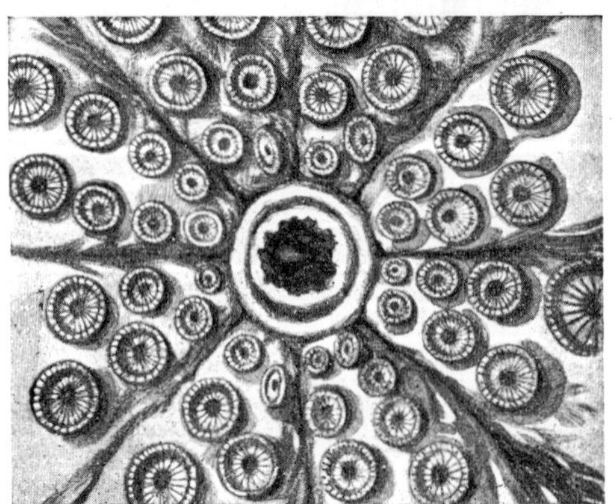

The mouth of the Octopus surrounded by the web and bases of the arms, with some of the suckers.

In conclusion, I would say to any reader who imagines that I have overstated the case as to the

[By: Theo. Carreras.

THE OCTOPUS OR DEVIL-FISH.

Different aspects of the Octopus are given in this picture. The figure in the foreground show the funnel through which water is ejected to force the Octopus in a backward direction as shown in the figure in the right-hand top corner.

hideousness of the Octopus, go to an aquarium and gaze upon the writhing, dirty-grey group of Octopi there for a few minutes. I shall then be honourably acquitted, I know.

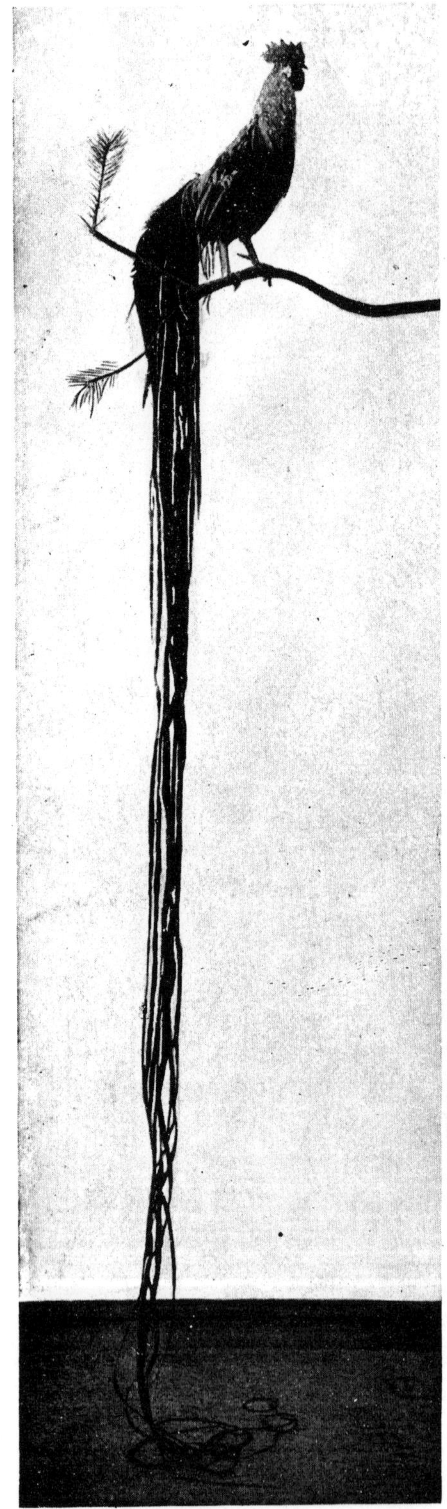

THE YOKOHAMA FOWL.

It has been proved that these beautiful birds were kept in Japan as far back as 600 B.C.

LONG-TAILED FOWLS OF YOKOHAMA

BY H. SEPPINGS WRIGHT

THE long-tailed cock, technically known as "The Yokohama," is a special product of Japan, where it was kept as far back as 600 B.C. It presents a pleasing and unique contrast to the general run of fowls. The fact that it is of Japanese origin accounts for the proud and beautiful bird we have in the Yokohama, for, no matter what section of art the Japanese touches, there his artistic soul finds full expression, whether it be in the much-admired masterpieces carved in ivory, or quaintly drawn in water-colour. Many interesting stories are told of the history of this remarkable bird.

The "Antient Almanack," a chronicle not unlike our "Domesday Booke," tells us:

"That once upon a time the Province of Tosa" (where the Yokohamas had always been extensively bred) "selected a feather for its crest, and required all retainers to attach a feather to their spears as a pennon, or emblem, on grand occasions and in state processions. Competition sprang up amongst the Samurai, or retainers, each individual striving to obtain the longest feather."

The feathers from the tails of the Yokohamas were chosen; in consequence there arose a great demand for the longest plumes. Breeders therefore bred for the greater length of tail. The price demanded for such feathers was enormous, and finally birds were raised with tail feathers measuring as much as eighteen or twenty feet in length, showing what could be done by scientific selection and special cultivation. Of course, the modern breeder has neither the same facilities, nor can he devote so much time to the art; consequently he is well satisfied with a tail of twelve or thirteen feet, such as the subjects of our illustrations.

The birds are always kept in high, narrow cages, lit from the top. If the bottom were lighted the bird would probably descend, and then, alas! for the beautiful tail! The bird remains on its perch all day, and to protect the long tail from dirt or damage two other perches are placed further back in the cage, over which the tail falls. In public gardens possessing an aviary, this bird is always a great attraction. About every other day he is taken out of his cage for a walk—an attendant, holding the precious tail up and well clear of the ground, following his wanderings. As a great treat the bird is washed in warm water about once a month and carefully dried by being exposed to the sun in some elevated spot—the ridge of a house for preference—so that the long tail may rest on the slope of the roof.

Rice is the chief article of diet, varied with occasional green vegetables. Plenty of water must be given him to drink.

When travelling, a special long box is made, fitted with a separate compartment for the tail; into this he is neatly packed, the tail rolled up in the second division. Once the tail is damaged or the feathers pulled out, it never reaches anything like its former magnificence.

The bird stands and looks much like a gamecock, somewhat heavier in make. The chief points are: small head and neck; ear lobes, oval or almond-shaped; small comb, single but small and even; beak strong and well curved; wattles must be small and neat; very bright eyes, full of life; long neck, well curved; long flowing hackle feathers; body nicely rounded and tapering towards the tail, which can be any length, with abundance of side hangers; in general appearance it is like a pheasant. Its colour is that of a game-fowl, but all the black feathers must have the beetle-green sheen. The white variety is very beautiful, with its snow-white plumage, bright-red face and eyes, and beak and legs of golden yellow.

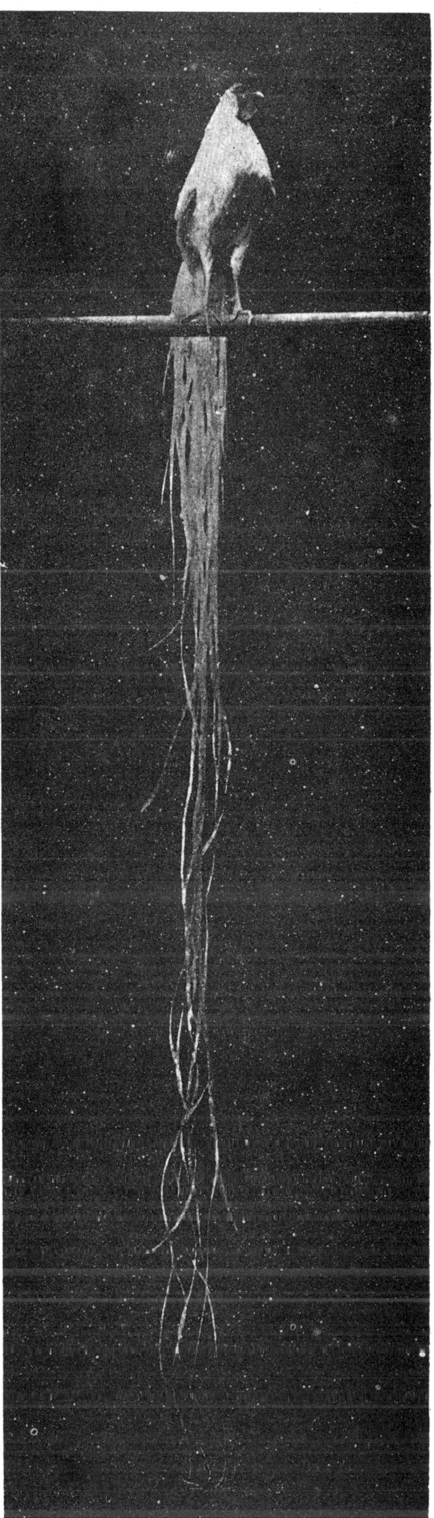

A WHITE LONG-TAILED FOWL.

The white birds are extremely beautiful, with their snow-white plumage, bright red face and eyes, and beak and legs of golden yellow.

COAL AND ITS STORY

BY E. A. MARTIN, F.G.S.

THE growth of the use of coal has been distinctly phenomenal. Probably known to the few for centuries, the knowledge of its existence was slow in travelling from place to place, and the primitive folk who lived in close proximity to its outcrops at the sides of the hills and mountains, found quite sufficient for their purpose, without any digging or other labour necessary for its extraction. Even the early Briton who waged war on Nature with flint implements found that he could warm or dry himself with a fire of coals, for he has left his axes imbedded in coal at Stanley, in Derbyshire.

But to-day the yearly output of coal in this country amounts to the inconceivable total of two hundred and sixty-three millions of tons, and the growth of output can be at once seen by a comparison with that of 1820, when it was estimated that only twenty millions in all were raised. How long will our coal last? This is a frequently-asked question. The Royal Commissioners reported in 1905 that at depths of less than four thousand feet there were one hundred thousand millions of tons of coal available in the United Kingdom, which means that at the present rate of output there will be enough coal at home to last for about four and a half centuries more. On the other hand, we must remember that there is untold wealth of coal abroad, and it is not likely that the world's supply will ever be exhausted.

A coal-seam is really a most insignificant part of the coal-measures in which it is found; for instance, a South Wales coal-field gives us twelve thousand feet of vertical thickness, and of this total, all the coal-seams put together would give us but one hundred and twenty feet of coal. Great masses of shale and sandstone intervene between the seams, and the cost of coal-getting is brought about to a great extent by the difficulties and dangers which they entail. A seam of coal less than a foot in thickness is, as a rule, not worth the working, but many of the best seams are only a few feet thick, and it will be seen how small a place they play in a series of measures such as we have instanced.

But why are the seams so thin? The mystery of their origin is wrapped up in this question. Coal is the compressed product of many years' growth of forests that were in existence in ages that have long passed away. If the whole of the growth of the thickest forest which is now living were suddenly to be overwhelmed and precipitated as coal, we might get as a result a thin seam of coal, but it would be unworkable, for it would be but an inch or two in thickness at the most. A single workable seam of coal represents to us the steady annual sheddings of trees extending over a space of time only to be reckoned by centuries, and these becoming buried deep down in the crust of the earth, the great pressure and heat brought to bear upon the deposits have combined to bring about a substance which now takes the form of coal. Dawson stated that he felt safe in asserting that every foot of coal implied the growth and fall of at least fifty generations of sigillaria trees, and that this necessitated many centuries of undisturbed forest conditions. We are thus able to form a slight estimate of the enormous mass of crushed vegetation which has gone to form our familiar coal. It is possible, too, to form an idea of the amount of time necessary to enable those changes to take place which have brought about the finished product. Suppose we stir up the black mud which is found at the bottom of our ditches and stagnant pools and canals, we shall find there all that is necessary

[By J. Teklenburg.

FOSSIL-TREE OF THE COAL PERIOD.

Found in position in the coal beds as it grew.

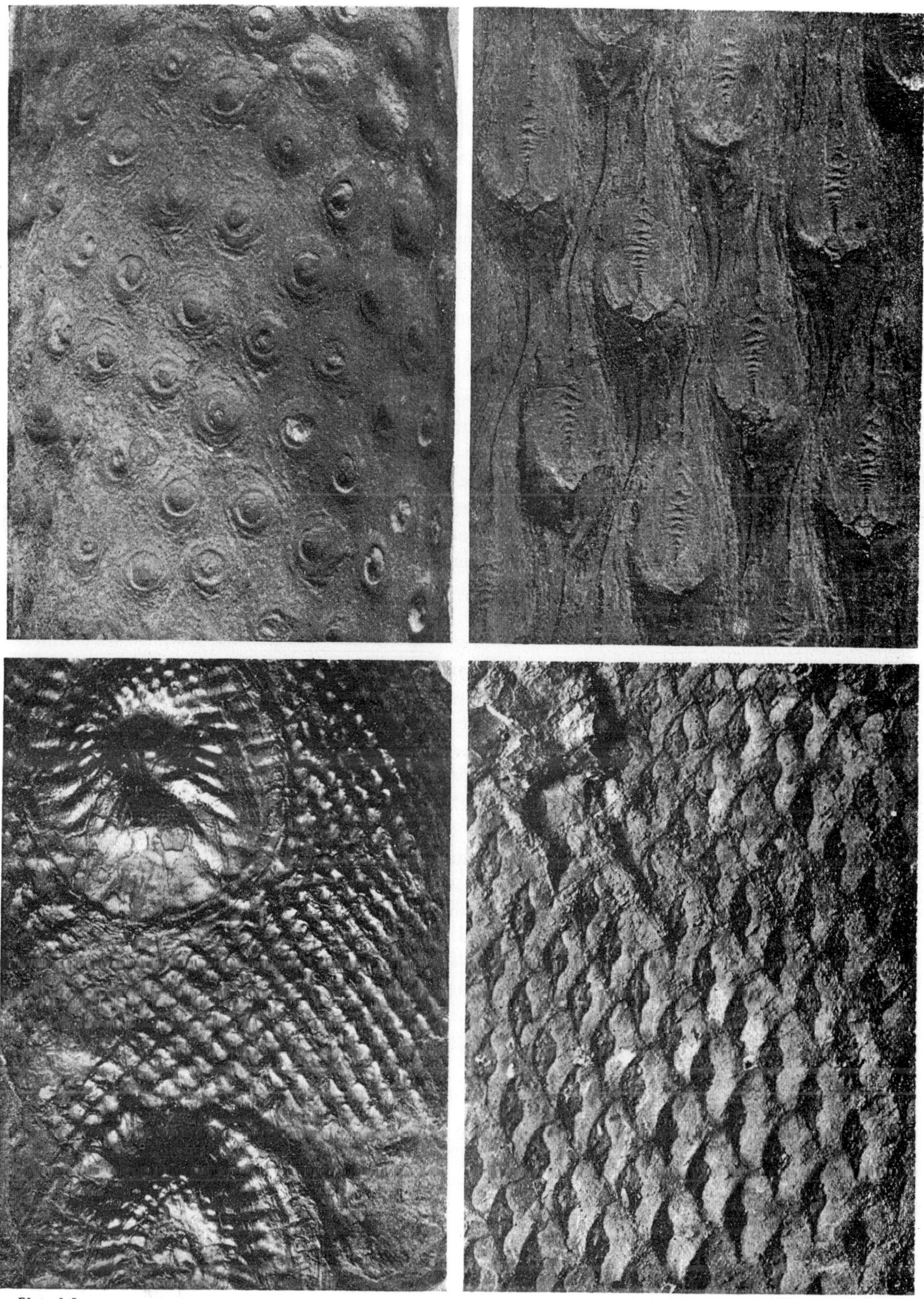

Photos by] [*E. A. Newell Arber.*

VARIOUS FORMS OF GIANT CLUB-MOSSES.

The top left-hand example shows the surface of an underground stem from which the rootlets extended at right angles. The others show the varied shapes of the scars made on the tree-trunks by the attachments of the leaf-stalks, and these are particularly beautiful and graceful in some species. The two large scars in the lower left-hand picture probably gave rise to cones.

FOSSIL-FERN FOUND IN COAL.

This fern is very common in coal, and its fossilized parts sometimes stand out very clear and distinct, as a film of black carbon.

ANOTHER COAL-FERN.

The "wedge-fern" has lobed leaflets, and the thick stem is particularly well preserved. This specimen was from the sandstone roof of a coal-seam.

to form a future seam of coal, provided the necessary physical conditions followed. But we can see no evidence of such conditions as being likely to follow. The country would have to be let down to the bottom of the sea; deposits of sand and clay would have to be laid down upon it; long periods of heat and pressure, and consequent chemical action, would also have to follow before our black mud would become coal.

Making allowance for differences in the forms of vegetation, a peat-bog forms a very good example of a bed ready to be changed into coal. But it would not become coal so long as it was exposed to the action of the atmosphere. If the peat deposits of Ireland were to disappear beneath the sea, and be acted upon by all the agencies which accompany marine sedimentation, there might be, in ages to come, rich deposits of coal.

The vegetation of the modern peat-bog, which might in future ages become coal, is very different from that of the old coal-forest growth. We look in vain in coal for our familiar trees or for the wild flowers of the country-side. Flowers with pretty petals there were none, as the insects, such as they were, had not learned to be attracted by highly-coloured appendages. The nearest approach to our forest trees which had then appeared were some cone-bearing trees. Most of the forest trees of those far-away times have, in succeeding ages, dwindled down to the horse-tails of our marshes, to the little creeping club-mosses of our mountains; these, with many fern-like structures, form the bulk of the coal-forming plants.

The modern club-moss closely resembles the giant coal-plant except in size, for the fossil genus in many cases reached one hundred feet in height. When the fruiting has taken place, the spore-cases open, and the microscopic spores are scattered far and wide, to provide for future generations.

Similarly, in those ages that are long past, the giant club-moss, with the enormous area which it covered, must have scattered its spores literally in showers, and we know that certain coals of the best kind are made up almost entirely of these spores and their spore-cases, which are saturated with resin and make a good blazing coal. Again, when we pass to consider the early horse-tails, we find instead of a growth of but a few feet high, fossil calamites commonly reaching to sixty or even a hundred feet in height.

The remarkable point about them is, as in the one previously mentioned, that their method of fruiting remains, in the modern plant, with minor differences, almost exactly the same as that which obtained in the times which we are considering, perhaps some two hundred millions of years ago. These plants cannot be satisfactorily examined in the coal itself, but in the shales and sandstones they have undergone but little change beyond carbonization, and their methods of fruiting have been microscopically examined, and wonderful results have been obtained. Much remains yet to be discovered, and palæobotany offers yet many prizes to the searcher.

The fern-like structures of the coal are many, and have received many fanciful names, generally based on some outwardly noticeable form. One, illustrated on the top of this page, must have been as plentiful in the coal-forests as is the common bracken of our own times. The coal-ferns, as we may call them for convenience sake, present many beautiful forms, and any waste-heap from a coal-mine will yield specimens.

To become coal, vegetation must pass through a number of stages, all dependent on the extent to which the gaseous constituents of vegetation have been allowed to pass away. The result is that coals vary according to the extent of the purity of the carbon which is left behind. Peat in this respect is but little removed from ordinary vegetation, its carbon being estimated at between fifty and sixty per cent. Lignite, or brown coal, has more carbon, or about sixty-seven per cent. The various coals have from eighty to ninety per cent of carbon, but when we come to anthracite we find that so much of its volatile products have disappeared that ninety-three to ninety-seven per cent. of carbon is found. Graphite has some ninety-five per cent. of carbon, and if we but knew how to do it at reasonable cost, we might turn our coal-scuttle into a receptacle for diamonds, for the diamond is itself almost pure carbon. But then we should value it the less.

Photo by] [*E. A. Newell Arber.*

FOSSIL-FERN OF THE COAL-BEDS.

A graceful form of which numerous species are found faithfully reproduced in the shale, which can be observed to be flaking away.

Photo by] [*E. A. Newell Arber.*

MICROSCOPIC SECTION OF A COAL-PLANT.

The section was made across the stem of a fern-like flowerless plant, which had no pith, the central tissue being woody.

SOME WONDERFUL BIRDS' NESTS

BY FRANK FINN, B.A., F.Z.S.

THE most remarkable nests built by birds are to be found in the tropical regions of the earth ; not only is there a greater variety of bird-types inhabiting those zones, involving greater variety of habits, but also the infinitely larger number of animal types generally implies the presence of more varied enemies of bird-life, thus stimulating the inventive powers of the birds, if the expression may be allowed. Thus we find arboreal beasts and reptiles much more in evidence in hot countries where the vegetation is so luxuriant, and these are the chief foes with which bird-life has to reckon. Hence

[By H. Grönvold.

A FLAMINGO ON ITS NEST.

Formerly these nests were thought to be sugar-loaf shaped, and so tall that the sitting bird dangled her long legs at the sides as she sat on top. It is now known that the mound-like nest, though very large, is too low for this, and that, as shown, the bird doubles her legs under her.

there is in such regions a wonderful development of that hanging type of nest which gives security by rendering it unsafe for foes to venture on an attack. One of the most curious, perhaps, is the long-celebrated structure of the Tailor-bird of India, one of the commonest small birds of the East. The bird is very like a wren, but olive-green instead of brown in colour. It selects a low shrub or plant with large leaves, and, according to the size of the leaf, either sews the two edges of one leaf together, or attains the same object, that of making a cup, by sewing two or more leaves together, if the leaves are smaller. The thread is usually the cocoon-silk of caterpillars, but the bird will gladly pilfer any ends of thread it may find lying about ; the needle is, of course, its slender bill, and though it does not, as has been asserted, make a knot in the end of the threads, it nevertheless fixes them securely by the resistance of the material. Inside the cup the nest is built, of

[By H. Grönvold.

THE NESTS OF THE SOCIAL WEAVER-BIRD.

The Social Weaver-birds of South Africa build their nests in an Acacia-tree, and unite to construct a common roof of Bushman's-grass. This is enlarged from time to time to accommodate new nests, the old ones not being used a second season

vegetable down, and is given additional security by parts of it being pushed through holes made in the leaves and fastened off like the stitches. Sometimes an over-pierced or unhealthy leaf will wither and break away from the twig but still remain attached to the nest.

As the brood of the tailor-bird is small in number and the materials of the tiny nest light, the strain on the frail supports is not so much as might be imagined, and the bird generally succeeds in rearing its brood without mishap. These nests are very difficult to find, and though the birds nested in the grounds of the Indian Museum, where I lived, I never could discover one *in situ.*

Many curious nests belong to the felted type, of which the beautiful structure of our chaffinch is an example ; but, of course, much more skill is required to make a nest of this kind when it has to

[By H. Grönvold.

THE EDIBLE NESTS OF A SWIFT.

These edible nests of the Swift are chiefly found in Indo-Malaya, and are constructed almost entirely of the birds' saliva. The Chinese use them as material from which they make a glutinous soup, this dish being esteemed a great luxury in the Far East.

be suspended than when it is a mere supported cup. A very celebrated felted nest is that of the Cape Tit, which was well figured by the French traveller, Levaillant, over a century ago; but he mistakenly attributed it to one of the grass-warblers. The real architect is a tiny, insignificant-looking little bird, and is commonly known to the Boers as the Cotton-bird. It does, indeed, make its nest naturally of plant-down, but since the introduction of sheep the little creature has been quite alert enough to take advantage of the better wool which they provide, just as the tailor-bird takes to our sewing-thread. The nest, which is supported by several twigs of a bush, round which it is built, is entirely made of this wool, vegetable or animal, and seems to be composed of a sort of cloth, so well is it felted together and interwoven, in several separate layers. It is domed over and rainproof, and the entrance is a short tube near the top, below which is a sort of watch-pocket—the cock bird's bed-room, according to native report. The late Dr. Stark observed that when the hen left the nest she sometimes shuts the door, as it were, by pinching together the edges of the entrance-spout, and he thought that this was done to regulate the temperature as well as to baffle enemies; that the latter result was fairly probable is evident from the fact that on one occasion the little bird herself had some difficulty in getting in when she came back after fastening up the spout-hole.

The ordinary nests built by the beautiful little Sunbirds of the Eastern tropics are of a felted type, consisting of scraps of fibre, dead leaves, and suchlike dry materials bound together with the webs of caterpillars and spiders; they hang from the end of a twig, and are oval in shape, with an entrance-hole in the side, which is overhung by a porch, no doubt intended to screen the head of the sitting bird, if she looks out, from the view of an enemy above. This is the more necessary as there is no attempt made to conceal the nest, which, after all, so often looks like a mere lump of casual rubbish that it might easily pass undetected for what it really is. The nest of the species shown, which comes from the Philippine Islands, is composed almost entirely of fibre, and so has a less "casual" appearance than those of the Indian species I have seen.

Very beautiful little pensile felted nests are those built by the Flowerpeckers of Eastern Asia; they are on a similar plan to those of the sunbirds, but have not the eave over the entrance, and possess a broader base of attachment. They are also much neater, the material being of more uniform character, and the foundation of plant-down and cobweb so well compacted that they can be folded up like little purses. Being very tiny, and placed at the tops of trees, they are not

[B H. Grönvold.

A NEST WITH A FALSE OPENING.

This nest of a Weaver-bird shows a long vertical tunnel leading to the nest proper. The false opening ha'f-way up the structure is intended to mislead any possible enemies.

likely to attract much notice ; the extreme fearlessness in captivity of the little builders shows that these birds have been little used to molestation by large animals, whatever their foes may be.

The woven nests of the Weaver-finches of Africa and South-eastern Asia may really be regarded as the most wonderful of all, inasmuch as real skill is needed to construct them, weaving being a much more complicated process than mere felting. A weaver's nest of the most perfect type, as exemplified by that of the Baya weaver, one of the few Asiatic species, consists of a long suspension-cord attached to a twig or palm-frond, and holding up a bulb expanded on one side into a nesting-chamber and prolonged downwards on the other into a long entrance-spout, up which the bird darts. The young and eggs are kept from tumbling down the spout by a strong partition which separates the nest-cup from the spout, and, being built before either, serves as a perch for the bird working inside the dome.

[*By H. Grönvold.*

AN OVEN-BIRD AND ITS NEST.

The Oven-bird of South America builds a quaint nest of clay on the horizontal bough of a tree, with a long narrow opening in one side. The Brazilians call the bird "Johnny Clay."

[*By H. Grönvold.*

THE AUSTRALIAN MOUND-BIRD.

The Australian Mound-bird makes a nest of dead leaves and grass, on which its eggs are laid. They are then covered up to a depth of two feet, and the fermentation of the vegetable matter hatches them. The mounds sometimes measure forty-five feet in circumference.

The most celebrated of all weavers, the Sociable Weaver of South Africa, collects in flocks of dozens of pairs to make a massed nest-foundation of grass in the branches of a tree, in the under-side of which the individual pairs each construct their separate nesting-hollow. The bird is plain and sparrow-like, whereas in most weavers the males, at all events, show much brilliant yellow.

Nests may be not only felted and woven together, they may be gummed by the saliva of the birds ; and this is sufficiently remarkable, as the saliva in birds generally is not conspicuous by its abundance. It is, however, particularly copious and sticky in the swifts, whose rude nests are usually gummed together, some tropical species constructing really elaborate nests in this way. The most remarkable is the strange nest built by a Central American swift, which is on the principle of the best weaver-bird's nests, with long entrance-spout and covered nesting-chamber at the top. It is, however, hung from a rock instead of a twig, and formed entirely of downy plant-seeds gummed together by the bird's saliva. There is a sort of false entrance half-way up the spout in the well-known British Museum specimen.

The little Eastern swifts of the *Collocalia* group build in caves, gumming their small half-saucer-shaped nests of moss, etc., mixed with saliva, against the wall It is in this group that the salivating habit is carried to its logical conclusion, for one species makes the nest entirely of dried saliva; and it is this nest that enjoys the unique distinction of being the "edible bird's-nest" so well known as a Chinese delicacy. It fetches a very high price, and the coast caves in the East Indies where these birds breed are regularly farmed out in consequence; but, as might be expected, there is very little flavour in the isinglass-like product of the little swifts.

The name "edible swallow" formerly given to this swift, reminds one of the true swallows, so well known as mud-builders even in our own country; but mud-masonry is practised in many countries by other birds than the swallow, such as the Oven-bird, a soberly-coloured, light brown bird of South America. The bulky nest of the oven-bird is placed on some solid support, such as a branch or post, without any attempt at concealment; as a matter of fact, the nest is a difficult one for a robber to negotiate, as the entrance at the side of the domed clay structure, which gets burned nearly into brick by the sun, does not communicate direct with the interior, but with a porch partly shut off by a firm clay partition from the inner room where the eggs are laid.

In the Zoological Gardens, too, has more than once been built a nest which is incubator as well, that of the Australian Brush-turkey. This bird, which belongs to a family distinct from the true turkey, scratches up a mound of any sort of vegetable rubbish it comes across, seizing the material in footfuls and throwing it back until a heap of several yards in circumference and several feet high has been formed. All this work is done by the male, and when the hen or hens come to the mound to scratch holes in it and deposit their eggs therein, he fills in the holes as soon as they have finished laying. Thereafter, he attends to the mound, in which the eggs are being hatched by the heat engendered by the decaying vegetation, often raking it over, apparently to keep the surface loose against the emergence of the chicks, which after six weeks' hotbed-incubation emerge from the shell full-fledged, and begin a perfectly independent existence.

[*By H. Grönvold.*

THE LITTLE HERMIT'S NEST.

The nest of this Humming-bird is made of silky plant-fibres, bound together with spider's web and attached to the end of a leaf.

[*By H. Grönvold.*

NEST OF THE LANCEOLATE HONEY-EATER.

This bird constructs a deep basket of grass and wool, and hangs it from a slender twig.

Photo by] [*W. Hope Hodgson.*

A WATERSPOUT.

In parts of the tropical seas this remarkable phenomenon of Nature may be witnessed. A dark column descends from a cloud, and makes connection with the sea. Up the column rushes many thousands of tons of water, and goes sailing away as a cloud to carry moisture to thirsty lands.

WATERSPOUTS

BY FRANK T. BULLEN, F.R.G.S.

GIGANTIC and never-ceasing as are all the ocean's activities for the well-being of the terrestrial life of the globe, there are none of them that are more obvious and direct, to even the meanest savage capacity, than the provision of fresh water, without which we all perish. A vast quantity of the amount needed is daily drawn into the air by evaporation, a steady unspectacular process which goes on continually wherever the sun is shining. But in order to realize how vast and rapid, as well as astounding, are Nature's processes, it is necessary that we should witness the birth, life, and death of a waterspout, a history which is comprised within a few minutes. Within the debatable area of the tropical seas between the limits of the north-east and south-east trade winds is the most favoured place to witness this stupendous alchemy of Nature. The wind has died into fitful airs from all parts of the compass, the sky is overcast with massive clouds and the atmosphere is tense with electrical energy. Suddenly there will appear in the sea, without any apparent proximate cause, a tiny whirlpool, if one may so call a rapid circular movement of the sea, whose centre *rises* instead of being depressed.

On looking upward, immediately above this movement of the waters will be seen a corresponding excitement in a cloud, from which will presently descend a dark tube, much like the trunk of some mighty elephant, swaying about as it lengthens, until at last it joins the mound of water beneath and establishes a connection between sea and sky. Now, in obedience to who knows what centrifugal impulse, the water of the sea rushes upwards with a roar which may be heard five hundred yards away. The cloud above grows black, menacing, monstrous. In a few minutes some mysterious operator has signalled that the one hundred thousand tons or so of water received above is sufficient for the cloud to bear, and immediately the connecting tube begins to dwindle, until soon it ceases to exist and its attenuated length is absorbed into the black bosom above.

Away sails the vast reservoir to thirsty lands far off, often letting fall at the breaking of the connection between cloud and sea a mass of water which is perfectly *fresh*.

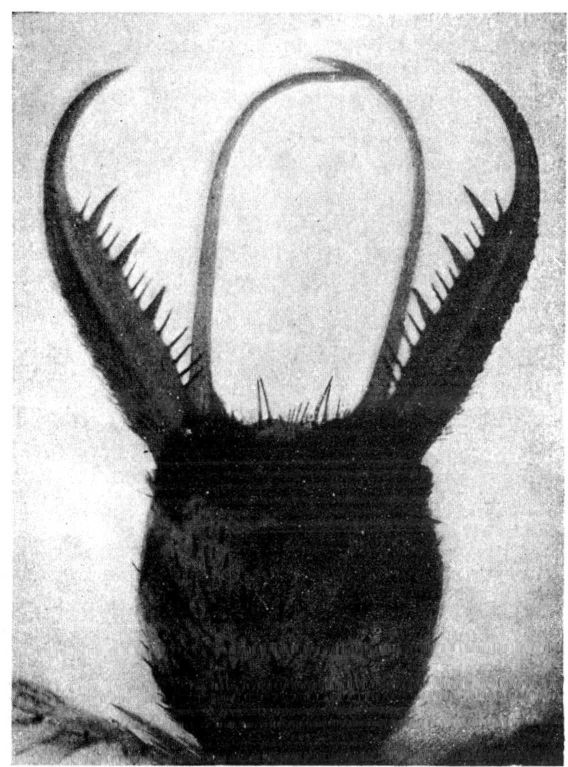

Photo by] [*J. J. Ward.*

THE ANT LION'S JAWS.

The head of the Ant-Lion as seen under the microscope. There is no true mouth, but the jaws are perforated, and when the points are fixed in its prey its juices are sucked through them.

THE ANT-LION AND ITS TRAP

BY JOHN J. WARD, F.E.S.

THE perfect Ant-Lion might easily be mistaken for a dragon-fly. Its gauzy, brown-blotched wings and similar-shaped body give it a considerable resemblance ; the short, clubbed antennæ, however, readily distinguish it. The fly itself is of little interest, but in its larval stage it is one of the most

Photo by] [*H. Main, F.E.S.*

THE ANT-LION. UPPER SIDE.

A photograph taken under the microscope, enlarging the creature to about eight times the natural size.

remarkable insects known, and has been a source of interest to entomologists for the past two centuries. It commences life from an egg deposited on the sand by the parent insect during her evening flight. When the larva is hatched, it immediately begins to construct a little pit in the sand, at the bottom of which it hides. Later on, we are better able to understand the true nature of this hiding-place.

The full-grown larva is a somewhat weird animal. It is about half an inch in length, with a body broadly oval in shape, and beneath it are six diminutive legs ill adapted for walking purposes ; indeed, it can only shuffle along, and, strange to relate, its shuffle is always in a backward direction. Its head is very conspicuous on account of its formidable jaws, or mandibles, these organs serving to capture and hold its prey, also to suck its juices ; for it is devoid of a mouth, properly speaking. It has six eyes on each side of its head. The prey that this curious grub seeks is small active creatures, such as ants, spiders, centipedes, wood-lice, etc. ; it demands living quarry, and lays out its plans accordingly for their capture.

Photo by] [*H. Main, F.E.S.*

THE ANT-LION. UNDER SIDE.

The legs are weak and ineffective for running purposes ; but the Ant-Lion mostly shuffles backwards.

Taking all things into consideration, the grub has many natural disadvantages to contend with. Ants, which constitute its chief prey, are extremely active, and, in open chase, the ant-lion's success is hopeless. Then, the grim aspect of its huge mandibles does not inspire confidence even in inquisitive ants, but the wily grub overcomes all its natural shortcomings. It is the larva that is properly called "ant-lion"; for the insect in its winged state does not prey upon ants.

The methods adopted by the grub are very remarkable. Selecting a dry and sandy situation beneath a tree, or sheltered by a wall, it proceeds to trace a boundary line to its proposed trap by depressing the end of its abdomen into the sand and walking backwards in a circle. The circle having been completed, with a diameter of from one to two inches, the larva then buries its

Photo by] [*Hugh Main, F.E.S.*

THE WINGED CONDITION OF THE ANT-LION.

When the Ant-Lion has passed its allotted time beneath the sand, and the intervening chrysalis stage, it emerges with much the appearance of a dragon-fly, with long, slender body and expansive gauze-like wings.

abdomen in the sand, in this manner obtaining greater power for its next movement.

Then by means of its fore-leg nearest the centre of the circle, it loads its broad, flat head with sand, which it suddenly hurls outside the circle in an astonishing manner. So it pushes backwards round and round inside the ring, continually throwing out the sand as it travels. When a circle is completed, it changes its position and excavates the next furrow in the opposite direction, thus working its legs on each side alternately, but always loading its head with sand by means of the inside fore-leg.

[By Theo. Carreras.

THE ANT-LION'S TRAP.

In this illustration the Ant-Lion's pitfall is shown in section, with the insect buried in the sand at the bottom.

Small stones are frequently a source of trouble during the excavation, and if one is too large to throw out the larva gets beneath it, lifting it on its back, and adroitly balancing it while it climbs the side of the pit and until it can deposit it on the outside. In due course, after many decreasing circles have been excavated, a conical pit is formed, and at the bottom of this the energetic grub proceeds to bury itself all but the end of its mandibles, which are allowed to project.

In this manner it patiently awaits the approach of an unsuspecting or inquisitive ant. The larva can endure long fasts, but sooner or later its patience is rewarded. An ant, while seeking out new feeding-grounds, meets with this smooth pit and its strange central object, and never having seen anything like it before (for an ant usually pays but one visit to the ant-lion's pit), its curiosity is aroused—it may mean food ! At all events, the naturally inquisitive ant desires to know, and so it proceeds warily to investigate.

It looks down into the hollow of the pit for some considerable time, as if expecting something to move. Apparently the curious central object has a fascination for it, and after a few minutes the ant seems determined to gratify its curiosity, and it proceeds to test the smooth and steep banks by means of its fore-legs.

The moment the ant steps over the edge of the pit the treacherous sand gives way, and the startled ant stumbles while endeavouring to regain its foothold, dislodging still more sand, with the result that it usually finds itself precipitated to the bottom of the pit, perhaps right into the hungry jaws of the ant-lion grub. Sometimes, however, it happens that the ant has so cautiously alighted on the sides of the pit that it detects the danger of the unsafe ground on

[By Theo. Carreras.

THE ANT-LION'S PITFALL.

The Ant-Lion excavates a funnel-shaped pit in the loose sand. Creeping insects looking over the brink lose their foothold and slip to the bottom, where the jaws of the concealed Ant-Lion are waiting to receive them.

which it is walking, and decides at once to return. Now, although it may with caution travel downwards over the sand spread so smoothly by the larva, yet to turn round, or climb upwards, is a very different matter, for the sand immediately begins to roll down the banks.

The moment the sand commences to fall, the ant-lion is alert, quickly hurling showers of sand up the banks, and before the ant realizes what has happened, it finds itself slipping down the sides of the pit. It is at once seized by the formidable jaws of the grub, from which there is no escape, for these organs act like suction-pumps, absorbing the juices of the prey while they hold it.

A large spider, or a beetle, when entrapped, proves a more troublesome captive, and if, before it is seized, its struggles bring down sufficient sand to allow it to climb out of the pit, it may escape; it is, however, a lucky insect that survives a visit to the ant-lion's den.

When the meal is finished, the ant-lion pitches the shrunken skin of its victim well beyond the borders of its excavation, probably to prevent future victims from taking alarm. It then proceeds to repair the damage to its retreat, and again goes into ambush.

The common species here described is found both in Southern France and in Switzerland, but does not occur in Britain. There are some three hundred species of ant-lions known to science, of which America claims at least thirty; so that they constitute a well-distributed group of insects. It should be mentioned, however, that all the species do not dig pitfalls, and, also, that many of them progress in the usual direction when walking. Probably when more is known of these insects the backward movement will be found to be associated with the pit-making habit. Those species which construct no pits have to capture their prey by hiding in suitable spots until a victim approaches.

The duration of the life of the larva depends largely on the food supply, and it may spend two years as a hunter in the grub state. It then spins a circular cocoon in the sand, in which it changes to a pupa, or chrysalis. In due course it bites a hole in the cocoon and emerges above ground as a winged insect, much larger than the original grub, and, as previously observed, resembles a dragon-fly, from which it can be readily distinguished by means of its clubbed antennæ.

Photo by] [*E. Step, F.L.S.*

THE VEGETABLE CATERPILLAR.

The caterpillar of a New Zealand moth falls a victim to a fungus which kills and feeds upon it, then shoots out a long stalk upon which its spores are produced.

VEGETABLE INSECTS

BY EDWARD STEP, F.L.S.

UNDER the name of the Vegetable Caterpillar, one of the products of New Zealand has long been regarded as among the marvels of natural history; but when its real nature is understood, it is seen to have rather less claim to be so considered than many everyday phenomena that escape attention. Of course, the imagination is fired by the thought of a large caterpillar walking about with a plant three times its own length growing out from behind its head, like the hart that Baron Munchausen shot with cherry-stones and met years later with a cherry-tree growing from its head. As a matter of fact, the caterpillar does not walk about with this encumbrance, for by the time the plant has developed the caterpillar is dead.

Everybody knows the rather large pale moth, called Ghost-moth, that flits about our gardens and fields at dusk, and whose caterpillar spends its days underground feeding on the roots of grass and other plants. There is in New Zealand an allied moth, whose caterpillar falls a prey to a species of fungus known as Cordiceps. The probability is that it eats the microscopic spore of the fungus whilst taking its usual food and the spore germinates inside it. Its growth fills up the whole of the caterpillar, upon whose organs it draws for nourishment, and whilst retaining the caterpillar form, converts a soft, juicy creature into a hard, wooden corpse. Then a shoot emerges just behind the head and rises into the air, the shoot attaining a length of six inches, of which the upper third consists of a thickened mass of spores which fly off into the air to destroy other caterpillars. It is

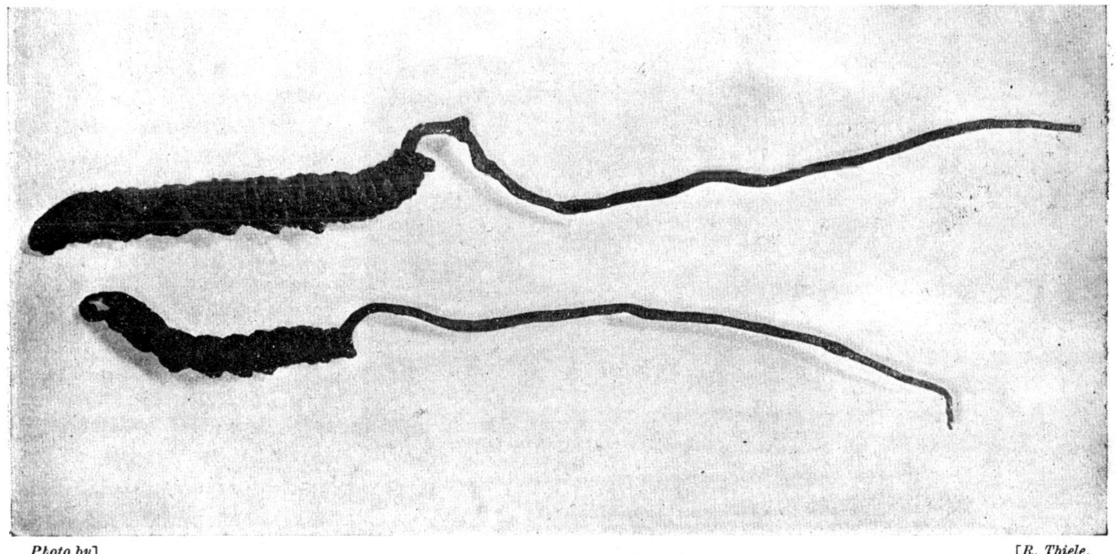

Photo by] [*R. Thiele.*

THE VEGETABLE CATERPILLAR.

As shown in this photograph the insect has been misunderstood by many, who have supposed that it walks about with this encumbrance attached to it.

this New Zealand species that is shown in our photographs: that on the present page showing it as though walking, but the correct position is given on page 48, where it is seen below ground with the fungus reaching up into the air.

A similar species occurs in China, where it is known by the poetical name of "the summer grass of the winter-worm." The Chinese attribute medicinal virtues to it, and no Chinese drug-store is without a stock of it tied up in little bundles. Other species are found in different parts of the globe, three of them in Britain. The last are much smaller than the New Zealand species; but one, often found springing from underground chrysalids, makes up for its smallness by its bright red colour.

At Guadeloupe a similar fungus attacks the wasp, and the wasp is said to have been found actually living with the encumbrance attached, though evidently in the last stage of existence. Others attack beetles, crickets and cicadas. Our common house-fly falls a victim in large numbers to a fungus of quite another sort. In autumn the flies may be seen on windows or walls, apparently only resting; but when an effort is made to dislodge them they are found to be fixed to their support by a fine feltwork of silky threads, with which also the fly's body is filled.

THE LEAF-CUTTING ANT

BY EDWARD STEP, F.L.S.

In tropical America there dwells an ant about the same size as the Red Ant of British pinewoods It is known in different localities as Sauba-ant and Coushie-ant. Most naturalists who have travelled in that part of the world have told us of their admiration at the sight of long processions of these ants streaming down the trunks of trees and along definite tracts through the forest to their great nest-heaps, every ant holding in his jaws a circular piece of leaf. The ants are almost hidden by these umbrellas, and the procession looks more like a stream on whose surface the leaves are floating. It was for long a puzzle what the ants did with these leaves, and it was variously suggested that they were used for food and for lining or roofing the galleries of their nests. The facts of the case, as usual where Nature is concerned, are far more wonderful than the theories. The ant packs these leaves into its cellars, where they rot and form a soil for the cultivation of a fungus which serves as food for the ant-grubs. When the fungus has drawn all the nutriment from this leaf-mould, it is carefully cleared out of the nursery galleries into spare-chambers, where it is eaten up by the grubs of certain beetles, such as ants always encourage, or at least tolerate, in their nests. Shafts lead from the galleries where these fungi are grown to the outer air, and the temperature and humidity of the gallery is regulated by opening or stopping up the shafts as circumstances may require.

[By T. Carreras.

THE COUSHIE OR LEAF-CUTTING ANT.

A single ant is here shown on its downward passage along a branch with the portion of leaf it has just cut. The nearer insect is a membracid that in colour and shape closely resembles the leaf-laden ant.

An unprotected plant-sucking insect, known as a membracid, that occurs in the same region, has developed a green hood, which covers its back in such a way as to closely resemble a Coushie-ant with its cut leaf, as shown in the illustration above. It has been suggested that this is a case of protective mimicry, and that the ants pass the membracid under the impression that it is one of themselves, and similarly engaged.

The underground galleries of this ant are very extensive. Bates tells that they had infested the Botanic Gardens at Pará so much as to arouse the indignation of a French gardener there, who set to work to try to exterminate them, or at least to reduce their numbers. His plan was to discover the main entrances to and exits from their nests, and over these to burn sulphur, and blow the heavy fumes into them by means of bellows. In this he was so successful that the destructive vapour entering at one point was seen to be coming out of other holes far away from the burning sulphur Bates measured the distance between the entrance of the vapour and one of its exits, and found it to be no less than two hundred and ten feet. And all the intervening space was tunnelled by a net-work of the ants' galleries. Truly a prodigious piece of engineering for creatures so small.

[By Theo. Carreras.

THE COUSHIE OR LEAF-CUTTING ANT.

These ants, which are here shown of the natural size, swarm in the forests of Tropical America, building large nest-hills in which they cultivate a species of fungus for the food of their young. To make a suitable soil for the fungus they cut leaves from the forest trees and carry them home in their jaws as shown.

STARS OF THE EARTH

BY EDWARD STEP, F.L.S.

THE Earth-stars are a group of fungi allied to the puff-balls ; but whereas the puff-balls are tolerably common and, therefore, well-known, the Earth-stars are comparatively rare and consequently little-known. In their earlier condition they might easily be mistaken for little puff-balls if they happened to be seen, but in this stage they are usually buried just under the surface of the soil. Getting to their full size, they push the soil aside and come into view. Then the outer layer splits from the apex into several pointed segments which turn away from the centre and give the whole plant a star-like appearance. In the centre of the star there stands a more or less globular body which, if cut through early, will be found to be of the consistence of crumb of bread, but later it breaks up into soft cobwebby threads, and amidst these are millions of the microscopic spores by which the plant is propagated. To set these free an opening appears at the summit of the globular body, or spore sac, and imperceptible changes of temperature will so affect the walls of the sac that many of the spores will issue from time to time in the form of a tiny puff of thin smoke. Even the temperature of the hand when examining one of these Earth-stars is sufficient to cause a demonstration of this.

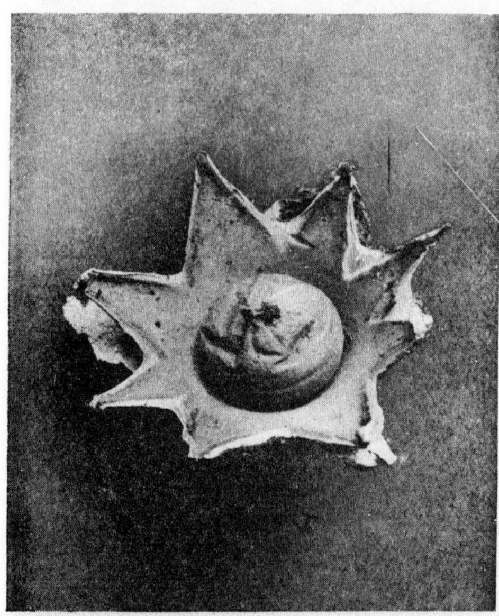

COMMON EARTH-STAR.

The outer coat splits into triangular lobes, which give the starry appearance to it when they have turned outward.

When the atmosphere is charged with moisture, the segments of the star rapidly absorb it, and by their expansion turn outwards and downwards, their tips often pressing upon the earth with such power that they force up the centre, and detach its connection with the soil. In consequence of this action the plant is often caught by the wind and sent bowling along for some distance, scattering its spores as it goes.

Photos by] [*E. Step, F.L.S.*

COMMON EARTH-STAR.

The round body in the centre of the star contains a powder like snuff, which consists of millions of microscopic spores.

There are a number of different species of Earth-stars, of which no fewer than a dozen are found from time to time in this country, but they are all very similar in character to those of which we give illustrations. One of these, however, will be seen to differ from our description in the fact that the spore-sac is smaller and is borne upon a stalk. This particular species is known as Bryant's Earth-star, and there are several others that share this possession of a stalked spore-sac. The opening of the spore-sac in Bryant's Earth-star is developed into a distinct fluted nozzle, whilst beneath the sac there is a sharp rim. Altogether the general form and graceful curves of this species are such as to fit it eminently for the purposes of the decorative artist. As already indicated, in their earlier underground condition there is little to distinguish them from Puff-balls. As a matter of fact, they were all at one time reckoned as Puff-balls, the Earth-stars being distinguished as Starry Puff-balls; but it will be seen that the outer jacket constitutes a real distinction.

Photo by] [*E. Step, F.L.S.*

BRYANT'S EARTH-STAR.

A rarer form, in which the spore-receptacle is seen to be stalked, and to have a distinct nozzle. The specimen in the background has distributed its spores and is now drying out of shape.

THE GIANT REPTILE OF WYOMING

BY R. I. POCOCK, F.R.S.

It is a common supposition, true in some cases, but not in all, that animals of past times greatly exceeded their modern representatives in size; but no one whose acquaintance with reptiles was limited to crocodiles, tortoises, snakes and lizards could dream of the extent to which some reptiles of the Mesozoic period, well termed "the Age of Reptiles," surpassed in bulk the living representatives of the class. These animals were not closely related to any existing species, except the crocodiles, but belonged to a special division—the Dinosauria; and strange as it may seem, it is none the less true that this group, containing species colossal alike in size and stupidity, is the stock whence birds took their origin.

One of the most wonderful, if not quite the largest of these huge dragons was Diplodocus, skeletons of which have been found in Wyoming in the United States. Even at the shoulder this animal was as high as the largest living elephant, while the height at the hind-quarters was not many inches short of thirteen feet. To support a body over twelve feet long and of huge bulk the limbs were massive and upright, and the broad, five-toed feet were armed with three strong, curved claws, the two outer toes of each foot being bluntly rounded. The neck was long, like that of a swan, but more robust and less flexible, and supported a comparatively small head, measuring not more than about two feet in length. The eyes were set high up, the nostrils were apparently just in front of them, as in most birds, and the jaws were armed in front with slender teeth set close together, like those of a comb. The head and neck measured nearly twice the length of the body ; but the tail, which tapered away to a slender point and must have been as flexible almost as a whip-lash, was close upon fifty feet long. It is difficult to imagine what purpose this tail can have served, unless, like that of the modern monitor lizard, it was used to slash such enemies as Diplodocus may have had amongst contemporary predacious Dinosaurs. It is supposed that this reptile was amphibious in habits, and fed upon plants it uprooted from the bottom of the water. Perhaps the claws were used for this purpose, and the formation of the teeth suggests that they may have played the part of a sieve, allowing the water to escape from the mouth in front and retaining at the same time the food within it. It may be that Diplodocus took refuge from its enemies in lakes ; and since no

RESTORATION OF THE DIPLODOCUS

This photograph is taken from one of the remarkable restorations of prehistoric animals in Carl Hagenbeck's Park, near Hamburg. It gives a good idea of the size and proportions of the animal contrasted with modern vegetation.

reptiles can breathe under water, it has been suggested that one of the uses of the extraordinarily long neck of this species was to enable it to raise its head to the surface, keeping the body deeply submerged. Even with all four feet resting on the ground, the animal could have thrust its eyes and nostrils into the air in water five or more fathoms in depth, while if it stood up like a kangaroo, supported on the hind legs and the muscular portion of the base of the tail, some six or nine feet would have been added to its stature.

It is impossible to know the causes which led to the extermination of Diplodocus and other Dinosaurs. Perhaps, however, some climatic changes dried up the swamps and lakes and destroyed the vegetation upon which they depended. With the close of the epoch which saw their disappearance, the carnivorous Dinosaurs, which probably preyed upon them, disappeared also.

[*By C. E. Swan.*

THE EXTINCT GIANT REPTILE OF WYOMING.

This huge reptile—known as Carnegie's Diplodocus—stood nearly thirteen feet high at the hind quarters, its body proper was twelve feet, the head and neck nearly twice that length, and the tail not far short of fifty feet long.

MARS AS A HOME FOR MEN

BY E. WALTER MAUNDER, F.R.A.S.

"WHEN the earth had to be prepared for the habitation of man, a veil, as it were, of intermediate being was spread between him and its darkness, in which were joined, in a subdued measure, the stability and insensibility of the earth, and the passion and perishing of mankind.

"But the heavens, also, had to be prepared for his habitation.

"Between their burning light,—their deep vacuity, and man, as between the earth's gloom of iron substance and man, a veil had to be spread of intermediate being;—which should appease the unendurable glory to the level of human feebleness, and sign the changeless motion of the heavens with the semblance of human vicissitude.

"Between the earth and man arose the leaf. Between the heaven and man came the cloud. His life being partly as the falling leaf, and partly as the flying vapour."*

This is a poetic definition of the conditions of habitability for a planet. But the scientific definition, though expressed in very different language and making use of different terms, in essence corresponds to it closely.

The falling leaf and the flying vapour are the tokens of a habitable world; habitable, that is to say, in the only sense in which we can properly use the term, by beings resembling men in their capacities of body and of mind. A world in which the bare rock stands naked to the sunlight, with only dry air between, can be no fit home for organic life in any highly-developed form.

* Ruskin, "Modern Painters."

A rough idea of the relative sizes of the planets nearest the Sun and—on, necessarily, a different scale—their relative distances from it. The comparative sizes of Mars and our satellite, the Moon, should be noticed, so greatly do they differ from the appearance of these bodies in the heavens.

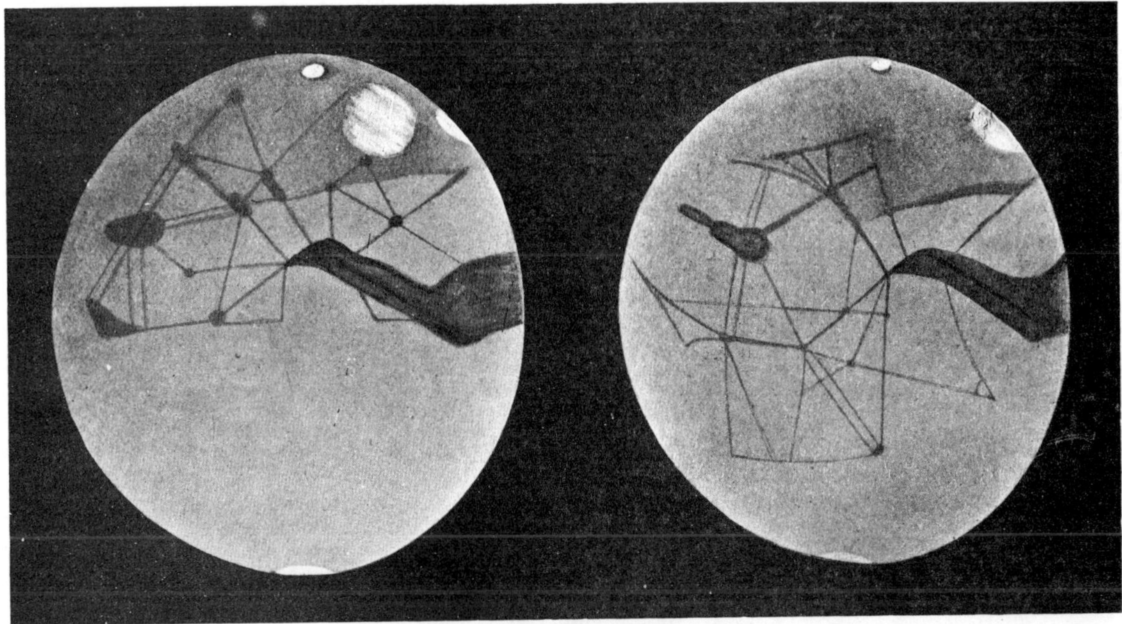

THE FIRST FROST OF THE SEASON.

Drawings of Mars made at the Lowell Observatory, Arizona, U.S.A.; that on the left by Professor Lowell, on 1910 November 18, and that on the right by Mr. E. C. Slipher, on 1910 November 22.

The essential quality of life is the power which the organic body possesses of taking dead matter into itself; of assimilating it and breaking up its structure, and using it either for its own growth or for the maintenance of its energy. The difference between the life of the plant and of the animal lies in the greater directness with which the plant can lay hold on inorganic matter; while the animal requires for its sustenance matter that has already undergone organic change.

Animal life, therefore, presupposes plant life, and plant life the presence of water, the one great neutral solvent; and of water as a liquid. Abundance of water, abundance of vegetation, are necessary if the higher, more developed forms of life are to flourish.

But water exists in the liquid form only within comparatively narrow limits of temperature. On the earth water freezes at 32°, and boils at 212°, but as the mean temperature of the earth as a whole is about 60°, and for the equatorial regions about 80°, there is a wide margin of safety on either side, and over the greater part of the earth's surface, water is normally in the liquid state. In the two frigid zones the mean temperature is 32° or lower. Here, therefore, water is normally found in the solid state as ice or snow. Life exists, indeed, in these two zones, but it exists because for half the year the temperature is above freezing-point, and because there is a continual influx of forms of life from the warmer zones.

The first question, therefore, to be asked with regard to the habitability of Mars is "What is the temperature of the planet?" We know from its position in the solar system that it receives from the sun, surface for surface, only three-sevenths of the light and heat that falls upon the earth. We can judge roughly of the effect of this by noticing that a zone on the earth slightly outside the Arctic Circle will, from the effect of its foreshortened presentation to the sun, receive for the unit of surface about three-sevenths of the light and heat falling on the Equator. But the mean temperature of such a zone will be fully 50° lower than that of the Equator. We may reasonably suppose, therefore, that the mean temperature of Mars is 50° lower than that of the earth, or, in other words, that it is about 10° instead of 60°. But a mean temperature of 10° is 22° below freezing-point; in other words, water on Mars is normally in the solid condition; it exists chiefly as frost, snow or ice.

But a world eternally covered in frost, snow or ice should sparkle under the sun's rays like a diamond, and Mars is the " ruddy " planet and has as poor a power of reflection as the Moon. Is it possible, then, that Mars is destitute of water, and is simply, like the Moon, naked rock ? No, for as the accompanying illustrations show, a white cap crowns each of the poles of Mars ; a cap so brilliant that it has been seen in the telescope to shine like a star of the first magnitude when a mist or fog has entirely hidden the body of the planet. And this cap, like snow upon the earth, increases in area as winter draws on, and diminishes with the approach of summer. Such changes are impossible to bare rock ; it is but reasonable to ascribe them to ice or snow. For as the white cap that is feeling the summer sun shrinks, so the region round it becomes dark, as if the snow had indeed been melted and become turbid water.

But the ice-caps of Mars are not relatively more extended than those of the earth, even in winter-

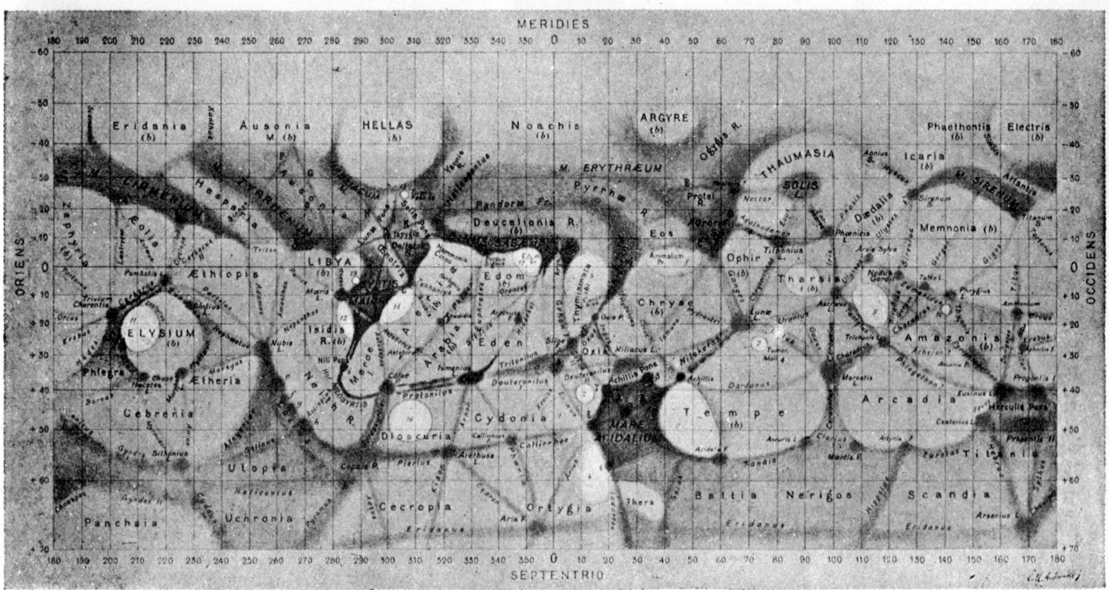

THE MOST COMPLETE CHART OF THE PLANET MARS.

This map was drawn by M. Antoniadi, and is based upon most recent discoveries. It is supposed that the dark parts of the planet formed the seas in ages gone by, as their names signify (vide, Mare Sirenum Mare Acidalum, etc.). The narrow lines connecting the darker spots are the much-discussed canals, accepted as such by some astronomers and rejected by others. The former suggest that the darkness of the lines is caused by vegetation which clusters round the darker spots, or "Luci," as they are technically called.

time, and in the summer they shrink and shrink, until in some years one of them has been known to disappear entirely. Is not this evidence that Mars is warmer, not colder, than the earth ?

This might be so, if we saw every part of the planet to the same advantage. But the Mars that we examine in the telescope is not the average Mars. The little planet, true to his warrior reputation, puts the best face on things ; we see him only under his happiest conditions ; he presents to us most fully the summer noonday of his tropics, and hides from us " the winter of his discontent."

The mean temperature of the earth is taken as 60°, but the mean *noontide* temperature of our tropics in summer-time will probably turn 100°. A similar difference would, therefore, give us 50° as the average temperature of the centre of the disc of Mars as we see it. We need to know not only the mean temperature of the planet, but the range of temperature, the range from midday to midnight, the range from midsummer to midwinter, the range from equator to the poles.

And this must be very great—greater than we experience under similar conditions here. We infer this from what has been ascertained as to the force of gravity at the surface of Mars. This we know to be about three-eighths that of the earth ; so that whilst a body falls sixteen feet in a second at the surface of the earth, it will only fall six feet at the surface of Mars It follows from this that the atmosphere of Mars is arranged in a different manner from that of the earth. On the earth, at a height of 3⅓ miles, half the atmosphere is left beneath us, and by an ascent of another 3⅓ miles half the remaining atmosphere will be passed through, and so on. On Mars we should have to ascend

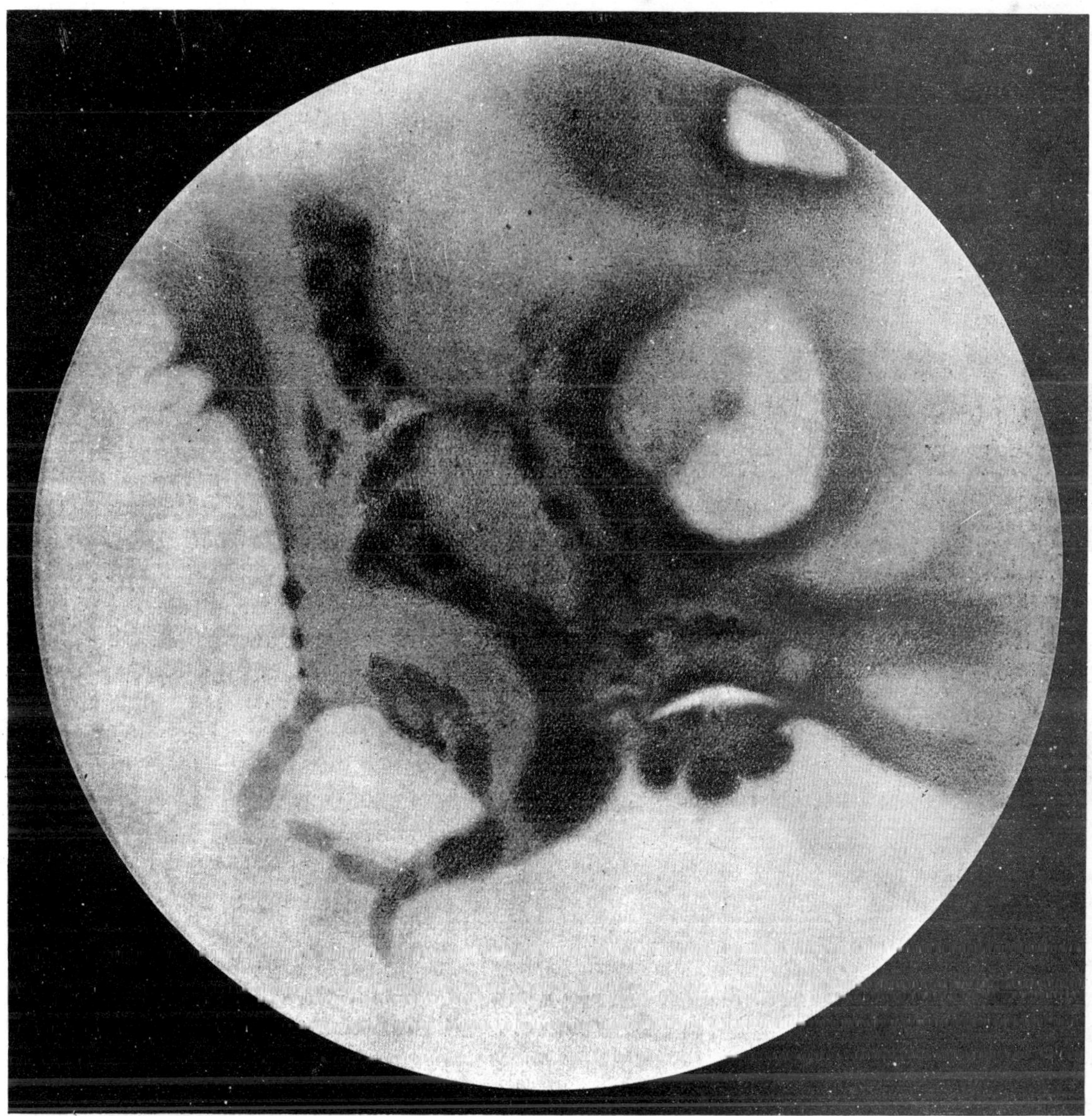

MIDSUMMER ON MARS.

Drawing of Mars, made by M. Antoniadi, on 1909 September 20—the day of the summer solstice of the Southern Hemisphere—with the 33-inch refractor of the Meudon Observatory. The great bright region in the lower part of the disc is "Aeria," one of the red lands that whiten under a low sun. The black marking, in the form of a cornucopia, that borders Aeria is the Syrtis Major, and midway between Aeria and the polar cap is the large round island "Hellas" that also grows white when near the edge of the disc.

8¾ miles to reach the level of half pressure, and 17½ miles to reach that of quarter pressure. If the pressure at the surface of Mars was as great as it is on the earth, we should be looking down through nearly three times the amount of atmosphere that we possess, and in all probability we could not discern his surface at all. As it is, the surface markings on Mars are almost as distinct as those on the Moon, a clear proof that we are looking through a thin and shallow medium. We cannot suppose the atmospheric pressure on Mars to be more than a seventh that upon the earth, probably it is much lower. But it follows from the rarity of the air on Mars that the vicissitudes of temperature there are far greater than upon the earth.

The atmospheric circulation of Mars must be very languid. This follows as a consequence of the small amount of heat received from the sun and of the feeble force of gravity. Just as a stone on Mars will only fall six feet in the second, instead of sixteen as with us, so a mass of heated air will rise with proportionate slowness. Some writers have indeed spoken of Mars as a "wind-swept planet," and have referred to its "hurricanes," "tornadoes" and "dust-storms." No such tumults can ever disturb its calm, clear, thin atmosphere. They are the creatures only of an unscientific use of the imagination.

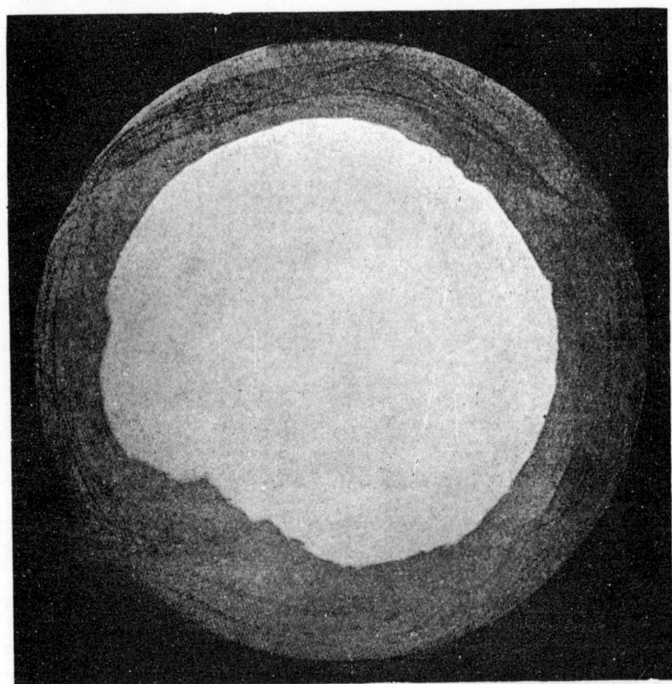

THE SOUTHERN HEMISPHERE OF MARS.
As observed at the Lowell Observatory in 1905, showing the extended area of the South Polar cap during the Southern winter.

If it be objected that the snow-caps of Mars are actually seen to melt away under the influence of the summer sun, a twofold answer is at hand. The rapidity with which the polar caps melt is the clearest possible proof that the sun's rays are not working upon great masses of ice. There is, indeed, no sufficient evidence that any considerable fall of snow ever takes place on Mars, or could take place. In the extreme temperature of the Martian winter "at a temperature vastly below any ever felt in Siberia, the smallest particles of moisture will be condensed into what we call hoar frost, and will glisten with as much whiteness as actual snow." And, on the other hand, though the mean temperature of the planet is low, it is not impossible, nor even improbable, since the extreme daily and annual range must be greater than anything we can experience here, that in mid-day of the tropics under a vertical sun—that is to say, in the region of the planet most favourably presented to our scrutiny—the temperature may rise

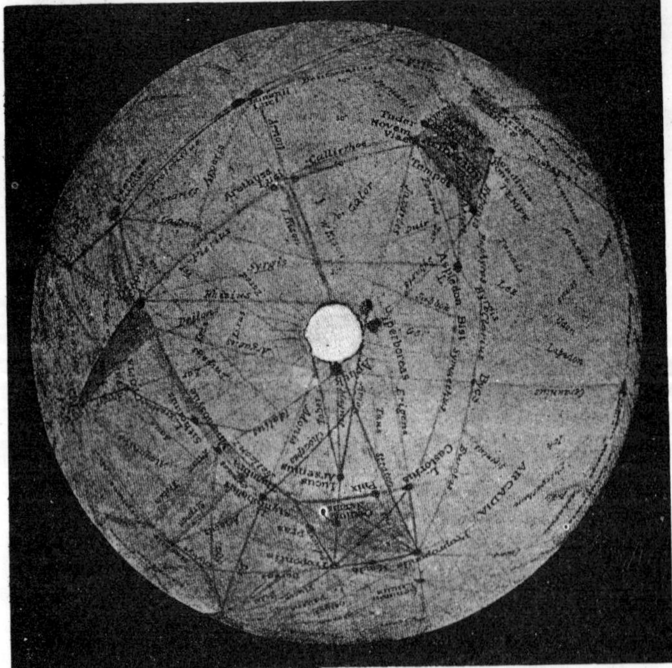

THE NORTHERN HEMISPHERE OF MARS.
As observed at the Lowell Observatory in 1905, showing the diminished area of the North Polar cap during the Northern summer.

for an hour or two in each day almost to the same height as in similar regions here. And as on Mars, owing to the low pressure of the atmosphere, water will boil between 100° and 125°, instead of at 212° as with us, there need be no surprise that the thin films of hoar frost are not only melted, but probably evaporated as well.

This view of the condition of the planet is strengthened by an observation that has frequently been made. In a beautiful drawing made by M. Antoniadi with the great thirty-three-inch telescope of the Meudon Observatory on September 20, 1909, when the southern hemisphere of the planet was at its summer solstice, the bright region in the lower part of the planet represents a ruddy district, called by Schiaparelli, "Aeria." It borders a dark marking having the form of a cornucopia, the darkest marking on the planet, and formerly known as the "Hour-Glass Sea," but now as the "Syrtis Major." It is a feature of Aeria and of several other of the ruddy lands of Mars that when near the edge of the disc—that is to say, early in the morning of the region or late in its afternoon—it becomes white instead of red, pointing to a deposition of frost in some form which disappears under the noonday sun and forms again shortly before he sets. In this case there can be no question of anything more than a thin film, whether of snow or of hoar frost, and as such regions sometimes go far to rival the polar cap in brightness, it is probable that it does not materially differ from them in character.

Under such conditions plant-life must find it difficult to maintain the struggle for existence, even in its lowest and hardiest forms. But for life in any of its highest forms, we may dismiss Mars as quite unsuited. For this a more genial climate, a denser and moister atmosphere, a less violent range of daily temperature are required, and these imply the conditions of a larger, denser planet, placed nearer to the sun; the conditions of our earth, mantled with vegetation and veiled in vapour, richly dowered with the cloud and with the leaf.

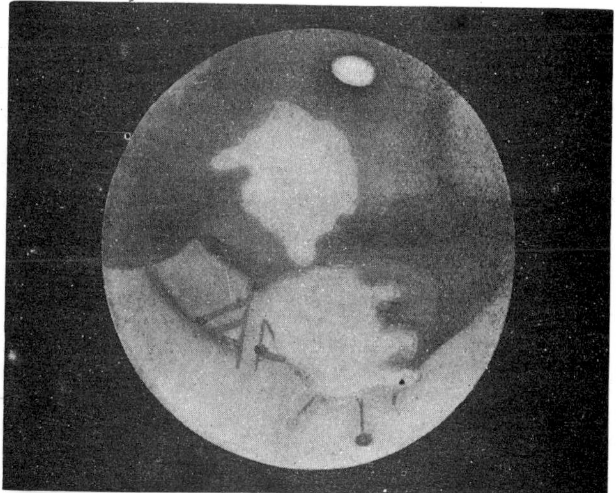

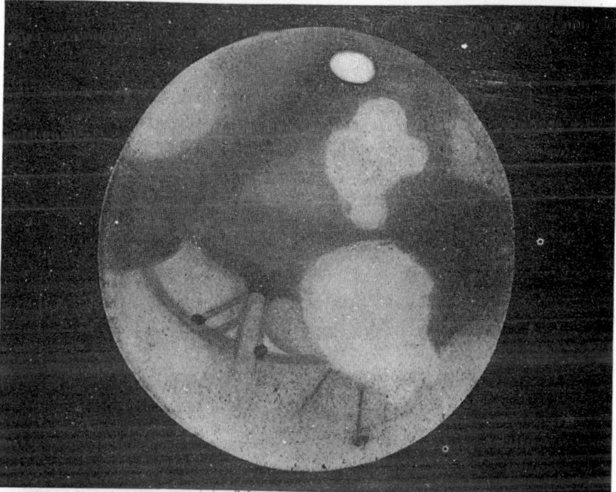

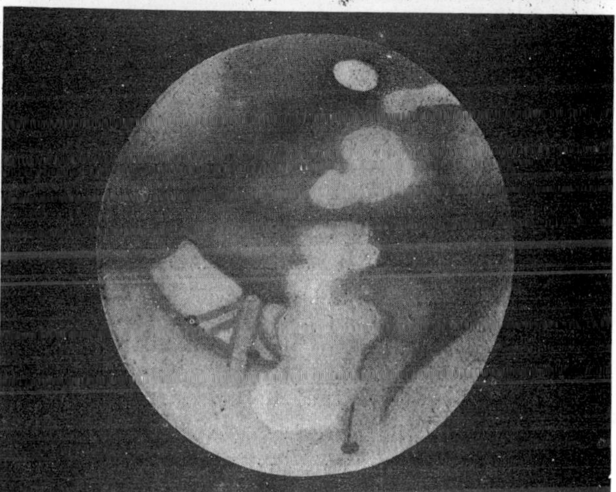

CLOUD-FORMS IN MARS.

These forms were observed and drawn by a German astronomer, Schröter, who noticed the changes of their mass more particularly round the equator. He therefore considered them to be clouds, and if this be so it demonstrates the existence of atmosphere and therefore a possibility of life on the planet.

THE LARGEST FLOWER IN THE WORLD

BY EDWARD STEP, F.L.S.

THERE are many plants whose flowers are so minute and inconspicuously coloured that one may pardon those who are not botanists for not knowing of their existence; but when we come to consider a flower that measures a full yard across it is rather astonishing to learn that it remained entirely unknown until the nineteenth century.

In the year 1818 Sir Stamford Raffles, with his wife and Dr. Arnold, were exploring the interior of Sumatra, when, suddenly, at the call of their Malay servants, they came upon a gigantic flower growing on the long twisting roots of some species of fig or vine. There were no leaves, the flower and flower-buds springing direct from these roots. Dr. Arnold, who was the first European to set eyes on this floral monster, confesses that, had he been alone, he would have hesitated to send home particulars of the flower's dimensions, from fear that it would be received as one of the old type of traveller's tales. Fortunately for his fears, Sir Stamford and Lady Raffles were within call and witnessed his measurements and particulars. The central portion of the flower containing the stamens and nectary was a foot across, and each petal, which was almost round, measured twelve inches in length. They were not thin and delicate like the petals of most flowers, but fleshy, and a quarter of an inch thick in the thinnest part, increasing to three-quarters of an inch at the thickest portions. The central cavity was computed to have a capacity of a gallon and a half, and the weight of the entire flower was calculated to be about fifteen pounds. The ground colour of the flower is flesh-tint or yellow upon which are heavy mottlings of a dull purple.

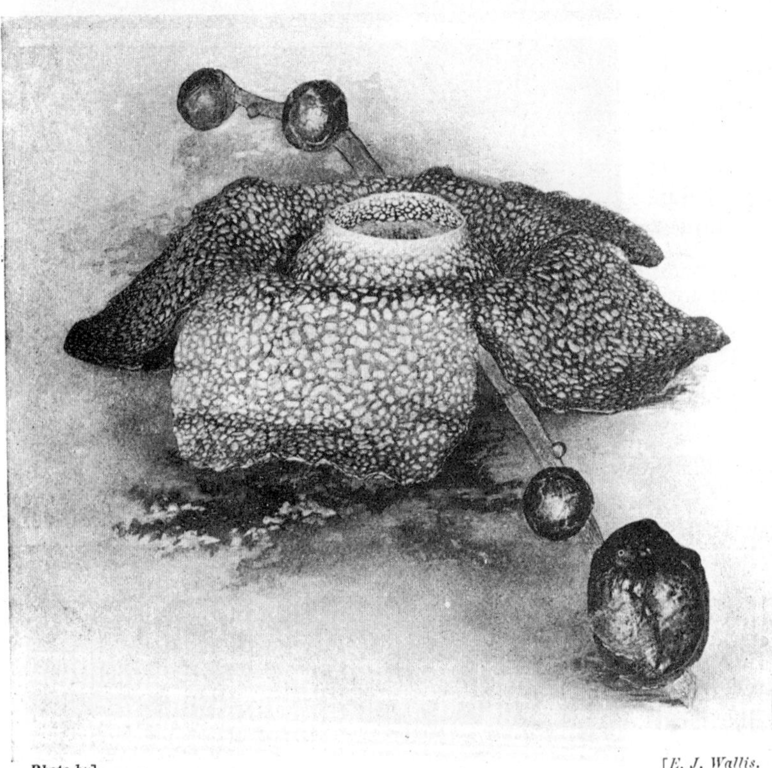

Photo by] [*E. J. Wallis.*

ARNOLD'S RAFFLESIA.

A parasitical flower in the forests of Sumatra. Four buds are also shown in various stages of growth, breaking forth from the roots on which the plant is a parasite.

The unopened buds are in form like cabbages. They take about a month to come to full size, and then are open only for a day or two, after which they rapidly decay. Resembling carrion in their general colouring, they also give off a carrion-like odour, which attracts swarms of flies. This is a device for securing the conveyance of pollen from the male to the female flowers.

Two or three other species of Rafflesia have been discovered since, but none so large as Arnold's.

[*By Arthur Twidle.*

ST. ELMO'S FIRES.

Sailors know these electrical manifestations by the name of "corposants," and erroneously regard them as **presaging bad weather.** Unlike many other manifestations of electrical force in Nature, these fires are unattended by any injury to **persons** or **property.** They are, however, considered ill omens by sailors, and consequently greatly dreaded.

ST. ELMO'S FIRES

BY FRANK T. BULLEN, F.R.G.S.

Ever since the dawn of seafaring, seamen have had their superstitious fears excited by the appearance on salient points of their vessels' spars of these extraordinary agglomerations of static electricity. Many and various are the allusions made to them in literature under differing names, of which St. Elmo's Fire is perhaps the most poetical. Sailors, however, know them best by the name of "corposants," an obvious corruption of the Italian appellation *çorpo santo*, or holy body. Even this word corposants has been corrupted by careless users into composants and other forms of the same kind.

Photo by] [J. J. Ward.

THE WATER SPIDER.

The spider has captured a worm, and is taking it to her home.

The popular idea that their appearance, glowing, lambent, and delicate, at a vessel's yard-arms or mast-heads presages worse weather has no foundation in fact. They merely indicate that the air is at the time surcharged with electricity, which, having no discharging factor at work, simply accumulates upon some favourable point and becomes visible as a soft, elusive, but never quite stationary flame. There can be little doubt that this light is similar to that of the Will-o'-th'-wisp, or *Ignis fatuus*, although not proceeding from the same source; the latter being probably luminous marsh gas.

If anything would convince a superstitious person of the harmlessness of St. Elmo's Fire it would be the fact, so frequently demonstrated, that it often covers like a halo the head of a seaman engaged in work aloft, and I myself have several times seen it streaming from my fingers when holding them up for the purpose. I cannot help confessing to a curious feeling of the uncanny on witnessing this phenomenon.

Photo by] [J. J. Ward.

The spider is diving with a load of air to her diving-bell.

Perhaps a belief in the supernatural origin of these most beautiful manifestations of the electricity in the air may be more readily understood than any other of the sea-superstitions now so rapidly dying away. Only appearing on the blackest of nights, moving from point to point without apparently passing through the intermediate space, unaffected by fiercest wind or heaviest rain, and insusceptible of being touched or moved, St. Elmo's Fires form what is probably the most mysterious and lovely of all the wonderful phenomena belonging to the ocean.

THE WATER SPIDER

BY R. I. POCOCK, F.R.S.

MOST people regard the diving-bell as exclusively a product of human invention. That is a mistake, for, thousands of years before man discovered this method of breathing and keeping dry under water, there lived a species of spider which had solved precisely the same difficulty; and this spider, probably in virtue of his clever invention, has survived to this day in our ponds and ditches. He is commonly known as the Water-Spider, and technically by a Latin name which means the "aquatic silver-swimmer." It is the appearance of this spider under water that has earned him this name, for the hairs which cover the hinder-part of his body are long and hooked at the end, and have the power of entangling the air when exposed to that element, so that the animal, partially enveloped in a bubble, looks like a ball of quicksilver as it swims beneath the surface.

The bell, which the spider makes for its home, is a thimble-shaped case of silk, anchored by threads to water-weeds, with the orifice turned downwards. To empty this bell of the water which fills it, the spider disentangles the bubble of air inside the aperture and this, rising to the top of the bell, displaces a certain amount of water. Another bubble is then fetched from the surface and liberated in the same way; and the operation is repeated until the bell is filled with air. This provides a water-tight air chamber, and exhausted oxygen is made good by the same process of bubble liberation.

In this bell two batches of eggs are laid, one in the early summer and one in the autumn. The summer eggs hatch quickly and the young soon start making diminutive bells on their own account; but after depositing the autumn batch, the mother closes up the mouth of the bell and passes the winter in a dormant state with them, waiting for the warmth of spring to hatch them. At the time of courtship, the male spins a bell alongside that of the female and makes a silk passage from one to the other, so that he can pass into her bell without getting wet. Water-Spiders feed upon aquatic insects and upon flies or gnats which alight on the surface of the water.

Photo by] [*J. J. Ward, F.E.S.*

THE HOVER-FLY.

THE HOVER-FLY GRUB AND ITS RAVAGES AMONGST THE "GREEN BLIGHT"

BY JOHN J. WARD, F.E.S.

To distinguish between insect friends and foes in the garden is not an easy task. The average gardener kills indiscriminately when insects cross his path, and especially if they happen to assume a grub-like form. Nevertheless, a little time devoted to discrimination between them would be time well spent; for his wholesale killing often results in the destruction of his greatest friends; indeed, he often kills much better killers than himself.

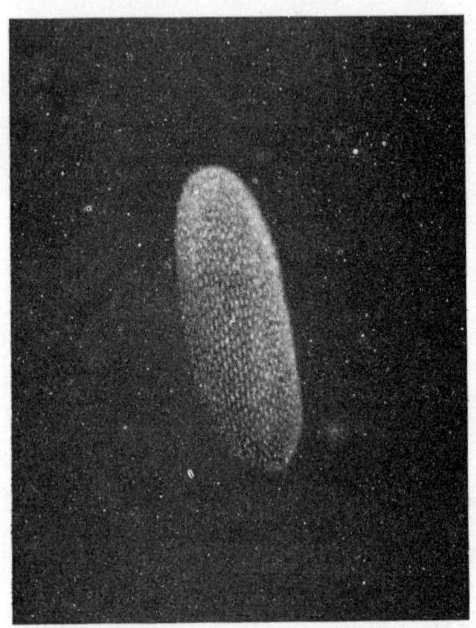

Photo by] [*J. J. Ward.*

EGG OF HOVER-FLY.

The egg here shown is magnified to sixty times the natural size.

My friend the Hover-fly has perfectly protected my sweet-peas from the ravages of the "green-flies," or "blight," during the extraordinary summer just passed. Although at first the green-flies existed in more than usual abundance, and their attacks were even more persistent than ever, yet the Hover-fly, and his near relatives, acted so promptly, and supervised things so thoroughly, that, later on, I could scarcely find a single green-fly.

It happened like this. By the side of my sweet-peas there is a bed of mint which quickly produced its flowers. Now of all things in the garden on which Hover-flies love to feast, probably mint flowers stand first. So these black and pale-yellow banded flies came in large numbers, and of various species, to the banquet, poising themselves above the mint flowers in their characteristic manner, held there by the rapid vibrations of their wings, and swooping down in hawk-like fashion to feast as opportunity occurred.

There was merry-making and love-making amongst them during every hour of sunlight; in due course, too, there was marriage, and also families. It was then that my sweet-peas had become badly attacked with the "green blight," which, like the Hover-flies, were thriving apace during the abnormally hot weather. It was then, too, that some of the female Hover-flies became deeply interested in my sweet-peas.

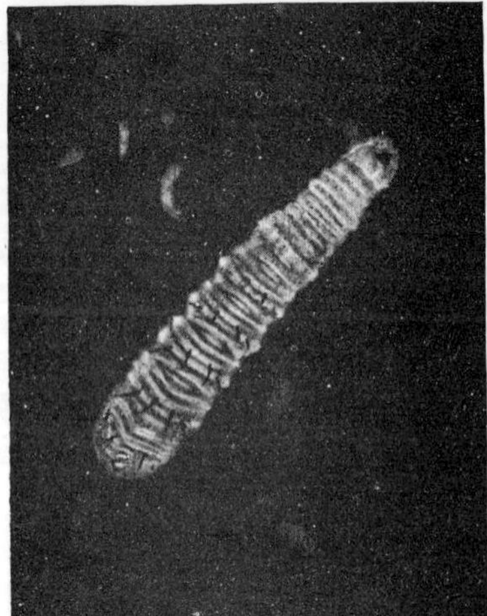

Photo by] [*J. J. Ward.*

GRUB OF HOVER-FLY.

The grub, photographed soon after emergence from the egg, is magnified about sixty times.

Day after day they were hovering before the flowers and leaves, alighting upon them and carefully searching them over. The moment they met with a green-fly, they would deposit one of their eggs near to it, and then fly to another flower or leaf for the same purpose.

Now, as everyone who has tried to cultivate plants will know well, green-flies possess extraordinary powers of reproduction. A single green-fly will bud out twenty or more young in one day, and these are all females, which quickly develop to maturity, and are almost at once able to produce young in the same manner as their parent, and which are again all females. Also, this budding process of reproduction continues for twenty or more generations during the summer months, without any males appearing. Males are eventually forthcoming in the late autumn broods, and after their appearance eggs are produced

A full-grown grub of the Hover-fly has captured a green-fly by spearing it with the three-pronged fork at its mouth; it then elevates it in the air by holding to the flower with its broad hinder part—for it has no feet—and in this position

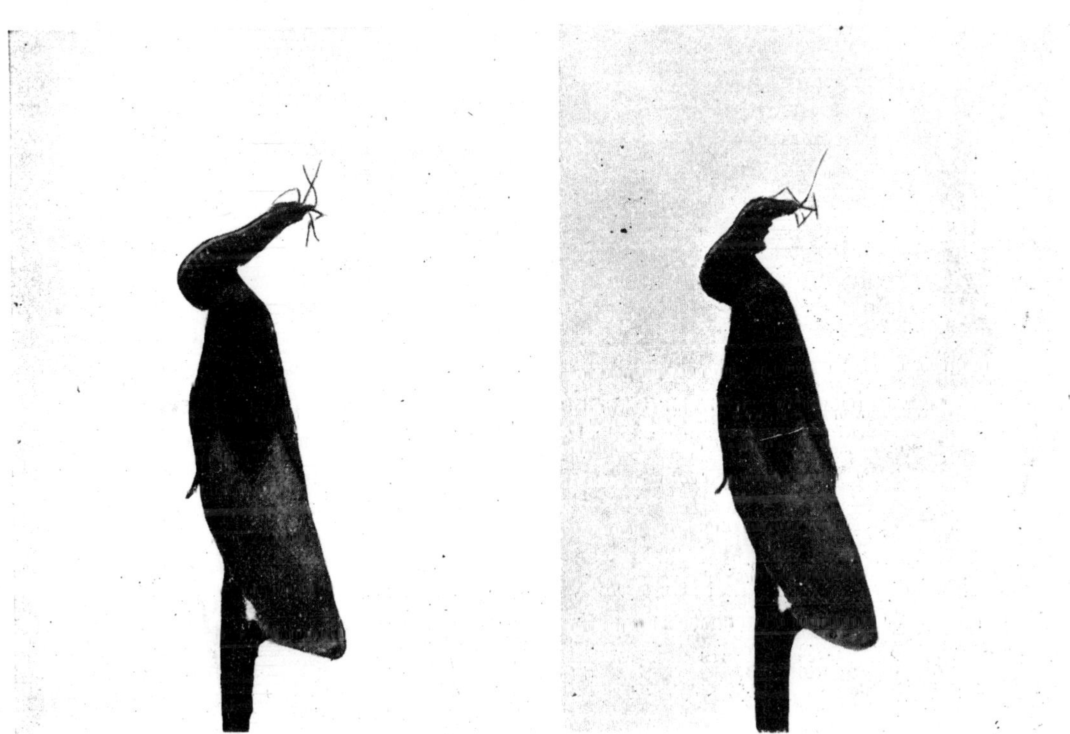

Photos by] *J. J. Ward, F.E.S.*

sucks the green-fly until only an empty skin is left. This is dropped, and another green-fly is at once sought, when the process is repeated over and over again.

HOW THE HOVER-FLY GRUB FEEDS.

instead of living young. When winter has passed, these eggs provide the first brood of budding females.

It is obvious, then, that every "green-fly" seen during the spring and summer months can become a mother. Réaumur has estimated that a single green-fly may be progenitor to no fewer than five thousand nine hundred and four millions nine hundred thousand individuals during the two or three weeks of its life.

The mother Hover-fly seems fully acquainted with this state of things, for the moment that she detects a green-fly on a leaf or stem, she is content to place an egg there; probably she is instinctively satisfied that the sight of a green-fly is a sufficient guarantee that ample provision will be forthcoming to give her offspring a good start in life.

Photo by] [*J. J. Ward, F.E.S.*

BUD OF SWEET-PEA ENLARGED.

In this enlarged photo the forms of the green-fly are clearly seen. Note the egg of the Hover-fly on the upper part of the flower.

In an illustration on this page the egg is shown as placed by the parent insect amongst the green-flies, and more in detail on page 66. Three days later a tiny, yellowish white grub emerges from the egg, which when fully extended is not more than one-sixteenth of an inch in length; but its courage and strength are as marvellous as its voracity.

The grub under observation, when hatched, travelled down the stem until it came to a leaf, on whose surface it met with its first green-fly. One of the legs of the latter touched the grub as it moved towards it. Instantly the grub thrust its head forward, and a moment later it was standing upright on the tail-end of its body with the green-fly elevated in the air—a magnified photograph of the actual incident is shown on the page opposite.

In this attitude the little grub held the green-fly for over an hour while it sucked its juices, and this in spite of the fact that its victim was much larger than itself and continually struggling to escape. It then cast away the empty skin of the green-fly and rested for a while. During the first day of its existence it captured and ate two half-grown green-flies and two smaller ones; and day by day, for ten days, its appetite increased in an astonishing manner.

It grew rapidly, and by the end of the second day had become quite an expert in manipulating green-flies. It is interesting to note that the grub is blind; it finds its prey by rapidly thrusting its head here, there, and everywhere as it travels. Sooner or later its nose comes in contact

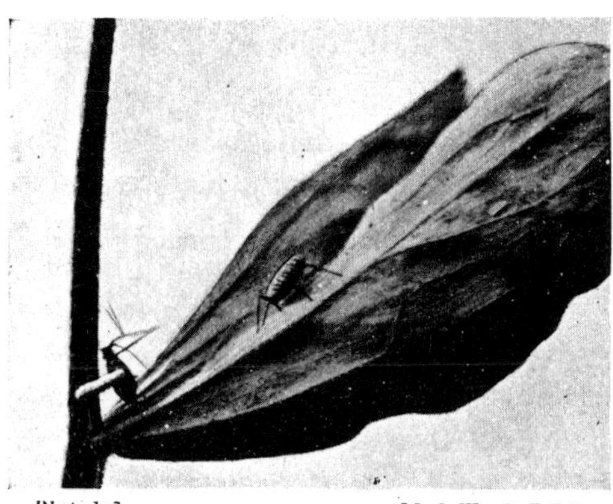

Photo by] [*J. J. Ward, F.E.S.*

HOVER-FLY GRUB.

The grub, two days old, has seized a green-fly. Four times the natural size.

with a victim, when a three-pronged fork, or trident, at its mouth is immediately thrust into the green-fly's body; so it is elevated into the air.

The method of feeding of the full-grown grub is clearly shown in the illustrations on page 67. The grub is of a green colour, with a white stripe down the centre of its back, which is tinged with purple at the base. When hungry, after searching for prey, these grubs are very ravenous, consuming green-flies at the rate of more than one per minute. As they feed during the night as well as the day, the enormous amount of destruction they work amongst these garden pests is incalculable; indeed, man's insecticides and fumigating devices scarcely count against the persistent onslaughts of these grubs.

The grub's method of locomotion is curious. It is not provided with feet, but adheres to the leaves by means of the rough edges of the skin at the sides of its body. It consequently has some difficulties to overcome when on the move; nevertheless, it travels at a remarkable pace.

It stretches out its head—which is little more than a tapering proboscis bearing at its apex the trident previously referred to—as far as it can reach, then, gripping the leaf by means of its trident and immediately drawing up its hinder parts until its tail-end almost touches its nose, appears as if about to turn a somersault. Just as you expect to see it perform that movement, however, its head is suddenly thrust forward again; so it moves along in this loop-by-loop fashion. It always appears to be in a big hurry, both when travelling and in making its thrusts around in search of prey.

At the end of the ten days' feeding period, the grub's enormous appetite declines, and it attaches itself to a leaf or stem by the trident at its mouth. There it hangs for another ten or eleven days, its skin hardening and becoming of a golden brown colour. After that time has passed, this pupa or chrysalis is suddenly burst open, and we find that the voracious grub has been transformed into a shining black and yellow-banded Hover-fly.

It is obvious that the Hover-flies are amongst the gardener's most valuable allies. When we recognize the enormous rate of increase of which the green-flies are capable, together with the knowledge that, if their multiplication remained unchecked, every green plant or tree on the surface of the earth would soon become converted into living green-flies, and that, in consequence, the food supplies of all terrestrial animals would quickly terminate, we can only marvel at the efficiency of Nature's sentinels which are ever on guard and ready to strike. When the green-flies become too abundant, Hover-flies and other natural foes of the green-flies quickly appear, and the work of destruction then goes on persistently until the normal balance is restored.

Photo by] [*J. J. Ward, F.E.S.*

HOVER-FLY GRUB ENLARGED.

A portion of the previous photo enlarged to sixty times the natural size.

Photo by] [*J. J. Ward, F.E.S.*

CHRYSALIS OF HOVER-FLY.

The grub having killed many a green-fly, has reached full size and turned into a chrysalis. Enlarged to four times the natural size.

Photo by] [*J. J. Ward, F.E.S.*

THE HOVER-FLY.

Newly emerged from the chrysalis, the Hover-fly is waiting for the full expansion and hardening of its wings. Four times the natural size.

THE SPECTRE OF THE BROCKEN

THE tourist who makes the ascent of the Brocken—a mountain in Germany—and reaches the summit about the time of sunset, may be startled by observing towards the east a terrific apparition among the clouds. Continued inspection will reveal a remarkable resemblance in outline between the spectre and himself, and that all his movements are imitated by the gigantic form. It is really a shadow of himself thrown by the horizontal rays of the setting sun upon a bank of mist that rises in the neighbourhood, and the size of the shadow depends, of course, upon its nearness to, or distance from, the place where the tourist stands. The same phenomenon may be seen, of course, at sunrise by looking in the opposite direction to that indicated for sunset. A traveller who had often been disappointed in his efforts to raise the spectre, thus describes his ultimate success :

" About a quarter past four I looked around to see whether the atmosphere would permit me to have a free prospect to the south-west, when I observed, at a very great distance towards the Achtermannshohe, a human figure of a monstrous size ! A violent gust of wind having almost carried away my hat, I clapped my hand to it ; and in moving my arm towards my head, the colossal figure did the same.

" The pleasure which I felt at this discovery can hardly be described ; for I had already walked many a weary step in the hope of seeing this shadowy image, without being able to gratify my curiosity. I immediately made another movement, by bending my body, and the colossal figure before me repeated it. I was desirous of doing the same thing once more, but my colossus had vanished. I remained in the same position, waiting to see whether it would return ; and in a few minutes it again made its appearance on the Achtermannshohe. I then called the landlord of the neighbouring inn, and having both taken the position which I had taken alone, we looked towards the Achtermannshohe, but did not perceive anything. We had not, however, stood long, when two such colossal figures were formed over the above eminence, which repeated their compliments by bending their bodies as we did, after which they vanished. We retained our position, kept our eyes fixed on the spot, and in a little time the two figures again stood before us."

THE SPECTRE OF THE BROCKEN.

A startling apparition of giants in the sky, witnessed by visitors who climb to the summit of the Brock[illegible] near the hours of sunset or sunrise.

POLYCYSTINS

BY RICHARD KERR, F.G.S., F.R.A.S.

An account of the "Marvels of the Universe" would be incomplete without illustrations and descriptions of some representatives of the Polycystina.

They are all marine creatures, and it is impossible to form any idea of their marvellous structure without the aid of a microscope. To the unassisted eye they appear as fine as flour sprinkled on a glass slip; but under a strong magnifying power the beauty and delicate tracery of these wonderful organisms is brought out in a marked degree. They are always a fascinating study. All students of microscopic forms of life are agreed in the opinion that the elegant shells or skeletons of these creatures are in the front line of beautiful and wonderful structures. In fact, any microscopist taking for his studies a wide range of nature would consider any collection of microscopic slides deficient if the Polycystins were absent.

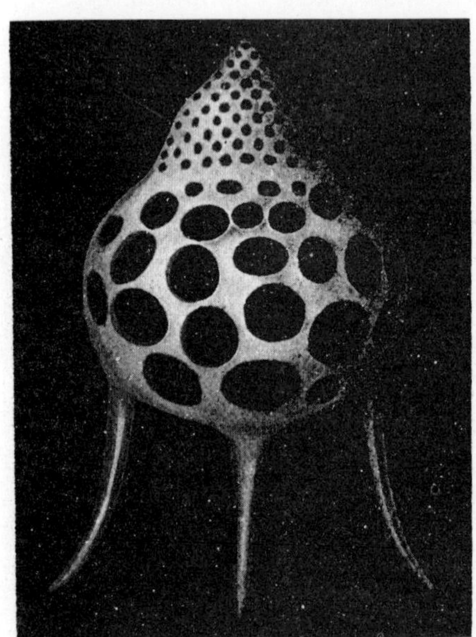

This Polycystin might be used as an art design for a brazier.

Inasmuch as they belong to the Rhizopods, which form one of the divisions of the Protozoa, they are classed among the lowest forms of animal life.

If the element of beauty could be included as one of the conditions necessary in the classification of animals, the Polycystins would occupy a much higher position in the list of denizens of the oceans. But, so far as research shows, they display very little of organic parts, their internal economy is simple, and the functions of life are performed under very elementary conditions. Therefore, they must remain in their humble position until microscopists detect the possession of organs which will warrant their promotion.

A nucleus is present, some yellowish bodies are present which may be green plants, and there is a very small particle of oil which enables the organism to rise in the water. Some observers state that there are also four nuclei corresponding with four lobes of the living protoplasm.

The naturalists who examined the living creatures as they were taken up from the ocean saw them in their gorgeous colours of crimson, olive, purple, and other tints which baffle description. The varied forms of the skeletons of the Polycystins amount to several hundreds. There appears to be no limit to the varieties of design in these specks of flint.

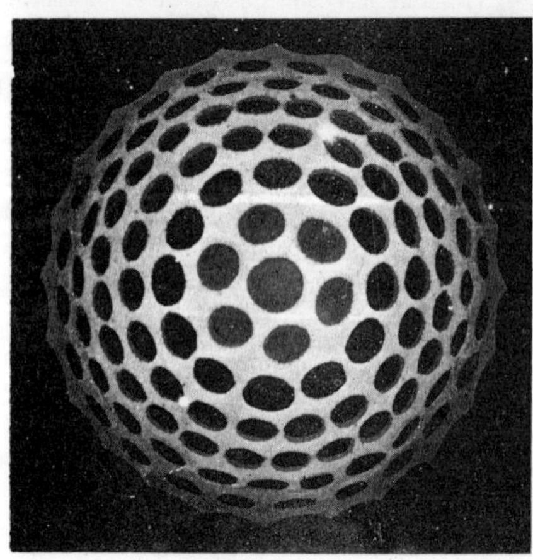

This form of Polycystin might be suggestive to the designer of golf-balls.

They appear to be more numerous than their partners in microscopic marine life—the Foraminifera; also, they excel them in the beauty and complexity of their shells or skeletons; and this is saying a great deal, for the Foraminifera are noted for their attractive and varied forms.

The Polycystina differ greatly from the Foraminifera in the fact that the skeletons of the former are of pure flint and of crystalline transparence, solid, each skeleton invariably composed of one continuous, unbroken piece, perfectly rigid, and unaffected by any chemical agents except those which are known solvents of flint; whereas the shells of the Foraminifera are composed of carbonate of lime. Occasionally flinty Foraminifera have been found, but these have been changed. Their original composition was carbonate of lime, but owing to the action of water containing silica, the chalky material was replaced by flinty matter.

There is still another important point of difference between these two great groups of marine organisms. The Foraminifera are divided into two sub-families—the Perforate and the Imperforate. But in the Polycystina there is no distinction afforded by Perforate or Imperforate, since all the Polycystina are Perforate. There is no exception to the rule that the flinty skeleton is so pierced by apertures in every direction and in every portion—excepting the spinous projections,

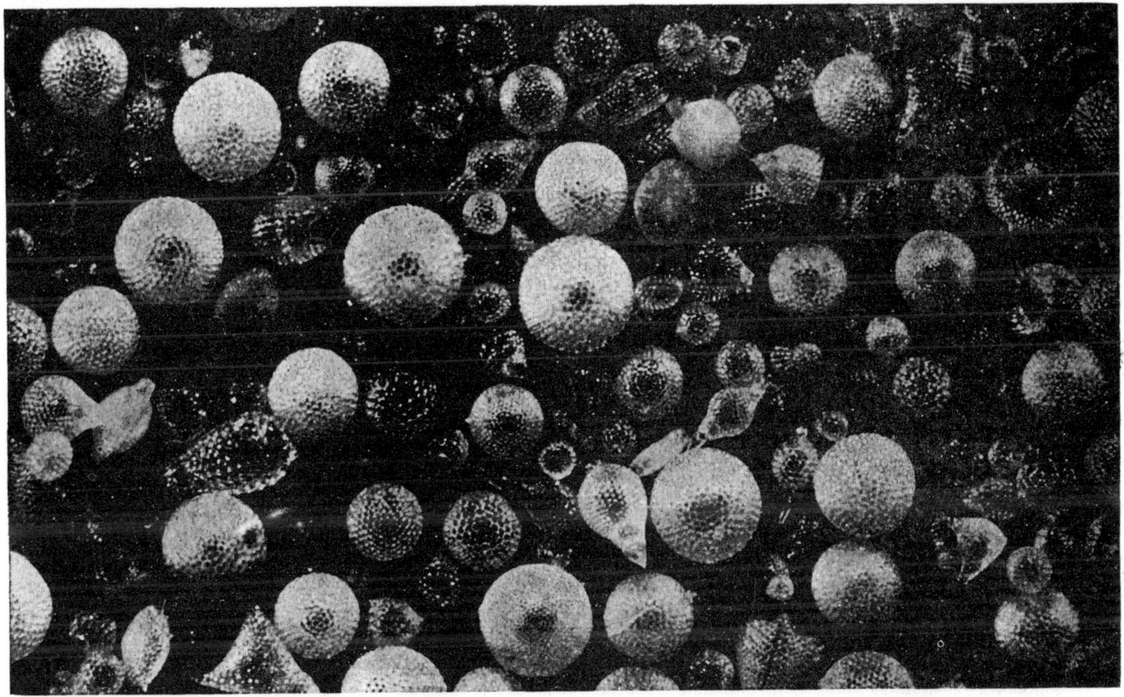

POLYCYSTINS.

This illustration shows how the Polycystins occur in Nature, the various forms mixed up together. The entire group shown would represent a mere speck of dust before they were magnified.

which are so but rarely—that the most complete communication may be said to exist between the contents of every chamber and the investing film of protoplasm on the exterior.

The flinty skeletons of the Polycystins are of many and curious shapes. Some are like baskets; others are spheres covered with spikes; others are cone-shaped; others seem as if they were wreathed into crowns; others are pyramidal, while many of them resemble the famous ivory ball-within-ball ornaments fashioned by the skilled Chinese artificer.

All of them, however small, are latticed. Even those whose diameter is the hundred and fiftieth part of an inch have their "windows," or apertures, where it would seem impossible for such to exist.

In those specimens which consist of concentric balls there are rods of flint supporting each sphere in its place, so that the complete structure is very strong.

In certain of the Polycystins the jelly-like body ramifies into all parts of the internal spaces as well as the external parts of the flinty skeleton. In others the creatures can retreat into the upper domes of the skeletons. Spikes or spines project at regular intervals from the surface. In life the creatures send out innumerable feelers for food. For this reason they are placed among the Rhizopods, or root-footed creatures.

Ehrenberg (1795-1876), a specialist in microscopic research as applied to the minute forms of marine life, seems to have paid more attention to the Polycystins than any of his predecessors had done.

Up to his time information about these organisms was limited to those found in the Island of Barbados. He believed that they had a much more extensive range than the West Indies, and he was encouraged in his belief by finding specimens of Polycystins at Cuxhaven, at the mouth of the Elbe. Being a great traveller and an experienced naturalist, as shown in the twenty-four works which he published, he discovered in the Nicobar Islands, which lie nearly in the same latitude with the Barbados, but in the East instead of the West Indies, at a height of two thousand feet, clays, marls and sandstones rich in Polycystins. In examining one hundred species under his microscope he found them partly identical with three hundred species found in the rocks of Barbados.

POLYCYSTINS.

Three more remarkable forms of these minute wonders. It is easy to imagine these as designs for articles of human workmanship.

He also found that the islands of Car Nicobar and Camorta are specially remarkable in this respect. On the latter there is a hill of three hundred feet, throughout which the Polycystins are in abundance.

Since Ehrenberg's time, geologists have shown that the Polycystins have a still wider range—that they form a large percentage of the composition of many of the rocks of the globe. Haeckel has proved clearly that the Polycystins of the rocks of Barbados are closely allied to the Radiolarians of the deepest parts of modern oceans. The Polycystins are a constituent in the composition of the jaspers of Siberia, of the tripoli of Richmond, Virginia, of the shales of Saxony, of the marls of Sicily, of the Cambrian strata of Thuringia, and of many other geological formations.

Sir Wyville Thomson states that when deep dredging in the Atlantic, the dredge brought up a large number of extremely beautiful Polycystins.

The illustrations will, it is hoped, induce the young Nature student to examine the actual specimens for himself. Prepared slides may be purchased at any of the establishments where microscopes are sold. But it is much more satisfactory to prepare the material for observation from a small piece of Barbadian rock, which may be obtained from any mineralogist.

A GROUP OF POLYCYSTINS.

These Polycystins are the fossil remains of minute creatures which have their modern representatives in the very deep waters of the Atlantic to-day. They were so numerous in past ages that many rocks are composed largely of their flinty skeletons

The method of procedure recommended by the expert in microscopic mounting, Martin J. Cole, is as follows :—" Take a piece of the rock not larger than a small marble, crush it and boil it in a strong solution of washing soda. Strain off the liquid, retaining the sediment. Wash the sediment several times, boil it in a test-tube containing nitric acid for about half an hour. This will remove any trace of the carbonate of lime. Wash the sediment repeatedly in water to clear it of the nitric acid."

The sediment when dry is ready for mounting. Some microscopists heat the dried Polycystins on a hot shovel in order to give the specimens a white opalescent appearance. They may be mounted on a dark background or as transparent objects are mounted.

The specimens represented in the illustrations vary in diameter from the hundredth of an inch to the two hundredth. Over three millions of them would occupy only a cubic inch of space !

The method just suggested of preparing and mounting collections or groups of Polycystins will

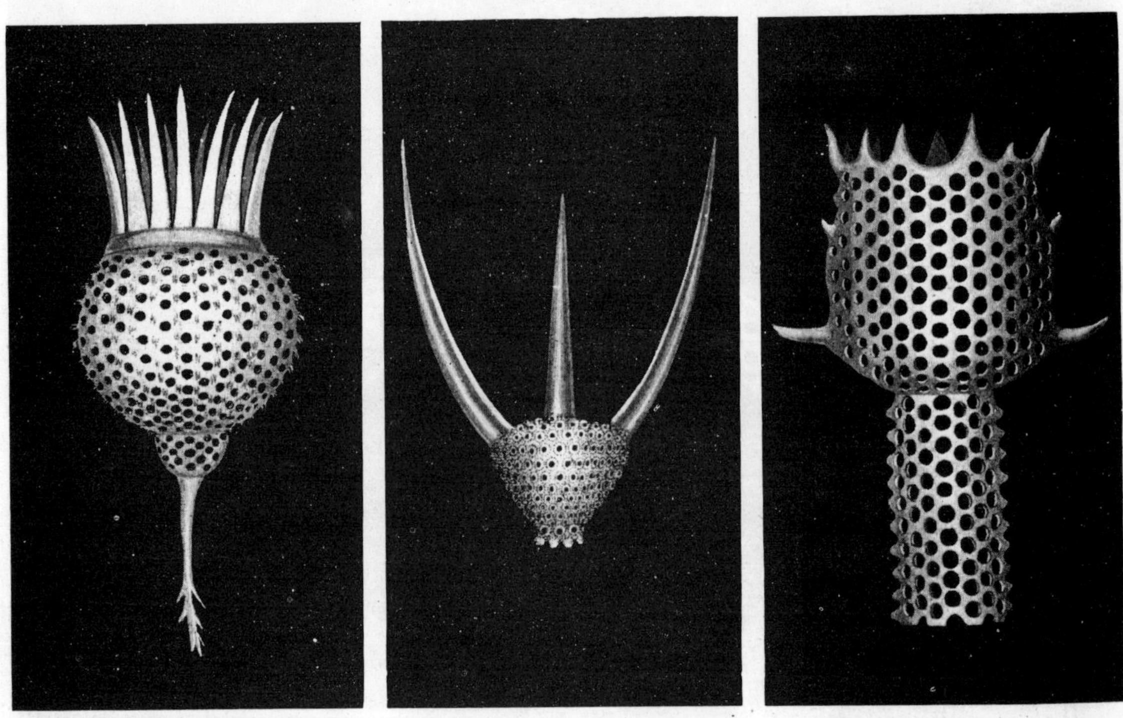

POLYCYSTINS.

In spite of their varied and graceful forms and delicate tracery, in actual size these Polycystins are like grains of fine sand.

hold good for those which form the chief constituent of friable rocks, but many of the Jurassic rocks, such as jasper, flint and chert, are too hard for treatment in this manner. It is necessary with these more compact rocks to grind them down to the thinness which will render them semi-translucent.

A slice of rock so treated is fastened to a glass slip by some transparent adhesive. Then when a strong light is focussed on the under side of the slide the Polycystins will appear and will make an attractive and valuable addition to the collection.

One or two geologists will not admit that Polycystins are found in the Atlantic Ocean, or that their range in Nature is so extensive as is commonly accepted ; but the late Thomas Rupert Jones, F.R.S., one of the editors of the famous " Micrographic Dictionary," and an experienced geologist and writer on geology, informs us that Ehrenberg found Polycystins in the Antarctic seas ; Bailey described them from the depths of the Atlantic ; Müller studied them from the Mediterranean, Haeckel in the Adriatic, Wallich in the Indian Ocean ; Wyville Thomson, Carpenter and Gwyn

A Polycystin that might serve as model for an artist's fishing float.

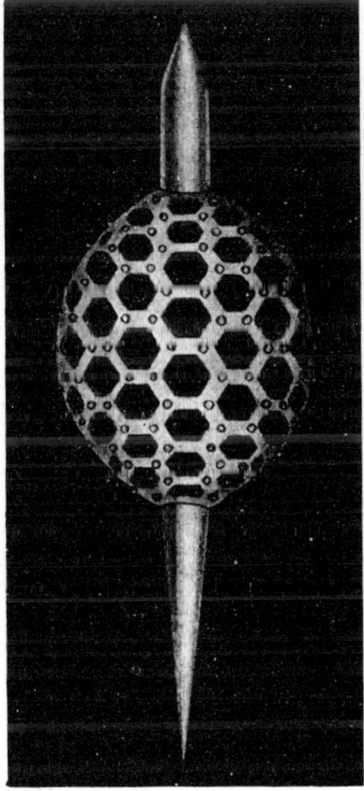

A Polycystin of spindle form, the swollen portion pierced with many six-sided openings.

Jeffreys noticed them in the deep-sea soundings of the North Atlantic.

He further informs us that the siliceous shells accumulated in thick deposits during the last geological periods, and that myriads of these exquisite microscopic forms may be obtained from many strata in Greece, Oran in Africa, and Bermuda. He suggests that the Polycystins are best examined as opaque objects, and, if minute details are required to be seen, as transparent bodies.

Although the Polycystins are classed under the Rhizopods, as stated, they are, with some exceptions, looked upon as fossil Radiolarians. Haeckel, to whom the vast collections of micromarine life dredged up during the *Challenger* expedition were given for the purposes of classification, etc., was of this opinion, and he divided the Polycystins into two groups, those whose fundamental form was spherical and latticed and those whose form was originally egg-shaped and perforated. He pronounces those of the Barbados as a deep-sea deposit, and that many of the Polycystins are to-day extant and unchanged in the ooze of the deep Pacific Ocean, that the deep-sea mud of the Central Pacific Ocean known as "Radiolarian ooze," consists for the most part of well-preserved shells of Polycystins.

J. Müller, another great naturalist, (in 1858) distinguishes the Polycystins as an animal enclosed in a perforated flinty shell. These shells are very minute and of many forms. Spherical, conical, egg-shaped and star-shaped kinds are common and in most cases are flinty prolongations of the shell, which are symmetrical and curved, or angular, or branched. The perforations of the shell are large and cause it to resemble a flinty network rather than a test. The prolongations of the flinty skeleton are not hollow, but consist of transparent solid silica which is developed during the digestive processes of the animal.

A quaint work on Nature, published

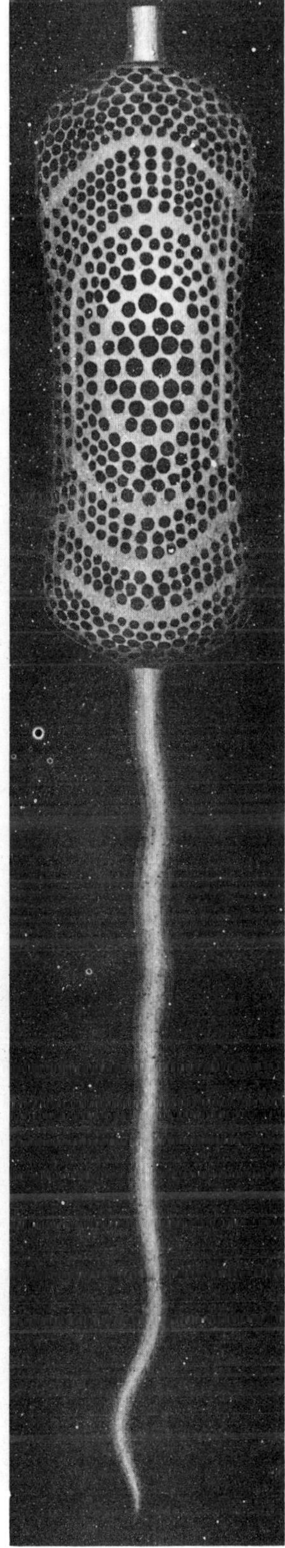

A cylindrical Polycystin

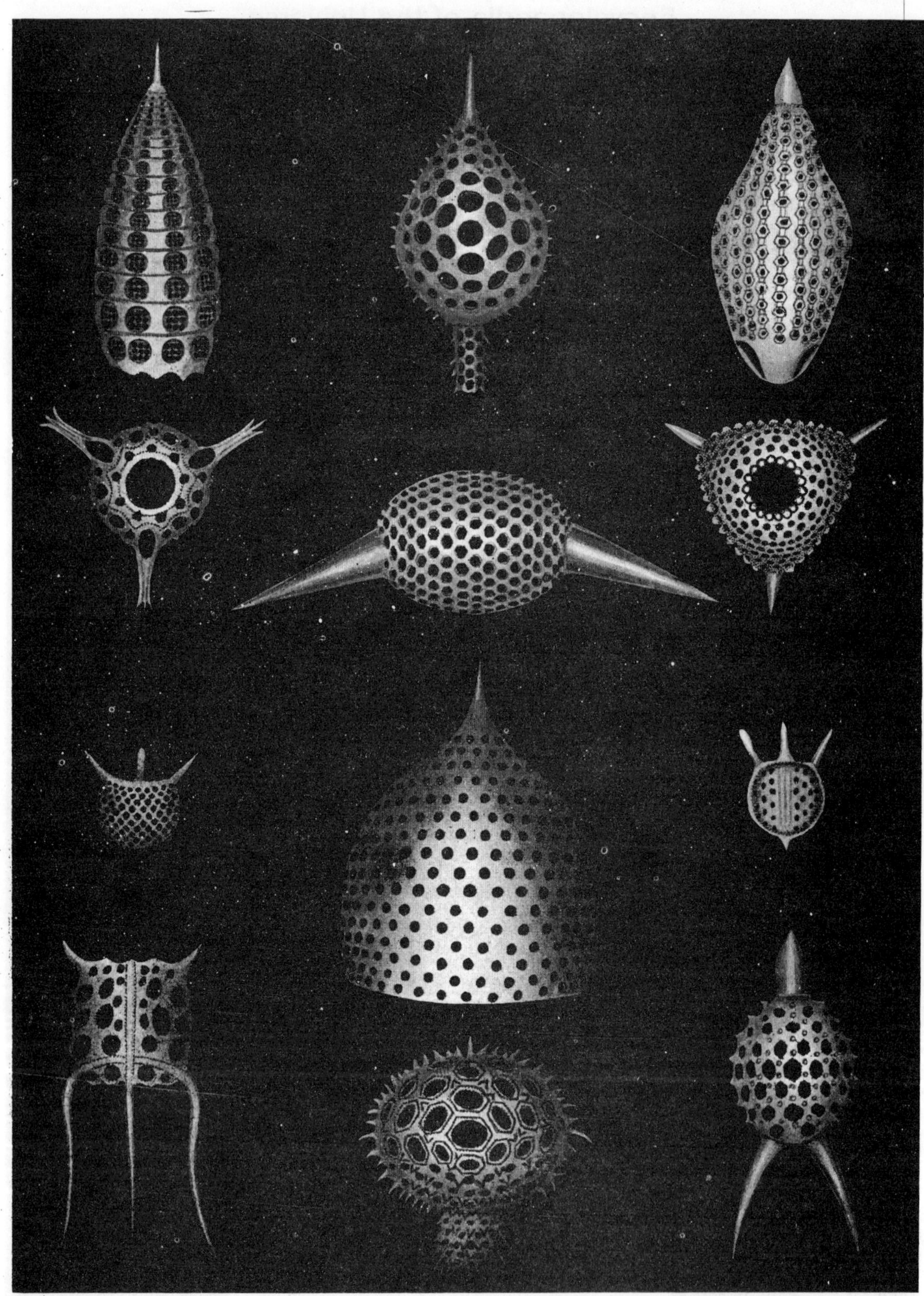

A POLYCYSTIN GROUP.

This second group of these marvellous organisms contains forms quite distinct from those on earlier pages. The entire collection of illustrations, representing thirty-six species, will serve to give a clear idea of the rich variety both of form and ornament to be found among these minute fossils

in 1873, entitled "A Catechism of Zoology," written by the Rev. J. F. Blake, M.A., etc., contains some useful particulars about the Polycystins, in the form of question and answer :

"Describe the Polycystina ?

"They form beautiful, siliceous shells, covered with large openings, out of which their straight pseudopodia are put, and often having spines and other ornaments.

"Why are they so called ?

"Because many of them have two or more shells, one inside another.

"How are they distinguished from the Foraminifera ?

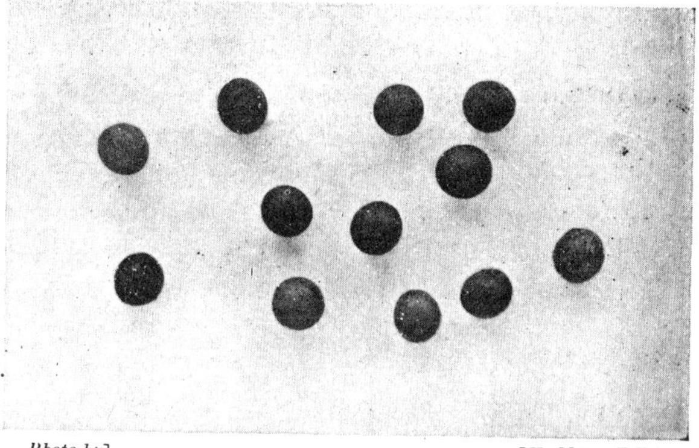

Photo by] [*H. Main, F.E.S.*

EGGS OF THE GLOW-WORM.

The Glow-worm's eggs, here shown eight times larger than the natural size, are luminous like the parent.

"Their shells are siliceous and not calcareous. Their openings, or foramina, are larger and their pseudopodia are straight and thick, instead of being thin and interlacing."

We give one more quotation from Haeckel, which is the wish of all lovers of Nature. It is the final sentence in the preface to his great work on the Radiolaria :

"May it be the means of inducing many naturalists to study more deeply this inexhaustible kingdom of microscopic life, whose endless variety of wonderful forms justifies the '*Natura in minimis maxima.*'"

THE GLOW-WORM AND ITS LIGHT

BY EDWARD STEP, F.L.S.

THE word "worm" is one of the most abused terms in our language. With our forefathers, snakes were worms, and they appear to have applied the word also to any creeping thing that was long and more or less cylindrical. Thus, among caterpillars we have "silk-worm," "palmer-worm," and similar designations. The Glow-worm is neither a caterpillar, a serpent, nor a worm, properly so-called, but a beetle. It is true that the female, which alone has any considerable power in her glow-light, has little resemblance to the majority of beetles, she being without the pair of wings and the horny wing-covers that are the predominant badge of the beetle tribe.

It is certain that few of those who have half-admiringly, half-fearfully, watched the weird greenish light showing among the herbage of the hedge-bank on summer evenings, have been bold enough to take up the little light-bearer and examine it. Even of those who have done so, the percentage is very small who have ever seen the Glow-beetle, to give the male a distinctive name. Yet, in those districts where the female, or her light, is frequently seen, it is not difficult to make acquaintance with the male. An open window and a light inside is the only preparation needed. Mr. Glow-worm

Photo by] [*H. Main, F.E.S.*

GLOW-WORM GRUB.

The grub or larva which issues from the Glow-worm's eggs is here shown about four times larger than the actual size. It also gives off light.

will fly in and probably alight on the table, after flying around the light, and can then be examined at leisure.

Though the light has its highest development in the mature female, it is present in the insect at all stages of existence. The small spherical eggs are luminous from the moment they are deposited, in the second half of July, until they hatch some four or five weeks later. The grub too, is luminous, and the same may be said of the chrysalis.

The Glow-worm is a life-long enemy to snails and slugs ; in fact, these molluscs constitute the sole food of this insect. It would appear to be a somewhat difficult matter for a Glow-worm to attack so sensitive a creature as the snail, who quickly withdraws his very soft body well into his shell at the slightest alarm. But from the moment he is well out of the egg-shell the grub seeks for the nearest snail and opens the attack, though he may have to crawl half into the shell in pursuit. He thus gets the snail into a corner from which there is no retreat, and he eats steadily on until the snail-shell is quite empty. That is the explanation of the majority of empty and unbroken snail-shells you may find under almost any hedge. The others probably have become vacant through the death of the snail from old age—if ever so-persecuted a creature does die a non-violent death.

Photos by] [*H. Main, F.E.S.*

THE MALE GLOW-WORM.

These enlarged photos show the upper and lower surfaces of the beetle in its usual attitude with head drawn in under the shield. The upper figure shows the head extended.

How is the Glow-worm's light produced ? and what is its purpose ? These questions have not yet been answered with entire satisfaction. Several explanations have been put forward. It was formerly the custom to speak of all luminosity produced by animals and plants as "phosphorescence." That term has been dropped because to most people it implied that the light was due to phosphorus. But what is the exact nature of the luminosity has yet to be determined. Its purpose has been stated to be to enable the winged male to find the wingless female ; but though this is probably one reason there must be others, or else the eggs and the grub would scarcely be provided with the same light. Moreover, scent appears to play a more important part than light in the mating of insects. A *hidden* female moth newly emerged from the chrysalis is often employed by the entomologist to attract a crowd of males to the exact spot, where of course they fall victims to their cupidity. Scent alone is the guide in this case.

The grub of the Glow-worm, as we have said, differs but little in appearance from the mature female. On referring to our photograph of the grub, it will be seen to resemble a caterpillar flattened out from above. Like the caterpillar of a butterfly or moth, the body is built up of rings or segments, jointed together and so giving a considerable power of movement in any direction. In the fore-part will be seen the three pairs of legs which are the common possession of all insects. From the hind-

Photo by] *H. Main, F.E.S.*

GLOW-WORM CHRYSALIS.

The chrysalis shown is a male, as indicated by the rudiments of wings hanging from behind the head.

most segment of all the grub can protrude at will a bunch of white fleshy points. The brush so constituted has a double use: when the insect has got smeared with the tenacious slime poured out by slugs and snails it uses this brush to cleanse itself with. The second use is in climbing over the smooth, slippery shells of snails in order to attack the gliding snail from behind the head. Its fine thin legs do not afford it a very powerful hold in such situations, but by pressing this brush against the shell it obtains additional support and is prevented from slipping back. The head of the grub is not visible in the photograph, for, except when it is feeding, it keeps its head retracted under the shield which forms its first segment.

When the grub has reached the full size it becomes inactive, and lies on its side in a somewhat contracted condition. Then the skin begins to split along the side just behind the head, and a creature of somewhat different form is disclosed. By alternate contraction and expansion of its segments it withdraws itself from the grub skin, and shows that it has reached the chrysalis stage. In the chrysalids of most insects there is no apparent distinction between the males and the females: the external sexual characters are not revealed until the final (usually winged) stage is reached. But owing to the great difference, already indicated, between the mature male and female Glow-worms, a similar difference is shown by the chrysalids: the male chrysalis, as seen in our photo, having the rudiments of wings in their wrappings showing plainly at the sides behind the head.

The Glow-worm remains in the chrysalis condition only about a fortnight. At the end of this time, during which a wonderful process of development has been going on within the chrysalis skin, the latter splits and there walks out the complete insect. If the chrysalis was that of a male, the change is very great, for the jointed upper surface visible in the grub is now concealed by a pair of semi transparent wings, which are folded up neatly, and in turn hidden when the insect is at rest by a pair of thin, horny wing-covers. These at first appear to be too short to cover all the soft body, but this soon contracts until the wing-covers cover all.

[*By Theo. Carreras.*

THE GLOW-WORM'S MYSTIC LIGHT.

A view in a Surrey lane, with the hedge-banks lit up by Glow-worms that have climbed the herbage. In some favoured places they are specially abundant.

The Glow-worm as an obvious beetle must be sought in appropriate localities in the month of June. At that time we have known a lamp with a white globe, exposed to open windows in the country, to attract the male Glow-worms in hundreds. They flew in and circled round the lamp—possibly mistaking its light for that of an extra brilliant female—and then dropped upon the table-cloth and, perhaps, called themselves names for being so easily deceived.

Photo by] [H. Main, F.E.S.

THE FEMALE GLOW-WORM

The female Glow-worm has no wings, and differs but slightly from the grub. The lighter patch towards the tail is the luminous area, which is more brilliant in the female than in the male.

SEA HORSES AND SEA DRAGONS

BY W. P. PYCRAFT, A.L.S., F.Z.S., ETC.

THE " Sea-horse," and its near relation the Sea-dragon, are both cousins of the " Pipe-fish," and all these are among the most remarkable of fishes, since their bodies are encased in a bony armour, and they swim in an upright position, propelling themselves by rapid vibrations of the back-fin ! But these peculiarities of structure and locomotion are soon forgotten in the contemplation of their shape and the conditions of life which have determined it.

The Pipe-fish must be taken as our starting-point. The long, tapering shape and the green colour of the body play an important part in the life-history of this creature, for it lurks in beds of long, grass-like sea-weed, the blades of which it so closely resembles as to be practically indistinguishable therefrom. And thus it escapes the eyes of prowling enemies, for escape by flight it cannot, and it is too feeble to be able to offer active resistance.

Nature has required to do but a very little modelling to fashion out of the body of the Pipe-fish the still stranger " Sea-horse," which is remarkable among fishes in being able to use its tail as a finger, to grip weeds and thus to come to anchor at will ; and when not thus in use, this wonder-

SEA HORSES AT HOME.

The remarkable development of the snout and breast of these singular fishes, combined with their accustomed habit of remaining erect, produces a wonderful resemblance to the head and shoulders of a horse. The tail-fin of other fishes is absent, and the tail is coiled around the stems of sea-weeds.

Photo by] [*W. Saville Kent.*

SEA DRAGON.

The Sea Dragons are related to the Sea Horses. Frills and lobes of skin from various parts of the body look like sea-weeds, and give the fish a very weird appearance.

ful tail is coiled up like a watch-spring to aid in keeping the body upright when swimming. To this new use of the tail we may ascribe the loss of the web, or " fin," which forms the characteristic termination to the tails of all other fish ; and it is easier to see how this end of the body got its shape than to imagine the causes which brought about the striking change in the position of the head, which is now bent at right angles to the long axis of the body. Partly on this account, partly from the shape of the head itself, the horse-like aspect of the fore-part of the body is due.

This fanciful likeness to a horse can be of no service to this fish, and it is not clear, exactly, what advantage is derived from this strange shape. Among weeds, however, the body looks like nothing in particular, and certainly does not suggest fish. Probably, therefore, potential enemies are put off the scent and evaded just on account of the creature's nondescript character.

The " Sea-dragon " stands in another category. A very slight examination of its general shape will be sufficient to show that it must be a kind of " Sea-horse," but one strangely transformed by extravagant trappings. These are donned, however, as no mere ornaments, but as very material aids in lessening the struggle for existence. For these same trappings are portions of its own skin, loose and flowing, and cut, so to speak, so as to look like fronds of the sea-weed amid which this strange creature lives. So disguised its personality is lost, so that the ravening wolves of the sea may pass it a dozen times a day and yet, save by the merest accident, they would never discover that they were passing not sea-weed, but flesh and blood, the object of their search ! Naturally, of course, the coloration of this odd-looking body matches its surroundings not less perfectly than its shape.

Though the powers of locomotion of these fishes are extremely limited, yet, strange to say, they have contrived to spread themselves practically all over the world : as may be gathered from the fact that the Pipe-fish and Sea-horse extend their range into our own seas, while to find the strange Sea-dragon we should have to search the seas of far Australia. It may be that the Sea-horses owe their wide distribution to accident; that is to say, they have been drifted hither and thither in the course of time on floating sea-weed, and getting into ocean currents have journeyed far until their bit of weed has got entangled in a weedy jungle on another shore.

But besides their shape and coloration, and many peculiar anatomical features which do not come within the scope of the present article, the fishes of this strange family have quite peculiar methods of guarding their young, which, unlike the young of most fishes, are nursed; and this, too, by the male parent. In the Pipe-fish and Sea-horse the eggs are carried in a pouch under the tail or embedded in the skin of the breast; and in the Sea-dragon a groove serves the same purpose. After the eggs have hatched the pouch serves as a nursery for the young, after the fashion of kangaroos.

The family to which these strange fishes belong is much larger than most people would suppose, since it includes no fewer than one hundred and seventy-five species. Nor are those which we have selected for special mention the only members deserving such notice, for there is an Australian species of Pipe-fish which has adopted a quite extraordinary mode of life, since it lives inside the sea-cucumber! This, however, is not the only fish which has found the interior of this animal a desirable dwelling-place; for the curious Fierasfer has long been known as addicted to this strange custom of haunting living caverns! But that is a story that must be reserved for telling on another occasion.

Photo by] [*W. Saville Kent.*

A SEA DRAGON.

This Sea Dragon is a native of Australia. It is so richly ornamented with loose and flapping ribbons of skin that it reminds one much of the fearsome creature that has been adopted as the Chinese national emblem. There can be no doubt that in the neighbourhood of the sea-weed jungles where it lives, these ribbons look so exactly like the sea-weed fronds that their fishy nature is not suspected by possible enemies. Its mouth is so small that it is unable to protect itself by ordinary methods.

BEAVERS AND THEIR WORK

BY R. LYDEKKER

BEAVERS have the distinction of being the only mammals that, under certain circumstances, permanently alter the configuration of the districts they inhabit. Such changes are produced in the first instance by the dams of timber, boughs, mud, and stones which these rodents construct across rivers, in order to extend the expanse of smooth water, or to obtain sufficient depth for their evolutions, and to enable them to enter their subaqueous burrows when the surface is frozen. Such dams often have a length of five hundred feet or more, with a height of twelve or thirteen feet, and a basal width of twenty feet ; the thickness tapering from the base till it becomes only about a yard at the summit. Rarely the dam may be about a mile in length.

Photo by] [*New York Zoological Society.*

THE BEAVER.

The broad flattened tail is well shown in this photograph. It was formerly thought that the beaver used it as a trowel.

Above the dam the water is headed up so as to form a pond or lake. In the course of centuries the lake silts up, by the aid of mud and vegetable growth, till eventually a plain of rich soil is formed, which, when clothed with grass, is called a beaver-meadow, and forms a new feature in the landscape. As an instance of the extent of these meadows and their drainage canals, it may be mentioned that the name of the Indian village which formerly occupied the site of Montreal means beaver-meadow.

Photo by] [*New York Zoological Society.*

BEAVER.

This beaver has started off on the search for further materials for lodge building.

Beavers are confined to Europe, Northern Asia, and North America, and in prehistoric times inhabited the English fens. Nowadays, they are almost exterminated in Europe, although pairs or colonies survive on the Elbe and Rhone, and in parts of Scandinavia. Occasionally these construct dams, but they generally live in burrows. In Canada, although their engineering skill has been overrated, they fell trees of considerable size for their dams by gnawing through the stem a foot or so above the ground ; such trees generally leaning over the water, and therefore falling in the right direction. From the subaqueous entrances to their burrows they frequently build mounds of brushwood and mud, which are known as " bank-lodges " ; in other cases, the structure

Photo by] [*New York Zoological Society.*

A BEAVER'S DAM.

In order to ensure a sufficiency of water around their lodge the Beavers construct a dam across a stream by piling branches and small trees, weighed down and cemented by mud and stones. These may be twenty feet thick at the base and twelve or thirteen feet high.

Photo by] [*W. A. Dimock.*

BEAVERS ON THEIR DAM.

In this photograph a pair of beavers are seen actually at work constructing their dam across a rapid stream.

is built far out in the water and is then termed a " lodge." Despite stories as to their complex structure, such lodges merely consist of heaps of brushwood and chips with a large central chamber in which the occupant may dwell secure from foes. The story that the Beaver uses its flattened, trowel-like and scaly tail to plaster down the mud of its dam is a pure fiction ; the whole of the work being done by the paws and teeth. The real use of the horizontally flattened tail is to assist the animal in coming rapidly to the surface to breathe or work.

Like squirrels, rats, and rabbits, Beavers have a pair of strong, chisel-like teeth in the front of each jaw, which grow throughout life, and always keep a sharp edge. It is by these that tree-felling is done. A peculiarity of the Beaver is the presence of an additional claw on the hind-foot, thus producing a kind of comb-like structure, probably used in cleansing the fur.

It is, however, not only on account of its marvellous engineering works that the Beaver has a special claim to interest ; for, in proportion to its size, it has a commercial value which is exceeded by few other animals. Its value at the present day is, however, somewhat less than was formerly the case, owing to the fact that one of its products is now held in little esteem. Here it may be well to mention that to the ancient Greeks the Beaver was known as *castor*, a name said to be derived from *gastro*, the stomach, and to have been applied to the animal on account of its paunchy appearance. Be the derivation what it may, the name has been adopted as the first half of the scientific designation of the beaver.

From *castor* is derived the word *castoreum*, applied to a resinous and strong-smelling secretion contained in a pair of pouches in the groin of the Beaver. Formerly castoreum was highly esteemed both in medicine and perfumery ; and not many years ago from eight to ten dollars a pound was the current price for this commodity in Canada. Nowadays, however, its use in medicine has, I believe, entirely ceased, while its employment in perfumery has been largely curtailed by more delicate vegetable scents.

In the old days, when "top-hats" were made of beaver-fur, the slang name for these decidedly cumbersome articles of head-wear was "castor"; this being exemplified by an old story current in the days when Chief Baron Pollock was at the head of the Court of Exchequer. On one occasion, when a brother judge was leaving a house after dinner, he was asked whether a certain hat was his "castor," to which he wittily replied: "No, it is Pollock's" (Pollux).

Photo by] [*New York Zoological Society.*

A BEAVERS' LODGE.

These lodges are frequently built right out in the water, but here advantage has been taken of a tree to build against. Sometimes a lodge will be built against a high bank.

Beaver-fur, or, rather, the soft under-fur, was employed in the manufacture of hats at least as early as Queen Elizabeth's time, the material being worked up into felt, which had a rough, hairy coating on the outer surface; and when the Hudson Bay territory was opened up, beaver-skins formed by far the most important item in the local trade. Indeed, so valuable and important were they considered, that they formed a kind of token coinage, represented by what were called "beaver-tallies"; the value of other commodities being reckoned in beaver-skins.

Eventually the old fashioned beaver-hat became—so far, at least, as men are concerned—replaced by the modern silk hat, which is, of course, made on an entirely different plan, the base being of what is called hatter's cloth, upon which is fastened a

Photo by] [*Reinold Thiele.*

THE ANGLER, OR FISHING FROG.

The Angler lies on the weedy or muddy bottom with his enormous mouth wide open. A passing fish that happens to touch the shred of skin that hangs like a bait from the long, curved fin-ray, is at once secured by the trap-like snapping of the well-armed jaws.

cover of silk plush, finished after the fashion of the old beaver, but showing an infinitely smoother and more glossy surface.

When France had the monopoly of the Canadian fur trade, the export of beaver skins attained enormous dimensions, the number of skins received at La Rochelle in the year 1743 being no less than one hundred and twenty-seven thousand and eighty. Despite continued persecution of the animal in the intervening years, the numbers showed no very great falling off a century and a half later, the Hudson Bay Company having sold sixty three thousand four hundred and nineteen skins in January, 1891.

Beaver skins, which may be used either with or without the long outer " water " hairs, are dyed black or brown, and sometimes furnished with white hairs to imitate the valuable sea-otter. They are usually employed for trimmings, cuffs, or muffs, according to the fashion of the day.

AN INFANT ANGLER.

Like most other fishes the Angler leaves the egg with a very different form from that of its adult stage. Note the remarkable length of the bones of the breast-fins, which later become short and paddle-like.

THE FISH THAT FISHES

BY EDWARD STEP, F.L.S.

THE Angler, or Fishing-frog, is one of the most remarkable of our fishes. Built more on the lines of a tadpole than a fish, two-thirds of its length (not reckoning the tail-fin) consists of head and mouth, and the remainder is so narrow in comparison that it is not to be wondered that it is regarded as mere tail. What has always been considered the most remarkable point in its structure is the possession of several long rod-like bones, which stand up on the upper surface of the head. To one of these is attached naturally a strip of skin-like substance which the imaginative naturalists of half a century and more ago described as worm-like and glittering. The Angler was said to manipulate this rod in so skilful a way that the appendage was as good as the artificial minnow which the human angler spins for the snaring of real fishes. The description was copied from book to book as the custom was—and still is to some extent—and nobody appears to have taken the trouble to look at a living Angler-fish to see how far description and fact agreed. There is really no need for invention in the matter.

A glance at the photographs will show that the Angler is not built on clipper lines. It would be impossible for him either to escape from an enemy or to secure his dinner by virtue of swift going in the water ; his lines are more those of the round-bowed barge. So the Angler has taken to a sedentary life on a weedy sea-bottom, where he can hide the greater part of him among the weeds

Photo by] [*E. Step, F.L.S.*

A SIDE VIEW OF THE ANGLER.

The side-fins are developed into paddles, by whose aid the Angler hoists himself along. The huge mouth extends back almost to these paddles.

or mud, open his capacious mouth at the edge of the jungle, and dangle his bait just above his lower jaw. His motto is the dubious saying, " Everything comes to him who waits." To a creature like the Angler, content with a life of stagnation so long as his bodily wants are satisfied, such a motto is sufficient. A fish of inferior size prowling around the jungle sees the bait dangling in the current and goes to investigate. A touch is sufficient. The nerves that clothe the rod-like spine flash the news to the creature's brain that a dinner is " within the range of practical politics," the lower jaw closes with a snap, and the dinner is secure within the bag-like mouth.

The breast-fins of the Angler are not like fish-fins at all, but are more akin to the flappers of the seal. They are thick and powerful, their bony structure reminding one of the human wrist, and the Angler uses them to hoist its heavy body slowly along. The skin of the entire body is soft and scaleless, so that a large specimen is a difficult burden to handle. The lower jaw and a line continuing therefrom along each side are fringed with a series of fleshy lobes which bear a close resemblance to the olive wrack among which it lies, and which is shown beneath it in the smaller photographs. This line of lappets marks the division between the brownish grey upper side and the white underside. The Angler grows to a length of five or six feet.

The Angler's " rod " is really the first bone of the first back-fin. Its base is attached to the snout, and the extremity is curved so as to enable the bait to hang loosely. It most resembles a decaying stalk of seaweed arising from a weedy boulder, and there is little doubt that it is so regarded by fishes who brush unceremoniously against it, and so set the terrible lower jaw in instant motion. This lower jaw is longer than the upper. Both are liberally set with several rows of long, tapering teeth, their points sloping towards the fish's throat. Cod a couple of feet long have been taken from the Angler's stomach, and in one case over seventy herrings, so fresh that the fisherman sold them without question that they had been taken in the usual way. This latter case looks as though a school of herrings had passed his way and the Angler had been able to take several gulps at them.

A story is on record of an Angler that was found dead on the shore with a fox hanging out of his mouth. It was conjectured that he had been stranded by a wave and had been unable to drag his soft but heavy body down to the water. The fox had come along in quest of crabs and had inadvertently touched against the trigger of the spring-trap, which had snapped to and put an end to his career. He has been found choked through attempting to swallow the cork float by which the position of a chain of crab-pots was marked; and others have been caught whose stomachs contained a sea-gull, a northern diver, or other marine diving bird. These instances, which might be extended, tend to show that the Angler's jaws act almost mechanically, and that little brain-power is used up in the discrimination of what they seize. The extent of these jaws, as shown in the photograph below, rather indicates that quantity is of more importance than quality.

The female Angler lays about a million and a half of eggs, and the number indicates that in early life the Anglers fall a prey to many other fishes, for which the rapacity of the elders can only be regarded as a reprisal—or *vice versa*. The young Angler passes through larval stages in which it is active and much more of a swimmer, its form being more upright and consequently more fish-like than it becomes in later life; the illustration of the last of these stages, given on page 91, shows the true nature of the "fishing-rod"—there it is clearly only one of the spines of the back-fin. In spite of the numerous eggs, the young Anglers appear to be scarce. The eggs adhere together in a vast glutinous sheet or ribbon, much after the fashion of the spawn masses of the frog, and float at the surface of the sea.

Cookery books may be searched in vain for any mention of the Angler as a food fish. The fish is too hideous in appearance to allow of its admission to the kitchen under its proper designation; but it is said that at certain fishing ports the body is separated from the head and sold under the name of "croan" and "dory." The latter is clearly a false pretence, for dory is the name of a distinct fish of quite other shape and habit.

Photo by] [*E. Step, F.L.S.*

A FULL VIEW OF THE ANGLER'S TRAP.

The Angler is not adapted for swimming, and lies almost concealed by its colour harmonizing with the sea-weeds. With its well-armed jaws open and its bait dangling above them, it waits patiently for any passing fish to come within range.

FLYING DRAGONS

BY R. LYDEKKER

Despite their title, the Flying Dragons of the Malay countries and the Philippine Islands are harmless little lizards, only a few inches in length, remarkable for their brilliant colouring and the expansion of the skin of the sides of the body into a large parachute-like structure, which enables them to take long flying leaps from bough to bough, or even from tree to tree. In the extent of these leaps they are quite unique among reptiles, and are, therefore, truly "marvellous animals," although they do not fly in the proper sense of that term; that is to say, they have not wings like those of a bird. In consequence, their flight, or rather their "glide," is always of strictly limited extent, and cannot be prolonged beyond the range of the original impetus given by the starting-point.

Photo by] [*W. Saville Kent.*

THE FLYING DRAGON.

In this photograph the remarkable dewlap is shown extended in front of the head. It is coloured brilliant orange or blue.

The parachute is supported on each side of the body by an elongation of the last six or seven ribs, and when the animal is at rest, can be folded up in fan-fashion. The tail is fully twice as long as the head and body, so that the total length of the creature is fully eight inches. There are about a score of different kinds of these beautiful little lizards, whose colouring is so brilliant and forms such complex patterns, that exact description is well-nigh impossible. Generally the upper surface displays bright metallic colours of varying tints, upon which are numerous dark blotches and bars, producing a marbled effect. The throat has a kind of dewlap, capable of expansion under the influence of excitement, and in one of the species is brilliant orange in the male, but bluish in the female.

[By Theo. Carreras.

THE FLYING DRAGON.

The Flying Dragon is a lizard of Malaya, where it lives entirely in the tree-tops. By a wonderful adaptation of its ribs it is able to take long flying leaps from tree to tree, balanced by its extensive tail, which is twice the length of its body.

Flying lizards, as they ought to be called, are strictly arboreal reptiles, dwelling high up in palms and other trees, and, therefore, seldom seen, as it is but rarely they descend to the ground. When the parachute is expanded, they look like gorgeous butterflies, or even brilliant flowers ; and when the parachute is suddenly opened preparatory to flight, the contrast to the appearance of the creature when at rest is really startling. Much the same, in fact, as when a moth with brightly-coloured under-wings suddenly starts to fly. For when a flying lizard is at rest, clinging tightly to a bough, the mottled colouring of the back harmonizes so exactly with the lichen-clad bark, that detection at a short distance is almost impossible, although the creature's presence is often revealed by its brilliant eye. Unlike many kinds of lizards, Flying Dragons appear to have no power of changing the colour of their skin in accordance with surrounding requirements.

When moving along the branches in search of the small insects which form their prey, Flying Dragons generally progress by a series of short hops or leaps ; and it seems that it is only when they have exhausted the food-supply of the branches on which they happen to be situated that they sally forth in flight in search of pastures new. They, of course, always fly from a higher to a lower bough, and when they have got too low down they are consequently compelled to reacquire a sufficient elevation by climbing the trunk. They appear to have no power of altering the direction of their flight.

THE STING OF A NETTLE

BY EDWARD STEP, F.L.S.

THERE are probably few persons who can say they have never experienced the irritation produced by the Stinging Nettle having come lightly in contact with their tender skin. We say " lightly," for there is truth in Aaron Hill's old jingle :

" Tender-handed stroke a nettle, and it stings you for your pains ;
Grasp it like a man of mettle, and as soft as silk remains."

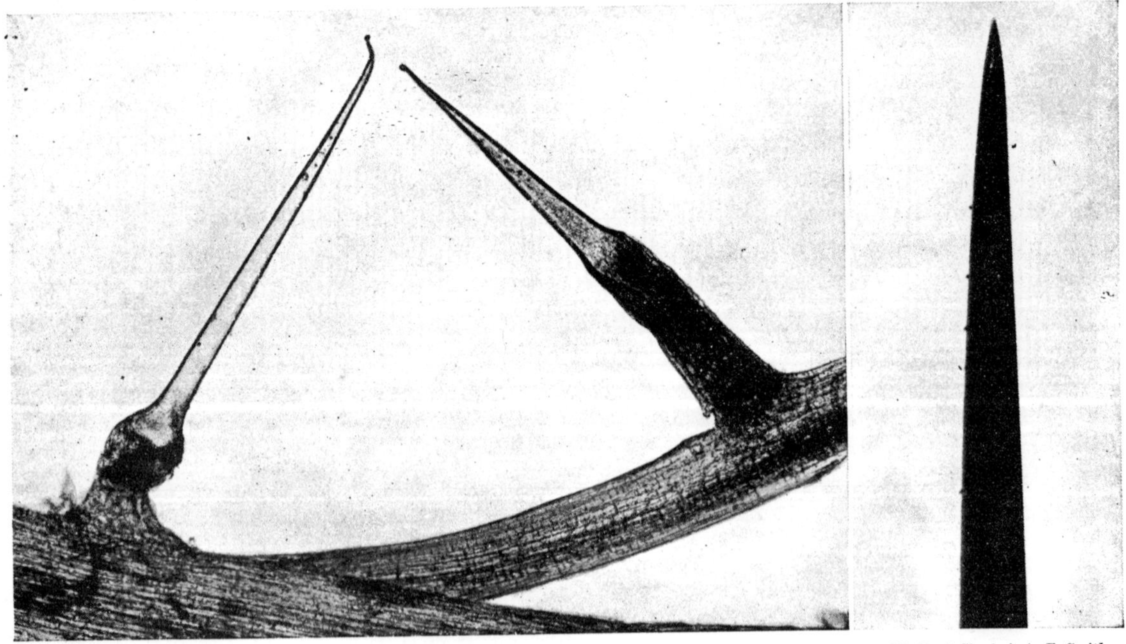

Photos by] [*F. Noad Clark & A. E. Smith.*

THE STING OF A NETTLE. A NEEDLE-POINT.

The sting of the Stinging Nettle is compared with the point of the finest needle made—a number 16. Note the little knob guarding the fine point of the sting, but which is broken off by the slightest contact. Both are enlarged to fifty times the actual thickness

Common as is the experience of being stung by this common hedgerow weed, few of its victims trouble to inquire by what means so delicate a contact can cause so much annoyance and irritation. Yet the sting of the Nettle is worthy of a little examination, for though its mechanism is so simple it is wonderfully efficient. If, with an ordinary pocket-lens, we examine the leaf or stem of the Nettle we shall discover that it is closely covered with very fine hairs. If we go further and submit one of these hairs to the more powerful lens of a microscope so as to increase its apparent size fifty times we shall see the hair as it is represented in our photograph. The hair, it will be seen, has a bulbous base from which it tapers to a fine point, where, however, it bends and slightly swells into a knob. In the enlargement this looks like a design to prevent the sting acting, and to be much like the button on the point of a fencing-foil. But just below the little knob the hair is exceedingly brittle, and the slightest touch will snap it across, leaving the now unprotected point free to enter the skin.

The hair, fine as it is, is pierced throughout its length by a tube which connects it with the swollen base, which is really a bag filled with poison. The pressure necessary to drive the point through the victim's skin also exerts itself upon the base, with the result that a minute drop of the poison is forced through the hollow hair into our blood, to be followed by the local irritation which sets up the small white swelling around the puncture. The fineness of this hair is made readily apparent to the eye by comparing it with an equal length of the point of the finest needle that is made, magnified to precisely the same degree—fifty times its thickness. Beautifully polished as the needle is, the microscope reveals a few burrs upon it, such as are not to be found on the Nettle's sting.

Photo by] [*P. H. Fabre.*

THE SCARAB'S HARVEST.

The Sacred Beetle is a dung-feeder; and here we have a party of Scarabs attracted to the feast, and busy rolling it into balls that they may push it away to the underground burrows they have prepared.

THE SACRED SCARAB

BY H. ST. J. K. DONISTHORPE, F.Z.S., F.E.S.

THE sacred beetle of the Egyptians was one of the insects best known to the ancients on account of its curious and interesting habits. Before dealing with the facts relating to its life-history, it may be as well to give a brief account of its more ancient history and the superstitions regarding it. It was worshipped and regarded as sacred by the Egyptians, and its figure is to be found in abundance on their paintings and amulets and carved on their tombs and monuments. Small figures of the beetle in jewels and stone were made in large numbers, and often interred with the

Photo by] [*P. H. Favre.*

SACRED BEETLES.

The Scarab having made a fine ball of food is joined by a brother who, under the guise of a helper, aims at appropriating the tempting sphere: but the owner repulses the false friend.

bodies of the dead to drive away evil spirits. The insect itself was often placed in the coffins, and sometimes it was embalmed. The Scarab was adopted as a religious symbol to represent the motions of the earth and sun and a unique birth; on account of their rolling balls of dung, and of their sudden appearance, often in great numbers, on the margins of the Nile, after the annual rising and falling of the water. They were considered symbolical of the world on account of the globular form of their pellets of dung, and from an odd notion that they rolled from sunrise to sunset. As a symbol of the sun, because of the angular projections from the head resembling the rays and the ball being rolled backwards with their feet whilst the beetles themselves look forward. As the sun appears to proceed in the heavens in a course contrary to the signs, thus the Scarabæi turn the balls towards the west, while they themselves continue to move towards the east; the first of these motions was considered to exhibit the diurnal, and the second the annual motion of the earth and the planets. They chose the insect as a symbol of the moon, which divides the year into months of twenty-eight days each, because they believed the pellet of the Scarabæus remained

Photo by] [*P. H. Fabre*

THE SACRED BEETLE.

The Scarab pushing backwards is the maker and rightful owner of the ball, which he is rolling along to his burrow. The other is trying to alter its direction with a view to getting it to his own burrow.

in the ground for a period of twenty-eight days. As there were supposed to be no females of this species, it was chosen as the type of Phtah, the creative power, and of Regeneration or Reproduction, because it was the first animal observed after the subsidence of the waters of the Nile. It was also supposed to represent Fecundity, and these beetles are even yet eaten by the women of the country to render them prolific.

Now, however, let us turn to safer ground, and consider the beetle from the more substantial basis of known facts.

The Scarab, which is found in Provence in France, in the south of Europe, in Egypt, Barbary, etc., is a large beetle, black in colour, with long legs and awkward action. It possesses ample wings, concealed beneath the wing-cases when on the ground, and can fly readily. The head is armed

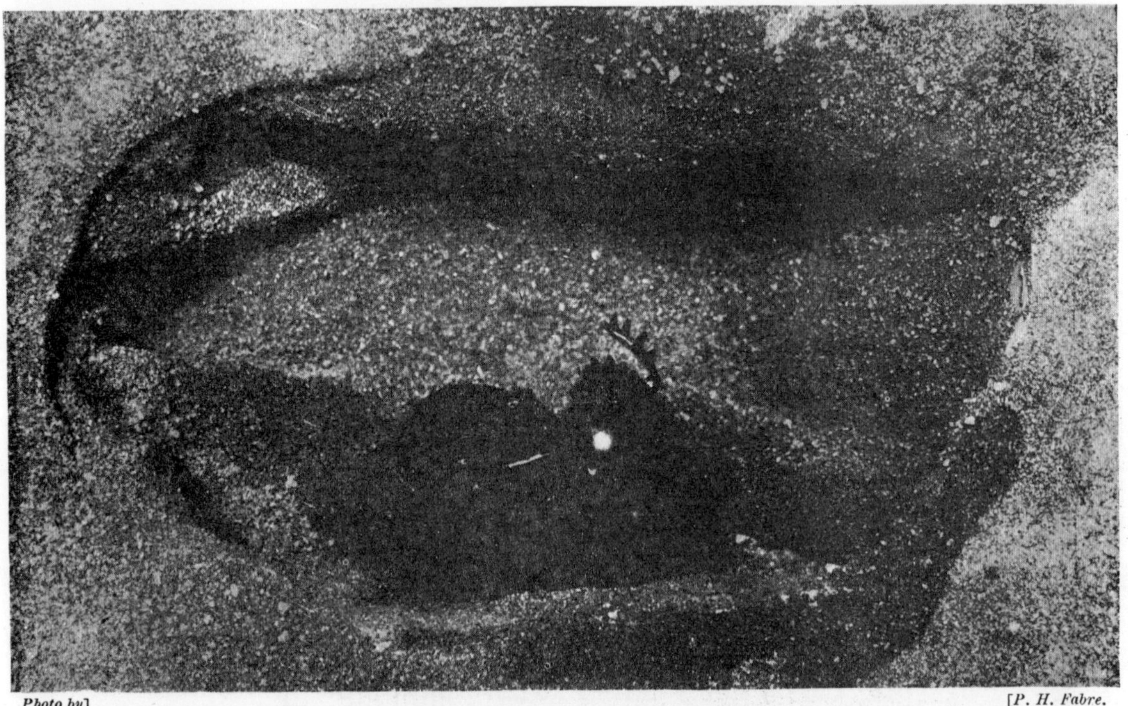

Photo by] [*P. H. Fabre.*

THE SCARAB'S CELLAR.

The underground crypt to which he has succeeded in getting his treasure. Here he can eat in peace after his prodigious efforts, or store up the food to be eaten by the Scarab grubs.

with teeth, with which it digs, and the back legs are long and bent round, which enables them to grasp the ball of dung. Most beetles, and, indeed, all insects, possess six feet, usually composed of five joints each; but in this species the front pair of legs do not possess any feet at all. In other of the dung-feeders which do possess front feet, these feet are often to be found broken or injured by continual digging in hard ground, and can be of little, if any, use to the beetles. Darwin suggests that the want of front feet in this species is due to the effects of long-continued disuse. The male beetle is said to stridulate (*i.e.*, to make a noise by rubbing hard parts of its body together in the same way that grasshoppers, crickets, and also some other beetles call to each other, or cry out when alarmed), to encourage the female when working, and to express distress if she is taken away from him.

The Scarab acts as a scavenger by breaking up and burying the droppings of cattle and other animals; it will thus be seen to play a very important part in the economy of nature. They collect balls of dung, which they roll and eventually bury in the ground. It was supposed that these balls contained the eggs of the beetle, and it is so stated in most of the works on natural history which deal with this insect. This, however, is not the case, as has been shown by the great French observer, J. H. Fabre, to whose patient investigations are owed most of the facts known concerning this interesting creature.

For two months, in June and July, the beetles are busy collecting and rolling balls of dung, which they devour. When the weather gets too hot they bury themselves in the cool ground and reappear in the spring. Their attention is then taken up with the continuation of their race.

The beetle finds by scent the droppings of some animal, to which it flies. It collects masses of dung with its front legs, which are bent, and passes them under the body to the long circular back legs, which shape the masses into a ball. The ball is turned underneath the beetle and rolled into a perfect sphere. In a very short time a large ball is collected, sometimes as large as an apple. The beetle then proceeds to roll the ball backwards with its hind legs. All the time the position of the ball is changed with the hind feet, so that every part of it comes in contact with the earth, and it is thus rolled hard and round. The object of the beetle is to move the ball away to a suitable spot where it can dig a hole, bury and devour it at leisure. It shows the greatest patience and perseverance in pushing the pellet over steep and difficult ground, or when it is obstructed by a root or stem. Should its pellet slip and roll down a slope, it will push it up again and again until the summit is achieved. Another discomfort often awaits our insect. Another beetle will not unfrequently join it, and proceed quite unnecessarily to help it push the ball along. The newcomer always has its face towards the ball, to which it fastens itself. Its object is to either rob the rightful owner of its property, which it promptly does if an opportunity is offered it, to decamp with it, or at least to share the feast, without having had the trouble to collect a meal

Photo by] [*P. H. Fabre.*

TUMBLE-DUNG BEETLES

Two other members of the same family as the Sacred Scarab, that share its habit of burying dung in tunnels they have dug.

for itself. It was often thought that when two beetles were found pushing or devouring the same ball, they were the male and female; and that if a beetle was unfortunate enough to allow its ball to fall into a hole out of which it could not remove it, it flew away and brought back with it other individuals to help. Monsieur Fabre has shown that neither of these ideas is correct. He found that if he fastened the ball to the ground with a long pin, the head of which was buried in the ball, the beetle would, after much pushing and pulling, lever the ball off the pin by raising it with its back, standing as high as it could on its feet. Could it not reach quite high enough to force the ball off the pin, it would make use of little stones he placed for it. If he did not help it, when the beetle could not reach high enough, it would fly away and leave the ball suspended on the pin. If, also, he placed the ball in a hole out of which the beetle could not raise it, after a time it flew away. In neither case, however, did it ever return with other beetles to help it out of its difficulties.

Photo by] [*P. H. Fabre.*

THE SACRED BEETLE'S FEAST.

Having rolled away a large ball of food, the Scarab sets to work to eat it, and will eat steadily on until he has consumed the entire mass.

When the beetle has reached a spot which it considers suitable, it proceeds to dig a hole with its armed head and powerful front legs. It forms a chamber large enough to receive the ball, and blocking itself up in this temporary dining-room, it begins to devour its prize. It never ceases to eat until the whole of the pellet has been consumed, taking from a week to a fortnight over it. When all is finished, it comes out of the ground again, flies away, and proceeds to find a fresh dropping. When one beetle has come to the assistance (!) of another, and a place has been selected to bury the prize, the original owner starts to do the digging. The treacherous friend now takes the opportunity to run away with the coveted morsel.

Special chambers are prepared in the spring for the eggs. The female beetle lines it with large quantities of dung arranged in different layers. The centre layer, which contains the egg consists of a creamy paste for the newly-hatched grub to feed on. The outer layers become coarser and coarser to be more suitable to the grub as it grows larger. The grub changes to a chrysalis when full fed, and when hatched the perfect insect breaks out of the earth and flies away.

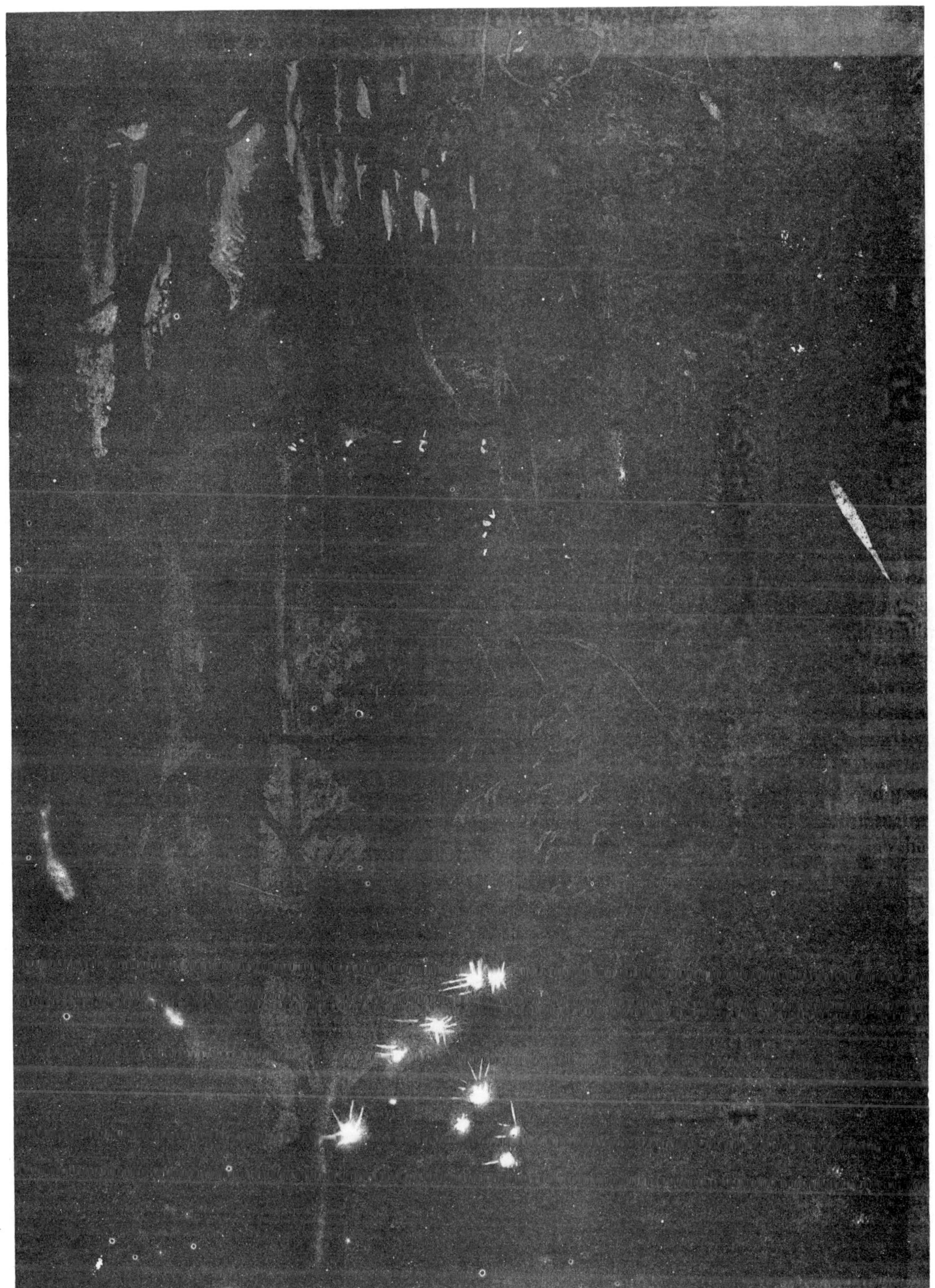

[*By A. E. Knight.*

LUMINOUS FUNGI IN A BRAZILIAN FOREST.

Certain toadstools are luminous, and give out a soft but brilliant light. It has been thought this is a device to attract certain insects who may be instrumental in the dispersal of their spores.

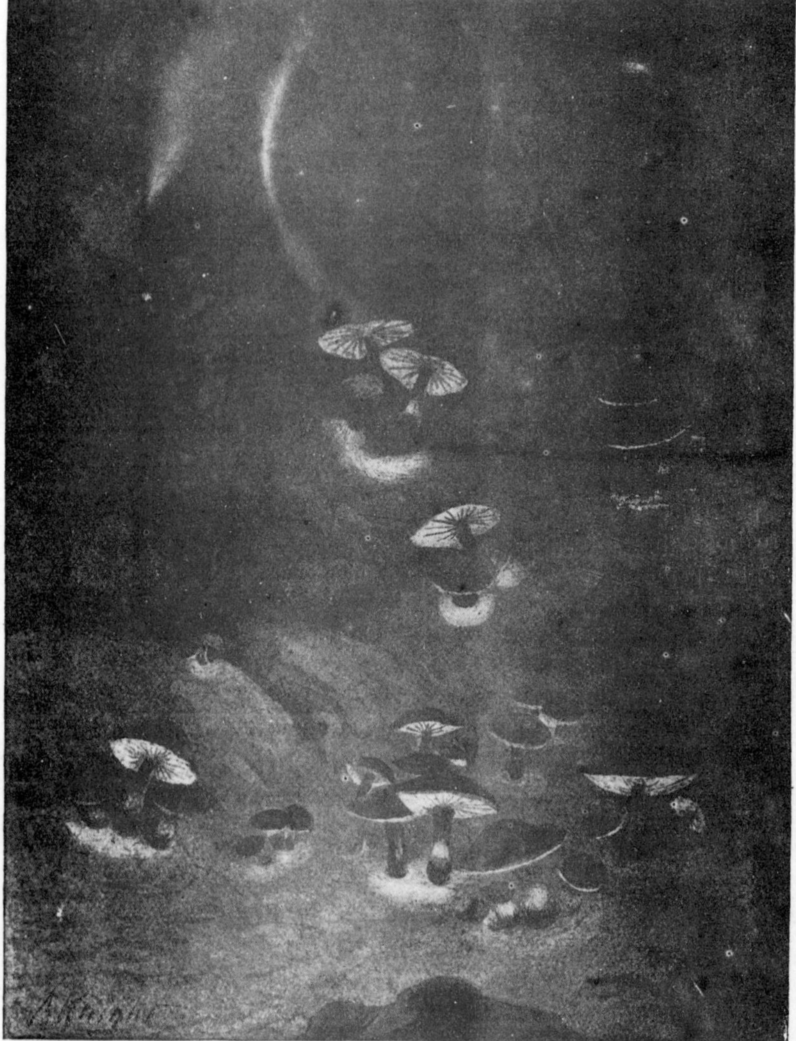

[By A. E. Knight.

LUMINOUS FUNGI.

In some places travellers have recorded that these give out light sufficient to read a book by.

LUMINOUS FUNGI

One remembers that in the stories of school-life of a generation or so ago it was one of the favourite "properties" of the writers to tell how a soft new-comer was terrified by an arrangement of touchwood, which glowed after dark with a strange, mystic light, forming a sentence of warning or doom. The touchwood consisted of rotten wood whose softened tissues were permeated by the delicate threads of which the earlier stages of the fungi consist. In the case of some of the large fungi that attack timber, these delicate threads develop into thick interlacing cords that become luminous in the dark; and in caves and mines similar cords run down the walls. But perhaps more striking is this luminosity when it is displayed by the mature fungus, the stage in which we can recognize it as a mushroom or a toadstool, whichever we may choose to call it. Certain of the large, leathery fungi that grow out from our trees are said to exhibit this luminosity; but the majority of cases on record relate to species that bear gills on the under-surface, like the common mushroom. One such group of tree-growing mushrooms is known as Oyster-mushrooms from the shape of the cap, which has a short stalk at one side instead of in the middle.

As early as the year 1840, Mr. Gardner, travelling in Brazil, met with such a mushroom which grew on decaying parts of a palm. In fact, he was led to discover it by finding the boys of a village playing after dark with this fungus. The plants attacked by the fungus gave out from every part a bright phosphorescent light much like that produced by the South American fireflies. Dr. Collingwood has recorded similar experiences in Borneo, which are supported by the observations of Mr. Hugh Low, who says that, when crossing the island by a road through the jungle, he saw the jungle all in a blaze of light, by which he could see to read. Investigation showed him that the light was produced by a mushroom. Then, soon after, Mr. Drummond reported a similar experience of fungi growing from tree-stumps in Western Australia. He gathered a specimen and laid it upon a newspaper, and found that by the light emitted he could read the newspaper easily. One such luminous fungus measured sixteen inches across, stood a foot high, and weighed about five pounds. A mushroom that grows on the olive-trees in the South of Europe is well known to give out a similar light, and other species with the same power have been reported from various parts of the world.

BIRDS AS BOWER-BUILDERS

BY FRANK FINN, B.A., F.Z.S.

THE remarkable erections of certain Bower-birds have long been familiar in the illustrations of natural histories ; but the purpose of, and the variety of construction in, these are not so well known. The Bower-birds belong to that class of bird that indulges in social dancing and posing as a part of their courtship. Not content with having a meeting-ground, the Bower-birds prepare it for the reception of visitors and even build and decorate playhouses.

In general habits and form Bower-birds resemble Jays, and build quite ordinary nests, the bower having nothing to do with nesting. Like Jays, also, they hop when on the ground, so that the attitudes given to them by some taxidermists, in which they are represented as running like Partridges, are quite misleading. The simplest form of playground used by any Bower-bird is that of the Tooth-billed species, which simply clears a space and covers it with large green leaves. The others all build some structure or other on the play-place, the best known being that of the Satin-bird of Australia, so-called from the satiny purple plumage of the old male. The hens and young

Photo by] [*W. P. Dando, F.Z.S.*

THE BOWER-BIRD AND ITS NEST.

The Satin Bower-bird is here shown building its arbour-like nest, both sexes putting the materials into position and adding decorative touches.

cocks are of a dull grey colour, and purple birds are so scarce that some Australian ornithologists look on the purple dress as a sort of anticipatory mourning plumage, put on by very old birds shortly before their death. The Satin-bird is a common species, and usually on view in Zoological gardens, where it freely builds its bower. This consists of an avenue of twigs set in a mat of the same, and round and in it the male shows off, constantly jerking up one wing and dancing absurdly with a leaf in his bill. His song is also amusing in the extreme, sounding like water gurgling down an obstructed pipe. Like Bower-birds generally, this species decorates its bower, but uses anything it can get hold of, such as shells, pebbles, bones and feathers. Other Bower-birds specialize more in their decorations: the Spotted Bower-bird, for instance, not only builds a larger bower than the Satin-bird, but uses more decoration, and prefers bones to most things for this purpose. This species, however, is also so fond of bright metal objects that it will even invade isolated houses near

Photo by] [*L. Medland, F.Z.S.*

A BIRD'S PLAYGROUND.

The Satin Bower-bird has built this arbour and laid out the surrounding ground with leaves, stones, beads, marbles, and anything else that appears attractive, and struts about its playground with evident pleasure.

the bush in order to steal them. Its plumage, as the name implies, is spotted, with buff on a brown ground, and the male has a collar of the brightest rose-pink.

Neither Satin nor Spotted Bower-birds can, however, compare in beauty of plumage with the cock Regent-bird, which is clad in golden-yellow and velvet-black; this bird's bower, however, is small and chiefly decorated with snail-shells. But by far the most extraordinary bowers are built by two species which favour flowers as decorations—Newton's Bower-bird of Queensland, a brown and yellow species, and the Gardener-birds of New Guinea, which are brown, with orange crests in the males. The Gardener-bird whose bower has longest been celebrated, builds a conical hut round the stem of a sapling, and in front of it lays out a lawn of moss on which are laid the flowers, buds and gay-coloured insects with which the bird makes the garden. As these decorations fade they are thrown away at the back of the hut and replaced by fresh ones. An allied species, whose constructions differ somewhat in detail, especially favours black objects, such as blackbeetles, as decorations.

The habits of Newton's Bower-bird, though comparatively recently made known, are, if anything, more extraordinary. This species raises heaps of twigs round two adjacent tree-stems, a yard

[By H. Grönvold.

NEWTON'S BOWER-BIRD.

This is, perhaps, the most remarkable of all the Bower-birds in its habits. It piles up sticks and twigs against two neighbouring trees, adding to them until they meet. In old erections one pile is several feet higher than the other, and may be nine feet high; all the work of bird no bigger than a thrush.

[By H. Grönvold.

THE REGENT BOWER-BIRD AND ITS BOWER.

This Bower-bird erects a smaller bower than some other species, and the most conspicuous decorations it uses are shells.

or so apart, adding to the piles year by year until they meet. In old houses one pile is always higher than another, and the higher one has been found three yards high—a good-sized edifice for a bird no larger than a thrush; this species being small for a Bower-bird. It is the higher pile or wall of the V-shaped arcade that the birds decorate with flowers, especially favouring white orchids and rock-lilies, while ferns and moss are also used. Quite a number of birds will frequent a large old bower, and they are quaintly touchy about having their decorations moved, so much so that the old males, who seem to be the artistic spirits of the community, will fight each other over the implied criticism of an altered decoration. Such a state of affairs seems to mark the limit attained by the evolution of bird social displays, and we may hail Newton's Bower-bird as the most highly evolved creature outside man in the matter of refined and intellectual amusement; indeed, considering the employments in which many of us humans find recreation, it is doubtful whether the bird has not the advantage of us.

The nearest allies of the Bower-birds are the Birds of Paradise on one side and the Crows—especially the Jays—on the other; and in both these groups we see indications of the instincts which strike us as so wonderful in the developments described above. The Birds of Paradise have the habit of assembling on certain trees to display; and at least one species, Lawes' Bird of Paradise, which is very like a Bower-bird in form and habits, has its dancing-place on the ground. The Crows, on the other hand, display a great fondness for collecting bright-coloured objects; at any rate, their habit of stealing these is well known to those who have kept or watched these birds in captivity. It is true that they only hoard these treasures up, but it may be that this seemingly senseless habit is the starved remainder of an instinct for collecting and displaying bric-à-brac which their primitive ancestors may have possessed, to gradually wear away in the struggle for life experienced by their descendants in the course of their wide distribution over the world. It is significant that the Bower-birds, with their wonderful habits, are confined to Australasia, a region where life seems to be more secure, judging by the survival there of so many forms of life which, in structure at all events, are undeniably primitive, pointing to a less acute struggle for existence.

THE LACE-BARK TREE

BY RICHARD KERR, F.G.S., F.R.A.S.

THE vegetable kingdom supplies many instances of the marvellous in Nature. Even with our limited knowledge of the mysteries of plant life we have abundant evidence that many plants are as wonderful in their way as members of the animal kingdom.

We may take as an illustration the inner bark of certain trees, which seems to be valued by native races more than by white people. This, however, may arise from necessity.

Doubtless many people have seen and inspected the curious and beautiful Lace-Bark of Jamaica, but few have any special acquaintance with specimens of the leaves and flowers or with the living plant as a whole.

It was permanently introduced to Kew Gardens, and in fact to Europe, in 1844. (A young Lace-Bark tree was introduced to the Kew Gardens by Rear-Admiral William Bligh in 1793, but it was soon lost.)

Trees of this species seem to be limited to the districts of Vere, Elizabeth and Clarendon in Jamaica. Mr. Wilson, a former curator of the Bath Botanic Gardens, Jamaica, procured seeds, and young plants, four inches high, which he sent to Kew in 1844, and thus was the means of introducing this rarity to our shores. In 1849 the plants were from eight to ten feet high, and one of them produced for the first time flowers and fruit copiously in the summer and autumn of that year. The authorities at Kew Gardens looked upon this as a very satisfactory event.

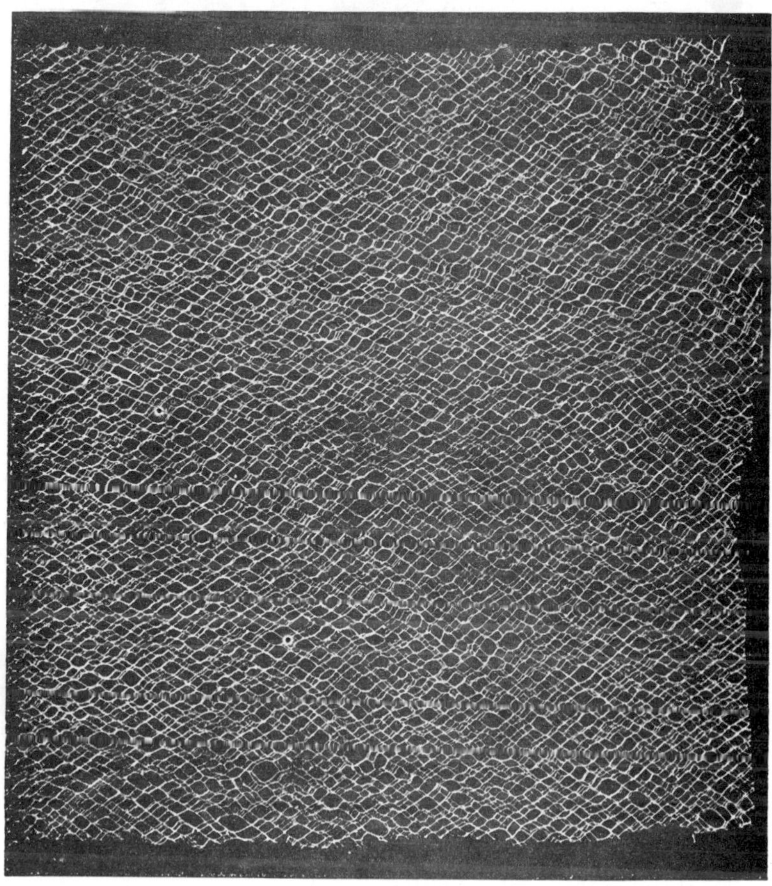

NATURAL LACE.
This is a photograph of a small piece of the inner bark of the Lace-Bark Tree, to show the structure of the lace. This being slightly enlarged, its mesh appears somewhat coarser than is actually the case.

The inner bark of this tree consists of layers of reticulated fibre, exactly resembling well-prepared lace; and its nature is best exhibited by cutting a portion from a branch, soaking it in water, and tearing down the bark. It is then separated by the hand into as many layers as that portion of the tree is years old. This last statement is interesting, as it answers the usual question: "Do these layers correspond with the years of growth?" The layers may number twenty or more. Dr. Lunan observes:—"The ladies of Jamaica are extremely dexterous in making caps, ruffles, nets, bonnets, veils, and complete suits of lace with it.

"In order to bleach it, after being drawn out as much as it will bear, they expose it (stretched) to the sunshine, and sprinkle it frequently with water. It bears washing extremely well, with common soap, and acquires a degree of whiteness equal to that of the best manufactured lace. The wild negroes have made apparel with it of a very durable nature, but the common use to which it is applied is rope-making. The Spaniards are said to have worked it into very strong ropes; and the Indians employ it in a variety of different fabrics.

"Sloane relates that Charles II. had a cravat made of the bark of this tree, which was presented to him by Sir Thomas Lynch.

"In the days of slavery the negro-whips were commonly made of the branches of this tree. A portion of a branch was taken, and the wood was removed from the part that was intended for the lash. The remaining portion of the branch formed the handle. If it was desired to ornament the handle, a portion of the bark at the end remote from the lash was unravelled until it formed a tassel of spreading layers of lace."

Photo by] [*E. J. Wallis.*

THE LACE-BARK TREE.

A photograph of a young tree growing in Kew Gardens. The tree grows on dry soils in Jamaica, and the layers of lace are found on removing the rough outer bark.

In the "Hortus Kewensis," second edition, it is stated that "when Mr. Wilson, the curator, sent us the young plants, he informed us that it is invariably found growing in very dry situations on marly limestone hills, where there is not a particle of earth to be seen. The young plants grow in the crevices of the rocks, and in order to obtain them with roots, a hammer or large stone is required to break away the porous limestone." He further adds, "that the soil for growing them in should be composed of one-third marl or lime-rubbish, for I am persuaded that pure loam will kill the plants."

Mr. John Smith, who collaborated with Sir W. Hooker in the production of *The Botanical Magazine*, enlightens us as to the methods adopted at Kew Gardens for the care of these young plants from Jamaica. " We are always most desirous to pay attention to information as regards the native habitats of plants, but in cases like the present one, we have found that when too strictly adhered to, successful cultivation does not always follow. In our experience we have never found any plant thrive by retaining it in its native soil, or in soil too closely resembling it. If we could also imitate all the various influences of climate that modify and control the growth

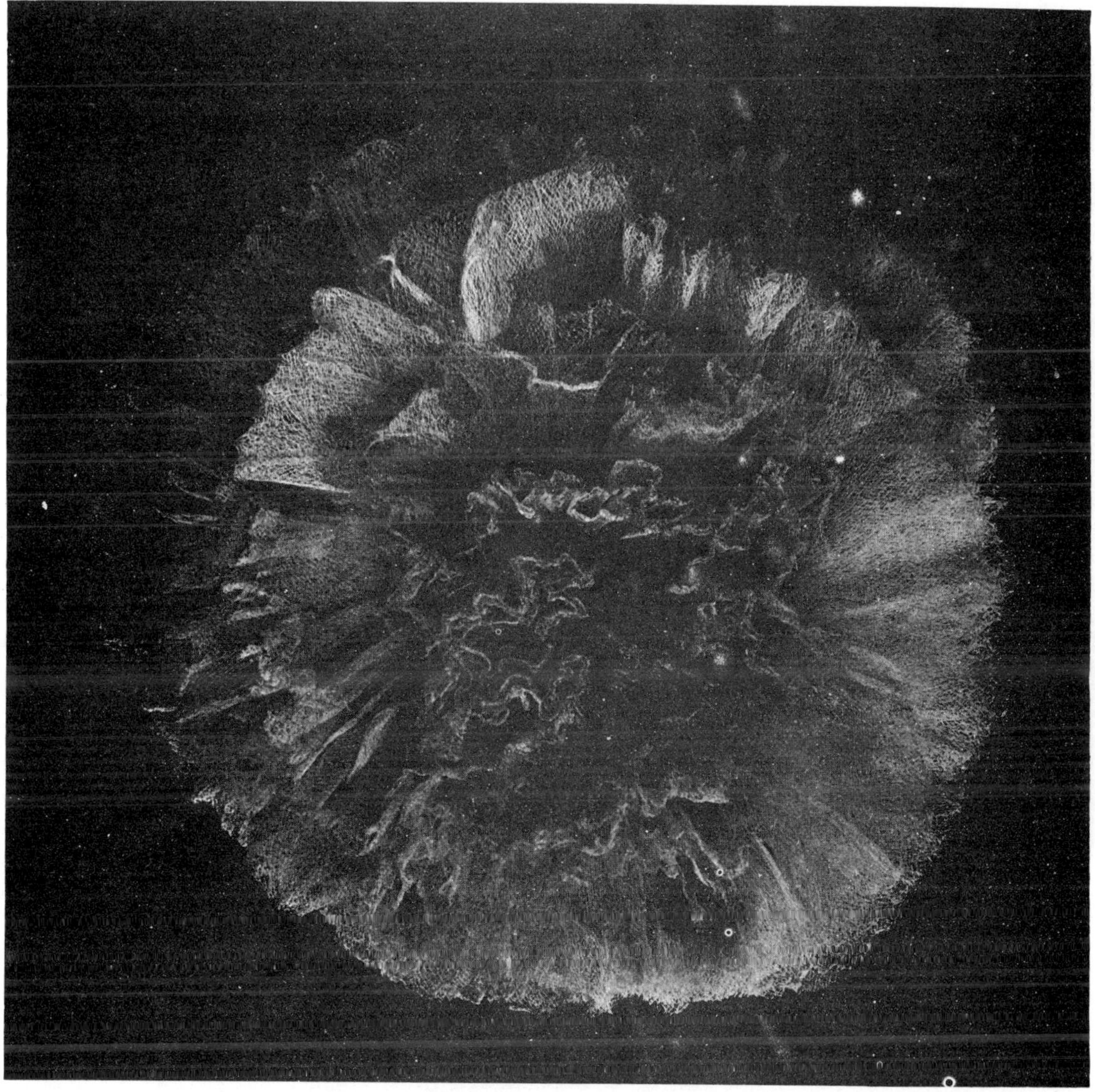

NATURAL LACE OF THE LACE-BARK TREE.

A reduced photograph of the lace around a branch about two inches thick, the layers being loosened to show their number and softness.

of plants in their native localities, it might then be proper for us to cultivate the Lace-Bark tree in marly soil, like limestone; but our plants afford evidence that such soil is not required when they are grown in an artificially heated atmosphere.

" We have used good yellow loam, mixed with a little leaf mould and sand. In this way they have attained a height of eight feet, and continue in a perfectly healthy state. In their native place the leaves are deciduous, falling off in the dry season. But the health of a general collection of tropical plants grown in a hot-house, will not allow us to put them under the influence of their natural dry season; we therefore find that some individuals change their habit and become evergreens. This has been the case with the Lace-Bark plants."

The flowers closely resemble those of the lily-of-the-valley.

Sir W. J. Hooker's description of the tree may be of interest to the botanist:

"A tree from twenty to thirty feet high, with branches too straggling and foliage (though of good size and glossy) too sparse to form a striking object, though really handsome when in flower.

"Leaves alternate, on rather short stalks, which are jointed on the branch; hence the leaves readily fall off in drying.

"Flowers pure white, or, in bud, greenish white, arranged in spikes which are solitary and terminal on a main branch, or on short side branches. . . . . The fruit is a smooth oval drupe."

In the museum of economic botany in Kew Gardens, the following articles made of lace-bark are exhibited: An early Victorian bonnet, a long purse, two small purses, a lady's collar mounted on satin, a pair of slippers, a peaked cap, and a whip as used in the days of negro slavery.

NATURAL LACE.

This photo represents, on a smaller scale and taken sideways, the same natural cluster of lace as shown on the previous page.

THE LUMINOSITY OF THE SEA

BY F. T. BULLEN, F.R.G.S.

WHILE on the land we have such feeble light-bearers as glow-worms and fire-flies, which bear about with them their own illumination, and are only decorative in their effect, the conditions of life in the ocean are so very different that another scheme of lighting is essential, and this is entirely dependent upon the light-giving power of living organisms. Even the tremendous power of the tropical sun-rays can penetrate the ocean only to the depth of a mere skin of water, never exceeding three or four hundred feet, and very feeble then—a sort of dense twilight. Yet we know now that the remotest depths of ocean have light, as efficient and as full as the surface of the sea at night with the appearance of which so many of us are familiar.

Yet although there is undoubtedly this universal system of lighting by many different organisms their power of emitting luminosity is not by any means constant in its effects, being acted upon by some unknown quality outside of themselves and apparently independent of their own efforts. How far this affects the light-giving fish of the depths we have no means of knowing or of learning; apparently the conditions under which life is lived in the abysmal depths being a sealed book to us. But it is only reasonable to suppose that the same influences, whatever they are, which cause the glow of the *Noctiluca*, that microscopic spot of jelly like an atomic brain, to wax and wane, will have a similar effect upon the highly-developed fishes of the profound depths which carry about with them so complete a system, apparently, of ordered illumination. If not in degree, certainly we should say in kind, judging from what we may observe of these fluctuations upon the surface.

Of all these fluctuations, that which is known to seamen by the feeble name of a "milk sea" is certainly the most amazing. It is indeed a stupendous manifestation of that unknown power which affects the lighting of the deeps. Thrice during my seafaring I have witnessed it, but the most striking appearance of the three was in the Bay of Cavité in the Philippine Islands. So intense was the glow of lambent light from the sea that the brilliancy of the scintillating stars was entirely paled, and they appeared like points of dead white against an absolutely black sky. There was not a cloud above the horizon, and the blackness was entirely due to the resplendent light from the sea. It washed against the vessel in wavelets of flame, amidst which the tracks of fish were shown by more vivid bands of light.

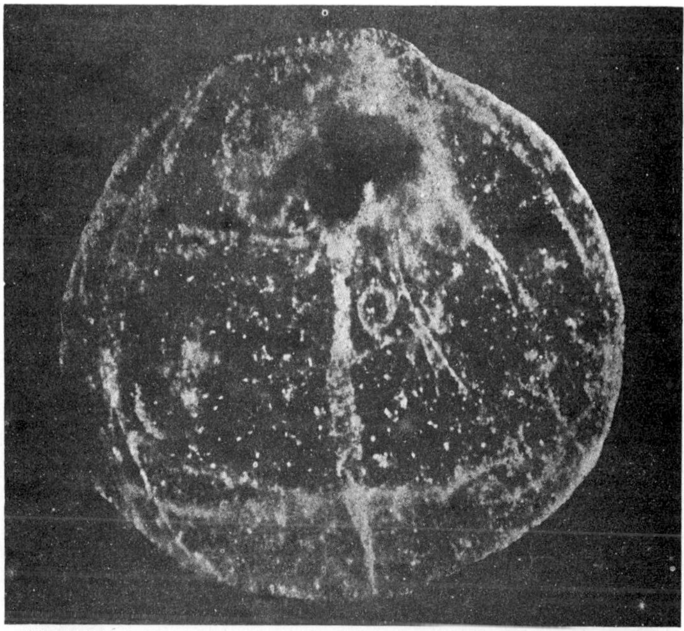

Photo by] [*W. Bagshaw.*

THE NIGHT-LIGHT OF THE SEA ENLARGED.

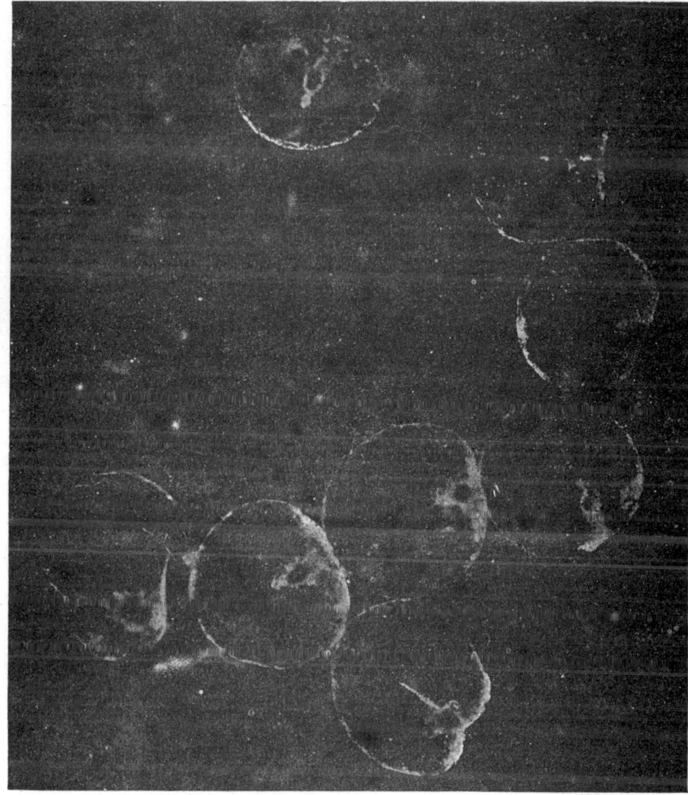

Photo by] [*W. Bagshaw.*

NIGHT-LIGHTS OF THE SEA.

Everybody knows how brightly the sea shines at night, occasionally. The luminosity is caused by these little spheres, which are one-twenty-fifth of an inch across.

COFFER-FISHES

BY W. P. PYCRAFT, F.Z.S.

THE Coffer-fishes are among the most remarkable of the fish tribe, and this, not only on account of their many structural peculiarities, but also on account of their singularly brilliant—one might almost say garish—coloration. But vivid as this coloration appears to be when seen in a museum or freshly taken from the water, it is by no means so conspicuous when seen in its natural surroundings. For the Coffer-fishes live amid a perfect riot of colour, and they have to " live up to it " ; thereby they secure a measure of protection from their enemies, since they harmonize so completely with their general surroundings that movement alone betrays them !

[by W. Saville Kent.

COFFER-FISH.

The Coffer-fishes are among the most brilliantly-coloured fishes known. Living in the immediate vicinity of the coral reef, this striking coloration harmonizes with the colours of the coral animals and makes the fishes inconspicuous.

The " Aracanas," which form the subject of our coloured plate, will give a good idea of how extraordinary this coloration may be But to appreciate this fully one would have to travel to the Great Barrier Reef of Australia, where, in some deep pool on the fringe of the reef they would be found feeding upon the coral polyps and shell-fish, biting off lumps of coral and crushing shells with their parrot-like beaks. These polyps, the shell-fish, star-fish, sea-cucumbers and sea-worms, on every hand, are arrayed in colours of indescribable vividness ; as if each were trying to outdo the other. And, indeed, this is actually the case, for a dull-coloured animal would be conspicuous in such a brilliant gathering, and so would speedily fall a victim to the wolves of the sea !

The Aracanas wear, under this vestment of colour, a bony armour, leaving only the fins and tail free. Hence swimming has to be performed by the rapid vibrations of the dorsal fin and that immediately under it on the under side of the body. The tail is used for steering and the breast fins as fans to draw currents of water through the mouth for breathing purposes, since, owing to the development of the armour, there are no movable gill-plates.

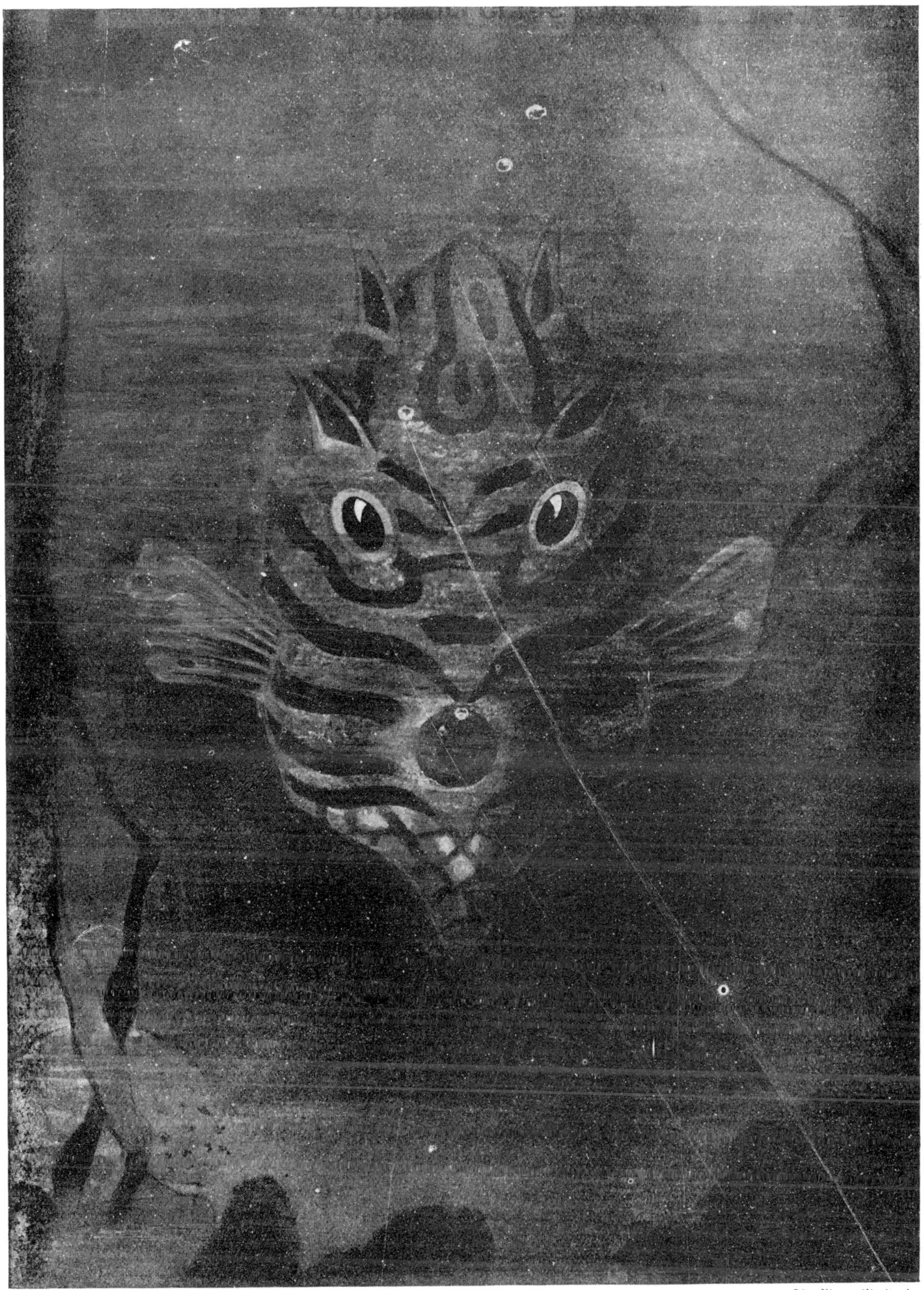

[By W. Saville Kent

COFFER-FISH.

A view of this remarkable fish as seen when coming towards you through the water. The appearance is then most grotesque and not at all fish-like.

MIMICRY IN SPIDERS

BY R. I. POCOCK, F.R.S.

MANY insects, as well as other animals, are protected from enemies by some quality such as a hard shell, a nasty taste or a repulsive smell which makes them unpalatable. After a certain amount of experience, varying in extent according to the species and individual, insectivorous birds and mammals learn to know these uneatable species by sight, and thereafter leave them unmolested. Clearly, therefore, a tasteful species will be greatly benefited if it resembles a distasteful one sufficiently closely to deceive its enemies. Resemblances of this kind are very common in Nature, and the phenomenon is called " mimicry."

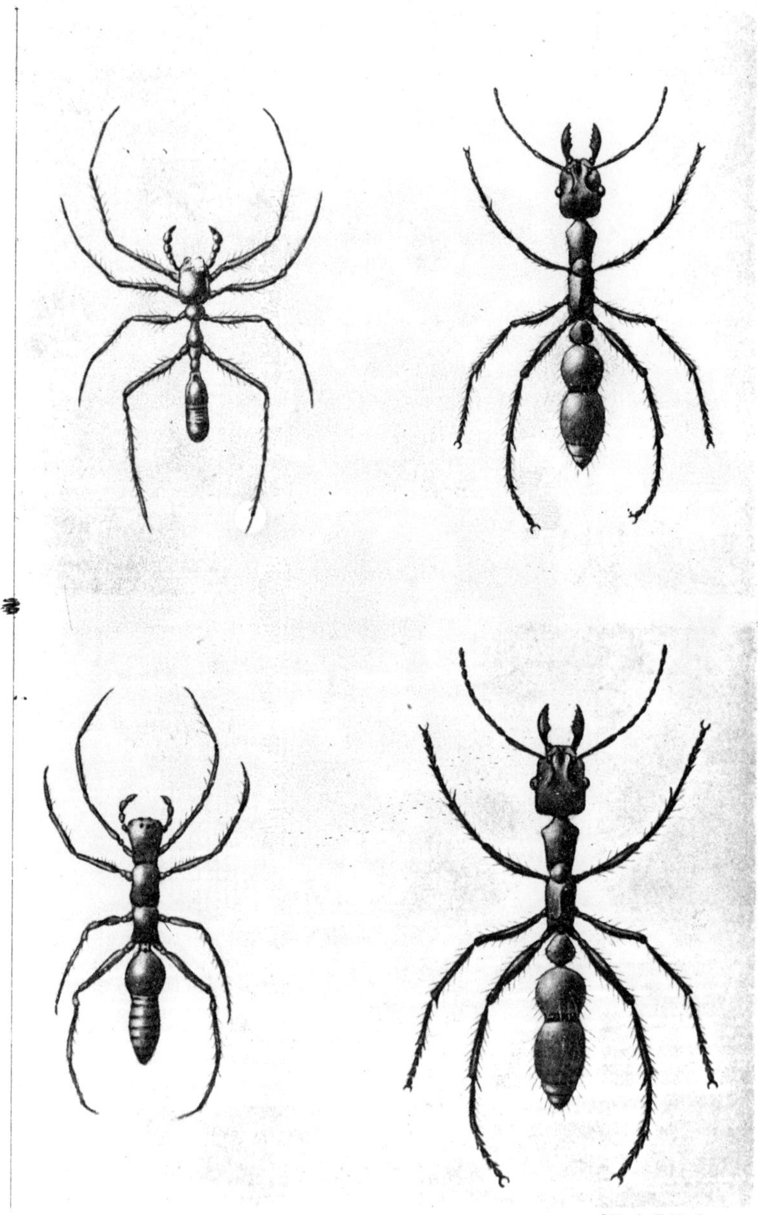

[By J. Teklenburg.

SPIDER MIMICRY.

The two spiders shown on the left are South-American species that both mimic one Ant, in order to protect themselves from the attack of a Mason-wasp that stores up Spiders for the food of its grubs, but leaves Ants alone. The Ant mimicked is shown on the right.

Now the majority of spiders are comparatively defenceless, and being highly palatable are extensively preyed upon by other animals. But the worst enemies they have to fear are the digger or mason wasps which, to feed their larvæ, store away in their burrows and mud-cells vast numbers of spiders, stung to a state of coma and so preserved against decay. Persecution by many enemies, and especially by these wasps, has reacted upon the instincts and structure of spiders in various wonderful ways, perhaps the most wonderful being the mimetic resemblance it has brought about between some of them and protected insects, more particularly those that the mason wasps do not attack. There are indeed one or two instances known of spiders accurately mimicking small snails which are useless food for wasp-larvæ and are neglected by small insectivorous birds. Other spiders for the same purpose imitate in colour and shape little shining hard-shelled beetles. Others, again, belonging to the same group as the common garden-spider have short legs, convex bodies and spotted coloration, so as to be exactly like the little beetles known as ladybirds, which are so nauseous on account of their acrid juices that very few animals will eat them. But the most perfect instances of mimicry in spiders are found amongst the species which imitate ants.

These belong to several distinct families, but the modifications which bring about the resemblance have followed almost the same lines in every case. Ants have a large head carrying a pair of feelers, and joined by a narrow neck to the long body, the fore-part of which bears six legs and is attached to the oval hinder part by a slender, flexible waist. Now spiders have no feelers and no head distinct from the body. They possess, however, eight legs and to imitate feelers they lift the legs of the first pair and, stretching them forwards, wave them in the air like feelers, thus leaving six legs for walking, as in the ant. The semblance of the ant's neck is produced by a deep groove on each side near the fore-part of the body, and this groove is filled with a patch of white hairs, which has the same optical effect as a piece of the body being cut away.

Thus is the detached head of the ant represented. Other likenesses to the ant are produced in analogous ways; and even the restless zigzag gait of the insect is unconsciously copied by the spiders to complete the deception. The mimicry is indeed often so close that the one animal can scarcely be distinguished from the other.

The reason for this particular kind of mimicry becomes perfectly clear when it is explained that out of the very large numbers of species of mason wasps that prey upon spiders, not one provisions its nest with ants. Some of these wasps, in fact, are known to have the greatest fear of ants.

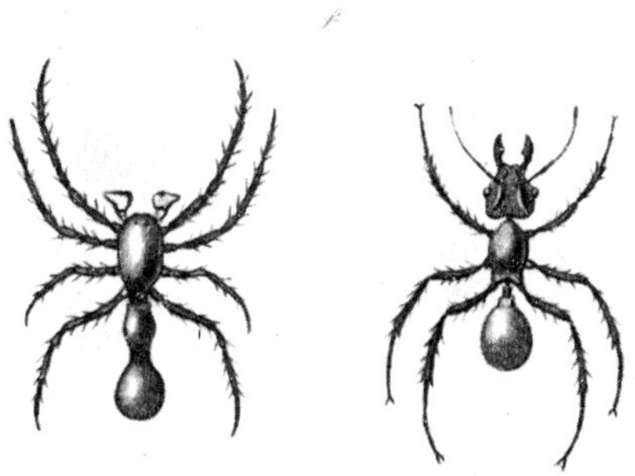

Another South-American Spider of which only the male, here shown on the left, mimics the Ant on the right.

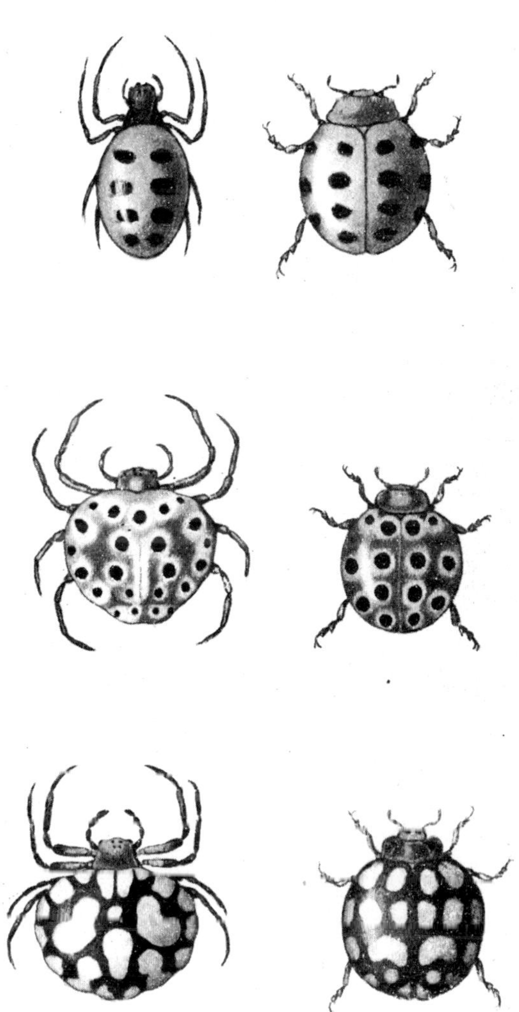

[*By J. Teklenburg.*

SPIDER MIMICRY.

The three Spiders of remarkable form and markings mimic Ladybirds, which have nauseous flavours and are therefore avoided by birds. They are orb-spinners, like our garden-spider, but come from Africa. The Ladybirds resembling them are shown to the right.

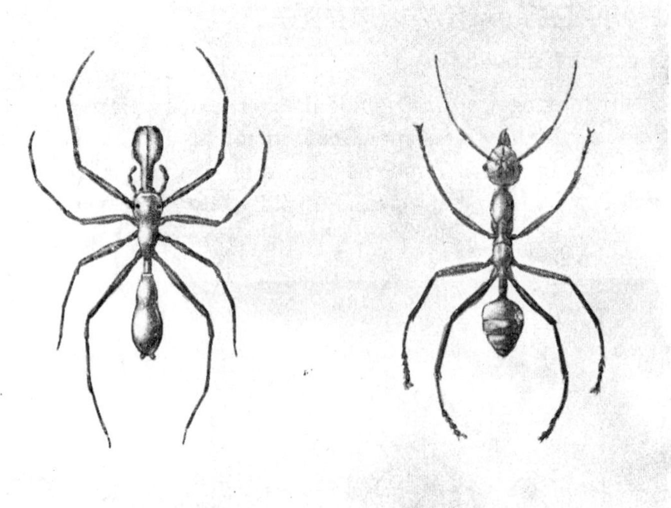

[By J. Teklenburg.

SPIDER MIMICRY.

The Indian Jumping-spider on the left mimics the Leaf-cutting Ant on the right. The jaws of the Spider represent the head of the Ant and two black spots look like eyes.

THE FATA MORGANA

The Italian title which appears at the head of this paragraph has always been held to be a synonym for those somewhat familiar appearances over marshy and malarious places which are known among us as "Will-o'-the-Wisp," "Jacky-my-lantern," etc. And as all investigation of their character and origin has confirmed us in our belief that they are simply agglomerations of luminous gases hovering over the places of their production, the romantic and superstitious suggestion contained in the Italian name has gradually faded. Few, indeed, are the country-folk in our midst to-day who would believe that the weird glow hovering over a dangerous swamp was intentionally being exhibited by a frivolous fiend bent upon luring them to destruction.

But in other parts of the world, notably Southern Europe, many of the old superstitions still hold undisputed sway over the minds of the people, who apparently cherish them as if they bore nothing but comfort and solace with them instead of being full of terror and distress. Fishermen and coasting seamen generally are among the most prone to quaint, disturbing beliefs of this kind, and the appearance, during a highly electrical condition of the atmosphere, of the phenomenon known as St. Elmo's fires (dealt with in an earlier page) is sufficient to terrify them greatly.

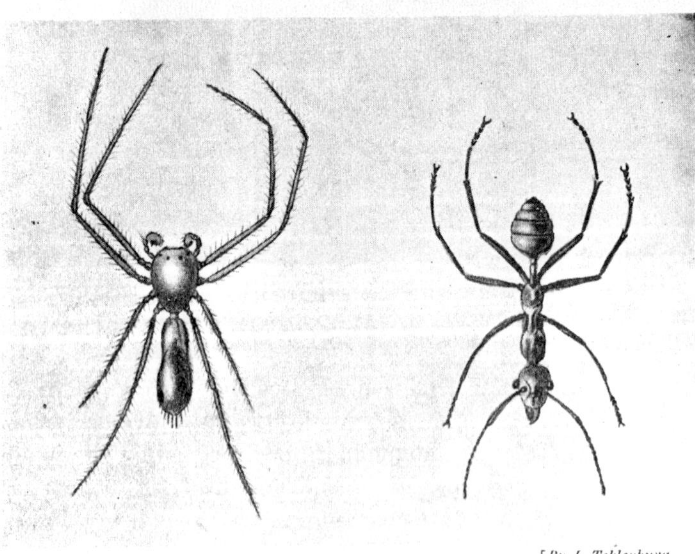

[By J. Teklenburg.

SPIDER MIMICRY.

Another Indian Spider (left) that mimics the same Ant (right), but only when it is going the other way. Note the spots at the rear of the Spider which look like eyes and produce this likeness.

But in addition to these well-known phenomena, there may be observed off the coasts of Southern Italy and Sicily in the proximity of marshy lands mysterious-looking lights wandering over the sea near the shore, having no reference to human agency, and acting entirely differently from the better-known but no less feared St. Elmo's lights. It must be admitted that among ignorant folk such appearances are sufficiently disturbing to warrant any trouble of mind, especially as the people are not in the least likely to hear any rational explanation of the sight, or, it must be confessed, to entertain it if they did.

[By H. Whanslaw.

FATA MORGANA.

The Fata Morgana is a kind of mirage seen over the Straits of Messina under certain conditions of atmosphere and sea. Houses and moving objects appear on the surface of the water and pass away in turn with movements of the current. Sometimes these reflections rise to a height of thirty feet above the water.

A fluffy, inchoate mass of light, itself indeterminate in shape and power, but appearing to be castles, palaces and what not, glowing and fading, rising and falling apparently without reference to the direction of the wind, but really acted upon by the lightest current of air; rebounding from the sea-surface as if it were a luminous bladder; what could any primitive folk make of that? I very much doubt whether, if such a phenomenon were to be seen on our shores, we should not have considerable terror among our fishermen and longshoremen, hard-headed and unromantic as we know so many of them to be. So curious is this particular vagary of the Fata Morgana, that it affords sufficient excuse for including it among the Marvels of the Universe.

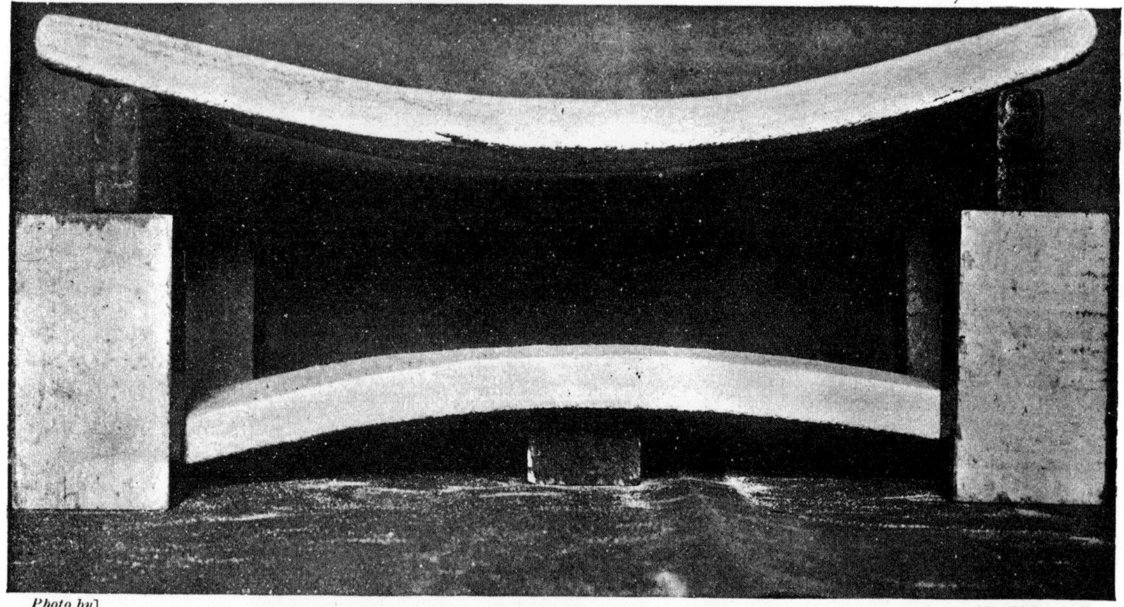

Photo by] [*R. F. Gwinnell.*

TWO SLABS OF FLEXIBLE SANDSTONE.

The upper slab supported at its ends sags in the middle; the lower slab is supported in the middle and droops at each end: thus showing its elastic character.

FLEXIBLE SANDSTONE

BY RUSSELL F. GWINNELL, B.SC., F.G.S.

"As firm as a rock" is a common mode of expressing stability, but here we have a case of a rock which will bend like indiarubber. If a thin slab of this sandstone be hung downwards, the free end can be made to swing to and fro like a pendulum while the upper end is held immovable. Not only so, but as shown in the illustration, a slab will bend by its own weight when supported only at the two ends or in the middle.

The two specimens photographed are each about two and a half feet long and they show a "sag" of the same number of inches. Their thickness is about an inch and a half. These specimens from Kaliana in the Punjab, India, are in the Geological Department of the Imperial College of Science.

Flexible rocks are distinctly rare, and the few known to occur are mostly sandstones; an exception is the Magnesian Limestone of Durham, which very infrequently exhibits slight flexibility, and is the only British rock possessing this character.

Georgia and the Carolinas (U.S.A.) possess flexible sandstones, but the best known example comes from Brazil, being found at Itacolumi, in the State of Minas Geraes. It is a porous yellow rock made of grains of quartz (the chief constituent of most sands and the clear, hard mineral found in granite, etc.), together with flakes of mica, a substance familiar to many people owing to its uses in the arts—in gas-fittings, electric insulators, etc.—where it is often miscalled talc. Now mica is a highly flexible mineral (thus it is bent to form lamp-chimneys), and it was formerly supposed that the flexibility of Itacolumite was due to the presence of this mineral in it. But many cases occur of micaceous sandstones where no flexibility is exhibited, such as the well-known flagstones used for street pavements ; while even in the Brazilian rock this peculiar property is only found in a small part of the whole mass. Associated with the sandstone—or, more properly, " quartz-schist "—of Brazil, are " mica-schists " which are far richer in mica, but quite inflexible.

Flexibility in the sandstone, then, seems to be due to the quartz and not to the mica. The grains of quartz are angular and interlock at their junctions, but yet leave many empty interspaces. The porous character of the rock allows movement in the spaces, while the hinge-like joints between the particles hold them together in spite of the shifting to which they are subjected when a slab is bent. If the reader clasp his hands together so that the fingers interlock, and then rub the palms together, he will illustrate roughly the action described. The Brazilian Itacolumite has an added interest in being the probable source of the diamonds which are locally so abundant.

PELORUS JACK

BY F. T. BULLEN, F.R.G.S.

It is very refreshing to find in the midst of so many manufactured stories about the lesser-known animals of our globe an item that, in spite of its romance and apparent fiction, is not only verifiable, but has been verified by so many competent authorities that its truth is beyond all shadow of reasonable doubt. Such a fact is found in the account of the Grampus which frequents the vicinity

Photo by] [*Muir & Moodie.*

PELORUS JACK.

The famous pilot of French Pass, New Zealand, which has been specially protected by Government. Although commonly spoken of as a pilot-*fish*, it is really a Grampus, one of the whale family

of the French Pass, on the direct route between Wellington and Nelson, New Zealand, and is popularly known as Pelorus Jack from his favourite cruising-ground, Pelorus Sound.

Briefly and baldly stated, the facts about this curiously-mannered animal, as verified by myself, are as follows. It is a Gray Grampus of about eighteen feet in length, which for a considerable number of years has been in the habit of meeting vessels using the French Pass at a point about five miles south of that narrow current-tormented strait, and accompanying them for about half an hour, rubbing itself against the bows in the rush of the boiling foam like an affectionate cat against the legs of its owner. The speed of the vessel in which I made the passage was fully eighteen knots; but the little whale seemed to maintain its position with the same absence of effort as does the Albatross over the masthead of a swift steamer—any movement of fin or tail was impossible to detect. Occasionally with the same utter ease it glided ahead for a ship's length, then dropped alongside again, until, its visiting time having expired apparently, it sheered off at right angles and disappeared.

Photo by] [*Muir & Moodie.*

PELORUS JACK.

The Grampus is here shown as seen from the deck of the vessel it is piloting through the French Pass, New Zealand.

The causes of its visiting ships like this are entirely mysterious, since a ship can supply nothing the creature needs; yet it is absolutely faithful in its attendance, and so general is the interest aroused by it in New Zealand, that a resolution of the Legislature has been unanimously passed requesting all persons to refrain from molesting it. This should not be necessary, but we must sadly admit that it is.

Around this most interesting mammal has concreted a mass of legend, as, indeed, was inevitable. There are not wanting old *identities* of New Zealand prepared to swear that they have known the ways of this whale for half a century, that it was once wounded from a ship, and ever afterward avoided that particular ship, and so on, and so forth. The Maori stories are highly picturesque and romantic, as might be expected from this strange people; but they all agree upon one point, which is that this whale is a minor deity well affected towards a particular tribe, which, indeed, it seems to have taken under its special protection.

THE SABRE-TOOTHED TIGER

BY R. I. POCOCK, F.R.S.

It is a great mistake to think that high organization necessarily involves success in the struggle for life. Specialization of structure means adaptation to special conditions; and if the conditions fail, the organism fails too, because of its inability to adjust itself to a changed

[By C. E. Swan.

THE SABRE-TOOTHED TIGER.

One of the monsters of Prehistoric times. Similar in some respects to the Tiger of the present day, its unlikeness is made evident at once by a glance at the pair of terrible tusks, sharp-edged and blade-like, projecting about six inches from the jaw.

environment. In this truism probably lies the explanation of the extinction of the Sabre-toothed Tiger, in some respects the most highly organized of all the cat family.

The word "sabre-toothed" explains itself, and alludes to the tremendous development of the canine teeth of the upper jaw, which were compressed, sharp-edged and blade-like, and projecting in some species six inches or more below the gum, constituted the most formidable stabbing instruments within the limits of the carnivora. The term "tiger," however, can only be applied to this great cat in a very general sense to indicate its equality in size and strength to that well-known species. In no other particular can it be supposed that close resemblance existed between them. Judging from the skeleton, the Sabre-toothed Tiger (*Machærodus*) was a short-tailed, heavily-built beast, with massive fore-quarters, suggesting very limited leaping powers and little speed of foot, while the immense fangs and the muscular development required to give force to their stroke affected the shape of the head in a remarkable degree, especially by lessening the space between the ear and eye. The cat-like physiognomy must have been still further altered by the massive development of the chin. Since there are reasons for supposing these animals inhabited more or less open country, it may be inferred that, like lions, they were self-coloured, and analogy suggests that their cubs were perhaps longitudinally banded.

In spite of their final extinction these animals held their own for long ages—that is to say, from Oligocene to Pleistocene times—and were widely distributed. Remains of many species have been discovered in England and other parts of Europe, and in India as well. Probably from Asia they made their way into the United States and passed thence into South America. How the very long-fanged species killed their prey and fed has been a much debated question. That they did not bite in the ordinary way seems certain from the weakness of the lower jaw and from the impossibility of its clearing the points of the fangs, at all events to any great extent, when open. Very possibly they used their fangs for piercing the carotid arteries of their victims, and, like weasels, subsisted largely by blood-sucking.

WASPS AND THEIR TOOLS

BY JOHN J. WARD, F.E.S.

In the early history of a wasp city there are two kinds of individuals only—the queen and her daughters. The latter resemble their mother somewhat considerably, but are diminutive in size and sexually imperfect. Consequently, the queen-wasp is the only perfect female in the city.

Photo by] [*J. J. Ward.*

THE THREE CLASSES OF WASP.

This group includes the three types of Wasp found in each nest. In the centre is the female or queen, noticeable for her superior size. On the left is the male or drone with his longer feelers; and on the right is the active little worker

The queen's husband never sees the thousands of citizens of which he is the father; indeed, he never sees the beginning and early stages of a city at all. He appears at the end of the summer, when the city is at the zenith of its glory, and when certain marriageable daughters have been born to the community from which he can select a wife. His function in life then having been fulfilled, he perishes, together with all the busy workers of the city in which he was born. Only the young queens are able to survive the winter and become the mothers of new communities.

Just how the above events come about need not concern us here, as this article is preliminary to another, where I shall be able to consider the habits and home-life of the wasp in detail. My present consideration is of the wasp individuals themselves, and, as we have seen, they consist of (1) the large-bodied female known as the queen, (2) the small-bodied imperfect females known as the workers, and (3) the slender-bodied males or drones with long feelers.

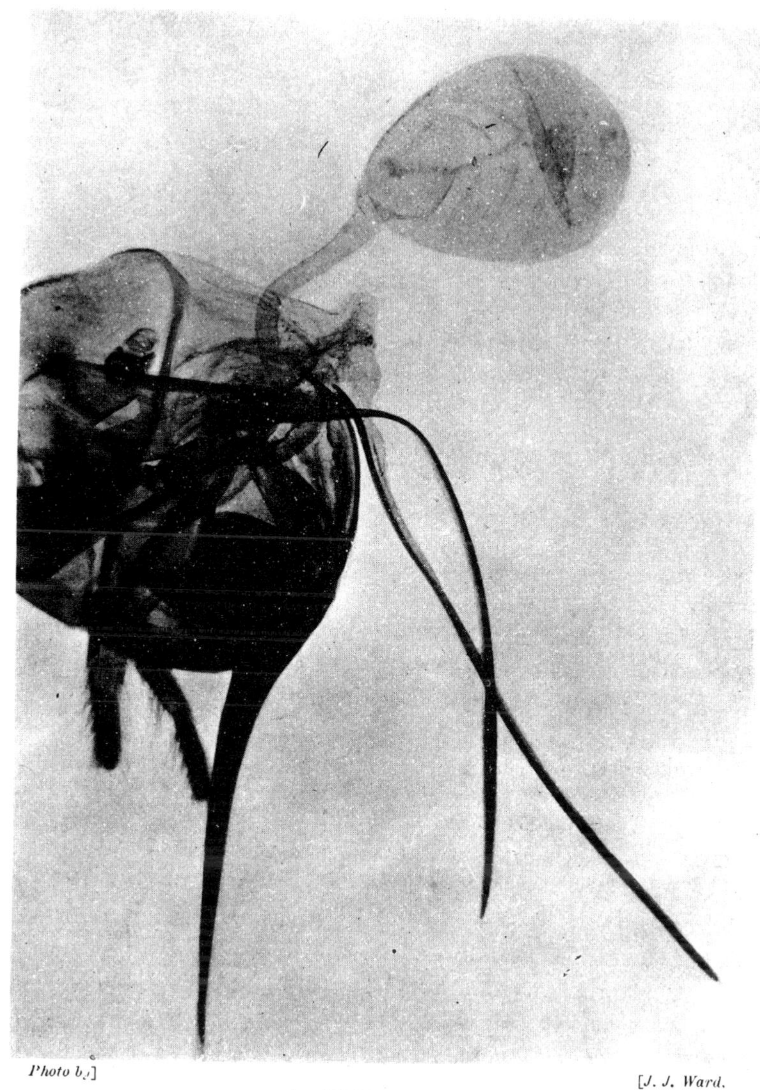

Photo by] [*J. J. Ward.*

A WASP'S STING.

The wound is made by the dark object to the left of the picture. But this is only a sheath in which moves the sharper sting, here shown removed from the sheath and to the right of it.

Both the queen and the worker-wasps are endowed in a marvellous degree with the maternal instinct. They are veritable slaves to nest-building and the rearing of the young. Consequently, Nature has provided them with some wonderfully-made tools to carry on these functions.

When we see a wasp on a wooden paling biting and tugging at its fibres, it is difficult to realize the strength that it is exerting in its efforts to bite and tear off a shred of the wood. Even if we look closely at its mouth-parts we can only observe its tiny jaws working persistently upon the wood, until at last a clean portion is revealed, showing that some of the fibres have been torn away, as shown in our photograph.

That portion of wood is then worked up in the complex machinery of the wasp's mouth and mixed with a gummy secretion, until at last a paper-pulp is produced, which is then spread out into a thin layer, and added either to the outer walls of the city or to the tiers of cells within. Just how the paper is manufactured, and what tools are employed in its production, we cannot quite understand; but a glance at the mechanism of the mouth-parts through a microscope reveals

WASP-STING AND NEEDLE

To show the fineness of the wasp's sting it is here photographed threaded through the minute eye of a No. 16 needle.

the complexity of the machinery used in the process.

The accompanying photo-micrograph shows the open mouth of a worker-wasp with the various organs in position, and which are so unlike the mouth-parts of man and other higher animals that it is almost impossible to make any adequate comparison between them.

The large horny jaws, or mandibles, seen on each side work sideways, not up and down like our own, and serve alike to rasp wood-fibres for nest-building purposes, to bite into the choicest plum or pear in the orchard, or to cut off the wings of a captured fly before it is carried off to the nursery. When the wasp is biting hard substances, however, the transparent tongue and other delicate organs of the mouth are withdrawn into a receptacle within the lower part of the head.

The upper lip, if we may call it such, is a simple scale-like body, while the lower lip (to complete the comparison) is of a much more complex structure. It is provided with a broad transparent tongue, very different from that of a bee, a butterfly, or a moth, for it is but few flowers that the wasp visits in search of their nectar.

Attached to the sides of the delicate four-lobed tongue, which is covered with minute hairs and bears a large gland at the tip of each lobe, is a pair of small four-jointed feelers, which serve as organs of sense. Outside the tongue and its feelers are two other complex organs, each bearing sensory hairs and a six-jointed feeler, and we may conventionally call them the modified lower jaw, the mandibles representing the upper jaw.

Whatever work the wasp has to perform, whether it is carrying a little stone out from the building site of the nest, biting through strong roots, manufacturing paper-pulp, robbing the orchard, or stealing tiny portions of meat from the butcher's shop—which it sometimes prefers to its more lawful prey, the flies—these are the tools which it employs. Just what function the delicate feelers and sensitive hairs serve, and which are associated with the senses of touch, smell, taste, etc., science is as yet quite unable to explain ; especially as they are probably connected with senses which are not analogous with our own.

The head, too, bears a pair of conspicuous feelers, and these again possess complex sensory functions. A glance into a thriving nest leaves no doubt in the mind of the observer that these are the organs by which communication is made one with another ; probably, too, they serve the sense of hearing. The large compound eyes, with their numerous lenses, so unlike our own, are likewise puzzling organs, especially as on the top of the head between these large eye-masses three simple eyes are to be found. Why the wasp should need three single eyes and two large compound eye-masses with thousands of lenses, each of which can receive an image, is, judged from man's standpoint, an inexplicable problem.

The wings of a wasp are also worthy of study. It will be observed that when a wasp alights and closes its wings that, though larger, yet they are not nearly so conspicuous as those of a house-fly. The explanation is that the two fore-wings are folded in a crease near the centre; in this manner their area on the body is reduced. Also, under a lens the lower wings will be seen to bear on their upper edges a row of minute hooks, and these are received into a groove on the lower edge of the fore-wings. By this means the two wings are united to form one strong wing. It is a curious fact that the male always has a less number of hooks, and, as he is the weakest flier, the inference may be made that they have some relation to the power of flight.

Other useful tools in the daily work of the wasp are its spiny legs, which serve it not only for locomotion, but to hold and manipulate its prey and to clean its feelers and wings; indeed, the legs can be used for all its toilet purposes.

A WASP'S FACE.

If a Wasp were as big as a kitten this is the appearance that it would present if we gazed at it face to face. The lower portion of the "face" consists of a pair of powerful cutting jaws.

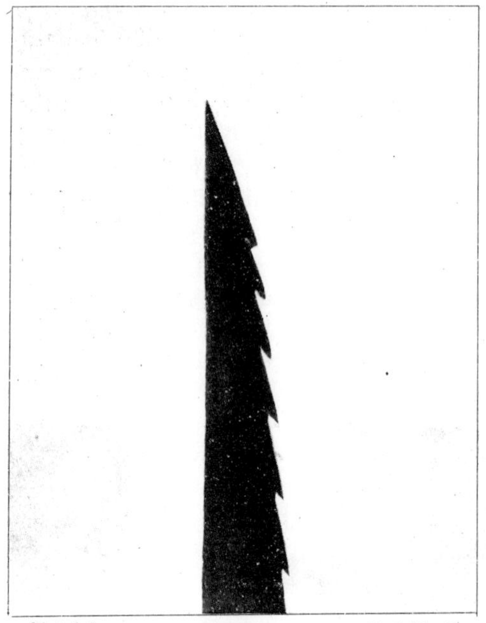

Photo by] [*J. J. Ward.*

POINT OF WASP'S STING.

A small portion of the point of a Wasp's sting is here much enlarged to show the barbed edge.

Finally, there is the smooth body clothed in a flexible suit of mail, perfect in fit, and rhythmic in action. Attached by its slender waist, the abdomen can be twisted about in a wonderful manner, until that terrible organ the sting is directed against an assailant.

What a complex weapon the sting really is may be appreciated from the accompanying photo-micrographs. As seen with the unaided eye, it appears as only a protruding point which might penetrate the skin just as the point of a needle or thorn would do; but under the lens it is revealed as a much more diabolical tool. The penetrating point becomes only a sheath for two darts barbed at the tips and furnished with powerful muscles at their base, and which can be projected into the wound, at the same time opening the way to a poison duct, from which, by the same action, formic acid is pumped into the puncture.

The queen wasp is not nearly so aggressive as the workers, and much less likely to sting; the barbs on her darts are often very imperfectly developed. A worker-wasp may lose its life while stinging an enemy, owing to its darts adhering, but the queen, in the interests of the race, is more protected, and uses her sting for its true function—namely, egg-depositing. The ovipositor of the worker-wasp, having largely lost its original function, has now developed into a complex weapon of offence, and is also used for paralysing living prey by means of its poison. It is scarcely necessary to add that the male wasp is harmless.

THE CANDLE-TREE

BY EDWARD STEP, F.L.S.

THE remarkable tree shown on the page opposite is a native of Central America. In the Isthmus of Panama, especially in the valley of the River Chagres, there are forests composed almost entirely of this tree, which from the peculiar shape of its fruits is known as the Candle-Tree. Dr. Seemann tells us that to enter one of these forests is like walking through an old-fashioned tallow-chandler's shop, for from the older branches and the trunks of the trees hang what appear to be long candles from two to four feet in length and of a yellow wax colour. These are the fruits of the tree, and they contain seeds that are much like lentils in appearance. Though these fruits resemble wax-candles, they are more akin to tallow, for they are said to yield as much as sixty per cent. of fat, from which a lamp-oil is prepared. They also form a capital cattle food, and their peculiar apple-like odour is imparted to the flesh of the animals.

Another species, known as the Cuajilote, is used as food by the people of Mexico, where it is a native. In both cases the fruit is preceded by large white bell-shaped flowers, and the trees in bloom have a fine appearance.

These trees are related to the Calabash-tree, of whose fruits bottles and other vessels are contrived by extracting the pulp from the cut end, but leaving the rind otherwise unbroken. Thus dried in the sun, they become hard, tough and watertight.

Photo by] [*International Press Photo Co., N.Y.*

THE CANDLE-TREE.

A remarkable tree that grows in the valley of the Chagres, Panama. The so-called "Candles" are the fruit of the tree and measure from two to four feet in length. They look like large wax candles and resemble them in colour.

Photo by] [*L. E. Adams.*

NESTLING MOLES.

This photograph of an opened Mole's nest shows the young Moles like little sucking-pigs. The position of the nest in the Fortress will be seen on reference to the picture on next page.

THE LIFE STORY OF THE MOLE

BY LIONEL E. ADAMS, B.A.

THOUGH perhaps no animal has been the subject of so many fantastic stories as the Mole, the true life story of a quadruped that spends practically its whole life underground is sufficiently strange to interest the lovers of marvels in Nature.

During May the Moles are born in a warm nest of dry grass under a mound of earth, and closely resemble small sucking-pigs, less than two inches long, pink, naked and blind. They increase in size at the rate of a fifth of an inch a day. On the tenth day they turn from pink to lead colour; in another four days they are clothed with a short velvety coat of the same leaden hue, which grows longer and darker till they emerge from the nest about the end of their fourth week, and commence life on their own account. They now resemble their parents in all respects, except in size, which is about three-quarters their full growth. These inexperienced juveniles come to grief in many ways, and may be found dead on the country roads during the summer. Towards the end of the year each Mole prepares a habitation for himself, which has aptly been called his Fortress, and merits special description. First, a cavity about the size of a large cottage loaf is excavated an inch or two below the level of the ground, the loose earth being ejected through the top and forming a mound. This cavity is to contain the nest of grass and leaves which the Mole brings in from above. Then, in order to protect the nest still further, he brings more earth by tunnels coming from the field outside and piles up an immense heap on the top of the original one, sometimes two feet high. These tunnels, being continued into the accumulating mound, usually remain intact, and were erroneously supposed by the old naturalists to be galleries arranged systematically to enable the Mole to elude his enemies when surprised in his Fortress. One such tunnel, however, does exist in every Fortress, leading downwards from the bottom of the nest cavity and then turning upwards into one of the tunnels leading away from the nest into the field, and termed the "Bolt-run." In this Fortress, then, the Mole lives solitary, sleeping curled up in the nest and emerging at intervals of every few hours to run along his tunnels, which extend in all directions from the Fortress into the open country, in search of worms, which form his usual diet. Occasionally, however, the fare is varied, a dead mouse or a slug, or even a live frog or a young bird, being devoured by the hungry Mole. Now, considering the strenuous life the animal is obliged to lead to find sufficient worms to support him, and considering the small amount of nutriment in a pint of worms, there is no mystery in his voracity. Hunger, moreover, induces bad temper, and when Moles meet they usually fight to the death—the vanquished forming a meal for the victor.

In March and April the males seek the females, but it is doubtful whether they ever occupy a joint Fortress as husband and wife. Certainly the breeding female makes a separate Fortress as a nursery, somewhat similar to that of the male, but smaller and without a Bolt-run.

During the time the young are in the nest the fields are bush-harrowed, and the Fortresses, as well as the ordinary small molehills, are levelled and scattered. This, however, does not harm the young in the nests below; indeed, in one way it is their salvation, since all visible signs of their existence have been cleared away and they are left unmolested. If these breeding nests were properly marked down and dug out, the race would be much reduced, as the mole only produces one litter a year—but then the mole-catcher's occupation would be gone!

During the summer the Mole leaves little trace of his existence, either foraging in his tunnels or on the surface of the ground when the worms come forth at night.

Mole workings are beneficial to heavy soil, and the heaps when scattered form good top-dressing; but on a sacred golf-green . . . !

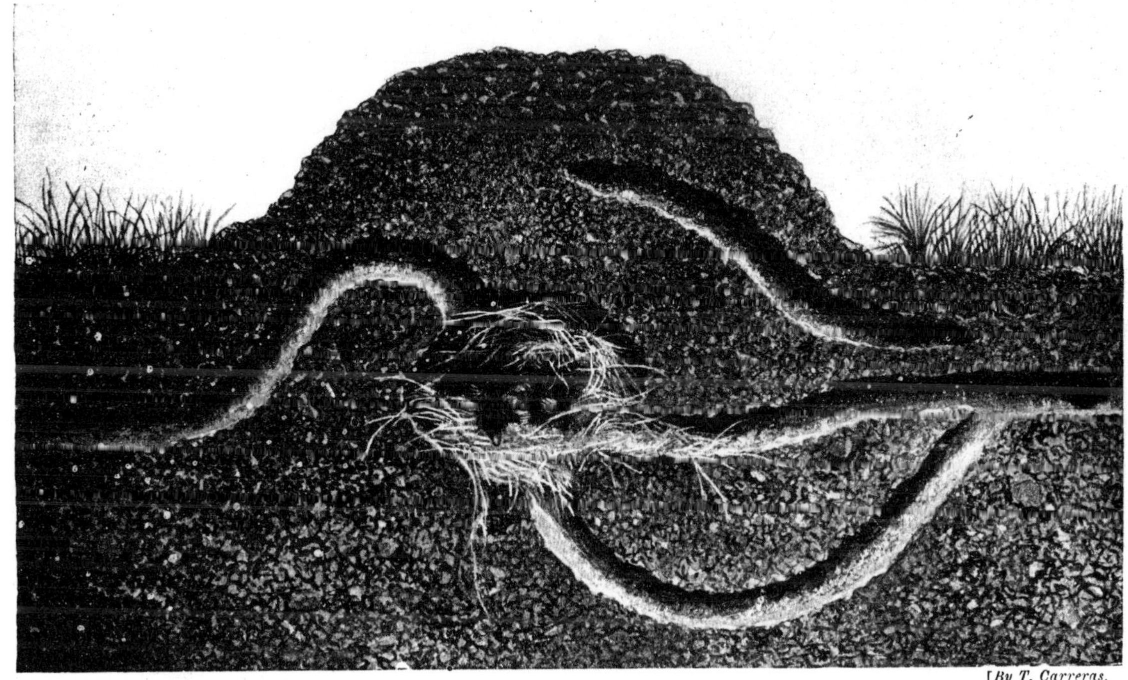

[*By T. Carreras.*

THE MOLE'S FORTRESS

A section through the fortress of the Mole, showing the nest wherein the young are born and reared. It will be seen to be far less complicated in its tunnels and galleries than was formerly supposed to be the case

BEEF-STEAKS THAT GROW UPON TREES

BY EDWARD STEP, F.L.S.

In the autumn and winter chiefly the rambler in the woods is almost sure to notice remarkable growths from the trunks of trees, mostly out of reach. These take the shape of semicircular brackets, often of considerable size, and of corky or woody consistency. Among them, however, in the earlier part of the autumn, will be found one of much softer character, dark red on the upper surface and flesh-coloured below. It may be a foot or more across, and weigh several pounds.

This is the Vegetable Beef-steak, or Beef-steak Fungus. In reality it looks more like a mass of liver when viewed from above; but when it is cut through, its streaks of lighter and darker red afford considerable justification for its popular name Another feature that supports the name is its juiciness. When cut, a thin red gravy exudes freely and plentifully from it. If this is tasted it will be found to be astringent and slightly acid.

Photo by] [E. Step, F.L.S.

THE MOLE.

The Mole may occasionally be seen above ground, and then makes desperate efforts to escape observation by digging a new burrow.

The Beef-steak Fungus is almost restricted to the oak-tree, but rarely it may be found upon other trees. Where there are oak woods of any size, it would be an easy task, during September, to gather a hundredweight of vegetable steaks in the course of an hour or two. Ordinarily, it grows about six or seven feet from the ground, but in dry seasons it will be found much nearer the roots, or actually on them. In the phenomenally dry year of 1911 most specimens were to be found low down, the trees evidently affording more moisture to the fungus a little above the roots. There can be little doubt that the astringency of its juice is due to tannin it has extracted from the oak.

It is, perhaps, unnecessary to explain that the Vegetable Beef-steak is a parasite. All the year its threads are extending farther and farther into the woody tissues of the oak, robbing it of nutriment, and breaking down its cellular structure, so that its timber becomes worthless. But it is not so speedy in its work of ruin as many other fungi, whose brackets we may find in the woods ; and, as

Photo by] [*Edward Step, F.L.S.*

A dish of Vegetable Beef-steaks cut into fillets preparatory to cooking, here reproduced to show the meat-like appearance of its interior.

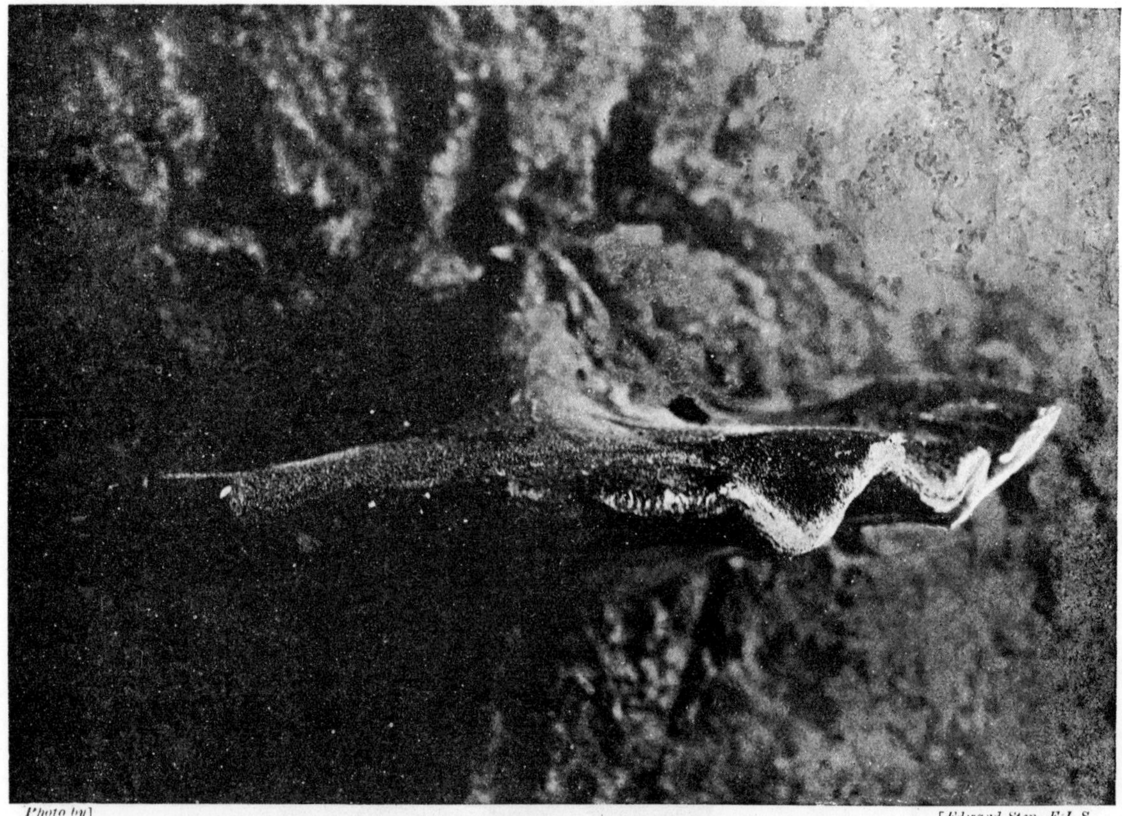

Photo by] [*Edward Step, F.L.S.*

THE BEEF-STEAK FUNGUS.

The Vegetable Beef-steak is here shown as it grows from the trunk of an oak-tree. The margin is often flat and even, though the form shown is quite characteristic.

a consequence, if we gather beef-steaks from a particular oak-tree one autumn, we may go back to it in several succeeding years at a similar date with the certainty of gathering another crop.

Our photographs not only show the Vegetable Beef-steaks as they grow, but also a dishful of " fillets " to give an idea of the meaty appearance of the flesh. There is really something in the name, for this strange fungus is—like many others of its tribe—a wholesome article of food. For my part, I think there are many fungi that provide better eating, but this is entirely a matter of personal taste, and there have not been wanting epicures who declare that the Vegetable Beef-steak can scarcely be beaten in this class of food-stuffs. It lends itself to varied treatment, or can be cooked almost without any preparation. For the simpler process, it needs only to be lightly washed to rid it of any woodland particles that may have adhered to its moist surface, then broiled like any other steak—or with another steak of the orthodox kind if it is to make a meal for one who harbours doubts as to the possibility of such things as fungi being palatable by themselves. Some stew it, adding seasoning of various kinds and yolk of egg when it is all but ready to be served up. It may be sliced and fried, either by itself or with animal steak accompanying it. Some there are who do not care for it as a solid, though they enjoy its flavour ; these have it stewed down to make a stock which can be used in various ways, whilst the " steak " itself may be rejected. Again, it has been treated like beetroot (which the flesh also much resembles), by being cut into thin slices and added to the salad-bowl.

The first appearance of the Vegetable Beef-steak is that of a more or less rounded, pale-red knob bursting through a crevice in the bark of the oak-tree and gradually spreading out into a somewhat flabby fan, thin at the edge and thick at its base, as may be seen in the photo of the " fillets."

A GIANT SILK-MOTH

BY W. F. KIRBY, F.L.S.

THE Cecropia Moth is the largest and handsomest moth found in the Eastern United States and Canada, where it is far from uncommon. It is one of the Eyed Silkworm Moths, of which the only British representative is the Emperor Moth, which is a fine insect, as British moths go, but is not more than half the size of the North American moth, which often measures six inches in expanse; though this again is equally surpassed by the great Indian Atlas Moth, which belongs to the same family, and is one of the largest moths known, sometimes expanding to nearly a foot.

Nearly all the moths of this group have short, stout, downy bodies, and a double row of appendages on the feelers, which are very large in the male, but are much shorter in the female. Such feelers are called pectinated, or comb-like, and, as in the present case, almost equally resemble a feather when these appendages are well developed. The feelers are the seat of some sense (probably analogous to smell) which enables the males to seek out the females at a distance (in some cases) almost of miles.

In most moths of this group each of the ample wings is ornamented in the centre with a conspicuous spot, which is sometimes round and opaque, as in the British Emperor Moth, and sometimes crescent-shaped and opaque, as in the Cecropia Moth; while in many species it is as transparent as glass or talc; irregular in shape in the Atlas Moth, or round in various other species.

The Cecropia moth is dark brown, with an irregular white band and white crescents on the

Photo by] [E. Step, F.L.S.

THE GIANT SILKWORM.

This North American caterpillar is four inches long, coloured with shades of green and blue The thick spines are red, blue and yellow, with sharp black bristles from the knobbed ends.

Photo by] [*E. Step, F.L.S.*

COCOON OF GIANT SILKWORM.

The full-grown caterpillar spins a large cocoon of white or brown silk in which it changes to a chrysalis and spends the winter and spring.

wings, all bordered with red; the body and base of the fore wings are vinous red, and the outer portion of the wings shades into grey, marked with brown lines, waved on the fore wings, and straighter on the hind wings; between the white and red band of the fore wings and the outer margin is a large roundish black spot towards the tip, and often others below it.

The moth appears to be less common in the Southern United States than further north. It was described and figured by the earliest writers on American butterflies and moths, and the caterpillar was described by Abbot as feeding on wild plum, and in captivity on apple. It really feeds upon a great variety of trees and shrubs, though manifesting a predilection for the rose family, such as apple, pear and plum; it also feeds on lilac, willows, maples, and the like.

Like others of the family, this moth is accustomed to cling to the under surface of a branch and let its wings droop down, in which position it might easily be mistaken for a swaying leaf.

The eggs of the Cecropia Moth are of a chalky white, streaked with brown. The newly-hatched caterpillar is nearly black, but undergoes several moults, during which it becomes a beautiful pale green, shading into blue on the sides, with black, red and yellow tubercles. When full grown the caterpillar measures four inches in length, and is green washed with blue on the back, and ornamented with blue, red and yellow firm tubercles on the body.

As in the case of most butterflies and moths, the Cecropia Moth is liable to the attacks of various parasites, several species of ichneumon flies laying their eggs on the caterpillar, when their grubs feed upon the contents of its body, and finally destroy it by the time that it attains its full growth.

But if no such untoward accident should happen, the caterpillar becomes a chrysalis, or pupa. This is enclosed in a large brown or white cocoon like that of the common silkworm, with an outer covering of coarser silk which is securely attached to the branch of a tree. Abbot says that this silk has been carded, spun and made into stockings, and that it will wash like linen; but as the cocoon is open at the end, the silk is of less value and less easily unwound than that of the mulberry silkworm, and I am not aware that much use is made of the silk of the Cecropia Moth for commercial purposes at present, though an allied Californian species, the Ceanothus Moth, a smaller and redder species, is regularly reared for its silk.

There is no doubt that there are many natural products which might be put to commercial and other practical purposes which have not yet been made use of; but it is possible for our zeal to carry

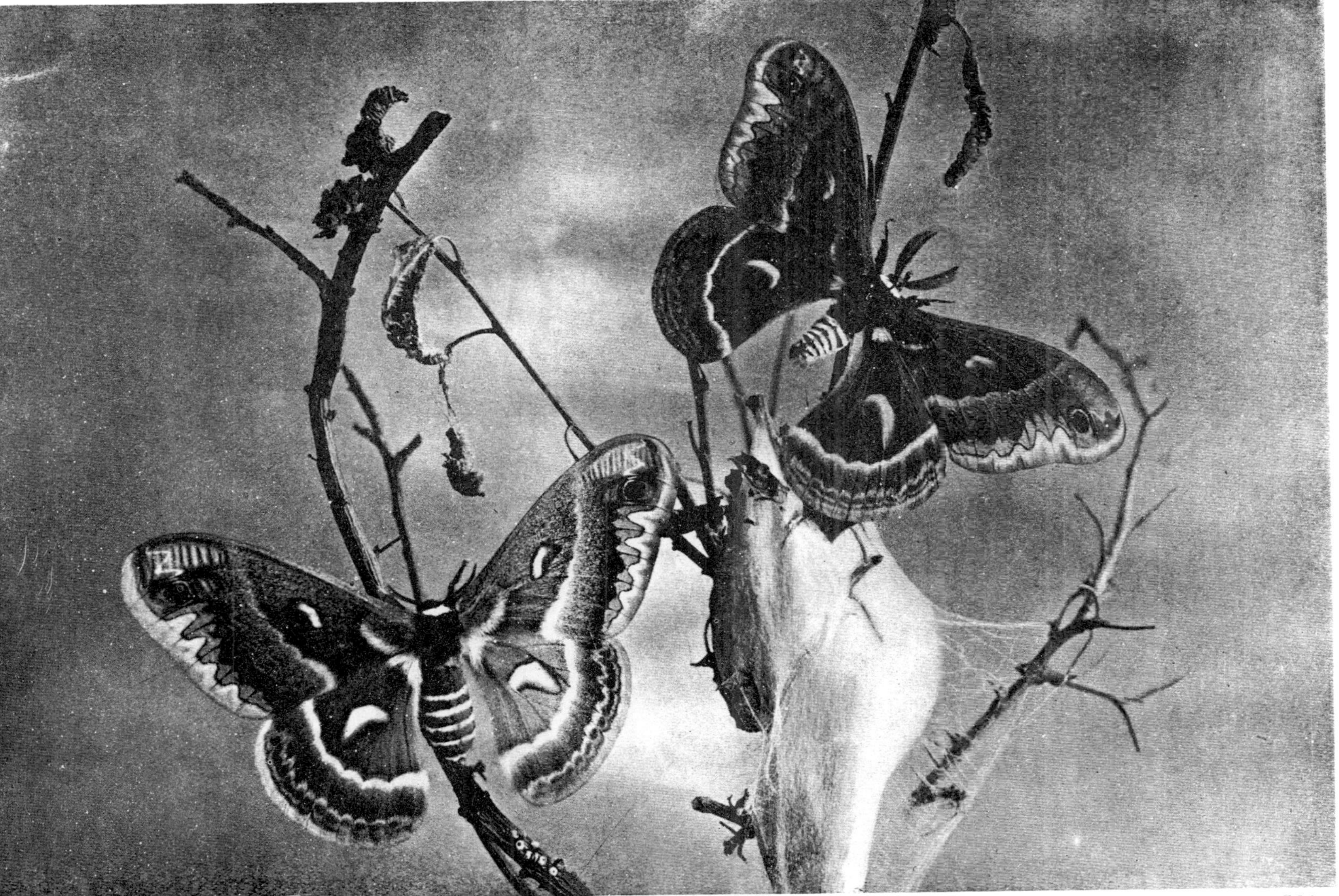

Photo by] [*Edward Step, F.L.S.*

GIANT SILK-MOTHS.

In this picture the two sexes of the perfect insect are shown. The male may be distinguished by the finer feathering of his feelers. The female has a stouter body more strongly marked by white and red bands.

Photo by] [*E. Step, F.L.S.*

THE GIANT SILK-MOTH.

The female moth is here shown of the natural size. A few of her eggs may be seen on the branch near the lower edge of the picture.

us too far. Especially is this the case when we attempt to introduce animals and plants into a new country. Some years ago a Swiss entomologist who was settled in the United States engaged in a series of experiments in breeding moths which he thought might be useful as silkworms. Among others, he received some eggs of the European Gipsy Moth, and these were blown out of the window into his garden; and the insect succeeded in establishing itself in the neighbourhood, and became a terrible pest, costing thousands of dollars annually in vain attempts to exterminate it.

When an injurious insect is introduced into a new country, it generally leaves its natural enemies behind, and is therefore enabled to multiply for a time without a check, if general conditions are favourable to it. On the other hand, conditions may not be favourable, and it will then disappear of itself. The Gipsy Moth formerly inhabited England, but was almost confined to the Fens; and when they were drained it disappeared, like various other insects. But though abundant everywhere abroad, and recently so terribly destructive in America, it was never generally common in England, and even when introduced, has died out more than once, though amid what appeared to be favourable conditions. There seems to be something in our climate inimical to this particular insect; but its introduction into this country has now, very properly, been forbidden, in common with several other dangerous insects which have been proscribed.

Several other species greatly resembling the Cecropia Moth, but differing somewhat in colour, are found in various parts of the United States, Canada and Mexico, to which they are restricted.

THE LUNAR WORLD

BY CAMILLE FLAMMARION.

At the hour of dusk, when all Nature seems to rest after the fevered activity of the day and to yield gently to the calm of the evening, a spirit of tender and high-souled thought comes over the thinker. It is the spirit of the vast skies and, above all, of the moonlight, whose rays, as soft as they are brilliant, spread little by little over the sleeping world.

The sun has disappeared behind the western horizon ; and with him the day has fled. The earth, a revolving sphere, is carried forward in its daily course, bearing us onward to the east and enveloping us more and more in darkness. But soon a pale light invades the firmament. Towards the east the silver orb of the full moon, the radiant Phœbe of ancient mythology, ascends into the heavens, where she reigns as supreme ruler until the dawn. Often she is the unwitting enemy of astronomers, whose observations of the stars, the planets or the comets she nullifies as she eclipses with her exceeding beauty the rest of the Universe. She is the star of the night *par excellence*, the star of mystery and of dreams, the first halting place of the human mind as it hastens forward to the conquest of the skies.

Ever since there have been eyes to see and brains to think upon the earth, the moon, of a surety, has held captive the first looks and the first thoughts. Her positions on the celestial globe, the changes of her appearance, her presence on certain nights and her absence on

[*By Arthur Twidle.*

THE EARTH AND THE MOON

In this picture we are supposed to be standing on some neighbouring planet and viewing the Moon from the side, thus getting an idea of its true form. As the same end is always presented to the Earth the idea we gain that the Moon is round from the terrestrial view-point is necessarily misleading.

others, then the uninterrupted succession of her phases and their exceeding regularity, all these particularities attracted the special attention of the inhabitants of the earth to this mysterious star.

The result of a long period of thought was the conclusion that these phases were due to the movement of the moon round the earth ; that the lunar sphere, our faithful satellite, did not possess any light of herself, and that the extent of her illumination depended on the position she occupied in relation to the sun, from whom she derived her light. When she is situated above the earth and is in conjunction with the day-star—that is to say, in direct opposition to the sun—the half of her here which is turned towards the earth receives the direct flood of the solar rays ; that is the period of the full moon. When she is lit obliquely or at right angles only a part of her surface appears brilliant, which varies according to the angle at which the moon receives the sun's rays. When the western half of the disc is visible that is the first quarter ; fifteen days later the eastern half is in its turn illumined, and that is the last quarter. At the moment when the moon passes between the earth and the sun she is totally invisible, because the half of the sphere receiving the light is directed towards the greater luminary, and the other half, which possesses no light in its own nature, remains too dark to be perceived. That is the new moon. But about two days after the conjunction of the sun and the moon the latter reappears to the west in the beautiful guise of a crescent delicately defined as if it were a sickle of light suspended in the darkling sky of evening. Sometimes during this phase the part of the sphere which is not directly lit is outlined in grey against the background of the sky. This *shadowed light* is the reflection of the *earth light*—that is to say, of the light of the sun reflected by the complete hemisphere of the earth and again reflected to us by the lunar disc. Our satellite, the moon, takes twenty-seven days, seven hours, forty-three minutes, eleven and a half seconds to reach the same position in the sky as regards its relation with the stars and to accomplish what is termed the sideral revolution, but it is necessary for her to continue her journey for about two days longer in order to come back to her position between the earth and the sun, who, in the interval of lunar circuit, have been displaced, the first in very fact, by the movement of the annual journey, the second in appearance only, owing to an effect of perspective resulting from the displacement of our globe, which brings the synodic revolution of the moon to occupy a total period of twenty-nine days, twelve hours, forty-four minutes, three seconds ; and this fact is of the utmost importance to us, for it determines the exact time of the moon's phases.

Photo by] [*C. Flammarion.*

THE MOON'S SURFACE.

"I possess upon the Moon, towards the centre of the visible hemisphere, a property of considerable dimensions, the Circle of Flammarion, with a diameter of 33 miles."

Photo by] [*The Yerkes Observatory.*

A MIGHTY CRATER, COPERNICUS.

This is one of the largest of the Moon's craters. The immense depression has a diameter of forty-six miles and is bounded by the circular range of mountains 12,000 feet high. The volcanic cones in the centre of the crater tower 24,000 feet above the lava bed.

That is what the thinkers amongst men have succeeded in finding out after several centuries of observations. But our curiosity is far from being satisfied by these astronomical preliminaries. As we contemplate the moon shedding her pure light upon sleeping Nature, other reflections cross our minds and the delicate crescent of silver light appears symbolic of a magnificent interrogation mark which the heavens have set before us.

The moon is a world, and it is the nearest to the earth. It receives from the sun the same rays which create the life upon the surface of our planet. Besides, everything in Nature has an end—a goal. Is it not, then, very natural for us to ask what part our satellite plays in the laws of the Universe and what the conditions may be which prevail in the lunar world whose destiny is so closely bound up with our own?

Photo by] [*The Lick Observatory.*

THE MOON

A beautiful photograph taken at the Lick Observatory when the Moon was fourteen days, one hour, old.

As soon as an explorer on this planet discovers a country that no civilized being has trodden before him, whether it be some region of burning desert or the rocky peak of a mountain at last scaled, or some frozen tract in the polar zone, the matter that interests us most to know is the exact quality of the locality and consequently to discover its exact characteristics.

We know that even upon our own planet the soil, the flora and the fauna are as infinitely varied as are the countries, and even in the most deserted regions the careful observer can always find out something which is worthy of remark. By the same method of reasoning the differences between one world and another must be greater still, and the true interest of astronomy is exactly to study the special characteristics of each and to consider its condition from the standpoint of life.

Moreover, the lunar world, in spite of its relative proximity, is very different from our own, and if an inhabitant of the earth could transport himself to its surface he would feel himself very out of place. Unhappily, none of us could prove these new impressions; not that the 238,610 miles (which is about the measure of distance separating us from this star of the night) would be too far for a terrestrial traveller, for many men, by boat or by railroad, have made as long a journey as if they had been to the moon. An employée on the Metropolitan Railway travels by the electric car as much as 238,000 miles in eleven or twelve years. But by the laws of Nature and by the mechanics of earth the crossing of the space between the planets, which is void of air, is for ever forbidden to us. So it is that we make excursions upon our satellite by the aid of the telescope and only receive news of her through the rapid waves of light, which, travelling at the rate of 186,772 miles a second, arrive on our planet from the moon in a little less than a second and a quarter.

The perfected methods of modern telescopic and photographic exploration can supply us with such precise facts about the earth's companion that our geographers and land surveyors might envy the astronomers, for the lunar topography is perhaps better known than that of the terrestrial globe. It must be understood, however, that reference is now made to the hemisphere turned towards us, which alone we can see; the other half of the globe remaining perpetually hidden from our eyes from the fact that the moon always presents the same surface to us.

Many millions of centuries ago the earth and the moon formed one single confused mass revolving rapidly on itself. This enormous fluid sphere was subject, owing to the attraction of the sun, to formidable tides. Men who have made the science of the formation of worlds their study are of the opinion of Sir George Darwin: viz., that under the combined influence of centrifugal force caused by very rapid rotation and by the solar tides a portion of the surface of the terrestrial globe was detached and constituted a single body in space in the form of the link of a chain, or a mass of material which afterwards congealed into a ball and became the lunar globe. This event—the birth of the moon—took place about fifty-six million years ago. According to Mr. W. H. Pickering, the profound depression of the Pacific Ocean should mark the site of the cradle of the moon. Should such be the case, one is then forced to admit that at the moment of separation the earth was already covered with a solid crust, a condition which appears doubtful, although the theory of mathematics teaches us that a viscous globe possessing even a thin skin can experience tides as intense as those prevailing when it is in a state of complete fluidity.

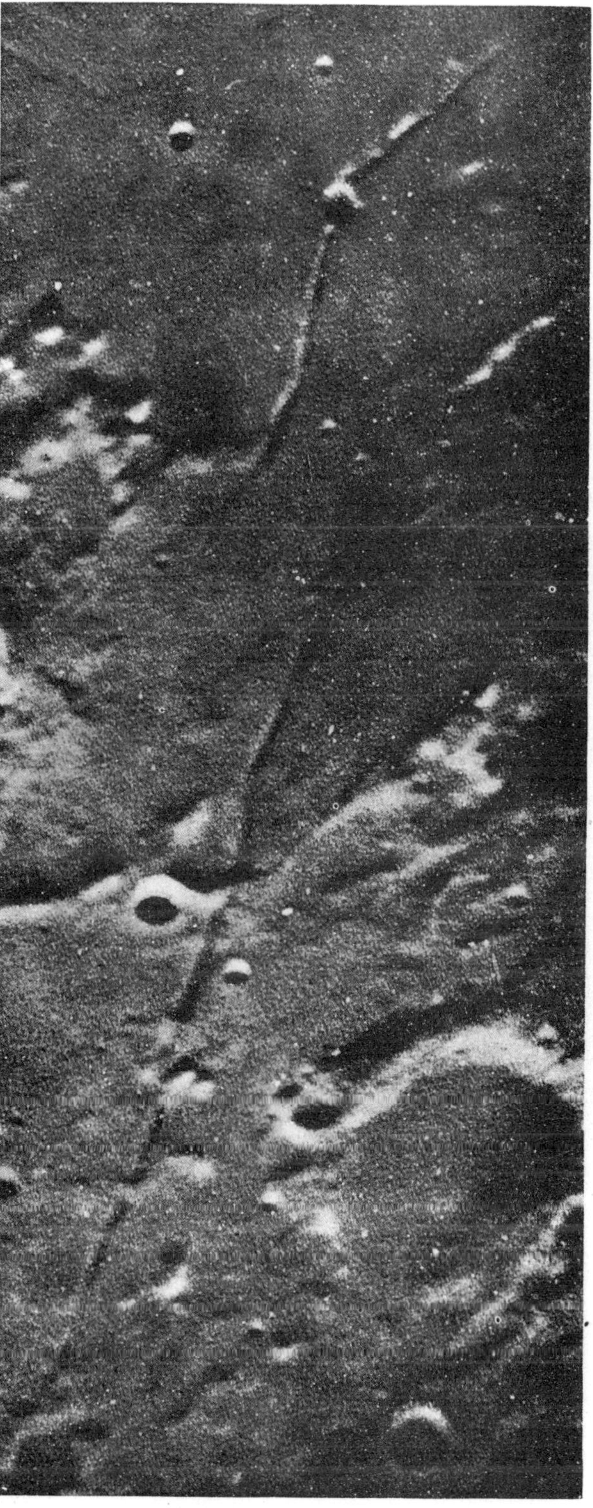

VIEW ON THE MOON.

This view represents the district known as Hyginus. Note the long diagonal depression, which is thought by some astronomers to be a formation similar to the supposed canals of Mars.

At this epoch the month and the day of the earth and of the moon had a common duration of five hours thirty-six minutes. Our satellite continued to revolve round the mother planet; gradually, however, getting farther away and lessening the speed of her revolutions. Her surface was troubled by immense tides provoked by the excessive attraction of the earth, which caused the formation of a great bubble. The lunar rotation caused this excrescence to veer round to the east, but the all-powerful attraction of the earth brought the prominence into line with herself. The swelling acted like a bridle upon the daily movement of the moon and caused the speed to relax, until the lunar day became at length equal in duration to the present lunar month. The tides have ceased, suppressed somehow by the action of the earth; but they have left the mark of their action in the lengthening of the hemisphere turned towards us, and which gives the moon the shape of an egg.

Photo by] [*The Yerkes Observatory.*

VIEW ON THE MOON.

One of the great grey depressions which cover a great part of the surface of the Moon. For want of a better term they are called seas. The one here shown is the important Sea of Serenity, possessing on the north-west of its borders two volcanoes—Posidonius and Caquet—which exhibit signs of activity.

So it is that the moon is of terrestrial origin. But it is less dense, for although the constituents of its mass are those of the terrestrial surface, they have, nevertheless, a density only twice as great as that of water, while the mean density of the earth is five and a half times as great, and in its central depths the latter must attain the density of iron and heavy metals, which brings its total density to at least seven to ten times that of water. Above all, the lunar land surfaces appear to be composed of formations of chalk, broken by a vast number of fissures and very light. The extreme brilliancy of their reflections creates the impression that the whole surface is coated with marble or a carbonate of chalk. During the day this reflection is equal to that of the whitest clouds.

On first thoughts it would seem that a vast atmosphere should envelope this little globe broken off from our own, and that a great deal of vapour would be found thereon. Certainly the moon at the outset was possessed of both water and air, but she has not retained them. Her volume is only a fiftieth part of that of the earth, her mass only an eighty-first part and the weight of her surface about one-sixth. A pound of matter transported from the earth would weigh no more than two and four-fifths ounces in the moon. So her feeble powers of attraction have not been able to retain an atmosphere, for she allows every molecule thrown up vertically to escape at a speed of at least one and a half miles per second; and, further, according to the kinetic theory of gas, this speed is frequently surpassed by the majority of gaseous particles.

As to the waters which first covered her mass, they have been quickly absorbed by the soil, which is as porous as a sponge.

Owing to its small size, the moon has cooled much more rapidly than the earth, and possessed a solid crust long before the latter body. The gases and vapours enclosed beneath this crust have played dire havoc with the walls of their prison. Under the enormous pressure of their expansive force, the surface of the moon has suffered the extremes of tension, to which the appearance of lunar territories to-day bears ample witness: the ground bristles with mountains, whose summits are worn away by the agency of the contraction of ages; there are numerous deep beds of rock, landslips of various natures, and corresponding upheavals.

Photo by]. [*G. E. Hale*.

THE MOON'S SURFACE.

Showing the innumerable craters which witness to the vast disturbances that have befallen our satellite.

Lunar topography differs essentially from that of the earth. While on this planet the higher masses of the crust are arranged in mountain chains, upon the moon they always take a circular form, from the outlines of the greatest "seas" to the smallest craters. To explain this characteristic configuration we must remember that the lunar soil, as it became colder and solidified, enclosed a considerable quantity of vapour and gases. These gases, seeking an escape from the interior, have brought great pressure to bear on the exterior surface of the moon. One can perceive evidences of these forces with the naked eye, but it is possible by means of photographs of the full moon to examine them in the long white tracks which spread out from the mountain of Tycho, and which appear especially brilliant. These white tracks seem to have been

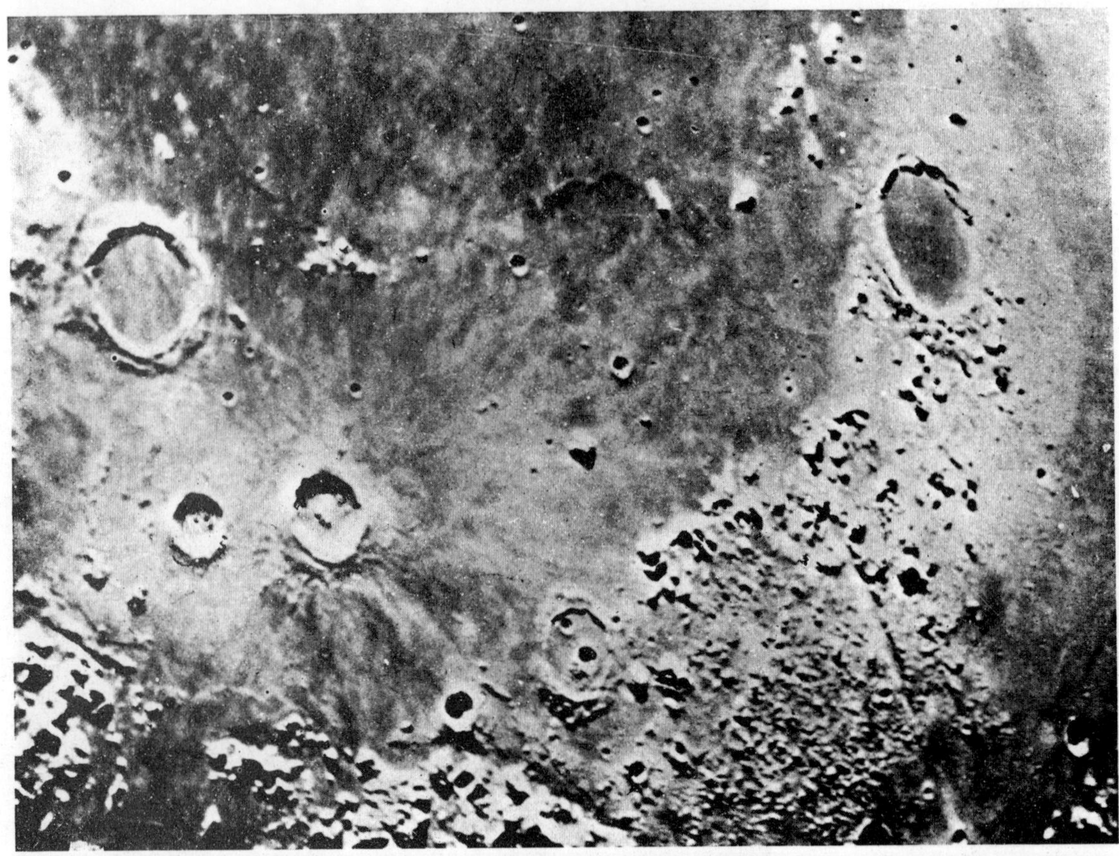

PLATO

The large crater-ring on the right hand of the picture is that of Plato. Within this vast district other craters have appeared and passed away, or, as is sometimes the case, have changed their position, and these phenomena support the belief that *all* the craters on the Moon are not yet extinct.

fissures across which floods of light-coloured lava have flowed, filling up the crevasses and solidifying. As for the mighty circles, it is possible that they were gas bubbles which rose to the surface and broke, producing a complete circular depression. The greatest formations must be the most ancient, the extent of the disturbed surface diminishing as the expansive force of the gases gradually lessened. The truly colossal dimensions of several of the circles and their surrounding heights give an idea of the forces which were in action in primitive times: the gaping mouth of Tycho measures about 52 miles and 1,435.8 yards in diameter, and the ramparts attain a height of about 3 miles and 552.7 yards. The circle of Clavius is 130½ miles wide, and has an elevation of 4 miles 615.4 yards. One can count upon the moon's surface about twenty circles whose diameter exceeds 60 miles, and about as many peaks of a greater height than Mont Blanc;

[*By A. Twidle.*

THE BIRTH OF THE MOON.

The Pacific Ocean is the possible cradle of the Moon. Attracted by the solar tides, it is supposed that a great nebulous mass was drawn away from the Earth, whose crust was insufficient to retain the semi-gaseous matter. The new-born Moon, however, could not escape the superior attraction of the Earth, so that these dual influences moulded it to its present shape and defined its present course.

Photo by] [The Lick Observatory.

THE YOUNG MOON.

Aged seven days, three hours

Photo by] [P. Puiseux.

THE MOON NEARING HER ZENITH.

the Mountains of Leibnitz rise to a height of 5 miles 222 yards, the Rocky Mountains to nearly 5 miles, the crater of Newton to 4 miles and 888.81 yards. A most curious feature is that the interior well of these mountain links is always at a very much lower level than the land on the further side of the enclosing mountains.

It is not possible that these circles owe their origin to volcanoes such as we are cognizant of, and the mountains surrounding them are not composed of material thrown up by these craters. They are districts of overturned soil caused through the same agency—a gaseous explosion coming from the interior. The eruptions have diminished in power, but they are still to be found almost everywhere, and have even formed little circular marks upon the ramparts of the larger circles, while their numbers are legion. The beautiful photographs taken from the Observatory at Paris show that over vast stretches of country a veritable deluge of mud finally invaded and submerged the lowest districts, often blotting out almost completely even the vestiges of these craters and burying the hollowed ground.

There are yet to-day very small craters which give off intermittent vapours. The little crater Linnæus appears to be one of these. For two and a half centuries it has been watched attentively, and sometimes it can be seen quite clearly, very deep and possessing a shadow; sometimes it has disappeared completely and has been replaced by what appears to be a very brilliant white cloud.

In the same lunar region, to the north-west of the disc, on the borders of a vast depression, which is circular and of a grey colour and has been named "The Sea of Serenity," there are two other craters a considerable distance one from the other —Posidonius and Caquet—which afford similar phenomena; that is to say, they appear at times to be filled with a white vapour which effectually hides their openings. Other points on the surface of the moon appear to be affected by similar changes. There is also Messier, whose variations are so strange.

It is possible that during the long and cold lunar night, which lasts about fourteen days, during which time the temperature of the hemisphere not illuminated is extremely low, water-vapour or carbonic acid gas is yet given off from the depths of these craters, and condenses in the form of hoar-frosts or snow on the interior of the bounding walls. The first rays of the rising sun dissolve these deposits; the ice melts, the snow evaporates and becomes invisible, and the crater reappears in its normal condition. This is precisely the impression created by the present (1911) observations of Caquet.

During the three hundred and fifty-four hours of sunshine which constitutes the lunar day, the surface of the soil must reach a point of temperature a good deal higher than that of our most burning desert or of the equatorial regions; for the heat must reach very nearly to one hundred degrees Centigrade and even more, perhaps even to nearly two hundred degrees in those parts which receive the solar rays perpendicularly; for they are never lessened by clouds nor the presence of fog, nor even of air, since the atmosphere which around the earth acts as a filter to strain the rays coming from the sun, and as a shield to hinder them from escaping and being lost in space, is so rarefied upon the moon that its effect is almost nil.

This great white globe, turn by turn too hot or frozen, whose great spots, dark and grey, mark the beds of dried-up seas; this world without water and with little air; this neighbouring habitation appears by no means hospitable.

Nevertheless, I possess upon the moon, towards the centre of the visible hemisphere, a property of considerable dimensions, the Circle of Flammarion, with a diameter of 33 miles; but I very much fear that it is rather a sterile property.

At the epoch when our world was yet only a mighty chaos, and was preparing itself slowly by the incessant travail of its elements for the future sojourn of humanity, the moon, more advanced in its planetary evolution, had probably given shelter for millions of years to living forms fitted by nature for the special conditions of life on her world. But she has grown old more quickly than our terrestrial habitation.

But it is not necessarily to be supposed that

Photo by] [*P. Puiseux.*

THE MOON ON THE WANE.

Photo by] [*The Lick Observatory.*

THE MOON IN THE LAST QUARTER.

Aged twenty-three days, eight hours.

we are dealing with a world completely dead. It is possible that beings of a different order to ourselves people the subterranean regions (or should we rather say sublunary ?), the last vestiges of a life that once on a time was vigorous and flourishing.

The moon is not only a celestial mirror which reflects the light of the sun to us. She relates, for our eyes to see, the history of an earth; a history the more appealing in that she is united to our own world by ties which can never be broken.

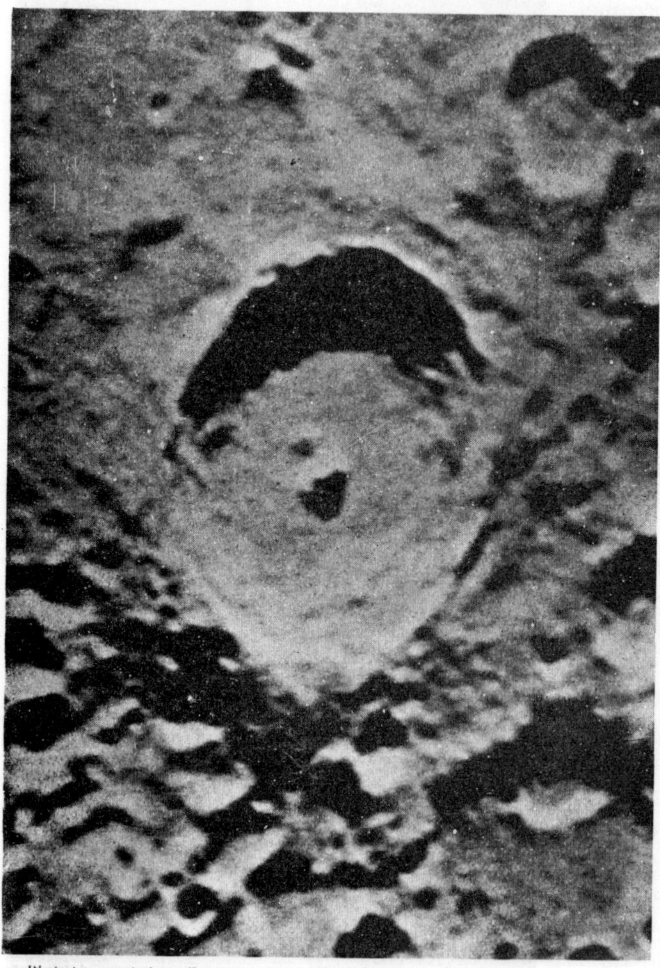

Photo by permission of] [*Yerkes Observatory.*

MOUNT TYCHO.

Tycho is, perhaps, the best-known crater on the Moon; but a "crater" in the Moon is a different formation to that of the Earth. It is the centre of an enormous bubble which has burst and left a circular chain of mountains to mark the place of the upheaval.

VEGETABLE SHEEP

BY EDWARD STEP, F.L.S.

So numerous are the instances afforded by plant life of counterfeit resemblances to animal and other forms that a book might easily be filled with an account of them. In quite recent pages of the present work we have already dealt with several of them —vegetable steaks, vegetable candles —and there are vegetable cows that give milk, vegetable butter, vegetable bread, vegetable putty, and so forth. Some of these names are given on account of the vegetable in question serving as an actual substitute for the animal product; in other cases they merely indicate an appearance that is deceptive, and may even at times give rise to feelings of annoyance with Dame Nature for having so successfully hoaxed us. Of this nature is the Vegetable Sheep.

The shepherd on the upland pastures of New Zealand may see afar off, resting on some rocky ridge, a sheep or group of sheep that refuse to come at his call when he is shifting the flock to a new feeding-ground. He climbs a long distance after them; but it is not until he has got quite close that he discovers he has been deceived by appearances. The supposed sheep is nothing more than a dense white clump of vegetation—the Raoulia, or Vegetable Sheep.

The plant is a member of the huge daisy family, and is near akin to that section of it known as everlasting flowers. In order to counteract the influence of the winds that sweep over its usual exposed places of growth, its branches divide into innumerable short twigs, and when these become clothed with small woolly leaves, you have a dense tuft of tightly packed vegetation as shown in our photograph. Our example, separated from its natural surroundings, might be mistaken at first glance for a mass of rock, it is so compact. If

Photo by] [*E. J Wallis.*

VEGETABLE SHEEP

A detached mass of this curious plant, which is a member of the daisy family. In the left-hand corner is a part denuded of flowers and leaves, showing the dense structure of the branches, which form a scaffolding sufficiently strong to withstand the most boisterous weather.

Photo by] [*S. Leonard Bastin.*

THE VEGETABLE SHEEP.

This plant grows in exposed and bleak places. In order therefore to overcome the disadvantages of its home it branches out in a dense tuft of woolly leaves and white flowers, which give the appearance of a sheep's coat.

looked closely into, however, the leaves will be seen, and a great number of tiny flowers. At the lower part of the left-hand side a portion of the plant is denuded of leaves and flowers to show how the stems branch again and again, and so build up a strong scaffolding upon which the leaves and flowers are produced in such a way that no wind can tear off portions.

A TUBE WORM

BY EDWARD STEP, F.L.S.

THE observant rambler on the sea-shore meets with many strange tubes on the rocks, on empty shells, and sometimes standing up from the sand. Some of these appear to be constructed of porcelain, some of fragments of broken shell; others of grains of sand cemented together. They are all the work of different kinds of sea-worms, which, having delicate bodies such as fishes and other predatory neighbours would delight to make a meal of, are constrained to build some sort of a house for their protection. For similar reasons these worms are all of a retiring nature; and he who is content merely to look at these tubes and wonder will never learn anything of their architects. Yet if a shell or a flake of rock supporting one of these tubes be placed in a glass of clear sea-water and watched, the observer will be rewarded by a beautiful sight.

When everything is still, a brush, like a camel-hair pencil, begins to be pushed out cautiously from the mouth of the tube. Then the filaments of which it is composed begin to separate, first into a funnel-shape and then more widely, until as the light falls across them you find that each thread is really a plume, the shaft giving off delicate transparent branches. This beautiful object consists of the breathing organs, or "gills," of the worm. Through these its blood circulates and absorbs the oxygen of the sea-water. They also serve as a fine net, by means of which microscopic food is gathered from the water and directed to the creature's mouth. A shadow falls across the creature or a step on the floor causes an almost imperceptible tremor, and at once the widespread plume folds up and glides back into the safety of the tube.

The animal which builds this tube, and of which we have only seen a small portion, has a long body built up of a series of rings, and around each ring there is a circlet of bristles, by whose aid it climbs up and down its tube. In building this tower it employs a cement of its own production, which hardens into a kind of silk to which fine particles of sand or mud adhere, and the whole sets into a firm tube. Some species build solitary tubes ; others combine in colonies of scores or hundreds of individuals each in its own tube.

THE PORTUGUESE MAN-O'-WAR

BY FRANK T. BULLEN, F.R.G.S.

THAT very delicate and beautiful denizen of the deep seas which is known to seamen by the above slightly sarcastic title is eminently worthy of a little detailed attention, perhaps all the worthier because in a Universe of Marvels its value is purely decorative, having not the least commercial use.

In some one or other of its hundred or so varieties it abounds in every tropical sea, exhibiting the delicate sapphire tints of its "sail" and the lovely shades of colour in its "oars" only when the ocean is sufficiently placid to allow so tender an argonaut to venture upon the surface. Then on some quiet evening, when the westward-hastening sun has flung along the level circle a succession of colour schemes each more wonderful than the last, a pigmy fleet of these timid voyagers may be

Photos by] [*Hugh Main, F.E.S.*

THE SABELLA AND ITS HOUSE.

The Sabella is one of the marine worms that protect themselves from their enemies by building a tube-shaped house of silk and mud or sand. In the first photo its closed gills are just protuding; in the second they are spreading out.

seen swaying in accord with some motion of the unstable sea that is imperceptible to man, dipping and rolling indeed as if making believe very much that the weather is " quite rough, you know."

Many people have confused this little creature with the more highly-organized mollusc known as the Nautilus, with its exquisitely beautiful shell. The Portuguese Man-o'-War has no shell or bone, or anything about it but what will melt under the fervent rays of the sun if out of the water. Its " sail " is but an air-sac, filmy almost as a soap-bubble, inflated by who knows what mysterious impulse from the little colony beneath. For herein lies another mystery: The Portuguese Man-o'-War is manned, if the term may be forgiven. It is not an individual, but a colony of individuals, which is subject to continual birth, growth, and death; but which appears to be actuated by a common impulse; one law which governs the whole without friction, knows no disobedience or insubordination.

[*By J. Teklenburg.*

THE PORTUGUESE MAN-O'-WAR.

This most brilliantly coloured member of the tribe of Jelly-fish makes a brave show in the waters of the Mediterranean.

Few things are more startling to witness, even by one ignorant of the organization of these colonies, than the way in which, as if at some imperative signal, all those sapphire sails are suddenly furled and the toy fleet disappears. Such a disappearance is usually only momentary; even while the onlooker is lamenting the absence of such a wonderfully decorative adjunct to the plain sea-surface, they will all reappear, nodding and rolling as bravely as ever.

Yet it must be noted that, beautiful and helpless and innocent as these miniature ocean voyagers appear, they possess a weapon that, while it is powerless to defend them from destruction, can, and does, inflict terrible punishment upon any aggressor. I allude to the extraordinary quality possessed by all of them of scarifying the flesh of anything that touches them almost as badly as vitriol does. It is probably an exceedingly potent digestive fluid secreted by the creature; but whatever it is, its effects are, in some cases at least, maddening. I can never forget seeing a pet pig of ours in a ship of which I was second mate writhing in agony, which culminated in our having to destroy her. She found a Portuguese Man-o'-War lying on deck, whither it had been washed in a gale the previous night, and essayed to eat it, with lamentable results. And I have seen a man temporarily bereft of reason through just touching the tentacles of a Portuguese Man-o'-War while swimming.

But, after all, this innocent and faery-like ship of the sea forewarns us of its dangerous qualities; for the bright colours it displays are given to it for the especial purpose of a danger signal. Whoever, therefore, becomes a victim to this potent sting must take the onus of his predicament on himself and not blame this beautiful denizen of the ocean.

Photo by] [*Paul H. Fabre*

MANTIDS.

There are many varieties of the grotesque family of the Mantids; but perhaps the best-known species is that of the Praying Mantis. Here female Mantids are shown quarrelling, and the war will be one of deeds not words, for the Mantids fight desperately.

Photo by] [*Paul H. Fabre.*

THE MANTIS LEAVING HER EGG-CASE.

This egg-capsule is deposited by the Mantis on the branch or twig of a tree. From it the young emerge, swinging by a delicate thread, after the fashion of spiders, till they reach another destination.

THE PRAYING MANTIS AND ITS RELATIVES

BY JOHN J. WARD, F.E.S.

IT would be difficult to find a more saintly-looking insect than the Praying Mantis. For hours together it will assume an attitude of innocence and quietness, its fore-legs raised as if in supplication. Not until an unwary fly approaches it do we realize the object of its quaint posture. It then instantly becomes obvious that its purpose is not to pray, but to *prey*.

In Britain we have no representatives of its group, its nearest relatives in this country being, on the one hand, cockroaches, and the other, grasshoppers and crickets. The Praying Mantis is the common European form, and it first begins to appear in Southern France, becoming more abundant further south. In general habits the European species is characteristic of its race; but its tropical relatives have evolved some extraordinary features.

In the first place, the Mantis is carnivorous. It requires living prey—flies, grasshoppers, caterpillars, etc.; but being slow of foot, it has to capture its quarry by craft.

The method adopted is to sit amongst the grass or leaves with its head held loftily erect, its thorax being developed into what resembles a long neck, thus extending its field of vision. Its colour may be green or brown, harmonizing with its surroundings so perfectly that it can only be recognized with difficulty. In front of its head its fore-legs are raised into what has been regarded as a praying attitude; hence the names "praying insects" and "soothsayers," which are sometimes used in connection with these insects. It then awaits the approach of a victim, and sometimes will remain motionless for several hours.

If the Mantis detects prey a little way off, but not near enough to effect its capture, it steals slowly and almost imperceptibly towards it. When within striking distance its raptorial fore-legs are thrown suddenly forward, and in an instant the victim is grasped clasp-knife-fashion between the toothed shanks and thighs of these powerful legs, after which it is ravenously devoured. If prey is abundant, however, it more often happens that fresh victims are continually captured and partly eaten, as if it enjoyed the sheer love of killing.

Such are the feeding habits of this Mantis of saintly mien. Even the Greeks endowed it with supernatural powers. In Nubia, the Hottentots regard a Mantis with the greatest reverence, and should one alight on an individual he is recognized as selected of heaven, and at once becomes a saint; but woe unto the native that kills a Mantis; he becomes no longer a skilled hunter, but is ill-fated until the end of his days. Turks and Arabs contend that these insects are in constant prayer, with their faces ever turned towards Mecca. France, Portugal, America and many other countries provide similar superstitions, all of which have doubtless arisen from the quiet demeanour and devout attitude which have served to disguise the highly predaceous character of these insects.

It is interesting to note that the Chinese have better realized their true character. I am informed that they are there kept in bamboo cages and matched like fighting cocks. Such a thing is quite possible, for should two female insects meet, a fight is tolerably certain. By means of their powerful fore-legs they are able to strike tremendous blows, and with these weapons they engage in combat, the result often being that the weaker of the combatants gets one or more of its limbs severed by one successful stroke of the enemy, or even his body may be cut through.

The male Mantis has a difficult task on hand when wooing his mate. She is both larger and stronger than he is, and should his advances not prove favourable, he rarely has the option of retreating; for while she will not accept him as her lover, yet she takes care that he does not become the husband of another of her species—her refusal takes the form of killing and eating him. Cannibalism is

Photo by] [*Paul H. Fabre.*

THE MANTIS.

The female Mantis destroying her unfortunate wooer. Wooing is a hazardous affair amongst the Mantids, for the rejection of a proposal means instant annihilation of the "proposer."

quite a common feature with these insects, the larger individuals frequently attacking and devouring the smaller ones.

The female Mantis deposits her eggs about September, the young being hatched in early summer. They resemble the adults, but are without wings in their early stages, when they prey upon greenflies. The numerous eggs are not deposited singly, but in a chambered capsule, which at first appears as a bubble of frothy foam, afterwards becoming hard and brittle, the capsule being attached to a twig or stem. When the young emerge, they drop down from the egg-cocoon on delicate strands, reminding one of young spiders.

It is, however, in the warmer parts of the earth that Mantids are seen at their highest development. In the tropics many species are to be found, all of which offer some wonderful features of interest. As we have seen, the Praying Mantis somewhat resembles its surroundings, but its foreign relatives carry such simulation to an extraordinary degree. Their leaf-like wing-covers so perfectly mimic the foliage which surrounds them that, when removed, they are almost sufficient to persuade a practical botanist into the belief that he is examining a vegetable structure, so faithfully are the veins and fine nervures of the leaves copied. Also, many of them have developed upon their legs irregular pieces of leaf-like integuments, so that, when waiting amongst the foliage for their prey to approach, they are completely disguised. It is remarkable, too, that as the season advances they slowly change in colour from green to brown, in harmony with the leaf changes.

Photo by] [*Paul H. Fabre.*

THE MANTIS PRAYING.

Not really praying, but preying in deadly earnest; for in this position the creature waits for its unwary victim, then the powerful fore-legs unfold and the unfortunate insect is grasped tightly between the toothed shanks and thighs.

Other species go further than merely harmonizing with their surroundings, and simulate some object which is of even greater utility to them. These, instead of resembling leaves, are brilliantly coloured like fine flowers, their attitudes and means of attaching themselves to the plant-stems combining with their external contour to heighten the effect. In this way their prey does not fall into their clutches merely by accident, but is directly lured towards them by their flower-like form. One species inhabiting India has its two hind pairs of legs flattened into broad pink or white plates, shaped like the petals of a flower, and when seeking for prey it spreads out these organs, two on each side, its fore-legs being almost hidden beneath its fore-parts. So butterflies and other flower-loving insects are attracted towards this apparent flower, which is really a death-trap for them.

Photo by] [*Paul H. Fabre.*

THE ATTACK OF THE MANTIS

Very terrifying is the attack of the Mantis as it raises itself to its full height to swoop down upon its enemy. There is little doubt as to the end of the conflict, for there are few insects which can withstand the onslaught of this rapacious creature.

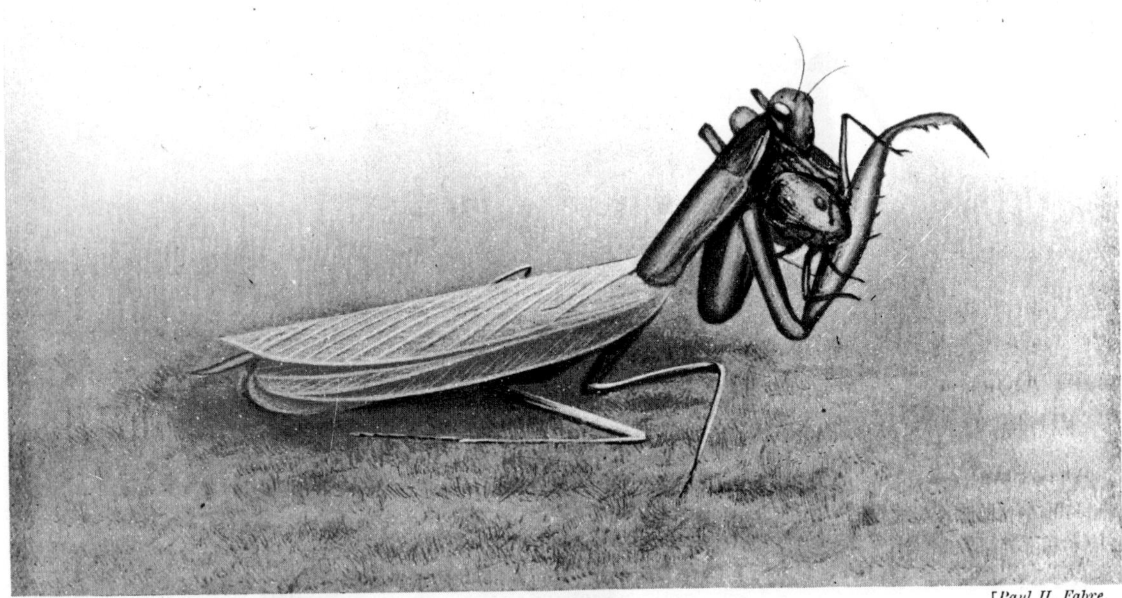

Photo by] [*Paul H. Fabre.*

THE VICTORIOUS MANTIS.

Once within the clutches of the powerful limbs the victim is immediately sacrificed to the rapacity of its cruel victor, for even if sport has been good and hunger appeased the Mantis will not leave its captive alive. Here the Mantis is devouring a grasshopper.

Still other forms have evolved in entirely other directions, having become mimics of the excreta of birds for the purpose of attracting flies.

A naturalist (Annandale) has recorded a remarkable instance of a pink Mantis found in Siam, and which was observed in the midst of some rhododendron-like blossoms. This insect being too large to successfully simulate part of a single flower, is provided with a green band, which divides its pink colouring into two sections, thereby assisting it to resemble two blooms. Its legs and the front part of the body are pink, and also the under-side of the end of its abdomen, which it turns up over its back. At the tip of the abdomen there is a black patch on which minute flies are fond of alighting. Thus we have the illusion of a flower with tiny flies feasting upon it.

The tiny flies are too insignificant for the Mantis to prey upon, and are only part of its device for attracting its larger prey. It was further noted that when the abdomen was turned down and the upper surface exposed, brown lines were revealed, which suggested the appearance of the rhododendron-like blossoms when they are withering.

GLASS SPONGES

BY RICHARD KERR, F.G.S., F.R.A.S.

THE Glass Sponges, more especially those known as the "Venus' Flower Baskets," are, without exception, among the most exquisite of all forms of marine life known to us. They have their abode on all ocean floors, generally at great depths. In size they range from a few inches in height to nearly two feet. They are attached to the bed of the sea or ocean by long wisps of flint, like spun glass. These flinty prolongations are part of the creatures' own structures.

The skeleton, the part with which we are most familiar in our museums, is a lacy fabric of flint in which the delicately-arranged lattice-work is in perfect order, based on a regular system which conduces in the best possible manner towards producing great beauty combined with wonderful strength.

In the living state all this intricate and beautiful network of flint is filled in with the flesh proper of the sponge-creature. It resembles the white of egg and it contains the life principle. Perpetual movement is the order of its existence, so that currents of water, laden with food and other requisites for its sustenance, are always passing in through the tiny openings in the lattice-work. The water in its passage, deprived of these vital products, passes out through the larger openings on the upper part of the sponge.

Numberless wonderful lashes, quite below the power of human vision, whip the water along through every portion of the sponge from the numerous entrances to the large openings.

But this is not all. On bringing the microscope to bear upon the fleshy substance, we find multitudes of exceedingly small rods of flint of many shapes. They are too small to be recognized by eyesight, however keen. Anchors, double anchors, spears, prongs, tridents and many other forms are present. All are composed of flint and appear to form a coat of mail, possibly for giving support to the delicate flesh of the creature.

The account of the vicissitudes which one of the first Glass Sponges brought from Japan to Europe underwent forms interesting reading. To be very brief, it was looked upon as a clever piece of Japanese skill and was given a prominent place in a collection of Japanese curios by the great naturalist Ehrenberg! Later on it was pronounced to be a dual work of Nature, *i.e.*, a natural glass coil, or "wisp," combined with a coral. It was then removed from the cabinet of Japanese art and placed among a collection of natural history specimens. Here, however, it was put in an inverted position! It is only fair to the deceived naturalists to explain the true condition of things. The Japanese dealer had ingeniously "made up" the so-called composite animal by uniting filaments of the "wisp" to a

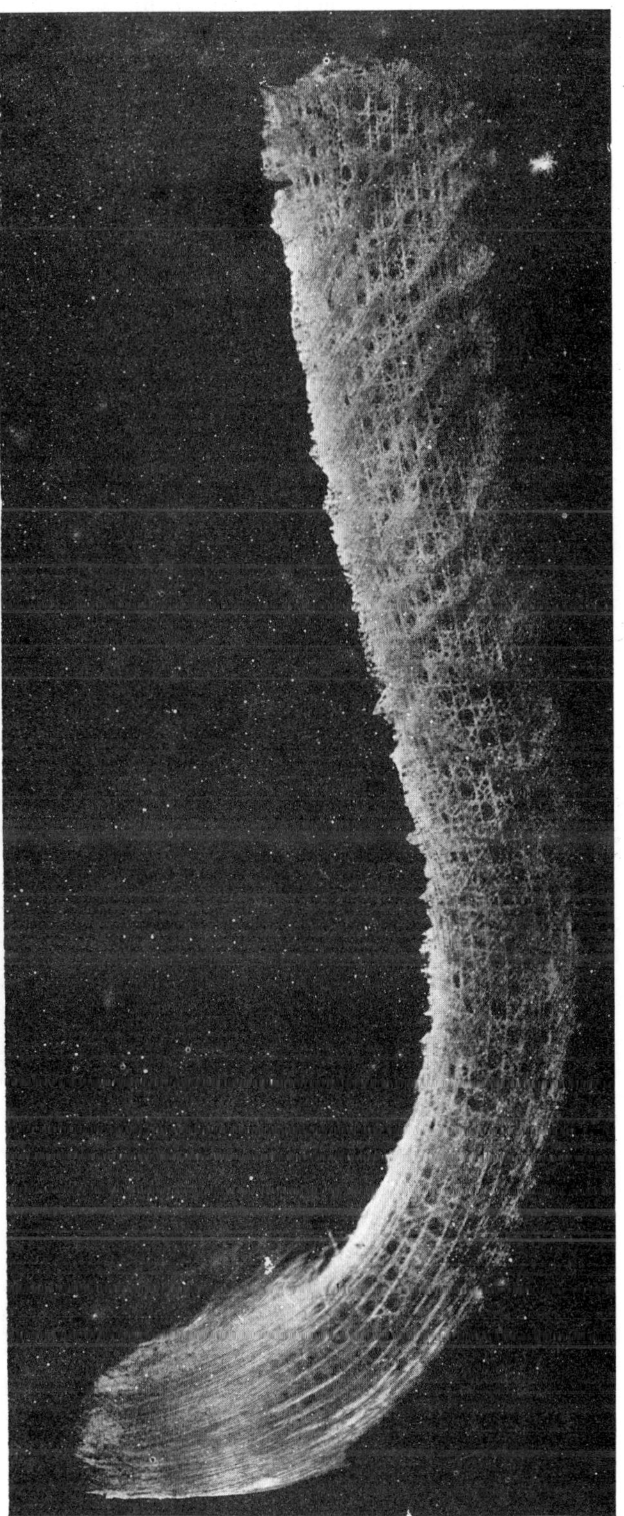

Photo by] [*W. Bagshaw.*

VENUS' FLOWER BASKET.

The species here shown is the most beautiful of the Glass Sponges.

piece of coral. One would have thought, however, that those engaged in its examination would have seen how impossible it would have been for any creature belonging to the lower forms of marine life to take in flinty matter for one part of its structure and lime for the other part.

It remained for Professor Lovén, of Sweden, to show that the specimen, so far as one portion was concerned, was a true Glass Sponge, and that it had been described in an inverted position. The other part was coral.

The first Venus' Flower Basket to reach England was sold for thirty pounds, one account says fifty pounds. At the time, its true history was not known. Attached to it was a small piece of shrivelled, brown, leathery substance. When this was softened and examined microscopically it displayed the spicules of flint, like harpoons, etc., already referred to. This at once indicated its animal origin. It was dredged up from the waters near the Philippine Islands, and was first

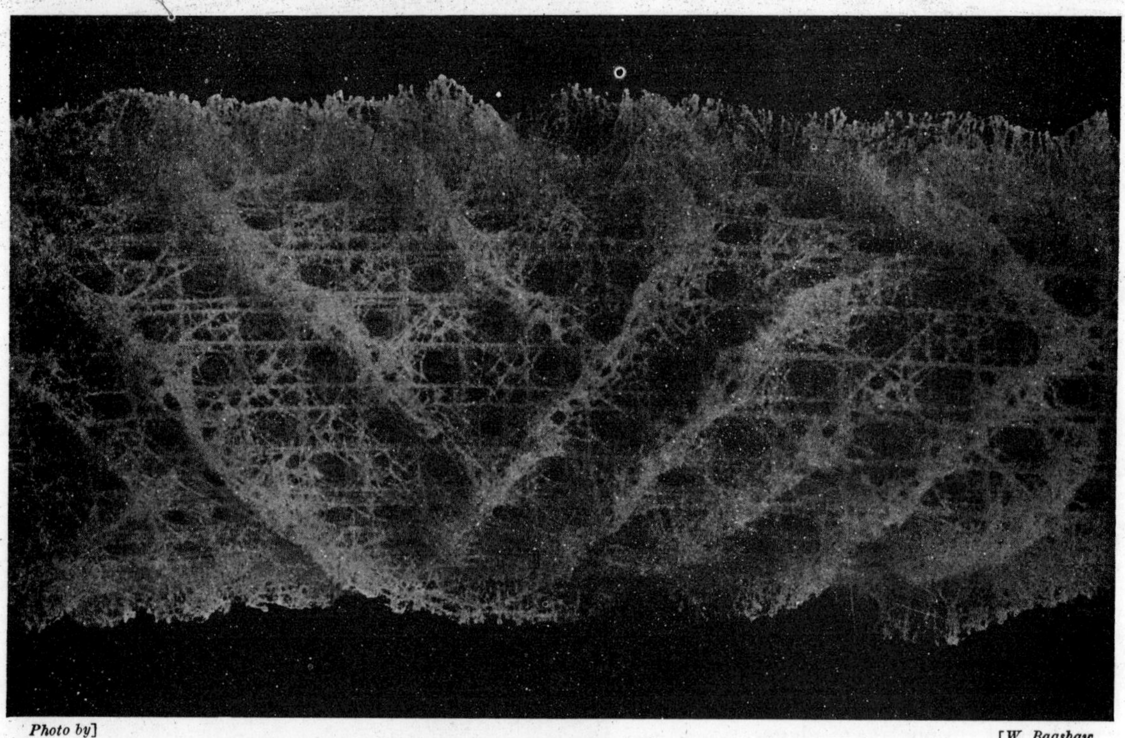

Photo by] [*W. Bagshaw.*

VENUS' FLOWER BASKET.

A section of the beautiful basket showing the slender threads and delicate webbing of spun glass. This structure gives free access to the particle-laden water which forms the sustenance of the inhabitant of the fairy palace.

described in a paper, in 1841, by Sir Richard Owen. From that date onwards more than one hundred differently-formed Glass Sponges have been brought up from ocean floors during the various expeditions made for marine exploration. But the most elegant of all the Glass Sponges is the cornucopia-shaped Euplectella, the "beautifully-woven" sponge. Its texture is square-meshed, yet every angle is rounded and ornamented by fine filaments which form a natural web and woof, soft, beautiful and fascinating. Delicate ruffles of spun glass in zigzag and spiraloid directions add to the complexity of the structure, while they evidently assist the stability of the wonderful skeleton. It has been frequently said that no man can make a bird's-nest so beautifully as the bird. With equal truth it may be said that no artificer in spun-glass handicraft can fashion a frame so marvellous as that involuntarily produced in the living Glass Sponge.

It will be noticed in the illustrations of the cornucopia specimens that as the tube gradually decreases in size so the square meshes are proportionately graduated with exquisite precision. The

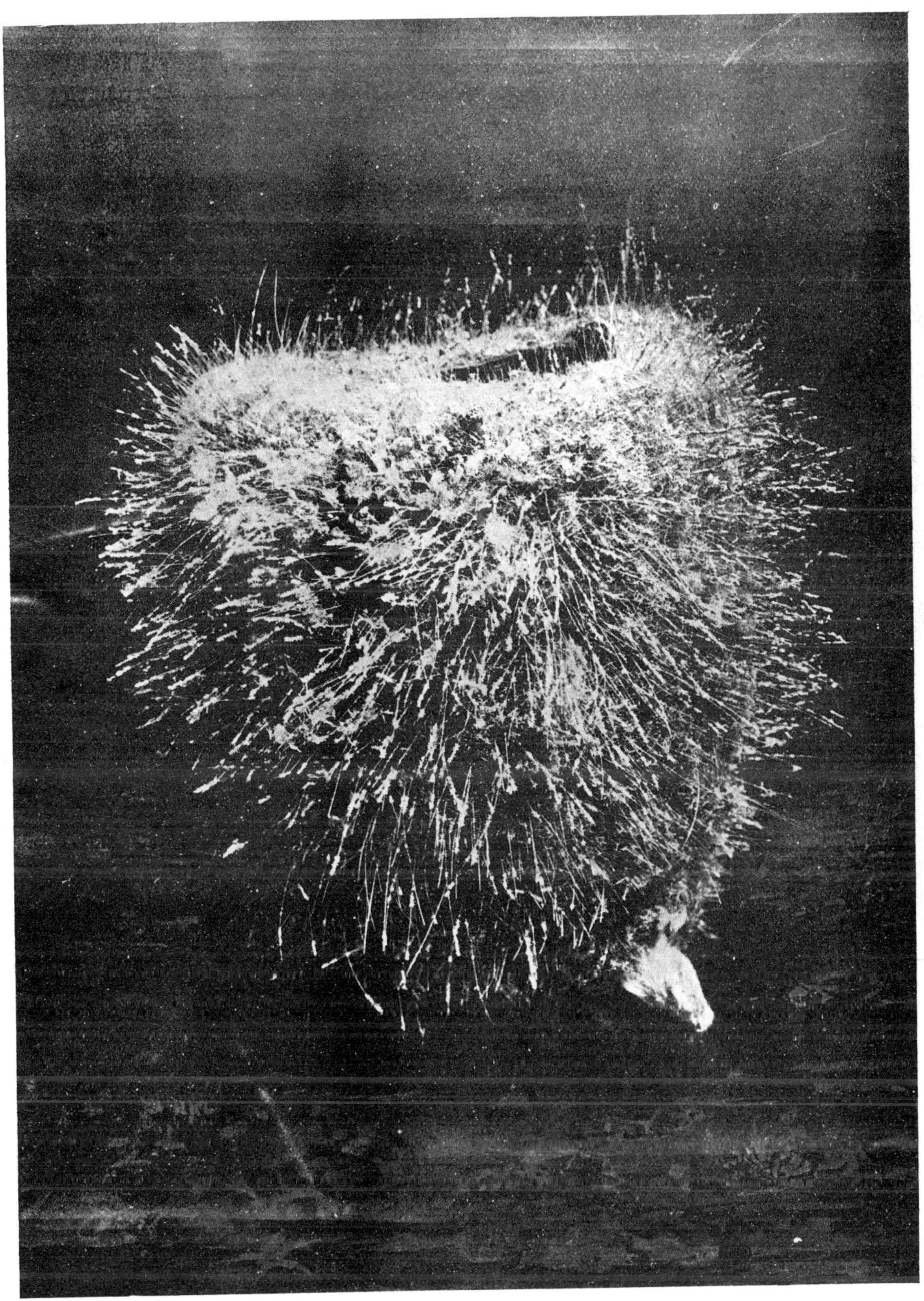

A GLASS SPONGE.

Over a hundred different specimens have been fished up from the "ooze and bottom of the sea," of which this example is one of the most striking. The filaments of these sponges are not transparent glass hairs, but threads of a pure and lustrous white substance, of which one of the principal constituents is flint.

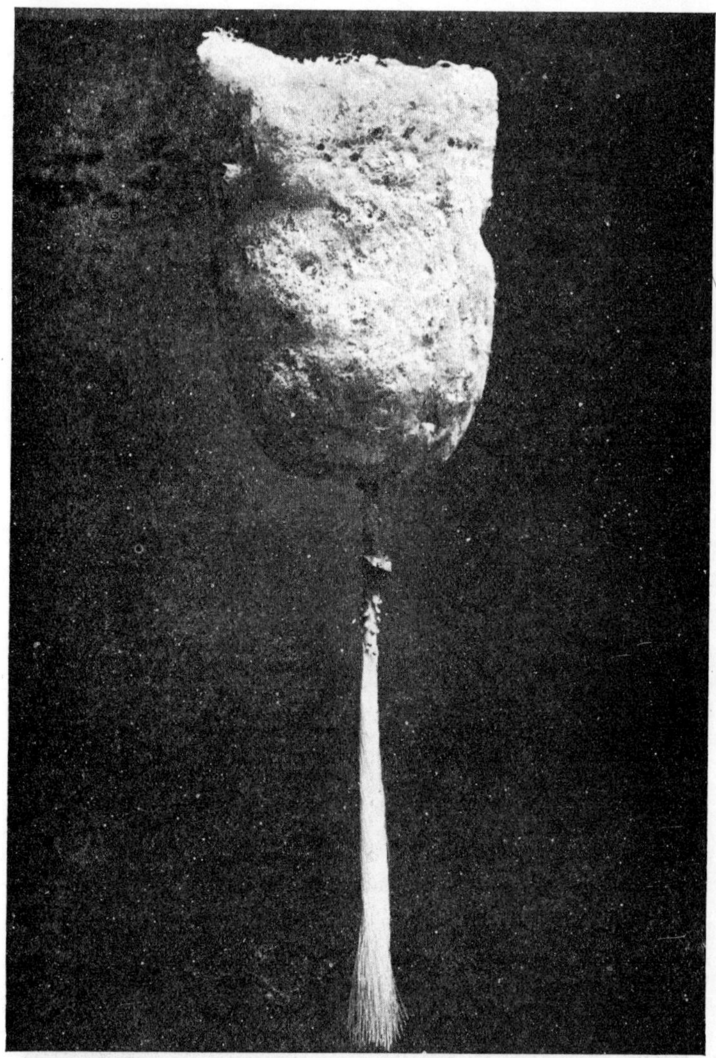

A GLASS SPONGE.

The actual sponge formation of this species is more evident than in those hitherto illustrated. The long coil of fine threads serves the sponge-creature as an anchor to hold it to its sea-bed.

mathematician would find it no easy matter to ascertain by what amount these squares should be reduced right down to the end of the tube so that harmony should be the result and not disorder. Yet all is easily accomplished in that life principle of the humble creature we call a "Glass Sponge."

The upper part of the sponge structure, known as the lid, was compared by Sir Richard Owen to the rose of a watering-pot. Through its meshes the waters flow outwards after traversing the ramifications of the animal.

The late Lady Brassey brought home several specimens of Venus' Flower Baskets from the eastern seas. They were greatly admired when exhibited in London. This was the first opportunity the public had of seeing so many. Even naturalists had only seen a few specimens.

Some years ago the writer of this article was fortunate in making the acquaintance of Mr. Cruikshank, who had resided for a long time with the late Rajah Brooke in Borneo. This gentleman's experiences in fishing for Glass Sponges may be of interest:

"My man and I spent many days every year in hooking up Venus' Flower Baskets from the mud of the sea-bed off Sarawak and other parts of the coast near Borneo. We only went out on very calm days. The man would row very gently while I sent down a fish-hook attached to a fine line, weighted with lead. On feeling the slightest resistance, the line would be carefully drawn up. Frequently a Glass Sponge would be captured. Contrary to the usual way of matters above ground, the creature in its living state does not look nearly so inviting as its skeleton looks when deprived of all the flesh and the micro-spicules which it contains. The grey substance enveloping the skeleton, which is the flesh proper, has to be washed away very soon, as our climate is hot. The skeleton is bleached and is then practically imperishable and uncommonly strong, considering its frail appearance and the light material of which it is built up.

"There is one more point worthy of notice which is not mentioned in any of the printed descriptions of the Venus' Flower Baskets which I have seen. In pulling up the line and hook, thinking we have made a capture, we find we have torn away a piece of the sponge, including the attached part of its skeleton; for you must remember that the long flinty filaments which are continuous with the sponge structure act as so many anchors by which the creature is fixed to the sea-bed. We keep the portion of the skeleton thus torn away. In about a year we are again fishing for Glass Sponges in the same locality. Among the specimens brought up, it happens at times that a sponge has involuntarily repaired itself where a breach had been made either by us or by others fishing in the same waters. This may seem incredible, but it is no more wonderful than the fact that other marine creatures can grow a limb in place of one lost.

"It is not easy to remove a hook from the sponge without making a rent, so that very few Glass Sponges fished up in this primitive manner, are free from damage. This fact is taken advantage of by the dexterous Oriental. He takes some of the long glass anchoring threads, and with them darns up the opening made by the hook, because he is aware that sponges repaired by the sponge-creature, if we may so put it, fetch a higher price than those not so repaired. It needs caution, therefore, on the part of the intending purchaser of 'repaired' Glass Sponges."

The late Professor H. N. Moseley, in his "Notes of a Naturalist on the *Challenger*," gives us an insight as to the methods adopted by the natives of Cebu for obtaining Glass Sponges:

"The special interest of this place lay in its being the locality from which the well-known delicately beautiful siliceous sponge called Venus's Flower Basket was first obtained. The sponge is dredged up from a depth of about one hundred fathoms in the channel between Cebu and the small island of Mactan.

"The fishermen use, to procure the sponge, a light framework made of split bamboo, with two long straight strips, about eight feet in length, forming its front and meeting at a wide angle to form a point, which is dragged first in using the machine. The long straight strips have fish hooks bound to them at intervals all along their length, the points of the hooks being directed towards the angle of the machine.

"The whole is very ingeniously strengthened by well-planned cross-pieces and is weighted with stones. It is dragged on the bottom by means of a light Manila hemp cord, which is attached to the angle. A stone attached to a stick is fastened just in front of the angle to keep the point down on the bottom. The hooks creeping over the bottom

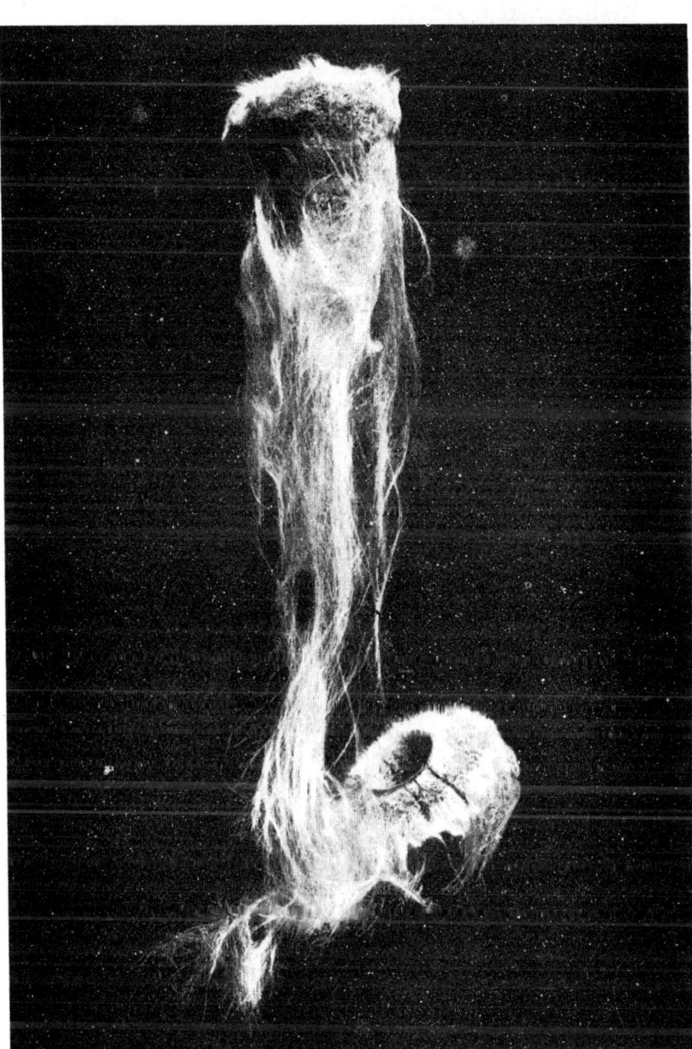

"THE BIRD'S NEST."

A curiously-shaped member of the Glass Sponges, which were once such a puzzle to naturalists. In fact, one of the first of these marvellous structures, which was brought from Japan, was placed by the great Ehrenberg in a cabinet of—Japanese art treasures!

and sweeping an area nearly fourteen feet wide catch in the upright sponges and drag their bases out from the mud. These sponges, once so rare and expensive, were a drug in the market at the time of our visit to Cebu. They were brought off to the ship in washing-basketsful, and sold at two shillings a dozen."

A very curious and inexplicable circumstance appears in the life-history of the Venus' Flower Baskets. Within the vase formed by the skeleton two small crabs are to be seen in several of the specimens. They are prisoners in an exquisite prison-house. They are not parasites, for instead of living upon their hosts, they live, as it were, at the same table with them. The incurrent streams of sea-water bring food to the crabs as well as to the living sponge. The crabs in this instance would be termed "commensals," or chums. This, however, is a department of nature study which is sure to be noticed in a satisfactory manner in a future part of "Marvels of the Universe."

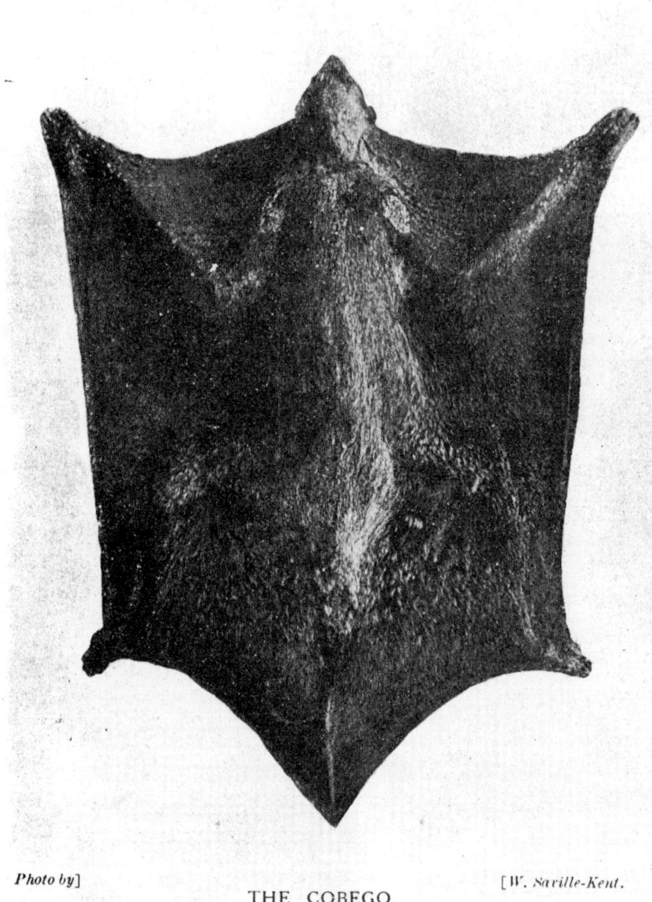

Photo by] [*W. Saville-Kent.*

THE COBEGO.

This strange animal is a native of the forests of the Malay Archipelago and neighbouring islands. It is enabled to fly, not with wings, like the bat, but by means of the fold of skin stretched across from foot to foot.

A writer in *Scribner's Monthly* (1875) echoes opinions held by many students of marine fauna: "The glass hair of which these sponges are woven is not transparent, as might be imagined; it is of pure and lustrous white, giving, in certain lights, an opalescent play of colour. The texture is like frost-work, phantom flowers, the finest and filmiest of the real Shetland lace; it is so exquisitely delicate and lustrously white as to beggar description and to make one turn disheartened away from analogy. In all the world there is perhaps nothing so fairy-like as these wonderful fabrics built up by this formless, structureless life in the darkness and stillness of the deep-sea waters."

AN ANIMATED AEROPLANE

BY W. P. PYCRAFT, F.Z.S.

FEW animals have puzzled naturalists so much as the Cobego, Kaguan, or Flying Lemur, a native of the great forests of the Malay Archipelago, Sumatra, Borneo and the Philippine Islands. This remarkable creature has been placed with the Bats and with the Lemurs, but nowadays it is regarded as most nearly related to the Insectivora—the insect-eaters: a group which contains such familiar creatures as the hedgehog, the mole and the shrew. But even here it seems somewhat incongruously placed, inasmuch as it does not eat insects, but leaves and fruit! However, what an animal eats is not nowadays regarded as the key to its genealogy. But we may leave this matter of

Photo by] [*W. Saville-Kent.*

THE COBEGO.

When at rest the Cobego adds to its chances of security by its method of hanging itself up with its feet placed close together on the bough of a tree and its head tucked away between its fore-legs. In this way it mimics some large brown fruit and escapes the notice of its natural enemies

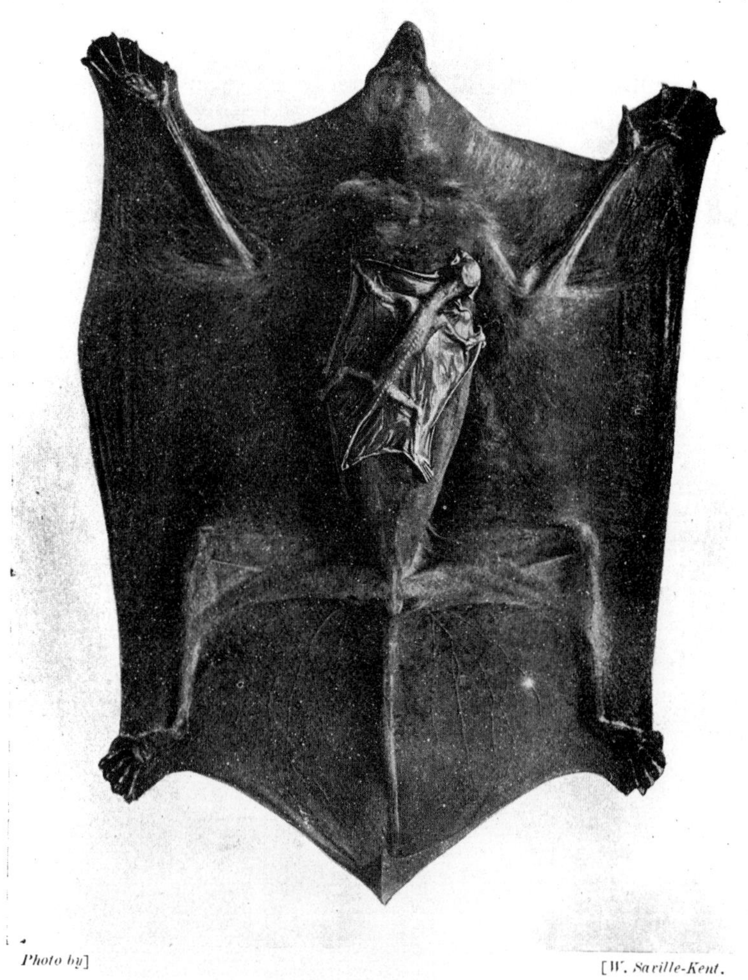

Photo by] [*W. Saville-Kent.*

THE COBEGO.

This illustration shows the mother Cobego carrying her young, a little hairless animal tightly pressed against her warm, furry breast. Owing to the difficulty of this means of transport the Cobego has but one young one at a time.

classification to those whom it may concern and turn to the animal itself.

This extraordinary creature, which is about two feet long, may be described as an animated aeroplane. It does not fly, like the bat, but by means of a fold of skin extending down the side of the body, so as to include the legs and tail, it is enabled to take flying leaps into space from one tree to another, descending, of course, as it goes.

As much as seventy yards can be covered in a single leap. As it approaches its goal the "flight" takes an upward sweep, and at the moment of alighting, the great hands, armed with powerful claws, are spread out, and a sudden grip is taken of the bark of the tree. Then the ascent of the trunk is begun by means of a series of jerky leaps, with occasional pauses for rest.

At such times the creature is conspicuous enough, as all moving bodies are. But when at rest it is far from easily discovered; for at such times it hangs suspended, back downwards, the feet placed close together, and the head thrust in between the fore-legs, so that it looks for all the world like some great fruit, such as abounds in these forests. And this semblance to fruit is the more complete owing to the light spots on the dark-brown skin. Thus disguised, it rests in peace, for prowling hawks and other carnivores pass it by unnoticed.

The Cobego has but one young one at a time. This limitation of the family is imposed by the fact that the mother has to carry her offspring with her wherever she goes, and even this must considerably hamper her freedom of action when necessity compels a leap into mid-air in search of fresh food. The youngster, which is at first quite naked, is carried across her breast, as shown in our illustration, and must be borne for a considerable time; until, indeed, it is able to leap with safety after its parent.

As we have already remarked, the Cobego lives upon leaves and fruit. But there must be something more in its feeding habits, or in the nature of its food, than we are yet aware of. And this, because of the very peculiar character of its teeth, and more especially of the incisors, or cutting teeth of the lower jaw, which are shaped like the teeth of a comb. That they serve some important end none can doubt; but no one has yet been able to throw any light on this problem. No other animal has similar teeth. It has been suggested that they are used for cleaning the fur, but this is hardly probable. One has been seen, when eating a banana, to suck it through its front teeth, using them as sieves; but since unripe fruit, and especially unripe cocoanuts, are eagerly devoured, this peculiar fashion of disposing of a banana cannot be regarded as the sole end and aim for such a remarkable modification of the lower incisors. In this connection it is interesting to note that the outer incisor of the giraffe has a notch in its cutting edge, which is used by the animal in tearing off leaves from slender twigs, the twig being drawn over the notch in the tooth. These teeth of the Cobego may, again, be compared to the "pectinated" or "comb" claw of many birds, which also is of unknown function.

That the Cobego is certainly one of the most contradictory of all animals we have already shown; but an additional illustration of this fact is afforded by the eyes; for though the animal is nocturnal, these are not large, as is the general rule among nocturnal animals. But then, it may be pointed out, the bat is nocturnal and might almost be said to be eyeless!

Although we have spoken of *the* Cobego, it must be pointed out there are two species. The larger, which forms the subject of our illustrations, is *Galeopithecus volans*; the smaller, *G. Philippinensis*, is, as its name implies, confined to the Philippines, and differs from its larger relative chiefly in point of size and the form of the cutting teeth of the upper jaw.

[By A. Twidle.

THE GIANT-TAILED MOTH OF MADAGASCAR.

This great moth measures eight inches from wing to wing, while its length, to the tip of its slender tails, is nine and a half inches. It has been suggested that these tails allow the moth a chance of escape when seized by any insect-eating bird, who, in their pursuit, would snatch at these easily-broken appendages, while the moth itself would escape uninjured.

THE GIANT-TAILED MOTH OF MADAGASCAR

In previous pages we have given particulars of the Giant Silk-moth of North America. The subject of the illustration on this page, and one of our coloured plates, is an even larger insect. It is a silk producer, like the American species, and the silk is declared to be very good; but whether it can be spun is, perhaps, another matter. Experiments in this direction are being made at the present time, the insect being bred for the purpose.

It is a native of Madagascar, and the two first specimens to reach this country were captured by a missionary in the south-east of the island. It was from one of these two original examples that our illustrations were made. Since then several others have been received at the Hon. Walter Rothschild's museum at Tring. Our example is a male, as shown by the beautifully plumed feelers. The conspicuous feature of the moth is the exceedingly long and slender tails to the lower wings. The utility of such an item in the adornment of the moth can only be guessed at ; but it has been suggested in similar cases that they serve to draw the assaults of insect-eating birds, who, in pursuit, may seize these tails—the nearest part—whilst the moth escapes with his vital portions uninjured.

This Madagascar moth measures eight inches across the expanded upper wings, whilst taking its length in the other direction we find it to be nine and a half inches.

[*By W. B. Robinson.*

THE FLYING FISH.

The Catalina Flying Fish, which is found on the coast of California, is very like a herring, save that its breast-fins are of abnormal length, reaching back to the root of the tail-fin. The fish, as a rule, measures ten to twelve inches in length.

FLYING FISHES

BY R. LYDEKKER.

THAT animals which swim at a great speed near the surface of the water, such as whales and fishes, should occasionally shoot themselves clear of that element, to take flying leaps in the air, is only what might be expected ; and we find the habit common to both groups, even such mighty creatures as spermwhales throwing their gigantic carcases completely out of the water. In the case of whales the leap is, however, solely dependent upon the original impetus ; but in certain groups of fishes it is converted into what may properly be termed a flight by the great expansion of the front, or pectoral, fins, which present a large surface to the air and thus serve the purpose of wings.

This power of flight attains its maximum development in two perfectly distinct groups of fishes, namely, the true flying fishes—or flying herrings, as they ought properly to be called—and the flying gurnards.

True flying fishes, of which the Catalina flying fish, from the coast of California, and the Celebean flying fish, of the seas around the island of Celebes, are shown in the accompanying illustrations, are very like herrings in general appearance, except for the great length of the pectoral fins, which reach at least as far back as the root of the tail-fin. As a rule, a flying fish measures from about ten to twelve inches in length, although specimens of eighteen inches are not very uncommon. They are beautiful silvery fishes, whose skimming flight as they rise in shoals from the crests of the waves just ahead of an ocean liner is one of the most beautiful sights in nature. They are essentially fishes of the tropical and subtropical seas ; but a specimen, perhaps driven to our seas by the unusual hot summer of 1911, was taken in a mackerel-net off the village of Wyke Regis, near Weymouth, in August of that year.

That the flight of flying fishes originated from an effort to escape the attacks of predatory fishes by leaving the water for a brief period and again falling into that element some distance ahead is admitted on all hands. But for many years there was a great difference of opinion among

[By W. B. Robinson.

FLYING FISHES.

The skimming flight of these fishes as they rise from the crest of the waves is very beautiful. Their scales glisten like silver, and the rapid vibration of wings creates the appearance of a glittering film surrounding them. They invariably inhabit tropical seas

[By W. B. Robinson.

THE FLYING GURNARD.

Except for the extraordinary elongation of the breast-fin these fishes differ but little from the ordinary red Gurnard. They, like the Flying Fishes proper, inhabit the warmer seas.

naturalists as to whether the pectoral fins, or wings (as they may conveniently be called), of these fishes act merely like those of a monoplane, and serve only to prolong the impetus of the initial leap from the water, or whether they really vibrate like the wings of a bird and thus impart a new and independent motion to the fish during its passage through the air. Some years ago the monoplane theory of the flight of flying fishes was almost unanimously accepted by naturalists. But as time went on several independent observers—myself among the number—who had watched the flight of these fishes from the bows of ocean steamers confidently expressed the opinion that there is a distinct vibratory movement of the wings. The death-blow to what may be called the aeroplane theory was, however, given by Col. C. D. Durnford, in two articles published in 1906. It was there pointed out that, as a flying animal, the flying fish is equipped with wings of a fractional sailing power as compared with those of a soaring bird ; and also that if the wings were many times larger, so as to place the fish on an equality in this respect with the bird, it could only sail with the bird's limitations as regards direction of the wind and with the bird's frequent assistance from soaring flight.

On the other hand, the true explanation appears to be that, in commencing its flight the fish starts with a tail-impelled and wing-assisted jump from the water to a height where the wings can work freely. When this is accomplished, the flight is continued by an intensely rapid and laboured wing-movement, which to most observers appears merely as a blurr. There are, however, short periods of slowing down, during which the movements of the wings become visible; these "slows" often preceding a special spurt, as when the fish has to top the crest of a wave. Finally, there is either a sudden cessation of wing-movement and a

[By W. B. Robinson.

THE FLYING GURNARD.

Showing the method of progress under water when the greater part of the now useless fin is laid flat against the side of the fish.

consequent immediate drop into the sea, or a short slow-down into visibility previous to such drop.

And yet, in spite of this clear and apparently conclusive statement, we are told in a recent work, purporting to be up to date, that flying fish fly in aeroplane fashion without any movement of their wings.

Very different are the flying gurnards, of which the common species is shown in the illustration. These fishes, which run to about eighteen inches in length, differ from the ordinary red gurnards of the fishmongers' shops mainly by the elongation of the pectoral fins into wings. They inhabit most of the warmer seas, and their habits are very similar to those of ordinary flying fishes, with which they probably agree in their mode of flight.

In addition to the above, there is a small fish, known as *Pantodon*, inhabiting the rivers of West Africa, and a second, *Gastropelecus*, from South American rivers, both of which likewise enjoy the power of flight.

Photo by] [*Russell F. Gwinnell*

FOSSIL SUNSHINE.

This stone was once part of the shore of a salt lake. The climate was tropical, and the scorching sun dried up the briny waters at the edge of the lake, so that the salt deposit formed into little crystal pyramids.

FOSSIL WEATHER

BY RUSSELL F. GWINNELL, B.SC., A.R.C.SC., F.G.S.[1]

THERE is perhaps no feature in connection with the study of rocks that is more wonderful than this: *it is possible to determine the climate of past ages.* Thus the fossils—the buried remains of animals and plants once living above ground—clearly indicate that London (or, rather, the site on which London now stands) once basked in tropical sunshine. Palms and fig-trees flourished, with here and there a prickly cactus, while turtles and snakes abounded, and in the sea dwelt crocodiles and sharks, besides such tropical shellfish as the cowry, now abundant on African coasts and used there as money. Temperate and, later, Arctic conditions set in, until the few plants were mosses and dwarfed trees, like the Arctic birch and dwarf willow, while Baltic and Icelandic kinds of shellfish lived in the adjacent sea, with the staircase-shell of Greenland. All of these variations

Photo by] [*Russell F. Gwinnell.*

A "THUNDERBOLT."

This "stony meteorite" fell in Yorkshire in 1881.

of climate were experienced within comparatively modern times, when plants and animals were already largely of kinds living at the present day in various parts of the world.

Now, not only *climate*—that is, the general conditions of temperature, rainfall and so on experienced throughout all the seasons—but also the *actual weather experienced on a certain day* may sometimes be traced in the rocks. These may, for instance, record the occurrence on a certain day, millions of years ago, of a short but sharp shower of rain, succeeded by brilliant sunshine, with a south-east breeze blowing rather strongly. In another case, a thunderstorm takes place and lightning flashes through the sky.

Such "FOSSIL WEATHER" is here illustrated. In the first picture, specimens A, C, E and F show the effect of *a showery day at the seaside.* The rain must have been very brief, as shown by the small number of rain-pittings on the mud or fine sand of the beach; sharp, also, because of the depth of the pits.

Specimen C is not the actual surface on which the shower fell, but the later deposit of sand which was washed over it, burying and preserving the rain-marks. It is thus a sort of "negative" or "cast," showing the pittings as little mounds, among which are ridges representing cracks due to the drying and shrinking of the sandy beach in the great heat of the sunshine which succeeded the shower.

In some cases the great depth of the pittings suggests that *hail,* instead of rain, produced them. Specimen A was formed recently at Cromer, and the regular shape of the pits here, and in F (from the Bay of Fundy), show that the rain fell straight down, so that evidently no wind was blowing. But in the case of E, formed ages ago in Co. Down, in Ireland, the direction of the wind is unmistakably indicated by the fact that all the rain-pittings are deeper on one side than on the other: the rain fell obliquely, driven slantwise by a strong wind. The actual direction of that ancient breeze is found by noting the position which the slab of stone occupied in the sandstone quarry whence it came.

Photo by] [*Russell F. Gwinnell.*

A "THUNDERBOLT."

An iron meteorite, weighing forty-four pounds, that fell in India. Notice the surface markings, a sure sign of the action of heat.

Permanent evidences of *thunderstorms* are rarer, though occasionally a flash of lightning, striking a sandy surface, will melt the grains, so that they run into one another and form a hollow tube, or pipe, of glass-like material. Such a "lightning-tube," from Starczynow, in Poland, is shown by B in the photograph. One small piece of it is viewed endwise, so as to show the central hollow. Much commoner are the cigar-shaped objects shown in D, and popularly known as "thunderbolts." These are hollow at one end only, and pointed at the other, and have in reality nothing whatever to do with thunder or any other condition of weather. They are the stony internal guards of a sort of ancient cuttlefish.

The nearest approach to such a thing as a "thunderbolt"—in the sense of something fallen from the sky—is a meteorite, or shooting-star. But these may fall in any

Photo by] [*Russell F. Gwinnell.*

RAIN, WIND AND LIGHTNING.

The "pitted" sections of sandstone bear witness to the storms of ages passed away. A and F show that the rain fell vertically, but in E the marks are deeper on one side than the other, showing that here the rain fell slantingly, driven by a strong wind. B shows the bore of lightning as it buried itself in the sand and fused the separate grains together, and so formed a tube. The cross-section to the left shows the actual channel bored by the lightning. D are the internal guards of a species of cuttlefish, but popular belief has erroneously regarded them as thunderbolts.

weather, and have no special preference for thunderstorms, although their arrival may occasionally happen to occur in thundery weather.

Unlike the fossil cuttle-bones, which may be best described as cigar-shaped, meteorites are usually more or less round, certainly never elongated. They are sometimes of stony matter, sometimes of nearly pure iron. Both kinds are shown on page 174. The mushroom-shaped object with radiating furrows is a "stony meteorite," which fell at Middlesbrough, Yorkshire, in 1881. The squarer lump, with characteristic surface-marking and a lustrous appearance, is made of nearly pure iron. Weighing forty-four pounds, this "bolt from the blue" fell near Vizagapatam, India, in 1870.

Photo by] [*Russell F. Grinnell.*

FOSSIL WIND.

The action of wind upon the wavelets lapping the sea-shore. The wind was blowing obliquely, so that the waves came on shore one way and rebounded in the opposite direction. The moving water rippled the sand and left on it this beautiful pattern.

Returning to *fossil wind*, the accompanying picture shows a slab of sandstone, which, in long-past ages, formed part of a sandy sea-shore. A gentle breeze rippled the water as it lapped on the beach, and the moving water rippled the sand beneath it. The little wavelets should have travelled landward, leaving the slender ridges marking their course; but the wind was blowing obliquely and, as a result, the ripples came ashore in a slanting way, and therefore rebounded in the opposite direction; hence the cross-rippling of the sand, so beautifully exhibited in the specimen. But in the illustration on page 177, the wavelets have come straight ashore, and consequently have fallen back in a parallel series of outgoing ripples, and there is no possibility of distinguishing between the effects of incoming and retreating waves, one series of corrugations being alone produced.

Fossil sunshine is further illustrated in the picture on page 173. The long slab of stone was once the muddy shore of a salt-lake, where now the Manchester Ship Canal exists. The climate was that of a tropical desert, and the scorching rays of the sun caused the briny water to evaporate and deposit all over the shore little crystals of salt, like dwarf pyramids in shape. The same process is happening now on the shores of the Dead Sea, n Palestine, and the Great Salt Lake of Utah. The salt-crystals are scattered in all positions, some on edge and some down flat on their square bases, with the sloping edges of the pyramid looking like the letter X.

The upper specimen on page 177 shows slender suncracks as well as ripple-marks. Many millions of years have elapsed since the wind rippled that water and the blazing sun dried and cracked the sand after the ripples had been impressed upon it, on an Irish shore. In the lower specimen, the sun had been more powerful, or else had had more effect owing to the finer, muddy nature of this Gloucestershire beach. The resulting sun-cracks are very broad and deep. Here, as in a previous specimen, the slab of rock is a " negative " or " cast " of the actual surface affected by the sun, and hence the cracks appear as ribs standing up from the surface. This is the mud which covered the original cracked surface, forming a protecting blanket upon it.

[All photographs illustrating this article are by the writer, some from private specimens and some, reproduced by permission, from specimens in the Imperial College of Science and Technology.]

Photo by] [*Russell F. Gwinnell.*

FOSSIL WIND.

Here an on-shore wind was sending the wavelets straight ashore, and so making only parallel ripples.

THE BIRD WITH HANDS

BY W. P. PYCRAFT, F.Z.S.

THAT strange native of British Guiana, the Hoatzin, Governor Battenberg's Turkey, or the Stinking Pheasant, as it is variously called, is one of the most remarkable birds living. Its whole life-history, from the cradle to the grave, is remarkable. But we are concerned here only with the earlier phases of its growth. Hatched in a rough nest of sticks placed on the bough of a tree overhanging the water, the young Hoatzin displays a singular activity from the moment it breaks the walls of its shelly prison. Normally this would mean an enormous mortality, but there is nothing normal about the Hoatzin, even in the nestling stage. Unlike all other nestlings, it is an expert climber, using beak and wings and feet with amazing dexterity in scrambling from branch to branch. Many birds, like parrots, employ the beak and feet when climbing ; but no other birds, not even the Hoatzin when adult, can use the wings for this purpose. But in the young Hoatzin the thumb and first finger are armed with long claws, and are very prehensile, serving at this stage of growth as fingers. More than this, the hand at this stage is exceedingly long, and to preserve the efficiency of the claws till the flight feathers have

Photo by] [*Russell F. Gwinnell.*

"A BROILING SUN."

Very fierce must have been the heat which poured down upon the shore and caused it to crack so deeply. Other deposits of earth filled up these cracks and bore the reverse impression shown here.

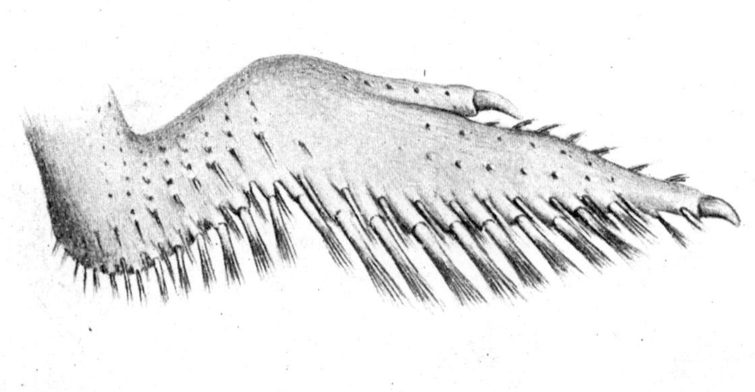

[By H. Grönvold.

THE BIRD WITH HANDS.

The wing of the young Hoatzin, showing the thumb and forefinger armed with the long claws with which it is provided.

developed far enough to break the force of a fall, should such an accident happen, the growth of the outermost quills is retarded lest they should hamper the grip of the claws. But as soon as the wing area is large enough to support the bird in mid-air, the quills held in restraint begin to grow, and with their growth the claws are gradually absorbed and finally disappear. That on the thumb is the first to go. By the time the normal flying wing has been developed, the growth of the forearm has exceeded that of the hand, and for ever afterwards the hand becomes the shortest, instead of the longest, segment of the wing. Should one of these youngsters fall into the river below, it promptly strikes out for the shore, and seizing hold of some overhanging bough trailing in the water, it speedily scrambles back to its fellows!

Curiously enough, the wing of the young game-bird reveals incontestable evidence of former similar climbing feats, thus showing that once on a time these birds also dwelt in trees, and reared their young there. In such wings—and the point can be verified in the chick of any barn-door fowl—the growth of the outer quills presents the same phenomena of development; but to see the claws we must examine the embryo. Thus it is evident that the days of arboreal nurseries must be remote indeed, for structures which appear and disappear during embryonic life can have played no active part in the life-history of their possessor for countless generations.

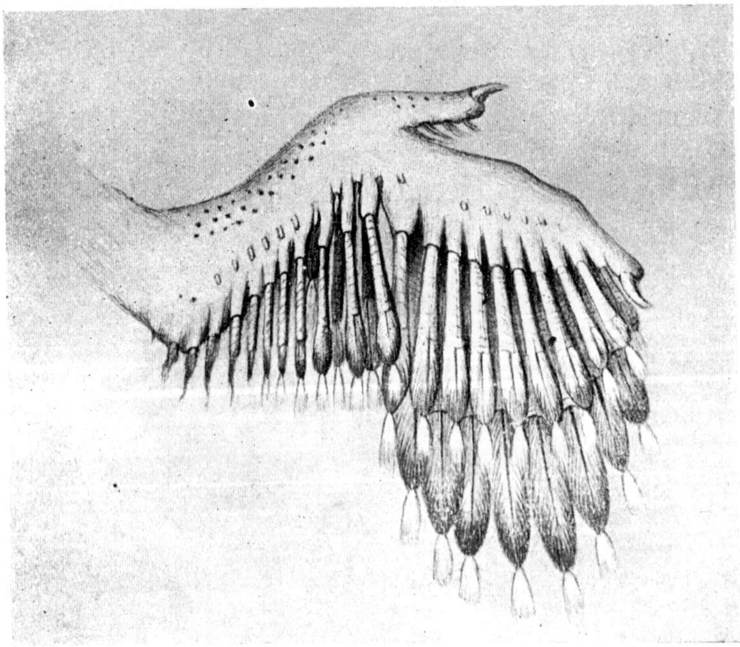

[By H. Grönvold.

THE BIRD WITH HANDS.

The wing at a later stage of development. Notice how the growth of the outermost quills is retarded so as not to impede the use of the claws.

THE NESTING OF CAIMANS

BY R. LYDEKKER.

MUCH confusion exists in the popular mind as to the distinction between crocodiles and alligators, Anglo-Indians almost invariably speaking of the representatives of these reptiles found in India as alligators, whereas they are really crocodiles. To point out the distinctions between the two groups is impossible on the present

THE BIRD WITH HANDS.

The Hoatzin inhabits British Guiana. The young bird is furnished with claws on its wings, and so is enabled to climb trees with great dexterity, until such time as the wings are long enough to sustain it in flight, when the now useless claws disappear.

Photo by] [*W. P. Dando, F.Z.S.*

THE CAIMAN.

A Caiman is neither a crocodile nor an alligator, though more nearly allied to the family of the latter. Their skin is deep black, with yellow markings.

occasion; but it may be stated that while all alligators have broad, stumpy snouts, most, although by no means all, crocodiles have narrower and more pointed muzzles. True alligators have, however, a much more restricted distribution than crocodiles; one species inhabiting the Mississippi valley and the other the valley of the Yang-tsi, in the heart of China. In tropical America there is, however, a nearly allied group of five species, often spoken of as South American alligators, but whose proper designation is caimans, or jacares. Several of these are more brightly coloured than crocodiles or true alligators, having the skin deep black, with yellow markings.

Crocodiles, alligators and caimans, like the majority of reptiles, produce their young from eggs, which in this group have hard shells and are about as large as those of a goose, although more elongated, and with the two ends nearly equal in size. In laying their eggs among dense vegetation the tropical American caimans differ from the crocodiles of the Old World, which make their nests in holes in the sand, and cover their eggs with a thick layer of the same, incubation being effected by the heat of the sun's rays. From twenty to sixty eggs may go to form a clutch. In the case of both crocodiles and caimans, the female parent takes up her station in the neighbourhood of the nest; and when the young reptiles are ready to emerge, they attract her attention by uttering shrill cries, those of young crocodiles being audible through a couple of feet or so of sand. Thereupon the mother crocodile digs down to the eggs, the shells of which are then chipped open by their occupants, which, like young caimans, are armed with a special temporary tooth for this purpose; but the female caiman merely has to remove the layer of brushwood with which they are covered.

THE COMMON DIVING BEETLE

BY HAROLD BASTIN.

A BEETLE is one of the chief marvels of the universe; it is so wonderfully equipped for the life which it leads, so liberally endowed with unerring instincts. This statement applies to beetles in general, but it is especially true of the common Diving Beetle, a creature which may be readily studied, since it frequents pools and ditches throughout the kingdom, and will soon make itself at home in a roomy aquarium. Let it be said at once that the Diving Beetle is carnivorous and that it possesses a voracious appetite. It is, indeed, a veritable fresh-water shark, and the dispatch with which it accounts for small fishes, newts, frogs, tadpoles, grubs, and other under-water dwellers has earned for it the equivocal title of " water devil."

Being an insect, the Diving Beetle is furnished with six legs. It has also four wings ; but while the lower pair are well adapted for sustained flight, the upper pair consist of hard plates which cover and protect the more delicate functional wings when the latter are not in use

A beetle is planned quite differently from a backboned animal, such as a rabbit or a fish. It possesses no internal skeleton for the attachment of its muscles. Instead, its skin is thoroughly permeated with a peculiar substance called chitin, which is remarkably hard and durable, well calculated to resist the weather, and to sustain the wear and tear of an active life. Thus, for practical purposes, we may regard the beetle as living within a suit of strong, light armour, to the inner walls of which its muscles are attached. This suit of armour is always divided into three compartments. The first encases the head, the second—from which spring the six legs and the wings—protects the " chest," while the third envelops the hind body.

This type of structure is common to all beetles. But in the case of the Diving Beetle we shall notice certain modifications which fit the creature in a marvellous manner for its aquatic life.

The Diving Beetle is not a mere surface swimmer. It is specially designed for under-water voyages. Its body is like a submarine—oval, somewhat flattened, having a highly-polished surface, and with its several parts so closely adjusted as to present practically a continuous outline. This smoothly elliptical contour and polished surface serve to lessen friction, which, as every naval architect will testify, is so serious a hindrance to speed.

YOUNG CAIMANS EMERGING FROM THE EGG.

These eggs are about the size of those of a goose, though more elongated in shape. The shell is quite hard, but the young Caiman is armed with a special temporary tooth with which he can break down his prison walls.

Let us now examine the means of locomotion, the mechanism which propels this boat-shaped beetle through the water. The two front pairs of legs, especially in the female, are used for crawling among weeds, for seizing prey, and for anchoring the insect when it desires to remain in one position beneath the surface. In the male the front pair of limbs is specially modified to enable him to seize and retain the female. One of the joints is flattened and furnished with sucker-like discs and cupules, well shown in the figure on page 183. But the larger hind limbs have undergone great modification and are, in effect, a couple of powerful paddles. By a peculiar keel-like prolongation of the chest, they are set far back below the centre of the insect, and their arrangement and attachment enable them to be brought at right-angles to the body, thus providing for a strong, wide sweep. Moreover, the whole form of what we may term the foot—the five-jointed terminal portion of the leg—has been flattened and clothed with stiff hairs until it is transformed into a perfect reciprocating oar-blade. These paddles are operated by powerful muscles. They generally strike the water simultaneously, driving the beetle rapidly forward in a straight line; but they are also capable of being used singly for the purpose of turning in a confined space, just as an expert sculler will use one oar to manœuvre his skiff.

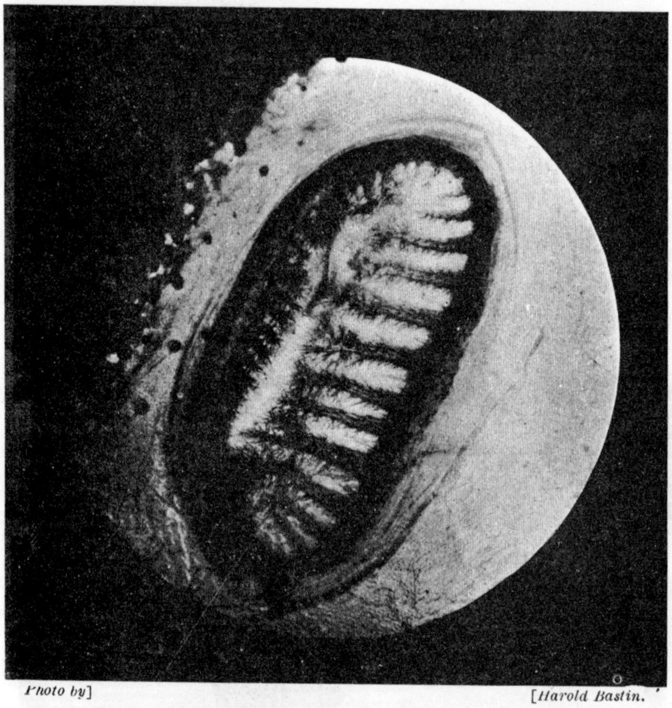

Photo by] [*Harold Bastin.*

THE "SPIRACLE" OF THE WATER BEETLE.

The Diving Beetle has no lungs, nor does he breathe through his mouth. He is supplied, however, with air-holes in his side and upon his back, each of which is guarded by a network of stiff hairs from the intrusion of any dirt or dust.

While the Diving Beetle is thus admirably fitted for progress through the water, its habits are by no means exclusively aquatic. True, its performance when afoot amounts to little more than an undignified scuttle, but its wings are large and well developed, and the beetle frequently makes long aerial excursions, especially at night, travelling from one pond or stream to another. It is, indeed, a wonderful colonizer, and there can be few suitable sheets of water, from one end of the kingdom to the other, which it has not made its own. This beetle has been known to pitch precipitately upon the roof of a greenhouse, probably mistaking the glint of the glass in the moonlight for water. Moreover, when imprisoned in an aquarium, it is liable to effect its escape and turn up in the kitchen or parlour, much to the consternation of the lady of the house.

The manner in which the Diving Beetle takes in its supply of oxygen is very remarkable. In order fully to appreciate it we must first fix in our mind the facts that insects have no lungs and that they do not breathe through their mouths. They take in air through holes in their sides, known as spiracles, each spiracle being guarded by a sort of quickset hedge of stiff hairs, which serves to keep out dust and dirt. Just within each spiracle, or breathing-hole, there is an ingenious little valve which opens and closes periodically by a muscular contrivance. The air which enters through the open valve is conveyed direct to every part of the body through a system of minute tubes, the final ramifications of which are so delicate that they penetrate the most distant extremities of the body. These air-tubes are soft and within them run spirally-coiled threads of hard material

Photo by] [*Hugh Main.*

THE LEGS OF THE DIVING BEETLE.

Few insects have such elaborately designed legs. The first in the page is the front leg of the male, with the wonderful pad which enables it to hold to a polished surface. The second is the leg designed for anchoring the insect and for crawling amongst pond weeds. The third leg is the powerful paddle, armed with stiff hairs, which propels the beetle through the water.

similar to that which constitutes the insect's external armour. They are, in fact, exact counterparts in miniature of our wire-lined rubber gas-tubing, and their structure is a safeguard against "short circuit." The elastic inner coil keeps the tube open even when it is subjected to pressure, as by the bending of a joint through which it passes; at the same time, the flexibility of the tube is preserved. Now, in the case of the adult Diving Beetle, the breathing-holes do not occupy their normal position along the *sides* of the body. They open upon the *back*, beneath the wing-cases. Moreover, the two terminal pairs of breathing-holes are much enlarged, while the wing-cases fit so perfectly upon the abdomen that an air-tight space is formed between them and the back of the insect.

Photo by] [*Harold Bastin.*

THE DIVING BEETLE AND GRUB.

The appetite of the grub is insatiable: it is always in search of prey. Its flat head is supplied with sickle-shaped jaws, which are traversed by a hollow tube leading to the closed mouth. By this means the blood is sucked out from the victim and goes to nourish the victor.

When the Diving Beetle wishes to take breath, it simply poises itself in the water with the hinder part of its body protruding slightly above the surface. It then curves the tip of its abdomen somewhat, when the air rushes into the four big breathing-holes, and fills the space between the wing-cases and the back. The chink is then closed tightly, and the beetle is once more ready for a diving excursion. Experiments made by Dr. Sharp, the eminent authority on insect life, revealed the fact that a male Diving Beetle comes to the surface to take breath once in every eight minutes and twenty seconds on an average, and remains poised for about fifty-four seconds. The longest interval that it was observed to pass without breathing was nineteen minutes, although the female insect, being less active in habit, rises less frequently than the male.

We have still to speak of the Diving Beetle's head. Its most important features are the eyes, the mouth, and the feelers. The latter are delicate, jointed appendages, which, so far as can be gathered, may be likened to the whiskers of a cat in that they serve the beetle's sense of touch, and are possibly of special service at night, or when the insect is foraging in murky water. The mouth consists of a whole set of useful implements. First, there is the upper lip—a hard plate hinged to the solid armour of the head. Immediately behind this are the jaws, which move sidewise, and not with an up-and-down motion, as is the case with the higher animals. After this comes a second pair of jaws, less powerful than the first, but highly sensitive. Each is made up of several parts, each carries a jointed feeler. So that these secondary jaws are well suited for holding food, examining it, and passing suitable portions towards the gullet. Indeed, they may be likened to a pair of

Photo by] [*Hugh Main.*

When the Grub is about to change into a chrysalis it constructs a rude cell at the margin of the pond, where—

Photo by] [*Hugh Main.*

it lies quietly for a few days, then wriggles out of its last grub suit—

dexterous hands, set just under the mouth. Moreover, their labours are ably supplemented by the lower lip, an organ which really represents a third pair of jaws intimately fused together. All these mouth-parts work in unison with amazing precision and address.

I will now ask the reader to glance for a moment through a microscope at the Diving Beetle's eye. The surface of this amazing sense-organ consists literally of thousands of facets. That each of these facets is a perfect lens may be readily perceived by looking through a magnified eye at some object, such as a postage-stamp. Every facet forms a tiny image of the stamp, and on page 187 we give a photograph of a single stamp with part of the beetle's eye used as the camera lens, the result being that the stamp is greatly multiplied in the photograph. That the beetle actually sees all the images is hardly probable. We must remember that each of our own two eyes produces an image, while only one picture is registered by the brain. The most plausible explanation of how the beetle really sees with its compound eyes was suggested by Johannes Müller early in the last century. The gist of his theory is that the beetle perceives with each facet not a complete image, but a tiny section thereof, and that these sections are pieced together to form one mosaic picture. Thus we must assume that the Diving Beetle sees only one image of an object at which it is gazing, but because its faceted eyes cover a large area of its head we may believe that the insect enjoys quite a panoramic outlook—seeing simultaneously to right and left as well as above and below.

So much for the physical equipment of the adult Diving Beetle. Let us now sketch briefly its life-story. The mother beetle makes incisions in the submerged stems of rushes or pond weeds with her sharp egg-tube, and inserts her white, slightly curved eggs. She acts thus, not because her youngsters will feed upon the vegetable substance, but as a precaution

Photo by] [*Hugh Main.*

and is transformed into a white mummy-like creature, very different from the perfect insect, but—

Photo by] [*Hugh Main.*

careful examination will show the several parts of the future Beetle packed closely together in this strange shroud.

against the egg-eating denizens of the pond. In about three weeks the young insects hatch. At first they are minute and pale, but they begin to feed almost as soon as they leave the egg, and rapidly attain a considerable size. They are now seen to bear a strange superficial likeness to shrimps, and one scarcely wonders that the old naturalists, who knew little about the amazing metamorphoses of insects, actually classed them with animals of the crab and lobster kind. Yet the young Diving Beetle is really very different from a shrimp. It has a long, spindle-shaped body, with six legs, and a broad flat head armed with sickle-shaped jaws. These jaws are marvellously contrived feeding-organs. Each is traversed through its length by a hollow tube leading to the closed mouth, so that when the insect fastens on its prey it can suck all the blood from the body without once relinquishing its hold. This, indeed, is its habitual way of making a meal. More-

Photo by] [*Harold Bastin.*

THE DIVING BEETLE.

The adult male and female in their native element. Notice the sweep of the oar-like legs, and the comparative size of the Beetle's eye.

over, its appetite is insatiable and its activity tireless. So that we may imagine it as skirmishing among its native pond-weeds and pouncing upon one hapless victim after another, each of which in turn is first sucked dry and then cast ruthlessly aside. It is, therefore, not without some show of reason that juvenile students of pond-life know this grub by the name of the water-tiger.

This programme of slaughter continues for a period of from four to six weeks. During this time the insect steadily increases in bulk, and repeatedly outgrows its skin—a little difficulty which it as often surmounts by wriggling out of its old suit and appearing in a new one of more ample proportions which has been perfected beneath. At length, having attained to its full size, the beetle-grub prepares for its further change to the chrysalis condition. At this stage in its career it crawls from the pond and constructs a rude cell high and dry within the bank. In this chamber

Photo by] [*A. E. Smith.*

SEEN THROUGH A BEETLE'S EYE.

The appearance of a postage-stamp when photographed through a Beetle's eye. This eye is "compound"; that is, composed of innumerable lenses, each lens focussing the object viewed and sending a reflection to the brain. The effect, it is supposed, is not that of seeing a multiplication of the object, but rather the lenses afford their owner a panoramic outlook.

Photo by] [*Stanley Mylius.*

THE GIANT TORTOISE.

Few animals live to a greater age than these huge, ungainly creatures. The photograph here shown is of a patriarch at Matara, which is computed to be at least 150 years old.

it changes into a chrysalis, and lies for some three weeks, taking no food, but displaying some power of movement if disturbed. At length it emerges as a perfectly-formed beetle, differing in every way from the grub, or "water-tiger," which has so far claimed our attention. Apart from the deep-seated physical changes which take place unperceived, this strange transformation is accompanied by two skin-castings. When the grub has formed its cell, and lain quiet for a few days, it wriggles out of its latest grub suit and appears in quite a novel guise. If we open its cell at this juncture, we find within a white, mummy-like creature, unrecognizable at first glance either as the grub which we knew or the beetle which we anticipate. Yet if we make a closer examination, we may trace the several parts of the coming beetle all folded together. The wings, in particular, are packed up in little bags, and folded forward among the legs.

Then comes the second skin-changing—a somewhat slow process, during which the wings and wing-cases "fill out" and assume their normal position above the insect's back. When all is over the pupa has been changed into a perfect beetle—complete in every detail save that it is soft and limp, while its colour is yellowish white. How great is the transformation may be appreciated by a comparison of the accompanying photographs.

Not long after attaining the adult state the beetle emerges from its cell and enters the water. It is still pale and soft, and fully a week will elapse ere it gains the colour and solidity of armour characteristic of its kind. But it soon begins to "hustle," and to indulge its voracious appetite to the full, with the result that in ten days' time it is as perfect a representative of the Diving Beetle fraternity as heart could wish.

GIANT TORTOISES

BY R. LYDEKKER.

ALTHOUGH in the absence of definite information as to the maximum age attained by elephants it is impossible to affirm with certainty that Giant Land-tortoises are the longest-lived of all animals, yet they have undoubtedly a very strong claim to be regarded as the Methuselahs of the animal kingdom, and are, therefore, clearly entitled to rank as marvellous creatures. At earlier epochs of the earth's history distributed over all the continents, Giant Tortoises, with the exception of one African species, which should perhaps be included in the group, are now restricted to the Galapagos Islands, off the west coast of South America, and the Seychelle and Mascarene Islands of the Indian Ocean, although many individuals from the last-named islands have been carried to various parts of the world.

As to the age attained by the Galapagos Tortoises, there appears to be no information ; and in the case of the species whose home is the islands of the Indian Ocean the records relate, at all events in the main, to exported individuals. One of the most interesting of these is the monster represented in the illustration on the previous page, the existence of which I had the good fortune to make known to naturalists. A year ago this Tortoise was living at Matara, in Ceylon, to which place it is stated to have been brought at the end of the eighteenth century ; so that, at a moderate estimate, its age may be put down as at least one hundred and fifty years. Till the year 1894, when it died after removal from its old haunts, Ceylon also possessed a second

Photo by] [*W. S. Berridge.*

THE GIANT TORTOISE.

The better-known Greek Tortoise has been put on the back of this great creature in order that its huge size may be the better appreciated

Giant Tortoise, which was found at Colombo when the island was taken over by the British in 1796 At the date of our occupation it was probably at least fifty years old; its age at the time of its death was doubtless not less than a century and a half.

Another record is formed by the celebrated Marion's Tortoise, which, till recently, at any rate, was living at the Artillery Barracks at Port Louis, Mauritius. This monster was brought to Mauritius in 1766, and as it had attained its present dimensions in 1810, it may be presumed to have been at least fifty years old when imported; so that its present age can scarcely be much, if at all, short of a couple of centuries; and there is another claimant (recently living at Tring) to a very great age.

Photos by] [*Hugh Main.*

SEED PARACHUTES.

These members of the Dandelion family bear their seeds clustered together on the top of their stalk. Each seed is situated at the base of a thin hair, whose head branches out into a fairy parachute, which is designed to carry the seed away from the parent. The first picture is of the puff of a Colt's-foot, and the last two show the unripened and ripe heads of the ordinary Dandelion.

SEED DISPERSAL BY PARACHUTES AND LIFE-BUOYS

BY SIR HERBERT MAXWELL, F.R.S.

In no class of living organisms is the evidence of design more convincing than in plants. Leaving out of account bacteria, diatoms and other vegetable micro-organisms, plants are denied the power of locomotion. They are provided, therefore, with an exceeding variety of appliances for the dispersal of their seeds, so that each race may have a fair chance of colonizing fresh ground. One of the readiest means of transport is the wind; but to take advantage thereof the plant must equip its seeds with mechanical aids to locomotion.

There are many plants which rely upon parachutes for dispersal of their seeds. Familiarity is apt to breed contempt, or at least indifference, in our regard for the crown of seeds in the common dandelion; but the fact that it is so common is in itself a warrant for the efficacy of this mode of dispersal. Nobody inspecting it through an ordinary pocket lens can fail to marvel at the exquisite

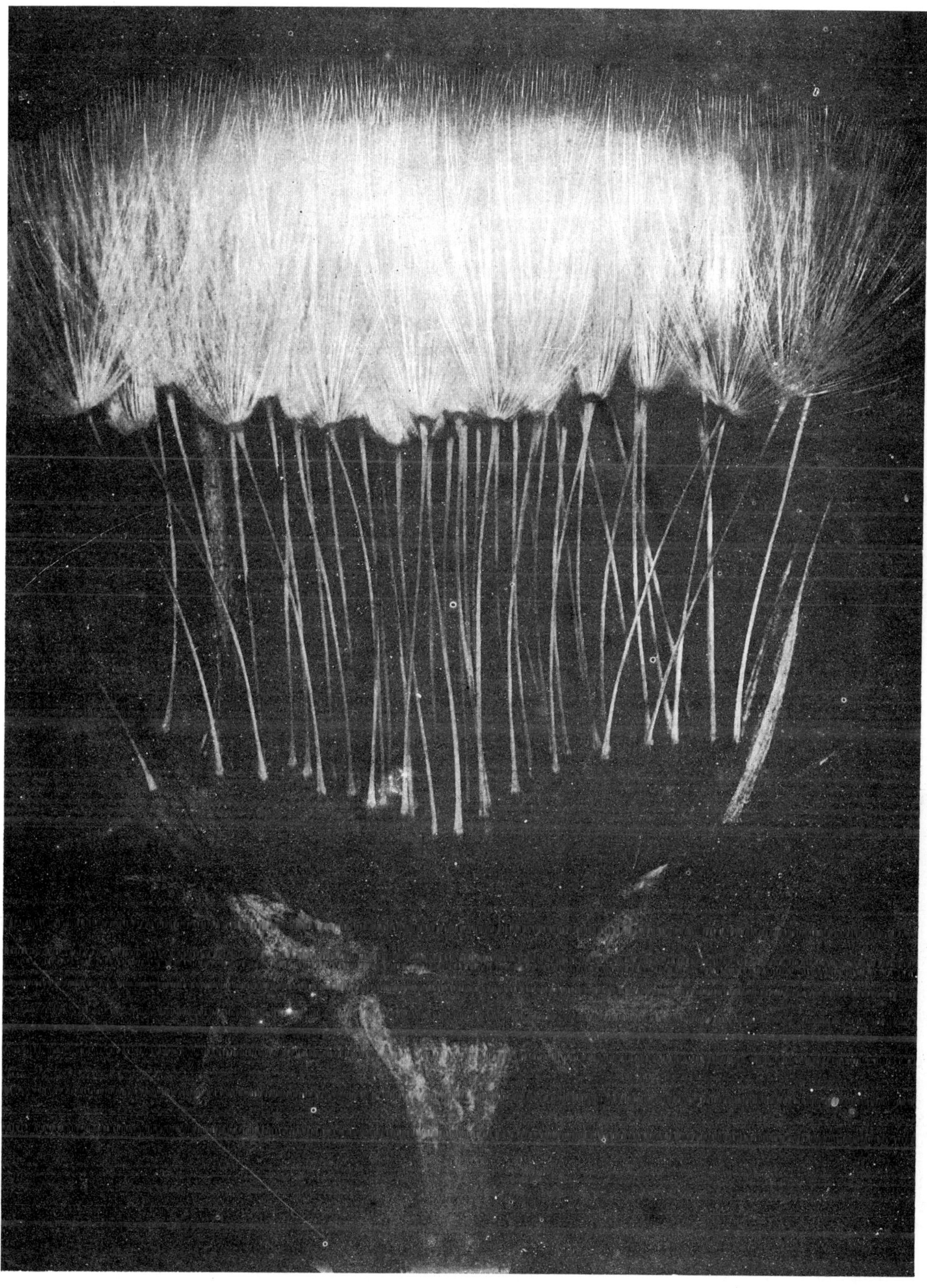

A DANDELION PUFF.

A photograph of a section of the puff showing the manner in which the seeds are disposed in the seed-vessel waiting for the wind to catch the feathery parachutes which will carry them off.

Photo by] [*Hugh Main.*

TRAVELLER'S JOY.

Each one of the little feathers that form the chief beauty of this familiar plant is destined to wing a ripe seed to distant soil.

Photo by] [*F. Noad Clark.*

THE PARACHUTE OF THE GOAT'S-BEARD.

The parachute of the Dandelion-seed is only the simple skeleton of such a one as this, for it has none of this delicate inter-tracery between its central threads

symmetry of the whole, the delicacy of the silky plume upon each seed, and its perfect adaptation for aerial transport. Nearly all plants of the Composite family produce similar parachutes; there is hardly any limit to the distance thistledown may be carried on a favouring breeze, and every gardener knows to his cost how vain is the attempt to exclude groundsel from his borders. In some of the Ranunculus family, such as the Traveller's Joy and certain species of anemone or wind-flower, aviation is contrived by the development of a long silky and feathered filament from the top of each seed. The common reed, which grows almost all over the world, and the pampas grass of our gardens owe the beauty of their plumes to the silky hairs provided for the wafting of the ripe seeds. These are still more conspicuous, owing to their snowy whiteness and length, in the cottonsedge of our moorlands. Many a time the deer-stalker has recourse to these in approaching his game, flinging a tuft of the down of cottonsedge in the air to ascertain the exact direction of a light wind. The cotton-plant is commercially more important than any other plant in the world, which is owing entirely to the length, tenacity and uniformity of the silky white hairs that cover the seed and are intended to facilitate its transport by wind. The art of the cultivator has been exerted to exaggerate this feature as much as possible, and the amount of this feather-weight material annually put upon the market can only be reckoned in millions of tons. Our native willows, though of a fa. different natural order from the cotton-plant, imitate it in a modest manner. The seeds lie in a tangle of fine white hairs, which bear them away on the breeze when the ripe capsules open. The willow-herbs, of which some species have beautiful rose-coloured flowers, are so named in English from a superficial resemblance in their seeds and leaves to those of willows; but in the willow-herbs each seed carries its own crown of hairs. A very similar arrangement is found in the milk-weeds, which are natives of the United States.

The mention of the willow-herb brings us to water as a vehicle for seed-dispersal; for if the seeds of the great willow-herb fall on the water, they may germinate afloat and take root when stranded. Of course, there are many other land plants the seeds whereof, falling in the water, float and may be transported to a suitable resting-place; but very few British species develop special structures for flotation. The white water-lily is one exception and the sedges furnish others. The fruit of the water-lily is a large berry, containing a number of small seeds. Each seed is girt with a spongy belt containing air-chambers. When the berry bursts under water, the seeds are borne up to the surface, and float about until their lifebelts rot, when they fall to the bottom of the water. In the sedges, the seed is very small, only one-sixteenth of an inch long, but it is cased in a sack four times as big, which enables it to float until it arrives at a suitable resting-place.

The most notable instance of fruits specialized for floating is the important one of the coconut. In this country we are only familiar with it after it has been divested of its smooth white skin, between which and the shell is packed a dense mass of the exceedingly tough fibres from which our coconut mats are made. This envelope gives the nut sufficient buoyancy to float for hundreds of miles, so that uninhabited coral islands become colonized with this most valuable food-plant.

A still larger example of this type is found in the Sea Coconut, or Double Coconut, of the Seychelles, which, growing upon a lofty palm, are washed out on the Indian Ocean as far as the Maldive Islands. Until comparatively recent times only the fruit was known from examples picked up at sea by sailors, the tree producing it being discovered much later.

Plants which cannot, or at least do not, equip their seeds either for flight or swimming, adopt an extraordinary variety of expedients to secure transport. Some of them actually educate their seeds to creep along the ground.

The common Stork's-bill owes its popular name to the long beak which forms in each flower as the seeds ripen. These seeds are five in number, each being armed with reflexed bristles and having a long tail, or "awn," proceeding from the top. These awns form the

Photos by] [*H. Main, F.E.S.*

FLOWERS AND FRUITS OF CRANE'S-BILL.

The five seeds are contained in five pouches. As they ripen the central pillar splits into five strips, and these, curling up, lift the pouches into the air until they are upside down, when the seeds drop out.

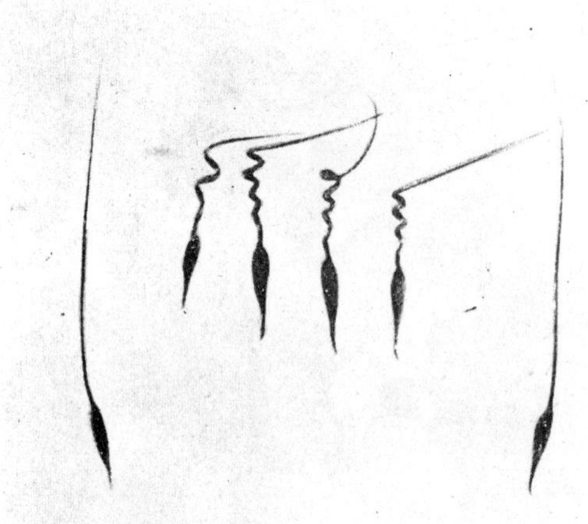

Photo by] [*H. Main, F.E.S.*

STORK'S-BILL FRUITS.

When the seed has reached the ground, the contraction and expansion of its tail drives the seed into the soil.

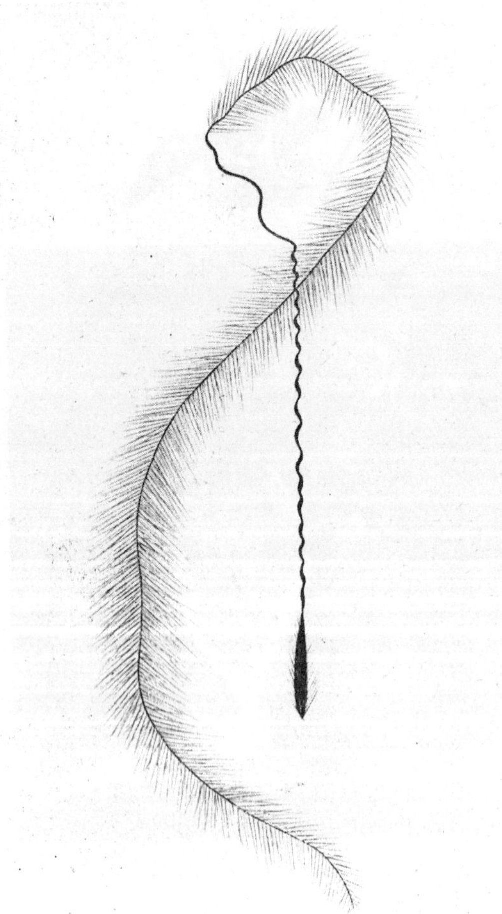

[*By J. Teklenburg.*

FEATHER-GRASS SEED.

central column or beak. When the seeds are ripe they move outwards, and are freed by the awns splitting off from the column. And now the true purpose of the awn becomes apparent. It is hygroscopic—that is, it curls spirally in dry weather and uncoils in damp. The alternation of these movements causes the seed to move along the ground, till it comes to a soft, unoccupied surface, when the awn by its wriggling drives the sharp base of the seed into the soil, and the reflexed bristles on the seed prevent it being withdrawn.

A great number of grass species rely for their dispersal on the hygroscopic properties of the awns and bracts with which they accoutre their seeds.

An extreme instance is furnished by the Feather-grass, a native of Russia, and, fortunately, not naturalized as yet in Great Britain. The seed of this species has a long awn, bent at an angle, exceedingly sensitive to drought and damp, and terminating in a very long feather-like tail which would enable it to fly off some distance from the parent plant before coming to the ground. It is also barbed with bristles like the seed of the Crane's-bill. When the spiral becomes damp it unwinds, with the result that the awn is lengthened, and, as the feathery portion is probably entangled in surrounding weeds, lengthening can only take place in a downward direction; so the seed is driven into the ground. At the next dry period, when the shaft again becomes a spiral, the seed cannot be dragged out of the soil because the barbs prevent it. At each atmospheric change the seed will get driven a little further in until it is deep enough for growth to begin. When this awn attaches itself to the wool of a sheep by winding and unwinding itself, it drives the hard, sharp point of the seed into the animal's skin, whence the barb-like bristles prevent it being withdrawn, thereby causing intense irritation and, in some cases, actually killing the sheep. It is a remarkable fact that in three other parts of the world, widely separated, namely, North America, New Caledonia and Queensland, there exist as many different species of grass which, by similar means, exact upon sheep the same

THE MILKWEED'S PLUMED SEEDS.

The Milkweed is an American plant whose seeds are crowned with a tuft of silky hairs which acts as a parachute and bears the seed to some distance on a gentle wind. This method is very similar to that adopted by the Dandelion and the Thistle.

fatal vengeance as does the Russian Feather-grass. This is only one of innumerable instances of the indifference shown by Nature to the suffering of individuals in her solicitude for the perpetuation of various species.

Photo by] [*J. J. Ward, F.E.S.*

THE DOR-BEETLE.

The Dor-beetle is the thick-set blue-black insect that frequently blunders against us as it flies. Owing to the mites that swarm upon its lower surface, it is often called the Lousy Watchman.

THE DOR-BEETLE'S MITE

BY JOHN J. WARD, F.E.S.

THE Dor-beetle, in its polished armour of what looks like burnished steel, is a familiar creature during the autumn evenings. Since, however, the similar story of its near relation, the Sacred Scarab, has already been told in this work, I need write but little here concerning it. My present consideration is of the host of tiny guests that usually accompany the beetle.

When, in its hurried flight, the beetle has collided with some object, it is not infrequently found on its back on a hard road or path, kicking its very hardest with its six legs, and yet quite unable to get its feet upon the ground. The metallic blues and purples of its underside show to perfection as it struggles, and it is at such a moment that we are tempted to pick up the insect. How often, though, have we immediately dropped it again—on account of the crowd of little, pale yellow "parasites" that are almost invariably found clinging to its body and legs?

One day I removed from a small-sized beetle forty-three of these mites—one of which is shown in the accompanying photo-micrograph—but that is a comparatively small number, for sometimes they literally swarm over the beetle. Yet, strange to relate, it never seems the least bit inconvenienced on their account; there is, I think, a plausible explanation of that curious fact.

In the first place I would like to call attention to the clean state in which the beetle is always found. Considering the noisome substances which it manipulates, and into which it completely burrows, it is rather surprising that it should not carry with it some trace of its labours; but it does not even retain the odour of its surroundings. The latter is the more remarkable when we take into consideration the fact that the ill-savoured burying beetles make their presence known in no unmistakable manner; but with burying beetles the Dor-beetle's mite apparently has no connection.

Since the beetle's body and legs are clothed in a horny armour, we naturally wonder on what the mites subsist; for it seems impossible for such tiny creatures to penetrate so hard a surface as that of their host. Also, they never appear to be sucking the beetle's blood. More often they are seen to be quietly travelling over its anatomy, or, when still, clinging to one of its hairs by means of one or both of their mouth-pincers—as shown in the photo-micrograph.

We have, therefore, on the one hand, a beetle which, apparently, is not the least incommoded by what, normally considered, would be an embarrassing number of "parasites" preying upon it, and on the other, a host of "parasites" which seem to possess no desire to prey upon their host.

Probably the explanation is that the mites are the beetle's allies, and not its parasites. It may be that the beetle owes its spick-and-span and odourless toilet to the regular services of these little attendants. Many similar instances are known of apparent parasites which, on investigation, have proved to be useful allies, even in relation to such removed animals as fishes and birds.

In cases of this kind the services are usually mutual, and it is possible that the little beetle-mite requires in its economy a diet similar to that of the beetle, together with an occasional aerial conveyance to new grounds for mating purposes. In consequence, it has probably acquired the habit of engaging the Dor-beetle as its motor omnibus, reciprocating the service rendered by diligently cleaning down the machinery of its locomotion, the product of which labour may serve it as food.

A feature which tends to point to that conclusion is the fact that these mites are frequently found away from the beetles, hiding under stones, and in damp places ; especially is this the case during the winter months.

At the advent of the cold weather, most, if not all, of the old beetles perish, and apparently the beetle-mites leave the beetles and wait until the next generation of its hosts appears in the winged state. It is obvious that they are able to live for months together without any assistance from the beetles, and just what advantages they derive from the association is difficult to understand ; unless, as I have previously suggested, they have some relationship to the functions of mating.

I am inclined to think that the mites are reared on the ground, and that when a family has reached maturity the individuals betake themselves to those quarters where the beetles will be sure to visit, and so they find a host. In this way they probably get carried to new grounds far from home, where they will meet with mates from other stock.

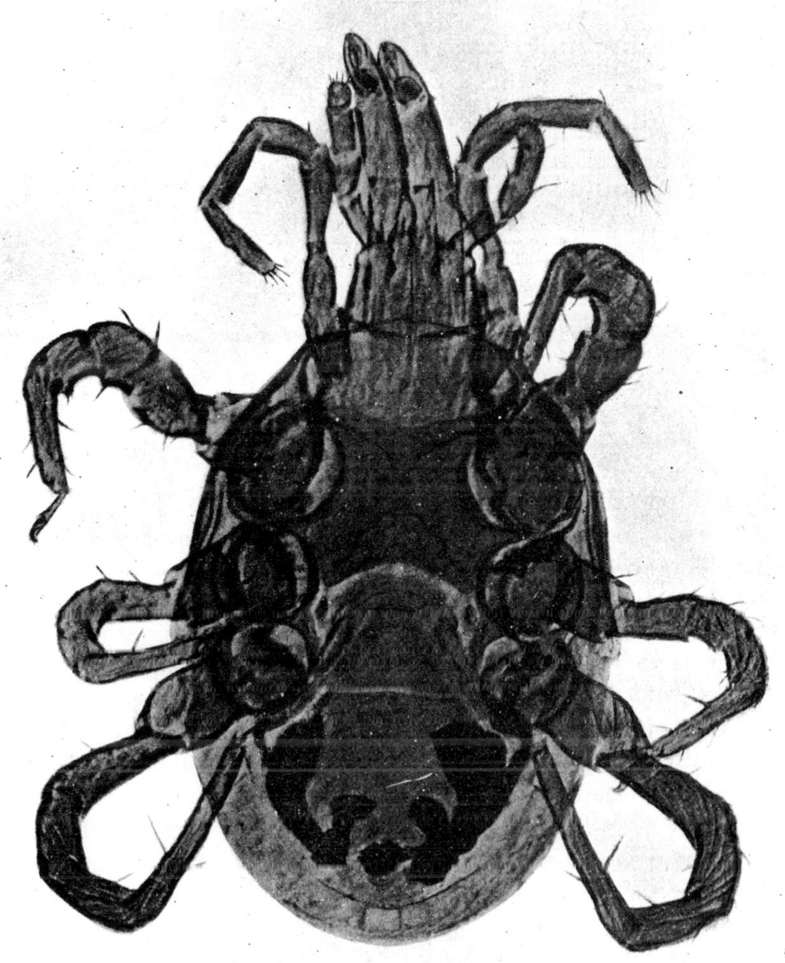

Photo by] [*J. J. Ward, F.E.S.*

THE DOR-BEETLE'S MESSMATE.

This is a greatly magnified portrait of the mite that runs over the surface of the Dor-beetle. A single beetle may be host to over a hundred of these mites, which apparently act as scavengers.

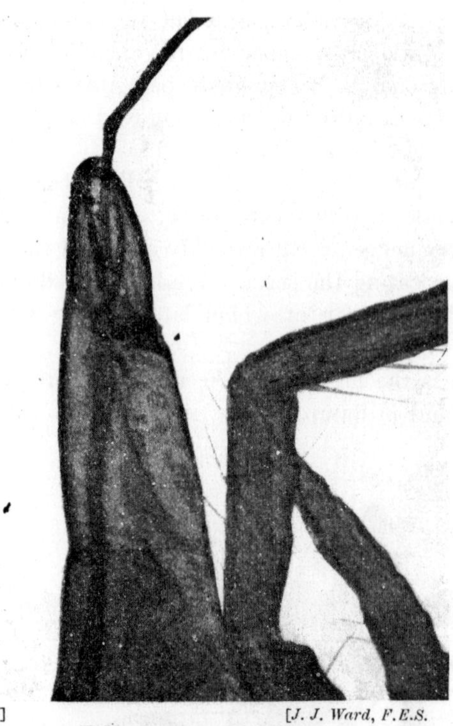

Photo by] [*J. J. Ward, F.E.S.*

THE MITE'S CLAW.

One of the claws of the Beetle's mite is here shown much more highly magnified. It is grasping a hair, and this serves to explain how it clings to the Beetle when the latter flies through the air.

In this connection it is interesting to observe that the legs and feet of the mites are adapted for locomotion on moist land, being provided with claws and suckers, and that they have no eyes. It would be a distinct advantage, therefore, when an individual had travelled a long distance to suitable ground to find a mate, and yet had been unsuccessful, to get quickly conveyed to another similar situation; indeed, it may be that the beetle itself is the profitable hunting-ground for finding their mates.

In any case, it is certain that the mite obtains some advantages from its association with the beetle, and as these on the whole do not appear to be of a parasitic character, perhaps the suggestions offered here will tend to further investigation regarding the habits of these curious little organisms.

The mites themselves are queer little animals, and may be regarded as degenerate spiders. In their early stages they have only six legs, but when their development is complete, another pair appears. Their projecting mouth-pincers can be either partially or wholly withdrawn into their bodies.

THE WINE PLANT

BY EDWARD STEP, F.L.S.

THE American Aloe, which has been extensively introduced into Southern Europe and may frequently be seen in English gardens, is not really an aloe at all, but an agave. Gardeners used to call it the Century Plant, from an erroneous idea that it did not flower until it was a hundred years old. It is a native of Mexico, and is one of the largest of the herbs. Its great leaves, whose form is well shown in our photograph, are thick and hard, and their edges are liberally armed with spiny teeth which make them look like a coarse-toothed saw. They form a huge rosette, the older leaves bending their tips to the ground, and the flower stem rising from the heart of the rosette to a height of thirty feet, branching like a candelabrum and bearing thousands of flowers.

Just before the flowering stem is due to make its appearance, the Mexicans cut out the heart of the plant in such a manner that it leaves a cup-shaped hollow into which exudes the sap that should have gone to nourish the flowering stem. This, the natives draw up into gourd-bottles and transfer to great skin-bottles, as shown on the donkey's back. This juice, which is described as having " a very agreeable sour taste," contains both sugar and mucilage, and the Mexicans long ago found that by fermenting it they could convert it into a sort of intoxicating wine, something like cider, to which they gave the name of Pulque. They also prepare from it a very strong brandy which they distinguish as mexical, or *aguardiente de maguey*. Pulque, however, has a very unpleasant odour—that of putrid meat—so that the newly-arrived European who wishes to get intoxicated by it has first to overcome his aversion to such odours. If he succeeds in that, he is said to prefer pulque to any other liquor. The Mexican Government draws a big revenue from pulque.

Photo by] [*R. J. Carmichael.*

THE AGAVE, OR AMERICAN ALOE.

The young flower-spike having been cut out of the centre of the plant, a sweet sap is poured into the hollow in such quantity that the Mexicans draw it up into gourds and then transfer it to pigskin bottles, such as are shown on the donkey. The juice when fermented becomes a sort of wine known as Pulque.

Photo by] [*H. C. Shepstone.*

THE AGAVE IN FLOWER.

The flower-shoot of the Agave grows to a height of thirty feet, with side branches bearing clusters of flowers, as seen in this photo.

But the juice, if evaporated, makes a good substitute for soap, and the fibres of both leaves and roots are valuable for various purposes. The old Mexicans made paper from these fibres much as the ancient Egyptians did from the Papyrus stems.

The agave is a very common plant in Tropical America, and is found in the mountains even at a height of ten thousand feet above sea-level.

THE FAIRY SHRIMP

SOME of the most beautiful forms of life are to be found in the most unlikely, even the most unlovely, of places. One has only to refer to the illustrations of pond-life a few pages later for confirmation of this statement; but here we have further evidence in the portraits of two young Fairy Shrimps. The marvels of pond-life are mostly found in water that is far from being drinkable, but the most likely places in which to find the exquisite and delicate Fairy Shrimp are the deep ruts near woods that were made by the broad wheels of timber wains and that have filled with rain-water. In such a restricted world, liable to be dried up by the sun, without the presence of water-weeds to purify the water, and to all appearance nothing in the way of food, the Fairy Shrimp lives and grows and enjoys its life.

The nearest relations of the Fairy Shrimp live in absolute brine in the salt-pans where sea-water is evaporated to obtain impure salt, so that it becomes even more surprising to find the Fairy Shrimp itself in such isolated and restricted quarters as woodland ruts afford.

The Fairy Shrimp is all but transparent, and the principal parts of its internal organization are visible from without. Like all the crab, lobster and shrimp family, to which it belongs, its body is built up of a number of rings, and each ring has its pair of appendages. Eleven pairs of these are swimming legs, which are also breathing organs, for every movement of them—and they are scarcely ever at rest—sets fans in motion which cause water to rush in and out over the gills through which the blood circulates and is purified.

The transparency of the Fairy Shrimp is one of the chief elements in its beauty, but this clearness is not so thorough as to produce monotony. The basal joints of its feelers are coloured bluish-green, and the tips are red. The termination of the tail also is red; and in the female the back is tinted with blue. Our photograph represents a couple of shrimps not fully grown, so that in some points they have not a full development of ornamentation. When mature they measure from an inch to an inch and a half in length. They always swim with their backs downwards as shown in our photo.

In spring—March or April—the female deposits eggs, minute globules bristling with spines, which probably enable them the better to adhere to the walls of their little pool. It is probable that these do not develop until the autumn, for ruts are mostly dried up in summer time. It is really most remarkable that any creature not possessed of the means for migration, in the form of wings or walking legs, should choose pools of such temporary character as these woodland ruts are.

The movements of the Fairy Shrimp are very quick, and it is by no means an easy matter to capture a specimen, though to watch its movements in a clear glass vessel is well worth the attempt.

Photo by] [*A. Leal.*

THE FAIRY SHRIMP.

The Fairy Shrimp is almost transparent. It is little more than an inch in length, and lives in puddles and cart-ruts. It always swims on its back.

ANIMALS WHICH GROW BACKWARDS

BY W. P. PYCRAFT, F.Z.S.

NATURE teems with contradictions, real or apparent, and surely nothing more contrary to our notions of the laws of growth can be found than is furnished by the history of the common fresh-water eel on the one hand, and of a curious South American frog on the other.

As touching the eel, it is now almost a matter of common knowledge that the adult eels leave the fresh water when they have attained maturity and make their way down to the sea. Here they make for the abysses of the ocean, spawn and die; for no eel ever makes the return journey. The young, orphaned before they are born, do not at first bear the least likeness to eels. As may be seen in our illustration, they are, in the first place, much flattened from side to side, not cylindrical; in the second, their bodies are transparent, and they have white blood and no fins. As time goes on they ascend towards the surface of the ocean and towards the light, and at the same time make their way landwards. All the while they are growing bigger (see photograph). But at last a stage is reached when a radical change of food is required, necessitating new jaws and teeth, and this change cannot be effected without closing for repairs, so to speak. While the new mouth is growing no food is taken. As a consequence, the material of the body itself is drawn upon to sustain life during the fast; so that, by a process of slow absorption, the body becomes gradually reduced in length and changed in shape, the latterly compressed form giving place to the more familiar eel-shape. While this transformation is taking place the blood is taking on its characteristic red colour, and the fins are developing By the time these tiny migrants have reached the rivers up which they are to ascend in vast hosts they are of about the size of darning-needles. In what enormous hosts these tiny eels, or "elvers," as they are called, invade the rivers of the west coast of England may be gathered from the fact that in the Severn they were caught in cartloads, and pressed into a "cake" known as "eelfare." Happily this wasteful practice has ceased. How they make their way from the deep sea off the west coast of Ireland into the rivers, not only of Ireland, but also England, west, south and east, is a mystery. A few find their way into Scotch rivers, but these are comparatively neglected. No less wonderful is the manner in which these tiny travellers ascend weirs and locks, and make their way overland to isolated pools and ponds.

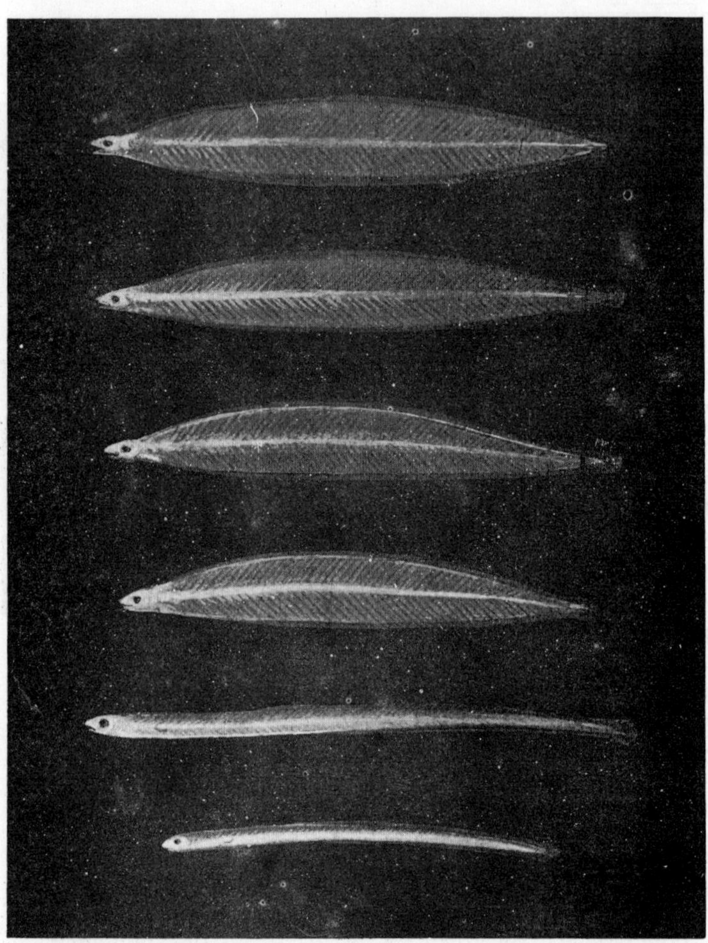

[By W. B. Robinson.

TRANSFORMATIONS OF THE EEL.

The Eel when newly-hatched is transparent and ribbon-like, as shown in the first drawing. It gradually gets smaller and thicker, until it becomes small and round as in the lowest figure, which is known as an Elver. From this point it again increases in size until it becomes a full-grown Eel.

Such journeys are always made by night, when the grass is wet with dew. Sooner or later these tiny travellers come to the end of their toilsome journey, and comfortably settle down till, maturity attained, the hunger for procreation seizes them, and they return to the salt sea which gave them birth, there to lay their eggs and die.

Until quite recently the facts relating to the birth and death of eels was one of the profound mysteries of Nature. Difference of sex could not be discovered; elvers known to develop into mature eels were known to ascend the rivers in millions every year; mature eels were known to descend the rivers to the sea every year, but never known to return. In consequence many wild fancies explaining the origin of eels were in general circulation, not one of which could be regarded by the naturalist as possessing an atom of probability.

The frog of which ment on has been made is the Paradox Frog of South America. In the tadpole stage this creature attains the enormous length of ten inches, and of this no less than seven inches is taken up by the tail, which, as may be seen in the photograph, is of huge size. As the animal nears maturity this tail slowly shrinks, so that by the time the adult, tailless stage is attained the total length of the animal does not exceed two and a half inches, and commonly not more than two inches! Thus while the backward growth of the eel is but a temporary phase, in the frog it is permanent.

[By W. B. Robinson.

THE PARADOX FROG AND ITS TADPOLE.

The extraordinary tadpole here shown attains to twice the size of its portrait; and yet as a completely developed frog it will have dwindled to a length of not more than two and a half inches. A remarkable case of growing backwards.

Photo by] [*W. A. Bentley.*

SNOW CRYSTAL.

Here the six rays have taken on leaf-like expansions.

MARVELS OF THE SNOW

BY WILFRED MARK WEBB, F.L.S.

IF inanimate objects lack the special interest which appertains to living things and do not excite quite the same sympathy, they are yet oftentimes exceedingly beautiful. No art of man could improve the appearance of many groups of crystals, and even if these are not alive, they grow. They may come into existence slowly and gradually from the tiniest beginnings, which arise in some suitable medium or by the introduction of a tiny crystal into a saturated solution, the whole of which will become a solid crystalline mass suddenly before our eyes.

Water comes into play in this case; but with a low enough temperature it will itself crystallize and give rise to all the beauties which we associate with ice and snow in the winter landscape; moreover, if we are not content with these we can bring to our aid the microscope, and in the individual crystals of the snow find such marvellous symmetry and such delicacy and intricacy of design as is difficult to imagine and well-nigh impossible to find in any other products of nature.

It will be easy to appreciate that snow-crystals require very delicate handling, and if we want to examine them we must do it under conditions which will ensure that they are not injured and are kept at a temperature and in surroundings which prevent them from melting; while to see their full beauty it is necessary to collect isolated examples. When photographs have to be taken as records or to demonstrate the wonderful variety of forms which they present, infinite trouble and much skill have to be devoted to the task. It is not surprising, as will soon be appreciated, that very few workers have been found who will make the necessary sacrifices. Of these none have been more successful than Mr. Wilson A. Bentley, of Jericho, U.S.A., who has gained a world-wide reputation. He was probably the first to seriously take up the study, and he has now devoted a quarter of a century to it. He modestly attributes much of his success to the favourable position in which he lives; but when it is remembered that the photo-micrographs of the snow-crystals have to be taken at a temperature of 32° Fahr., and that the photographer has to stay out in the open air often during the whole of a snowstorm in order to find one or two perfect crystals, the enthusiasm of Mr. Bentley must be considered the most important factor in the success of his work.

Photo by] [*W. A. Bentley.*

SNOW CRYSTAL.

The expansions of the six rays are developing into six-sided plates.

Two thousand photo-micrographs in all have been collected by him, which means an

Photo by] [*W. A. Bentley.*

SNOW CRYSTAL.

Here the spaces between the six rays have been filled up, and—

average of about eighty in a year, and this fact, taken in conjunction with what we have just said, will give some little idea of the time that has been devoted to this labour of love. The crystals are caught on a black receiving-board, and it requires very careful examination to detect the perfect ones.

Photo by] [*W. A. Bentley.*

SNOW CRYSTAL.

—a complete six-sided figure is the result. Here six intermediate rays have been formed.

These may, as has been indicated, be very few and far between, and after a whole day has been spent in the search only one or two good examples may have been secured, for a great many of them come into collision and get broken, or they are merged together and therefore spoilt. Those who are photographers will know the many adjustments that have to be made in the case of ordinary objects of which a picture has to be obtained, but if the microscope lens has to be brought into play difficulties will be greater, and when it is remembered that the snow crystals are tiny ephemeral things that may melt away in a moment, it will be recognized that the whole process has to be carried out "against time." It is true that there may be more excitement, as there is the danger of losing some treasure before its form has been recorded, but the strain on the worker will be inevitably the greater.

Practically speaking, no two crystals are quite alike. Water, as is well known, may be obtained in six-sided prismatic crystals, and snow consists of beautiful stellate groupings of such crystals, while the six-sided arrangement always appears, and it is truly wonderful that so many changes can be rung upon this plan. The crystals may be like flowers, like wheels, like studs ; but there are always six branches. They lend themselves excellently to photo-micrography, because their growth takes place to a great extent in one plane, and they are themselves no thicker than a piece of thin paper. Mr. Bentley's photographs, which are reproduced here, will show better than words can describe, the full beauty of form which is to be met with. It is interesting to note that if

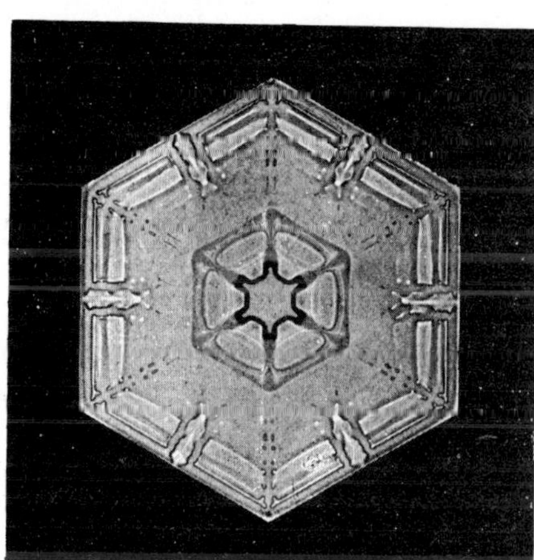

Photo by] [*W. A. Bentley.*

SNOW CRYSTAL.

These two lower figures have followed—

Photo by] [*W. A. Bentley.*

SNOW CRYSTAL.

—much the same lines as the upper ones.

the two thousand crystals which he has photographed were put into one solid mass they would only occupy a cubic inch.

Roughly speaking, we may distinguish feathery and solid crystals. The first grow quickly and are characteristic of low-lying storm clouds which are not very cold. In the most open forms there are the six distinct rays just joining at the centre (see the first illustration on page 204), each of which otherwise develops independently, but symmetrically, so that we get what are practically six crystals growing together to form new elaborate and delicate shapes. When the centre is at first a little more solid and the connection between the rays at their bases greater, we get probably the most effective of all the types of snow crystals such as are shown most beautifully in the full-page illustration. The compact crystal, on the other hand, is said to come, in some cases, from a height as great as eight miles, where great blizzards arise, and grows slowly as it is blown about, perhaps for days, adding gradually to its size and weight, until it finally comes to earth.

Some examples of the very solid kind of crystal are given in the illustrations on page 205. They

Photos by] [W. A. Bentley.

SNOW CRYSTALS.

Three further variations. The central example shows the six rays but slightly decorated.

may grow and continue to retain their very simple though accurate hexagonal outlines, or their angles may develop. In this case two things can happen; either the solid type of crystal will be reproduced in their branches, as shown on page 208, which are occasionally quite elaborate, or they may depart from their original method and become open and more delicate. In this way we may get the forms which we have already alluded to with the solid centre and lace-like branches shown in the plate.

Not the least beautiful feature of the crystals, apart from their form, is the internal structure caused by the inclusion of tiny air-bubbles, and these also enable us to see how the growth of the crystal has gradually taken place.

The marvellous regularity with which the crystals increase in size in the six directions is shown by these bubbles. Take, for instance, the lower right-hand crystal shown on page 205. Within each side is a curious and complicated figure, at the top of which is a little four-rayed object with a dot in the centre. Each of the six is practically the counterpart of the other. It will be seen, however, that there is a pair of processes running from the inner hexagon towards the figures, twelve in all. Four of these join up with the figures, but the rest stop short. Here there is a little discrepancy;

Photos by] [*W. A. Bentley.*

SNOW CRYSTALS.

A comparison of the forms on this page and those previously given will show how great a range of variation is possible with strict adherence to the same fundamental plan. That plan is six straight rods radiating from a common centre, the interspaces variously filled up in the numerous examples given.

Photos by] [*W. A. Bentley*.

SNOW CRYSTALS.

An assemblage of varied forms. The central example differs widely from the others, but its adherence to the six-rayed type is manifest, though the rays are not equidistant.

but, judging from the small projections which are left on the figures, originally all of them were accurate.

As regards the branches of the crystals it will be noticed that, as a rule, these come off in pairs, but that sometimes towards the tips of the rays, as seen in the topmost crystal in the plate, only the branch on one side may develop. It is interesting to notice, too, what happens when branches coalesce; and though in the majority of our illustrations they point outwards, in certain forms there is a tendency for some of them to grow towards the centre. This is indicated to some extent in one or two of the more complicated crystals.

To the nature-lover and the artist, the perfect symmetrical forms have the greatest charm, but to the student of snow crystals some of the results which occur through one side of the crystal developing in a different way from the other have a very special interest, while very curious structures sometimes occur as a result of accident.

In certain cases parts of the crystals seem to stop growing, and we get apparently three-rayed structures, though examination will show, as in the case of the example in the centre of the illustrations above, that the six rays are there. Then, again, there may be no branches, or very few, and these may occur on one part of a solid crystal, while the other angles remain unaltered, and occasionally we get quite an irregular form showing little of the usual beauty of the snow crystal. We have mentioned the stud-like forms, and these consist of a columnar hexagonal crystal with two plate-like crystals, one at each end.

One aspect of the subject must not be overlooked. Lace-makers and designers of all kinds are often desirous of obtaining new patterns, and they could not do better than to examine and take advantage of Mr. Bentley's pictures of snow crystals. No doubt they would find some of them too elaborate to reproduce, but others could be modified, while the simpler ones in many cases might be adopted as they stand.

The whole subject of inorganic growth is at present of particular interest owing to the strange forms which have recently been obtained by the depositing of inorganic matter out of solutions and the suggestion which has been made that, after all, the bodies of plants and animals owe some of their shapes and development to the laws which govern inorganic matter and are not dependent upon the intervention of what we call life, though its meaning is still beyond our ken.

SATURN—THE RINGED PLANET

BY T. E. R. PHILLIPS, F.R.A.S.

In the early morning of September 17th last two bright planets were very close together in the sky—so close that they could be seen in the same field of the telescope. They were Mars and Saturn—the swiftly-moving, blood-red star of war, and the leisurely planet to which the old astrologers assigned the character of lead. And of all the planets these are the two which appeal most strongly to men's minds to-day: Mars, because it is the planet of romance—it is a near neighbour; we know a good deal about it, and it *may* be inhabited by beings something like ourselves. Saturn, because of the unique beauty of its system, and the old enigma of its rings. Prior to the discovery of Uranus, by Sir William Herschel, in 1781, Saturn was the most distant planet known. It revolves round the sun at a mean distance of more than nine and a half times that of our own globe, and at its very nearest approach to us it is still seven hundred and forty-four millions of miles away. The globe of Saturn, like that of Jupiter, is soft and plastic and very hot, and the dense clouds enveloping it are disposed in a number of more or less parallel streaks. Occasionally definite spots or other irregularities are seen, and from the study of these we learn that the great globe—seven hundred and sixty times as large as the Earth—is spinning round in between ten and eleven hours. But of course, it is in the rings by which the planet is girdled that our interest is chiefly centred, and the story of their discovery and explanation furnishes a remarkable chapter in the history of astronomy. It was Galileo Galilei who, in the year 1610, first applied a telescope to the study of the heavens. It was only a very small and imperfect instrument as compared with the giant

[*By T. E. R. Phillips, F.R.A.S.*

THE PLANET SATURN.

The glory and mystery of Saturn lie in its wonderful series of rings that greatly puzzled Galileo and other astronomers. The composition of these rings baffles even modern telescopes to determine; but they are believed to consist of loose particles of matter.

telescopes of our day, but it revealed to him many things that the eye of man had never seen before. The mountains in the Moon, the four chief satellites of Jupiter, thousands of stars studding the region of the Milky Way—all these were quickly found, and then Galileo turned his attention to Saturn. How astonished he was and how puzzled by what he saw! What could those little globes be, almost in contact with the planet, one on either side of him? Surely the planet is triple, he thought; and then to his amazement he watched them dwindle away, till, in about two years' time, they disappeared entirely! The men of those days had laughed at Galileo's alleged discovery, and now even he himself began to wonder whether in some strange way his eyes and glass had not played him false. But the day of his triumph was at hand, for he soon saw the little bodies reappear in their old positions, and he watched them grow till

TWO VIEWS OF SATURN.

These are two of the phases of the planet that have puzzled observers. In the first the rings appear to be changed into a straight rod passed through the orb; in the second the rod-like character is lost, and only a dark bar across the orb is seen. This is really the shadow cast by the ring, the latter being unseen in this phase.

they were no longer like globes, but handles attached to the body of the planet through which the dark sky could be seen. What a mystery Saturn was to the astronomers of those days is shown by their curious drawings, and it was not till 1655 that Huygens found the solution of the riddle. The planet, he said, "is girdled by a thin detached ring, inclined to the ecliptic," and this statement afforded a complete explanation of those strange disappearances and reappearances which had been so great a puzzle. As the planet revolves in its orbit and comes into different positions relatively to the Earth the rings are seen in a variety of phases. Sometimes they are presented edgewise to view and become invisible, owing to their remarkable thinness; sometimes they are open widely enough to look in small telescopes like loops or handles, as Galileo saw them in 1616; and sometimes—when their inclination to the line of sight is greatest—they are seen to project above and below the planet's globe.

[By The Abbé Moreux.

SATURN RISING.

The appearance of the planet which would be presented to us if we could stand upon Saturn's nearest satellite and watch its rising above the horizon of jagged peaks. Owing to its nearness the planet would appear to fill a great part of visible space, as here shown.

But the discovery that Saturn is surrounded by a ring, only solved a part of the strange puzzle. There remained the further problem—What *is* the ring? Is it solid, liquid or gaseous? Is it stationary, or is it in perpetual revolution round the planet? Such questions could only be answered by mathematicians, and amongst the great names associated with the story at this point must be mentioned that of James Clerk Maxwell, who, in 1857, found that the ring must be composed of a great number of little separate bodies, each one pursuing its own path in accordance with Kepler's laws of planetary motion. Of what size these bodies are no man knows, for they are far beyond the reach of the most powerful telescope we can ever hope to make.

Saturn's ring is far more complex than was at first supposed. Thus, amongst other details, there is a strong dark line or gap dividing it in two, and then there is a very mysterious-looking, faintly-luminous ring commonly known as the "crape" ring, between the bright one and the ball. This latter is partly transparent, so it is likely that the particles composing it are less densely packed than elsewhere, and permit the body of the planet to be seen between them.

Saturn is accompanied by ten satellites, the most distant of which adds yet more to the mystery of this marvellous system by revolving in a direction contrary to that of all her fellows.

The coloured plate shows us how Saturn would look could we be transported to one of his satellites and view the globe and rings at close quarters. Another of the illustrations gives us a picture of the rings as seen from above the planet's clouds at midnight in the summer time. Not on earth is it given to man to behold so glorious a spectacle as that gigantic arch spanning the sky, and illuminating the cloudscape below with its soft radiance.

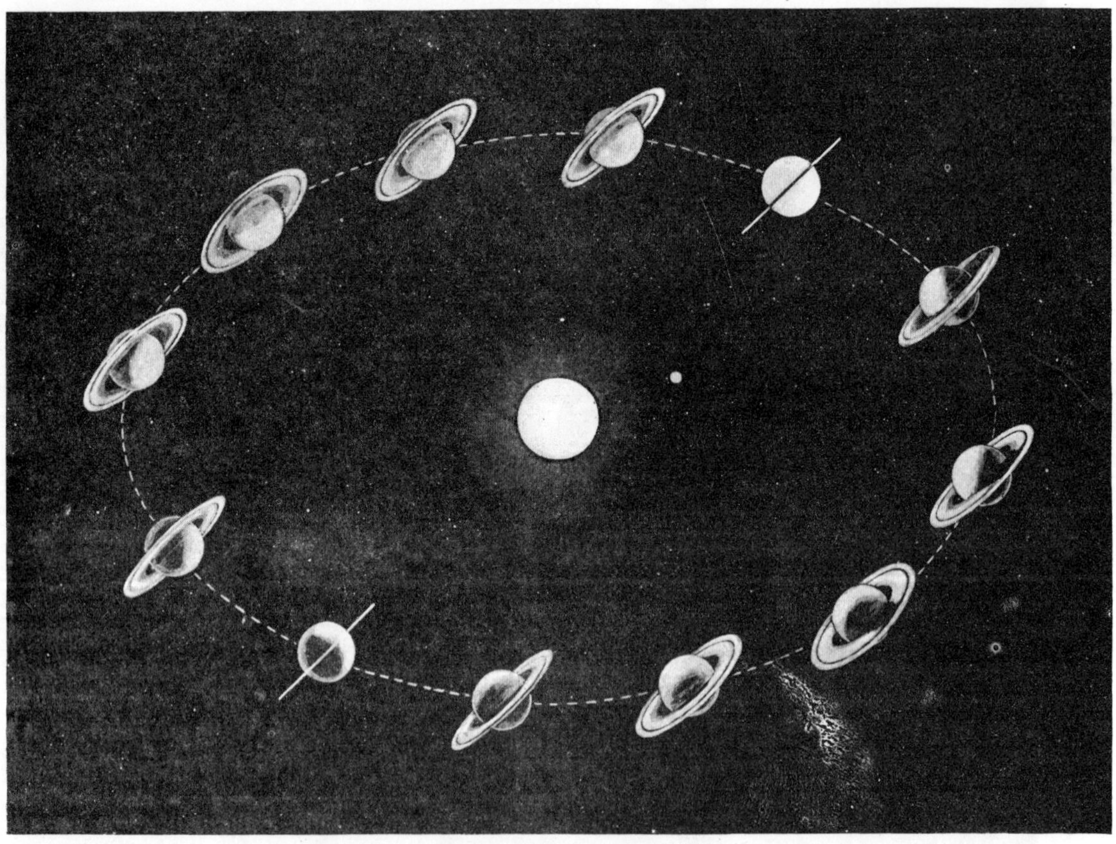

SATURN'S PHASES.

In this diagram the different appearances of Saturn as seen from the Earth are explained. The planet follows the course indicated by the dotted line. The central sphere is, of course, the Sun, and the Earth is a little higher and to the right.

[By A. Twidle.

SATURN'S RING ABOVE THE CLOUDS.

Saturn is hidden by clouds, but the Sun's light from below throws the planet's shadow across the ring. Above, one of the planet's satellites is seen at the full.

SEA CUCUMBERS

BY JOSEPH SINEL.

It is doubtful if any person outside the ranks of the zoologists would be able to trace any sign of relationship between the leather-clad, sausage-shaped animals known as "Sea Cucumbers" and the familiar Star-fishes and Sea Urchins of our shores. Yet the relationship is there, and they are all included in the same division of the animal kingdom.

Diverse as are these forms, they have more than one bond of unity. In their early stages they are all alike, except in microscopic and technical details. They are at first little barrel-shaped organisms that, except in a few cases, swim freely in the sea, and in these stages undergo many wonderful transitions as they advance towards their ultimate and adult form.

Even in adult life a common plan pervades them all. They are symmetrical and built on a five-cornered plan. This plan, masked, varied, and in many cases all but suppressed, can be traced by close study.

All, with the exception of the swimming form shown on page 214, have an arrangement of plates or spines of lime either upon or within their skins. Even in their internal anatomy what at first appears contradiction can be seen to be but variation.

In the Sea Urchins the plates are jointed together to form the well known prickly "shell." In the Sea Cucumbers the plates are separate and scattered within the substance of the skin.

As means of locomotion they possess a curious arrangement of what are termed "tube feet." These are simply tubes of thin elastic material, which can be protruded through holes in the skin and into which water is forced by means of a series of canals termed the "water-vascular system." These canals drawing in water from the sea through a portion of the skin specially modified to act as a filter, forces it into the tubes and thus renders them turgid and enables them to act as locomotive appendages. A little sucker at the tip of each tube allows of a hold upon the surface travelled over, so that by this means the creature can progress at a fair rate.

In the Star-fishes and Sea Urchins these tube-feet are arranged in five rows, but in the Sea Cucumbers they are not as a rule so regularly disposed. In the latter group some of the tube-feet are so placed as to take no part in locomotion, and so modified as to become mere sense organs, as in the two shown on page 216; whilst in the one that lives at the sea-surface they are modified into swimming appendages and resemble the tentacles of a Jelly-fish. In fact, in all members of the family these tube-feet have other than locomotive uses. Romanes has shown from experiment that even in the Star-fishes, where they are situated on the under side, they act as the organs of smell and that it is by their means that the animal is enabled to find its food.

A SWIMMING SEA CUCUMBER.

Although this creature has more resemblance to a Jelly-fish, its organization shows it to be a near relation of the Sea Cucumbers such as are shown on the opposite page. A muscular web stretches between the "arms," and by its expansion and contraction the animal makes its way through the surface waters.

The eggs of the Sea Cucumbers in most cases are set free in the water, hatching and development there taking place quite independent of the parent; but in some the eggs are hatched and the young developed within special brood-pouches possessed by the mother. The eggs of the Saffron-coloured Sea Cucumber become attached to the back of the mother, where the young hatch out and remain until they have attained a considerable size, as shown on page 215.

The Sea Cucumbers have around the mouth a circle of ten tentacles, which can be expanded or withdrawn. In some species these are small and simple; in others they are elegant branched appendages; whilst in others they are like little fern fronds, and are well shown on the opposite page.

These tentacles assist in locomotion, and it is by their means that food is secured; they are also believed to be sense organs.

Most Sea Cucumbers burrow, worm-like, in the sand or mud. Some creep over the ocean floor; others lurk in rock crevices and in coral reefs; whilst one species—already mentioned—swims like a Jelly-fish, at the surface of the sea.

About five hundred different kinds of Sea Cucumbers are known to science, and of these some fifty are found in British waters.

[*By Theo. Carreras.*

SEA CUCUMBERS AND THEIR YOUNG.

The Saffron-coloured Sea Cucumber is abundant at the Falkland Isles, where it adheres by its suckers to the big seaweeds. The branching expansions around the mouth are feelers. By these food is caught from the waters and conveyed to the mouth. The young attach themselves by their suckers to the back of their mother, and as each has its own expanded feelers the general effect is rather remarkable.

A SEA CUCUMBER THAT RESEMBLES A SEA SLUG.

A form that differs remarkably from the ordinary Sea Cucumbers. It glides over the ooze and mud at the sea-bottom and feeds upon this material.

In size they vary from a length of three or four inches to a couple of feet or more. Their food consists of anything, living or dead, which comes within their reach and which is not too bulky to enter the mouth.

The limy plates which occur in the skin of some species are often of very beautiful patterns—St. Andrew's crosses, curious little stools, and wheels with spokes, being among the common forms; whilst they also occur as exquisitely shaped little anchors, each anchor accompanied by a perforated supporting plate.

Mr. Moseley, in his " Notes by a Naturalist on H.M.S. *Challenger*," tells how " the King [of the Sandwich Islands, Kalakau] took the liveliest interest in the work of the *Challenger*, and was almost the only distinguished visitor of the many to whom I had exhibited microscopical objects during our voyage who recognized the well-known anchors in the skin of these creatures and named them at first glance. These anchors stood us in good stead at all the ports visited and were described in all the colonial newspapers as belonging to the ' Admiralty worm,' supposed to be the most wonderful of the deep-sea discoveries of the expedition." On page 218 is given a photo-micrograph of some of these.

Some of the Sea Cucumbers are highly prized as an article of food under the names of *Trepang* and *Bêche de Mer*; they form the subject of extensive fisheries in various seas, notably in the Malay Archipelago, the New Guinea coast and California.

The Chinese appreciate them highly and rank them with the gelatinous " bird's-nests " as an ingredient for soups.

A SLUG-LIKE SEA CUCUMBER.

A deep-water species that feeds on the ooze at the sea-bottom. In form it closely resembles one of the Sea Lemons among the Sea Slugs.

When taken from the sea the Sea Cucumbers are split open and cleaned; they are then boiled in sea-water, flattened out and dried on wicker frames—first in the sun, and then over a wood fire, where they are slightly smoked. They are then packed for market and command a good price. About a dozen sorts are thus employed, their value varying according to district and to species, those with the fewest limy plates being considered the best.

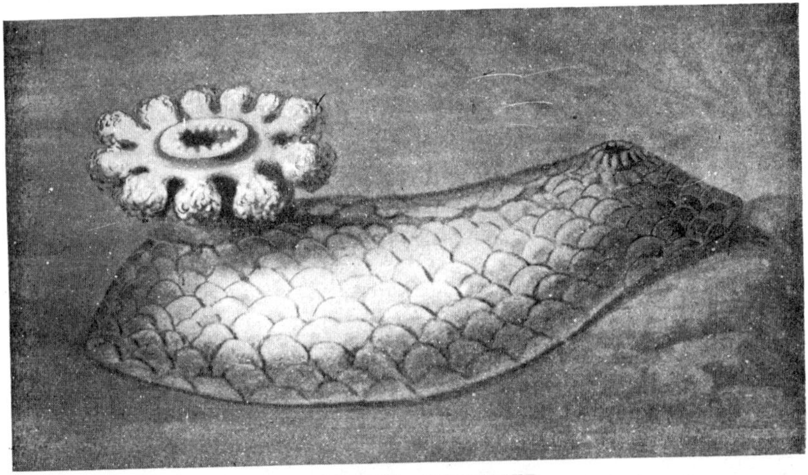

SCALY SEA CUCUMBER.

Another slug-like form from the Kurile Islands. The upper side is covered by large scales, and the mouth with its circle of feelers looks like a chimney.

The term *Bêche de Mer* (literally French for "Sea Spade") is said to be taken from the shape of the dried article, which somewhat resembles a spade; but some authorities say it is derived from the Portuguese *bicho*, a worm or grub. The term *Trepang*, which is also applied to the commercial article, is simply the Malay name for the animal, Tripang.

A number of different kinds are indiscriminately sold under the name of *Bêche de Mer*, so that it is not advisable to refer to any particular species as being the delicacy in question. The probability is that, after they have received the curing treatment described, there is no great difference perceptible, and, after all, it is their glutinous character that recommends them chiefly to the Chinaman's palate.

Sea Cucumbers have not been found as such in a fossil state. All that has been found to show that they did exist in remote periods are the anchors, gratings and wheels, such as are shown from present-day examples on page 218. The oldest strata in which they have been found are those of the Carboniferous limestone.

Certain of these Sea Cucumbers have for so long climbed over the rocks that they have at length developed an upper and under side, the former supplied with three lines of sucker openings, the latter with only two; and this arrangement clearly shows the origin of this two-sided form, especially as the two digestive openings are still situated at the two extremities of the body. This transformation shows itself particularly clearly in the species which is illustrated at the top of this page, where the mouth is thrown back upon the upper side at the end of a vertical chimney, and this gives the creature the appearance of a tiny locomotive without wheels, but with smoke issuing from the funnel in the shape of branched tentacles.

A SPINY SEA CUCUMBER.

This species conforms more to the shape which has given these animals their popular names, but it differs in the fact that it is covered with spines, and the feelers are not branch-like.

But to see still greater variations in the appearance of these Sea Cucumbers it is necessary to go to the species which live in the greater depths of the sea. Here there are many whose former symmetry is made manifest by the presence of a mouth on the inferior surface of the body. Sometimes this is a simple opening, and sometimes it is a veritable snóut. In such forms the under side is entirely different from the upper, either owing to its flattened form, or to the nature of its appendages. These may consist of little openings, sometimes of long tentacles, which, however, are of no use for locomotion. There are some which have no appendages of any sort, and some which crawl about on the ocean-bed after the manner of slugs.

Besides these species, there are in the great depths some which live all but entirely buried in the mud, where they remain almost completely motionless. On this account their bodies have been absolutely changed as to symmetry. In the examples shown on page 220 it has taken the form of a U, so as to bring the two digestive orifices outside the region of the mud.

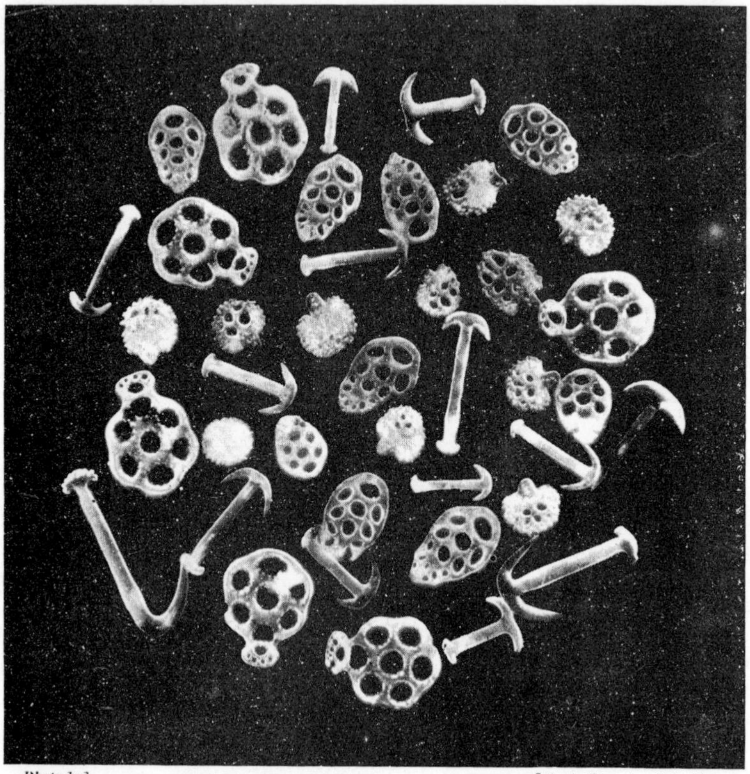

Photo by] [*H. S. Cheavin, F.R.M.S.*

ANCHORS AND GRATINGS.

These remarkable objects—which are here shown much enlarged—are found embedded in the skin of certain members of the Sea Cucumber family. Each particular species has its own peculiar forms.

The Sea Cucumbers of which we have spoken have very peaceful manners. But there is one which has dreamt of vast spaces and the greatest liberty. It has, therefore, become a swimmer, or, rather, a sea-navigator. It gets to the upper seas and balances itself gently on the waves. Different conditions of life have altered the structure of its body. The species illustrated on page 214 is the only one at present known that has taken to the Jelly-fish's form and habits.

There is a remarkable circumstance connected with the Sea Cucumbers, though it is a story in which they take but a passive and probably unwilling part. The hinder opening of the Sea Cucumber's body contains extensive branching organs known as the respiratory tree, because within it circulates the creature's blood and absorbs oxygen through its walls. To keep up this supply of oxygen, sea-water is drawn in at short intervals, and expelled when the oxygen has been extracted from it. Taking advantage of these openings, a small fish, known as the Fierasfer, is said to enter the interior, and, turning, remains there with its head at the entrance ready to pounce upon any suitable food that comes by. The story has been doubted by some, and probably it stands in need of corroboration. It is a curious story, for it is neither a case of parasitism or of what is called commensalism, such as those examples of partnership described in relation to the Hermit Crabs and the Ant-nest Beetle. The Fierasfer takes nothing from the Sea Cucumber ; neither does he confer any benefit upon it : therefore, he is neither a parasite nor a messmate.

The Fierasfer is a thin, ribbon-like fish, about six inches long.

Photo by] [*W. Saville Kent.*

SNAKE-LIKE BÊCHE DE MÉR.

The Bêche de Mer, or Trepang, is a Sea Cucumber much in request as food by the Chinese and other Eastern peoples. They are dredged for by Malays, who cure and dry them for the market. Some of them are two or more feet in length.

Photo by] [*W. Saville Kent.*

PRICKLY BECHE DE MER.

A Sea Cucumber that would probably be rejected by the Bêche de Mer merchants, because of the hard spines with which it is covered.

BURROWING SEA CUCUMBERS.

These are two distinct kinds that agree in their habit of sinking their bodies into the ooze, whilst keeping their mouths above it. Although bearing no resemblance to cucumbers, they may be likened to certain forms of gourd, which are the vegetable cucumber's relatives.

A MESSMATE OF ANTS

BY H. ST. J. K. DONISTHORPE, F.Z.S., F.E.S.

PERHAPS one of the most interesting beetles which occur in Britain is the subject of this sketch. It is an ants'-nest species, and passes its whole life in the home of its host—the large blood-red Robber Ant, our only slave-making species. The writer was fortunate enough to witness one of its slave raids when at Bewdley Forest, in Worcestershire. Only two other observers, Charles Darwin and Frederick Smith, have seen this before in Britain. The worker ants march in a body from their nest to a colony of a black species of ant, which they have previously located. They attack the latter, kill any that offer resistance, and steal the cocoons, which they carry back to their own nest. When the black ants hatch out from these cocoons they act as slaves to their captors.

To return to our beetle. In 1710 a single specimen was taken by Sir Hans Sloane on Hampstead Heath. A second was found by Dr. Leach, when travelling in the mail coach between Cheltenham and Gloucester, about 1820. Both these specimens are now in the British Museum. It was not found again for many years, and was considered doubtfully indigenous to Britain. On May 25th, 1906, the writer discovered it again when investigating a nest of its host, the slave-maker, at Woking, and since then has found it in numbers in the nests of this ant there. On one occasion over sixty individuals were found in a single nest.

The beetle is a true guest to its host; that is to say, it is fed by the ants, and supplies them in return with a sweet secretion which exudes from small orifices in the segments of the hind body. It is a reddish-brown creature, with tufts of golden hair, which cover the places from whence the sweet fluid springs. When it wishes to be fed it taps an ant with its feelers, in the same way that one ant does to another when asking for food, and the latter feeds it from its own crop. A near relation of this beetle, which is also a dweller in ants' nests, has gone a step further. It not only uses its feelers when supplicating its host, but also strokes the ant's cheek with one of its front feet, which is also the procedure of an ant under these circumstances. The beetle can also feed itself, as when kept in observation-nests in captivity it has been seen to bite at dead ants and suck caterpillars and other creatures given to the ants as food.

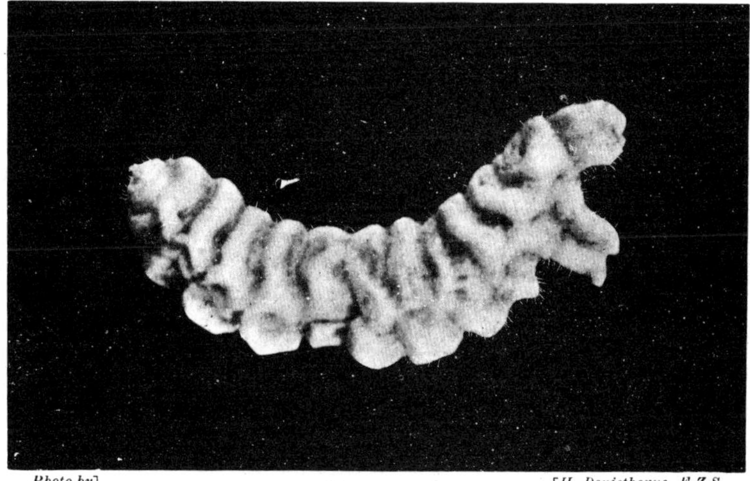

Photo by] [*H. Donisthorpe, F.Z.S.*

ANT-BEETLE'S GRUB.

This grub is fed by the ants, even with their own young. It is very voracious, and its presence leads to a great reduction in the number of ants in a nest.

This beetle is not very ant-like in appearance when examined by itself, but when it sits among a lot of its hosts—and it is always to be found where the ants are thickest—it becomes practically invisible. The reason for this appearance is that the light which is reflected from the concave sides of the " chest " of the beetle appears to the eye like the narrow back of an ant, and the rolled-up hind body of the beetle reflects the light in the same way as the rounded hind division of a fat ant. When their hosts change their nest, or move from place to place, these beetles move with them. They can also fly, possessing ample wings wrapped up under their short wing-cases.

Photo by] [*H. Donisthorpe, F.Z.S.*

ANT FEEDING ANT-NEST BEETLE.

The beetle is treated as an honoured guest and fed by the ants. When it wishes to be fed it taps an ant with its feelers as here shown, and the ant feeds it with food from its own crop. In return it supplies the ants with a sweet secretion from its body.

The courtship takes place in May, and the female lays her eggs on the egg-masses of the ants. The egg, which is very like that of an ant, is exceedingly difficult to detect. The young grub hatches from the egg in a very short time, and proceeds to feed on the brood of its host. The ants not only feed it by mouth (this has been proved by feeding the ants with honey coloured red, when the red colour can be traced in the intestinal canal of the little grub through its transparent skin), but they place it on their own grubs. The beetle grub is very like that of an ant, and though it possesses six short legs, it does not use them, and imitates the behaviour of an ant grub. The ants pay it the greatest attention, and when danger threatens the nest they carry it first into safety. It is extremely voracious, and devours large quantities of the ants' grubs. In nests where the beetle has been a dweller for some years a decrease is produced in the number of worker ants. Now, as is well known, ants create females by feeding their grubs on special food, and to try and make up for this loss of worker ants they try to turn grubs which they have started to bring up as females into workers. The change, however, is too late, and the result is what are called false females. These are neither females nor perfect workers, but of an intermediate form which does not work or bite. These nests are the centres from which the beetle spreads to other nests. At first the beetle is kept in check by the ants digging up its chrysalis and carrying it about as they do their own cocoons, which of course kills a delicate beetle chrysalis; but as more false female ants are produced, fewer chrysalids are disturbed, till at last the destruction of the ant colony is brought about.

This beetle is spread over the whole of Europe and North and Central Asia as far as Thibet.

Photo by] [*H. Donisthorpe, F.Z.S.*

BEETLE FEEDING THE ANT.

In this photo the ant is seen licking up the sweet substance that exudes from small openings on the beetle's back.

THE SULPHUR-BOTTOM WHALE

BY FRANK T. BULLEN, F.R.G.S.

Prodigious in size, even among mammals that are noted for their vast proportions, the Sulphur-Bottom Whale (to give it the trivial name conferred upon it by whalemen) is more enveloped in mystery than any other of the whales, for various reasons, some of which will here be shown. But first of all, a brief description of the mighty animal is necessary. It is a toothless whale, having in place of teeth a series of plates of baleen (whalebone), dependent from its upper jaw. But these two rows of fringed plates are very short and ineffective compared with the enormous screens with which the Greenland whale is provided. The reason of this is obvious if we examine the throat of the creature. For whereas the Greenland whale, because of the tiny aperture of its throat, must carefully sift only the smallest of marine organisms (brit) from the surface inhabitants of the sea, the rorquals, of which the Sulphur-Bottom Whale is the chief, can swallow huge masses of fish,

Photo by] [*H. C. Shepstone.*

THE SULPHUR-BOTTOM WHALE.

The Sulphur-Bottom is the largest of the Whales. The example photographed measured eighty feet in length and weighed eighty-seven tons. It is one of the toothless Whales, teeth being replaced by fringed plates of "whalebone," which strain off the comparatively small fishes upon which it feeds from the water.

Photo by] [*H. C Shepstone.*

THE SULPHUR-BOTTOM WHALE.

This Whale owes its name to the colour of the under surface, which is pale yellow, and thrown into furrows and ridges extending from end to end. In these furrows many sea-creatures find firm hold, as they cling to the Whale and get carried about the oceans by it.

such as cod, mackerel, herring, etc. Not that it has any more discrimination than a trawl; its only anxiety is to get enough.

Feeding in such a different manner, upon lively food, all the rorquals have developed speed and agility far beyond that of the Greenland whales and the cachalots. Their forms are more elegant, their bodies longer in proportion to their size. Both of these qualifications—*i.e.*, speed and elegance—culminate in the Sulphur-Bottom, which has an authentic maximum length of one hundred and fifty feet, with a maximum girth of little more than thirty feet. As far as size is concerned, this may be compared with the great "Bowhead" of the Behring Sea, which, with a length of seventy feet and a median girth of fifty feet, may very well be the larger creature of the two.

There is another feature of the Sulphur-Bottom Whale which differentiates him from all the rest of the deep-sea mammals, his enormously high vertical spout. One of the rudimentary lessons which the budding whale-fisher must learn is the ability to distinguish between whales by their spouts, which are visible several miles away. Or perhaps I should say that this knowledge *was* necessary before the present equipment of whalers, which makes comparatively easy prey of any whale, was invented. In those bygone days, then, from the crow's-nest of a South Seaman even the veriest novice could not fail to detect a Sulphur-Bottom by the tall column of vapour rising vertically into the pure air for a matter of twenty to thirty feet, and hold his peace about it, since it belonged to an animal that was of no earthly use to anybody.

It really seemed as if the Sulphur-Bottom knew of his immunity, since he cultivated quite a friendly acquaintance with ships in the lonely seas which are his habitat, often lying by them for days in a calm and occasionally rubbing his barnacle-infested belly against their bends, sending disagreeable tremors through the whole fabric and fostering superstitious fears among the sailors, who could not understand this colossal familiarity.

The trifling reason for the popular name given to this monstrous mammal is to be found in the light yellow hue of its corrugated belly. All other rorquals have the belly corrugated longitudinally from the vent to the verge of the lower lip, but the colour is universally a dirty white with this one exception. The depth between these ridges of blubber, speculation as to the uses of which is at present mere waste of time, is from two to four inches, and the elasticity of the whole fabric so great that I have seen a blanket-piece from the belly of a humpback when skinning the whale extend from the whale to the futtock shrouds (just beneath the main-top) under the strain, and when cut off fly up into a space only ten feet long—this from a stretch of forty-five feet.

This arrangement of the belly blubber is a perfect boon to the limpets, barnacles and other parasites which infest these whales, for within those snug folds the shellfish find a cosy abiding-place from which nothing but the death of their host can dislodge them. No one who has not seen the condition of the underside of a great rorqual can credit how terrible are the ravages made by these parasites upon the peace of the gentle monster compelled to give them lodging-place, or seeing, could fail to pity.

Alas ! the day of the Sulphur-Bottom, as of all the harmless rorquals, is practically over. Many whaling companies, armed with the latest lethal weapons, which no whale can resist—or escape—are established in most of those desolate outliers of the world which, since the seals were exterminated therefrom within the last century, have been of no use to any. Now a new war of extermination is being carried on against the whales. It will not last so long as the one did against the seals, for several reasons, principal among which is the highly modernized character of the weapons, the speed of the vessels, and the length of the period necessary for reproduction of whales as against seals ; while other factors are the same, such as the absence of all control over such far-distant operations, and the want of knowledge of what the extermination of these vast sea-creatures and the consequent disturbance of the balance of Nature will mean.

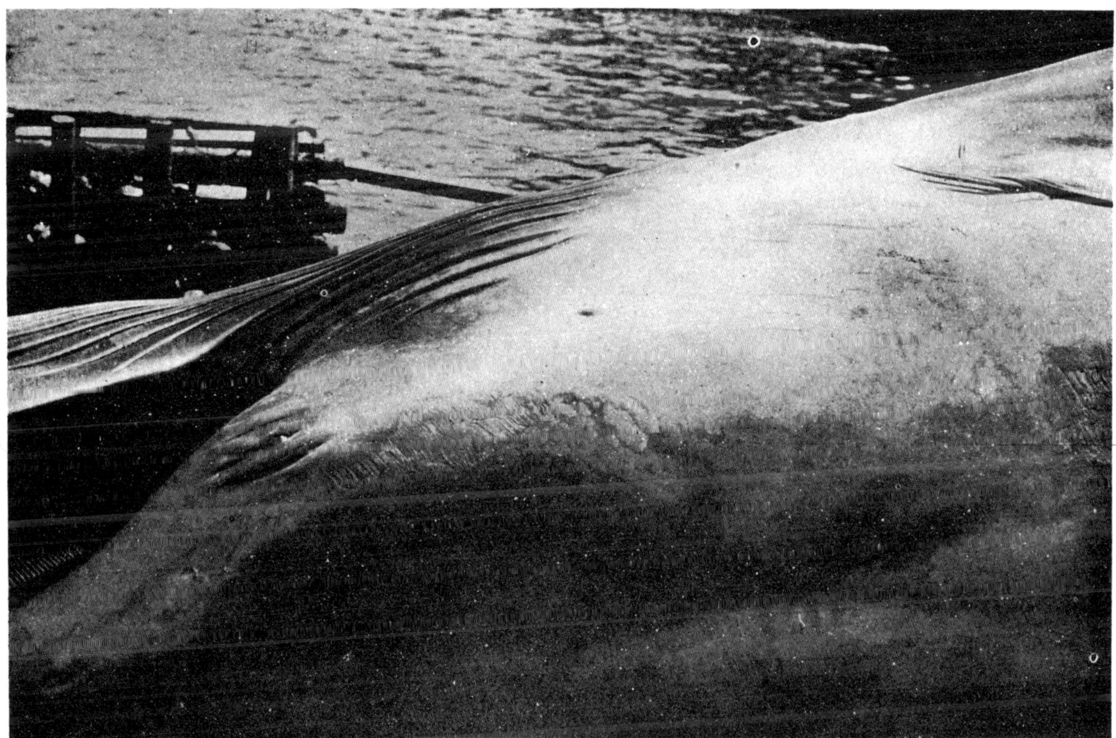

Photo by] [*H. C. Shepstone.*

THE SMALL EYE OF THE SULPHUR-BOTTOM WHALE.

Here a portion of this huge creature is photographed to show the comparatively small size of the eye (to left of photo) and its great distance from the ear—shown to the right in the upper part of the picture.

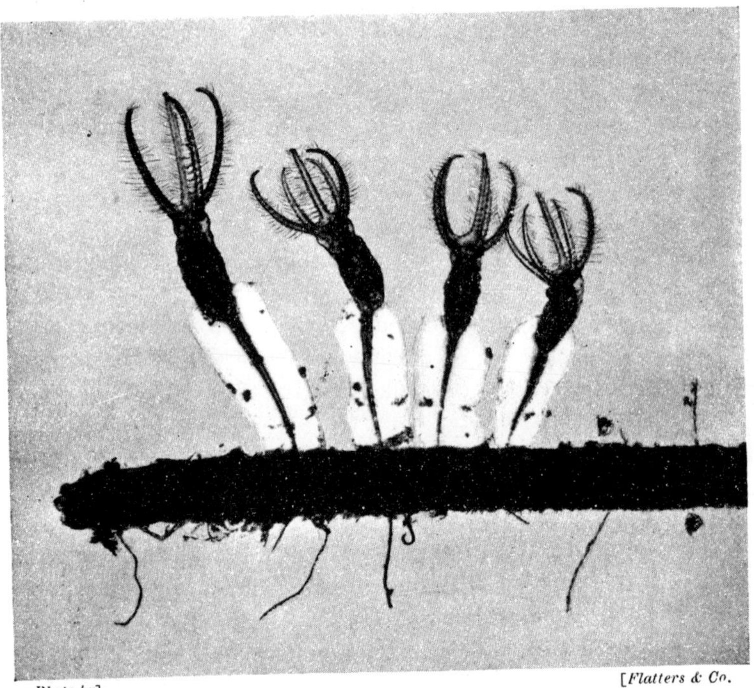

Photo by] [*Flatters & Co.*

THE CROWNED HORN.

This beautiful creature lives in a transparent bag, from which it extends itself to catch food. The mouth is surrounded by the crown of five arms. These arms bear rows of fine hairs whose movements create currents which bring food to the mouth. About thirty-six of these creatures placed end to end would measure one inch.

MINUTE MARVELS IN THE POND

BY WILLIAM WEST, L.D.S.

AMONGST the many marvels of the universe there are none more wonderful than the minute forms of both vegetable and animal life to be found in the water of ponds and ditches, which need not of necessity be stagnant to be prolific. The most interesting forms inhabit clear ponds and running streams, in which aquatic plants grow. If we take a sample of water from one of these ponds and examine it carefully we shall probably see minute green spheres revolving over and over in their progress through the water. They are known by a name which means Revolving Globes, and are claimed by the botanist as plants and by zoologists as animals. They were by the earlier microscopists supposed to be single organisms ; but they are now proved to be made up of innumerable pear-shaped cells united into a spherical network. From each of these cells project two minute lashes, which by their regular movements propel the compound body through the water. In the interior of the network sphere are seen smaller spheres, which are set free by the break-up of the outer and larger one. In their interior may be seen eight still smaller globes, so that we often get in the one sphere several generations visible. About twenty of the larger globes placed side by side would measure one inch.

Let us examine a piece of weed from the same pond, and perchance we shall see a small brown, rod-like body, about the twentieth of an inch in length, projecting at right angles to the green weed. This is the house of a marvellous little animal called the Little Brickmaker. The animal, as soon as it is hatched from the egg, attaches itself to a piece of aquatic weed and sets to work to build a tube around itself to live in. To enable it to do so, it is provided with four beautiful fan-like lobes, which radiate from the central mouth. These lobes have their edges fringed with fine transparent hairs capable of rapid movement, which fall and rise in rotation, one after the other, so rapidly that their motion produces the appearance of four revolving wheels. The object of this action is to create a vortex in the water the centre of which is the animal's mouth, so that any particle of food coming within the current is drawn down the miniature whirlpool and swallowed. The Little Brickmaker seems to have the power of discrimination, for what it does not require for food it rejects, though sometimes it is captured again for brick-making. The recaptured particle is conveyed to a little receptacle situated at the base of the fans on the outer surface. The particles are drawn into this chamber and made to revolve rapidly, until they become solid enough for the brickmaker's purpose ; when this is so, he swings his body over to the side and deposits and cements the newly-made brick on the edge of his tube ; then proceeds to make another. As he grows taller, so he builds his house higher to enable him to retreat into it when alarmed. The slightest vibration of the microscope is sufficient to alarm him, and he requires a little time to gain confidence enough to emerge from his retreat. The time occupied in making a brick is about three minutes ; but it depends, of course, upon the amount of material at his disposal. To see the process well, it is advisable to put a little carmine in the water ; he quickly utilizes it and makes carmine bricks. I once gave a Little Brickmaker carmine, and after he had laid two or three rows of red bricks around his tube, I put some chalk into the water. He then made some layers of white bricks. Next I supplied him with indigo ; the result was he built his tube of the national colours, red, white and blue, in successive layers.

Photo by] [*A. Leal.*

THE CREEPING PLUMES.

Jelly-like material found coating water-weeds are seen under the microscope to protrude delicate tubes whose summits spread out into exquisite plumes, which are fringed with still more delicate hairs. These serve as a net for capturing food, and for refreshing the blood.

Another interesting creature may be found attached to the same weed ; it is called the Beautiful Floscule. This charming animal is somewhat funnel-shaped, and fixed by its foot (the narrow end of the funnel) to the weed ; but it builds itself a clear transparent tube, which enables us to see its internal economy and all the processes adopted for its well-being. This tube, like the Brickmaker's, is open at the top, which enables the creature to retreat entirely inside when alarmed. If we watch carefully for a few minutes we shall see a long brush protruded from the end of the tube. This brush consists of innumerable hair-like appendages, so fine as to be scarcely visible singly. When it has come out sufficiently these hairs are gradually separated and thrown wide apart. We then see that they are attached in five groups to the margin of the crown-like edge of the animal's head. This head is shaped somewhat like a beautiful glass vase with a deeply scolloped edge. Or we may liken its appearance to a flower with five petals, and on the apex of each petal a little knob to which the hairs are attached. By means of a slight vibratory tremor smaller creatures are drawn in and devoured. The entire Floscule is so transparent that we can see the whole process of mastication taking place. The eggs —as many as nine in a batch— are laid and hatched in the crystal tube.

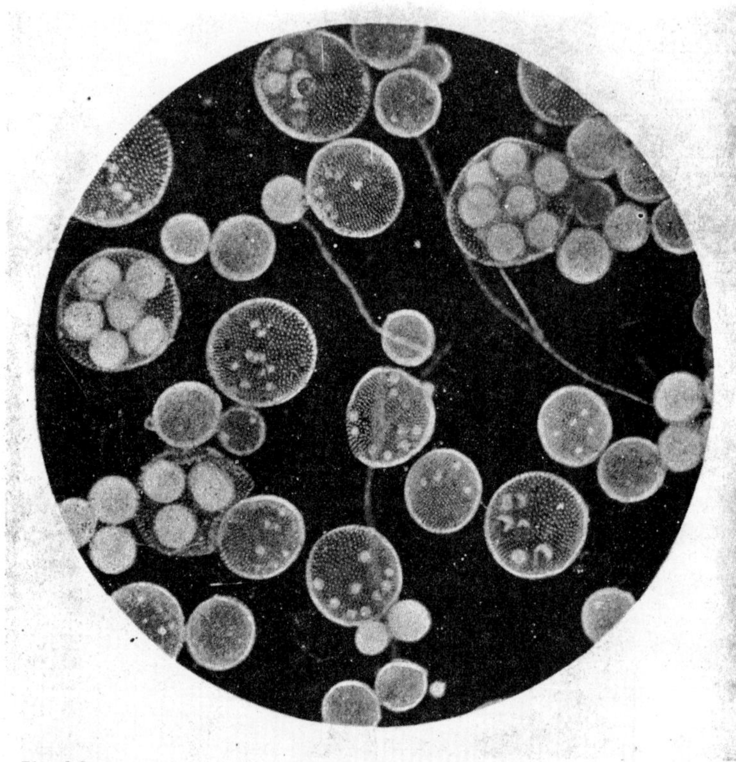

Photo by] [*A. Leal.*

THE REVOLVING GLOBES.

These bright green globes are about one-twentieth of an inch across, hollow, with younger globes revolving within them. They are of singular beauty, and are claimed by botanists to be plants and by zoologists as animals.

There is another charming tube-dweller probably to be found on the same piece of weed, and known as the Crowned Horn. In this case the crystal vase is closed around the neck of the animal, and when the latter withdraws the tube is folded in on itself. It is not smooth and straight, like that of the Floscule, but appears creased, as if it were too large for its inhabitant ; this creasing enables it to fold in and become shorter when the creature withdraws in alarm. The head of the animal is ornamented with five horns, with tufts of short fringe arranged along them. These horns are curved inwards and form a beautiful crown ; hence the name of the animal.

In placing our piece of weed into the bottle to convey it from the pond we probably disturbed some interesting forms which have fallen to the bottom ; so let us examine some of the sediment. Here is one of the lowest forms of animal-life, the Amœba, a transparent speck of jelly-like substance, capable of assuming any form it likes. It captures and devours its food by extending its margin into finger-like processes, and so enveloping the desired tit-bit, literally getting outside it.

Another form, called from its shape the Sun Animalcule, possesses the same power of taking in its nourishment at any part of its body ; but to enable it to procure food it is furnished with innumerable long hair-like extensions of its body, which radiate all around and give it the appearance

Photo by] [A. Leal.

THE LITTLE BRICKMAKER.

This remarkable creature not merely builds the tower in which it shelters, but also the bricks of which it is constructed. The size of the bricks may be estimated by the fact that the tower is about one-thirtieth of an inch in height.

of sun rays. It multiplies by a contraction across the centre of the sphere, and eventually cuts itself into two, and so two individuals are produced in place of one.

Upon the submerged roots of trees surrounding our pond and on the stems of aquatic weeds may be seen a mass of a jelly-like substance, which is a colony of animals of a higher order than those already mentioned, and known as the Creeping Plumes. When the piece of jelly is undisturbed we shall see cautiously projected from it numerous feelers arranged in horse-shoe form around the head of the animal, and together forming a most beautiful plume. Each of the feelers is fringed with fine lashes, which vibrate in rotation up one side of the feeler and down the other, so that they appear to be travelling along. The internal arrangement is well seen through the transparent material of the body. A somewhat similar form is the Crystal Crest-bearer shown on page 230.

In the coloured plate that shows a large number of these microscopic marvels there is a figure of the Bell-animalcule, which has the form of a wine-glass with a very long stalk which constantly coils into a spiral and then straightens itself again. The rim of the glass or the bell, whichever we choose to regard it as most like, is fringed with pellucid lashes that rise and fall rhythmically and create the illusion that the bell is whirling around. The movement creates a vortex, as in the case of the Little Brickmaker, carrying food particles to the creature's mouth. Many variations of this form of obtaining a livelihood are to be found among these pond marvels. A group of them, known as Green Trumpets, are shown on this page. Here it will be seen there is a modification of form, by the bell-shaped end being tapered gradually to the base, so that the whole becomes trumpet-shaped instead of wine-glass shaped.

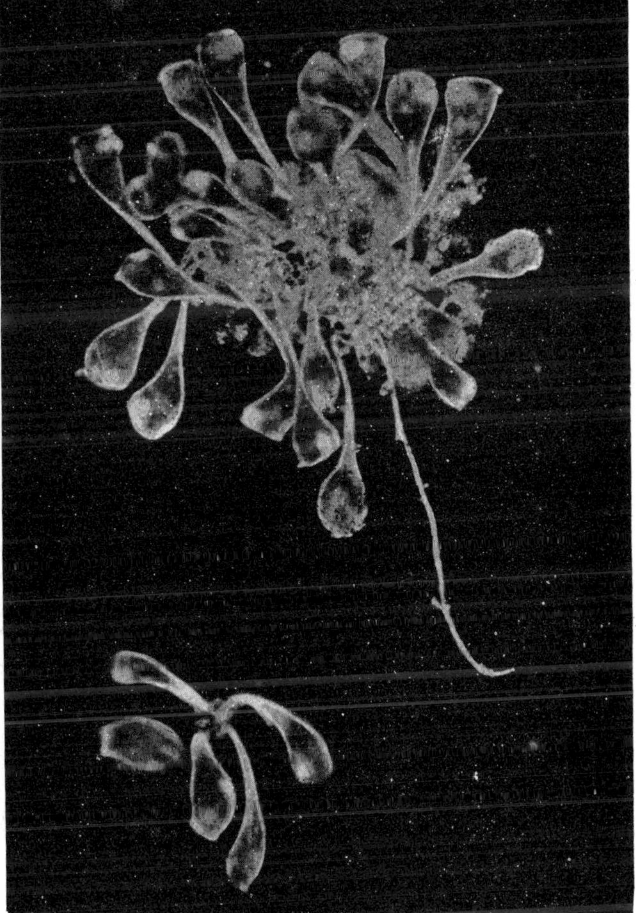

Photo by] [A. Leal.

THE GREEN TRUMPETS.

Each of these brilliant Green Trumpets measures about one twenty-fifth of an inch. They may be found under the leaves of Duckweed. They constantly change their shape from a round knob to a trumpet and back again. They increase in numbers by throwing out buds like plants.

HERMITS AND THEIR MESSMATES

BY EDWARD STEP, F.L.S.

In the article upon a Tree-climbing Crab we spoke of the Robber Crab as a Hermit Crab that had abandoned the habit of protecting its unarmoured hinder parts by thrusting them into an empty shell. The true Hermits may be found commonly upon our own shores, and the commonest of them must be very familiar to all who have spent a seaside holiday away from the crowded resorts, and where there are rocks and rock-pools. In the fascinating employment of watching the varied life in one of these natural aquaria, we are sure to see one or more baby Hermit Crabs scuttling across the bottom, the crab almost hidden by the red or yellow shell into which it has curled its tender tail. As soon as the young Hermits have passed through their infant swimming stage, and assumed the form of their parents, they look about for an empty sea-snail's shell of the right size, and thereafter they spend their days in dragging a shell about wherever they go. They grow and, like other crabs, have to throw off their natural armour as it gets too small for their expanding bodies. With each casting of their natural shell comes the necessity for obtaining a more roomy snail's shell for their hinder parts. The suitable shell may be found; often it is found by two Hermits at once, and both decide that it is the only possible shell worth having. In that case the onlooker will be amused at the conflict that will ensue between them, each in turn gaining a temporary advantage of the other and momentary possession of the shell. At length one of the claimants to the property will feel that he has no chance against the superior tactics of his opponent, and possibly will come to the conclusion that there are much better houses in the market—more solidly built, more sanitary, or more artistically designed and ornamented.

Photo by] [*A. Leal.*

THE CRYSTAL CREST-BEARER

In general structure the Crystal Crest-bearer is much like the Creeping Plumes. The base of the exquisite plumed crest takes the form of two horse-shoes.

At every change of the natural shell this sort of thing goes on; but when the young Hermit has made several changes into more commodious residences, he finds that the rock-pool is too small a world for his ambitious ideas, and one day when the risen tide has come far above his pool, and

Photo by] [*E. Step, F.L.S.*

THE COMMON HERMIT-CRAB AND ITS ANEMONE MESSMATE.

In the photo the task of dragging the whelk-shell and its tall anemone appears a hopeless one, but as the anemone consists mainly of water its weight in the water is not very great. Of old the anemone was thought to be a parasite; it is really a messmate.

Photo by] [*E. Step, F.L.S.*

THE HERMIT-CRAB AND ITS WORM MESSMATE.

A beautiful, many-coloured sea-worm, eight inches long, dwells in the upper part of the Hermit's house, and keeps it clean. At meal-times it creeps along the Hermit's back, and helps itself to particles of food.

Photo by] [*E. Step, F.L.S.*

HERMIT AND SPONGE.

This little Hermit, which lives in deep water, induces a sponge to grow over it, into which it can retreat when necessary. The growth of the sponge does away with the necessity for finding a larger house every time the Hermit casts his natural shell.

made it for the time being a mere little hollow of the boundless ocean, he walks out and down to deeper water. Here he grows apace, and soon we find him full-grown and occupying a big whelk-shell, perhaps four inches long and weighing more than two ounces.

At this stage in his existence we shall find that he is not leading the solitary life usually attributed to human hermits. Our Hermit Crab is a sociable being, though his leaning may not be to his own kind. We are almost sure to find perched up on his house a broad, tall column of flesh surmounted by a spreading nimbus of broad flat rays. This living tower is a sea-anemone, to which the erroneous name of Parasite Anemone was given in days when friendship among the lower animals was not understood, and if one species attached itself to another it was regarded as a case of parasitism. But this anemone is more in the way of being a paying guest than the expensive hanger-on implied by his name.

The Hermit is rather a nervous fellow, and at the slightest alarm pulls himself back into the whelk shell with a jerk. If you pick him up now you will observe what you may not have noticed much before—that his right hand with the pincer-claws is at least twice the size of the left, and that now when all else but the tips of his long walking legs is inside the shell, this big right hand is folded across the opening and forms an efficient street-door. But though the street-door may be sufficient to keep off minor foes, it could not save the Hermit from a big-mouthed fish whose palate teeth would be able to crush the whelk-shell. It is here, probably, that the services of his anemone friend come in. If you handle this anemone you will find your fingers covered with an evil-smelling slime, which is only to be got rid of after much expenditure of soap. It is probable that its odour is objectionable to crab-loving fishes, who are thus warned off by the presence of the anemone. On his part, he would benefit by getting tit-bits when the Hermit is cutting up his meals. Another advantage to the anemone consists of frequent change of scene without any exertion on his part, and the certainty of being conveyed to places where meals are to be had instead of waiting patiently for something to turn up.

But the Parasite Anemone is not the only companion of the Hermit. In the hinder and otherwise empty part of the whelk-shell, a beautiful creature, six or eight inches in length, and known as a Nereis, winds his richly-tinted, iridescent body round the spiral chamber. When meal-time comes this Nereis, smelling the feast, pops out his head over the head of the Hermit and helps himself to a portion. This Nereis is one of a group of sea-worms that form most deadly baits for the salt-water angler, no fish being able to resist it. It is, therefore, a good thing for Nereis to be provided with safe quarters and a share of the Hermit's meals ; and in return it is surmised that he acts the part of a scavenger and keeps the house clean. Just inside the Hermit's doorway may also be found a minute zoophyte, related to the corals, whose exact importance in the household is not so clear, though he may capture more minute and insidious foes that seek to creep in and attack the Hermit's unprotected parts. This zoophyte is a fixture, attached by a stony skeleton to the whelk-shell; but the anemone removes when the Hermit changes his cell, and seeks another host when the Hermit dies. As the stony base of this zoophyte runs up into hard spines at intervals between the soft bodies of the animal, it is clear that it helps to fill up the mouth of a shell that is too large, without blocking the access of sea-water to the interior. A much enlarged portrait of this guest will be seen below.

Prideaux's Hermit is another large species, that has a similar friendship for a different anemone, known as the Cloaklet, from its habit of wrapping itself round the Hermit's cell, commonly in this case a shell of the large Top. Some years ago I had the remarkable experience of obtaining two specimens of this Hermit that, having escaped from crab-pots in harbour without their stony houses, had taken refuge between the stones of the harbour wall, and because the holes were too large, had induced or permitted Acorn-barnacles to build on their heads and arms, and so help the crabs to fill up the doorway. A photograph of one of these remarkable specimens is reproduced. The fact that a Hermit Crab placed under unnatural conditions had met the difficulty in such a way would have been remarkable; but that two individuals should do precisely the same thing under those conditions was rather startling. Were it not that psychologists tell us that animals so low down in the scale of organization can have no glimmering of intelligence, one would be inclined to think they had both followed one line of reasoning, and acted upon it! The photograph on page 235 does not show one of these intelligent Hermits in the cave of its choice. From its situation

THE HERMIT'S ZOOPHYTE COMPANION.

This greatly enlarged picture shows part of the zoophyte colony that lines the mouth of the Common Hermit's portable house. It may also be found entirely covering the shell in which the Downy Hermit is housed—see photo on next page.

Photo by] [*E. Step, F.L.S.*

THE DOWNY HERMIT AND ITS AUGER-SHELL HOUSE.

The Downy Hermit is a small, deep-water crab. The Auger-shell it has selected for a house is mainly covered by a zoophyte which builds hard spines, and these help to protect the Hermit from being eaten by fishes. There are also the stony tubes of certain sea-worms upon the shell, but their presence is probably accidental.

this was not possible at the time; but the attempt has been made by the author to reconstruct the conditions with one of the actual crabs concerned. How long they had been in residence when my attention was called to them is not known, but the fact that the Acorn-barnacles were so fully developed is proof that they had been there some time. It is worthy of note that the anemone ordinarily found wrapping the house of Prideaux's Hermit must secrete some acid, for I have obtained numerous examples in which the hard limy material of the shell had been dissolved away, and only the animal matter was left, like layers of tissue paper, but retaining the shell form.

A smaller species from deeper water is the Downy Hermit, of which a photograph is given, living in the large Auger-shell, which is here encrusted with the spine-like bases of polyps, similar to those described as lining the mouth of the Common Hermit's borrowed shell, and which is shown on page 233. It is as often coated by a small species of compound sea-anemone, which dissolves away the Auger-shell, but continues its shape, so that the Hermit's cell becomes an entirely living one. Several of these smaller species of Hermits might be mistaken at first sight for young examples of the Common Hermit, but that they occur in deeper water. Another of these, shown in a photograph on page 232, induces a little sponge to cover its hinder parts, and the sponge constantly growing does away with the necessity for the crab changing its home. Yet another deep-water species—the Hairy-handed Hermit—is covered by a sea-anemone, which has never been found growing apart from the crab, and the crab has never been found apart from the anemone. This is one of the most perfect forms of animal comradeship known.

The examples cited—and they might be increased in number—are all drawn from British waters, but all over the world the race of Hermit Crabs exhibits the same characteristics. Whether it was that Nature was niggardly towards the first of the family and omitted to provide their hinder parts with the shelly covering proper to crabs, or that the founder of the dynasty took to thrusting this part into shells and set a fashion that has led by disuse to the absence of crab-shell, must remain for the present a mystery. But in general habits, wherever we may find them, the Hermit Crabs have had to adopt similar methods to meet their present-day necessities. Some have entered into partnership arrangements under which an anemone provides them with a kind of blanket, without the interposition of whelk or other shell. The Robber Crab that has taken to a life on land, is only a huge Hermit, and as such he is compelled to protect his tender hind-body by thrusting it into a hole in the ground. Long persistence in this habit of thrusting the hinder part of the body into a spiral shell has brought about a corresponding twist in the body itself to make it fit comfortably. With this change several of the hinder feet have disappeared from several of the species, and the final pair have been modified so that they now form a hook by which the crab maintains a hold on its borrowed shell, and is able to drag it about to all appearance as though it were a part of itself. Of course, with such a creature there are at times exhibitions of strange likings for out-of-the-way fashions in houses. For some reason Hermits have appropriated to this use such things as glass bottles, and clay tobacco-pipe bowls, that have dropped overboard from ship or fishing-smack and fallen to the sea-bottom. It cannot be said that the form of these substitutes for spiral shells is at all convenient for the purpose. My private impression is that they have not seriously been adopted with a view to lengthy residence in them, but merely tried on experimentally to see how such an arrangement would work. They have probably been found by the naturalist in the act, and it has been assumed that they had been living under such uncomfortable conditions for some time.

Whatever the psychologists may say to the contrary, it is difficult to avoid seeing in these habits, and these business-like arrangements with their partners, some faint beginnings, at least, of the intelligence that is seen in its higher developments in the dog, the horse, and man.

Photo by] [*E. Step, F.L.S.*

PRIDEAUX'S HERMIT-CRAB.

This Hermit usually lives in a shell that is wrapped around by a sea-anemone—the Cloaklet—but this one has lost his house and had to take shelter in a hole inshore. Note the Acorn-barnacles he has allowed to build on his head and arms.

Photo by] [*J. H. Crabtree.*

THE HOUSE-FLY.

Although one of the commonest objects of our everyday life, there are really few persons who know anything about the structure and habits of this dangerous pest. The accompanying article and photographs will help to supply the deficiency.

THE HOUSE FLY

BY E. J. SPITTA, F.R.M.S.

CONSIDERABLE attention has of late been directed to the possibility, if not the probability, that this household pest can be a most dangerous means of carrying infection.

Before going further, it should be at once mentioned, that, although we have spoken of *the* House Fly, as if one type only existed, there are at least four different kinds that are popularly regarded as the same. Of these, three are very similar, only differing for the most part in size and slight modifications in the wing; such differences being only understood and appreciated by the expert. Each of them has a suctorial mouth, too, that is to say, has a tubular tongue for obtaining its food, whereas the fourth and largest variety of all, the Stable Fly, is of quite a different construction—at least, so far as its head is concerned; being provided with an apparatus for biting as well as for sucking. This consists of lancets and a tongue. The first thing, then, to be kept in mind is that the House Fly mostly met with does not bite, because it cannot, and that it is only the much rarer Stable Fly which is provided with instruments of torture. This statement, it should be mentioned, is not common knowledge, nor, indeed, very possibly, is it in agreement with the opinion of many, who, on the contrary, very confidently assert that House Flies *do* bite, *especially towards the end of the season.* But it is in this last sentence that the controversy is explained. The Stable Fly only puts in an appearance in houses towards the end of the summer, and its visits last for a short period in comparison with those of the House Fly: hence, as this Stable Fly can, and does, bite, and is usually mistaken for the House Fly, biting powers are attributed to the latter which only belong to the former.

All varieties have much in common to be explained, but we will first discuss the differences, commencing with a description of the sucking apparatus of the House Fly. It will be seen extended in the photograph on page 245, which represents the entire female insect, photographed from a specimen prepared as a slide for the microscope, and therefore one compressed as flat as possible in the mounting. Although this is the only way to get the whole insect into focus for the purpose of photography, still it has the disadvantage of showing the proboscis in a manner that suggests to the student it is a direct prolongation of the body. This, however, is not the case; it really hangs down at right angles from it. [In another picture the lancets of the Stable Fly are shown with their sheath; these hang down in just the same way and serve to explain the arrangement.]

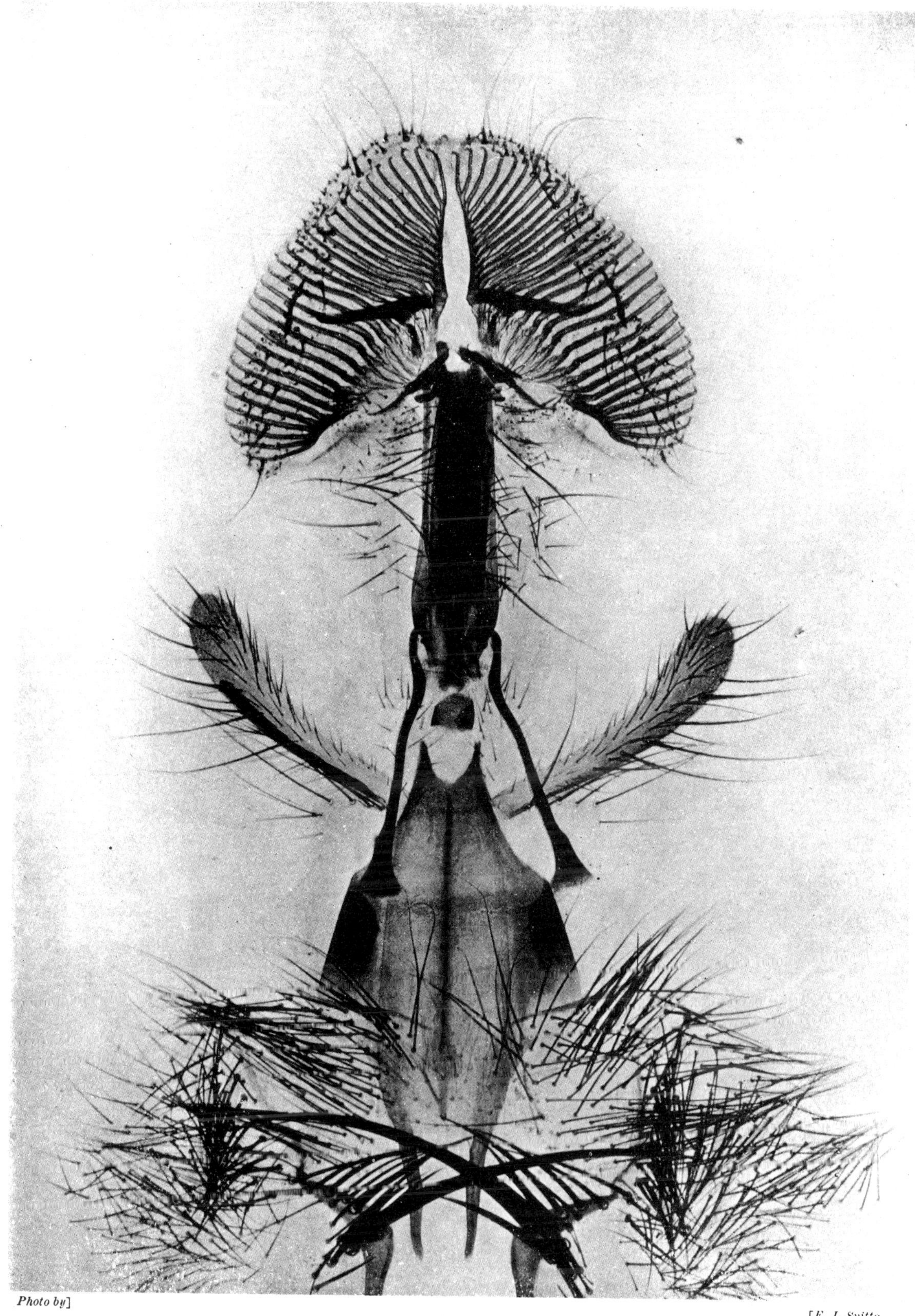

Photo by] [*E. J. Spitta.*

SUCKING APPARATUS OF HOUSE-FLY.

This is the thick process beneath the head of the Fly with which it makes continual dabs at our food as it walks over the table. It is here flattened out for examination under the microscope, which shows it to consist of many delicate tubes opening out into two larger ones. Through these tubes all the Fly's food is sucked up, and passed through the larger tubes to the mouth.

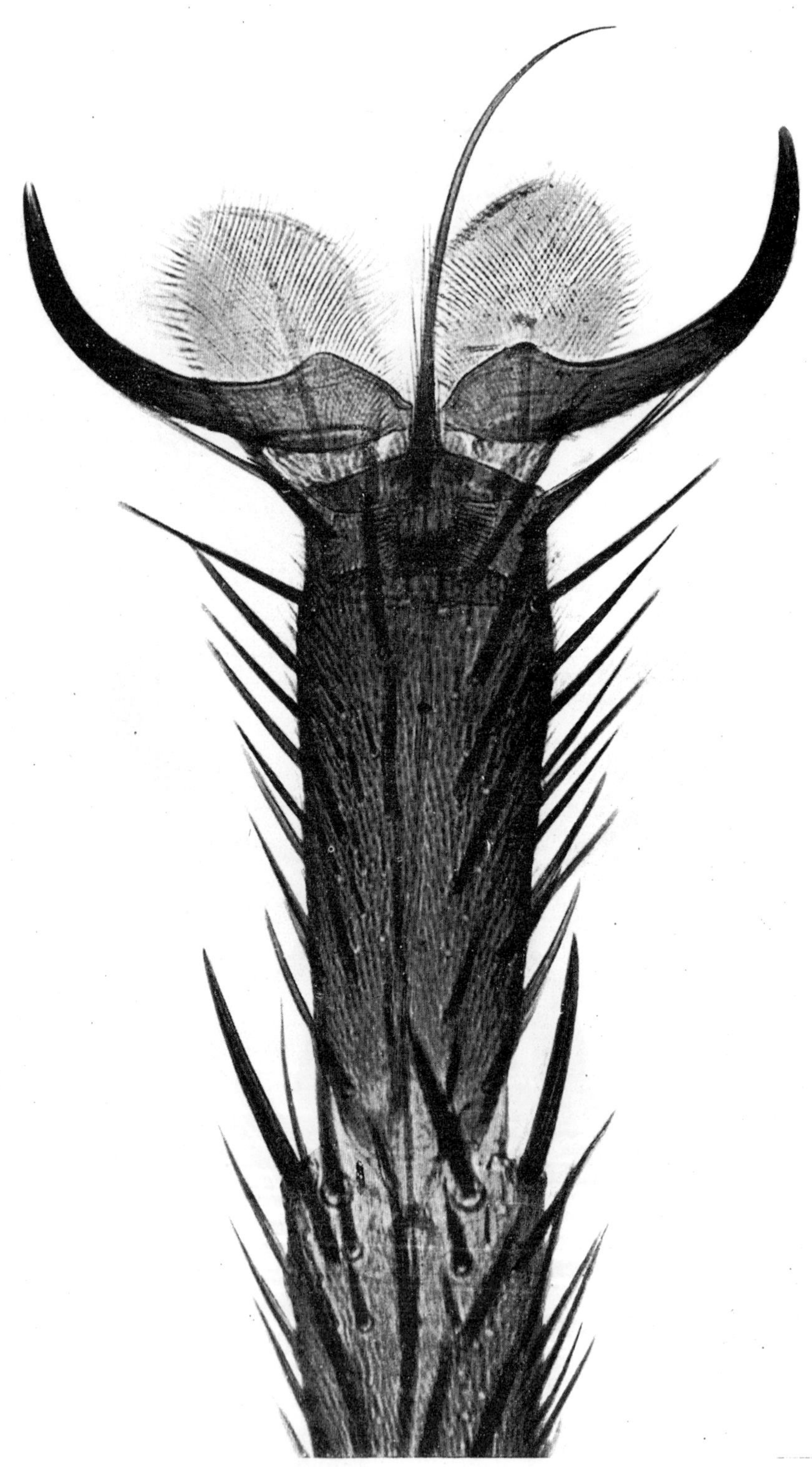

Photo by] [*E. J. Spitta.*

THE FLY'S FOOT.

The foot of the Fly consists of two horn-like projections attached to the end of the leg, at the base and sides of which are seen two pads. It is by means of these pads that the insect is able to hold on the ceiling or upon glass, even when upside down. The details of these pads, more highly magnified, are shown on later pages.

The use of this position in Nature, becomes apparent with but little consideration. It is for a two-fold purpose; first to allow the fly to commence taking its food at once, without change of posture, and also for the purpose of allowing room for its eyes to look around in all directions to warn it of approaching danger.

The eye of these insects, being compound—by which it is meant it is composed of numerous facets (something like four thousand), each furnishing an image—can see simultaneously in almost any direction; which explains why it is so very difficult to catch a fly.

The tongue, as a whole, can be studied by examining the photograph, where it is shown lying in front of and between the two clubs. It is a two-lobed organ, both lobes simultaneously acting as one, and being capable of closing or collapsing like the shutting-up of an umbrella, and of withdrawal, as a snail retires into its shell. The entire organ is really enclosed in an exceedingly fine and delicate membrane, which for some reason is not easy to see or to photograph in the case of the House Fly. It is better shown in the tongue of the Gadfly, where the delicate membrane, broken away in parts, is distinctly visible covering the top of the lobes. The object of this delicate covering undoubtedly is to prevent dirt getting into the suctorial tubes, which are shown very plainly in the more highly-magnified photograph of the same organ. All these tubes, it will be seen, empty into two main collecting tubes, one for each lobe, by which the nutriment enters the system of the fly. Before proceeding further, it should be noted that the two black wire-like-looking objects, one curving to the right and the other bending to the left, at the lower part of the lobes, are *not* tubes, but the supports of the tongue—like the framework of the umbrella—and which serve to assist in the opening and closing of the little organ. The extended organ is well seen in another photograph, which was prepared by a special method *without* pressure.

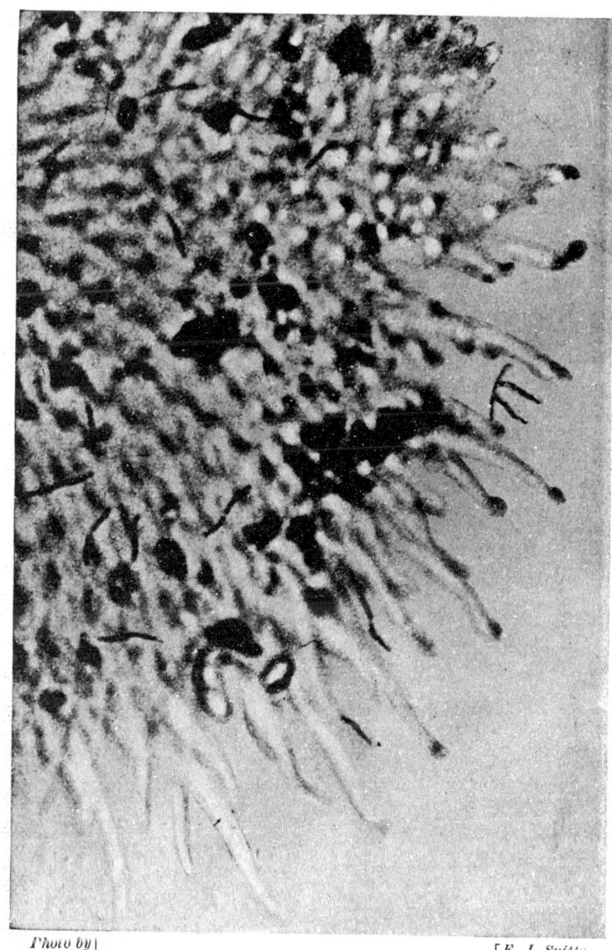

Photo by] [F. J. Spitta.

PART OF FLY'S FOOT WITH BACTERIA.

A portion of the pad of a Fly's foot showing how this insect conveys Bacteria, often of a pernicious kind, to our food. The black patches and streaks are the Bacteria.

The tubes are ingeniously made, and merit a little close attention. Even a cursory glance at the photograph shows that they are not solid, like pieces of brass tubing, but that they are formed by a series of juxtaposed rings or ribs. These are believed to be formed of a gristle-like substance so as to be somewhat elastic. Further, close inspection of the same figure will reveal the fact that each rib (in the larger tubes) is *not* a complete ring, but only about three-quarters of the same—like a horse-shoe, in fact. To convert the string of incomplete ribs into a continuous tube, a membrane of exceeding delicacy not only covers the spaces between the rings, but stretches across their open ends. A completed tube then, in section, would appear like a horse-shoe, where the heels have been

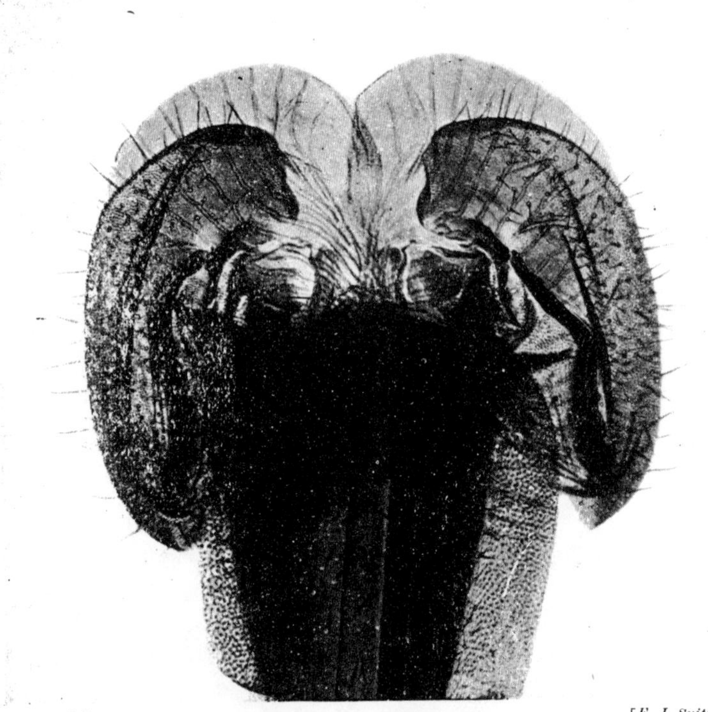

Photo by] [*E. J. Spitta.*

THE BASE OF A GADFLY'S TONGUE.

This photo shows the delicate envelope of the tongue, which is not easy to see or to photograph. It prevents dirt getting into the fine tubes.

joined together by a membrane. But this is not all. The free ends of the gristle-rings are curiously fashioned, for whilst one appears to end closely resembling a pitchfork, the termination of its neighbour looks like nothing else than a spike. The object of this arrangement is believed by some to be as follows: It must be recollected that the tongue has to be opened and closed with great rapidity; there must not be any hitching, as sometimes occurs in opening an umbrella, for the insect might lose some food thereby. Now when the tongue is closed, it is easy to understand in the collapsed state of the organ, the membrane that we have explained as forming part of the tubes must of necessity be folded over into pleats by the compression, thereby effecting economy of space. It is thought, then, that as in compression the spikes of the tubes of one side all the way down drop into the forks of the other side so as to pack away closely, in opening the former serve the additional purpose of rapidly and somewhat forcibly freeing the membrane, should any of the folds accidentally attempt to stick together. This is a beautiful thought and a beautiful arrangement.

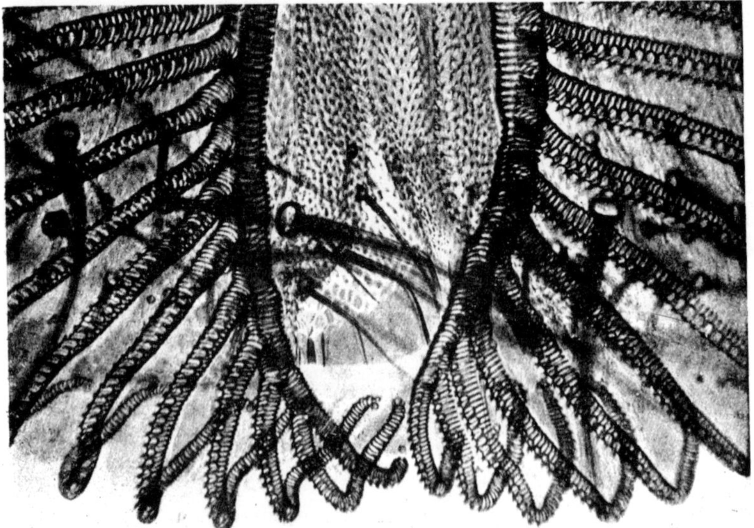

Photo by] [*E. J. Spitta.*

THE LOWER EDGE OF THE FLY'S TONGUE.

This beautiful photograph shows the ends of the fine branch tubes through which the Fly's nutriment is sucked up, and the two main tubes into which they all connect.

As the second and third varieties of House Flies have each sucking mouths of a similar nature to the common kind, we can now pass on to the consideration of the biting apparatus of the Stable Fly. The entire fly is shown, but in life the apparatus is seen *in situ* hanging from the head of the fly at right angles to the body.

Two very delicate lancets, so transparent as to be exceedingly difficult to photograph, can just be seen leaving the sheath near the lowest part

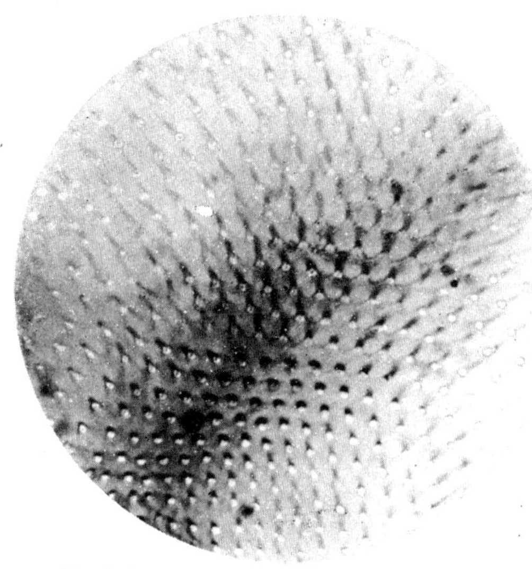

Photo by] [*E. J. Spitta.*

PART OF FLY'S FOOT.

This photo shows a portion of the Fly's foot under the microscope. The dots are the tubes with sucker ends which enable the insect to walk up glass or over the ceiling.

of this photograph; they are better seen, however, in the photograph which is on a larger scale and has been made with the special object of showing the sheath and lancets alone. In the actual use of this apparatus, the bite is made with the lancets, which cut out a hole large enough in diameter to allow the exit of sufficient blood for the sucking tongue to be placed into.

Space will not allow us to deal with an explanation of the clubs seen on each side of the tongue, which in all the varieties of flies are probably sense organs; neither will it permit much being said concerning the wings, although passing mention should be made of the fact that differences in the position of the veins (as the linear markings are called), their arrangement and general appearance, offer to the expert a means—in addition to that of size—of distinguishing the House Fly from the others that are frequently confused with it. Mention might be made here also—for want of a better place—of the curious fungus disease so fatal to the House Fly, which causes them to adhere to window-panes and such like. When closely examined, the fly appears as if invested with a white substance which not only covers the body more or less completely, but even spreads to the wings and to the glass adjacent. A portion of a wing highly magnified is shown, in which this creeping fungus can be seen spreading itself as a white, transparent network, winding its way amongst the microscopic hairs, reminding one of clover in the way it pushes its wire-like fibres between the blades of grass upon the lawn.

The wings move with great rapidity in flight. Messrs. Kirby and Spence quote the substance of an anonymous writer, who says that he calculates that a House Fly in ordinary flight makes about six hundred strokes—which carries it about five feet—every second. If alarmed, however, the velocity can be increased to about thirty to thirty-five feet per second. In this space of time, the writer adds, "a racehorse would clear only ninety feet, which is at the rate of more than a mile a minute." The little fly, then, in his swiftest flight, goes, in the same space of time, about one-third of a mile. In conclusion, they further add: "A little calculation seems to show that did the fly equal the racehorse in size, and retain its present powers in the ratio of its magnitude, it could traverse the globe with the rapidity of lightning."

The breathing apparatus of the fly much resembles that of other insects. One of the chief anatomical differences between a mammal and an insect is that in the former the muscles lie outside the bones—as in the human subject, for example; whereas in the insect the internal organs are enclosed in a sac-like body, made, in many instances,

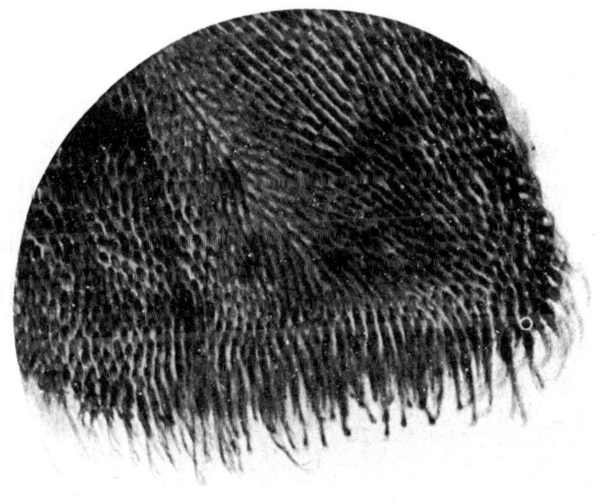

Photo by] [*E. J. Spitta.*

FLY'S FOOT.

Portion of the cushion of a Fly's foot, less highly magnified than the previous photo. It is in between the little prominences that Bacteria are carried about.

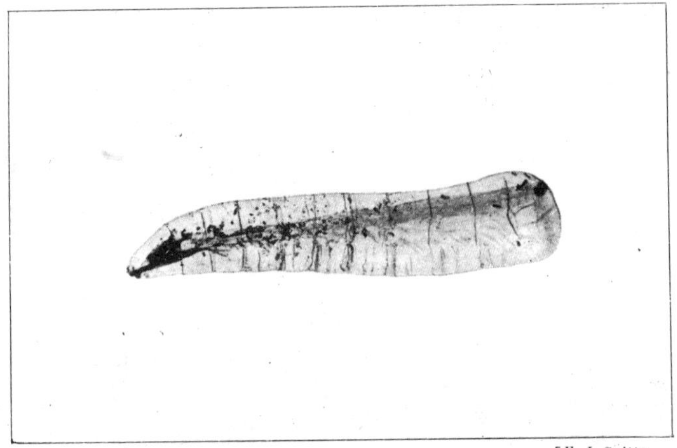

Photo by] [*E. J. Spitta.*

THE GRUB OF A FLY.

This much enlarged photo of the "maggot" which issues from the Fly's egg shows the true form. The mouth is at the pointed end and the digestive tract is shown as a darker band.

of a horny substance closely resembling chitin. In the mammal air is taken in by the mouth, and after being led through tubes (the bronchial tubes), enters the lungs, which may be described as consisting of an abundance of little sacs, capable of opening and closing, about which the small blood-vessels encircle. The air, in passing through these air-sacs, re-oxidizes the blood in the adjacent blood-vessels by simple contact, and thus purification of that fluid is effected. In insects, however, the arrangement is modified. Holes in the sides of the body of the insect or between some of its segments (called spiracles) exist, which lead into tubes that pass in all directions, and are all of different sizes, throughout the entire body. The blood, however, is not contained in any vessels, but is loose, and circulates freely all about the contents of the sac, and playing over these air-tubes, wherever they may be, causes re-oxidation of that fluid. The spiracles, or little holes, are more highly developed and easier seen in the Water Beetle than in the case of the Fly. One of these spiracles was given in an earlier page to show how Nature adopts an ingenious method for keeping out dirt or any foreign matter from the tubes by stretching across the entrance numerous twig-like little branches, closely associated together, and which act as a kind of mechanical filter. The tubes from these holes, which pass throughout the body in all directions, are beautifully made, consisting of spirally wound fibres, the spirals themselves (figure of 8 in form in the large tubes) being designed to be of a somewhat spring-like nature, so as to assist in the distribution of the air as the insect draws in and lets out its breath.

We have never seen the result of any experiments for ascertaining the rapidity of breathing in the case of the House Fly, but in that of the Bee it has been estimated at about forty when at rest, but when pushed, to reach from one hundred and ten to one hundred and sixty times a minute. It is not improbable the breathing of the Fly is much the same.

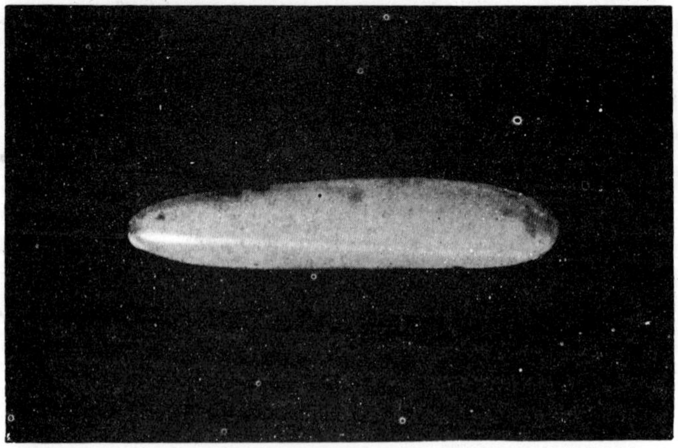

The egg-laying apparatus of the Fly (called the ovipositor) is a curiously-shaped arrangement, the object of the different parts not being accurately known. It is somewhat long and stout, being especially made for the purpose of pushing aside bits of straw, etc., as the insect deposits its eggs deep down in the warm parts of a heap of stable manure. Here they hatch. On this page is shown the egg of a House Fly, whilst on the opposite page is a group of eggs which, although often sold as those of the House Fly, are really derived from some other species. The eggs of the Stable Fly

Photo by] [*E. J. Spitta.*

EGG OF HOUSE-FLY, MAGNIFIED.

The egg is very long in proportion to its width, and quite devoid of ornamentation. It hatches in twenty-four hours, producing a maggot like that shown above.

Photo by] [*A. E. Smith.*

GROUP OF A FLY'S EGGS.

These are the eggs that are commonly shown under the microscope as those of the House-fly, but they are those of another species, as will be clearly seen by comparing them with the photo on the page opposite. These eggs are much more beautiful than those of the House-fly, owing to the wing-like expansions and the honeycomb pattern.

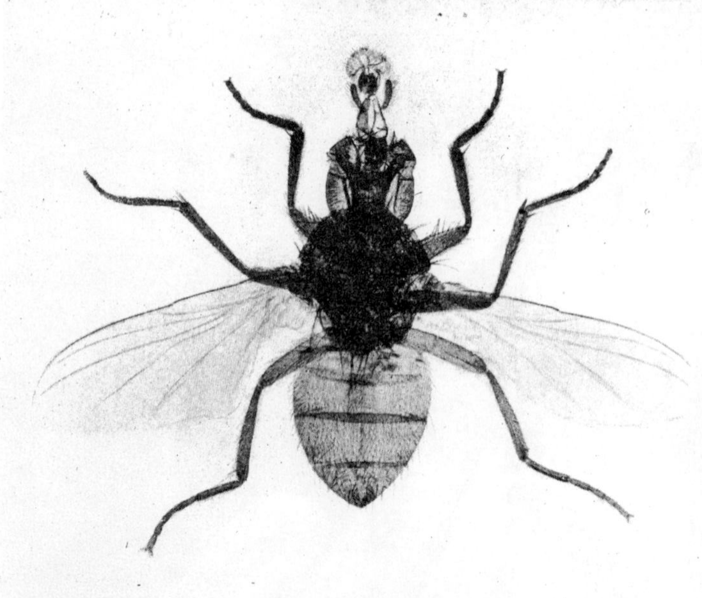

Photo by] [*E. J. Spitta.*

THE HOUSE-FLY.

The Fly is here shown sufficiently magnified to make its parts distinct. Note the spread-out "tongue" and compare it with the corresponding part in the photo below.

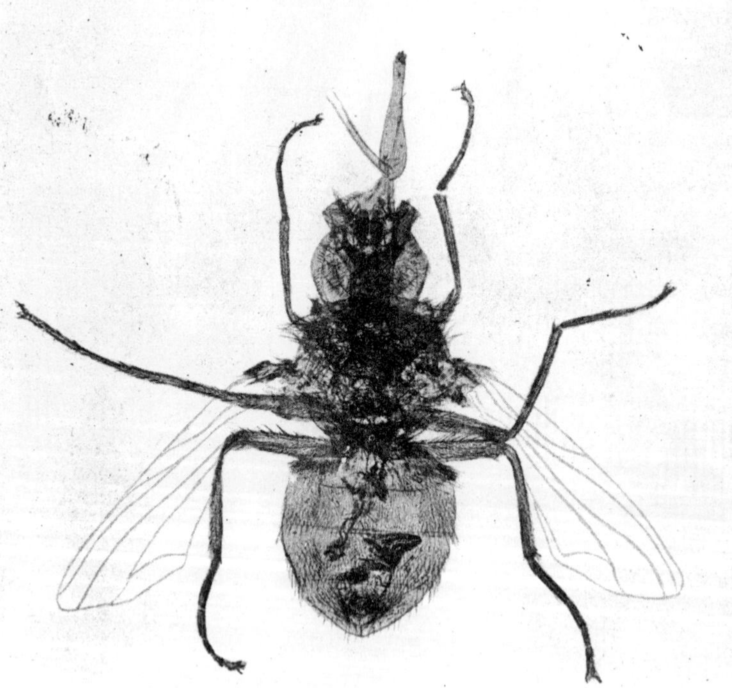

Photo by] [*E. J. Spitta.*

THE STABLE FLY.

The great difference in the "tongues" of the two Flies that are so frequently confused is here made very clear. The lancets with which this Fly stings are shown a little to the left of the tongue.

are difficult to get, for when the Fly is obtained, it is not easy to keep alive, as it must have fresh blood upon which to live.

The egg of the House Fly hatches in about twenty-four hours, when the grub puts in an appearance. It remains as such about seven or eight days, changing its skin several times as it increases in size, until reaching finality when about half an inch in length. The grub now changes into a chrysalis, which is smaller than the grub, being only about a quarter of an inch in length. Having remained three or four days, the case cracks, and out emerges the complete Fly. Mention should here be made that it is a misconception to suppose that flies grow. No growth takes place in their life save at the grub period, when they change their skins, as already explained; hence the smaller variety of fly, occasionally and rather rarely seen upon the window-panes, is not the House Fly in the act of growth, but a smaller species.

Before quitting the subject, we must not omit to mention the marvellous foot of this little insect, shown on page 238. Here it is seen to consist of two horn-like projections attached to the end of the leg, at the base and sides of which are seen the two pads. It is the pads which are so wonderful, and their construction merits more than a passing attention; for it is by means of these little arrangements that the insect is able to hold on the ceiling or upon glass, even when upside-down. Each pad is like a cushion, and from it project numerous small cone-like-looking objects, each of which ends in what at first sight appears to be a minute hair, in some instances surmounted by a little pad.

In times gone by it was thought the Fly exuded from its *pad* an adhesive fluid, which glued it on the ceiling or pane of glass; but this is believed now not to be quite correct. Since more perfect microscopes have been available, it has been found that each hair is really a tube; and hence each may be considered as a perfect suctorial arrangement, complete in itself. Experiments, however, of somewhat recent date most certainly show that it is highly probable flies can at will pour out some sort of adherent material—very possibly from the tubes themselves; for they have been, and indeed frequently are, found with their feet only firmly adherent to the glass, although quite dead, when all suctorial powers must necessarily have ceased to exist.

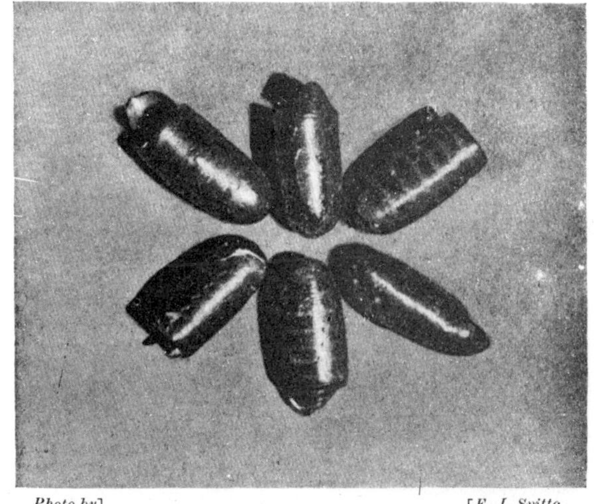

Photo by] [*E. J. Spitta.*

THE INTERMEDIATE STAGE, MAGNIFIED.

When the grub is full grown its skin hardens into a brown oval case in which the grub changes to the chrysalis. A few days later one end is thrown off and the complete Fly emerges.

The foot, however, has unfortunately one more very special interest—a painful one to the human subject; for it is by this means that it certainly can carry germs of infection. The bacteria of disease are known to be exceedingly small, so small, indeed, that many thousands can be comfortably located on the point of a needle. Hence, if the Fly alights on any poisonous matter, it is quite conceivable how bacteria can be lodged either upon the hair-like suckers themselves or between them. A photograph has been devised to show this, which depicts part of a pad magnified at five hundred diameters, upon which a photograph of the bacilli of consumption (for convenience of representation magnified twice as much) has been cast. The number of germs that could be carried even by a single pad of one foot of a House Fly is therefore painfully evident. It shows, indeed, that the House Fly is a living terror to mankind, and it cannot be wondered at that the Medical Officers of Health in all districts are now making such a stir for the extermination of one of the most dangerous pests that can occupy our houses.

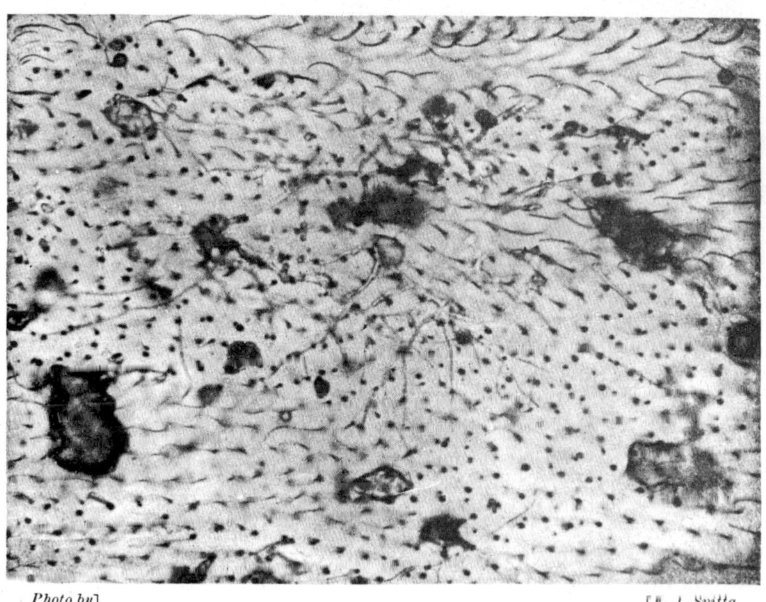

Photo by] [*E. J. Spitta.*

PORTION OF FLY'S WING, HIGHLY MAGNIFIED.

The pale threads and dark patches in this photo show the minute fungus which attacks Flies in the autumn and kills off thousands of them.

A MUSHROOM PARASITE

It is not an uncommon thing for a fungus of one type to attack one of a distinctly different type. When any of the larger mushroom or toadstool forms begin to decay they are frequently overrun by one of the numerous species of mould, such as attack most decaying matter and help in the final break-up of its tissues. But it is quite a rare occurrence for one of the mushroom type of fungi to attack another species of the same group. Yet there are three closely-related species, known as Russules, which appear to be the exclusive victims of hereditary feud, for there are two other and smaller species of mushroom, known as Nyctalis, which will grow nowhere else but on one of these three Russules.

Photo by] [*E. Step, F.L.S.*

MUSHROOMS GROWING ON A MUSHROOM.

The clusters of small white Mushrooms are of a totally distinct kind from the larger Mushroom upon which they grow as parasites. They do not grow upon the ground, and have only been found upon this and one or two other Mushrooms.

The Nyctalis, which is quite small when compared with its four-inch-topped host, is of a light grey colour and issues in clusters from the upper side of the Russule's cap. Sometimes, however, it hangs from the gills on the lower surface of the cap; and in the example photographed they are growing from both parts, whilst another is sprouting from just above the base of the stem.

The Nyctalis is a distinctly rare fungus, and the fungus-hunter who comes upon it in the woods for the first time will be very pleased with himself. It gives out a very distinct and penetrating odour, which, once smelt, may be instrumental in tracking it down on other occasions.

THE LONG-NOSED MONKEY OF BORNEO

BY SIR H. H. JOHNSTON, G.C.M.G.

Few monkeys present such an extraordinary appearance as the Long-nosed Monkey of Western and Southern Borneo. In the female and young of this species the nose is much smaller in its projection from the face, and in the immature forms of either sex the short nose turns up like the caricature of a snub-nose among humans. Now there is a genus of "snub-nosed" monkeys closely allied to them, with ridiculous turned-up noses, projecting not more than an inch from the muzzle. These range from Eastern Tibet and Central China to Annam. At some time or other a species of this group must have found its way to Borneo when the huge island was still connected by land with the Malay Peninsula, for the more striking forms among the beasts and birds now existing in Malaysia may be traced back to an evolutionary area in the Eastern

Photo by permission of] [*Sir H. H. Johnston.*

LONG-NOSED MONKEY.

This extraordinary example of the Proboscis Monkeys is a native of West Borneo, where its bare pink face has led the natives to call it "The White Man." The long tail and the lower part of the back are white, which affords a strong contrast with the chestnut and yellow-brown of the upper parts of the body.

Photo by] [*Fuller & Osborne.*

THE GLOBE FISH

This fish of remarkable form is brilliantly coloured, and covered with spines. At ordinary times it is much more fish-like, but when danger threatens it puffs itself out and erects its spines.

Himalayas, Assam, China, or even Tibet. Here it underwent isolation and an independent development, which produced in the male not only a very bright and peculiar coloration, but exaggerated the projection of the nose until it grew into the drooping proboscis which assumes such a quaint form in the unusually fine specimen of a fully developed male shown in the accompanying illustration—a specimen from West Borneo in the Natural History Museum at Munich (Bavaria).

Nature is constantly repeating herself, and seems ever and anon to try experiments in peculiar developments which lead to nothing—or to nothing of much apparent use—until exactly the right thing is hit upon ; and then the new idea runs away with Nature and is carried out in the fullest possible manner. The human genus is a bundle of Nature's plagiarisms of her earlier manner. The upright position and bi-pedal walk have been tried in reptiles and in birds and in many types of mammals ; notably in the far-back lemur ancestors or collaterals of man. The projecting nose (reaching its highest development in the white man) is no new idea, but possibly also showed itself in certain extinct lemurs, and has attained a comical distinctness (if incomplete and without any cartilaginous bridge) in the East Asiatic Proboscis Monkeys.

The male Long-nosed monkey is very remarkable for his coloration. The bare face and drooping proboscis are pink, like the face of a white man ; and so we find this monkey is called " White Man " by the natives of West Borneo. The top of the head is bright chestnut, the whiskers, shoulders and chest are ginger-colour, the hair of the arm-pits being white and that of the upper arm chestnut, while the lower arm is grey fringed with white. The hind limbs are grey, and the hair of the hind parts is reddish, and that of the back deep, glossy chestnut-red. In vivid contrast with these tints of chestnut, yellow-brown and ash-grey are the lower back and tail, which are pure white. The Long-Nosed Monkey lives mainly on leaves ; and the adult male grows to a considerable size—as large as an average-sized baboon.

THE GLOBE OR PARROT FISH

BY FRANK T. BULLEN.

AMONG all the multitudinous mysterious denizens of the sea with whose habits we are acquainted, only to become more mystified than ever as to the meaning of those habits, none are more curious than the numerous varieties of a fish whose popular name is given above. It is exceedingly widely distributed, being found in both tropical and temperate zones ; but in its greatest profusion frequents tepid waters and the vicinity of coral reefs.

Its peculiarities are many, but one of them may be taken first as the most important, viz., that it is both poisonous and innocuous according to its habitat, this fairly well-known fact preventing its ever becoming a food fish. As a general rule, the appearance of a fish will determine its acceptance as an article of food, some of the sea-creatures being so repellent in their appearance that it needs considerable courage or pressure of hunger to attempt eating them. To this rule, however, there are several important exceptions, such as the John Dory and the Flat-head, of Australia. Indeed, it is difficult to assign reasons sometimes for the rejection of fish as food. I once caught some flying gurnard in St. Vincent which were almost the most beautiful fish I have ever seen ; but even the negroes refused to eat them, without knowing why.

Photo by] [*Fuller & Osborne.*

THE GLOBE FISH.

The appearance of the distended fish when viewed from the front. The distention of the body causes the protrusion of the four large teeth, which then somewhat resemble the beak of a parrot.

Another fairly safe rule which is followed, or used to be, among seamen is that of rejecting as food all very gaudy fish. I need not say in the tropics, because the temperate seas are notable for the sober colouring of their fish. This simple rule is generally observed by natives, from whom, indeed, seamen have obtained it, and it effectually cuts off the Globe Fishes, because they are nearly all of very brilliant hues. How brilliant may be gathered from Mr. Charles Darwin's remarks in "The Voyage of the *Beagle*," that when handled, one of these fish emitted from the skin of its belly a most beautiful carmine-red stain, which remained permanently bright and irremovable from the fabric upon which it fell.

Another feature of this most curious fish calling for wonder is that some of its varieties are covered with spines almost as thickly as a hedgehog. These spines are quiescent as long as the fish is unmolested; but if it is caught and necessarily handled, they then come into play. But only in connection with another peculiarity, which is perhaps the greatest of all shown by this fish. I allude to the power which it has of self-inflation. On many occasions I have caught this fish and thrown it on deck, when immediately it has emitted gurgling sounds such as you hear when holding an empty bottle, open neck upwards, under water. Coincidently with this, the body of the fish gradually became globular, the spines stood erect, and having attained its full distension, the creature remained thus. In fact, the passing of its life did not affect its appearance; after death it still appeared distended and prickly.

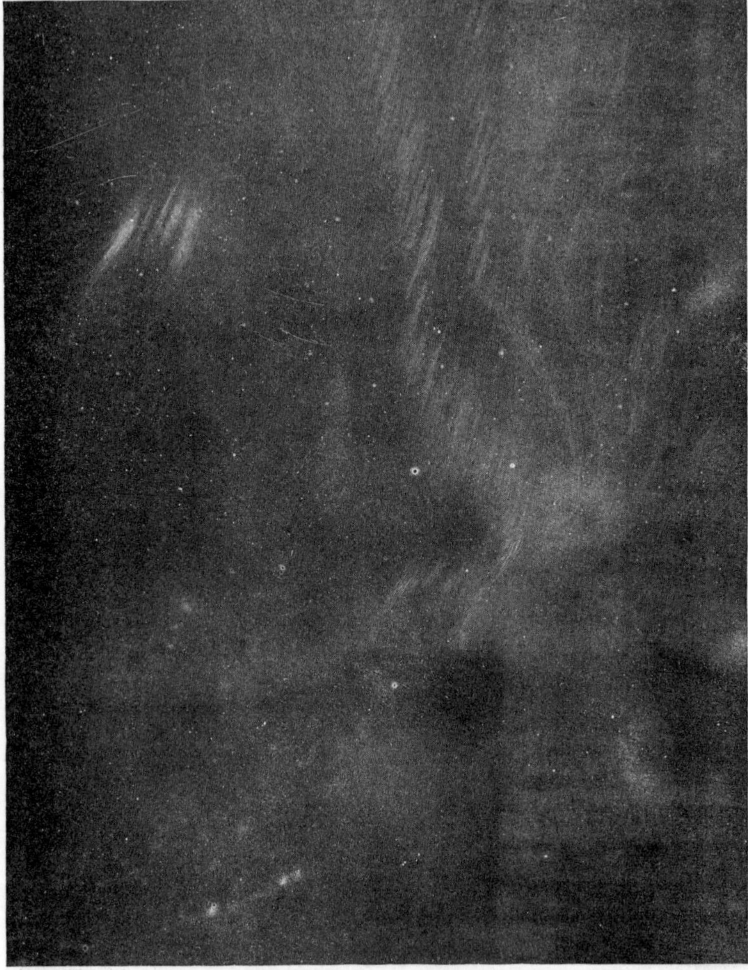

[*By Max Raebel.*

NORTHERN LIGHTS.

The variation in the colours of these Lights is probably due to the breaking up of the rays in the aqueous vapour.

The trivial name Parrot Fish is really unsuitable, because it only derives from the beak-like appearance of the teeth when the fish is inflated. But those teeth are the common inheritance of many tropical fish, and are identical in their appearance with varieties entirely distinct. They are like human incisors, but far more powerful; and I have known them to sever copper wire.

THE AURORA BOREALIS, OR NORTHERN LIGHTS

BY MAX RAEBEL.

MOST of what we now know of the nature and origin of the Aurora Borealis we owe to the Norwegian scientist, Professor Birkeland, who has succeeded, aft r years of observation and experiment, in reaching definite hypotheses on the subject.

In this paper I shall content myself with giving a short account of the scientific research on the subject of the Northern Lights, adding a few particulars of my own observations.

[By Max Raebel.

THE AURORA BOREALIS AT TRONDHJEM

The Northern Lights reach only the more rarefied upper air of our atmosphere, and have the property of condensing aqueous vapour. Their colour is generally green, but sometimes violet and red.

[By Max Raebel.

NORTHERN LIGHTS.

These were formerly thought to foretell a hard frost, but M. Raebel's observations lead him to quite the opposite conclusion: in most cases milder weather follows immediately after a display.

According to the latest views, the Northern Lights waves are negative electric rays which the sun, and especially the solar spots, send out into space. These are what are called "kathode" rays. They bear a strong resemblance to radium rays, and may therefore be regarded as a kind of immense electrons. It has long been known that these rays are closely connected with the sun-spots, as we find in both the same eleven-year periods. When a ray of this kind comes into the neighbourhood of a magnetic expanse, as, for instance, on the earth in the vicinity of the magnetic pole, the ray will be, so to speak, sucked up by it, as Birkeland expresses it. If the course of the ray, on the other hand, is not directed towards such an expanse, it will be diverted from its course, and in this way the most varied forms arise. As our earth is very small in comparison with the universe, great expanses may very well be affected at the same time by the magnetic disturbances, and Birkeland has discovered, by placing instruments at Potsdam and Bosekop in the north of Norway, that the variations in the magnetic elements were alike at the two places at the same time, only that the oscillations were more intense in Norway.

The Aurora Borealis causes magnetic disturbances which are called magnetic storms. The theory has been propounded that the earth derives its magnetism from the Northern Lights.

We owe to Professor Stormer the mathematical calculation of the course of the rays and of their length—a portion of the subject in which the French scientist, Poincarré, has also done good service.

In one of his experiments which consisted in passing an electric current through a bottle filled with rarefied air, Birkeland has produced light-waves (corkscrews) similar to what may sometimes be seen in the Aurora Borealis.

The Northern Lights reach our atmosphere, but only the more rarefied upper air. They have the property of condensing aqueous vapour, and there are clouds whose origin is attributed to the Northern Lights. At the South Pole there exists a similar light, called the Aurora Australis.

This light is visible, not only in the immediate vicinity of the poles, but sometimes as far as forty or fifty degrees away from them. For instance, the Magnetic North Pole is situated at a latitude of 78° north, and at a longitude of 71° west of Greenwich (1900). Thus the limit of the Northern Lights zone in the East (Asia) is almost at the 40th parallel, while on the other hand, in the West (America) it extends nearly to the 20th.

At a latitude not too far north—for instance, Scandinavia—the Aurora Borealis is generally green, but sometimes also violet and red. The violet colour always appears at the extreme edge, or underneath the light wave. The variations in the colour are probably due to the breaking-up of the rays in the aqueous vapour, and probably the nature of the air also contributes, for the sun also has different colours at different times, and the Northern Lights seen in daylight in spring are rosy red, but gradually change to green as the twilight comes on.

It is a popular saying that hard frost follows a vivid display of the Northern Lights. This is not correct, for during the last fifteen years I have observed about four hundred such displays, and have come to the conclusion that, in ninety cases out of one hundred, warmer weather follows almost immediately. On the other hand, it is correct to say that bad weather often follows the Northern Lights, and this is easily explained. As the light waves can form clouds, it is quite natural that cloudy weather results. In winter this generally makes the air milder, and the change of temperature causes wind. Thus we have the bad weather.

The Aurora Borealis is not affected by moonlight or clouds. It can sometimes be seen in summer, but only in the twilight zone between latitudes 58 and 54. It is most vivid in October and March, or when sun-spots are in process of formation.

When one has first discovered the most active side of the sun, one may be fairly sure that vivid Northern Lights will appear after the lapse of twenty-seven days, when the sun turns the same side to the earth.

And now some particulars of the circumstances under which my drawings were made. I have always been interested in the Northern Lights, and wished to possess some pictures of them, but as these are scarcely to be found outside scientific works, there was nothing else for me to do than to set to work myself. First I tried photography. Our cameras

By permission of] [*The Royal Society.*

SOUTHERN LIGHTS.

This photo and its companion, both taken in the Antarctic Regions, fittingly accompany the drawings of M. Raebel, as they demonstrate the essential similarity of the Northern and Southern Lights.

are probably good enough, but the plates are not nearly sensitive enough to catch the rays, which are always changing. On the other hand, it is possible to photograph landscapes by the light of the Aurora Borealis, and on one single occasion I succeeded in photographing a rayless Aurora Borealis arc. These large arcs rise but slowly in the heavens, and either dissolve into rays (when they approach each other) and disappear south of zenith, or else return as they came.

As I did not succeed with my camera, I made an attempt with pastel drawings. I often stood for six hours at a time in frost and snow on the summit of Graakall Hill, sketching with my pencil the forms of the lights. By the help of a good memory I completed the drawings next day with pastel crayons, and criticized them at the next appearance of the lights. In this way I made good progress, and after working day and night for four or five months, I think I can say that, as far as the phenomena of the Northern Lights are concerned, my drawings are fairly correct and true to nature.

It may be noted that a landscape lighted by the Aurora Borealis throws no shadows, as a single ray in zenith annuls all shadows. There is a strong resemblance between the light of a vivid Aurora Borealis and that of the full moon.

By permission of] [*The Royal Society.*

AURORA AUSTRALIS.

This photo, taken during Capt. Scott's Antarctic Expedition, shows how closely the Northern and Southern Lights resemble each other.

Last winter I made the remarkable discovery that the light of the zodiac has the same origin as the Aurora Borealis, viz., the solar spots. About every twenty-seventh day last winter there appeared a vivid Aurora Borealis at the same time as a strong zodiacal light was seen at other places. I may mention January 25th, February 21st, March 19th and April 14th. Of these, the lights on the 19th of March were the most vivid, and displayed such unusual forms that the like may perhaps never be seen again. Some of these forms are reproduced in my drawings. Among these was a corona which appeared at 9.30 and 10.45 p.m., and which could scarcely be depicted satisfactorily by any painter. A cinematograph would have been necessary. From the north-west sky there came with great velocity a wave of light in the form of a hollow cone, and consisting of very regular green rays. As soon as this cone reached zenith, there appeared round its lower edge an intensely bright violet line, and from this green and violet flames were projected with great velocity towards and through each other, as may sometimes be seen when the searchlights of several men-of-war are brought

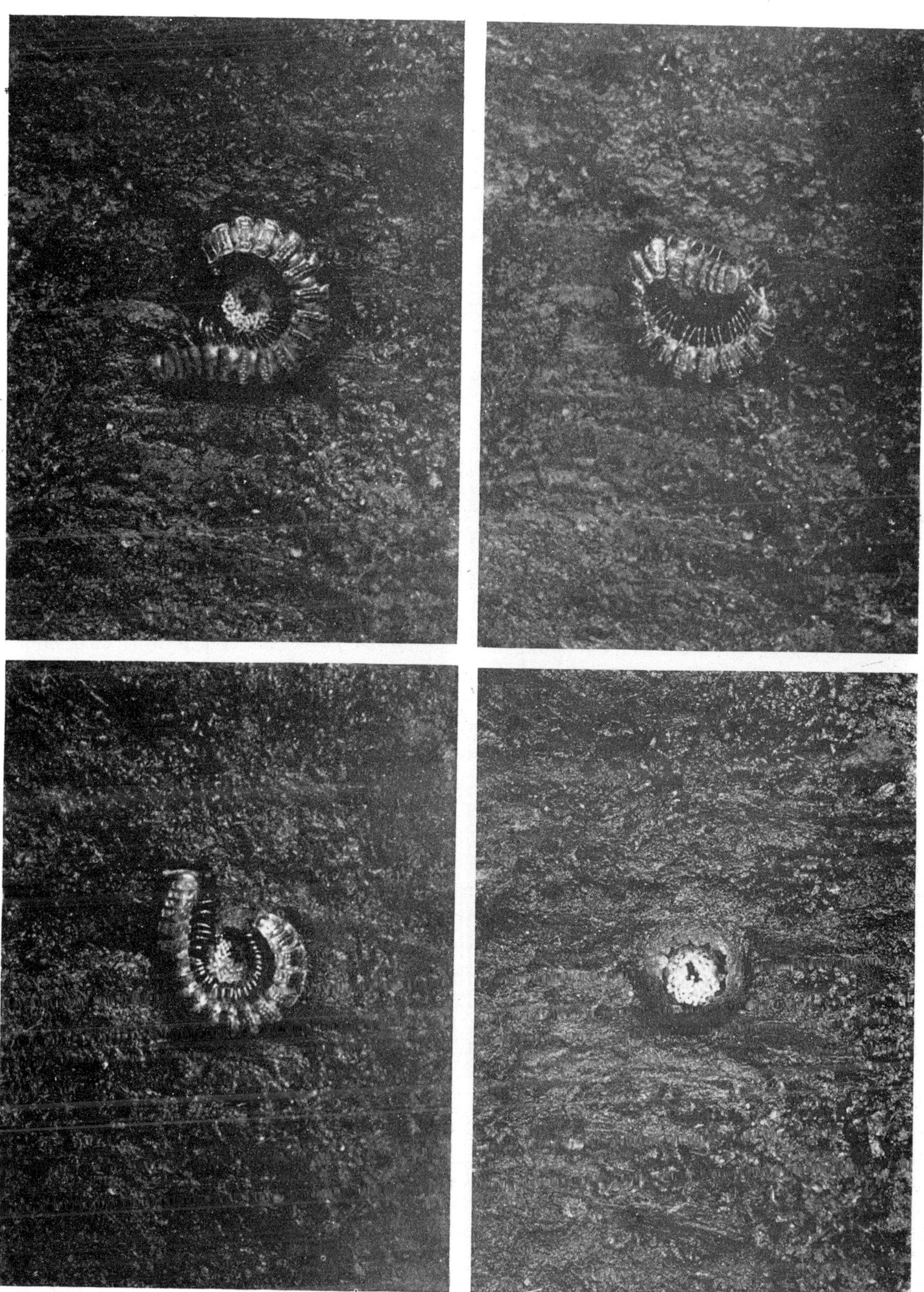

Photos by] [*Hugh Main, F.E.S.*

THE NEST-BUILDING OF A MILLIPEDE.

These four photographs represent successive stages by which the female Millipede fashions the nest that is to protect her numerous eggs, and at the same time adds to the layers of eggs inside the walls. In the last of the four pictures the Millipede has been removed to show the half-built nest more clearly. These and the following photographs are all twice the natural size.

into play at the same time. This hollow-cone form seems to point to the fact that magnetic currents and eddies have the same form as hurricanes and whirlwinds, and revolve round a centre. Perhaps an Aurora Borealis ray struck such an eddy in the centre and was divided in its circumference. After the latter had sucked up the rays, they were probably drawn in towards the centre, and after some minutes died away.

In any case, a vivid display of the Northern Lights is a magnificent and never-to-be-forgotten spectacle, and no one who has the opportunity should fail to visit the North of Norway towards the end of the winter. It is not necessary to travel farther than to Trondhjem, but it must not be much further south.

Photo by] [*Hugh Main, F.E.S.*

A pair of Millipedes at home selecting a site for the nest or egg-store.

Photo by] [*Hugh Main, F.E.S.*

THE MILLIPEDE'S NEST.

The female has commenced to construct the lower walls. A number of eggs can already be seen in the centre.

A MILLIPEDE'S NEST

BY HUGH MAIN, B.SC., F.E.S.

THE Millipedes, or Thousand-legs as they are commonly called, are of a retiring disposition, hiding under stones and logs of wood. If we turn over one of their shelters the little creatures at once endeavour to hurry away out of sight. There is something repugnant to many people about their appearance, so on the whole they are generally left severely alone.

A closer acquaintance with them, however, unfolds a most interesting story. With the help of the illustrations we can follow a small part of it which does not appear to have been described before, viz., the way in which the female builds her nest and lays her eggs in it.

A pair of Millipedes were found, in the month of March, hiding under a piece of wood lying on the ground. The wood was quite damp and had a rough, thin layer of earth on it. Wood and Millipedes were taken home and special precautions adopted to prevent the wood drying up and to give the creatures the shade they like. They are vegetarians; but although supplied with various things to eat, they seemed to find sufficient nourishment among the earth on the wood.

They lived together very happily, but evidently disliked being exposed to a bright light, always seeking the shaded side of the piece of wood when exposed to view.

On looking at them on the morning of April 17th, the female was seen to be in the act of making her nest. The wall was about half built, and she continued her work all the rest of the day and through the night, and it was finished early on the next morning. A second nest was made, unobserved, shortly afterwards. On May 7th, about 11 a.m., a third nest was seen to be just started. We shall follow the building of this nest. All the photographs are twice the natural size, so a good idea can be obtained of the actual dimensions of the different parts.

When first seen, the wall was only raised a little above the level of the wood, and it contained an incomplete ring of spherical, white, semi-translucent eggs, one layer deep, and irregularly three or four eggs wide. The female rested on the top of the wall and worked round it, generally in the direction of the hands of a watch, with the feet on the side of the body towards the centre holding on to the upper edge of the nest. She extended the anterior segments of the body over the interior of the nest and passed out the eggs, two at a time, from the openings of the oviducts, which are situated just behind the second pair of legs, *i.e.*, not far behind the head. After completing the ring of eggs, she proceeded to raise the wall of the nest about the height of the eggs themselves, and then deposited another layer of eggs as above described.

The method of raising the wall was very curious. She foraged all round the nest, as she continuously revolved on the top, and ate up the small particles of earth within reach. The material thus obtained was passed out at the hinder end of the body in a prepared condition, and was moulded into position by the large flaps at the extremity. As the wall was raised it was made dome-shaped, and egg-laying alternated with wall-building. Only once did she stop work, and that was when brought into a stronger light about 4 p.m., and in the third photo on page 255 we see how the work had progressed. The mouldings are very noticeable, and the mass of eggs. She resumed her task, and at 5.40 p.m. the opening at the top of the nest was reduced to the size seen

Photo by] [*Hugh Main, F.E.S.*

Eleven hours from the start saw the dome completed all but the chimney, which took three hours more to erect.

Photo by] [*Hugh Main, F.E.S.*

THE NEST HIDDEN.

The nest is finally covered over with earth, which completely disguises it, and prevents the eggs from drying up.

Photos by] [*Hugh Main, F.E.S.*

THE NEST VIEWED FROM THE SIDE.

To the left is seen the Millipede's nest completed with the chimney; to the right the same nest after its builder has covered it with earth to hide its form and make it appear a mere lump of soil.

in the next photo. It was not till 10.15 p.m. that the dome was finished, only a small opening being left at the top. She then proceeded to add a tube or chimney, forming a passage communicating with the interior of the dome, and this was completed at 1.15 a.m. of the next day.

From the experience of the first nest, I knew that the next step was to cover over the whole structure with small rough pieces of earth, right up to a level with the top of the chimney. As, however, I wished to secure a photograph of the nest before this was done, I gently drove the female away, covered up the nest with a pill-box, and went off to bed.

In the photographs on page 257 we get a view of the completed nest uncovered, and the same nest quite hidden by small pieces of earth, and the female resting alongside it. The two nests appear in elevation in the final photographs, and show how wonderfully the nest is protected from observation by the covering of earth. This coating also helps to keep the nest damp, thus preventing the eggs within drying up. The female either resented my interference with her plans or forgot her nest, as she never paid any more attention to it. On May 17th, the eggs in the first nest hatched and the little white, six-legged creatures climbed out of the hole at the top of the chimney. It is not till later in life that the young Millipedes get their full number of legs.

TWO DRAGONS OF THE PRIME

BY SIR H. H. JOHNSTON, G.C.M.G.

TENNYSON, like some other poets, had the prophetic instinct strongly in him. He realized and wrote of things in the first beginnings of invention and discovery long before the conception of them could have grown up in the mind of humanity at large. And when somewhere about 1850 he wrote of "Dragons of the prime that tare each other in their slime," he certainly seems to refer to the extinct gigantic Dinosaurs, which were not fully revealed by the discoveries of geologists till some years afterwards. The Dinosaurs originated, first of all, probably in the Old World, in what is called the Triassic Period at the beginning of the Secondary Epoch, from a group of generalized reptiles represented by the living Sphenodon of New

Photo by permission of] [*Sir H. H. Johnston.*

A "DRAGON OF THE PRIME."

A restoration of the huge herb-eating reptile known as a Stegosaur. Its remains were found in Colorado, but very similar forms have been found in this country. Its enormous crest consisted of bony plates.

Zealand. The earliest Dinosaurs were certainly carnivorous, and this flesh-eating type developed in course of time into formidable reptiles of from fifteen to thirty feet in length, though never such monsters in bulk as grew up in the course of ages out of the herbivorous or seaweed-eating Dinosaurs.

The vegetable-eating Dinosaurs reached an extravagant size, and evolved many strange growths of armour in course of time. It is popularly supposed that these marvellous monsters occurred mainly in North America; but although the biggest-known Dinosaurs are found in the later Secondary formations of the United States, there were others almost equally great inhabiting England, Belgium, France, Germany, East Africa and India. Some of these monsters seem to have ranged, with very little variation in species, between Western Europe and North America. Of

Photo by permission of] [*Sir H. H. Johnston.*

A DINOSAUR.

The Dinosaur here represented is one of the flesh-eating kind. It will be noticed that it is a much smaller reptile than the huge Stegosaur on the previous page, which is a vegetable feeder.

such was the Stegosaur of Colorado, here illustrated, which reappears in the closely-allied *Omosaurus*, in the formations of Oxfordshire and Dorsetshire, or as *Polacanthus* in the Isle of Wight. The picture which is given of a Stegosaur should be considered, perhaps, an American subject; the huge armoured reptile which has been restored with such wonderful vividness and accuracy by Herr A. Pallenberg, for exhibition in the gardens of Carl Hagenbeck's famous animal park at Stellingen, being almost certainly the American Stegosaur. The carnivorous Dinosaur, with its spinal crest of bony plates, is intended for a Megalosaur. The Megalosaurs originated in Europe, and are most commonly met with there in the formations representing the first three quarters of the Secondary Epoch, but they afterwards extended their range to western North America in time to attack and perhaps help to extinguish the last preposterous Stegosaur.

Photo by] [E. Step, F.L.S.

THE WORM-EATING SLUG IN A RESTING STATE.

THE WORM-EATING SLUG

BY EDWARD STEP, F.L.S.

To the average gardener, amateur or professional, the statement that all slugs are not his enemies will be received as rank heresy. But it is a fact which even he would be compelled to accept if he were a little more observant and less inclined to treat all creeping things as specially designed to thwart his efforts. The slug to which special reference is now to be made is one that few people who are not naturalists have ever seen. It passes the greater part of its life underground, only coming to the surface when its underground runs are flooded, or on moist nights when its special prey, the earthworm, is also out for an airing.

The Worm-eating Slug is purely carnivorous, and would, figuratively speaking, turn up its nose at the choicest bed of seedlings that were ever grown. In nearly all respects it differs from the well-known ordinary slugs. The latter, it will be remembered, are thickest at a little distance

Photo by] [E. Step, F.L.S.

THE WORM-EATING SLUG EXTENDED.

In this condition the Slug enters the burrows of the earthworm and eats the worms. A hard shell on the tail end protects this Slug from a similar attack from one of its own kind that may follow it down the same burrow.

behind the head, where they are covered by a flap with loose edges and known appropriately as the mantle. On the right margin of this mantle is an opening to the gill-chamber, where the slug's blood is purified. Within the substance of the mantle are a few grains of shelly matter that, in a poor way, represent the spiral shell of the snail. From this stoutest part of the body the common sorts of slugs taper away to the rear.

In the Worm-eating Slug much of this is reversed. The stoutest part of the slug is right behind, and it tapers forwards to the very slender head. We were once told by a friend who had discovered one of these slugs in his garden that he had witnessed the extraordinary sight of a slug gliding backwards. He was an observant man, or he would never have taken note of the fact that slugs ordinarily go with the thickest part forward; but not sufficiently observant, or he would have noticed that this remarkable slug had a rather prominent shell. The Worm-eating Slug wears quite a small mantle covering its air-chamber, which is situated at the extreme hind-end of the body, and this mantle is almost covered by the small, thick shell. An ordinary slug has its mouth armed with jaws for cutting the stalks of the juicy seedlings, and a wonderful ribbon-shaped tongue studded with flinty hooks for rasping the said seedlings into pulp. The Worm-eating Slug has lips, but no jaws. It has a long ribbon-shaped tongue like other slugs, but the hooks upon it are barbed.

Photo by] [*J. F. Cooper.*

THE SLUG'S TEETH.

A portion of the tongue or tooth-ribbon of the Worm-eating Slug. There are fifty rows of teeth, and each row consists of fifty-one teeth.

These differences are all beautifully related to the peculiar habits of the Worm-eating Slug. Its great mission in life is to restrict the undue multiplication of the beneficent earthworm. Its small head and tapering body, capable of extension to a length of three inches or more, enable it to enter the earthworms' burrows like a ferret in a rabbit's run. In so doing its body fills the space, so its air-chamber must be at the extremity of its body to be provided with air. Now these slugs not only consume worms and subterranean beetle-grubs, but are not above making a meal off their own kindred. If one of them were followed by another in the same boring, the second one might begin to eat the first; so to guard against accidents of this kind each wears over his air-chamber the hard little shell which is proof against the tongue-hooks of his kindred. A portion of this wonderful tongue-ribbon, much enlarged, is here shown. Throughout its length it bears about fifty rows of the curved and barbed teeth, and in each row there are fifty-one teeth. The ribbon is partly thrust out, the hooks are pressed into the worm's flesh, and he is dragged into the slug's gullet.

THE TRAVELLER'S TREE

EVERYTHING that has life is in one way or another a marvel: but we are too apt to consider as such only those things that are huge or ugly, or to which sentiment has given a fancy name. The Traveller's Tree is a case in point. A palm-like plant, somewhat nearly allied to the banana and

THE TRAVELLER'S TREE.

This fine palm-like tree grows in the island of Madagascar. The lower parts of the huge leaf-stalks are hollowed so that they catch and hold a large quantity of water. If the leaf-stalk is cut across, a stream of water gushes down to refresh the traveller who may need it.

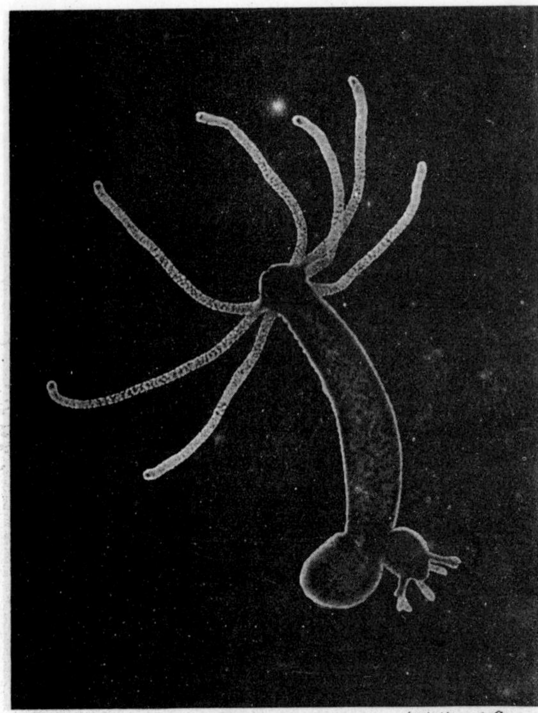

Photo by] [*Flatters & Co.*

THE BROWN HYDRA.

The Hydra is here shown fully extended, with a bud which will drop off its parent and begin a separate existence.

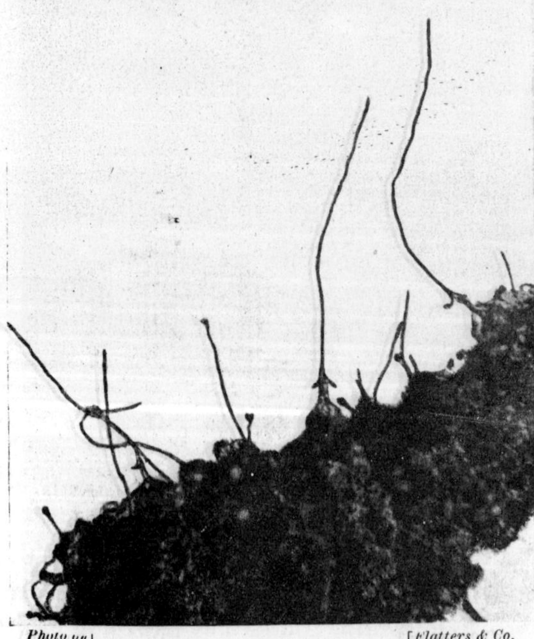

Photo by] [*Flatters & Co.*

THE HYDRA'S STINGING THREADS.

A small portion of the Hydra is here greatly enlarged in order to show the stinging threads with which it catches and kills its food.

plantain, it is one of the most gracefully beautiful of such growths. But one seldom hears a word of its beauty, only of its supposed usefulness to the thirsting traveller who is struggling across the parched desert. The early travellers told how Providence had planted this tree in such places for their special benefit, and many a pious lesson has been founded upon this " fact." Even when at the last gasp, with swollen tongue clinging to the dry palate, and making utterance impossible, when every throbbing vein was crying out for water, you had only to come upon the Traveller's Tree to find all the liquid refreshment desired. Cutting the leaf-stem across, a stream of pure, cool water gushed out, and all was well.

The plentiful supply of liquid is a *real* fact; and no doubt an exhausted traveller would not be too critical as to its flavour, said by later investigators to be disagreeable. But the weak part of the wonder-story is that the Traveller's Tree will not grow in the desert, and insists upon having its roots in well-watered soil!

As may be seen from our fine photograph, the gigantic long, flat leaves are arranged in two opposite rows above the woody stem; and these leaves are used for thatch, for dishes, and small portions folded serve for spoons. Its comparatively small flowers are clustered at the base of the leaf-stalks. The plant is a native of Madagascar.

THE FRESH-WATER HYDRA

BY K. G. BLAIR, B.SC., F.E.S.

To make acquaintance with the Fresh-water Hydra all that is necessary is to go and collect a small bottle full of the floating duckweed or the thread-like green weeds that are found in almost any pond. If this material be emptied out into a bowl of clean water and allowed to stand for an hour or two in a good light, the Hydras that are almost sure to be present will have had time to recover from the shock of the disturbance, and many will probably be found to have attached themselves to the sides of the vessel and thus be more readily visible. They may be seen as a slender green stem, about one-sixth of an inch long, standing out from the side, each with a crown of about eight still more slender tentacles waving about in the water. If carefully watched, they may be observed to move about, which they do either by bending over and taking hold of the surface with their tentacles, then letting go with the other end, and attaching it again close by their new hold, or by gliding over the surface on their end of attachment, much in the same way as a snail crawls Care must be taken not to jar the vessel or to disturb the water, or the Hydra will draw in its tentacles and contract its body to a mere globule of dark green jelly and fall to the bottom.

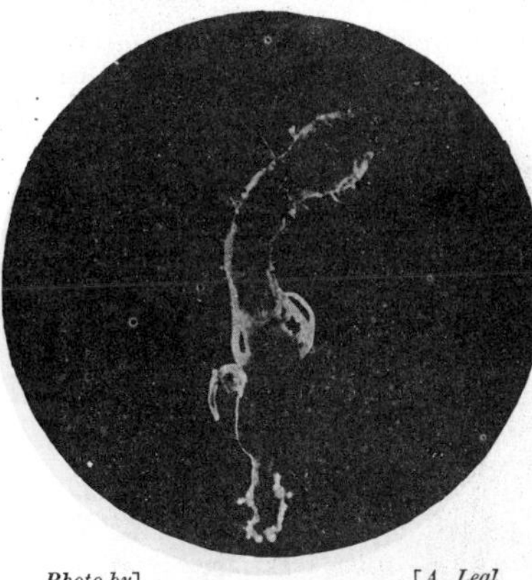

Photo by] [*A. Leal.*

Hydra swallowing a water grub.

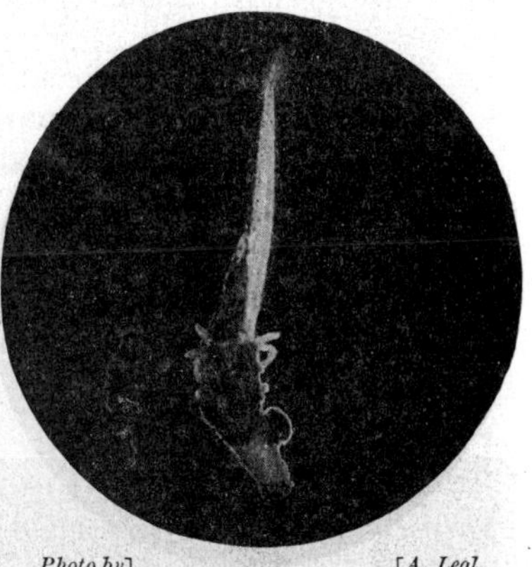

Photo by] [*A. Leal.*

Hydra swallowing a young fish.

In spite of its extreme tenuity, the whole animal is hollow, the cavity occupying the thread-like body sending a branch up each of the tentacles; its only opening is the mouth, which is situated in the centre of the circlet of arms. Further, the body-wall of Hydra is composed of three distinct layers; the middle one of these is of a jelly-like consistency, the layers on either side of this, both within and without, being composed of cells. The cells of the inner layer project into the central cavity, which serves as the stomach of the animal, and their secretions act upon the small animals captured and swallowed by the Hydra, so that they can be digested. The cells of the outer layer are considerably smaller, and are mainly muscular in function, effecting the elongation and contraction of the animal, though among them are various kinds of special cells, such as the reproductive cells and the stinging cells. The latter are arranged in little groups, or batteries, mainly upon the tentacles, and upon receiving the requisite stimulus, such as is afforded by the touch of any passing creature, they shoot out a fine thread-like dart that pierces the body of the prey and appears to exercise a paralysing effect upon it, so that it can be seized by the tentacles and swallowed without further struggling. These darts, highly magnified, may be seen in the photograph on page 264. Water-fleas and other minute animals form the staple food of the Hydra.

A Hydra will very often be observed to have one or perhaps two young ones growing out like buds from the main body, at about one-third of its length from the base. The buds remain attached to the parent for some time, and put out their tentacles and catch food on their own account, though their central cavity still remains in communication with that of the parent. They may even put out secondary buds of their own, so that we find one individual consisting of three generations, all enjoying a common stomach which branches up between them. The Brown Hydra found on the undersides of water-lily leaves and such-like places is more prolific in this manner than the Green Hydra of the duckweed; it is also rather larger and more slender, and a well-budded

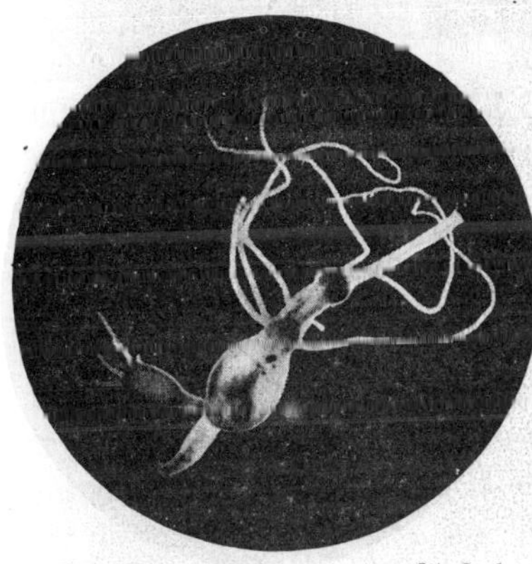

Photo by] [*A. Leal.*

One Hydra swallowing another.

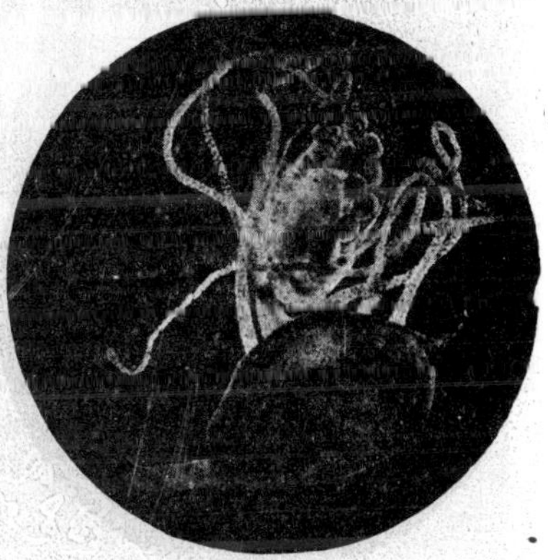

Photo by] [*A. Leal.*

Hydra capturing a water-flea.

specimen with the long, thread-like tentacles of all its members waving gently in the water affords an extremely beautiful sight. Sooner or later, however, the buds become detached and commence housekeeping on their own account.

There is also a sexual method of reproduction, in which, instead of a bud being formed, an egg takes its place. This, after fertilization, secretes round itself a hard coat, and drops off, and in due time the shell splits and a young Hydra emerges.

Like many other animals that reproduce their species asexually—that is, by budding, or some similar process—Hydra is able to replace any lost parts; but in Hydra this power is developed to such a degree that even if cut up into pieces, each piece can replace all the rest and develop into a complete individual.

Though such an inconspicuous creature, Hydra presents us with a number of most interesting phenomena. For instance, its very colour—green or brown, according to species—is due to a minute, single-celled plant that lives within the cells of the digestive layer of the body-wall. The plant is not a parasite, as it causes no harm to its host; indeed, it is probably useful to it, both by the fact of its colour rendering the Hydra less conspicuous and less easily found by its enemies, and also by its vital processes yielding products that are useful to the Hydra. In return, the plant doubtless receives considerable protection in being enclosed within the body of the Hydra, and probably obtains its food from waste excretions of its host.

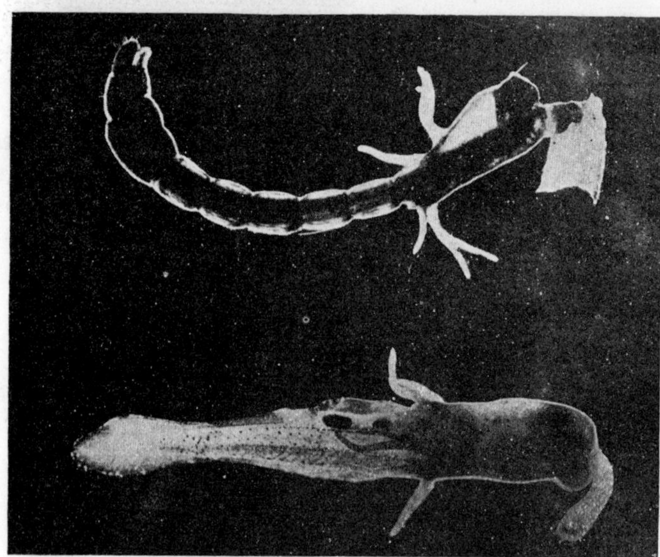

Photo by] [*Flatters & Co.*

THE HYDRA AND ITS VICTIMS.

In the upper figure the Hydra has captured and is swallowing a worm; in the lower it is a young fish that has fallen a victim. The distinction between captor and prey may be traced by means of the tentacles, which surround the mouth.

In spite of its frail appearance, and the benevolent part it plays as host, the Hydra is a creature to be reckoned with in pond-life. Its capacity for nourishment is enormous and vastly disproportionate to its size. Many are the water-fleas and other small fry that the Hydra can account for, and occasionally even very minute fishes find their way down that capacious gullet. The victims are caught and stung to death by the thread-like tentacles shown on page 264. Even its own kind are not outside the limits of its voracious appetite. In the lower left-hand photograph on page 265 an instance of this flagrant cannibalism is shown. The vanquished is already becoming part of the victor.

When we come to examine the affinities of the Hydra we find that it stands alone as an inhabitant of fresh water, all its relatives being marine. It is closely allied to the jelly-fish, so beloved of nervous bathers, and less closely to the sea-anemones; indeed, it can hardly be doubted that Hydra itself is descended from marine ancestors, but in adapting itself to fresh-water conditions it has, as is usually found in such cases, become considerably modified, especially as regards the development of the young. In Hydra this modification is evident in the suppression of the free-swimming stage in the life history that is found in its marine cousins; in the comparative rarity and simplicity of the sexual method of reproduction; and in the fact that the buds sooner or later drop off and start life as new individuals, instead of remaining attached to form a branching colony.

By permission of] [*Sir H. H. Johnston.*

UMBRELLA BIRDS.

Two distinct kinds of Umbrella Birds are here shown—the Pendulum-bearer to the left and the Bare-necked Umbrella Bird to the right. They are both from different parts of Tropical America. The remarkable appendages from the throat are connected with the vocal organs, and increase the depth and mellowness of their flute-like calls.

UMBRELLA BIRDS

BY SIR H. H. JOHNSTON, G.C.M.G.

SOUTH and Central America are the home of several remarkable groups of Passerine birds of relatively large size, and of more primitive structure than the Singing-birds which make up such an overwhelming contingent in numbers of the whole bird class as it exists at the present day. Amongst these there is, perhaps, no more extraordinary form to be seen as regards the development of plumage than the Umbrella Birds. The best known of these was introduced to European collections as early as the eighteenth century, coming from the forests of the Amazon in Northern Brazil. It had been seldom seen alive by white men, as it lived in the higher branches of forest trees, where it subsisted on wild fruits. But the natives pursued it for its extraordinary plumage and obtained specimens by the use of their blow-guns. The great naturalist-explorer, H. W. Bates, gave an excellent description of the peculiar vocal powers of this Umbrella Bird. "It drew itself up on its perch, spread widely the umbrella-formed crest, dilated and waved its glossy breast lappet, and then, in giving vent to its loud, piping note, bowed its head slowly forwards."

Further research amongst the forests and mountains of Equatorial America produced still more remarkable developments of these birds in the way of plumage and appendages. Of such are the two species here illustrated—the Pendulum-bearer from Ecuador, and the Bare-necked Umbrella Bird from Costa Rica. In the first-named the chest appendage or lappet grows to a length of as much as eight inches in the male; in the females of all forms of Umbrella Birds the eccentric developments of the plumage do not attain the same proportions. In the Bare-necked Umbrella Bird the appendage is shorter, almost devoid of feathers, and is covered with a brightly coloured red and yellow skin. It was shown by Bates that these appendages were pouches communicating with the breathing and vocal organs, and increased the depth and flute-like mellowness of the loud call.

Photo by] [*W. Saville Kent.*

A GIANT CLAM.

An empty shell of a Giant Clam on an Australian coral reef exposed at low water. The pure white of the interior contrasts strongly with the dull-coloured and weedy exterior.

Photo by] [W. Saville Kent.

A GIANT CLAM.

In this photograph the Giant Clam is alive, and between the partly open valves of the shell the "mantle" which lines and produces the shell is seen. The weight of the shell keeps it in position among the rocks.

THE GIANT CLAM

BY SIR H. H. JOHNSTON, G.C.M.G.

THE Giant Clam, or Tridacna, like the familiar oyster, is a bivalve, and is the supreme development in size known amongst all shells. The specimens illustrated on page 270 measure about three feet across at their widest; but according to the late W. Saville Kent, probably the largest may attain to ten feet in width off the coast of Queensland, and the entire animal weighs as much as two thousand pounds. These Clams are evidently among the latest developments of bivalve shells, since they cannot be traced back farther geologically than the Miocene Period or middle of the Tertiary Epoch. The muscular power developed in these huge Clams is very great. When undisturbed this mighty Clam usually remains with the shell partially open, but if a man were so foolish as to insert his hand or foot between the interlocking valves they would at once close on such an intrusion, and it would be impossible to open them without recourse to the forcible application of crowbars or axes. The Clams of to-day—or, at least, those of enormous size—come mainly from the shallow parts of the Pacific Ocean, amongst the coral islands; and Mr. Saville Kent, in his monumental work, "The Great Barrier Reef of Australia," alludes to the loss of life occasioned among the men who wade or dive among the coral reefs in search of pearls or *bêche-de-mer*, by their inadvertently inserting a foot between the gaping valves of the Giant Clam. The outer surface of the shell in the older clams is so disguised by a parasitic growth of weeds, coral and sponges that they are not easily distinguishable from their inanimate surroundings, and the "mantle" of the Giant Clam, which may be seen between the open valves, is not brilliantly coloured like that of the smaller species, but is only light brown, with stripes of darker brown. The fisherman—black, white or yellow—once caught by the Clam, is held there till he drowns, unless a comrade comes to his assistance and prizes open the shell. The Giant Clam, however, merely rests on the rocks, anchored by its weight, and does not attach itself permanently by any bundle of filaments or cartilaginous plug.

The picture on page 269 (an original photograph by the late Mr. Saville Kent) shows the Giant Clam as and where it lives, and illustrates very well its remarkable development of shell-teeth.

The powerful muscle which enables this Clam to close its shell is said to be good eating, and this use of the Giant Clams of the Pacific was noted by Captain Cook in his voyages of exploration.

By permission of] [*Sir H. H. Johnston.*

GIANT CLAM SHELLS

In this photograph both inside and outside of the enormous shell is shown. The high ridge at the back of one valve fits into a corresponding groove in the opposite one, and in conjunction with an elastic ligament constitutes the hinge upon which the valves open and shut.

THE GIANT CRAB OF JAPAN

BY SIR. H. H. JOHNSTON, G.C.M.G.

THE Giant Crab of Japan, or Kämpfer's Crab, is supposed to be the largest of the crab tribe in existence. The male can span between its outstretched legs as much as eleven feet. This is a moderate estimate, for there are records of specimens so large that they could span as much as eighteen feet. It is the opposite extreme in its class to the minute " water-fleas " found in every pond, which are barely a hundredth of an inch in total length. The Giant Crab of Japan belongs to the section known as the Spider Crabs, which are represented in the seas of Europe as well as in the Pacific Ocean. The largest European Spider Crab is the Spiny Spider Crab, but this is not so large or remarkable as the monstrous crab which was discovered in the waters of Japan and named after the German biologist, Kämpfer. The specimen here illustrated, perhaps the largest in any

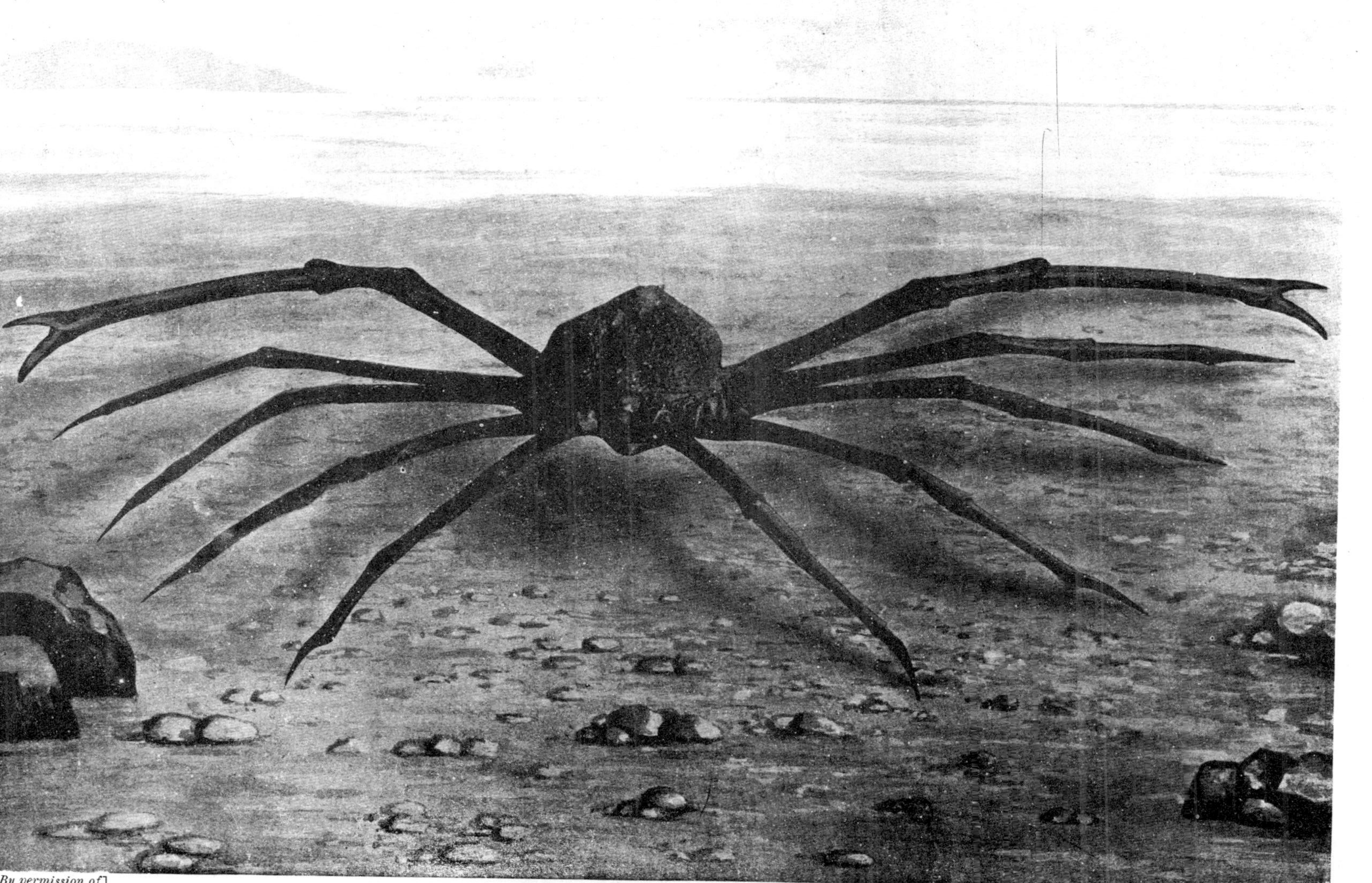

By permission of] [*Sir H. H. Johnston.*

THE GIANT CRAB OF JAPAN.

Kämpfer's Spider Crab is here seen in retreat towards the sea, where naturally it remains; it is not a shore-haunting crab, but remains in fairly deep water among the rocks and weeds, as shown in the coloured plate. **Between the tips of its first pair of legs it can span eleven feet or more**

collection, was brought from Japan to the Natural History Museum of Munich by Professor F. Doflein.

The Spider Crabs live on the sea-bottom, at varying depths from the surface. They clamber about amongst the weeds and rocks, and often show in their structure or adornments a remarkable resemblance to their surroundings. Some species actually plant upon their spiny shells anemones, sponges, or seaweeds, causing them to take root there. But in spite of all attempts at disguising themselves and of their sometimes enormous size, they are constantly preyed on by the octopus and many fishes.

"ORGAN-PIPE" CORAL

BY EDWARD A. MARTIN, F.G.S.

AMONGST the great variety of forms assumed by the hard skeletons of the coral-polyps none is more beautiful than the group known as the " Organ-Pipe Corals." The chief living form has the remarkable peculiarity which has given it its popular name, and bears what is really a remarkable resemblance to a number of pipes of an organ clustered together, as though placed in position by the hand of man, rather than a normal natural growth. The pipes are arranged in fairly parallel upright rows, with spaces between them, and they are seen under the microscope to be made up of an immense number of chalky spicules fused together by the animal. All these pipes are held together by transverse plates, or platforms, placed at distances of about an inch apart, as though dividing the whole into floors or stories, with the different floors supported by a large number of hollow

Photo by] [*E. Step, F.L.S.*

THE ORGAN-PIPE CORAL.

This coral, which is of a fine though dull red colour, forms colonies of a foot or more across. It attaches itself to other corals, and can be found at low tide in the pools near the edge of the coral reefs. The little platforms connecting the tubes are the stage from which new growths begin by budding from the older ones. The tubes are here slightly enlarged.

Photo by] [*W. Saville Kent.*

THE ORGAN-PIPE CORAL.

This photograph of the living coral, taken through the water, shows the polyps expanded. They are the builders of the "organ pipes," which are more clearly shown in the other photograph from the dead and empty coral. The tentacles of the polyps are coloured a bright emerald green, making a fine contrast with the red of the tubes.

pillars. It is from these floors that new tubes are begun by the budding off of new polyps from the older ones.

Both the tubes and the platforms are of a striking crimson colour, the polyps themselves and their enclosing flesh, with the exception of their tentacles, being of the same colour as their hard skeleton. The tentacles, which are eight in number, are of a fine emerald green. When living, each of the tubes encloses within it a little coral polyp, and the effect of hundreds of green tentacles projecting from hundreds of crimson tubes is marvellous in the extreme. It must be remembered that each polyp is essentially the same as a sea-anemone, although much smaller, and also that the hundreds of rosetted polyps in a single lump of coral are all united together and form one compound colony. The crimson layer of fleshy material covers the outside of all the tubes and the platforms, and the hard chalky " coral " is the skeleton of the animal. It is the flesh which secretes from the ocean the lime which goes to make up the coral, and imparts its own crimson colour to it ; and when the tubes have departed from the upright sufficiently to allow room for another tube between them, a fresh tube will commence to be built from one of the transverse platforms by a process of budding from the parent stock.

The colony is fed in the same way that a sea-anemone obtains its food. The polyps cannot go in search of food, so they have to wait until it comes within reach of their tentacles. Then, like the anemones, they grasp at the tiny particles of food which are wafted towards them, and the food is conveyed by the tentacles to the mouth, which is in the centre of each cluster of tentacles. Not the least remarkable thing in connection with these corals is the fact that all the polyps of a colony are in organic connection with each other, and thus the food which reaches any one of the numerous mouths goes to support the whole organism.

The Organ-pipe Coral does not form reefs or islands as some corals do. The masses in which it is found are comparatively small, usually not much more than a foot or so across ; and it is most frequently attached to other corals. It is found in comparatively shallow water, in the pools at the edge of the coral reef at low-water.

THE SCORPION AT HOME

BY K. G. BLAIR, B.SC., F.E.S.

SCORPIONS are essentially tropical creatures, though a few species are found in the south of Europe. They are nocturnal in their habits, hiding away by day under stones, and, owing to their retiring disposition, and the fear and disgust in which they are usually held, until recently very little was known of the details of their life-history. Of recent years, M. J. H. Fabre, an eminent French naturalist, has investigated the habits of the larger of the two common European species, and to the delightful account he has given us we are indebted for many of the following particulars.

This Scorpion, known as the Scorpion of Languedoc, attains a length of about three and a half inches, and is of a pale brownish-yellow colour. It inhabits dry hilly regions almost devoid of vegetation by reason of constant exposure to a scorching sun, and is to be found under the flat stones and slabs of rock that are scattered everywhere about. If a stone be turned over, a circular hole, several inches deep, will be found, often with the Scorpion at the entrance, with claws extended and tail curled over his back, ready for all comers. On a hot sunny day the Scorpion loves to lie thus immediately beneath the stone, heated right through by the sun; but on a cold or wet day it retires into its burrow.

Photo by] [*P. H. Fabre.*

THE END OF THE MALE SCORPION.

The lady Scorpion takes the government of the household into her own hands, and has her husband served up for breakfast.

The large pincer-claws are used as weapons of attack and as purveyors of food to the mouth; they are endowed with a very delicate sense of touch, and when the animal is walking are held well forward, slightly spread, and sway gently up and down; for though provided with four pairs of eyes, these are all directed so much sideways that the Scorpion is not able to see straight in front of it, and relies upon the delicate perceptions of its claws for information as to where it is going. The tail, armed at the end with the poison dart or sting, is usually carried curled forwards over the back, or when at rest curls to one side of the body.

When attacking its prey, the Scorpion approaches quietly and seizes it with its pincers; if the fly seized remains quiet, it is simply brought to the mouth and held there while devoured at leisure, without the sting being brought into play at all; but should it struggle violently the Scorpion, which above all things dislikes a disturbance, curls the tail forward over the back, holds the sting poised, as it were, for a second, then pricks the victim once or twice to keep it quiet. Though so deliberate in all its movements, the Scorpion does not seem to care much how it holds the prey or in what region of the body the dart pierces it.

The effect of the sting varies very much according to the victim, and experiments were tried, by inducing the Scorpion to attack different animals, to determine this effect. Spiders curled up and died at once; most insects, especially those of more delicate organization, were at once paralysed, while death followed more or less rapidly; on the other hand, a large centipede survived

Photo by] [*P. H. Fabre.*

THE HOME-COMING OF THE BRIDE.

Having at last found a cavern that promises well as a dwelling-place, the male Scorpion leads his bride into it. Next morning the cave has only one tenant. See photo opposite for the sequel.

for several days. Strangely enough, though an insect such as a moth or a beetle very quickly dies after being stung, their grubs seem to be quite immune, and even after their soft bodies have been pricked many times they appear quite unaffected by the poison, and in due time produce the perfect insect as though nothing had happened. A beetle reared from such a larva was stung to see whether it would now be immune, but it succumbed as rapidly as one that had never been inoculated. Unfortunately these experiments were not extended to a human subject, but from other less exact reports it would appear that, while the sting of a Scorpion has been known occasionally to prove fatal, it does so probably only when the blood of the patient is very much out of order.

Photo by] [*P. H. Fabre.*

THE FINISHING STROKE.

If its victim struggles the Scorpion curls its tail over its back and pricks its prey once or twice to quiet it.

The love affairs of the Scorpion are most entertaining. A male Scorpion out for a walk one warm evening in April or May meets a lady, usually somewhat larger than himself. The two stand face to face with claws extended; and, resting the fore part of the body firmly on the ground, each raises the hinder part, as if trying to stand on its head, until their tails cross, when they rub them affectionately one against the other and hook the tips round one another. This performance is apparently the usual greeting in polite Scorpion society, though in the case of a young couple thinking of matrimony the actions are more ceremonious and partake of the nature of a caress. If the lady be agreeable, he seizes her two hands, one in each of his, and walking backwards, draws her gently after him. They may perambulate in this manner for some hours, but eventually he finds some crevice under a stone that he considers suitable, enlarges it by scratching up the earth with his feet and sweeping it out with his tail, all the time maintaining his hold of the lady, and finally pulls her in after him. If she is satisfied, well and good, but if not they come out and wander round to find another home. Sometimes during their perambulations a bigger male than the first will put in a claim to the lady: they do not, as might be expected, engage in mortal combat, but each takes hold of one arm and pulls until one obtains the advantage, when the vanquished abandons the lady, and sets off to find another. When, however, the course of true love runs as it should, the happy couple at last find a house to suit them and take shelter there late that

Photo by] [*P. H. Fabre.*

MOTHER AND FAMILY.

The young Scorpions are about one-third of an inch long when they leave the egg. They climb upon the back of their mother, and crowd together there without food for a week.

night. Next morning, if the stone be turned over, the female is found to be in sole possession, with possibly a fragment of her spouse beside her. The details of the awful tragedy are not fully known, but it appears that the lady, who has hitherto remained perfectly passive, suddenly takes the government of the household into her own hands and has her husband served up for breakfast. Exactly how she manages to overpower him, especially considering that he had hold of her hands, is not known, but the result is always the same.

Whatever may be thought of her conduct as a wife, the female Scorpion forms an exemplary mother. About fourteen months after the tragedy of mating she deposits her eggs. It was formerly believed that Scorpions brought forth their young fully formed, but M. Fabre's researches have proved that they are really deposited as eggs, though the eggshell is an extremely delicate

Photo by] [*P. H. Fabre.*

A FRIENDLY GREETING.

Two Scorpions meeting stand face to face with claws extended. Each raises its hinder part, as if it were trying to stand on its head, until their tails cross, when they hook the tips round each other.

membrane, from which the soft, helpless young are unable to free themselves without the aid of the mother. As soon as the eggs are laid, the mother proceeds to tear open the membrane with her pincers and carefully detach it from the young; as soon as all are free she collects the débris and swallows it. The little creatures are now about one-third of an inch long, white, and almost helpless; they climb with difficulty to the back of the mother, where they lie closely packed together for a week. At the end of this time, during which they have taken no nourishment, they undergo a kind of moult, but instead of the old skin being left intact, as happens at later moults, it peels off in rags and tatters, which remain as a kind of blanket over the mother. The young Scorpions are now about half as long again as before, and become more lively, climbing about the mother and sometimes starting to wander off by themselves, only to be caught and brought back again. In another week they are able to catch food for themselves; the mother now ceases to take any interest in them, and the family soon becomes dispersed, each member wandering off to fend for itself.

THE COCHINEAL INSECT

BY HAROLD BASTIN.

THE Cochineal is one of the very few insects that are directly serviceable to mankind ; and nowadays the beautiful innoxious colour for which it is famed is rapidly falling into disuse, having been largely displaced commercially by the derivatives of aniline. The fact remains, however, that for centuries man depended for this dye, and for several of the pigments, such as carmine, used by artists, upon the product of a tiny creature closely allied to certain of the worst pests with which the modern agriculturalist has had to combat. For the Cochineal Insect is a near relative of the dreaded "fluted cushion scale," which, in a single year, reduced the Californian orange crop from eight thousand to six hundred car-loads by its ravages. Nor is it far removed in affinity from the "green-fly" which infests our roses and broad beans.

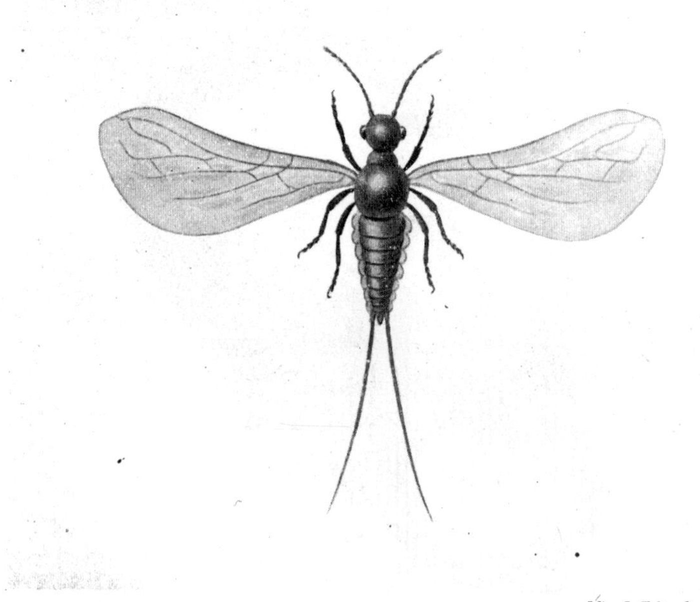

[By J. Tektenburg.

THE COCHINEAL INSECT.

The male of the Cochineal Insect when mature is provided with a pair of wings, but he is short lived and has little opportunity for using them. The figure is greatly enlarged: in actual size the insect is little larger than the common Green-fly.

The Cochineal Insect feeds, by sucking the sap through its delicate, tubular mouth, upon several kinds of cactus of the "prickly-pear" group. Its favourite food-plant is known as Nopal. A native of Mexico, it was introduced into the Eastern hemisphere, after several unsuccessful attempts, about the year 1827, and was established with more or less success on the borders of the Mediterranean.

It was in the Canary Islands, however, that its acclimatization was most successfully effected, and here it was formerly the object of an extensive commerce. Forty years ago the imports of Cochineal from the Canaries to Britain were valued at over £800,000.

The life-story of the Cochineal Insect is very remarkable. The young, when first hatched from the egg, are mite-like creatures, with six legs, which move rapidly about upon the stems of the cactus until they find suitable anchorages for their sucking-trunks. Then each one settles down to a sedentary existence and subsists upon the sap.

The female Cochineal Insect develops backwards—if I may be allowed this hibernicism. It begins life, as we have seen, as an active, six-legged creature ; it ends as an inert grub, so degraded in form that most of its external organs and appendages cannot be distinguished. But when once it has got its sucker through the skin of the cactus into the juicy tissues beneath, the sucker is never withdrawn. The insect enjoys a continuous feast, for which it pays with loss of the power of movement. The male, on the contrary, undergoes a complete metamorphosis, and eventually attains to the dignity of two delicate wings, and the liberty to use them. Its body is red and its wings white. It has also large eyes and a pair of wonderful ten-jointed feelers; but it lacks a mouth, and is so quite unable to feed. Needless to say, its life is of short duration.

The Female Cochineal Insect becomes, at the close of her life, a veritable egg-sac; and it is at this period that her commercial value is at its zenith, as her tissues then contain the greatest amount of colouring matter. The insects are brushed or scraped in their thousands from the cactus plant and fall upon sheets which have been placed upon the ground. The strange harvest is then killed, either by immersion in hot water or by the heat of a specially constructed kiln or oven. It is afterwards subjected to a careful drying process, at the conclusion of which it is ready for the market. For commercial purposes, Cochineal is separated according to its quality and place of origin into a number of grades, some of which command far higher prices than others. Cochineal was formerly used medicinally, by the way, chiefly as a specific for whooping-cough. It has been estimated that a pound of the dried commercial Cochineal contains about seventy thousand insects. From these about ten per cent. of pure dye is obtained.

Photo by] [*Harold Bastin.*

COCHINEAL INSECTS ON CACTUS.

The Cochineal Insects are here seen closely attached to the Cactus and spending their lives in sucking its juices. The dried bodies of the insects as shown here constitute the Cochineal of commerce.

If the female insect is left undisturbed upon the cactus plant—as, indeed, some of them must be in order that the species may continue—she deposits her eggs beneath her body. A few hours later she dies, and her body subsequently dries up and hardens, forming a kind of concave scale, or shell, which protects the eggs, and from beneath which the young eventually creep.

It is this scale-like form of the female of the Cochineal Insect and its allies that has earned for the group to which it belongs the general designation of Scale Insects. Any amateur gardener who has a greenhouse in which a collection of cacti is growing may find either the Cochineal Insect or one of its near relations somewhat disfiguring his plants by their brown scales attached firmly to the skin. As a rule these are not suspected of being insects, but are regarded as marks of decay.

ENCYCLOPAEDIA OF THE UNIVERSE

ENCYCLOPAEDIA OF THE UNIVERSE

A POPULAR WORK ON THE MARVELS OF

THE HEAVENS :: THE EARTH :: PLANT LIFE :: ANIMAL LIFE :: THE MIGHTY DEEP

WITH 1000 FINE ILLUSTRATIONS

Vol. II

Editor

WALTER HUTCHINSON

B.A., F.R.G.S., F.R.A.I., Etc.

WITH AN INTRODUCTION BY

LORD AVEBURY, P.C.

AND WITH CONTRIBUTIONS BY LEADING SPECIALISTS, INCLUDING :

Sir Harry Johnston, G.C.M.G., K.C.B.
Richard Lydekker, F.R.S.
Frank Finn, F.Z.S.
Edward Step, F.L.S.
E.W. Maunder, F.R.A.S.
W.P. Pycraft, A.L.S., F.Z.S.
Russell F. Gwinnell, F.C.S.
Frank T. Bullen, F.R.G.S.
Tickner Edwardes
J.J. Ward, F.E.S.
C. Flammarion
Lionel E. Adams, B.A.
Harold Bastin, F.E.S.
S. Leonard Bastin
K. G. Blair, F.E.S.
Sir Herbert Maxwell, F.R.S.
E. B. Poulton, F.R.S.
G. F. Chambers, F.R.A.S.
H. St. J. K. Donisthorpe, F.E.S.
Richard Kerr, F.G.S.
W.F. Kirby, F.L.S.
Hugh Main, F.E.S.
Ernest Marriage
E.A. Martin, F.G.S.
T. E. R. Phillips, F.R.A.S.
R. I. Pocock, F.R.S.
Dr. E. J. Spitta, F.R.M.S.
E. W. Swanton
J. Sinel
W. Mark Webb, F.L.S.

DISCOVERY PUBLISHING HOUSE PVT. LTD.

NEW DELHI-110 002

Published by:
Namit Wasan

DISCOVERY PUBLISHING HOUSE PVT. LTD.
4383/4B, Ansari Road, Darya Ganj
New Delhi-110 002 (India)
Phone : +91-11-23279245; 23253475; 43596065
E-mail : discoverybooksindia@gmail.com
discoverypublishinghouse@gmail.com
namitwasan9@gmail.com
web : www.discoverypublishinggroup.com

Edition: 2020

ISBN: 978-81-8356-873-9 (Set)

Encyclopaedia of the Universe

Printed at:
Infinity Imaging Systems
Delhi

CONTENTS OF VOLUME II.

Vol. II

PEARLS AND PEARL OYSTERS

BY CHARLES DE BOISMAISON.

WHEN the Pearl Oyster was very young, it left the parental mantle and floated along, a mere film, on the ebb and flow of the tide. Each day it grew a little bigger, a little heavier, a little more important, while its increasing size protected it more and more from the voracious appetites of one or other species of the innumerable hosts of marine life that lie in wait for such minute organisms. As it grew in bulk and weight, it became less active, till at last it sank upon "the ooze and bottom of the sea" and fastened itself by its beard, which grew near the hinge uniting its shells, to the most convenient rock that chanced to project from the gravelly soil. There it grew still larger, still heavier, still more important, opening its two shells to let the water flow in, for the water carried the minute life which nourished it; and so a year or two passed by. One day a microscopical creature ventured too near these open doors and was swept in by the current, but not near the expectant mouth. It was caught in the tentacle-fringed mantle that was moving with a gentle undulation along the lip of the shell. This very movement carried it still further into the mantle of the oyster, and caused the tiny victim to break the filmy skin that envelops the edge of this mantle and carry a fragment with it into the interior of the expectant oyster. Now is the biter bit. The oyster cannot swallow his disappointing visitor, for that visitor is nowhere within reach of his mouth; and while he is himself denied the expectant feast, the little creature who was to form that feast is causing intense irritation to his nervous system. Time, however, and his own constitution help to ease him of his annoying intruder. His nervous disorder causes him to secrete the delicate film with which he has lined his shell and which has hardened into mother-o'-pearl.

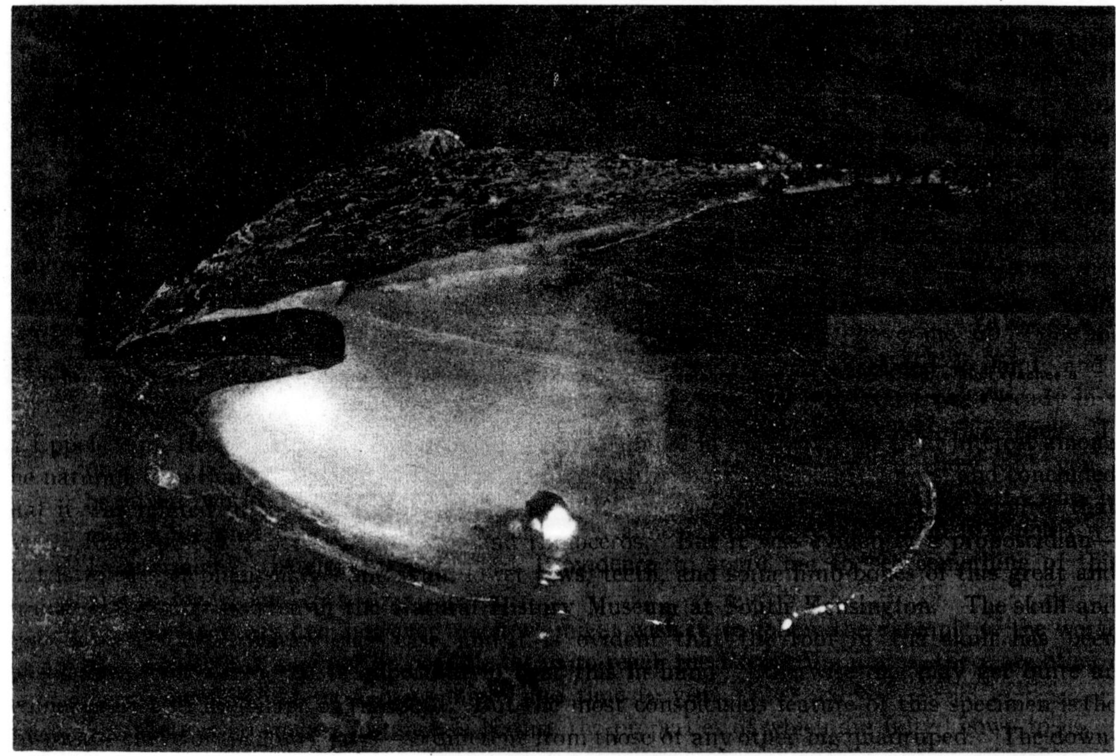

Photo by] [*W. Saville Kent.*

THE PEARL OYSTER'S SHELL.

The open shell of the Pearl Oyster after the death of the animal, showing a large pearl. This was probably formed in a free condition and globular in form, but afterwards became attached to the shell.

With one film after another he coats the now lifeless cause of his annoyance. Still he laboriously excretes the delicate substance, till at last he possesses a treasure that makes him the envy of vain woman and the desire of covetous man—a pearl, a pearl of price ! Now comes the diver and wrenches him from his rocky bed. He is bundled into sacks with thousands of his unfortunate companions ; with them he forms a heap of dying and dead oysters piled on the shore. It is not long before he, too, is dead, and the two shells gape open, disclosing the sheen of his beautiful inner home. He is then dismembered, and his little green-grey pulpy body is cast with those of his kindred into the great tub standing ready to receive them. Now these murdered innocents take a late revenge on their captors. Their bodies decay and liquefy into an evil-smelling, disgusting " broth." Did I say evil-smelling ? The term is too light for the foul stench that is given off by hundreds of decayed oysters. The putrefying remains are left for a year or so, and then the liquid is drained away. At the bottom of the tub will at last be found the pearl which has caused the destruction of the unhappy oyster.

By permission of] [*Charles de Boismaison.*

THE ENGLISH RIVER MUSSEL.

It is this shell which is the chief pearl-producer of the Scots fisheries.

By permission of] [*Charles de Boismaison.*

THE GIANT STROMBUS.

This splendid shell has occasionally produced pearls of a pink hue, and of good quality, though apt to fade.

That is the life history of the pearl oyster and the jewel it secretes. As many as one hundred and fifty pearls of varying sizes have been found within a single shell.

But the common pearl oyster is not the only shell that bears pearls ; specimens have been found in the great univalve shell known as the Giant Strombus, a photograph of which is given on this page. This shell, which has an inner lining of coral pink, bears a pearl of the same pink hue, which, however, is apt to fade with age ; nor are these gems very often found in Strombus, though eagerly sought for on account of their good commercial value. Again, there is the river-mussel which inhabits the fresh-water streams of Scotland and the North of England and which was once noted for its pearls, though of late they have greatly decreased in commercial importance, as they are of no great size and have a dull, lifeless sheen which is comparable to that of glazed porcelain.

Pliny relates that a red-coloured pearl was in his day taken from a shell named Myes, which inhabited the Thracian Bosphorus. Another shell in which pearls have been found is the hammer-headed oyster, but such specimens are prized more on account of their rarity than for any peculiar quality of the gem itself. The common edible mussel and some species of the wedge shells must also be added to the list of pearl-bearers. These names are necessarily taken as representative of certain families, many members of which are to be found throughout the world, all bearing the same specific quality of pearl production. For instance, the river-mussels of England and Sweden have

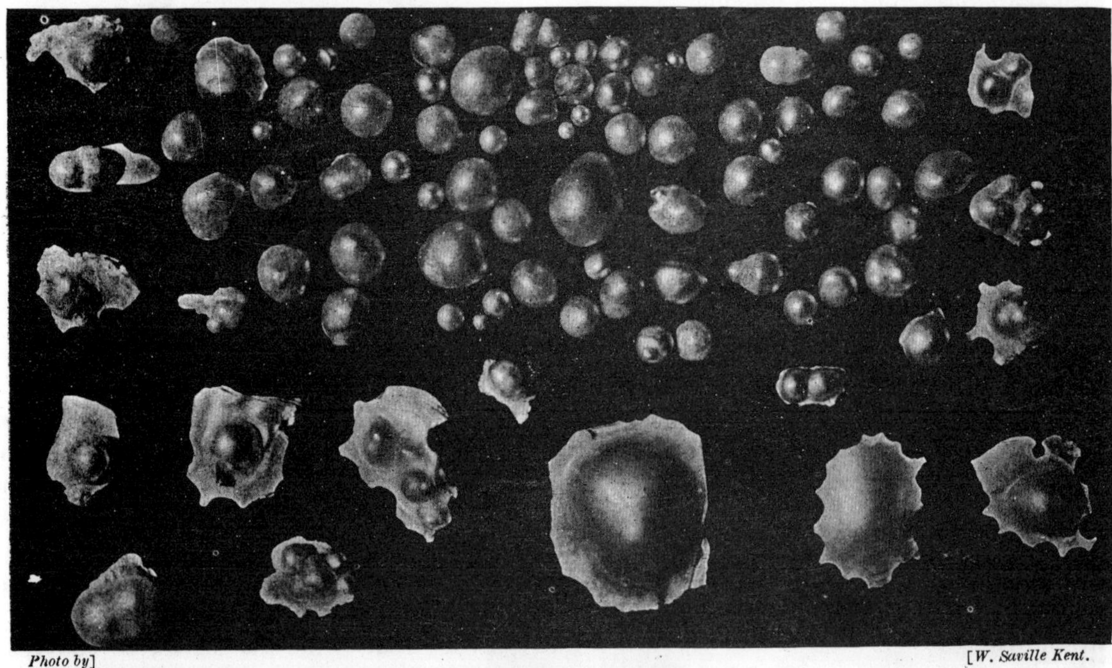

Photo by] [*W. Saville Kent.*

PEARLS

A photograph showing the several varieties of pearls. In the higher part of the picture are more or less perfect pearls, amongst which, however, can be seen several "baroque" pearls. To the left are pearls immersed in a nacreous matrix, while at the bottom of the picture are simple "blister" pearls.

their counterpart in those of America, which are often furnished with a nacre of great depth and beauty and sometimes contain excellent pearls.

A variety of colours are to be found in pearls—some are dull white, like those of the English rivers; many shells bear pink pearls; while uncommon shades, such as apricot, green, deep rose, purple and black are sometimes found. There was in the famous Hope collection a pearl found in American waters, which glowed with the fiery colours of the opal, tempered with that soft lustre which is the peculiar attribute of this gem. Mr. Saville Kent mentions in his book, " The Naturalist in Australia," the finding of several pearls the colour of gold, and a case was brought before the notice of the writer of a man who possessed a set of studs headed with pearls that could not easily be distinguished from solid gold. But the pearls of greatest beauty and renown are those taken in the fisheries of the Persian Gulf or Ceylon. They are of the true Orient variety—creamy, mellow and lustrous. This lustre results, not from any peculiar quality of the nacre, but from the slightly uneven surface of each coating, which causes the globule to catch and diffuse the light. To what circumstances the coloration of pearls is due is a question not yet completely settled, though the most acceptable theory ascribes the phenomenon to some mineral substance held in solution in

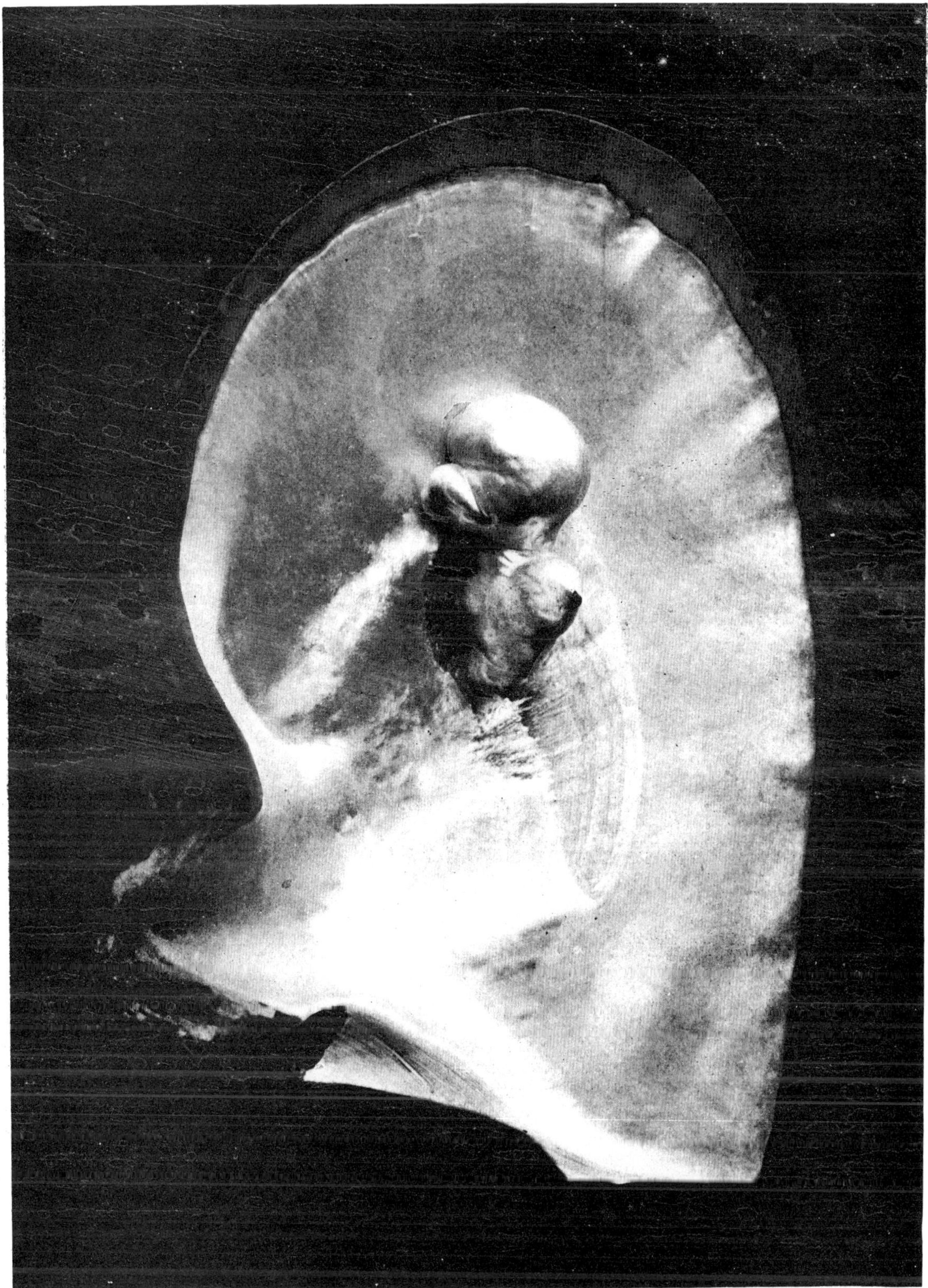

Photo by] [*W. Saville-Kent.*

A WESTERN AUSTRALIAN PEARL.

From the collection of the late W. Saville Kent. This remarkable pearl is just two inches in its largest diameter, and bears a singular resemblance to a human head.

the waters which furnish the food supply of the mollusc. The theory that the colour is due to external influences rather than to an inherent quality of the oyster itself is supported by the fact that a pink and a black pearl have been found in the same shell. Such a change of colour could have been brought about by a change in the tidal currents, bringing at one time one mineral substance, at another material of a different nature; but even here the data are insufficient to form a conclusive proof.

All pearls are not perfect, for only those secreted in the fleshy parts of the oyster attain the regular shapes that are essential in the true "virgin" pearl. These shapes are either globular, pear-shaped, or egg-shaped. Some pearls adhere to the shell of the oyster, and becoming shaped like half-globes, are known as "button" pearls; others are formed round a core of irregular shape, and themselves become irregular. These are known as "baroque" pearls. A pearl is easily skinned, layer after layer of the precious substance peeling readily off. Baroque pearls are sometimes subjected to this treatment, for it may happen that a perfect virgin pearl will be found secreted in the misshapen globule.

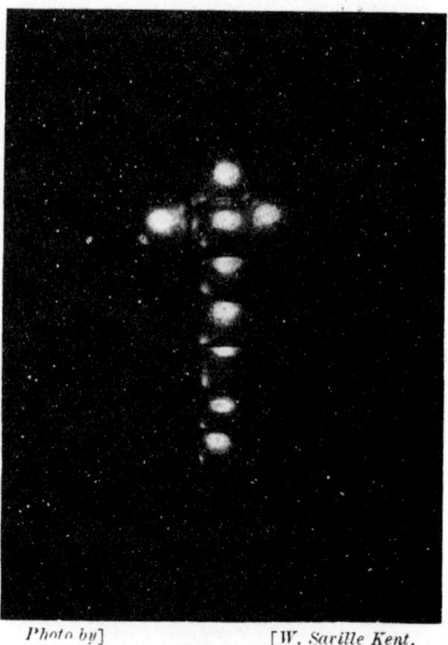

Photo by] [*W. Saville Kent.*

THE SOUTHERN CROSS.

This remarkable Western Australian pearl has been valued at £10,000. It is composed of nine amalgamated pearls.

Several such specimens are shown in the lower part of the illustration on page 282, while a particularly fine example attached to the mother-shell is shown on page 283. In the sixteenth and seventeenth century such pearls were greatly in demand for making into grotesque pendants; and famous jewellers have employed their skill in devising settings to show off these quaint gems. Several good examples of such work are to be found in the collection of jewellery lent by Mr. Pierpont Morgan to the Victoria and Albert Museum, South Kensington. Yet others are fashioned by the mollusc to repel the invasion of some boring parasites; they are hollow, and are therefore termed "blister" pearls; some of these hollow pearls are not even composed of the same material as the pearl proper, but partake of the nature of the mother-of-pearl, and to these is given the technical name "coq de perle."

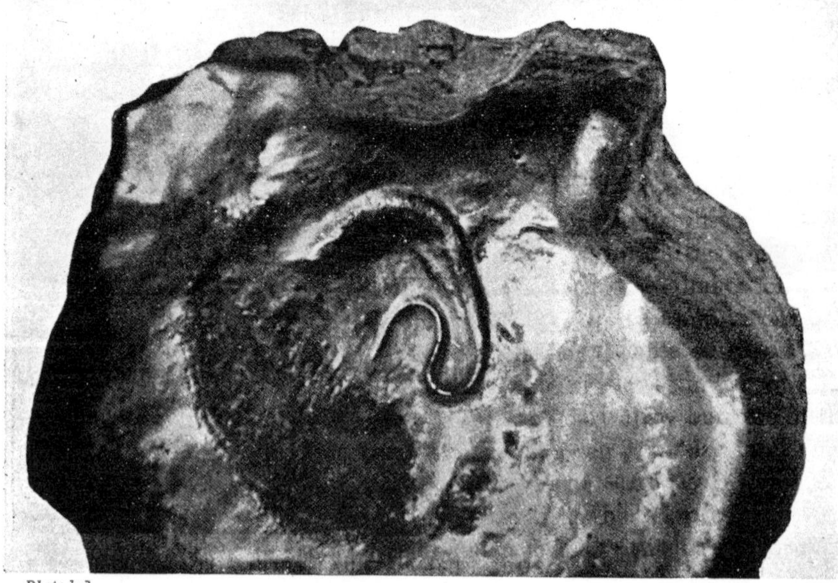

Photo by] [*W. Saville Kent.*

A STRANGELY-FORMED PEARL.

A small fish, Fierasfer, has here entered the shell of the mollusc and has been entombed in the nacre deposited by his host

There is only one certain test for pearls, and that is trying them between the teeth. An imitation is easily detected by its hard, slippery surface, for the real gem always feels gritty.

Mention has already been made of the primary cause of pearl formation—the embedding of a minute organism in the body of the mollusc. Researches have lately been carried out to ascertain the exact nature of this organism. It has been found that all species of pearl-bearing shells are not attacked by the same parasite. Some, indeed, owe their pearls to the desire to rid themselves of the irritation caused by a minute grain of sand or particle of wood. A photograph is given on page 284 which shows a small fish that has entered the shell of a mollusc and has been entombed in nacre by the oyster. But as the Ceylon oyster fisheries produce the finest pearls, the parasite of the Ceylon oyster is most worthy of investigation. It has been proved by Professor Herdman to be the grub of a tapeworm. This grub becomes embedded in the oyster, dies, and is entombed by its host in fold on fold of nacre; while the form of the grub, which is globular, influences the shape of the pearl.

So it is with a dead worm in its little shining tomb that the woman of fashion adorns herself. It is the possession of numbers of little dead worms hidden in lustrous globes that makes her to be envied of her fellows. "Vanity of vanities," saith the preacher, "all is vanity!"

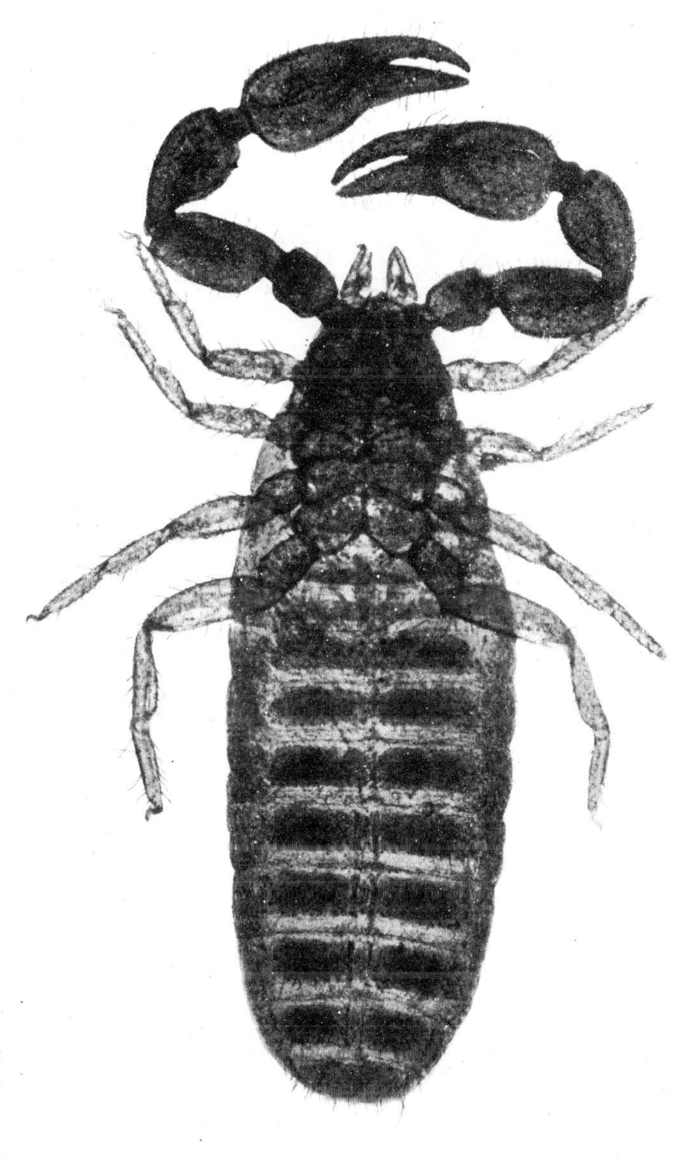

Photo by] [*Ernest Marriage*

A FALSE SCORPION.

Such is the name given to this formidable-looking creature, which, however, is here represented about forty-seven times larger than the natural size. With its great nippers it attaches itself to the leg of a fly, using its host as a means of conveyance.

FALSE SCORPIONS

AMONG the surprising adventures of Sinbad the Sailor, it is related that he once found himself surrounded by precipitous cliffs up which it was impossible to climb. After fruitless efforts, he at last made his escape by tying himself to the leg of a "roc," a gigantic bird, which flew out of the valley carrying Sinbad. Possibly a somewhat similar service is rendered by the house-fly to the small animals known as Chelifers, or false scorpions, one of which is represented in the accompanying photomicrographs. The whole Chelifer is shown about forty-seven times, and the illustration below about sixty times the natural size. If one looks closely at flies, especially in warm weather in the autumn, now and again a small object will be seen attached to one leg of a fly, generally one of the last pair; it is pretty sure to be one of these creatures. One day in September I caught six of them in one room (not in an old house, by the way), all held on by one claw only, and in most cases the grip was taken high up on the fly's leg. The one from which the illustrations were made, a particularly large specimen (though its length of body was increased in the course of preparing it as a microscopic slide), was attached to the foot of a fly, and it was seen in this position with some hours' interval between observations. The flies seem to suffer a certain amount of inconvenience, which they show in endeavouring to get quit of their passenger by rubbing their legs together. The Chelifer apparently uses the fly, as Sinbad did the roc, as a means of changing his quarters. Autumn is a time of migration, and he possibly feels the need of a warmer place of abode for the approaching winter.

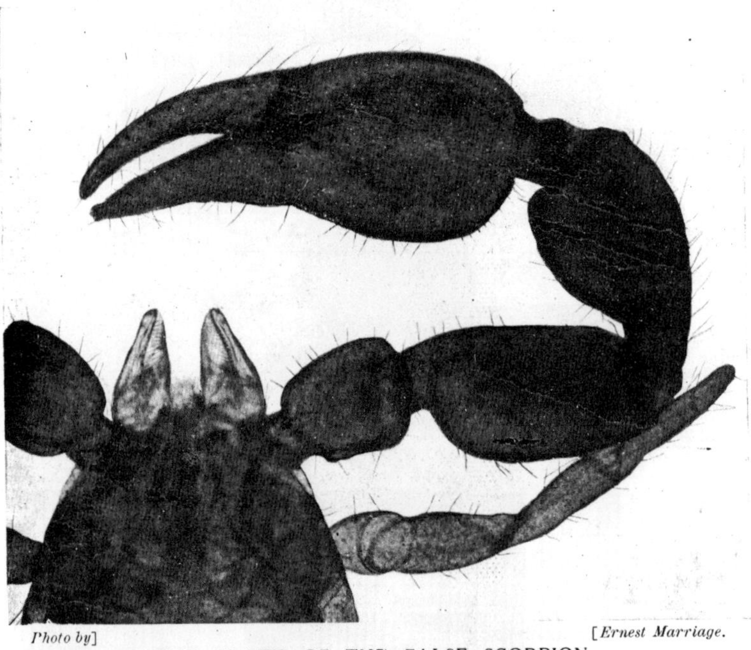

Photo by] [*Ernest Marriage.*

THE CLAWS OF THE FALSE SCORPION.

These are attached to the head of the creature, which is further armed with two pairs of powerful jaws with which to suck the juices of its prey. The creature is also known as the Book Scorpion, as one of its favourite haunts is an old and disused book.

On the other hand, these false scorpions may actually be parasites on the flies. In the larger illustration, two pairs of jaws armed with long teeth are shown, which are probably used for gripping and at the same time sucking the juices of its prey after the manner of the real scorpions. Thus the Chelifer seems well equipped for a parasitical existence, and the absence of eyes in the creature points in the same direction. However, as far as my personal observation goes, these little animals are never attached to the body of the insect. Again, their connection with flies may be purely accidental. Both for hibernating and for laying their eggs, flies frequent places which are likely haunts of the Chelifer; the latter may seize the fly's leg in mistake for its regular prey, and once fastened, it certainly clings with great tenacity.

In colour the Chelifer is a reddish-brown. Besides the formidable lobster-like claws already mentioned, which are attached to its head, it has eight legs, a characteristic of the spider tribe to which it belongs. It bears in several respects a remarkable resemblance, though of course in miniature, to the true scorpions (fortunately not natives of the British islands), to which it is akin. Anyone comparing these photographs with drawings or photographs of real scorpions will see how similar are the claws and head; the false scorpion, however, lacks the formidable tail with the sting at its tip. They are also called Book Scorpions, because in olden, and probably dirtier, times they were to be found in libraries. They are not uncommon in store-rooms of old houses, in piles of dusty and neglected papers, and in stables; also under the bark of trees, in rotten wood or moss and under stones. Though the Book Scorpion looks a fearsome creature when seen under a

[By A. Twidle.

THE HORNBILL ON HER NEST.

The method of "nesting" of the Hornbill is very peculiar. The hen makes her nest in the hollow of a tree, and when incubation commences the cavity is filled up by the male bird, with an aperture only for the beak of the sitter. He then feeds his mate assiducusly until the eggs a.e hatched.

microscope, it is probably more useful than harmful, and helps, by killing and eating mites and small insects, to preserve rather than destroy books.

By permission of] [*Sir H. H. Johnston.*

THE RED HORNBILL.

It is supposed—and there is little reason to doubt the supposition—that the Hornbill is the original of the fabled Phoenix of the Ancients.

HORNBILLS

BY

SIR HARRY JOHNSTON, G.C.M.G.

THE Hornbills, which range in size from a bird no larger than a crow (about eleven inches in length) to one that exceeds a large turkey in size and measures five and a half feet from the tip of the bill to the end of the tail, belong to the Picarian order of birds, of which they are the largest representatives. Together with the Todies, Kingfishers, Bee-eaters, Rollers and Motmots, they constitute a division of this order, which is characterized, amongst other features, by a joining together for half of their length of the third and fourth toes. Probably their far-back ancestors had their toes turned two and two like those of the Trogons, during which process the third and fourth toe grew close together. Then the second toe was once more turned forward, producing the type of foot which is now characteristic of the Hornbills and Kingfishers.

But the casual observer will be more struck by the extraordinary beaks of these grotesque birds than by their peculiarly fashioned feet. The beak is always very large, though perhaps it does not in any species attain to such a disproportionate size as in the American Toucans (whose resemblance to the Hornbills is merely accidental and not due to near relationship). In some Hornbills not only is the beak very big, but it is surmounted by a casque or helmet, commencing on the forehead and growing on the top of the upper mandible. Thus certain Hornbills possess a really marvellous appearance, and undoubtedly attracted so much attention from the ancient Aryan invaders of

India and Ceylon that they originated the myth of the Phœnix: though how they came to be associated with fire and immortality I cannot say. But the Arab and mediæval Italian accounts of the "Phœnix Bird" of Ceylon and Malaysia can all be traced back to an exaggerated description of one or other species of Hornbill; and there is said to be still amongst the Crown jewels of the Sultan of Turkey a cup which is carved out of the solid helmet of a Malay Hornbill, and is described as being made "from the beak of a Phœnix." In this species both beak and casque are solid and have the appearance of ivory; but in all the other Hornbills the swollen part of the upper mandible, and, above all, the helmet or casque, are hollow; that is to say, the structure is a series of cells with light pithy walls, so that the monstrous beak and its excrescence are much lighter to carry than they appear. Another species of Hornbill found in India and Malaysia has a double casque on the top of its beak divided into two horns, while a kindred form here illustrated has the beak and casque of a bright brick-red colour, the casque being curved like a crescent and consequently apt to be manufactured into an ornamental sword-hilt by the natives.

The big Rhinoceros Hornbill, which is here illustrated in two forms, has the beak yellow-white in colour. The casque, which is coloured red, orange and black, is curved crescent-shaped in the Indian form, and in another species or sub-species found in the Malay Islands it is very large, and straight along the top. In the great Ground Hornbills of Africa—which have developed long legs and bear a misleading resemblance to turkeys—the casque is abruptly truncated at the end as though it had been chopped in half by an axe, and the sides are curiously undulating.

But less than half the species of this extraordinary group (which amongst other peculiarities develops thick black eyelashes) have no casque on the top of their beaks, though there may be a

By permission of] [*Sir H. H. Johnston.*

THE NIPAL HORNBILL.

This bird is remarkable for the peculiar notches in the beak. Although they have the appearance of teeth they do not seem to be used for the purpose of mastication, as all food is swallowed whole.

slight ridge instead. In the absence of a helmet, the exuberance of development about the head is manifested in some cases by the growth of a crest or mane. Mane is almost the word which might be applied, for example, to the great Red Hornbill of Nipal. In the still more remarkable-looking Maned Hornbill of Sumatra and the Malay Peninsula the head-feathers grow into a thick crest which projects over the basal part of the beak in delicate white spikes.

It will be observed that the beak of the Nipal Hornbill is strongly toothed, the teeth, of course, being merely due to deep notches in the edge of the mandible. This is a feature developed in not a few Hornbills; but it is difficult to see how it can be useful to the bird, since the Hornbills do not seem to crush or masticate their food, but to swallow it whole. The teeth, like the coloured ridges in the beaks of some Hornbills, may be only an additional "swagger" development, intended, like the crests, to strike terror into their enemies. Curiously enough, in confinement, Hornbills—especially the Ground Hornbill of Africa—are amiable, good-natured birds, that never seem—at least, that is my experience, and I have kept or observed not a few of them in captivity—to make use of their formidable beaks in an aggressive manner or with the spitefulness of parrots, and they probably look much fiercer than they really are, and depend for protection more on their exceedingly nasty-tasting flesh than on their clumsy and not over-strong beaks. The black and white Hornbills of Africa have developed an extremely loud trumpet-call, which is sometimes like the braying of an ass, sometimes like the loud cries of a woman in distress. It certainly serves to scare away from the Hornbills' haunts not only the native hunter, but also monkeys, baboons, wild cats and civets, which might contemplate any tearing open of the nest and devouring of the callow young.

Photo by] [*W. S. Berridge, F.Z.S.*

THE BLACK HORNBILL.

In this variety the great horn is supplemented with a black mane. This horn, it should be remembered, is hollow and constructed of a series of cells with light pithy walls, which minimizes the apparent weight.

Hornbills have very peculiar nesting habits. A hollow is excavated in a tall tree, generally in the fork between two branches. The bottom of the nest is covered with chips and dust dropped into the excavation. On this surface about four eggs are laid—sometimes more, sometimes less. As soon as the last egg of the clutch has been deposited, the female applies herself to incubation, and the male then, using his beak as a trowel, plasters up the orifice of the nest with clay in such a way that there is only a small hole left for the protrusion of the beak and head of the female, who is thus

By permission of] [*Sir H. H. Johnston.*

THE MANED HORNBILL OF SUMATRA.

In this species the casque of the Hornbill has disappeared and has been replaced by a great crest or mane, which in the species here shown is white.

immured until the eggs are hatched. During this period the male feeds her assiduously with insects, carrion, small reptiles and fruit. When the young are hatched, male and female tear away the dry plaster of mud, and thenceforth one or other of the parents mounts guard over the callow young until they are large enough to leave the nest. Hornbills are, of course, monogamous, and I should think paired for life.

The Black Hornbill of Western Equatorial Africa resembles certain Asiatic Hornbills in the difference of coloration between male and female, the adult male being black in plumage, while the female has the head, back and breast of chestnut-brown. For the most part, however, the coloration of the large Hornbills is mainly black and white, the black being often glossed with tints of iridescent green and blue. When there is any third colour in the plumage it is generally the chestnut-red referred to. But these birds are often rendered still more picturesque by brightly-coloured spaces of bare skin round the orbits of the eyes, or on the throat.

By permission of] [*Sir H. H. Johnston.*

RHINOCEROS HORNBILLS.

The beak of these birds is yellow-white; the casque is coloured red, orange and black. Notice that the casque is, in the upper bird, which comes from the Malay Islands, very straight-edged, while the lower Indian form is sharply curved.

Two Black Hornbills of West Africa develop a wattle on the throat, which is leaden-blue in colour, contrasting in the male with the bright-red and purple-blue spaces of wrinkled skin at the base of the beak and round the eyes. The Maned Hornbill of Sumatra, here illustrated, has the naked spaces on the face almost a cobalt blue in tint when the bird is alive. In other forms the cheeks are lemon-yellow.

THE KING CRAB

BY R. I. POCOCK, F.R.S.

THE King Crab is a wonderful creature for many reasons. In the first place, he holds amongst existing species in the animal kingdom a position which is almost unique for its isolation. Although commonly called a crab, because he lives in the sea, breathes by means of gills and uses the basal segments of his legs for chewing his food, he is not a crab at all in the sense in which that term should be properly understood. Indeed, there is far less of the crab about him than there is about a wood-louse, or a sand-hopper, or any of that great and heterogeneous assemblage of aquatic organisms popularly designated shrimps. It would be much nearer the mark to call him the King Scorpion, for his kinship to the Scorpion, although admittedly remote, is not now doubted by any competent judge. So long as this relationship is clearly understood and remembered, no harm is done by adhering to the popular name of King Crab for these animals. But we have to trace their lineage a long way back in the geological records before finding animals obviously related to them, which, by their superficial as well as fundamental resemblance to Scorpions, carry conviction as to the truth of the kinship above claimed. Unlike the King Crabs, these animals, the Eurypterida, were narrow, long-bodied, active swimmers, with the integument jointed as in Scorpions. In the King Crabs, as shown in the illustrations on this page, the body is exceedingly broad and short, and possesses only one large transverse joint, which divides it into an anterior portion and a posterior portion. The edges of the former are produced sideways, and in front into a great downwardly bent, roof-like expansion, completely concealing the stout locomotor limbs which are tucked away beneath it. Of these there are five pairs, all except the last pair ending in powerful nippers. The mouth is placed in the middle of the lower surface of this region of the body, between the basal segments of the limbs, which are armed with spines, in the place of teeth, for tearing to

Photo by] [*W. S. Berridge, F.Z.S.*

AMERICAN KING CRAB.

On the upper side the King Crab is entirely protected by the two large armour-like shells. Only the long slender "tail" is exposed, but this is covered, likewise, by hard shell.

Photo by] [*W. S. Berridge, F.Z.S.*

THE KING CRAB.

The animal is here photographed on its back to show the jointed character of the under surface, and the five pairs of limbs. Four pairs of these are provided with nippers. The mouth is between the bases of these limbs.

pieces the worms and other soft-bodied animal organisms upon which the King Crab feeds; and in front of the mouth there is a pair of nipper-shaped jaws used for pushing food into the mouth. The part of the body behind the waist-like joint is produced sideways into a shelf-like ridge armed with spines, and it is hollowed out beneath to form a chamber for six broad foliaceous appendages, the last five of which carry two clusters of delicate plate-like gills. Loosely, but strongly-jointed to the end of the body, there is a long, spiniform tail.

From this description it will be seen that, except for the fusion of all the segments of the body to form the two areas above described, the King Crabs have a strong resemblance to those very ancient creatures called Trilobites, which occur in the earliest fossiliferous rocks, but became extinct no one knows how many millions of years ago. Like the Trilobites, too, they have a pair of large kidney-shaped eyes, each composed of multitudes of ocelli, on the upper side of the body.

King Crabs are essentially shallow-water species, usually living on sandy or muddy bottoms at a depth of not more than six fathoms. In the early summer they come to the shore in great numbers to spawn; and advancing with the spring-tides they leave their eggs in the mud near high-water mark so that they are exposed to the heat of the sun between the high tides.

Existing King Crabs, of which there are several well-marked species, have a curious distribution. They are found only on the western coast of the Atlantic, where they range from about the latitude of New York to the Gulf of Mexico, and on the western shores of the Pacific, from Japan in the north to Borneo and the adjacent islands in the south. That they formerly lived in European waters is shown by the occurrence of fossil species both in Tertiary deposits and in the chalk and other rocks of the Secondary epoch; and their wide geographical separation at the present time seems to be only intelligible on the supposition that there were formerly, although not necessarily contemporaneously, shallow water coast-lines between Europe and North America on the one hand and between Europe and Eastern Asia on the other, and this shows what far-reaching inferences as to the past history of the earth may be drawn from the study of one single small group of animals.

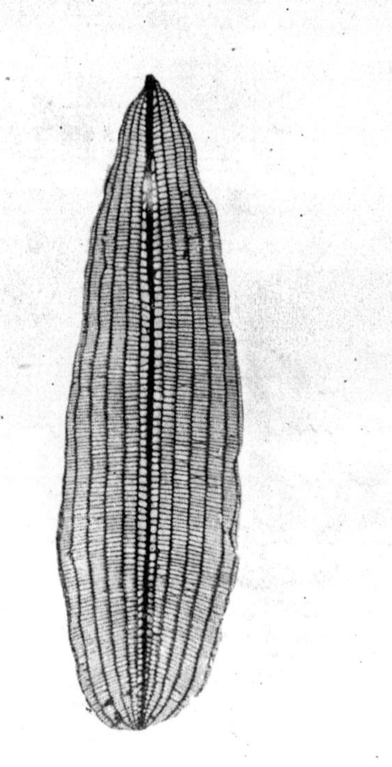

Photo by] [*W. Plomer Young.*

LATTICE LEAF.

The light portions of the leaf are clear spaces between the nerves. But for its colour and freshness it might be mistaken for one of the leaf-skeletons found in winter among the grass.

A LIVING SKELETON LEAF

BY EDWARD STEP, F.L.S.

It is a noticeable peculiarity of many water-plants that grow with their leaves submerged that the latter are thin and slender as a whole, or are divided up into a number of thread-like parts. A familiar instance is afforded by the common Water-Crowfoot of our ponds and ditches, which has leaves of two sorts, neither having the slightest resemblance to the other. The one kind of leaf is more or less ivy-shaped, feeling greasy to the touch and persistently floating on the surface of the water; the other kind is split up into a large number of dark green threads, and is always spread out at some distance below the surface. The floating leaf acts as a buoy, enabling the flowers to rise into the air. The submerged leaf is so greatly divided in order to expose as great a surface as possible to the water, and so

Photo by] [*E. J. Wallis.*

THE LATTICE LEAF PLANT.

This plant is one of the greatest botanical curiosities of the world. Its leaves are so completely riddled with holes that they appear at first sight not to have sufficient substance to carry on the ordinary functions of the leaf, but the object of this arrangement is to expose as large a surface as possible to the water in which it lives. It is a native of Madagascar.

enable it to extract the dissolved carbonic acid gas from the water. Many of the smaller seaweeds and their fresh-water allies are similarly fashioned. The Water Yam's leaf, commonly known in this country as the Lattice Leaf, unfolds from its bud with an unbroken, or almost unbroken, surface, but as it increases in size and the parallel veins get wider apart, the spaces between become perforated, and the soft green portion finally becomes visible only as a covering round the veins. By this mode of development, it will be seen, an enormous surface is exposed to the water.

The Lattice Leaf is about a foot long, but only a few inches broad, with a stalk twice or thrice as long, which enables it to float free of its neighbours. The flower-stalk, which rises above the water, divides into two or more branches, each bearing a large number of tiny flowers.

THE LUCERNARIAN

THE beautiful Lucernarian, as it has been often called, is an animal that in popular language may be said to be a kind of go-between 'twixt the jelly-fish and the anemone, possessing some of the attributes of each, but not exactly resembling either.

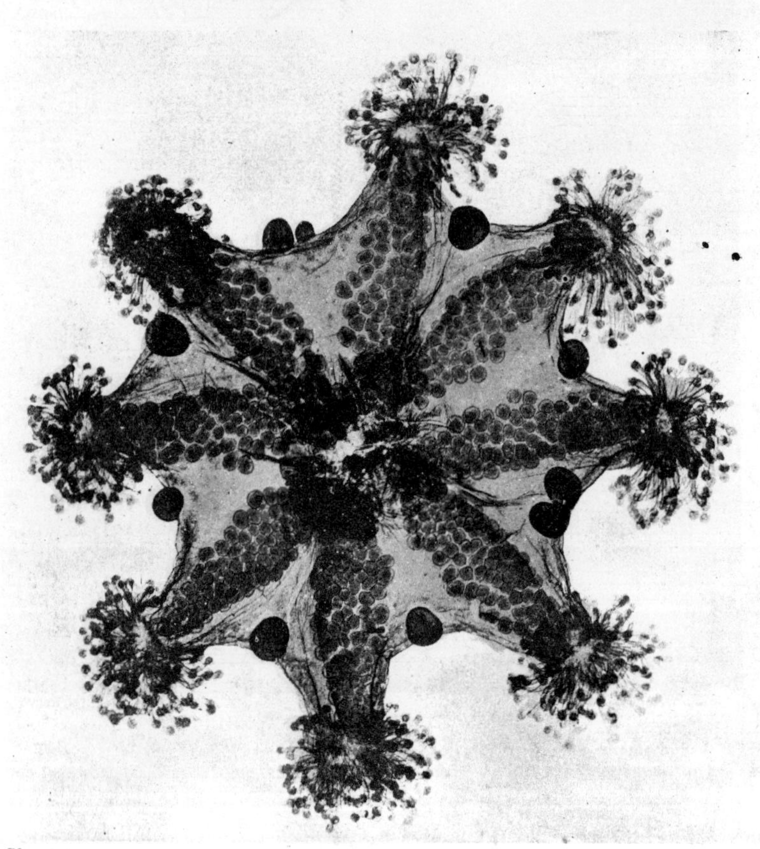

Photo by] [*E. J. Spitta.*

THE LUCERNARIAN.

This beautiful little sea-creature takes rank between the jelly-fish and the sea-anemone. It is about the size of a shilling, and is often to be found on pieces of floating seaweed.

This little bell-shaped organism—about the size of a shilling—is frequently found adhering to pieces of floating seaweed or to other objects. The body consists of a *single* individual form, and is not, like nearly all of the allied species, of a compound mass including several. It attaches itself to any object by a short pedicle, or foot-like process, well shown in the illustration on the opposite page; but it can detach itself at will, and swim *rapidly* about by contracting and opening its umbrella-like form. It can also move about *slowly*, like a leech, by fixing the tentacles, to be mentioned shortly, and dragging its body forwards till fixed again by its foot or stalk.

The Lucernarian is provided with eight of these tentacles arranged round the margin of the umbrella, as seen much magnified in the accompanying photograph; and each tentacle is composed of numerous increments—we have counted over seventy in one tentacle—each little tufted detail being a complete suctorial agent in itself; hence the whole tentacle taken collectively is a powerful organ.

The margin of the disc is eight-sided, the eight little processes supporting the tentacles radiating from the central, or mouth portion, which is really a four-sided arrangement. When food is taken, the whole of the umbrella portion rapidly folds itself over the prey, if such be *too large* to enter entirely the four-sided mouth ; hence in these circumstances, when the animal is feeding it assumes all manner of curious shapes, which for the most part resemble that of the object contained within.

Before closing this short account, notice should be made of the curious-looking little black objects—almost circular in shape, sometimes single and sometimes double—that lie on the margin of the disc of the animal between the tentacles, for their actual use has been much discussed. Some have thought them to be organs for stinging the prey—paralysing or killing it—although such is difficult to believe, seeing that no secreting glands or delivery tubes can be made out. Others believe—and this is the opinion mostly held—that they are simply additional suckers used in some way to assist the animal in making its *slow* gliding motion already referred to, and hence have been appropriately named " anchors."

THE ICHTHYOSAUR

BY SIR HARRY JOHNSTON, G.C.M.G.

WÜRTTEMBERG, in Southern Germany, is quite a small kingdom, only some seven thousand five hundred square miles in area ; but from the point of view of the science of extinct animals it is perhaps the most wonderful country in the world. This arises partly from the fact that during long periods of its ancient history it has been the great gulf of a shallow sea—a sea interspersed with islands and peninsulas. Together with the western parts of Bavaria it has produced some amazing links in the chain of life, many of which are referred to in the course of the present work. The sea which covered so much of the kingdom of Württemberg in the Secondary Epoch was no doubt connected through the valley of the Danube with the ancient sea of Hungary, the Black Sea and the Caspian—the " Sarmatian Sea," as it is sometimes called by zoologists. This sea swarmed with fish, and consequently attracted to its waters an incredible variety and number of fish-eating reptiles of those

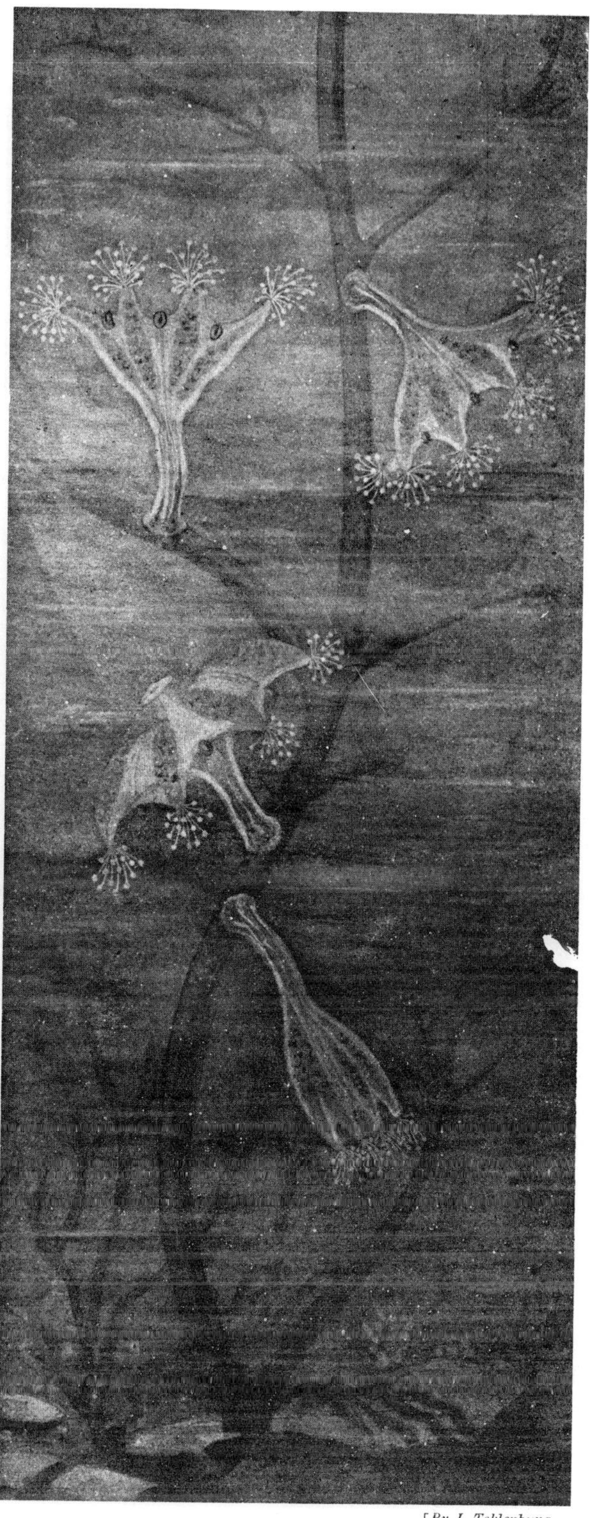

[*By J. Teklenburg.*

THE LUCERNARIAN.

Various phases of activity are here shown; for though the Lucernarian can swim rapidly, it can only crawl slowly, and is chiefly occupied in throwing out its tentacles for food.

distant days. Ever and again some catastrophe, such as a storm blowing them on shore, or a sudden landslide of the cliffs, immured these fish, together with giant amphibians and still more gigantic or extraordinary reptiles, in mud. In course of time the mud hardened into natural cement and rock, and was covered in its turn by other layers of deposit ; and thus large portions of Württemberg have become one of Nature's great museums, in which specimens of antique fish, reptile, bird and beast have been locked up for ages away from the discovery of men of science.

In some cases the ancient clay, hardened into rock, has made such a complete mould of the dead animal, that by filling up this mould with plaster-of-Paris an actual cast has been obtained, showing the shape of the dead creature. This has gone far to help zoologists to determine—with more or less accuracy—the appearance of some of the Ichthyosaurs when they were alive.

From these records of the rocks in Württemberg, in England, and in North America, we are able to understand that the Ichthyosaurs were a remarkable order of reptiles developed, far back in the Earth's history at the close of the Primary Epoch, from some very primitive ancestor of all the reptiles, which was at the same time closely related to the preceding frog-like creatures. In the course of

By permission of] [*Sir H. H. Johnston.*

A SKELETON OF AN ICHTHYOSAUR.

The members of this race of monsters were first of all frog-like creatures, who afterwards developed into huge water-reptiles, with great back-fins and a vertical tail. Examination of this skeleton will show that the tail end of the spine turned downwards in order to strengthen the lower lobe.

ages we see these Ichthyosaurs growing into huge water-reptiles, with something of the appearance of whales and dolphins. They evolved enormous eyes, protected by plates of bone, with which to scan the waters for their prey. Fins like those of a whale grew out of the back, and the tail developed in course of time two large lobes, which, however, were vertical, instead of horizontal like those in the warm-blooded whales. At first it used to be thought that most of the Ichthyosaurs had died with a broken end to their spine ; but through the investigation of Professor Fraas, of Stuttgart (some of whose photographs are here reproduced), it was shown that the tail-end of the spine turned downwards to strengthen the lower lobe of the tail. It is also evident from these records of the rocks that the Ichthyosaurs did not produce their young in the form of eggs which had to be hatched, but developed them inside the body until they could issue forth—three or four in number, perhaps—as living creatures, ready at once to take up the battle of life.

The fore-limbs of the Ichthyosaurs gradually lost the appearance of five-fingered hands, and became paddles filled with many small bones and covered by a leathery skin and much tissue, while the hind-limbs degenerated into mere flaps on either side of the vent. All the Ichthyosaurs of Europe retained their large number of sharp-pointed, uniform teeth ; but those which attained an extreme development in the shallow seas of North America had, ere the order became extinct, lost their teeth

[*By Charles Knight.*

THE ICHTHYOSAUR.

This restoration of the creature shows the great eyes protected by plates of bone. All the Ichthyosaurs of Europe had a number of pointed teeth, but later generations inhabiting the shallow seas of North America became toothless like so many of the whales.

By permission of] [*Sir H. H. Johnston.*

THE ICHTHYOSAUR.

This fossil reveals to us some of the life-history of the giant lizard-fish. We learn that the young Ichthyosaur was not hatched out of an egg, but was born fully developed.

and become toothless, like so many of the whales These strange water-monsters died out, as far as we know, before the beginning of the Tertiary Epoch and before they were displaced in the waters by the mammalian whales.

THE PLANET JUPITER

BY REV. T. E. R. PHILLIPS, F.R.A.S.

THOUGH surpassed by Venus in lustre, by Saturn in beauty, and by Mars in general interest, Jupiter easily holds the palm amongst all the planets in point of size. In fact, he is larger than all of them put together and exceeds our own little Earth more than thirteen hundred times. It is true we have every reason to believe that his globe is greatly expanded by heat, and that were he in the same physical state as the smaller planets his bulk would be much less, but all the same the revolutions of his satellites tell us that he must be more than three hundred times as massive as our own globe. This giant orb revolves round the sun at more than five times the distance of the Earth, and requires nearly twelve of our years to complete one journey.

If we look at Jupiter through a good telescope we see at once that he is not a true sphere, but flattened like the Earth at the poles, only to a much greater extent. This is because he is spinning round on his axis very fast indeed, and the same circumstance causes the dense vapours surrounding him to assume the form of parallel streaks and bands. These markings are by no means uniform, but irregular and spotted, as will be seen from the accompanying drawings.

Now it might be thought that by watching the spots as the planet spins round it would be easy to find out exactly how long the planet's day is. But this is not so ; partly because the markings we see are not solid or permanent features, but cloud-like forms which are constantly changing their shapes and places, and partly because the time it takes a spot to go round and come back again into view depends on its distance from the equator. This is very interesting, for it means that Jupiter, as we see it, does not rotate "all of a piece" like the Earth or any solid globe, but that the length of the day varies with the place on the planet's surface. If the Earth behaved like this, of course no permanent maps would be possible, for continents and islands and oceans would be perpetually changing their shapes and positions. In the case of Jupiter no part takes so long as ten hours to rotate, and equatorial spots perform their daily journey in about nine hours, fifty and a half minutes, though the time is somewhat variable. It follows that whereas a spot on the Earth's equator is carried round at a speed of about seventeen and a quarter miles a minute, a spot similarly placed on Jupiter has a velocity of nearly four hundred and seventy miles a minute, or well over twenty-eight thousand miles an hour !

The belts of Jupiter often show very beautiful and diverse colours. Reds and browns and purples and blues are not uncommonly to be found, and these, like the forms of the markings, are subject to gradual change.

But although change is the law of this planet, yet from time to time certain objects appear which last for many years. The most famous of these is the Great Red Spot, which was so conspicuous in 1878 and the years following, and which, though subsequently becoming very faint, has lately shown marked signs of revival. It is perhaps hidden at times by overlying clouds, but it is quite likely that it is identical with objects seen at various times during the last century, and perhaps with one seen by Hooke so long ago as 1664. Naturally many speculations have been made as

[*By T. E. R. Phillips, F.R.A.S.*

JUPITER.

Such is the appearance of Jupiter when viewed through a powerful telescope. It is so luminant that it shows only a bright disc and no spherical shape. Four of its eight attendant moons are visible, but are so far away that they appear only as small dots.

to what so remarkable an object can be—as, for instance, that it is a volcano, or a part of the true globe projecting through the clouds—but no one knows what it is. It is quite certain, however, that although it has lasted so long, it is not a fixture on the planet, for it is constantly moving both in longitude and latitude. In the large coloured drawing this famous object is seen much as it has appeared in recent years, lying in the hollow it has made for itself in one of the dark belts. Another long-enduring object in Jupiter is an outbreak of dark matter between the two chief belts in the southern (upper in the drawings) hemisphere. This disturbance, which first appeared as a comparatively small object in 1901, is well shown in two of M. Antoniadi's beautiful drawings. It is in the same latitude as the Red Spot, and, drifting much more rapidly, overtakes and, flowing round it, passes it every two years. In the first of M. Antoniadi's drawings the Red Spot is seen disappearing towards the left, and in the other coming on the disc, surrounded by the dark material of the great disturbance.

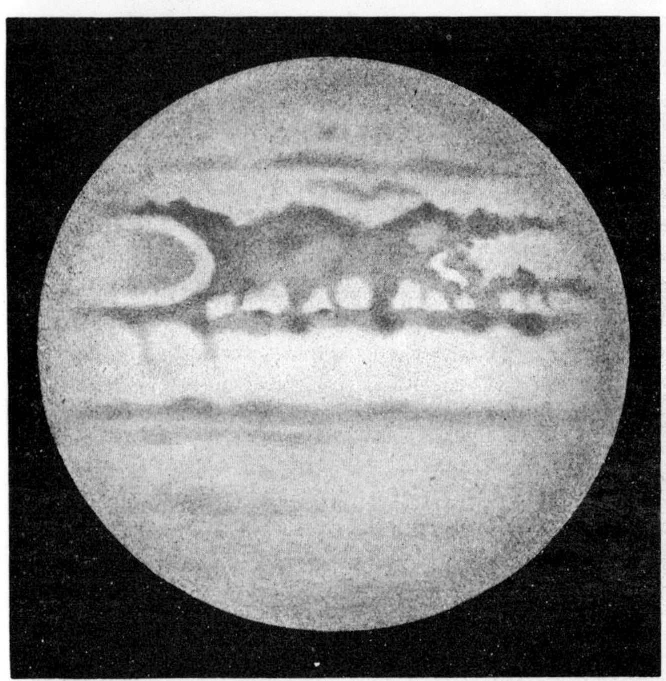

JUPITER.
Jupiter, from a drawing by M. Antoniadi in May, 21 days, 21 hours, 57 minutes

From what we have seen, it is clear that we cannot suppose Jupiter to be inhabited by creatures possessing animate life such as we know it on earth. Its hot, and perhaps molten, state renders it unfit for their abode; nevertheless Jupiter is intensely interesting, because it illustrates so well the theory of the gradual evolution of worlds. We know that although it is not hot enough to shine by its own light, it has many points of resemblance to the Sun, and thus it falls naturally into line between that giant globe of fierce heat and dazzling light on the one hand, and the smaller, colder and more solid bodies of our system, such as the Earth and Mars or the "Dead World" of the Moon, on the other. We can hardly doubt that in the remote ages of the past our own Earth was soft and plastic as Jupiter is to-day, and that earlier still it was incandescent like the sun, but being so small a body, it has cooled down comparatively quickly and has long been fitted to be the abode of animal- and plant-life.

As already mentioned, Jupiter is attended by satellites—eight are so far known, and four of these were discovered by Galileo in 1610. These four are so bright that they can be seen with a good pair of field-glasses, but the others are very faint indeed, and three of them were discovered and are best studied photographically. The most remarkable of all is the outermost, found by Mr. Mellotte on photographs taken at Greenwich in 1908. It is about sixteen million miles from the planet, takes between eight and nine hundred days to complete a revolution, and, like the most distant satellite of Saturn, revolves in a contrary direction to that of all the others.

In the large coloured drawing one of the satellites is seen just passing in front of the planet, and the black spot to the left of it is its shadow. We should get a somewhat similar effect if we could stand on our own Moon during a total solar eclipse, for the Moon's shadow would be seen as a round spot passing swiftly across the surface of the Earth. The other dark spot in the drawing is another

JUPITER IN RELATION TO THE SOLAR SYSTEM.

An imaginary view of Jupiter and the nearer planets revolving on their orbits round the sun, showing their relative distances and as far as may be their relative sizes. Jupiter in reality is thirteen hundred times the size of the Earth.

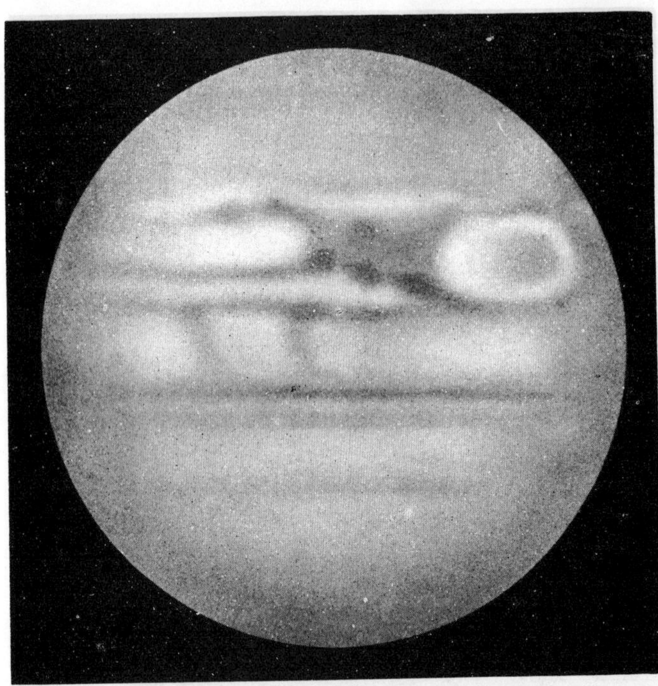

JUPITER.
In June (18 days, 22 hours, 9 minutes), by M. Antoniadi.

of the satellites, which, though it shines brightly against the sky, looks dark by contrast with the greater brilliancy of the background near the centre of the planet's disc.

A BEE ARTISAN

Bees, like men, exhibit varying degrees of civilization. Some species, of which the hive bee is the best-known example, dwell in vast communities, cherish one "queen mother" and labour together for the common good. But there are other kinds of bees which shun the society of their fellows. Each one toils alone to promote the well-being of its little family of grubs, and seeks no assistance from its friends and neighbours. These are the so-called "solitary bees," the most interesting of which in their habits are the Leaf-cutters. There are seven British species of these insects, and they all resemble one another in their great skill and ingenuity. In the following paragraphs, however, I shall deal only with the life-story of the particular Leaf-cutting Bee which commonly frequents gardens, both in the country as well as in the London suburbs. I have even seen the signs of its toil in a small garden close to the Holloway Road. These signs, by which the presence of the bee in a given locality may be readily known, take the form of circular and lozenge-shaped gaps in rose-leaves, showing where pieces have been cut away.

JUPITER.
In June (20 days, 21 hours, 16 minutes), by M. Antoniadi.

How and why these cuts are made I shall shortly describe. But in the first place we must consider the origin of the bee's housekeeping. The male of the species is rather smaller than his mate, and his front feet are beset with dense fringes of yellowish hairs —for what purpose I have never been able to discover. But save for the fact that he is the nominal master of the little bee household (in which, by the way, he seems to take not the slightest interest), he scarcely counts, for he is a lazy fellow who passes most of his time sipping nectar from wayside flowers or in revelry with his good-for-nothing companions.

The female Leaf-cutting Bee deserves a more careful introduction. She has been aptly likened to a stout hive bee in form and general appearance. She is both larger and more robust than her partner, and this is only natural, her life being passed in a round of arduous labour. Indeed, it is her habits rather than her appearance that call for our close attention, for she is one of the most skilful artisans of the bee-world, and is expert in at least three callings.

She makes her début in the early summer, and as soon as her nuptials are accomplished she at once turns her attention to house-hunting. Instinct guides her to rotten wood, such as an old gate-post, the defective beam of a shed or green-house, or the dead heart of a pollard willow. Here she commences to drive a tunnel, following, as a rule, the line of least resistance—*i.e.*, the particular layer of wood which has become softened by decay. She rasps industriously at the wood with her jaws, rakes the chips beneath her body by means of her legs, and then, by walking backwards, pushes the débris out of the hole.

The bee continues to demonstrate her skill as a carpenter for some days, until she has made her tunnel long enough for her purpose, or is stopped by a layer of wood too hard for her jaws to cope with. She then performs a brief toilet, carefully combing the dust from her abundant brown hair, and flies off to a rose-bush. She is now about to prove her dexterity as a pattern-cutter. It is not an easy matter to catch her actually at work; but if, during the sunny morning hours, we stand patiently near to a likely bush, we shall sooner or later be rewarded. The sight is well worth waiting for. The bee hovers in doubt for a moment. Apparently she is making up her mind as to which leaf she will attack. Suddenly her decision is made. She alights upon the leaf, grasping

The male of this Bee is smaller than the female, and is a drone leaving all work to his partner.

The Bee at work, leaf-cutting. (Natural size.)

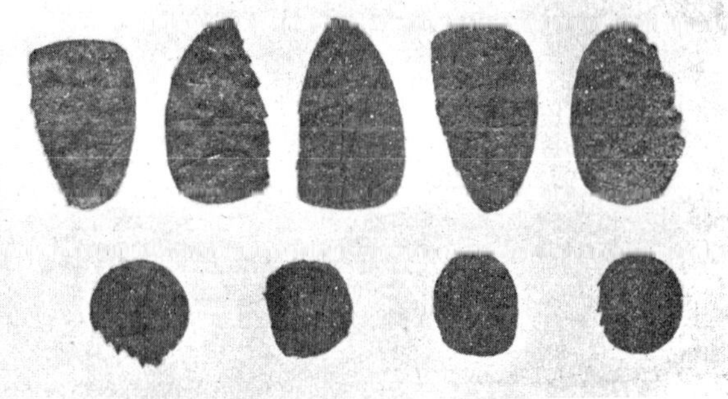

Photos by] [*Harold ... stin.*

THE LEAF-CUTTING BEE.

Nine of the eleven pieces of leaf which go to form ... e cell of the Bee. (Natural size.)

with her legs the piece which she desires to cut away. Then with her jaws she rapidly cuts the leaf-tissue, and just as one imagines that she has cut away the last strand of her support and will fall to the ground, she sets her shining wings in motion and darts off, carrying the severed piece of leaf tightly gripped beneath her body.

The marvel is that the bee can arrange so exactly as not to fall when the last attachment is removed. She must " have her wits about her " in a way which human beings (save, perhaps, the most skilful acrobats) can scarcely comprehend, while her nerves and muscles must work together with absolute precision. Still more wonderful is the manner in which the pieces of leaf are cut, each to a particular shape, with unfailing accuracy.

The reader must know that each piece of leaf as it is cut is carried to the bee's tunnel and there used to build up a thimble-shaped cell or nursery for a potential bee-grub. The number of leaf-fragments required for the formation of each cell varies somewhat with circumstances, but there are commonly seven lozenge-shaped and four circular pieces—the former constituting the

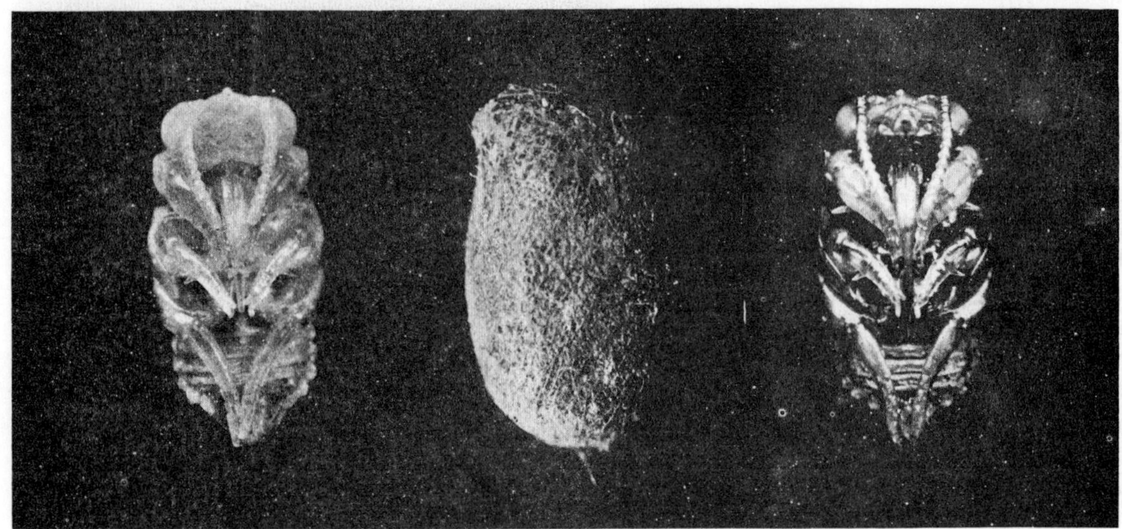

Photo by] [*Harold Bastin.*

THE LEAF-CUTTING BEE.

In the centre is the cocoon spun by the full-grown grub. Outside the silk is coarse, inside it is fine and soft. To the left is the new chrysalis, looking like some carved Chinese ivory. To the right is an older chrysalis, which grows darker as the insect nears the perfect state.

sides and lower end of the cell, the latter the wad or cap with which it is closed. The reader will appreciate this description after glancing at the accompanying photographs. The remarkable point about the whole business, however, is the fact that not one of the leaf-fragments exactly resembles another in shape. Each is cut specially by the bee to cover a particular area as the work proceeds. How does she accomplish this amazing feat ? The parts of her mouth and legs are equivalent to a box of useful tools, but she has nothing that can be compared to a two-foot rule or a pair of compasses ; certainly she has no notebook in which to record figures. We are left to assume that she carries her measurements " in her eye," as the phrase goes, on each occasion that she leaves her burrow and visits the rose-bush. Certain it is that she brings home just the right-shaped piece of leaf each time and never evinces a sign of doubt or confusion.

As soon as the Leaf-cutting Bee has completed the thimble-shaped portion of a given cell she changes her trade for the third time. We have admired her successively in the callings of carpenter and pattern-cutter. She now becomes a cook—and a very skilful cook, too, for she concocts a dish which is at once appetizing, nourishing, and perfectly adjusted in quality to the requirement of the eater. She gathers nectar and pollen from flowers, pours these ingredients into her thimble-

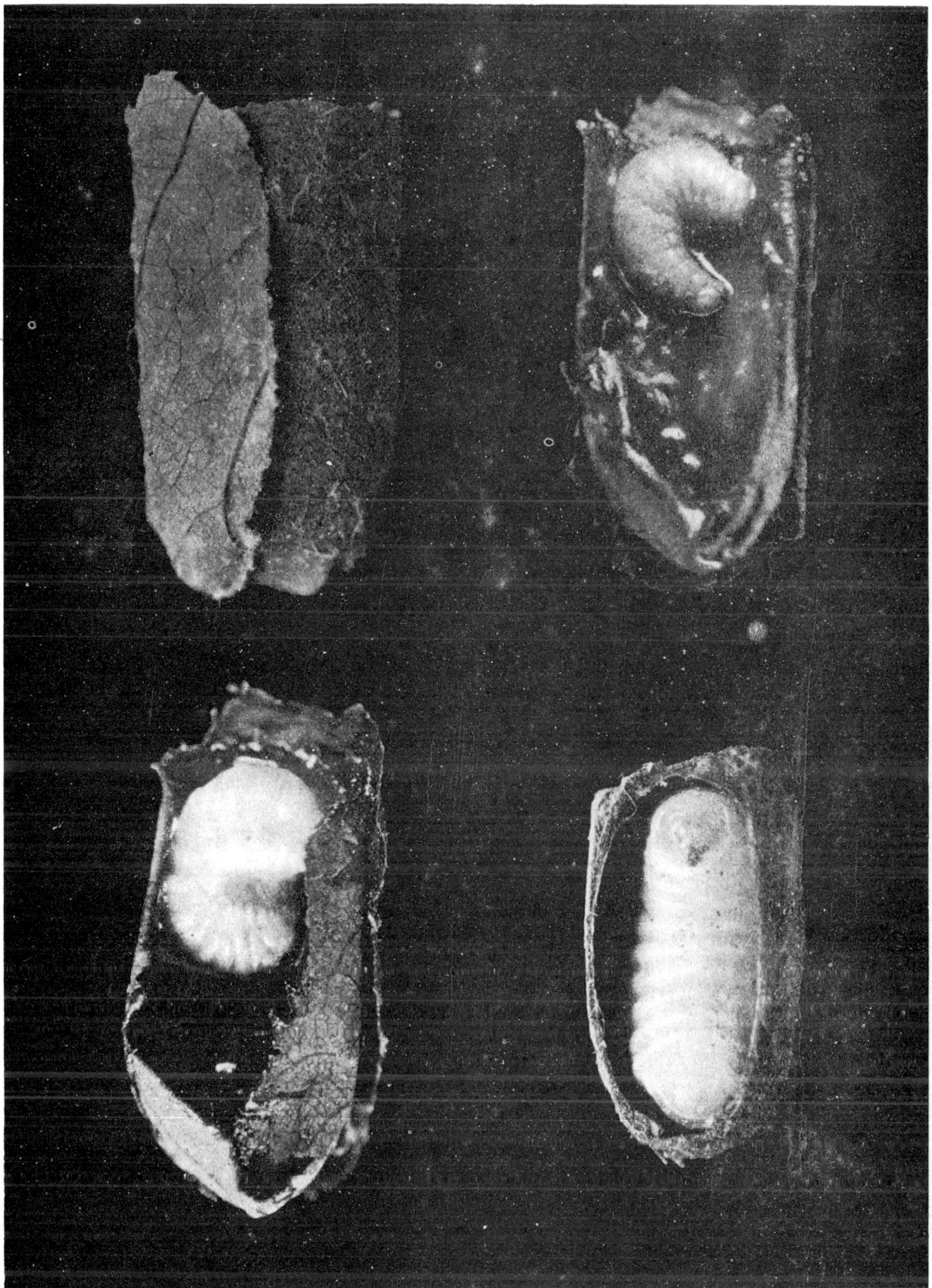

Photo by] [*Harold Bastin.*

THE GRUB STATE OF THE LEAF-CUTTING BEE.

First is the leaf cell in which the mother Bee lays an egg, and deposits food. Second is the newly-hatched grub lying on the top of the pudding. In the third stage he is quickly devouring his food supply, and in the fourth he has eaten all his provisions and is ready to spin his cocoon.

Photo by] [*Harold Bastin.*
THE NEST OF THE LEAF-CUTTING BEE.
Showing the tunnels bored in the decayed wood by the industrious mother, and the neat cells piled one on top of the other, each containing a single egg.

shaped cell, and carefully mixes them in accordance with a recipe which is an ancestral secret. When the cell is more than half filled with the sweet pudding, the bee lays an egg in it. She then makes further excursions to the rose-bush, cuts her four circular pieces of leaf, and uses them to close up the cell's entrance.

Having made all snug, the industrious bee proceeds at once to construct and provision another cell (not forgetting to lay an egg in it), and so continues until her tunnel is almost filled, the last inch or so of space being plugged with wood-chips to keep out intruders.

In the accompanying photographs I show a female Leaf-cutting Bee and underneath nine of the eleven pieces of leaf which go to the formation of a single cell. This will give the reader some idea of the task which the industrious insect performs. But it must be remembered that to the labour of leaf-cutting and leaf-carrying is added the gathering of provisions for each cell that is made, not to mention the work of tunnelling which is undertaken in the first instance. It seems impossible to estimate the average number of cells that a bee constructs in the course of the season, for the reason that when one tunnel is filled the insect may, and sometimes undoubtedly does, make and fill a second tunnel. Moreover, two or more bees not infrequently drive their tunnel close together in the same piece of wood, and it becomes difficult to decide exactly where the work of one little artisan ends and that of another begins.

Let us now turn to what I may call the nursery life of the Leaf-cutting Bee. The egg in each cell hatches a few days after it is laid, and the tiny white bee-grub begins to feed upon the good things provided by its mother. It literally wallows in the sweet pudding and slowly absorbs it ; so that, while at first there is a large quantity of pudding and a very little grub, at the end of a few days there is a big grub and no pudding at all. For the mother bee is such a wonderful cook that she puts exactly the right quantity of food into the cell. There is absolutely no waste ; and when the grub has licked up the last atom of sweet pudding its desire for food ceases quite suddenly. Its babyhood is now over. It spins within its leaf-cell a cocoon of silk, coarse without, but very fine and smooth within, and with a kind of cap at one end, which is readily cut away when the grub ultimately becomes a perfect bee.

Strange to say, the further transformation of the insect is delayed until the succeeding year. When the grub has spun its cocoon it becomes quite inert and lies enshrouded throughout the whole winter, only changing to the pupa in the late springtime, a few weeks prior to its assumption of the perfect winged state. The newly-formed pupa of a bee is one of the most beautiful natural objects with which I am acquainted. It looks exactly as though it had been carved out of pure white ivory by an Oriental craftsman. The wings, legs and antennæ are folded close together upon the breast. Gradually, as development proceeds, the various organs grow darker, colour appearing first in the eyes. Finally a delicate membrane which envelops the whole insect is cast aside, and the perfect bee appears.

Photo by] [*Harold Bastin.*

THE LEAF-CUTTING BEE PARASITE.

This wasp-like creature seeks to lay its eggs in the cocoon of the Leaf-cutting Bee grub. If it is successful, its progeny feed on the sleeping creature, and—

It is a remarkable fact that the worst enemies of insects *are* insects. This is well illustrated by the case of the Leaf-cutting Bee, for one of her most persistent foes is a near relative. It is known as a Cuckoo Bee, because its habits are not unlike those of the cuckoo among birds. It evades the responsibilities of parentage by means of a most dastardly trick. The mother Leaf-cutting Bee forms her cell, provisions it, and lays her egg. She then goes off to cut the circular leaf-fragments which are to cap the cell ; and the Cuckoo Bee, who has been waiting about with an out-of-work sort of mien, suddenly becomes alert. She hurries into the leaf-cutter's tunnel and lays one of her own eggs side by side with that of the rightful owner. The usual result is that the Cuckoo grub devours most of the food, while the grub of the Leaf-cutter pines and dies ; but a few instances are recorded in which both grubs have developed into perfect insects. In either case, the Cuckoo Bee gets her offspring reared " free, gratis and for nothing."

Another insect enemy of the Leaf-cutting Bee is a tiny wasp-like creature which contrives to lay its eggs within the cocoon spun by the full-fed grub ; and this time it is the fat grub itself, not a lion's share of the sweet pudding, that is the objective. The minute grubs of the parasite fix upon the big sleeping bee-grub like so many leeches, and suck, and suck, and suck, until nothing but a dry and empty skin remains. Each is then transformed into a tiny pupa. The accompanying photograph is of a bee's cocoon with part of the side cut

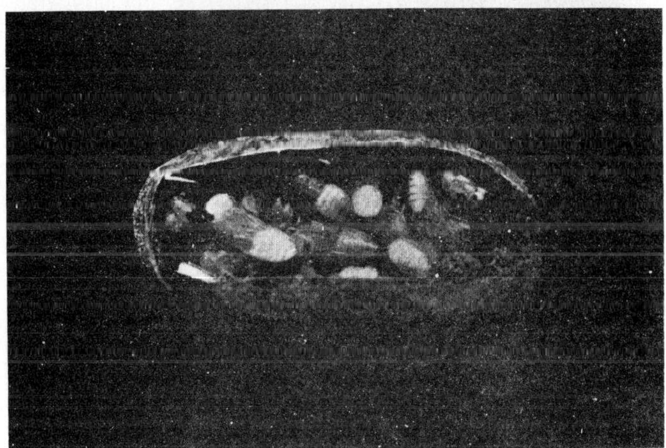

Photo by] [*Harold Bastin.*

—instead of the delicate chrysalis that should inhabit the cell, only the tiny chrysalides of the robbers are to be found.

away to show the interior packed with the parasite chrysalis. One cannot help speculating upon the probable feelings of the industrious Leaf-cutting Bee could she take a peep into such a cocoon. Happily she is spared all such distractions. Her work over, she perishes. Her offspring she never sees, nor can she, in the nature of things, be prescient of the dangers which beset their adolescence.

VEGETABLE HORSEHAIR

PECULIAR to Tropical America and the adjacent West Indies there is a family of aloe-like plants, many of them with beautiful flowers, which grow in crowded rows upon the branches of forest trees. These plants, named after Tillands, the Swedish botanist, are often grown in our hothouses for the sake of their flowers and ornamental foliage. But among the great number that have this upright, somewhat stiff habit, which reminds one of their relation to the pineapple, there is one of entirely different appearance, and known as Spanish Moss. It attaches itself, as they do, to the branches by means of suckers rather than roots, and hangs down in dense tufts, so that it largely hides the foliage of the trees upon which it grows and produces the effects illustrated by our photographs of a grove in Florida. These swaying tufts are often many feet in length, and appear to be clad in silvery scales. The slender stems and branches of which the tufts consist bear two rows of thread-like leaves, two or three inches long, and rounded sheaths almost as long, which clasp the stems.

From Stereo copyright by] [*H. C. White Co.*

"SPANISH MOSS."

Though a true flowering plant, allied to the Pineapple, this "moss" is so called from its resemblance to the Beard-moss lichen which is plentiful in some oak and pine woods in this country.

The name is quite a misnomer, for the plant bears tiny green flowers ; but it has no doubt been designated a moss because the effect produced by its hanging stems is very similar to that of the long silvery grey lichen that hangs freely from the oak and pine trees in some of our forests, and is known as Beard-moss. In both cases the hanging plant gives its host a weathered and venerable, hoary appearance.

So far we have not given any indication of the reason for the title of this note ; but if a long string of the Spanish Moss be pulled from the tuft and peeled, it will be found that it has a tough fibrous interior or core. It may be presumed that the Spanish conquerors of America discovered this fact and turned it to account, if any value attaches to the name of Spanish Moss. At the

From Stereo copyright by] [*Underwood & Underwood.*

"SPANISH MOSS."

This view shows a forest grove in Florida, with the branches heavily draped with the long silvery trails of the so-called "Spanish Moss," whose fibrous core is used as a substitute for horsehair. The plant is not a parasite as generally supposed, but obtains its nourishment entirely from the atmosphere.

[*By Theo. Carreras.*

SEA-SLUGS.

The two Sea-Slugs here represented of the natural size are the Bushy-backed Slug (upper) and Homberg's Triton (lower). The branching plumes that ornament their backs are extensions of the liver. Those of Homberg's Triton closely resemble the polyps of Deadman's Fingers, upon which the Slug feeds; and the Bushy-backed Slug might be mistaken for a cluster of red-brown seaweeds.

present time the plant is gathered in quantity as having a commercial value. It is steeped in water until the outer tissues decay and can be washed off with ease. Then it is dried and used as a good substitute for horsehair in the stuffing of cushions and mattresses.

We have referred to the upright-growing aloe-like species. There is an interesting point concerning them which is worth mentioning. The bases of their leaves are dilated, so that, growing in rosette fashion, they form vases in which, during rainy periods, they collect a good deal of water, which helps to tide the plants over the dry seasons. A water-weed, one of the bladderworts—of which several species are well-known in this country for the traps they bear for the capture of minute pond life—has found its way to these elevated cisterns, and by throwing out runners, connects several of these plants together, and there preys upon the small insects that fall into the water.

The marvel here is, that a small weak plant whose nearest relations inhabit ponds and ditches should presumably have changed its habitat so greatly as to have become, so to speak, an air plant whilst remaining an aquatic plant.

BEAUTIFUL SLUGS OF THE SEA

BY EDWARD STEP, F.L.S.

SOME of the most beautiful forms in which life manifests itself are utterly unknown to the vast majority of mankind; and strange to say, one may search in vain in the ordinary natural-history books for any adequate representation or description of these subjects. The Slugs of the sea are in this category. Even the great Darwin, who was struck by the quaint beauty of these creatures and thought it must bear some intimate relation to their life and habits, could not obtain from the

Slug specialists of his day any information that would support the idea ; on the contrary, they gave him reason to believe that their beauty was " probably of no service to these animals." A closer acquaintance with them as living animals rather than as museum " specimens " demonstrates clearly enough that both form and colour have special relation to their mode of life and are of the greatest service to them as a protection against their natural enemies.

The Sea-slugs must be sought among the weeds and corallines that cover rocks and stones at extreme low water of spring tides and at lower levels. But they are not creatures that can be discovered at a glance, for they harmonize so well with their surroundings that even the largest members of the race may be but a few inches before our eyes and yet be unseen. One of the best known of the British species is familiar in books under the name of Sea Lemon, and it is really a good name, for, alike in form, size and colour, it closely resembles the half of a lemon that has been cut lengthwise. As depicted in natural-history books it is a very conspicuous object ; but when it is engaged—as it commonly is—in feeding upon the similarly-coloured Crumb-of-bread Sponge that encrusts our submerged rocks, it is no longer half-lemon or Sea-slug, but part of the sponge. The low wart-like irregularities of its back then appear to be the pores in the substance of the sponge ; and so the sponge is doubly valuable to it ; for it provides a meal and security from interruption whilst it is enjoying that meal. These encrusting sponges, it should be remarked, give off a strong unpleasant odour, which, in all probability, makes them unacceptable as food to other marine creatures, though not to the Sea Lemon.

But the Sea Lemon is too plainly dressed a Sea-slug to be taken as a pattern of the family, though there are several near relations that are as little adorned. One of these is only about an

[By Theo. Carreras.

SOME SEA-SLUGS.

The Sea-Slugs here represented are all enlarged to about twice their natural length. The figure to the left is white with a rosy tinge. The one above it, covered with tubercles, is yellowish ; the rosette on the lower portion is the breathing organ. The upper figure to the right, with the long snaky appendages, is coloured red with variations towards orange. Its appearance is much like that of a Sea-Anemone.

[By Theo. Carreras.

THE GREEN ELYSIA.

This Sea-Slug is little more than half an inch in length. Its green body is slender, but along each side there is a fin-like expansion. It varies its colour according to the hue of the weeds upon which it is found.

inch long—the common Sea Lemon measures three or four inches—and bright scarlet in colour; consequently, we find it feeding upon a red sponge. Johnston's Sea Lemon is creamy white in hue, and it feeds upon the paler varieties of Crumb-of-bread Sponge, which varies from white to orange. Most of the Sea-slugs have a pair of tentacles, or sense organs, standing up on the back behind the head, and these often bear ring-like plates which appear to be the sensitive portions. At the other extremity of the body there is mostly a flower-like expansion, which consists of the gills spreading around an orifice. In a large number of species there are in addition a varying number of filaments, which wave over the back and produce beautiful effects. Shakespeare speaks of one wearing his heart upon his sleeve for daws to peck at; many of these Sea-slugs perform a similar physiological feat by wearing portions of their liver on the outside of their bodies. These filaments are really offshoots or branches from the liver, and their often beautiful colours are due to their food or to the presence of bile. There is a double advantage about this arrangement, for whilst the practice of wearing the liver outside gives more room for the expansion of the stomach within, the colour of the food consumed shows up in these appendages and so helps to make the slug less visible. For example, some years ago I experimented amongst other species with a pale form of the Plumed Eolis by putting it successively into glasses of sea-water, each containing a sea-anemone of a distinct colour; and as it fed upon these in turn it changed colour accordingly. Its back is covered with several hundreds of filaments resembling the tentacles of the common Beadlet Anemone; and when it is at rest after a meal, with the body curled round, it looks exactly like the anemone it has last eaten—having adopted that anemone's colour.

Homberg's Triton is a very large species—four to eight inches long—and the outgrowths from its sides take the form of circlets of plumes, each of which reproduces a small seaweed, whilst the warts on its back resemble the unexpanded polyps of the compound animal known as Deadman's Fingers, upon which the slug habitually feeds! The Bushy-backed Slug, which is much smaller—about an inch and a half long—has flat side ornaments cut and jagged into a close resemblance to the red-brown seaweeds among which it lives; whilst the Crowned Sea-nymph, which feeds upon zoophytes and corallines, has the same parts got up to closely mimic the class of animals upon which it preys. Indeed, most of the Sea-slugs, different as they are in form, colour and ornament, are all more or less "got up" on a mimetic basis.

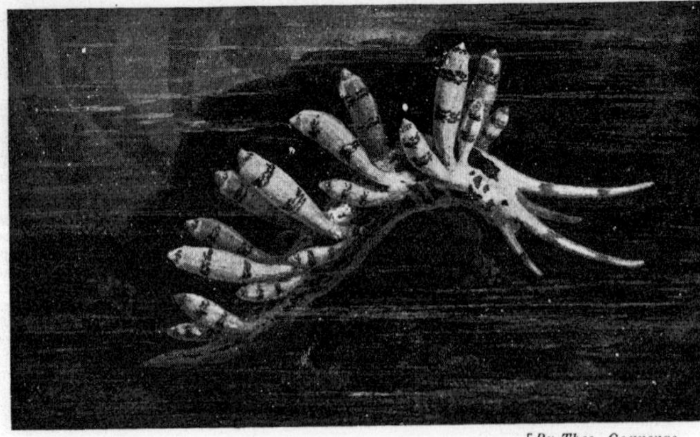

[By Theo. Carreras.

LITTLE EOLIS.

This Sea-Slug measures less than a quarter of an inch, but its transparent yellowish-white, marked with brown and olive-green, makes it a pretty object. It lives on seaweeds, and mimics certain zoophytes

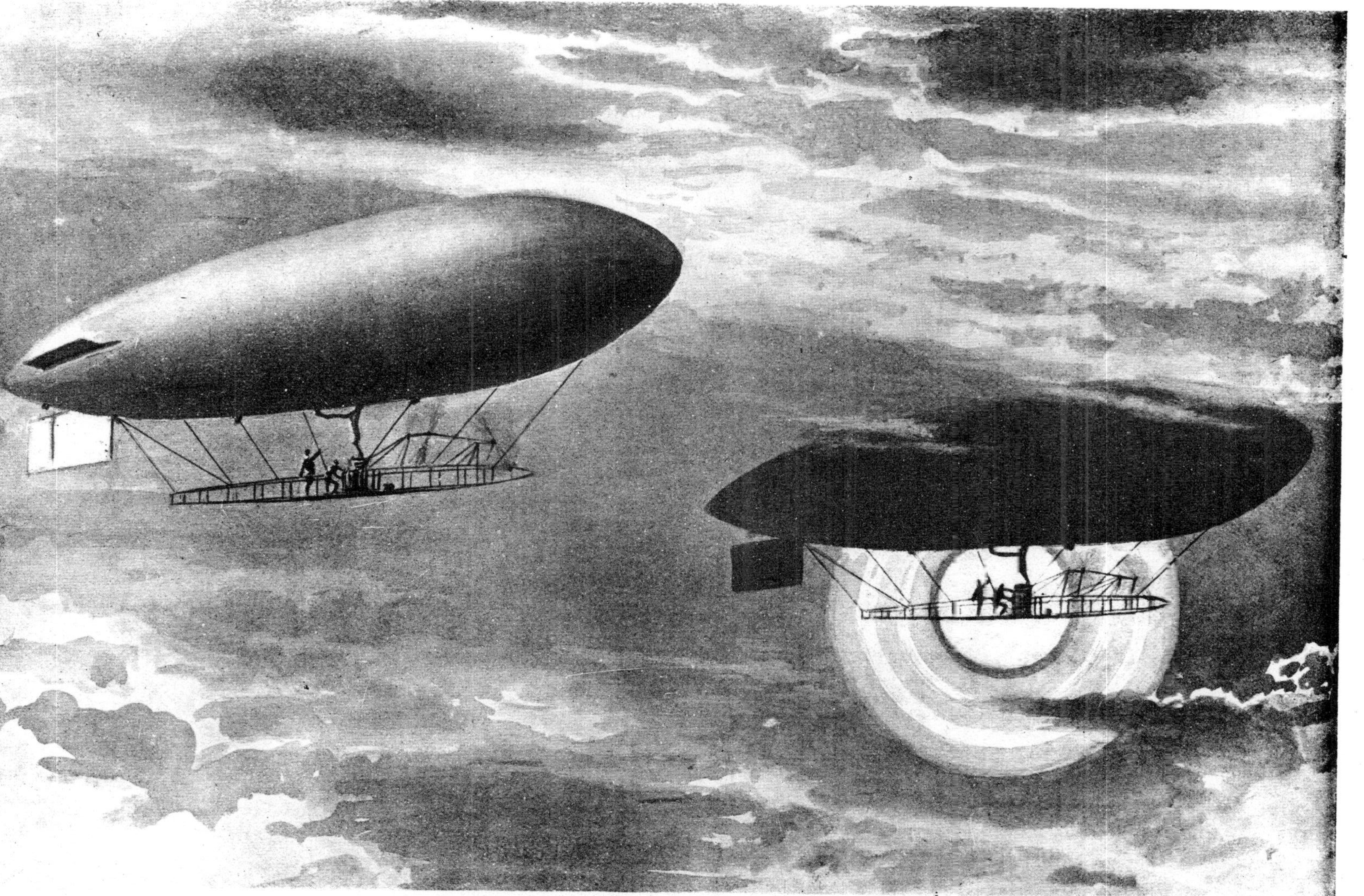

[By A. Twidle.

ANTHELIA.

This peculiar phenomenon is caused by the rays of light slanting so obliquely that they throw the shadow of the observer on an opposing bank of dense mist, while the circular halos are caused by the diffraction of light by the water globules of which the mist is composed.

By permission of] [*Sir H. H. Johnston.*

A GORGED PYTHON.

This enormous creature has just swallowed a goat weighing 52 lb., whose horns, however, were removed before the carcass was thrown into the den. The jaws of the Python are capable of almost limitless distention, while the action of swallowing is mechanical.

ANTHELIA

SUCH is the name given to a phenomenon which is only rarely observed. Certain climatic conditions are necessary for its appearance. There must be a great deal of humidity in the atmosphere and the sun's rays must fall obliquely—that is to say, the sun itself must have only just risen. The usual form which the atmospheric moisture takes is that of a bank of fog, and as the condensation of vapour takes place more rapidly in high latitudes, Anthelia are most likely to be observed in mountainous regions. Suppose the reader has access to such a locality; let him climb some mountain and wait for the break of day. Looking to the west, a great white bank of fog hides the neighbouring hills, while the sun's rays are just piercing the eastern mist. Should fortune favour him he will see before him his own self silhouetted against the white background; but Nature has

Photo by] [*W. Saville Kent.*

THE SUCKING FISH.

This is one of the laziest fishes on record; for while equipped with every means for obtaining its own livelihood, it prefers to attach itself by means of the strange oval sucker on its head to some object that will carry it within reach of its prey.

endowed him with more than earthly beauty. Round his head appears a complex halo formed by ring after ring of rainbow hues, shading from an outer rim of crimson or violet to an inner circle of fine green, while the interspaces between the rings shade to yellow. The outermost of these four concentric rings is white and fades away till it is lost in the cloud, but only occasionally does this evanescent circle appear at all distinctly. Sometimes the range of colouring is reversed, the red light forming the inner shade and green the outer; but the exact reason for this contrast has not yet been accurately ascertained; it is possibly due to some law of diffraction which has still to be discovered. The general course of the phenomenon, however, is simply diffraction, the sun's rays being reflected at a peculiar angle by the minute globules of moisture of which the fog is composed, while the reason of the apparition being noted by the observer is that the sun's rays, streaming from the east, are in exact contact with his centre of vision. It is owing to this circumstance that his shadow is thrown on the western bank of clouds.

As may be supposed, the appearance of the perfect concentric rings is only possible in wide atmospheric spaces.

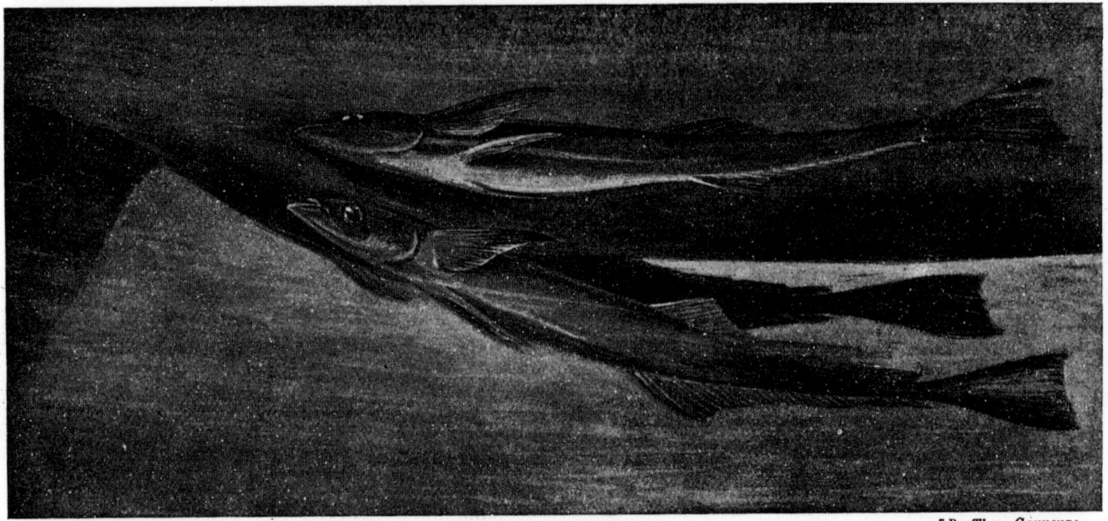

[By: Theo. Carreras.

THE SUCKING FISH.

Showing the method of its attachment to the belly of a shark. The sucker is so formed that no amount of pulling backwards will detach the creature—this prevents the "wash" of the larger fish breaking its hold—while the slightest jerk forward sets it free. The fish is thus provided with means for quickly chasing its prey.

THE PYTHON AND ITS PREY

BY SIR HARRY JOHNSTON, G.C.M.G.

OUR interesting photograph represents an enormous Python which has just swallowed a goat weighing fifty-two pounds. The horns of the dead goat were removed before the carcass was thrown into the reptile's den. This was done as an experiment, for it was scarcely believed that the Python in question would be able to swallow such a large morsel, although seven days previously he had in the course of four or five hours swallowed a smaller goat of twenty-one pounds and a small deer of forty-three pounds, only a short time elapsing between the two efforts. In the case of the fifty-two pound goat, the Python attacked it without hesitation, and had half swallowed it before the photographer arrived; but he had to make considerable efforts, nevertheless, to get it down, and from time to time emitted deep groans.

Pythons, and their near allies, the Boas of South America (the range of the Old-World Pythons extends from Africa to the Malay Archipelago), are without poison-glands, and their structure is slightly more primitive than that of poisonous snakes; for they possess minute vestiges of hind legs, observable externally in the form of two claws near the vent. They dart at their prey open-mouthed, and usually seize it by the head with their strong, sharp teeth. Immediately afterwards they coil their strong body round and round their victim, using the tail as a fulcrum and attempting to crush the creature they have seized into a sausage shape. As soon as the Python believes his victim to be dead, he commences the process of swallowing, which is almost mechanical, and once having been begun cannot be stopped. The jaws distend like an indiarubber bag to an extent which seems well-nigh incredible. Mr. Hagenbeck, to whom we are indebted for the photograph, states that he has seen a python swallow a goat which weighed ninety-three pounds in the course of an hour and a half.

THE SUCKING FISH

BY FRANK T. BULLEN.

FROM a sociological point of view, the study of the Remora is full of intense interest. For here we have a creature that to all outward appearance is fully equipped for the battle of life, in no way handicapped by weakness or inferiority of any organ ; who has deliberately, and apparently of malice aforethought, chosen to live upon the energies of others, to expend the minimum of effort to obtain that sustenance which is imperatively necessary for his support.

To the attainment of this much desired end the Sucking Fish in all its varieties has devoted all its energies until now its equipment for the purpose is perfect. The top of its head is flat and oval in plan, while upon it is formed a strangely conventional apparatus which is an efficient sucker, though it looks more like a piece of regularly designed grating than anything else, or perhaps the sole of a tennis-shoe. All the other organs of the fish are normal; indeed, it is rather better equipped for swimming than many fish.

Its procedure in the furtherance of its desires is to attach itself by means of this arrangement on top of its head to the underside of any smooth, level surface, whether animate or inanimate does not matter, which is likely to carry it into the vicinity of food, always preferring such a situation as promises that the food will float towards its mouth as it lies closely attached to its host ; for which purpose it always fastens itself with its head in the direction in which its host is progressing. But this is not the only reason for the creature choosing such a position: the arrangement of the sucker is such that while a direct pull backward only seems to increase its hold upon whatever it is attached to, so that it may be torn asunder rather than relinquish its grasp, a very slight push forward will instantly release it.

THE SUCKING FISH.

A good view of the strange sucker, looking for all the world like the sole of a tennis-shoe, with which the fish has attached itself to the glass of the aquarium.

But of all the places in which this lazy fish loves to lurk, there are none so greatly to its mind as the roof of the mouth of some large shark, or some kinds of whales. Here, with its head pointing in the direction in which its host is going it can, and does, take toll of all food that enters, which, as it is omnivorous, like the shark, never comes amiss to its palate. It is able to perform all the natural functions while in this position ; and so it reaches at once the summit of its ambition, that of obtaining all its needs with the minimum of effort.

The ingenuity of the Chinese fisherman, which has enabled him to utilize the voracious cormorant for fishing purposes, has also seen the possibilities in the Remora, and the unwilling creature has been pressed into the service of man in a remarkable manner. In waters where turtle abound, a boat will put forth, having a number of Sucking Fish attached to her bottom, every one of which has a ring fitted round its body in front of its tail, which has a long thin cord fastened to it. Arriving at the fishing-grounds, the fisherman detaches the suckers from his boat by a gentle push forward from a bamboo, and with the same instrument prevents their return. They are thus forced to swim about until they find a turtle, to which they immediately attach themselves, and are then hauled in, bringing their temporary host with them.

Photo by] [*J. J. Ward.*

A BIRD-EATING SPIDER.

BIRD-EATING SPIDERS

BY R. I. POCOCK, F.R.S.

It would not be quite the truth to say that "Bird-eating Spiders" are creatures of romance, because it is undeniably true that large spiders exist in the tropics which eat small birds when they can catch them. Nevertheless, the popular belief that they subsist upon birds is equally undeniably false, and can only be ascribed to the proneness of man to accept and remember the marvellous and forget all about the commonplace. The erroneous idea in question was unwittingly set on foot by a lady, Madame Merian, who wrote a history of the insects of Surinam in the eighteenth century and published therein a sensational figure of a great spider eating a humming-bird; but although she carefully explained in the text that the spider feeds mostly upon ants and only hunts for birds when the insect-supply runs short, popular sentiment was so shocked at the idea of the most loathsome creature in existence being permitted to prey upon the most beautiful and charming, that all else regarding the spider was forgotten. Linnæus, too, contributed to the promulgation of the error by naming this spider *Avicularia* or the "bird-catcher."

Photo by] [*E. Step, F.L.S.*

WEST INDIAN BIRD-EATING SPIDER.

Big spiders of this kind occur in various tropical countries. They have long thick legs, covered with a dense coat of long hairs. They are nocturnal in habit.

The comparative rareness of the occurrence recounted by Madame Merian may be gauged by the circumstance that apparently only three travellers since her time have observed a similar thing; and the observations in at least two of these cases were made upon widely different species of spiders, one being an inhabitant of Borneo, a second of Martinique, while the third, which possibly belonged to the same kind as the true *avicularia*, came from Brazil. This shows that there is more than one of these monstrous spiders to which the term "bird-eating" is applicable, and since it is probable that all the spiders related to them that are capable of eating small birds now and again practise this habit, just as they now and again eat small lizards and mice, the title in question, misleading though it be, has been extended in popular parlance to embrace at all events the larger members of the entire group.

Now this group is very widely distributed. It ranges in America from the Southern States of the Union to Argentine and Chili; in Africa it is found everywhere south of the Sahara; while in Southern Asia it spreads from Bombay to New Guinea, and even occurs in Australia. The number of species represented in these countries has never been counted, but it certainly runs into hundreds. All are nocturnal, and their principal variation in habits is found in the nature of retreats in which they lurk during the day. Some dig burrows in the soil; others shelter in rock-crevices, holes in trees, forked branches or rolled leaves, but whatever be the nature of the retreat, it is always lined with silk. The true *avicularia* is one of the species which lives in trees. Another with similar habits is now and again imported to England in bunches of bananas from the West Indies. It is probably amongst the tree-dwellers that the bird-eating propensity is most prevalent, because coming out at night in search of prey, opportunity is afforded them of taking birds unaware on their nests or of pouncing upon them roosting in the branches.

In actual bodily bulk these giants of the spider-world are not so large as one would suppose, the greatest length hitherto recorded being only about three inches; they are, however, stoutly built, and the length and thickness of their long, hairy limbs, which in the largest kinds may span over eight inches, make them look far bigger than they are in reality, and give them the appearance of equalling a rat in size. They can run with astounding speed; but for the most part they are extraordinarily lethargic, although they will fight fiercely if interfered with. Their weapons of offence and defence are a pair of strong steel-like piercing fangs, which may be half an inch long and, like the fangs of a rattlesnake, are perforated at the tip to emit the poison secreted by a gland lodged in the jaw. The virulence of this poison may be guessed from the fact that a frog was quickly killed by the bite of a comparatively small South African spider of this group, and that a kitten was made violently sick by a wound from an Australian spider, also of small dimensions.

Photo by] [*E. Step, F.L.S.*

SPIDER AND HUMMING BIRD.

These big spiders do not, as their name implies, subsist upon birds, but they are known occasionally to catch and kill humming-birds as well as lizards and mice. They kill their prey—mostly insects—by means of a pair of sharp, poisonous fangs, half an inch long.

But these spiders are protected by other attributes than their poison-glands. Although the long hairs which clothe the body and limbs are seemingly soft and silky, they are nevertheless exceedingly penetrating, and since they set up irritation in the human hand which has been described as maddening, it is evident that to the sensitive skin of the palate and tongue they must cause excruciating pain. A cat, indeed, has been known almost to die from the effects of eating a portion of one of these spiders.

Now, animals provided with poison-glands, sharp spines or irritating bristles commonly possess some special means of self-advertisement by which they make themselves known to possible enemies by appealing to their sense of sight or hearing. The rattle of the rattlesnake and porcupine, the hood of the cobra and the black-and-yellow colouring of humble-bees and wasps are illustrations of this. Many Bird-eating Spiders are doubly endowed in this way. They are often conspicuously

[*By A. Twidle.*

THE SPERM WHALE.

There is something positively appalling in the contemplation of his colossal head, which appears, when the animal is poised upright in the water, like some huge flat-topped column of black marble against which the waves beat vainly.

coloured in one way or another on the upper or under sides, the tints of the under-side being exposed to view when the animals stand up with their fore-limbs raised and fangs extended in their habitual attitude of defence. Formidable, nevertheless, as these spiders are, mortality amongst them, at all events when young, must be very high, judging from the large numbers of eggs they produce, as many as 2,000 having been counted in a single cocoon. Probably tropical ants destroy numbers of them. Even the full-grown spiders fall victims to large solitary wasps; and it is said that the contests between these redoubtable antagonists always result in the spider's death.

THE CACHALOT OR SPERM WHALE

BY FRANK T. BULLEN, F.R.G.S.

MONSTROUS as are the proportions of this amazing whale, in mere size he is far surpassed by two of the whalebone whales, the Sulphur-Bottom and the Bowhead or Arctic Right Whale; his utmost size being about seventy feet in length, forty-five feet in greatest girth, and weight about one hundred and ten tons; tail one hundred and twenty square feet in area, and utmost speed fourteen miles per hour. And yet for several reasons he has an entirely indisputable right to be considered the Monarch of the Peoples of the Sea. First of these reasons comes the great fact that, while never aggressive, like all true heroes, when attacked he will turn and fight, and unless very speedily disabled, becomes the most terrible antagonist imaginable. Secondly, in majesty and terror of his appearance. There is something positively appalling in the contemplation of his colossal head, which appears, when the animal is poised upright in the water, like some huge flat-topped column of black marble against which the waves beat vainly. Adjunct to the head is the immense lower jaw, a tapering shaft of bone, sometimes twenty feet in length, with a double row of massive back-curving teeth, between fifty and sixty in number and several

[*By A. Twidle.*

A SCHOOL OF SPERM WHALES.

One huge bull will convoy a harem of cows and young bulls, which latter fight furiously among themselves at intervals, until one of them, victor over all the rest, challenges the great leader and ousts him from his position.

inches apart. This jaw hangs down at an angle of about forty-five degrees when the animal is in its normal position, revealing the cavernous throat and comparatively insignificant tongue. The interior of the mouth is livid white, which gleams palely against the blue of the sea and effectually dispels by one glimpse of it any idea of the whale not being able to swallow anything much larger than a herring. But then, this idea arises from the popular ignorance of the great number of different kinds of whales which exist in the sea, and the concentration of all the scanty knowledge of the subject upon their Arctic representative, the great Bowhead.

Now, in very truth, there could hardly be a greater number of fundamental differences between two members of one family than there are between the Sperm Whale and the Bowhead or Arctic Right Whale. In the first place, despite his thick covering of fat, the Sperm Whale is found all over the oceans where the temperature does not fall much below fifty degrees, even in the almost hot waters of the Persian Gulf; but never by any chance in those frigid waters so necessary to the well-being of the Right Whale—they appear to act as an impassable wall.

Photo by] [*H. C. Shepstone.*

WHALE SPOUTING.

The so-called "spouting" of the Whales is a column of vapour from the blow-hole or nostril, due to the hot expired breath condensing in the cooler air.

As might be expected from the appearance of its teeth, it has no screen or sieve of *baleen*, or whalebone, in its mouth, this alone hinting that its food is of a more important character. Yet its teeth are confined to the lower jaw; the upper jaw has a corresponding double row of sockets into which the teeth fit. This, of course, points to the fact that great as may be the prey upon which the Sperm Whale feeds, it must be of a soft and easily divided nature; also, that mastication is out of the question.

But perhaps the most wonderful and mysterious characteristic of the Sperm Whale is the great reservoir of liquid spermaceti which he carries in his head—the "case" of the whalemen. I have myself assisted to bale out a case which yielded between two and three tuns of the bland, snow-white grease—at least, it became snow-white grease upon exposure to the air; within the case it was colourless as pure water. This case fills the whole top of the vast head, and through it runs the single breathing-tube which feeds the lungs with fresh air from the spiracle at the extreme forward top end of the head. This creamy substance, or "head matter" as the whalers call it, has absolutely no connection with the small brain. No exact knowledge of the use of this tank of spermaceti exists; but it is generally supposed to act as a buoy to bring the massy body more swiftly to the surface for air, by reason of its specific lightness in comparison with the water in which it floats.

There is another curious point of difference between the Sperm and all other whales, which is the disparity of size between the sexes. In all other whales there is a general equality, or where there is a difference it is in favour of the female. But among the Cachalots the male grows to twice the size of the female on the average, and in particular cases to fully three times as large. They are polygamous, a single pair never being seen; but one huge bull will convoy a harem of cows and young bulls, which latter fight furiously among themselves at intervals, until one of them, victor

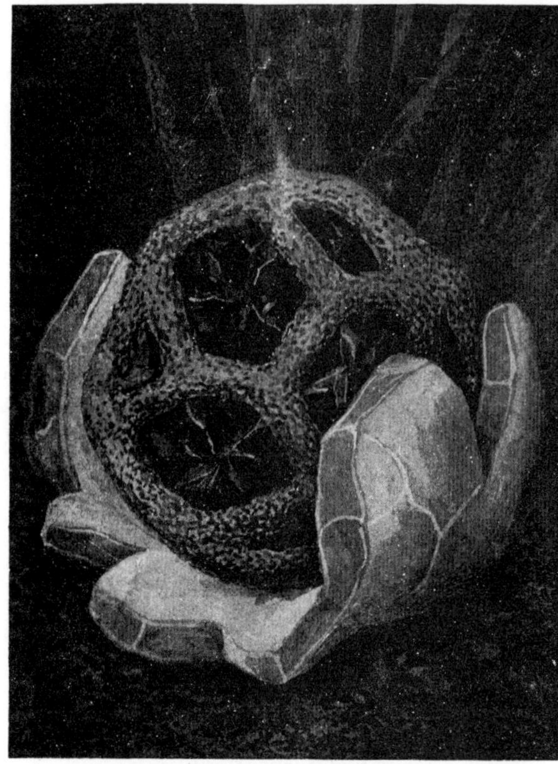

[By Theo. Carreras.

THE LATTICE FUNGUS.

At first a soft egg, the fungus bursts its skin and enlarges to a great extent, and ultimately appears like a network of red coral. It has a disgusting odour.

over all the rest, challenges the great leader and ousts him from his proud position. Thenceforth the ejected bull roams the seas alone, never to join another school, but becoming more morose, vicious and difficult to attack. It is a pathetic sight to see how, when the leader of a "school" of Sperm Whales is attacked, his faithful wives will gather round him, regardless of danger, though unable to help. Many fantastic stories are told by whalemen of this self-sacrificing propensity on the part of the cow-whales, but they have no place here.

THE LATTICE FUNGUS

BY EDWARD STEP, F.L.S.

One of the most remarkable of the varied forms of fungi is the Lattice Fungus, whose earlier condition is shown on this page, and whose full development appears on the coloured plate. Emerging from the earth as a gelatinous ball, its expansion causes the rupture of its envelope. At this stage the spaces in the lattice are filled and the lattice-bars covered with an olive slime, in which the spores are embedded. As the hollow sphere increases in size this slime gets thinner and reveals the red colour of the lattice-work, which finally stands clear and open like a network of coral, and measures three or four inches across. It varies in colour from bright vermilion to dingy red. In the South of Europe it is a tolerably common fungus, but in this country it is rare and of very limited range. Its habitats are woods and gardens; but however desirable it may appear to have such a thing of beauty coming up in one's garden, it is not so in fact: it gives off an intolerable odour which produces headache.

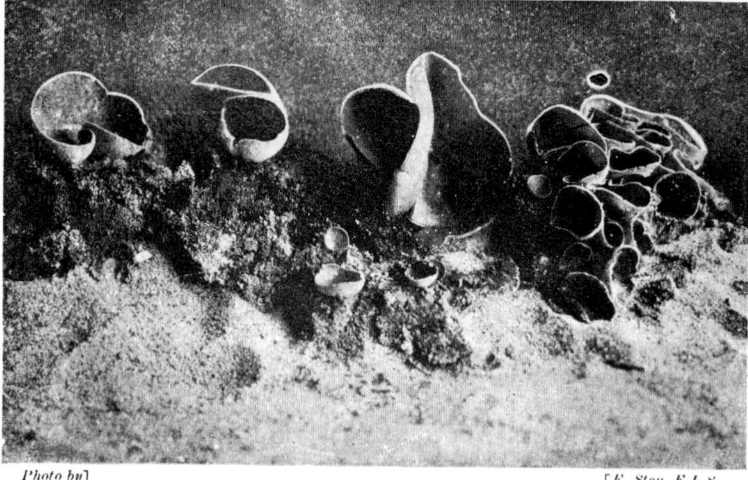

Photo by] [E. Step, F.L.S.

THE ORANGE ELF-CUP.

This brilliant little fungus is here shown only half the natural size. The cups are at first almost closed, but as they expand they also become shallower and, finally, almost flat. They are wax-like in appearance, and brittle.

THE ORANGE ELF-CUP

The lower part of the coloured plate in which the Lattice Fungus appears is filled with the wax-like cups of another

By permission of] [*Sir H. H. Johnston, G.C.M.G.*

THE ARCHÆOPTERYX.

A restoration of the earliest winged animal that could properly be termed a bird. The wings, it will be seen, have fingers attached; and the tail-feathers are arranged along the sides of a lengthy extension of the spine, instead of being clustered around a stump, as in modern birds.

beautiful fungus, the Orange Elf-Cup. It is a representative of a very numerous family of fungi, of which many are quite minute; but most of them are beautiful. Certainly no one would deny that epithet to the present species if he came upon a colony of it on the margin of the wood or on a low bank with the autumn sunshine upon it. It may be met with in scores or a very few of the cups, in crowded groups or scattered individuals: small spherical ones with almost closed tops, fairly open deep cups, and widely expanded saucer-shapes. Their substance is firm, wax-like and brittle. They may be almost white with a pink tinge, but are usually of some fine tint of lucent orange. The spores of this group are contained in microscopic cylinders embedded in the flesh of the cup; and the careful observer may see as he approaches minute puffs of smoke or vapour given off. In reality it is a discharge of spores from these pits. The Orange Cup has a pleasant though faint aroma, and it is said to be edible, though it is too small to be worth cooking.

THE ARCHÆOPTERYX

BY SIR HARRY JOHNSTON, G.C.M.G.

"How did birds originate?" is one of the unsolved problems of evolution. They are, of course, nothing but a branch of the reptile class—exquisite reptiles which have achieved in a supreme degree the art of flying (though not a few of them have since taken to an exclusively ground or water life), with warm blood, highly-organized brains, and a skin covering of feathers (feathers and down being only a modification of the same substance as produces nails and scales). The Archæopteryx is the earliest bird type as yet discovered in the records of the earth's history, and although it retains many reptilian features, it is still an out-and-out bird. Its genus may have developed forms as large as a raven, but the fossil skeletons of the two species found in the slate or "lithographic" rocks of Bavaria, and dating from the Upper Jurassic (the middle part of the Secondary Epoch), indicate birds of the average size of a rook. The Archæopteryx still retained teeth in the jaws, mainly in the upper jaw. It had large eyes, supported by bony plates, and feathers attached to fore-limbs which were like those of the modern bird in having only three fingers. But the thumb was large, and no doubt freely projected beyond the feather contours. It was furnished with a large claw, as were also the two joined-together fingers. These in their terminal joints seem to have been uncovered by the wing surface, and with the thumb probably enabled the Archæopteryx to use its fore-limbs occasionally for climbing up rough surfaces, just as is done by the young of a very remarkable bird still living in South America, the Hoatzin (see page 177).

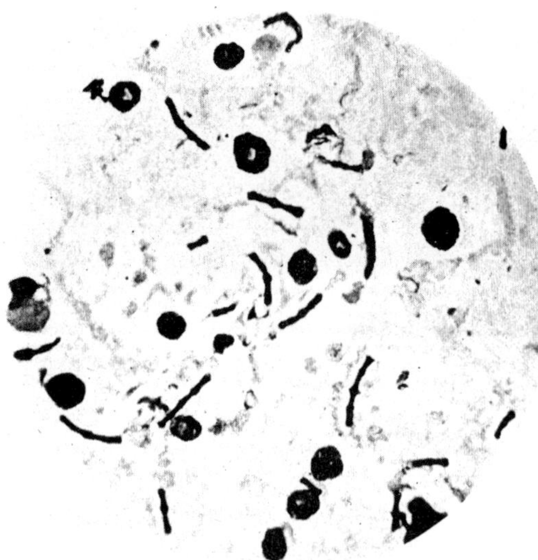

Photo by] [*E. C. Ash.*

BACILLI OF ANTHRAX.

In this photo the rod-shaped objects are the bacilli. They produce spores which are able to retain their virulence for twenty years. Magnified about 500 times.

Besides this greater freedom of the fingers, the Archæopteryx differed from modern birds in its long and lizard-like tail. We are accustomed to regard many living birds as being "long-tailed." This only means that they develop long-quilled feathers growing out of the tail-stump; but, as a matter of fact, all existing birds are very short-tailed, the vertebræ being much reduced in numbers. The Archæopteryx, however, retained the long tail of the reptile so far as its bony skeleton was concerned, and the quills, instead of growing in a bunch from the end of the tail, as they do in modern birds, grew out in pairs from the sides of the twenty vertebræ. It is probable that the body of the Archæopteryx was covered with down, and that well-developed feathers only grew on the wings and tail. The legs were scaly, like those of modern birds, except the thigh part, which was either feathered or downy.

BACTERIA

BY EDWARD C. ASH, M.R.A.C.

It is a most surprising fact that man's most powerful and most merciless enemies are minute plants —organisms of so small a character that they are only visible under the highest-power lenses of the modern microscope. And yet they are responsible for most, if not all, the "ills and pains" to which humanity is subject. Every day wars between the most highly-developed, the most perfected of all animals—man, and the lowest and the most rudimentary plants—Bacteria, continue.

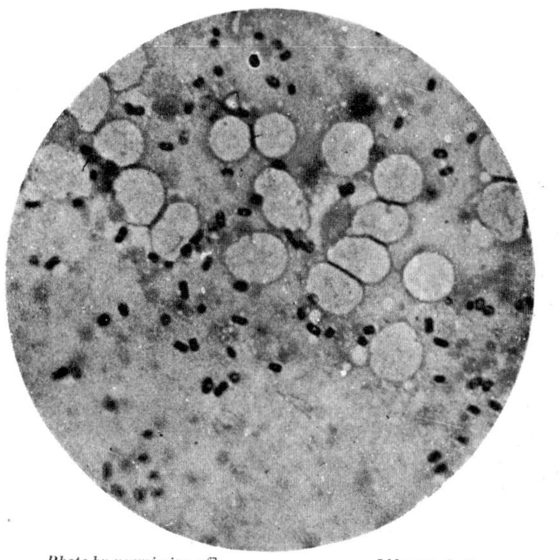

Photo by permission of] [*Messrs. Leitz.*

BACILLUS OF THE PLAGUE.

The large objects in this photo are blood corpuscles. The small dark bodies are the Plague germs. They are magnified about twelve hundred times.

The word "Bacteria," to the average mind means "enemy." But this idea is a great mistake; for if it were so, human life could not exist. Man would have to retire and leave the world to the inhabitants of the microscopic kingdom; and before long the results of all his improvements would collapse and things would return to a condition of wilderness nature.

Bacteria are everywhere: the air we breathe, the ground we tread, and even our bodies, are inhabited by countless numbers of these minute plants. Bacteria, like human beings, have each their allotted tasks, and each species plays an all-important rôle in the microscopic kingdom, and indirectly in the kingdom of the higher animals. With the aid of the microscope the scientist peers into a world inhabited by millions of organisms. The average man lives practically unaware of the existence of a world which is only to be seen under lenses; and it is fortunate that he is able to do so, for even the strongest nerves could not long withstand the effect of seeing swarms of organisms covering everything. Who, for example, could enjoy his or her meals, or even a mouthful of food teeming with all kinds and conditions of microscopic life?

What are Bacteria? Bacteria are low forms of plants. They are roughly divided into main groups respective to their shape. Thus we have rod-shaped Bacteria or Bacilli, spherical Bacteria (Cocci), and many others. The majority have no means of locomotion. Some, however, are furnished with swimming organs, known as flagella. These hair-like appendages appear either on one or both sides of the organism, or are sometimes present on all sides, and can be clearly seen in the photograph of the Typhoid Bacilli.

It is reported that the first man to see Bacteria in motion became wildly excited, under the impression that he had witnessed plants changing into animals. Mobile Bacteria do not seem to have any idea of direction, but travel onwards until they meet an obstacle, and then their subsequent direction depends solely on the angle of impact.

Whilst some species of Bacteria spend their lives in the work of destruction, others are equally concerned in building up compounds. Some, again, are colour-producers, in fact, artists of great merit, and occasionally cause much surprise to those interested in milk—blue milk, red milk, yellow milk, and even black milk, being the results of their labour. Then, again, others have learned that by "uniting we stand and by dividing we fall."

This statement may seem rather Irish, because, it will be seen later, Bacteria are characterized partly by their method of propagation, which is by division.

Bacteria have found that co-operation gives beneficial results ; hence, certain species seek partners of honourable intentions. The partner chosen may be a higher plant, such as a tree, thus many billions of times greater than the micro-organism. Or the matter may be satisfactorily arranged between Bacteria of different species.

Bacteria are singularly fortunate in the rapid and easy method they have of reproducing their species. This method is by fission—that is to say, each Bacillus divides into halves. The parts so formed rapidly mature and divide again. Hence we have the astonishing theory that under optimum conditions one bacillus may have the honour to be the ancestor of two hundred and eighty billions of its kind in forty-eight hours. Under normal conditions, however, the multiplication is much less rapid. From an experiment carried out a few years ago twenty-one Bacteria in three days produced four hundred and ninety-five thousand—certainly a large family.

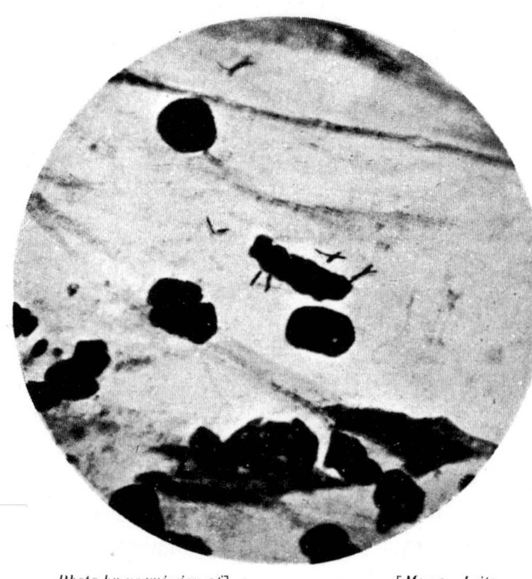

Photo by permission of] [*Messrs. Leitz.*

TUBERCULOSIS.

The germs of what has been called the White Scourge, the consumption that destroys so many thousands every year, is here shown magnified about twelve hundred times.

They have also a further method which can hardly be considered reproduction. This is by spores, in which form the organism is able to remain dormant for lengthy periods. Spores are egg-shaped, transparent bodies, singular for their great resisting power, being able to retain their vitality under the most adverse conditions.

Although for years it was generally believed by scientists that many diseases were caused by some microscopic organism, it was not until a comparatively recent date that it was understood how these organisms obtained their fell results. Many theories were advanced, such as that the arteries of the sufferer were congested with the personæ of the Bacteria. But to-day it is known that Bacteria are able to produce a liquid, and that it is this liquid which is responsible for the disease. Fortunately, man, owing to an army of hard-working—but never paid, nor usually even thought of—allies, is able to protect his life against his would-be destroyers.

To recapitulate, Bacteria are microscopic plants ; their habitat is most varied ; they occur in every conceivable position, and are tenants, welcome or unwelcome, even in the mouth and alimentary tract of the higher animals. Some spend their lives in bitter antagonism to man and his works, destroying all he produces, interfering with his domestic animals, his crops, and lastly, but not of least importance, ruining his health and persistently attempting to destroy his life. Others of a more friendly disposition, for reasons best known to themselves, help man directly and indirectly. They render willing assistance in the production of some of his foods, such as cheese, butter, vinegar, etc. Others help in the manufacture of his clothing—for example, in the tanning of leather; and some, again, are of the greatest economical importance in the prevention of disease by destroying refuse. This last characteristic has been utilized by man, and so to-day we have a bacteriological method of sewage disposal, thus partly solving a most perplexing and difficult question. It is hardly necessary to mention that micro-organisms require food for their nutrition and for the production of energy, and are as dependent upon a food supply (except during the spore, or resting, stage) as are the higher animals. Most of the flowering plants rely upon the atmosphere for their carbon, which they obtain by means of the chlorophyll (green-coloured matter) present in their leaves. Bacteria, however, have no cells containing chlorophyll, and so, being totally devoid of this substance, must necessarily obtain their carbon elsewhere. It therefore follows that Bacteria are seldom able to lead a "free life," depending to a greater or less extent on living or dead structures for their food. Hence, the majority of Bacteria are parasitic (*i.e.*, preying on living matter) or saprophytic (living on dead matter). It was partly this total absence of chlorophyll that led many to think that Bacteria should be placed rather in the animal than in the vegetable kingdom. But in totalling up their respective animal and vegetable-like characters no doubt remained that Bacteria were more like a very low form of plant. This conclusion caused much astonishment to those well-informed people who had considered Bacteria a species of cheese-mite.

Photo by permission of] [*Messrs. Leitz.*

TYPHOID.

The germs of Typhoid as they appear under a high power of the microscope. They are magnified about twelve hundred times.

Although about four-fifths of the air is nitrogen, yet the higher plants are continually suffering from a nitrogen famine, because they have no means of "tapping" the atmosphere; and they depend, except in a few cases, on the soil for all the nitrogen they obtain. Certain species of Bacteria, however, have the power of obtaining nitrogen from the air, and do so either independently or else in partnership with plants, such as beans and peas, little aware that by so doing they are signing their own death-warrant. For the plant, in obtaining the nitrogen from the Bacteria, is unscrupulous enough to devour the rightful owner. It is easy, therefore, to understand the vast importance of the nitrogen-fixing organisms to the plant world, and indirectly to man.

It is interesting to know that Bacteria vary greatly in size, from .3 of a micron to 30 or more microns in length; a micron ( ) being one twenty-five-thousandth of an inch. In a few cases the individuals of the same species are not constant as to size. The Typhoid Bacillus, although usually a rod 2 to 3 μ long and .6 μ broad, occasionally occurs as a much larger and broader organism, 10 or even 20 μ long and 2 μ thick. The Comma Bacillus, so called because it somewhat resembles a comma, is a minute curved rod one twenty-five-thousandth of an inch long, and is responsible for the disease known as Asiatic Cholera. It is interesting to know that the Comma Bacilli are able to live for several months in polluted water, whilst in distilled or very pure water they rapidly die; in fact, seldom remain alive for longer than twenty-four hours.

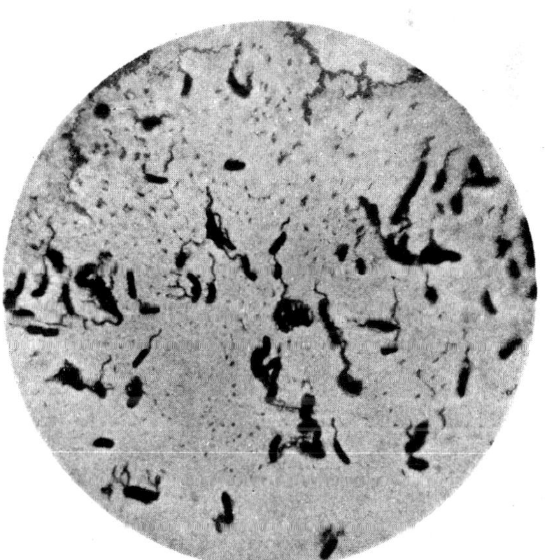

Photo by permission of] [*Messrs. Leitz.*

ASIATIC CHOLERA.

This organism is commonly known as the Comma Bacillus from its often assuming a curved form.

The Tubercle Bacillus, the germ responsible for consumption, is from 3 to 5 μ long. It is difficult—in fact, impossible—to realize what vast or exceedingly minute sizes represent. A simple

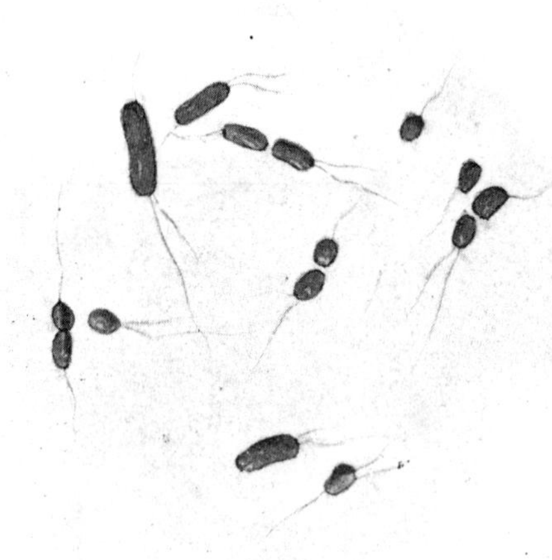

[By A. E. Knight.

BUBONIC PLAGUE.

The Kitasato Bacillus shown above is so small that many thousands of them could be picked up on the point of a fine needle.

experiment will give the reader a good idea of " bacterial " size. If you take an ordinary pin, and lightly prick a piece of paper, the hole you have made will be of sufficient area to accommodate three hundred and thirty-three comparatively large Bacteria in a column of single file end to end, and notwithstanding, there will be plenty of room to spare between the ends of the column and the sides of the circle. If, however, you prefer to use a comparatively small Bacillus for the experiment you may make a hole half as large as the small pin-prick and you may still place three hundred and thirty-three of the chosen organisms in a similar column as in the last example without any fear of crushing. Those interested in comparative sizes may work out endless examples giving most amusing results.

There are in existence even smaller organisms, organisms which are not to be seen with the most powerful lenses, although we may have absolute proof that they are present. These minute bodies are able to pass through porcelain filters and are members of the ultra-microscopic kingdom.

Anyone with an average amount of intelligence perusing a daily paper will have noticed that an occasional case of plague causes much uneasiness. It is not surprising, when one considers the rapidity with which plague spread in former years, that the occurrence of a single case should be a matter of general interest. To-day, with scientific sanitary methods, an isolated case has little or no chance of producing an epidemic, as it would have been likely to do a few years ago.

THE NIGHT-FLOWERING CACTUS

Photo by] [W. Saville Kent.

A NIGHT-FLOWERING CACTUS.

This is Macdonald's Cereus, a native of Honduras, which creeps along the ground. Its pure white fragrant flowers are fourteen inches across.

To those who believe that all beauty in Nature exists for the gratification of man's esthetic sense, the night-flowering species of Cactus present a difficulty, for the flowers do not begin to open until late in the evening, and by the morning they have collapsed into limp rags devoid of beauty. In the meantime the night-moths, for whom alone their beauty and fragrance have been evolved, have visited them and pollinated them.

The group of Cacti to which the magnificent night flowers belong are distinguished by the name of Cereus, which means pliant, and has reference to the long slender stems, which are regularly grooved, whilst the ridges between the grooves are adorned at intervals with clusters of terrible spines. Many of them climb by leaning against trees, others run along the ground. The Large-flowered Cereus shown on this page has flowers about a foot across. The unopened flower-bud gives no hint of the glorious interior, for the

Photo by] [*H. C. Shepstone.*

THE NIGHT-FLOWERING CACTUS.

A reduced photograph of a West Indian species which measures a foot across, and is highly fragrant. The flower does not open until late in the evening, and by the next morning early it has closed for ever. Like most other fragrant white flowers that bloom at night, it is fertilized by moths.

outside of the sepals is dull brown ; but when these separate and turn back, their inner faces are seen to be bright yellow, which has the effect of accentuating the purity of the numerous white petals and of the vast number of long, thread-like stamens in the centre. To this chaste beauty is almost superfluously added a powerful and delicious fragrance calculated to lead the night-moths from afar to the exact spot. This plant is a native of the West Indies, and is the one most frequently seen in English greenhouses.

A somewhat similar species, but with larger flowers, comes from Mexico; but the finest is Macdonald's Cereus, from Honduras, whose flowers are frequently fourteen inches across, with red and orange sepals and white petals. This species creeps along the ground, whereas the other two mentioned are climbers.

THE CICADA

BY K. G. BLAIR, B.SC., F.E.S.

THE vocal accomplishments of the Cicada have attracted attention to their possessors from very early times. They were well known to the ancient Greeks, by whom they were held in great esteem as songsters; indeed, Anacreon, in an ode to this insect, addresses it as "almost like a god," though the attributes that he ascribes to it as a foundation for this extravagance—"earth-born, musical, impassive, and without blood"—are hardly sufficient to make the likeness very impressive. The Greeks, moreover, recognized that the male Cicada alone is the musician, for another poet, Xenarchus, sings: "Happy the Cicadas' lives, since they all have voiceless wives."

The Cicadas are mostly tropical insects, though about twenty species are found in Southern Europe; one is a native of this country, though rare and seldom found outside the New Forest. In habits they vary considerably; ours and many of the European species sing only in the hottest sunshine, while in the tropics they are usually silent during the day and commence singing only when the sun goes down. Once started, however, they are in no hurry to stop, and the power of their voice, the pertinacity with which they continue throughout the livelong night, and, above all, the immense numbers of them all around, all singing at once, make the night hideous in districts where they occur. As to the nature of the sound they produce opinions differ immensely; probably in great part this difference is due to the various species that have been the subject of observation in different parts of the world. The Greeks, as we have seen, highly appreciated the sound, though how much of this was due to the effect of a poetic imagination upon stories received from others is open to question; modern observers in the same districts say that the sound in itself is not unpleasant, but that the monotony and the pertinacity of it make it perfectly distracting; others in other parts of the world say that for shrillness and intensity it is like nothing so much as a locomotive whistle; others again compare the noise with that made by a threshing machine.

Photo by] [*P. H. Fabre.*

CICADA EMERGING.

The winged Cicada having all but emerged from the chrysalis skin, hangs limply while its wings are expanding more fully and its new skin hardening sufficiently to enable it to take hold with its feet.

The object of the noise is a matter of conjecture. The vocal organs consist of a complicated series of chambers, stretched membranes and mirrors, all concealed by a flap found on the underside of the body, on either side, behind the base of the last pair of legs. As these are found in the male sex only, it is to be presumed that their purpose is more or less sexual; but whether it is to attract the female to the performer, or whether it is a case of rivalry, each male trying to outdo his brothers, is uncertain. Since no auditory organs have been detected in either sex, with the possible exception of a membrane that has been supposed to be auditory in function within the vocal apparatus itself, it may be that if the vibrations set up are sufficiently intense they may be perceived merely in a general sense by their intended recipient.

Photo by] [*P. H. Fabre.*

CICADAS HUSTLED.

The Cicadas having pierced the shrub to obtain the sap, ants, beetles and other insects parched with thirst hustle them away in order that they may benefit by the Cicadas' industry.

The Cicada belongs to an order of insects known to scientists as "bugs," an order of which the creature to which this appellation is popularly assigned is a somewhat degenerate member. Though the perfect insect lives in trees and bushes, its early stages are passed underground. The female deposits her eggs within the stems of plants. With her strong saw-like ovipositor she pierces the stem and drives it in parallel with the fibres of the plant, forcing them asunder, and in the cavity so made lays about a dozen eggs. Then, climbing a short distance further up the stem, she makes another slit, fills that with eggs, and so on until she has made in all thirty or forty such chambers, either in one long stem or in several. This operation takes place about the middle of July. On a hot, sunny day in October the eggs hatch. The young grubs are white spindle-shaped bodies, with the first pair of legs enclosed in a common sheath and slightly movable, forming a lever by means of which they work their way to the mouth of the chamber; all their other limbs are fastened down to the side of the body, and thus safeguarded from any injury they might receive in the passage to the exterior. Having proceeded so far, they hang for a while by their tails in the sunshine; soon the skin splits and the normal grub wriggles forth and falls to the ground. Here it wanders about for a time, looking for a suitable place, and then burrows into the soil. It is not content with merely burying itself, nor does it concern itself with looking for food, but it burrows to a considerable depth, probably three or four feet or more; having reached the required depth, it seems to remain quiet without taking any nourishment until the following spring. In this respect however, as indeed in others, what may be true of one species is not true of another; for though in its general outlines the scheme of their life-history is the same throughout the group, in details of habits there may be considerable variety according to the species.

The Cicada grub is now a soft, white little creature, with powerful legs, short and broad, well adapted for burrowing in the earth. Instead of the biting jaws that are usual amongst insects, the Cicada, in the early state as in the perfect state, has its mouth-parts modified, like the rest of the "bugs," to form a long, sharply-pointed, tubular proboscis, with which it stabs the stem of some plant and through which it sucks up the sap. The Cicada grub feeds in this way upon the roots of various plants and trees; it does not remain fixed in any one spot, but burrows here and there in the earth, tapping the roots first in one place, then in another. In its wanderings it often penetrates a considerable distance from the surface; ten feet, even twenty feet down they have been found. The Cicada passes by far the greatest portion of its life in the earth in this manner; our British species lives three or four years in this condition, while the common species of the Eastern States of North America spends sixteen years as a larva, successive broods thus appearing every seventeen years. Although this is the average length of life of this species in the Northern States, in the Southern States it becomes reduced to thirteen years, and as there are naturally intermediate stages, it must not be supposed that the insect is seen in its perfect state only once every seventeen years. In one district or another it turns up every year, but in any given district the numbers in the intermediate years are quite insignificant, whereas in the year when a brood is due they come in their millions and there is no escaping from them.

Photo by] [*P. H. Fabre.*

CICADA EMERGED.

The newly-emerged Cicada, having sufficiently hardened, sits on the empty chrysalis-skin basking in the sunshine before taking its first flight. (Natural size.)

It may be wondered how a soft-bodied creature like a Cicada larva can burrow in hard dry earth, even though provided with fairly powerful legs. As a matter of fact, it moistens the earth before attacking it, making use of the supply of sap in the tree-roots for this purpose. Digging its proboscis into a root, it gorges itself with the sap, proceeds to the working face of its burrow, which may be some little distance away, and voids the superfluous liquid on to the hard earth, which is then comparatively easily removed. The same tactics are familiar to all in the case of its little relative, the Frog-hopper, or Cuckoo-spit Insect, whose larva in a similar manner pierces the stems of our garden plants, and with the sap, after it has passed through its alimentary tract, forms the well-known froth, or cuckoo-spit, that is so conspicuous upon the plants.

When full grown the larva casts its skin once more; but instead of being soft and white, it is now covered with a firm, hard brown skin, though still very similar in form. This condition is almost equivalent to the chrysalis of a butterfly or moth, but there is the very important difference that the Cicada nymph, as it is called, can still feed and still has the full use of its limbs. This state lasts only a few days, but in that time the creature makes careful preparations for its forthcoming brief period of existence in the air and sunshine. Coming to within a few inches of the surface, it selects a suitable root as a reservoir from which it may draw supplies of sap for future operations; it then proceeds to drive a tunnel vertically upwards to the surface of the ground, returning from time to time to the root at the bottom for liquid to moisten the earth as the work progresses. The damp earth it works up into a paste or mortar, with which it lines its tunnel. Some American species having reached the surface, continue the tunnel as a round, hollow tower some inches high;

Photo by] [*P. H. Fabre.*

CICADA AND CHRYSALIS.

Here a chrysalis has climbed up the heath-shrub in order that the influence of sunshine may assist it to burst its skin and emerge in the winged condition. Two other chrysalids are represented by the empty skins, showing how, except for the slit in the back, they are intact. A Cicada that has emerged is waiting for its wings to harden.

By permission of] [*Sir H. H. Johnston.*

THE BLACK COCKATOO.

This imposing bird is mainly a glossy greenish-black in colour, relieved by a band of scarlet across the outer tail feathers.

others in the same way form a covered way some little distance up a tree-trunk. The object of these structures is unknown; it has been suggested that they are to keep water out of their burrows in case of a flood, but this hardly seems likely to be the true explanation.

One bright sunny morning in May or June the nymph comes up out of its burrow, climbs up some tree-trunk, or stem of a plant, ready to enter upon its final state. It waits a short time in the sunshine, and then by strong muscular effort it bursts the skin down the middle of the back and forces its way out of the slit. Gradually it comes further and further out, until the tail only is gripped by the edges of the slit as the hard, stiff nymph-skin tends to resume its normal position. Then it bends right backwards and hangs down, soft, limp and powerless, as though the effort were too much for it.

Very soon we notice the wings begin to grow—at first short, thick and stumpy, bearing little resemblance to the large transparent wings with the sharp black veins of the full-grown insect—they begin to swell out at the base, and, as the blood is forced into the veins from the body, these expand, stretching out as they do so the folded and crumpled wing. Having attained their full dimensions, they are still limp and soft, and the insect remains suspended by the tail while its teguments harden. Then a sudden stiffening and straightening of the body brings it once more upright; it seizes the plant with its legs, folds the wings along the body, and sits basking in the sunlight until it acquires the strength and its wings the rigidity to enable it to take to flight.

It will be remarked that in our second illustration several other insects—ants, beetles, etc.—are almost as much in evidence as the Cicadas. Thanks to its sharp, strong proboscis, the Cicada enjoys a considerable advantage over other insects, in that while they, in common with most animal and vegetable life, are parched with thirst and weak and drooping in the hot, fierce sunshine, the Cicada can obtain a plentiful supply of cool, sweet sap from within the bark of the branches of trees. Other insects of all kinds, eager for a taste of the refreshing liquid, come up and wait for a chance.

Ants are not so patient and try to hustle the giant away, crawling underneath it, tugging and pulling, even seizing its beak and trying to draw it by force from the wound. The Cicada, whose motto seems to be "Anything for a quiet life," flies off to another branch and leaves the marauders in possession. It will thus be seen that the relations subsisting between the ants and Cicada (or grasshopper, as the Greek name of the creature is translated) are just the reverse of the case as related by Æsop, where, we are told, the Cicada, after spending the summer in singing and dancing, on the approach of winter went to the ant and begged for a morsel of food. The ant, however, replied that it should work during the summer and lay up its own food for the winter, and refused to relieve it.

THE BLACK COCKATOO

BY SIR HARRY JOHNSTON, G.C.M.G.

It is curious in the coloration of birds what a tendency there is for closely-allied species to pass from white to black or dark-grey, possibly because such birds may have been derived from ancestors of grey plumage. Thus we have not only white swans, but black swans, and, curiously enough, the cygnet of the white swan is born of a grey colour, whereas the cygnet of the black swan is almost white in tint; and whilst there are white Cockatoos—those which are best known and commonest—there are also black Cockatoos. In point of beak development and body size the Black Cockatoos are perhaps the most noteworthy of this remarkable sub-family of Parrots.

Photo by] [*H. Main, F.E.S.*

THE MOLE CRICKET.

Being like the Mole a burrowing animal, the Mole Cricket is remarkable for having its fore-limbs shaped much like those of the Mole, and equally efficient as digging appliances.

When adventurous Italians and Portuguese began to penetrate to Further India in the fifteenth and early sixteenth centuries, they noted with great interest in the stories of their travels the "white" parrots which they saw in the possession of Malays. This is the first mention of Cockatoos in literature, but the birds referred to had really been brought by ships from much farther east than the Malay Peninsula or Sumatra. The Cockatoo sub family is restricted in its range to the Australian Continent and the Australasian Islands of New Guinea, the Moluccas, Timor, Celebes, etc., with one outlying representative in the Philippine and Sulu Islands. There are legends collected by trustworthy German travellers of gigantic Black Cockatoos living in the unexplored interior of New Guinea. These may some day come to light and make the example of a Black Cockatoo here illustrated paltry in comparison. But Banks' Black Cockatoo is a sufficiently remarkable bird in appearance, and was the first example of the genus to be discovered by, or through, that most noteworthy pioneer explorer of the British Empire, Sir Joseph Banks, the companion of Cook. Banks probably obtained his first examples of this Black Cockatoo from New

Guinea, not on the occasion of his own explorations of the coast of that huge island, but from persons who accompanied Cook on his second expedition to the South Seas. This imposing Cockatoo is mainly a glossy greenish-black in colour, with a large crest. The sombre tints of its plumage are relieved by a band of scarlet across the middle of the inner side of the outer tail feathers. Unlike what occurs in most parrots, where the sexes can scarcely be told apart, the female of Banks' Cockatoo differs from the male in having the head and part of the back spotted with yellow and the under surface barred with yellow and russet. Banks' Black Cockatoos feed (appropriately enough) on the seeds of the banksia tree and on the cones of the casuarina. They fly heavily and much less willingly than the White Cockatoos, and their cry is less of a scream than of a low whine.

THE MOLE CRICKET

By permission of] [*Messrs. Leitz.*

THE MOLE CRICKET WALKS ABROAD.

The Mole Cricket likes to have his burrow in light soil and near water. He issues forth only at night, and sings a low continued trill. As in the case of the Cicada, the females are mute.

It is a trite observation to say that when Nature sets any of her subjects to do a given piece of work, she provides them with efficient tools for the purpose; but one cannot help being struck by the fact that two of her excavators, as far removed in relationship as are the poles from the equator, are fitted out with similar appliances and with a general similarity of body form. The Mole Cricket is a near relation of the merry chirper that likes to be in close proximity to our kitchen fireplaces, where there are many scraps of food to be found as well as cheerful warmth. The Mole Cricket, however, prefers to make a home of his own in loose soil, where he can hunt for his food. He sinks a more or less vertical shaft, and works much like the mole, scraping the soil with the claw-like edges of his digging feet and throwing it behind him. In so doing he comes across tender roots, which he cuts through and eats; likewise worms and the grubs of beetles, which also serve to appease a rather voracious appetite. From his main shaft he bores horizontal galleries right and left, to give him a greater extent of hunting-ground. His wife likes a special chamber just below the surface, in which she deposits three or four hundred eggs, and watches them until they are hatched. She is even said to guard her grubs and feed them until their first change of skin, afterwards letting each fend for itself. The young are much like their parents in miniature, except that they have no wings. As in all the Grasshopper tribe, to which they belong, these come gradually; but it is not till about three years after the eggs were laid that the new generation has become fully fledged. Like the House Cricket and the Cicada, the male Mole Cricket "sings" at night by rubbing his upper wings one over the other, producing a soft continuous trill. Though common in Central and Southern Europe, the Mole Cricket is only found locally in the south of England and is entirely absent from the north.

Occasionally he turns up in garden-ground and is then regarded as a great pest by the gardener owing to his habit of cutting through roots and eating them.

Photo by permission of] [*Sir H. H. Johnston.*

THE LARGEST PTERODACTYL.

The large creature is the Pteranodon, a flying monster whose wings measured twenty feet across, and it is engaged in stalking the smaller flying reptile that is perched on the rock-ledge above. Its fossil remains upon which this restoration is based were found in the United States; and the photograph was taken from the sculpture of Herr Pallenberg at Hagenbeck's Tierpark.

By permission of] [*Sir H. H. Johnston.*

A PTERODACTYL

The Pterodactyls were winged reptiles which flourished in the Secondary Epoch. Their wings were not like the wings of birds, but rather like those of the bats, a membrane being spread from the body to the greatly lengthened fifth finger.

THE PTERODACTYLS

BY SIR HARRY JOHNSTON, G.C.M.G.

ONE of the unsolved mysteries in the evolution of reptiles is the ancestry of the Pterodactyls, or winged Saurians. Though they offer an example of most elaborated development for the special purpose of flying through the air, their fossil forms appear quite suddenly in the rocks dating from the middle of the Secondary Epoch (the Lower Lias). Some features in the skull suggest relationship to a primitive order of reptiles, of which the Sphenodon of New Zealand is the only living representative [this order will be illustrated in a later page], and other points in the skeleton recall similar structures in the Crocodiles. The brain and the bones in many particulars offer a striking resemblance to the brain and bones of birds. But this, as so often occurs in Nature, is very misleading, being due not to any real relationship, but to similar effects produced by a similar mode of life. The wing in these flying reptiles consists of an immense membrane growing out from the sides of the body and limbs, and depending mainly for support on an extravagant development of the fifth or " little " finger of the hand. This fifth finger, instead of being the smallest in the series as it is in so many land vertebrates, is about six or seven times the length of the other fingers. Three of these (the first, or " thumb," is missing) remain quite outside the wing, instead of being covered by its structure as is the case with modern birds. In complete contrast to the Pterodactyls, the bird's wing consists of very little membrane, its flying surface being made up by feathers—exaggerated scales—which grow from the edges of the flipper formed by the second and third fingers. The Pterodactyls retained four fingers and five toes, but lost all vestige of the thumb;

the ancestors of the bird retained the thumb but lost all traces of the fourth and fifth fingers and of the fifth toe.

The Pterodactyls proper, the remains of which are commonest in Germany, were quite small flying reptiles, the largest having a spread of five feet across the expanded wings, while the smallest, with a body about the size of a lark, flew with wings measuring ten inches from tip to tip. But the supreme development of this order during the last part of the Secondary Epoch in the United States gave to the world flying monsters with wings *twenty feet* across! In this type—the Pteranodon—the muzzle has developed into a huge beak without any teeth; but the true Pterodactyls had teeth in their jaws, which in some instances grew into ferocious fangs.

The accompanying illustration of an immense Pteranodon stalking a smaller flying reptile is photographed from the model erected by Mr. Hagenbeck at Stellingen after the designs of Herr Pallenberg.

FROST CRYSTALS

BY RICHARD KERR, F.G.S., F.R.A.S.

WHEN the air of a room charged with moisture deposits its load of vapour on the cold glass windows in hard, frosty weather, beautiful designs are wrought in solid ice. The ice-trees are formed! All the panes of glass become clothed with frosted-silver robes as if by elfin craftsmanship. Each pane displays a fairy like scene wondrously pure.

Photo by] [*G. Parkin.*

FROST CRYSTALS ON THE WINDOW-PANE.

A delicate moss-like pattern woven on the window.

One cannot but have noticed the similarity which many of these patterns bear to several forms of vegetable life, including different kinds of seaweed.

Here is the frond of a fern ; there is a frost-picture which might be taken for the negative of a photograph of a landscape. Close by we have the miniature counterpart of one of the giant trees of the Western States of America. Here are radiate branches spreading out like certain tropical plants, and which remind one of the fluted stonework which forms the inner support of the roof of a cathedral. On this pane is a fairy glen of mimicked foliage which forms a barrier to all intending intruders.

The finest and costliest lace, the result of years of thought and patiently-applied skill, is not equal to the choice design fashioned on a window-pane in a single dark and frosty evening. Even to the unaided eye it is beautiful, but it shows still more wonderful details of ornamental structure under the magnifying-glass. Yet with what apparently elementary materials the designs are fashioned !

Photo by,] [G. Parkin.

FROST ON THE PANE.

A delicate formation of the Frost Crystals on the window-pane, apparently due to a small amount of moisture in the air. The long thin growths closely resemble some of the trailing mosses.

The late Dr. Maddox, a skilful microscopist, in his warm room, breathed on the window-pane during frosty weather, then, focussing his microscope towards the glass, he had the gratification of seeing a great number of minute crystals, which he photographed by attaching his camera to the microscope. Scores of frost-crystals occupied a space about the tenth of an inch in diameter !

Frost-crystals seem to prefigure living forms; crystal clusters upon crystal until arborescent forms spring forth with great rapidity.

It may be said that these resemblances are only matters of coincidence. On the other hand, there may be an underlying principle common to the vegetable and to the mineral which would account for the uniformity of design in both. Or, to put it in another way : Are the innate forces which govern both the plant and the mineral identical ?

Possibly structural forces which have an agreement lie latent in plant and in mineral.

So far as water is concerned, there is a marvellous crystalline force which not only shows its results in all the exquisite branching frost-foliage of the window-panes, but which accounts for the beautiful molecular architecture which belongs to ice, and which, though latent, comes into play when the water is sufficiently cooled.

Mosses and tree-forms like those on the window-pane are to be seen in many specimens of agate and chalcedony. These dendritic shapes add greatly to the value of the gems and are composed of iron and manganese.

We are not strictly accurate in our use of the word " crystals " in connection with these various vegetable forms on the window-pane, or those in the stones named.

None of these curious and beautiful shapes belong to the recognized systems of crystals. They have not the regular exterior form of true crystals. They seem to occupy an intermediate position between shapeless matter and perfect crystals.

Photo by] [*J. Leadbeater.*

A WINDOW-PANE STUDY OF FROST CRYSTALS.

In this photograph the Frost Crystals have built themselves up into beautiful curved shapes which remind one strongly of the fronds of some of the smaller Spleenwort Ferns. The design is complete, covering a large area very fully.

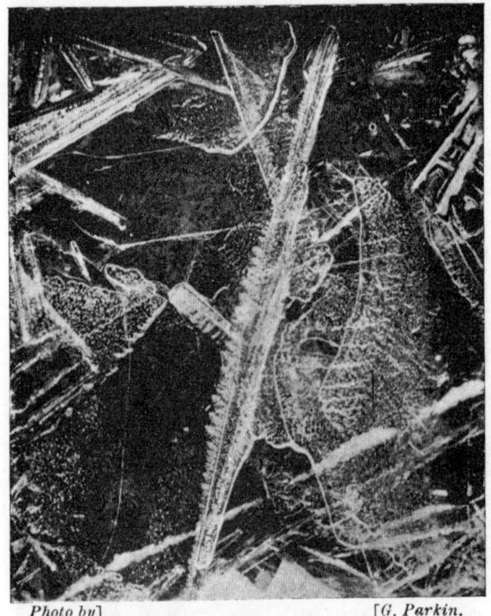

Photo by] [*G. Parkin.*

ICE CRYSTALS.

These Crystals forming in water are the first stage in the production of solid ice.

It has been said that foliage-forms assumed by hoar-frost are the result of the markings of the surface on which it is deposited. To a very small extent these may have some effect, but they do not account for the beautifully-curved lines and mimic forms of plant-life displayed by hoar-frost. The window-pane designs are but another variety of hoar-frost, and as the glass is smooth and completely devoid of any guiding lines, their forms must depend upon some other cause.

Most of the curved patterns on the window-pane can be accounted for by the currents of air charged with moisture in the room. These currents can be detected by smouldering paper or any such light combustible. The air is never still or motionless unless all the air in the room is of one temperature.

Cold, the absence of heat, is one of the chief agents in the condensation of vapour which the atmosphere always holds in larger or smaller quantities, whether in a warm room or out of doors.

When a current of air passes over stretches of water or the damp surface of the earth it becomes charged with moisture, and as it is only capable of retaining a certain quantity, much less when cold than when warm, it follows that warm air, when so saturated, must surrender some of its moisture on cooling. The moisture thus liberated is deposited as snow, rain, or hoar-frost. This enables us to see what takes place in a warm room during a frosty night. The currents of air passing along in curves part with their fog-like particles of water, which are frozen into some of the various forms already referred to.

So far we have said but little about the chief agent in the formation of frost-figures. Water holds the pre-eminent place among all the substances so extensively used in the operations of Nature on our globe. It is composed of invisible gases. It is almost colourless; it is brilliant, without taste in its pure state; it has no smell, and it is transparent. It assumes the gaseous, liquid or solid states with great facility. It is the largest ingredient in the composition of animals and plants, and it forms an important part in several rocks of the globe.

It is not generally known that the crystals and window-pane frost-forms have the power of reproducing lost parts. Paget, in his work on "Pathology," shows that crystals, as well as certain animals, possess the remarkable power of repairing their own injuries. His statement is as follows: "The ability to repair damages sustained by injury is not an exclusive property of living beings, for even crystals will repair themselves when, after pieces have been broken from them, they are placed in the same conditions in which they were first formed."

Beauty of form and of colour are both embodied in the window-pane frost-foliage, but the colour is only

Photo by] [*G. Parkin.*

ICE CRYSTALS.

Another group of the simple, first stages in the formation of ice on water The Crystals are seen to be long, thin and spear-like.

to be fully detected by the use of the polariscope.

There are innumerable examples of beauty to be found in Nature. Those persons who are fortunate enough to have their powers of observation so educated as to see and admire the exquisite designs apparent in most natural objects ought to find the world less sombre than those who pass unnoticed Nature's lavish illustrations of symmetry and ornate beauty.

The following account of a phenomenon which occurred many years ago in Sweden regarding warm air highly charged with moisture and the extraordinary ways in which the moisture is liberated when brought into contact with cold air, will bear out some of the statements just made.

A large company had gathered in a ballroom. The night was icy and starlit. The air in the ballroom had evidently become highly charged with moisture. The heat was so oppressive that several ladies fainted. An officer present tried to open the window. The frost-crystals had done their work, for they had extended to all the crevices, so that the window was frozen fast to its casement. As a last resort he broke a pane of glass; the cold air rushed in, and it *snowed in the room.* The warm air could sustain its great load of moisture no longer when mixed with the cold air from the outside. This is vouched for by Professor Dove, of Berlin.

As children we were delighted whenever the curious frost-figures decorated our bedroom windows. In our innocence we may have thought that the ferns, the seaweed, and the branches of trees had been traced out in some mysterious way. So they had; and in a measure the full secret of their beauty is still a mystery to us. Then, when out of doors we saw that all the blades of grass and all the twigs were glistening, our wonderment was greatly increased.

A fan-shaped pattern in which the fans overlap.

Photos by] [*J. J. Ward.*

FROST CRYSTALS.

Here the Crystals have grouped themselves into long ribbon-like forms, this being probably due to the trickling of moisture down definite lines.

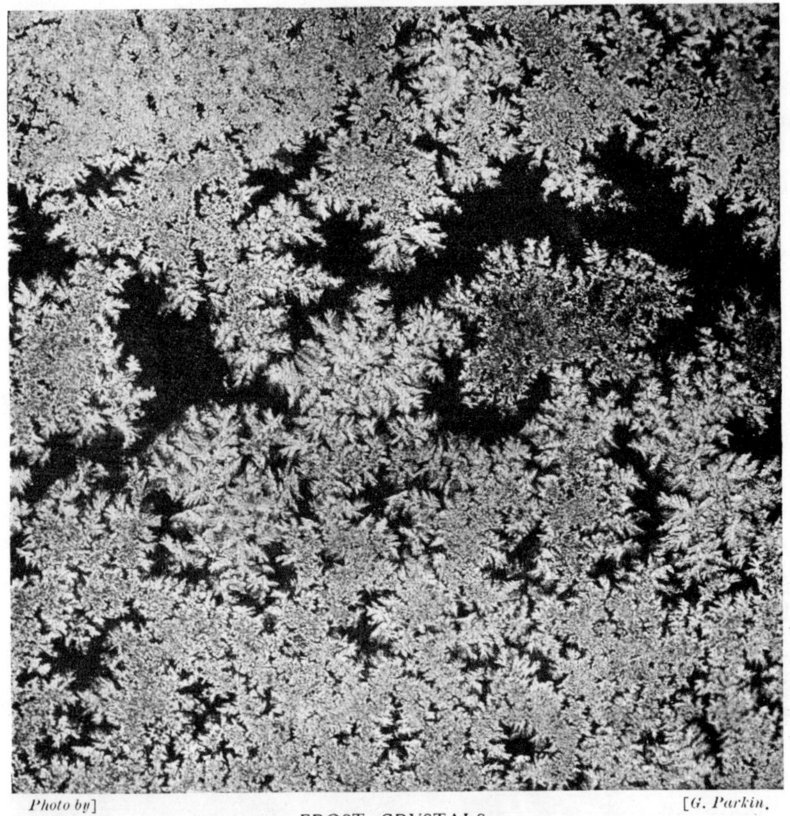

Photo by] [*G. Parkin.*

FROST CRYSTALS.

In this example the form is most like the pattern woven by lichens on a rock, when growth proceeds by an extension of the fringed margins to be made more solid later.

A TAILLESS LEMUR

BY SIR HARRY JOHNSTON, G.C.M.G.

NATURE, in the development of living forms, is constantly trying experiments. One of these has been the fabrication of a mammal which might walk about erect on its hind legs and use its hands for other purposes than locomotion. In the perfect carrying out of this idea Man came into existence, and really became the marvellous creature that he is by using his hands and fingers for manufacturing and for executing many a behest of the brain. This idea of walking about on the hind legs first made its appearance ages ago in reptiles, and brought the Bird into existence, as well as some of the strangest Dinosaurs. It crops out again in the Marsupials (phalangers, opossums, kangaroos), and again makes its appearance more markedly amongst the Lemurs, or Half-apes. We know now that there were Lemurs once existing in Madagascar nearly as big as men in size and probably walking about erect on their hind legs, one of Nature's numerous unsuccessful attempts to make a Man, and probably extinguished finally by the coming of real men to Madagascar. But amongst the existing Lemurs in that strange island is the one here illustrated; the Indris, largest of living Lemurs, which frequently walks and leaps on its hind legs only. With these powerful limbs the Indris is able to jump from branch to branch over a distance of as much as thirty feet! When the tailless Lemurs are on the ground they do not progress on all fours, but stand erect and move by a series of leaps, keeping their balance by holding the arms above the head. The first toe is large, as it is in Man, but it is opposable to the others, as in monkeys. The tail is only a stump four or five inches in length. The thick fur is divided rather vividly into black and creamy-white areas. The range of the Indris seems to be restricted to the forests of the east coast of Madagascar, where it lives in small troops of four or five, probably a family of father and mother and grown-up children.

THE SQUID

BY EDWARD STEP, F.L.S.

UNDER the general term of Cuttle-fish at least three distinct groups of related animals are included. These are the Octopus, the Sepia or true Cuttle, and the Squid. The Octopus has been described already (see page 26). The Squid differs from the Octopus in that its body is cylindrical instead

By permission of] [*Sir H. H. Johnston, G.C.M.G.*

A TAILLESS LEMUR.

The Indris, here represented, is the largest of the Lemur family. Its tail is reduced to a short stump, and its thick fur is **mapped out into black and creamy-white areas.** A native of the Madagascar forests, it is able to take flying leaps from branch to **branch as much as thirty feet apart.**

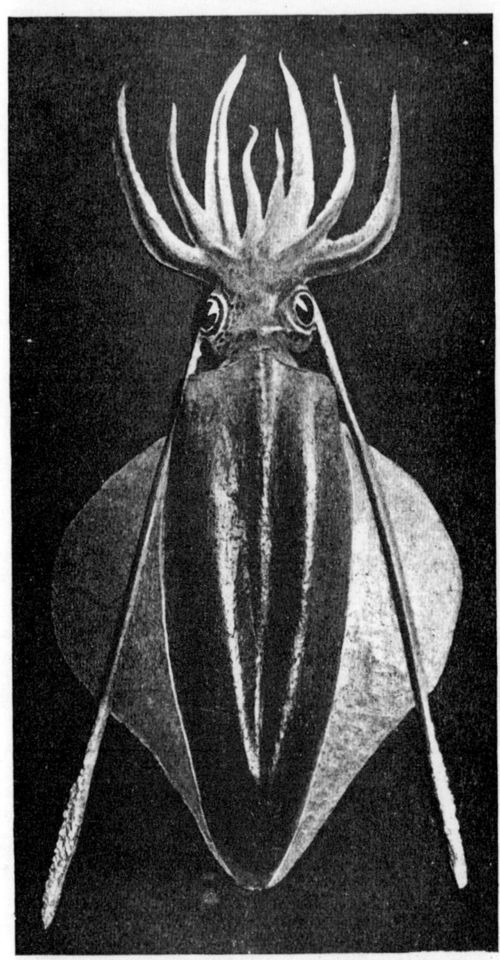

Photo by] [*W. Saville Kent.*

A TASMANIAN SQUID.

This Squid differs from the European kind mainly in the rounded margins of the fins.

of being globular, and that in addition to the eight arms of the Octopus it has two tentacles three or four times the length of the arms. Also, the hinder upper part of the body is provided with a pair of fins, which enable the Squid to swim in the middle or upper waters, whilst the Octopus is compelled to crawl over the bottom. The Octopus has merely the rudiments of a skeleton, represented by a couple of small rods embedded in his outer integument. In the Squid the rods are developed into what is known as the pen, a long, flat band of clear, thin gristle, in adult specimens fifteen to eighteen inches in length. To give it additional strength this pen is keeled along its under side. It is not easy to get a perfect example, as it is very fragile; tender and easily torn when fresh, brittle when dry. One would imagine that to this "sort of a backbone" the muscles of the Squid would be attached, but this is not the case: the pen lies quite loosely in the mantle. A reduced photograph of two of these pens will be found on page 351.

Like the Octopus, the Squid has a water-siphon, through which it can eject a very forcible jet of water, which has the effect of sending the Squid rapidly backwards. It may sound Irish, but this is its mode of progression, modified by the action of its fins. Into the siphon there opens a tube connected with the ink-bag. Though the ink-bag is small, its contents are so highly concentrated that only a very small quantity poured into the outgoing current of water is sufficient to produce a dense black cloud in the surrounding sea, enabling the Squid to shoot back out of sight and sink to the sands, where its colouring renders it inconspicuous.

As in the Octopus, the eight arms, as well as the two tentacles at their broad extremities, are covered on their inner face with powerful suckers, by which they attach themselves to their prey whilst the parrot-like beak is tearing it to pieces. Fish is the principal food; and let us not exclaim, "Poor fish!" for there are few things fishes appreciate more in the way of food than the flesh of the Squid. It is generally a matter of which is the bigger; if the fish, then the Squid may be eaten; if the Squid, then the fish may be the victim. There is no better bait for sea-fishing than the flesh of Squid, and it is one of the principal foods of the sperm whale.

The eggs of the Squid are laid in long capsules of a gelatinous character, each capsule containing a number of eggs, and many capsules being attached together to form what fishermen know as a "sea-mop," which will contain about forty thousand eggs.

The Squid is the largest of the Cuttles found ordinarily around the British Islands. Without reckoning the arms, its length is about two feet. Occasionally we have had visits from much larger species. The largest-known Cuttles swimming on the surface with arms extended have no doubt upon occasion served as the great sea-serpent. One of these was cast upon the shore of County Kerry a couple of hundred years ago, and was ascertained to have been nineteen feet in length, plus the length of its tentacles. These, unfortunately, had been mutilated by rolling and beating on the shore ; but what was left of them measured eleven feet. Forty years ago two large Squids of this kind were stranded on the shores of Newfoundland and these measured respectively forty-seven and forty feet long.

Our full-page illustration shows the Squid in mid-water pursued by a conger, and resorting to its accustomed tactics—the submarine equivalent of " throwing dust in the eyes " of its opponent. The ink-bag is brought into requisition and a black aqueous fog is interposed between its pursuer and itself, under whose cover it will make off backwards. A very little of this ink will go a long way, as may be learned by cutting up a Squid for bait in fishing. The unskilful amateur fisherman who has hooked a Squid for this purpose and does not know how to manage his capture may have his face and clothes spoiled by the profuse discharge. The experienced fisherman contrives to keep the Squid's head turned away from the boat. The ink has a burning, smarting effect on the skin, and its stains are not easily removed.

When the Squid is at rest it lies on the sandy bottom, where it is well concealed, for its dirty white flesh is peppered with darker dots, and is thus very similar in appearance to the sand around it.

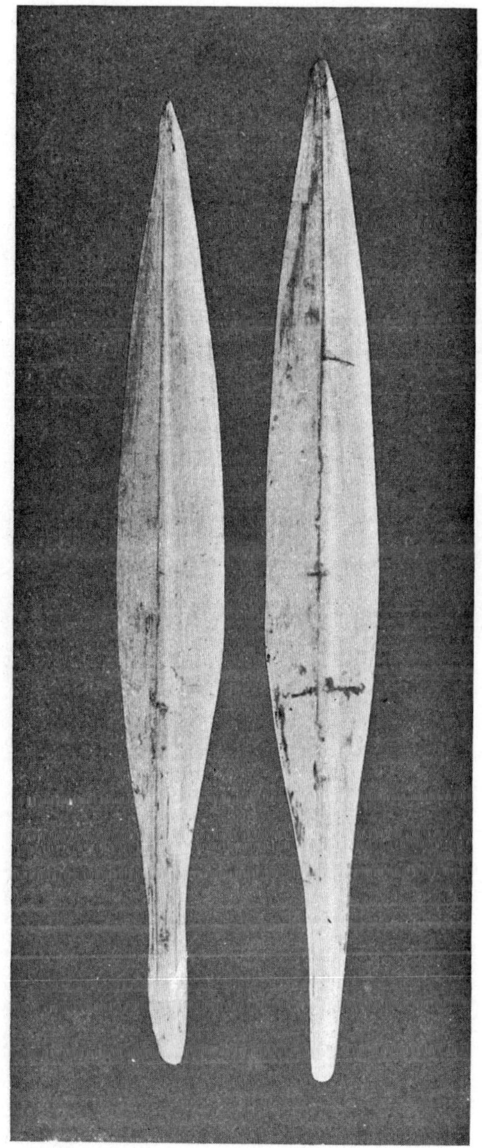

Photo by] [*E. Step.*

SQUIDS' "PENS."

As a sort of a backbone the Squid bears in its mantle what is known as a "pen." It is thin, transparent and gristle-like. The photograph is only one-third of the actual size.

HOW NATURE PROTECTS INSECTS

BY EDWARD STEP, F.L.S.

INSECTS have a multitude of natural enemies, including some of their own class, whilst to many birds, reptiles and some small mammals they form the staple of their food supply. The constant onslaught upon them that this implies is no doubt one of the reasons for their marvellous fecundity. At the same time it must have led to the gradual destruction of the unfit, and the development of natural defences. The unarmed and the undisguised would fall victims wholesale and have little chance of leaving descendants. It is thought that those that varied in form and colour in the direction of harmonizing with their surroundings would stand a better chance of being overlooked by their enemies, and, therefore, of transmitting their advantage to their progeny. In a similar way stings might be developed from organs originally intended and used for piercing plants in the process of egg depositing, and the throwing off of waste substances from the body might have made certain insects unpalatable.

However the advantages may have arisen, we find that many insects are to-day effectively protected by stinging or biting apparatus, the discharge of corrosive or otherwise offensive fluids, a general unpalatableness accompanied by conspicuous colouring that

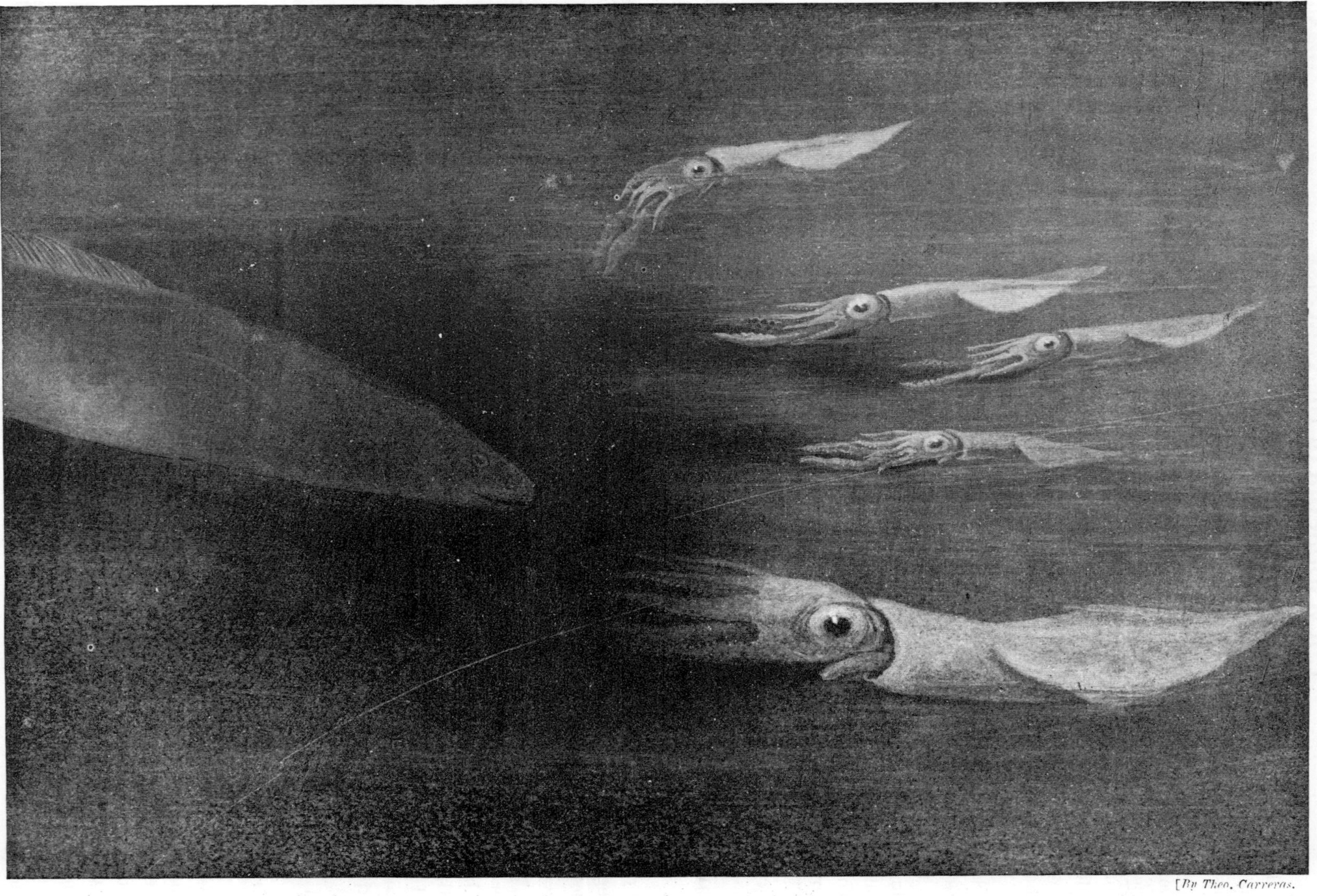

[By Theo. Carreras.

SQUID AND CONGER-EEL

The Squid is esteemed as food by the larger fishes. In this picture a Squid is in danger of attack by a Conger-eel, but by pouring a little of its ink into the stream of water it is forcibly ejecting from its siphon, a dark cloud is interposed between attacker and attacked, whilst the latter sheers off backwards into safety.

advertises the fact to all whom it may concern ; or a scheme of colour that so admirably harmonizes with their accustomed surroundings as to render the insect all but invisible. Then, again, there is a class of defences in which the whole external form has been so modified as to make it bear a very close likeness to some other and uneatable natural object.

These various defences constitute too wide a subject to be dealt with in a short article. We must content ourselves for the moment with calling attention to a few non-combative defences adopted by insects ; not those in which the jaws, the sting, the poison or the terrifying aspect is relied upon, but what we may fitly describe as the self-effacing type of protection. To properly appreciate the efficiency of protective coloration we must go to living nature. The cabinet drawers of the entomologist will exhibit clearly all the diversities of form and colour by which the naturalist distinguishes one species from another, and may well fill us with admiration for the beauty of colour and design ; but the fullness of that beauty can only be appreciated when we study those same insects in the places where they live, and learn the true significance of every tint and pencilling. The butterflies in the cabinet are chiefly admired for the rich tints with which the upper surface of the wings is painted. But this, as a rule, is mere sexual show ; the marvel of the butterfly's wing is mainly to be found on the underside. We may watch the apparently joyous flight of the butterfly in the sun, and feel delight in its rich colouring, for it is one of the most brilliantly conspicuous things in Nature. A cloud passes over the sun, and the butterfly drops upon the nearest herbage. We hurry up to view it at leisure ; but

Photo by] [*F. Enock, F.L.S.*

STICK-LIKE CATERPILLAR.

In this photograph a caterpillar of the Swallow-tailed moth is attached to a dead shoot of ivy, and is scarcely to be distinguished from the dead twigs.

Photo by] [*F. Enock, F.L.S.*

CATERPILLAR ON ROSE-SHOOT.

The caterpillar of the Willow Beauty moth is here shown among the shoots of a rose-tree. Its colouring, form and attitude are such that it is easy to overlook it as one of the branches.

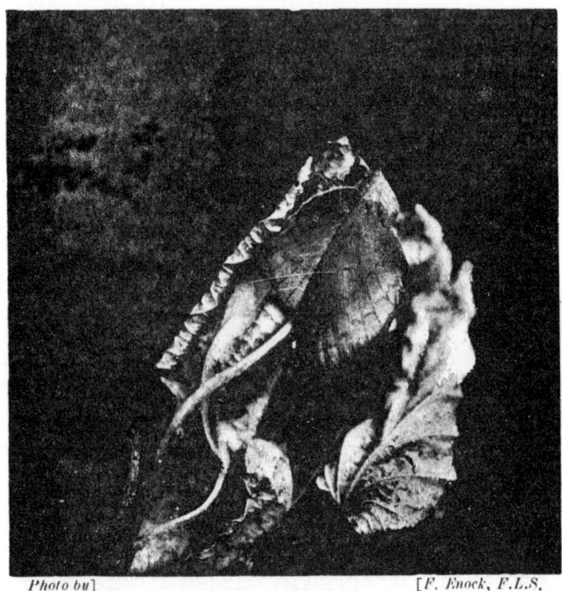

Photo by] [*F. Enock, F.L.S.*

LAPPET MOTH.

The Lappet moth when at rest so carries its wings as to give itself the appearance of a cluster of dead leaves, which it resembles in colour.

Photo by] [*E. Step, F.L.S.*

GREY ARCHES MOTH.

The Grey Arches moth sits by day on the trunks of trees, and its colour and markings so closely resemble the lichens and irregularities of the bark that it is seen with difficulty. In the photo the moth is about one-third less than the natural size

our butterfly has completely disappeared—though probably it is sitting on a leaf or a flower within a few inches of our eyes. What, has happened is that the wings have closed together over the back, completely hiding their brilliant upper surface and bringing the more subdued ornamentation into view. This, as a rule, so harmonizes with the immediate surroundings that the butterfly is no longer recognizable. One of our photographs shows the conspicuous little Orange-tip butterfly that has alighted upon a flower cluster of Jack-by-the-hedge. Its orange-tips are no longer prominent, but the green mottling of the underside is turned to us, and so well does this match with the green and white parts of the flower cluster that, even in the photograph, it is difficult at first to detect the butterfly.

I have said "even in the photograph," because, as a rule, it is far easier to detect a disguised insect in a photograph than it is in Nature, for the area that the eye has to wander over is small, and in most cases the photographer almost mechanically centres his chief object. Another beautifully protected butterfly is the Green Hairstreak. In this, only the underside of the wings is coloured green; but when the butterfly settles on the new foliage of the beech, as it likes to do, and turns up its wings to expose the green side, it is seen with difficulty. The Comma butterfly, whose wings have a very ragged margin, is bright brown of various shades above, but beneath it is clouded with duller, darker browns, and when it is at rest and showing this side, it might pass for a dead leaf. The Lappet moth, of which also we give a portrait, is, when at rest, like nothing so much as a bunch of dead leaves. The Hook-tip moth is much like the Comma butterfly in the effect it produces when at rest, for it has the butterfly habit, not common among moths, of throwing its wings up vertically. The Dotted Border, on the contrary, spreads its wings half open and lets the edges come into close contact with a dead leaf. If this is a wintered oak-leaf, as frequently happens, the colour matching is almost perfect; in any case, it is so close as makes the detection of

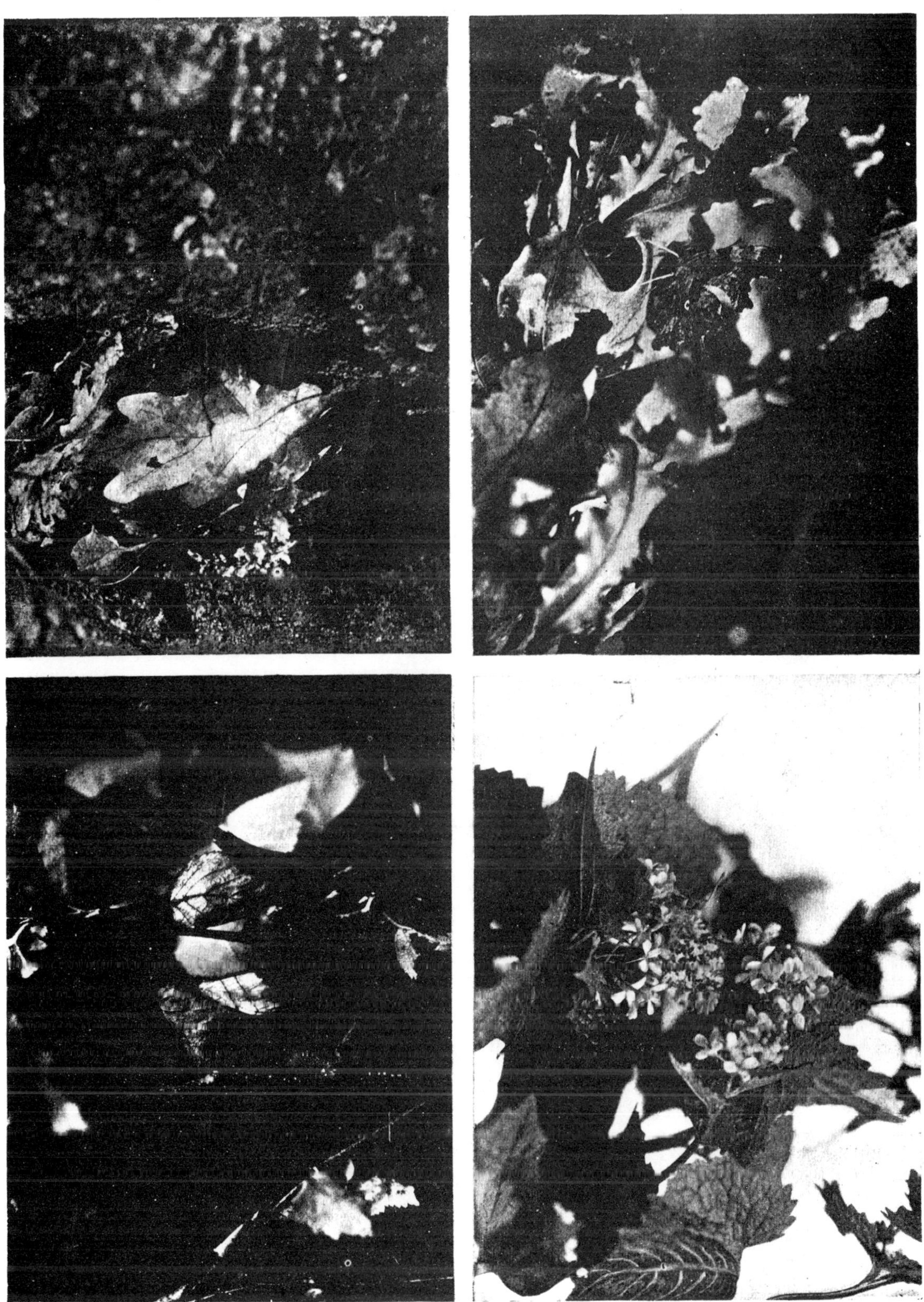

Photos by] [*F. Enock, F.L.S.*

PROTECTIVE COLOURING OF INSECTS.

These four photos show four living insects in their natural resting attitudes. The first is the Dotted Border moth on a dead oak-leaf, similar in colour, and therefore not easily seen. The second shows the Comma butterfly among dead leaves, with which its brown underside harmonizes. The third is the Hook-tip moth, looking like a withered leaf, a resemblance helped by the vein-like lines across its wing. The last photograph shows the Orange-tip butterfly at rest on the flowers of Jack-by-the-hedge. All are reduced from natural size

the moth most difficult. Even if the moth settles on a tree-trunk the resemblance is still to a dead leaf, only now it looks like a thin leaf that has been carried by the wind and has caught on the rough bark in its passage to the ground.

The last-mentioned moth—the Dotted Border—belongs to the large Looper family, so called from the habit of the long slender caterpillars, which in walking draw the greater part of the body up into a loop; the legs being situated near to the head and a pair of clasping organs at the other extremity, this is the only method of progression open to them. We shall have a few words to say of some of these caterpillars before we have done, but our present concern is with the moths that develop from them. They mostly rest with their wings spread flat and almost as widely as the "set" specimens in the collectors' cabinets, and their wings are marked with patterns that fit in wonderfully with the cracks and crevices in the bark of trees or with the grain and weather streaks in the wood of fences and posts. The Waved Umber moth, both in colour and pattern, matches the oak palings on which it was sitting when photographed. Most of such moths spend the whole day in these positions, and appear to be unseen by any enemies, or the trick would soon be abandoned, for moths so openly exposed would become extinct. Some of these Looper moths seen in the cabinet are conspicuously coloured, a broad expanse of white, for example, being contrasted with a patch of black or dark brown. But see such a moth on tree trunk or broad leaf and you pass it unnoticed, for it is then to all appearance a bird-dropping.

Photo by] [*E. Step, F.L.S.*

CATERPILLAR OF BEAUTIFUL YELLOW UNDERWING MOTH.

This caterpillar is coloured dark green chequered with white, and is scarcely distinguishable from the branch of heather upon which it is feeding.

A somewhat similar type of coloration—though intended to produce a different illusion—is adopted by some of the smaller of the thick-bodied moths. In this class of moths the upper wings are folded when the insect is at rest, over the lower ones, as shown in several of our photographs. The Marbled Coronet moth is one of these, and its light markings may appear to be merely white or grey lichen spots on the bark of the tree it is resting upon. Another instance of the same type is afforded by the photograph of the Marbled Green moth which shows a portion of a wall in a much frequented thoroughfare of one of our popular South Coast resorts. The particular patch of wall was within a few inches of the shoulders of the numerous passers-by, yet to our knowledge those two moths remained unseen and unmoved for at least six hours in the height of "the season." There seems to be a kind of consciousness in the moth that, so long as it remains immovable, it will be passed as merely one of the lichen-spots that ornament these walls. An interesting point about this moth is that the caterpillar feeds on these same lichens, and constructs for itself a chamber of moss and lichen—unless it can find a hole in the masonry—in which it hides during the day.

A larger example is afforded by the Grey Arches moth, shown on page 354. This moth spends the day on tree trunks in or near woods, and it will be seen how admirably the streaks and out-

Photo by] [*E. Step, F.L.S.*

THE MARBLED CORONET MOTH AT REST ON A BEECH TREE.

This beautiful moth has its grey wings marked with white spots which strongly resemble the lichen-spots that are found on the bark of most old trees.

lined spots of its grey-brown wings fit in with the crevices of the bark. So, too, with the lighter Grey Dagger moth, which is often found on lichen-spotted trunks, though our example was photographed on the author's fence, where it was much less conspicuous than in the narrow limits of the photograph. It will, indeed, be found that in almost every case these sitting moths are invisible save to the trained eye of the entomologist.

And much the same may be said of caterpillars. If they are not marked with warning colours, or clothed with long hairs or spines, they are coloured to harmonize with some part of their food-plant. The majority are green, but if they feed only at night and spend the day in or on the earth near the root of the plant they will be dirty brown in hue. If day feeders in the open their green may be streaked or spotted in some fashion that will break up their outline. An example is afforded by the photograph of the Beautiful Yellow-underwing moth on page 356. This caterpillar feeds openly on heath and heather, and its green back is chequered with diagonal white spots in such fashion as to repeat the effect of light on the small glossy leaves of its food plant. Even the huge

Photo by] [*E. Step, F.L.S.*

THE GREY DAGGER MOTH ON A FENCE.

Most frequently this moth would be found on a tree trunk where its dagger marks would fit in with the cracks in the bark.

caterpillars of several of the Hawk moths, though they may be as thick as one's finger, are so marked with diagonal stripes of white and violet that in some cases these look like the light reflected from the edges of leaves, or nerve branches on a curled-up leaf.

In the case of most of the Looper caterpillars already referred to, the general form and habit is to resemble a twig, the caterpillar holding only by its hind claspers, standing rigidly at an angle corresponding to that of the real twig. Examples will be seen in the caterpillar of the Swallow-tailed moth on an ivy stem and in that of the Willow Beauty moth on the rose, as shown on page 353.

These few examples could be multiplied *ad libitum*, but for the present sufficient has been said to call attention to very remarkable phenomena that exist, here, under our eyes. The reader, if he will keep his eyes open upon his daily walks, even between home and the railway station will find further illustrations upon the familiar walls and fences.

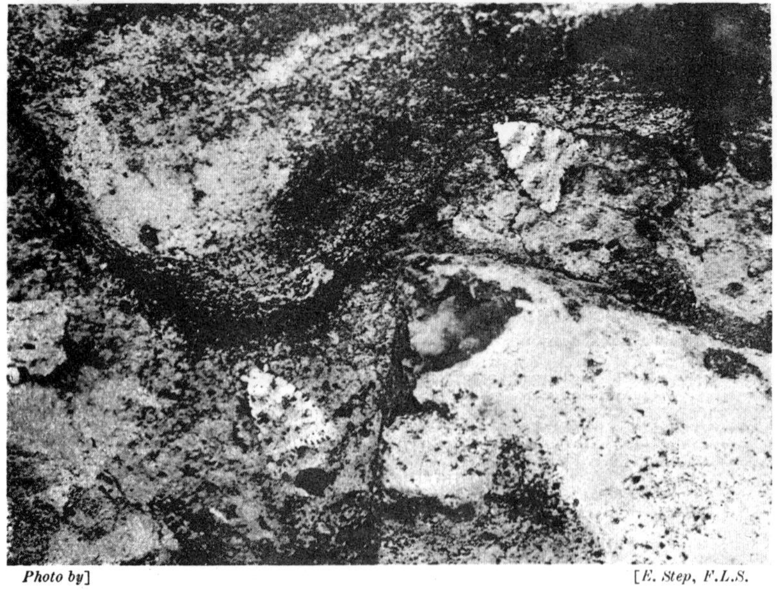

Photo by] [*E. Step, F.L.S.*

THE MARBLED GREEN MOTH.

This moth rests on exposed lichen-covered walls and rocks. So close is its resemblance to its surroundings that the specimens photographed were passed by hundreds of persons without being seen.

BIG SNAILS

BY LIONEL E. ADAMS, B.A.

THERE are people who regard snails as noxious objects to be collected in a flower-pot and stealthily heaved into the next garden, or else crushed beneath the foot. If our British snails attained the size of the great tropical Agate Snails, the former method of riddance would be dangerous and likely to lead to a breach of the peace, and the second method too horrible to contemplate! If, however, the not unnatural antipathy to slime can be overcome, these handsome creatures can be made most interesting pets—even though, perhaps, a trifle unresponsive. They have been called Agate Snails on account of the high polish and curious variously-coloured markings which adorn some members of the tribe. In captivity, when the young emerge from the bird like eggs, they should be separated from their parents or they will certainly be devoured, as most snails are cannibals. They possess a powerful weapon of offence, a long, flexible ribbon set with numerous rows of minute but serviceable shark-like teeth, the rasping of which can be plainly heard when the animals are feeding.

Photo by] [*E. Step, F.L.S.*

MOTH ON OAK FENCE.

The Waved Umber moth, alike in its colour and its markings, fits into the general appearance of the oaken fence so well that it is scarcely perceptible.

There are some one hundred and sixty different kinds of Agate Snails, mostly inhabiting the tropics ; only a single representative of the tribe being found in Britain. This species, however, is so peculiar in its habits and structure that it is quite worthy of mention among the Marvels. Not exceeding a fifth of an inch in length, the highly-polished, transparent and colourless shell is known from its slender form as the Hairpin shell. It is inhabited by an equally colourless animal, which is so transparent that the food it has eaten may be seen within. This absence of colour in animal and shell is due to the fact that it lives three or four feet underground and has never been found alive on the surface. This subterranean habit of life has also deprived it of the power of sight, though it possesses perfectly-formed eyes at the ends of its "horns," or tentacles. Its food is likewise mysterious, no one having yet been able to determine whether it consists of vegetable or animal matter ; but the fact that it is specially attracted by buried bones gives colour to the belief that it is carnivorous by preference. There are large species of Hunger Snails that run the Agates rather close for size. These also are mostly tropical, the species photographed (page 361) being from New Caledonia. Our only British representatives of the group are tiny snails found on trees and rocks, of which the larger of the two measures only two-thirds of an inch.

Our largest British land snail is known by various names. Archæologists, who used to find it on the sites of old Roman camps and villas, supposed it to have been introduced into this country as a dainty by wealthy Roman officers, and called it the Roman Snail. But though this (or a very similar species) was in high favour with the Romans as a nutritious delicacy, the shells may be found in earthy deposits which were formed long before the Romans appeared in these islands ; and it is merely a coincidence that the Roman camps and villas were built on chalky soil, which is the natural habitat of this snail. Another name, the Apple Snail, has also been bestowed upon it erroneously. This term is very appropriate as regards its shape, but its scientific name, *pomatia* is derived from a Greek word meaning a "trap-door," and not from the familiar Latin *pomum*—an apple. The ancient Greek naturalist gave it its name because he observed that on the approach of winter the snail forms a thick calcareous covering over the opening of the shell, both to keep out the cold and as a defence against the attacks of beetles and other enemies.

The value of this snail as food is patent to anyone who wanders through the markets on the Continent, where the gathering of the snails and fattening them in "snail-farms" forms an important industry. In certain districts they are fattened on vine-leaves and are known as Vine Snails. At one of these farms in Switzerland as many as eighty thousand are fattened annually; their wholesale price varies from ten to seventeen francs per thousand according to their quality. The Roman Catholic Church permits their use as food during Lent, in which season Paris alone is

A MONSTER SNAIL.

One of the huge Agate Snails of Africa. They are the largest land-snails known, and different species of them are found in several of the warmer parts of the world. The one shown is a native of Nigeria, whence living specimens have been brought to this country. Their eggs are as large as those of pigeons.

said to consume fifty tons daily! In fact, the increase in the consumption of this snail caused a necessary restriction for fear of its extinction, and the authorities of Côte d'Or, in France, have established a close season from April 15 to July 15. The Continental export trade extends to both North and South America; indeed, in Buenos Ayres the species is now firmly established. Their import to this country also is considerable, one hundred thousand being consumed annually in London, where they are utilized extensively in thickening certain soups and in giving a glaze to various kinds of pastry.

It is our British custom to decry as "foreign messes" many really excellent Continental dishes; though why prejudice should extend to this most nutritious mollusc, while the indigestible whelk and the dubiously wholesome mussel are tolerated, is hard to understand.

The largest land Snails in the world are the subjects of the illustrations on this and the opposite pages. The family to which these giants belong is not only exceedingly numerous, but is represented in most parts of the tropics and the temperate zones—two quite small and inconspicuous forms alone being found in Britain. The special genus or group to which the Giant Snail belongs is confined to South America and the West Indian Islands. These giants, which sometimes attain the length of six inches, are said to be still sold as human food in the Brazilian markets.

The eggs of a specimen of the Oblong Giant Snail (shown in the illustration on this page), which came from the Island of Trinidad, were ascertained to measure as much as thirty-one millimetres in length and twenty millimetres in breadth; or, in more familiar terms, one and a fifth of an inch in length and four-fifths in breadth. Perhaps the marvellous size of these eggs may be best realized if we compare them with those of our Common Thrush, to which they are, roughly speaking, equal. The young, when they eat their way out of the shells, are about the size of our common periwinkle, though of a more elongated form. These eggs differ from those of most land snails inasmuch as they are hard and brittle like those of birds, instead of tough and leathery like those of turtles and crocodiles. When freshly laid the white is used extensively as cement for mending china and glass.

SNAILS' EGGS.

The egg of a Trinidad Snail is here shown slightly enlarged. The lower example shows that the snail's shell is already formed before the young leaves the egg.

ROXBURGH'S FIG

THE Fig family of plants is a large and varied one. Few of those who grow the familiar india-rubber plant in their windows or conservatories realize that it is a near relation of the tree that in the Mediterranean region produces the fruit that comes to us flattened and slightly pressed in neat little boxes. Neither do they associate it with the banyan of India, that will cover acres of ground, sending roots down from the far-spreading branches, which roots ultimately be-

Photo by] [*E. Step, F.L.S.*

THE SHELL OF A HUNGER SNAIL.

The Hunger Snails are so-called in Greek by the naturalists on account of their voracious appetites. The species here shown of the natural size is a native of New Caledonia. Our British representatives of the family are not more than two-thirds of an inch long.

come stout props and trunks supporting a leafy roof, under which an army could shelter. There are other Figs in Burmah, South America and other warm regions which take advantage of other trees as supports up which to climb, and end by strangling their supporter and occupying its place.

Roxburgh's Fig, which forms the subject of our photograph, was discovered in Silhet seventy-two years ago, and named in honour of Roxburgh, the famous botanist, who added so much to our knowledge of the flora of India. Its leaves have little resemblance to the seven-lobed leaves of the ancient fig-trees that luxuriate in old gardens in the South of England (where they are believed to have been originally introduced by the Romans and re-introduced at a later period by Cardinal Pole). Roxburgh's Fig, apart from its fruit, might be taken for a mulberry-tree, for its foliage is very similar in shape. Like the Common Fig, it grows to a height of about twenty feet only; but the remarkable point about it, which is clearly shown in our photo, is the manner of producing its large fruit. In the Common Fig the fruits are produced singly or in pairs from the base of the leaf-stalk; but those of Roxburgh's Fig come in clusters from the lower branches, the main trunk, and even those parts of the stout roots that are above ground. They are much larger than the European Figs, shaped much like a turnip, but ribbed and covered with coarse down. Our photograph, for which we are indebted to Major Gage, the superintendent, was taken in the Botanical Gardens at Sibpur, Calcutta.

Photo by] [*E. Step, F.L.S.*

THE ROMAN SNAIL.

This is the largest of the British Snails, and is here shown of the natural size. By many people—especially on the Continent—it is esteemed a greal delicacy. It is found only on chalky soils.

THE LOVE-STORY OF THE STICKLEBACK

BY JOHN J. WARD, F.E.S.

THE Stickleback in yonder pond is a gay fellow indeed. There is no creature either in the pond or the brook that is so gorgeously attired. Even the male newts, with their flowing crests and gaudy orange and black hues, cannot compare with his exquisite suit of iridescent sapphire blue and glowing red. Also, he is as courageous as he is brilliantly coloured.

He has built a tub-shaped nest from particles of decayed vegetation, well cemented together, and hammered into shape by means of his head. Now that it is complete and almost hidden from view (being partly covered by mud and sand), he is very proud of his accomplishment; and his

Photo by] [*Major A. T. Gage, I.M.S.*

ROXBURGH'S FIG-TREE.

This photograph from a specimen in the Calcutta Botanical Gardens well illustrates the tree's peculiarity in producing an abundance of fruit not only on its lower branches and trunk, but even on the above-ground portions of the roots.

Photo by] [*W. B. & S. C. Johnson.*

THREE-SPINED STICKLEBACK.

The Three-spined Stickleback is here photographed the full natural size. In the breeding season it is known to juvenile anglers as the Redthroat, from the resplendent bridal tints it assumes.

present business in life is to lure a lady Stickleback into that structure, so that she may assist in stocking it with eggs.

Presently one is seen approaching; and then the eyes of this gay fellow flash fire-like glances towards her as he manœuvres round and about his household in ecstatic delight. His colours glitter with extraordinary brilliancy as his rapid movements continually expose them to the sunlight. Then he makes a headlong dive right into the entrance of his nest, almost immediately backing out again and once more encircling his preserve. Lady Stickleback, however, apparently regards him with a very indifferent air. Suddenly his prospective bride darts to one side, and, like a flash of light, a third Stickleback has appeared. In an instant there is a whirl of colour: silver, blue and red are immediately merged into a blur of grey. In truth, we are witnessing a desperate duel for a lady fair.

The colours of the combatants promptly reappear again, for they have quickly separated, the owner of the nest instantly diving to guard its entrance. His adversary, however, is not yet beaten, for he also has made a rapid plunge for the nest. That manœuvre, though, was a fatal mistake, for the anger of the rightful owner was at its height, and, by a dexterous turning movement, he has charged the lower part of his foe with his formidable spines.

His rival has retreated mortally wounded, but the victor glows more brilliantly than ever. Lady Stickleback had been quite forgotten during the fight; but, to the great delight of her valiant suitor, she has now entered his home. After depositing four eggs therein, and pushing her way out through the opposite side, leaving a clear passage through the nest, she at once swims away.

[*By J. Teklenburg.*

TEN-SPINED STICKLEBACK.

This species is less common than the Three-spined Stickleback. It is a thinner fish than that species, and its adornment in the breeding season is a rich black, which has earned for it the familiar name of Tinker.

Such are the scenes which may be witnessed during the merry month of May when the Stickleback's "fancy lightly turns to thoughts of love"; for he is a confirmed polygamist, and requires many consorts to fill his nest with eggs, which he faithfully guards until the young fry appear. Even then he vigilantly protects the offspring, and promptly returns them to the nest by means of his mouth when they go astray.

COMETS

BY G. F. CHAMBERS, F.R.A.S.

GREAT Comets which appeal ostentatiously to the public are so rare that probably the announcement made in the newspapers in September, 1911, that no fewer than seven Comets were then under the observation of astronomers in differents parts of the world fell rather flat on the minds of the uninitiated. In reality, Comets are very common, especially now that so many giant telescopes are available for picking them up and following them.

Photo by] [*J. J. Ward, F.E.S.*

STICKLEBACKS AND NEST.

In this photograph a male Three-spined Stickleback is guarding the nursery it has built against the attack of a marauding Ten-spined Stickleback (upper figure). (About one-third less than natural size.)

In bygone times, when there were no telescopes available, of course it was only the specially brilliant ones which attracted notice, and as these only appeared at rare intervals of, it might be, twenty or thirty years, they came to be looked upon as portents, and portents of evil rather than of good to the nations. Shakespeare, therefore, was no more than abreast of the times when he wrote :

"Comets importing change of times and states
Brandish your crystal tresses to the sky,
And with them scourge the bad revolting stars
That have consented unto Henry's death."

Allusions in the same spirit may be found throughout the literature of the world from classical times down to the last century, and, indeed, belief in the mischievous influence of Comets is not even yet extinct. On several occasions during 1911 I was asked whether the intense hot weather was not due to the prevalence of Comets ; but the answer was bound to be in the negative, and it may as well be stated once for all that there is not the slightest ground for the opinion that Comets exercise any influence whatever on mundane affairs, except it be to frighten nervous people.

According to a more scientific level, it may be pointed out that nearly all Comets at their first discovery need to be termed "telescopic" Comets ; that is to say, are only visible in telescopes, and the great majority of them never become anything else. Whilst some Comets are picked up by chance, there has been carried on now for many years, especially in America, a systematic search for them by sweeping over parts of the heavens in the neighbourhood of the sun immediately after sunset and before sunrise. In other words, the best chance of finding a Comet is when, after a long journey from the distant realms of space, it approaches the sun—the normal place of assembly for such Comets as come within the ken of us who live on the earth.

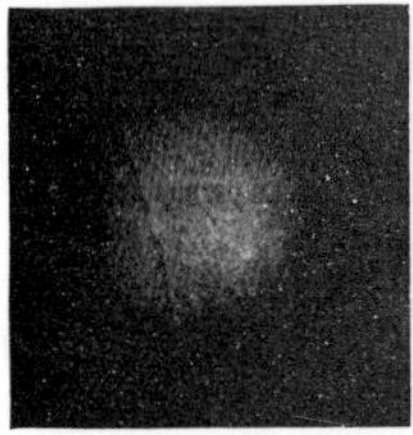
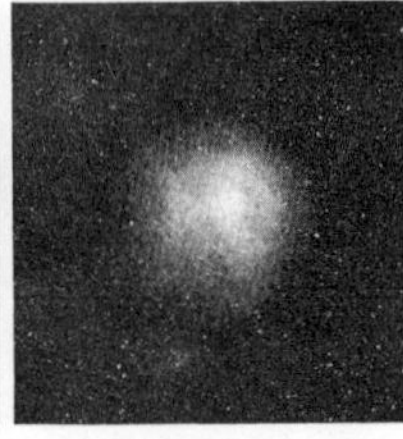

TELESCOPIC COMETS.
Without nucleus. With a nucleus.

When a Comet can only be seen through a telescope it has no tail. Comets visible without the aid of a telescope always have tails.

When first discovered by means of a telescope, a Comet is usually a small speck of light, perhaps no bigger than a star of the twelfth magnitude, or even less. As it comes nearer the sun it will grow both in apparent and also in real size, and will present the appearance of a more or less circular patch of grey light. It may or may not have a central condensation of light, which will give it the appearance of a star surrounded by a sort of halo or cloud of light. When such is the case, the star-like condensation is technically called the "nucleus," and the surrounding portion the "coma." The question of a tail then arises. Does the Comet or does it not exhibit a tail? This is a matter of uncertainty. It may be taken as a general rule that Comets which are only visible in a telescope have no tails, but that the Comets visible to the naked eye always have tails. This latter fact accounts for the popular idea that unless a Comet has a tail "the man in the street" does not regard the object as a Comet at all, and takes no interest in it. To the rule just stated, that naked-eye Comets always have tails, I cannot recall a single exception; but the other part of the rule that telescopic Comets never have tails is not so universally true. But in the case of a telescopic Comet, whilst there may be no tail, properly so called, visible, it will sometimes happen that a coma with its nucleus acquires an elongated shape, which may or may not indicate that a tail will soon appear.

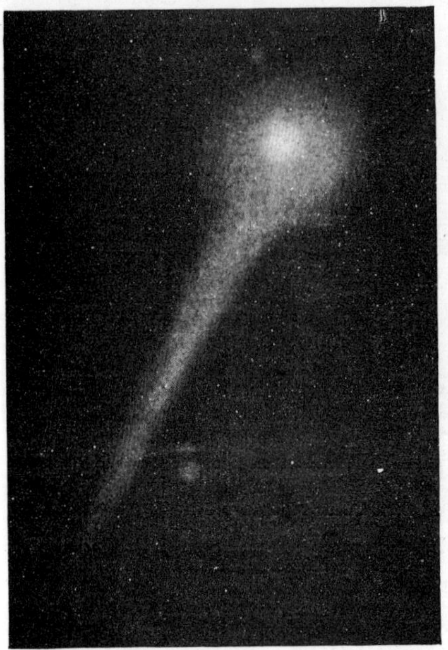

RORDAME-QUÉNISSET COMET 1893.

This was a particularly beautiful Comet. The light of these celestial bodies is due chiefly to electric discharges.

The general nature of Comets and their tails is a matter not easily dealt with. Now and again a Comet appears in which it will be noticed that luminous matter is persistently ejected from the nucleus, indicative of action closely resembling in appearance, and perhaps in actual nature, the ejection of water from the jet of a fountain under pressure from below. On the other hand, with many Comets there is no appearance of the tail being the result of internal forces; but the tail simply grows as from nothing, and this perhaps is the more common case, though probably to say that it grows from nothing is quite unphilosophical.

At this point, perhaps, the reader will ask the question: "What is a Comet?" "What is it made of?" I am bound frankly to confess that we do not know; that these questions do not admit of a straight answer.

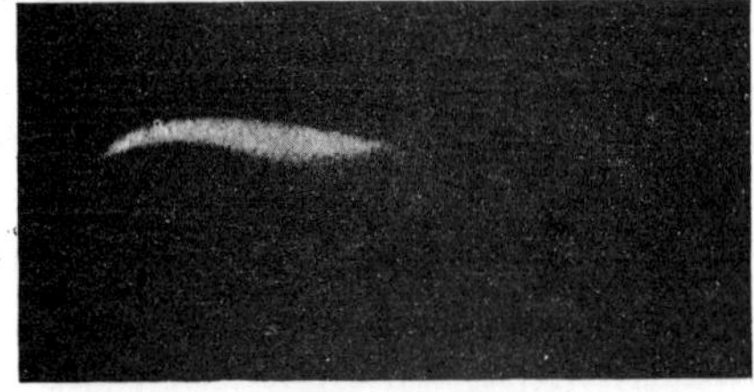

DONATI'S COMET.

This, one of the most brilliant and picturesque celestial sights ever seen, appeared in 1858, and was visible for 112 days.

The internal structure of tails is an interesting matter of inquiry. Whilst we can give a reason for their being curved (because, being more ethereal than the heads, they lag behind the more dense heads), we cannot pronounce definitely as to what is the cross section of an average tail. Judging by the fact that its outer margins are more luminous than the centre, it may be inferred that the tail is cylindrical in section. Whilst it cannot be said that we really understand how the tail of a Comet is formed, or why it should be formed, it may at least be conceded that

[*By Arthur Twidle.*

SOME NOTABLE COMETS.

In this picture we have brought together some of the most notable Comets of recent times. On the left-hand side the upper figure shows Donati's Comet (1858) passing in front of Arcturus. Below it is Coggia's Comet of 1874, and at the foot are two views of Holmes' Comet of 1892. The upper centre is occupied by the many-tailed Comet of 1744, and below it Sawerthal's Comet of 1888. In the upper right-hand corner are two views of Biela's Comet (1846), below it the Great Comet of 1882, and at the foot the Comet of 1862.

some manifestation of an electric nature may be involved. To this it may be added that some recent Comets, especially Morehouse's of 1908, have exhibited transformations in their tails of a kind not generally noticed or recognized in the past. These transformations have included a development of swellings in the tail not unlike knots in a rope; whilst in a somewhat larger number of cases it has been noticed that a Comet has, as it were, shed portions of its tail, which have fallen off and become distinct fragments. Even beyond this there are a few cases on record, such as that of Biela's Comet, where Comets have broken up, have become disintegrated and scattered through space to form the objects which we call luminous meteors.

As regards their motions, some Comets are very regular and others very erratic. The Comets which are regular in their movements are technically termed "periodic Comets," and circulate

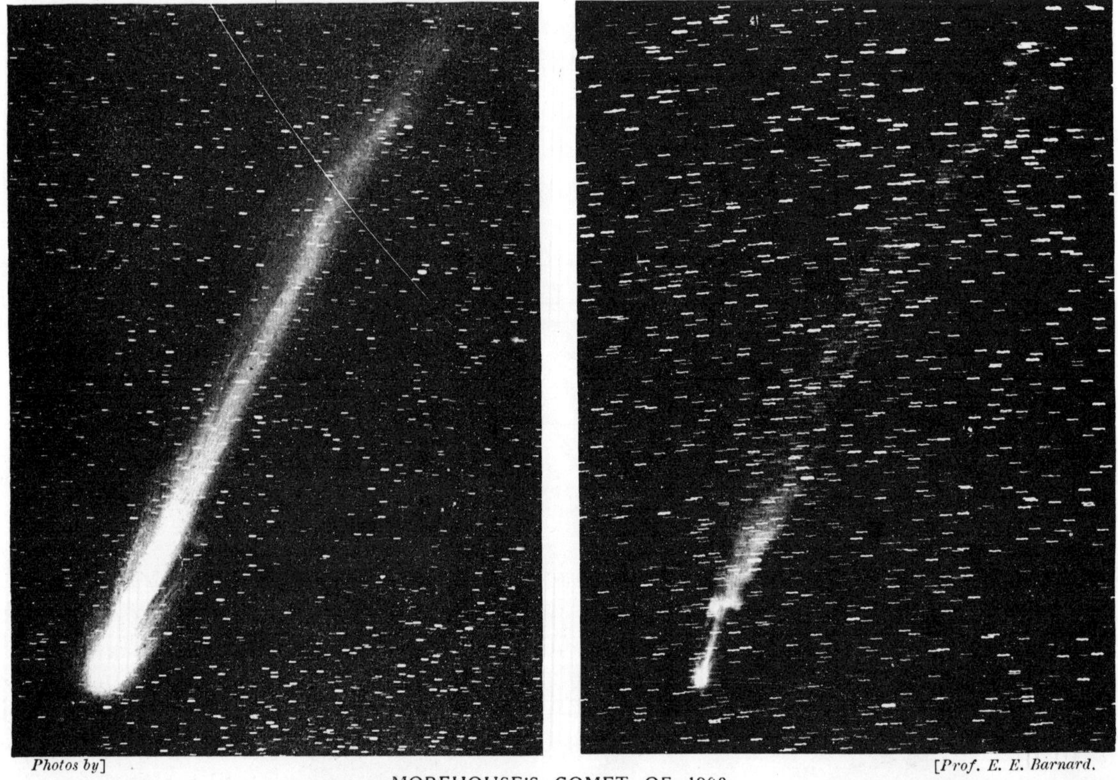

Photos by] [*Prof. E. E. Barnard.*

MOREHOUSE'S COMET OF 1908.

These two photographs exhibit remarkable developments in the tail of this Comet which had not been generally noticed previously. These changes have included the formation of swellings in the tail not unlike knots in a rope, as shown in the second photo.

round the sun in regular or closed orbits. The erratic Comets come up to the sun and become visible to us, having travelled from whence we know not, and pass on after they leave us, whither we know not.

The periodic Comets may be grouped in two classes. The first class includes those Comets which visit us with great regularity at stated intervals of from three to a dozen years, or thereabouts; whilst there is also a small group which seem to belong to our system, and may be expected to reappear about every seventy years. Of these the well-known Comet of Halley is a noteworthy example. The second class, which we verily believe are really periodic, judging by the calculated form of their orbits, have periods supposed to amount to many hundreds of years. Therefore, they have not proved to us yet their periodicity by paying us a second visit, though they may do so perhaps some centuries hence, if the world lasts so long.

Something should be said about the numbers of the Comets. The total number recorded from the earliest times down to the end of 1911 is nearly eleven hundred, and the orbits of about five hundred of these have been calculated. It is obvious, however, that the number which have appeared during the last two thousand years is vastly more than the eleven hundred just mentioned, because before the invention of the telescope none of that class of Comets which we call "telescopic" were observed, although, doubtless, hundreds of such came into our system and went away again. This fact will be readily grasped when I state that 1911 produced eight Comets, of which no more than probably one would have attracted the notice of the few astronomers or historians who chronicled Comets in pre-telescopic times. Even the number of eight belonging to the year 1911, large as it may seem, is not unprecedented; for 1898 produced ten Comets, 1886 produced nine, 1858 produced eight, as did 1881, and, not to multiply instances, there have been in several recent years as many as seven Comets each year.

A subject of some interest is the absolute dimensions of Comets as regards both heads and tails, measured in miles. This is a matter in which we meet with great extremes.

The nucleus of the Comet of 1806 was estimated to be only thirty miles in diameter, whilst the nucleus of the third Comet of 1845 had a diameter of eight thousand miles. So likewise as regards the heads, or comæ. The fifth Comet of 1847 had a small head, only eighteen thousand miles across, whilst the first Comet of 1811 had a diameter of more than a million miles. But it is in the case of the tails of Comets that we come across the most stupendous figures. Whilst these may show such moderate dimensions as only hundreds of thousands of miles, the tail of the Great Comet of 1843 was calculated to be two hundred millions of miles long.

As a last word, let me say that cometary astronomy is a branch of the science well worthy of the attention of amateur students of nature; for Comets, as will be readily understood from what has gone before, are constantly coming and going; the changes of size and form which they often go through are very curious and diversified, and a large number of Comets may be studied with the aid of the telescopes of moderate size, which are now so common, and comparatively cheap, considering the prices of half a century or more ago.

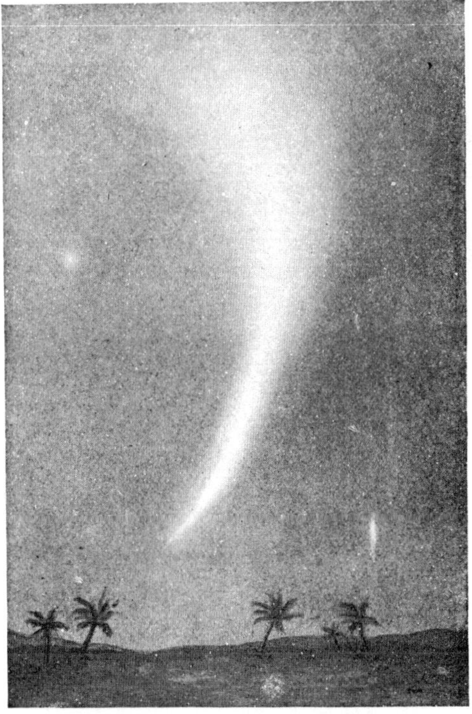

THE DAYLIGHT COMET.
As seen in North Africa in January, 1911.

HALLEY'S COMET.
Edmund Halley suggested the identity of the great Comets of 1531, 1607 and 1682, and predicted a reappearance at intervals of seventy-six years.

ENCKE'S COMET.

A tail, or a "nucleus," is not an essential detail of a Comet, as its nature is determined by the shape of the orbit in which it travels. This Comet has a particularly short period of 1,204 days.

MAMMOTHS AND EXTINCT ELEPHANTS

BY SIR HARRY JOHNSTON, G.C.M.G.

THE Elephant order came into existence somewhere in North-East Africa, probably in Egypt, arising from a generalized type of ungulate related to the surviving Hyrax, of African rocks and forests, and even to those strange aquatic mammals, the Manati and Dugong, which still remain as eaters of seaweed and waterside vegetation in the seas and estuaries of the tropical world. The earliest type of Elephant was a creature not much larger than a pig, which developed a very long head and snout, no doubt for the purpose of groping and grubbing after roots along the banks of rivers and devouring on a large scale riverside vegetation. With these habits it gradually developed its front incisor teeth into formidable tusks, and at last became of the type which we know in a general way as "Mastodon," from the hilly, or nippled, appearance of the surface of its large molar teeth. The Mastodons grew to an enormous size, possibly exceeding in height anything since attained by the True Elephants, and they developed very remarkable tusks. The earliest forms had tusks in the lower, as well as the upper, jaw. One branch devoted itself entirely to the development of its lower tusks and became the Dinotherium, a creature which may have survived in Tropical Africa down to the human period, its fossil remains having recently been discovered in East African strata of relatively recent origin. Some people think a Dinotherium still lives in the inaccessible marshes of Lake Bangweulu and the region of the Upper Zambezi, perhaps also in Congoland, in all of which countries there are traditions of a "Water Elephant."

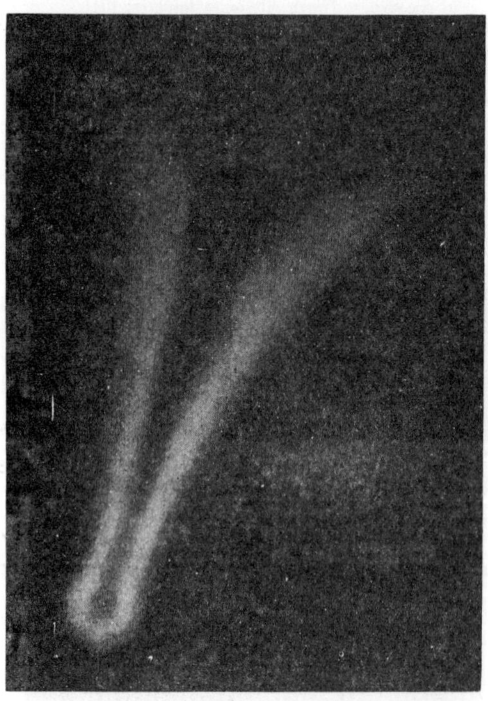

THE COMET OF 1811.

This Comet was notable for its extreme brilliancy. It is a curious fact that stars which the slightest fog would conceal shine clearly through millions of miles of the unsubstantial stuff which makes up a Comet.

The more elephant-like type—Mastodon—spread from the Old World to the New, and grew to a great size in North and South America. This creature may also have lingered in America down to the coming of Man. In Africa, Mastodons (we now know), extended their range as far south as Cape Colony. In Eastern Asia, from out of the Mastodon type grew the True Elephant, which soon attained, like the bigger Mastodons, an almost world-wide range, only the continent and islands of Australasia and Madagascar being without Elephants. The chief region of development for the Elephant genus was, of course, the Old World, where it exhibits two leading types, which might

[*By Sir H. H. Johnston, G.C.M.G.*

MAMMOTHS.

The Mammoth resembled structurally the Indian Elephant of to-day, but had much longer and more curved **tusks, and its** body was covered thickly with hair and wool. Mammoths have been found preserved in the ice of Siberia, **and their flesh has** been thawed and eaten within the last few years.

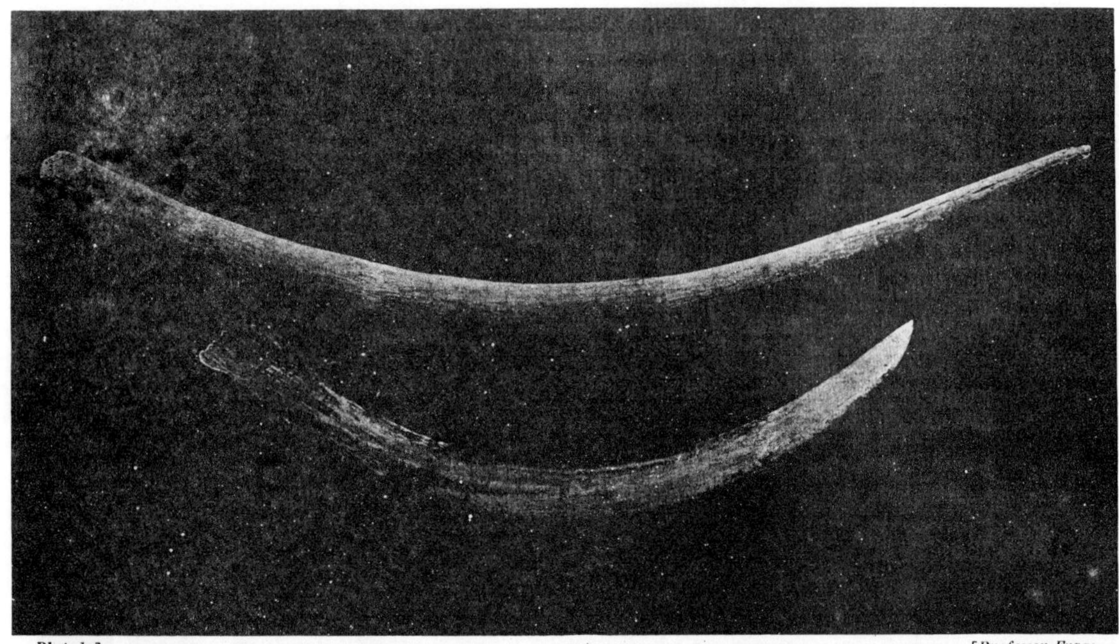

Photo by] [*Professor Fraas.*

ELEPHANTS' TUSKS, PAST AND PRESENT.

The upper tusk in this picture is that of an extinct "African" type of Elephant of about half a million years ago. The lower figure is the largest known tusk of the African Elephant of to-day. It will be seen that the ancient species, which inhabited Europe and North Africa, had tusks half as long again as those of the modern Elephant.

be roughly called the "African" and the "Asiatic," the African being the less specialized of the two. African Elephants once inhabited (even in the human period) the greater part of Europe and Western Asia, as well as the African Continent. One of their extinct species (*antiquus*) developed tusks exceeding anything we know of in modern developments, as is displayed in one of the illustrations to this article. An allied type—*ganesa*—in India (as may be seen by the example in the British Museum) probably reached a climax in tusk development which has never been surpassed. But the Asiatic group of True Elephants—typified by the Indian Elephant of to-day—was not far behind in the matter of tusks. It gave us, amongst other extinct forms, the Mammoth, which is little more than an "Indian" Elephant, with somewhat more specialized molar teeth, and with a thick covering of hair over its skin. The Mammoth did not perhaps attain to a greater height at the shoulder than the largest specimens of Indian Elephant known to-day (and the biggest of these—twelve to fourteen feet high at the shoulder—have been recorded from Siam and Sumatra), but it certainly had tusks of a size and length (ten feet) never attained by any existing species of Asiatic Elephant, and leaving even the biggest known Africans behind. These tusks were, as a rule, much more curved than those of the African group of Elephants, and even tended to be spiral.

The Mammoth, which probably first assumed its specific type in Asia, spread across Bering's Straits (then, no doubt, dry land) to North America, and traversed the New World down to Northern Mexico, developing several sub-specific types of great size, one or two of which, no doubt, co-existed with Man. In fact, in Alaska and North-West North America it is probable that the Mammoth only became extinct a short time before the arrival of the European explorers. Northwards and westwards from Central Asia the Mammoth spread to the shores of the Arctic and Atlantic Oceans. It permeated all Europe as far south as the Danube, the Rhine, Northern Italy and Northern Spain. Its range extended westwards as far as Ireland, where, likewise, it may have lingered down to the coming of Neolithic Man with improved weapons. Speaking generally, however, the Mammoth became extinct in Europe not long after the last of the Glacial periods had melted into a more genial climate. Its extinction, however, was not brought about by a rise in temperature, but by the growing boldness and skill of Man as a hunter.

We know much more about the Mammoth than other extinct Elephants, because specimens of it have been preserved whole in the ice of Siberia, so that men of science have actually had the creature's stomach to examine with undigested fragments of its food, and anyone who has been able to visit the museums in Russia has been able to see the actual skin of a Mammoth coated with shaggy hair, while there have even been dinners given in Russia at which frozen Mammoth flesh has been thawed and cooked for the delectation of a few scientific enthusiasts.

The very fine illustration of a Mammoth's skull here given is from a photograph kindly lent by Professor Fraas. This depicts an unusually fine specimen of this extinct Elephant recently exhumed in the wonderful little kingdom of Württemberg, which has provided museums in modern Germany with so many marvels of vanished beast, bird and reptile.

THE SPINY SPIDER

THE family of spiders is a very large and comprehensive one, but a still greater number are said to exist in the Southern hemisphere. Some are extremely forbidding in appearance, the terrible Tarantula being only surpassed by the repulsive Scorpion Spider. Others, however, approach things beautiful, and one of those especially attractive is certainly the Spiny Spider, a native of Trinidad.

About the size of a sixpenny-bit, it rejoices in having a series of six spines—each covered with minute hairs—arranged around its body, and these, together with the legs, as well as the head and its appendages, are coloured a kind of bright brick-red. The body, however, is almost pure white, which offers a great contrast to the spots dotted all over its surface, which have much the same colour as the spines, outer jaws, feelers and head.

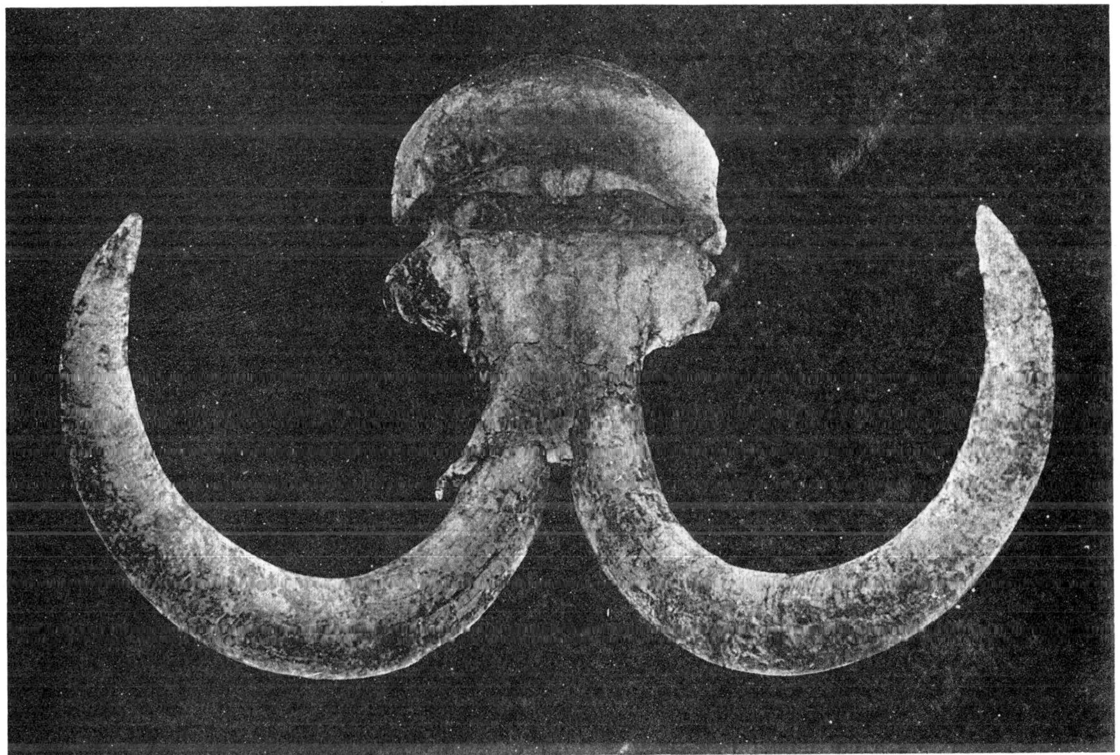

Photo by] [*Professor Fraas.*

A MAMMOTH'S TUSKS.

This photograph shows the skull and tusks of an enormous Mammoth discovered in Württemberg in 1911 by Prof. Fraas. Note the considerable curvature of the tusks, which are also unusually thick for a Mammoth.

A PLANT THAT SETS WATER TRAPS

BY JOHN J. WARD, F.E.S.

I HAVE already introduced to the readers of this work some strange plants which catch and eat insects, and in that article I pointed out that this habit was the natural outcome of those plants living in boggy and wet situations, because in such places the soil lacks a sufficiency of the nitrogenous matter so necessary in the plant economy. Here I want to introduce another curious plant with similar habits, but in this case one that has disposed of its roots entirely and become completely aquatic. The plant in question is the English Bladderwort (of which there are three recognized species) of the deep pools and sluggish waters. In the first place, it is obvious that the Bladderwort by becoming aquatic has made an advantageous change. Its roots were more or less useless for feeding purposes, owing to the fact that there was no material on which they could feed. Therefore, their natural function was lost, and they, in consequence, have disappeared. At whatever remote period in its evolution the Bladderwort lost its roots, we may assume that it then grew in watery surroundings. That being allowed, we can imagine that the next thing that would happen would be that the plant would find itself afloat. From thence its possibilities of development would improve. In floating about it would doubtless come in contact with decaying animal and other matter, which would provide it with those nitrates which had hitherto been its main want. And, as in the case of the insectivorous plants previously considered, its primitive glandular structures would scarcely lose the opportunity of making the best use of this all-important material. That, possibly, was the beginning of things, and then, by a process of natural selection combined with what the philosopher Bergson defines as "the internal push that has carried life, by more and more complex forms, to higher and higher destinies," the Bladderwort would evolve the complex mechanism that we find it possessing to-day. To-day, indeed, it is an exceedingly successful water-trapper. It floats near the surface, with its traps all set for minute insects, water-fleas, the small fry of fishes, etc. The traps are popularly

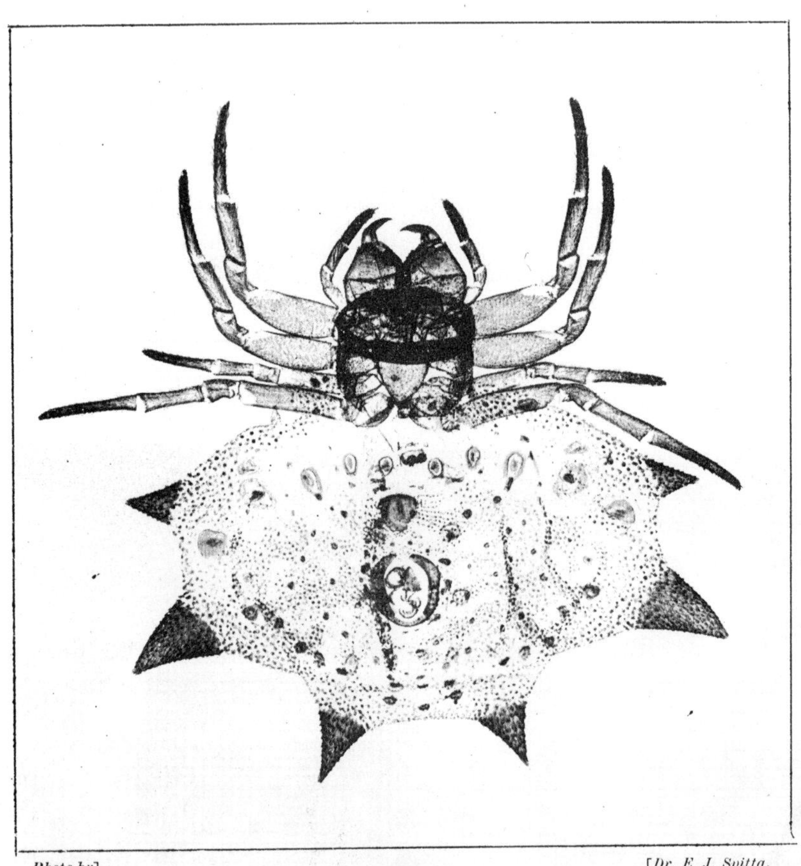

Photo by] [*Dr. E. J. Spitta.*

THE SPINY SPIDER.

A view from the underside of a remarkable Spider which is a native of Trinidad. It is three-quarters of an inch across. It has a white body, spotted with red, and the six spines, as well as the legs and jaws, are also red.

[*By J. Teklenburg.*

THE BLADDERWORT.

The Bladderworts are small aquatic plants that float in ponds and ditches. The large leaves consist of many branching fine threads which bear little pear-shaped vessels with spring doors which allow minute water animals to pass in but prevent their exit. The dead bodies of these little victims are used for the support of the plant.

termed "bladders," and are borne amongst the leaves, of which they are modified portions.

In the accompanying photo-micrograph they will be seen to be somewhat pear-shaped structures supported on short footstalks. The entrance is at the narrow end, and is closed with an elastic valve which opens inwards. When the prey is scarce the valves act very promptly, giving way the moment they are touched, and precipitating the inquiring water-flea, or other animal, into the interior, and then immediately closing again. It is quite impossible for the captive to escape, as the valve cannot be opened from the inside. Around the entrance of the bladder are several branched bristles, which probably serve to conduct the prey to the vicinity of the valve.

The bladder often contains several water-fleas or similar organisms, which will sometimes remain alive for two or three days, but eventually they perish and begin to decay. Inside the bladders, too, are numerous living animalculæ which seem to thrive on the decaying prey. Whether these are to be looked upon as parasites on the plant's preserves, or whether they in turn serve as further nourishment for the plant, is a difficult point to decide. On the inside walls of the bladders are innumerable minute glands (which are visible as little points in the photo-micrograph), and it is these organs which supply to the plant its nitrogenous food. They absorb the material after it has reached the condition of a liquid manure, and do not, so far as Charles Darwin's experiments showed, digest their prey in the manner of the Sundew and Butterwort previously considered.

Photo by] [*J. J. Ward, F.E.S.*

THE BLADDERWORT.

A portion of the plant, natural size, showing the position of the bladders in which minute aquatic animals are entrapped and done to death.

Photo by] [*J. J. Ward, F.E.S.*

THE BLADDERWORT'S TRAPS.

A few of the bladders that constitute the traps by means of which the plant catches its food, on a scale of magnification which shows them twenty times larger than life size.

What is perhaps the most curious point regarding the Bladderwort is what induces the tiny prey to go so readily to its slaughter. It may be the naturally inquisitive characteristics of these organisms for penetrating crevices that lead them into trouble; but the valve of the bladder is transparent, and possibly that may be the attraction. It may create a bright spot which acts as a lure, and so draws the victims to the entrance. Then, should they stumble against the valve, they are immediately entrapped.

THE AUSTRALIAN SPINY LIZARD

BY R. LYDEKKER

AMONG the many strange and weird creatures whose home is Australia, none is more grotesque or more suggestive of a nightmare than the eight-inch-long reptile to which the colonists give the name of Spiny Lizard, or Thorny Devil. From the latter title, and still more from its scientific designation of Moloch (borrowed from the evil Babylonian deity), the Spiny Lizard might naturally be regarded as a reptile as much to be dreaded and avoided as a cobra or a rattlesnake. But, as a matter of fact, it is perfectly harmless, inhabiting dry and sandy districts in Southern and Western Australia, and subsisting apparently to a great extent on ants. What, then, it may well be asked, is the object of this wonderful panoply of horny spines with which the entire upper surface of the head, body, limbs and tail is so thickly covered? But it must be remembered that spines may serve just as well for defensive as for offensive purposes; and it seems quite reasonable to suppose that those of the Spiny Lizard are designed to prevent any hungry bird from making a meal of what may be a very toothsome creature. And in confirmation of this, it may be mentioned that, according to the meagre accounts of their habits we possess, it is by no means uncommon to see two or three of these lizards lazily sunning themselves on the top of a sand-dune, where, in spite of a certain limited power of changing their colour to harmonize with their surroundings, they can

Photo by] [*W. Saville Kent.*

THE AUSTRALIAN SPINY LIZARD.

The Spiny Lizard is a perfectly harmless reptile, inhabiting dry, sandy districts in Southern and Western Australia, where it appears to live chiefly upon ants. The photograph shows it about one-third less than the natural size. Its spines are defensive, not offensive.

scarcely fail to be more or less conspicuous. Generally, however, these lizards partially bury themselves in the sand when in repose.

The Spiny Lizard, which is the sole representative of its kind, is a relative of the Stellion Lizard of Southern Europe, which, although not unlike the smaller British Lizards in general appearance, represents a separate family. From all the others of its own family the Australian species is broadly distinguished, not only by its protective armour of spines, but likewise by the very small size of its mouth, and by the wall of teeth which crown the summit of its upper jaw being inclined inwards instead of standing vertically.

Possibly some of my readers may be led to imagine that the Spiny Lizard is a diminutive survivor of some of the giant reptiles of former epochs of the earth's history, which were likewise armed with formidable spines, or spines and vertical plates, on their heads, bodies and tails. But there is one important difference (to say nothing of others) between the two : namely, that whereas in the giant extinct species—Dinosaurs, as they are called—the spines consist of bone, covered presumably in life with hollow horny sheaths, like those of cattle, in the Spiny Lizard they consist of horn alone.

Photo by] [*W. Saville Kent.*

SPINY LIZARDS.

It is no uncommon sight to see two or three of these Lizards sunning themselves on the top of a sand-dune. They have a limited power of changing colour to harmonize with their surroundings. They partially bury themselves when in repose.

There are, however, in the warmer parts of North America certain small lizards, locally known as Horned Toads, which present a remarkable resemblance in general appearance to the Spiny Lizard, having their heads, bodies and tails similarly covered with horny spines—albeit somewhat differently arranged. So close, indeed, is the resemblance between the two that it might be imagined that they were near relations ; the shortness of the head from which the American species derive their name of toads being a feature of little importance. But the resemblance is only skin-deep, for the Horny Toads have teeth attached to the sides of the jaws, in the manner characteristic of the almost exclusively American family of Iguanas [I don't want to be told that iguanas are found in India and Africa, as, in spite of popular opinion, they are not], and it is to that group that they belong.

The resemblance between the Spiny Lizard and the Horny Toads is thus in a certain sense accidental (although nothing in Nature is really so), and may be due to the same special need for protection from birds and other enemies.

Another feature common to the Spiny Lizard and the Horny Toads is the extreme flatness or depression of their bodies, this being a characteristic of lizards which dwell on the ground, as distinct from those dwelling in trees, which have very thin or laterally compressed bodies, as is

[*By A. Twidle.*

DEEP-SEA FISH.

In this group of remarkable forms of Deep-sea fish the upper example is the Angier-devil; a front view of the same species is shown in the lower right-hand corner. The middle figure is the Sturgeon-like Gurnard, and the two spherical fishes below are the Tiny Globe-fish, only an inch and a quarter in length.

the case with the great majority of the members of the Iguana group. The depression is, however, carried to a very unusual degree in the Spiny Lizard and the Horned Toads, and is in accordance with their habit of squatting close down on the ground.

If a Spiny Lizard be placed in a saucer of water, the fluid will be rapidly absorbed by the skin. This is very remarkable in a creature inhabiting arid districts, in which it would be natural to suppose that there was some special means to prevent the fluids of the body from drying up.

[By A. Twidle.

THE ILLUMINATED ANGLER.

Fins and tail are enormously developed in this fish, but the mouth, though still immense in its gape, by comparison with the body, is normal in shape and position. The creature seems to be well adapted for pursuit, and to have abandoned the lurking habit altogether.

DEEP-SEA FISH

BY FRANK T. BULLEN, F.R.G.S.

CONSIDERABLE trepidation must be felt by anyone who approaches the task of a popular description of the exceedingly strange fish depicted in the accompanying drawings. For several reasons, chief among which are (1) they have no popular names, the titles given them by their discoverers usually running into nine syllables of mixed Greek and Latin, and (2) that those specimens which have been dragged to the surface from the darkling depths where they lurk can hardly be considered representative. In fact, except to a favoured few they are and must remain unknown from the very nature of the case.

Few among us will consider this a hardship when we contemplate their weird and gloomy shapes, and most of us will hope that the few diminutive specimens that have been secured in the most accidental manner are full sized—we hardly like to think of such awful creatures as growing big. But

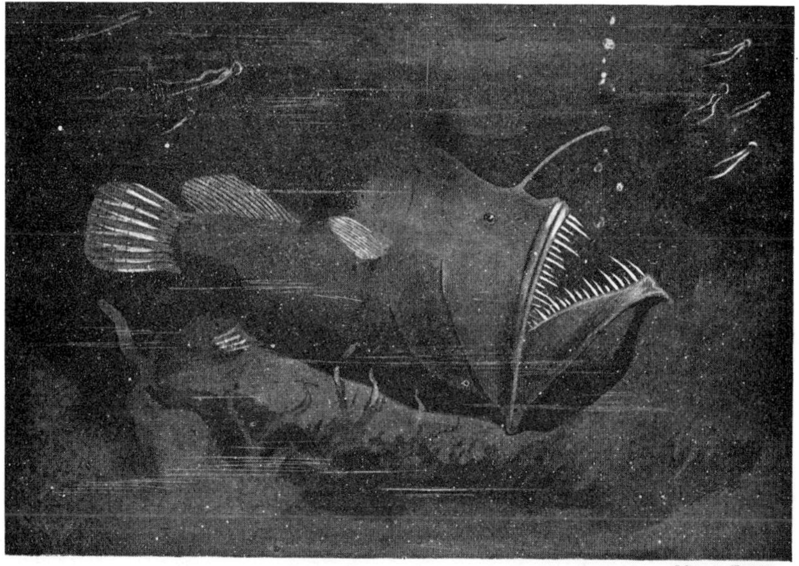

[By A. Twidle.

THE INVERTED ANGLER.

Judging by appearances one would suppose that this Angler carried his stomach on his back. He is a degenerate Angler, for his fishing-rod is contemptible, while every other organ except his mouth is only just in evidence.

we do not know, and by a very little exercise of the imagination we can people the abyss with stealthily gliding forms that materialize the very worst figments of a severe nightmare.

Amidst a world where all is paradoxical and unexplainable, we find what we might expect, that colour is absent; for, with hardly an exception, these extraordinary shapes are either black or dirty-grey, in keeping with the gloom of their native surroundings. But there any universal characteristic ends. Species apparently allied most closely in all other respects will differ as widely as possible in some most important detail. As, for instance, one will have eyes like pin-points; another eyes like saucers. One will have a mouth like a coal-sack, with a body behind it like a whip-lash; another has a mouth like a pin-hole and a body like a porpoise. Or one light, slender, almost elegant in shape, has an immense propelling organ or tail; another with a body like a porpoise, has practically no tail-fin at all: its body tapers off to a point. But we must pause in this brief introduction to consider a few of the creatures here depicted, popularizing, as far as possible, the terrible appellations given them.

The Snipe-beaked Ribbon-fish shown on p. 383 has been found, as usual with the deep-sea creatures, in widely separated districts, and has been brought up from depths varying from two hundred to two thousand fathoms; but the possibility of specimens of little-known fish having got into the trawl on its way down must always be borne in mind. Especially so in

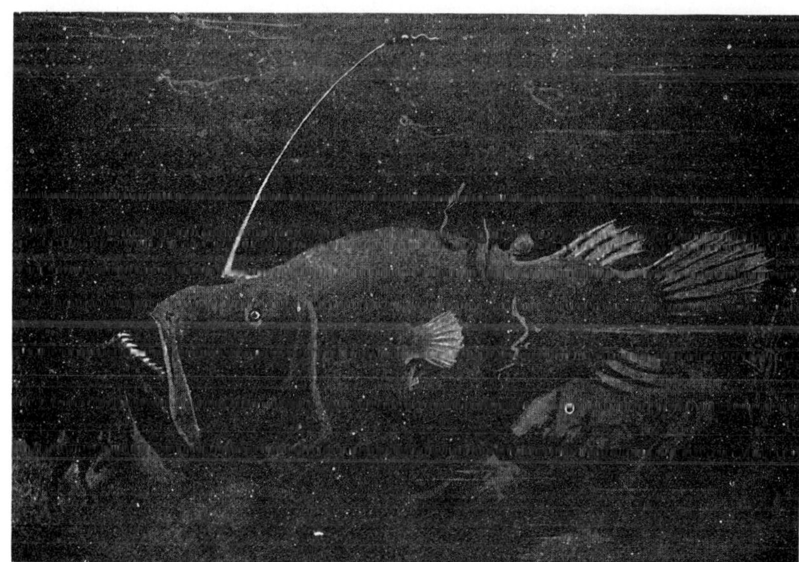

[By A. Twidle.

THE LURKING ANGLER.

Like most of the Anglers, this fish is ill-adapted for swimming. The best description that could be given of it is that of an animated stomach. It is only about two inches long, and it has been found in water seventeen hundred fathoms deep.

this case, as a specimen was brought up by a trawl from the *Fish Hawk* in two hundred and fifty-two fathoms off Newport, which was clinging by its long jaws to the outside of the net. The largest specimen yet found was brought up from a depth of one thousand eight hundred and eighty-eight fathoms, and measured six hundred and fifty millimetres, or, roughly, twenty-five inches in length. But again, in measurements of queer fish dredged from great depths, it must always be remembered that, owing to the necessarily small size of the trawl, the possibility of securing any but the smallest and feeblest specimens is very small.

" A sack-throated fish with a whip-like tail " is a very free translation of the scientific name of our next subject, which is one of the most paradoxical creatures of all that perverse population.

[*By A. Twidle.*

THE SACK-THROATED WHIP-TAIL.

A side view of this extraordinary fish is shown in the opposite plate. Here the view s from above. It is one of the most paradoxical creatures of all the deep-sea population. Its principal characteristic is an astounding gluttony. Its entire length is over six feet.

Its principal characteristic may be inferred from its appearance—an astounding gluttony. Only five specimens have been taken, three of which were choked by attempting to swallow fish many times larger than themselves. These three were found on the surface ; another specimen was taken in nine hundred fathoms. The largest individual had a body twenty and a tail fifty-eight inches long.

The Angel Fish must be passed over promptly, as it evidently, in spite of its extraordinary appearance, belongs to a surface-frequenting species, and, moreover, only one specimen seems to have been taken, four inches long.

Why our next specimen should have been provided with such an amazing suit of protective armour is another of the mysteries of ocean. It does not appear to have any value, yet judging from the character of its spines, conical flinty protuberances covering the whole of its body, each of which

[By A. Twidle.

DEEP-SEA FISH.

In this group the upper fish is an Angel-fish whose great development of fins indicates that he has strayed from the upper waters. The extraordinary form below it is the Sack-throated Whip-tail, and beside it is the Snipe-beaked Ribbon-fish, whose form well bears out the name. At the bottom of the plate is the Armoured Globe-fish.

in turn is studded with needle-like flexible prickles, it must be a morsel to destroy the stomach of any fish, however omnivorous, that ventured to swallow it. So far as is known, it does not inhabit abysmal depths, only one hundred and twenty-nine fathoms being the deepest, and its size is not even hinted at in the scientific descriptions. But probably it is very small.

Our old familiar friend, the Gurnard, is a quaint-looking fish in its normal appearance, but when it goes in for decorations and protective covering it becomes fantastic, if one may safely apply the term. Our picture represents a curious kind taken from depths not exceeding two hundred fathoms, and in spite of its habitat still retaining its vivid colouring, the best specimen obtained being a bright crimson. As usual, the size of this specimen is small, only three hundred millimetres in length—or, roughly, about a foot.

Angler-fish abound in the great depths of an amazing variety, some of which are shown in our illustrations. From most of the external characteristics of these bizarre fish one would be led to the conclusion that they were essentially liers in wait, dependent not upon speed but on finesse for the supply of food which their general appearance indicates that they require in fairly large quantities. But here again that condition of paradox confronts us. One specimen shown is certainly ill-adapted for swimming, its fins and tail being of the feeblest, most rudimentary kind; the latter, indeed, appearing as if intended more for a feeble defence than for propulsion, owing to its spikiness and exiguity of membrane. And the pectoral fins, whereon it must depend for its balance, are so tiny as to appear ludicrously inefficient for any practical purpose.

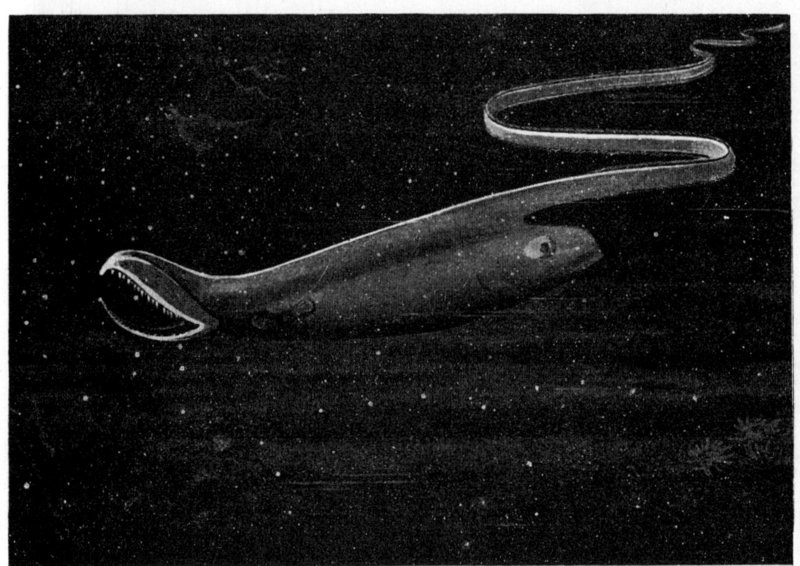

[*By A. Twidle.*

SACK-THROATED WHIP-TAIL.

In this view the fish has recently made a hasty meal. The fish eaten was larger than itself, but the stomach has stretched to accommodate it. The eye of the victim will be seen showing through the skin of its captor.

Moreover, its eyes are mere pin-points, rudimentary, in fact, like most of its other organs, the exceptions being the very efficient lure, or fishing-rod, and the enormous mouth. Perhaps the best concise description that could be given of this fish is that of an "animated stomach," and to its prey it should be a satisfaction that its size is so insignificant, only about two inches long. It has been found in one thousand seven hundred fathoms.

When we turn to another specimen, however, we find fins and tail enormously developed; indeed, the back and belly fins are each as long as the entire creature, while the pectorals, or chest fins, now extraordinarily near the tail, have increased in size until they are nearly as wide as the body. The fishing-rod has dwindled to an insignificant rudiment; the mouth, though still immense in its gape by comparison with the body, is normal in shape and position; the body is more elegant and fish-like, and altogether the creature seems to be well adapted for swift pursuit and to have abandoned the lurking habit altogether. Another peculiarity possessed by this species is that its back is covered with luminous filaments, which, when all fully excited, must give the creature a dazzling appearance.

THE EARTHWORM

BY K. G. BLAIR, B.SC., F.E.S.

EARTHWORMS are perhaps the commonest and most generally known of all our native fauna, yet it is truly astonishing how few people are acquainted in any detail with their habits or realize how great are the results brought about by their work.

They prefer to live in moist earth, into which they penetrate partly by thrusting in their pointed heads and pushing away the particles on either side, but mainly by swallowing the earth as they go along, passing it through their bodies, whence it emerges at the tail end in the state of a semi-fluid, fine comminuted mud. A considerable quantity of this material is made use of by the Worm to form a smooth, firm lining to its galleries, an operation in which the flattened tail plays the part of a trowel; the rest is brought to the surface and there ejected in the form of the well-known worm-casts.

The body of the Earthworm is of fairly simple structure, consisting of one long tube within another. The inner tube is, of course, the alimentary canal, the greater part of which is formed by the intestine, whose function it is to absorb nutriment from the materials passing through it. In many animals, ourselves for instance, the intestine is long and coiled in order to increase this absorbent area; in the Earthworm it remains straight, but the same object is attained by a great inward fold along the back. This arrangement will perhaps be better understood after a glance at our fourth illustration, which represents a section through the front part of a Worm. The outer tube is formed by the skin and muscles, while the space between the two is crossed by a large number of transverse partitions, and it is these partitions, seen from without, that give the Worm its characteristic ringed appearance.

Its food consists mainly of decaying leaves, which it pulls partly into its burrow;

Photo by] [*H. Main, F.E.S.*

THE COMMON EARTHWORM.

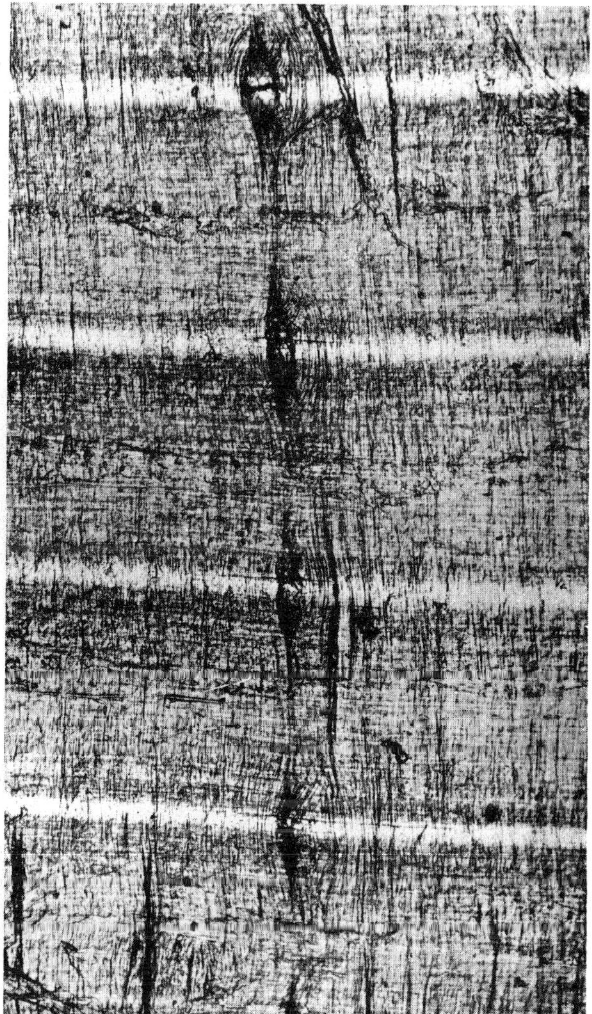

Photo by] [*J. H. Crabtree.*

PART OF AN EARTHWORM'S BACK.

In this photo of skin from an Earthworm's back magnified about thirty times the natural size, the rings and the apertures are shown clearly. From these pores a fluid is poured out that protects the worm from the attacks of bacteria.

but, in addition, the earth swallowed by the Worm in making its burrow is acted upon by the digestive fluids in its passage through the body, and the organic matter contained in it extracted.

It is, of course, only at night that Worms become active ; during the day, especially if the atmosphere be warm and moist, they like to lie close to the mouth of their burrows, concealed by a plug of leaves or heap of castings, and they frequently line this part of the burrow with bits of leaf or small stones, presumably to keep their bodies from direct contact with the earth. They seldom entirely leave their burrows, but reach around as far as they can, keeping the tail in the burrow ready for instant retreat. For this purpose the rows of short, strong, retractile hairs along the body, that are readily felt on pulling the Worm backwards through the fingers, give it an excellent hold upon the sides of the burrow. Though they usually confine their workings to within about a foot or two of the surface, yet in very cold weather, or in times of drought, they will penetrate to a depth of four or five feet, and there construct a little chamber, which they line also with small stones ; here they lie coiled up, often two or more Worms together in a knot, until more favourable conditions are resumed.

Photo by] [*H. Main.*

Egg-capsule of Earthworm. Each capsule contains four eggs, but only one produces a young worm.

Owing to the immense numbers in which Worms exist in the ground, the effect that they exercise is considerable. In ordinary agricultural or pasture land the whole of the upper foot or so is being continually passed through the bodies of Worms and brought to the top, thus, as it were, being repeatedly though slowly " ploughed." The effect is not at once evident ; but when it is considered that as castings are brought to the surface, burrows are formed below which sooner or later collapse, and that this process is being continuously repeated, it will readily be perceived that in time the effect will be considerable. Darwin, indeed, quotes numerous instances of a layer of top-dressing of sand or lime being laid on a field and a number of years later a pit being dug revealing this layer some distance below the surface all round the pit. A large number of observations show that this burial takes place at an average rate of rather over one-fifth of an inch per annum. The same explanation applies to objects left on the ground—stones, coins, etc.—being gradually buried, and shows why a stony field regularly ploughed remains stony, but when left for pasture becomes soft, even turf, and why the floors of old Roman houses and other objects of antiquarian interest have lain protected beneath a foot or so of earth.

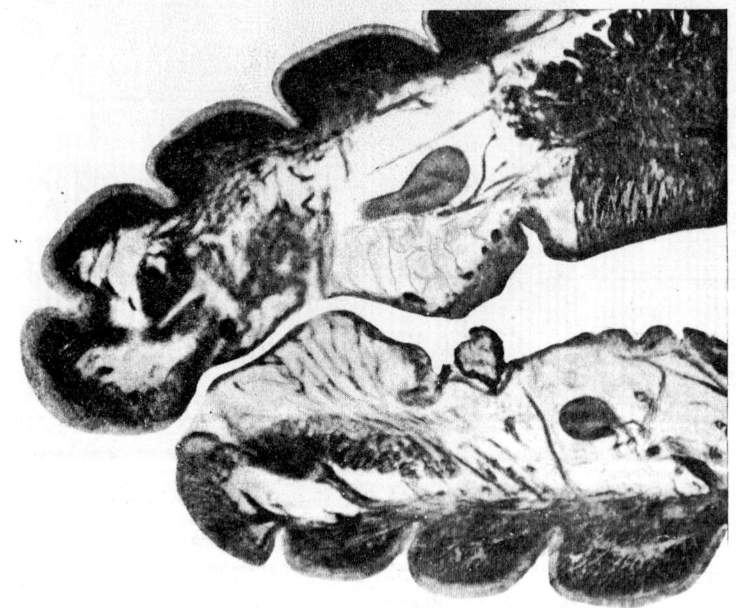

Photo by] [*J.H. Crabtree.*

SECTION THROUGH "HEAD" OF EARTHWORM.

The first four rings of the Earthworm are here magnified about twenty-two times, to show the simple mouth. The two dark patches, one above and one below the gullet, are nerve centres.

THE DOVE FLOWER.

These are the flowers of a tree orchid from tropical America whose central organs bear a remarkable likeness to a fluttering white dove. The Spanish-Americans of Panama know it as *El Spirito Santo*, or flower of the Holy Ghost. The flowers are slightly enlarged in the photograph.

THE DOVE FLOWER

One of the most remarkable phenomena in plant life is the resemblance that certain flowers, or parts of flowers, bear to animal structures. We need not go far to find examples of this, for the Bee Orchis, the Fly Orchis and the Butterfly Orchis supply really good illustrations; yet one cannot say what is the purpose of these resemblances, for so far as at present ascertained they appear to have no distinct relation to the habits of the plant.

Still more puzzling is the remarkable likeness the inner organs of a Central American orchid bear to a dove hovering with half-expanded wings, as shown in our photograph. In this case it is not a resemblance of the whole flower, which might be seen from a distance and deceive some animal. Until one makes a close inspection of the flower and peers right into its heart, nothing striking is visible; further, the mimic is on a scale so small that no one could mistake it for a bird. Yet, there is the likeness.

The plant is one of those species of orchid that grow upon the trunks and branches of trees in the tropical forests, where the air is so laden with moisture that the roots of these orchids, hanging down from the tree but far from the earth, are able to obtain all the water the plant needs. To tide it over the dry seasons the plant stores its food-material in false bulbs as large as swans' eggs. When the growing period returns, leaves a yard long and six inches broad arise from these bulb-like stores, and with them a flower-stem five or six feet in height, of which the upper third bears the creamy-white flowers, which are about an inch and a half across. They give off a strong, sweet scent. The Spanish-speaking people about Panama call it *El Spirito Santo*, or Holy Ghost flower, from this resemblance to the conventional art representations of the Holy Spirit.

Photo by] [*J. J. Ward, F.E.S.*

PAPER NAUTILUS.

This exquisitely beautiful structure is a false shell, for it is not inhabited by the animal, but produced solely for the purpose of holding her eggs. In earlier days many disputes arose between naturalists as to the true nature of this shell. The photograph is slightly larger than natural size.

[*By Theo. Carreras.*

THE PAPER NAUTILUS.

The female Nautilus holding the wonderful shell she has constructed for the reception and care of her eggs. She is represented as going backwards, impelled by the discharge of water through the siphon. The small, shell-less creature below is the male. The scale is one-half the natural size.

THE PAPER NAUTILUS

MANY an idle tale has been spun around the Paper Nautilus, and learned folk give the doubtful honour of their origin to the Greek sailors, who, in crossing and recrossing the Ægean Sea, became familiar with these strange little fairy vessels and their faithful pilots. "Argonauts" is the name that has been given to the species, and therein is to be found the germ of the first legend; for the word means "the sailing-ship," and it would not be long before the charming but fallacious romance obtained credence that the Paper Nautilus sails upon the ocean with numerous tentacles stretched out from the delicate shell, two of which are raised aloft and spread out into thin, filmy sails to catch the wind that shall propel it through the waves, while the remaining arms are used by this quaint navigator after the fashion of oars. Vastly wonderful, but vastly untrue!

Again, it was stated with all possible circumstances of verity, and supported by the most elaborate and ingenious argument (aye, and believed even as lately as the beginning of the last century), that the Nautilus was a parasite who, by fair means or foul, had obtained possession of his host's shell and sailed therein—a theory equally wonderful and equally untrue.

Yet ours would be a sorry task if we had to repudiate such extravagant theories and make of our romantic ship nothing but a small and insignificant ocean-dweller. Such, however, is not the case. The sober facts of the life history of the Nautilus corroborate the statement that Truth is stranger than Fiction. The Nautilus is not a pilot, she is a nurse; her shell is not a ship, it is a cradle—or, as Professor Huxley more aptly described it, a perambulator. It is only the female of the species that bears the shell; the male is a small creature not more than an inch in length:

A MICROSCOPE SLIDE.

In the black ring are some Diatoms of the natural size, like specks of fine dust.

and the race of which they are members is that of the cuttle-fishes. Imagine a cuttle-fish clasping with two long arms a large oval and not very wide snail-shell which looks as though it were made of crinkled tissue-paper, with the greater part of its body and the remaining arms hanging outside the shell, and you will have a very fair idea of the appearance of the Paper Nautilus.

The first actual observations recorded of their habits are those of Mme. Jeannette Power, who studied these creatures in her vivarium at Messina. She states that the young Nautili are naked when they issue from the parent shell, but that in the females the shell begins to form within ten or twelve days. Its formation is due to a secretion exuded from the two large tentacles which sweep backwards from the creature's beak over the bag-shaped body. Should anything happen to this delicate construction the owner is able to repair it in the same way that any other mollusc can repair his home, except that the tools here used are different; for most molluscs obtain their secretion from the mantle which lines the interior of the shell, while the supply of the Nautilus is derived entirely from her two tentacles. These tentacles fulfil another important purpose, for they act as spring clasps to this frail cradle (now become the treasure-house of the eggs of the Nautilus), and press it against the bag-shaped body, which entirely fills up the opening of the shell, so that the eggs may be securely packed in their appointed vessel. To better serve the purpose of security, each tentacle is supplied with a fine membrane, which spreads out like a web from the central ridge and can be made to fold over the greater part of the cradle, adhering closely on account of its elasticity to the ridges and depressions in its surface.

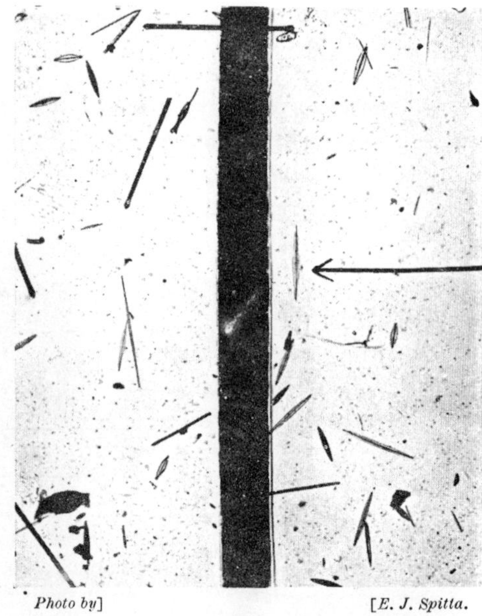

Photo by] [*E. J. Spitta.*

A DIATOM AND A HUMAN HAIR.

The centre of the slide is here shown enlarged eighty times, with a human hair stretched across. The black band is the hair, and beside it is a Diatom, to which an arrow points. The same Diatom is shown at the foot of the page on a larger scale.

There are several species of the Paper Nautilus, all of which are found in the warm seas. The most common is that found in the Mediterranean, such as is depicted on page 389, while the first illustration shows a very beautiful variety, noticeable for its network ridges and the bolder toothing along the crest of the shell.

Photo by] [*Max Poser, F.R.M.S.*

A BEAUTIFUL DIATOM.

The central area of this graceful form is marked with fine cross-lines. Only fine lenses of high power can separate these lines. The original photograph from which this block was made was on a scale of 3,600 times the natural size; here it is reduced to half that scale, or 1,800 times

DIATOMS: WHAT ARE THEY?

BY E. J. SPITTA, F.R.M.S.

DIATOMS, or Brittleworts, are minute algæ. But what does that mean? Plants belonging to the class algæ grow in water, whether salt or fresh; seaweeds are algæ. All of this group, speaking popularly, are separated into three grand divisions: olive-coloured, red-coloured and green-coloured. One of

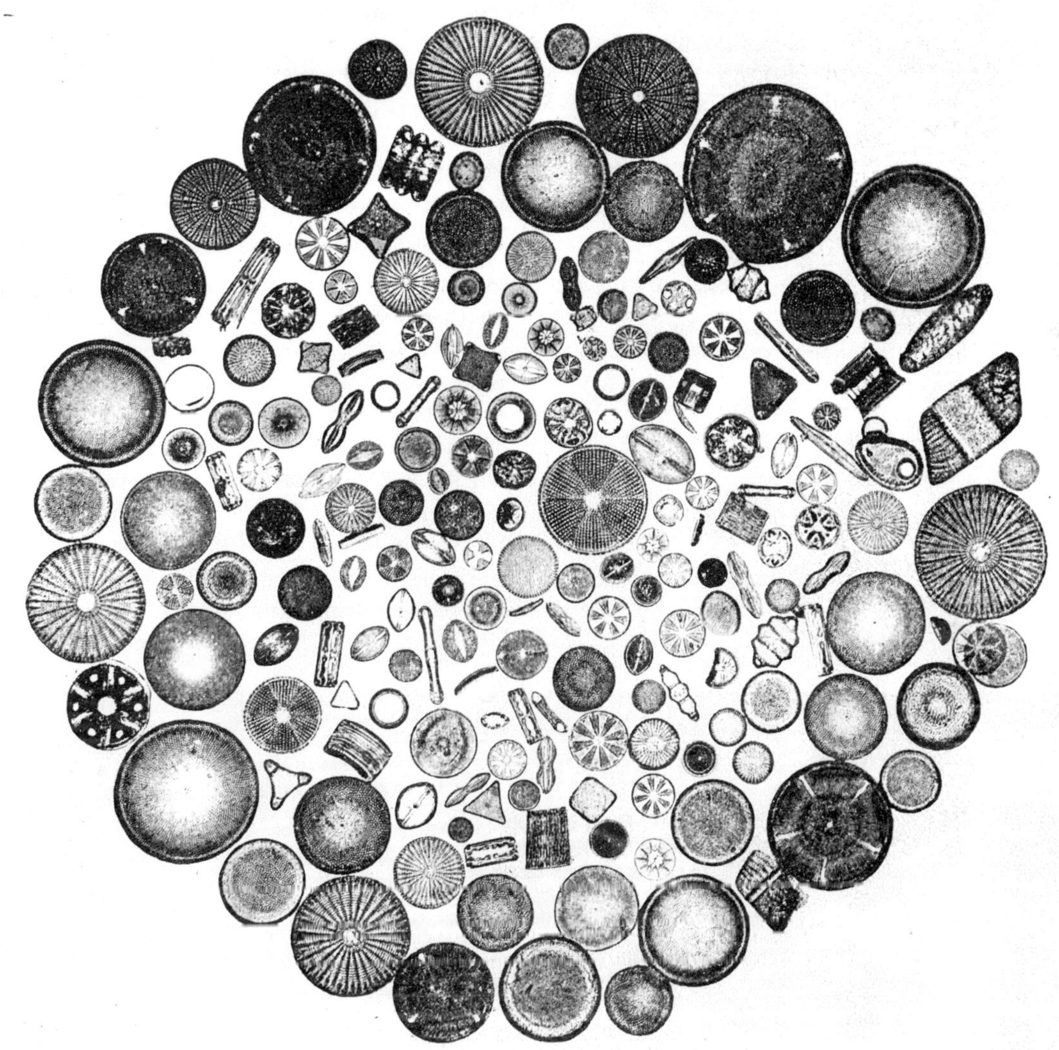

Photo by] [*A. E. Smith.*

A GROUP OF DIATOMS

Each of these microscopic plants, in spite of its minuteness, is delicately sculptured, each species with its own special pattern. They are here shown about sixty times larger than the actual size. Several of them will be found, much more highly magnified, on other pages.

the sub-divisions of the green-coloured is peculiar in that its members all possess a flint-formed base, which supports what is oft-times called the fleshy exterior or envelope. These are Diatoms. So hard is this flinty matter, that if any Diatoms are placed on the tip of a knife and held in the flame of a spirit-lamp to burn off their envelopes, the pure flint framework remains perfectly intact; in

Photo by] [*E. J. Spitta, F.R.M.S.*

PART OF A DIATOM.

This photo shows a small portion of the Diatom shown at the foot of page 390, but under a much greater magnifying power of the microscope. Here the fine lines of the former photo are seen to consist really of minute dots. The scale of magnification is 1,000 times the actual size. It is the summit of the photo-micrographer's ambition to thus resolve the lines into these dots.

fact, this is one way of cleaning them. It is these " remains " that are of such interest to the microscopist, for they are found in every variety of form, whether boat-shaped, circular, spindle-shaped, or like a strip of tape, shaped as a basin or as a three-cornered hat. Sometimes forms are met with twisted as a portion of a spiral. They all vary, too, in size, some only requiring a moderate-power microscope to render them visible, whilst others need the highest of magnifications. But this is not all. If the highest possible magnifying power be used, and every optical construction and mechanical adjustment of the microscope be employed at its best, fine markings are revealed, called " secondaries," or " secondary markings," which were invisible before. These consist of minute lines ruled close together, dots or pearls and cavities. The way in which these minute markings are executed, their exquisite symmetry of design and their regularity of arrangement, surpasses man's comprehension, both as to the method by which they have been sculptured, as well as to comprehend or even speculate upon any possible purpose for which they can have been formed. It must

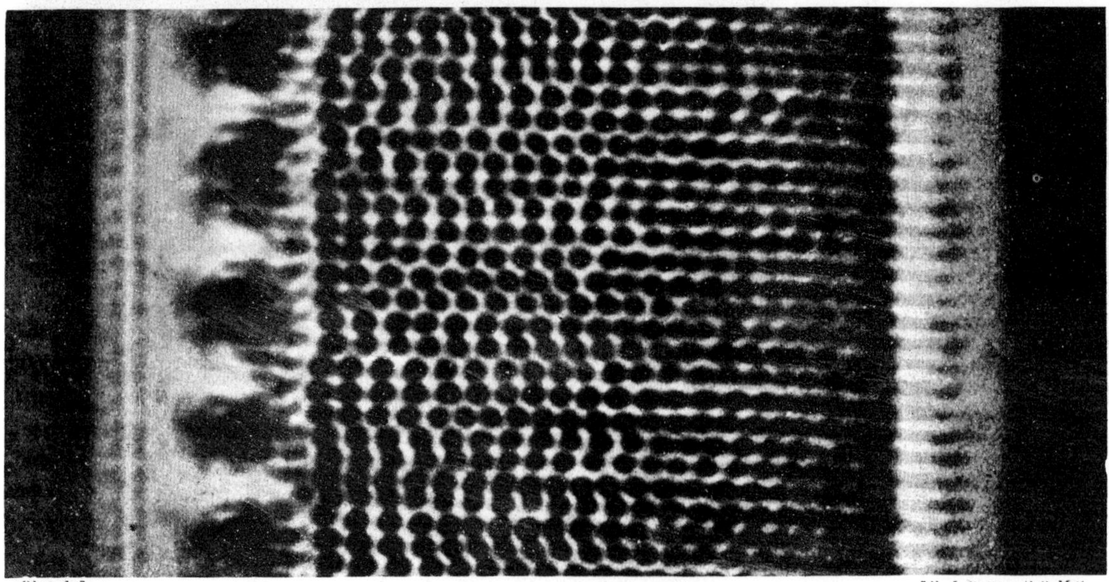

Photo by] [*E. J. Spitta, F.R.M.S.*

PORTION OF A DIATOM.

This photo shows a smaller part of the Diatom shown above, but still more highly magnified. In consequence the dots are seen to be cavities—perforations of the valve. The scale of magnification here is 3,200 times the actual size of the portion shown.

be remembered, too, that these markings are not themselves visible until the cloak of green integument is removed.

The mind is appalled also when one comes to think that the dimensions of these markings lie in the region of thousandths or even hundreds of thousandths of an inch! Then, too, unless one is familiar with such small quantities as microscopists are accustomed to, the mind seems unable to grasp what magnification is required to be able to see objects of these dimensions.

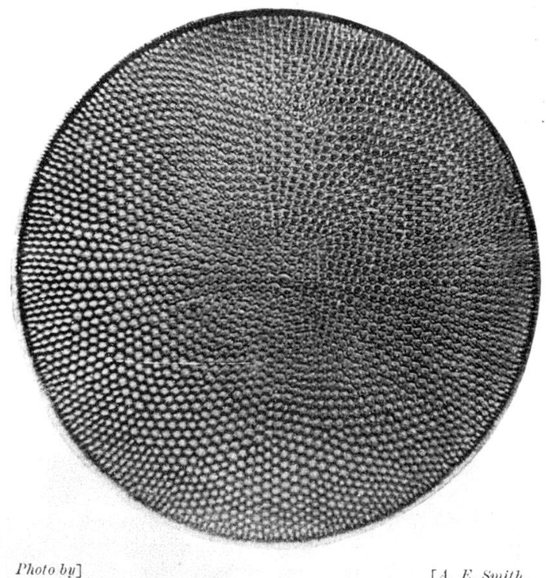

Photo by] [*A. E. Smith.*

A CIRCULAR DIATOM.

This photograph is nearly fifty times larger than the natural size of the Diatom, which is shaped like a watch-glass, and beautifully sculptured, every dot under a still higher magnification resolving itself into a six-sided figure.

To make this clear the following attempt has been carried out. Our first illustration on page 390 is a photograph of an ordinary microscope "slip"—a small piece of glass three inches long and one inch wide. In the centre is seen a ring of black varnish which attaches to the slip a little "cover-glass" made of extremely thin glass about .17 millimetre in thickness. Under this have been placed a little scattering of Diatoms; but they are quite invisible to the naked eye. The only appearance capable of being seen under the cover-glass is what looks like a little accumulation of dust more in one part than another.

In the second photo we are looking at a portion of the space enclosed by the black ring, but more magnification has been employed, and we now see a little, long and thin Diatom lying parallel to a black band running from the top to the bottom of the circle. This band is a human hair of the head. It has been so placed to educate the mind to appreciate the size of the Diatom in question. The exact Diatom is indicated by an arrow pointing to it.

Apply considerably more magnification and we now see, as at the foot of the same page, the same little Diatom much amplified in size, and we notice lines are now visible running across from side to side the whole way down its extent. If now we come to higher magnification still, using microscopic objectives of the finest quality and employing the utmost refinement of the microscopist's art, we find that these lines are really capable of being resolved into dots (see upper photo on page 392), and, moreover, we discover if we measure them that they are so small as to require one hundred thousand to be placed side by side to reach across a halfpenny! The mind literally reels at an assertion of this kind—it is a statement that is at once staggering and appalling. The

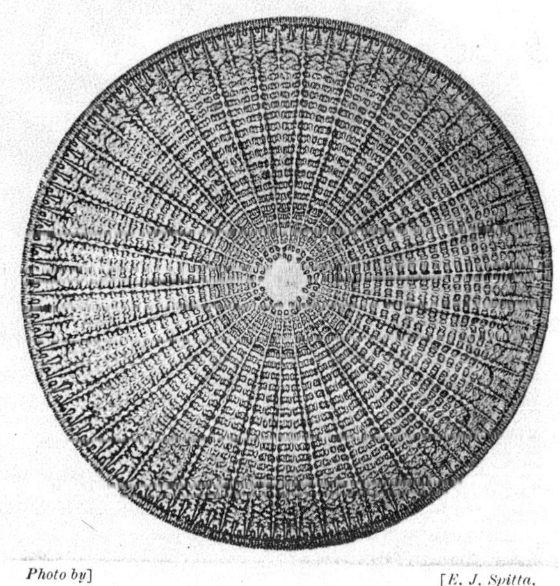

Photo by] [*E. J. Spitta.*

ANOTHER CIRCULAR DIATOM.

In this form the ornamentation is entirely different from the above, the markings making more than twenty concentric circles cut by thirty radiating lines from the clear centre. (Nearly fifty times larger than life size.)

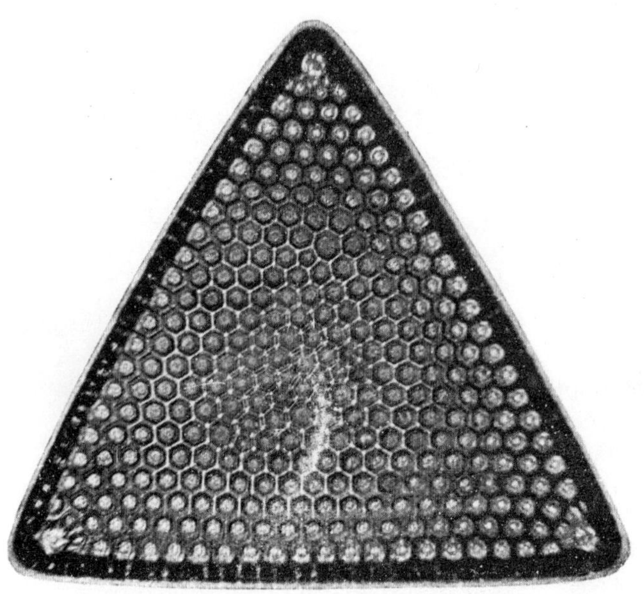

Photo by] [*A. E. Smith.*

A TRIANGULAR DIATOM.

This Diatom is here enlarged to 500 times its natural dimensions, a scale of magnification which resolves the central dots into six-sided openings.

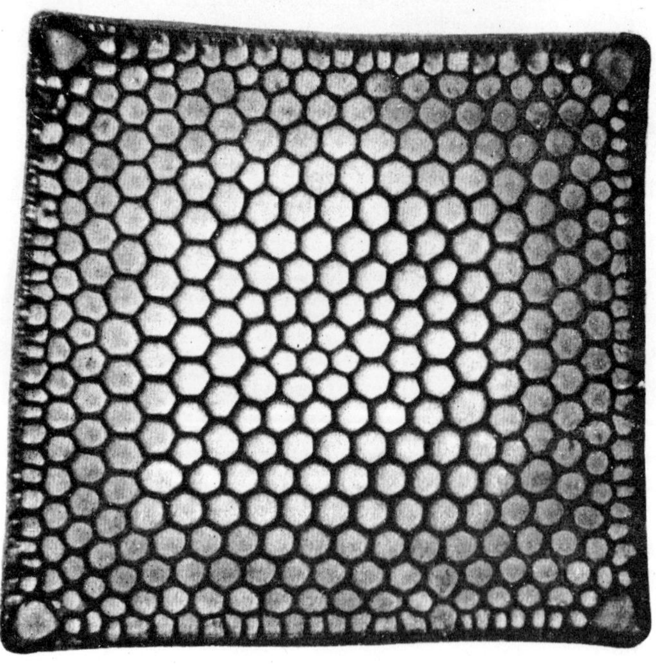

Photo by] [*A. E. Smith.*

A CUSHION-SHAPED DIATOM.

A nearly square Diatom, closely related to the one shown above, magnified 440 times.

magnification that is required to effect this resolution is very great, in the original negative amounting to about three thousand eight hundred diameters, an amount which, if applied to a man, would raise his stature to over four miles, and if a halfpenny were magnified the same amount would extend it to over three hundred feet.

These interesting little bodies really consist of two valves (united by a kind of band), one overlapping the other much as a lid of an ointment-box covers the box itself. They are distributed in a living state all over the world, in ditches, ponds, lakes, rivers and seas, having even been found in the Arctic ice; and as fossil remains they are found sometimes in great heaps in many parts of the world; the city of Richmond, Virginia, for example, being built on a layer of fossil Diatoms over fifteen feet in thickness. Diatomaceous earth—as the material is sometimes called—is to be found in Great Britain at Peterhead, Dolgelly, Lough Mourne, Raasay, Cantyre, Mull, and other places. Fossil Diatoms, too, have been found in volcanic dust.

Some of the comparatively large varieties are composed of very marked dots as secondaries, one being shown in the illustration at foot of page 392. Others are circular in form, as seen in the illustrations on page 393, the lower reminding one of a spider's web. Boat-shaped varieties are furnished by later illustrations, whilst the honeycomb or cavity type of secondary structure is displayed on page 396, in which the little holes are so small that fifty thousand would have to be placed side by side to reach across a halfpenny.

In the foregoing remarks we have dealt with the Diatoms chiefly from the microscopist's point of view, that is to say with the hard flinty skeleton or support for the living gelatinous envelope. It is the envelope that has chief interest for the botanist, for *that* is the only part that has life, and by its vital activity it extracts the dissolved flint from the water and deposits it in a solid form within its own simple body. We say simple body advisedly, for the Diatoms are among the simplest of all the plants—those that consist of a single cell. They never reach a higher development than this, and though a Diatom that is multiplying by division may be said to consist temporarily of two cells, it does not remain long in this condition. In certain species a number of Diatoms may be seen united in a chain, but each link is a separate and complete plant.

More than ten thousand distinct species of Diatoms have been named and described. As already indicated, they are ubiquitous. They occur in such vast numbers in the living state as to form a brown scum on stagnant water. They sometimes cling in thousands to the stems and leaves of higher freshwater plants and to seaweeds. They may even be collected from the face of wet walls and rocks.

One more remark should be made before closing this brief account of one of Nature's marvels. Some of these plants possess a power of movement which has baffled all attempts at explanation, several theories having been put forward; but all seem open to certain objections. To examine this movement it is only necessary to gather a little mud out of a pond, when several varieties will be sure to be obtained; and no little surprise will be felt at seeing some of these minute organisms floating across the field of view of the microscope in a slow but elegant manner.

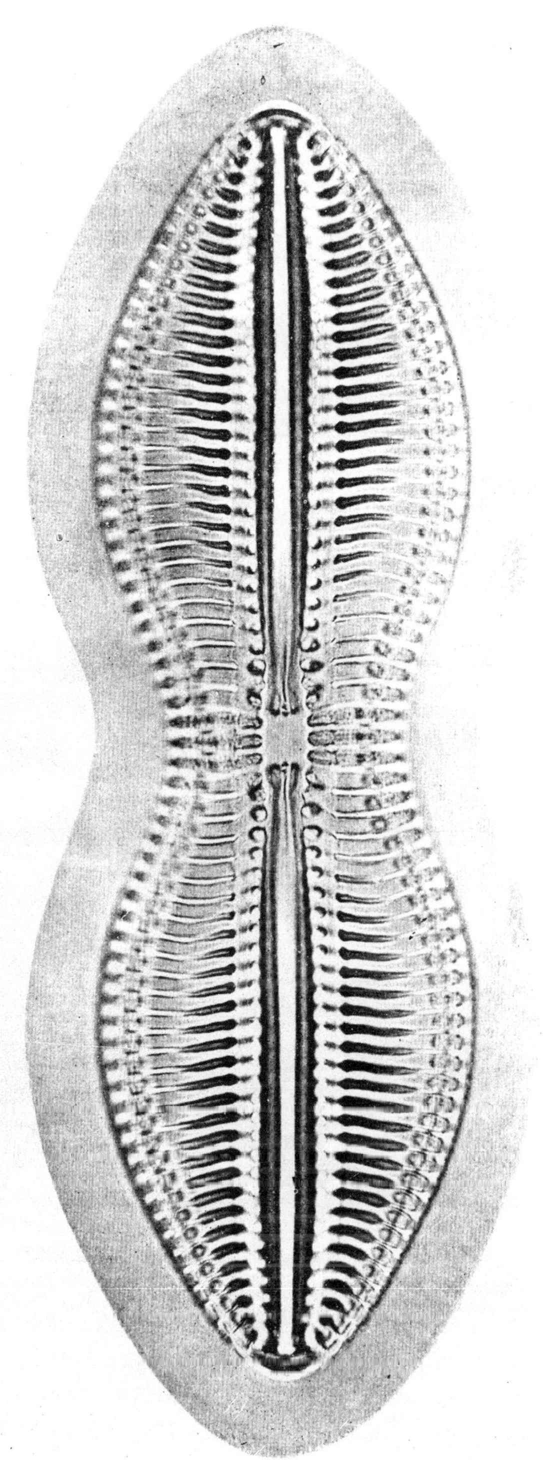

Photo by] [*Max Poser, F.R.M.S.*

A DIATOM.

This photograph of one of the "Little Ship" section of Diatoms is about 750 times larger than the natural size.

Photo by] [*T. F. Smith.*

PART OF A DIATOM'S SHELL.

This small portion of the flinty shell has been magnified about two thousand times.

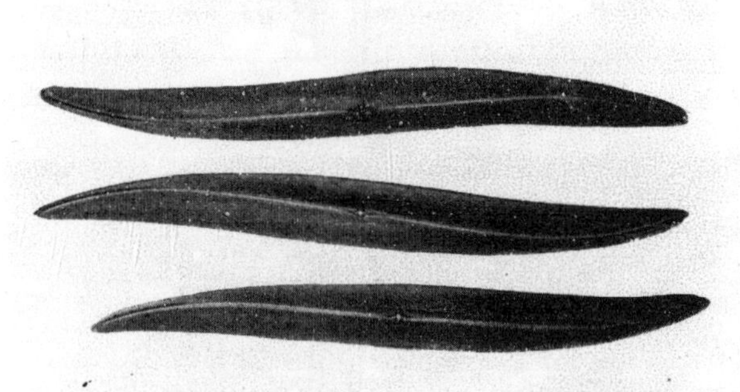

Photo by] [*Flatters & Co., Longsight.*

SPINDLE-SHAPED DIATOMS.

Enlarged about six hundred times.

Photo by] [*E. J. Spitta.*

HONEY-COMBED DIATOM.

The spaces in the valve of this Diatom are so minute that fifty thousand of them would be required to reach across a halfpenny.

THE GREAT SEA EAGLE

BY SIR HARRY JOHNSTON, G.C.M.G.

THE Sea Eagles belong to a genus which (with two subgenera) is found almost all over the world, and is invariably addicted in the main to a fish diet, though its members do not disdain other forms of flesh food if they are within reach.

These Eagles are noteworthy also for their loud screaming cries and for their bold, unusual, and often handsome coloration.

Some species found in Australia are quite small birds—not larger than a buzzard—and, like the screaming Fish Eagle of Tropical Africa, are white, chestnut-brown and black. But the Fish Eagles of the Pacific coast of the Americas and Eastern Asia are very large birds, almost, but not quite, the largest of the birds of prey. The most striking of these Fishing Eagles is Steller's Sea Eagle, of the Pacific coasts of North America, Alaska, Kamschatka and Japan, a bird which was first discovered by a Russian naturalist of that name in the eighteenth century. It is a favourite subject in Japanese art, and is, in fact, a great lure to the artist for both form and colour. The big, narrow, deep beak is a bright lemon-yellow, and the plumage is dark sepia brown and snowy white. Some of the finest examples of these magnificent birds measure about four feet in length from the tip of the beak to the end of the tail, and have a spread of over six feet across the wings.

Steller's Sea Eagle differs from the more typical Sea Eagles by having fourteen tail feathers instead of twelve. Its food consists of young seals, foxes and grouse, besides fish.

HOME LIFE IN WASPLAND

BY JOHN J. WARD, F.E.S.

IT was a bright, sunny day in early November and a young queen Wasp had just returned from her honeymoon tour. By some means she had lost her husband; the loss, however, did not cause her the slightest anxiety. As a matter of fact, as the sun retreated her spouse had felt advancing the chilled air which foretold an early frost, and an hour before he had alighted and crawled beneath a log of wood that rested against a wall. In all probability, too, he would die there; for the male Wasp rarely lives for many hours after his wedding, and if a frost follows it almost certainly spells his doom.

The queen herself was also now feeling the cold, but what she felt more was a drowsy feeling that seemed to overcome her. She had alighted on the sunny side of the bole of an old elm tree and was slowly crawling over the bark. Presently she came to a split in the tree, where the bark was loose, and through which she made her way, eventually finding herself in the heart of the tree. For a minute or two she crawled round and about the inside; then, selecting the driest spot, she opened wide her strong jaws and took hold of a projecting portion of the hard wood and gripped it firmly, folded her wings round her body and immediately fell into a sound sleep.

Later on she awoke, feeling a little dazed and stiff; but through the opening by which she had entered the tree she could see some rays of sunlight penetrating and she steered towards them, coming out on to the bark; and there she sunned herself for nearly an hour. At the end of that time she appeared to have got thoroughly warmed, and then she started on her toilet. After cleaning her jaws very thoroughly and briskly rubbing her face, she brushed down her wings and body by means of her legs. Finally she drew each of her feelers, one at a time, between her folded forelegs, cleaning them from base to tip. Her toilet then being complete she immediately spread her wings and pushed out into the air.

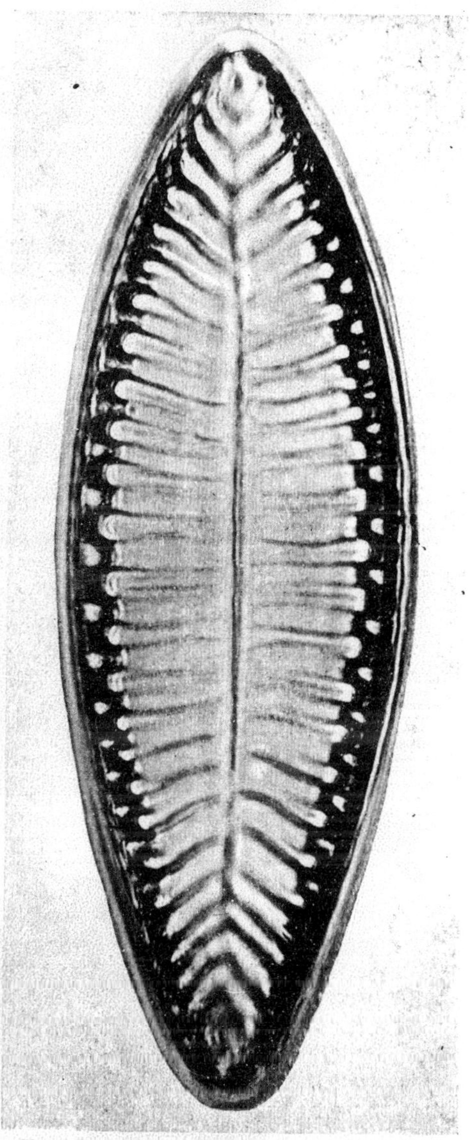

Photo by] [*A. E. Smith.*

A CANOE-SHAPED DIATOM

The movements of these minute plants through the water naturally suggest that this species is canoe-shaped. The markings are of a stronger character than in some others here illustrated. As shown, it is about eight hundred times larger than natural size.

This queen Wasp did not, of course, know anything about dates; had she have done so she would probably have been greatly surprised to know that her sleep had lasted from November 9th until April 19th. For more than five months cold and wet days had come and gone while she had remained dormant. Now her hour of victory had come: she was to become the mother of a new Wasp community; indeed, she was destined to be the queen of a thriving city with perhaps as many as fifty thousand inhabitants—all her children.

THE GREAT SEA EAGLE.

The species represented in this photograph is the enormous Sea Eagle of Japan and the northern coasts of the Pacific. It has a bright lemon-yellow beak, and the dark parts of its plumage are sepia-brown, whilst the rest is white. The spread of its wings is over six feet, and its entire length about four feet.

Her flight did not continue very far before she suddenly swooped down, alighting upon an old and broken fence, where she flew from post to post as if at play ; but queen Wasps have no time for play ; there is probably no animal that lives which works so hard and so persistently. She had discovered the building material for the city which she was to establish and she was making herself acquainted with its exact location.

Presently she ascended in the air again, and, after flying along the whole course of the fence, she suddenly turned and flew in a circular direction towards a high hedgerow bank. There she spent twenty minutes or more, running in and out of the crevices amongst the sandstone and soil of the bank. In one of these holes (which probably was once the home of a pair of field-mice and their family) she spent a great deal of time. Just what she was doing inside was difficult to say, but it was fairly obvious that she was cleaning up the place a bit, for every now and again she would come to the outside of the hole and on each occasion she carried something in her mouth which she dropped outside. Sometimes it would be a scrap of dried nut-shell, sometimes a bit of root or a seed, and not infrequently a little pebble. These actions made it quite clear that she had commenced operations for building a Wasp city and was clearing the ground.

Photo by] [*J. J. Ward, F.E.S.*

NEST OF A TREE WASP.

The species of Wasp that builds in trees and bushes is different from those that construct their nests in the earth ; but the internal arrangements are the same.

After nearly two hours at this work she came outside very dusty and dirty, but she quickly arranged her toilet (which is always an important matter with a Wasp) and then she flew towards the old fence, following the circular route by which she came earlier in the day, and meeting the fence at the end just where she had left it. Afterwards she made some hundreds of visits to that fence, but she always followed the same course, although the distance would have been much shorter had she flown direct to her building-site instead of following the long, circular route. The fact is, if she had gone any other way she would have got lost ; for Wasps have to find their way by familiar landmarks, and if the latter are disturbed they become like a ship without a rudder. If, however, it should discover its nest by accident during its wanderings it will find it again and again afterwards, but always by the last route which it discovered.

In this connection the writer, during the past summer, had a live nest of Wasps under observation in his photographic studio, and in the ordinary way these Wasps flew in and out the open doorway ; but in the early morning, before the door was opened, the Wasps that had just emerged from their cells learnt their way out through a triangular ventilator immediately over the door. Now, although the ventilator was only separated from the open door below by a strip of wood about one and a half inches wide, yet those wasps which had left by the ventilator always returned that way. If the ventilator was closed they alighted on the outside and were quite lost. Occasionally, in trying to find their way in, one would fly within the doorway, and in a moment it would recognize its whereabouts and fly to the nest. When, too, the nest was removed two yards to the opposite side of the studio the Wasps were equally lost, and when they returned would alight upon the spot on the table where the nest should have been, accumulating there in large numbers and becoming very angry with each other.

These experiments, then, tend to show that the Wasp followed the circular route because that was the only way to her building-site which she had learnt. So for days together she continued to make this journey. Each time that she alighted upon the fence she tore off shreds of the wood fibres by means of her strong jaws, and these she masticated into a pulp and carried back to her home in the bank. The first use of this material was to form a stout little *papier-mâché* pillar attached to a projecting portion in the roof of the chamber—for the Wasp builds the roof of its home first. Then to the end of this footstalk a cap of the same material was added about half an inch in diameter. To the under side of this cap four cells were attached ; and in that manner the first stage in a new Wasp city was built.

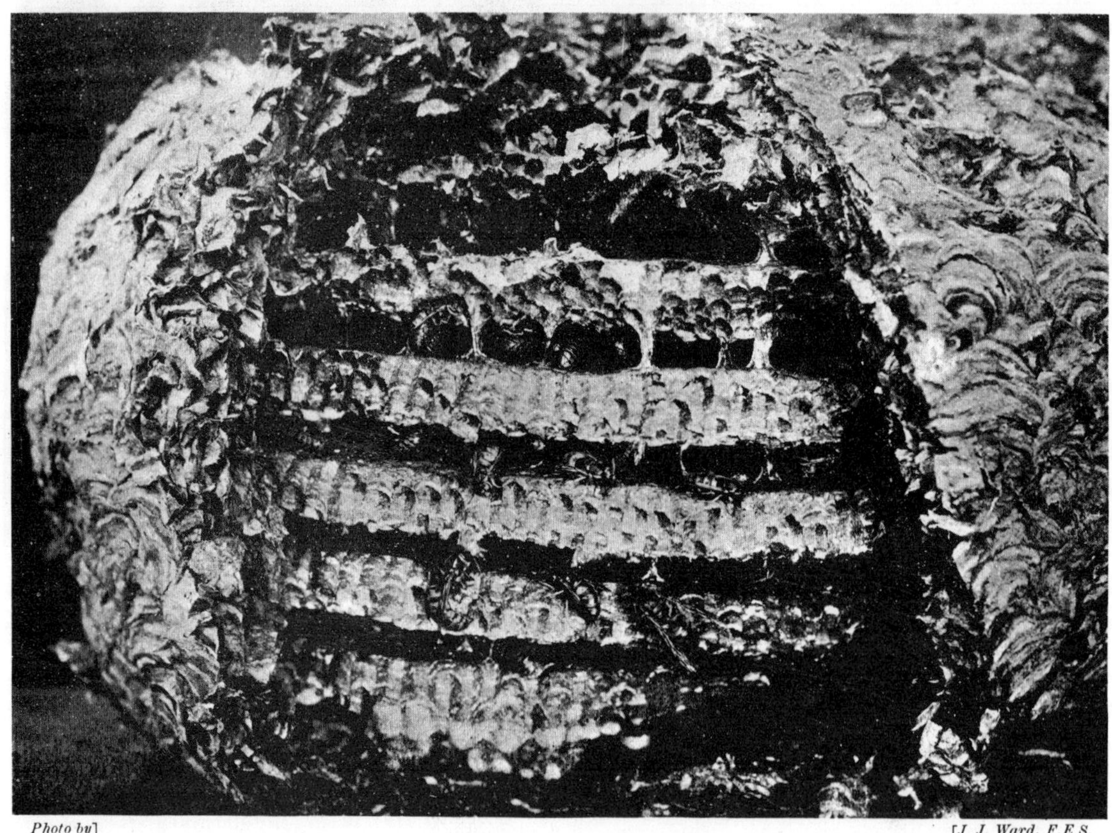

Photo by] [*J. J. Ward, F.E.S.*

INTERIOR OF A WASP'S NEST.

By cutting away part of the paper walls the arrangement of the combs is seen. It will be noticed that they hang one from another on paper pillars, and that the cells all open downwards. Only just sufficient space is left between the combs to enable the Wasps to crawl about, and attend to the grubs.

In each of those cells she then placed an egg. From those eggs, too, would come the first four citizens of the new city. She then added other cells by the side of the first four, extending the edges of the covering cap as she built them. In this manner a layer of cells soon appeared and in each one she deposited an egg as it was completed. Meanwhile the eggs which she laid first hatched out their grubs, and this queen—in name only—then had nursery matters to attend to. The young grubs had to be fed on chopped insects and vegetable food, which she had to find and prepare. Still the Wasp-mother kept pace with it all, building new cells between the intervals of feeding her offspring.

In the course of three weeks the grubs which emerged from the first four eggs deposited had grown so large that their fat, white bodies almost completely filled up their cells, and they themselves then proceeded to seal up their cells with a white cap; after which they troubled their mother no more. Ten days later the cells were cut open from within by the sharp jaws of their inmates and four worker Wasps emerged.

Photo by] [*E. Step, F.L.S.*

WHERE THE WASP OBTAINS MATERIALS.

This photo shows a small portion of a fence that is a favourite collecting ground for Wasps engaged in building nests. Each light streak shows where a shaving of wood, the same size, has been scraped off by a Wasp to be worked up into paper building material.

That was a glad day for the mother Wasp, for her powers of paper-making had declined. Here, however, were four young and strong daughters, less in size than their mother, and sexually imperfect, but like her in all other respects, for they could build new cells and were just as capable as she in tending the young. Day by day new workers appeared, and soon the queen Wasp did nothing but deposit eggs in the cells as rapidly as the workers could build them.

So the city grew apace; new tiers were suspended to those above by short pillars, until eleven floors had been made. The central and largest tier alone contained more than four thousand cells, and some of these were used for the rearing of at least two broods.

At the end of the summer the Wasp city was at the height of its glory; food was abundant and the thousands of busy citizens were extremely active. About this time, too, some very large cells were built in the lower tiers and extra food was supplied to the grubs that hatched within them. Then there came a terrible blow to the community—the vitality of the queen Wasp became exhausted and she could no longer deposit eggs. At first the worker Wasps did not realize that anything was wrong, but later on they found themselves idle, having no new grubs to feed.

Then the large cells were cut open from the inside and from them emerged a host of young queens, while at the same time from other cells appeared another host of slender-bodied Wasps with very long feelers. The latter were the male suitors of these queens of the future. In the course of a few days they selected their mates and left the nest never to return. What eventually happens to these males and queens has already been told.

Many of the workers still remained in the nest, but they appeared to have all suddenly gone mad. Their sole occupation was that of pulling out the half-developed grubs from their cells, dragging them to the entrance of the nest and leaving them there to perish on the ground. Many ridiculous solutions of this action have been advanced. It is not infrequently attributed to a merciful instinct on the part of the workers, who apparently realize that, with cold weather approaching, they will not be able to rear their charges and consequently they mercifully terminate their suffering. The true explanation is, I think, one of sanitation. An instinctive impulse guides them to clear the nest of the starving grubs, just as they would rid it of any other matter which threatened the health of the community. Then, however, arises the question: Why should they clear a nest which every hour is becoming more and more deserted? Until late in the year there are still queens and males to emerge, and the last service the workers render to their race is to remove the grubs they cannot rear, so that these sexual individuals may mature in a healthy atmosphere.

Photo by] [*J. J. Ward, F.E.S.*

WASP PAPER.

Here are shown, natural size, samples of three qualities of paper made and used by the Wasps in the construction of their nests. The centre sample is that of the Common Wasp, characterized by its shelly patches. Above and below are shown paper from the nests of two species of Tree Wasps, which is made in long layers.

Afterwards, when all the grubs are removed from the cells, the young workers forsake the nest, and, having no home ties or young to feed, they become freebooters, giving themselves over to orgies in any warm kitchen where they can scent savoury food being prepared. Their orgies may revive their spirits somewhat for the time being, but even though they may escape from the hands of the cook, yet their time is short; for, having deserted their nest, they now have no home, and as night approaches they seek shelter perhaps beneath a leaf or in some similar situation, and so, sooner or later, wet and cold overtake them.

Occasionally the Common Wasp will vary its building site; perhaps the commonest place other than underground is in the roof of a building. A monster and beautiful nest, which was built on a beam in the roof of an empty house in Warwick during the extraordinarily hot summer of 1911, is

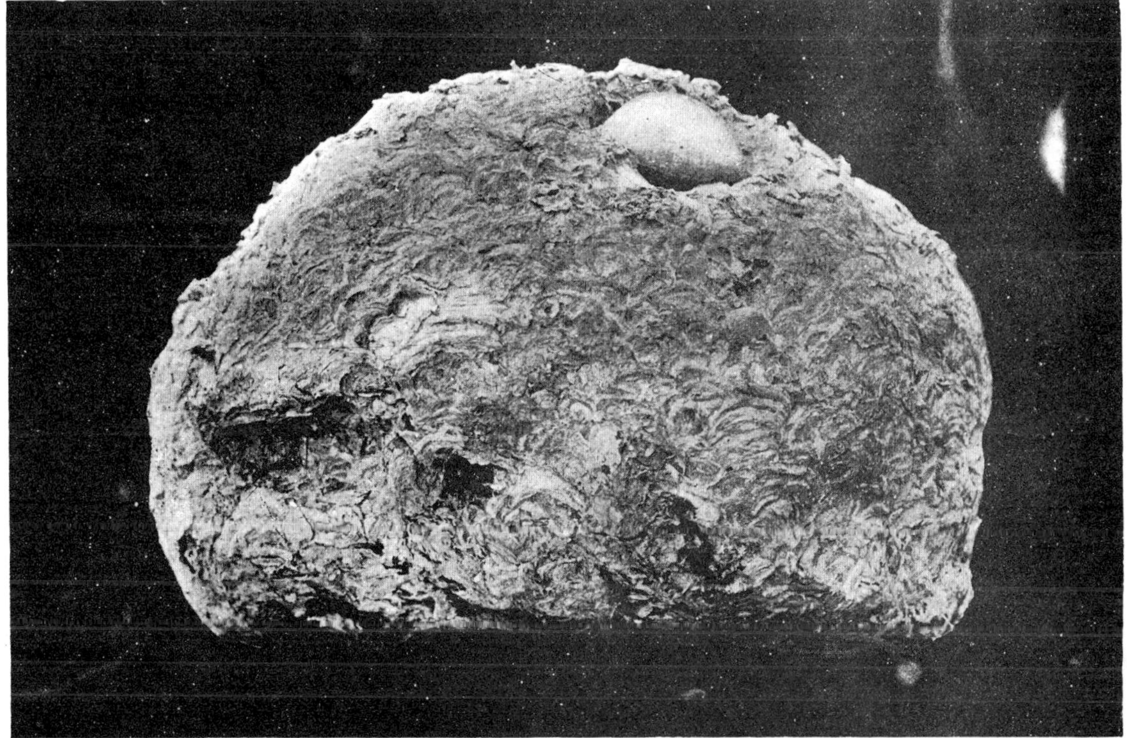

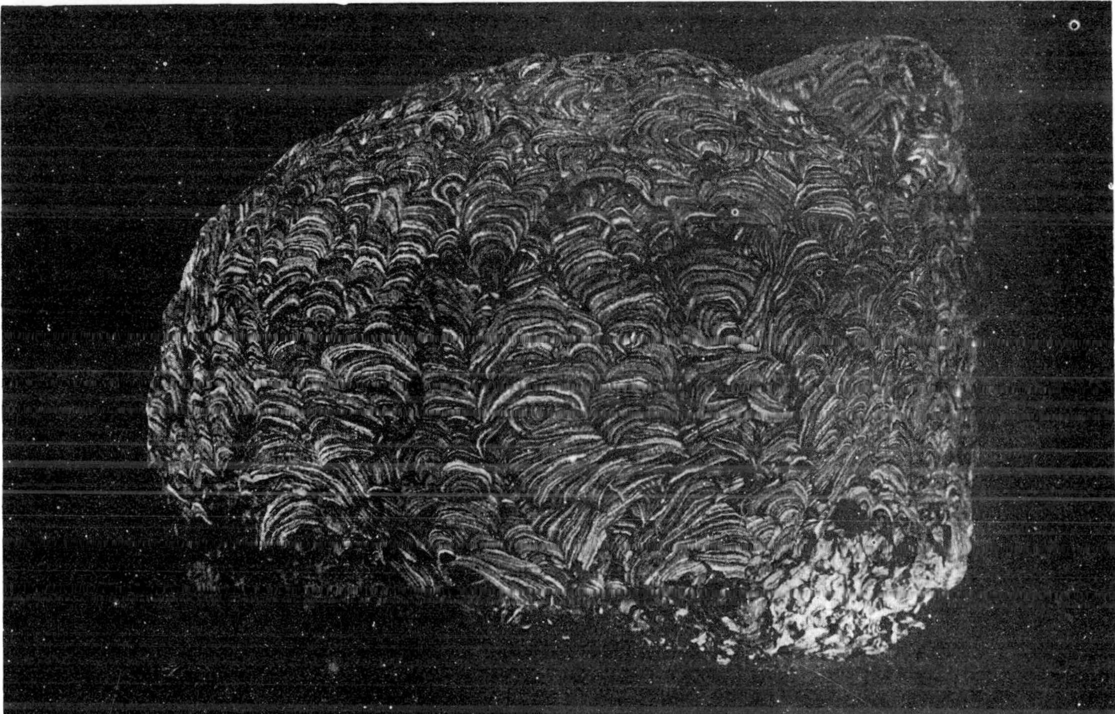

Photos by] [*J. J. Ward, F.E.S.*

TWO WASPS' NESTS.

The upper nest shows that in excavating the Wasps met with a stone too large for removal, so the nest was built around it. The lower photo is of a monster nest of the Common Wasp, built in the roof of an empty house, and measuring just five feet around it.

shown amongst the illustrations. The nest was of an oblong form and of extraordinary size, measuring five feet in circumference, with a diameter of eighteen inches. How many tiers it contained is uncertain, as it was preserved whole for a museum specimen. Judging from its measurements, it would contain at least sixteen tiers. Another remarkable feature of this nest was its pale biscuit-colour; each little band of masticated paper added to the shelly patches by the individual workers could be distinctly seen, as shown in the photograph.

The British Tree Wasps construct their nests on almost identical principles with those of the Common Wasp here described, excepting that they are suspended amongst the branches of the trees. Each of the three species found in this country favours different trees: one selects thorny hedgerows, another holly bushes, while a third usually places its nests upon the branches of larch-trees. These nests often surprise country people when discovered for the first time.

There are also three British species of Ground Wasps, not including the large species known as Hornets, which sometimes build underground, though more often in a roof or in a rotten tree; but the Wasps themselves require a practised eye to distinguish. The outer covering of the nest is, however, the best guide to their recognition, as each species seems to have its own particular method of paper manufacturing. In the case of the Ground Wasps and the Tree Wasps the distinction is at once obvious (see illustration), but between the individual species it is less easy to recognize.

In this country the nests of the Tree Wasps do not reach such large sizes as those of the

Photo by] [*Harold Bastin.*

A BRAZILIAN WASP'S NEST.

This nest was built in an orange-tree. It will be seen to differ from the nests of British Wasps in form as well as in its smooth surface. The paper substance of the outer case is as thick and durable as cardboard. It is known as the Dutchman's Pipe.

Ground Wasps, but in tropical America and Central Africa very large examples are found; sometimes they attain three and four feet in height. Amongst the illustrations is shown an example of a nest from Brazil, which is popularly known as the Dutchman's Pipe, its shape somewhat resembling a large pipe-bowl with a small mouth, while the portion of the branch on which it was suspended serving as the stem, completes the resemblance.

The builders of these nests are known as Pasteboard Wasps, as the walls of their nests are thick and tough, like strong cardboard. In addition, they have a beautifully polished surface, which makes them impenetrable by rain and mist. The resemblance to cardboard is so good that it would almost deceive a paper manufacturer, and to others it certainly appears as an artificial production.

These nests are not infrequently placed on orange-trees, and their structure is rather curious from the fact that a central orifice passes through all the combs, commencing at the entrance.

Photo by] [*J. J. Ward, F.E.S.*

THE DEVELOPMENT OF A WASP.

In this photo the lower cells of the comb contain eggs and young grubs; above the cells are capped with a papery material elaborated by the full-grown grubs before becoming chrysalids. In several the cap has not yet been made, and the grub is seen filling the cell. On the left a Wasp has just emerged from the chrysalis. (Twice the natural size.)

Having regard for the facts we have sketched in this and an earlier article, we may claim that in its social economy the Wasp is certainly as marvellous as the Bee.

A PLESIOSAUR

BY SIR HARRY JOHNSTON, G.C.M.G.

FROM the earliest reptiles, in the latter part of the Primary Epoch, were evolved not only the Fish-lizards or Ichthyosaurs (which adopted a wholly aquatic life and had an appearance something like whales or porpoises), but also a remarkable order known as Sauropterygia, or Saurians with paddles. These, perhaps, had some affinity with the ancestors of the tortoises and turtles, though the resemblance in structure may be mainly due to a parallel line of development. The Sauropterygia are generally summed up for the unscientific public by one name—Plesiosaur. [This really means "more lizard-like," because to its discoverer the long-necked sea-reptile seemed more like a lizard in shape than did the contemporaneous Ichthyosaur.]

Fifty or sixty years ago, when interest was first aroused in the marvels amongst extinct forms of life, there was no illustration more popular in works dealing with the past phases of the earth's history than one which depicted a fight between a Plesiosaur and an Ichthyosaur. Both these groups of aquatic reptiles flourished during the Secondary Epoch of the earth's history; and it is possible, of course, that in their common pursuit of fish they may have quarrelled and fought.

The extraordinary Plesiosaur which is here illustrated, both in Professor Fraas's picturesque and accurate restoration and in its skeleton form, is known scientifically as *Thaumatosaurus*, and was discovered in the Upper Lias formations of Holzmaden, in Württemberg, by the well-known German professor of palæontology at Stuttgart, by whom both illustrations have been kindly lent.

Photo by] [*A. E. Tonge, F.E.S.*

NEST OF THE TREE WASP.

This nest of the Tree Wasp was built under a sloping roof. As photographed it was in a somewhat early state, but was gradually enlarged until it measured six inches across the middle. The photograph is twice the actual size.

This Plesiosaur was a large animal, about fourteen feet long. Its jaws were armed with long, sharp fangs, but its neck, in striking contrast to the Ichthyosaurs, was very long and slender.

Yet in most of the Plesiosaurs the neck must have been stiff, almost rigid, except for lateral movements—to judge from the conformation of the joints. In another very remarkable form, also discovered in the Upper Lias formations of Württemberg, the end of the tail spread into two broad flukes.

One of the illustrations to this article is photographed from the carefully-sculptured restoration of a Plesiosaur in Carl Hagenbeck's Animal Park, near Hamburg. It is the work of a German artist, Herr A. Pallenberg, who has developed, under the guidance of Professor Fraas and other German naturalists, a great talent for realizing and reproducing the appearance of extinct reptiles, birds and mammals.

But it is questionable, except on some seaweed-covered rock nearly on a level with the water, whether any Plesiosaur was able to resort to land. These marine reptiles seem to have spent their lives entirely in the water, and would have found it as difficult to move on a solid surface as a porpoise or a turtle.

Yet apparently they did not produce their young alive, as did the Great Fish-Lizards. Perhaps, like the existing turtles, they laid their eggs on sandy beaches or amongst the smooth rocks along the margin of the sea.

Photo by] [*Carl Hagenbeck.*

A PLESIOSAUR.

A restoration, in Hagenbeck's Tierpark, of one of the great Saurians with paddles, known as Plesiosaurs, which were contemporary with the Icthyosaurs shown in earlier pages.

The food of Plesiosaurs, besides fish, consisted of cuttle-fishes and crustaceans; and the types of later development show by the fossilized contents of their stomachs that they were in the habit of swallowing stones to assist them in grinding up the harder parts of shellfish.

The largest known members of this order of sea-reptiles with paddles attained to a length of twenty feet. Plesiosaurs of various genera and species frequented the shallow seas all over the globe down to the verge of the Tertiary Epoch. They must have looked very like the average sketch of a sea-serpent, with their long necks, turtle-like bodies and short, thick tails. In fact, one is tempted to believe that there are still Plesiosaurs which have survived down to the present era in the great oceans, and whose occasional appearance gives rise to the stories of a great sea-serpent.

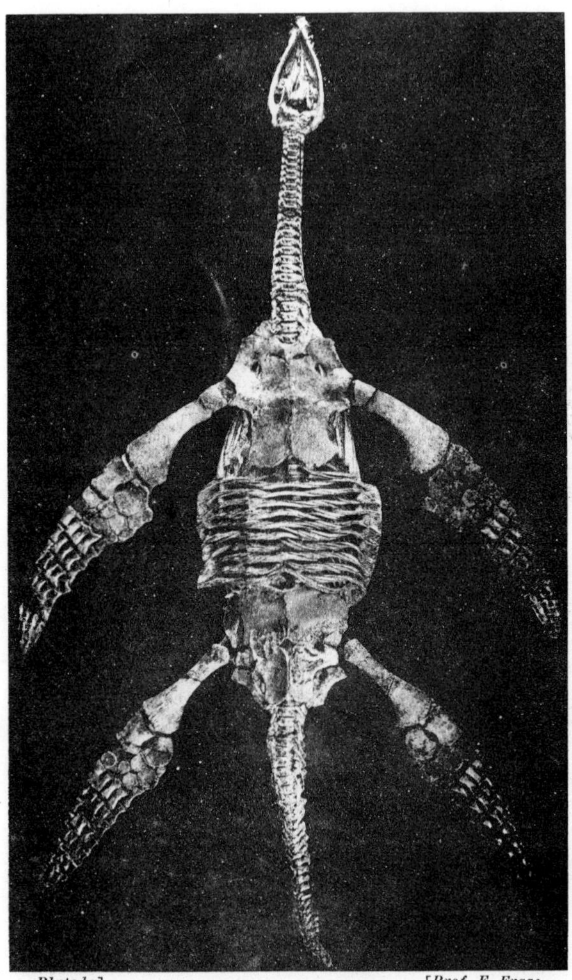

Photo by] [*Prof. E. Fraas.*

A GIANT PLESIOSAUR.

This photo represents a perfect skeleton of a large Plesiosaur which was found in the quarries of Holzmaden, in Württemberg. It is a fossil of the Upper Lias formation.

THE HAND FLOWER

ONE hundred and twenty-five years ago there was discovered near the town of Toluco, in Mexico, a tree about thirty feet in height and covered with large leaves, much like those of the Plane-tree in shape.

The Mexican Indians regarded this tree with feelings of great veneration: they believed it to be of great age—as, no doubt, it was—and that it was the only tree of its kind in existence.

But the chief reason for their reverence lay in the peculiar structure of the flowers, from which a long and skinny hand, with claw-tipped fingers, appeared to protrude. So great was their interest in this tree that as soon as its flowering time arrived they flocked to it and gathered every flower; so that there was little chance of its producing seeds.

But although it was thus prevented from propagating itself in that neighbourhood, the Toluco example was not the only tree of its kind; for about fourteen years later a traveller in Guatemala discovered its probably true home, where forests of it existed. It is therefore surmised that in much earlier times some wandering Indians, being struck by the remarkable flowers, carried off a young plant to Toluco and thus established the cult of the Hand-flower in Mexico. They gave it the terrible-looking name of Macpalxochitlquahuitl, and beside this the Greek name given to it by the scientific botanists looks quite pale and simple.

The flower is about the same size as our photograph, and consists of a cup-shaped calyx of leathery consistence and a rust-red colour, deeply cleft into five parts. The likeness to a hand is furnished by the five long stamens, which are united towards their base and of a brighter red than the calyx. There are no petals. At the bottom of the cup there are five cavities of a bright yellow colour, which are filled with a sweet fluid which is said to have the flavour of toast and water. This, no doubt, is to make a palatable drink for some insect whose services are required for carrying the pollen from flower to flower, and so to help in the fertilization of the future seeds.

THE ARCHER-FISH

BY W. P. PYCRAFT, F.Z.S.

THE Archer-Fish well deserves its place among the marvels of Nature, for it is the only creature save man which brings down its prey by shooting! Further, it has this advantage over man, that

[By Prof. E. Fraas.

PLESIOSAURS OF THE UPPER LIAS.

In this picture Prof. Fraas has restored the probable life appearance of the Plesiosaur whose fossil skeleton is shown on the opposite page.

its supply of ammunition is inexhaustible. Briefly, it lives largely on insects which haunt the margins of streams and rest on leaves and twigs overhanging the water. Prowling along at the surface of the stream, the hungry and eager fish sooner or later marks down a luscious-looking fly, and then, with careful aim, brings it down with a drop of water squirted from its tubular mouth.

The Malays set great store by the Archer-Fish on account of the entertainment which it affords them. They call it the "Ikan Sumpit," and keep it in small aquaria the more easily to watch its wonderful marksmanship.

How did the Archer-Fish discover its ability to shoot? Another remarkable feature of this strange phenomenon is the fact that the fish can see its prey sufficiently clearly to enable it to aim surely, for it must be remembered that the eyes of fishes are adapted to see clearly only under water. Thus we must assume that the eyes of the Archer-Fish differ from those of other fish. That this is a matter demanding investigation is shown by the fact that the only other fish which takes an interest in the upper world is that known as the "Four-eyed Fish," or Anableps. This creature has bulging eyes and habitually swims at the surface, so that the upper half of the eye is above the level of the water. This portion is divided by a sort of Plimsoll mark from the lower half, and is modified to meet the requirements of terrestrial vision, while the lower half agrees with that of fishes generally. The lens of the eye does not undergo any special change; by a modification of the iris, or coloured portion of the eye, two distinct pupils are formed, one above the other!

The Four-eyed Fish is a native of the rivers of tropical America. The Archer-Fish enjoys a much wider distribution, being found in the fresh waters of the East Indies, North Australia, Polynesia and New Zealand.

Photo by] [*E. J. Wallis.*

THE HAND FLOWER.

This is the flower of a Central American tree, whose five red stamens, united towards their base, resemble a long, thin hand with clawed fingers—a resemblance that has caused the Mexican Indians to regard it with veneration. It is here shown about the natural size.

THE BECKONING CRAB

BY EDWARD STEP, F.L.S.

THE Crab family presents us with a great variety, not merely of form, but of habit also. Even the Crabs of a single country—our own, for example—make this diversity plain; but when we take the Crabs of the world into consideration we are astonished by the many ways of life they have adopted. Some, save for an annual trip to the sea, spend their lives on land away from the shores. Others, though keeping close to the sea, live mostly out of water; and among these is the quaint-looking Crab of our photographs. Represented by different species in several parts of the world, the Beckoning, or Calling, Crabs all have similar habits and the same grotesque appearance—in the males, at least—due to the disproportionate development of one of the pincer claws. Whilst one of these is scarcely larger than the walking feet, the other is larger than the little body

[By W. B. Robinson.

THE ARCHER-FISH.

The Archer-Fish is a native of the East Indies, where it is reputed to catch insects from the riverside vegetation by shooting drops of liquid from its mouth, with such good aim as to bring them down into the water, where they are, of course, snapped up by the Archer. The Malays call it "Ikan Sumpit," and keep it in aquaria in order to watch its marksmanship.

Photo by] [*W. Saville Kent.*

BECKONING CRABS AMONG MANGROVE ROOTS.

The Beckoning Crabs are so called from their manner of carrying the relatively huge pincer-claw, which makes them appear as though beckoning to a friend at a distance.

to which it is attached, and which appears always to be in danger of being overturned by the superior weight of the big fist. As they run along the sands with this big claw held up—as though they were afraid of falling over it if carried in front of them—they appear to be beckoning. One species that occurs in the Far East is named by the Japanese *Siho Maneki*, which is said to mean "beckoning for the return of the tide."

Latreille, the French naturalist, who defined the characters of the group, gave them a scientific name which signifies laughable or ridiculous.

But in spite of their grotesque demeanour, the Beckoning Crabs are a very useful people.

We acclaim certain individuals as empire-builders. The Beckoning Crabs are more than that: they are world builders—industrious creatures that, whilst looking primarily after their own interests, are helping to wrest new land from the dominions of the sea, and to convert barren shores into rich soil.

The Beckoning Crab is a burrower. Like the benevolent earthworm, they excavate tunnels a foot or more deep, and perhaps an inch in width. Bringing up pellets of damp sand from the depths, they carry it to a distance of three or four feet from the mouth of their hole before depositing it, and when their tunnel is completed they carry down bits of seaweed and other vegetable *débris* to be consumed as food. Some species are said to make stores of food in a terminal chamber of their tunnel. The burying of vegetable matter by hundreds of thousands of these little burrowers has an appreciable effect. Mr. H. O. Forbes, who observed their work on the coral islands of the Eastern Archipelago, says: "By the inner margin of some of the islands, and forming lagoonlets in some of them, there are soft, limy mud-flats, which are gradually becoming land, mainly by slow elevation and by crustacean agency. One of the largest of these is in West Island. Its lagoonward portion near the entrance conduit, which is submerged at high tide, is tenanted by two, if not three, species of Beckoning Crab. They live in narrow, corkscrew burrows, round the tops of which there is always a little mound, just such as is seen about an earthworm's; and, indeed, they are most perfect worm-substitutes. I counted one hundred and twenty of these holes in an area of only two feet square, and as there were many square acres in the ground of which I speak, some idea of the number of this busy army may be obtained. They were incessantly active during the recess of the tide, and even during high water, which is generally perfectly still, in carrying down twigs of trees or fucus leaves, scraps of coconut shell and seeds, laying the foundation of the future land."

Forbes further tells us that the entry of a man into this region is the signal for a stampede of the Crabs to their holes, where they wait within the entrance, watching the intruder with their long-stalked eyes over the folded big pincer claw, which serves as a sort of street-door. It is just the position of affairs that is shown in the photograph below.

A distinct species of the same group, which is known as the Boxer, from its attitude when alarmed or disturbed at its work, is a native of the southern parts of the United States. It has been described as occurring in millions on the seashores and along the tidal rivers of Carolina, where they carry out similar operations to those of their relatives in Southern seas. They are the prey of otters, birds, turtles and alligators, but though these enemies destroy vast numbers of them, the fecundity of the race is such that no appreciable difference in their numbers can be noticed. Another American species, known as the Fiddler, owing to the peculiar movements of its big arm, lives in salt-marshes and on fresh-water shores. It constructs a sort of porch of mud at the mouth of its burrow, and sits there on the watch.

Photo by] [*W. Saville Kent.*

THE BECKONING CRAB AT HOME.

When alarmed these Crabs scurry into their holes, blocking the entrance with the big pink-coloured claw, over which the long-stalked eyes keep watch upon the intruder.

THE FAN-CORALS

BY EDWARD A. MARTIN, F.G.S.

THE four examples of deep-sea life which we here show are not usually recognized by most persons as Corals at all; the word "Coral" as a rule conveying a meaning as of something hard and stony, and bearing various pretty markings which are at once recognized as having been the clefts in which the polyps lived and died. But they are no less true Corals than are the Madrepores and the Brain-Stones which are so familiar as ornaments in the cabinet or the drawing-room. As a matter of fact, they belong to the same great order of Nature as that to which the beautiful Red Coral belongs,

Photo by] [*H. J. Shepstone.*

A FERN-LIKE HORNY CORAL.

Like all the horny Corals the branches of this species are covered, when living, with a rind of flesh, of which the polyps form part. The branches start off from the stems in two directions only, so that they present a fern-like appearance.

although even this has to be polished before it assumes the brilliant appearance which the article possesses when it is placed on the market.

The Fan-Corals—or, as they are sometimes called, the "sea-shrubs"—are scientifically known as the Gorgoniæ, although most of them are too pretty to even suggest any resemblance to the Gorgon's head of ancient myth. Nor is the title "sea-shrub" a happy one, as they are no part of the vegetable world at all. Like many other wonders of the sea, the name was given them because of outward resemblance only, and it remained for science to place them in their right place in the scheme of Nature. All Corals are distinctly animal, and if only we could watch the polyps working under a microscope we should at once see clearly their animal nature. Sometimes in museums we find a specimen of Coral with all its flesh intact, whilst peeping from out of the flesh the snow-white polyps have been caught in the act, so to speak, and preserved in that condition in spirit.

Photo by] [*E. Step, F.L.S.*

THE SEA-FAN.

Here the skeleton is all in one plane In life the " Fan " was coated with a thin covering of flesh in which the polyps were **embeddeu.** The photo is about one-third less than the natural size.

Photo by] [*H. J. Shepstone.*

AN AUSTRALIAN SEA-FAN.

In this species—quite distinct from that shown on the previous age—the connections of the branching network are maze-like. It is flexible like the others, and covered in life by a layer of gelatinous flesh.

Let us see for a moment what a Fan-Coral is like.

The same description will hold good for the familiar Red Coral. As shown to us in the illustrations, we only see the skeleton of the animal. Just as much as human beings are made up of flesh and bone, so also are the Corals, the difference being that in some cases the " bone " is not composed of lime or calcareous matter, but is sometimes made of horny material; whilst in other cases the hard skeleton is made up alternately of pieces of limy and horny material. On the next page we can see in the centre of the picture some portions of stems which seem to be alternately light and dark. This is where the two kinds of material take part in turn to make up the stem. Such a stem, it will at once be seen, has a great advantage in seas that are troubled with submarine swell, since the deposition of pliable horn at intervals enables the creature to bend to and fro in a sea that would be fatal to other forms not so well adapted to it. This is, indeed, a wonderful adaptation in Nature, and it may seem strange that other forms have not in the course of time become similarly adapted. But it will be at once seen that the other forms we illustrate above and on page 415 have other means of support in the shape of cross-growths of hard material, which bind the whole of the structure together. These cross-pieces of the Coral skeleton remind us of the cross platform-like pieces which we saw in the red Organ-pipe Coral in a previous article; but the shape of the skeleton is altogether different.

In these two forms the " Fan " is all in one plane. We can imagine that Nature wished to prevent the straggling appearance shown on the next page, and betook herself to cross-junctions in order to steady the whole colony. The most economical way to do this was to press all the branches out into one flat plane, and thus the cross-bars became of the shortest possible length.

Now we see that however much the young shoots may cluster around one another at what answers for the root—or, rather, we should say, the place of attachment to the rock—they quickly flatten out as they grow older. Just imagine for a moment all the horizontal pieces of Coral removed, and the loose stems would be swept all over the rock around. In fact, they have adopted in lowly animal life a system which, if adopted by the higher forms of vegetation, might prevent many a noble tree from biting the dust after a storm of great severity. We can see, also, that although the Fan-Coral is able to sway forward and backward fairly freely owing to its horny build, it is given additional strength and rigidity from the fact that no part of it can bend in any direction individually without affecting the whole. And its fan-like build causes it to be almost unaffected by currents or sea-bed swells which strike at right-angles to the principal plane.

To see these creatures in a living state, we must imagine them as being completely clothed with a thin layer of vivifying flesh. The flesh is external and the bone internal, and this is, as far as possible, the complete opposite to the arrangement obtaining in the crabs and other crustaceans, in which the flesh is inside and the skeleton outside.

The flesh of this order of Corals covers the inside hard corallum as the bark of a tree covers the woody trunk. It is the living principle of the animal. You will see no clefts in the hard part in which the polyps might have lived had they been like the Madrepores, for the clefts were, in the latter, actually in the skeleton itself. All these Gorgonian Coral-polyps were seated in little clefts which opened out from the flesh, so that when the flesh had decayed there remained nothing to show where they had formerly lived. We have only the internal axis left behind.

The order to which the Fan-Corals belong has an additional claim to beauty over and above the order of the Brain-Corals and the Madrepores, in that every one of the eight tentacles which always surround the mouth of each of these little sea-anemone-like polyps is beautifully fringed. The tentacles are always eight in number. The whole organism is a compound one, and all the polyps intercommunicate by means of a network of canals which permeates the fleshy bark and is joined to each of the polyps in succession.

Nature is not slow to cover up man's work, and so when a broken submarine cable is brought up it is found to be generously coated with these flexible corals among other things.

Photo by] [*H. C. Shepstone.*

THE SEA CANE.

In this species the slender flexible branches are unconnected by cross branches, and are very long and like canes.

THE SUPERB LYRE BIRD

BY SIR HARRY JOHNSTON, G.C.M.G.

It will scarcely be denied that one of the world's marvels is the Lyre Bird of Australia, and that being the case, and Australia having been colonized by people of Anglo-Saxon race, the readers of this work will not be surprised to learn that every possible attempt is being made to exterminate this largest and most remarkable of Passerine birds.

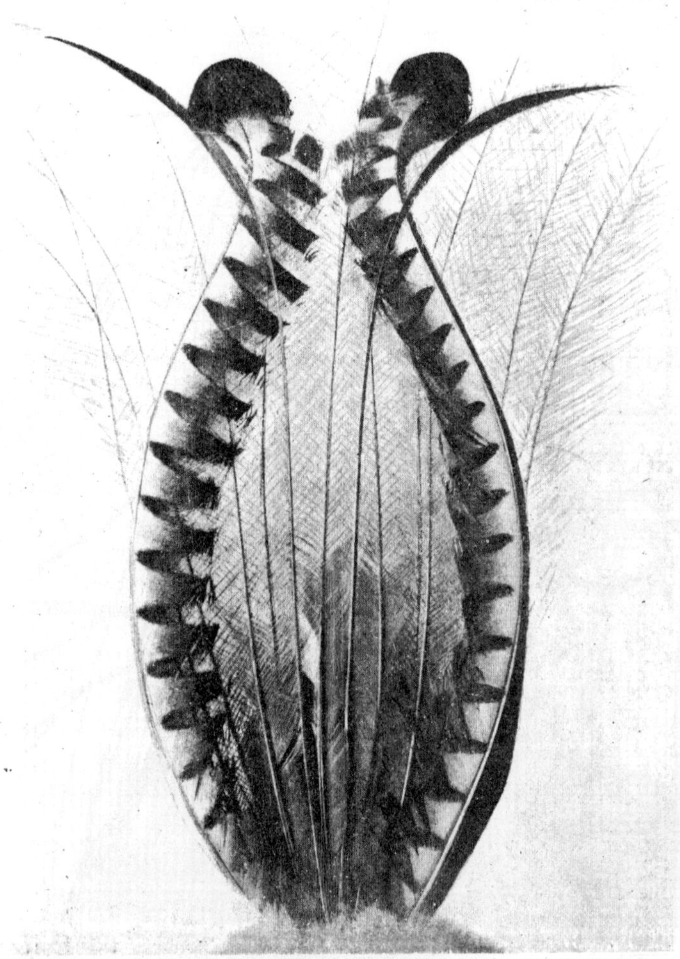

Photo by] [*W. Saville Kent.*

TAIL FEATHERS OF LYRE BIRD.

When the Lyre Bird walks or runs this exquisite arrangement of feathers is carried horizontally; but when showing off to fascinate the hens, the tail is erected as shown in the photo.

There are three species of Lyre Birds, all of them restricted in their range to the eastern and southern parts of the island-continent of Australia. The largest of these three species, and the one here illustrated, is the Superb Lyre Bird of New South Wales, which is thirty-three inches long in the adult male. The colour is a dark umber-brown, which becomes russet on the throat, while the back has a tinge of purplish-grey. The naked space round the eyes is lead-blue in colour, and the two superb lyrate feathers of the tail are a rich chestnut-brown, marked on the broad inner side of the web with notches, which are semi-transparent. Though the Lyre Birds are unmistakably members of the great Passerine order, they exhibit features in their anatomy and habits rendering them peculiarly interesting as survivors of early types, which were connecting links assisting to bridge over the interval between the birds of more or less gallinaceous type, on the one hand, and the cuckoos, woodpeckers, and singing-birds on the other. Lyre Birds live mainly on the ground, and prefer to run and jump rather than to fly, though quite able to use their wings if they wish. It is said that with their strong legs they can leap to a height of ten feet from the ground at a single bound. They carry the tail horizontally when they run, and with their strong feet scratch amongst fallen leaves and dust, much after the fashion of pheasants and domestic fowls. But although the male Lyre Birds make the utmost use of their strange tails in showing off for the fascination of the female, they are apparently husbands of only one wife, before whom they strut and whirl and dance with erect tail and drooping wings—pecking at the ground as they do so and uttering loud gurgling sounds. The nest is built with a dome or roof, and is constructed either

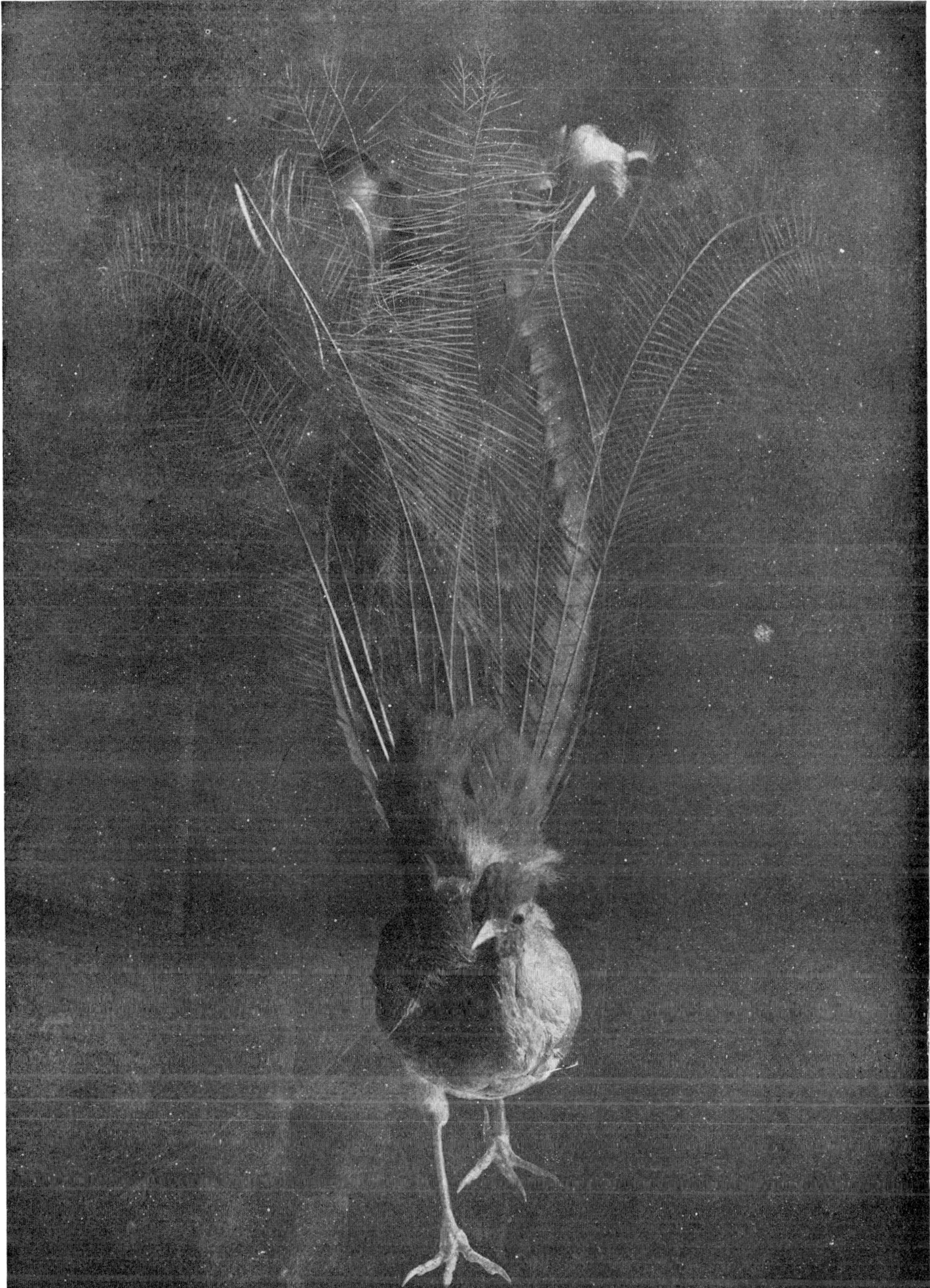

By permission of] [*Sir H. H. Johnston, G.C.M.G.*

THE SUPERB LYRE BIRD.

The Lyre Birds are all restricted in their range to the south and east of Australia. The Superb Lyre Bird, shown in this photo, is peculiar to New South Wales. It is a male, that sex alone having this magnificent adornment.

on a ledge or in a cranny of the rocks, or in the fork of a low tree, or amongst fallen logs. The materials they use are sticks, bark, grass and leaves, and the inner part of the nest is lined with moss, fibres and the bird's own feathers. Only one egg seems to be laid, which is a purplish-brown in colour, marked with spots and streaks of a darker tint. They are very clever mimics of the sounds they hear around them, and are birds which might easily be tamed, domesticated and bred in captivity. It is little else than a disgrace to the young nation of Australia that so marvellous a form should be brought so near to extinction.

THE DRIVER ANT

BY LOUIS P. BOWLER, F.C.I.

THE large black species of Ant common to the African jungle is generally known as the Driver Ant, for the reason that every living thing makes room for these ants to pass; for if by any chance a mouse or an elephant stands in their road, it is forced to speedily move on or remain and be eaten up. The Driver Ants will not budge an inch from the direction they are bent on travelling; no obstacle will prevent their advance but fire or water. They reign supreme as conquerors amidst the inhabitants of the jungle, and are universally respected in the insect and animal world, commanding admiration for the resolute manner in which they carry out the

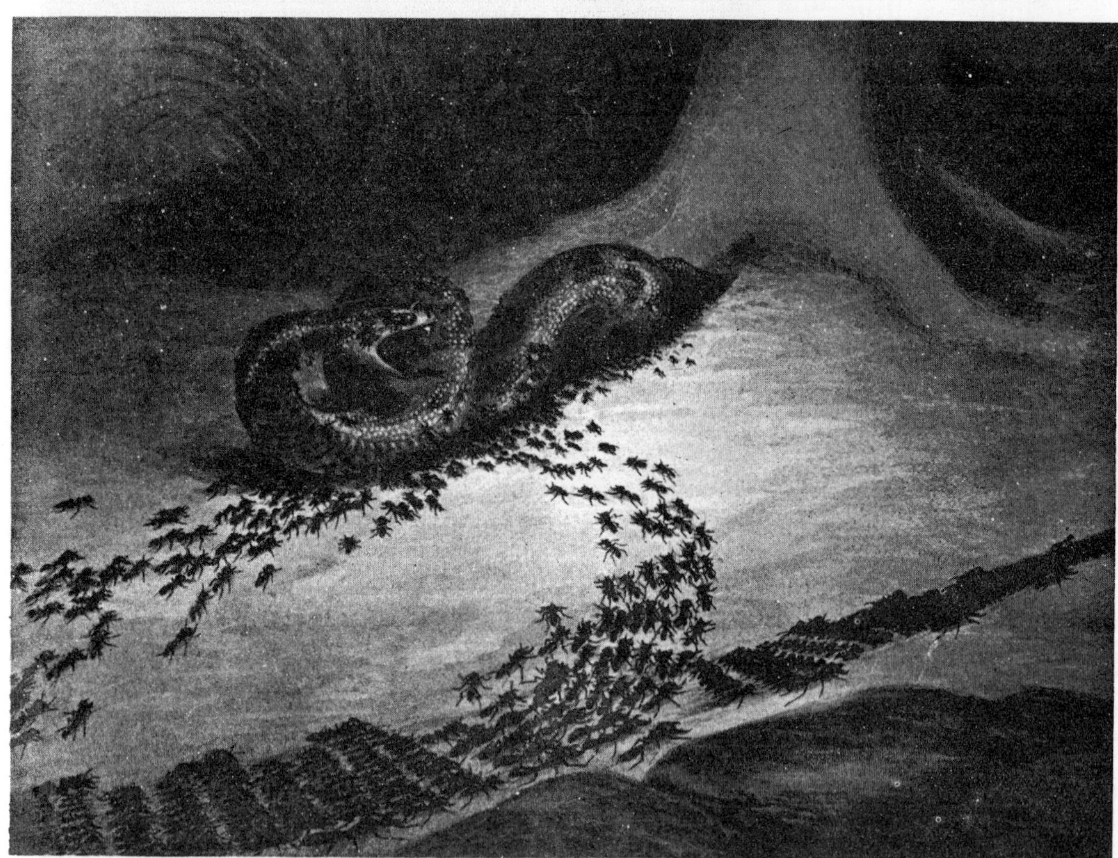

[*By Theo. Carreras.*

DRIVER ANTS ATTACKING A SNAKE.

The snake is the Horned Viper, and it was attacked whilst casting its skin. The Ants covered every portion of its body, hanging on by their pincer-like jaws. The snake writhed and struggled for a quarter of an hour, but in the end was killed and finally eaten by the Ants.

[By Theo. Carreras.

ANTS TUNNELLING A DRY WATER-COURSE.

These Ants being accustomed to take a certain route, which was liable after rain to be crossed by a stream of water, took the precaution to tunnel this portion and so safeguard themselves against being swept away by a sudden rush of water.

scavenger's duty allotted to them in Nature's burden: a work they share with the Red, White and Sugar Ants as sextons and sanitary inspectors, checking the spreading of disease in the animal and vegetable kingdom, and adding enormously to the improvement of the soil.

In Africa, just before the rainy season sets in, it is a common occurrence to see long lines of Driver Ants on the march. They rapidly advance in perfect formation, in lines two to three inches wide, and often extending for nearly a mile, with an organization and military precision that is truly marvellous, as shown by their well-drilled battalions led by their officers. It is interesting to watch the various sections pass at the double, first a column in close formation, consisting of warrior Ants possessing formidable-looking nippers, led by one or more very large Ants, twice the size of the others; behind this column follows a mixed mass of smaller Ants. These are the carriers, carrying various loads: some bring eggs, others their young, bits of leaves and sticks, grass, berries, dead bodies of insects, and general goods and chattels from their storehouses and towns.

All trot along with a determined swaggering air, as if conscious of their irresistible strength by sheer force of numbers. They tear along, tumbling like a cataract over obstacles, and run over one another or not, as the procession winds in and out of the paths in the forest, looking, especially in the moonlight, like a long black snake unfolding itself from the jungle growth.

Their sagacity and engineering aptitude are truly remarkable, as is seen by the wonderful tunnels they construct, generally under the exposed portion of the road on their line of march, where there is any danger of a torrent washing down through a sudden downpour of rain. I came across one of these tunnels at Subilsu, near Tarkwa, which measured thirty-two feet in length, and within this length I found several perpendicular air, emergency, or escape shafts, measuring eight, fourteen and eighteen inches deep, sunk below the surface. They were placed at a little distance from a position where a possible wash might occur.

The carrying strength of the Ant is out of all proportion to its size. I have often seen an Ant struggling along with a beetle a dozen times its own weight; but this feat sinks into insignificance with the extraordinary powers they possess of communicating with one another. In watching their movements, I have noted the rapidity with which the Ants forming the rear sections of a procession become aware of any interruption to their advance column a long way ahead. They appear to have a system of instantly telegraphing the news up and down the line; for all the fighting Ants become excited, and rush forward to give battle or to investigate the cause of the interruption.

[By Theo. Carreras.

A BALL OF ANTS.

A huge black mass of sleeping Ants hanging to the roots of a tree in the shade, seen by the author in Africa.

I tested this on one occasion by placing a friend one hundred yards ahead in their procession, and told him on a given signal to break their line, while I timed with a watch how long it took before a commotion transpired amongst the Ants passing where I stood.

In ten seconds from the time my friend called out, all was bustle and excitement; the carriers halted, but the warrior Ants rushed forward, showing clearly that they were aware of trouble ahead.

While I was travelling a bush-path near Bensu, in the Gold Coast Colony, during the dry season, my natives suddenly warned me to quit the path and make a detour to avoid an open, dry, sandy spot that lay in the road, saying, "that a big palaver of Ants was taking place, as they were electing a new king on the throne" ("stool," they called it), "and if we disturbed them they would flog us bad" (meaning "bite"). The ground in front of me was black with Ants, and streams of them were pouring in from all sides. The noise they made rustling over the leaves and ground sounded like the swish of a scythe. In the centre of this living mass was a small open space, encircled with rings of Ants, all facing outwards with their nippers held up. There were several clearly defined rings,

[*By Theo. Carreras.*

A GENERAL ASSEMBLY OF ANTS.

This "big palaver," as the natives termed it, was seen by the author in the Gold Coast Colony. A small open space in the centre was encircled by rings of Ants, all facing outwards with their nippers held up. They appeared to be guarding the centre, where there were a few Ants of a lighter colour and evidently of greater importance.

quite stationary, the Ants appearing to be guarding the centre, in which were half a dozen larger Ants of a lighter colour than the rest. These evidently were important individuals. I became so absorbed at the sight of this magic circle that I failed to notice the hundreds of these creatures who were investigating my body; some penetrated to the calves of my legs, others climbed to my neck, and then at a given signal they all bit at once. I had often heard of this peculiar habit of biting all together, but never experienced it before. Needless to add, I beat a hasty retreat and tore my clothes off, to find these tigerish creatures had buried their pincers into my flesh, and had to be scraped off.

On another occasion, at midday, I noticed a huge black mass hanging to the roots of a shady tree. An inspection proved this to be a mass of Ants. They were clinging to one another in this bunch, fast asleep.

I once witnessed a fierce fight between a cerastes (horned adder) and a swarm of Driver Ants. The snake had evidently been attacked while shedding its skin, and was struggling frantically amidst a dense mass of Ants, who covered every portion of its body. They were furiously hanging on with their sharp pincers, and every second a fresh Ant fixed its grip. The snake writhed and squirmed about amongst its implacable enemies for fully a quarter of an hour, and then its struggles became feebler, and it stretched its length out on the ground; in a very few minutes its body was covered two inches deep with Ants tearing and cutting away its flesh.

[*By Theo. Carreras.*

TWO IRISH ANEMONES.

The larger of the two species here shown is Richard's Anemone, with a pale green column studded with rounded warts, and white tentacles. The smaller species shown on the rock above and also to the extreme left is Haddon's Anemone, with flesh-tinted column, and pale tentacles with dark base. (Natural size.)

SOME SEA ANEMONES

BY EDWARD STEP, F.L.S.

THERE is a pretty constant tendency to name the newly-known according to its resemblance to the well-known, and much confusion has arisen from the practice. As a case in point, take the wonderful Sea Anemones. The earlier observers were impressed by their superficial resemblance to certain flowers; so when they admitted them into their schemes of classification and had to invent a name for the group of hitherto nameless things they manufactured from the Greek lexicon the word *Anthozoa*, which means "flower-animals." As an English equivalent they invented Sea Anemones, which served to fix the floral idea, and different species were named with the names of particular flowers. Minor poets wrote of them as flowers of the sea, and so, even to-day, there lingers the idea that they are as much vegetable as animal, if not more so.

SEA ANEMONES. [*By Theo. Carreras.*

The species in the upper part of the picture is the Gem Pimplet, a flesh-tinted Anemone with pellucid tentacles barred with black and grey. The lower figure is the Dahlia Wartlet, a large and showy species, common on our shores. (Natural size.)

If the Sea Anemones, with their beautiful tints and petal-like tentacles, are not flowers, what are they? To give an idea of their true nature in popular language, we may describe them as coral-polyps that deposit no coral. Coral-polyps are united to form colonies of many individuals, but Sea Anemones (with a few exceptions) are individually separate.

The majority of visitors to the seaside know only one kind of Anemone, the kind that may be found in plenty on the rocks between tide-marks, and as the waters have receded they are seen only in a closed and more or less flaccid condition. To see the Anemones in their full expanded beauty we must peer into the clear water of the rock-basins that are left full when the tide goes out. Some of them never venture so far up the shore as to get into

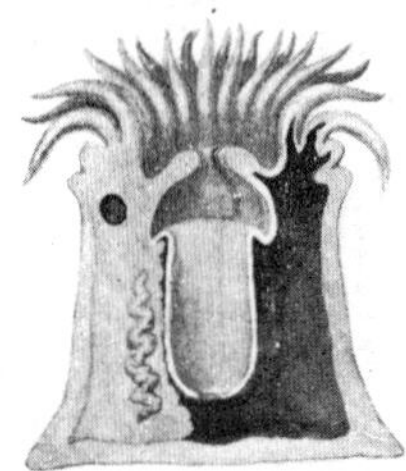

A SEA ANEMONE.

An ideal section to show the general plan of the interior parts.

these pools, and must be sought on the rocks that are never uncovered by the tide ; but there are quite enough of different varieties found between tide-marks to afford a correct idea of their beauty, their structure and their habits. But before glancing at a few of these forms, let us consider what the Sea Anemone is in general.

Roughly speaking, it is a hollow cylinder of soft flesh, closed at both ends, with a spreading base which adheres closely to the rock, much as a slug's body adheres to its support. Within the cylinder a number of plates of varying breadth jut out from the walls into the cavity, dividing it into a number of open compartments. When the Anemone is fully expanded, the top of the cylinder is seen to have a slit-like opening, which is the mouth of the animal, and around it, spreading out much like the petals of some flower, are the tentacles in a single row or several rows one within another. These tentacles serve the Anemone as hands. They close upon any small living creature that comes in their way, and by the emission of barbed microscopic threads from their surface kill and secure their prey at one operation. Pushing it towards the mouth, the lips part and the victim is received into the gullet, which is a bottomless sack opening into the general cavity. As the food is digested, it falls into this cavity, and is absorbed by its walls, the undigested portion being ejected through the mouth.

The common kind of Anemone to which we have already referred as the best-known species is named the Beadlet, from the fact that just below the outer row of tentacles is a series of bead-like spheres of a brilliant blue. A narrow line of blue also runs round the base of the column. Commonly,

[*By Theo. Carreras.*

THE ARROW MUZZLET AND THE CUP-CORAL.

The only British Coral is shown in the upper half of this picture, and to the right of it are seen two specimens of the limy corallum left on the death of the animal. Below is the Arrow Muzzlet, which buries its column in the sand. (Natural size.)

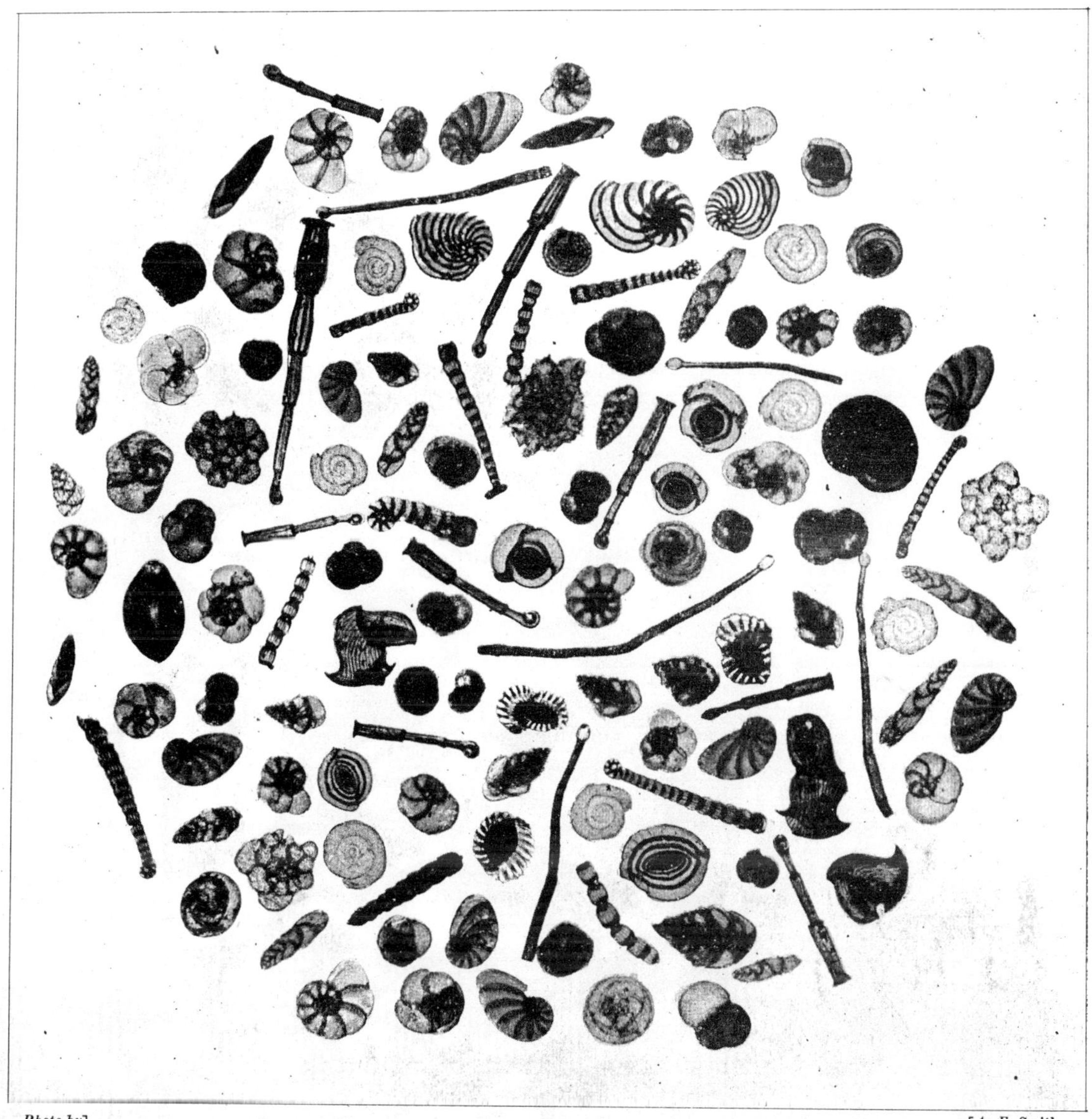

Photo by] [*A. E. Smith.*

A GROUP OF FORAMINIFERA.

This group of minute shells under the microscope shows a few of the numerous forms of Foraminifera. They are magnified about twenty-five times. The entire group would thus go comfortably into a space no larger than the uncut end of a drawing-pencil.

the Beadlet is liver-coloured, or dark red, but it is one of the most variable of species, so far as colour is concerned. The specimens that are found on the rocks when the tide has receded are not full grown; in later life the Beadlet prefers to migrate below tide-marks, where it may always be submerged. The tentacles are hollow, and so it is easy by withdrawing water from them to pack them into a small compass; the same, indeed, may be said of the entire animal. Children know this fact in their natural history well, for they like to amuse themselves by touching these exposed Anemones in a way that irritates them and causes them to squirt out water that they may shrink into a smaller space. In some of the lighter-coloured species, such as the Gem Pimplet and the Dahlia Anemone, this hollowness of the tentacles may easily be seen, for they often keep their young in them until they are sufficiently developed to be cast out into the waters.

These infant Anemones are exceedingly graceful and active creatures, constantly changing their forms and positions, roaming about on the rock until they have found a spot to their taste upon which to settle. They are alternately spherical and worm-like, with all sorts of intermediate shapes. The worm-like shape is assumed for purposes of locomotion, the base retaining its hold until the tentacles have laid hold at a distance ; then the base disengages itself, the column is drawn up to the tentacles, and may again lengthen and take hold in an opposite direction.

The Opelet, which is one of our largest species, is a very inconstant Anemone—always shifting its quarters, and found as much, perhaps, on the broad fronds of the oar-weed as on the rocks. At least, this is the case when it has attained to full size, and it may frequently be found floating tentacles downward, and its broad base hollowed out like a boat. This is a very graceful species, with

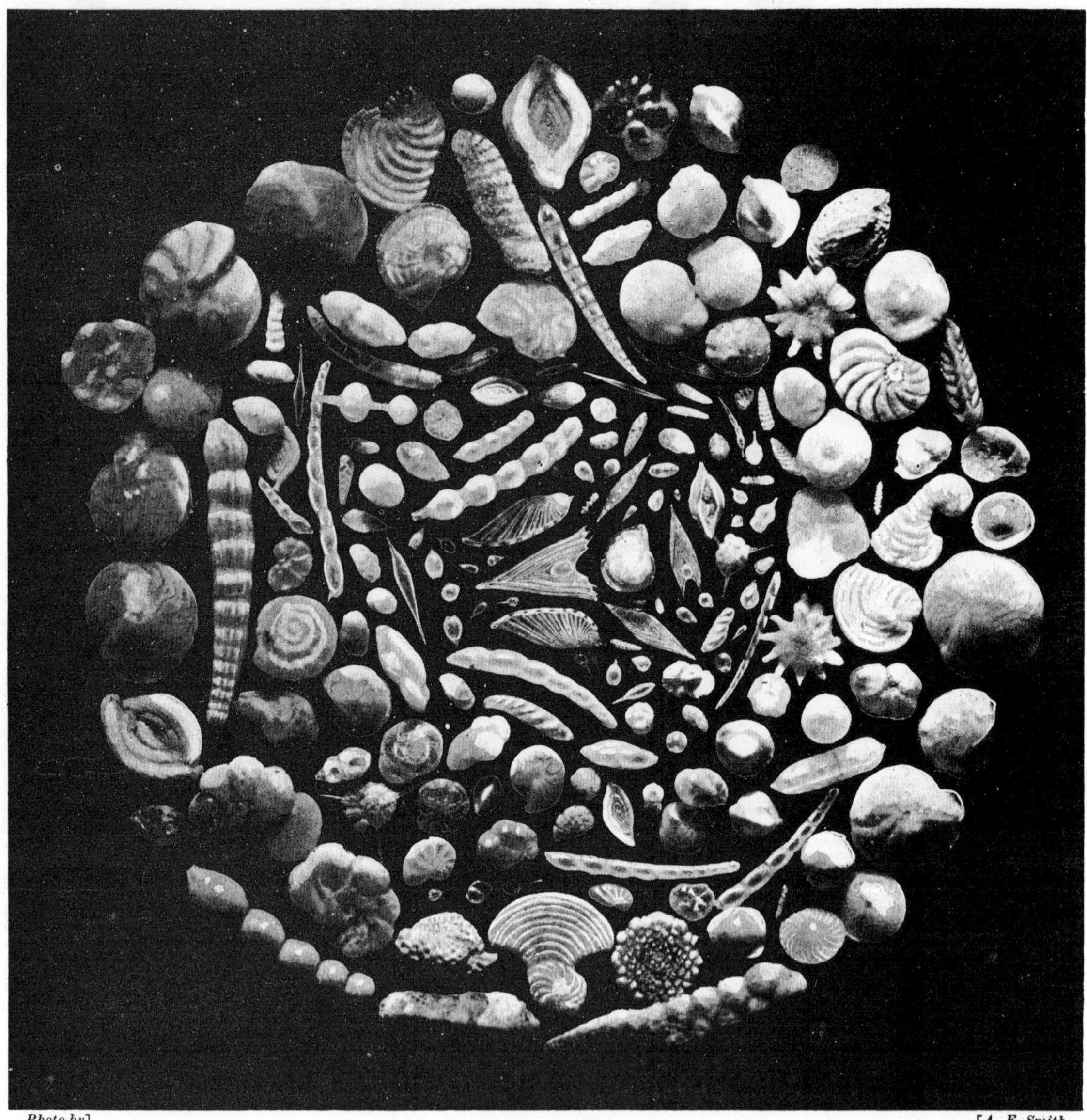

Photo by] [*A. E. Smith.*

FORAMINIFERA.

Another group of varied forms of these wonderful examples of minute life. The scale of magnification is about twenty-five times larger than the actual size—that is to say, the real size of the entire group is one-fifth of an inch across its broadest part.

a multitude of long, tapering tentacles. The broad-based column is flesh-coloured or lilac, whilst the tentacles are mostly bright green, with pink or crimson tips. In Provence this Anemone is converted into a table delicacy known as "Rastegna."

The Dahlia Wartlet is our most showy species, its banded thick tentacles, with their clear tints, making a strong contrast with the crimson disk. The column is mostly hidden, for it is covered with suckers to which bits of shell and small shingle adhere, so that when the tentacles are withdrawn the Dahlia looks like a small rounded heap of gravel. There are many interesting species we should like to mention, but space forbids. For an idea of some of these we must refer the reader to the illustrations in these pages, and to the coloured plate.

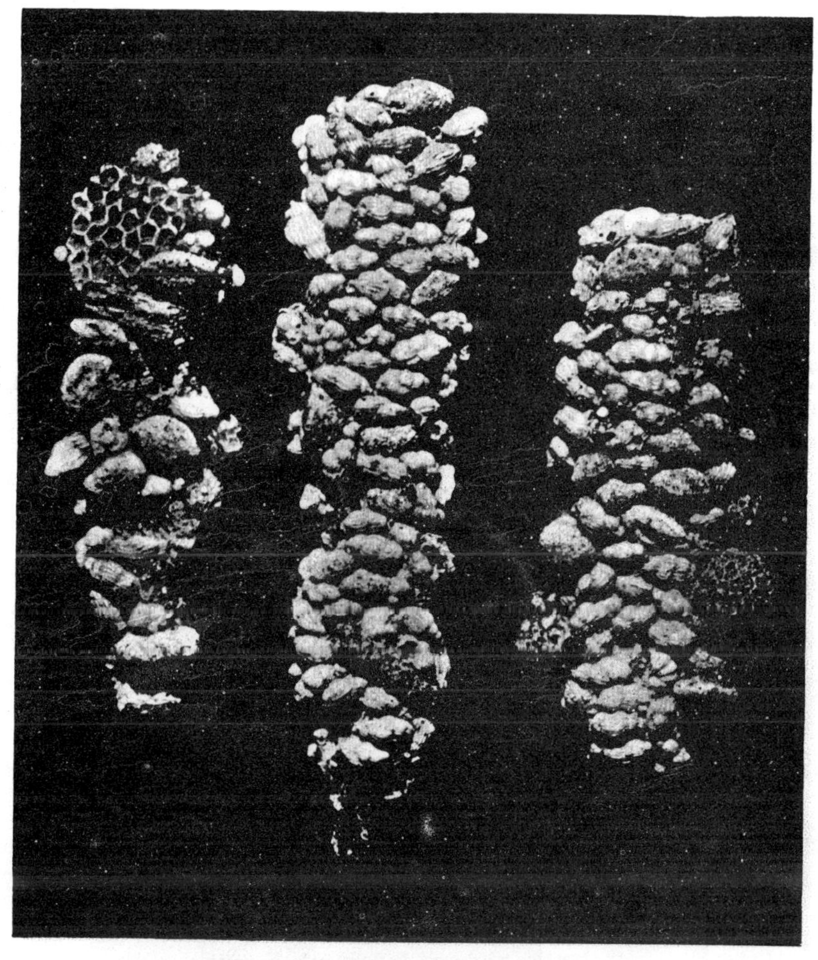

WORM TUBES BUILT OF FORAMINIFERA.

These Foraminifera have been picked up by a minute worm and cemented together to form a secure retreat. They are shown twenty-five times larger than the natural size.

FORAMINIFERA

BY RICHARD KERR, F.G.S.

THESE are small marine creatures, which possess very beautiful shells composed of carbonate of lime. The shells, however minute, are perforated throughout to allow numbers of tiny feelers to protrude. These feelers may be used both as a means of locomotion and for drawing in food particles. Because of the minute perforations in the shells the name Foraminifera is given to them as a whole, however varied their shapes may be.

"About the size of a pin's head" is a common expression, used so frequently to describe the size of any very small object that it has almost become a sort of mental unit of measurement. It may very well hold good for some of the shells under consideration, but for myriads of them the pin's-head dimensions would convey an exaggerated idea of their actual size. Suppose the pin's head to be hollowed out, it would readily form a receptacle for a score or more of some forms known to students of micro-marine life. This shows us that the powers of the microscope must be employed to enable us to see them to advantage by unfolding the beauty of their structure. We must not lose sight of the fact that there are several great families of Foraminifera much larger than the pin's head, such, for example, are the nummulites, or coin-shaped fossils, to which reference will be made.

Some of the shells consist of a single compartment ; others of a number of compartments arranged one after the other in a single line, which may be straight or rolled on itself as a spiral. There are others with cells or chambers arranged in double or triple rows. Others have their minute compartments fashioned around invisible axes, if we may so term them. Now if we imagine the first cell to be an exceedingly small one and the next a little larger, and so on as the gradual additions are formed, until the most recently added cell is the largest, we shall have an idea of a structure that resembles a pyramid with its graduated divisions. This is a very characteristic form among the " Forams." We might proceed in this manner to describe different structural arrangements of the cells, and we should find at the end of a long period of time that we had by no means exhausted all the different types and shapes assumed by these tiny marine organisms.

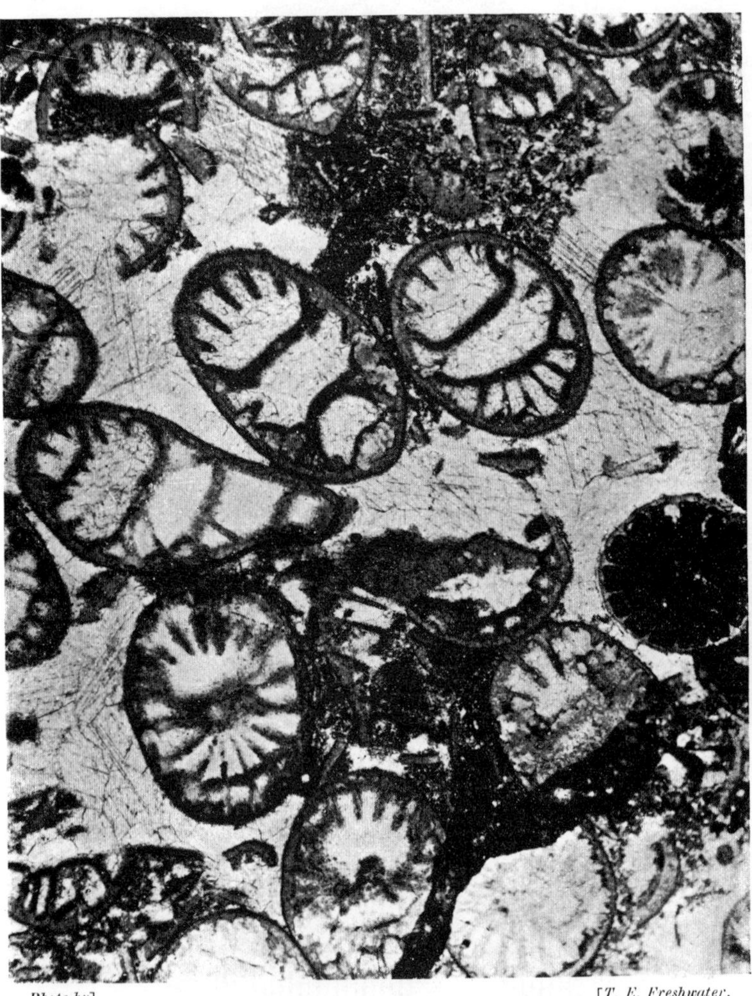

Photo by] [*T. E. Freshwater.*

FOSSIL FORAMINIFERA.

This is a section through a piece of limestone largely consisting of Foraminifera. Many important buildings are erected of such stone, and the Pyramids of Egypt are formed of nummulites which are a large form of Foraminifera. (Magnified about fifty times.)

With very few exceptions, which are doubtful, the body proper of the creature is continuous through all the different compartments in each shell. This is a fact which is very marvellous when we consider that scores of cavities or chambers are present in nearly all these little shelly homes.

In the many wonderful processes involved and agencies utilized through countless ages in the building-up of this great world, the " Forams " occupy a prominent position, owing to the extensive results of their existence. They are a tiny people, but they have left behind them, and their modern representatives are still leaving, a record on the side of the elevating influences probably second to no other forms of life. Their work goes on, and is likely to go on as long as the world lasts. In point of actual numbers, it is not at all unlikely that they are the most numerous agents that are employed in the construction of the earth.

Even if we omit all reference to the geological formations of America, Asia and Oceania, we shall find abundant evidence in the rocks of Europe and Northern Africa of the tremendous record

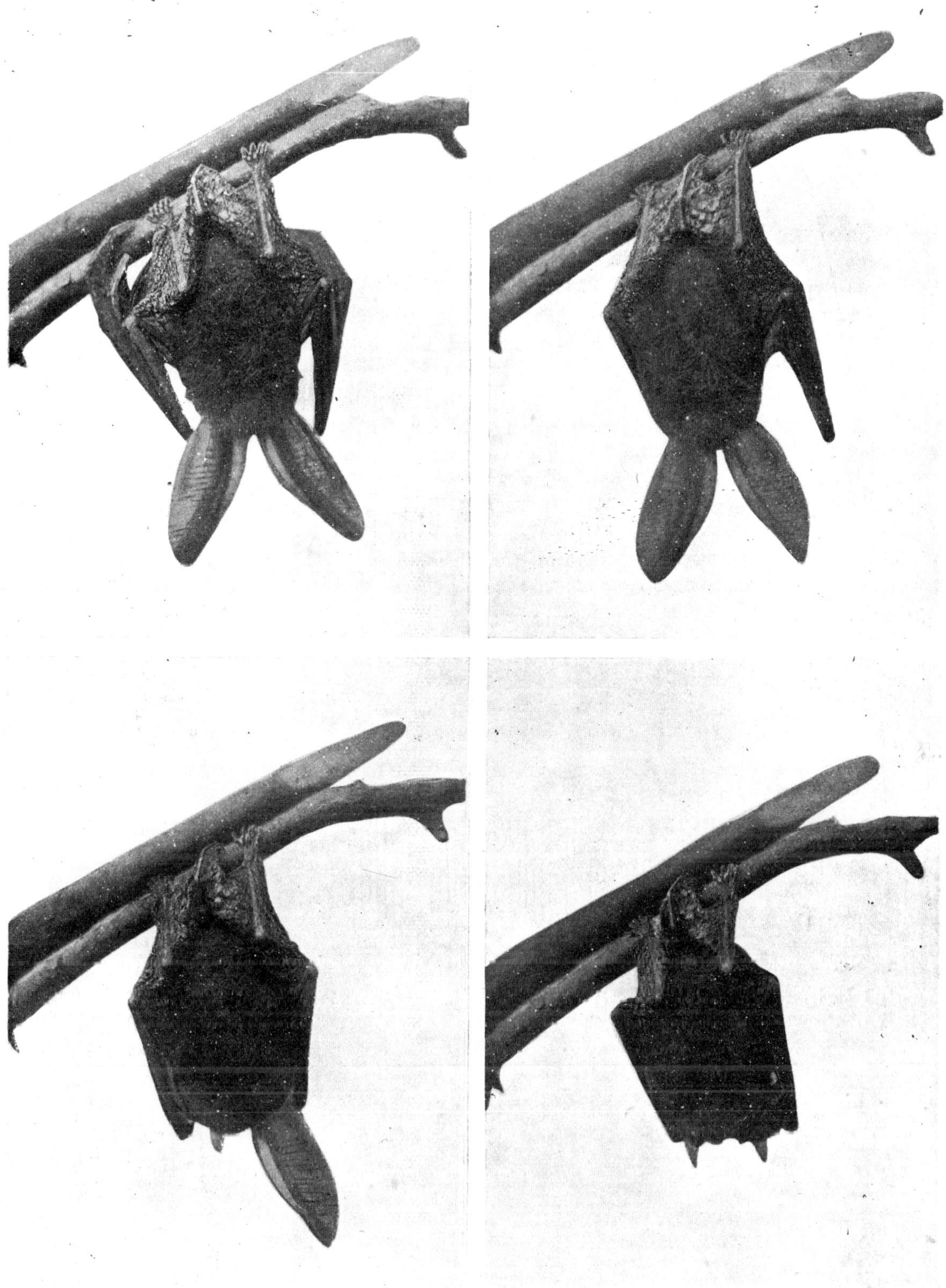

Photos by] [*J. J. Ward, F.E.*

THE LONG-EARED BAT PREPARES FOR REST.

Having taken secure hold, head downwards, the Bat folds up his wings and tucks them closely to his side. Then first one long ear and then the other is likewise folded and tucked away, until little resemblance to a Bat is left.

of the building powers of the Foraminifera. A superficial list will be sufficiently convincing: The chalk formation extending along the south coast of England, hundreds of feet in thickness; the similar formation on the French coast; the Tertiary rocks of the extensive geological formation known as the "Paris basin"; the corresponding strata of Austria and Italy; the calcareous deposits of Russia, and the rocks comprising the Pyramids of Egypt consist, for the most part, of shells of the Foraminifera! They are present in vast numbers among the sands on the seashore. An ounce of sand, taken at random from the shores of the Adriatic Sea, contained six thousand beautifully formed micro-shells quite as attractive as that of the Pearly Nautilus. D'Orbigny, closely examining the sands on the shores of the Antilles, calculated that there must have been nearly four millions of these shells in a single ounce! A row of figures representing the number contained in a cubic yard would baffle our powers of comprehension. That being so, we must admit that the Foraminifera form a population gigantic beyond all means of intelligible expression; for they are found on all seashores; the successive accumulations of their shells form banks that are dangerous to navigation, and they assist in conjunction with corals in building up reefs and islands in tropical seas. They constitute the chief organic deposit on the floor of the Atlantic Ocean. This was proved when the first cable was being laid to America. Samples of "ooze" consisting almost entirely of shells and fragments of shells of Foraminifera were brought up. Under microscopic examination they were found to bear a close resemblance to the fossil forms of which the chalk cliffs of Dover and other localities are composed.

With regard to the rocks which form the principal building material of the city of Paris and the neighbouring towns and villages, we are told by D'Orbigny that a cubic inch of this stone contains fifty-eight thousand shells of Foraminifera!

[*By H. Whanslaw.*

THE BOMBARDIER AND THE GROUND BEETLE.

The little Bombardier Beetle, pursued by a Ground Beetle with evil intentions, repulses the enemy by ejecting an acrid fluid which becomes converted into a vapour by contact with the air. The emission is accompanied by sharp little reports.

THE LONG-EARED BAT

BY JOHN J. WARD, F.E.S.

In the whole history of life on the earth probably no animal has possessed proportionately larger ears than those of the Long-eared Bat, which is one of the most grotesque of animals. It is only occasionally seen in flight during daylight. Then it becomes a most curious object indeed, for during flight it is quite unlike any other creature of the air. Its huge ears are erected and lowered continually in a most extraordinary manner, and being almost as long as its body, they give it a very striking effect when seen against the light. In the natural order of things, however, this Bat does not fly until very late at night. This habit, I think, throws some light on the purpose of its huge ears.

I am inclined to think that it hunts by sound, and that its ears serve as huge ear-trumpets, by means of which it can detect the faintest sound-vibrations, even those of the flight of gnats and other small insects on which it preys.

Photo by] [E. Step, F.L.S.

INDIAN STICK INSECT.

In form and colour this "Stick" is much like a small branch of a bush. It is one of the wingless species, and is here shown not quite the natural size. The remarkable eggs shown on the next page almost invariably produce female insects.

While holding this species of Bat in my hand I have heard it screech with terror, and its voice is a very razor's-edge of sound; so shrill is it that it is almost inaudible to the human ear. Yet with that voice it calls its mate from afar. Probably, too, its late habits in the silent night adapt it the better to detect such delicate sounds. Thus, when it rests it folds up its ears that it may not hear.

THE BOMBARDIER BEETLE

Among the methods adopted by insects for the purpose of discouraging enemies who would make a meal of them, is the production of objectionable odours and fluids. Thus, large numbers of the great Bug family evolve the disgusting odour that reaches its highest development in the repulsive Bed-bug. The pretty beetles known as Lady-birds, if molested, exude from their joints a yellow, oily fluid that is objectionable to birds, and advertise the fact by their red and black coats.

Various species of Ants eject a corrosive fluid known as formic acid, and the group of small beetles known as Bombardiers have a somewhat similar defence.

The last segment of the Bombardier's body bears a couple of glands which secrete an acid fluid into a pair of little bladders. When the Bombardier is pursued by a larger beetle or other enemy, muscular pressure appears to be exerted upon these bladders, with the result that there is a cracking report and a little of the acid is ejected and turned to a bluish vapour on contact with the air. The action may be repeated several times if the first explosion does not serve the purpose of repulsing the enemy.

The Bombardier is a small beetle, found chiefly along the shores of rivers, where it is often pursued by the larger Ground Beetles which are carnivorous, and therefore the Bombardier stands in some need of special protection. Such a pursuit and repulse is shown in our illustration, where both Beetles are shown twice the actual size.

Photo by] [*F. Noad Clark.*

EGGS OF STICK INSECT.

The eggs are shown about nine times larger than the real size. They have a close resemblance to seeds. Each is provided with a lid, which the newly-hatched "Stick" pushes off, as shown in one of the upper figures.

STICK INSECTS

BY K. G. BLAIR, B.SC., F.E.S.

STICK Insects afford a beautiful example of those animals which in their bodily structure, combined with the habit of remaining motionless, bear so close a resemblance to some inanimate object of their natural surroundings that they very readily escape notice, and consequent consumption, by their enemies. As their name implies, they resemble sticks and twigs of the foliage in which they usually dwell; if disturbed, the resemblance is made yet more close by their habit of folding the limbs close to the body with the forelegs straight out in front and feigning death; it is then difficult to realize that one is dealing with a living insect and not with a piece of dead twig. In connection with this habit should be noticed the shape of the front legs. At the base they are very slender, but about the middle of the first joint they thicken rather suddenly and then taper away to the foot. When they are held straight out in front in the proper "stick" attitude, the head fits exactly into this hollowed part and the eyes are able to look out over the base of the leg, while the continuity of the stick-like body remains uninterrupted.

Stick Insects are found throughout the warmer parts of the world, but the kind here illustrated comes from India. Though many of its relatives possess wings, this species is wingless in all stages. They are nocturnal in their habits, remaining motionless among the twigs of their food-plant during the day. They walk rather slowly, and before coming to rest have a peculiar habit

THE SOLAR HALO.

The silvery sheet of Cirro-stratus covers the sky, and a shining circle of light is formed by the sun's rays passing through the ice-crystals which form the cloud. The sun forms the centre of the circle, and is just under the edge of the dark cloud.

Photos by] [*G. A. Clarke.*

THE THUNDERCLOUD.

The Cumulus-nimbus is the grandest and most massive of all clouds. The one shown is commencing to dissipate in a heavy shower, as may be seen at the left of the picture.

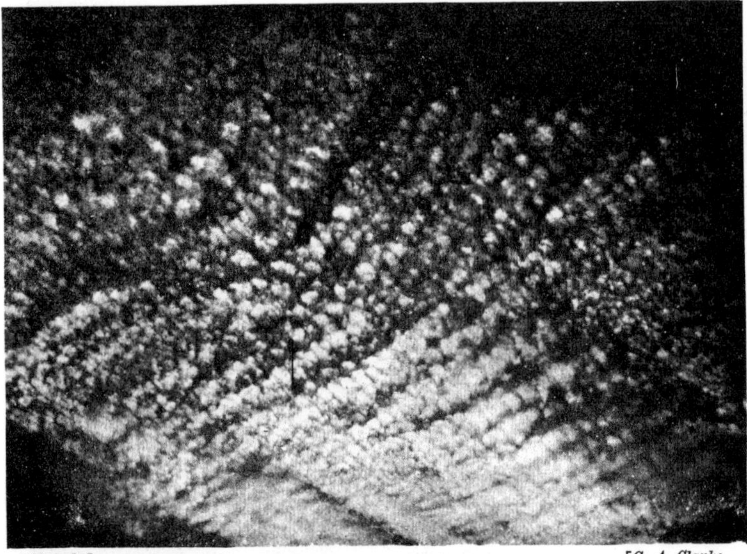

Photo by] [G. A. Clarke.

CIRRO-CUMULUS CLOUD.

The marvellous beauty of the Cirro-cumulus is well shown here. It is, as a rule, an accompaniment of fine weather.

of swinging gently from side to side. It is a curious fact that the male of this species is excessively rare (indeed, the writer knows of only one example), but the propagation of the species seems to be in no way impaired thereby; generation after generation may be reared, but all are females.

The eggs are simply allowed to fall to the ground. They are no less curious than the insects themselves, and bear an exceedingly close resemblance to the seeds of some plant of the gorse tribe. They are little shortly oval bodies, with a lid-like structure at one end, in the middle of which is a little yellow button. When the egg hatches the lid is pushed off and the young Stick Insect drags itself out, slowly straightens its back and limbs—for in the egg it was all curled up—and presently wanders off to look for something to eat. It is very like the adult in form, but not so long-drawn-out, and does not at first assume the stick-like attitude. It feeds on privet, rose and a few other plants, and after moulting several times becomes mature in about five or six months.

While young they possess the power of growing new legs in case of accident, but the process occupies three or four moults, and if the loss occur too late in life no attempt to replace it is made.

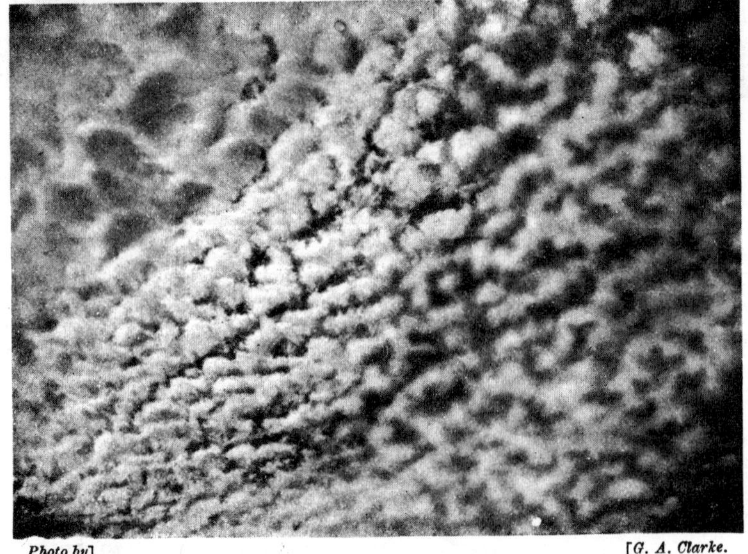

Photo by] [G. A. Clarke.

ALTO-CUMULUS CLOUD.

This is another variety of the Cumulus clouds shown in the previous illustrations.

THE CLOUDS

BY GEORGE AUBOURNE CLARKE.

Of all the marvels that are to be found in the universe, probably none can surpass the clouds for beauty of form and colour; and whether it be the scarlet pageantry of the sunset or the purple gloom of the thunder-cloud, there is always in the picture before us a sublimity that compels our admiration.

Even the most casual observation cannot fail to bring to our notice that the forms assumed by the clouds are many and various; that, for instance, the swiftly changing and passing shower-clouds of springtime are quite different from the beautiful wavy bands which we call a " mackerel sky." The natural association of certain cloud-forms with certain kinds of weather demanded some means of distinguishing between these different forms ; and, though space does not permit of our here entering into the history of cloud nomenclature, it will be sufficient to say that in 1891 an International Meteorological Conference authorized the publication of an atlas and classification of cloud-types for the guidance of observers. This classification has been generally accepted, and the names here mentioned are in accordance therewith.

Clouds owe their origin to the sun. The sun's rays heat the ground, and the air lying immediately above the ground thus becomes warm and charged with evaporated moisture, and this

Photo by] [G. A. Clarke.

LENTICULAR FORM OF CIRRO-CUMULUS.

It is to the Lenticular form of the Cumulus clouds that we owe many of our most gorgeous sunset effects.

warm, moist air, being lighter than the colder air above it, naturally rises upward. When this ascending air-column reaches a height where the temperature is sufficiently low, the water-vapour which it contains condenses into very minute drops of water, and these then become visible as a cloud. A simple illustration of this process is to be seen in the familiar puffs of steam from the exhaust of a locomotive. This condensation is repeated at successively higher levels, and consequently we meet with clouds at greatly differing heights.

Highest of all float those delicate filaments resembling feathers or wisps of hair, popularly known as " mares' tails," but in cloud classifications called *Cirrus*. Twenty-nine thousand feet is the average height of such clouds, practically the same as that of Everest's snowy peak. At such a great altitude the temperature is far below freezing-point, and these clouds are therefore composed of ice-crystals. Sometimes the Cirrus becomes sufficiently extensive to cover the whole sky with a pearly or silvery sheet, and is then known as *Cirro-stratus*. When this occurs, we frequently see bright rings of light encircling the sun or moon at a radius of about twenty-two degrees. These are called *halos*, and are caused by the ice-crystals refracting the sun's rays. Halos are popularly regarded as portents of coming storm or wet weather, owing to the fact that large quantities of Cirro-stratus are usually found in the front of an approaching atmospheric depression.

Next in order of height come the *Cirro-cumulus* and *Alto-cumulus* types, and as a rule they accompany fine weather. The general characteristics of these two types are similar ; but while the Cirro-cumulus may be as high as twenty-three thousand feet, the Alto-cumulus may be found as low as ten thousand feet. These clouds are extremely beautiful when stretched across the sky in long bands and fleecy masses, exhibiting a wavy, rippled structure, or marshalled in parallel lines of tiny cloudlets, and the fancied resemblance of the ripples to the markings on a mackerel has earned for

Photo by] [*G. A. Clarke.*

LENTICULAR FORM OF ALTO-CUMULUS.

This again is a form of cloud much in evidence during fine sunsets. The edges of the cloud reflect intense colours, which form a fine contrast with their denser parts and produce those wonderful effects that are so fascinating to the beholder.

them the name of " mackerel sky." One particular form of these clouds—the *lenticular* one—gives rise to our most beautiful sunsets ; for, when the last rays of the sun impinge upon their rippled under-surfaces, the whole sky seems transformed into a sea of shining gold or fiery crimson.

Still lower down, at about six thousand five hundred feet, is found a similar but much heavier cloud, called *Strato-cumulus*. This is a type frequently to be seen, and is sometimes heavy enough to cover the whole sky with its dark, wavy masses.

About the ten thousand feet level there is occasionally seen a uniform, greyish sheet called *Alto-stratus*, through which the sun shines dimly. It is the " watery sky " of popular weather-lore, and is almost always followed by rain. Owing to its uniformity it cannot be photographed effectively.

The *Cumulus*, or " wool-pack " clouds, are probably the best known of all clouds, and consist of

Photos by] [*George Aubourne Clarke.*

CUMULUS CLOUDS.

Of this fine set of Cloud photographs the first shows the dark wavy masses of Strato-cumulus covering the sky; the second and third present us with Cumulus in two of its finest aspects: while in the fourth a Cumulo-nimbus, or Thundercloud, may be seen piling up its enormous mass.

dome-shaped masses of condensation with flat bases. When they pass over the sun they appear as dark masses with bright edges. The average height of their base level is about four thousand feet, and that of their summits about six thousand feet. They are quite plentiful during summer, and are a fine-weather type.

Unfortunately this last remark cannot be applied to the *Cumulo-nimbus* clouds, for they appear at their best during thunderstorms, and likewise give rise to the sharp hail- and rain-showers of springtime. These clouds are really an over-development of the Cumulus type and have a similar base level; but they continue to heap up mass upon mass of cloud, till their summits reach almost as high as the Cirrus. Imagine a pile of cloud three or four miles high, and containing twenty or more cubic miles of condensed vapour! It is fascinating to watch the growth of their fantastic forms, which continues until conditions become unstable, when the whole huge fabric collapses in a torrential "thunder-shower."

Photo by] [*E. J. Wallis.*

THE TELEGRAPH PLANT.

The two small leaflets, at the base of the large leaflet, are in constant movement, revolving on their stalks. The large leaflet moves up and down only, like the arm of a railway signal.

The *Nimbus*, or rain-cloud, is a mass of cloud without definite form, from which steady rain falls. It is sometimes accompanied by small detached pieces of cloud which float below the main mass, and to which sailors have given the name of "flying scud." Its height is six thousand feet or thereby.

Stratus, the lowest of all clouds, is simply a sheet of lifted fog, found anywhere up to about three thousand feet; and, like the Alto-stratus, does not lend itself to photography because of its uniformity.

THE TELEGRAPH PLANT

THE telegraph referred to in the name of this plant is the old semaphore apparatus with which, before the application of electricity, messages were signalled from hill to hill. It survives in the signal-posts and arms of the railways by which sections of the line are indicated to be free or blocked according to whether the arm is depressed or horizontal. The suitability of the name will be manifest to anyone who watches the behaviour of the plant. It is a native of India, Ceylon, etc., and belongs to the great pea family. Like so many other members of that family, its leaves are broken up into three leaflets, of which one is enormously larger than the other two, which look like mere stipules, one on each side of the leaf-stalk. These small leaflets are in almost constant movement, and this quite independently of any movements in the air. In this country, of course, they can only be grown in hothouses, and there, where the air is quite still, the movements can be seen to best advantage during the hours of sunlight. The leaflets revolve, so to speak, on their stalks, their movements describing an ellipse. Sometimes they move very gradually, sometimes by jerks; sometimes repeated up-and-down movements take the place of the elliptical progression. The movement of the large leaflet is not so obvious, as it maintains a more or less horizontal position during the day; but at night its tip always points towards the earth.

Now these movements are perfectly spontaneous; that is to say, they are not due—as in the case of the Sensitive Mimosa—to external stimulus or irritation. It is safe to say that they serve some important purpose in the economy of the plant. You may try to influence their movements in the ways that are adopted towards the Mimosa, but the Telegraph Plant pays little or no attention to your efforts. The plant bears upright spikes of violet flowers, much like those of the lupine.

A BEAUTIFUL SHRIMP

BY E. J. SPITTA, F.R.M.S.

THE Crustaceans are an important division of animals, to which crabs, lobsters, crayfish, shrimps and prawns belong, and which are characterized by having an external skeleton of shell in many pieces. They are mostly aquatic, and all, even the terrestrial forms, breathe through the integument, or through the projections of the integument which are known as "branchiæ," or gills. Their integument, or covering, is made of a peculiar substance called "chitin," which is also the chief ingredient of the wing-cases of insects. It is of great hardness, resisting almost any ordinary means of cutting or biting, as witnessed by the hard shell of the lobster or the tough skin of the shrimp or prawn. Its actual formation as a covering skin is a mystery at present, and its power of resisting alcohol, ether, water,

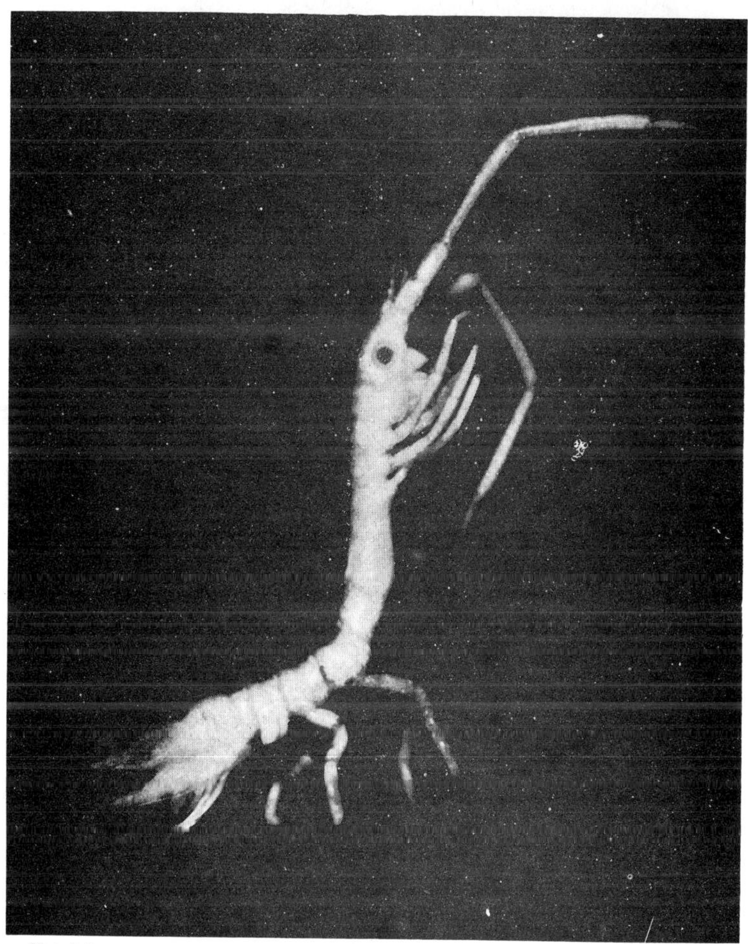

Photo by] [*E. J. Spitta.*

A BEAUTIFUL SHRIMP.

Arcturus, as this little creature has been named, is noted for its beautiful pink colouring and its bright blue eyes. It is here shown in the characteristic attitude which it takes up on a branch of coralline, after the manner of the Mantis on a land plant, and with a similar purpose.

acetic acid and alkalies single it out as a very curious substance. So resisting, indeed, is the material that if shrimps or prawns, for example, be softened out by treatment in various chemical solutions, the original form of the little animal is elegantly and perfectly maintained notwithstanding.

There are an almost endless number of animals belonging to this group, and some have different habits from others. Their formation, too, varies, some having eyes projecting on little stalks, like lobsters, whereas others possess eyes, called "sessile," those which are attached to the body without any sensible projection or visible support.

Our present illustration is an example of a curious little crustacean, known as Arcturus, to which one or two interesting peculiarities are attributed, although it must be distinctly acknowledged they are not universally accepted as facts. The Arcturus is seen in the first illustration reared up on its hind legs, and it is said that this is the position in which it watches for its prey; whereas another account attributes to this little member of the sea a peculiar habit of fretting such things as the wood of piers or harbour supports for a purpose only known to itself.

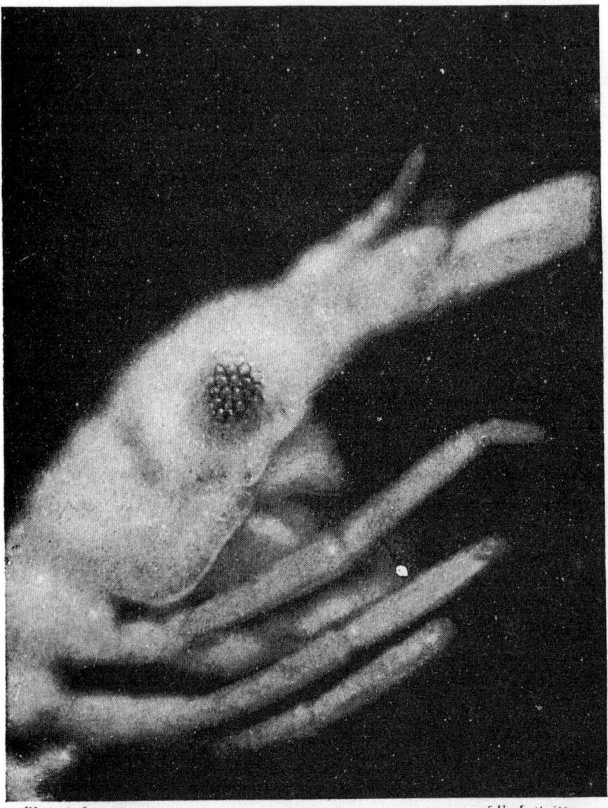

Photo by] [*E. J. Spitta.*

A BEAUTIFUL SHRIMP.

Some of the Crustaceans, such as the Lobsters, have eyes projecting on little stalks This Shrimp has "sessile" eyes, that is to say they are set close to the body without any appreciable projection.

The most interesting features, however, of this little animal are to be found really in its beautiful pink colour, its shining skin, and its remarkably beautiful blue eyes, each being more like a mulberry than an organ of vision, and one of which is shown in the second illustration considerably magnified.

THE DODDER

WHEN wandering over heath or downland in summer one may often come across a furze-bush over which, to all appearance, someone has been industriously twining and entangling a few hundred yards of crimson thread. Here and there upon the threads are thick knots, half an inch or more across. This tangle of threads is really a living plant, leafless but not flowerless, for what appeared at a little distance as a knot is a dense cluster of tiny bell-shaped blossoms. This is the Dodder.

Now there is one thing that will at once strike the observer who is told that the Dodder is a plant, however ignorant of botany he may be, and that is, that it is remarkable for a plant that bears flowers to have no leaves; yet search the interminable threads from end to end and you will find no trace of them. Another thing worth notice: trace any of the lower threads of the tangle downwards and you will find that it has no roots—it does not even reach the ground. Then, perhaps, you conclude that, like mistletoe and many tropical orchids, it began life on the trunk or branch of its host, where a seed may have caught in a crevice. But that is not so.

The Dodder seed must fall to the ground before it germinates and becomes a seedling. Most seedlings emerge from the seed provided with a pair of fat leaves of different shape from those they

Photo by] [*H. C. Shepstone*

FLOWERS OF THE DODDER.

In this photo, which is three times larger than nature, several clusters of the flowers are shown. The two stout upright stems must not be taken for those of the Dodder: they belong to its victim. The Dodder stems are the much more slender, twining ones.

Photo by] [*E. Step, F.L.S.*

THE DODDER.

The fine threads are the leafless stems of this plant-parasite. In this common species, which attacks heath and furze, the threads are crimson, in others they are pink or yellow. (Natural size.)

put forth later. These seed-leaves are, in effect, a couple of pockets filled with as much food as the baby plant requires until it has developed the beginnings of a root system and a leaf or two that can gather food from the atmosphere. Not so with the Dodder. It comes of what the Eugenists would call " degenerate " stock, the thriftless, unemployable class, doomed from their birth to be parasites and criminals. The Dodder is one of them. For ages all its ancestors have got their living at the expense of the industrious plants and never done a hand's-turn to provide their own food. Of course, if you go far enough back you will find honest and industrious ancestors.

The Dodder seed, on germination, produces a mere thread without any food-provision. That thread bends from its base and sweeps its limited horizon, its tip boxing the compass against the sun. If during these revolutions it chances to touch against a suitable plant, its future is assured ; if not, it dies within two or three days of its birth. But if it touches the stem of a furze- or heath-plant the thread at once begins to twine round it, from north to west or south to east, and makes its hold secure by means of little suckers. Then it drives root-like tubes into its victim, and through these sucks as much of its victim's juices as it requires. It revels and riots in the good things of plant-life, without troubling to inquire whether the poor furze can spare it all. Its threads lengthen rapidly, and before long it indulges in a display of flowers.

These flowers, if examined curiously, will reveal the honourable stock from which the Dodder's degenerate ancestors came. It is the beautiful family of the Bindweeds—the Convolvulus of our cornfields and hedges, the Morning Glory of our gardens—for the flowers, though diminutive, bear all the family features.

SEA GOSSAMER

BY RICHARD KERR, F.G.S., F.R.A.S.

" SEA GOSSAMER " is an apt description of these frail creatures which float in the waters and spread out their fairy meshes of fine, transparent threads. So beautiful are these tubular jelly-fishes, and so unlike any other living organism, that they defy comparison, though the description of an expert on the Sea-Comets approaches nearly to the truth, and may be considered typical of many of these jelly-fishes. "Imagine, if you can, a spiral axis of flexible crystals, about which is a garland of hundreds of diaphanous pearls of a vivid red colour and an infinity of pendants full of life and light and motion, glistening among the waves like some crowning triumph of the lapidary's art—a royal sceptre endowed with vitality and consciousness—and you have at best a very slight and imperfect idea of the Sea-Comet."

These spangled webs of crystallized dew can elongate and contract amazingly, and a thread of polyps which would stretch out to about three feet can compress itself into a finger's length. Another characteristic is their frailty. The Gossamer cannot be gathered with the hand from the water; it will break into an infinite number of particles, and even when "floated" into a convenient receptacle and raised out of the sea in its confined state it soon loses its beautiful filaments and shrinks to an insignificant size.

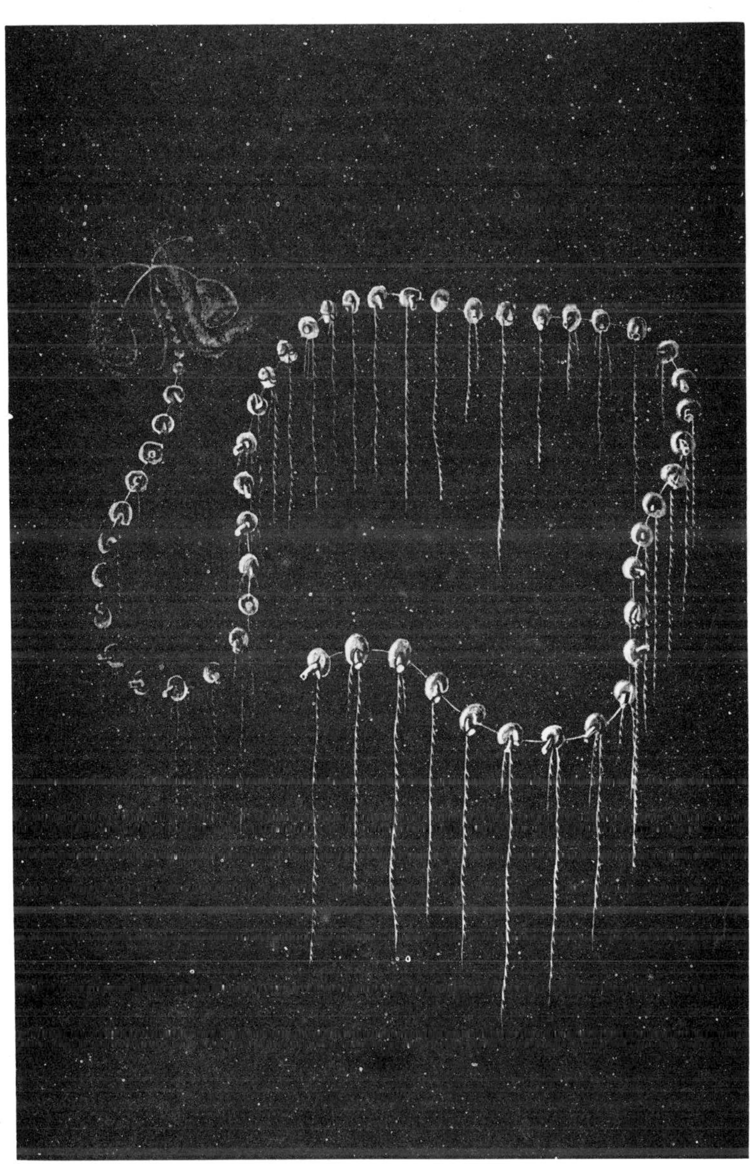

THE PRAYA.

This thread-like Jelly-fish is named after its favourite haunt, the Porta della Praya, Cape Verde Islands. The thread uniting the many polyps is so contractile that three feet of "gossamer" can compress itself into a finger's length.

The principle by which these creatures obtain their food-supply is that of the lasso. As the Sagartians in the Persian army captured their enemies' horses and riders by entangling them in the noose of the skilfully-thrown rope, so these slender jelly-fish catch and seize the minute life of the ocean by means of often-times invisible threads, which, besides, have the power of stinging their prey. Their power of movement, too, is peculiar and is situated in the usually larger swimming-bladders. Roughly speaking, it is a cup-shaped bag lined with muscles, by the contraction of which the water is expelled and the animal in the reaction is urged in the opposite direction.

The members of a large division of these thread-like jelly-fish have been called by naturalists "siphon bearers," on account of their possession of suctorial tubes which serve for the reception and digestion of food, and which are the most important and the most constant appendages of their structure.

All siphon-bearers are free-swimmers and, therefore, each must have a swimming apparatus. This is formed of two different organs—the swimming-bells and the swimming-bladders. The swimming-bell resembles the umbrellas of the larger jelly-fish. The swimming-bladder is an air-chamber. We may consider these creatures as colonies of bells. The bells of each colony are joined together in a great variety of ways and positions. The individual bells appear to be able to act as separate creatures, although united ; and they seem to perform their part in life on the division of labour principle. The duty of one set of bells is to steer or push forward the whole craft ; another set looks after the food supplies, as if commissariat officers ; others produce eggs so necessary for the continuation of the species ; others control the stinging tentacles ; while all take part in the floating of the colony.

The siphon-bearers are possessed also of curious formations which alternate with the polyps along the polyp-stem. These are the " tasters "—flask-shaped and with a long filament attached to the base. What the function of these " tasters " may be has not been determined, but whatever it is, it has nothing to do with " tasting."

THE "FLOATS" OF THE SEA-COMET.

These beautiful clusters of transparent tissues are ranged at intervals throughout the length of the polyp-stem. Each cluster consists of little polyps, "tasters," and reproductive bodies.

Most beautiful of all the attributes of the Sea-Gossamer, however, is its quality of giving out a soft phosphorescent light, illuminating the stem and the congregated polyps with their waving filaments, and making a miniature auroral curtain floating in the sea.

A brief description of the four creatures represented in the illustrations will suffice for the purpose of this article.

The Sea-Comet is one of the siphon-bearers. Its Latin name signifies that its feelers, or tentacles, are devoid of stinging power, and these are single tubules. It is one of the largest of the thread jelly-fish ; its polyp-stem alone attaining to a length of two or three feet. It is composed of transparent tissues, and its system of float-bearers, illustrated on an enlarged scale on this page, is peculiarly beautiful.

The Golden Helmet is not more than three-quarters of an inch long ; it is found in the Mediterranean, the Indian Ocean, and off the coasts of Greenland and Norway. Unlike most of the siphon-bearers, it only possesses one pair of swimming-bells. The polyps are attached to the stem that branches from the junction of the bells, and to each polyp is attached a capturing tentacle.

The Bladder-bearer works upon the same principle as the airship. It keeps to the surface waters of the ocean by means of an air-bladder situated at the extremity of a narrow tube, which supports the creature in any position. Underneath there are two rows of bells which contain water and which pulsate rhythmically, with the result that the animal is moved forward as the water is pressed out. The siphons for drawing in food particles and the tentacles which capture and convey the food to them form a mass of slender filaments below the bells.

The Praya, as its Latin name implies, possesses two swimming-bells. Between the two swimming-bells the end of the long tubular body of the animal is attached by slender tubes,

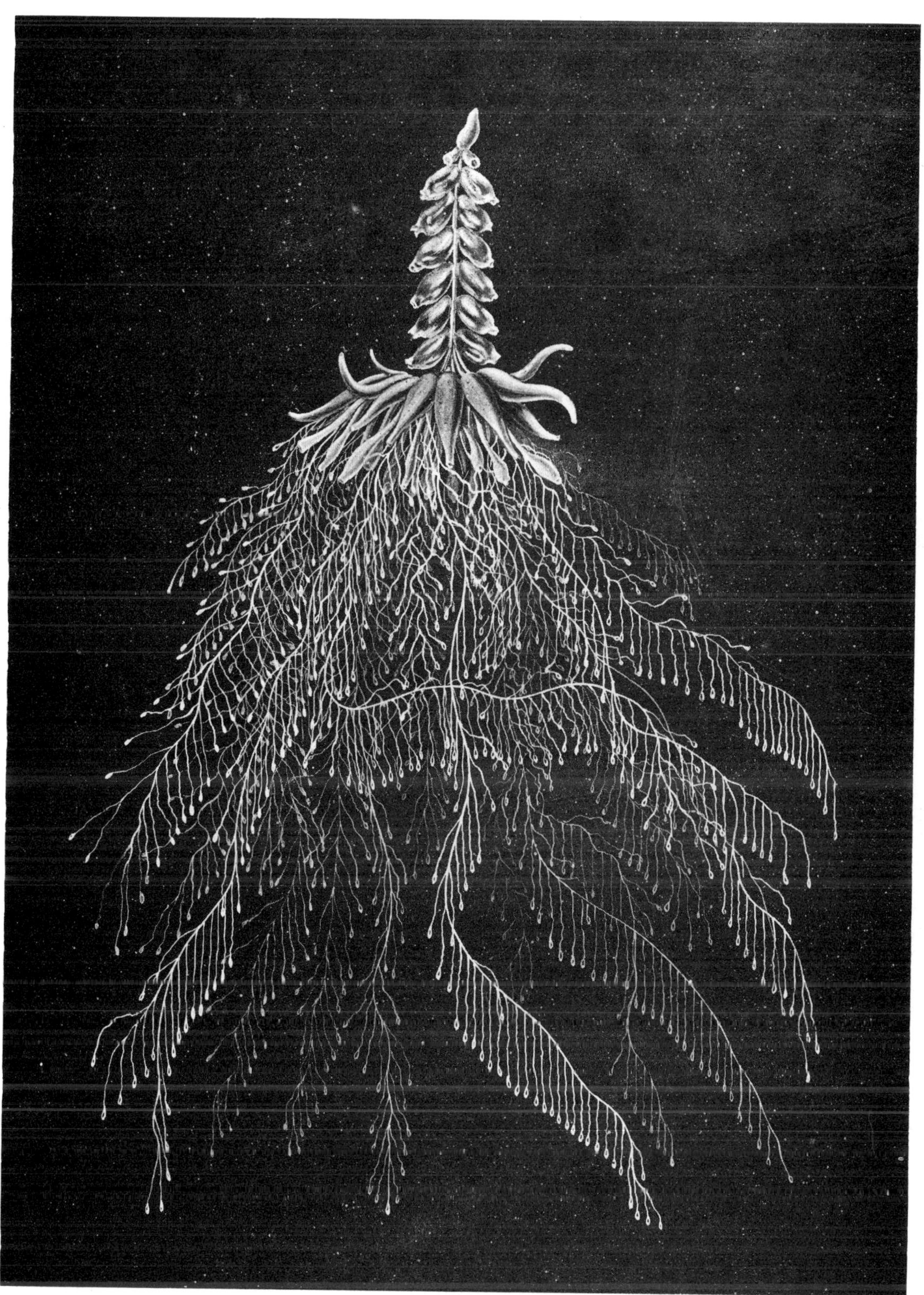

THE BLADDER-BEARER.

The Bladder-bearer, in common with its kin, is literally a host in itself. It is a host of polyps or swimming-bells united to a common system of swimming-bladders. Different functions are assigned to different sets of polyps; some steer the whole craft, others look after food supplies, while others again produce eggs or control the stinging tentacles.

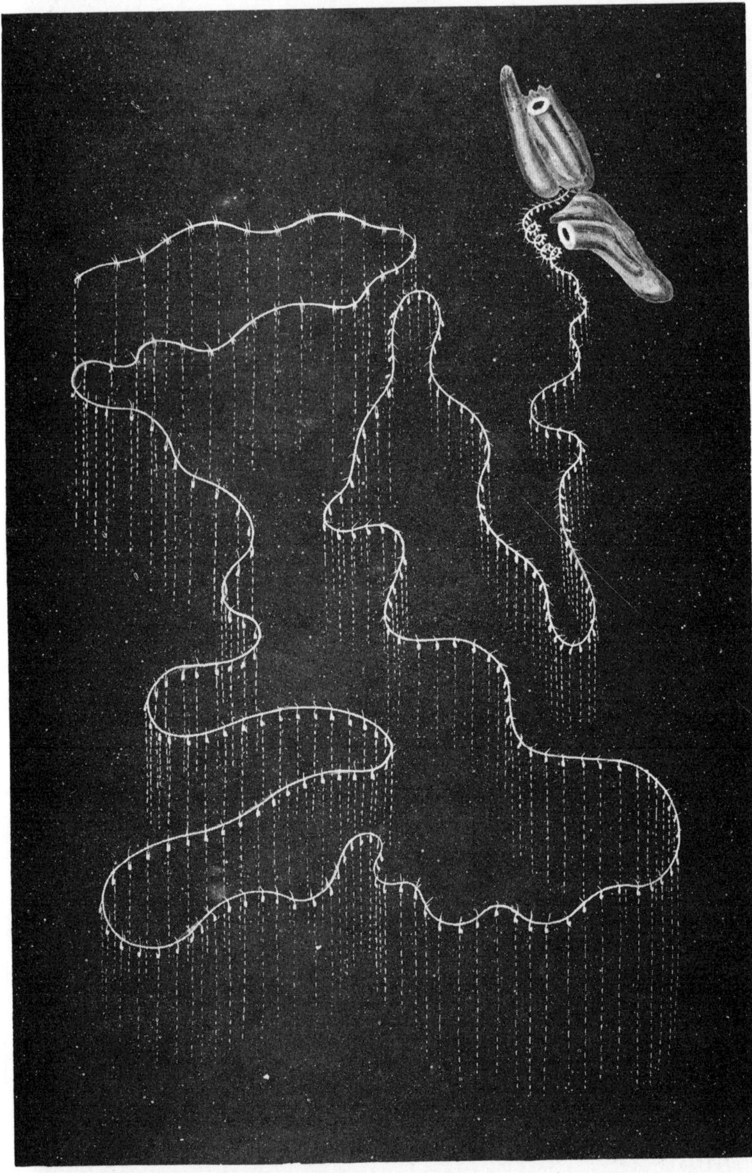

THE GOLDEN HELMET.

So called from the two large swimming-bladders which serve to propel the whole colony of polyps attached to the principal "thread."

through which a nourishing liquid passes along canals into the bells. The movement forward is accomplished by the alternate absorption and ejection of water, so that the bells advance head foremost and regulate the motion of the whole compound animal. The bells also perform the duties of rudder. When they act together the creature advances straight ahead ; but when the right bell is alone in action it goes to the left, and so on.

TOUCANS

BY SIR H. H. JOHNSTON, G.C.M.G.

THE Toucans are a group of birds which have developed very fantastic and, in some cases, most disproportionately large beaks : in fact, the Toco Toucan, which is here illustrated in colour, is probably the culminating marvel in big beaks, proportionately to the size of the bird. In their distribution they are entirely Tropical American, and even here their range is somewhat limited. No Toucans are found, for example, in the West Indian Islands or to the west of the Peruvian Andes. They are mainly confined to the Equatorial region of north-eastern South America and the forested parts of Central America. Like so many Tropical American animals, in fact, their geographical origin seems to have been from the direction of West Africa, and their ancestors probably passed over to Brazil from Africa, together with the parent forms of so many beasts and birds, at the close of the Eocene Period, before the land bridge or archipelago of islands which once connected the Guinea Coast with Guiana or Brazil had broken down.

The Toucans with their big beaks made a great impression on naturalists, and on travellers in South America, during the eighteenth and early nineteenth centuries ; so much so, that when the natural history of Africa was first investigated and explorers encountered a very common type of African bird—the smaller hornbills, or " tocks "—they were wont to style them " Toucans."

This error is very prominent in the earlier travel-books of the great explorer Stanley. He and others frequently mentioned the "Toucans" of the Congo Forest. But these are really small hornbills, and although they bear a strong superficial resemblance to Toucans, there is absolutely no near relationship between the two groups.

Toucans have, as their nearest affinities, the barbets—a family of small and often brilliantly coloured birds, with rather large and serrated beaks, which are found throughout the tropics of the Old and New Worlds, and most abundantly in Equatorial Africa and Asia. Not much more distant in relationship are the woodpeckers; the jacamars and puff-birds of South America; and perhaps the honey-guides of Tropical Africa and Asia. All these groups agree in having their toes placed two and two, instead of three and one. The beaks of Toucans are not only large and sometimes very serrated at the edges (for the purpose of masticating food), but they are usually conspicuous either for their striking coloration or their ornamental carving, or for both these adornments. Yet, despite their great size, they are of cellular construction and very light. The beak in the species illustrated in the coloured plate is not only the largest proportionately in all the group, but is really a beautiful object in its vivid orange colour, with a large blackish spot on the lower mandible. The beaks of other Toucans that are not so huge are more elaborate in their coloration, the upper mandible being a vivid black and orange, and the lower mandible pure white. The beaks of other types may be white, red, green and black, while the plumage of their owners includes many brilliant tints of green, scarlet and orange, besides deep black and pure white. But no Toucan has any bright blue in its plumage; the nearest they get to it in one or two species being a bluish-grey. On the other hand, the bare spaces of skin which appear in some forms on the face may be a brilliant

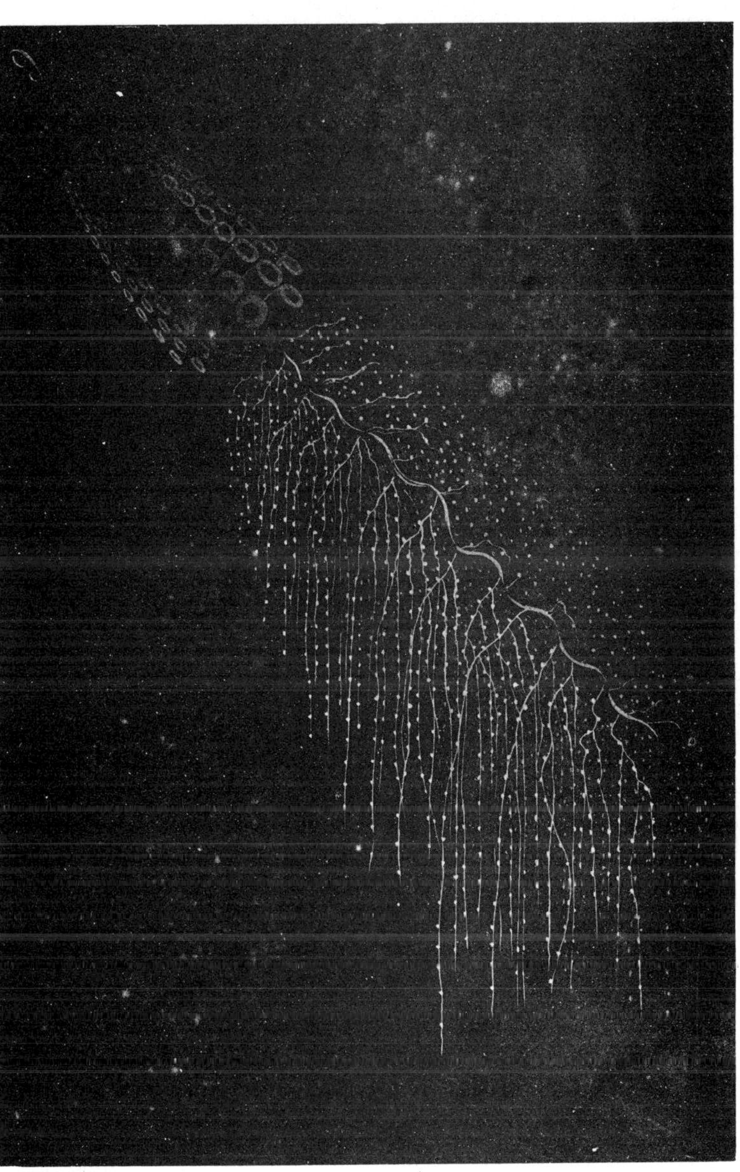

THE SEA COMET.

Very beautiful is this inhabitant of the Mediterranean. Its spiral axis of flexible jewels is often more than forty inches in length, and is garlanded with hundreds of diaphanous red pearls and an infinity of pendants full of light and motion.

cobalt blue, and this is the case (though not, perhaps, to the same extent) with the Toco Toucan of my painting.

The Toco is about the largest of the family, and may be as much as two feet long from the tip of the beak to the end of the tail. But ordinarily the Toucans are small birds, ranging in size from that of a thrush to a turtle-dove. They live mainly on seeds and fruit, but add insects to their dietary, especially during the breeding season, and apparently will also eat small mice and lizards if they can get them. But unlike the hornbills, they do not swallow whole the larger morsels of their food for subsequent digestion in the crop and the muscular gizzard or stomach. On the contrary, they make much use of their beaks in mastication. Sometimes, when the fruit or grub has been swallowed at a gulp, it is regurgitated from the crop after an interval and masticated between the serrated edges of the beak.

Photo by] [*A. S. Rudland & Sons.*

THE CURL-CRESTED TOUCAN.

This curious beak enables its owner to get within reach of the fruit hanging from the branches of the trees, which are too slender to bear the whole weight of the bird. This species derives its name from the feathers on its crest, which have the appearance of steel or iron filings.

The eggs of Toucans appear to be always white, with a glossy surface, and they are laid—two in number only—in a nest which is formed in the hollow or hole of a tree.

THE PELYCOSAUR

BY SIR HARRY JOHNSTON, G.C.M.G.

ALLUSION is made in this work to the order of reptiles represented at the present day by the Tuatera "lizard" of New Zealand. This last is a most remarkable animal closely related to extinct reptiles that inhabited Scotland, Germany and India in very remote periods—at a guess, some six millions of years ago. At one time the Tuatera was fairly common in the North Island of New Zealand, but owing to ceaseless persecution by the stupid colonists (quite unable to appreciate the marvel they had amongst them), it has become almost extinct, and the few survivors only linger at the present day on islets of the Bay of Islands in North New Zealand. The order to which the Tuatera belongs is particularly interesting, because it is rather basic in character and shows us a somewhat generalized type of reptile from which several very divergent orders may have been in time developed, such as, for example, the Tortoises, the Dinosaurs and the Crocodiles. Reptiles of this order often retain in their skulls the traces of an opening in the cranial bones on the top of the skull. Though there is some doubt as to the meaning of this opening and the gland it exposes, it seems on the whole most likely to be the vestige of a third eye which may have been

Photo by permission of] [*Sir H. H. Johnston, G.C.M.G.*

THE PELYCOSAUR

Notice the enormous upward development of the spines, which form a distinctive feature of this extinct reptile

developed in early types of amphibian and reptile and have been of use to them as they basked on the sand or mud and wished to be aware in some way of what was going on immediately above their heads.

A very interesting group of the extinct reptiles of this order was that known as the Pelycosaurs, in some forms of which the spines of the vertebræ were enormously developed upwards, till they gave to the body the huge shield which is so well shown in my illustration—a photograph of the restoration by Herr Pallenberg, exhibited in the grounds of Mr. Hagenbeck's park at Stellingen.

THE OTUMPA METEORITE.

This Meteorite, which derives its name from the place of its fall, is an enormous mass of iron weighing 1,400 pounds.

METEORITES AND AEROLITES

BY EDWARD A. MARTIN, F.S.S.

THAT meteoric stones actually reach our earth from outer space is a fact which everybody now accepts, but we have only to go back a century to find that even scientific people were loth to believe in the possibility of such an occurrence taking place. Those who said that they had seen a meteorite fall were laughed at for being so easily deluded, as it was thought, and this in spite of the fact that from very early times such falls had been seen and recorded. Much superstition had come to be connected with these early falls, and science was not inclined to aid it by recognizing the truth concerning the stones which, as was then thought, had come down from heaven. We alter the term nowadays, and are content to say that they came down from the heavens, including in that term the whole realm of the star-lit sky. An inquiry into certain reported falls which was made early in the last century resulted in the acceptance by Edward Howard of the principle of extra-terrestrial meteorites, and whilst the French philosophers were still discussing the matter a fall, in 1803, of some two or three thousand stones took place at L'Aigle, Orne, and this was followed by an exhaustive report by the physicist Biot, which compelled the whole scientific world to recognize the fact that such stones did actually fall from outer space.

For the earliest record of a meteorite we have to go to the Chinese, for in some of their early manuscripts it is mentioned that falls have again and again been seen, the earliest of these dating from as far back as 644 B.C., but so far as we know the stones themselves have not been preserved. Of those that can still be seen the oldest is that which has been for centuries an object of venera-tion, if not worship, of hosts of Mohammedans, and which is still to be seen built in the wall of the holy Kaaba, at Mecca. Its history goes back to the seventh century. There appears to have been another which fell in Thrace about 470 B.C., and which, five hundred years later, was stated by Pliny to have been still in existence, although all further record of it has since been lost.

Amongst the more famous of recent falls we have that which fell at Ensisheim, in Elsass. A document which is still preserved in the church shows that it fell in 1492, and it was accompanied by a prolonged noise as of a crash of thunder. It was seen by a child to strike the ground, where it made a hole more than five feet deep. It is now kept in the Rathhaus. It weighed no less than two hundred and sixty pounds.

Another of the most famous falls is what is known as the Pallas-iron. Of this a piece weighing about seven pounds is preserved in our national collection at the Natural History Museum. But the whole mass weighed as much as fifteen hundred pounds, and the greater part is now to be seen in the St. Petersburg Museum. This had been seen by a Cossack, at Mount Kemirs, in Siberia, as early as 1749, and was recognized by Pallas in 1772. We are told that it was regarded by the Tartars as a holy thing fallen from heaven. It has a very irregular surface, with large sponge-like pores in which olivine is found.

Many meteorites have a very smooth and almost polished exterior, and of these two massive specimens are to be seen at South Kensington. One, which comes from the Cañon Diablo, Arizona, weighs about two hundred pounds, and it is remarkable that many other meteorites have come from the same neighbourhood, and in some cases they have contained minute diamonds. The

THE CRANBOURNE METEORITE.

This is a giant among Meteorites, and weighs three and a half tons. The rusting of its surface is remarkable, which is caused by corrosion from exposure to the atmosphere.

surface, though smooth, is very irregular, and pieces appear to have been gouged out of it as it was flying through space, reminding us on a large scale of the frost-chippings which are noticed upon many flints. The second meteorite referred to is from Otumpa, in the Argentine, and this weighs even more than the former, turning the scale at fourteen hundred pounds.

These smooth-surfaced specimens are very different in appearance to others in which the surface has readily corroded from exposure to the atmosphere. Such an one is that enormous mass which fell at Cranbourne, near Melbourne, which weighs no less than three and a half tons. The rusting of the surface is remarkable, and small pieces appear ready to flake off at any moment. Unlike the dark though shining appearance which characterizes so many of the meteoric irons, this one is of a reddish colour. It stands in a special case at the Natural History Museum. One can only imagine what were the concomitants of the fall of a mass such as this. It must have been considerably larger when it first sailed into the thin atmosphere a hundred miles above the earth's surface. Then, as it rushed along with a commencing velocity varying from ten to forty-five miles a second, we can see its surface becoming scorched by the great heat evolved by friction with the atmosphere, giving rise to numerous detonations as it compressed the air in front of it into great wave-like disturbances. Then, finally, it became so retarded by the resistance of the air that by the time it reached the earth its initial velocity was lost, and it merely fell, as would an enormous stone dropping from a position of rest on the outskirts of our atmosphere. The air thus acts as a veritable armour, protecting us from bombardment from the outer spheres. But even the striking of the earth at a velocity of but a mile a second might be enough to do considerable damage, and a stone which fell in Finland, in 1901, at a rate, it has been calculated, of five hundred and eighty-four feet a second, made a hole in stiff loam of about twenty-two inches. On the other hand, several stones which fell on the ice, a few inches thick, at Hessle, near Upsala, in 1869, rebounded without damage either to the ice or to themselves.

Photo by] [*H. J. Shepstone.*

THE SELMA METEORITE.

The present weight of this Meteorite is 306 pounds; but it must have been considerably greater when it first sailed into the thin atmosphere one hundred miles above the earth; then its velocity would cause friction that would scorch and wear away its surface.

The greater the trail left behind as a meteor sails across the sky, the greater has been the fusion of its surface. This fusion appears to affect but the thinnest possible layer of the meteor, and with the exception of this thin layer it is intensely cold. It has no heat whatever as it sails through space, and the few seconds of friction with the air is insufficient to heat it to any great extent, since allowance has to be made for the heat carried away from the mass by radiation into the cold air. But the sudden generation of heat causes the fusion and luminosity of the outer layer, and as one layer is carried away as a luminous haze, another layer is acted upon in the same way, and in some cases the trail may develop into a luminous cloud, from which the meteorite is seen

A GROUP OF METEORITES.

The curious shaped Meteorites at the top of the page are excellent examples of the corrosion caused by the fusion of their surface in the transit through the air. The centre Meteorite is of iron, and shows the structure of the mass. The left-hand lower specimen is the Imilac Meteorite, with olivine enclosed, and the right-hand is a typical example of a smooth Meteorite.

apparently to emerge. When the case is one of a stony meteorite, the unequal heating must result in chipping and flaking, and fragments may fly off with a series of small explosions.

Great as the Cranbourne meteorite is, its size has been considerably exceeded by one found in western Greenland, and by another which is still remaining at El Ranchito, in Mexico. The former mass, which measures eleven feet long, weighs thirty-six and a half tons, and was brought from Melville Bay to New York, where it is now preserved. The Mexican mass, which is thirteen feet long, has been estimated to weigh no less than fifty tons. These two meteorites are, in fact, the largest known.

Photo by] [*H. J. Shepstone.*

A STONY METEORITE.

With a Meteorite of this nature, the fusion caused in its passage to earth is unequal, and results in much chipping and flaking of the surface, and these fragments may fly off with a series of small explosions.

When we consider such enormous world-fragments as they were before reaching our atmosphere, we cannot help regarding them as comparable with small planetoids, travelling, as we now know, in an exceedingly elongated orbit around the sun. There is no doubt that they are made up of the same chemical elements as those which have gone to form all other members of our solar system. No new elements have been found in them, whereas thirty-two of our own elements have been recognized in them. Speculation as to their origin has been to a great extent set at rest by the aid of the spectroscope, and there is little doubt that meteorites, shooting-stars, comets' tails and the rings of Saturn are made up of the same materials, whilst the two latter consist of enormous numbers of meteorites whirling around their respective orbits, and possibly giving rise to luminous haze by constant collision and the heat engendered by it.

Careful scrutiny of the paths of well-known comets has shown that they are identical with those of well-recognized meteor-showers. The inference is at once seen that meteor-showers will occur when the earth passes through a comet's tail or the path along which a comet has passed. This

Photo by] [*Hugh Main, B.Sc., F.E.S.*

THE BINSITTA MOTH.

This beautiful moth, with glistening white wings dappled with brown, emerges from the remarkable chrysalis shown below.

is known to be the case. Apparently the tails of some comets have gradually lagged behind until portions of them extend along the greater part of their orbit, and when the earth crosses that orbit shooting-stars are seen.

It would be fruitless to catalogue here the many falls that have occurred. Our national collection has nearly six hundred specimens, of which a dozen fell in the British Isles. The latest was that which was seen to fall at Crumlin, Antrim, in 1902. This weighs eight pounds. Most of them were stony meteorites, or as they are called, aerolites, consisting for the most part of silicates. The meteoric irons are known as siderites, and consist chiefly of nickel-bearing iron. Others contain both iron and silicates in large proportion, and these are called siderolites.

It is scarcely to be wondered at that these messengers from space have been regarded with awe and veneration in the past. Even with the cold light of science thrown upon them we do not regard them as any the less wonderful. Our knowledge of them has increased, but with this increase of knowledge has come a more thorough appreciation of the marvels of our solar system, which may itself be perhaps but a portion of a greater system which at present we have not been able to fathom.

THE SNAKE'S HEAD CHRYSALIS

BY HUGH MAIN, B.SC., F.E.S.

When a butterfly or moth is in the chrysalis stage of its existence it is usually very inactive and thus unable to escape by any movements from its enemies, such as birds or lizards, who would gladly make a meal of it.

We find that the caterpillar often buries itself in the ground, or hides itself away in some crevice, or spins a cocoon more or less strong, in which the next change may take place, and where the chrysalis can remain secure from the weather and protected from observation.

In other cases where the chrysalis is not thus hidden away, it has been noted that it frequently bears a resemblance to some other object, and so, for one reason or another, escapes being taken for what it really is.

Such resemblances, which also occur

Photos by] [*Hugh Main, B.Sc., F.E.S.*

THE SNAKE'S HEAD CHRYSALIS.

The chrysalis, to protect itself in its inactive state, assumes the appearance of a bird-eating snake. The black eyes, the nostrils, snout, mouth and chin are all faithfully reproduced.

in all the various stages of an insect's life, may be classified into various groups, one including such cases as those in which the insect becomes practically indistinguishable from its surroundings, and another in which it is quite conspicuous, but looks more like some easily recognized noxious creature than something quite harmless and perhaps good to eat.

It is difficult, without a good deal of patient observation, to gather from their actions what are the views of a bird or lizard upon such resemblances, but many examples are readily appreciated by human beings.

In the second class above mentioned many examples could be given, but one of the most remarkable on record is shown in the illustrations on page 457 of the chrysalis of a moth. This was found attached to the branch of a small bombax tree in Upper Burma. It bears a wonderful resemblance to the projecting head of a small snake, and not only that, but to a particular bird-eating tree-snake which occurs in the same part of the world.

The staring black eyes, the nostrils, snout, mouth, chin, are all there, and certainly give the chrysalis a most forbidding aspect.

The moth itself which emerged from the chrysalis is very beautiful, the wings having a lovely white, silky sheen with patches of darker scales.

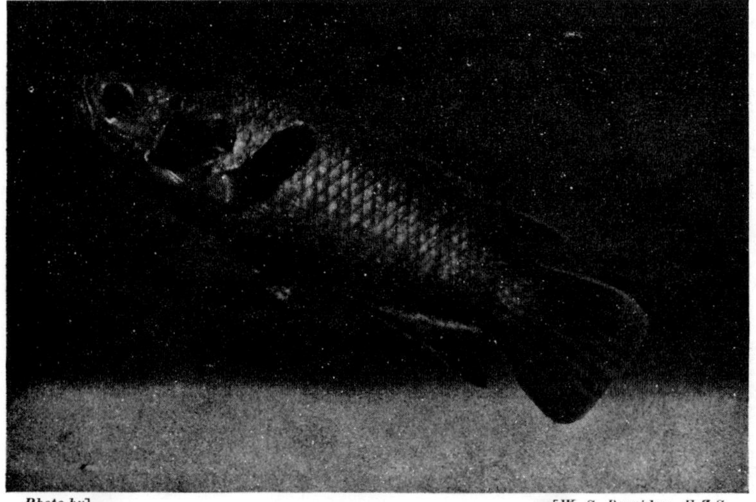

Photo by] [*W. S. Berridge, F.Z.S.*

THE CLIMBING PERCH.

The first of these curious fishes was found by Lieutenant Daldorf, in 1797, in a crevice of the bark of a palm.

THE CLIMBING PERCH

BY W. P. PYCRAFT, A.L.S., F.Z.S., Etc.

It is always the unexpected that happens in Nature. Doubtless he who coined the phrase "Like a fish out of water" as a vivid means of expressing extreme discomfort amid unusual surroundings, felt confident that no fish could leave its watery world and live. But as a matter of fact there are many species which possess this power, and the wonderful Anabas is among this number. Stranger still, it can not only live without discomfort ashore, but it can both crawl and climb! And this by means of the stout spines which arm the hinder pair of belly-fins, the unpaired fin which runs along the body between these paired fins and the tail, and the stout spines on the gill-cover. In crawling, it thrusts the spines of the gill-cover into the ground, bends its tail and inserts the belly-spines, then straightens out the body, which is thus thrust forward, when the same movements are repeated. Thus the fish wriggles along at a fairly rapid rate. This fish was first discovered in India, and was reported to be able to climb low trees, but confirmation of this feat is still wanting. It can, however, as our illustration shows, climb up reeds to enjoy "a place in the sun," or to gain the sloping bank for the purpose of making excursions on land. The latter journeys are probably only undertaken when, by a premonition of impending drought and the probable drying-up of the stream in which it is living, it desires to find a more secure abiding-place.

Now it is obvious that these peregrinations would be impossible but for some special breathing apparatus contrived to enable the creature to breathe the upper air. This apparatus takes the form of a pair of large cavities, on either side of the head, above the gills. This cavity opens down-

[By W. B. Robinson.

THE CLIMBING PERCH.

For a long time the curious habits of this fish were considered to be fable, but that the fish does climb out of the pools it inhabits has now been incontestably proved.

Photo by permission of] [*Gerald N. White.*

THE NEST OF THE TRAP-DOOR SPIDER.

This photograph is taken looking down on the Spider's nest, and shows the admirable way in which the insect hides its living-place.

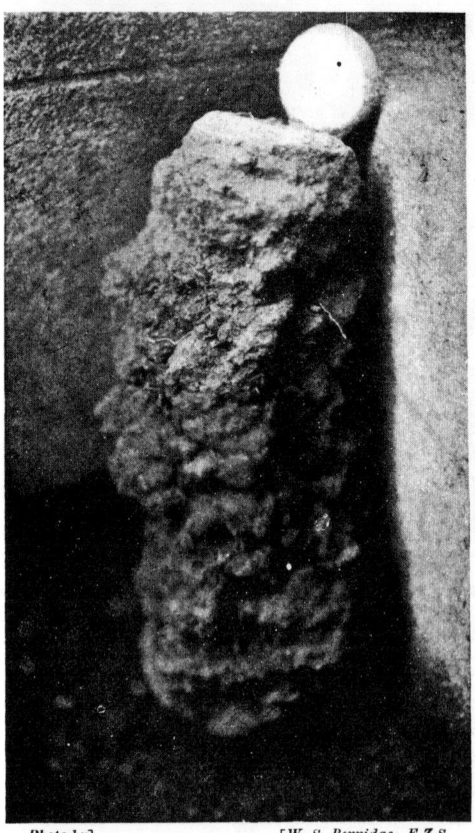

Photo by] [*W. S. Berridge, F.Z.S.*

THE NEST OF THE TRAP-DOOR SPIDER.

This nest has been dug out of the ground, and shows the long tube excavated by the Spider for her eggs. The whole tube is lined with silk.

wards into the gill-chamber, and is divided up by thin, scroll-like plates of bone covered with a delicate skin, over which ramifies a network of blood-vessels for the purpose of absorbing oxygen.

The Climbing Perch is a freshwater fish found in India, Burma and the Malay countries, while other species are found in Africa.

Nearly related to the Climbing Perch is the "Snake-head," which is found in the grassy swamps of India, China and the Philippine Islands. Like its more famous relative, it is also able to live long out of the water and is provided with a similar apparatus for breathing atmospheric air, save that the scrolls of bone are wanting. Thus the breathing surface is much less in extent than in the Climbing Perch, and its sojourns on land can only be of comparatively short duration. Nor does it possess the power of locomotion on land, so that it would seem that this provision is only for the purpose of tiding over a temporary shortage of water or of breathing atmospheric air when, owing to the drying-up of the stream, the water which remains is too foul to breathe by means of the gills.

TRAP-DOOR SPIDERS

BY R. I. POCOCK, F.R.S.

NOT without reason have the burrows of Trap-door Spiders been called "triumphs of protective architecture." The essential feature distinguishing these burrows from those of other spiders—indeed, from those of all other animals—is the presence of a movable lid hinged to the edge of the aperture, so that it can be opened and closed at will. Moreover, the outer surface of the lid is covered by the Spider with particles of moss or other vegetation growing in the vicinity of the burrow, so that, when closed, it exactly matches its surroundings and becomes practically invisible. By thus concealing the burrow the Spider secures the double purpose of hiding from his enemies and surprising his prey.

When on the watch for food he takes up a position at the top of the burrow beneath the lid, which is raised just sufficiently to allow the occupant within to peep through the narrow chink. Then woe betide any beetle or fly that passes within striking distance. Like a flash the Spider is upon it and drags it underground, the whole performance being accomplished with such marvellous speed and precision that it is all over in less than a tick of time, too swiftly, indeed, for the human eye to follow the consecutive details of the act. The closing of the lid is effected automatically, because the Spider cunningly lays on the threads composing its hinge in such a way that they are on the stretch when the lid is open and thus tend always to keep it shut. But why, it may reasonably be asked, is the Spider at such pains to conceal the burrow and to capture prey with such lightning speed? Because, like all spiders nearly, he lives in mortal dread of those dire enemies of spiders, the mason-wasps, which all through the spring and summer are relentlessly ransacking every nook and cranny in the search for spiders wherewith to provision the nests of their young. Against these powerful, venomous and armoured insects the Trap door Spider has no chance, even when fighting within the narrow precincts of the burrow.

Photo by permission of] [*Gerald N. White.*

THE NEST OF THE TRAP-DOOR SPIDER.

The same nest as that shown on the opposite page, but here the trap-door is open, balanced on the delicate silk hinge which, when the Spider has emerged, closes the lid down again over the aperture.

The persecution of these insects, carried on generation after generation for untold thousands of years, has had a wonderful influence in fashioning the structure of the burrows made by some species of these Spiders; for quite commonly the burrow is not a simple tube, but is complicated internally by side-branches or secondary chambers, usually shut off by internal doors from the main part which opens to the surface of the soil; and it is interesting to note that in nearly all cases these additional passages for escape are found in burrows which are closed by what have been aptly called, from their delicacy and thinness, " wafer " lids, to contrast them with the thick, heavy " cork " lids, with a bevelled edge, which fit tightly into the slightly expanded aperture of the burrow. Nearly all the cork-lidded burrows are simple tunnels, but many of those furnished with wafer-lids are more complex; and perhaps the most perfect burrow, from a defensive standpoint, that has yet been discovered is that made by a species of Nemesia which occurs in South Europe. From a point a short distance within the main burrow a second passage, mounting obliquely towards the surface, is excavated; and to the upper edge of the orifice an inner door is hinged in such a way that it can be swung upwards so as to block the main passage, or pulled over the aperture of the side tunnel. The Spiders thoroughly understand the use of the doors. If conscious of interference with the main door from without, they quickly mount to it, and grasping its inner surface with their fangs and driving their claws into the walls of the silk-lined burrow, hold on with might and main to prevent it being forced open. If unable to resist the strength of the marauder, the fate of the occupant is sealed

Photo by] [*W. S. Berridge, F.Z.S.*

THE NEST OF THE TRAP-DOOR SPIDER.

A view of the opposite nest *in situ*. When the Spider has entered and the door of the nest is shut, the inmate holds on to the inside of the door by the mandibles and the first two pairs of feet, while the third and fourth pairs are pressed against the walls of the tube.

unless he can fall back upon a second line of defence. Here comes in the use of the inner door made by the species of Nemesia above mentioned. Abandoning the outer door, he plunges deeper into the burrow and thrusts up the inner door so as to block the passage. If the wasp forces this, the Spider still has a chance, and retreating up the side passage, pulls the door after him. If the wasp then descends to the bottom of the burrow, and finding it apparently unoccupied, quietly departs, well and good; but if he detects the inner door, now flush with the wall of the main tube, and attempts to force it, the Spider's last hope lies in the strength with which he can hold it shut.

Trap-door Spiders are distributed all over the warmer parts of the world, and are akin to the great Bird-eating Spiders. They vary greatly in complexity of structure, the most highly-organized having the jaws and forelegs armed with rows or clusters of strong spines to help in digging. They live in all kinds of places, some, indeed, building their silk burrows on the surface of tree-trunks; but to show, in conclusion, that their activities are guided entirely by instinct, it may be mentioned that the newly-hatched little ones make trap-door burrows as perfect at the start as those of their parents after months, perhaps years, of practice.

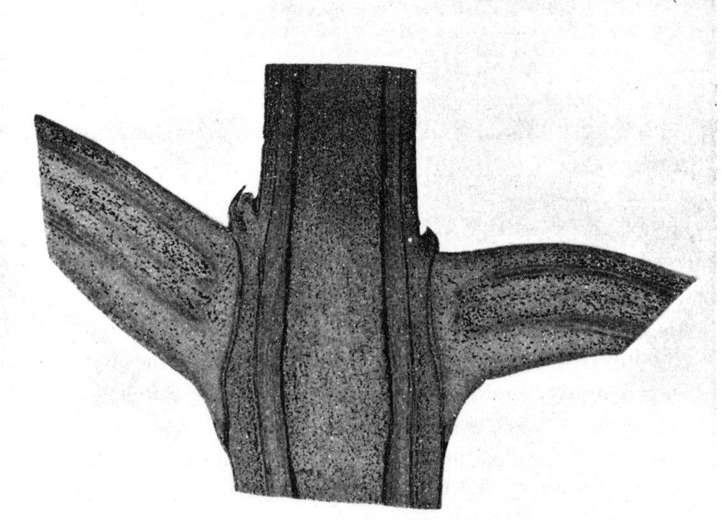

Photo by] [*E. J. Spitta, F.R.M.S.*

WHY A LEAF FALLS.

Here is a slightly magnified photograph of a section cut along the stem of a Horse Chestnut showing the junction of the leaf-stalks with the main stem of the plant. Notice the growth of the cork cells at the junction of the left-hand stalk with the main stem.

WHY A LEAF FALLS

BY E. J. SPITTA, F.R.M.S.

THIS curious and wonderful phenomenon is brought about by the growth of an intercepting wall of cells which cuts off the flow of the sap from the leaf at the junction of its stalk with the stem of the plant. It is called the "cork" layer. To understand this fully, resort must be had to the examination microscopically of a longitudinal section through the stem of the plant at its point of union with the stalk of the leaf. A photograph of this is seen slightly magnified in the illustration on this page. It was made from the Horse Chestnut.

The stem of the plant has two leaf-stalks attached to it, one on the left and one on the right as looked at in the photograph. It is the right one we first examine. A portion in this situation is shown more magnified in the full-page plate, where it will be seen coming from the lowest part of the stalk (at its junction with the stem) and reaching almost across its entire width, a finger-like projection of cells all piled one on the other, and all closely approximated and packed together. The top or tip of this finger is illustrated at a great magnification in the last illustration, where it is readily seen to be blunt-ended and larger than the part which supports it, as the head of a pin is larger than the pin upon which it is carried. This is supposed to be for the purpose of pushing the cells of the stalk aside, and to make room for the intercepting wall to grow and increase in *thickness*.

The sap from the mother tree, of course, usually has free access to the leaf by passing through the stem and then the stalk; but this cork layer, which commences to grow in the early autumn

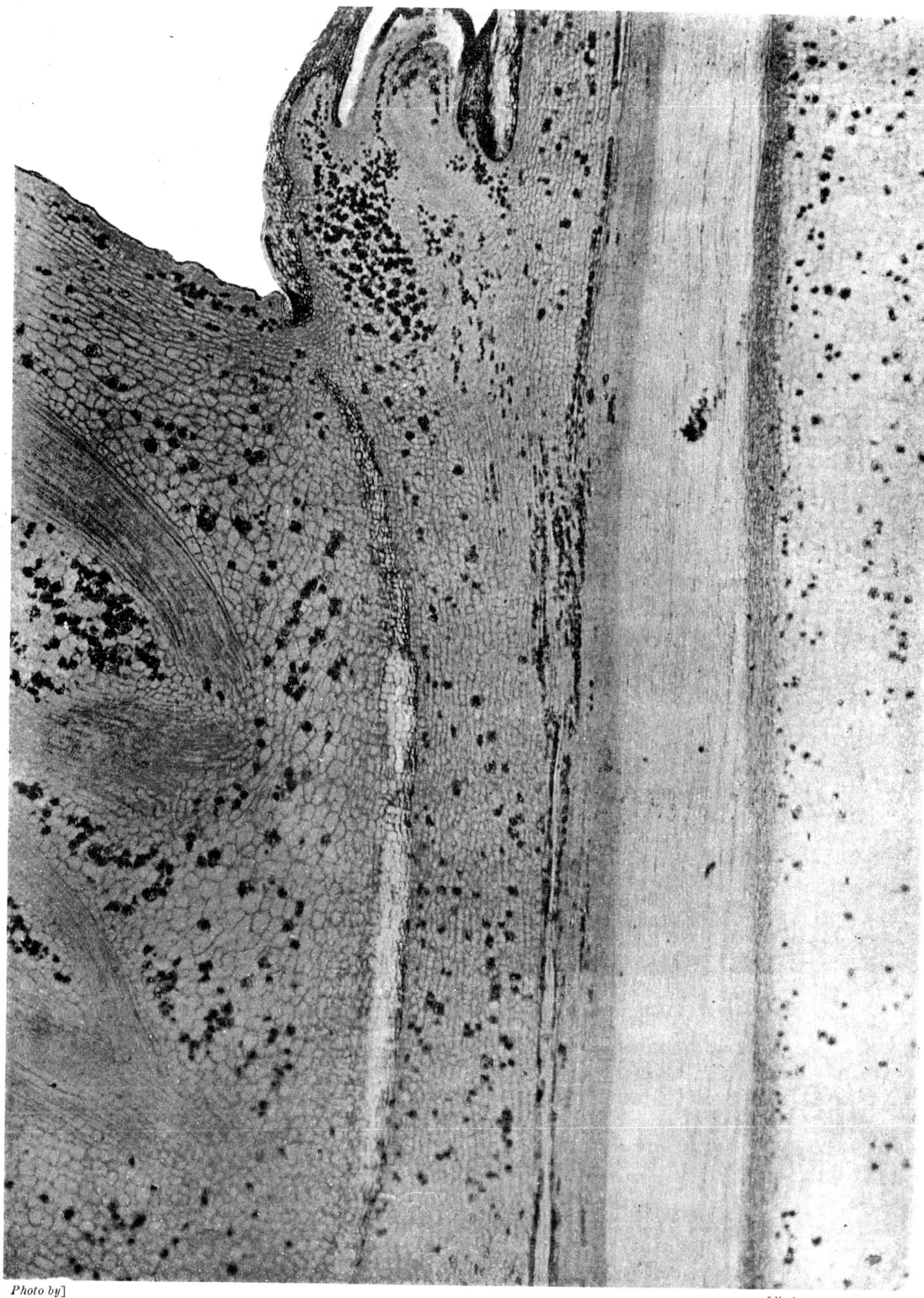

Photo by] [*E. J. Spitta, F.R.M.S.*

WHY A LEAF FALLS.

In this photograph the point of junction has been more highly magnified to show with greater distinctness the lighter coloured cork cells packed closely together. As autumn advances these cells will stretch right across the connecting space till they reach the base of next year's bud, when the leaf will fall off.

and reaches the other side of the stalk towards the end of that season—although this varies in different plants—acts like shutting the gates of the harbour against the incoming tide. In other words, it deprives the leaf and stalk of its life-giving sap. Nourishment being thus cut off, the leaf and stalk soon commence to shrink like a piece of orange-peel dried in the sun (see the left stalk in the photograph on page 462, which is shrinking), and this act of shrinking slowly increases, eventually loosening the attachment of the stalk at its union with the stem, so that the first puff of the autumn winds completes its severance and the leaf falls to the ground.

It should be noticed, however, that this cork layer has another function, for it also seals effectually the point of severance, so that the plant, after the fall of its leaf, shall not spring a leak in this situation.

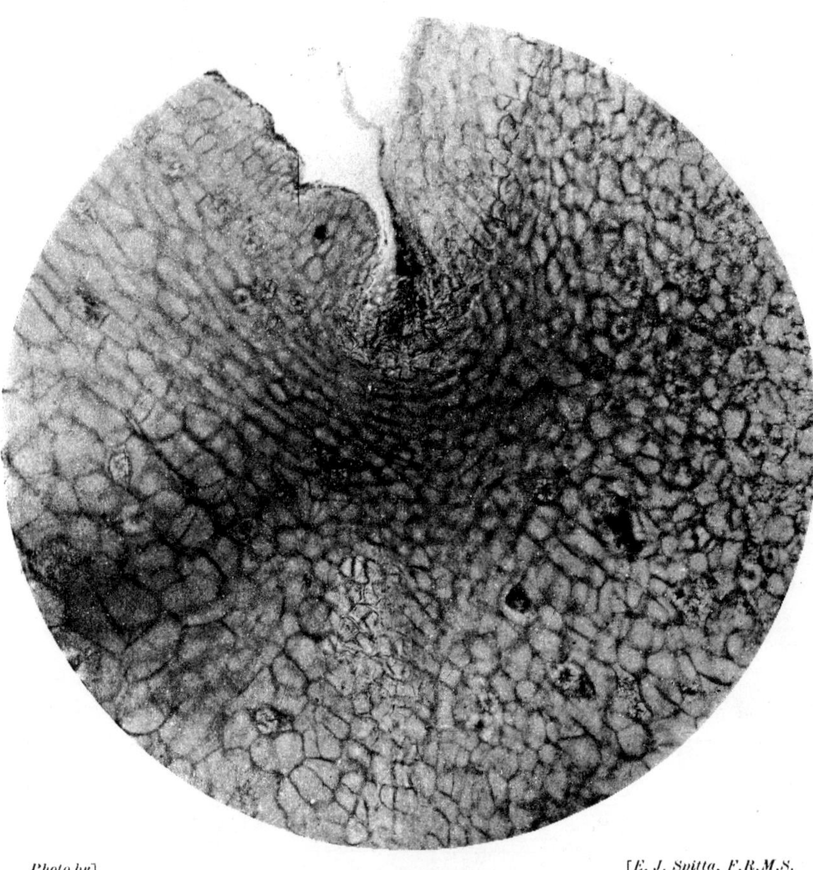

Photo by] [*E. J. Spitta, F.R.M.S.*

WHY A LEAF FALLS.

The foremost cells of the growth of cork are here magnified to show that they are "blunt-ended."

THE DEVIL'S COACH-HORSE BEETLE

BY JOHN J. WARD, F.E.S.

He had just come from under a heap of stones and was travelling at a rapid pace along the dusty road, when I approached. I do not think that he heard me coming, but probably the vibration from my footstep warned him of my presence, and, being a warrior bold who knew no fear, he instantly halted. He set his legs wide apart and grasped the ground well with his feet, while his jointed feelers quivered excitedly for a moment, and then became still—as shown on page 466.

Now a Devil's Coach-horse Beetle that is not prepared to meet a foe and fight until death, if need be, is no credit to his race. This fellow was no coward, he evidently had no intention of running away; indeed, his attitude seemed to distinctly invite a quarrel.

While he was awaiting events, there was ample opportunity for viewing this gentleman of the road, and it did not take long to decide that he was undoubtedly the ugliest insect that one could expect to meet on an English roadside. He looked somewhat like a large and exaggerated earwig without tail-pincers, and with a big head bearing a pair of cruel-looking eyes. His colour, too, helped his hideousness, for, quite unlike, I think, any other animal in the British fauna, he was a funereal, dead black from head to tail.

I pushed the toe of my boot towards his tail-end. That was more than his irritable and aggressive nature could tolerate, and in an instant he turned completely round, fully prepared for battle. He held his head low down and half raised his tail, and seemed all ready to make a sudden and ferocious charge—as shown in the first photograph on page 467. Probably that manœuvre scares away many of his foes; his tactic, however, is to invite the charge rather than to make it.

Photo by] [*J. J. Ward, F.E.S.*

THE DEVIL'S COACH-HORSE.

There is no member of the insect world so pugnacious as this Beetle. He is here shown attacking the stalk of a leaf which has been used to annoy him.

He, therefore, adopted his usual procedure and once more became motionless in his changed attitude. Another movement of my toe instantly produced another and more desperate posture. He rapidly changed the position of his legs, raised high his tail in scorpion fashion, opened wide his huge jaws (which open sideways), and positively glared with his cruel-looking eyes in a most savage manner—as the second photograph on page 467 shows. Then from a pair of glands at the tip of his tail he discharged two globules of a volatile fluid with a horrible odour.

Brave would have been the insectivorous foe that dared to tackle such a meal. Doubtless, in the course of his life he had scared away many birds, hedgehogs, toads, lizards, and similar enemies by these same manœuvres, but he had never before met a wandering naturalist bent on seeing what fighting powers a Devil's Coach-horse Beetle really possessed. Here, therefore, was a new venture for this fellow of menacing attitudes. After the one discharge of the evil fluid, that means of defence was for the time being exhausted. Its next defence is its jaws, for although its tail assumes a stinging posture, yet it is quite incapable of stinging. So its wide-opened mandibles would be the foe's next consideration. Can it really bite? the unsatisfied attacker would naturally inquire.

A small folded portion of coarse brown paper would probably represent a fairly tough natural foe. So with that object in one hand and the camera release ball in the other I made some experiments. I gently pushed the folded paper towards him. He at once savagely seized it, much after the manner of an angry bull-dog, attacking it not only with his mandibles, but also with his legs. He held it so firmly that I was able to lift him up from the ground (see photograph), but this movement only made him still more desperately angry—as shown in the next photograph. So he held on literally with "tooth and nail."

Photo by] [*J. J. Ward, F.E.S.*

THE DEVIL'S COACH-HORSE.

So furiously has he laid hold of the cause of his annoyance that he has been lifted up bodily from the ground hanging to the leaf-stalk entirely by his jaws.

After a few minutes I laid him and the paper down, and after the latter had been quiet sufficiently long to justify its release, the beetle let go. He made no attempt to move away, but doggedly waited for the paper to move again.

When he had rested for about five minutes I laid a leaf-stalk beside him. His blood being up, and apparently having laid one enemy low, he was ready for the next, for he instantly pounced on this, taking the end of it firmly in his jaws. My touching the other end of it made him very angry, causing him to get his legs well over it—as the photograph on page 465 shows.

I then lifted the leaf-stalk, when to my great astonishment he raised his tail and slowly turned half a somersault, and in that attitude he held firmly on, his mandibles bearing his whole weight (see photograph), until I laid him and the leaf-stalk down again.

Photo by] [*J. J. Ward, F.E.S.*

THE DEVIL'S COACH-HORSE.

Already in his rambles the ferocious insect has realized the approach of a prospective enemy and is prepared to do battle.

An examination of the folded brown paper showed that his mandibles had neatly pierced it. Were those jaws strong enough to pierce the human skin? To decide that point, after he had rested I offered him the tip of my forefinger. Again he was ready to attack, and time after time he endeavoured to get a grip, but the roundness of the finger prevented this. I could distinctly feel the nip of his jaws, and the amount of pressure which they exerted was really astonishing. Suddenly I felt that he had accomplished his purpose, when I instantly withdrew my finger. The skin was firmly gripped and he was holding on. He had, however, made a mistake in biting so hard on this occasion, for his wedge-shaped jaws had penetrated only the surface skin; but so thoroughly had they been driven in that he was not strong enough to withdraw them again.

Having performed his part so well, I carefully released him and placed him on the ground. After cleaning up his mouth-parts and waiting sufficiently long to see that no further enemies were abroad, he pursued his path on the roadside, quite ready to attack his prey in the same fearless manner that he had dealt with his supposed foes. He is a most useful insect, and should never be killed; his prey consists of almost any insect with which he meets, and a slug or a worm does not come amiss to him; he will even attack a young toad; also he has ample wings, is a bold flier, and is common throughout the British Isles.

THE WONDERFUL SEA-URCHIN

BY EDWARD STEP, F.L.S.

THE Sea-Urchin, or Sea-Hog, is so called because it resembles the hedgehog, or urchin, by being covered with protective spines that keep most enemies at a distance. With a remark to the effect that this is a curious correspondence between a land-beast and a sea-creature, the average person

Photos by] [*J. J. Ward, F.E.S.*

THE DEVIL'S COACH-HORSE.

In the two upper photographs the creature is shown in his characteristic attitudes of defiance. Like the scorpion he raises his tail to threaten his enemy, but unlike the scorpion he is innocent of sting. In the lower pictures he is shown attacking a piece of brown paper "tooth and nail."

Photo by] [*H. J. Shepstone.*

THE COMMON SEA-URCHIN.

The Urchin has been denuded of its spines to show the arrangement of the plates in bands extending from summit to base, with their bosses for the spines and perforations for the water-tubes.

is inclined to pass the Sea-Hog by as having no further interest for him. Let me say that this is a great mistake. There are few creatures existing that are so wonderfully constructed and organized. Let us consider it for a while.

The Sea-Urchin is shut up in a stony box which cannot be opened without killing the inmate ; it is entirely without any limbs, yet capable of walking, not only over horizontal surfaces, but also up vertical rocks and even under the roofs of submarine caverns, with as much ease as on the level. The box in which the creature is encased is a wonderful structure ; and wonderful are the thousands of spines which suggest to a big-mouthed fish that the Urchin, if swallowed, might cause some serious stomachic troubles, and that, in consequence, the Urchin had better be left alone. It is not only on broad, solid surfaces that the Urchin can climb ; he can ascend the slender bars of the crab- and lobster-pots, enter at the summit just as the crab does, and help to dispose of the bait. Naturally, the crabber is annoyed at the Urchin's presence ; he looks upon the Urchin as standing in the room of a good-sized crab, and his feelings do not allow him to spend time in musing on the marvels of Sea-Hog structure. He is determined that the particular Sea-Hogs he found in his crab-pots this morning shall not trouble him again ; so he brings them to shore in his boat and then dashes them to pieces against the rocks ! In one acquainted with the Sea-Hog's history, this act of vandalism gives rise to feelings such as the china-maniac might experience on seeing some cunningly-designed piece of his treasured ware shattered before his eyes. But no human

Photo by] [*H. J. Shepstone.*

"SHELL" OF A SEA-URCHIN.

The spines have all been detached to show clearly the polished bosses upon which they move. Each big knob is encircled by smaller ones which support small spines.

pottery, fashioned out of a lump of plastic clay, can compare with this natural box of eggshell-ware built up of nearly six hundred pieces, all fitted together edge to edge with such nicety that there is not a hair's space between them. And yet this ware is only about one-twenty-fourth of an inch in thickness!

A remarkable feature of this structure is seen in the five-sided character of all the little plates that go to make up the cupola-like box. The Sea-Hog, in truth, is built up on a pentamerous system. Well nigh everything about it is arranged in fives or some multiple of five. Turn it over, and you will see that its mouth —which is lowermost—is guarded by the tips of five white teeth, which meet in a common point. Then the plates of which the box is composed are arranged in bands according to their size and character; these are of three kinds and there are ten bands of each. The smallest of these plates are perforated with a large number of minute holes which are arranged in pairs, and there are three pairs in each row. Other plates bear spines, which are attached by cup-and-ball socket-joints to solid polished bosses on the plates, the spine and the boss being held together by a muscular band which moves the spine in any direction. When an Urchin has been denuded of its spines these little bosses will be seen standing out in regular rows from the shell.

Photo by] [*E. Step, F.L.S.*

THE COMMON SEA-URCHIN.

Here represented about half the natural size. Common on all our shores, it is esteemed as food in the countries round the Mediterranean, where it is cooked like an egg.

In order to manipulate these spines, and for other purposes, the outside of the box is coated with a layer of thin flesh, which is very sensitive. Touch it gently with any slender instrument that will pass between the spines without touching them, and immediately every spine in that area will turn its point towards the spot touched. At

Photo by] [*E. Step, F.L.S.*

FLEMING'S SEA-URCHIN.

This is, perhaps, the most beautiful of the British Urchins, its long sharp spines being particularly graceful. They are an inch and a half long, whilst the "shell" is three inches across, and nearly as high.

the same time you will become aware of the presence of active living bodies among the spines that at first you will regard—quite reasonably—as little worms with jaws like the beak of a bird, except that some have three bills instead of two. These remarkable bodies were formerly considered as parasites, and named and classified as a distinct group of animals. They are really part of the Urchin organization, and apparently modified spines, for they have a solid, stony base, and a pliant, fleshy " neck " between it and the beak. The probability is that when you touch the delicate flesh of the Urchin that covers its box your instrument will be in the neighbourhood of one of these bodies ; if so, it will turn in the direction of the annoyance, snapping its beak and probably taking a grip. Their principal purpose is apparently to keep offensive refuse from accumulating between the spines, for they may be seen to pick up such particles and convey them as far as their length permits, and so handing on the duty to others. But they also serve as weapons of defence, for they have been observed to repulse Star-fishes by biting their sucker-feet.

Photo by] [*H. J. Shepstone.*

TWO JAPANESE URCHINS.

These are in life covered with bristles, similar to those of the Heart Urchin. The pores for the water-tubes are all on the upper surface and form the outlines of a flower's petals.

The Urchin has sucker-feet also. The minute perforations in the five-sided plates of which the box is built up give exit to little fleshy tubes, each ending in a sucker! It is by means of these apparently weak but really powerful suckers that the Urchin climbs up rocks or travels over the sea-bottom in search of food. The tubes are filled with water and the suckers are worked by water power ; a special system of water-vessels inside the box supplying the necessary pressure. In our Common Sea-Urchin these tubes extend from the base to the summit, so that by their aid, should the Urchin get turned upside down, it can easily right itself again. Some of its relatives, however, are not so liberally provided. The Heart-Urchin shown on page 473, for example, has them only on the upper and under surfaces, not down the sides.

Certain of these organs, distinguished by ending in a clear round head, are noticeable on the upper surface of the Urchin. The comparative robustness of these heads is due to the fact that each of the three jaws bears on its exterior a transparent pouch, filled with liquid, from which two tubes run nearly to the tip of the jaw. The liquid in the pouch is poison. When the jaws are closed

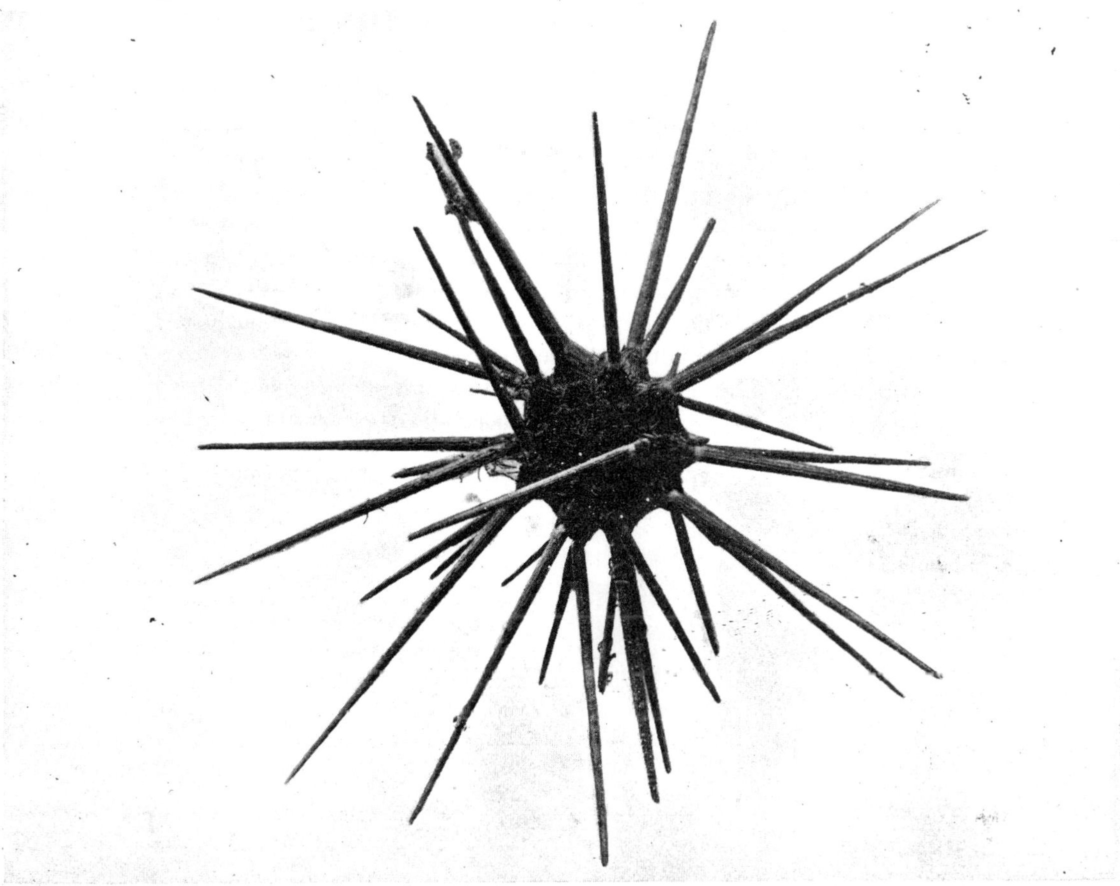

Photos by] [*H. J. Shepstone*

TWO SEA-URCHINS.

In both these forms the spines are comparatively few. They serve to give an idea, with the other illustrations, of the great **diversity of form there is in this family.** In the upper figure, though the spines are few, they are long and so disposed as to keep **an enemy at a safe distance.** In the lower photo they are very stout

upon any available part of an enemy their bite is a poisoned one. Experiments made to test the virulence of the poison have shown that a single one made to bite a frog has caused the victim's heart to stop beating. The three-jawed organs, which are by far the most numerous, open quickly, and close with a snap. One of their special functions is to seize the swimming forms of parasites that are ever seeking to establish themselves on the Urchin's delicate exterior. A third kind end in a pair of toothed jaws that give them a close resemblance to the head of a snake. Their function appears to be to catch small animals as food, and hold them until the tube-feet can bend that way and convey the captive from one to another until it reaches the Urchin's mouth.

The Urchin's box is full of wonders, although the Urchin is an animal of a low grade in the scale of life. It is a creature without a head or a brain, a heart or circulatory system. There is a nervous system, but it is not of a high order. The complicated masticatory apparatus, which is known as "Aristotle's Lantern" (shown on page 474), is worked by a ring of muscles; the five large teeth, whose tips show externally in the centre of the underside, form the chief part of this apparatus.

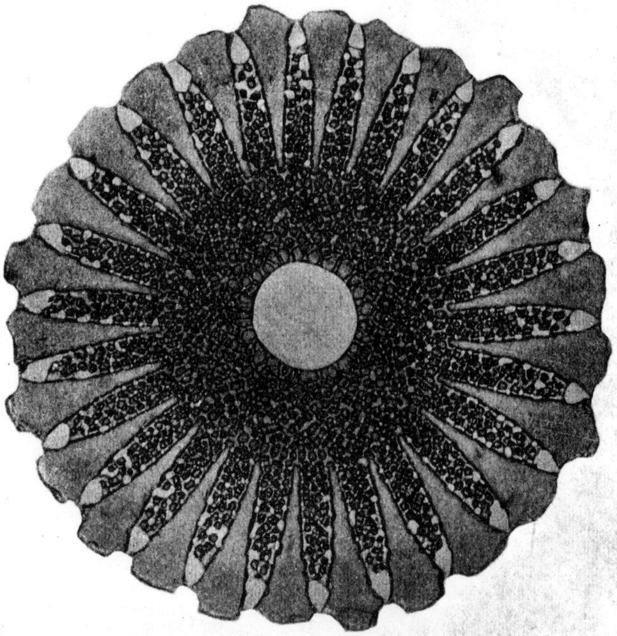

Photo by] [*E. J. Spitta, F.R.M.S.*

AN URCHIN'S SPINE.

This is a highly magnified view of a section cut across an Urchin's spine. Each species has its own peculiar design, which corresponds with differences in the grooving of the exterior.

The chief food of the Common Sea-Urchin appears to be the leathery brown fronds of the great Tangles and the numerous small animals that live upon them, among other things small crustaceans and worms. The fact that it is frequently found in crab-pots, attracted thither by the pieces of fish provided as bait, shows that it is only restricted to smaller animals naturally as food by the fact that it cannot capture larger ones. One of our rarer Urchins, with purple spines, when living upon exposed limestone rocks, especially noticeable on the west coast of Ireland, excavates holes in the rock by the rotation of its spines. These hollows are sufficiently deep and wide to accommodate the Urchin and save it from being washed away by heavy seas. Recent observations, however, seem to show that it does not start with solid rock, but takes advantage of an existing crevice and enlarges it to suit its own circumference. These internal matters, however, we must leave, and devote a little space to the forms of Urchins represented by our photographs.

The Common Sea-Urchin, which may be regarded as the typical form, bears about four thousand spines, each a wonderful piece of workmanship. If one of these be microscopically examined it will be seen to be a thing of great beauty, whether we examine it as a whole or in parts. Its full length under a low power of magnification reveals itself as a beautiful tapering shaft, fluted in a manner characteristic of the species; the cup-like hollow at its base, which revolves so freely on the polished boss of the shell, is itself polished perfectly. If we cut the spine across and examine a thin slice of it we shall see that in spite of its apparent solidity it is really a symmetrical network of ribs and hollows. We have said that there is not a hair's space between the edges of the living plates

Photo by] [*E. Step, F.L.S.*

THE HEART-URCHIN.

This species burrows in fine sand and mud. It is covered with fine bristles (cleaned off the middle specimen), and has the pores all on the upper side. (Slightly reduced from natural size.)

that make up the Urchin's box, and this is true ; yet the plates are not in actual contact, for a film of living animal matter separates them. It is by the vitality of this film that the box increases in size, for it is always depositing fresh particles of lime equally on all the edges of all the plates, so that increase takes place simultaneously in all parts. The Urchin multiplies by means of eggs, and the young go through preliminary stages of existence in which they present not the faintest resemblance to their parents. But the adult form in little gradually develops within the infant body, which is ultimately thrown off.

In the breeding season our Urchin contains five large masses of eggs, and in some localities on our own coasts these make it a desirable article of food. On the shores of the Mediterranean also Urchins are esteemed, and at Naples they are known as *frutta di mare,* or fruit of the sea. So great is their fecundity that it is calculated that our Common Sea-Urchin produces twenty millions of eggs in a single season.

There are throughout all seas many variations from this typical form, and a few of them may

Photo by] [*E. Step, F.L.S.*

THE WHEEL-URCHIN.

An African species of Cake-Urchin which burrows in the sand. The spines have been cleaned off. One half the margin is indented to form teeth like the cogs of a wheel. (Natural size.)

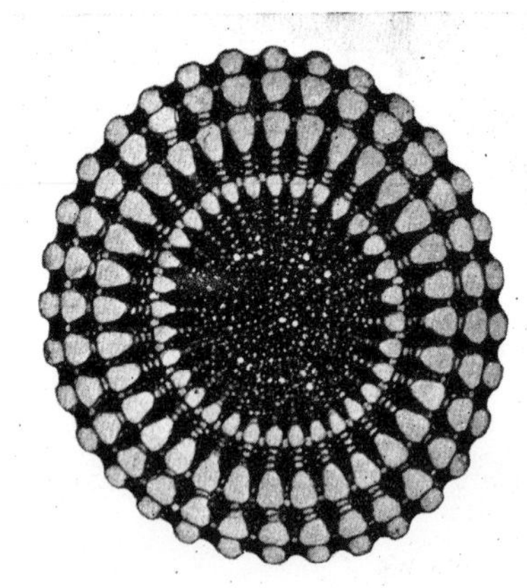

Photo by] [*E. J. Spitta, F.R.M.S.*

SECTION OF SPINE.

This magnified section of an Urchin's spine presents an entirely different pattern from that shown on page 472.

be obtained in the sea around our own islands. In these differences, which constitute distinct species, we find some with shorter or longer spines, each with a constant and characteristic grooving and internal pattern by which the species may be distinguished. They differ also in colour, some deep red, some green, others purple. Some, of which the Heart-Urchin shown on page 473 may be taken as a type, are clothed with silky bristles, all pointing one way, instead of the spines, which point in all directions. These are burrowing Urchins that live in sand or mud, and, in consequence, are not so well known to the public. These will be seen to be oval instead of round, and as they live upon substances that require no mastication they are without teeth; instead, the mouth is a simple open scoop situated at the front of the lower surface, so that the mere act of burrowing pushes the sand or mud into it, and the nutritive material mixed up in the sand is then digested out in the internal organs.

Another group comprises the Cake-Urchins and Shield-Urchins, which are very flat and thin. Like the Common Urchin, they have the mouth in the centre of the underside, and it is provided with apparatus for masticating the food; but in the matter of spines they come nearer to the Heart-Urchins. One of these is the well-known Sand Dollar, of North America, and the little-known Wheel-Urchin, from West Africa, which is represented by the lower photograph on page 473. The specimens have been denuded of their spines. A remarkable point about the Wheel-Urchin is the character which has suggested the name—the cog-like indentations upon half the edge of the disc. The pores through which the suckers emerge are all on the upper side, and they run off from the little five-sided button in the centre like the petals of a flower. This button, by the way, should be mentioned as one of the marvels of the Sea-Urchin and Star-fish organization. It is porous and a natural filter. It is the intake for the water-vessels that work the thousands of sucker tubes. When one regards the minuteness of the pores in the walls through which these tubes pass, it must be manifest that the freedom of the water filling them from any particles of solid matter is an absolute necessity for the working of the system. This porous plate is Nature's plan for overcoming that difficulty. No solid matter can pass through it, and all the water for working the tubes does pass through it and is kept separate from what passes through the mouth. The Cake-Urchins having two almost flat surfaces, require support for the roof, and so we find the interior is strengthened by limy partitions connecting the upper and lower surfaces.

"ARISTOTLE'S LANTERN."

The above name was given by the ancients to the masticatory apparatus of the Sea-Urchin, which consists of five strong teeth, of which only the polished tips show outside the mouth.

The Sea-Urchins are a very ancient family, and their fossils are among the most abundant of the many forms of life that have left evidence in the rocks of their former existence. They appear as far back as Silurian times, for fossil forms are found in the rocks formed at that period. Certain beds of the Upper Chalk are full of fossil Sea-Urchins. The fossils known as "fairy loaves" in chalk districts are among the most plentiful of these, and until recently that particular type of Urchin was supposed to be

[*By W. B. Robinson.*

THE TARPON.

The Tarpon is noted for its huge bulk and prodigious agility. Its appearance is that of a gigantic herring, but it can be distinguished by its enormous scales (more than four inches across), the long filament from the hinder margin of the back fin, and the unique possession of the large single plate of bone on the throat.

extinct, but a closely allied living form was dredged by the *Challenger* from a depth of nearly three miles in the North Atlantic. Our Common Sea-Urchin is found as a fossil in the Coralline Crag beds of the Pliocene Period. Other forms in plenty are found in the Greensand, the Carboniferous Limestone, the Oolite and the Devonian rocks.

Truly, we may say that our despised Sea-Urchin is a box of marvels, both within and without.

THE TARPON

BY W. P. PYCRAFT. F.Z.S., A.L.S., Etc.

THE Tarpon by common consent occupies a very conspicuous position among living fishes. In the first place, he is of extremely ancient lineage, preserving in his person many tokens of this fact:

Photo by] [*J. A. Dimock.*

THE TARPON.

Very remarkable are the jumping feats of the Tarpon, especially when it is remembered that the creature is some seven feet in length and exceeds two hundred pounds in weight.

though these must be sought for, if not " with forks and hope," at any rate, with the anatomist's knife. On this account the ichthyologist makes much of him. In the second place, he displays a huge bulk and prodigious agility, and may be caught with rod and line by those who are strong enough to hold him, and thence he is held in high honour among fishermen.

Most of us must deem ourselves fortunate if we succeed in finding a specimen in some museum, since to meet with the Tarpon alive one would have to cross the " herring-pond " and make for the south-eastern coasts of North America, whence it could be followed southwards to the West Indies and Brazil. For choice the Tarpon fishers select Florida. Though strictly speaking a " sea-fish," and, as it were, hugging the shore, it frequently ascends rivers, and even enters lakes far inland, drawn thither in the chase of the mullet, its favourite food. At times the Tarpon behaves after a very extraordinary fashion, leaping high into the air, and descending again with a resounding splash, the body bent like a bow and the great gill-covers wide open, revealing the crimson gills. Why such acrobatic feats are performed no one seems able to say, at any rate when these are what we may call spontaneous movements ; for similar agility is displayed when desperate efforts are being made to break the line or dislodge the hook of an all-too-successful angler, and often with success, for the inside of the mouth of the Tarpon is almost " armour-plated."

To really appreciate what these leaps mean, it must be remembered that a Tarpon may measure some seven feet in length and exceed two hundred pounds in weight. The careful angler is never in a hurry to strike when Tarpon fishing ; hurry is fatal to success, but it requires considerable experience and self-control not to strike the moment the bait is taken. If the lure is to achieve its purpose it must be gorged. There are exciting moments while this is being done, but they become much more so when the victim suddenly realizes that something is checking his freedom of action. Then, with a mighty rush he makes for the surface and leaps six feet or more into the air. At the downward plunge he races on at a furious speed till checked by the line. Then follow more leaps and more wild rushes, till after an hour, perhaps, or more, resistance fails and the gallant fish is slowly drawn to the side of the boat and gaffed. The seine fishermen in Florida have a wholesome dread of the Tarpon. For, while dragging their nets, as in the Pensacola and Homosassa rivers, one of these giants is not infrequently caught. Then trouble in real earnest begins, for one or more of the unwilling captors have before now been seriously injured, and even killed, by the tremendous leaps which the enraged captive indulges in, in order to make his escape. If the men escape, the nets are ruined. Occasionally, however, and much to the relief of all concerned, the enclosed monster leaps clear over the edge of the net.

But though regarded as one of the greatest of game-fishes, no one used to eat Tarpon. The fishing was purely for sport, and all that was saved from the carcase was the scales, which are of huge size, measuring about four inches across. They find a ready sale in the curiosity-shops in Florida, while the rest of the body was either eaten by vultures or used for manure for the

Photo by] [*J. A. Dimock.*

In angling for this giant the expert endeavours to gorge his victim, and does not strike the moment the bait is taken,—

Photo by] [*J. A. Dimock.*

—for if the Tarpon realizes that something is checking his freedom he plunges and leaps frantically until exhausted with the unequal struggle.

Photo by] [*J. A. Dimock.*

THE TARPON.

"At times the Tarpon behaves after a very extraordinary fashion, leaping high into the air, and descending again with a resounding splash, the body bent like a bow and the great gill-covers wide open, revealing the crimson gills."

orange-trees. Improved methods of cooking, however, seem to have changed this, earning for the victim a more glorious end, since it is now described as most excellent eating.

The general appearance of this magnificent fish can be well seen in our illustrations. Bearing an unmistakable likeness to a gigantic herring, it differs therefrom not only in the matter of the great scales, but also in the long filament which grows from the hinder margin of the back fin. Yet, closely examined while still in the flesh, another curious point would be noticed—to wit, the large single plate of bone on the throat, which is found in no other living members of what are called the "bony" fishes—to distinguish them from sharks, sturgeons, lung-fishes, and so on. The coloration of the body is like that of a herring—blue and silver.

Until recently, comparatively speaking, but little was known about the Tarpon, and not the least remarkable fact which a study of its life-history has revealed, is that concerning its early period of development. Briefly, there is every reason to believe that Tarpon retreat to the deep

Photo by] [*J. A. Dimock.*

THE TARPON.

"Though strictly speaking a 'sea-fish' it frequently ascends rivers and even enters lakes far inland, drawn thither in the chase of the mullet, its favourite food."

sea to spawn. And this because the young Tarpon enter the world in a "larval," or incomplete state of development, bearing not the least likeness to their parents. On the contrary, they closely resemble the larval stage of the eels, which we have already described in these pages, and, like them, as they grow older they grow smaller—up to a certain point. Then the tide turns; the frail, anæmic, scaleless morsels become gradually transformed into the giants so eagerly sought by the fisherman.

This peculiar larval state is met with in three near relations of the Tarpon—the Ox-eye, the Lady-fish, and the "Ten-pounder," or Elops. But none of these attain the magnificent proportions of the Tarpon. The Ox-eye is an Indian species of about five feet long, and in marked contrast with its larger relative, is most excellent eating, and is, in consequence, kept in ponds or tanks for table purposes. The "Ten-pounder" and the Tarpon frequent the same waters, but another species of the same genus is found along the West Coast of Africa, spending a part of the year, at least, in rivers.

Though the jaws of the Lady-fish are but feebly armed with teeth, there are large crushing teeth

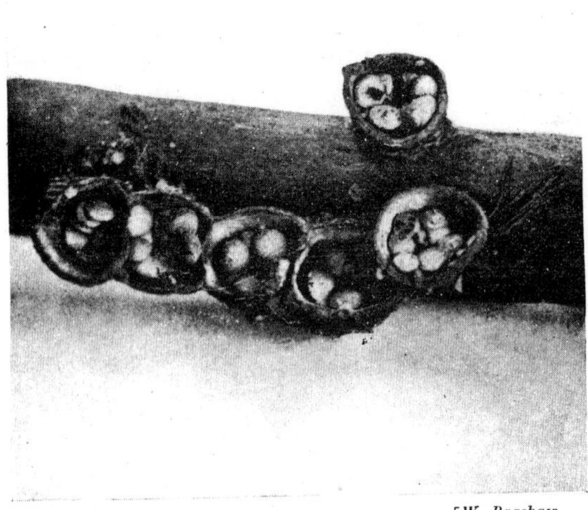

Photo by] [*W. Bagshaw.*

THE BIRD'S-NEST FUNGUS.

A group of this quaint little fungus growing on a dead stick. Each "egg" or spore-case is attached to the nest by an elastic thread. (Natural size.)

within the mouth, which are used for smashing shellfish. Opinions differ as to its palatability, but it is held in high regard by anglers on account of the sport it affords. Compared with the Tarpon, it is a pigmy not exceeding thirty inches in length and a weight of ten pounds when full grown.

Mention has been made of the long and rather delicate filament or thread which projects from and far beyond the hinder margin of the back fin; and it has occurred to some of my readers, no doubt, to ask what purpose this may serve. So far, this question remains unanswered; but it is possibly, as are the similar fin-threads of so many other fishes, of use as a special sense-organ—an organ of touch in various ways. At any rate, it is significant to notice that when a shoal of Tarpon is sighted it is always by the great commotion in the water which they make, either breaking the water sharply, or showing their backs in the slow and regular movements of the porpoise, or swimming with the dorsal-fin and its queer whip-like thread thrust out above the surface. The latter action suggests the wetting of the finger and holding it up to learn the direction of the wind, a plan often adopted by the weatherwise. The fish may similarly gather information by the play of the wind on this fin-thread. It is a pity that it has not occurred to Tarpon fishers to investigate this matter. The above observations as to the movements of shoals of Tarpon were made by a particularly keen naturalist, Mr. Rowland Ward, and are recorded in some detail in his delightful book on Tarpon-fishing in Florida, published some years ago. But he makes no suggestion as to the use of this "feeler," as we may assume it to be.

THE BIRD'S-NEST FUNGUS

Photo by] [*W. Bagshaw.*

BIRD'S-NEST FUNGUS.

Two specimens are here shown enlarged to twice the natural size.

One of the prettiest of the smaller fungi that are not of microscopical smallness is that depicted in the accompanying photographs. It is also one of the few for which the novice has not to hunt far to find its name. The first sight of it springing in a cluster from a rotten root at the copse-side, or—as the specimens photographed were—growing from decaying wood in the garden—suggests that it is the nest of a fairy-bird, complete with eggs. That is the view one gets of them from above; from the side they are vase-shaped or bell-shaped, which does not fit in with our notions of ordinary nest-shapes.

Our forefathers appear to have overlooked this neat little fungus, or we should have had some quaint legend to account for its form, and probably some wonderful medicinal or magical powers attributed to the "eggs." These nests are really the fructification of the fungus, whose vegetative parts exist as mere threads ramifying among the rotting wood and leaves upon which it feeds. Then at a certain point in its life, little knots are formed upon the threads and gradually assume a short club-shape. The thicker end is closed by a thin veil, which soon breaks ; the edges of the bell turn over, and the eggs are revealed. If these eggs are shaken out they will be found to be too thin and flat to keep up the egg illusion More-over, each one is attached to the walls of the interior by an elastic thread from its under surface.

What are these " eggs," and what their purpose ? If we are satisfied that our nest is some form of plant-life we shall at once come to the conclusion that the eggs are seeds, and in this we shall be near the truth, though not absolutely reaching it. The plant, being a fungus, produces no seeds, but

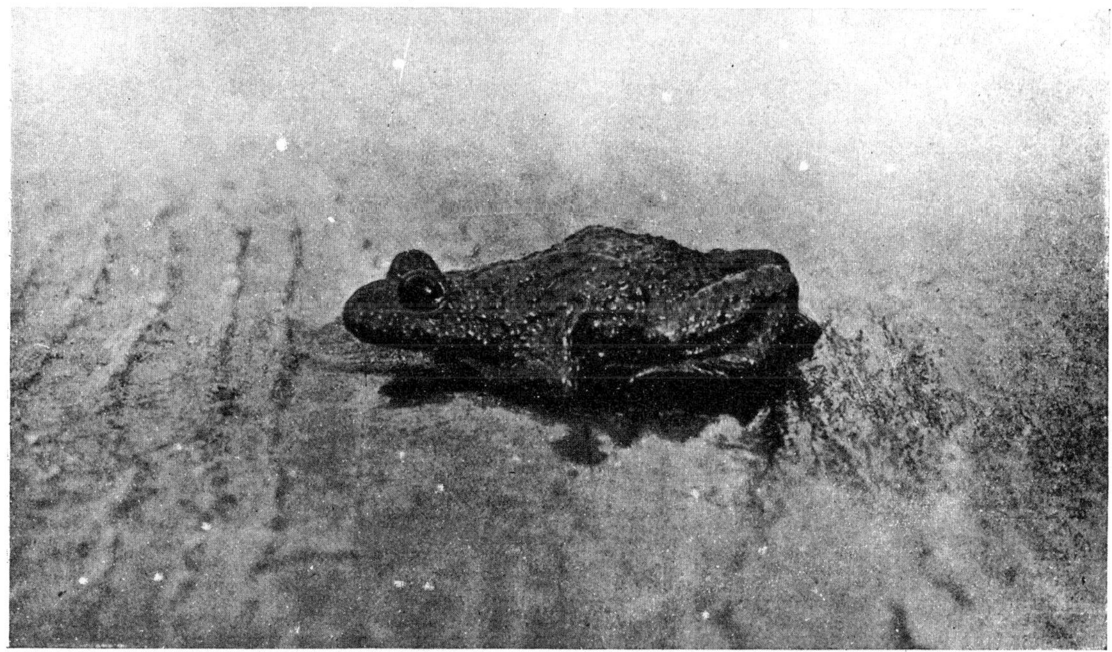

Photo by] [*W. S. Berridge, F.Z.S.*

THE FIRE-BELLIED TOAD.

This Toad is common in many parts of the Continent. It takes its name from the colour of its under surface, which varies according to locality, the yellow variety inhabiting streams at considerable elevation, while the orange form confines itself to the ponds and rivers of the plains.

reproduces itself by means of spores, which are very different in character, though they attain the same end. The egg is a special lentil-shaped body, in which the spores are produced.

Our photographs represent the Bird's nest Fungus of the natural size and enlarged to twice the size.

THE FIRE-BELLIED TOAD

BY R. LYDEKKER.

THE Fire-bellied Toad is a representative of a small family of the frog tribe characterized by the disc-like form of the tongue, which may be either free, or adherent to the floor of the mouth ; the latter condition obtaining in the Fire-bellied species. Although unknown in the British Isles, this Toad is common in many parts of the Continent, whence it extends eastwards into Asia. In addition to the peculiar conformation of its tongue, it may be recognized by the warty character of the olive-brown skin of the back, which may, or may not, be marbled with black ; and by the smooth skin of the under-parts, which may be either yellow or orange, with black marblings.

Whether the ground-colour of the under surface of the body is yellow or orange depends, it seems, on what naturalists call station; the yellow-bellied phase living in streams at a considerable elevation in the mountains, while the orange-bellied form inhabits ponds and rivers in the plains. What may be the meaning of this difference in colouring has not yet been ascertained; but its existence can hardly be regarded as a reason for including this Toad among marvellous animals, as such "dichroic" (*i.e.*, two-coloured) species are by no means uncommon.

Neither can I consider the death-feigning instinct possessed by this Toad in a high degree (as shown in the second illustration) as entitling it to a position among the marvels of the world, for this trait is shared by quite a number of animals—from the American opossum to the English click-beetle and dor-beetle—and is also exhibited by the common British Toad, which, as every gardener knows, will often, when thrown on the surface in the course of digging a piece of ground, lie sprawling on its back with its limbs as rigid as though in the grip of death.

Photo by] [*W. S. Berridge, F.Z.S.*

THE FIRE-BELLIED TOAD.

Many Toads possess the instinct of death-feigning, but this species shows this curious method of self-protection in a high degree.

One peculiarity of the Fire-bellied Toad is the great size attained by its tadpole, which is further characterized by the unusual development of its tail-fin.

And this leads me to point out that in one respect all frogs and toads are really some of the most marvellous of all animals; for what can be more wonderful than the development of the purely aquatic, vegetable-feeding, gill-breathing, limbless, long-tailed tadpole into the amphibious, carnivorous, lung-breathing, four-legged, tailless frog or toad? It is, indeed, a metamorphosis paralleled only by the development of the creeping caterpillar into the winged butterfly; the quiescent pupa-stage being, however, missed in the case of the frog. Most of us have probably noticed the manner in which tadpoles or frogs, after their emergence from the spawn, develop first the hind and then the fore limbs, while they exchange external for internal gills, and when both pairs of legs are fully formed gradually absorb their long, fish-like tails, till they finally leave the water as miniature frogs.

But this ordinary process is only one of many marvels connected with the development of frogs and toads. Many kinds of frogs, for instance, spend their lives in trees, where there is no opportunity for the tadpoles to live in water; but Nature is not to be beaten by trifles of this kind, and we accordingly find that the eggs are much larger than usual, thus permitting the whole of the tadpole-time being passed within the egg, and allowing the young to make their appearance in the world as full-blown frogs. Again, a frog inhabiting dry districts in the Solomon Islands, lays eggs of the size of marbles in rocky situations, from which also emerge in due course perfectly-formed frogs. Other tree-frogs take advantage of the moisture contained in the cavities of the boughs or leaves of

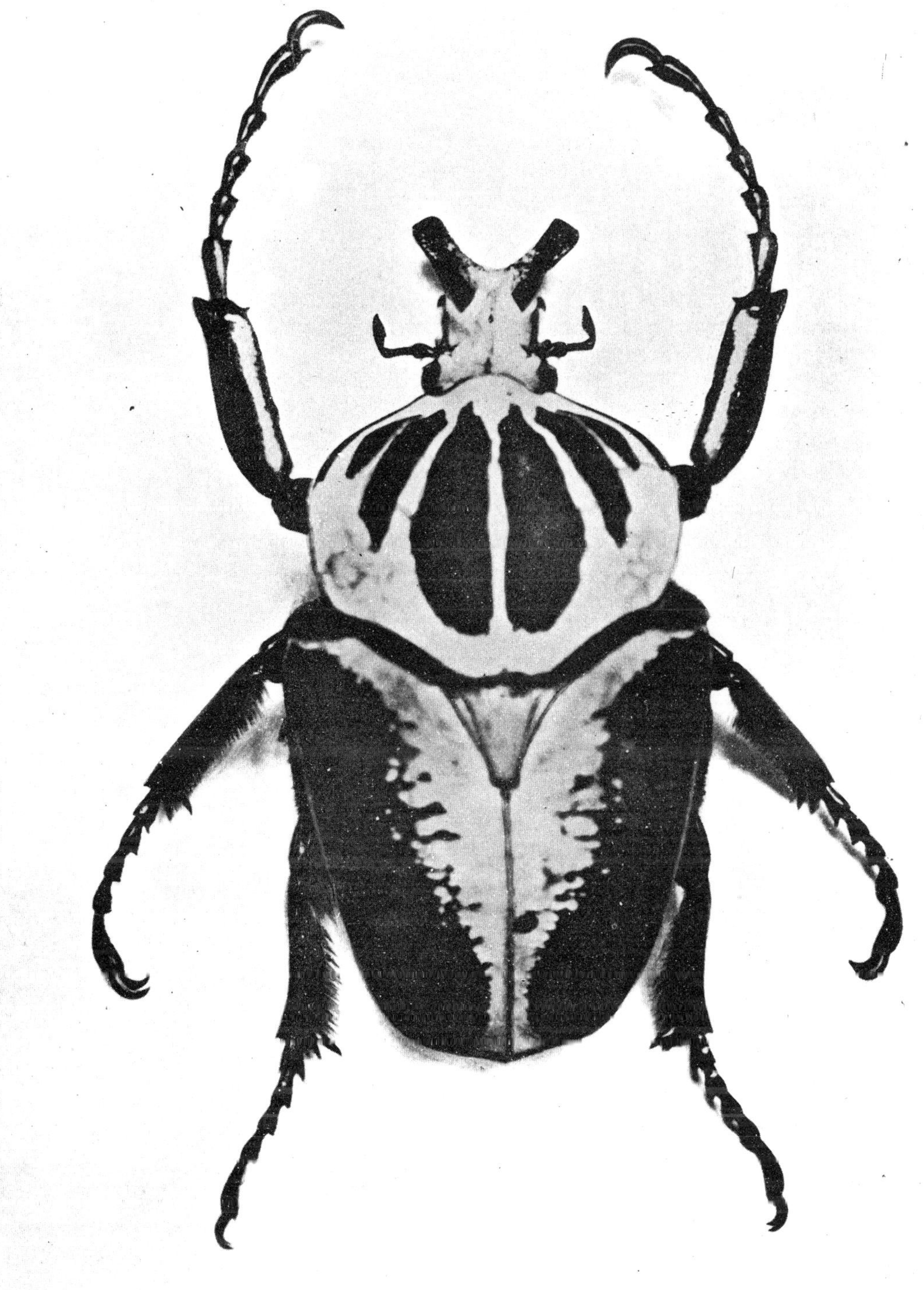

Photo by] [*J. J. Ward, F.E.S.*

THE GOLIATH BEETLE.

This is probably not only the largest Beetle but the largest insect known. It is of a chocolate-brown colour marked with creamy-white, while the large wings glitter with a beautiful green lustre when in flight, and is shown here about natural size.

THE GOLIATH BEETLE.

The horns on the head of the male Beetle are said to be weapons for fighting its adversaries, but they are also of use in tearing the bark of certain species of vine from which the insect sucks the sap.

the trees in which they dwell; so that the tadpoles may undergo their development either in little pools or in masses of froth like that of our well-known "cuckoo-spit" insect.

Stranger still is the development of the so-called marsupial frogs of South America, the females of which have a long pouch in the hind part of the back, wherein the fifteen or sixteen eggs are placed by the male, there to hatch and produce tadpoles which undergo the usual course of development. Even this is surpassed in the case of Darwin's frog of Chile, the males of which take the eggs into a special chamber in their throats, there to remain till they develop into fully-formed frogs. The nursery arrangements of the far-famed Surinam Toad will form the subject of a special notice.

THE WORLD'S LARGEST BEETLE

BY JOHN J. WARD, F.E.S.

AMONGST land animals greatness in bulk has not proved a success. Many huge extinct animals are now represented by quite diminutive forms. Indeed, development to a large size is, as it were, a blind alley which leads nowhere. As a consequence, insects, which have always remained small, have become the most abundant of living land animals, their species outnumbering those of all other groups put together. The Goliath Beetle is not only the largest among living beetles, but possibly the giant of all insect life—past and present.

The full-fed grub constructs an oval-shaped cocoon of mud, with very thin walls, which are strengthened round the central part by a thickened belt. This cocoon varies in size according to the amount of food consumed by the individual, and some are as large as a swan's egg.

In habits the Goliath Beetle is very like our Rose Chafer, and especially so during its flight, for it does not extend its horny wing-covers at the same time as its membranous wings. They are only momentarily raised to allow the flying wings to project from the sides, but are not separated.

While our Rose Beetles feed largely on the nectar of flowers, the chief support of the Goliath Beetle is the thick juicy sap of some species of vines, which the male obtains by wounding the bark with the horns on the front of his head, and which he works at very persistently, the female coming to the feast when the bark is broken. The horns on the head of the male are also said to be used in fighting his adversaries.

Dr. Bennet states (in the *Transactions of the Entomological Society of London*, 1899) from his own observation of one of these insects in Bululand, West Africa, that, while soaring around a vine hanging from an immense tree, it gave forth " a sound not unlike that of steam emerging from a small safety-valve." He also states, its " large wings glitter and scintillate with a most beautiful greenish lustre." Its movements were described as being so rapid that it was only the " strange sound " that first arrested attention.

These huge beetles are chocolate in colour, marked with creamy-white, and have a surface soft as velvet ; but their remarkable appearance and giant size are a menace to their species, as they are eagerly sought for by collectors, and their ultimate extermination from this cause is a possibility. In addition, they are used by the natives as ingredients for soups and fricassees.

THE GNOME-CRAB

BY EDWARD STEP, F.L.S.

AMONG the numerous varieties of form to be found among Crabs, the family represented by our photograph is singular in the fact that its members are almost globular, and, therefore, look very uncrablike. Unfortunately, the species figured has no English name. Perhaps the name of Gnome-

Photo by] [*E. Step, F.L.S.*

THE GNOME CRAB.

The Gnome belongs to a family of Crabs that are peculiar for their globular form, and is remarkable besides for his thick coat of short hair, like shabby velvet; so that when he tucks in his claws he looks much like a dark pebble. He further hides his identity under a sponge attached by means of a special pair of nippers to his back. (Natural size.)

Crab would best fit its appearance as in the photograph. In ordinary his form and proportions are not so clearly displayed, for it is his custom to carry a mass of living sponge upon his back; or, if the sponge is not there, its place is taken by a growth of polyps of various kinds.

The Gnome is best known as a native of the Mediterranean and the Adriatic, but it also occurs somewhat rarely on our own shores; though this latter fact was not known to British naturalists until the year 1824. The first recorded British example was discovered in Billingsgate Market, among some oysters that had come from Whitstable. Later it was found off the Sussex coast, and again off Plymouth. The individual who sat for the portrait on page 485 came from Cornish waters. In addition to his globular form, the Gnome is remarkable for his close pile of short hairs

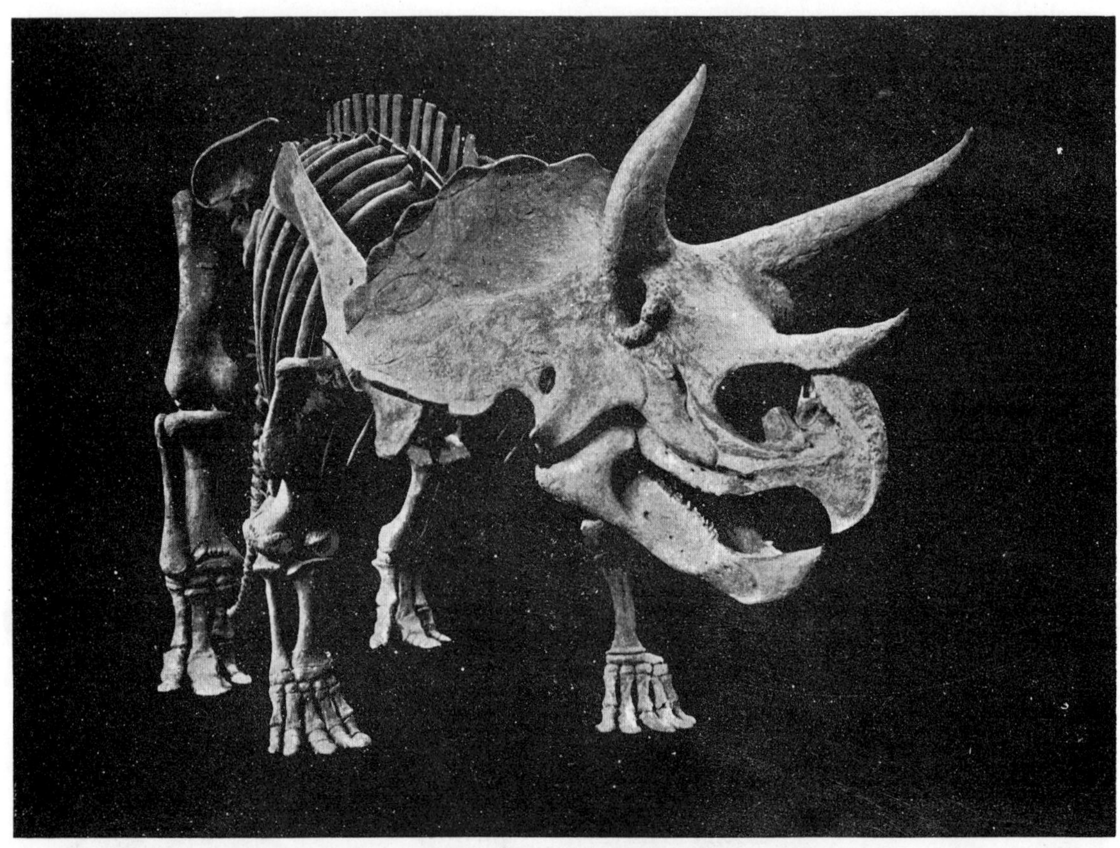

THE SKELETON OF A TRICERATOPS.

This strange creature was one of the group of vegetarian Dinosaurs, which developed during the latter part of the Secondary Epoch, and which resumed progress on all four legs after its ancestors had for some ages hopped about like kangaroos on the hind limbs only. In full-grown specimens the enormous skull was from seven to eight feet long.

of a dull brown tint which looks like shabby velvet. He is moving lazily in the picture, but when quite at rest he tucks his limbs (or most of them) under him, and then looks much like a dirty, dark-coloured pebble. The exceptions to this general covering of velvet are the claws and the pink nippers, which appear to have been exempted on the principle that "cats in gloves catch no mice"; crabs in gloves might also find it difficult to catch their prey. The two hinder pairs of legs have their attachment almost on the upper side of the crab, and are usually—the fifth pair always—laid on the back, and provided with nippers.

There is a reason for the last-named departure from rule in the case of crab-feet. The nippers are a special development for holding the sponge we have referred to as usually concealing the Crab. Bell, who wrote on the British crabs and their allies, in 1853, strangely confused cause and effect

Photo by] [*H. J. Shepstone.*

THE TRICERATOPS.

Besides the three enormous horns, from which the animal derives its name, the Triceratops was possessed of a curved beak at the end of the upper jaw. The skin was probably armed with knobs and plates like those of the crocodile, and the skull developed behind the head into a toothed bony frill.

in this matter. He says he has received a number of young specimens of this Crab from the Mediterranean, and finds that in every case they have been overgrown by a sponge, which even hides the last two pairs of legs and presses them against its back, rendering them immovable. The truth is, of course, that the Crab has adapted its legs to take this uncomfortable-looking position in order that it may securely hold its sponge-messmate as a disguise, and under cover of this inoffensive mantle approach its prey without arousing fear or suspicion.

Some of the foreign relatives of this marine Gnome hold bivalve shells and leaves over their backs in the same manner and with the same object in view.

HORNED, FRILLED AND ARMOURED DINOSAURS

BY SIR HARRY JOHNSTON, G.C.M.G.

VERY marvellous amongst the most amazing group of reptiles ever evolved in past ages (the Dinosaurs) was the Triceratops genus, which came into existence in Western North America in the last period of the Secondary Epoch, at a time when many groups of reptiles were "over-doing it," becoming, that is to say, so fantastic, so huge, or so fettered in their movements by extravagant developments of armour, length of tail, or head-covering, that they must at last have found it very difficult to cope with active enemies or to reproduce their species with sufficient rapidity to keep pace with the growth of mammalian rivals, whose later development of bulk and limb they were anticipating.

Photo by] [*Sir H. H. Johnston, G.C.M.G.*

THE TRICERATOPS.

This is a remarkable restoration of the outward form of Triceratops, a vegetarian Dinosaur which existed at the close of the Secondary Epoch in the United States. Herr Pallenberg, the designer of this group at Stellingen, has shown us a male, female and young one.

It has already been related in these pages how the sub-order of the Dinosaurs arose, addicted to a purely vegetable diet, and described as "bird-limbed," because they walked mainly on their hind legs, which became very long, and used the shorter fore-limbs for grasping. But a section of these bird-limbed Dinosaurs relapsed into going about on all four legs, and although their hands and arms continued to be a little smaller and shorter than the feet and legs, they grew so thick and substantial as not to exhibit much disproportion. Prominent for its extraordinary appearance amongst these vegetable-eating Dinosaurs of very mammalian appearance was the monster called Triceratops, from the three bony projections or horns growing from the upper surface of the skull, one pair just above the eyes and a single horn from the top of the nose. There were probably no front teeth, but the fore parts of the upper and lower jaws were fashioned into something like a hooked beak. But there remained a substantial set of grinding teeth in the back part of both jaws.

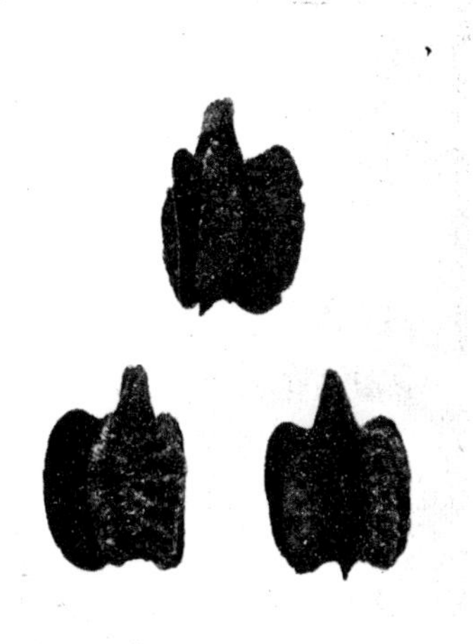

Photo by] [*Hugh Main, B.Sc., F.E.S.*

THE EGG-CAPSULE OF THE LEAF INSECT.

The mimicry of this insect is apparent throughout its life-history, while the eggs have all the appearance of ribbed seeds. They are here shown twice natural size.

It is quite possible that these extravagant Dinosaurs had ceased to lay eggs, and, like some living reptiles and other extinct forms, had taken to developing the young within the body and giving birth to them fully alive and able to take care of themselves. And no doubt, as occurs with most very specialized forms of vertebrates, the number of young produced at a time may have become fewer and fewer, until at last the Triceratops, so admirably reproduced in sculptured form by Mr. Pallenberg at Stellingen, may have had only one calf at a time. Through the enterprise of Mr. Hagenbeck, a wonderfully interesting group, which we may suppose to be father, mother and baby, of this Horned, Frilled and Armoured Dinosaur, is set forth for exhibition amid very natural surroundings of wood and water.

LEAF INSECTS

BY SIR HARRY JOHNSTON, G.C.M.G.

MIMICRY in insects is a subject of vast complexity, deep interest, and one which still arouses fierce conflicts of opinion amongst students of natural history. Whether it arises to some extent from the deliberate exercise of will power on the part of the insect—which, being conscious of its resemblance, partial or even extraordinary, to its surroundings, deliberately makes use of this similarity for purposes of eluding the observation of enemies or disguising itself in order to prey on unsuspecting victims—or whether, on the other hand, the mimicry has been begun and has become perfected by a series of accidents resulting favourably for the species, are points not yet determined as universal articles of faith on the part of biologists. To the observer with a normal mind it would seem to be an inevitable conclusion that these insects, like birds, fish, reptiles and mammals, are completely conscious of their resemblance in colour or form to their surroundings and make full use of this advantage, deliberately choosing to place themselves where the resemblance may be most effective. When travelling through the forested regions of the Uganda Protectorate

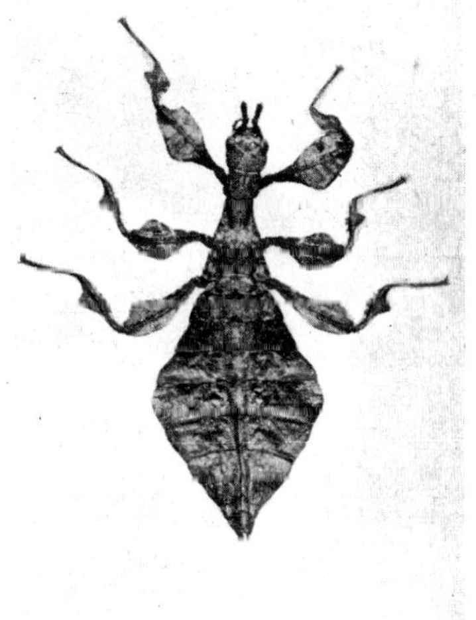

Photo by] [*Hugh Main.*

THE YOUNG LEAF INSECT.

The eggs hatch out into reddish-coloured and active Leaf Insects, who become green soon after they begin to feed upon leaves. (Three times natural size.)

I have on more than one occasion been startled by the marvellous mimicry of certain butterflies. Unfortunately they were so wary that they eluded my net, and have perhaps remained uncaptured and undescribed down to the present day, since I have met with no specimens of the "Greenleaf Butterfly" in any collection. These butterflies, when they closed their wings, were absolutely indistinguishable, even to a sharp eye, from bright green leaves. Had it not been that they were not completely confident of their assimilation and flew away from the twigs on my approach, I should never have guessed that they were anything but a collection of green leaves. Most people are familiar with the butterflies and moths that resemble in almost every detail the dried or dead leaf. Other moths exactly simulate bark or lichen. There are mantises in Central Africa which imitate to a nicety large pink flowers with a green stalk and a green sheath to the calyx. But perhaps the most spectacular and marvellous of insects in mimicry of leaves are to be found amongst the Carabid Beetles; and in a widely different group, the Stick and Leaf Insects of the great order Orthoptera—the same order which includes the mantids, the cockroaches, earwigs, crickets and locusts.

Photo by] [*Hugh Main, B.Sc., F.E.S.*

THE LEAF INSECT.

This Leaf Insect is a native of Seychelles. Together with the rest of its family (which is very abundant in Assam), it is a remarkable instance of protective resemblance. Leaf Insects range across the islands and continents of the Indian and Pacific Oceans from the Seychelles to Fiji.

The Leaf Insects in the accompanying photographs only occur in the tropical regions of Asia and the islands of the Indian and Pacific Ocean. They do not extend their range either to Africa proper or to America, yet they are found as far west as Mauritius and the Seychelles, and as far east as the Fiji Islands, or even (it is said) Tahiti. Curiously enough, they seem to flourish particularly well on islands rather than on broad continental areas; but the Leaf Insects here illustrated would probably come from Assam, several species being found in India and Malaysia. These remarkable creatures are about an inch and three-quarters long. The resemblance to a leaf is perhaps more marked in the female than in the male, the female having very short antennæ, while these appendages of the head are relatively long in the male. The chief resemblance to leaves arises from the expansions of the limbs and the coverings of the wings. In the female, which is far more leaf-like than the male, and is consequently much more stationary, the hind wings are not present, or, at any rate, are only represented by a minute rudiment. The female is unable to fly. On the other hand, the male of this genus has merely rudimentary front wings and large hinder wings, and is able to fly, at any rate, to some extent. The abdomen, also, in the male is not nearly so large, flattened, or leaf-like as it is in the female. Both sexes, however, are very leaf-like in their colouring;

Photo by] [*Sir H. H. Johnston, G.C.M.G.*

LEAF INSECTS.

This is a puzzle-picture from life, and has been photographed in the collection of Mr. Carl Hagenbeck at Stellingen. It shows several live Leaf Insects feeding on a yoang oak-tree.

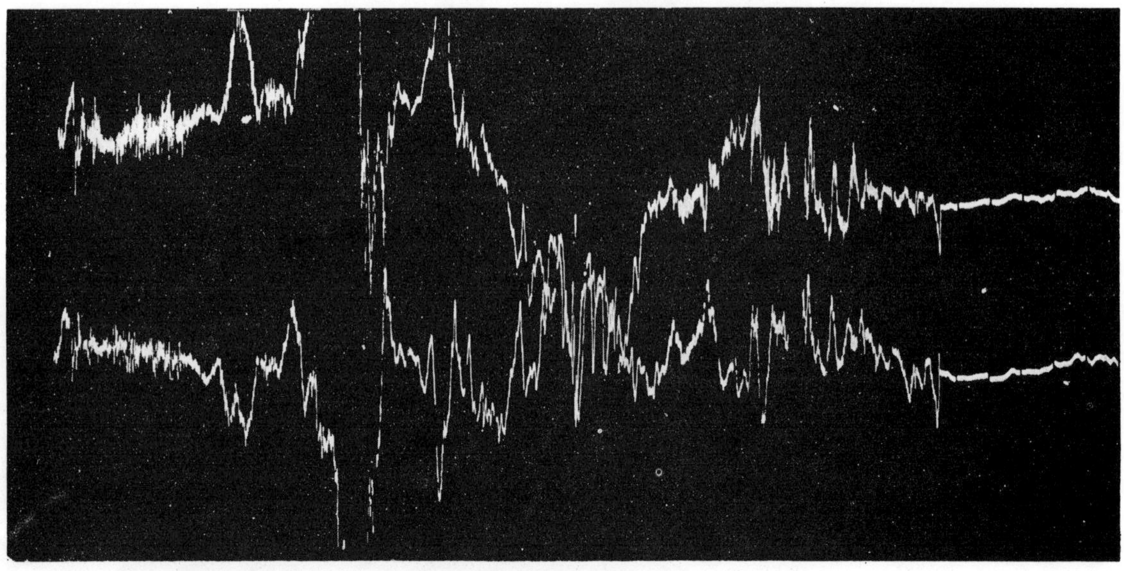

THE TRACING OF A MAGNETIC STORM.

Automatic register of the movements of two of the magnetic needles at Greenwich Observatory during the "magnetic storm" of February 13-14, 1892. The beginning of the trace shows the usual character of the magnetic movements, but at 5h. 30m. a.m. on February 13 a sudden vibration was shown by the needles in all the magnetic observatories of the world, after which their oscillations became very violent and rapid.

but the range of colour is considerable, even apparently in the same individual and in normal health. Those from which my photographs have been taken (with the kind permission of Mr. Carl Hagenbeck), and which are exhibited at the Insectorium in the Tierpark at Stellingen, are not merely emerald-green or grass-green ; but some of them are yellow or reddish-yellow, especially when in contact with slightly fading oak-leaves. Others, again, are bright sulphur-yellow or russet-green. The resemblance in general tone to the foliage of the oak (on which these insects were feeding at the time of my visit) was so extraordinary that, although I possess very good sight, I thought at first that the Curator was merely holding up for my inspection bunches of oak-leaves ; and it was only when I saw the insects begin to move that I realized they were there. The likeness to vegetation begins with the very extrusion of the egg from the body of the female, the eggs being of large size and enclosed in still larger capsules exactly resembling ribbed seeds. So close is the resemblance that it might even deceive the eye of a botanist, who would believe that he had picked up a seed-capsule enclosing a seed. These eggs fall to the ground and in due time hatch out into minute, reddish-coloured, active young Leaf Insects, not unlike baby locusts in appearance. They at once proceed to crawl up the stems of plants, and soon after they have begun eating the leaves they become green in colour. They do not at first possess to the full the remarkable developments of the front or hind wings (according to sex), or the leaf-like expansions of the legs. But these soon appear, and until they have come to perfection the insect is protected from observation by the artfulness of its attitudes and its " colourable imitations " of leaf or twig.

The food of these Leaf Insects is purely vegetable. They eat nothing but leaves, and can actually be deceived into nibbling each other in the belief that they are eating leaves ! But it is said that they do not inflict mortal injuries on their companions, as they merely nibble away the appendages to the limbs, which soon grow again. In fact, by their rapid assimilation of vegetable matter they really become, in course of time—as the natives of the countries in which they exist believe them to be—walking leaves, their very colouring matter being hardly more than the assimilated chlorophyll of the plants on which they live.

SPOTS ON THE SUN

BY E. WALTER MAUNDER, F.R.A.S.

"THERE are spots on the Sun" has passed into a proverb, with the pessimistic meaning that there is no character so bright but that an attentive search can detect some slight blemish or imperfection in it. But the spots really on the Sun add, without doubt, to its importance to us, and certainly to its interest.

What are these spots, and what is their significance? The Sun, as examined under good conditions with a telescope magnifying from one hundred to two hundred diameters, shows a very brilliant surface of nearly uniform brightness. Nearly uniform, not quite; for a close examination shows that the whole is covered with a very delicate network. Everywhere there are minute points of light a little above the average in brightness; everywhere in the intervals between these bright points there are dots and lines that are a little less bright than the average. A photograph of the Sun, taken on a sufficiently large scale to bring out this delicate structure distinctly, looks like a highly magnified picture of an irregularly woven fabric. The unvarying brilliancy of the Sun's surface, as seen by the naked eye, really covers a very intricate structure, and considerable ranges of brightness; but these are all details in high light; there is no question of real darkness on the Sun, there are only differences of degree in brightness.

In general these differences are not great, and might even be overlooked on a hasty glance, or if the "seeing" were not good; and in any case, this "granulation," or "mottling," or "photospheric network," as it has been termed, is too minute, and changes too quickly in its details, for much to be made out from it. The Sun would, therefore, be a very uninteresting object for telescopic scrutiny if larger and more definite features were not recognized from time to time, and such features are found in the "spots" and the "faculæ."

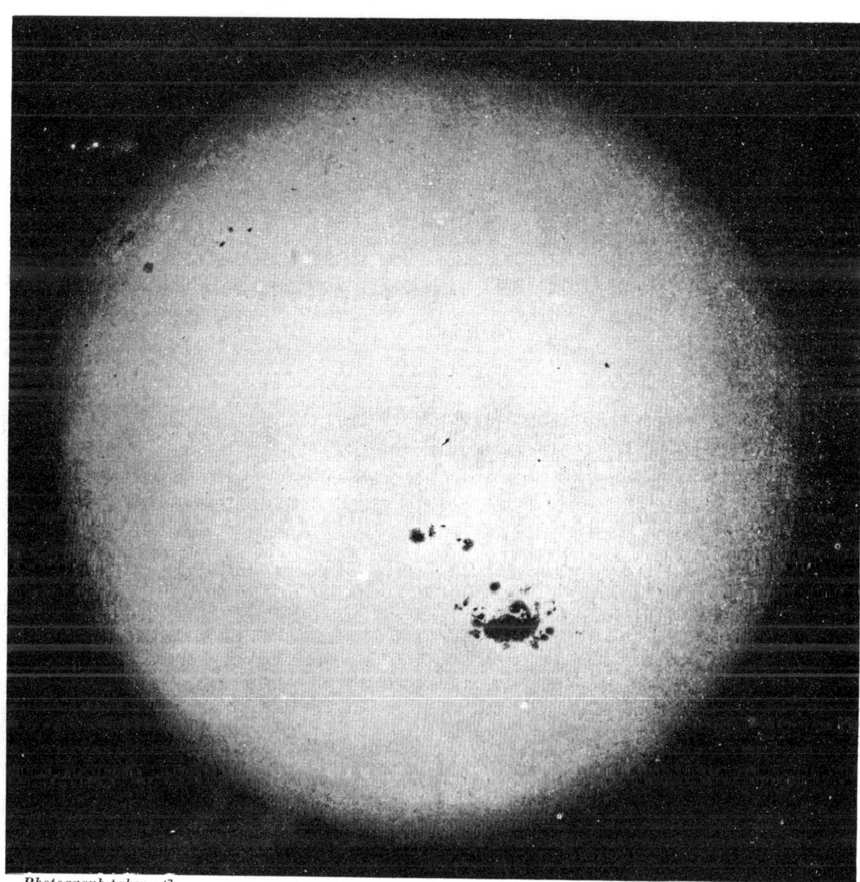

Photograph taken at] [*Greenwich Observatory.*

SPOTS ON THE SUN.

Photograph of the Sun taken on February 13, 1892, showing in the southern portion one of the largest spots ever seen upon the solar surface. The total area of the great spot was 3,000 millions of square miles; including the small spots close to it, the entire area of the whole group amounted to 3,500 millions.

The "spots," though really bright

objects, appear dark, sometimes very dark, through contrast with their surroundings ; the " faculæ " are very bright—bright even as compared with the brightest parts of the Sun's surface. The " spots " appear to be exaggerations of the least brilliant interstices of the " mottling " ; the " faculæ " of the most brilliant ; but are so much larger, more definite and stable than the fine details of the photospheric network as to make them quite distinct classes of the solar details. But the gradation is complete from the largest Sunspot down to the minutest detail of the granulation, and it is not always easy to distinguish between a very small, faint " pore," as the smallest spots are called, and some of the larger, more distinct, of the interspaces of the " mottling."

Spots vary much in size, appearance and history, but the type most frequently seen begins usually after the " photospheric network " in some region has been observed to be apparently perturbed, and to take on a coarser appearance than usual. Then two small " pores " appear near to each other, and begin to grow and move apart. One of the two soon shows itself as a round, sharply-defined object, especially on its advancing side, and before long it becomes the leader of the group, moving forward on the Sun's surface away from the rest of the group, often at the astonishing speed of three hundred miles an hour. Meanwhile the other member of the original pair has been growing, possibly even more quickly, but without showing the same symmetry and definiteness. The two are now widely separated in consequence of the rapid forward motion of the leader ; but new spots have broken out in the gap, and the entire group consists of a long chain, the largest members of which are the original pair, now at the two ends of the chain. Then the general bright surface of the Sun—the " photosphere "—which is brighter than usual, and apparently piled up in front of the leader spot and on either side of the chain, begins to flow over and invade, first the smaller middle spots, and then the rear spot. The accompanying photograph, taken on July 31, 1906, shows a fine but short stream of spots at this stage ; the small middle spots being almost hidden by this boiling over of the photosphere, while the rear spot is crossed by many bridges or invaded by streams of bright photospheric material, actually rushing down into its depths. In the photograph taken three days later this bright material has all but covered the hinder portion of the group, while the leader spot has still retained its outline ; its

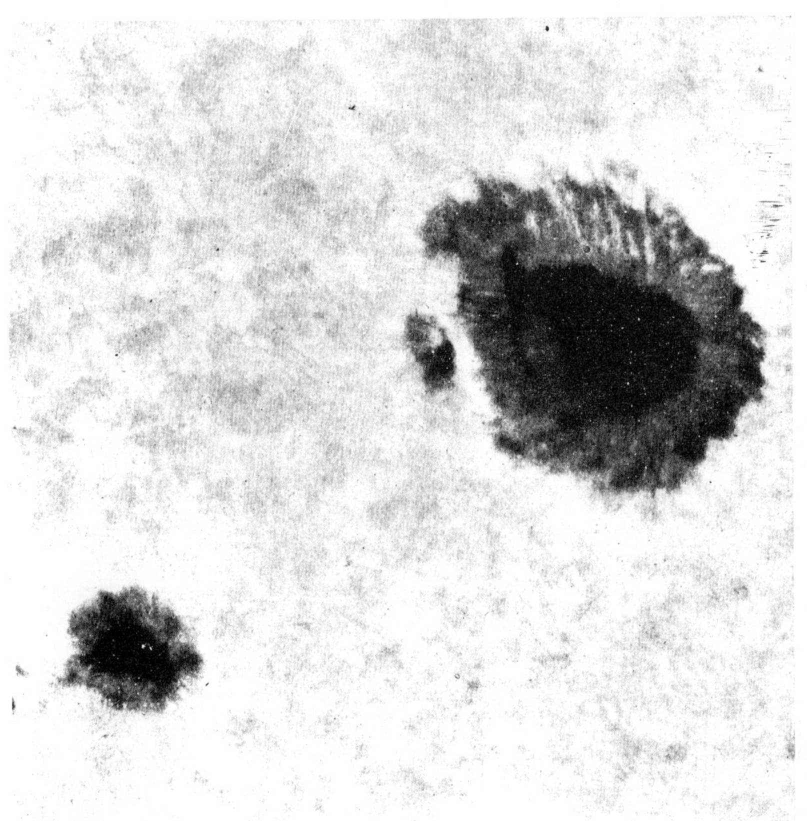

Photograph taken at] [*Greenwich Observatory.*

SPOTS ON THE SUN.

The larger spot is about 800 millions of square miles in area, or 32,000 miles in diameter. It shows a spot in its stable condition, this spot lasting for more than three months. The smaller spot is only about 120 millions of square miles in area, and is a member of a group that lasted only a few days and changed rapidly.

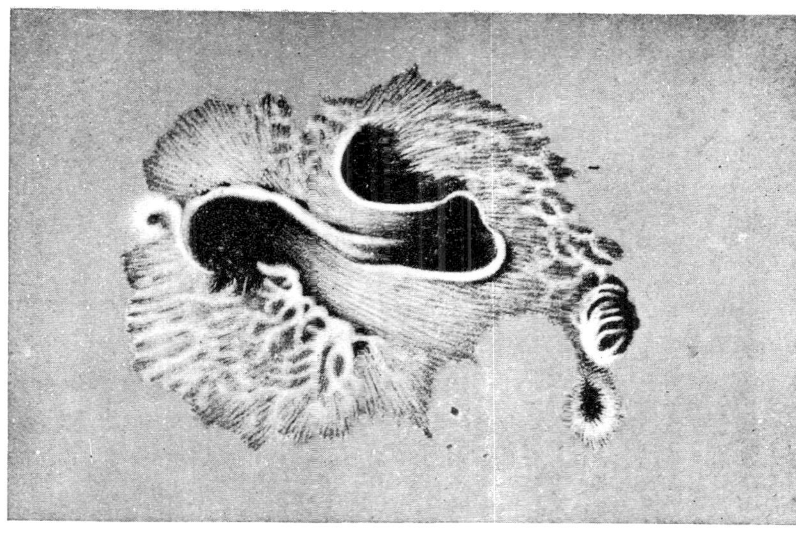

SPOTS ON THE SUN.

Drawings of Sunspots made at the Observatory of Harvard College, U.S.A., illustrating the manner in which bright "bridges" form across spots, and the bright material of the general solar surface overflows and hides, or pours down into and fills up, spots when in the stage of decline.

apparently elliptic form is due to the fact that the group is now seen near the edge of the Sun, and is, therefore, strongly foreshortened, as it is being looked at nearly sideways. The next phase in the history of a group of this type is for the whole of the hinder portion to disappear under the photospheric material, leaving the leader spot alone. As an isolated round spot, it may last for some weeks, or even for two or three months, without undergoing much change beyond a slow contraction. Sometimes this process goes on until the spot has filled up ; at other times it breaks up into a number of small fragments before its disappearance.

This particular stream was a short one for its size, being *only* about fifty thousand miles in length. Some groups of Sunspots have been known to stretch out to a length of nearly two hundred thousand miles ; other spots, like the great one shown on the photograph taken on February 13, 1892, have been less drawn out and more compact, but have covered a greater total area. The principal group on that photograph covered an area of three thousand five hundred millions of square miles, more than eight hundred times the area of Europe ; seventy worlds as large as our own could have lain side by side within its bounds. We do not know what the depth of the spot was ; without doubt the depth was very shallow as compared with its area ; but this would still be the case if it were some few thousands of miles deep. The actual dimensions of Sunspots are, therefore, very often very great indeed, as compared with that of the earth ; worlds as large as our own could lie in such a spot as peas in a saucer.

But, as compared with the Sun itself, Sunspots are very small. It is very seldom, even when they are largest and most numerous, that one five-hundredth part of the Sun's surface is covered by them. And as the very darkest part of the darkest Sunspot is really very bright, and the faculæ

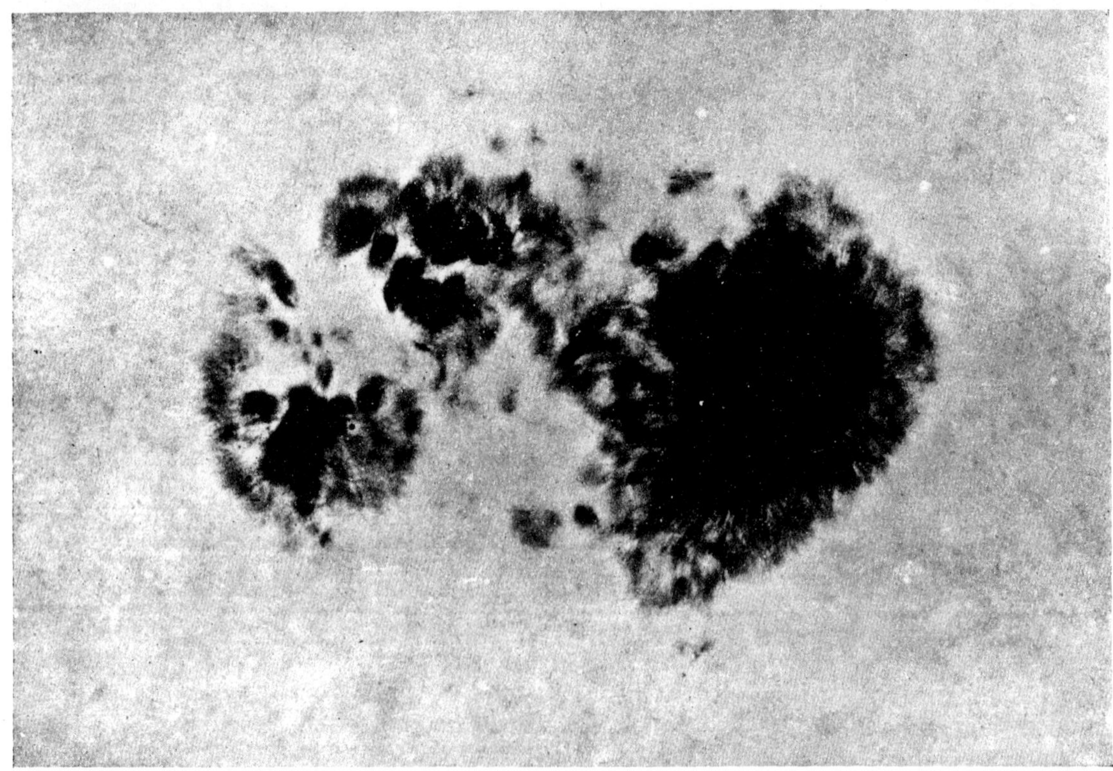

Photograph taken at] [*Greenwich Observatory.*

SPOTS ON THE SUN.

A compact stream of spots, about 50,000 miles in greatest length and 30,000 in greatest breadth, and covering 2,500 millions of square miles. No fewer than fifty worlds as large as our own could lie side by side within its borders. The group is beginning to decline, and the bright material of the Sun's general surface is overflowing the following half of the group.

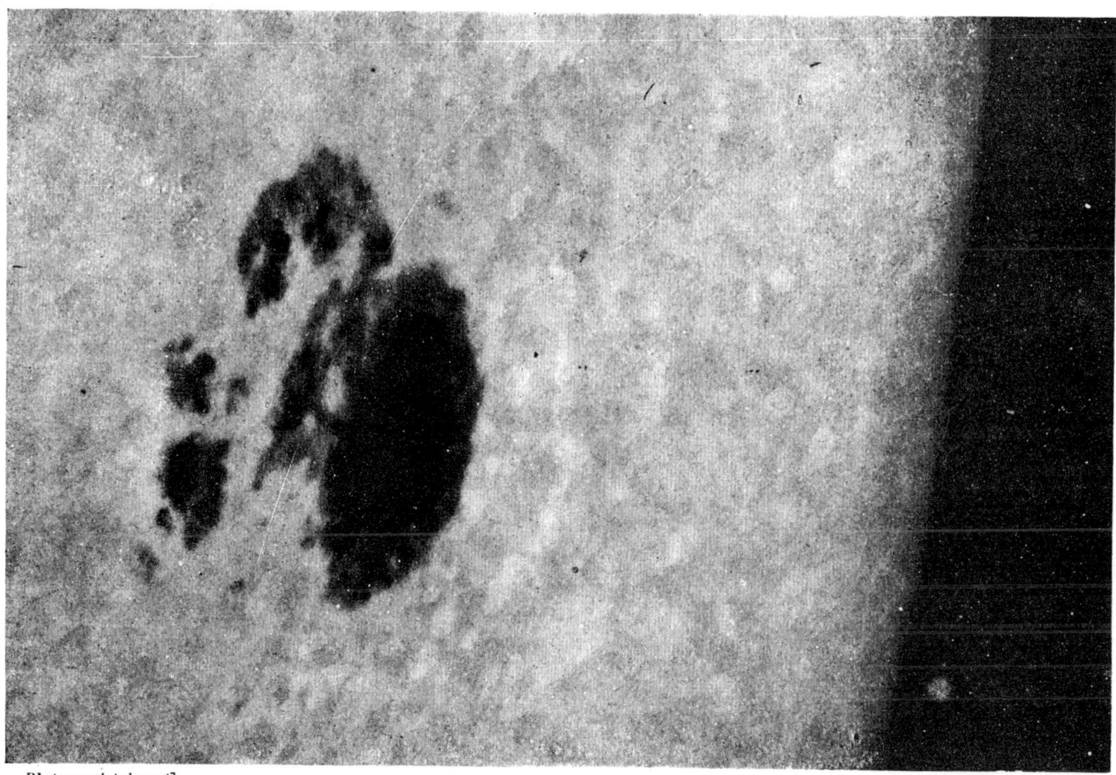

Photograph taken at] [*Greenwich Observatory.*

SPOTS ON THE SUN.

The same group of spots as on page 496, but seen three days later, when the rotation of the Sun had brought it near to the edge of the disc. It is therefore seen much foreshortened. The bright material of the Sun's general surface has overflowed, and almost entirely hidden the following half of the group, but has left the leading spot almost untouched.

which always accompany a spot are considerably brighter than the Sun itself, there is very little loss of light or heat, if any, even during the greatest outbreaks that have ever been witnessed. Up to the present, therefore, it has been a moot point whether the Sun sends us most light and heat in years when spots are large and numerous, or in years like the present, when they are almost entirely absent. At one time Sir Norman Lockyer, and other leading solar observers, held that the total radiation of the Sun was greatest when there were most spots ; at the present time the opposite opinion appears to be gaining ground ; but the evidence in either case is very slight and conflicting.

But what are the spots ? The Sun itself is held to be a globe of intensely heated gases, surrounded by a thin shell of luminous clouds where some of these gases are partially condensed as they come into contact with outer space. This shell of intensely brilliant cloud is the photosphere, and the granulations of the surface are the regions where such condensation is greatest and the clouds thickest and most brilliant. A spot, then, is a region in which these bright photospheric clouds are not able to form ; and it is very instructive to watch how the photosphere spreads itself above and over a spot when this is in its declining stage.

Many attempts have been made to trace a connection between the numbers of spots on the Sun and the changes in terrestrial weather. Rainfall, hurricanes and cyclones, Indian famines, have all been made the subjects of inquiry ; so far without any result that can be described as satisfactory or conclusive. In this country we never have a hot summer, or a very wet or cold one, without some attempt being made to foist the responsibility for it upon the solar changes. But seasons of such pronounced character have happened almost indiscriminately at times when Sunspots were specially numerous, and when there were none at all.

In one direction, and as yet in one direction only, has a distinct and unmistakable connection been traced between the Sun's spot activity and terrestrial changes ; and this connection is one that, as a general rule, does not affect the interest or comfort of the world at large. It is, of course, generally known that the magnetic needle does not point directly to the true geographical north, and that the amount of the deviation changes slowly from year to year. It undergoes also a slight change in the course of each day, moving a little westward in the afternoon, and eastward again during the night and morning. The amount of this daily movement varies with the time of the year, being greatest in summer-time ; it varies also from year to year, being greatest in years when there are many spots. But besides this daily range the magnetic needle is also subject to sudden disturbances of a very remarkable character, and these also are connected with some action by the Sun, an action which is often closely connected with the outbreak of a great spot. Thus on February 13, 1892, when the great Sunspot shown in the photograph on page 493 was on the disc, the needles in the magnetic observatories all over the world suffered simultaneously a sudden twitch, after which they began to move rapidly backward and forward, instead of following their usual slow, dignified oscillation once in the twenty-four hours. A month later the rotation of the Sun upon its axis brought the same spot round a second time to the same apparent position upon the disc, and immediately a fresh convulsion seized the magnetic needles. More recently still a yet greater disturbance occurred on September 25, 1909, when for several hours the submarine cables all over the world ceased working, and this coincided with the presence of a very beautiful and active spot, though not one of the largest size. Such disturbances are due to streams of minute invisible particles shot forth from restricted areas of the Sun—most frequently areas in which Sunspots have formed or are forming—and driven out in long rays through space. From time to time one of these rays will overtake the earth in its revolution round the Sun, and strike it on the sunset side, producing instantaneously a disturbance of the magnets all over the world.

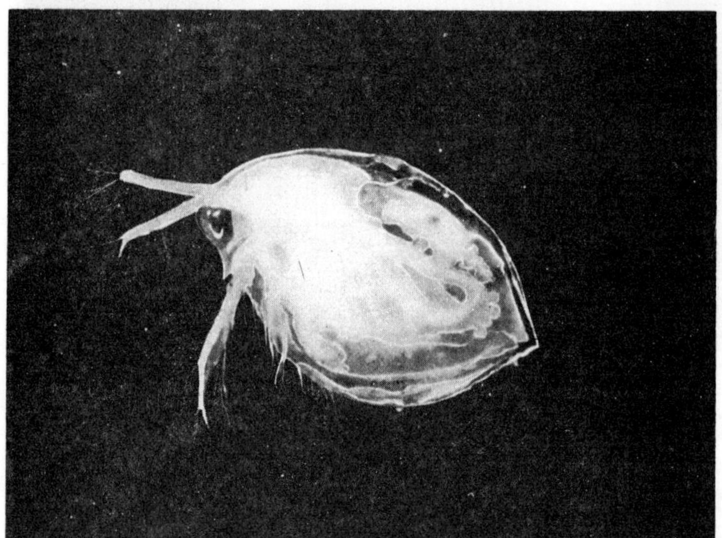

Photo by] [*J. H. Crabtree.*

A WATER FLEA.

Water Fleas are relatives of the lobsters and crabs, and, like them, are clad throughout in armour. This species is Daphnia, which prefers the deeper, clearer space of the freshwater pond. Its shell is so fine that the whole structure of the little creature is distinctly seen.

WATER FLEAS

BY K. G. BLAIR, B.SC.

If a sample of the water contained in almost any pond be taken home for examination at leisure, careful inspection in a good light, even without the aid of a magnifying glass, will almost certainly reveal a number of small organisms, the largest not so big as a pin's-head, swimming about, especially against the sides of the vessel containing them, in a series of sharp, jerky movements. Amongst these are sure to be the minute crustaceans—relatives of the lobsters and crabs—popularly known as Water Fleas, and a good pocket lens should enable the observer to recognize one or more of the kinds here illustrated. Of these Daphnia usually prefers the deeper clear spaces of the pond, coming

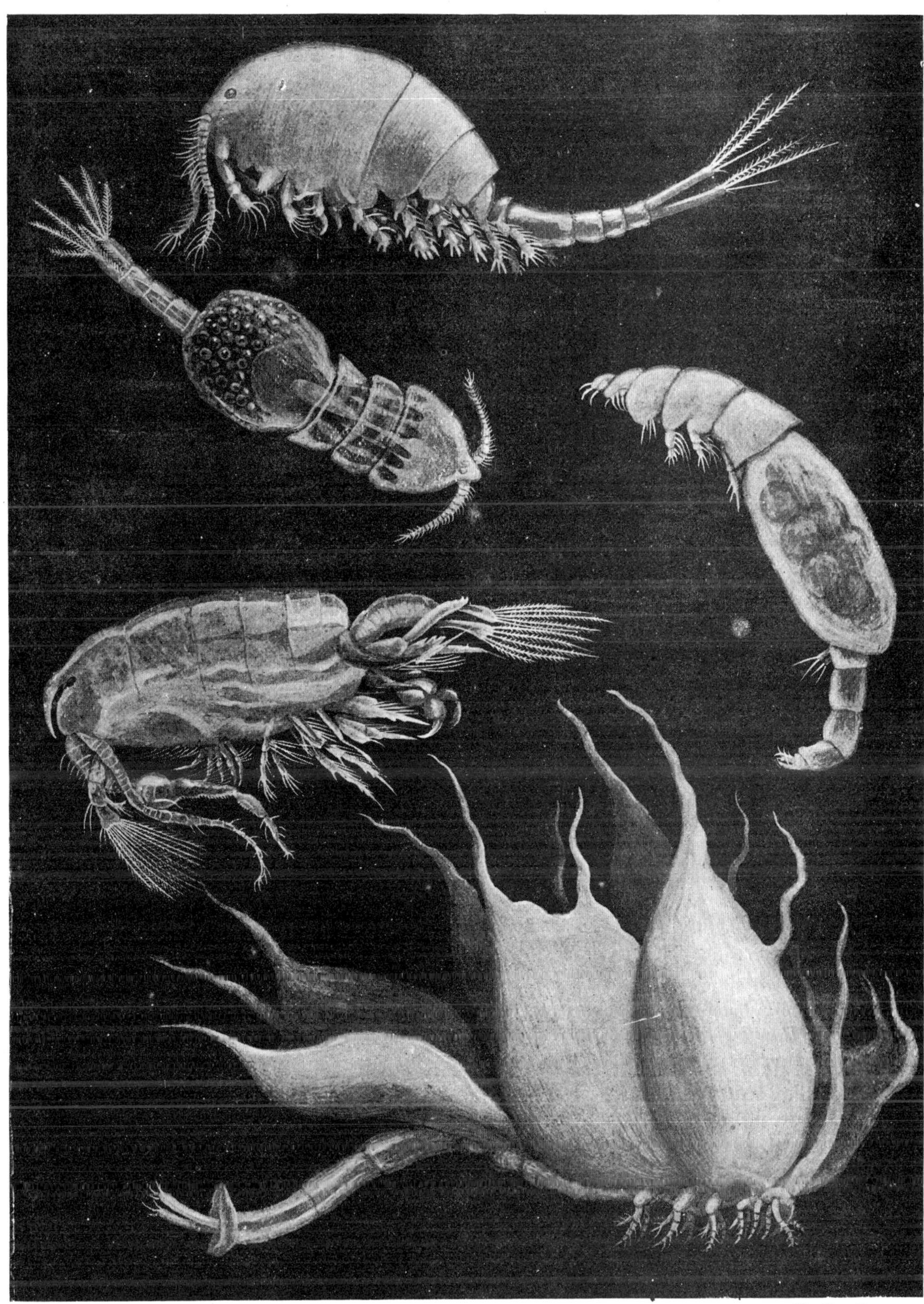

[*By Theo. Carreras.*

WATER FLEAS.

There are many curious forms amongst the Water Fleas of marine life, and some are here shown. Notice particularly the Butterfly Flea at the bottom of the page, with large wing-like bladders. All are largely magnified

A WATER FLEA.

Cyclops, the one-eyed, is very abundant in fresh water and inhabits the muddy ooze of the pond. The single eye is placed in the middle of its forehead. Note the curious egg-pouches of the female. With the rest of the Water Fleas, it is a very minute creature, about the tenth of an inch long.

near the surface on dull days, but retiring to the lower layers in bright sunny weather; sometimes on looking down into such a pond a pale pinkish cloud will be observed, and on bringing some of this up in a net or bottle, it will be found to consist entirely of a dense struggling mass of Daphnia. Cyclops, on the other hand, so-called from his having but a single eye situated in the middle of his forehead, like the giant of Greek mythology, prefers a soft muddy bottom in shallow water, while the active, fussy little Cypris prefers to paddle about water-weeds.

In Daphnia the body is enclosed in a shell-like carapace from which only the head and the large branched swimming-arms project. The latter are not true feet, but are more comparable to the "feelers" of lobsters; the feet of Daphnia are enclosed, like the rest of the organs, within the shell; they are used to promote a constant stream of water towards the mouth, and as this current brings the minute organisms which form the food of Daphnia, we may say that its feet are used, not for locomotion, but merely to kick the food into the animal's mouth. Cypris is even more completely enclosed within its shell, which in this case is hinged along the back, so that it is really bivalve, like the shell of a cockle, and, like it, can be shut completely when at rest. In both Daphnia and Cypris the shell is minutely sculptured with very fine lines. Should it touch the surface of the water, the latter is at once thrown off as if off a duck's back, and the unlucky swimmer is unable to get back beneath the surface until it reaches some object down which it can climb and forcibly drag its shell down. In Cyclops the carapace is not so large and the body and limbs are free.

In the case of creatures like these, whose whole world is often a small pool or puddle liable at one time to be frozen solid, at another, and more frequently, to be altogether dried up, conditions which may perhaps be compared with those in which the Martians live, active existence is liable to long periods of suspension. As an adaptation to these we find that the Water Fleas are extremely prolific, able to reproduce very rapidly while conditions remain favourable, while their eggs are well able to withstand desiccation, and may be blown about in the dust without any diminution of vitality. As soon as rain comes, and they find themselves in a puddle, they hatch, and a new period of rapid growth and reproduction sets in until the next interruption. It is in this way that their presence so soon after rain in even the smallest and most temporary pools is to be accounted for.

On the approach of winter Daphnia produces another kind of egg, enclosed in a hard-shelled, purse-like sheath. These, instead of developing rapidly like the summer eggs, remain floating in their sheaths throughout the winter, and do not hatch until the following spring.

The Water Fleas are by no means confined to fresh water, but exist in vastly greater variety in the sea, where they sometimes assume most fantastic forms. A peculiarity common to those that frequent the open ocean is the enormous development of spines and hairs, which are of service in keeping the animal suspended in the water, on the same principle that a feather or dandelion seed floats in the air.

Small and delicate though these creatures are, geological evidence of their existence in long-past ages of the world's history is by no means wanting; indeed, the very rocks that contain the bones of the gigantic reptiles about which we have already learned in these pages contain also beds in which the tiny shells of contemporaneous Cyprids are crowded together in vast quantities, proving that Cypris abounded in the shallow lakes and marshy pools of those days even as it does in similar situations to-day.

Photo by] [*S. L. Bastin.*

LIGHT AND THE PLANT.

The Eucalyptus is typical of plants that grow in places where the heat of the sun is intense. The effect of its scorching rays would tend to dry up the leaf, and these, for this reason, hang downwards and so escape the danger.

PLANTS WHICH SEEK AND AVOID THE LIGHT

BY S. LEONARD BASTIN

It does not take long for the student of Nature to discover that there are a great many things in this wonderful world which no one can explain. There is little doubt that some of the most perplexing problems arise in the case of plants, more especially where the question of relation to environment is concerned. No longer is it possible to regard the plant as a kind of living machine, where there is nothing to consider but a question of mere cause and effect. Such a point as this can scarcely be better illustrated than by the behaviour of plants in connection with the presence or absence of light.

Now, in a general way, it is quite true that plants desire as much light as possible. Everyone has noticed the manner in which the window geraniums of the cottager grow. Unless the housewife takes care to turn the specimens from time to time, the development will be all on one side—the foliage and, finally, even the stem of the plant leaning away from the darkness of the room towards the light quarter. Plants in shady places, such as at the bottom of hedges, grow tall and spindly, and so bring their foliage up to the necessary light. In a measure the phenomenon is explained by the fact that the extension of vegetable tissue takes place more rapidly where there is little light than in situations where the illumination is very brilliant. This will help us to understand how it is that the geraniums already mentioned grow to the window. On the shady sides of the leaf-stalks growth will be more rapid than on the sides which are well illuminated. Naturally this causes the stalks to bend over in the manner which has been noticed, although it should not be assumed that we have here a complete explanation. Even at the threshold of our study of these light-seeking and light-avoiding plants, we must confess that we do not know the whole story. Apart from the stalks, it is easy to see that the leaves themselves seem to twist round, so as to bring their upper surfaces into full illumination. The point is also well illustrated in the case of many climbing plants, such as ivy and the ampelopsis creeper. We may say that the tissue of the leaf responds to the action of the light, but this is only a roundabout way of putting that which we already know well.

Photo by] [*S. L. Bastin.*

LIGHT AND THE PLANT.

The New Zealand Veronica growing with the leaves one above the other to shield them from the sun

Some kinds of seedlings show an astonishing sensitiveness to light. Darwin has described the manner in which the first leaves of certain grasses will bend towards a distant lamp which emitted so little light that a pencil held close to it did not cast any shadow on a white card. The great

Photo by] [*S. L. Bastin.*

LIGHT AND THE PLANT.

Compare this young Eucalyptus with the photograph of the leaves of the fully-developed tree on page 501. It grows in the shadow of its parent, and is not injured by too much light and heat. Its leaves therefore spread out in the fashion here shown.

Photo by] [*S. L. Bastin.*

LIGHT AND THE PLANT.

A view of the New Zealand Veronica taken from the side, showing the parallel of the leaves.

scientist has also shown us how seedlings, kept in a dark place, and laterally illuminated for two or three minutes at intervals of about three-quarters of an hour, bowed towards the direction from which the intermittent light had come.

Sunshine is, of course, a necessity if plants are to carry on the formation of their green matter but it is possible to have too much of even this good thing. Such an effect is most commonly to be seen in the warmer countries of the world, and the outcome is peculiarly striking in Australia, where, of course, the sun is at times nearly overhead. The fierce rays pouring down upon the surface of the foliage are very harmful in their effects upon the delicate chlorophyll, and as well an undue transpiration of moisture is likely to take place. The situation has been well met by an arrangement whereby nearly all the Australian trees and shrubs present their leaves edgeways to the sky. The broad expanse of the organs is not exposed at all to the direct rays of the sun, save in the morning and evening, when the amount of light and heat is somewhat lessened. Those who have travelled in Australia say that the forests of eucalyptus and acacia cast a small amount of shade compared with that afforded by trees in more temperate regions. It is a singular fact that the foliage of the seedling eucalyptus trees is held in a horizontal position; not until the third or fourth year will the mature type of leaf be produced. Although the shade under the parent tree may not be very dense, yet the youngsters will certainly receive the sunlight in a very much modified form. In order to secure enough for their purposes it is necessary that they should expose their full surfaces upwards.

We need not, however, go to the other side of the world to find instances of plants which in some

Photo by] [*S. L. Bastin.*

LIGHT AND THE PLANT.

A seedling Veitch's Creeper has here been grown in a box to show that the leaves turn to the light while the tendrils turn away from the light.

way or another shun the light. We have seen that the ivy-leaves twist round in order to be fully illuminated; but in the case of the young shoots of this plant there is a very marked tendency to avoid the light. Examine a plant of ivy which is growing up a trunk or against a wall, and note how closely the growing shoots press against the support, striving their hardest to get away from the light. This avoidance of light is to be seen even more strongly in the case of the tendrils of the vine and the ampelopsis creeper. It is quite astonishing to watch the manner in which the delicate threads seem to search out the corners and crevices, just as if they had some definite knowledge of what they were doing. As a matter of fact, it is simply that they have been endowed with a tendency to grow away from the light. This is, of course, a valuable habit, seeing that it is in the cracks of the wall that they are most likely to be able to find a suitable hold.

Photo by] [*S. L. Bastin.*

LIGHT AND THE PLANT.

The Rose of Sharon flower turns upward to the light in order that the flower may be properly displayed, but—

Although it does not entirely explain the downward growth of the root of the seedling, there is small doubt that there exists a tendency on the part of all underground processes to get away from the light. A pretty little experiment, and one which is very simple to carry out, will prove this. Fill a tumbler with water, and across the top of this stretch a piece of fine muslin. The material may be readily fixed with a rubber band or a portion of twine. On the top sow a few seeds of white mustard, and put the whole thing under a glass shade. Place a cloth covering over the shade so that the seedlings will start life in absolute darkness. After a few days it will be seen that the shoots have grown straight up

Photo by] [*S. L. Bastin.*

LIGHT AND THE PLANT.

—when the seed vessel is almost ripe the stem gradually turns away from the light until the capsule bends towards the earth that shall soon receive the ripened seeds.

and the roots straight down into the water. Now by adjusting the cloth, allow the light to stream in from one side only through a narrow slit. In a few hours a remarkable change will have come over the little plants. The shoots will be leaning over to the direction from which the light has come, and the roots will be curving away towards the darkness under the cloth-covered part of the shade. Obviously the roots have a distinct tendency to grow away from the light.

One is at a loss to explain the causes of this light-avoidance on the part of plants, although the purposes which are served thereby seem in most cases to be fairly clear. There is a still more baffling problem before us when we come to consider the case of those plants which produce organs capable of both light-seeking and light-avoiding, according to circumstance. In two accompanying illustrations are shown the flowering points of the well-known Rose of Sharon. Now in the first picture it will be noticed that the stem producing the bud is curved very decidedly in an upward direction. This habit of growth towards the light ensures that the flower shall be properly displayed. It is not very long after the flower is over and the seed has been set that a striking change is seen to be taking place in the bearing of that which is now a capsule. Gradually the stem turns round, growing away from the light until it arrives at an entirely different position; instead of being curved upwards, it is bent down towards the soil. It now only remains for the capsule to open in order that the seeds may pour out on to the ground in such a way as would not have been possible without this change in the bearing of the stalk. We may find another instance of a similar nature in the case of the ivy-leaved toad-flax. This little plant, which is so abundant upon old walls, must be very familiar to everyone. If an example be examined, it will be noticed that when the small purple flowers are in their full beauty, the stem bearing them bends in a pronounced style towards the light. After the setting of the seed, should the case continue to be held in this fashion, there would be small chance that the contents would find a suitable resting-place. Most probably they would tumble to the ground, where they would be unable to grow into mature plants, on account of the coarse weeds and grasses with which they would be surrounded. But just at the time when the seeds in the capsule are getting ripe, the stem changes its habit and shows a strong tendency to grow away from the light. In this way the case is brought round into close contact with the wall, and by the time the actual dispersal takes place, it would not be easy for the seeds to do other than fall into some crevice. Thus, when the seeds germinate, the little plants will find themselves in an ideal situation for their peculiar habits of growth.

Photo by] [*S. L. Bastin.*

LIGHT AND THE PLANT.

Mustard-seed has here been sown on muslin stretched over a tumbler, and the light has been allowed to penetrate to the seedlings from the left-hand side. Notice that the leaves lean towards the light and the roots turn away.

Whilst admiring such beautiful arrangements as these, the student cannot well refrain from asking, How is this change of bearing brought about? It must be confessed that at present it is not possible to offer any solution which will in any way supply a sufficient reason. The great botanist, Kerner,

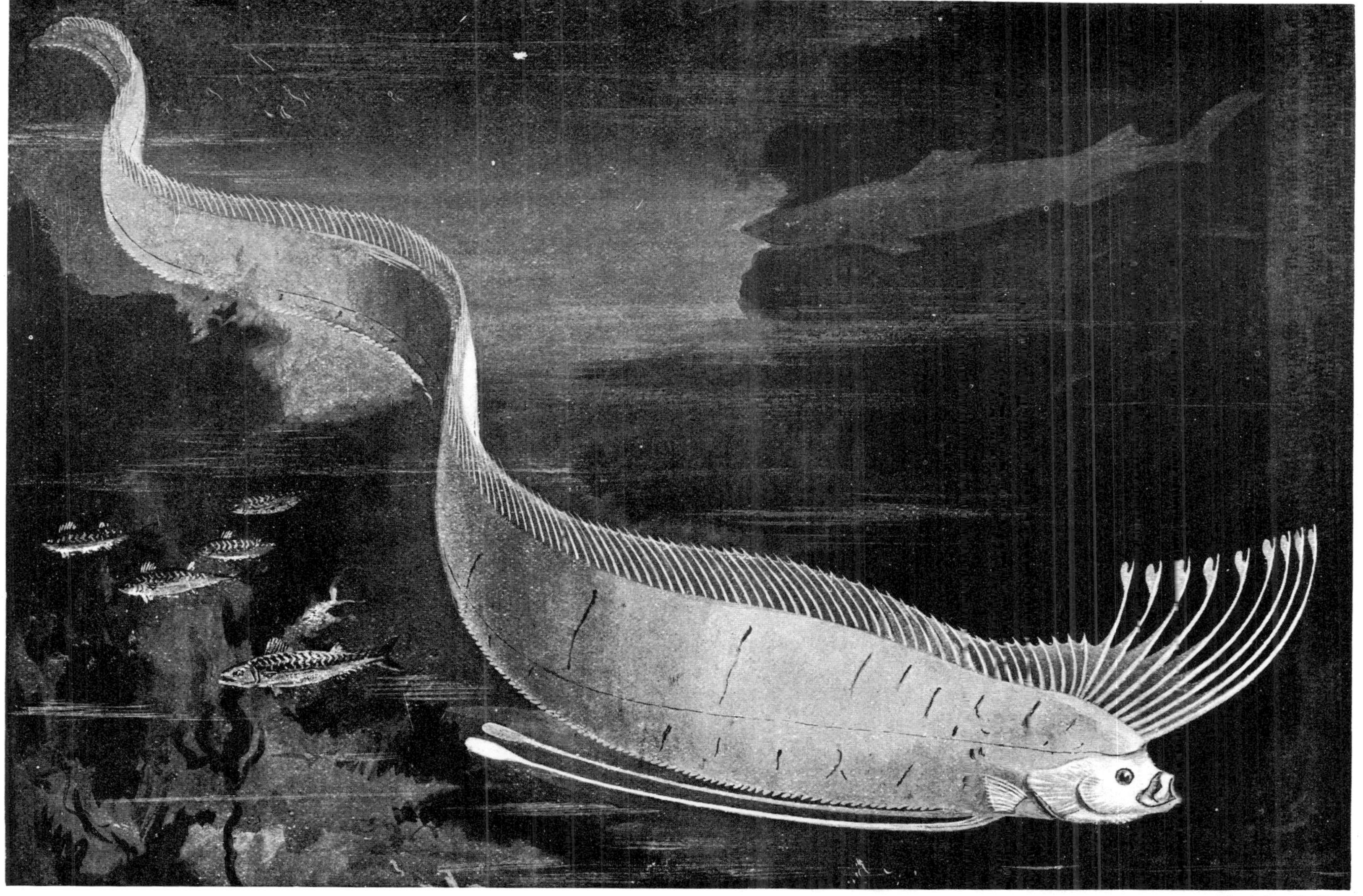

[*By Arthur Twidle.*

THE OAR-FISH.

This species of Ribbon-fish is not only of remarkable appearance but attains to great length, often as much as twenty-two feet; so that legend has often associated it with the Sea-serpent.

Photo by] [*E. J. Spitta, F.R.M.S.*

THE GRASSHOPPER.

A photograph of a specimen prepared for a microscopic slide. Notice the large flying wings, the difference between the four fore-limbs which are used for alighting and the long jumping legs; also the thick thighs of the Grasshopper.

has expressed an opinion that in such a case as that of the toad-flax there seems to be almost sufficient evidence for assuming that the plant possesses some kind of instinct. More recently Professor Francis Darwin has declared that he can discern the dawnings of consciousness in some of the more advanced cases of vegetable movement. Altogether, one feels that the subject of light-seeking and light-avoiding on the part of plants is one which presents a fascinating subject for research.

THE OAR-FISH

BY FRANK T. BULLEN, F.R.G.S.

STUDENTS of the Book of Revelation must immediately recall the Beast with seven heads and ten horns upon seeing a picture of this extraordinary inhabitant of the deep sea. Not only so, but those who have read and been interested in the many quaint accounts of sea-serpents, which, to the detriment of the general gaiety, have become so scanty in the newspapers of late years, will also find in the fearsome crest of this creature some justification for the lurid descriptions of the maned serpent of wondrous size so often reported. The connection between the imaginary monster and the real fish is closer even than might be suspected from just a glance at the picture, when it is realized that the Oar-fish has been found of a length of twenty feet, and is believed by ichthyologists to reach a much greater length in its mysterious deep-sea habitat.

Its trivial name has been given with less justification than usual merely from the blade-like expansion of its two ventral fins. It is unsatisfying, for one cannot but feel that so strange a deep-sea monster should have a far more striking epithet to distinguish it from the ordinary. Fishermen, because of its appearance having at times coincided with the arrival of the herring shoals, have called it "the King of the Herrings," but there is no other evidence to show that it has any connection with those valuable food-fish.

Like all abnormal sea-creatures, it is mysterious in its comings and goings. Certainly it inhabits the oceanic abysses, and when it is seen, it is evident that it has been compelled to visit the upper air by some convulsion of Nature. Yet, contrary to general experience of the deep-sea people, it is able to accommodate itself to the tremendous alteration of pressure between, say, ten thousand feet and the surface, a change of conditions which usually has the effect upon the hapless creature enduring it of turning it inside out!

It is not in any sense a beast of prey; in fact, it is as harmless as the great sturgeon or the mighty right whale. Its mouth is just a tiny opening well beneath the head, unfurnished with teeth, and what is more difficult for us to realize, its jaws are either so feeble as to be hardly noticeable, or do not exist. Yet as it glides along the surface of the sea at great speed with its huge mane of reddish-brown rays waving in the misty air and its wake trailing behind, easily encouraging the observer to double or treble its real length, it presents a sufficiently regal, imposing appearance to justify any ordinary observer describing it as a sea-serpent of appalling dimensions.

Unlike the great majority of abnormal or mysterious denizens of the great deeps, the Oar-fish has obligingly presented itself for inspection many times, having apparently, when on the surface, lost all control of itself and drifted ashore at some convenient spot for discovery. Dr. Günther, in the "*Challenger* Reports," mentions fifty-six specimens known to science and gives their distribution as world-wide. There are many varieties described, but the probability is that these are only local peculiarities, slight variations of the same creature.

Photo by] [*E. Step, F.L.S.*

THE GREAT GREEN GRASSHOPPER

This is a somewhat rare Grasshopper in England, although quite common on the Continent. It is very beautiful, of a brilliant green, with a slender body and long thin "jumpers"

[*By R. J. Lewis.*

THE HEARING ORGANS OF A GRASSHOPPER.

While it is the male Grasshopper who makes the shrill sounds peculiar to them, it is the female who possesses the best means of hearing the music. The apparatus here shown belongs to the female of a Natal Grasshopper.

GRASSHOPPERS

THE Grasshopper is a common insect, of which there are many kinds. In one respect it is a familiar as well as a common insect, for you cannot walk through the fields in summer without becoming aware of his presence beside you, seeing that he keeps up his shrill note all the way, and appears to be keeping pace with your walk. This, however, is only an illusion due to the vast number of Grasshoppers that are in the field. One calls, and another a foot or two away, answers, and so the continued shrilling beside you gives the impression that you have a musical attendant dogging your steps. It is chiefly in this way that the Grasshopper is familiar to people who are not naturalists. He is too lively and elusive to be seen much, save by those who are determined to watch his ways.

There are three points in the story of the Grasshopper that are worth pointing out as distinguishing him from most other insects. These are, his admirable jumping powers, his skill as a musician, and his remarkable organs of hearing.

Although the life history of the Grasshopper may be found in most books, still little or no mention is made, still less any explanation offered, as to its jumping apparatus and how it is used.. When one comes to think of the powerful manner in which the little orthopteran leaps—especially considering the size of the insect—the omission is rather curious.

Nature intends that the Grasshopper shall leap in a powerful manner, and accordingly, in the first place, modifies the internal arrangement of its body so as to lighten it. The hollow tubes which lead from the spiracles—or little holes in its body through which it breathes—are very numerous indeed, it being said that they are more in number and larger in size than in the case of most insects; but, besides this, certain air-sacs or air-bladders are found in various parts of the body, which are for the purpose of lightening it. But this, however, in no way explains the *method* of jumping and the special arrangement of limbs for so doing.

On page 508 the Grasshopper is shown as it appears when compressed as a specimen. Attention is attracted not only to the large wings used to fly with and to assist the insect in alighting, but also to the special form of the legs, their anatomical construction being of a highly interesting nature.

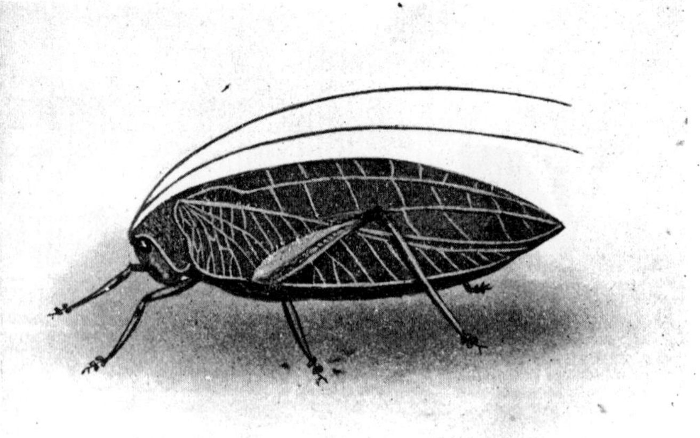

[*By R. J. Lewis.*

A NATAL GRASSHOPPER.

A beautiful insect, with green body and wings reticulated with yellow, bluish thighs and red jumping legs.

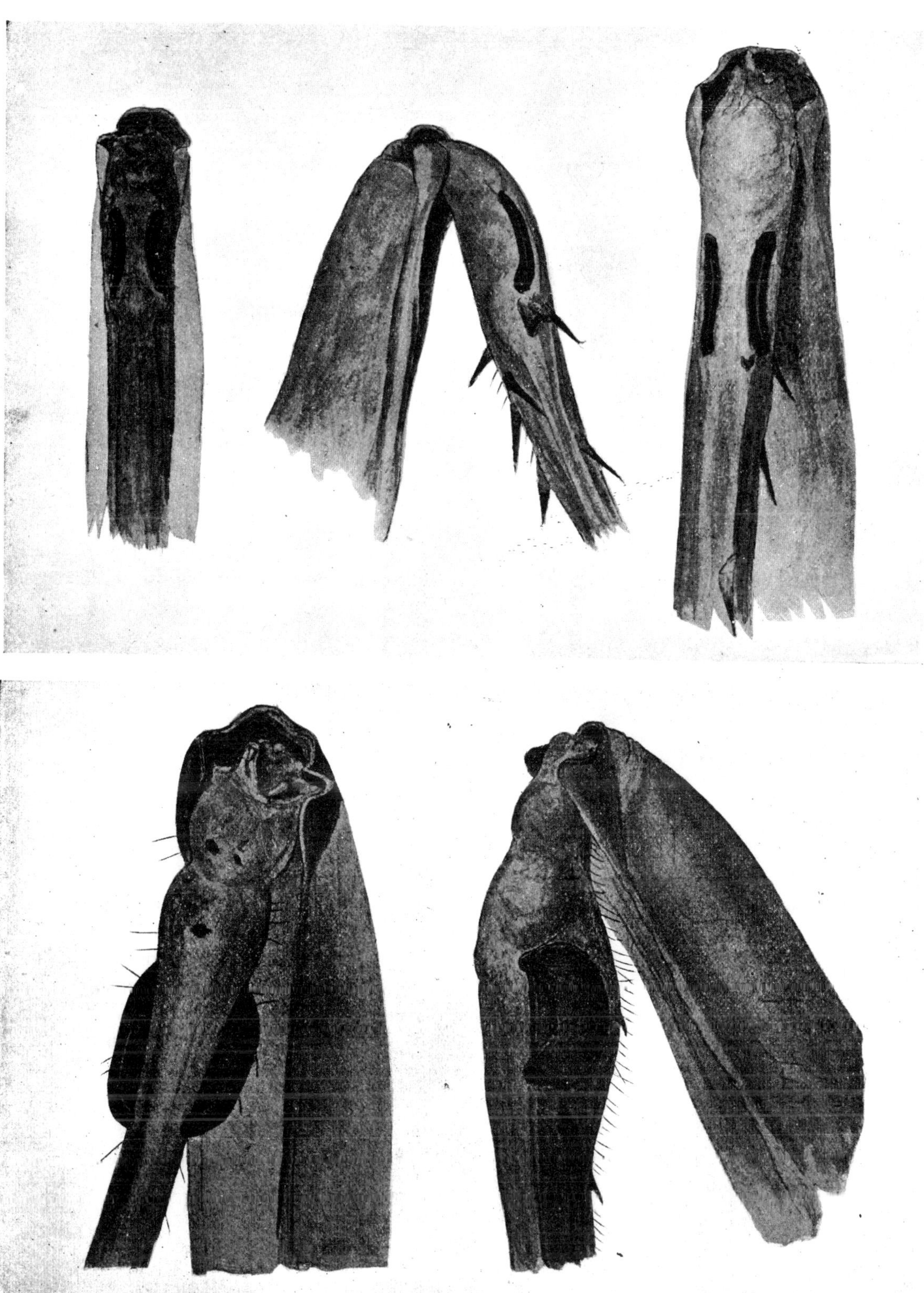

[*By R. J. Lewis.*

THE HEARING ORGANS OF THE GRASSHOPPER.

In the Short-horned Grasshoppers these take the form of two films stretched over little spaces on the back behind the roots of the wings; but in the Long-horned Grasshoppers they are in the shank of the leg, as the illustrations show. In the upper are the "ears" of the Great Green Grasshopper, in the lower those of a Natal species.

Notice the spikes on this hind leg of the Grasshopper, which enable the creature to obtain a good purchase before jumping.—

—The enlarged joints of the leg give greater power to the jumper,—

Photos by] [*E. J. Spitta, F.R.M.S.*

THE JUMPING APPARATUS OF THE GRASSHOPPER.

—while the fore feet are each provided with a strong pad to reduce the shock of alighting.

The two hind legs, or "jumping limbs," can be seen to be very different in construction, both as to shape and strength, from the remaining four limbs, termed "the alighting legs." Take them at their commencement at the trunk and note how plump these thighs are —possessing muscles of the strongest description. The joint between this thigh and the next portion—corresponding to our tibia, or lower leg—is of great development too; and careful examination of the central illustration on this page will show that this articulation is possessed of several ribs or strengthening bands, doubtless to give it additional rigidity. The tibia, a long and thin tubular-like limb which passes from the knee to the ankle, is provided with very pronounced spikes arranged in a double row on its posterior part and which appear very firmly attached to the main structure. The first photograph on this page shows these spikes which are used for the powerful leap, where the leverage required is very great. To bring this about, the insect places the tibia of each leg with its spikes on the ground, *well underneath its body,* so as to get a very strong grip before using all its full strength to make a powerful spring. It is obvious that the spikes must be exceedingly strong, for the strain on them at the moment of springing must be very sensible, and it would be fatal for these limbs to slip, for all purchase being gone, the insect would fail to move.

On turning now to the last picture on this page, and looking at these alighting legs, their difference is easily apparent. They are quite lightly made, and one would think not nearly strong enough for their object. This, indeed, would be the case, were it not that in the act of alighting the wings play such an important part. The end of each leg is the most attractive feature of the limb, for it will be seen, when highly magnified as in our illustration, that it consists of a double hook, and also what appears to be a very large and pronounced pad. Each pad is globular in form and so strongly stuffed as to resist any attempts at thinning or reducing in size whilst being prepared for a microscopical specimen, which so commonly occurs with other parts of insects whilst immersed in the first of the preparing fluids. It is quite obvious that this ball is for the purpose of acting as a concussion-relieving arrangement, one to assist in reducing the shock to the insect's body when alighting upon some hard object, such as a piece of wood or stone. But, in alighting after a powerful leap, additional assistance has to be brought into play, first by the *resisting power of the wings* at the moment of coming to rest (as a chicken will rapidly flutter its wings at the moment before coming to the ground), and secondly, by the hooks at the ends of the jumping feet which, coming into play, act as anchors.

It is a surprise to most people to be told that the shrill note of the Grasshopper and the Cricket is not produced by organs in the throat, as sounds are produced by man, also by the beasts and birds. As the insects do not breathe through the throat and the connected parts, but take in air through minute openings along their sides, it is evident that if they produce sounds it cannot be by the passage of air over vocal chords; so we must look elsewhere for the mechanism. This will be found on the two upper wings. The wing has a net-like structure of strengthening ribs, or "nerves," and between these is an expanse of transparent film. One of these ribs is stouter than the others, and ridged like a file, whilst over one of the clear spaces the film is drawn tight and tense, like the head of a drum. The insect, by a peculiar movement, draws one of these upper wings quickly over the other, with the result that one file passes over the other, which causes the drums to vibrate and, so to speak, magnify the sound. Whoever has kept the male of the Great Green Grasshopper as a pet—and an interesting pet he proves to be—will know that his shrill note of satisfaction after you have given him some tit-bit of food can be heard all over the house.

THE LONG-HORNED GRASSHOPPER.

The long feelers have given this creature its name. It is a near relative of the Great Green Grasshopper.

Now, as in the case of the Cicadas, already described in these pages, it is only the males that have the apparatus necessary for the production of these sounds; but it is supposed that the chief reason for the chirping is to give pleasure to the silent females, and so we find that, as a rule, they are provided with finer organs of hearing than the males.

These "ears," or auditory organs, are no less remarkable than the sound producing mechanism. In the Short-horned Grasshoppers and Locusts they consist of two clear films stretched over little spaces on the back just behind the root of the wings; but in the Long-horned Grasshoppers and Crickets they are situated in the shank of the first pair of legs, just below what we may term the "knee." In some groups of Grasshoppers there is a slit-like opening in the leg leading to the ear-chamber; in others there is a broader opening covered by a vibrating film; in others, again, there is a coarse shield, like an ear-trumpet, communicating with the ear-chamber. Examples of some of these forms of auditory organs are shown in the illustrations on pages 510 and 511.

There is great variety of forms and colourings in Grasshoppers. Perhaps the most beautiful feature is the wing which, when opened, reveals wonderful tints. Especially noticeable in this respect is the Flying Grasshopper shown in the coloured plate. The Natal Grasshopper is also a resplendent species, with bright green wings veined with yellow, purple thighs and reddish legs. Among the most peculiar forms are the Indian hooded species and the Proboscis Grasshopper of East Africa. The snout of this creature is prolonged to a great length and tilted upwards. Both of these examples are shown in the coloured plate.

A NATURAL CASE OF LOCK-JAW!

BY W. P. PYCRAFT

It has been contended that a common cause of extinction among animals is the excessive development of some one or other of their structural peculiarities. The "Irish Elk" is cited as a case in

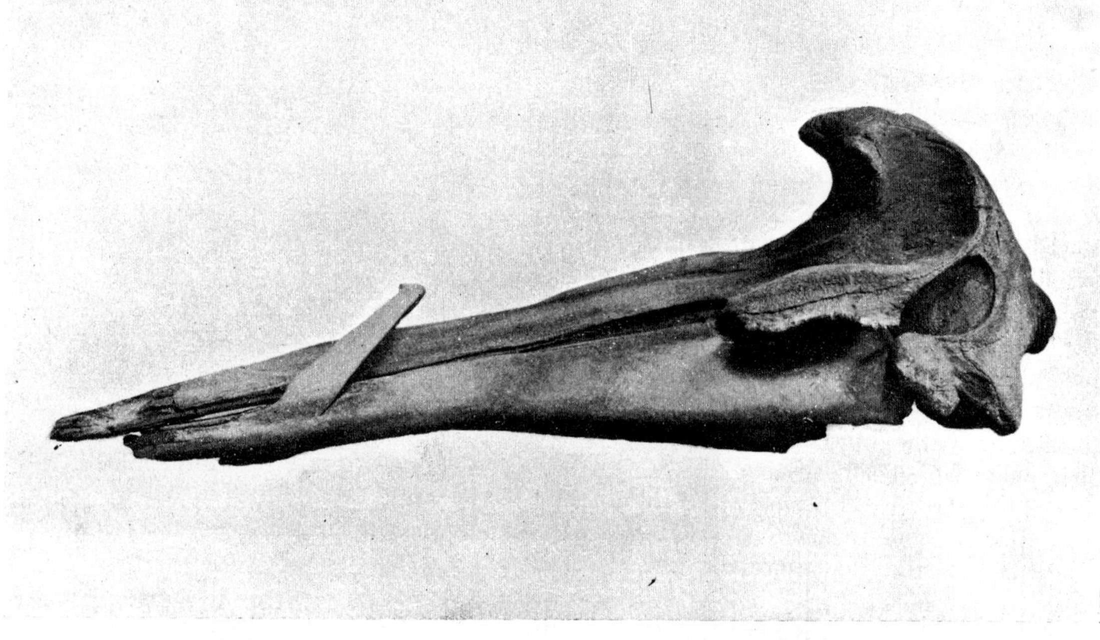

LOCK-JAW!

The excessive development of some structural peculiarity has resulted in the extinction of many species; and it is likely that Layard's beaked whale will in time be exterminated owing to its inability to obtain sufficient nourishment. The cause is the growth of the two teeth of the lower jaw, which curve over the upper and effectually close the mouth.

point, when, by the enormous development of the antlers, the weight became so great that the bearer became at last so hampered as to be unable to escape its enemies. The sabre-toothed tiger is made to afford another instance. In this animal, as everybody knows, the upper fangs were of enormous length, so that, even though changes in the nature of the hinges of the lower jaw allowed the mouth to be opened to an abnormal degree, the animal was so handicapped in securing its prey that ultimately the race became extinct. Layard's beaked whale, an exceedingly rare South African cetacean, affords an exactly contrary illustration to that of the sabre-toothed tiger. Herein all the teeth save two in the middle of the lower jaw have been lost. These two have a curious strap-like form, and curving upwards and inwards, arch over and meet above the upper jaw so as to effectually draw the jaws together, allowing, at most, the opening of a mere slit for the intake of food, which must therefore consist of extremely small crustacea and molluscs sucked into the mouth, perhaps by the action of the tongue. No other animal now living, or in times past, ever

[By Theo. Carreras.

THE DUCK-BILLED PLATYPUS.

This strange animal is a link between the reptiles and the mammals. The female lays a soft-shelled egg about as large as those of a whitethroat, in a chamber at the end of her burrow. The Duck-bill presents another resemblance to the lizard in the temperature of its blood, which is much lower than in most mammals

displayed such an extraordinary state of affairs. When first discovered it was supposed that these teeth were abnormal, but it is now known that such is not the case ; though the ultimate closing of the mouth may be only a feature of extreme old age.

THE DUCK-BILLED PLATYPUS

BY R. LYDEKKER, F.R.S.

If there be an animal in the world which specially merits the designation " marvellous," that animal is assuredly the Duck-billed Platypus, or Duck-bill, as it is better called, of Southern and Eastern Australia and Tasmania, for it is neither more nor less than a living survivor of a large group of

Photo by] [*W. Saville Kent.*

THE DUCK-BILLED PLATYPUS.

The beak of the Duck-bill is an adaptive structure designed for shovelling up the river shells and worms which form its food When out of the water the webs of the feet are folded back beneath the soles of the feet, to leave the claws free.

creatures that in former epochs served to connect mammals with reptiles. First brought to the notice of the scientific world by the naturalist Shaw, in the year 1799, the Duck-bill was described by him under the name of *Platypus*, in allusion to its splay front-feet, but as that name had been previously used for a group of beetles, it was replaced by *Ornithorhynchus* (the Greek for Duck-bill) ; but although the former name was abolished from scientific literature, it has survived as the popular designation of this strange creature. In spite of the long time the Platypus has been known to science, it was not till the latter part of the nineteenth century that we became fully acquainted with its most remarkable peculiarities.

This has been largely due to two things. In the first place the Platypus is one of the comparatively few species of peculiar mammalian types that has never been exhibited in the zoological gardens of Europe ; and, in the second, naturalists, after their custom, persistently refused to credit the reiterated statements of the Australian natives, that the female Platypus lays eggs at the end of a long burrow in the river's bank, from which the young are eventually hatched. To bring aquatic creatures of this kind alive to Europe would probably be very difficult; and the attempts hitherto made have all ended in failure.

One of the earliest naturalists to study the natural history of the living Platypus was Dr. George Bennett, who procured from the natives specimens of the young soon after they were hatched, but who nevertheless refused to believe the story of their origin. Thus matters remained till the 'eighties, when an English naturalist was despatched to Australia to investigate the whole question; the result being to confirm the native statements that the female lays a couple of soft-shelled eggs about as large as those of a whitethroat in a chamber at the end of her burrow. This important discovery was made in 1884. It had been previously known that the skeleton of the Duck-bill presents several most important resemblances to that of lizards; and when it was subsequently ascertained, not only that the young are hatched from eggs, but that the temperature of the blood of the adults is considerably lower than that of ordinary mammals, it became a practical certainty that this strange Australian creature was to a great extent a link between mammals and reptiles, although so much nearer to the former than to the latter as to be included by naturalists in the mammalian class.

Photo by] [Sir H. H. Johnston, G.C.M.G.

THE LESSER SUPERB BIRD OF PARADISE.

The male has a superb gorget of emerald-green shot with turquoise-blue. From the sides of the neck grow out huge plumes of velvet-black tinged with purple and sepia-brown. The female is only brown and grey, without any plumes.

Nor was this all, for some years after the settlement of the egg question it was found that, till nearly full grown, the Platypus has two pairs of true teeth of a peculiar type in the upper jaw, and three pairs in the lower jaw; and also that these teeth appear to have some resemblance to those of certain extinct mammal-like reptiles. When the animal becomes full grown, these teeth fall out, and are replaced by horny plates, which were long supposed to be the only tooth-like structures ever developed in the Platypus. One other discovery showing the low grade of the creature, is that the milk of the female, instead of being supplied to the young by means of a special nipple, or teat, is merely discharged on to the surface of the breast through a number of minute pores, and is then lapped up by the young.

But even these did not exhaust the list of errors and misunderstandings connected with the Platypus, for there is another which is perpetuated in every picture of the animal that has come under my notice. In all such illustrations it will be observed that when the Duck-bill is on land—

Photo by] [Sir H. H. Johnston, G.C.M.G.

THE STANDARD-BEARER BIRD OF PARADISE.

This remarkable form, discovered by Dr. Alfred Russel Wallace in the island of Jilolo, has two of the wing feathers prolonged into whitish plumes which can be erected like pennants.

Photo by] [*Sir H. H. Johnston, G.C.M.G.*

THE KING OF SAXONY'S BIRD OF PARADISE.

From the back of the head grow out plumes of the most extraordinary character to be found in the whole of the bird class. They are—in the adult male—twice as long as the bird's body, and are divided into enamelled lobes of a pale glistening blue.

to which it apparently but seldom resorts—the edges of the webs of the front feet are represented as projecting considerably beyond the claws, although it must be quite obvious that the creature could not walk on slippery ground, much less burrow into river-banks, if such were really the case. As a matter of fact, when the Platypus is out of the water, the edges of the front webs are folded back beneath the soles of the feet, so as to leave the claws completely free.

In appearance, a Platypus may be likened to a very large and very fat mole (its length is from eighteen to twenty inches), covered with coarse brownish fur, and having a duck's beak jammed on to its muzzle, while its feet are webbed, and its eyes, although small, well developed. So strange-looking is the association of a duck-like beak—at all events in dried specimens—with hair, that there is little wonder the skins first brought to Europe were regarded as samples of those "faked" animals so commonly manufactured in the Far East. To the man in the street the beak of the Platypus may appear its most remarkable feature; and in a sense this is undoubtedly the case, although the naturalist looks upon it merely as an adaptive structure designed to aid the animal in shovelling up the river-shells and worms which form its food, and in no sense comparable in interest with its teeth, the structure of its skeleton, and its egg-laying habit. Although in museum specimens the skin of the beak is hard, black and shrivelled, it is stated to be quite soft and sensitive in life, with the tip a pinkish flesh-colour. In the young the beak is short.

As incidentally mentioned, the Platypus is an essentially aquatic animal, dividing its time mainly between swimming in the water and reposing in its burrow, which is driven, with the entrance below water-level, for a distance of from twenty to forty, or even fifty, feet into the bank of the river or pond inhabited by these animals. There is usually an exit on land, and the eggs are deposited on a layer of grass in a chamber at the end.

Photo by] [*Sir H. H. Johnston, G.C.M.G.*

THE SUPERB BIRD OF PARADISE.

In this form, which has a vivid emerald gorget, the most striking adornment is a huge frill of feathers growing from the back of the head and sides of the neck, and capable of erection something like a peacock's tail coverts. The general tint of the plumage is deep black, shot with lovely purples, chestnut-browns and deep greens.

BIRDS OF PARADISE

BY SIR HARRY JOHNSTON, G.C.M.G.

All things considered, perhaps these are the greatest marvels of bird development and very likely the latest achievements in time as regards avian evolution; for the Paradise Birds are an outgrowth of the crow type, and this is one of the most highly-organized divisions of the Passerine or Perching Birds.

The crow-like character of the Birds of Paradise is evident even to the eye of an untrained observer in glancing at such forms as the Paradise Crows of New Guinea and Northern Australia. All round the equatorial belt of the earth's land surface there are crow-like birds which have further an affinity with the starlings, orioles, mainas, piping crows and hang-nests; and from a group of

Photo by] [*Sir H. H. Johnston, G.C.M.G.*

THE EMPEROR WILLIAM'S BIRD OF PARADISE.

This rare and delicately beautiful form comes from German New Guinea. The side feathers of the male are pure white, the rest of the plumage being chocolate-brown, straw-yellow and deep beryl-green.

these, probably in the large island of New Guinea, there arose, in recent times, the wonderful Paradise Birds and Bower Birds, which are restricted in their range to New Guinea and the adjacent islands on the west and to the northernmost portions of the Australian continent (where they are known as " Rifle Birds " as well as Bower Birds). The distinction between Birds of Paradise and Bower Birds is not a very great one as regards structure, but the Bower Birds have developed extraordinary features in regard to courtship —features which, however, are discernible here and there amongst the true Paradise Birds. The male of the Bower Bird (and of certain Paradise Birds) not only prepares a piece of ground on which he may dance and display his feathers before the female, but also erects " bowers," or tunnels, of twigs, either in connection with the nesting-place or separately from it. This characteristic has already been dealt with in this work (see page 105). Some gorgeous Paradise Birds are really Bower Birds, such as, for example, the Golden Bird of Paradise of New Guinea, which is a blaze of gold and orange, with black face, quills, and tail; or Macgregor's Bird of Paradise of the Owen Stanley Mountains, which, in addition to its golden-yellow and black plumage, is remarkable for a crest of three quaint golden plumes; or the yellow-crested species of Dutch New Guinea—a brown Bower Bird with enormous horizontal orange-coloured crest to its head. The true Paradise Birds also prepare dancing-grounds on which the male may perform many strange antics and exhibit his plumage in such a way as to show off its eccentric developments or its gorgeous colours. But a number of species rarely frequent the ground, and give their displays on bare branches of trees or on the smooth lianas which, in the dense forests of New Guinea, stretch like ropes from tree to tree.

In the semi-mythical accounts of Paradise Birds which were extant in Europe down to the beginning of the nineteenth century, it was stated that not only had these birds no legs, and, therefore, could never perch, but that they spent the whole of their time floating about in the air, the female laying and incubating her eggs in a hollow on the back of the male. Occasionally they rested by suspending themselves by their tail feathers from the branches of trees. No suggestion was offered as to how they obtained their food. Not only was this legless condition attributed to the Great Bird of Paradise (celebrated for its magnificent development of filamentous plumes on the side of the body, rising in a golden cloud above the outspread wings when the bird makes its courting display), but to such other of the Paradise Birds as reached Europe in this mutilated condition down to one hundred years ago. This conception arose from the way in which these skins were preserved for commerce by the Malay traders of the Archipelago, who, to save bother in preparing and packing, cut off the legs. As a matter of fact, Birds of Paradise are provided with very stout limbs and occasionally rather large feet, eminently adapted for perching and scratching, and the eggs, instead of being laid on the back of the male, are deposited (never more than three in number and sometimes only one) in nests built in bushes or in trees constructed, sometimes very roughly, of twigs, leaves, roots, the stems of lycopodium, moss and grass. The eggs vary in colour from a dark reddish-brown with darker blotches and streaks, or even black markings, to paler tints of grey or greenish-white, marked with spots and streaks of purple, brown, or black.

The food of Paradise Birds is very varied, and includes not only insects, snails and worms, but also small frogs and lizards and many kinds of fruit, berries, seeds and figs. Their diet in captivity amongst the natives of Malaysia usually consists of cockroaches, of which disgusting insect they are very fond, and papaw fruit or bananas. At least one species here illustrated—

Photo by permission of] [*Sir H. H. Johnston, G.C.M.G.*

THE "MAGNIFICENT" BIRD OF PARADISE.

The "Magnificent" or Golden-naped Bird of Paradise belongs to a group of small but beautifully decorated birds found in New Guinea and most of the great islands round about the western half. The plumage in the decorative parts is so metallic that it looks like gold, silver and gleaming copper.

the Purple and Gold Bird of Paradise—sucks the honey from flowers. Probably this is a trait to be found in other Paradise Birds possessing long and curved bills. This particular Paradise Bird has developed a long, flat, extensile tongue which is fibrous at the end and admirably adapted to loading itself with the nectar at the bottom of the flower calyx. The family, in fact, is sometimes divided into Paradise Birds with relatively straight, crow-like beaks, and those with beaks that are long, slender and curved.

Photo by permission of] [*Sir H. H. Johnston, G.C.M.G.*

THE SIX-WIRED BIRD OF PARADISE.

The forehead of this bird has a depressed patch of feathers like pure silver, contrasting vividly with the jet-black, the deep bronze, copper and beryl-green of the rest of the plumage. The extraordinary wire-plumes grow out from the sides of the head.

Nearly all these birds are noisy in their cries. Some of them are excellent mimics of other birds. Most of them utter bold, harsh noises, especially in the breeding season. The Purple and Gold Paradise Bird, for example, is sometimes called by the natives Kó kó kó, or Ka ka ka, on account of its loud cry. The King Bird of Paradise (a gorgeous little creature of cardinal-red and emerald-green) makes a sound like a cat's mew. Others utter loud whistles or flute-like notes, or harsh, grating, metallic sounds as ugly as anything a parrot can devise.

The fully developed, gorgeously apparelled male usually leads a somewhat solitary existence out of the breeding season, but the young of both sexes, together with the adult females and nearly adult young males, fly about in little flocks, frequenting chiefly hilly and well-wooded districts. Not a few of the Paradise Birds are restricted in their range to the high mountains of New Guinea, especially in the extreme north-west and south-east. Here, in the Arfak and Charles Louis Mountains on the one hand, and in the Owen Stanley range on the other, they frequent the forests up to heights as great as nine thousand feet. The very notable and beautiful species are restricted in their range to single islands, such as Jobi or Jappen, Waigiu (from which comes the Great Bird of Paradise), Salawati, Jilolo, Halmahera or Bachian (the most western limit of their distribution). Others are thought to be restricted to the Arfak Peninsula of north-west New Guinea, or to the Charles Louis or Owen Stanley ranges of mountains. But caution is necessary as regards conclusions of this kind, because in the gradual opening up of New Guinea—still, perhaps, the most unexplored portion of the earth's surface except the Arctic regions—it is found that species at one time thought

Photo by permission of] [*Sir H. H. Johnston, G.C.M.G*

THE BLOOD-RED BIRD OF PARADISE.

These specimens have been photographed by the kind permission of Professor Lampert from the Royal Collection at Stuttgart. The male possesses four elongated tail plumes instead of two; possibly a sport and not a permanent variety

Photo by permission of] [*Sir H. H. Johnston, G.C.M.G.*

THE SICKLE-BILLED BIRD OF PARADISE.

The general colour of this bird is black, velvet-black (the female is brown-black and grey); but the black of the great side wings of the male is enamelled with ultramarine and emerald, and these jewel tints recur in the tail.

to be peculiar to the mountains of the extreme south-east or north-west are now found to range discontinuously throughout the whole length of this immense island. Perhaps the two regions of New Guinea richest in forms of Paradise Birds, marvellous for their colour or strange developments of plumage are the Arfak Peninsula of the north-west (Dutch New Guinea) and the Owen Stanley Mountains of British New Guinea. Some of the most noteworthy members of the group are illustrated in colour or black-and-white in this article. They are as follows:

The Blue Bird of Paradise. This is *the* Blue Bird of the world. No other bird type exhibits in its plumage such a gloriously vivid gamut of blue as does this lovely member of a group of Paradise Birds which develops filamentous plumes at the side of the body (the same group which includes the well-known Great Bird of Paradise). The coloured plate will give some idea of this range of tints, extending from sea-blue and smalt to intense purple, blue-black, cobalt and azure. The female of this species (perched in the foreground of the illustration) is more brightly coloured than is the case with most other Paradise Birds, in which the female (in the gorgeous species) differs markedly from the male, and is usually coloured in shades of brown, grey or russet-black. The Blue Bird of Paradise is mainly, if not entirely, a British subject, as it has not hitherto been obtained outside British New Guinea, where it is found in the Horseshoe and Owen Stanley Mountains.

The Purple and Gold Paradise Bird is sometimes known as the Twelve-wired Bird of Paradise, as it has twelve long, curved, wiry appendages to the side feathers, which, except when distended for display, usually emerge from over the hidden tail. The side-plumes which produce these extraordinary wires are one of the bird's chief beauties, for they are prolonged in two exquisitely soft, rounded masses and are, like the under-tail coverts, of a pale maize-yellow, the rest of the bird's plumage being a vivid purple with gleams of emerald-green. This lovely creature fortunately stands confinement fairly well, and one or more species may be seen at the time of writing in the

London Zoological Gardens, though my coloured illustration is chiefly based on examples of the bird exhibited in Germany and Holland.

The Blood-red Bird of Paradise comes from the Islands of Waigiu and Batanta and is noteworthy for its vivid red side-plumes, the rest of the plumage being straw-yellow, intense beryl-green, old-gold, warm sepia and pinkish-fawn, while the two (occasionally there are four) prolonged plumes of the tail are like filaments of black steel. The Emperor William's Bird of Paradise is a lovely creature. It comes from German New Guinea and is, of course, related to the Blood-red and the Great Birds of Paradise. The side-plumes are of pure white and contrast exquisitely with the deep chocolate-red of the underparts of the body, the intense emerald-green of the throat, and the glistening straw-colour of the upper neck and of the belt between throat and stomach. The Superb Bird of Paradise, like the Lesser Superb Bird of Paradise, develops wonderful gorgets, breast shields and back-shields of iridescent verditer-blue touched here and there with smalt or emerald and contrasting with the intense brown-black of the rest of the plumage. The Standard-bearer Bird of Paradise has a curious combination of colours. The greater part of the plumage, the beak and the legs are a pale fawn-brown fading to an écru-white in the eccentric wing-feathers, which rise up like standards in the male bird. The crown of the head is a pale lilac, and—most striking of all—the gorget is of the most intense emerald-green in some lights, changing in other aspects to a vivid blue. The sides of the stomach are also tinted with emerald. The various species of *Diphyllodes* Paradise Birds are not large in size but are extraordinarily vivid in colouring, and are remarkable for the large shield of feathers growing up on the nape of the neck, and usually of a glistening gold or straw-white, the surface of which glitters as though it were made of the precious metal. Two large, curled plumes grow out of the central tail-feathers, and the tints of these birds are a mixture of gold or straw-white with equally glistening chestnut-red, black, orange, beryl-green and steel-blue.

Photo by permission of] [*Sir H. H. Johnston, G.C.M.G.*

THE ASTRAPIA, OR "PARADISE PIE."

This belongs to a group of long-tailed Birds of Paradise. The Astrapias, of which there are several species, have the plumage of the males nearing black, with iridescent lights. These are so bright that they appear (except when seen in shadow) to be vivid green, gold, purple and copper, the tail especially being an intense purple. The feathers round the head and neck are erected, with a velvet frill.

The " Paradise Pies " are large birds the size of a crow, with very long tails, and plumage which has intense black for its basis, but a black which is spangled, banded and shot with metallic tints of blue, golden-green, emerald, purple and copper. The Six-wired Bird of Paradise is, with other members of its genus, remarkable for a vivid patch of pure glistening silver on the forehead, the rest of the plumage being black with tints of bronze, purple, gold, green and blue. But the Queen of Saxony's Bird of Paradise, besides the silver forehead and the tints on the upper parts of the body of black, purple, emerald and gold, has fluffy side-plumes of pure white streaked with orange and vermilion.

THE TEETH OF THE IGUANODON.

Specimens of teeth from the Natural History Museum at South Kensington. They show various stages of wear, and denote that their owners were vegetable-feeders.

Perhaps, however, as regards plumage developments the most extraordinary member of the whole group is the small bird sometimes known as the King of Saxony's Bird of Paradise, which comes from the Amberno Mountains, near Geelvink Bay, on the north-west of New Guinea. This bird in its body is no larger than a thrush, and its colours are little more than a combination of brown-black and chestnut glossed with green and purple, while the under sides are ochreous-yellow. But from the back of the head grow out plumes of the most extraordinary character to be found in the whole of the bird class. They are—in the adult male—twice as long as the bird's body. These two streamers are divided into enamelled lobes, usually about thirty-seven in number. Each lobe on the outer aspect is a pale glistening blue, exactly as though it had been enamelled. The under side is dull reddish-brown.

It would be thought that before the twentieth century had been reached, educated men and women would have combined (after reading the works of Alfred Russel Wallace, F. H. Guillemard, d'Albertis, Bowdler-Sharpe, Salvadori, and others to have preserved as a very precious heritage for the human family the Paradise Birds of New Guinea and the adjacent islands. But not so. The destruction of these birds, chiefly for the purpose of decorating women's hats and the turbans of E[illegible] monarchs, or for making show-cases in the museums of eccentric Americans, is going on at such a terrible rate by native plumage hunters (financed by Dutch, French, British and German firms), that many species are now on the verge of extinction, if not already extinct, [illegible]st others are being rendered scarcer and scarcer and driven far from the haunts of men into [illegible]sible mountains, where their beauty can excite neither the envy nor the admiration of their persecutors.

Photo by] [*H. J. Shepstone.*

THE IGUANODON.

As it is restored in Mr. Hagenbeck's Tierpark. There was for a long time much doubt as to the appearance and nature of **this creature, but a discovery** of the remains of twenty-three specimens at Bernessart, in Belgium, has to a large extent determined its characteristics.

THE IGUANODON

BY R. I. POCOCK, F.R.S.

THE study of extinct reptiles has brought to light marvels only dimly guessed at half a century or so ago, and has proved the nearly world-wide existence during the Mesozoic epoch—so appropriately termed the age of reptiles—of large numbers of these cold-blooded vertebrates which, as compared with existing members of that class, were no less remarkable for their colossal size than for their grotesque appearance. But strange and interesting as these bygone organisms are in themselves, one of the most important and unexpected results their investigation has led to is the discovery that to them we must look for the parentage of the two most highly-organized classes of vertebrates, the Birds and the Mammals.

One of the earliest of the bird-like types to be revealed was the Iguanodon, whose bones, embedded in rocks below the chalk, were found some fifty years ago by Dr. Mantell in the Weald of Sussex. The name Iguanodon, meaning "Iguana-toothed," was suggested by the shape and structure of the teeth, the crowns of which were compressed and pointed, while their edges, front and back, were armed with small, close-set cusps like the serrations of a saw, thus resembling somewhat the teeth

THE SKELETON OF THE IGUANODON.

A splendid reconstruction in the Natural History Museum, South Kensington. The height of the Iguanodon was about fourteen feet and the length twenty-five feet. The tail is deep and laterally compressed as if for swimming.

Photo by] [*W. S. Berridge, F.Z.S.*

THE GREAT ANT-EATER.

The Great Ant-eater is as large as a collie-dog Self-confident of his enormous strength, he carries his huge tail in such a way as to make himself as conspicuous as possible. It should be noted that his mode of living has caused the Ant-eater to develop elongated and tubular toothless jaws.

of the modern Tropical American tree-lizard known as the Iguana. Now, since the Iguana is known to feed mainly upon fruits, it was natural to infer that Iguanodon was herbivorous likewise. This view, no doubt, is correct; but dismissing the teeth, the resemblances between Iguana and Iguanodon come to an end, for the structure of the skeleton proves conclusively that Iguanodon was certainly not arboreal and cursorial like Iguana, and equally certainly that he was not a lizard in the strict sense of the word. Nevertheless, a glance at the restoration of this giant reptile suggests that calling him "bird-like" is a mere misuse of language In a measure that is so; for unquestionably the front legs have no sort of likeness to the wings of birds, and no existing birds have a long, massive tail like that of this animal. All the same, the skeleton of the hip and hind limbs shows some very curious bird-like features, which have inspired the group-name Ornithopoda, or bird-footed, for this reptile and others related to him. In the first place, all birds walk upon two legs only, the body being supported upon three toes armed with claws, the middle toe of the trio being the longest. That this was the case with Iguanodon is shown by footprints left in the sandstones of the wealden rocks. Iguanodon, it is true, was short-footed and had the bones between the toes and the ankle separated; while birds are remarkably long-footed, and have a single bone below the ankle. But too much importance must not be attached to this difference, because it is known that the bone in question in birds is really composed of three fused together, each corresponding to one of the toes. There are also certain points in the structure of the ankle itself in Iguanodon which serve to bridge over differences between typical reptiles and birds. Again, in the hip or pelvis of Iguanodon the two bones below the socket of the thigh bone were long and slender, and projected obliquely backwards beneath the root of the tail. This is another character in which a condition intermediate between ordinary reptiles and birds is found; and since resemblances of this nature are interpreted as evidences of kinship, there is no avoiding the conclusion that Iguanodon was related to the ancestral stock of birds.

In general appearance, as he stood on his hind legs, supported by his great thick tail and with his comparatively light forequarters and small head carried erect or inclined forwards, he must have looked like a gigantic kangaroo; but his mode of progression, instead of being a succession of hops, was no doubt an awkward waddle, like that of a duck, with the tail trailing behind and the short front legs perhaps used now and again for additional support. These legs had five fingers, the fifth being somewhat, and the first, or thumb, considerably, divergent from the remaining three, giving a broader span to the paw, possibly for support when walking, possibly for grasping the branches upon which this creature fed. But the thumb was very peculiar in being composed of a single large, stout and pointed bone, probably covered with a hard, horny sheath. What the use of this may have been is not known, but when the bone was first discovered, detached from the paw, it was

Photo by] [*W. S. Berridge, F.Z.S.*

THE TAMANDUA ANT-EATER.

Black and yellowish-white are the markings of this short-haired species, which is about the size of a young fox. It uses the smooth tail to hang from the boughs of the trees amongst which it is chiefly to be seen.

supposed to have belonged to the head; and the original drawings of Iguanodon represented the animal as carrying a horn on his nose like that of a rhinoceros.

The teeth, described above, were arranged along the sides of the jaws only, the front part of the mouth being entirely toothless, and possibly encased above and below in horny sheaths, like the beak of a bird, with sharp edges suitable for cutting green vegetation and stripping the bark off trees.

ANT-EATERS

BY R. LYDEKKER, F.R.S.

SINCE ants of various kinds—inclusive of the so-called termites, or white ants—abound, under suitable condition of climate and soil, in all the warmer parts of the globe, it is only natural to expect that such a rich supply of easily assimilated food would not be neglected by the higher animals. And, as a matter of fact, members of several distinct groups of mammals have become specially

[*By Theo. Carreras.*

THE AARDVARK.

Such is the title, signifying ground-hog, given by the Dutch to the South African species of Ant-eater. By day they secrete themselves in burrows, which they can dig with amazing rapidity, while at night they raid the ant-nests and lick up the swarming inhabitants by means of the long, sticky, ribbon-like tongue.

Photo by] [*Flatters & Co., Longsight.*

THE TRANSFORMATION IN FLAT-FISHES.

When quite young a Flat-fish has the same appearance as the young of any other fish, but gradually a strange transformation takes place. It gradually leans over till one side becomes the under surface, and the fins on the back and belly take up a new position as side-fins. In the photograph the fish (considerably enlarged) is shown in its two stages.

modified and adapted for feeding on ants. In Tropical America we have, for instance, the true or typical Ant-eaters, which form the main subject of this article ; in the warmer parts of Asia and Africa these are replaced by the curious Scaly Ant-eaters, or Pangolins, which look like gigantic, animated spruce-fir cones, and are regarded by many persons as reptiles ; in New Guinea and Australia the group is represented by the Echidna, or Porcupine Ant-eater, a cousin of the Duck-billed Platypus, which it resembles in laying eggs ; in Australia there is also the little marsupial Banded Ant-eater, of the size of a squirrel ; while Africa is the habitat of the Great Ant-Bears, or Aardvarks, easily recognized by their donkey-like ears and long, thick tails.

To thrive in the ant-eating line it is essential that a mammal should have powerful claws to dig up the ants'-nests, a long, extensile, and more or less worm-like tongue, coated with a sticky saliva, to lick up the ants themselves as they swarm out when their dwellings are opened, and a tough skin to resist the bite and acid secretion of these ferocious insects. Moreover, ants do not need much biting, and many Ant-eaters, like the typical American group, the Pangolins, and the Echidnas, have accordingly discarded teeth altogether, and developed more or less elongated and tubular toothless jaws. On the other hand, the Australian Banded Ant-eater and the African Ant-Bears have retained teeth.

Most ant-eating mammals, being more or less harmless and defenceless creatures, either dwell in burrows or are of arboreal habits. Not so the Great Ant-eaters of Tropical America, which are fully as big as large collie dogs, and have enormous flag-like tails, carried in a manner to make them as conspicuous as possible ; for these animals are fully conscious of their great strength and power, which render them dreaded as much as the jaguar by the South American natives. Indeed, if a Great Ant-eater has but once got a man firmly gripped in its vice-like hug, there is said to be no chance for the hapless victim. And little wonder, when we remember that, exclusive of its tail, a Great Ant-eater measures fully five feet in length, and that the muscles of its powerful and strongly-clawed fore-legs are constantly exercised in digging out ants'-nests.

The extraordinarily elongated and tubular muzzle, coupled with the tiny opening of the mouth and small size of the ears, gives to the head of the Great Ant-eater a most remarkable appearance ; indeed, in the ordinary and obvious meaning of the term, these Ant-eaters are the most long-headed of all mammals, presenting in this respect a strange contrast to their compatriots, the sloths, which are nearly the shortest. Great Ant-eaters, of which there are two species, are exclusively denizens of the great forest tract of Tropical and Sub-tropical America, where they range from Honduras and Costa Rica in the north to the northern part of Uruguay and the Salta district of Argentina in the south.

Much smaller is the Tamandua Ant-eater, which is also a dweller in the tropical forests, where it is represented by several local races, but is of totally dissimilar habits and appearance. For these Ant-eaters, which may be compared in size to a three-quarter-grown fox, have short hair all over the body, and comparatively smooth cylindrical tails, which are endowed with prehensile power, and are, in consequence, used for suspending their owners from the boughs of the trees amid which they spend so much of their time. Owing to that perverse habit of travellers which always prevents them from telling what you want to know, it is not easy to ascertain whether Tamanduas descend to the ground in search of ants, or whether they find a sufficient supply of these insects in the tree-tops.

Although the muzzle is much shorter than in the Great Ant-eaters, the mouth has a similar small

Photo by] [*E. J. Spitta, F.R.M.S.*

THE TRANSFORMATION IN FLAT-FISHES.

The most curious part of the transformation is the change in the position of the eye. The bones of the head are twisted round in such a way that the second eye is transferred to the opposite side of the face.

tubular aperture, from which, when the creature is on the feed, the long, worm-like tongue is protruded and retracted with the rapidity of lightning. Although presenting considerable local or individual variation in the matter of colouring, Tamanduas usually show a large black patch occupying the greater part of each side of the body, with the rest of the hair yellowish-white.

The smallest and, at the same time, the most beautiful of the true or American Ant-eaters is the Golden, or Two-toed species, which is not larger than a rat, and has the body enveloped in glossy golden fur, and a prehensile tail longer than the body, with its under surfaces bare, so as to obtain a firm grasp of the bough around which it may be curled. The feet are completely modified into climbing organs, the front ones having only a single pair of functional toes, armed with hook-like claws, although four toes are retained on each hind foot.

The Little Ant-eater, as it is often called, is restricted to the very hottest parts of the Tropical American forest-region, where it is exclusively arboreal in its habits, obtaining even its food amid the branches.

Examples of the Great Ant-eater and the Tamandua are commonly exhibited in the Zoological Gardens in Regent's Park; but, being nocturnal creatures, do not attract the amount of attention that is their due.

Photo by] [*Flatters & Co., Longsight.*

SEA FEATHERS.

A natural group of these Hydroid colonies. In spite of the small size of this species the arrangement of the polyp-cells along the upper side of the branches can be plainly seen.

THE TRANSFORMATION OF FLAT-FISHES

NATURE has, in the course of the long life-history of our planet, produced creatures of a myriad different forms and variations as diverse in their structure and their uses as it is possible to conceive. While some of these forms have survived through many ages, others have passed or are passing away, and it is according to their observance or non-observance of one unvarying law that they owe their survival or extinction. This law is—Adaptability to Circumstances.

In some of her creatures the whole race has had to undergo a series of strange contortions before arriving at a form suitable to their environment, while in others, each individual is compelled at a definite period in its existence to pass through a curious process of transformation in order to arrive at the same end. Such a case is that of the Flat-fish.

As an example of Flat-fishes, let us take the Plaice. Any of our readers will remember it as a broad, flat fish, particularly thin, with a long fin running down the whole length of either edge. There is also a small fin just behind the gill. On the upper side the skin is a dark, mottled grey-brown, and underneath it is white. The two rather bulgy eyes are situated on this upper side.

But the fish was not always like this. For a short time after it was born it was much like the young of other fish, with two equal sides and an eye on each side of the head, as is shown in the illustration on page 532; but in a little while the tiny creature began to lean to the left side and

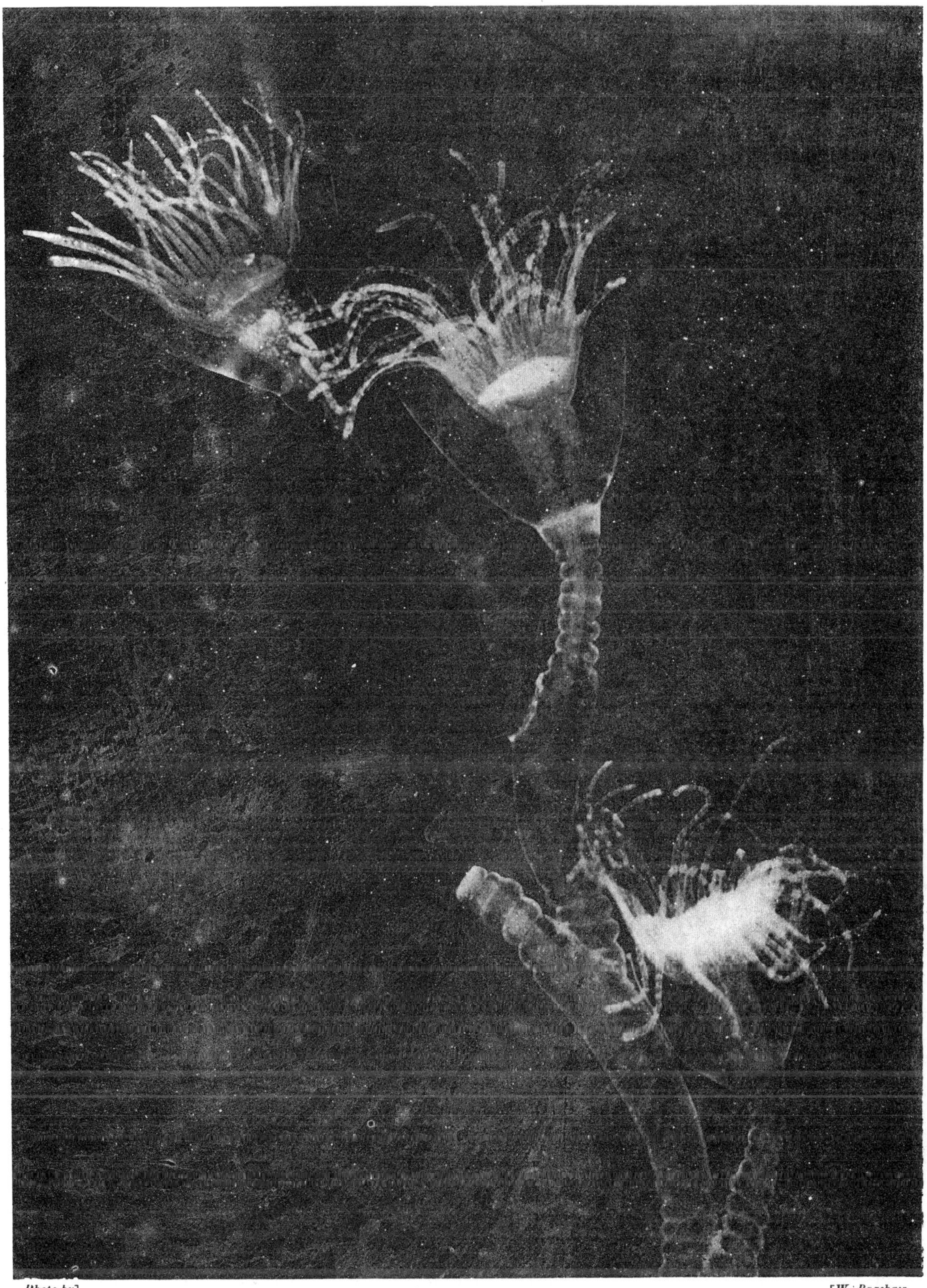

Photo by] [*W. Bagshaw.*

BELL CORALLINE.

A branch enlarged very considerably, showing the bell shape of the crystal cell. Like those of Obelia, the polype float off as minute jelly-fish.

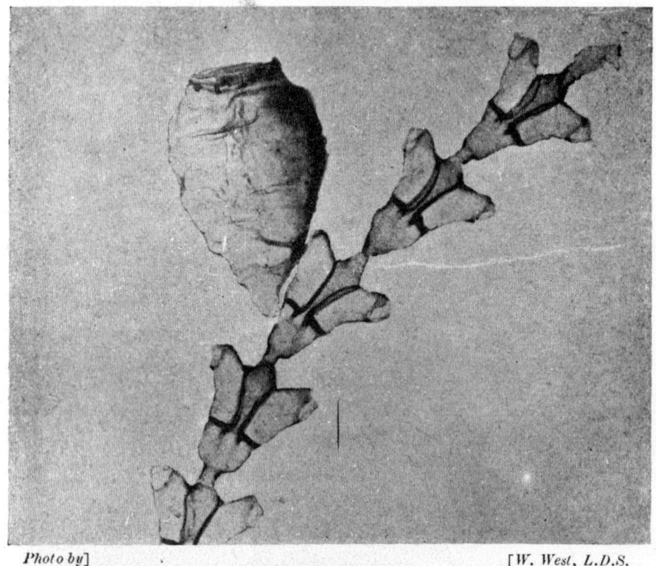

Photo by] [*W. West, L.D.S.*

SEA-FIR.

A portion of a branch of the species known as the Sea-Oak Coralline. The ordinary polyp-cells are here arranged in pairs. The large capsule is that in which the young are produced. (Enlarged about twenty times.)

to seek the sandy bottom of the sea. Here it lay half-hidden in the silt until the upper surface became brown and the left side developed into the under side, which retained (except in rare instances) its original white colour, while the fins on the back and belly took their place along the sides and the under eye moved to the upper side.

Flat-fishes are not really of different construction from other fish, as will be seen by the following homely illustration. Let the reader picture a fish —say, for example, a salmon—at rest in the ordinary position of swimming. Consider now pressure being applied to its two sides so as to squeeze them more and more together, until at last the fish becomes quite flat. The sides remain sides just as before ; and we have only to suppose the salmon turning over to swim on one of them like the plaice, and we have the condition explained.

The most remarkable part of this transformation is the change in the position of the eye. The turning over of the eye was for a long time considered a problem and was thought to be due to a positive shift of one eye to the other side of the skull ; but anatomists now show the whole thing is due to a twist of the bones of the face which slowly and gradually takes place during the fortnight.

It is now easy to realize the necessity which brought about the change. The fish is henceforth adapted to its mode of life. It is now able to use the second eye, which in its original position would have been useless. At the same time it can lie unnoticed on the sand owing to the similar coloration of its upper surface ; and also it will be able, when disturbed, to make use of its two large fins to flicker away with that quick undulating movement which is characteristic of Flat-fishes. This movement is essentially due to the peculiar character of the transposed fins, which are attached to the body in the usual manner of back fins, but not after the fashion of side fins, for these can be spread out from the body. It should also be noticed that the breast fin of the fish has since the period of transformation become useless owing to its peculiar position on what is now the

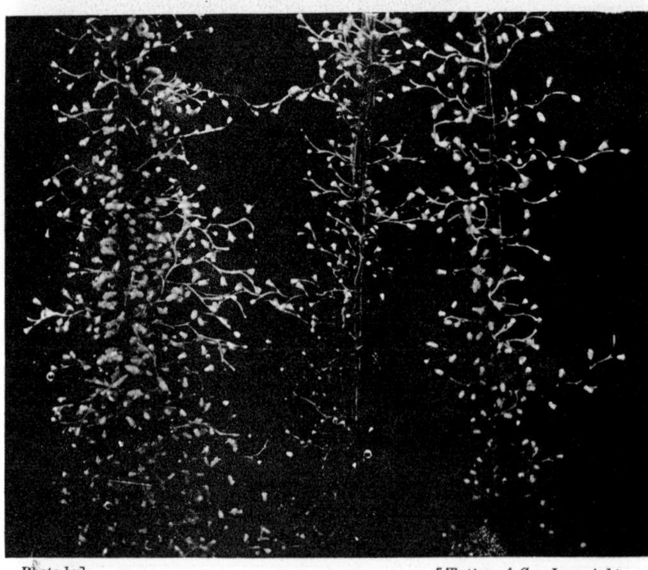

Photo by] [*Flatters & Co., Longsight.*

OBELIA.

The polyps are here arranged alternately on each side of a slender stem which runs over the leaves of Sea-grass. The polyps ultimately float off as minute free-swimming jelly-fish. (Slightly enlarged.)

upper surface, and has therefore failed to increase with the growth of the fish and become merely a rudimentary limb.

SEA-FIRS

BY EDWARD STEP, F.L.S.

THE occasional visitor to the seaside who is attracted by the beauty of the smaller seaweeds that he finds thrown up by the waves, and is induced thereby to make a small collection of them, is sure to get among them some plant-like specimens that are not seaweeds at all. If, after cleaning, separating and mounting them, he shows them to a more learned friend, he is astonished to be told that these branching growths are really of an animal nature. Then he discovers that, more or less thread-like though they be, they are of harder, more horny character than most seaweeds. Often, when on the seashore, a tiny tuft of one of these objects may be washed up whilst still attached to an empty shell. If this be dropped into a glass of sea-water and allowed to rest for a time, its true nature will be evident. These creatures are known in the mass by the popular name of Corallines—a vague term, for several stony-coated seaweeds are included under it—and one of the commonest of them is known as the Sea-Fir, because its stems bear some resemblance to the branch of a fir-tree. But most of them, being separately unknown to the public, have no English name by which to designate them, which makes it rather difficult to enter upon a popular account of them. Instead of using the scientific name by which they are known to naturalists, we have therefore used this term Sea-Firs to cover the whole group.

Photo by] [*W. West, L.D.S.*

OBELIA.

A small portion of the colony magnified about twenty times to show the character of the cells and the polyps.

Returning to our glass of sea-water, we shall now find that the branching colony of small animals has recovered from the shock of being rolled over by the waves. We find that both stems and branches are deeply notched along both edges, but in parts of the colony the notches are filled with broad crystal vases, whose edges are divided into fringes of waving tentacles. We now find that our Sea-Fir is a colony of tiny cells, in each of which lives a transparent animal, comparable to the Freshwater Hydra described in earlier pages of this work. As a matter of fact, the animals of the Sea-Fir colony and the Freshwater Hydra are not very distantly related, though

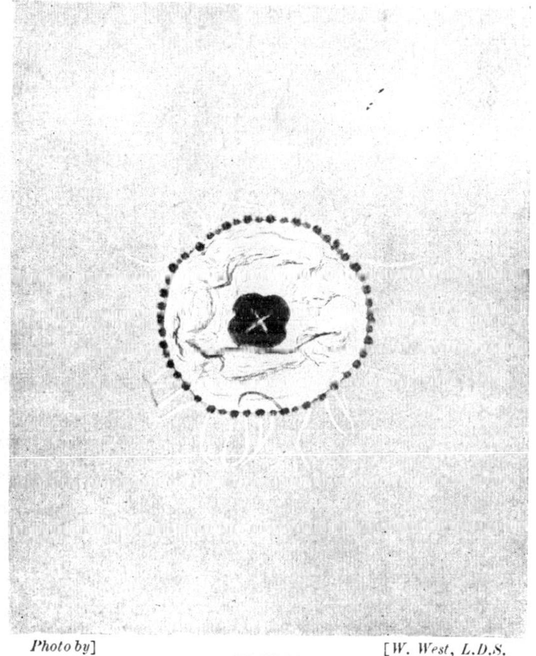

Photo by] [*W. West, L.D.S.*

OBELIA.

A single polyp from a colony such as is shown in another photo. It has now become a little jelly-fish, and is here viewed from the under side. (Magnified about twenty times.)

Photo by] [*W. West, L.D.S.*

SEA-FIR.

This species is known as the Sea-Hair Coralline. Its small cells and nursery capsules will be seen to differ in shape from those of the Sea-Oak Coralline. (Enlarged about fifteen times.)

the former is broader in proportion to its length. The Hydra, it will be remembered, is a little bag, closed at the lower end and with only the mouth opening at the upper extremity in the centre of the fringe of tentacles. Its walls consist of three layers of flesh, an outer and an inner coat with a jelly-like layer in between. From certain cells of its outer coat it is able to eject long stinging threads which kill or paralyze its prey. Now in a general way the "Hydroids," or Hydra-like polyps that live in the cells of the Sea-Firs agree in structure with the Freshwater Hydra, except that some of them do not possess the stinging threads.

The beginning of the Sea-Fir colony is an egg which, on hatching, produces a swimming larva, but this soon settles down and develops into a creeping, root-like tube attached to shell or seaweed or some other firm base, with a single Hydroid in its little cell. This produces a bud, which develops into another Hydroid, say, on its left side. The second Hydroid produces a bud on its right side; and so with this continual alternate budding we soon have a stem closely beset with cells and Hydroids on each side. At intervals a Hydroid buds on both sides, with the result that a branch stem starts off, and so the tree-like colony arises. If we look at the photograph on page 535, which shows a few Hydroids of a different family, as they appear under the microscope, we shall be struck with the general resemblance between these minute animals and the Sea Anemones, already described in a separate article. Hydra, Hydroids, Sea Anemones and Coral Polyps are constructed on much the same lines, and are regarded by naturalists as all members of one great class, though each of the four animals named represents a distinct family in that class.

Photo by] [*W. West, L.D.S.*

SICKLE CORALLINE

The stem and branches are spirally twisted, and the polyp-cells all stand on the upper side of the branches. The four extra large cells shown near the stem are the nursery cells in which the young are produced.

One of the species photographed (page 536) is known as the Sea-Oak Coralline, in which the regular arrangement of the horny cells on the stems and branches produces a resemblance to the backbone of a small fish, such as a sardine. The photograph represents only a small portion magnified, for the whole branching colony is only about an inch in height. Each cell contains, or has contained, a little polyp, which can, in the case of any alarm, withdraw itself completely into its cell. Some species have a sort of lid which closes automatically when the polyp retires. In winter all the cells may appear to be empty, but an element of life remains at the bottom, and in spring

Photo by] [*H. J. Shepstone.*

THE BOMMI FISH.

The local name by which this fish is known in the coast region of Liberia. It is also known as the Mud-skipper, and is prevalent throughout the tropical regions of the Old World. It is a remarkable instance of a sea creature adapting itself to a land existence.

redevelops into a new polyp. The whole colony may be the growth of only a few weeks, as may be proved by floating a plank or a cork buoy for a short time in the sea and finding it supporting a number of such colonies.

On many of these Hydroid colonies at certain seasons there will be found cells of urn-like shape, larger than the others, containing special Hydroids. The ordinary cell contains what we may call worker Hydroids, borrowing a name from the honey bee community. They are sexless individuals, unable to propagate their species ; but the special cells contain sexual individuals. In some species these remain attached to their cells and produce eggs, which float off and hatch into larvæ like that by which the colony was founded. In other cases—such as Obelia—they develop into minute jelly-fishes, which quit their cells and float away. They are very small—less than a quarter of an inch across—and transparent. Scarcely anybody but the naturalist ever sees them, though towards the end of summer the sea-water near the shore may be thick with them. A fine muslin net drawn through the water and then rinsed in a vessel of sea-water will yield scores or hundreds of them. In a few cases these develop into quite large jelly-fishes, such as may be noticed by the average seaside visitor. These jelly-fishes produce eggs, which are distributed in the sea as they swim about by pulsations of the little bell. One of our photographs (page 537) shows an enlarged representation of the jelly-fish stage of Obelia, after it has floated off from the colony, and in another photograph (page 535), of a different species, these jelly-fishes are seen still in their crystal vases ready to depart.

The Sickle Coralline, of which a small portion is shown under the microscope in the photograph on page 538, has its stems and branches spirally twisted, giving a distinct and pretty appearance to the plant-like colony. Another photograph (page 534) shows a magnified portion of one of the Sea-Feathers, so called because the branches are arranged equally on each side of the stem and so bring about a resemblance to a bird's feather. A number of different species of these Sea-Feathers may be found upon our own coasts, and they may be readily distinguished from other groups by

Photo by] [*W. S. Berridge, F.Z.S.*

THE BOMMI FISH.

The fish is said to prefer to lie with the extremity of its tail in water, and it is enabled in some way to oxygenate its blood through the tail-fin.

THE BOMMI FISH.

Apparently it is the other fishes that are the chief enemies of the Bommi, for its flesh is unpalatable, and it is therefore despised as food both by shore-bird and man.

the fact that their polyp-cells are always seated upon only one side of the stem or branch. One of the characteristics of this large and widely distributed family—a characteristic shared in a less pronounced degree by some other families—is the production of special polyps in special cells which possess neither tentacles nor mouth. They are set apart for the special duty of defending the colony and of catching food. They are armed with stinging-thread cells, and these threads are shot out on suitable occasions like the harpoon of the whaler. Other species have similar stinging-threads coiled up in microscopic cells of their tentacles ; and some, instead of having threads with barbed points, have clinging discs which adhere to their prey and enable the tentacles to haul it to the polyp's mouth.

Among larger forms of these Hydroids that are worth looking for are the Bottle-brush Coralline and the Lobster-horn. In the former the branches are arranged all around the stem, and after curving from their connection with the stem they take an upward direction, so that they bear a very striking resemblance to the useful kitchen implement whose name has been bestowed upon the species. In the Lobster-horn the stem is stouter and straight, and the short finer branches are similarly arranged in a series of whorls, so that the likeness to the long feelers of a lobster is by no means a fanciful one.

In all these Hydroids the stem with its cells, which are all composed of a thin, horny substance of the same nature as the hard parts of insects, is invested with a continuous layer of thin jelly-like flesh. When one of the polyps captures some microscopic creature and digests it in its interior cavity he alone does not get the benefit of that food. Although there is no circulatory system of blood-vessels, or any other perceptible method of transmission from one part of the colony to the others, there is an imperceptible system. This consists in the minute lashes, or cilia, with which the investing coat is covered. By the constant vibrations of these cilia the food digested by any one polyp is distributed equally all over the colony.

Sea-Firs and other Hydroids may be found by a little systematic searching on almost anything that is submerged in the sea. They cover the stems of many of the seaweeds ; they stand up like

Liliputian forests on the shells of many sea-snails ; on the walls of rock-pools ; on the upper valves of oysters and scallops, and even on the back and legs of many of the Spider-crabs, as well as on piers and piles and submerged timber in general.

THE BOMMI FISH

BY SIR HARRY JOHNSTON, G.C.M.G.

Bommi, of course, is only a local West African name, the title, at any rate, by which this " fish out of water " is known in the coast region of Liberia. Englishmen in West Africa also speak of it as the " Mud-skipper." It is necessary, perhaps, to quote its unwieldy scientific name—*Periophthalmus*—by which it is known to scientific men. It exists throughout the tropical regions of the Old World from westernmost Africa to the shores of New Guinea. Apparently, however, there is only one species.

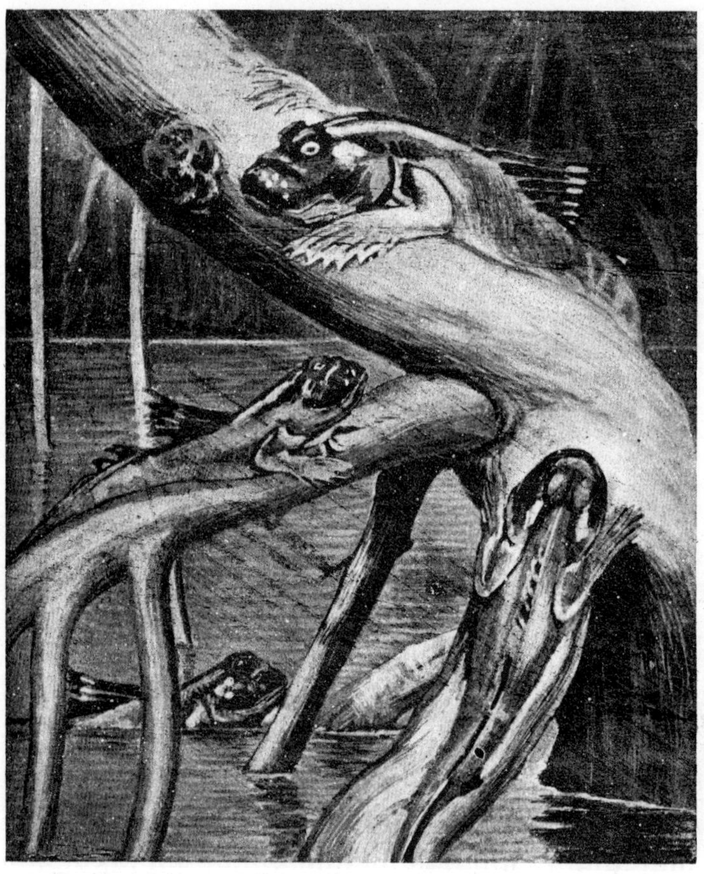

THE BOMMI FISH.

The Bommi has developed out of its fore-fins very serviceable arms and appendages somewhat like webbed fingers; while the hind-fins support the body for leaping, and are provided with a kind of sucker for the purpose of clinging.

The Bommi Fish is a member of the Goby family of marine or estuarine fishes, most of which are remarkable in one way or another for developing powers of walking or jumping over the ground rather than progressing through the water by the use of their fins. The Bommi Fish, indeed, is fast becoming a land animal. It resorts to the water for breeding purposes, but as soon as it is mature, its one desire seems to be to leave the water (though it always remains in the proximity of that element) and lie on the mud or sand, or even ascend the trunks or branches of mangrove-trees or pandanus. It is said that it usually prefers to lie with the extremity of its tail in the water, and it is enabled in some way to oxygenate its blood through its very vascular tail-fin, which in this way replaces the gills. But this hardly explains the ability of the Bommi Fish to remain for such a long time quite away from the water on the trunk of a tree.

Of course, this creature is not in any way on the original line of ascent which was followed by pristine types of fish that gradually developed into amphibians and from amphibians into reptiles. It is an entirely independent attempt (there are many parallel ones) on the part of modern fish to adopt a land life. The fish ancestors of the reptiles, mammalia and birds developed fore-limbs and hind-limbs from out of paired fore-fins and hind-fins, which had a very considerable interval between them ; but nearly all modern fish, including the Bommi, allowed the interval between the

Photo by] [*The Yerkes Observatory.*

THE GREAT NEBULA IN ORION.

Perhaps the most beautiful of all the gaseous nebulae. It is situated round the central star of the Sword of Orion, and seems to have been discovered by a Swiss astronomer, Cysat, in the year 1618.

CLUSTER STARS.
Once, owing to deficiencies of the astronomical instruments, classed as nebulæ.

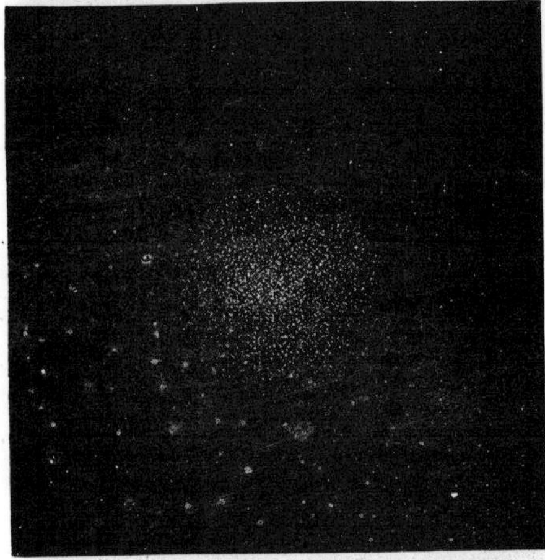

CLUSTER STARS.
In Aquarii, through Sir John Herschel's telescope.

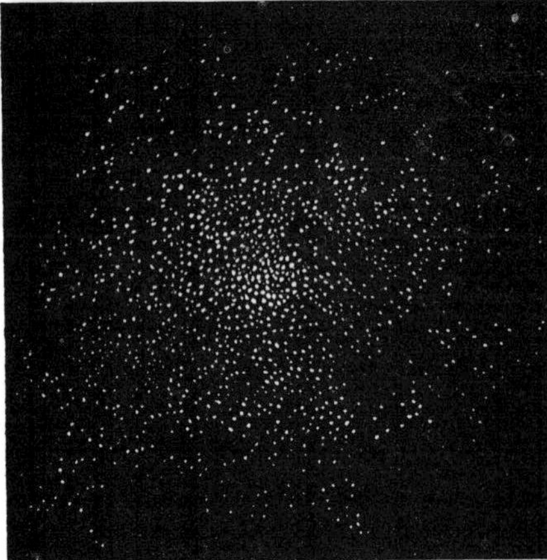

CLUSTER STARS.
The same group in Aquarii, through the Earl of Rosse's telescope, showing the numberless stars which go to form the mass.

fore-fins and the hind-fins to become less and less, until the two pairs are now often close together and in the front part of the body. The Bommi, like other Gobies, has developed out of its fore-fins, however, very serviceable arms with appendages exceedingly like hands with webbed fingers, but the hind-limbs simply become a sort of pedestal upon which the body rests. This serves as a fulcrum from which great leaps can be taken ; and also as a kind of sucker, enabling the Bommi to stick pretty closely to the bark of a tree up which it may climb by a series of deft jumps.

The rather ornamental fringe of back-fins can be erected, and no doubt because of its armature of spikes acts as a slight protection. The creature's real safeguard seems to lie in its nasty uneatable flesh, which causes it to be avoided by human beings, and apparently neglected by most of the shore-frequenting birds. Its enemies would seem to be other fishes in the water, before whose advent the Bommies may be seen jumping out of that element on to the mud and hopping well out of reach of anything that is unable to leave the water.

STAR CLUSTERS AND NEBULÆ

BY G. F. CHAMBERS, F.R.A.S.

WHEN one looks at the stars shining on a clear night the natural impression conveyed to the mind is that we are looking at so many single points of light. The telescope, however, shows us that the light from many of these single points comes from a pair of stars so close together that they ordinarily seem to be one star only. Further examination by the aid of increased optical assistance will disclose the fact that a certain number of these double stars are really triple, being three stars looking as one; whilst a certain number are quadruple, and so on. From such groups to the Clusters and Nebulæ, to be now considered, though the stretch is a great one, yet so far as the Clusters are concerned the difference is only one of degree. It is not difficult to realize that the question whether a group of stars is spread out so that the component stars can be individualized, or is so squeezed together as to appear a confused mass, is a matter entirely dependent upon the nearness or the remoteness of the observer's position from the objects looked at. To put it in another way, to accord with the facts as disclosed to us on the Earth, the distinctions just mentioned depend entirely on the telescopic power at the command of the observer, and this statement can be supported historically.

Towards the close of the eighteenth century, a painstaking French astronomer, named Messier, armed with a telescope of only two inches aperture, compiled a list of all the groups of stars and cloudy patches of light, one hundred and three in number, which he could see in the heavens as visible in the latitude of France. Though his telescope enabled him to satisfy himself that many of his objects were discrete groups of stars, yet there were others the stellar character of which was wholly undiscoverable by him. Jumping over a period of more than half a century, during which Sir W. Herschel resolved a certain number of Messier's patches of light into separate stars, we reach the epoch of 1846, when Lord Rosse, with his giant Reflector, carried a good deal farther than Herschel had done the separation of the so-called Nebulæ of Messier into individual stars. But a point was soon reached when Lord Rosse's instrument, powerful though it was, was powerless to resolve a large number of Nebulæ (some of Messier's, many of Herschel's, and many more of Rosse's own) into separate stars; and thus it came about that the study of Nebulæ, so far as knowing what they were was concerned, came to a standstill until the spectroscope was brought into use, and the discoveries of Bunsen and Kirchoff, about 1860, paved the way for the development by Huggins and others of discoveries which revolutionized to a great extent this particular department of sidereal astronomy.

The object of this article is to invite attention to some of the most famous, perhaps it would be better to say some of the most striking and conspicuous, of the Clusters of stars and Nebulæ which are to be found in the heavens. Messier's labours, limited though they were, owing to his very inadequate optical means, have borne great fruit, in so far that the objects which are entered in our catalogues with the letter "M." attached to them, to commemorate the fact that they were first catalogued by him, comprise practically all the Clusters and Nebulæ which may now be ranked as of first-class importance, excepting a few which were beyond Messier's ken because they are in parts of the southern hemisphere which are invisible in France.

The illustrations here given will bring under the notice of the reader a few of the chief of these objects, it being remembered that no distinction was, or could be, drawn by Messier between what we now call "Clusters" and what we now call "Nebulæ," he lumping all together under the phrase "Nebulosities and Masses of Stars" (*Nébuleuses, amas d'étoiles*).

The most familiar of the Clusters are undoubtedly

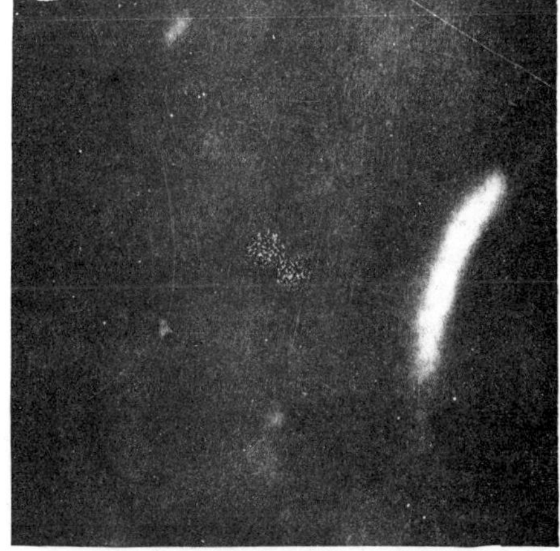

THE "DUMB-BELL" NEBULA.
As seen through an ordinary telescope.

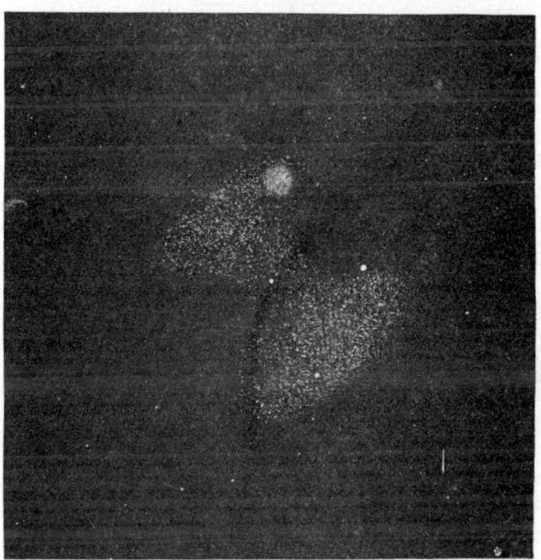

THE "DUMB-BELL" NEBULA.
Through Sir John Herschel's telescope.

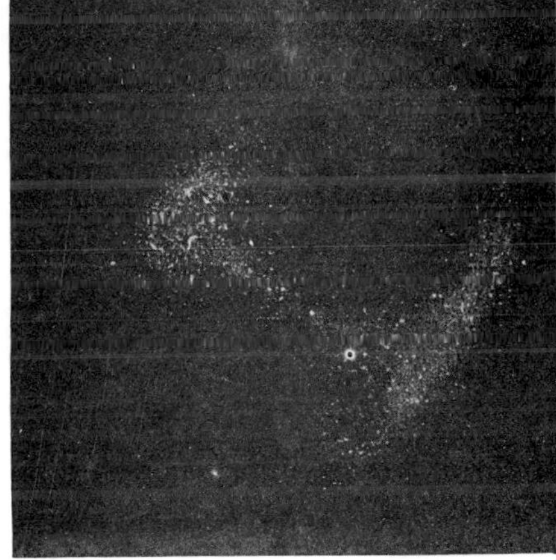

THE "DUMB-BELL" NEBULA.
Resolved into thousands of separate stars by the giant instrument of the Earl of Rosse.

those which from ancient times have borne the names of the "Pleiades" and the "Hyades" in the constellation Taurus and the "Præsepe" in Cancer, familiarly known as the "Beehive." These three objects though truly Clusters, are not usually included under that technical name, being no more than groups of stars gathered together within a limited area. Of Clusters properly so-called the most striking are those known as 5 M. Libræ, 13 M. Herculis, 15 M. Aquarii, and the Cluster surrounding the southern star ω Centauri. All these may be described as very bright and symmetrical in outline; hence they are commonly known as "globular Clusters." Most of Messier's other Clusters are of altogether irregular form, and one of them (11 M. Antinoi) was likened by Admiral Smyth to a flock of geese. Another very remarkable Cluster is that in the sword-handle of Perseus, the official designation of which is 33 H. VI Persei. This, and a closely adjacent group, were well termed by Webb as "gorgeous" and by Smyth were spoken of as "affording together one of the most brilliant telescopic objects in the heavens." I can confirm both these opinions from personal observation, as the reader can well do for himself, armed with a telescope of even the most moderate power. Indeed, the more moderate the power, the greater will be the sensation to the spectator, because the larger the field, which will therefore embrace many outlying stars which would be lost to view in a telescope of higher power and therefore more contracted field.

Photo by] [*H. J. Shepstone.*

A SPIRAL NEBULA.

Lying a little south-west of the bright star Eta in the "Great Bear." It is a more advanced form than the Orion Nebula, the gaseous matter condensing gradually into worlds.

Hitherto I have spoken under the designation of remarkable Clusters of objects which have a certain amount of concentration and symmetry, constituting, in fact, patches of light in the sky; but it must not be forgotten that the largest and finest group of stars in the heavens is the Milky Way, the very name of which suggests that its essentially stellar character was not realized by our ancestors, who gave it that name.

Turn we now to the Nebulæ properly so-called. But in doing so we are met at once by the preliminary difficulty of the uncertainty which has existed and to some extent must be said still to exist as to when an object can no longer be called a Nebula and must be ranked as a Cluster, and vice versâ. It may seem strange to put the matter thus, but the three pictures of the "Dumb-bell Nebula," in Vulpecula, will suggest the propriety of the statement. The topmost picture gives the object as seen in any small telescope. In other words, as a patch of light without trace of star. Seemingly it is a genuine Nebula.

Photo by] [*The Yerkes Observatory.*

THE NEBULA IN ANDROMEDA.

A Persian astronomer of the tenth century mentions this splendid Nebula. It is possible that it is a system in a fluid or viscous state, and that it "may be found to be the nearest celestial object of all beyond the solar system."

But if resort be had to telescopes of considerable power, especially such giant instruments as those employed by the Herschels, Lord Rosse, and, nowadays, at so many of the national observatories of the world, it will be found that this nebulous patch, as so seen in a humble telescope, is no more than thousands of separate stars, which are easily recognizable as such, but, it may be, with truly nebulous matter behind or around them.

Photo by] [*The Lick Observatory.*

THE TRIFID NEBULA.

So named from its peculiar division into three distinct nebulous masses.

Resort to the spectroscope in the observation of Nebulæ may be said to date from 1864, when Sir William Huggins brought it to bear on a Nebula in the constellation Draco. He found that the spectrum afforded by that object exhibited three brilliant detached lines, and he drew the inference that the object in question was not an agglomeration of separate stars, but a concrete mass of incandescent gas. This deduction created no inconsiderable sensation in the astronomical world, for up to that time it had been generally taken for granted that all Nebulæ were nought else but immense masses of stars too closely compacted together to be individualized —a supposition powerfully encouraged by Lord Rosse's success in showing to be unquestionably stellar numerous objects which, up to the time when his gigantic Reflector at Birr Castle was brought to bear on them, were recognized to be patches of nebulous matter, the physical nature of which was unknown.

We on the Earth associate with masses of burning gas a constant variation in the intensity and likewise in the visible outline of the supplies of gas in course of ignition, but there is an entire absence of recorded signs of any such changes in any of the Nebulæ which are now set down as gaseous in their constitution. This undoubted fact cannot but be regarded as tending to militate against the gaseous theory of Nebulæ, notwithstanding the strong proofs which are available to support that theory, which has met with general acceptance on the part of astronomers. It would lead me too far afield to go into the technical details of spectroscopy in connection with Nebulæ, but there is one very significant fact which cannot be passed over. It is that, whilst the very large and irregular-shaped Nebulæ are those which have especially yielded proofs of being gaseous in their nature, there is a class of small Nebulæ known as "planetary Nebulæ" which also give indications of being gaseous. These Nebulæ received their name from the fact that they have somewhat the appearance of planets, being as a rule very small and of circular outline.

The fact is also noticeable (though what its significance may be we know not), that these planetary Nebulæ are not distributed indifferently over the heavens, but are chiefly to be found in or close to the Milky Way.

Whilst the gaseous Nebulæ yield spectra of bright lines differing from one another, and the existence of the bright lines constitutes the proof of their gaseous character, the resolvable Clusters of stars give spectra which, though feeble, are continuous—a physical fact which carries a meaning of its own.

Photo by] [*Hugh Main, B.Sc., F.E.S.*

THE TORTOISE BEETLE.

The full-grown grub of the Tortoise Beetle (natural size). Notice the curious forked appendages.

THE GREEN TORTOISE OR HELMET BEETLE

BY JOHN J. WARD, F.E.S.

THE scene is a piece of waste ground with low herbage, chiefly consisting of clumps of thistles and flowering, white dead-nettles; and amongst those clumps you can find the Tortoise or Helmet Beetles at home.

The insects are inconspicuous when resting on the leaves, since they are green in colour and somewhat flat-bodied. There is, however, no difficulty in identifying them when once they are observed, for in a moment it becomes plain how they acquired their popular names. The broad and slightly convex shell which encases the beetle extends considerably beyond the body which it covers, and at once gives one the impression of the covering shell of a tortoise, although the beetle is only about one-third of an inch in length. It looks still more tortoise-like from the fact that its head is hidden from view owing to the curious development of the thorax, or central division of its anatomy, which has become dilated so as to cover the head of the beetle; hence, too, its name—Helmet Beetle.

Having discovered and recognized the quarry, the next thing is to inquire what purpose the beetles can have amongst the thistles and dead-nettles. Their occupation there is largely that of seeking their mates and the depositing of eggs by the female insects.

Photo by] [*Hugh Main, B.Sc., F.E.S.*

THE TORTOISE BEETLE.

Showing its remarkable method of self-protection, which is to hold its food-waste by means of these forks over its body and so conceal itself from its enemies.

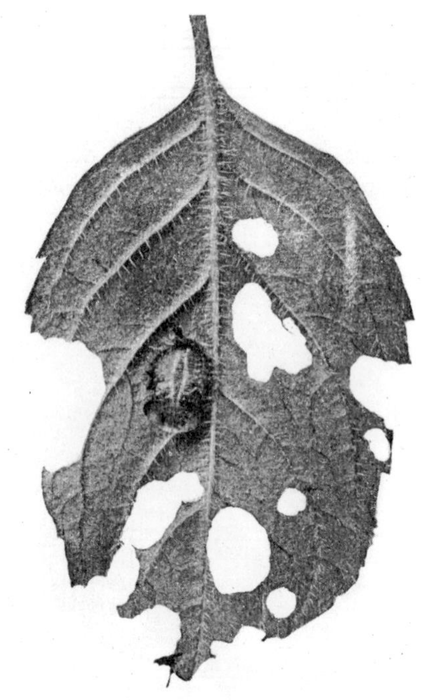

Photo by] [*Hugh Main, B.Sc., F.E.S.*

THE TORTOISE BEETLE.

Similarity to its surroundings and an appearance of rubbish also characterize the chrysalis stage.

Photo by] [*Hugh Main, B.Sc., F.E.S.*

THE TORTOISE BEETLE.

It is from the appearance of the perfect insect that the beetle gets its name. European specimens are chiefly dull green and brown, but foreign varieties are brilliant as jewels.

In due course, from the eggs the young larvæ emerge, and these are strange and weird-looking little animals of extraordinary habits. The structure of the beetle itself is curious, but that of the larva is still more so. Like its parents, it is green in colour, while its body is both broad and flattened. Its fore-parts are much larger than those which follow, and around its margin is a row of curious spines beset with tiny bristles. At its tail-end, which is slightly upturned, appears a forked appendage which turns over its back, extending nearly to its head. These details of structure are, of course, difficult to see in the earliest stages of the larva, but later on they become more obvious, and it is then that we begin to realize what a strange animal it is.

In the first place, it is apparent that these insects have in the course of their evolution been subjected to the attacks of insectivorous foes. The green colour of the beetle, its flattened form, and its consequent inconspicuous appearance, are undoubtedly features of a protective character. Then, in the case of the larva we find analogous characteristics. Here also the protective green colour is again utilized, and the bristly spines are a feature associated with unpalatableness. Furthermore, there is the curious forked apparatus at the tail, which is directed over its back; what useful purpose can that serve?

When the larvæ commence to feed, the function of that organ becomes obvious, for then we find it quickly converted into a dark-coloured umbrella, under which the larva is almost hidden from view. This canopy is the more remarkable when we consider how it is produced, for it is nothing more than the waste products of the body, which, instead of being rejected, are gathered and stored for this purpose, only occasionally being thrown off, when a new layer is quickly produced.

At first glance it appears that the larva is necessarily living under most unsavoury conditions; but we have to remember that the process of digestion in leaf-feeding larvæ is different from that of the higher animals. Only the juices of the leaf are required, and the digestion consists rather in the grinding and crushing of it to obtain its juices, than in actually digesting it. Consequently, the leaf particles which pass through the digestive system of the larva, if softened in water and examined by the microscope, reveal the leaf tissues almost in their entirety. Therefore, when exposed to the air, these leaf particles quickly dry and form a covering which is scarcely more offensive than crushed particles of dry leaves would be.

Photo by] [*E. Step, F.L.S.*

ACORN BARNACLES.

The common Acorn Barnacles that trouble the bare feet of the bather are here shown, about half the natural size, encrusting the rocks between tide marks. The dark patches are the roosting-places of limpets.

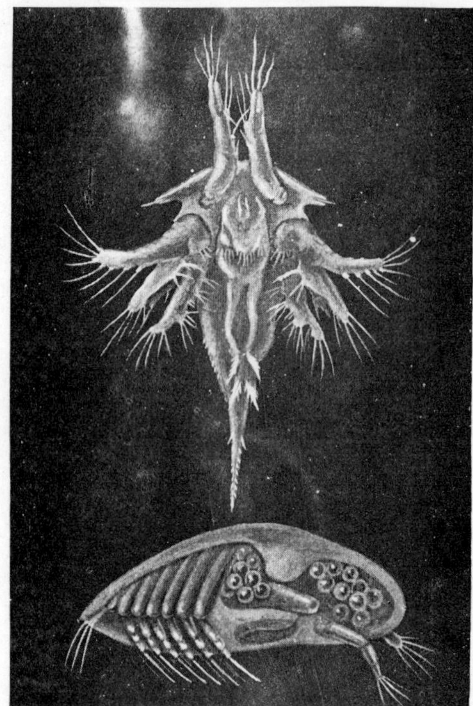

[By Theo. Carreras.

THE EARLY STAGES OF THE BARNACLE.

In the first stage it seeks its food. In the second it has no mouth.

[By Theo. Carreras.

AN UNUSUAL NAUPLIUS OF A BARNACLE.

The Nauplius is the name of the Barnacle in the first period of its existence. These long spines are supposed to support it in the water.

Allowing that point, we find that the larva has struck on a very marvellous protective device, for under its canopy it is largely protected from the eyes of its foes; for the leaves often bear such pellets of material dropped by the various leaf-feeding larvæ. It has been contended that the covering serves to protect the larva from the sun's rays; but I am inclined to think that it is purely an instance of protection by simulation of surroundings. Indeed, the larvæ of some species of these insects which inhabit tropical America carry above them a dense mass of long and curved fibres which resembles a miniature bird's-nest. These fibres are likewise produced from the waste products of the body; but in some peculiar manner scores of the tiny portions are joined together to form threads many times the length of the body. Probably in this case the simulation would be that of a leaf gall, or some similar structure.

We may, therefore, regard this strange habit of the larva as a means of protection from the eyes of birds, lizards, ichneumon flies, and other foes. Its green colour and bristly spines would seem to have lost their function with the advent of the canopy, but possibly they are still useful when occasionally the covering is shaken off and renewed.

When full-fed, this curious grub throws off its forked tail and changes into the chrysalis, but remains amongst the leaves; even then it is still surrounded with spines and looks a curious object while hanging to the leaf.

There are several species of Tortoise Beetles found in Great Britain, much alike in habits and mostly of a dull green or brown colour, which quickly fades on the death of the insect. Some tropical species, however, are very beautiful, retaining their colour so well that they are not infrequently mounted and used as jewels.

BARNACLES

BY BERNARD C. WHITE

ONE of Huxley's most popular sayings was his definition of the Barnacle: "A creature that stands on its head and kicks food into its mouth." But it is not to be expected that such a complex organism as the Barnacle can be completely summed up in so few words, for our subject assumes during its career three distinct forms, and in one at least of its stages it is entirely unattached. Again, in another stage it has only a rudimentary mouth, which is useless for the absorption of food; and we must, therefore, conclude that this saying of the great scientist should, as with all popular sayings, not be taken too literally, although it forms a good working idea for general use.

First of all, then, the Barnacle is cousin to the crab. In its early form it is very like the first stages of the crab, and even when it grows older and takes on its peculiar characteristics, it has many points in common with the members of its family. On page 552 is shown the young Barnacle, totally unlike its ultimate appearance. It is provided with three pairs of legs (of which each of the hindmost pair develops into two branches), two frontal horns, a single eye on the under surface of the head and a long upper lip. It is called a Nauplius. These characteristics are common to every species, but some possess besides long and horny spines, as in the Nauplius shown at the bottom of page 552. What the exact purpose of these spines may be is not yet determined, but it is possible that they assist the tiny creature to float on the surface of the water where it can obtain the minute ocean life that serves it for food. After several "moults," the Nauplius is ready to take on the second stage of its existence. It casts its coat once more and is this time completely transformed. It is encased in a two-parted glassy shell, shaped rather like a mussel; another pair have been added to its legs; instead of one eye, it has two, which have grown one on each side of the original member. A pair of its antennæ have gone and the remaining pair have developed into little tubes, each headed by a disc which acts as a sucker, and into each of these tubes opens a gland. This new individual has, besides, no mouth; all that it may require in the way of nutriment is stored up inside the shell, and there is further stored a cement, which will be used for the final development of the Barnacle. At this period the tiny creature—which is little more than a quarter of an inch in length—bears a strong resemblance to the water-flea, and is therefore named a Cypris, after one of the members of that group. A further succession of "moults" takes place, during which the posterior feet have increased in size, and in a little time the Cypris is ready to take on the nature and habit of a mature Barnacle. In its first existence it was forced to seek its own food and to paddle its own canoe. In the second it still paddled, but the canoe held all the provender necessary for its existence. Now it is determined either to settle down for the remainder of its life or to be carried by other means

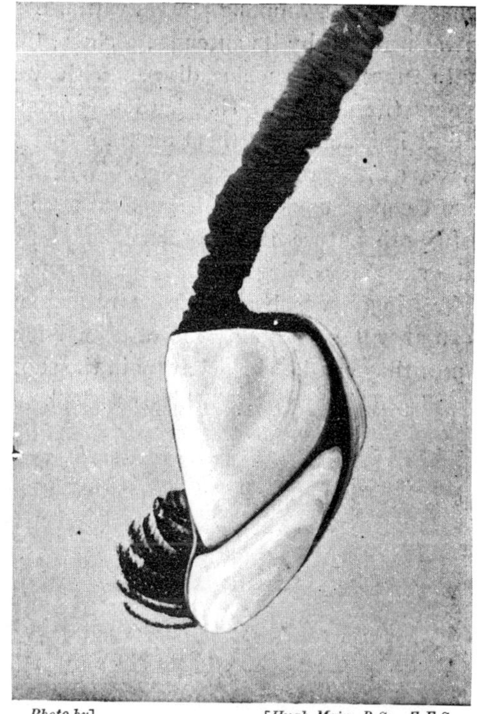

Photo by] [*Hugh Main, B.Sc., F.E.S.*

THE SHIP BARNACLE.

With tentacles half-expanded.

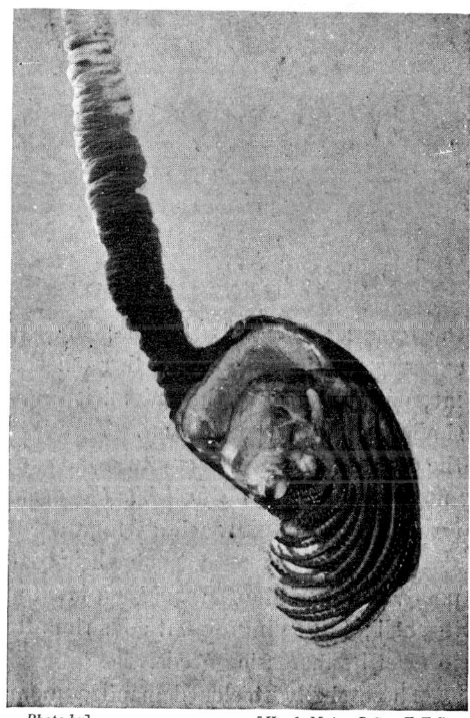

Photo by] [*Hugh Main, B.Sc., F.E.S.*

THE SHIP BARNACLE.

With one side stripped off to show the position of the creature in the shell.

than its own swimming powers within reach of its food. It therefore chooses a convenient spot and attaches itself by its antennæ. These, the reader will remember, were in the Cypris stage converted into tubes with sucking discs. With these the Cypris fixes itself, and through the hollow channel pours the cement. This hardens gradually into a stout membrane which forms the tube of the adult Barnacle. All this time the structure inside the two-parted shell has been altering. The organs have, as it were, turned round. With the succeeding moults the bluish-white shells, which are the most conspicuous part of the creature, have been forming, and when the Barnacle finally rids itself of Cypris nature it has become such an object as is seen in the two photographs on page 553. Of the six pairs of legs, each one has developed into two jointed branches bearing beautiful plumes, like fringes, which are constantly in motion, curving outwards and inwards, so that the Barnacle can at will spread out a fan of twenty-four delicate fronds, or retract them by curling them inwards, until they can be packed away in the snowy shell.

" But," says one, " this is a shell, not a crab," and certainly outward appearances do justify

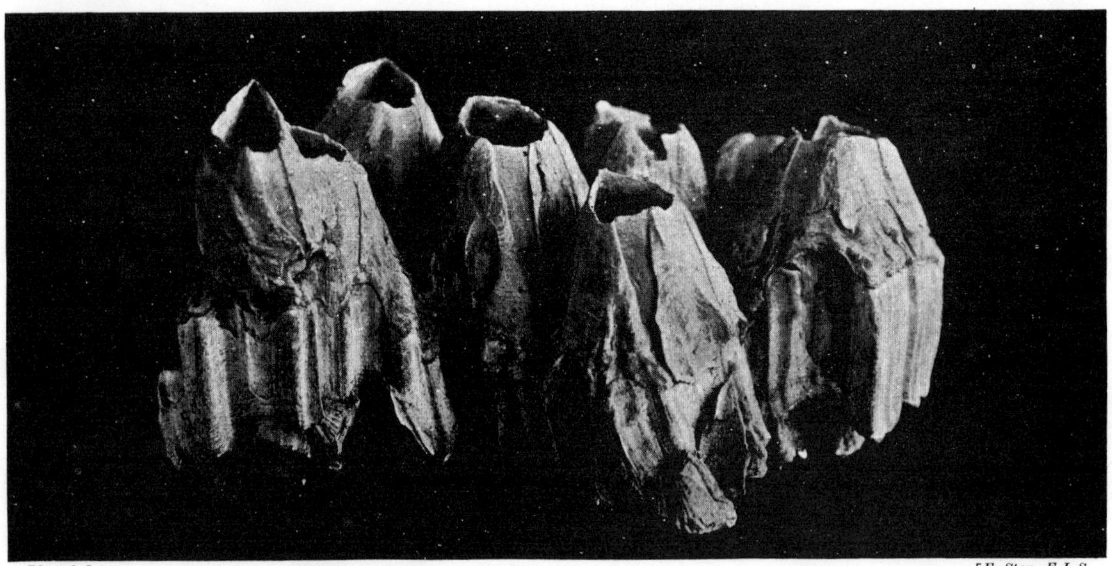

Photo by] [*E. Step, F.L.S.*

A BIG ACORN BARNACLE.

These large specimens of Hamer's Barnacle were attached to the bottom of a ship. The trap-doors are missing, but they are much like those of the common species that encrusts the rocks, as shown in another photo. (Natural size.)

the remark ; but it is the life-history of the Barnacle which has influenced the judgment of the expert. No mollusc passes through the three stages just described, while crabs do ; besides, the internal structure of the animal is crab-like, and if we regard the shell as a horny covering, the similarity is still more distinct. Turn it round so that the mouth occupies a similar position to that of the crab, and the whole creature falls into proper order. The mouth has complex jaws, the branched antennæ, " *kicking the food into the mouth,*" take the place of legs (and gills), while to complete the resemblance, the Barnacle at times casts off some of its covering (including the curled feet), just as the crabs throw off not only their coat of mail, but their inner linings as well. In one particular point, however, the Barnacle is a law to itself ; it never discards the curious shields which compose its shell ; but as it grows, it adds rim after rim to every part, and by this means enlarges its covering to meet its increasing needs.

These shelly shields demand a moment's notice. The parts of which they are composed are of such odd shapes : some are triangular, others have one straight edge and one curved, and some near the edge have one side raised high and gradually sloped away on the other till they reach the level of the remaining sections. Again, their number and the variety of shapes appear to differ

Photo by] [*Edward Archer.*

A GROUP OF SHIP BARNACLES.

This keg-buoy and spar, covered with Barnacles, was washed ashore at Greystones, in Ireland. It is a good example of the enormous masses in which Barnacles congregate together, and this species is typical of all the Stalked Barnacles.

Photo by] [W. B. Johnson.

THE ACORN BARNACLE.

A beautiful photograph of a group of the common species on a mussel shell. Each individual is displaying his feathery tentacles in search of food.

with every species, one variety, such as the Ship Barnacle, having only five component parts, while another may have as many as forty different valves; but, in reality, these smaller valves are only sub-divisions of the larger masses, and can be grouped into the larger divisions, and hold exactly the same relative position; so that even the valves in the Acorn Barnacle, which in appearance is totally unlike the Ship Barnacle, occupy the same position as those on its more conspicuous relation; while the forms and varieties of even a single species are numberless and of great divergence. For instance, a specimen of the common Acorn Barnacle has been found with the shells narrow and elongated and snowy white, about an inch and a half long, forming a roughly shaped tube, and totally unlike the usual form. Again, Darwin has grouped a number of the shells of the Amphitrite Barnacle which vary in every conceivable way, with colouring ranging from pink markings to purple or the appearance of onyx. Of this species Darwin wrote: "After studying such varying forms . . . it is difficult to avoid, in utter despair, doubting whether there be such a thing as a distinct species, or at least more than half a dozen distinct species in the whole gens of Acorn Barnacles."

Photo by] [E. Step, F.L.S.

SHIP BARNACLES.

This bunch of Barnacles is attached to the cork of an empty beer-bottle washed up on the Cornish coast. (Reduced to about one-half the actual size.)

This brings us to a consideration of the two distinct types of Barnacle: the Acorn Barnacle and the Stalked Barnacle. Their chief differences are in their general appearance and their method of living. The Stalked Barnacle has been better equipped by Nature for the development of its species than the Acorn. It usually attaches itself to floating objects, and is further enabled, by means of its long stalk, which will attain a length of as much as six inches, to reach its food, while the always moving tentacles create a current and bring in a constant supply of minute organisms for the voracious appetite.* The Acorn Barnacle, on the other hand, is stationary; and though he curl and uncurl his feathery fingers in the manner shown on page 556, he is unable to obtain so extensive a commissariat, and has to await the arrival of his prey instead of being carried out to "pastures new." Yet the Acorn Barnacle has adapted himself to and flourishes in his surroundings; every rock and post beyond high-water mark is covered with his kin. Many an unwary bather, when he has attempted to scramble on to some likely rocks, knows to his cost how numerous is the company of this species and how sharp are their shells.

Another point of interest in the Barnacle is that both sexes are contained in the same individual, so that each one can reproduce its kind. But Nature has further provided for the continuity and vigour of the species by means of external influence. Just as some plants whose ordinary flowers possess the power of fertilizing themselves are provided with other flowers which require pollinating from other individuals in order to invigorate the strain, so Barnacles are provided with complemental males which introduce fresh vitality, and so minimize any of the evil tendencies of heredity. These males are extremely minute—about the size of a pin's head—and attach themselves to the shield of the Barnacle.

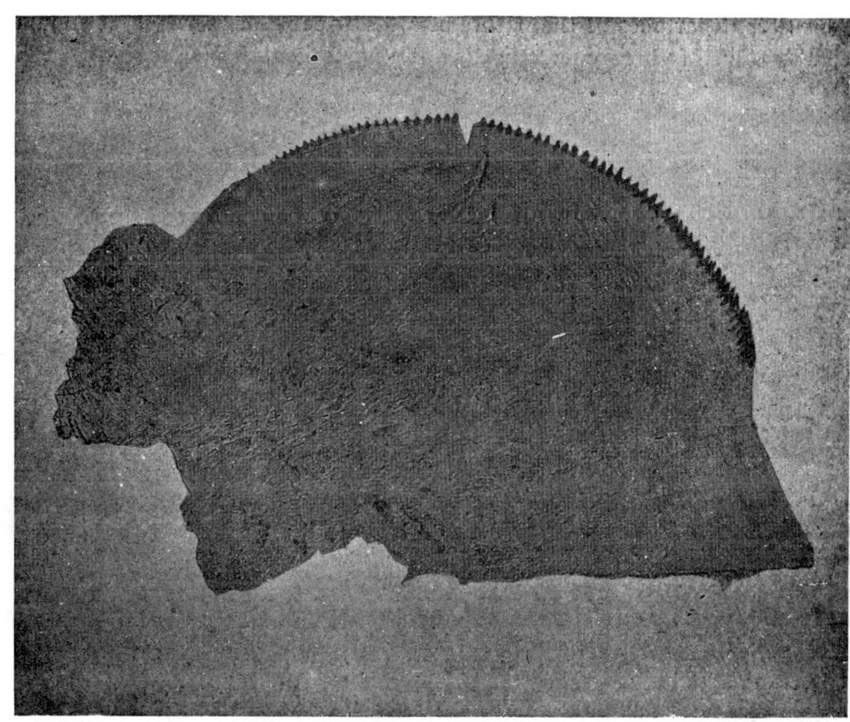

Photo by] [E. J. Spitta, F.R.M.S.

THE TEETH OF THE LEECH.

The tenacity of the Leech is proverbial, and its bite is performed by a complex arrangement of circular saws, one of which is here shown. (It is greatly magnified.)

There are many different species, both of Stalked and Acorn Barnacles, and they show a great variety both as regards the attachment tube and the shells of the shield. The best known variety of the former class is the Ship Barnacle. Another species, which is found in Australian seas, has the tube about three-quarters of an inch long, covered with fine, silky brown hairs, while the shells are as pointed as the claws of a mole; in fact, its general appearance is that of a brown mole's foot. In another the tube is covered with horny scales, while a third has the tube fine, flexible and semi-transparent, marked with white dots. It has the appearance of a snake's skin, and the shells of its shield are slightly pink. Its beauty justifies its title of "the Elegant Barnacle." The largest of the Stalked Barnacles is the Giant, of which a fine specimen has a shield of about three and a half inches in length, with a thick, scaly tube. In colour it is a dirty ivory.

Amongst the Acorn Barnacles there are several deserving of notice. The largest is the Parrot, a denizen of the South American seas, with purplish-pink shells, which attain a girth of between

* This may account for the higher development and the greater average size of the Stalked Barnacle.

five and six inches, and a height of over three inches. Its flesh, when cooked, has the delicate flavour of crab. Another curious specimen is the Diadem Barnacle, which burrows into the thick skin of whales or turtles. Its appearance is that of a chalky white tube, with raised rings ; but only the beak is visible, the rest of the animal being embedded in the hide of its host.

THE LEECH AND ITS BITE

BY E. J. SPITTA, F.R.M.S.

THE Medicinal Leech, it must be acknowledged, is a repulsive looking animal, but it has an arrangement for biting which is at once wonderfully constructed and marvellously adapted to the purpose.

It should first be explained that this bite is not a bite at all in the ordinary acceptation of the word, for it partakes of the nature of a "saw-cut" or, perhaps, it should be more strictly called a combination of three cuts radiating from one centre. This triangular incision is caused by the simultaneous action of the three so-called tongues—transparent-looking, semicircular bodies—one of which is shown on its side in the illustration on page 557, moderately amplified. If the teeth be very highly magnified upon a dark ground, they appear as in the illustration on this page, and are seen to be sharp-pointed and sharp-edged fangs—like the teeth of a circular saw. Their movement is to and fro, which is effected by the muscle upon which each tongue is set.

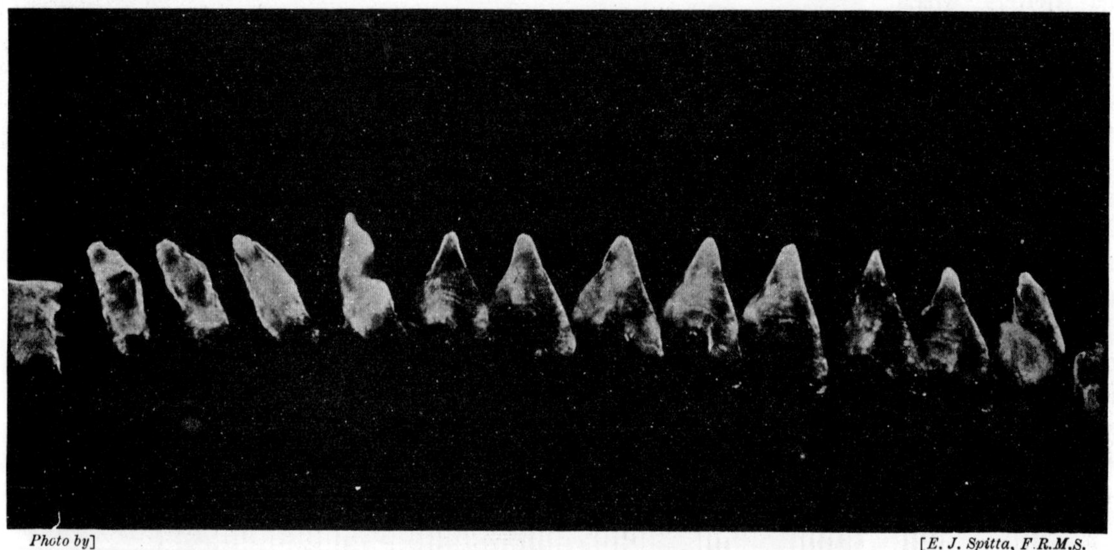

Photo by] [*E. J. Spitta, F.R.M.S.*

THE TEETH OF THE LEECH.

A portion of the "saw" has here been very highly magnified to show the teeth, sharp-pointed and sharp-edged fangs which border the saw.

LIANAS OR BUSH-ROPES

THE remarkable Lianas illustrated are very familiar objects in tropical forests. Especially noteworthy are they throughout the region of the Cameroons, also in Western Uganda, and in the forests of the Northern and Central Congo. In these districts they are the stems and branches of various kinds of Landolphia, a genus of rubber-bearing trees or shrubs, which furnishes nearly all the wild rubber derived from Africa. The Landolphia genus supplies most of these rubber "vines," so called because their habit of growth is very like that of a vine. They climb over the big trees, sometimes to a height of two hundred feet above the ground, hanging occasionally in

Photo by] [*Max Ferrars.*

A NATURAL BRIDGE.

A number of Bush-ropes have thrown themselves across a forest stream. One of them has sent down a twisted root towards the stream, which will develop into a stout trunk.

Photo by] [*Sir H. H. Johnston, G.C.M.G.*

THE RUBBER LIANAS OF THE UGANDA FOREST.

These Rubber "vines" climb over big trees, sometimes reaching as high as two hundred feet. The rubber is a sticky sap flowing between the outer bark and the inner core of woody fibre

great loops like enormous cables, loops on which it is very tempting for the native boy or the light-hearted traveller to swing himself, were it not that he never knows what his pull may dislodge on his unfortunate head in the way of cataracts of leaves and heavy branches, or swarms of biting ants.

The flowers of these Landolphias are deliciously scented. They are usually white, with a golden centre, and not unlike in shape the flowers of the periwinkle. The foliage which grows out of these Lianas is a glossy dark-green, and the fruits are large and sometimes brightly coloured. Not all the Landolphias yield valuable rubber, but most of them do. And this rubber is a sticky sap (known scientifically as *latex*), which flows from the lining between the outer bark and the inner core of woody fibre. The natives are usually too impatient now, to obtain the rubber by scoring the bark here and there and letting the latex drip into a little vessel. They prefer to cut the Lianas up in sections, strip off the bark and bast, and boil this till they extract the viscid latex. In one way and another these Lianas are being rapidly destroyed in the search for wild rubber, and very little effort is made to replace these creepers by sowing their seed. It is conceivable, therefore, that before many years are over these remarkable Lianas will cease to be a striking feature in the Equatorial forests of Africa.

Liana, however, is a common name for many quite unrelated plants, the native or traveller not troubling himself, as a rule, with questions of natural affinity. The feature that strikes him and leads to his scheme of classification is not the structure of the flowers and fruits, such as the botanist would look to, but the fact that their stems have developed into natural ropes and cables. There is a necessity for this style of growth imposed by the forest conditions. Every green plant must have light, some demanding more than others, and if they cannot get it, they die. A place in the sun is not a mere pleasure or luxury, it is a necessity; and there is in the forest—especially the tropical forest, with its continuous and rampant growth—a keen struggle for light and existence always going forward. This is a thing that most people find hard to realize. That animals should fight among themselves for mastery and sustenance

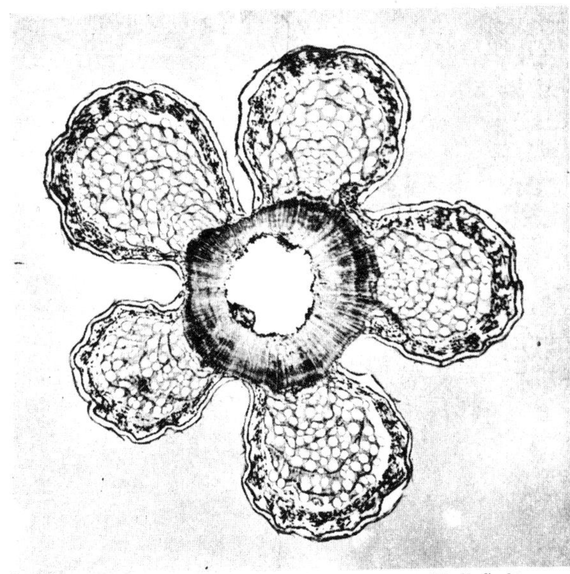

Photo by] [J. H. Crabtree.

A BRAZILIAN LIANA.

This section shows one of the remarkable ways in which these Bush-ropes add to their thickness without losing pliability. (Magnified.)

they can understand, but how the fixed and inanimate trees should or can do so appears to them impossible.

Go into a pine-wood or a beech-wood in our own country, and you may see things on a small scale that will help you to realize the position. In the pine-wood the straight, bare shafts of the trees rise to a great height, and the foliage is almost confined to their tops, where it forms a dense, unbroken canopy. There is plenty of light down below, but it is not active, life-sustaining light. Only fungi and a few mosses can live there, and the floor is covered with the monotonous carpet of dead pine-leaves. Such light as there is has had to pass through the chinks in the canopy and the life-giving violet rays have been filtered out and absorbed by the canopy.

In the tropical forest you will find, in addition to the great trees, a complete tangle of these Lianas, or bush-ropes, climbing up the trees and stretching across in lines and loops. Look at our photograph on page 559, where a stream rushing through the forest lets a little light in on either side. Many of the bush-ropes that feel its influence have worked out to it as being nearer than the light above the tree-canopy. Some of them have then attempted to rise; but being too slender, have fallen athwart the stream, and tangling and matting with others, have formed a natural bridge—the kind that probably gave man his idea for the first suspension bridge. Now these bush-ropes in many cases are not true climbers, like the hop or the convolvulus—their parents were trees—and they did not all start life in the ground. Some of these Lianas increase their strength without losing suppleness by developing longitudinal corrugations—as we strengthen our girders without adding materially to their weight—and this may be seen in the cross-sections photographed on this page.

One of these Lianas, the Clusia—known in South America as the Wild Fig, but not a fig of any kind—in some respects resembles the banyan. Despairing of its progeny reaching to the light from the forest floor, the Clusia has learned what sort of outfit to give its seeds to afford some of them at least a chance of producing a successful plant. It embeds them in a pulpy coat that is attractive to birds, and the birds, carrying them away, eat the pulp and leave the seed sticking to an upper branch, where, on germinating, its first leaves can get sufficient light to work and feed the seedling. The young Clusia sends its roots down the trunk of the tree,

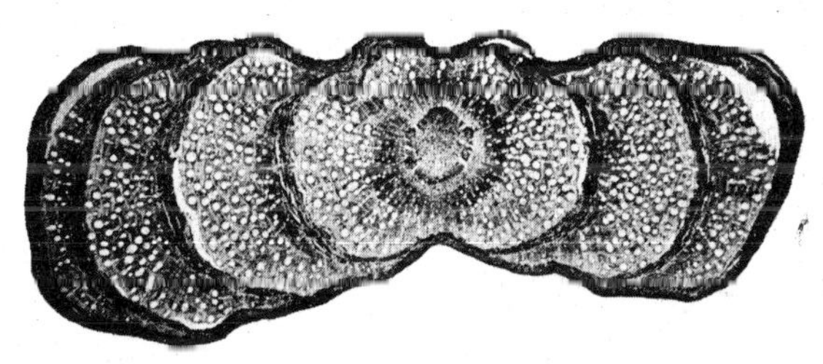

Photo by] [J. H. Crabtree.

A BRAZILIAN LIANA.

A section cut across one of these Bush-ropes and magnified. The circular centre shows that originally it was round in outline, but by pressure in twining it has grown in two directions only.

where they find trickling moisture and food material, and at last they reach the earth. They expand into thick trunks all round the supporting tree, and the true trunk at their summit waxes fat and puts out many leafy branches above those of its host, gradually shutting out all light from it, and finally killing it by suffocation. Then the dead tree rapidly succumbs to the disintegrating influences of boring beetles, termites and fungi, until it has mouldered right away and left the victorious Clusia perched up on its scaffolding of thick aerial roots.

So, it will be seen that trees can fight as well as beasts and men.

THE PROTEUS, OR OLM

BY R. LYDEKKER

PROBABLY not many of my readers have studied Sir Humphry Davy's "Consolations in Travel"; but those who are acquainted with that charming book will be mindful of the article on the Proteus, or Immortality, in which it is stated that the blind salamander inhabiting the subterranean waters of Carniola, Carinthia and Dalmatia is "a far greater wonder of nature than any of those which the Baron Valvasor detailed to the Royal Society, a century and a half ago, as belonging to Carniola." Why it was called "Proteus," seeing that it never changes its form, or why Sir Humphry selected it as an emblem of immortality—unless it be that it will live for years in confinement—is by no means easy to guess; but both these attributes have become practically a part of the creature.

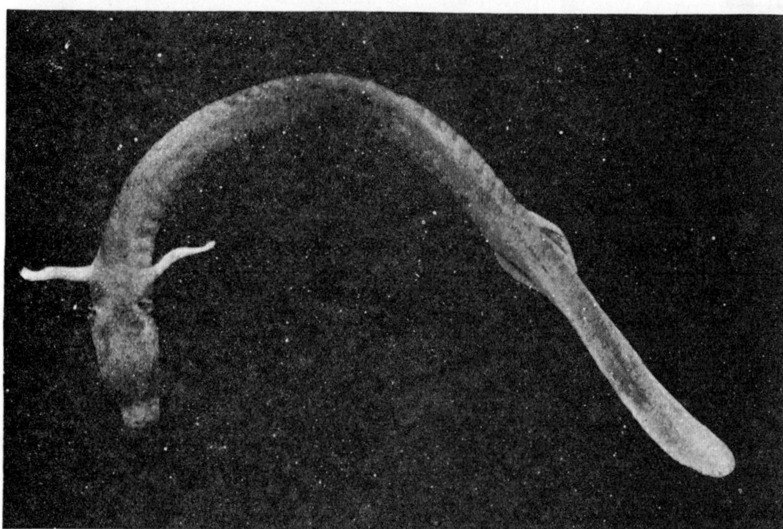

Photo by] [*W. S. Berridge.*

THE PROTEUS, OR OLM.

The fore feet of the Olm resemble hands, each with three fingers, while the hind feet are two-toed, and in old specimens almost abortive. The site of the rudimentary eyes is marked by two black beady points on the sides of the head.

Originally discovered in the waters of the great cave of Adelsberg, near Trieste, the Olm (to use its German title) is an eel-like salamander of about ten inches in length, of a transparent fleshy whiteness, with a pair of branching coral-red gills on each side of the neck, which may be snipped off with a pair of scissors without doing any apparent harm to their owner, who is also furnished with lungs. Its fore feet, as Sir Humphry Davy did not fail to observe, resemble hands, although having but three fingers apiece, and are too feeble to be of any use in grasping or supporting the weight of the body; while the hind feet are two-toed, and in old specimens are almost abortive. Two black beady points on the sides of the head serve to show that the Olm is descended from salamanders with functional eyes, which have become rudimentary from lack of use. Moreover, the fact that when the Olm is exposed to the action of light its skin darkens like an undeveloped photograph indicates that its ancestors were coloured after the fashion of ordinary salamanders. Whether when they are returned to darkness the colour fades out does not appear to be ascertained.

Sluggish and slow in their habits, Olms spend most of the time deep in the water; but if the latter be deficient in oxygen, they will rise occasionally to the surface to take a gulp of fresh air

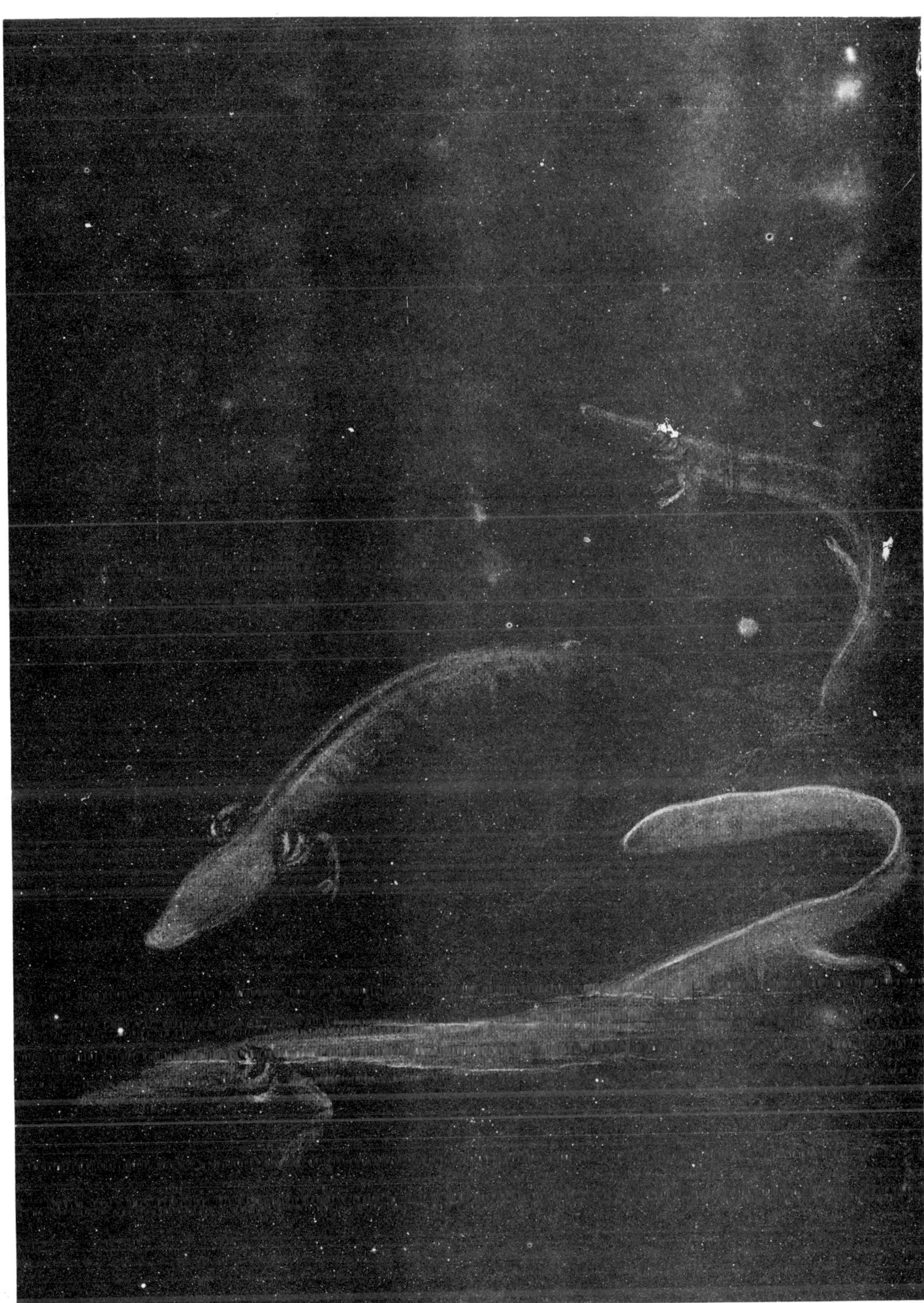

[By Theo. Carreras.

THE PROTEUS, OR OLM.

This queer inhabitant of subterranean waters was first discovered in the Adelsberg Cave, near Trieste. It is a blind salamander, about ten inches in length, of a transparent fleshy-whiteness with coral-red gills. These gills can be snipped off without injury to the creature, while exposure to the light will darken the skin.

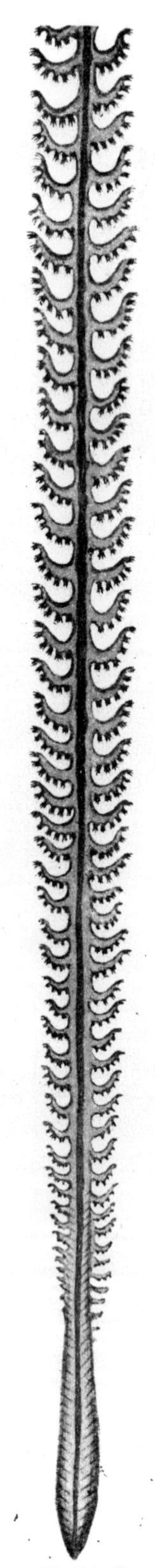

[By Théo. Carreras.
The Wonderful polyps.

into their lungs. As a rule, however, the gills suffice for their respiration; and when specimens are kept in a glass vessel, it is interesting to notice how the colour of the gills changes as they are alternately filled with pure and impure blood. Reproduction is effected by means of eggs; and the tadpoles resemble their parents, with the exception that the tail-fin extends forwards on to the body, and that the eyes are relatively larger,

For a long time the Olm was believed to represent an isolated type; but towards the close of last century a very similar creature was brought up in the water of an artesian well from a depth of one hundred and eighty-eight feet, near San Marcos, Texas.

THE SEA-PENS

BY EDWARD A. MARTIN, F.G.S.

A GLANCE at our illustrations will at once show how appropriate is the popular title of "Sea-pens" as applied to this family of animals. They are just as much members of the animal kingdom as are the corals, or the "Sea-fans," already dealt with in previous chapters. We are all, perhaps, too much in the habit of regarding an animal as a creature which of necessity has at least four legs, entirely overlooking the fact that any form of life must of necessity be animal or vegetable; and we know quite well that a Sea-pen is certainly not a vegetable. Maybe it looks like one, but we must not judge by appearances; and all who have read our previous articles will have come to recognize that many marine growths, although at first mistaken for vegetables by early investigators, have turned out after all to belong to the *animalia*. But the appropriateness of the title "Sea-pens" none will gainsay, for they were so named when people were in the habit of writing with quill-pens much more than they do now; and here we have veritable living quill-pens, so far, at any rate, as appearances go. It must be confessed that most of them would be but clumsy quills if one had to hold them in the hand; but one must not expect too detailed a resemblance where popular nomenclature is concerned.

In the common type of Sea-pen—and here perhaps it would be well to give a word of warning against comparing these with the "pens" found in certain molluscs of the cuttle-fish tribe—the

[By Theo. Carreras.
Another type of polyps.

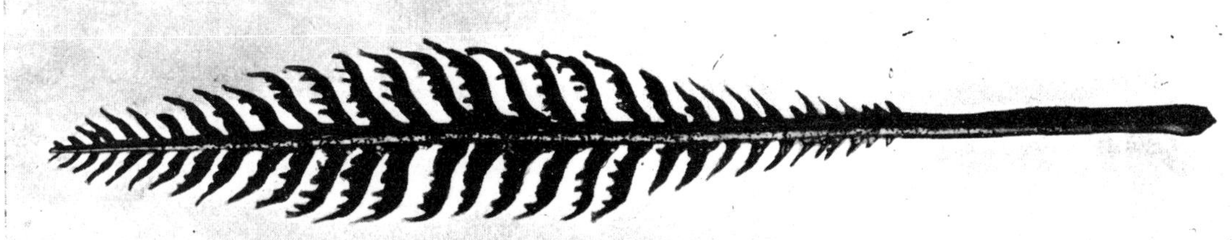

Photo by] [*Fuller & Osborne.*

A FERN-LIKE SEA-PEN.

The polyps are on the upper sides of the "barbs" only.

barbs are borne on the stem in one plane, just as in the case of the barbs of a feather; whilst to add to the resemblance there is a centre hard stalk, or corallum, of lime, which runs up the middle of the whole organism; and this, of course, gives it the support necessary to hold the structure together. That part of the animal which answers to the business-end of a quill-pen, as one might say, is really what serves as a root, by which it is embedded in the sea-bed. The creature seems to love a muddy or sandy bottom, and in this respect is in marked contrast to many other lowly creatures which love only a clean, hard bottom. The shaft of the quill is embedded more or less loosely in the mud or sand, and to that extent is not altogether immovable, but its powers of locomotion, if any, must be very small. The older naturalists attributed to it the power of flapping the *pinnæ* of which each barb is composed, after the manner of the fins of a fish, and so maintaining a certain amount of locomotion; but Lamarck, who, at the beginning of the nineteenth century, was engaged in his zoological researches, expressed his doubt of this theory, and it has since been proved that he was right, for "when placed in a basin of sea-water, the Sea-pens are observed never to change their position but to remain in the same spot and lie with the same side up or down just as they have been placed. They inflate the body until it becomes to a considerable degree transparent and only streaked with intercepted lines of red, which distend at one place and contract at another; they spread out their *pinnæ* and the polyps expand their tentacles, but they never attempt to swim or perform any process of locomotion."

Some of the Sea-pens are phosphorescent, and these present a magnificent sight in the darkness. Linnæus tells us that where they occur in numbers on the bed of a sea, they cast so strong a light that the fishes and worms which sport around them can be easily counted with their aid. Many of the coral-reefs owe a great part of their adornment to these graceful and brightly-coloured zoophytes, growing wherever the coral sand has collected sufficiently to give them a foothold. The

Photo by] [*Fuller & Osborne.*

THE HART'S TONGUE SEA-PEN.

The stem is capable of expansion and contraction.

possession of this attribute in a peculiar degree has given its name to one of the species which is to be found in British waters. This is the Phosphorescent Sea-pen. In repose its colour is dull purple-red, but when the delicate transparent polyps are expanded and the brilliant phosphorescence changes the dull hue to a deep purple it becomes an object of splendour and deserves its fame as the most beautiful of British Sea-pens.

When we come to consider the manner in which a Sea-pen is built up, we shall at once notice a great similarity to the corals. The barbs are full of separate individuals, each in its own little sac, from which it can protrude itself, or into which it can retract itself at will. They are, indeed, little coral polyps ; but instead of being able to create masses of coral-limestone reefs, all they do is to secrete from the sea-water just enough lime to form the hard, central rod, and this is almost the only thing the whole organism has in the way of a hard skeleton. In this it bears a close resemblance to the precious red coral, which is but a skeleton, and is covered with sarcode or jelly-flesh when living, and this contains the individual polyps seated therein. So with the Sea-pen, the central rod is hidden by a layer of flesh, sometimes of considerable thickness, and the barbs themselves are also flesh-covered. Then on the barbs are placed the little polyps, differently arranged in different species. These polyps rest in their separate little clefts, and, when living, protrude their rosettes of tiny tentacles and wave them in the water, endeavouring to pick up the morsels of nourishment which stray within their reach. They are veritable sea-anemones in miniature, and they always have eight tentacles surrounding their central mouth.

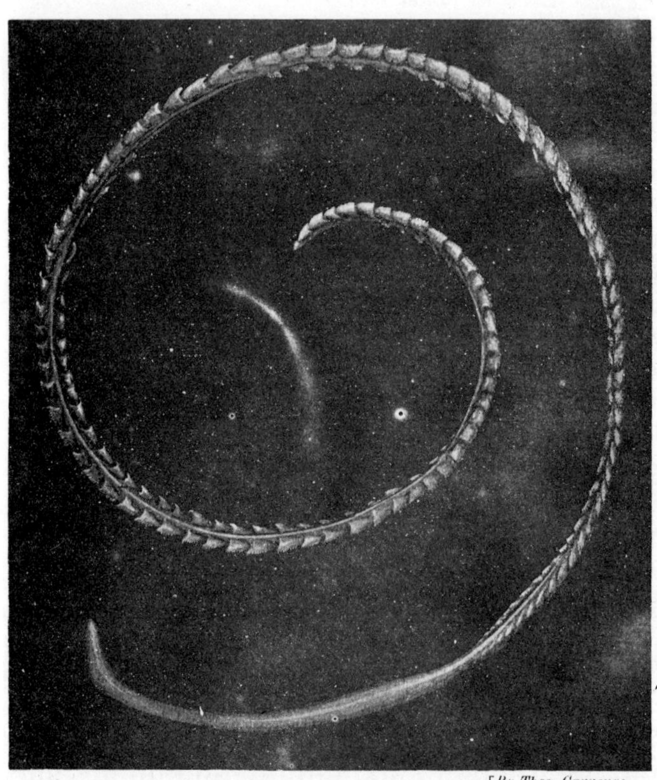

[*By Theo. Carreras.*

A "CHALLENGER" SEA-PEN.

A very beautiful and very intricate species. The polyps fringe the edge of each one of the curled fronds. While curled in this and the following illustration, Sea-pens are more or less flexible, though with a hard central stalk, which inclines to the perpendicular.

The typical quill-pen arrangement is departed from in some species, for the barb will be found at times so thick and bushy as to look like a fair-sized brush, of which the handle is formed by the naked lower portion of the stem, or rachis. The Phosphorescent Sea-pens of British seas, mentioned above, are of this kind, and in warmer seas they may be found a foot in length. The lower stem is of a creamy colour. One can well understand why this creature was long thought to belong to the vegetable world. First, the stem was planted in the mud of the sea-bed. Then there was a central soft stem, with a hard " pith," as one might call it. On each side of the stem small leafy branches were given off, and on the upper sides of these there grew a number of rosette-like flowers which we know now to be polyps surrounded by tentacles, each having the power of withdrawing within the particular cleft in which it lives. This latter characteristic, however, clearly shows its animal nature. Again, each polyp, though closely associated with its neighbouring polyps, is a single individual with a distinct mouth and body, and its own series of tentacles. Many thousands of them

[*By Theo. Carreras.*

THOMSON'S SEA-PEN.

The central column of this beautiful Sea-pen is pliable. The creature is a pulsating mass of life. Wavy movements pass from base to tip. The polyps are arranged in whorls surrounding the stem. It is a living note of interrogation.

[By Theo. Carreras.

THE THORNY SEA-PEN.

Each barb is divided into layers, which are separated by hard but brittle white needles. Between the needles, or spines, the rosetted polyps are protruded.

go to form the whole structure of the Pen, but although so numerous, they all act by one movement and have one central axis connected with a system of obscure circulation.

As far as can be ascertained, the separate polyps play no part in the production of the ova, which are formed in the cavities leading from the fleshy compartments of the extremity. These cavities were found by the late Charles Darwin to be " filled with a yellow pulpy matter which under a microscope consisted of round, semi-transparent grains aggregated together into particles of various sizes. All such particles, as well as separate grains, possessed the power of rapid motion, generally revolving round different axes, but sometimes progressive."

Some of the Sea-pens have been observed to suddenly withdraw themselves bodily into the mud when they have been slightly touched. On the other hand, the lower stem, or rachis, has considerable power of expansion in some species, and this part is always free from polyps. They have been dredged from great depths, and they are generally found in somewhat large numbers wherever discovered. This seems to show that they make but little use of the slight powers of movement which they are said to possess.

The name " Cock's-comb " is sometimes applied to the Sea-pen, and, seen in the right position, with a colour which does actually resemble the decoration on the head of a domestic fowl, the name is somewhat appropriate.

We have seen that in the sponges and corals those minute pieces of mineral known as spicules are invariably found. So in the flesh of the Sea-pens similar spicules are found, and these tend no doubt to give some slight support to the jelly-like flesh. We seem to see here the commencement of that solidification of the integument which afterwards enclosed the flesh inside a hard skeleton in the case of the crustaceans and the sea-urchins.

The straight internal rod or axis is longest in the genus of the Sea-Pens known as the *virgularia*, and one of the rods in the Natural History Museum measures no less than three feet long. The fossil rods of similar creatures have been met with in the London Clay of Sheppey, but they are extremely rare.

Many new species of Sea-pens were discovered during the *Challenger* expedition of 1873-76. Amongst these was the beautiful Tentacled Sea-pen shown on page 566. It was dredged up from a depth of ten to twelve fathoms off the coast of the Philippine Islands. Each frond or *pinna*, which spread out fan-shaped from the main stalk, is fringed with delicate, white, tentacled polyp-cells —hence its name. These cells, and the polyps themselves, are so minute that they can only just be detected in the illustration. The main stem and the bases of the branches are of a reddish hue. But perhaps the most splendid of all is Thomson's Sea-pen, named after the leader of the *Challenger* expedition, which was found at the immense depth of six hundred fathoms off the coast of Buenos Ayres. Its average length was about four feet. The thick swelling stalk with a smaller end bulb has a pointed free end which is buried to a considerable depth in the mud. The rather fleshy main stalk does not branch out into well-defined pinnules or fronds, edged with polyp cells, as do those on page 565, but the polyps themselves are ranged in a curved line round the stalk, and are thickly placed, eight or ten forming a single curved fringe. The polyps are entirely separate from one another even in their growth from the stem. They are very long, and the cells which are usually noticed and which form a distinctive feature of the Tentacled Sea-pen are absent. The polyps in this instance, however, are supplied with long tentacles which are seldom withdrawn, and careful observation has ascertained that these tentacles are themselves supplied with long threads or pinnules. Of the species found round the British coast perhaps the most interesting is the Wonderful Sea-pen. It is common in the waters to the East and North of Scotland and is to be found off the coast of Norway; but the extremely fragile structure of this Sea-pen makes a perfect specimen a rarity. It is one of the finest of marine polyp-systems. There are two series of half-moon shaped wings which are wrapped round the main stem and spread out after the fashion of branches. These series have the appearance of two ornamental ribbons folded in an inverse direction round the stem and develop ornamental wings alternately. These wings are waving, vandyked

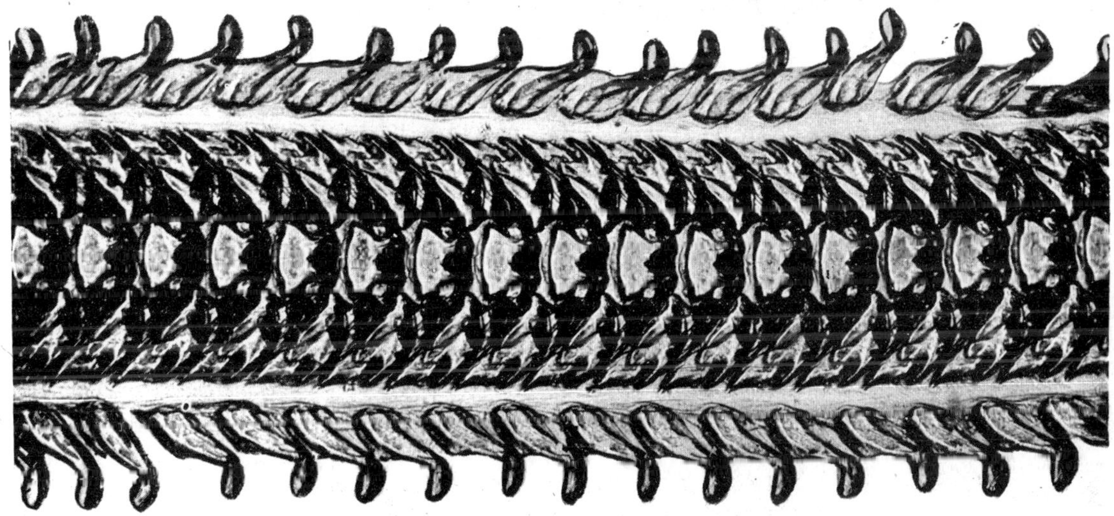

Photo by] [*J. E. Cooper.*

THE PERIWINKLE'S HOOKED TEETH.

They act as sickles to scrape off minute seaweeds for food.

and fringed on their outer edge, and in colour are a brilliant yellow. On the toothings of the edge are situated the polyps, which are white and semi-transparent. When the tentacles of the polyps are displayed the margin of each wing is irradiated with a series of frail silvery stars.

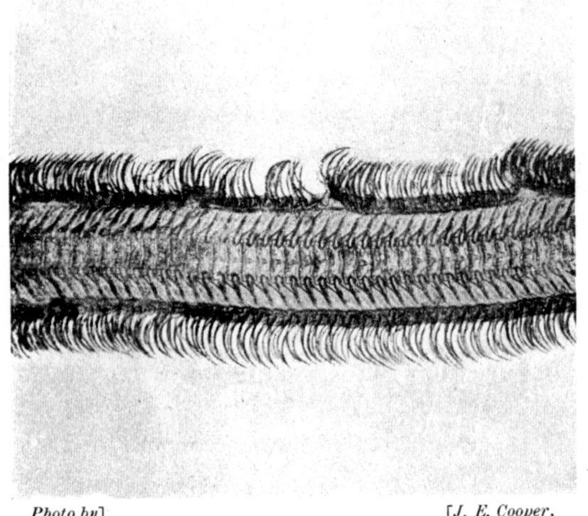

Photo by] [*J. E. Cooper.*

THE TEETH OF A MARINE SNAIL (TRITON).

From Cape Verde Island.

SNAILS' TEETH

BY J. E. COOPER.

TEETH are usually associated with biting, but in one natural sub-kingdom of animals it is exceptional to find them used for that purpose. The creatures referred to are scientifically called *Mollusca*—a term which includes Slugs, Snails, Whelks, Squids, Limpets and other shell-bearing animals.

Bivalve shells (*e.g.*, mussels and oysters) have no teeth; but practically all univalve shells possess them, generally in great numbers. They are employed for various purposes, some of which are here described.

In Slugs and Snails the " tongue," or lingual ribbon, is a slipper-shaped membrane, lying at the back of the throat. Its use is to break up the animal's food, for which purpose it is covered with a multitude of minute pointed teeth composed of chitin, a very hard and brittle substance. The arrangement of the teeth on the ribbon resembles that of the scales on an ancient coat-of-mail; the teeth themselves can be erected or depressed by special muscles at the will of the animal. The process of digestion in a Snail has been aptly compared to harrowing a field, the food being gradually broken up by the hundreds of minute rasp-like teeth with which the tongue is covered.

The Great Black Slug of our fields has about thirty thousand of these scale-like teeth, while the Common Snail manages to rub along with as few as one thousand five hundred.

In the Glassy Snails the individual teeth are considerably larger and fewer in number, though they serve the same purpose of pulping the food. The small Fresh-water Limpet, which abounds under stones in many of our streams, is related to the Snails; its teeth, however, are more loosely arranged on the tongue and are much narrower in proportion to their size.

In the majority of the carnivorous marine molluscs both the purpose and arrangement of the tongue are totally different from the foregoing. Here the tongue, which is long and narrow, is usually placed in a retractile proboscis. The teeth are fewer but more powerful, and are specially adapted for use as a drill to bore holes in other shells, chiefly helpless bivalves. When once the hole has been drilled the proboscis comes into play as a pump to suck out the victim's juices.

Photo by] [*J. E. Cooper.*

THE CROW-OYSTER'S TEETH.

They are quite harmless despite their formidable appearance.

All the Whelk tribe feed in this manner. From the commercial point of view the Whelks are of some importance. They all have a liking for mussels and oysters, when they can get them. Our

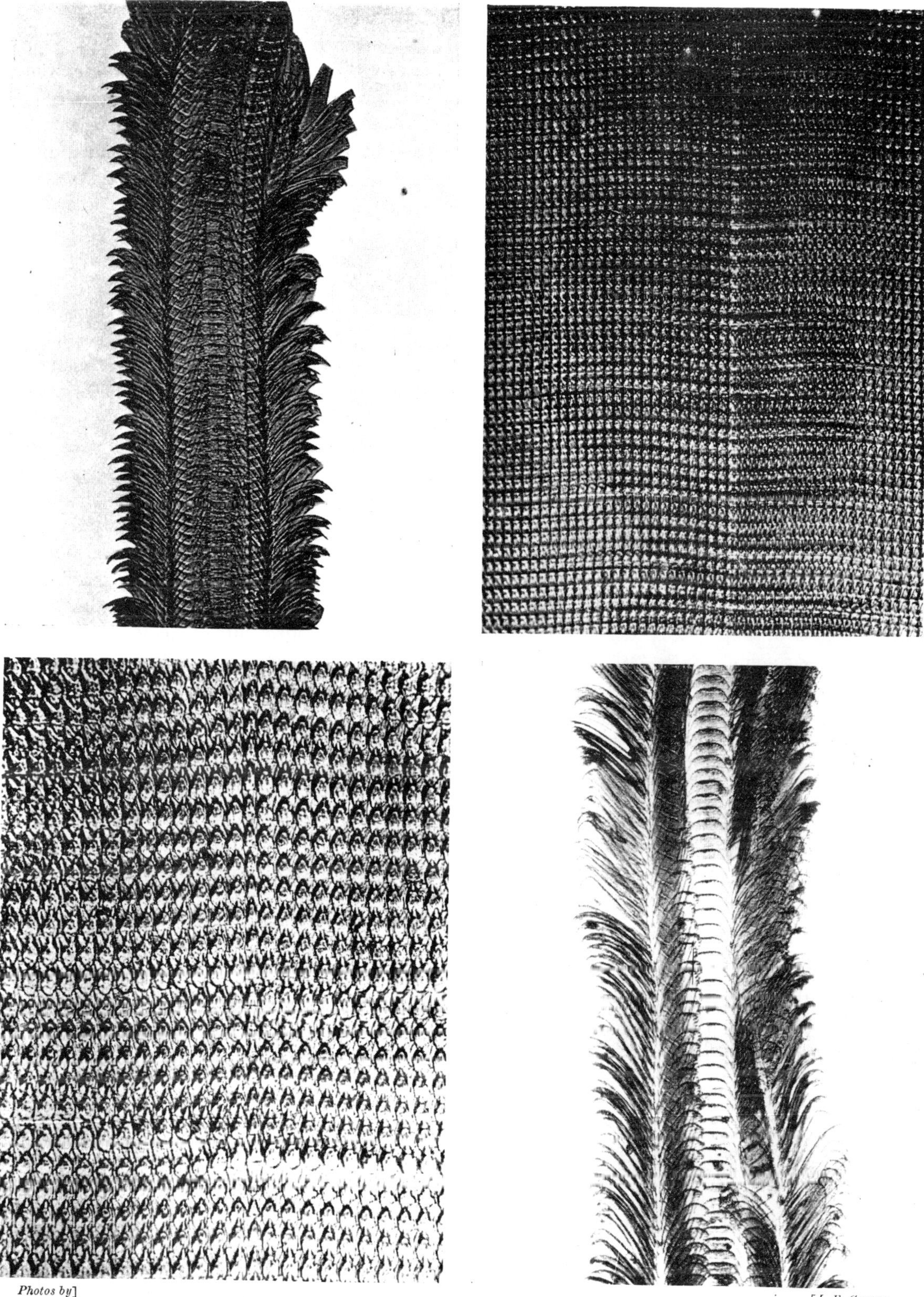

Photos by] [*J. E. Cooper.*

SNAILS' TEETH.

The top left-hand photograph shows the elaborate set of the River Snail; the right-hand shows some of the Black Slug's 30,000 teeth. The teeth of the Common Snail occupy the bottom left-hand corner, and on the right are the fan-like teeth of a species of Fresh-water Snail.

The Fresh-water Limpet has a remarkably uniform set of teeth,—

—and shows by these teeth that it is related to the Snail.

Photos by] [*J. E. Cooper.*

The Top-shells are a striking contrast to the Volutes. Their teeth are innumerable, and of three distinct patterns.

largest species is the Almond-whelk; some of the smaller kind, however, such as the Dog-whelk, appear to do more damage. At Burnham-on-Crouch the Dog-whelks are collected by the bucketful from the oyster-beds at low spring-tides.

The common Netted Dog-Whelk is a more useful member of this tribe. It swarms in crab-pots on our southern coasts, where it plays the part of a scavenger.

In connection with these carnivorous molluscs it is curious to note how often the attacker does a vast amount of work without reaping any benefit; it is not at all uncommon to find bivalve shells with several partially-bored holes in them.

A very degenerate form of the carnivorous type of tongue is found in the Volutes. They possess but one band of trident-shaped teeth, which cannot be of much use to their owner, as the total length of the tongue is three-eighths of an inch, while the shell itself is four and a half inches long. The tongues of marine shells that are vegetarians may be roughly divided into two sections: (1) those in which the tongue is used to collect food; and (2) those in which it is a digestive organ. Limpets and Periwinkles are good examples of the first section. They have very long tongues armed with comparatively few teeth. The tip of the tongue is protruded when the animal is feeding, and the teeth act as a rake or sickle to scrape off minute seaweeds, etc. As the teeth get worn by use, new ones are continually being pushed forward to replace them. The common Periwinkle can never be short of teeth, as its tongue is twice as long as the shell is high and is said to carry three thousand five hundred teeth.

In the other section the tongue's chief function appears to be digestive, as in the Snails. Our common Purple Top-shell may be taken as an example. The central teeth are rasp-like and, no doubt, serve to break up the food. The use of the numerous hair-like teeth at each side is uncertain, possibly they act as brushes to push food over the central teeth.

These remarkable hair-like teeth are also found in our two River Snails, the *Viviparas,* but in this case they are curved at the extremity. This would make them more effective as brushes.

The American "Crow-oyster," or "Slipper-limpet" (which is now well established in some of our estuaries), has a striking arrangement of teeth The teeth are harmless to other molluscs; but the shell is cordially hated by oyster-growers, as it is exceedingly prolific and crowds out the baby oysters.

A somewhat similar set of teeth is shown by the Tritons. The one figured is from a Cape Verde Island species.

These are but a few examples of the wonderful beauty and variety of molluscan tongues and teeth.

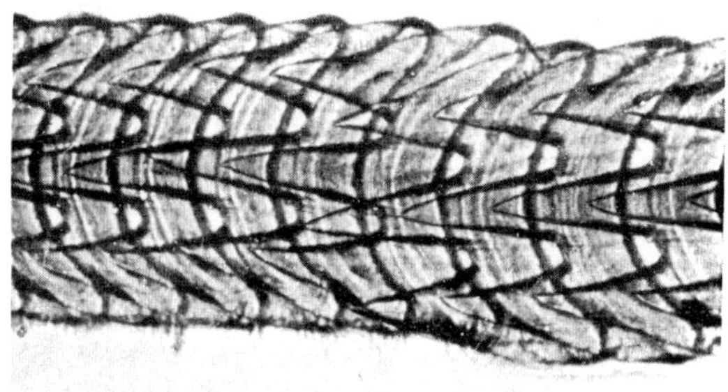

The Volutes have a very simple tongue, merely one row of three-pointed teeth.

FOSSIL SEA-SERPENTS

BY W. P. PYCRAFT, F.Z.S.

DESPITE the fact that the Mosasaurus represents the head of a most interesting and distinguished family of reptiles which have long since had their day and ceased to be, no one of them has, until recently, received what we may call public recognition, though *savants* have written much and speculated much about them since 1770, when the " head of the house " was brought to light. They have been thrust into the shadow by those old sea-dragons, the Ichthyosaurs and Plesiosaurs, whose remains are so much more common in museums, and have accordingly attracted so much more attention. Yet the Mosasaurs, as we may call the family collectively, are a most interesting tribe, and lose nothing by comparison with their more famous rivals.

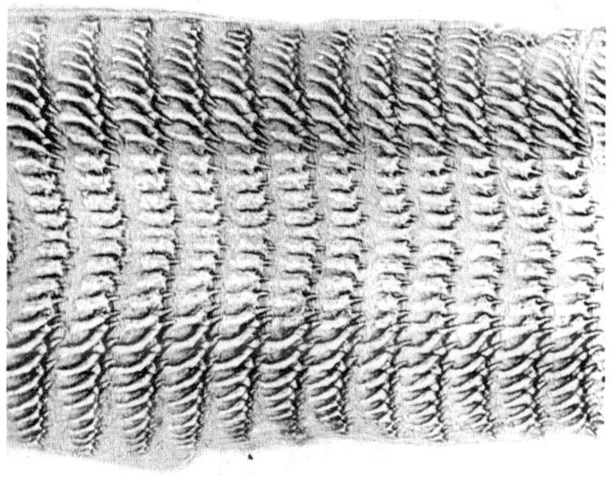

The delicate tooth-coils of the Glass-snail.

The first indication of their existence was obtained when some workmen, blasting the rock in one of the caverns near Maestricht, in the valley of the Meuse, exposed a pair of huge jaws, armed with formidable teeth. They were not long in conveying the news of their discovery to Herr Hoffmann, a Dutch military surgeon attached to a neighbouring fort, and known to be interested in such " curiosities." After laborious weeks he succeeded in detaching these remains, and the marvel thereof soon spread. His house was besieged by the curious. In a little while the canon of the neighbouring cathedral, marking the excitement, claimed the bones as his own, as lord of the manor, and bore them off in triumph to the cathedral, and poor Hoffmann died without regaining his prize. But time brings its revenges. In due course came the French Revolution, and the armies of the

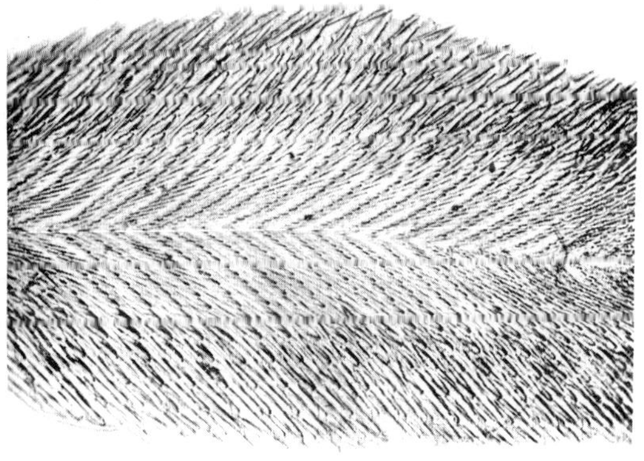

Photos by] [*J. E. Cooper.*

The fine teeth of the Worm-eating Slug already described in these pages.

Republic surged at the gates of Maestricht. The town was bombarded, and by the advice of the *savants* who accompanied the army to select their share of the plunder, the cathedral and its neighbourhood was spared from the fire of the artillery. When at last the city fell, a rush was made to secure the treasure. But the canon of St. Peter's, suspecting the reason of such peculiar forbearance, removed the specimen to a vault. He was, however, speedily compelled to disgorge his ill-gotten prize, which was triumphantly borne off and deposited in the Jardin des Plantes, at Paris, to be studied and named by the great Cuvier. He bestowed the name "Mosasaurus"—the Lizard of the Meuse. A cast thereof is one of the treasures of the British Museum of Natural History. For a long while discussion waxed furious as to what kind of animal these jaws really represented, for Cuvier's verdict was not universally accepted, though it proved to be right in the end, and it was not till other remains in other parts of the world came to light that the mystery was solved.

We now know that the Mosasaurs represented the last of the great sea-dragons; they attained their zenith as the old Ichthyosaurs and Plesiosaurs were passing away; that is to say, they throve during the Cretaceous age. By many, nowadays, they are described as fossil Sea-serpents; and it is certain that they are nearly related to the lizards of to-day. All the evidence goes to show that

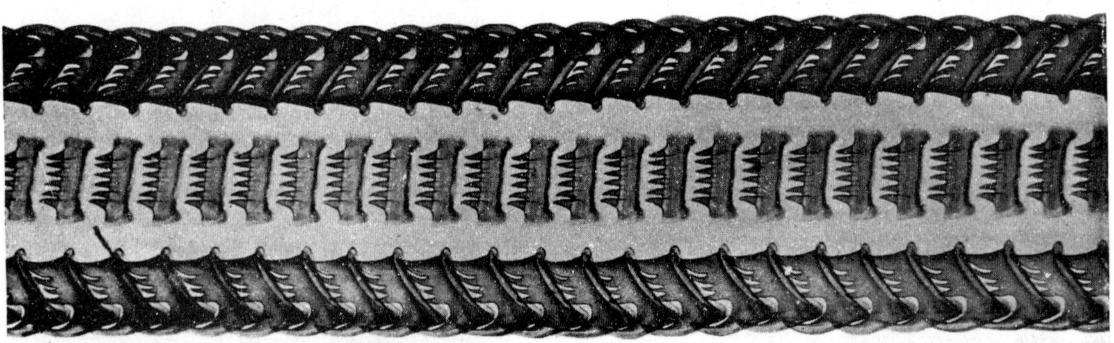

Photo by] [*J. H. Crabtree.*

THE BEAUTIFUL TEETH OF THE COMMON WHELK.

they, or their forbears, were land-dwellers. At first, we may suppose, like the Monitor lizards haunting river banks and taking readily to the water, they became gradually more and more aquatic, and in proportion their limbs gradually assumed a paddle shape, as did those of the Ichthyosaurs and Plesiosaurs before them, and whales and porpoises which have followed.

We may assume that this process of transformation began in what is now the European area, and that as they became more and more aquatic so they spread further and further afield, for fossil Mosasaurs occur both in North and South America and in New Zealand. Some of these were giants of fifty feet or more in length; some did not exceed fifteen feet. In some, as in the species of the genus Edestosaurus of the Upper Cretaceous of North America, the body had become excessively long and snake-like, and hence the association with the Sea-serpent. They must have swarmed in the old seas that once surged over what is now Western Kansas, for thousands of specimens have been obtained from this region.

Some of these, undoubtedly, had a long fin down the back, and all, the world over, have a remarkable feature about the lower jaw. This was hinged across its middle, and the two halves were not fixed at the snout end. Thus they resembled the jaws of snakes, from which we argue that they must have fed upon creatures of greater bulk than the width of the head. But once the victim was caught between those terrible backwardly curved teeth, escape was impossible.

[By C. E. Swan.

THE MOSASAUR.

A reconstruction of the last of the Sea-serpents. It is probable that the Mosasaur was first of all a land-dweller and then gradually developed aquatic habits, till finally the lizard-like limbs assumed a paddle shape and the species which at first had been (it is supposed) restricted to Europe was enabled to spread over the entire world.

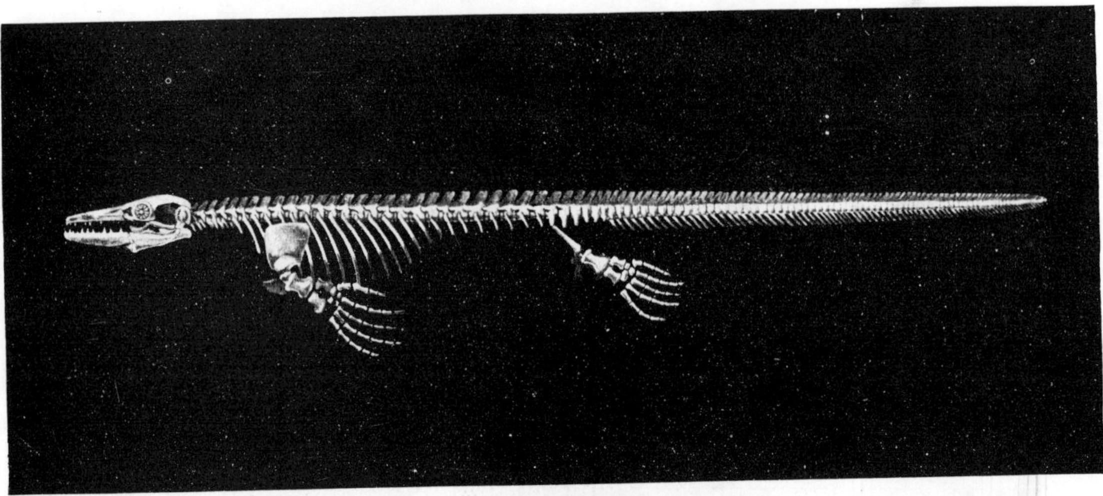

THE MOSASAUR.

A reconstructed skeleton of a creature which often attained a length of fifty feet or more. An interesting feature is the lower jaw, which is hinged across the middle and resembles the jaws of snakes, indicating that the prey of these animals was larger than the width of the head.

THE GIANT WATER-LILY

BY EDWARD STEP, F.L.S.

WE have already figured and described the largest known flower (Rafflesia), and by comparison with it the Victoria Water-lily is a rather small giant, but putting Rafflesia aside, the Victoria is of quite noteworthy proportions. Then, added to its magnificent flowers, the plant has several notable features which would give it first place in any popular competition between the two plants. The aquatic flower has beautiful tints ranging from pure white to rose; it gives forth a delightful odour, and it is set off by magnificent leaves that are in themselves entitled to rank as marvels.

Although discovered so far back as the year 1801, and recorded by several travellers in subsequent years as growing in various South American rivers, it does not appear to have been taken seriously until 1837, when Sir Robert Schomburgk found it in the Berbice River, British Guiana, and fully described it to the Royal Geographical Society. That was the year of Queen Victoria's accession, so what more natural than that the beautiful plant should be named *Victoria regia?*

From a thick, fleshy rootstock running in the mud of the stream the plant sends up its leaf-buds to the surface, where they expand into huge tray-like leaves which measure from two to four yards across. Rich deep green on the upper surface, the edge is turned up all around to a height of three or four inches, as though to show the crimson or purple tinting of its lower side. The leaf-stalk is attached to the centre of this underside, and from it radiate a number of very stout nerves with many thick branches connecting them, so that the entire underside is mapped out into a great number of cells. This conformation, it will be understood, at once adds greatly to the strength of the leaf and also to its buoyancy. It is worthy of note that it was the contemplation of this leaf-structure that gave Sir Joseph Paxton—then gardener to the Duke of Devonshire—the idea upon which he designed the Crystal Palace, iron-work taking the place of the nerves, and glass instead of leaf-tissue filling in the interstices. The nerves of the leaf as well as its stalk are hollowed by air passages to increase their buoyancy, and are beset on the exterior with sharp, stout spines to keep off fishes that might otherwise be inclined to browse upon the plant. The flower is about fifteen inches across.

ENCYCLOPAEDIA OF THE UNIVERSE

ENCYCLOPAEDIA OF THE UNIVERSE

A POPULAR WORK ON THE MARVELS OF

THE HEAVENS :: THE EARTH :: PLANT LIFE :: ANIMAL LIFE :: THE MIGHTY DEEP

WITH 1000 FINE ILLUSTRATIONS

Vol. III

Editor

WALTER HUTCHINSON

B.A., F.R.G.S., F.R.A.I., Etc.

WITH AN INTRODUCTION BY

LORD AVEBURY, P.C.

AND WITH CONTRIBUTIONS BY LEADING SPECIALISTS, INCLUDING :

Sir Harry Johnston, G.C.M.G., K.C.B.
Richard Lydekker, F.R.S.
Frank Finn, F.Z.S.
Edward Step, F.L.S.
E.W. Maunder, F.R.A.S.
W.P. Pycraft, A.L.S., F.Z.S.
Russell F. Gwinnell, F.C.S.
Frank T. Bullen, F.R.G.S.
Tickner Edwardes
J.J. Ward, F.E.S.
C. Flammarion
Lionel E. Adams, B.A.
Harold Bastin, F.E.S.
S. Leonard Bastin
K. G. Blair, F.E.S.
Sir Herbert Maxwell, F.R.S.
E. B. Poulton, F.R.S.
G. F. Chambers, F.R.A.S.
H. St. J. K. Donisthorpe, F.E.S.
Richard Kerr, F.G.S.
W.F. Kirby, F.L.S.
Hugh Main, F.E.S.
Ernest Marriage
E.A. Martin, F.G.S.
T. E. R. Phillips, F.R.A.S.
R. I. Pocock, F.R.S.
Dr. E. J. Spitta, F.R.M.S.
E. W. Swanton
J. Sinel
W. Mark Webb, F.L.S.

D P H

DISCOVERY PUBLISHING HOUSE PVT. LTD.

NEW DELHI-110 002

Published by:
Namit Wasan
DISCOVERY PUBLISHING HOUSE PVT. LTD.
4383/4B, Ansari Road, Darya Ganj
New Delhi-110 002 (India)
Phone : +91-11-23279245; 23253475; 43596065
E-mail : discoverybooksindia@gmail.com
discoverypublishinghouse@gmail.com
namitwasan9@gmail.com
web : www.discoverypublishinggroup.com

Edition: 2020

ISBN: 978-81-8356-873-9 (Set)

Encyclopaedia of the Universe

Printed at:
Infinity Imaging Systems
Delhi

CONTENTS OF VOLUME III.

THE AFRICAN JAÇANA OR LEAF-WALKER

BY SIR HARRY JOHNSTON, G.C.M.G.

THE Jaçanás—the name is derived from Brazil—are a group of plover-like birds, which ranges throughout the tropical regions of the world, having representatives in Africa, South America, India, Malaysia and China. It will be observed from the accompanying illustration, which is that of the largest and commonest African species, that this bird develops enormous feet, though the excessive apparent size of the feet is partly caused by the great length in the claws. Indeed, as a fact, these birds have proportionately rather a small hind toe, and have probably been developed from some plover-like type, in which the hind toe was in course of atrophy. This short length has been made up by the development of claw, which is about twice the size of the actual toe to which it is an appendage.

The object of these greatly lengthened toes and claws is to enable the birds to stand on floating river vegetation, which they haunt for the insects which swarm about such plants. They are frequently called " Lily-trotters," on account of their association with water-lily leaves. Many of these birds are very beautifully coloured. The species here illustrated, which has, perhaps, the most remarkable development of feet, is not quite so showy as the Brazilian and Chinese forms

Photo by] [*H. N. King.*

THE GIANT WATER-LILY.

The enormous leaves of this Water-lily may measure as much as twelve feet across. They are a rich deep green on the upper side with a crimson under surface, and are turned up all round to a height of three or four inches. The flower is white shading to rose, and is beautifully scented.

THE AFRICAN JAÇANÁ.

This is the larger of the two Jaçanás, or Leaf-walkers, found in Africa, and is also the commonest, ranging right across Tropical Africa from west to east. It is about the size of a small plover, and has blunt spurs on the wrist of the wing.

(some of which develop remarkable spurs on the wrist of the wing). Nevertheless, the common African Jaçaná is a pretty creature, with its plumage of rich chestnut-brown, black quills, and black top to the head and neck, and a face, throat and chest of pale golden-yellow, while the beak and the frontal shield, the big legs and feet are a pale leaden-blue.

THE LACE-WING FLY

BY K. G. BLAIR, B.SC.

WHILE among the countless hosts of the insect world most of the species are of no economic importance to mankind, yet there are, of course, vast numbers that do him considerable injury; and in attempting to exterminate these, or, at any rate, to keep them in check, large sums of money and an immense amount of labour are annually expended. One of the most successful methods of keeping down the numbers of one of these undesirable kinds of insects is to find out its natural enemies, and by using all possible means of favouring these and encouraging their multiplication to increase the efficacy of their work as a scourge upon the pest.

One very common insect pest, which, though not perhaps of great commercial importance, is yet a source of constant annoyance to gardeners, amateur and otherwise, is the green fly, aphis and its allies. Yet this little insect has a great number of enemies, especially amongst other insects, and by finding out these and what are their habits a great deal can be done to keep down the numbers of the aphis. One of these natural enemies that has already been treated of in these pages is the hover fly, another is the lady-bird, while yet another is the Lace-wing Fly. The latter is a rather large insect, the common species measuring about an inch and a half across the expanded wings. Its body is of a vivid green with some golden markings, and the eyes are rather large, prominent and of a brilliant golden colour. It has two pairs of large gauzy wings, each of which shows a delicate net-like tracery of veins, the feature from which the fly derives its common name. It frequents gardens, bushes, or rank vegetation—anywhere, in fact, where the green fly abounds; but it usually remains hidden away during the day and only flies for a short period about dusk. It is then easily captured, as its flight is slow and heavy, as though the expanse of wing were too great to be under the complete control of the slender body.

The Lace-wing Fly is to be found throughout July and August. The eggs are laid on the undersides of leaves, sometimes singly, but usually in groups. Instead of being laid directly on the surface of the

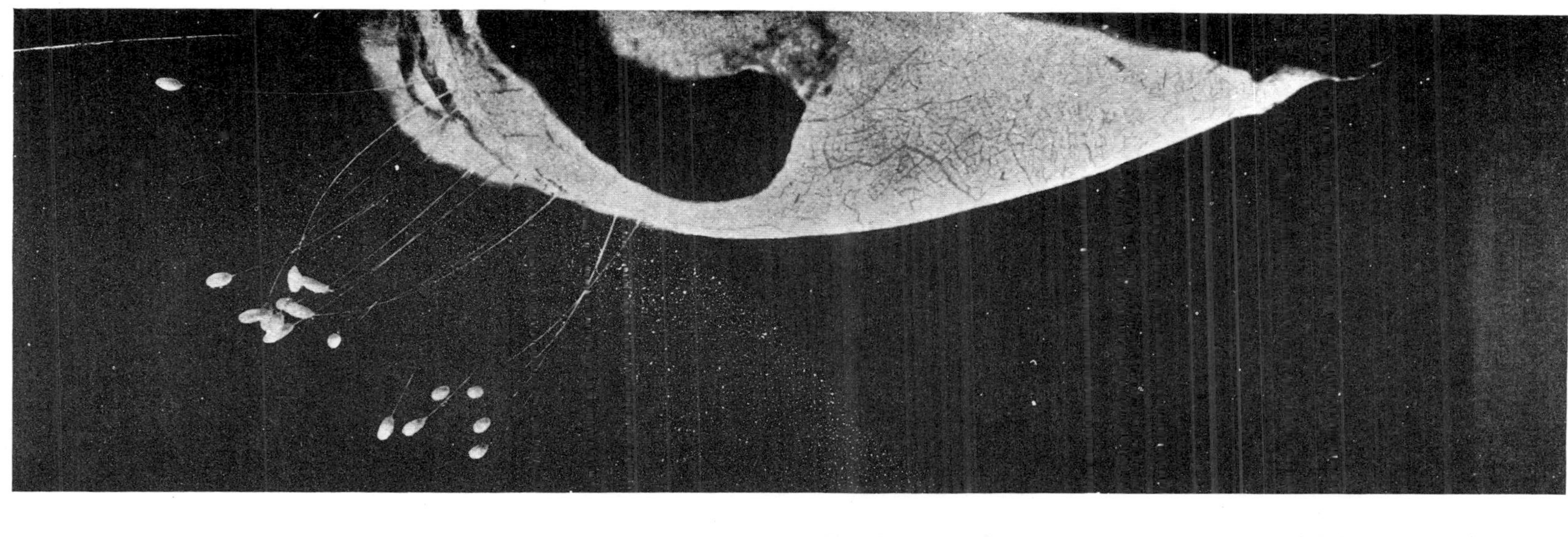

Photos by] [*Hugh Main, B.Sc., F.E.S.*

THE EGGS OF THE LACE-WING FLY

These little groups of stalked eggs were classed by some of the elder naturalists as mosses. In reality they are nearly as large as the finest pin's head and are supported on the slenderest of silken threads. Sometimes they are deposited separately on the underside of a leaf, and sometimes the stalks of the eggs are united in a common bundle as in the lower photograph.

Photo by] [*Hugh Main, B.Sc., F.E.S.*

THE GRUB OF THE LACE-WING FLY.

It is provided with two pairs of long, sharply-pointed, sickle-shaped jaws, each pair fitting together to form a hollow tube. With these jaws the grub has here seized its prey (an Aphis) and is sucking the juices out of its victim's body.

Photo by] [*Hugh Main, B.Sc., F.E.S.*

THE GRUB OF THE LACE-WING FLY

As soon as it has emerged from the egg it will attack and devour any unhatched egg or emerging grub within reach.

leaf, each is suspended at the end of a long slender stalk ; in some species all the stalks of a group of eggs are united in a common bundle, though more often each egg remains quite distinct. It is interesting to note that these little groups of stalked eggs were classed by some of the older naturalists as mosses !

When first laid, the eggs are green in colour, but gradually change, until just before hatching they are white with the exception of two little brown spots at one end ; these are the eyes of the enclosed larva showing through the shell. The little pearly-white larvæ appear after about a week, and are at once very voracious. Each will remain for a while on its own empty egg-shell and attack and devour any unhatched egg or emerging larva within reach ; so that in the cases where the eggs are all united in a bunch, it is only the first larva or the first few that stand a chance of survival. Throughout life, in fact, the larvæ are very liable to become cannibals if the least excuse of shortage of aphides be afforded them ; though if these are plentiful they may be kept together with a fair amount of safety. The larva is provided with two pairs of long, sharply-pointed, sickle-shaped jaws, of which the corresponding members of each pair fit together in such a manner as to form a hollow tube between them leading from the tip into the mouth. When it seizes the prey the larva merely nips it between the points and sucks up its juices into the mouth without chewing the body in any way. As soon as the aphis is sucked dry, the larva tosses away the corpse, at once spears another green fly, treats it in a similar summary fashion, and so on with one after another in rapid succession, soon causing havoc in the colony that it has attacked.

Another kind of Lace-wing Fly, instead of discarding the empty skins of its victims, carries them about on its back, so that it is soon completely concealed under a heap of débris, and until it moves is passed over as a mere lump of dirt in the middle of a leaf. In this case the main object of the concealment is, no doubt, to protect the larva from the attacks of insect-eating birds ; that it should enable the creature to approach its prey unobserved is here of little importance, since aphides are probably too sluggish to move whether they perceive the foe or not, though it may be noted that the species which practise this mode of concealment are found more

Photo by] [Hugh Main, B.Sc., F.E.S.

About three weeks after leaving the egg the grub spins a cocoon in some curled-up leaf—

often on beech and other trees on which the infesting green fly are more thinly distributed and more active in habits than those we find on the rose-bushes.

In favourable circumstances the Lace-wing Fly larva grows very rapidly, and, after moulting twice, becomes full-grown in three or four weeks after leaving the egg. It now seeks a retired spot in which to complete its transformation. In some crevice, or in a withered and curled-up leaf, it spins around itself a white silken cocoon formed of the secretion of glands developed for the purpose which open into the mouth. When complete, the cocoon is almost spherical, though slightly longer than broad, and very much shorter than the full-grown larva; the latter, indeed, has to lie curled up within it, with its head tucked under its tail. So closely is it packed away, that it is marvellous indeed how it can retain sufficient freedom of movement to complete the cocoon; for it must be remembered that the spinning is all done from the inside. In this state it remains throughout the winter; but early in the following spring it sheds its skin once more, and the pupa is disclosed. During its long rest it has decreased somewhat in bulk, so that it is now not quite so tightly packed and has room to indulge in more freedom of movement; even yet, however, the pupa is considerably longer than the cocoon in which it is enclosed, and both head and tail are curled forward over the breast. The long slender legs of the fly can now be made out, all folded against the lower side of the body, and the rudiments of wings are folded down between the second and third pairs of legs. The long, thread-like antennæ, or "feelers," are curled back behind the eyes, and thence down with the legs and wings to the lower surface. Unlike the pupa of a moth or butterfly, the limbs are not firmly soldered down to the body, and are not even lightly attached, as in the pupæ of beetles or bees, but each limb remains distinct. At first the difference between the pupæ of the Lace-wing Fly and that of a bee, for instance (see PART VII.), is not very marked, because, though each limb is

Photo by] [Hugh Main, B.Sc., F.E.S.

—and with marvellous skill he continues to spin until he is completely hemmed in.

Photo by] [Hugh Main, B.Sc., F.E.S.

The cocoon is much smaller than the grub, who has to curl himself up in the manner here shown.

Photo by] [Hugh Main, B.Sc., F.E.S.

When ready, the insect makes a clean cut nearly round the cocoon, lifts the lid and leaves his case empty as shown here.

The first transition which takes place in the cocoon, the grub changing to chrysalis.

The early chrysalis.

Photos by Hugh Main, B.Sc., F.E.S.

The mature chrysalis (uncurled), showing already the antennae of the perfect insect.

separate, they are, nevertheless, quite motionless for the very good reason that the muscles to move them are not yet developed. But as the fly within the delicate pupa shell approaches maturity, the muscles are formed, and when almost ready for emergence the limbs of the pupa become functional.

In this stage, too, the insect is furnished with a pair of powerful jaws, unlike those of the larva and unlike those of the fly—jaws, in fact, useful only to the pupa, and that for one special purpose. The muscles that work these, like the muscles that move the limbs, only become functional late in pupal life; but when the insect is ready to emerge, the first operation is for the pupa by means of these jaws to make a clean cut nearly round one end of the cocoon: it then crawls out of the cocoon, forcing up the lid to do so; as soon as it is clear, this falls back by the natural elasticity of the part left as a hinge, so that at first sight the empty cocoon still appears complete. The pupa now, but for its imperfectly developed wings, closely resembles the perfect fly, both in appearance and colour. It endeavours to find some upright rough surface, such as that afforded by a fence or tree-trunk, up which it can climb, in order that its final transformation may be successful. After a short rest the skin splits down the back, and through the rent the fly emerges. At first the wings are quite small and shrivelled in appearance, but are soon expanded by the forcible injection of blood into their veins; even now they are moist and limp, and are kept hanging vertically downwards while they harden. If the insect has not been able to find a support that enables it to assume this position, the wings are almost certain to be crippled and useless for flight. When sufficiently stiff, they are turned down roof-wise over the body, as seen on this page; but some little time has yet to elapse before they are ready for use. Early morning, and again the late afternoon, appear to be the favourite times for the emergence of the Lace-wing Fly, and about these hours the newly-emerged flies are common objects upon walls and fences.

COLOURED STARS

BY G. F. CHAMBERS, F.R.A.S.

With a few notorious exceptions the Stars which we see shining in the heavens on an ordinary clear night are devoid of colour, exhibiting, as a rule, a more or less pure white light. The exceptions to this rule are single Stars of a yellow or deep orange, or fiery red colour. These are not uncommon, but isolated blue or green Stars are very rare. Indeed, amongst the conspicuous Stars I can only recall one that is green, namely, β Libræ. The German astronomer Krüger has given the following as typical coloured Stars:

White.—Sirius (α Canis Majoris); Altair (α Aquilæ); Regulus (α Leonis).

Yellow.—Capella (α Aurigæ); Pollux (β Geminorum); Arcturus (α Boötis).

Orange-red.—α Herculis: Betelgeuze (α Orionis).

Colour in connection with Stars comes under our notice more directly

Photos by] [*Hugh Main, B.Sc., F.E.S.*

THE LACE-WING FLY.

Two photographs of the perfect insect, resplendent in its gauzy wings and beautiful green surtout broidered with gold.

in dealing with double Stars, many of which exhibit the curious and beautiful phenomenon of "complementary colours." Perhaps before going any further, I ought to explain the two phrases just used, namely, "double Stars" and "complementary colours."

When one gazes upwards at the sky, with the naked eye or an opera-glass, all the Stars we look at appear as single points of light; but a telescope will often show that we are looking at two Stars so closely side by side that inadequate optical assistance, or none at all, makes them appear to be only one Star. The larger the telescope, the greater the number of the Stars, apparently single, which are found to be double. As a matter of fact, several thousand Stars are now known to be really pairs of Stars. But this is not the whole case. A certain number of Stars, seemingly single to the naked eye, or in a small telescope, are found, when examined with a larger telescope, to be

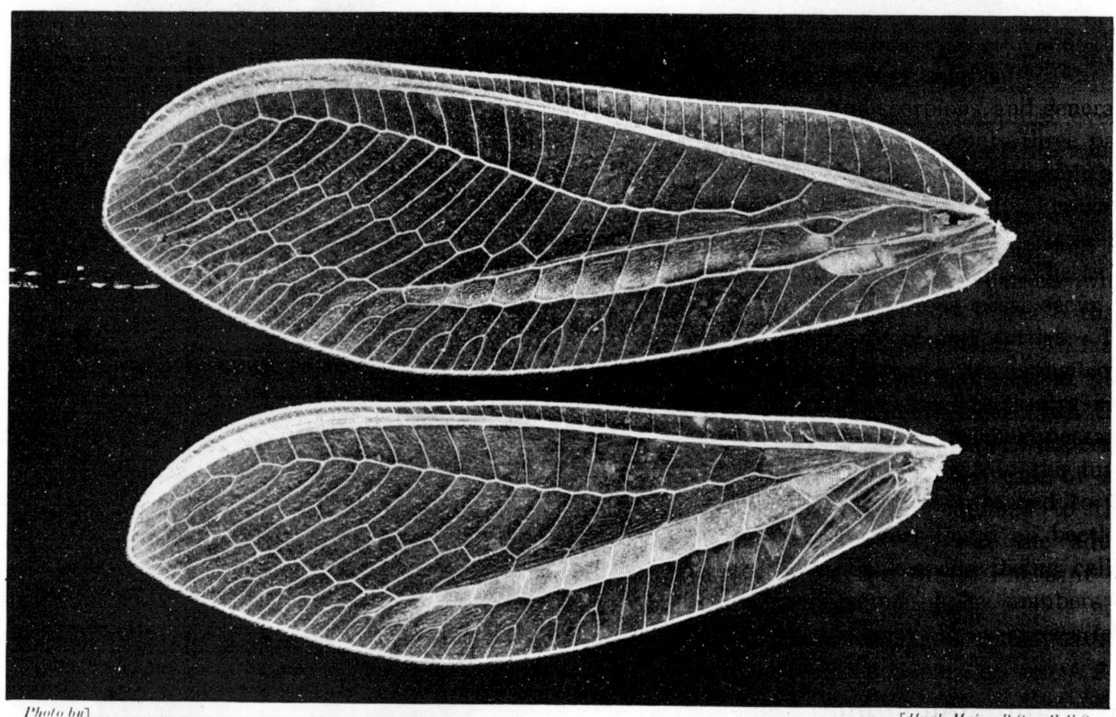

Photo by] [*Hugh Main, B.Sc., F.E.S.*

THE WINGS OF THE LACE-WING FLY.

A photograph of the beautiful wings, showing the delicate net-like tracery of veins from which the fly derives its name. They are about an inch long, but the flight of the insect is slow and heavy, as though their expanse was too great to be controlled by the slender body.

triple; that is to say, one becomes three, whilst there are such combinations as quadruples, quintuples, sextuples. After this we talk of them simply as "multiples."

I must not go too much into details as regards these matters, or I should get too far astray from the main point of this article, which is Coloured Stars.

The second definition which I said it was expedient to present was the term "complementary colours." In order to understand this, the reader must be reminded that if we take a source of pure white light to start with, and pass a beam of it through a prism, we find the emergent beam of light is transformed into a coloured beam, showing various gradations of colour between violet at one end of the emergent beam and red at the other end. The various intermediate hues pass into one another in beautiful succession, and the result is the well-known spectrum which we used to regard as made up of seven colours, but which scientific men, following Sir David Brewster, often now only regard as consisting of three colours, to be called truly primary, namely, yellow, red and blue; and that the other colours are no more than an admixture of these primary tints. If in white

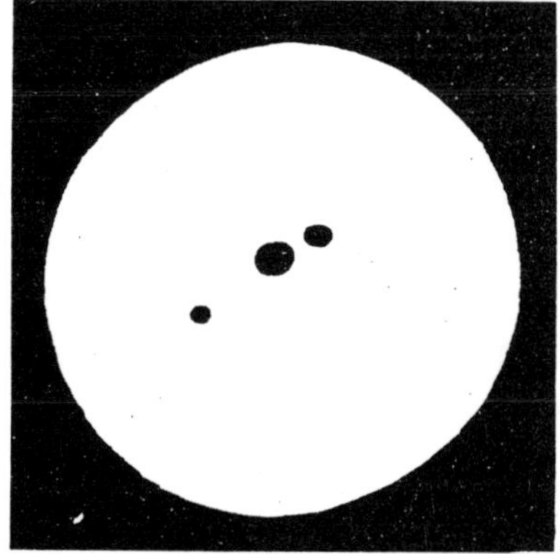

COLOURED STARS.

Colour is most frequently to be found in double or multiple stars which,—

light any colour is suppressed, a mixture of the remainder is called a complementary colour, for it is the colour required to restore the sensation of white light. Blue and yellow mixed produce *green*, and accordingly green is the complementary to red. In like manner, a mixture of red and yellow produces *orange*, which is complementary to blue. Similarly, indigo is complementary to yellow, and yellowish-green to violet. It may be convenient to have these facts put in tabular form, and slightly varied in detail, thus:

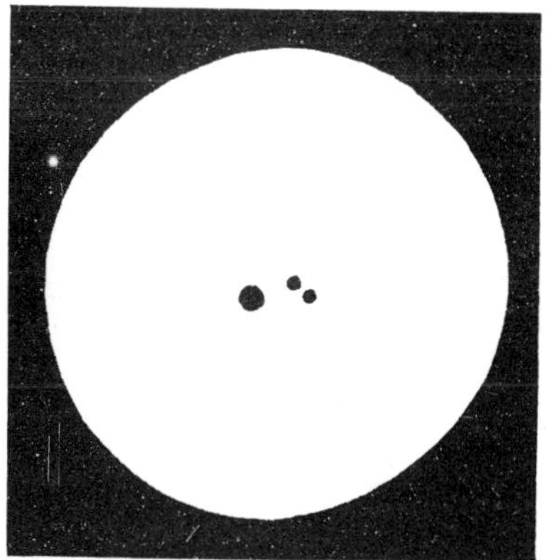

COLOURED STARS.

—although showing a white single light to the naked eye, can under a powerful telescope—

Red is complementary to	Bluish Green.
Orange	Sky-Blue.
Yellow	Violet-Blue.
Greenish-Yellow	Violet.
Green	Pink.

The situation of things is really this: that any two colours which are complementary as above give, when mixed in proper proportions on a rapidly revolving disc, a composite tint which is almost white.

We are now in a position to carry the matter farther as regards Double Stars, the two members of which exhibit complementary colours. In such cases, the larger Star is usually of a ruddy, or at any rate, orange hue, and the smaller one blue or green, or something between the two, say, bluish-green or greenish-blue, somewhat according to the fancy of the observer. If complementary colours are noticed in the components of a Double Star which are of very unequal size, the circumstance may be attributed often or mainly to the effect of contrast; yet it is permissible to suppose that sometimes the colour seen is actually the intrinsic colour of the Star.

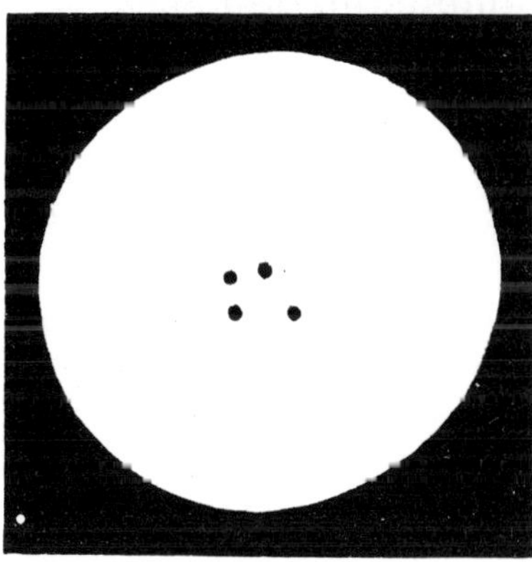

COLOURED STARS.

—be separated into two, three, or more distinct stars of different hues, and in many cases—

The number of Stars which exhibit various gradations of colour between crimson, on the one hand, and only yellowish on the other, are very numerous indeed. They may be counted by hundreds, not to say thousands, and are scattered over the whole heavens, and are of various sizes from the first to the eighth or ninth

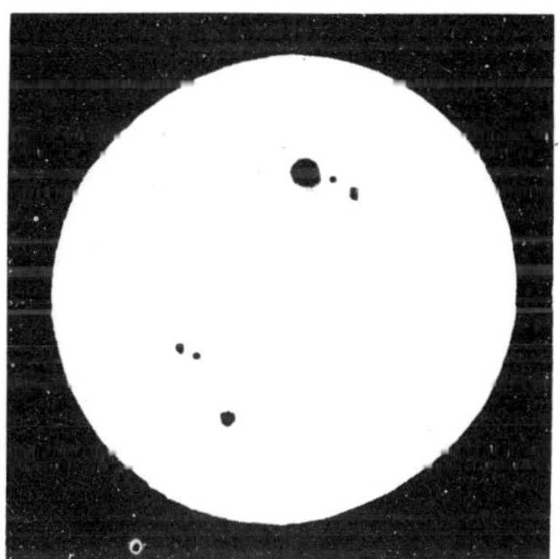

COLOURED STARS.

—these colourings are "complementary" one with another, so that a red star will combine with a green to produce the effect of white light.

magnitude. I prefer to draw the line there, because it is puerile in my opinion to pretend to trace, colour in Stars so small that they can scarcely be seen at all.

Much speculation has been indulged in as to the physical condition of the Stars which we see exhibiting the various gradations of colour, included for convenience sake under the general name of "red" Stars. It seems a simple, perhaps one might say obvious, theory that they are Suns in an incandescent condition; in other words, are celestial objects on fire. Certainly, in looking at some of those which have a very pronounced colour, this theory might seem to be the right one, but I hesitate to say that it is so. The fact that so many of the red Stars are variable in their light, or, to put it the other way about, that so many of the Stars which undergo variations of

Photo by] [*W. Saville Kent.*

A CHAMELEON.

Africa is probably the original home of the Chameleon, and even now its distribution is limited to Africa and the most southerly districts of Europe and Asia. Madagascar has reared both the largest species (which is about two feet in length) and the smallest—less than three inches.

brilliancy are often red or reddish, would seem to countenance the notion that changes of the nature of combustion periodically take place in them.

Various attempts have been made to devise some means of comparing the colours of Stars with some definite standard. I have often thought that if a scheme of shades of colour could be reproduced by little circles of coloured glass in regular gradation, the pieces of glass being mounted in a frame of convenient size to hold in the hand, such a contrivance would be a colour photometer capable of being made of practical usefulness in estimating the colours of Stars. Such a method of investigation would not be without its interest, and might add considerably to our knowledge of these jewels of the night.

STRANGE CHAMELEONS

BY SIR HARRY JOHNSTON, G.C.M.G.

THE Chameleons are a very peculiar and easily recognized sub-order of lizards, restricted in their distribution to Africa and Madagascar, Southern Spain, Syria, Asia Minor, Western and Southern Arabia, Southern India and Ceylon. They would seem to have originated in South or East

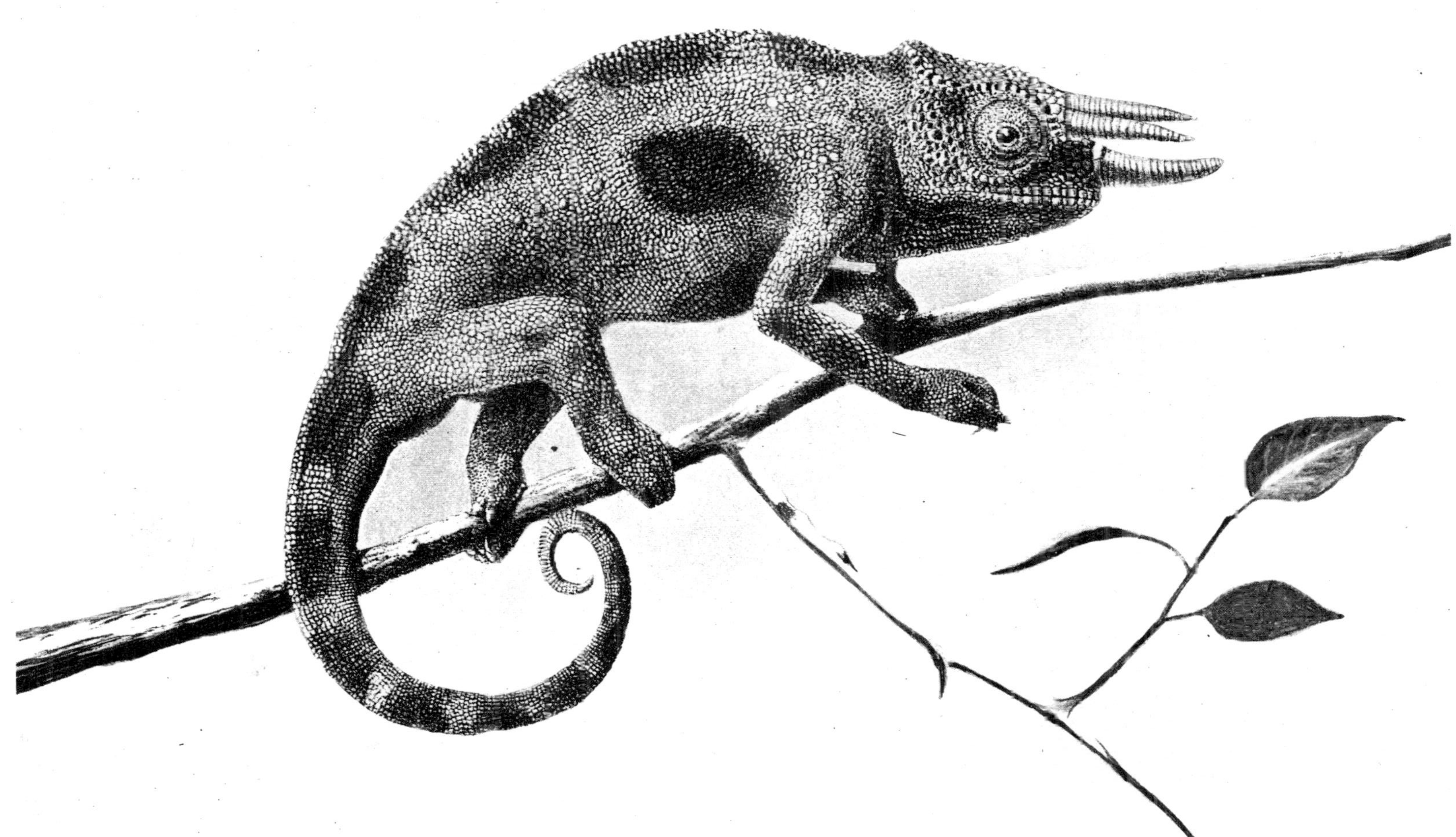

From a Lithograph] [*Executed for the Zoological Society of London.*

JOHNSTON'S CHAMELION.

This three-horned Chamelion was discovered in the forest region of Mount Ruwenzori (Equatorial Africa) by Sir Harry Johnston's Expedition in 1900. The illustration, originally drawn for the Zoological Society's Proceedings as here reproduced, represents the life size of an adult male. Only the male develops these three horns.

Head of female of Johnston's Chameleon. Note that the female has no traces of the three horns so prominent in the male.

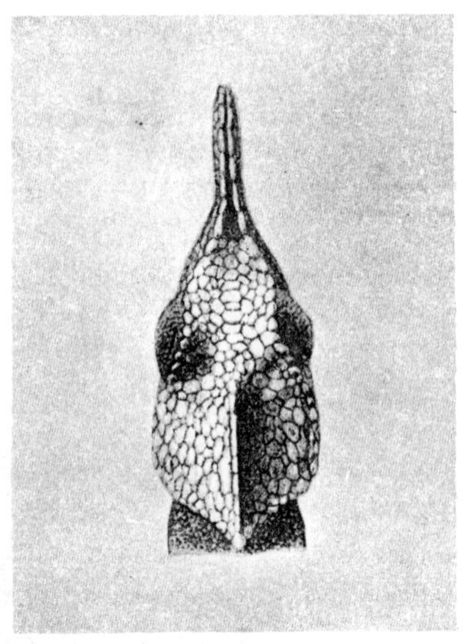

Top of the head of the Long-nosed Chameleon: a male specimen.

Africa, possibly from some group related to the Agama lizards of the Old World. They develop about twenty-five different species in Madagascar, where they are represented by two genera, one of which contains the smallest known Chameleons. One of these species is less than three inches in length. On the other hand, in Madagascar is also found the Giant Chameleon, which is two feet long. Only one species is found in Southern India and Ceylon, but it is sometimes stated that there is a second type of Chameleon in the last-named island. This is a very excusable mistake, arising from one of the most remarkable cases of mimicry known in nature. A genus of Agama lizards has acquired the most extraordinary external resemblance to Chameleons, having just the same shaped head, with excrescences like horns, the same serrated back and hunched-up body, but can, however, be distinguished from the real Chameleons by the fact that its tail is not really prehensile, and, still more, by the arrangement of the fingers and toes. In this imitative Agama lizard the thumb and the fifth toe are opposed to the other digits; but in the true Chameleons, not only is the tail very prehensile, but the fingers and toes are arranged in groups of three and two for the purposes of grasping, instead of four and one.

Another remarkable feature in the structure of Chameleons is the large revolving eyeballs, which can be turned independently one of the other, so that they can direct the sight of either eye simultaneously in all directions. The actual eye opening is small. Another strange development is in the immense club-shaped tongue, which has a long, narrow root of elastic fibres. The tongue is retracted in a telescope fashion into the capacious throat, but it can be most rapidly ejected to a length nearly equalling that of the animal's body. The club-shaped end of the tongue really consists of two flaps, which can open and

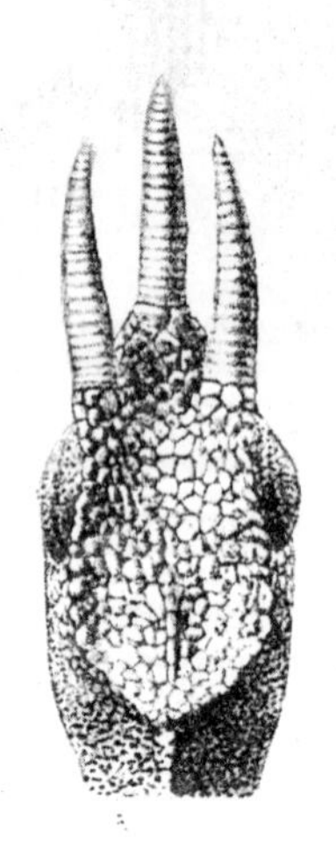

Top of the head of the male Johnston's Chameleon, showing the three horns.

Head of the female Long-nosed Chameleon, in which the strange projections of the nasal bones and forehead are only just indicated.

seize all or part of the insect at which the Chameleon is aiming. In any case, the saliva which covers the extremity of the tongue is so sticky that it adheres to any creature with which it comes into contact.

The Chameleon has become proverbial from its power to assume rapidly changes of colour and its obvious ability to suit its colour to its surroundings within certain limits. The mechanism by which these changes of colour are achieved is of a very complicated nature, and those wishing to understand it are advised to consult technical works of biology. It is due to a shifting of colour-cells backwards and forwards to the surface of the skin, and the colours range from white and black

From a Lithograph] [*Drawn for the Zoological Society of London.*

THE LONG-NOSED CHAMELEON.

This is the adult male of a strange-looking species of Chameleon discovered in 1900, on the eastern side of Ruwenzori, by Sir Harry Johnston's Expedition. These projections of the bones of the face and forehead, covered with skin and scales, are only developed in the male forms, and, like the horns in an allied species, are intended merely for show and not for any purpose of offence

to green, yellow, bluish-grey and reddish-brown (even brighter tints exist in a South African species). But this colour changing is not confined to Chameleons. It is almost equally developed in one or two American lizards and in Indian Agama lizards.

Two genera of Chameleons are found in Tropical Africa (apart from Madagascar). One of these is peculiar to the great forest belt stretching from the Cameroons to Uganda. This genus produces species almost equally small with the tiny Chameleon of Madagascar. In South Africa there is likewise a very small Chameleon, which is remarkable for producing its young alive. The species here illustrated, however, is a large animal, and is found in the forests of Ruwenzori. The male of one kind develops three remarkable horns, and of another an immense projection on the nose, and also over the top of the skull.

"ODD" AND "EVEN" IN NATURE

BY W. P. PYCRAFT, F.Z.S.

In the animal world, what the text-books call bi-lateral symmetry may be regarded as the rule That is to say, their bodies may be divided longitudinally into similar right and left halves. But among whole groups of animals this symmetry is conspicuous by its absence. As, for example, among the univalve mollusca—whelks, snails, etc.—where the shell is spirally coiled. Higher in the scale of animal life, from the fishes upwards, " bi-lateral symmetry " is the rule. But there are some striking exceptions. Among the birds we have the little Wry-bill Plover of New Zealand, in which the beak is sharply turned to the right, a device enabling the bird to pick out small crustacea from under stones. Among the owls there are some quite extraordinary cases of " asymmetry," for among certain species the aperture of the ear differs either in size, or shape, or both, on the two sides of the head. In the Long- and in the Short-eared Owl, for instance, this aperture is of enormous size, and is divided by a pleat of skin into two chambers. On the right side of the head, as may be seen in our photograph, the aperture of the ear lies low down in the lower chamber, the upper one being blind. On the left side this condition of things is reversed, the blind cavity being below, the passage to the inner ear above! In the rare Tengmalm's owl this asymmetry extends even to the skull itself, the bony walls surrounding the passage to the inner ear being strikingly different on the two sides of the head.

Photo by] [*W. Saville Kent.*

A CHAMELEON.

The Chameleon is proverbial for its ability to change colour. But this is only true within certain limits, and is due to a shifting of colour-cells backwards and forwards to the surface of the skin. The tints it can assume range from white and black to green, yellow, bluish-grey and red-brown.

FAIRY RINGS

BY EDWARD STEP, F.L.S.

Among the common objects of the country that are always somewhat of a mystery to most of those who have observed them are the well-defined rings that often make their appearance in meadows and pastures. They are of various sizes and differ in the degree of their completeness. The smaller ones are usually complete rings, but the largest may be only half, or a mere segment of a circle. These started as small circles some years ago, and increased each season, until they met with part of other

Photo by] [*W. Saville Kent.*

A CHAMELEON CATCHING A FLY.

This immense club-shaped tongue can be extended to a length equalling the animal's body, and can be retracted telescope-fashion It really consists of two flaps which open and seize the prey. It is covered with a substance so sticky that nothing can escape its adhesive powers.

Photo by] [*W. P. Pycraft, F.Z.S.*

THE EAR OF THE SHORT-EARED OWL.

Notice that the ear is divided into two cavities. On this, the right side of the head, the upper hollow is blind and the lower conducts sound to the inner ear, but on the left side the positions are reversed and the upper passage conveys sound.

rings and the contact broke both. Our ancestors could not understand them, and the natural phenomena they could not understand they attributed to the intervention of the devil or the fairies. If the result seemed evil to them, they debited the affair to the devil's account with humanity; but if some good were traceable to it, they credited it to the fairies. Now, along the bright green circular lines in the fields they found an abundance of small mushrooms, or toadstools, which were clearly gifts from the good folk—the fairies; so the circles were called "Fairy Rings," and a plausible theory was invented to justify the name.

Here is the approved theory of Fairy Rings as given to us by Keightley, who knew all about the good folk and their ways: "The Elves are extremely fond of dancing in the meadows, where they form those circles of a livelier green which from them are called Elfdans (Elf-dance): when the country people see in the morning stripes along the dewy grass in the woods and meadows, they say the Elves have been dancing there. If any should at midnight get within their circle, they [the fairies] become visible to him, and they may then illude him."

This is very pretty; and it is surely a thankless task to be called upon to explain the true nature of Fairy Rings, and so help to destroy another bit of the romance handed down to us by our fathers. But the botanist has this saving grace—that though he may shatter bits of the old belief, he has much more wonderful things to tell if you will only listen to his modern fairy-tales. He has been inside the charmed circle—not at midnight, but in broad daylight. The fairies that marked this circle have become visible to him, but they have not "illuded" him. He has caught them and put them under his microscope, and their ordinary invisibility has no longer served them.

In plain language the maker of the Fairy Ring was a minute speck of vegetable dust that floated in the air from some previously existing Fairy Ring. It was a little oval fragment of mushroom, thrown off from one of the "gills" which on the underside of the mushroom's cap radiate from the central stem. Such a fragment is known as a spore, and the length of its longer axis only amounts to eight microns—a single micron being $\frac{1}{25,400}$th of an inch.

The spore sank between the grass-stems and reached the soil, where the congenial dampness caused it to swell and burst its jacket, from which it issued as a filament slender as a fine hair. This entered the soil among the grass-roots, where it found decayed animal and vegetable matter, which it absorbed. Upon this food it grew, lengthening and branching, and the branches crossed and re-crossed until they formed a little patch of felted network. Then, having exhausted all the food under that little patch, all the new growths started away from it, and at the proper season each produced upon itself a tiny knob. And the little knobs all grew bigger and bigger, and waited just under the surface for a warm, dewy night. When it came, they sucked in all the moisture they could and swelled up, so that a little stem that had formed in the middle lifted the knob up into the air, where it expanded and spread out in umbrella fashion. And lo ! all these little umbrellas formed a little ring. They remained for some days, the umbrellas or caps spreading until they had flat tops an inch or so across, and you could see that underneath they had deep plates or gills set on edge and radiating from the top of the stalk. From the sides of these gills they threw off hundreds of thousands of spores, which floated away on the air, and a few of them dropped in the right places and burst their jackets as their parent did.

Having got rid of their spores, all the umbrellas—really little mushrooms—collapsed and decayed, and their remains went to feed the grass, so that it became more vigorous and of a brighter green all along that circular line. And people said, that is where the fairies took hands last night and danced in a ring. But the underground threads went on working and getting farther and farther away from the centre, so that each year the circle of bright green grass grew wider.

THE SENSITIVE PLANT

BY S. LEONARD BASTIN

EVEN amongst the higher plants there are many instances which make us feel that the dividing line between the two great living kingdoms cannot be very definitely drawn. Most of all one feels this in connection with the various species which evidence the power of movement, seen, perhaps, to the most striking effect in the case of the Sensitive Plant. A native of Tropical America, this singular plant is now naturalized in nearly all the warmer parts of the world and, indeed, is very commonly grown under glass at the botanic gardens in this country. The

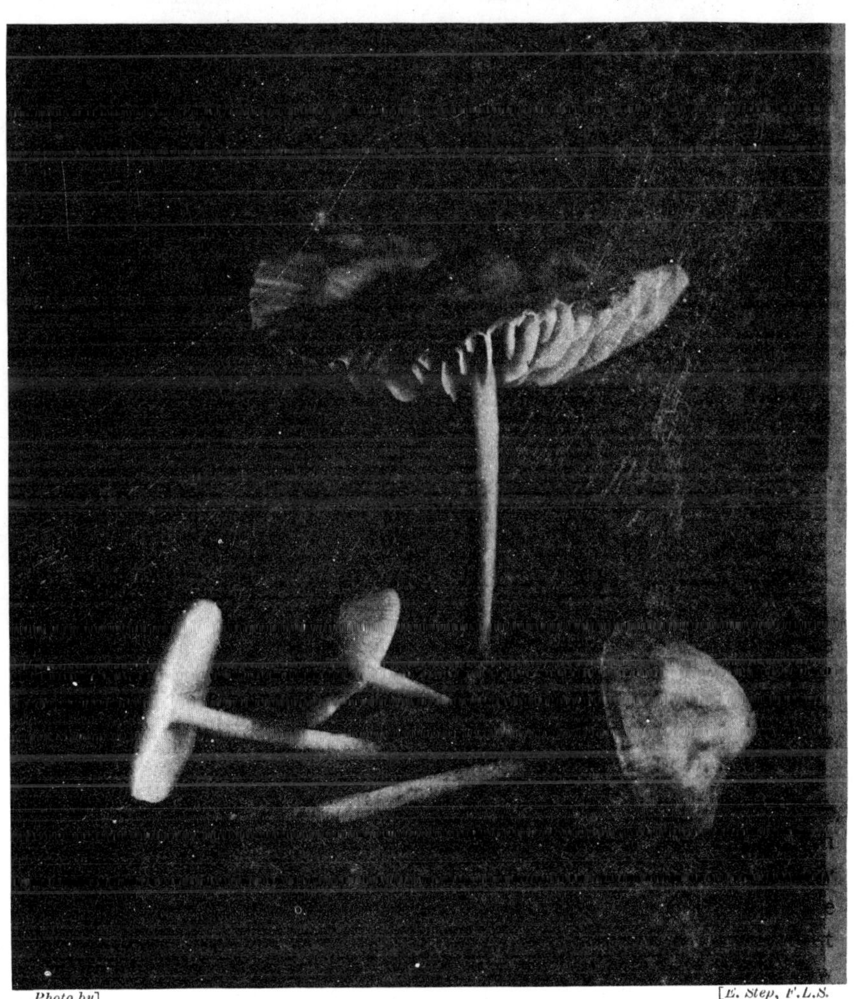

Photo by] [*E. Step, F.L.S.*

THE FAIRY RING MUSHROOM.

A few of the Fairy Ring Mushrooms taken from the ring in order to show them more clearly. The photo is twice the natural size. Note the spore-bearing plates or "gills" radiating from the stem.

most distinguishing feature of the Sensitive Plant is, of course, its habit of responding at once to any touch or injury. Should one of the leaf-tips be knocked never so lightly, the changes start to take place almost immediately. One by one the little leaflets begin to close up, next the secondary stalks bearing the leaf divisions fall downwards, and finally the main stalk collapses with a sudden jerk. It is more than likely that the leaves above the one which has been touched will also show signs of having received the shock. This is a very singular outcome, and one which seems to show that in its powers of shock transmission the Sensitive Plant has something which is not altogether unlike the nervous system of an animal. A sudden smart tap at the base of the stem of a plant brings about the collapse of the whole of the foliage.

Now our knowledge of the causes underlying these responsive movements on the part of the

Photo by] [*E. Step, F.L.S.*

A BIT OF A FAIRY RING.

Here is shown a small portion of a ring made by the true Fairy Ring Mushroom, better known to epicures by its French name, Champignon. Owing to their small size they are partly hidden by the taller grass, so that it is difficult to obtain a photograph of the whole ring.

Sensitive Plant is not by any means complete. It is certain, however, that the collapse of the stalks is due to changes in the distribution of water contained in the thickened part at the base of the leaf-stalk. In its normal condition the lower part of this process is distended in such a way that complete support is given to the stalk. After the reception of a shock the water passes from the lower cells to other parts of the thickened portion. As a result of this a certain flabbiness arises, and there is no longer the rigidity necessary to keep the leaf-stalk in an upright position.

Most of us, as children, have wondered in a fanciful way whether plants could feel pain. However this may be, the Sensitive Plant appears to be acutely conscious of things which cause suffering in animals. If a lighted match be applied to one of the leaves for only a fraction of a second, the organ shrinks back as if in great pain, so that one feels that such an experiment is positively unkind!

Photo by] [E. Step, F.L.S.

A RING OF ST. GEORGE'S MUSHROOMS.

This Mushroom gets its name from the fact that it appears in spring—about the period of St. George's Day. An idea of the size of the ring can be gathered from the statement that each of the Mushrooms composing it measured four or five inches across. They are very fleshy, and give out a strong starchy odour.

Photo by] [*S. Leonard Bastin.*

THE SENSITIVE PLANT.

Originally a native of North America, this plant has now been naturalized in all parts of the world. It obtains its name from its immediate perception of the slightest touch, when its leaves close up and the leaf-stalks collapse with a sudden jerk. It is here shown in an undisturbed state.

Like animals, the Sensitive Plant is powerfully affected by anæsthetics of all sorts. If a specimen with its foliage displayed in a normal fashion is exposed to the vapours of chloroform or ether, there is to be seen a gradual drooping of the leaves, for all the world as if the plant had passed into a state of unconsciousness. Electric shocks will agitate the Sensitive Plant to a great extent, and it has been observed that even a thunderstorm will bring about movements on the part of the foliage.

One very strange phase in the life of the Sensitive Plant is that, like animals, it seems to become "tired" if it is irritated too much. After being touched several times in fairly rapid succession the leaves fail to respond, and the power of movement is not restored until a period of rest has been allowed. On one occasion it was necessary to take some specimens of this plant by rail. During the early part of the journey the plants showed their disapproval by the collapse of their foliage and general leaf agitation. After a while, however, they seemed to become used to the novel and disturbing conditions, and finally the foliage was displayed in an entirely normal manner.

Towards the close of the day the Sensitive Plant begins to make preparations for the coming night. One by one the leaflets close together, until an observer feels, with Linnæus, that the plant must have gone to sleep. It is a wrong idea, however, to associate the evening movements of the Sensitive Plant with the taking of rest. More properly the change in position is to be regarded as a protective measure. In all parts of the world the temperature reaches its lowest point during the hours of darkness, and the leaf-closing of the Sensitive Plant tends to prevent an undue loss of warmth. In the case of specimens which have been forcibly prevented from closing their leaves, the foliage has suffered severely during the night and has been damaged to such an extent that it was of no further use. After all, however, one is bound to admit that there are a good many things in the behaviour of the Sensitive Plant which it is not at all easy to understand.

THE SURINAM TOAD

BY R. LYDEKKER, F.R.S.

In the article on the Fire-bellied Toad reference is made to the remarkable nursery arrangements which obtain in certain frogs and toads; and among these none is more peculiar than that characteristic of the Tropical South American species commonly known as the Surinam Toad, or sometimes, on account of its aquatic habits, as the Surinam Water-Toad.

An ugly-looking creature at the best is this curious toad, which is one of the larger members of the group, and is easily recognized by its depressed and triangular head, with small, beady eyes, and some protruding flaps or filaments of skin on the upper lip, at the gape of the mouth, and in front of the eyes. Very distinctive, too, are the star-like expansions on the tips of the front toes, which are quite separate from each other, and, likewise, the fully-webbed hind toes, each armed with a claw. The skin covering the back of the broad and depressed body, like that of the head, is dark blackish-brown in colour and dotted over with small tubercles, but on the under surface is whitish, sometimes with a brown line along the middle. Each of the tubercles, or papillæ, on the skin of both surfaces of the head, body and limbs is armed with a minute horny spine, some of them being also provided with a poison-gland at the base. There are likewise four rows of larger poison-glands on both aspects of the body.

If the mouth of one of these toads be opened it will be seen to be completely devoid of both tongue and teeth, although the place of the latter is taken in the adult by horny plates. The lack of a tongue—although not of teeth—is shared by an allied African family of frogs.

These Toads inhabit the hottest and dampest districts of the Guianas and Brazil, and never voluntarily leave the water, although during the dry season, when many of the ponds and pools give out, they are compelled to bury themselves in the mud. They awake, however, with the coming of the rainy season, and then commence the business of spawning in the flooded forests, where they are at this season exceedingly difficult to find.

In the spawning season the skin of the back of the female becomes very much thickened and softened, and serves as a receptacle for the eggs, which are stuck into it like currants on the top of a cake. It was formerly believed that spawning occurred on land, and that the male took each egg singly and buried it in the skin of his partner; but in the spring of 1896, two of these Toads spawned in the London Zoological Gardens, when it was seen that the process took place in water, and that, by a special arrangement, the female is enabled to deposit the spawn on the lower part of her own back. The eggs are then pushed forwards, one by one, by the male, and gradually pressed down so as to cause them to sink into the soft and yielding skin until they become completely buried. When development has taken place, the young toad becomes enclosed in a pocket-like cavity furnished with a thin lid, like the hole of a trap-door spider; the lid, which is of a shining horny

Photo by] [S. Leonard Bastin.

THE SENSITIVE PLANT.

The stalk has been touched and the whole plant is disorganized. Fire, an anaesthetic, an electric shock, or even a thunderstorm will cause the same effect, though continued irritation will exhaust its responsive faculties.

[By Arthur Twidle.

THE SURINAM TOAD.

The most remarkable characteristic of this ugly creature is the mode of nurturing and developing the eggs. These are deposited on the back of the female, where they sink into the thick yielding skin and each pocket-like cavity is in due time covered with a shining horny lid. After eighty-two days the young Toads lift up these lids and emerge, fully developed. The Toad spends most of its time in the water.

nature, being perhaps developed from a portion of the shell-membrane.

In these cells the young toads undergo their full development, not emerging until they are miniature replicas of their parents. As a rule, from sixty to seventy offspring are developed in the back of each female, but in some instances the number may be increased to so many as one hundred and twenty; the whole process of development occupying eighty-two days. When ready to emerge the young toads do not appear to require any assistance, each pushing off the lid of its cell by thrusting forth its head or a leg, and then proceeding to climb out. After the young have come forth, the outer layer of the skin of the back of the female dries up and is shed, while the honeycomb-like cells gradually close up, the position of each being indicated by a small, pimple-like elevation.

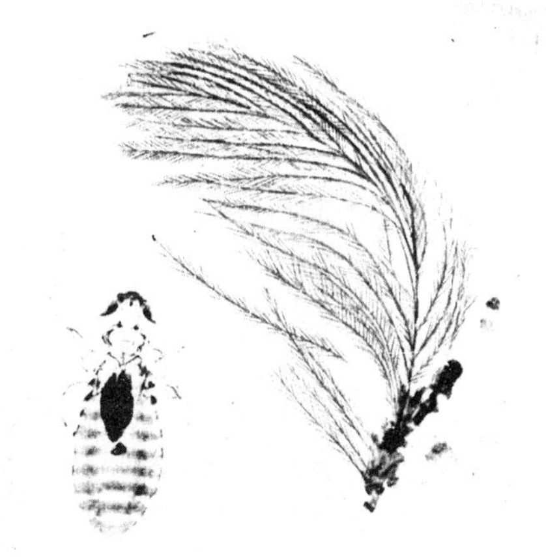

Photo by] [*H. S. Cheavin.*

A PARASITE OF THE PARTRIDGE.

Notice the method of attachment of the eggs to the feather.

Strange as is this mode of nurturing and developing the eggs, it is not absolutely unparalleled, for there exists in the island of Ceylon a species of frog belonging to the same group as the so-called Flying Frog of Java, in which the eggs, to the number of about a score, become attached to the skin of the under surface of the body of the female, on which, after hatching, they leave small, shallow pits. In this case, however, it does not yet appear to be known what becomes of the tadpoles, although it is believed that they do not undergo development in the parental skin. Neither is it ascertained by what means the eggs are implanted in the skin.

It may be added that those large and handsome brown Mauritian cowry shells in which the upper surface is marked by a number of yellowish spots are frequently called Surinam-Toad cowries, on account of a fancied resemblance to the back of a nursing female of the species forming the subject of the present article.

BIRD PARASITES

BY JOHN J. WARD, F.E.S.

FORTUNATELY, science has no blushes; otherwise, I might apologize to my readers for introducing this somewhat indelicate subject. All living animals, great or small, are pestered more or less by other animals specially adapted to prey upon them; it is possible that there are no exceptions to the rule.

Even the slug, whose viscid secretions might be regarded as an effective barrier to all such trespassers, has its own special parasite, which plunges through the exudations of its host with perfect freedom; while the frog proceeds upon his watery way accompanied by at least twenty kinds of these uninvited visitors. The bullet-proof hide of the rhinoceros and the leathery skin of the hippopotamus are subject to the attentions of the torturing inflictions of a tick.

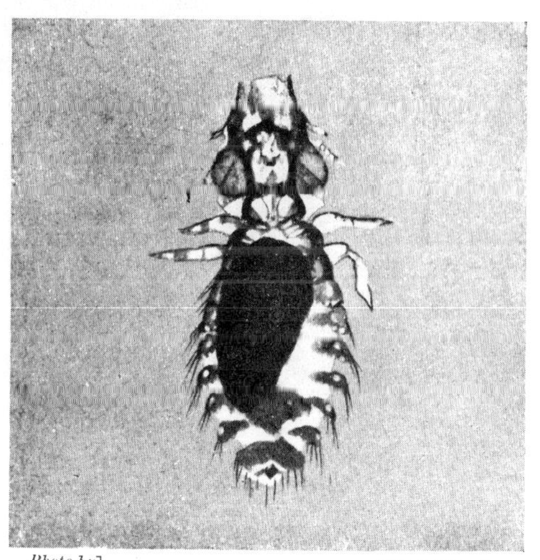

Photo by] [*H. S. Cheavin.*

A PARASITE OF THE THRUSH.

It will be seen that many birds possess their own particular species of parasite, which will die if separated from its particular host.

Photo by] [*H. S. Cheavin.*
A PARASITE OF THE DUCK.
Showing the egg-case attached to a hair.

Horny skins or integuments are, indeed, of no avail against the depredations of parasites armed with complete sets of surgical instruments for the express purpose of penetrating them. Fishes, and even the great whales of the deep, likewise, serve as hosts for many parasitic organisms.

Remarkable as it may seem, monkeys probably approach the nearest to freedom from external parasitic visitors. They are not infested by fleas; and pediculi, or lice, are but very rarely found upon them. That they often appear to be cleaning their fur (or perhaps more often that of their nearest neighbour) from these pests is true, but observation has shown that they remove only bits of scurf and similar secretions from the skin. Indeed, it is this very habit of picking particles from the coat of each other by means of their sensitive fingers and opposable thumb that has prevented parasitic organisms from establishing themselves upon them as permanent and specialized species.

Other animals, which have only been able to scratch and rub themselves when attacked, have become hosts to numerous species, some of which are so specialized as to become their own particular parasites, preying on no other hosts.

Birds are animals which present numerous instances of that character, many species possessing quite distinctive parasites; while other parasitic organisms will prey upon only nearly related bird species. On the other hand, canaries and fowls entertain mites and parasites of a more catholic taste, which often make excursions to the persons having charge of these birds.

It follows, therefore, that when a parasite is found on an animal it does not necessarily mean that it is the natural parasite of that animal; that is a point which needs careful study.

Not infrequently parasites lose their way, and having introduced themselves to the wrong host, they then are compelled to use the latter as a waiting-room, as it were, in the course of their travel until the train they require comes along. Very often they never again meet with their true host, and consequently they live but a short time.

Domestic poultry have perhaps more than a fair share of parasites; yet, as the accompanying illustrations show, distinct species patronize the fowl, the duck, the goose, and the pigeon. When we see a fowl taking a sand bath, we observe it adopting an alleviative against the vermin which are annoying it, the sand serving to dislodge its visitors. Also, the fowl is not by any means confined to one kind of parasite; it harbours at least eight nearly-related species of the organisms known as Bird Lice, which swarm among its feathers and are often left on the hands of those who have to do with it. Neither are they easily brushed off, for, in common with many parasites, they possess flat and highly-polished bodies which offer scarcely any resistance to touch, and they can also cling very tenaciously.

Photo by] [*H. S. Cheavin.*
A PARASITE OF THE FOWL.
The fowl, however, is the host of more than one species of parasite, harbouring eight varieties of Bird Lice.

Quite apart from these insect parasites, fowls are

often badly infested with a mite nearly related to that found on the Dor-beetle, which I described and illustrated in an earlier article; but in this case it is tolerably certain that their mission is not to attend the toilet of their host so much as looking after their own wants; for when they are hungry they are pale in colour, but when their digestive organs are full, they are of a purplish hue, a colour produced by the blood of their victim which they have consumed.

Nevertheless, it should not be assumed that all the parasites of birds are undesirable guests. There are insect and mite organisms which are designated "feather and hair eaters," and many of these render good service to their hosts by thinning their body-plumage during hot weather. The slender-bodied parasite of the duck shown amongst the illustrations, which may also be found on the goose, is an example of that order. Pigeons and doves also seem particularly favoured with this class of parasite; while other species patronize the fowl, teal, turkey, etc.

It is a remarkable fact, too, that when several genera infest one host they confine their attentions to particular localities. Perhaps some will inhabit only the feathers of the neck, others those of the breast, and still others those of the wing feathers. They can move with remarkable facility, and have a way of turning sideways round the quill of the feather when touched; and they can also travel as readily backwards as they can forwards. These features combined with the fact that their bodies are flat and polished, make their chances of being caught when amongst the feathers very improbable.

The mites (which are much smaller, and may be distinguished by their eight legs) are more nearly related to the spider family; the six-legged parasites being insects properly speaking. Almost every bird species has one or more kinds of these organisms allotted to it, and some of them are, indeed, weird looking animals when viewed through the microscope, the third pair of legs of the male sometimes being exaggerated into monstrous organs. Occasionally six-legged forms are discovered, but these are the mites in their larval stage; after the final moulting of their skin eight legs appear.

Although, as previously stated, some mites prey

Photo by] [*H. S. Cheavin, F.R.M.S*

BLACKBIRD PARASITES.

It is a remarkable fact that when several kinds of parasites infest the same host each species usually confines itself to a particular locality.

Photo by] [*J. J. Ward, F.E.S.*

THE PARASITE OF THE TAWNY OWL.

This "tick" is an exception to the rule that all parasites are slender-bodied. Notice that it possesses eight legs instead of the usual six in all insects

Photo by] [*H. S. Cheavin, F.R.M.S.*

A FOWL LOUSE.

The parasite is not always a pest; he sometimes renders good service to his host by thinning the body plumage during hot weather.

upon the blood of their host, yet it is probable that the great majority of them are, like the slender-bodied parasites of the duck and the pigeon, useful attendants on their toilet.

The tawny owl presents by way of contrast a parasite that is not slender-bodied (see photograph), but which exhibits an aldermanic outline suggestive of unearned increment. That organism, also, is seen to possess eight legs; it is, however, much larger than a mite, and is known as a "tick."

Ticks are not so common on birds as the bird-lice and the mites previously mentioned, but some bird species become prey for them; the swallow and the pigeon are hosts not infrequently selected. They are probably the most troublesome of all their parasites; they are certainly the largest feeders. Their bodies are of a leathery consistency, and their mouth parts are adapted for penetrating the skin and for blood-sucking, while their feet are provided with sucker-like organs.

Probably the birds acquire these troublesome visitors from the herbage amongst which they move, as they are vegetable feeders during certain stages in their development, and they also breed on the ground.

It is the female tick which has to be feared, as the abdomen of the male cannot be distended with blood like that of the female. The latter often pumps herself up to the size of a pea. Also these parasites are not at all particular regarding their host; indeed, in this respect they are most adaptable, and it does not by any means follow that a tick found upon an owl is an owl parasite; it would probably have travelled with a hedgehog just as readily if fortune had turned it that way. It is well known to country people that dogs and horses which are quite clean on starting out will often return after hunting with some of these pests about them. Even man himself is often annoyed by them, for they are just as likely to grasp his clothes by means of their hooked feet as they cling to the tall grass-blades as they would the coat of a passing sheep or a dog; they are especially annoying at hay-time. It is obvious, therefore, that birds which continually move in and out amongst vegetation while searching for their prey are most likely victims.

Photo by] [*H. S. Cheavin, F.R.M.S.*

A ROOK LOUSE.

All these examples of parasites are much enlarged.

Although the ticks do not breed on their hosts, yet it is obvious that some of their other parasites do, for attached to the quills of birds' feathers their numerous eggs may be found in abundance. These tiny eggs are often beautifully sculptured, and under a microscope they make most charming objects when well illuminated.

THE SUBTERFUGES OF SNAILS

BY BERNARD C. WHITE

PROTECTIVE resemblance is a method of self-preservation that widely prevails in the world of Nature. It is well known to be frequently adopted by beasts and birds, and, above all, by insects; but it is not so widely known that it is also in vogue in the world of molluscs.

A remarkable instance is mentioned by Karl Semper. The Snail was Xesta, which was found to mimic the markings of another snail, Helicarion. Now Helicarion possesses the power of shaking off the tail when seized or irritated, and so of escaping from the clutches of its enemies. The result is that these enemies, finding the game of hunting it not worth the candle, leave Helicarion unmolested. Xesta has recognized its neighbour's immunity from attack, and has adopted the appearance of Helicarion, although it is unable to shed its tail at will. However, as it lives associated with Helicarion, the deception is sufficient to safeguard it from attack.

Two splendid examples of protective coloration are furnished by the Common Limpet and the Blue-rayed Limpet. In the first, its colour exactly resembles the surface of the rock on which it lives, and when the edges of the shell are drawn down to fit tightly into its shallow hollow in the rock, the creature looks for all the world like a rough protuberance of the rock itself. In the case of the Blue-rayed Limpet, the young Snail feeds on the iridescent leaves of one of the larger sea-weeds, and is possessed of an oval, thin, semi-transparent shell, ornamented with vivid blue lines running from the beak to the margin, and can hardly be distinguished from its food-plant. But in later life it prefers the rounded stalk, and finally the roots of the seaweed, which are entirely devoid of iridescence. It therefore adapts itself to its newly-chosen surroundings. The shell becomes thickened and compressed at the sides; the hitherto smooth, sharp base assumes a shape which will accurately fit the stalk. The blue lines are no longer wanted—in fact, would make their owner conspicuous—and so disappear, and the shell loses all resemblance to its former self. The mature Limpet now scoops out a little cavity in the seaweed, and becomes actually part of the root or stem.

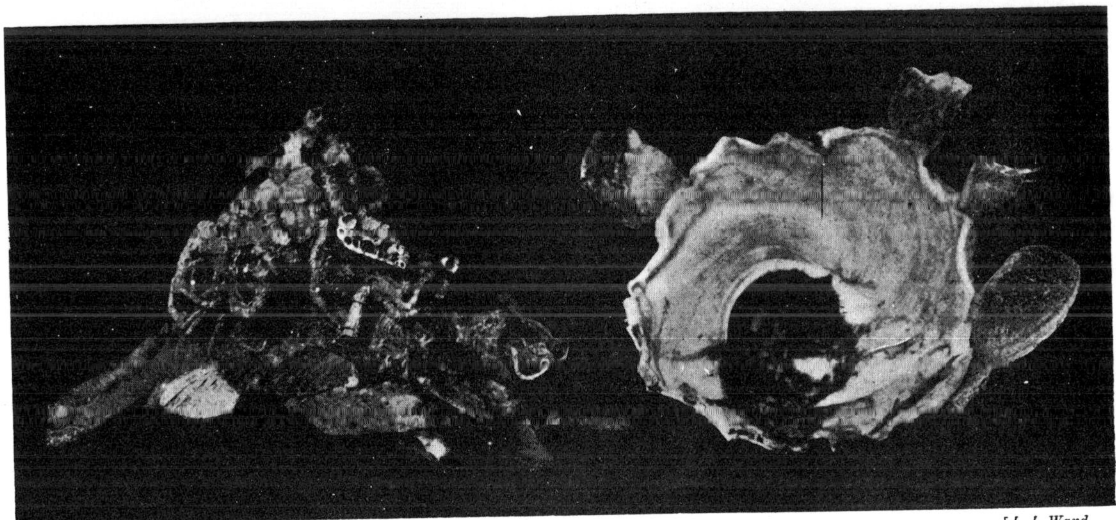

Photo by] [*J. J. Ward.*

CARRIER SHELLS.

In order to save himself from the attacks of enemies the Carrier Snail hides his shell under a coating of cement, to which he attaches any pieces of rock or shell that come his way. As a result his home has only the appearance of a mass of rubble.

In the foregoing cases the resemblances have been produced in the actual shells, but there are Snails which are known to adopt certain external aids to concealment. For instance, the Lesser Bulin, which has a glossy shell and feeds upon the immature mosses on the bark of trees, covers its shell with slime and drags it through the débris to be found in the crevices of the weather-worn bark, repeating the process until the shell is entirely concealed and the birds pass it by as a knotted excrescence of the bark. The Pellucid Glass Snail, again, has formed the habit of attaching to itself bits of leaves or soil, which entirely cover and conceal both shell and animal. But the Snail which has succeeded best in disguising his home is the Carrier Snail of Borneo. The denuded shell is in shape like a top-shell ; but the Snail secretes a cement by which he fixes all manner of stone, shells, and other débris to the top of the successive whorls. The result is that the whole creature is completely hidden under this conglomerate mass.

There are several varieties of Carrier shells, all with the same propensity, though while some

Photo by] [*J. J. Ward.*

CARRIER SHELLS.

The owners of these queer dwelling-places will take great pains in their adornment. In some specimens only a certain kind of shell or stone will be used (but this may be due to a limited choice of building material), and these particles will be inserted to outline the different whorls of the shell.

appear to favour the larger and rougher pieces of rock and to set them in the cement more carelessly, other Carriers choose minute particles of shell or stone, which are very carefully and regularly set in the cement and neatly outline the successive whorls of the shell. There is, however, more arrangement in this queer device than meets the casual eye, for the choice of material is not altogether haphazard, as certain Snails appear to have a preference for certain ornaments. For instance, a specimen dredged up during the *Challenger* Expedition is ornamented entirely with long and pointed shells ; others are set entirely with stones, and one large specimen, about six inches in diameter (also from the *Challenger* collection), bristles exuberantly with a species of oyster-shell. There is only one Carrier that is not worthy of the name. That is the Indian Carrier, who plasters his shell with cement alone ; but this is laid on in a very delicate and fine fan pattern. There is one other unornamented Carrier, but he neither cements nor adorns his house, and for a very good reason ; for the lower edge of the whorls branches out into flat spines, and these take the place of extraneous ornament.

Outward appearance seems to point to the fact that this species has evolved its peculiar

[*By Theo. Carreras.*

A STARFISH OPENING AN OYSTER.

One of the most persistent and destructive raiders of the oyster-beds is the Starfish. Having chosen his victim, he attaches his fingers to the shell beneath him by means of his innumerable suckers, and raises his body, exerting all his force to drag open the shell which the oyster has securely shut. It is a long struggle, but at length the muscular power of the oyster gives out and the shell gapes open. Then the Starfish devours his prey in the manner shown at the background of the picture.

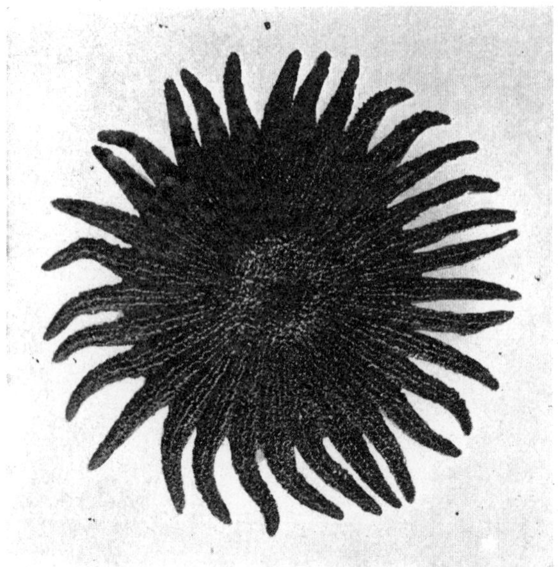

Photo by] [*H. J. Shepstone.*

MANY-ARMED SUN-STAR.

Of similar form to the Common Sun-star of our coasts, this species is seen to have more than twice the number of arms.

structure from the characteristic habit of its race. Whether this be so or not is a question hard to decide, but interesting to discuss.

STARS OF THE SEA

BY EDWARD STEP, F.L.S.

THE poet Montgomery has said :

> "The heavens
> [Are] thronged with constellations, and the seas
> Strown with their images "—

which is probably an unconscious paraphrase of Link, the Leipzic apothecary, who, a hundred years earlier, wrote : "As there are stars in the sky, so there are stars in the sea." But the analogy will only stand in the domain of poetry, for alas ! the matter-of-fact astronomer has proved that the stars of the heavens are not star-shaped, but mere spheres, whereas the Stars of the Sea are of true stellate form. The confusion caused in our language by the progress of discovery is really serious ; small wonder that old-fashioned folk resent the new-fangled learning that overturns so much of the knowledge painfully acquired in youth. Here, for example, we have a large family of animals called Starfishes, because of their conformity to what for ages was understood to be the shape of the stars. First comes the naturalist with his objection that their organization is entirely different from that of the fishes ; so half of the name is shown to be erroneous. Then the astronomer demonstrates by means of his telescope that stars are not stellate, and the other half is rendered misleading. There is nothing for us to do in this dilemma but to fall back upon the artist, who will tell us that the actual forms of things are of little account ; what really matters is the appearance presented to the unassisted eye, and no artist would dream of depicting a star other than as a radiate body. As for the fish element in our word, everybody knows that whatever lives in the sea is a fish, excepting the mermaid, and she is a compromise.

Photo by] [*H. J. Shepstone.*

THE BRILLIANT STAR.

A very beautiful Starfish found off the coasts of Mozambique, of a bright orange colour with red edges.

There are hundreds of different species of Sea-stars—even the waters that girdle the British Isles yield more than half a hundred of distinct kinds—and we must have a comprehensive vulgar name for them that will save our people from the necessity of speaking of them as asteroid and ophiuroid echinodermata. Let it be Sea-stars, then, a name that is sufficiently old, for Sir Thomas Browne speaks of the "Stellæ Marinæ, or Sea-stars."

Every seaside tripper is familiar with the common orange-coloured Starfish, or " Five-fingers," that is flung up on the beach by the waves—that is to say, familiar with its superficial appearance, as he has turned it over with toe or walking-stick, or viewed it in the pail of a treasure-seeking youngster. Not one in a thousand of such trippers or loungers on the sands has ever taken the trouble to consider this common object of the shore; but, in truth, it is one of the most marvellously contrived pieces of animated machinery that you can find. Turn it over again, and what do you see? There are the five spine-clad arms, or " fingers," all united by their broad ends; but no legs, no fins, or other apparent means of locomotion. It is even without a head, which most people would consider a *sine quâ non* for success in the struggle for existence. That Five-fingers *is* a success in the battle of life we may assure ourselves by hunting around the base of the rocks at low-water; we shall find him represented there by numerous brothers and sisters, to say nothing of cousins. If you are on a part of the coast where oysters are laid down in beds to make them fit for the market, inquire of the fisher-folk what they think of Five-fingers, and you will get a very bad character of him. They will tell you that he is one of the most troublesome enemies they have, and will probably add a little to the true facts of the case by relating to you how he inserts a finger into the oyster-shell and hooks out the luscious inmate to make himself a meal. Not only so, but that when they have with much labour baited their hooks and gone fishing, Five-fingers comes along and cleans off the bait. In addition, they will add, he crawls into their crab-pots and appropriates the bait that was intended to beguile the crab and lobster to their doom.

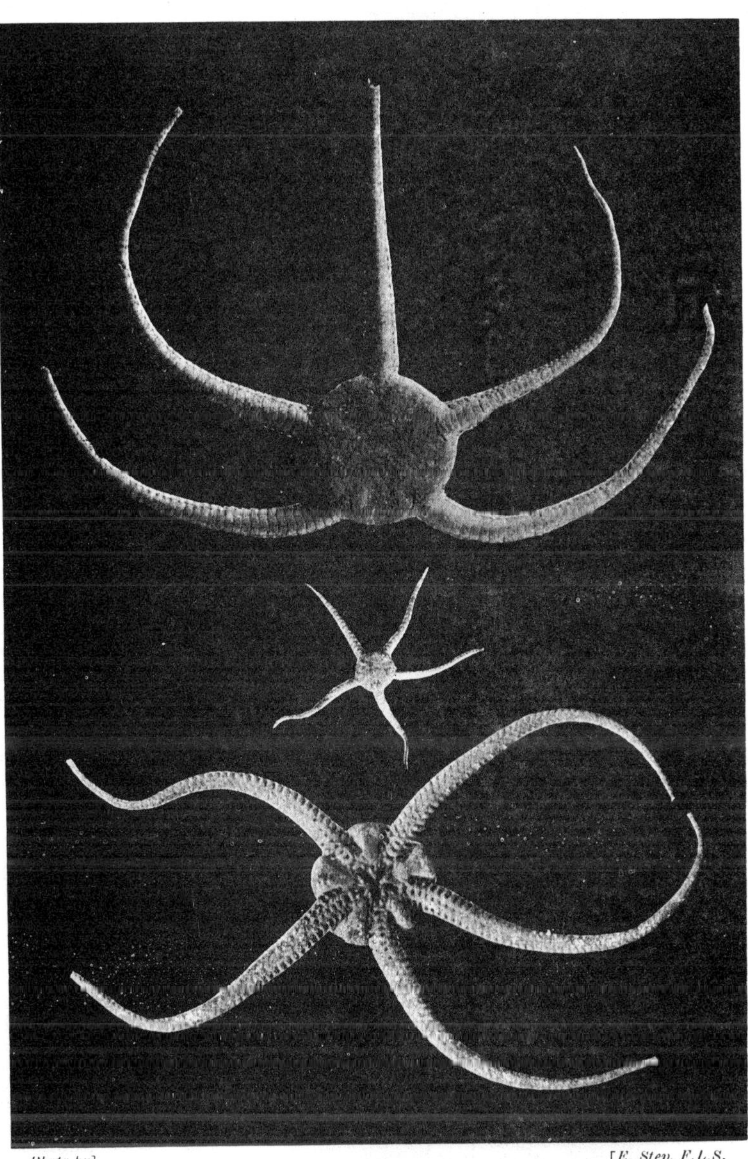

Photo by] [*E. Step, F.L.S.*

SAND STARFISHES.

Sand-stars and Brittle-stars are very touchy individuals, and manifest their annoyance at any interference by snapping off their arms as readily as a lizard snaps off his tail. By this means they are often enabled to escape from their enemies.

Most of this is true, for Five-fingers has a catholic taste, though his preference is distinctly for " shell-fish," but he cannot insinuate his " fingers " between the tightly-closed valves of the oyster's

Photo by] [*E. Step, F.L.S.*

SUN-STAR.

One of the most beautiful of the Sea-stars. Its crimson upper side, often varied with white, is studded with little bundles of fine spines. Its arms vary from twelve to fifteen. It is found in deeper water than Five-fingers. (About one-third of the natural size.)

shell; he has a better plan for accomplishing his nefarious ends. Like his cousin, the Sea Urchin, of which we recently wrote in these pages, Five-fingers is provided with tube-feet, which end in suckers and are worked by hydraulic power. If Five-fingers can attach a number of these to the valve of the oyster's shell, and then bunch his five arms together so that he stands, as it were, upon tiptoe, he can exert such a strain that the oyster's muscles cannot stand for long, and the shell gapes. If you turn Five-fingers over you will see that he has a very small mouth, and the eating of an oyster appears to be entirely out of the question. What he actually does is more wonderful: he passes out a portion of his stomach through the mouth, surrounds the oyster with it and digests the mollusc first and then swallows it. You may often surprise him among the rocks when his five arms are thus contracted, and if you can act smartly and turn him over, you can see the protruded walls of the stomach enwrapping a mussel or three or four small sea-snails.

What we previously said of the organization of the Sea Urchin applies to a large extent to the Sea-stars. The five-sided idea is manifested here in the five arms. There are the similar tube-feet, and the beak-like organs for defence and cleanliness, the stony filter that supplies the water-tubes with water free from grit, the same method of locomotion. But it will be noticed that there are no teeth to the mouth. Five-fingers cannot feed upon any hard substances that cannot be digested without previous mastication. It will be noticed, too, that the tube-feet are all produced from a deep groove, which runs along the centre of each arm on the underside only, and that these grooves all extend from the tips of the arms to the mouth, which is in the centre of them all.

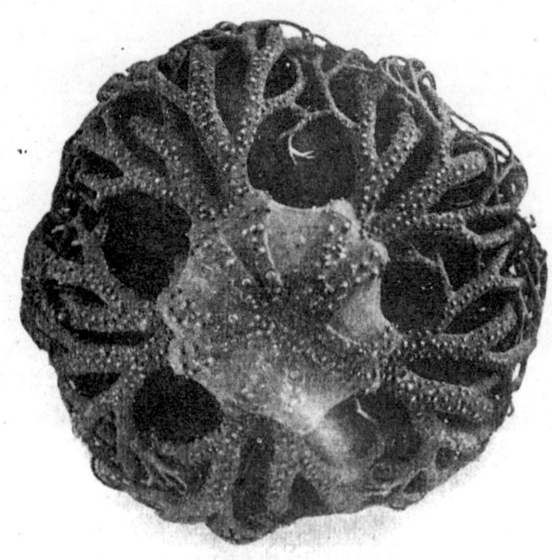

Photo by] [*H. J. Shepstone.*

THE SHETLAND ARGUS.

This remarkable Sea-star is a native of northern seas. It has only five arms, but these fork from the base, the branches constantly subdividing, so that the ultimate divisions are exceedingly numerous. The arms are here shown in a contracted state; when extended the Star measures more than a foot across.

The Sea Urchin had a skeleton of five-sided plates whose edges joined to make a complete spherical or cupola- or heart-shaped box, invested by a thin layer of flesh and armed with movable spines. The skeleton of the Starfish is not continuous and box-like, but more like scaffolding to support a covering which, though it appears thick and leathery, is a very delicate and sensitive material. There are spines, too, but they are comparatively few and short, as are the beak-like defensive organs. In some species, like the larger Spiny Starfish of our western coasts, they are both larger and more numerous; whilst in some other forms they are so short and close together that on a superficial inspection they may be regarded as non-existent. The photograph on page 608 shows a Starfish that has been cut through horizontally and the fleshy portions all dissolved away, leaving only the skeleton, which is seen to be a meshwork of little rods rather than solid plates, as in the Sea Urchins. By it is a

Photo by] [*H. J. Shepstone.*

THE GIANT STARFISH.

This species is found on the Pacific Coast of the United States, and attains a diameter of as much as two and a half feet. Notice the horny plates which cover the whole upper surface of the Giant.

Sun-star, from which also the flesh has been abstracted. Such treatment produces very beautiful and instructive specimens.

Starfishes do not all conform to the same shape, as the last-mentioned photograph will show. The Sun-stars have from eleven to fifteen arms, shorter than those of Five-fingers, and with the central disc much broader. Another photograph shows it with the flesh on. The spines are gathered into numerous bundles. Another form common under stones on our south and west coasts is the little greenish-grey Cushion-star, or Starlet, in which the angles between the five arms are almost filled up. Several modifications of this type occur, until we reach the extraordinary Bird's-foot Sea-star, which is so thin, it looks as though it had been under a steam-roller which had completely filled up the intervening spaces. The Five-horned Sea-star, from Mozambique, belongs to a family whose skeleton consists of p'ates rather than rods, and the spines are replaced by a granular roughness. Another form has very numerous arms, which is supposed to give it a resemblance to a sunflower.

But there is a distinct division of Sea-stars in which the arms are long and snake-like and less fleshy. They are collectively known as Sand-stars and Brittle-stars. They are more or less bony-looking and spiny, and their arms, instead of broadening to their bases to form the disc, are of more equal breadth throughout, and their bases are far apart when they unite with the more circular disc. The type is well seen in the Common Sand-star photographed on page 605, while the species of this order of Starfishes are numerous. There are no sucker-feet as in Five-fingers, though these are represented by undeveloped fleshy cones which serve as sense-organs but are useless for locomotion. The Sand-stars and Brittle-stars make their way over the sand and rocks by wriggling,

Photo by] [*E. Step, F.L.S.*

SKELETONS OF SEA-STARS.

Here the fleshy portions of the Stars have been dissolved away to show the beautiful and intricate bony scaffolding upon which the various forms are based

bending their arms into S-like curves after the fashion of a snake. Of course, if all the arms wriggled in the same manner at the same time, there would be no progress in any direction ; but a distinct order in their movements is observed. Two of the arms arrange themselves as nearly parallel as possible, and curve their slender tips away from each other, so enabling them to get a grip upon the sand. Then by forming another curve farther back simultaneously, they drag the disc forward. The three other arms, pressing with their tips upon the sand, straighten out their previously-formed curves and so *push* the disc forwards. By this correspondence in the movements of the arms the Brittle-star can wriggle along with considerable rapidity.

These Brittle-stars, as their name indicates, are very touchy individuals, and manifest their annoyance at any interference by snapping off their arms as readily as a slow-worm or a lizard snaps off its tail. This is often a means of saving their lives. A fish or a crab comes along and seizes our Brittle-star by an arm—for one arm or another is always nearer than the disc—and the Brittle-star cheerfully relinquishes it and makes haste to wriggle away while his enemy is eating up the discarded member. The Brittle-star, like Sea-stars in general, can easily grow another arm to replace the lost one. It is no uncommon experience to meet with individuals thus mutilated—in fact, if you are in want of perfect specimens for the cabinet, it is extremely difficult to obtain them, because if they are perfect when you determine upon the one you will have, the probability is that it will go to pieces before you have secured it.

Photo by permission of] [*R. Lydekker, F.R.S.*

THE BARK GECKO.

The Bark-Gecko is a native of Madagascar, and is about eight or nine inches in length. It has a large, flattened head, depressed body and short, trowel-like tail, with the toes terminating in large discs, which enable it to cling to the bark of a tree. When thus posed its markings and colour are identical with the lichen-clad bark, and so it is enabled to escape detection.

Our last example of the diversity of forms in these Sea-stars is the Shetland Argus, in which the disc is obscurely five-sided, with ten ridges on its upper surface. The five arms are very long, and immediately after they leave the disc they divide into two branches, and these again branch into two, and so on almost *ad infinitum*. These branches are very flexible, and cling round weeds and fishing-nets, or whatever offers. Owing to the great number of the ultimate branches, it is no easy matter to disengage an Argus when it has got hold of any of these objects. A superficial view of the Argus suggests a relationship to the Feather-stars.

THE BARK-GECKO

BY R. LYDEKKER F.R.S.

THAT many animals are coloured to resemble their surroundings, so that by means of this protective resemblance they may more or less completely escape detection, is a fact familiar to all persons who are in the habit of making an intelligent use of their eyes, and more especially those who have travelled in foreign countries. Ordinary grasshoppers, tree-frogs and many kinds of tree-snakes are, for instance, coloured like the grass or foliage which forms their home ; while desert-dwelling grasshoppers, snakes, and mice have a brownish livery, mottled with black and yellow, to harmonize with the sand and pebbles of their surroundings.

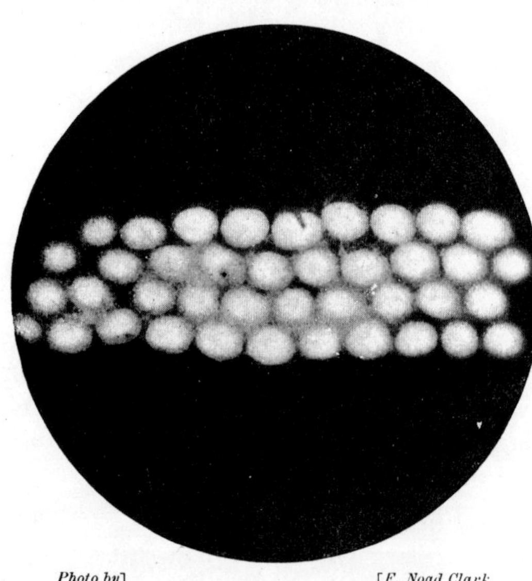

Photo by] [*F. Noad Clark.*

THE EGGS OF THE ARGULUS.

The Argulus will lay as many as 600 eggs at a time. They are deposited in rows, are oval and pale yellow in colour, while each one is glued to its neighbour with an albuminous material. They are here shown under water and magnified twenty times.

In some instances this general protective tone of colouring is intensified and enhanced by the development of special markings, as in the case of the woodcock and the snipe, whereby the resemblance to surroundings is made much closer than would otherwise be possible. One of the most remarkable instances of this kind of protective resemblance is presented by a lizard found on trees in Madagascar, belonging to the group typified by the little Gecko of Southern and Eastern Europe, and hence known as the Bark-Gecko. Measuring some eight or nine inches in length, these lizards are built somewhat on the lines of a miniature crocodile, having large and flattened heads and broad, depressed bodies, although the tail is peculiar in being short and trowel-like, while the toes terminate in the large disc-like expansions common to Geckos generally.

Such a form of body is, of course, admirably adapted to enable the creature to cling close to the bark of its native tree ; but in order to enable it to escape detection when thus posed, its colouring is almost precisely identical with that of lichen-clad bark, being a mixture of black and irregular streaks and blotches of white and grey. As may be seen in the specimen exhibited in the Natural History Museum, many of these light blotches are almost indistinguishable from the patches of grey lichen on the tree-bark ; and when viewed from a short distance, almost the only thing that attracts attention to the presence of the creature are its large, glaring, yellow and red-streaked eyes. Whether when alive the Bark-Gecko closes these betraying eyes at the approach of danger does not seem to be known ; but possibly they are less conspicuous in nature than in museum specimens.

There appear to be two distinct varieties of these remarkable Geckos, but whether these are restricted to different situations is another factor in the economy of the creature of which we are still ignorant. In the typical variety the tail is short and broad and the markings less distinctly like lichen-clad bark than in the second variety—or Lichen Bark-Gecko, in which the tail is longer and narrower.

Photo by] [*F. Noad Clark.*

THE ARGULUS ATTACHED TO A STICKLEBACK

The fish is here very greatly enlarged to show two adult Arguli attached to their host. The favourite place of attachment is on the gills, though they are frequently found on the fins and tail and occasionally on the eye itself. The fish when attacked struggles violently to free itself, but is seldom successful

THE FISH-LOUSE, OR ARGULUS

BY FRED. NOAD CLARK.

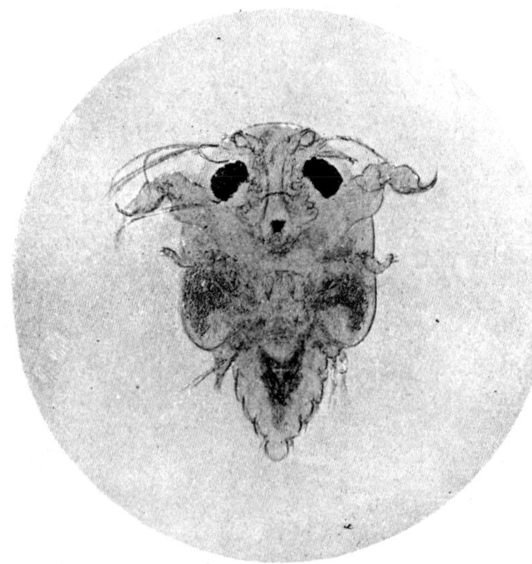

Photo by] [*F. Noad Clark.*

THE GRUB OF THE ARGULUS.

It is only just hatched, and shows the relatively large eyes and barbed hooks which take the place of the suckers that appear at a later stage. (Magnified fifty times.)

THIS creature with the unpleasant name, which one associates with another parasite not unconnected with the "genus homo," is a most interesting beast. Unlike the latter, however, it is not, technically speaking, an insect, but belongs to one of the smaller divisions of the great class Crustacea, previously referred to by contributors to this work. It is a parasite in the true sense of the term, for it derives its sustenance from the living body of a "host," which in the case of the Argulus may be the carp, roach, tench, stickleback, and other fresh-water fishes. It has been reported to be found on tadpoles and minnows, but it was from the stickleback that the specimens which are here illustrated were obtained. By reason of its exceptional structural character, the elaborate nature of its apparatus for attachment to its host, the interesting details of its metamorphosis, and the curious organs and mode of reproduction, a study of this obscure creature will well repay the observer.

The Argulus may be described as a transparent, jelly-like animal of a greenish hue, the bulk of whose body appears as an oval slightly convex shell, which entirely covers the various organs which lie on the under surface of the body, with the exception of the swimming feet and so-called tail. The female is larger than the male, and an adult measures about one-third of an inch in length. It may be distinguished from the latter by the presence of two sharply-defined black spots, or ovaries, at the base of the tail-plate. The eyes will be seen as two dark masses near the top of the body, one of which is shown on page 613, and shows its compound nature, similar in structure to others of the class, and analogous to the compound eyes of insects. There are about forty facets or lenses in each eye, and they are capable of a slight rotary twitching movement. A curious feature, however, occurs in the existence of a third or single eye, situated a little lower down on the middle of the back. On either side of the paired eyes are seen two curious appendages (page 613), tubular in form, with a delicately fringed margin and provided with muscles for producing a vacuum. At casual sight these might be taken for eyes, but are in reality suckers, by which the parasite is enabled to attach itself to the smooth surface of the fish, whilst feeding or when depositing eggs on stones or aquatic plants. In addition to these suckers, the Argulus is furnished with an armament of spines and hooks on the under surface of the body, which assist it in retaining its

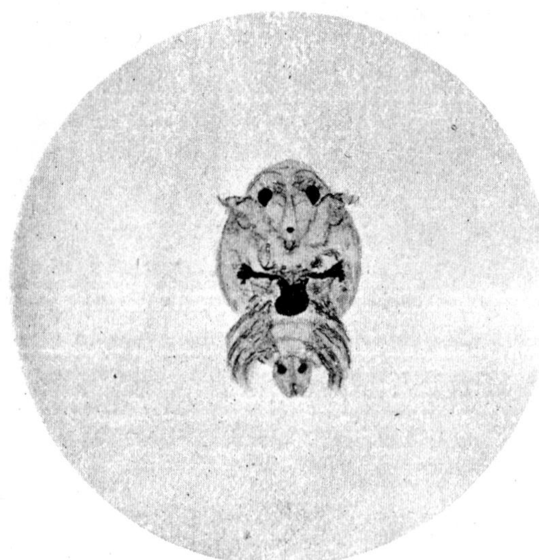

Photo by] [*F. Noad Clark.*

THE GRUB OF THE FEMALE ARGULUS.

It has attained the third stage of its existence and is rapidly assuming the form of the adult state. (Magnified twenty times.)

hold when at rest. A pair of antennæ, or feelers, or, as they are sometimes termed, " foot-jaws," lie just above the paired eyes. They are the organs of some special sense and probably of taste (page 614). The mouth and feeding organs are of a complex nature and lie in the centre of the body. An incision is made in the skin of the fish with a pair of minute serrated jaws, which lie within a pouch-like organ called the siphon, and a long telescopic proboscis, or poison-sting, is inserted into the wound. The reason for this is not very clear, but it is probable that an irritating fluid is injected, which induces inflammation and a flow of blood; at any rate, the blood of the fish forms the diet of the Argulus. The latter sometimes takes two or three days at a stretch to finish its meal, without once relinquishing its hold on the unfortunate host.

Photo by] [*F. Noad Clark.*

THE EYE OF THE ARGULUS.

One of the compound eyes, showing the kernel-shaped crystalline bodies and the dark pigment patches. (Magnified one hundred times.)

A most noticeable feature is the incessant motion of the swimming feet, in which the four pairs participate simultaneously. Each of these is divided into two branches of equal length, bearing comb-like hairs (page 614), which latter are again subdivided into delicate secondary hairs. This movement is never arrested during life, even when the creature is at rest on its host or when depositing eggs. The Argulus swims with a darting motion, frequently making a series of somersaults, and occasionally it is seen to maintain for an hour or more a whirling motion in the water. When about to attach itself, the current of water caused by the movement of the fish's fins gently carries it on to the fish. When once attached by means of the powerful suckers, all efforts of the fish (and they do struggle strenuously) to shake it off are in vain. On page 611 are shown two Arguli attached to a stickleback, giving some idea of their relative size. When searching for these Arguli, one finds about one in every four or five fish affected. The favourite place of attachment is on the gills, although they are frequently found on the fins and tail, and in some cases upon the eye itself. Let us hope that the sensitiveness of the fish's eye does not approach that of the human eye. They are evidently distasteful to the fish, and have often been observed to be taken into the mouth and shortly afterwards ejected uninjured, even when the fish are hungry. Between the suckers and the swimming feet is a pair of jointed legs, or foot-jaws, whose use is to assist the Argulus in attaching itself to the fish and for locomotion. The stomach of the Argulus is on a comprehensive scale, and branches of it ramify throughout the greater part of the body. In this way food is stored up and the creature is enabled to live for a considerable time away from its food supply. I have a record of an adult living a fortnight without food. Under the

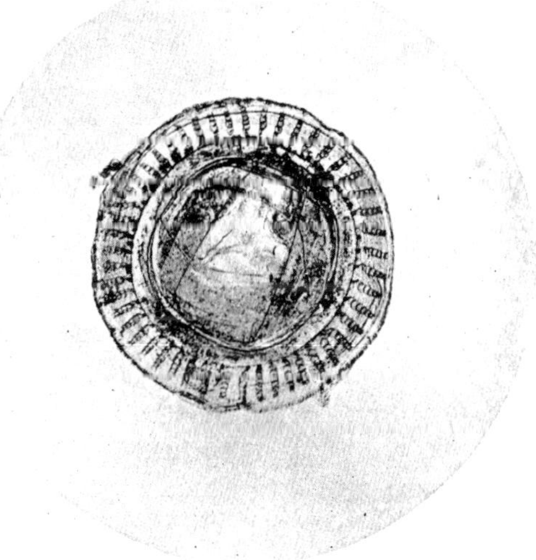

Photo by] [*F. Noad Clark.*

THE SUCKER OF THE ARGULUS.

Found only in the adult stage. By means of the many-jointed rays and the muscles the Argulus can produce a vacuum which enables it to adhere so tenaciously to its host. (Magnified about fifty times.)

Photo by] [*F. Noad Clark.*

THE FOOT-JAW, OR FEELER, OF THE ARGULUS.

These are only to be found in the grub stage. They take the place of suckers, to attach the parasite to the fish, and are also useful in cutting the egg for the liberation of the young grub. (Magnified fifty times.)

microscope it is interesting to watch the circulation of its blood; this is quite transparent, and contains small round globules, which traverse the stream in pulsations every second. Little is known of its means of respiration, but it is supposed that the constant motion of the swimming-feet previously referred to is associated with the function. On the other hand, the curious two-lobed tail-plate, as it is called, has been thought to be an organ of respiration, but its primary use is no doubt as a rudder.

In the female Argulus, previous to the depositing of her eggs, the whole centre of the body, from the mouth to the tail-plate, is seen to be full of eggs, to the number of four or five hundred. A distinguishing feature also of the sex is the presence of a number of star-like patches of brown pigment over the surface of the egg-sac, which have a pretty appearance under the microscope. Under natural conditions egg-laying commences about the end of April, and there are at least two broods during the year. Eggs laid late in the year remain over the winter and hatch in the spring; but the average time from the laying of the egg until hatching occupies twenty-five days. On one occasion an Argulus was observed to have laid over six hundred eggs. The operation of egg-depositing, as seen under the microscope, is an interesting one. The three first pairs of legs are in motion as usual, the fourth or last pair being held rigid, with their bases (which are rather prominent) pressed together over the entrance to the egg-sac and relaxed alternately as each egg makes its exit. This done, the Argulus raises herself higher up on the object on which the eggs are laid, and moving a little from side to side, places the eggs in regular order in ribbon-like rows of from three to six broad. The time taken for laying each egg is eight seconds. As seen on page 610, the egg is oval in shape and pale yellow in colour; each one is glued to its neighbour with an albuminous material. The eyes of the young Argulus before hatching from the egg are clearly visible, being relatively large at this early stage. The young larva then makes its exit, tail-end first, by cutting a longitudinal slit in the egg. Page 612 shows a highly magnified photograph of the larva just emerged. As will be explained later on, this tiny speck of active life bears but a slight resemblance to its later stages of existence.

Photo by] [*F. Noad Clark.*

THE SWIMMING FEET OF THE FEMALE ARGULUS.

The incessant movement of these swimming feet is never arrested during life; not even when the Argulus is attached to its host. The four pairs of feet move simultaneously (Magnified fifty times.)

It immediately sets forth in quest of a host, to which it attaches itself and commences feeding. In confinement under artificial conditions, these newly-hatched larvæ have been known to live five days without food. It may seem wonderful how

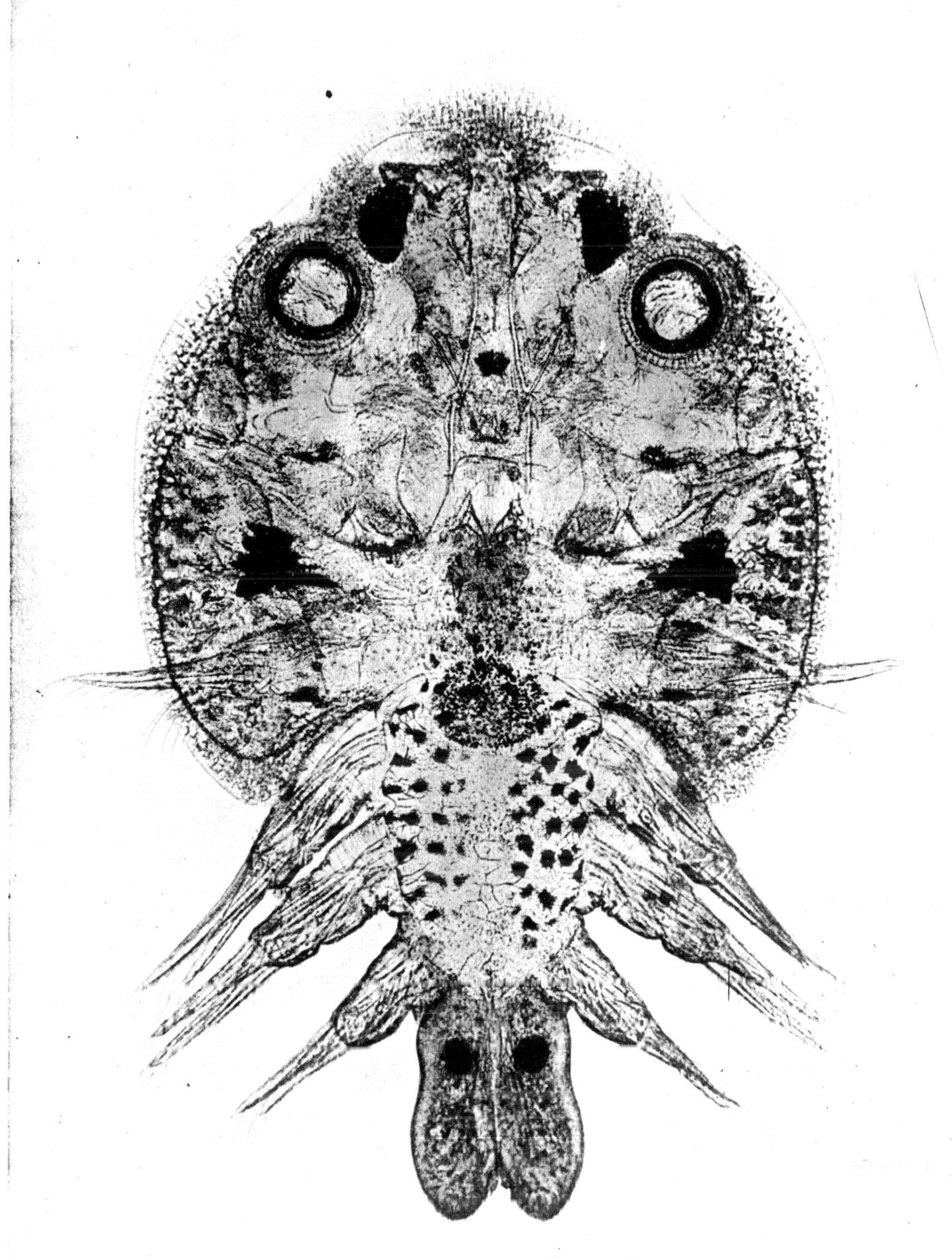

Photo by] [*F. Noad Clark.*

THE ADULT FEMALE ARGULUS

Note the two black spots on the tail-plate which are the distinguishing feature of the sex. A full-grown parasite measures about one-third of an inch in length. It is a transparent, jelly-like creature of a greenish hue, the bulk of the body appearing as an oval slightly convex shell.

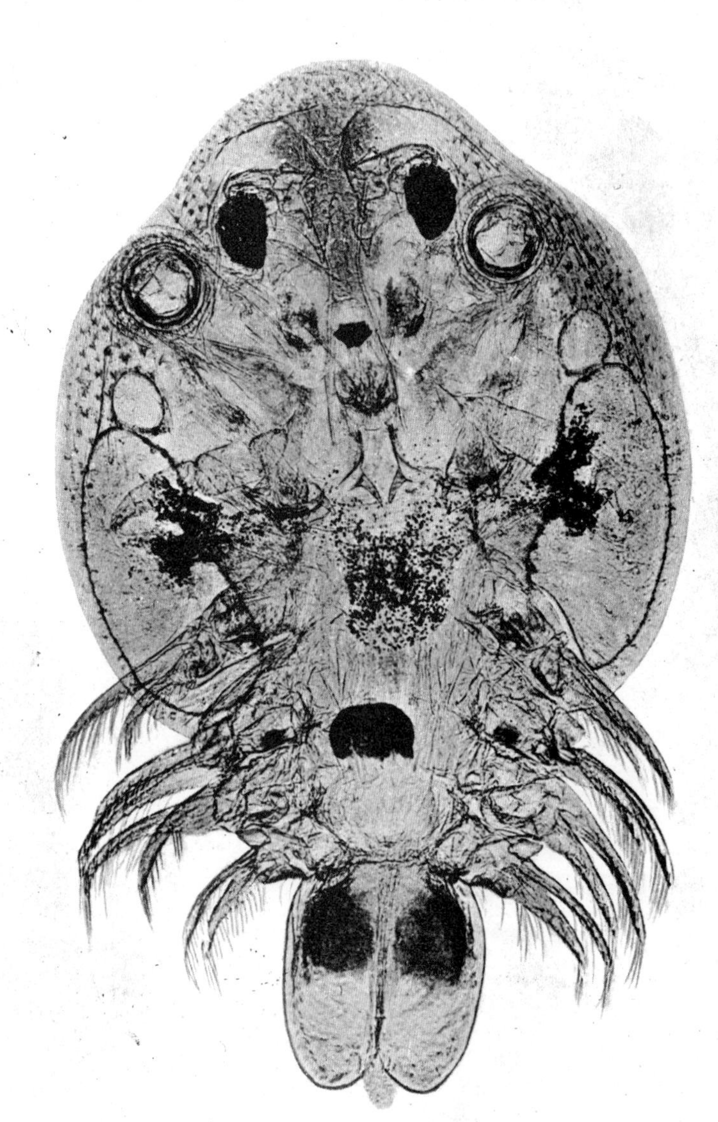

Photo by] [F. Noad Clark.

THE ADULT MALE ARGULUS.

He is not so large, but in other respects closely resembles his mate. The average length of life is probably over six months. Arguli have been known to live in salt water, though they are usually to be found on fresh-water fish.

such a minute creature is able to find its source of livelihood; but it must be remembered that the sticklebacks usually frequent ponds and still water and are generally plentiful; so of the many thousands of young hatched, there is every chance of some finding their goal, more especially as their abundance is probably due to their being distasteful to their enemies. At this stage the swimming feet are undeveloped, their function being undertaken by two pairs of bristle-like limbs. These impart to the larva a darting motion similar to the water-fleas. A powerful pair of triple-barbed hooks, or foot-jaws (which are replaced at a later stage by the suckers before mentioned), answer the purpose for attachment to the fish, and it was by their aid that the young larva cut its way out of the egg.

The young Argulus now enters upon its metamorphosis by successive changes of form and structure, to intimately describe which would not come within the scope of this article; but at each stage interesting and remarkable events take place in its life-history. On about the twenty-fourth day the Argulus has cast off the rudiments of its early stages of existence and reached the perfect state. This progress has been accompanied by a series of moults which continue to take place at intervals throughout life. The cast-skin is a replica of the body it has left, the minutest structures being retained in the cast.

About the middle of June, when fish are plentiful, as many as a dozen Arguli have been found on a single fish, and nearly all had one or more attached. In captivity, the Arguli, if numerous, will attach themselves to the number or thirty or forty on one fish. Needless to say, the host lives but a short time after. I once took a young roach with two parasites attached, which remained so during twenty-five days. They may often be captured in the free state by using a fine net in ponds or canals where sticklebacks congregate, and there is no doubt that they also pass the winter in the adult state. A case has been recorded of the capture of three Arguli in the Faröe Channel, thus showing that they will live in salt water. These presumably had been conveyed to sea by the medium of some fresh-water fish.

Their average length of life is probably over six months, and I have kept several such under observation. They are frequently victims of a filamentous fungoid disease, from which their hosts also suffer, and from their earliest stages are occasionally found to accommodate colonies of the well-known Vorticellæ, or bell-animalcules.

FLAMINGOES

BY SIR HARRY JOHNSTON, G.C.M.G.

THESE beautiful birds are considered to be related distantly to the geese and ducks and screamers on the one hand, and to the ibises and storks on the other. In the opinion of the late Professor Garrod they also show some affinity to the bustards and thick-knees. It is certainly curious that the young of the Flamingo, in their streaky brown and white plumage and in the shape and colour of the beak, do offer some resemblance to the stone-curlews, or thick-knees. Extinct forms living anciently in France and Northern Europe in which the bill was straight, the legs were shorter, and

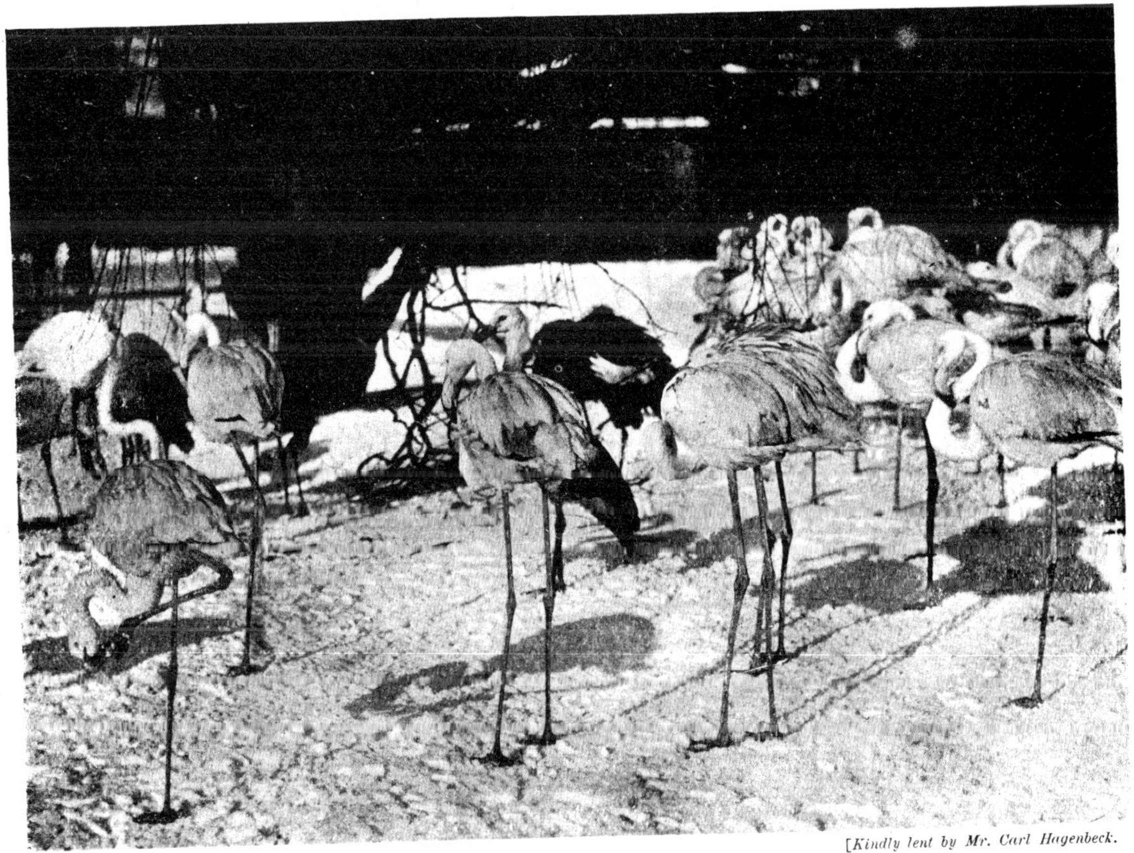

From a photograph] [*Kindly lent by Mr. Carl Hagenbeck.*

FLAMINGOES.

These curious and brilliant birds have been photographed standing on the frozen snow in Mr. Carl Hagenbeck's Animal Park. Flamingoes once frequented England, and it was only the ruthless destruction of their kind that drove them away from our fens.

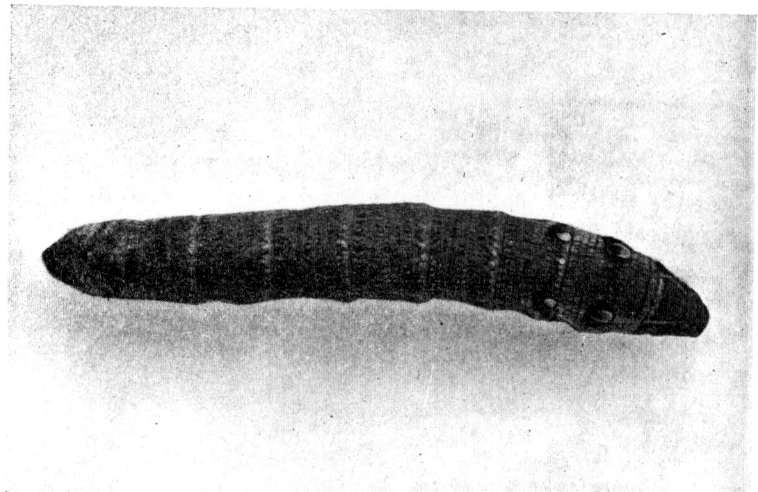

Photo by] [*Hugh Main, B.Sc., F.E.S.*

THE CONNAGH WORM.

This is an excellent example of the power of a harmless insect to make itself look like a fearsome beast. Here the Elephant Hawk grub is shown in its natural state.

the toes longer, seem to have connected the Flamingo group with the ancestors of the ducks and geese. In their structure, Flamingoes certainly resemble the duck and goose order more closely than any other at the present day.

No one can mistake a Flamingo for any other type of bird. It is the giraffe of its class, with the longest neck proportionate to size; in that respect exceeding the swan. In the larger species the legs are also perhaps of a greater proportionate length than in the case of any other bird. The short toes are webbed. The lower part of the beak is abruptly turned downwards, and inside the beak (which is short and straight in the young bird) there is a horny growth, serving the bird as well as actual teeth in breaking up its food, which consists of small crustaceans and fish, water-insects, and mud composed mainly of vegetable refuse. The Flamingo, like the duck, prefers obtaining its food through and in the water, passing the mud or water through its beak as through a sieve, and retaining in that way minute particles of food. In captivity, Flamingoes will subsist on porridge, on shrimps, pounded fish and pounded grain. They are able to swim as well as to wade in deep water, and can also fly as strongly as geese or swans; but like these birds, their powers of flight are lost for a short part of the moulting season in the autumn, when the old quill feathers are shed and the new ones not sufficiently grown. At this time Flamingoes are very helpless, as they can only wade or swim away from the human enemy hunting for their beautiful plumage.

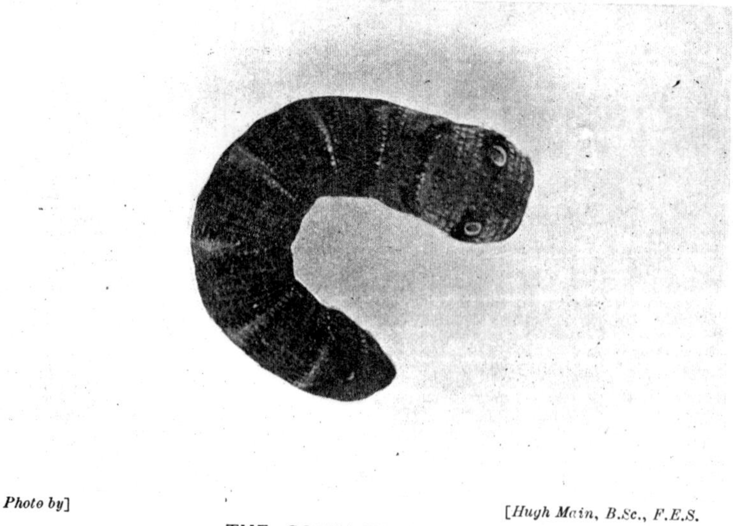

This plumage ranges from an exquisite blush-tinted white (with deep pink and scarlet wing and tail feathers and black quills) to a universal pale scarlet in some of the American forms. In all species except one the legs are a deep pink colour.

Their nesting habits are peculiar. They scrape up with their webbed feet little columns or mounds of mud. The top, which is often only a few inches above the level of the water, is slightly hollowed, and on this are laid a single, or perhaps two,

Photo by] [*Hugh Main, B.Sc., F.E.S.*

THE CONNAGH WORM.

Here the grub has been disturbed. He has compressed his head and the tapering segments until the extended part has become distended and the front oval patches look like two staring eyes.

FISHES THAT SWIM CURIOUSLY.

The Cat-fish shown in the upper part of the photograph swims belly uppermost, and this freak is provided for in its coloration, for whereas usually the back of a fish is darker than the belly, in this case the colouring is transposed. The Sun-fish shown below is built for sudden descent in the water, head-first.

THE NEEDLE-FISH.
This curious fish swims head downwards. Notice its peculiar shape and the tail fin.

eggs of bluish-white. The mother bird folds her long legs beneath her and squats on the top of the mound.

Flamingoes once lived in England and Holland, and at the present day, as can be seen by the accompanying illustration, they are able to exist quite happily in Mr. Hagenbeck's park, near Hamburg, even when the shores of their lake are covered with deep snow.

THE CONNAGH WORM

BY HUGH MAIN, B.SC., F.E.S.

THAT it is not necessary to leave our own shores to find examples of a harmless insect making itself look like some fearsome beast may be gathered from the accompanying illustrations.

The caterpillars of the Elephant Hawk Moths occur in many parts of the British Isles in July, August and September. They are two or three inches long when full grown, green or brownish in colour, with the front segments tapering to the small head, a short distance from which are two or four large oval coloured patches. That of the Small Elephant Hawk Moth is shown in its usual resting position on its food-plant, one of the bed-straws.

The green-coloured caterpillars are only noticeable among the bed-straw on careful examination, so closely does their colour resemble that of the plant.

On the contrary, the brownish caterpillar in the photograph is quite easily seen. If disturbed, however, it at once withdraws its head and the tapering segments into the succeeding segments, which are thus distended. This action also brings the front oval patches into a position where they look like two big staring eyes. The caterpillar then throws itself violently about by vigorously bending its now turgid body rapidly from side to side. Altogether its appearance and action give it a most uncannny character, sufficient to give even an entomologist quite a start on making a first acquaintance with the insect.

SOME MARVELS OF FISH LOCOMOTION

BY W. P. PYCRAFT, F.Z.S.

EVEN among fishermen it is generally believed that a fish, in swimming, is propelled by its fins ; yet, with certain exceptions to be described presently, this is never the case : the fins serve merely as balancers, steering organs and brakes. Locomotion is performed by more or less rapid side-to-side movements of the tail, which, be it noted, represents all that portion of the body lying behind the vent. The so-called "tail" is really the tail *fin*, which, though an important steering organ, does not, as we have remarked, serve to propel the body.

THE SUN-FISH.

Sun-fishes depend largely for food on prey caught at enormous depths; especially are they partial to larval eels, which can only be had by diving; hence their grotesque appearance and the adaptation of the dorsal fin for progression by means of vibration.

Now, as to these exceptions : though few in number, they are certainly remarkable in kind. Most of these swim by means of the dorsal or back fin, which by excessively rapid vibrations acts as a propeller, driving the body through the water. In many such cases the body is of a most extraordinary shape, being about twice as deep as it is long ; that is to say, the length from the snout to the fork of the tail fin is only about half the height ! In some cases, as in the Sail-fish, this depth is formed by the enormous development of the dorsal fin and its corresponding fin of the under surface of the body—the "anal" fin. In others, as in the Sea-bat, the body itself is expanded. Now if the details of these curious fishes are for the moment ignored, and attention be concentrated on their general outline, it will be found that it makes a rough approximation to that presented by the grotesque-looking Sun-fishes, which have the appearance of having undergone amputation of the hinder end of the body. They have been humorously described, indeed, as "cherub-fishes, because they are cut off behind"! This general semblance is not without a purpose: for all these fishes are dwellers at the surface of tropical seas, but apparently depend largely for food on prey caught at tremendous depths. The Sun-fish, we know, lives largely, during a part of the year, on larval eels, which can only be had in quantity by diving; and in all these cases the motive power is furnished by the vibration of the dorsal and anal fins. Thus, then, the common likeness in contour has been brought about by adaptation to a common end, this peculiar shape apparently being the most suitable for sudden descent head-first.

But even among fishes which normally swim by vigorous side-to-side, lashing movements of the tail there are some which, on occasion, make use of vibrations of the dorsal, or dorsal and anal fins, to attain special ends, generally stalking prey. The Pike is one of these. The celebrated John Dory is another. As everybody knows, the John Dory is exceedingly thin from side to side, and deep measured from the back to the belly. Thus, end on, he is a mere line in the water, and this fact serves him in good stead when stalking a victim—say, a sprat. Instead of striving to overtake his prey with a rush, he approaches his victim warily, moving by means of the vibrations just referred to. As soon as he is sufficiently near to ensure success, he shoots forward his wonderful tubular mouth, and in an instant the fish is engulfed and the telescopic mouth goes back to its place again!

Photo by] [*G. R. Balance.*

A YOUNG ICE FLOWER.

The rarefied air of the Alps holds the water-vapour in suspension, and the intense cold freezes it into crystals of ice. These are attracted one to the other, and so go to create the Ice Flower. Notice the little tufts of ice near the central bloom, which will shortly be incorporated in the flower.

The Pipe-fish of our coasts is another fish which, like the Sea-horses, swims by means of vibrations of the dorsal fin. But this fish moves, not horizontally, after the fashion of fish, but vertically. And the same is true of another and much more remarkable fish. This is the Tortoise-fish, so called because the scales of the body have become welded together to form a box-like case, as inflexible as the shell of a tortoise, only the dorsal fin and the extreme end of the tail being capable of movement. This shell is almost as flat as a paper-knife, and the ventral edge is almost razor-like, so as to offer the least possible resistance to the water. To secure the greatest possible rapidity of movement in this strange position, the dorsal fin has shifted to the extreme hinder end of the shell, and looks at first like the tail fin, which last has become bent round, as it were, to the ventral or under surface! A shoal of these fish seen swimming in an aquarium has an almost uncanny effect, for the slender bodies are almost transparent.

Photo by] [*Mrs. Aubrey Le Blond.*

A GARDEN OF ICE FLOWERS.

No words can convey the exquisite beauty of a field of Ice Flowers when the sun has just touched slantingly the sharply-pointed petals. It is as if a cart-load of perfectly-cut gems had been strewn around, star-like points of blue and red shining with dazzling brilliance.

Photo by] [*G. R. Balance.*

ICE FLOWERS.

These splendid examples of the frost's work were covering the rough surface of an old wall. Atmospheric conditions have been such and the air sufficiently still to allow the delicate petals to combine and form singularly perfect rosettes.

By way of contrast, we have fishes which swim belly uppermost! A well-known instance of this kind of topsy-turvy locomotion is furnished by some of the Cat-fishes. One of these, a native of the Nile, was well known to the ancient Egyptians, and was frequently depicted swimming after its peculiar fashion on their wall-paintings. But associated with this strange manner of locomotion is a no less striking peculiarity of coloration; for whereas, for protective purposes, in other fishes the back is much darker than the belly, in this fish the belly is darker than the back! Thereby, of course, the fish escapes the keen eyes of other predaceous fishes, for the dark upper surface (the belly) counteracts the effects of the strong light from above, while the white under surface (the back) neutralizes the blackness of the depths below. But locomotion in this case is performed, as in ordinary fishes, by side-to-side movements of the tail and tail fin.

Only one other group of fishes contains members which swim belly uppermost. These are the Globe-fishes. Normally they swim after the typical fashion. But when alarmed the fish, rushing to the surface, suddenly inflates the abdomen and turns turtle. The globular belly now projecting above the surface of the water, is caught by the wind and borne along. Thus are enemies avoided—partly by the dangerous armature of spines which stud the whole body, and partly by removal to a distant spot.

As a rule, probably, most of us regard the breast fins in fishes as the rowing organs; their use for this purpose is negligible. They serve as balancers and steering organs. But among the Skates the breast fins have been developed to form huge muscular wings, and these alone are concerned with locomotion, driving the body through the water by means of their peculiar wave-like undulations.

In most of the Skates, indeed, locomotion by means of the tail would be impossible, this organ serving merely for steering purposes. In some species, however, it has become modified to form an excessively long, whip-like organ, armed with a large and most poisonous spine, thus serving both as a rudder and a weapon of offence.

ICE FLOWERS

BY MRS. AUBREY LE BLOND.

PERHAPS the most beautiful feature of an Alpine winter is the Ice Flower, or monster hoar frost, which adorns the upland Swiss valleys.

In certain places, however, Ice Flowers are finer and more often seen than in others, the best examples being always met with on comparatively small lakes fed by large streams.

The Lake of St. Moritz, in the Engadine, is an ideal home of the Ice Flower, for the stretch of water, being of moderate size, quickly catches all over and offers a surface of smooth, black ice, while the river Inn, which feeds it, enters by a fast-flowing stream, which does not freeze, and leaves it over a cascade of open water.

Above these two pieces of moving water dense mist hangs at night, and gently drifts over the black ice. The mist is composed of tiny crystals of ice—for at the usual temperature, many degrees below freezing-point, water could not remain in suspension. Some of these infinitesimal rods of ice

Photo by] [*Mrs. Aubrey Le Blond.*

DETAIL OF AN ICE FLOWER.

The combination of crystal upon crystal can be distinguished in the photograph. The velocity of their impact causes them to melt at the point of contact; but the pressure ceases once they are united, and they freeze together.

fall on and are scattered over the shining surface of the lake. Others are magnetically attracted to them, and thus, petal by petal, the Ice Flower is built up.

"The different shapes in which crystals are found are classified into six systems, but nobody knows why a particular substance should adopt one form rather than another. The hexagonal form adopted by vapour of water, though apparently somewhat complicated, is, in reality, perhaps the most natural shape. A number of equal spheres placed in contact arrange themselves in such a way that the lines of cleavage form hexagons or equilateral triangles."

My old friend, Mr. Gilbert T. Walker, now head of the Meteorological Department in India, gave me years ago, the above note as applying to Ice Flowers.

As soon as the sun strikes the mist it begins to evaporate, but for some time the shining rods of ice floating about can clearly be seen. In perfectly calm air and during intense cold these

Photo by] [*G. R. Balance.*

A GIGANTIC ICE FLOWER.

The flowers do not usually much exceed two inches, but if neither wind nor thaw nor a fall of snow interrupts their formation they can attain a great size. This flower was a record growth, for it measured over a foot in diameter.

crystals are particularly large, and when neither wind nor thaw nor a fall of snow interrupts the building up of the Ice Flowers they reach a great size. I have measured petals three inches long on the St. Moritz lake.

Ice Flowers of great beauty are often found on weeds and grasses at the edge of steep, small streams. They form a delicate, feather-like border, and continue to increase in size till destroyed by thaw or snow. On trees and palings in the neighbourhood of mountain torrents we find similar growths; but often these are in valleys to the depths of which the sun never penetrates, and in that case much of their loveliness is absent.

No words can convey the exquisite beauty of a field of Ice Flowers when the sun has just touched slantingly the sharply-pointed petals. It is as if a cart-load of perfectly-cut gems had been strewn around, star-like points of blue and red shining with dazzling brilliance.

It will be recollected that Ice Flowers are merely glorified hoar-frost, but the hoar-frost we are accustomed to see in England is so minute in its details that the eye does not obtain the flower-like

Photo by] [*Mrs. Aubrey Le Blond.*

A PATCH OF LARGE ICE FLOWERS.

Ice Flowers are merely glorified hoar-frost, but the hoar-frost we are accustomed to see in England is so minute in its details that the eye does not obtain the flower-like general impression conveyed by the Swiss variety.

Photo by] [S. L. Bastin.

WINGED SEEDS.

The Ash produces its fruit in "keys," a bunch of long green wings, each carrying a single seed.

Photo by] [E. Step, F.L.S.

WINGED SEEDS.

In the Sycamore the seeds are in pairs, and are carried by a pair of wings.

general impression conveyed by the Swiss variety. The tracery on window-panes is more familiar to us in England, and in that also we can discern the crystallized form always taken by the vapour of water when it freezes.

In Tyndall's "Forms of Water" the magic circle of snowflake—ice—water—vapour—can be followed out. Under "Atomic Poles" he tells us how every tiny crystal which floats in mist—or, rather, into which at a certain temperature mist resolves itself—possesses two magnetic poles. One repels and the other attracts. As they lightly drift about, the attractive poles come together and the repellent poles retreat from each other. If pieces of ice are floating in water, when they touch each other they freeze together. Similarly, on a glacier, when the gentler slopes at the foot of an ice-fall are reached, crevasses close up. The ice now begins to present a uniform surface, for the same natural law is at work as that which binds the particles of hoar-frost which have been brought together by magnetic attraction, though, of course, in the case of a glacier the ice is brought in contact by other means.

The tiny magnetic rods, sorted in a six-sided pattern by attraction and repulsion, are, as it were, cemented together by regelation, though the simile must not be strained, for they actually grow together as one flower. Regelation, or re-freezing, is produced by pressure. Pressure actually *melts* ice, and directly the pressure is removed the ice freezes again. Now picture two sides of a crevasse pressed together, as the glacier which has been strained, distended and cracked over a steep bed reaches more even ground. First the pressure melts the surfaces as they come into contact, but afterwards, as the great body of ice settles down into an unbroken stream, the pressure passes away from special points and everything is again a uniform frozen mass.

With the petals of the Ice Flowers, if one may so describe these frozen, floating, magnetized atoms, the same thing occurs. They lock themselves together at the last moment with an appreciable velocity, and latent heat is freed which melts their surfaces as they come in contact. But the pressure ceases once they are united, and they freeze together. Thus we see exemplified, even in these fragile children of the mist, the natural law that the freezing-point of water is lowered by pressure.

Tyndall describes Ice Flowers of a totally different nature to those which adorn the frozen surface of a lake. These flowers actually lie beneath the surface of the ice. A slab of ice is placed in the path of a concentrated ray of sunshine. Part of this beam is stopped. Part goes through. "The former produces internal liquefaction; the latter has no effect whatever upon the ice. But the liquefaction is not uniformly diffused. From separate spots of the ice little shining points are seen to sparkle forth. Every one of these points is surrounded by a beautiful liquid flower with six petals . . . This is a reversal of the process of crystallization."

These Ice Flowers *within* ice are formed through the melting of the spot where the ray has been stopped by a crystal, the melting ice shrinking in bulk and leaving a vacuum. It is the surface of this vacuum, shining like silver in the sun, that forms the flowers. They can be seen any sunny day with the aid of a pocket lens.

There seems little doubt that the popular idea that more frost is needed in England to form ice of a given thickness than is required in the Engadine, six thousand feet above sea, has a scientific basis. The air-pressure at a great elevation is less than at sea-level, so water at six thousand feet above sea freezes at a temperature which would not freeze it at sea-level. The ice of high altitudes always seems more brittle than English ice, and a greater thickness is considered necessary before it is safe to venture on it. Perhaps this more rapid freezing may partly account for the greater size of the Ice Flowers in the Engadine than in places some thousands of feet lower.

"Water: Its Origin and Use," by W. Coles-Finch (Alston Rivers, 1908), is a volume which those interested in ice and snow, glaciers and rainfall, springs and salt lakes, will do well to consult. It has about two hundred illustrations of water in all its forms, and gives sufficient preliminary information to prepare the enthusiast for making personal observations.

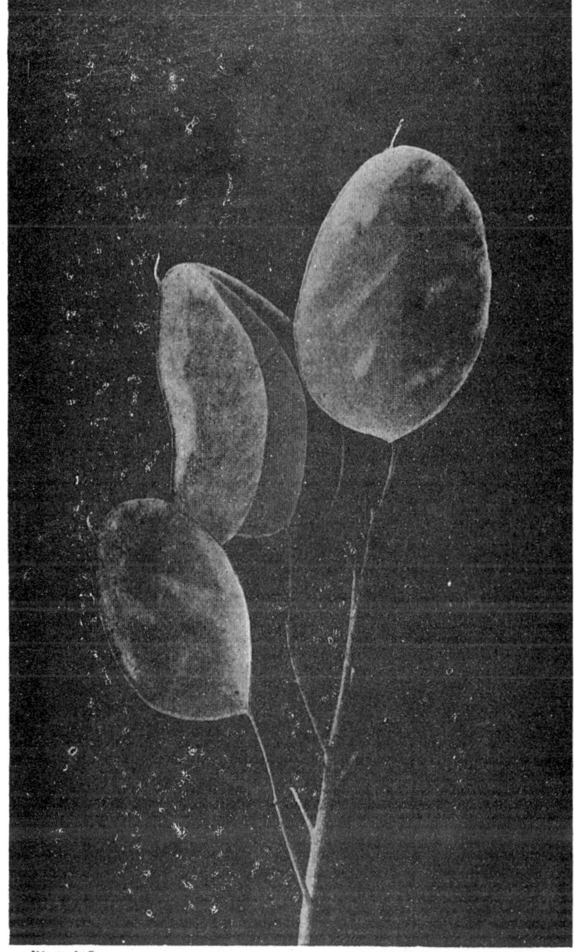

Photo by] [*Hugh Main, B.Sc., F.E.S.*

WINGED SEEDS.

Some plants, such as Honesty, supply their seeds with a papery border, which allows them to be carried far away.

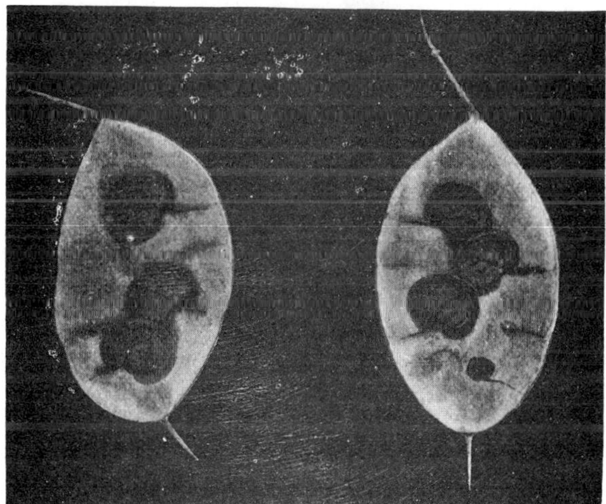

Photo by] [*Hugh Main, B.Sc., F.E.S.*

WINGED SEEDS.

Notice how flat the seeds are and how they are packed close to the thin membrane, which acts as a sail.

Photo by] [*Hugh Main, B.Sc., F.E.S.*

WINGED SEEDS.

The seeds and scales of the Fir-cone. Notice the marks on the scales showing the position in which the winged seeds are stored away, and the curved edges which prevent the unripe seed falling out from the cone.

Photo by] [*E. Step, F.L.S.*

WINGED SEEDS.

The seeds of Sabine's Pine. All seeds of conifers have no special envelope to protect them when they have escaped from the cones.

WINGED SEEDS

BY SIR HERBERT MAXWELL, F.R.S.

THE three necessary conditions for the development of plant-life are space, light and fresh ground in which the matured seed may germinate. To obtain these necessities the seed must travel some distance from the parent plant before its fall to the ground. One of the methods of travelling is by means of wings, and the seeds of many plants are, therefore, furnished with a strong membrane, which spreads out from the actual seed, and is large enough to take advantage of the wind to carry the precious freight beyond the baneful influence of the parent stem.

The race of pine-trees spend two or three years in building splendidly protective fortresses round their seeds; but if all the cones of a pine were to fall at its foot with the seeds within them, the mountain could never be clad with forest; wherefore each seed, escaping when ripe from the cone, is provided with a beautiful membranous wing, upon which it is wafted, sometimes to a great distance, on the chance of alighting on a vacant piece of ground.

In most species of pine the cone opens of itself and releases the ripe seeds; after which the seeds soon lose their vitality if they do not find a spot favourable for germination. But Bishop's Pine (***Pinus muricata***) forms a singular exception to this habit. Although this tree is a native of the hot coast of California, the sun has no power to cause the cone to open, nor will it do so until forced to gape by the heat of a forest fire. This may not occur except at long intervals; accordingly the cones persist on the branches for an indefinite number of years (I have seen living cones unopened after forty years), waiting for the necessary conflagration to enable the still living seeds to fly forth on their inch-long wings. In 1907, Mr. J. Trotter succeeded in raising seedlings from an old cone which he detached from the branch and caused to open by putting it in an oven.

Bishop's Pine exhibits another feature strongly suggestive of intelligent design. Most pine-trees, those, at least, which part soon and easily with their cones, are satisfied to protect the seeds by building strong, imbricated, wooden walls around them. But cones that may be exposed for twenty or thirty years to the attacks of rodent animals or strong-beaked birds require additional defence; consequently the cones of

Photos by] [*E. Step, F.L.S.*

PINE CONES.

Two of the largest cones produced by Pine-trees. That to the left is the cone of the Maritime Pine, which has proved so valuable in the reclamation of land from the sea in Brittany. That to the right is the cone of Sabine's Pine, a North American species. The first is reduced one-fifth; the second one-half.

Photo by] [W. P. Pycraft, F.Z.S.

THE "GILA MONSTER."

This is a lizard, some two feet long, found in New Mexico, and is remarkable for the fact that, of all known lizards, it alone has the mouth armed with poisonous fangs like those of a snake. This poison is sufficiently strong at times to kill even men.

Bishop's Pine carry a sharp, recurved spine on the external apex of each scale, rendering them extremely unpleasant to handle.

Pines, and all other conifers, produce naked seeds (although these are wedged in among the scales), a character which betokens a type of vegetation which held the field when the chalk of Sussex and the greensand of Surrey were in process of formation under the ocean. But in the course of unnumbered ages plants took to providing a special envelope for their seeds, a device which proved so successful that at the present day it prevails among the vast majority of species. All flowering plants and broad-leaved forest-trees protect their seeds in this manner. But they have not been so proud as to discard the device of equipping seeds with wings, which their forerunners had proved to be so successful during hundreds of thousands of years. Many plants of the modern type have adopted it.

Thus the ash produces its fruit in " keys," consisting of a bunch of green wings, each wing with a cell at the base in which is a single seed—wing and cell together measuring about an inch and a half. The maple family have improved on this arrangement by providing two seeds, each with a wing, and as the carpels containing the seeds are formed in pairs, they sally forth on a pair of wings. In our native field maple the wings are extended horizontally in a straight line ; in the sycamore (which is also a species of maple) they are set horizontally at an angle with each other. Presumably the sycamore's arrangement is the more effective, because, while the field maple does not spread rapidly, the sycamore, although not indigenous to Great Britain, establishes itself so quickly and widely from seed that, if unchecked by agriculture, etc., it would soon take exclusive possession of some districts. Herr Dingler found by experiment that, in still air, the winged seed of sycamore fell six feet in 5.6 seconds, whereas when he clipped off the wings the seed fell the same distance in 1.2 seconds.

Other plants, instead of wings, provide their seeds with a continuous papery border, which is equally effective in retarding their descent, thereby exposing them to the drifting force of wind.

The wych-elm, various lilies, bignonia and the common honesty, are good examples of this method. In Dingler's experiments, unmaimed seed of bignonia took 24.6 seconds to fall six feet; deprived of its membrane it took only 4.6 seconds.

Such are a few examples of seed dispersal by aeroplane. Others have recourse to balloons, such as the bladder-senna, a pretty shrub easily grown in British gardens, although our climate is seldom dry enough to render its balloons an effective means of transport. Its yellow flowers are followed by extravagantly-inflated pods, each about two inches long and containing a few hard seeds. It is a native of the Mediterranean region, and grows into the very crater of Vesuvius, where the pod dries to sufficient buoyancy to ensure it being wafted to a considerable distance from the parent plant.

THE "GILA MONSTER"

BY W. P. PYCRAFT, F.Z.S.

THE Gila Monster is a lizard, some two feet long, found in New Mexico and Arizona, and is remarkable for the fact that, of all known lizards, it alone has the mouth armed with poison fangs, like those of a snake. The venom which is furnished by glands at the base of each tooth, escapes into the wounds inflicted by a bite by means of grooves down the teeth. Though less virulent in its effects, as regards the human race, this poison is sufficiently strong at times to kill even men. As a rule, however, it is used only in the capture of its more active prey, such as frogs, worms and centipedes, and the eggs of birds and lizards are also eagerly sought for. It is a most repulsive-looking creature, the head being broad and flat, with small, wicked-looking eyes; while the general loathsomeness of its appearance is greatly increased by its coloration, the body being irregularly banded with black and orange, and the tail even banded or ringed with the same hues. Further, the upper parts are covered with curious wart-like prominences, recalling those of toads.

These colours, black and red, are what are known as "warning" colours, advertising the dangerous character of the wearer to all whom it may concern. As a rule the Gila Monster hides by day

Photo by] *Geo. Wharton James.*

THE "GILA MONSTER."

It is a most repulsive-looking creature. The head is broad and flat, with small, wicked-looking eyes. The body is irregularly banded with black and orange, w. ile the upper parts are covered with wart-like prominences These markings and colours indicate the dangerous character of the creature.

under the roots of trees, emerging at dusk in search of food. During the heat of the dry season it "æstivates," or passes into a state of coma, just as dormice, bats and lemmings, and many other animals, "hibernate" to escape the rigours of a prolonged winter. It is not, by the way, the extremes of temperature which induce this enforced rest so much as the shortage of food.

From "Extinct Birds" *[By permission of Walter Rothschild.*

THE DODO.

At least one living specimen was brought to this country; but the race, which was indigenous to Mauritius, was exterminated by the voracity of travellers.

WINGLESS BIRDS

BY W. P. PYCRAFT, A.L.S., F.Z.S.

To speak of flightless birds seems like a contradiction in terms. All the same, there are a number of species at the present day to whom flight is impossible; and when we extend our survey over the past history of the world, we shall find a yet more numerous company who had lost what would seem to us their greatest heritage. To enumerate all the known species in this predicament would require a volume in itself; it must suffice, then, to make a selection of the more remarkable instances.

To begin with, let us take the Ostrich tribe. Taking the fossil with living species, we find that this ancient group has contrived to distribute itself practically over the whole world; yet all save the Tinamous of South America are flightless. Six species still survive. Of these the largest is the African Ostrich, a full-grown cock standing as much as eight feet high. Though the wings are of considerable size, they are utterly inadequate to support the body in the air, but are used as sails to increase the speed when running; and it is exceedingly swift of foot. A dweller of the open plains, it contrives to avoid lions and other beasts of prey, and,

From "Extinct Birds"] *[By permission of Walter Rothschild.*

THE SOLITAIRE.

Although useless for flight the wings of this species were used as weapons, and were furnished with curious bony bosses on the wrist joints.

THE MOA.

This enormous bird stood about seven feet high. It is related to the Kiwi, and is the most notable type of the flightless bird. The stages leading to the wingless state are: first, the muscles of flight decrease in size; then the keel of the breastbone vanishes; the wings become smaller, degenerating into mere stumps, and finally disappear altogether.

perhaps as a further measure of safety, commonly associates with herds of zebras and antelopes, just as the South American Rhea herds with deer and guanachos. Not the least of its many peculiarities is the transformation which the foot has undergone, the toes having been reduced to two, and there are signs which seem to show that sooner or later the outer of these two will also vanish ; for it is much the smaller, and has been so far reduced in length as to suppress the claw. The Ostrich and the Tinamou alone among their kind possess a tail, or, rather, tail-feathers, and in the Tinamou these are hardly distinguishable from the rest of the plumage. These have degenerated, and in most cases vanished, in sympathy, so to speak, with the loss of flight.

Australia harbours two kinds of Ostrich : a Cassowary and an Emu, the first-named haunting the scrub, the last the open country. But in both the wing is far more degenerate than in the Ostrich or Rhea. In the Emu it must be carefully sought for among the feathers of the body. All that remains is a stump of a few inches long, and the same is true, as regards size, in the case of the Cassowary. Its presence is revealed, however, by a number of long, porcupine-like quills. These represent the barrels of once functional quill-feathers, but for some mysterious reason they have become solid, and the rest of the feather has vanished, though in young birds a thin terminal, vane-bearing portion of the feather is present. Later, it is shed. Such quills can serve no useful purpose, and their hypertrophied condition is inexplicable. Though flightless, these great birds have contrived to pass from the mainland of Australia into New Guinea and the adjacent islands. Clearly this migration must have taken place during the time when these now separated areas formed a continuous land surface. Curiously enough, the Australian species is dull-coloured, like the Emu, while the island species, which are numerous, have shed the feathers of the head and neck, and developed in their stead, over the bare skin, the most vivid hues of red, yellow, blue, and violet.

Photo by] [*H. J. Shepstone.*

THE EGG OF THE ÆPYORNIS.

The largest egg known. The Æpyornis is supposed by some to be identical with the "Roc" mentioned by the traveller Marco Paolo. An egg in the British Museum has a circumference of two feet two inches. It is here shown with an Ostrich egg on the left-hand and a hen's egg on the right.

To this peculiar decoration wattles on the neck and a great horny casque have been added. These brilliant colours and peculiar ornaments, set off by the jet-black plumage, have made the Cassowaries the most remarkable of living birds.

We pass now to New Zealand, the home of the strange Kiwi, the last of the living Ostrich tribe on our list. Several distinct species are embraced under the name of " Kiwi," but all are small, dull-coloured species, with exceedingly degenerate wings which can only be found after a hunt among the feathers.

The Kiwi, so-called on account of its loud piercing cry uttered at night, is remarkable among birds for the fact that its nostrils are placed at the extreme tip of the beak, where they can be of most service in scenting the smell of its favourite food; for the Kiwi, or " Apteryx "—a name referring to its almost wingless condition—feeds by night, and has but feeble powers of sight. It has been said that the Kiwi has a by no means keen sense of smell; but this statement was made after experiments with captive birds, and without allowing for the fact that the captives were seeking a familiar scent, ignoring other odours which, for them, were not associated with food.

[*From a restoration by W. P. Pycraft, F.Z.S.*

THE PHORORHACOS.

By far the most formidable of the fossil giants. It stood eight feet high, with a head bigger than that of a racehorse, and was armed with a hooked beak. The wings were ridiculously small so that the victims of the giant must have been run down.

But New Zealand has yet other flightless birds. Two of these, the Owl Parrot and the Weka Rail, must be mentioned here. These are birds not even remotely related to the Ostrich tribe; yet they have long since lost the power of flight, though their wings are by no means so degenerate as in the Ostrich. Why this should be we shall briefly explain presently.

Let us turn for the moment to the consideration of some of the flightless birds of the past. It is a very surprising fact, but it is nevertheless true, that flightlessness overtook at least one species of one of the most primitive of known birds. This was the Hesperornis, a giant fossil Diver, standing some six feet high, which lived so far back in time as the Cretaceous epoch, a time when the jaws of birds yet bore the teeth inherited from their reptilian ancestors. Hesperornis must have been the last of a long line of flightless birds, dating back to the very dawn of avian development, though there must have been a time when its yet more remote forbears flew, at any rate, passably well.

THE GREAT APTERYX

Another of the large flightless birds. The cause of their flightlessness appears to be over-abundance of food which can be obtained the year round without persecution from animals of prey. The birds under such conditions have no need to escape the rigours of climate, or of enemies; neither to hunt for their food nor to find roosting-places.

We measure, roughly, the period of the flightlessness of Hesperornis by the fact that all that was left of its wing was a spike-like remnant of the upper arm-bone; all the rest of the wing had vanished; not even the Ostrich tribe can show such complete reduction. Among sea-birds it was among the largest which have ever lived; its only rival, indeed, seems to have been a Penguin, which was equally large. But here the rivalry ended, for they never came into conflict, and for the very sufficient reason that they lived in different geological epochs, and in widely remote regions of the globe; Hesperornis remains having been found in the Cretaceous rocks of North America, and the remains of the Giant Penguin from the Eocene formation of New Zealand. Time was when New Zealand harboured great numbers of Ostrich birds, varying in size from turkeys to giants far larger than the African Ostrich. These are commonly known as Moas, and seem to have been finally exterminated by the Maoris. Some species of Moa retained vestiges of wings; but in others, not only the wing, but even the bony shoulder-girdle which supported it, had vanished completely. No other bird yet known to us has so completely obliterated all traces of what, but for the facts before us, we should have regarded as an absolutely indispensable organ.

Photo by] [*W. S. Berridge, F.Z.S.*

THE KIWI, OR APTERYX.

It is so named after its cry, and its race is peculiar to New Zealand. There are several species, but all are small, dull-coloured, and provided with exceedingly degenerate wings, which can only be found after a hunt among the feathers.

In the island of Madagascar, that vast preserve of archaic animals, there lived another giant Ostrich, whose eggs are recovered from bogs even to-day. They are the largest eggs yet known to us, as may be gathered from the fact that they had a capacity of about three gallons! The Æpyornis is supposed by some to be identical with the "Roc" mentioned by the traveller Marco Paolo. An egg in the British Museum measures two feet seven inches round its long axis, and has a circumference of two feet two inches. Yet, in spite of the enormous size of the egg, this bird was not so big as some of the Moas, since it does not seem to have stood much more than seven feet high. In the disproportionate size of its egg, the Æpyornis recalls the Kiwi, which has been described as "the bird which lays an egg bigger than itself!"

By far the most formidable of the fossil giants was the great South American Phororhacos, standing some eight feet high, with a head bigger than that of a racehorse; and armed with a most ferocious looking, hooked beak, he must have been a very terror by day to the creatures on which he preyed. The wings were ridiculously small, so that his victims must have been run down; they could hardly have been taken by surprise after a successful stalk, for with such a towering height stealth would avail nothing.

The term "giant" is a relative one. The two birds now to be discussed are pigmies compared with the eagle-headed monster just described; they are giants compared with their largest living relations. These two birds are the Dodo and the Solitaire, of the islands of Mauritius and Rodriguez respectively, and they represent the largest of the Pigeon tribe, or, rather, their remains mark the record stature attained by pigeons.

When the old explorers first visited these islands they found these quaint-looking birds in large numbers. It did not take long to discover that they were good to eat, and in a very little while, like the walrus and the carpenter with the oysters, they had " eaten every one ! " At least one living specimen of the Dodo, however, found its way to this country, and was exhibited in London. Later, its remains found a resting-place in Oxford, where they still remain. The story of the Dodo and the Solitaire is hardly apposite to our present discussion, and would take long in telling. Of the two birds, the Solitaire is really the more remarkable ; yet somehow its memory has been overshadowed by that of the Dodo. Both were flightless, though the wings were by no means so degenerate as in the Ostrich tribe.

Photo by] [*W. Green.*

FOSSIL SEA-LILIES.

These beautiful feathery lilies are really animals, nearly related to Starfishes. They were abundant at some periods of the world's history, and their fossil remains are numerous. This fine example was found in a bed of Lias.

Why did these birds, giants and dwarfs alike, become flightless ? The reason is not hard to find. Whenever and wherever the ground furnishes an abundance of food, which birds can obtain the year round without persecution from predatory animals, there is no longer any use of wings to escape the rigours of climate, or enemies. And, accordingly, flight speedily falls into desuetude. As a consequence, the muscles of flight, and the bony framework to which they are attached, gradually decrease in size. The keel of the breastbone is the first to go ; then the wings become visibly smaller, and eventually degenerate into mere stumps ; finally, as in the case of some of the Moas, to disappear altogether, when the evolution of the flightless birds becomes complete. With any material change in the environment, however, such as of climate, or the advent of predatory animals, their doom is sealed, for escape is impossible.

SEA-LILIES

BY BERNARD C. WHITE

THE " inviolate sea " is a realm apart. It is a world teeming with life; a world of infinite variety; but it is nevertheless a hidden world whereof the very conditions of its existence enable it to guard its treasures jealously. It has therefore escaped the notice of mankind in general, who have yet to realize the wonderful secrets of its vast organization. Forms of life, so beautiful, so complex, so outside the very borders of the imagination have been brought to light by the naturalist after ages of patient research that the mind reels at the conception of such a fairyland and of such strange folk.

The sea has its gardens, with the weeds and the flowers, much in the same manner as the earth is made beautiful with a thousand different forms of vegetation. But there is this difference between a sea-garden and a land-garden. Many of the sea's choicest "plants" are not plants, but animals: Sea-fir, Sea-anemone, Sea-cucumber, all fall under this category, so also do Sea-lilies.

These beautiful flower-like forms, known to science as Crinoids, occur in great beds at the bottom of the ocean; more especially those species which inhabit the shallower waters.

We have evidence in fossil remains that they were abundant in the very far-away ages of the world's history. Vast tracts of the ocean's bed must have been covered with the waving masses of their feathery plumes (some of them reaching to a height of seventy feet), for so thick is their débris in certain beds of limestone that the rock is known as Crinoidal Limestone. They were in existence long before the era of the great coal forest, and by a study of these fossils we can distinguish many and various changes in the structure of these species; in fact, the variety is much greater amongst the fossil Crinoids than amongst those known to exist to-day.

Our earliest knowledge of Crinoids was entirely obtained from the study of fossil remains; for it was not until about the middle of the seventeenth century that living Sea-lilies were discovered. It is wonderful to think that so much had been found out about them from the bare testimony of the

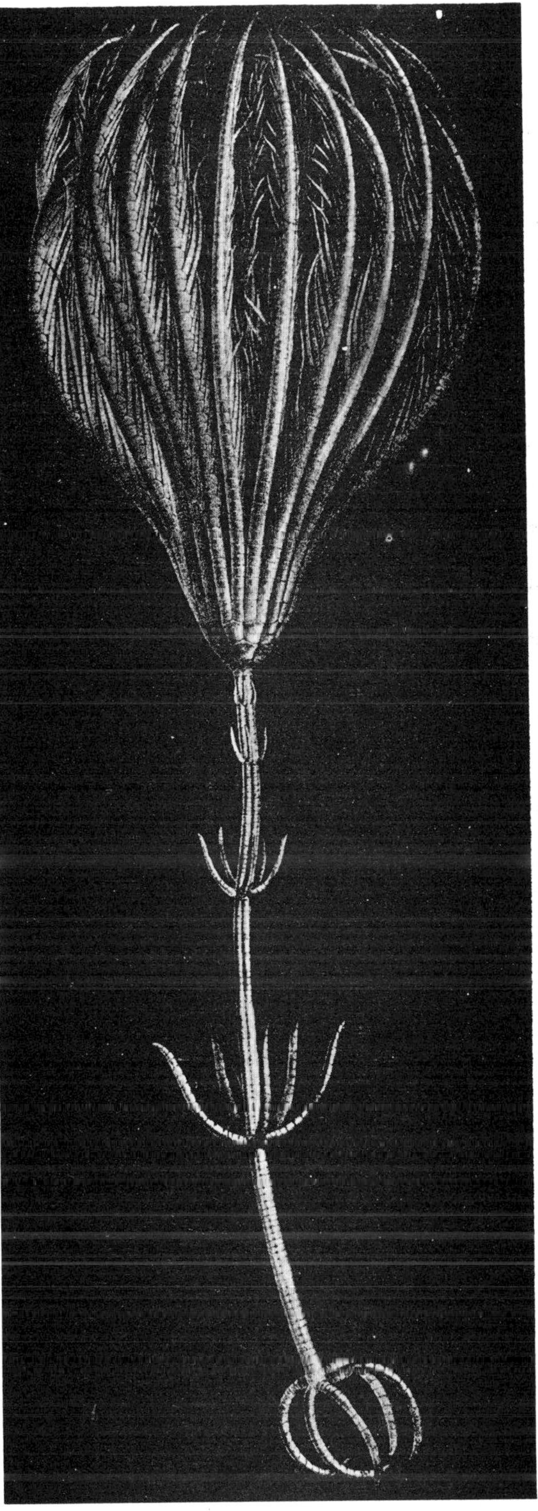

THOMPSON'S SEA-LILY

This Sea-lily is grass-green in colour. Specimens have been dredged up from a depth of over one thousand fathoms.

rocks, for already half a century earlier Edward Lluyd had demonstrated that these remains belonged to animals, and were not the petrified stems and flowers of plants, as they were first supposed to be. He even pointed out their affinity to the Feather-star, a species of Starfish, which, though not "stalked" in its perfect form, is rooted like the Sea-lily, and has the same system of arms supported by an internal skeleton of limestone joints. The discovery of the living specimen set all doubt at rest as to its animal nature, but it was long before its relationship with the Starfishes was acknowledged. The further discovery of a rooted species by G. O. Sars, in 1864, reawakened interest in the Sea-lily and was indirectly the cause of the *Challenger* expedition of 1872, which added very considerably to the present knowledge of this subject.

Lluyd has been proved absolutely right. The Sea-lily is an animal, and it is nearly related to the Starfishes, with which the Feather-star is the connecting link. Although low down in the scale of life, it is a very intricate organism. In common with all the members of its great family, the Crinoid has the power of depositing lime in its soft tissues. The deposit is most abundant in the skin, and, as it were, coats it, so that in time the creature has built up a continuous skeleton. In some cases, as in the Sea-pens, this skeleton is inside; in others, as in the Sea-urchins, it is to all intents and purposes formed on the outside; and so it is in the Sea-lilies. The chief part of the Crinoid skeleton is the cup situated at the head of the long stalk and supporting the feathered arms. It contains the most important organs of the creature's body. In the cover of the cup is the mouth, and radiating from this mouth to the branched arms are grooved channels which convey the food particles to the creature. This cup is composed of two circlets of five plates. In some species two or more of these plates will unite to form a single plate, but their pentameral origin is discernible. And here it should be noted that the whole creature is built up (like the Sea-urchins and Starfishes) on the principle of five. There are always five arms, however many times they may branch, and the internal organs can be roughly grouped into systems of five.

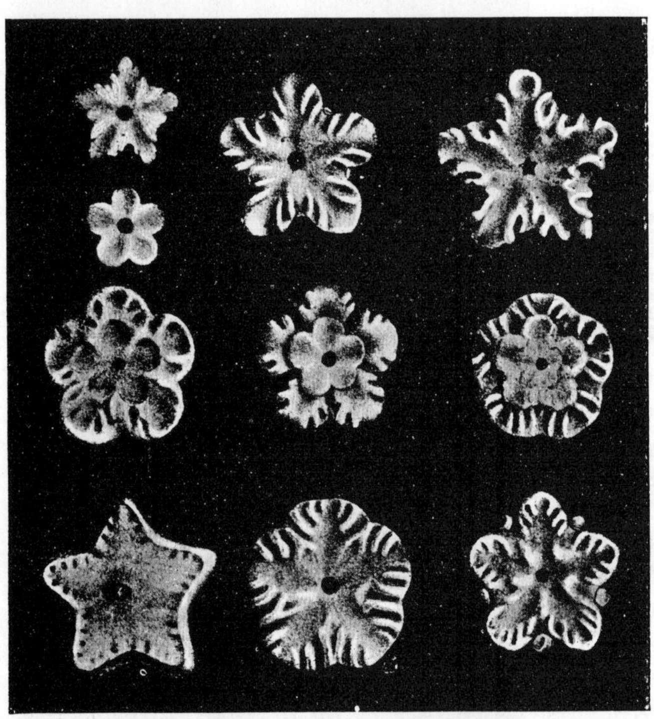

THE JOINTS OF THE SEA-LILY.

These joints are of exquisite workmanship. The sutures or grooves by which one segment fits into the other form varied and elaborate patterns. These segments, here shown "end-on" to give some idea of their construction, are all taken from one species of Sea-lily, that shown on page 643.

The arms of the Sea-lily are very sensitive and highly-organized members. On the inner surface they are grooved for the purpose of carrying the continual stream of water swept down by the feathered arms to the creature's mouth; for the water holds the minute organisms on which the Sea-lily feeds. Beneath this groove lies a system of water-vessels, which create hydraulic pressure and control the movement of the limbs. There is also a nerve-system running the whole length of the arm and a system of blood-vessels. The exterior surface of the arm is formed of the same jointed material as are the stalk and the cup; each segment fits closely into its neighbour, either by means of sutures or grooved markings, on the ball-and-socket principle, or simply by a cohesion brought about

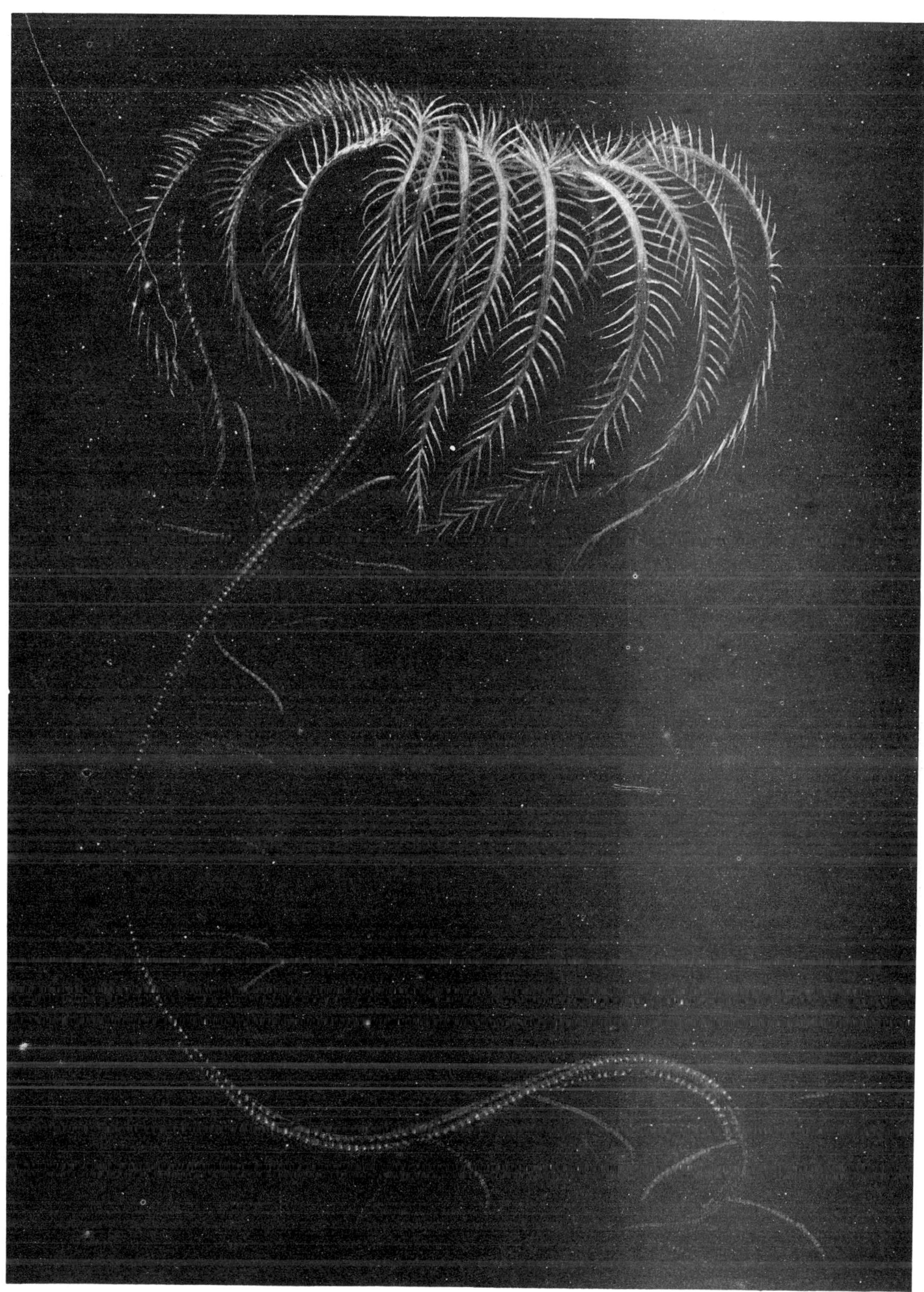

THE BEAUTIFUL SEA-LILY.

There is hardly anything more beautiful or more fragile in ocean-life. It is as brittle as glass, and when irritated has a habit of shedding the feathered arms. These arms are moved one at a time. Some of these Lilies are white, some purple and some yellow. Forests of them occur in the ocean depths, and they are particularly abundant off the coasts of some of the West Indian Islands.

THE HOLOPUS.

This strange form has no stalk and is only affixed to the rock by the enlarged base of the cup. The limestone "armour" in Holopus is much thicker than in any other of the Sea-lilies. The living creature is possibly white, but after death it takes on a greenish hue. It has only been found at great depths in the Caribbean Sea.

by the faultless contours of the joint. This latter principle, it will be remembered, is employed in the shell of the Sea-urchin, where the smooth edge of one division fits the smooth edge of the other, and the two are held together by little more than the marvellous accuracy of construction. From each segment of the arm of a Sea-lily grows a tiny stalk or pinnule, and it is these pinnules branching from every segment of the arms that give the feathered appearance to the Crinoid. In the later-developed Crinoids the pinnules have been arranged to the right and left of the central groove of the arm on alternate segments, for the obvious purpose of sweeping the surrounding waters more effectively, and so of obtaining a greater supply of nourishment. In fact, the whole structure is arranged with this purpose in view ; even the long stalk serves to raise the cup and arms of the lily out of the restricted area of the sea-floor to a larger field of energy. The pinnules are very sensitive organs. In them are stored the immature ova. They are very sensitive to touch. As soon as they are aware of a strange presence they close down over the groove of the arm ; then the arms which stretch out star-like from the cup close up much in the same way as a flower closes its petals. As soon as the danger is apparently passed, the arms once more expand and the ceaseless movement of the pinnules commences again.

But the variety in the forms and structures of Crinoids is immense. All are not stalked after the fashion of Thompson's Sea-lily, which is shown on page 643. The Holopus, for instance (see above), has no stalk at all, and is only fixed to the sea-bed by means of the enlarged basal joint of the cup. Some of the stalks are provided at the junction of every ten segments with five little jointed feelers, or cirri ; others, such as the fossil species known as the Pear Encrinite, have a perfectly straight stem. Others, again, have " roots "—that is to say, a bunch of branching cirri at the end of the stalk by which they support themselves in the mud ; while the majority fix themselves to the rock solely by the secretion of limestone at the basal extremity. Certain, though only a very select company, of the Crinoids were not attached. One of these was the Cup-Lily, which had an enormous feathered corolla. The comparatively weak stem was attached to a large hollow ball, and it is probable that this species floated in the water with the crown hanging downwards; another, the Many-armed Lily, which was in existence at a much later date, was also free-moving (only anchoring itself at intervals), with its stem tapering to a point; while in one or two species the stem itself is reduced to a mere knob. In the fossil Crinoids of the English chalk the stem is entirely absent, and it is supposed that they were free-swimming. As these variations from the stalked lilies become more marked the approximation of the species to the Feather-star becomes more apparent. Now, allowing for the different position of the mouth, the general anatomy of the Feather-star is essentially the same as that of the Starfish, and it was the discovery of the relationship that existed between the Sea-lilies and the Feather-stars that enabled naturalists to "place" these wonderful organisms in the animal kingdom. This discovery followed upon an investigation into the life-history of the Feather-star, when it was found that in its early stages the Feather-star grew on a stem, and that only in its maturity did it break off from that stem and "the Stalked Crinoid" become a true Feather-star.

When the young of the Feather-star leaves the egg it is about one-thirtieth of an inch long, shaped like a tub and divided into five segments by bands of long vibrating "cilia," or infinitely slender hairs, and a tuft of these cilia at the tail. Gradually minute limestone particles are formed within the creature's body and give it the appearance of the illustration at the top of page 648. Longitudinal and horizontal bundles of these limestone particles continue to form, and finally the plate at the end of the stalk appears, and before long the Feather-star is fixed to

[*By Theo. Carreras.*

THE ROSY FEATHER-STAR.

The Feather-star is a connecting link between the Starfishes and Sea-lilies, for in its early stage it is to all intents and purposes a Sea-lily, and is shown in that stage to the right of the picture It is to be found in all temperate seas.

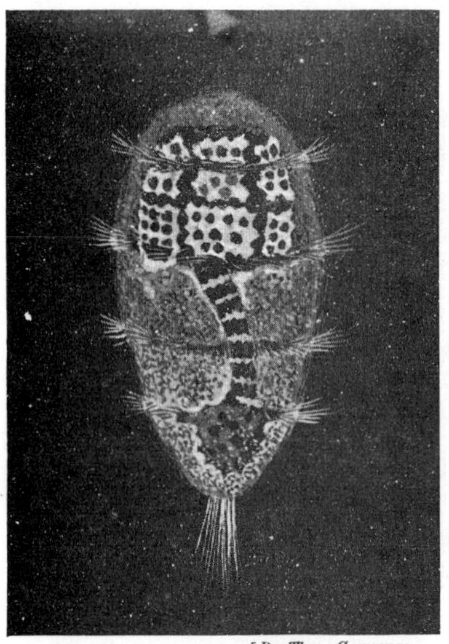

[*By Theo. Carreras.*

The first stage of the Rosy Feather-star after it has left the egg. The chalky skeleton is already beginning to form within.

some firm substance and has attained the second stage in its history (see the second illustration on this page). In the third stage the arms, with the pinnules, begin to develop, the petals of the bud open out, and the cirri are pushed out from the enlarged uppermost stem-joint. This stage is shown on the right-hand of the picture on page 647. As the Feather-star reaches maturity it drops from its stem and curls and uncurls its feathery rays in its progress through the water, while the cirri which appeared by the uppermost stem-joint now serve the Feather-star as hooks by which it can attach itself to the seaweeds.

THE BIRD OF PARADISE FLOWER

THE remarkable plants of which we give photographs are near relations of the Banana. Unfortunately, black-and-white gives but a poor representation of the flowers which owe their glory to their brilliant tints, which have caused people to see in them some resemblance to the beauty of the birds of paradise. There are four or five different species, all natives of South Africa. When the one illustrated was first brought to this country, about a hundred and forty years ago, George the Third was the reigning monarch; and as the plant was newly-discovered and had no name, it was considered a fitting compliment to his queen (Charlotte of Mecklenburg-Strelitz) to name the plant after her. Hence its scientific name from that day to this has been Strelitzia reginæ; much as, at a later date, the new great water-lily was named Victoria regia in compliment to our present King's grandmother. That first-found species is still considered the finest of its race.

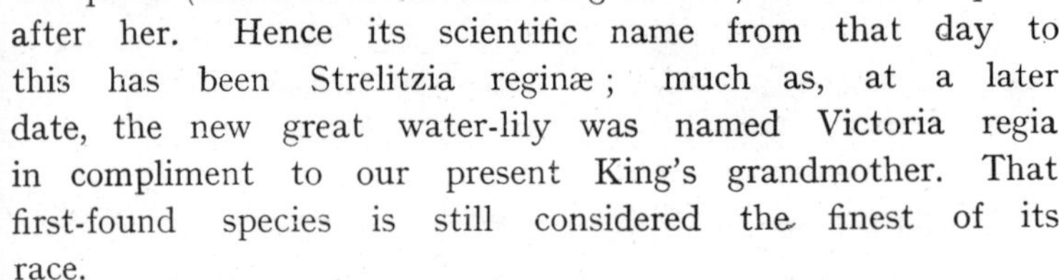

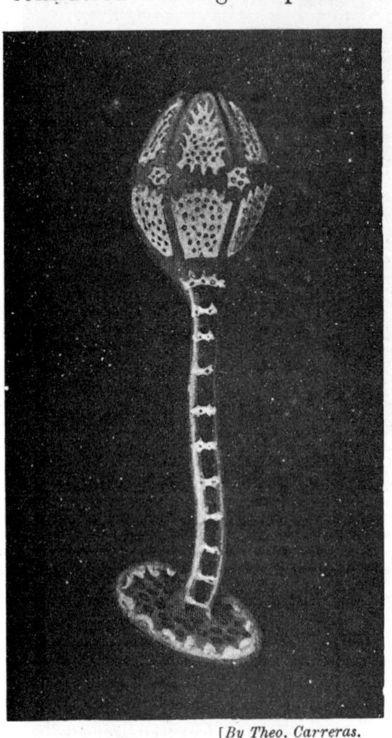

[*By Theo. Carreras.*

The second stage: here the thread-like tail has developed into the long chalky stalk, but the base by which it is affixed to the rock is very large.

It is a noble plant, about five feet in height, the long stalks of the older leaves ensheathing the newer and inner ones. They are not attached to any stem in the ordinary sense, all springing direct from the ground. There is, of course, what botanists call a contracted stem, but it is not apparent in most of the species. The flower-cluster is hidden at first in a boat-shaped, leaf-like wrapper, of similar nature to the hood of our Cuckoo-pint. When the flowers open they present a gorgeous appearance, for their slender sepals and petals are brightly coloured in orange and purple; so that in this condition the Bird of Paradise plant has been fitly described as one of the most splendid of all herbs. The seeds, which are used by the Kaffirs for food, are each surmounted by a tuft of orange-coloured hairs.

The Majestic Bird of Paradise plant makes a more imposing appearance before it is in flower, owing to its superior size, which is twice that of Reginæ; but its flowers, being white, do not compare for magnificence with those we have illustrated.

Photo by permission of] [*The Cape Government Railway.*

THE BIRD OF PARADISE PLANT.

A native of South Africa and allied to the Banana. Its flowers on bursting from their boat-shaped sheath present a beautiful display of colour—orange and purple.

THE GARDEN SPIDER

BY E. J. SPITTA, F.R.M.S.

When it is borne in mind that out of the six hundred kinds of Spiders to be found in Great Britain the Garden Spider and its immediate relations are the only ones that spin the beautiful web with which we are all acquainted, is it to be wondered at that we classify the little animal amongst the marvels and wonders of the universe? This little Arachnid being so well known, it is useless to append an illustration of the entire animal, and our article is therefore entirely illustrated with photographs of the less-known marvels of the creature's organization.

In popular estimation the Spiders are insects, an error largely due to the fact that they are small. In former days, to be small was considered to be of little account, and, therefore, all sorts of small

Photo by] [*H. J. Shepstone.*

BIRD OF PARADISE PLANT.

A nearer view of a single cluster of flowers springing from the sheath that protects the buds.

animals of great variety in structure and organization were bundled together in a sort of lumber room with a door labelled "Insects," and left for modern investigators to sort them out and label them more accurately. If a beetle, a bee, a fly, butterfly, or other fully-developed insect, be examined superficially, it will be seen to be distinctly divided (or *insect*ed) into three chief regions—the head, the thorax or chest, and the abdomen or hind-body. The head bears antennæ or feelers, the mouth-parts and the *compound* and *simple* eyes. To the thorax, or middle region, the six legs (*always* six) and the wings are attached. [Compare an insect with a Spider, and some radical differences will be obvious at a glance.] The Spider is divided into two regions only, the foremost being a consolidation, so to speak, of the head and thorax into one, which bears the mouth-parts, the more numerous but all *simple* eyes and the eight walking limbs. The hind-body of the insect is made up of a series of ring-like segments; but in the Spider it is one and undivided. Another, and more important, point as showing the essential separateness of the two classes represented by a Spider and an insect is development. The insect, after leaving the egg, has to pass through two intermediate stages before it attains its complete form. The almost microscopical Spider, when it hatches out, only differs from its mature mother in point of size.

There are, of course, internal differences to correspond with the external. The respiratory system of an insect consists of a series of tubes, with openings all along each side of the hind-body. These tubes are rendered non-collapsible by a continuous spiral wire of chitin, the horny material of which the hard parts of insects are formed, much in the same way as india-rubber gas-tubing is strengthened by a coil of metal wire. These tubes traverse every part of the insect, including the wings. In the Spiders we find some of these tubes also; but the principal work of respiration is carried out in two (in the Bird Spiders, four) chambers called lung-sacs situated in the fore part of the hind-body, and having their openings in the under surface just behind the legs. An insect, after it has attained adult form, does not cast its skin; but a Spider does.

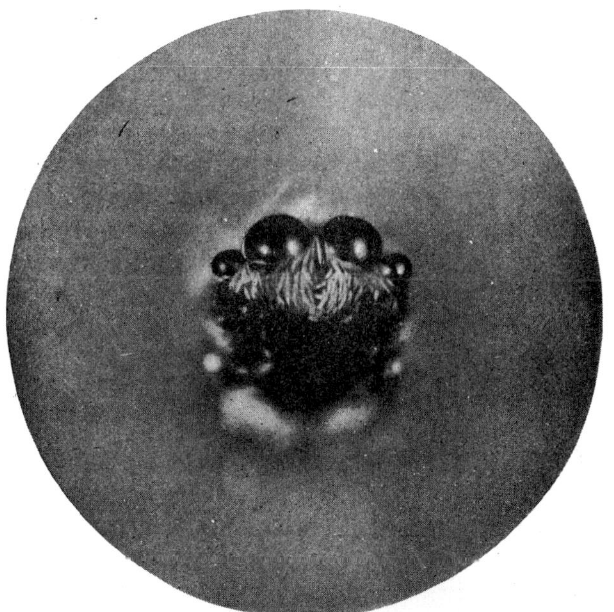

Photo by] [*E. J. Spitta, F.R.M.S.*

THE SPIDER'S EYES.

These are single, and in this way differ from those of insects, which are all composite.

The Spider whose huge snare frequently confronts us in the garden is familiarly known as the Garden Spider, sometimes as the Cross Spider, from the fact that it bears upon its back certain lines and spots that at a little distance have the appearance of a cross. The adult Garden Spider has a very hump-backed look, owing to the great development of the hind-body. Having constructed its snare on geometric lines—many "spokes" radiating from the centre to the boundary lines, and strengthened by a spiral of enormous length that runs round and round, knitting them all together—the Spider sits in the centre waiting for any small winged creature that may blunder into it. Her senses are so acute that she can tell the moment this happens, and in what part of the snare she may look for her victim.

The head of the Spider ends in two very prominent cone-shaped bodies, which perform the function of pincers and are called the "falces," the cone part of each falx being called the base, whilst the hooked extremity is called the fang. Each fang is hard, sharp and curved, being movable at its attachment to the base, which

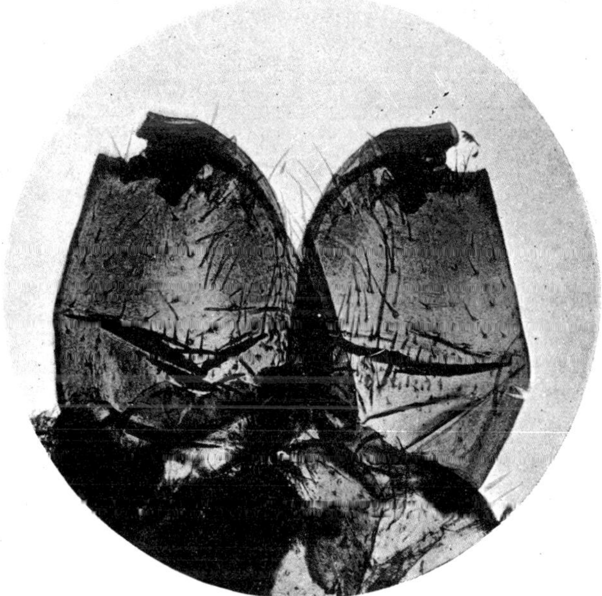

Photo by] [*E. J. Spitta, F.R.M.S.*

THE SPIDER'S NIPPERS.

The "falces," or nippers, are here shown with the fangs closed down, where they lie between the rows of teeth. Notice the curious jointing.

is provided with a groove, along the edges of which are set a row of terrible teeth (see the illustrations on pages 651 and 652). The falces are not used in the act of taking food, but are for the purpose of catching prey, holding and killing it. The last operation is brought about by means of a rapidly-acting poison injected into the victim through a little hole at the top of the fang. It must therefore be clearly understood that a fly caught by a spider is almost instantaneously killed (which fact grasped, lessens the feeling of animosity in the onlooker), this being proved by the fact that if it be immediately released, as the writer has often done, and the web around it entirely removed, it is found to be quite dead. Not so, however, in the case of spider victims, for the poison does not seem to affect them in a like manner; upon releasing one in a similar way to the fly, it has been found to recover itself rapidly and hasten away. Before leaving the subject of the falces, attention should be directed to the powerful muscles which actuate the fangs; in the accompanying illustration they are plainly shown as radiating lines passing downwards from the fang into the base. These pull the fangs downward upon the bases, so as to lodge them neatly between the rows of teeth, thus effectively preventing the escape of the victim. The fangs shut completely down are shown on page 651.

Photo by] [*E. J. Spitta, F.R.M.S.*

THE FANG OF THE SPIDER.

An enlarged photograph of the falx, showing the fang erect. Notice the powerful muscles which drag the fang between the teeth, and so secure the Spider's victim.

The eyes of a Spider are set in the front part of the head; they are not of the compound variety found in insects, but are of the single type, and called "ocelli." They vary in number and position, and are largely concerned in the classification of the different orders; in the photograph on page 651 they are shown specially dissected out.

The palpi—with little hooks on their ends—look like short legs. They are sensory organs of some kind, but their exact nature is not known, much difference of opinion existing between different naturalists. They are seen in the full-page photograph.

The mouth of a Spider is situated at the front of the head, in the neighbourhood of the base of the falces. It contains a kind of tongue and has two very pronounced lips, an upper and a lower, provided with teeth. The mouth apparatus is used for sucking purposes and not for devouring, as a Spider apparently never eats its prey; simply sucks it dry.

The spinnerets at the end of the body vary in number in different species; they are seen, in the upper photograph on page 654, to be conical projections, each being surmounted at the tip by

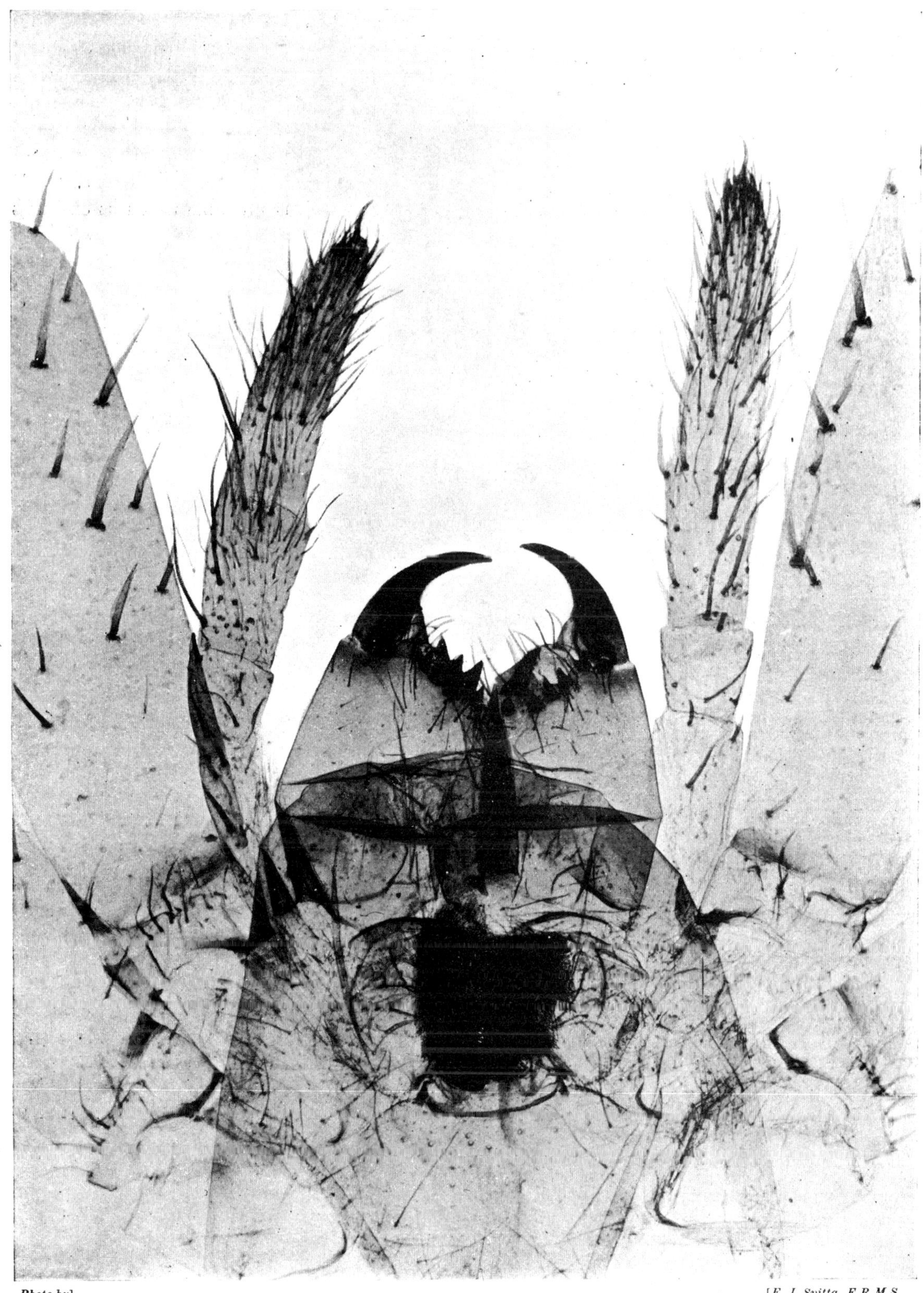

Photo by] [*E. J. Spitta, F.R.M.S.*

THE MOUTH OF THE SPIDER.

The mouth of the Spider is situated near the base of the falces. It contains a kind of tongue and two very pronounced lips, an upper and a lower, provided with teeth. The mouth is used for sucking purposes, not for devouring, as a Spider never apparently eats its victim; only sucks it dry.

Photo by] [*E. J. Spitta, F.R.M.S.*

A SPINNERET OF THE SPIDER.

These spinnerets vary in number in different species, and are conical projections with numerous hair-like appendages at the tips.

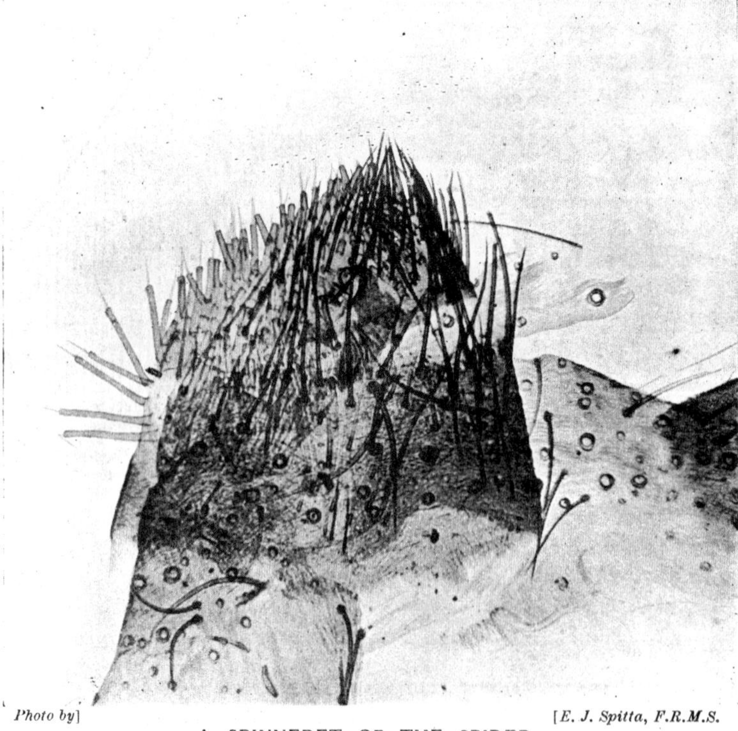

Photo by] [*E. J. Spitta, F.R.M.S.*

A SPINNERET OF THE SPIDER.

One of these projections more highly magnified.

numerous hair-like appendages. In the lower photograph one of these is seen more highly magnified, and it will be noticed it seems enclosed with a kind of black ring. It is probable that this exercises a sort of pinching action to stop the flow of the fluid web, for it is well known that Spiders can use all their spinnerets simultaneously or only part of one, as they wish.

On the next page the terminal tubes are well shown. It will be better seen in the photograph which is more magnified still, that each individual tube ends in a little delivery tube—much like the narrow brass ending of the fireman's hose—which is to give a greater projective power. Before passing on, note should be made of the very beautiful skin of the body. Magnified considerably in an accompanying illustration, it will be seen to be an object of great beauty in design.

The legs and feet next claim attention, the joints being of especial interest. In the bottom left-hand corner of the full page it is evident that a kind of protecting membrane surrounds the joint, the special object of which it is difficult to understand, but it is usually thought to be of a protective nature. Curiously, however, the envelope does not seem to be present in all species, or even in all the joints of the same animal. Owing to the legs of the Spider being required to have a very free movement, joints of a ball-and-socket type are in evidence, and one of these articulations is shown in the illustration on page 657. Apparently, just beneath a very long hair (which, of course, is really outside the joint) a little central band can be noticed, uniting the two parts of the joint together, reminding the anatomical student of the round ligament in our own hip-joint. This band is to assist in holding the parts in juxtaposition, although permitting free movement in all directions.

The foot, seen greatly magnified on the same page, is of wonderful structure, at a glance a comb can be plainly seen with several teeth for striking into the web, but on closer inspection, a second —out of focus from being so far behind—is just visible. A third and very bent hook is not always present, but one is visible in this specimen, being situated below the two great ones. It is said not to be used for the web at all, but to be of service in crawling up rocks or slippery surfaces, being capable of ejecting an adhesive fluid to render movement more easy by stopping all slipping.

If a claw is lost, it will be found to have grown again later on, but, curiously, does not seem to occupy its final place until the animal performs its next moult, every Spider changing its skin about five times from birth to maturity. Whilst waiting for this to occur, a new claw may sometimes be seen quite properly formed, resting within

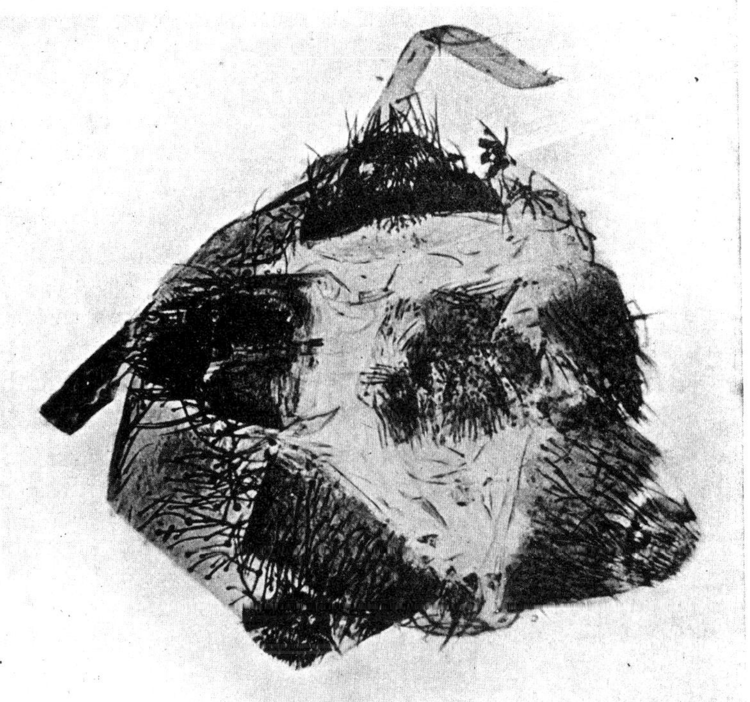

Photo by] [*E. J. Spitta, F.R.M.S.*

A SPINNERET OF THE SPIDER.

A direct view of the conical projection, showing more clearly this dark ring controlling the web.

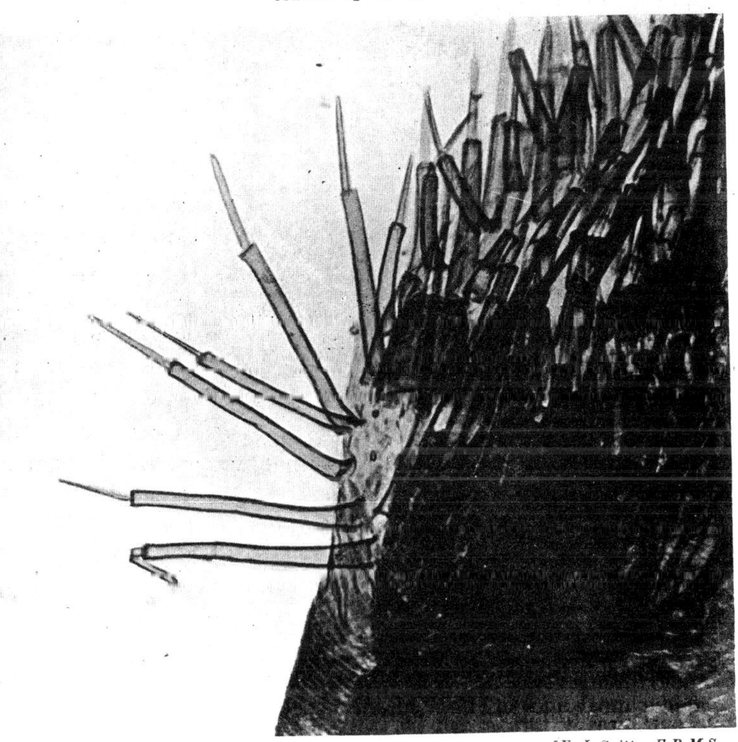

Photo by] [*E. J. Spitta, F.R.M.S.*

A SPINNERET OF THE SPIDER.

The delivery tubes, by which means the fluid thread is projected.

the leg, evidently in readiness to be pushed forward at the next moult; an illustration of this is given also on the same page.

The Garden Spider, like most of her relations near and remote, is an unsocial creature. Alone, she sits in the centre or at the side of her snare, where there is elbow-room for scores, but resents intrusion, even of her own kind. She is solely intent upon the capture of something that may be eaten. Even when one of the other sex comes courting she only tolerates his presence for a time; and he goes courting nowhere else. The allotted end of a male Spider is to make a meal for his wife. "There is no waste in nature!"

When the female Spider deposits her eggs, she spins a soft, fluffy cocoon, usually in the angle of the garden fencing; and there you may find her guarding it with her body spread over it. Some Spiders attach their egg-cocoon to their body and carry it about until the eggs hatch. The young of the Garden Spider, when newly hatched, do show some tendency to sociability, for they keep together—though there are six or eight hundred of them in one batch—for a time. They spin some exceedingly fine lines that are almost invisible, and cluster together in a mass in the centre of this composite web. This is apparently a defence against ichneumon-flies, which seek to lay eggs in them, but are deterred by invisible lines that may throw their wings out of action. If any of the lines by which this nursery is suspended be touched, the minute spiders will scuttle off in all directions; what seemed like a solid ball of them appears to evaporate. When all is quiet a few minutes—or seconds—later, they gradually return and pack themselves into a solid ball again. They appear to leave the egg with a sufficient reserve of food material stored up in their bodies to enable them to grow considerably before it is necessary to begin hunting for their food. When that time arrives the little fellows separate among the herbage, and henceforth each one looks after himself, and himself alone. Even looking after himself is not a sinecure; for so many are the dangers that lie in the way of spider development—in the form of birds, lizards and ichneumon-flies, for example—that very few of them attain to the full stature of their parents.

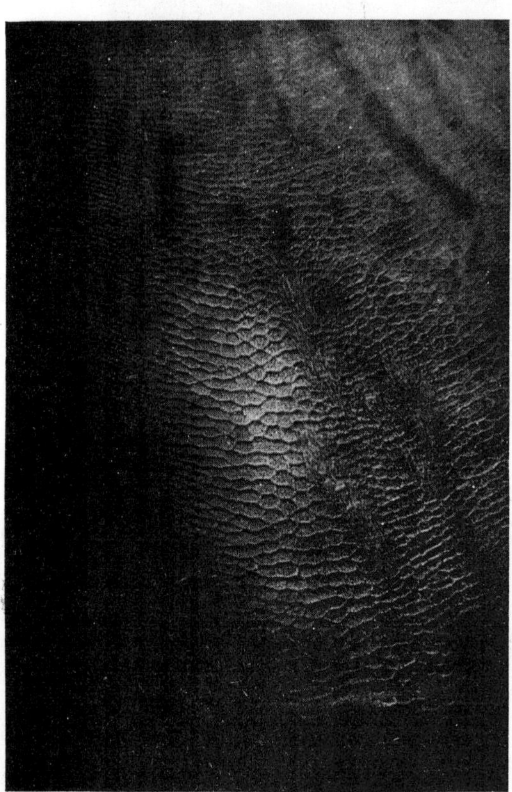

Photo by] [*E. J. Spitta, F.R.M.S.*

THE SKIN OF THE SPIDER.

A photograph which ably demonstrates how skilful study may discover hidden beauties in apparent ugliness.

SHIFTS FOR A LIVING

BY W. P. PYCRAFT. F.Z.S.

HOWEVER much of poetry we may weave into our conceptions of life, we are sooner or later brought face to face with the fact that all living creatures, including Man himself, must spend a more or less considerable amount of time in the purely animal struggle to find enough to eat. As a matter of fact, the importance of this business of eating can hardly be exaggerated, for the whole bodily structure affords eloquent testimony of the multifarious ways in which changes, now in

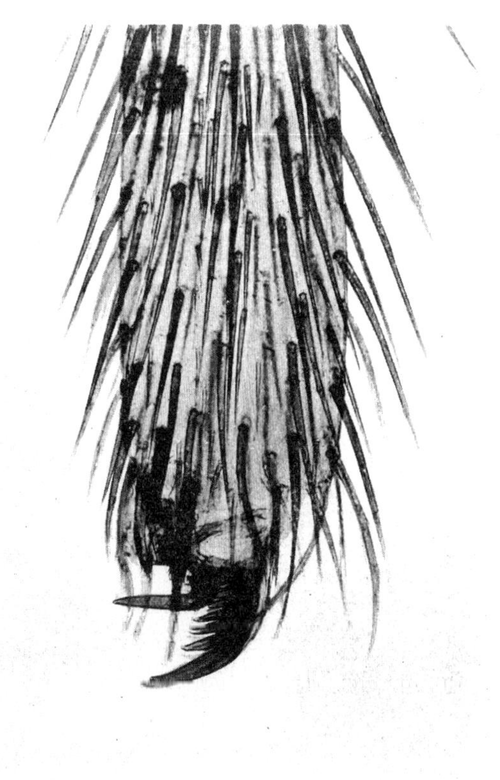

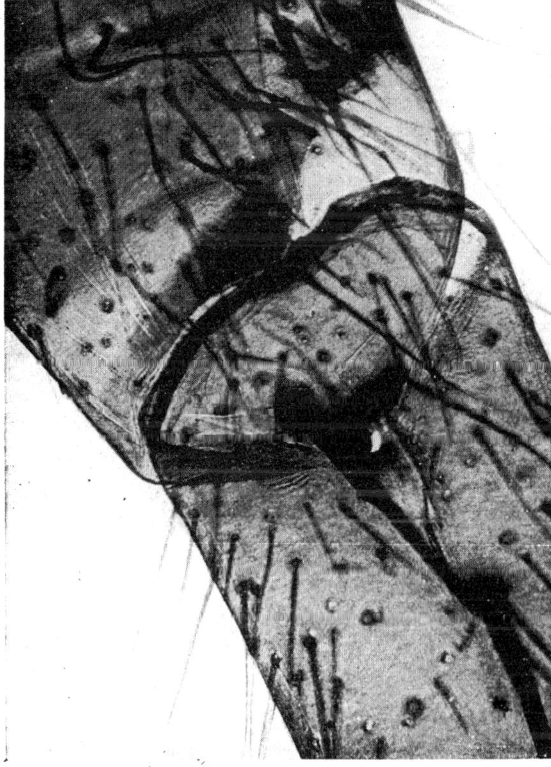

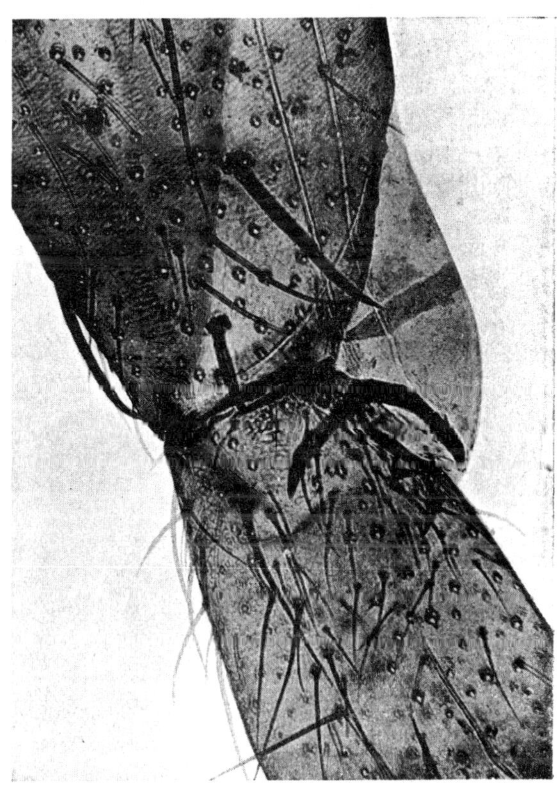

Photos by] [*E. J. Spitta, F.R.M.S.*

THE LEGS AND FEET OF THE SPIDER.

The upper photographs show the feet of the Spider. Notice in the left-hand picture one of the claws bent at right angles. This is supposed to eject an adhesive fluid to enable the spider to crawl up slippery places. The second photograph shows the growth of a new claw. In the lower pictures the leg joints are shown, with the peculiar protecting membrane surrounding them.

Photo by] [*E. Step, F.L.S.*

THE RED GURNARD.

A species allied to the Sapphirine Gurnard in the coloured plate. The Gurnards are "bottom-fish," searching for their prey in the ocean's bed rather than seeking it in the mid waters, for by their shape they are prevented from swimming quickly.

this part of the body, and now in that, have come into being whereby nearly-allied animals have been enabled to avoid competition by seizing hold of food-sources otherwise inaccessible. On the other hand, indeed, the most striking features of some animals are those which have been evolved to enable them to avoid being eaten!

The search for examples illustrating these facts need neither be long nor laborious; any animal we chose to select at random would suffice. Let us take the cases afforded by the fishes in the coloured plate. To begin with the uppermost figure, the Sapphirine Gurnard of our own seas. Observations on the habits of this fish in its native element are naturally well-nigh impossible, but could we trace its daily life we should probably find that its vivid hues were, in part, at any rate, due to the "avoidance of being eaten" to which reference has just been made. But we are concerned more particularly with the more direct evidence as to bodily modifications. Considerations of space forbid a minute examination of this strange creature; but we shall find enough for our purpose in the great breast-fins alone. Immediately in front of these will be noticed long, lean fingers. Now these have been formed by reducing the web of the fin so as to set free the first three or four of the fin-supports, which now serve partly as legs for crawling along the bottom of the sea, and partly as sensitive fingers, or organs of touch, wherewith to find the daily bread. Very well. A cousin of the Gurnard, one of the flying-fishes known as the Flying Gurnard, has retained all its fin-supports in their original position, and has further enormously increased the size of the whole fin so as to form a very efficient pair of wings, enabling its possessor to take considerable flights in mid-air—to avoid being eaten. The Golden-finned Parrot-fish—one of the Great Australian Barrier Reef fishes—is remarkable, not only on account of its gorgeous coloration, but also because of the peculiar armature of its jaws, which, like those of its numerous relations, have become modified to form a parrot-like beak whereby pieces of living coral may be bitten off and eaten. The vivid carmine, sapphire-spotted, Richardson's Plectropoma is, in spite of his beautiful appearance, one of the ravening wolves of the sea, as is plain from the great heavily-armed mouth ; his whole bodily framework speaks of rapine. The strange-looking Holoxenus is another carnivorous denizen of the sea, and one cannot help wondering what may be the meaning of its extraordinary coloration, for apart from its vivid hues the body is scaleless and presents the appearance of being covered with velvet pile. Our " Father Lasher " is a near relation of this strangely-coloured fish. Finally, we come to the curious little Jumping-fish, one of the Blennies, of which we have representatives in our own seas. Not much seems to be known of its life-history save that it haunts rock pools, and possesses prodigious powers of leaping, only exceeded by those of its cousin, the Bommi Fish, of which a vivid account has already been given in these pages.

RATTLESNAKES

It is the rattle that makes the snake, or, at least, it is to the peculiar coil of dried, horny plates that the snake owes its universal fame. America is the home of the Rattlesnake, and there are about sixteen species known to naturalists, though this number is sometimes increased by the inclusion of a nearly allied tribe, the members of which can be distinguished from the true Rattler by possessing four horny plates on the head and degenerate rattles.

Though " the rattle " is the most extraordinary feature of the snake, there are one or two points worth a moment's notice. One is the system of fangs. Through these fangs run slender canals, connecting with ducts that lead to the poison glands. They are " perfect hypodermic syringes," and when not in use can be folded back along the upper jaw and protected from injury by a fleshy membrane. There are two apparent fangs with which the snake wounds his enemy ; but at the back

Photo by] [*Doubleday, Page & Co.*

PRICE'S RATTLESNAKE.

The Rattlesnake is warming itself in the sun, for it loves the warmth. Notice how the markings of the creature blend with its surroundings. The rattle is particularly well displayed.

of each fang are several in reserve, ranging from one full-grown to the mere germ of one. On the fracture of a fang, the first one on the reserve falls into a socket by the side of the one holding the broken fang. It soon becomes securely fixed to the maxillary bone and ready for use, while the base of the broken fang falls out, and the now vacant socket is ready for a new instrument.

Another curious point about the snake is the deep pit between the eye and the nostril. What the purpose of this formation may be has set naturalists wondering, and they are still wondering.

The colouring of Rattlers is a matter not without its interest, for their particular scheme of coloration seems to combine the two functions of warning and of protective colour. The Diamond Rattler, for instance, has elaborate markings of a diamond shape in dark-brown, red-brown and yellow, which extend right down the back, and these hues are particularly warning hues. Others of the Rattlesnakes are coloured so similarly to their surroundings that they are difficult to detect, such as the one instanced by Mr. Backus : " Peering with eager eyes into the deep shadow of a thicket . . . I suddenly heard the shrill locust-like rattle of a snake. It sent a sickening chill through my heart, causing it to stop beating momentarily and then go violently pounding against my side. The rattle was so extremely loud that I knew I was right on him—and a big one. But where ? I could not see him. I looked down to the ledge on which I was creeping—no snake ; under a projecting rock at my side—still no snake to be seen, but this rattle always dinning in my ears and coming apparently from every direction. I lifted my head a little. There, about a foot away, on a flat rock at the height of my shoulders, was a big black Rattlesnake, coiled up amongst some dead and flattened pine-branches of almost the same colour and marking as himself."

Photo by] [*W. S. Berridge, F.Z.S.*

THE SNAKE'S RATTLE.

This is composed of a series of hard horn cups with heavy rims fitting into each other closely and secured in place by a horny "button." When the owner is angry he causes these rims to vibrate and produce the sound for which he is famous.

The rattle is made up of depressed rings, which overlap a smaller cup with something of a stalk terminating in a button. The button attaches the one cup in the hollow of the neighbouring cup. So cup fits into cup, and the series of rims form the rattle. A new rim is formed every time the Rattlesnake casts its slough ; but when the older joints become worn out and feeble, they drop off and new ones take their place. The number of rings, therefore, does *not* indicate the age of the snake—a fable that is too readily believed.

As to the use of the rattle there have been many theories. Some say it is used as a lure for insectivorous birds ; but unfortunately for this theory, the snake, when out hunting, is peculiarly quiet ; another that it is used to terrify its prey ; a third, that it is used as a musical instrument during courtship ; but this is hardly likely, for the rattle has been proved to be more vigorously used when the owner is in a state of anger. The most probable solution of the problem is, therefore,

[*By C. E. Swan.*

GIANT GROUND SLOTHS.

The Ant-eaters and Sloths of South America are apparently entirely dissimilar; they are nevertheless nearly related, for they are descended from a common ancestor, *The* Giant Sloth, who had the head of a Sloth, the body and limbs of an Ant-eater and a tail which was a compromise between Sloth and Ant-eater.

that the rattle is used, not as a danger-signal, but as a means of protection. The Rattlesnake is not the most venomous of snakes, and is supplied with only a limited quantity of poison. He therefore only uses it under great provocation and upon a persistent enemy. He has discovered that his poisoning propensity has made him dangerous and a creature to be feared. He therefore works on that fear by sounding the alarm. The more timid of his foes will immediately retreat and the Rattler be left undisturbed, without serious interruption of his sluggish habits and without loss to his cherished store of poison.

GIANT GROUND SLOTHS

BY R. I. POCOCK, F.R.S.

South America possesses many strange animals, but perhaps none more strange and grotesque than the Sloths which spend their lives suspended upside-down to the branches of the forest trees, and the Great Ant-eaters, which prowl beneath in search of the colonies of White Ants upon which they feed. Nor would it be easy to find amongst related animals a greater contrast than these two present. The Ant-eater has a snout of preposterous length, with a huge bushy tail to match, while the Sloth has a short and rounded head, and can boast at most a mere vestige of a caudal appendage. Yet that these animals are related, and tolerably intimately related too, is proved by the former existence in South America of immense beasts, whose structure may be briefly epitomized by saying that they had the heads of Sloths, the bodies and limbs of Ant-eaters, and tails which were a compromise between those of the other two in length. These are the Giant Ground Sloths, of which the best-known forms are the Megatherium, as large as a rhinoceros, and the Mylodon of about half that bulk.

Photo by] [*H. J. Shepstone.*

SKELETONS OF THE GIANT SLOTH.

The largest kind of Sloth was about the size of a Rhinoceros, and they are the skeletons of this species which are shown here. The smaller kind were about half the size of their giant kindred.

Judging from the structure of the mouth and teeth, and comparing them with those of the Sloths, which feed entirely upon leaves, there is no doubt that these animals were vegetarians likewise; but it is equally certain that they did not climb trees in search of their food, for few trees grow branches strong enough to support such a weight. They must, therefore, have lived by eating grass, browsing upon low herbage, reaching the lower branches of taller trees, bending down saplings, or by bodily uprooting the larger trees whose foliage was out of their reach. For this latter task their powerful fore-limbs, armed with strong claws, were well fitted, while their massive hind-quarters and short, thick tail were admirably adapted for supporting, tripod-fashion, the weight of the body when upreared to reach the branches overhead. It is likely enough, too, that the mouth was provided with a long, prehensile tongue like that of a giraffe, and that the upper lip and nose, at least in Megatherium, were prolonged into a proboscis resembling that of a tapir.

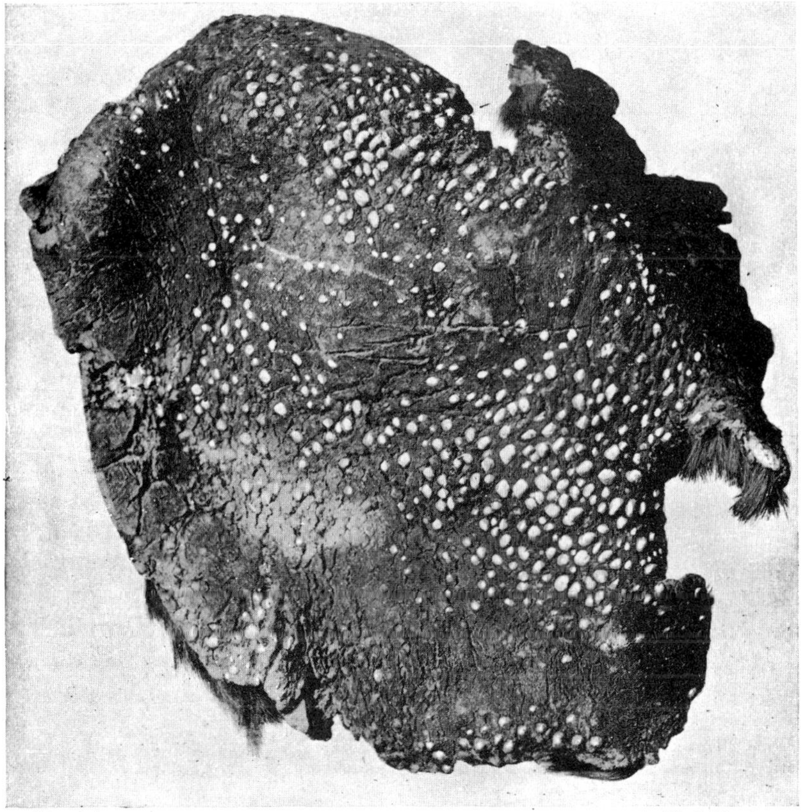

Photo by] [*H. J. Shepstone.*

THE SKIN OF THE GIANT SLOTH.

Showing the inner side, with the little bosses, or "nodules," of bone, close-set in the hide, which thickened and strengthened it till it was as impervious yet as flexible as chain-mail.

There were a great many different kinds of Ground Sloths in South America besides Megatherium and Mylodon, and both of these spread northwards into North America. Most of them, however, were restricted in distribution to South America, where they became extinct long before mankind made his way from his cradle in the Old World into the New. But not all of them. Some managed to hold their own down to comparatively recent times, and survived, beyond doubt, until Argentina and Patagonia were occupied by men of the same race as those that were found in those countries by the first European settlers that landed there. It is, indeed, quite possible that some of these settlers might have seen the Ground Sloths alive had they penetrated sufficiently far into the interior.

This possibility rests upon the comparatively recent discovery of the remains of species related to Mylodon in a cave on one of the inlets of the sea in Patagonia. Not only were bones here found, but such perishable portions of the animals as pieces of skin, stained with streaks of blood, covered with adherent hair, and, generally speaking, in such a remarkably fresh state of preservation, that, making all due allowance for the protection against damp and the destructive effects of weather provided by the cavern, it is impossible to believe they can have lain there for many hundreds of years. Mixed up with these remains, too, were bone-awls fashioned by man ; and many of the bones of the Ground Sloths had muscular tendons attached to them, and appeared to have been split by the hand of man for the extraction of their marrow. Nor was this all. There was a quantity of dried cut grass on the sandy floor of the cave, and pieces of dung, composed apparently of digested remains of this hay and believed to be the droppings of this Sloth, were also discovered. By

THE SHELL OF A TORTOISE.

The Tortoise can shut himself up in his fortress against all enemies. In this case the lower part of his shell is hinged so that he can draw it up and shut himself entirely in.

putting these facts together, it has been supposed —and there seems nothing extravagant in the supposition—that the Indians who made the tools of bone and chopped the hay, kept and fed these Ground Sloths to slaughter them, as required, for their own food!

Another interesting peculiarity about the pieces of skin above referred to is that they were thickened and strengthened by a quantity of close-set bones resembling small pebbles, showing that this Mylodon-like Ground Sloth was invested in a tough, bony coat almost as resisting and flexible as chain-armour. The zoological interest of this fact lies in the kinship it suggests between these creatures and the South American armadillos, which likewise have bony plates in the skin, and are known to be related to the Sloths and Ant-eaters.

Thus the study of this group of animals, known to zoologists as the Edentata, shows how the small and insignificant species, which lurk in the tree-tops or skulk on the ground, have outlived the huge and unwieldy forms in the contest for existence.

HOW THE TORTOISE GOT HIS SHELL

BY W. P. PYCRAFT, F.Z.S.

"FAMILIARITY," it has been well said, "breeds contempt," and this was certainly never more true than of the Tortoise. If this were as rare an animal, say, as an Okapi, we should all of us, long ago, have discovered that it was one of the most remarkable of living animals. As it is, some may wonder why it is accorded a place among the "Marvels of the Universe." Yet, of living creatures, none more thoroughly deserves it. In the first place, it stands alone among the vertebrates—or backboned animals—in having its skeleton outside its body, just like "a mere Beetle." How has this come about? In all other vertebrates, as in man himself, for instance, the backbone is embedded in a mass of muscle—which we in common speech call "flesh"; and the back is lithe, it

THE SHELL OF A TORTOISE.

In this species it is the upper part, so that he can draw it down and make himself secure.

THE SHELL OF A TORTOISE.

The shell of a Tortoise is here shown with (on the right-hand side) and without (on the left-hand side) its skin-covering. It should be noticed that the joins in the horny skin do not correspond with the joins in the bony plate, and this points to an entirely separate growth of these two parts of the Tortoise's shell. The grooves made by the joins in the skin show up as white grooves on the shell-bone.

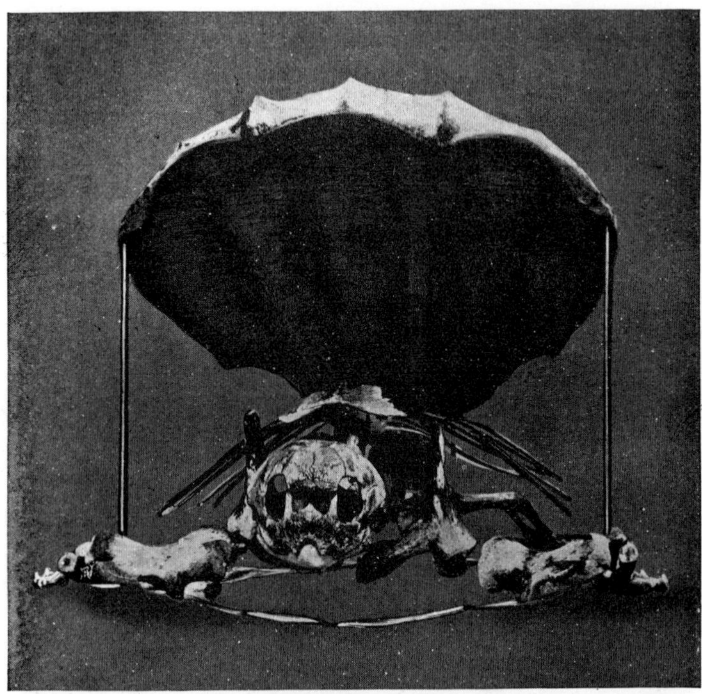

THE SHELL OF A TORTOISE.

Here the skeleton on a Leathery Turtle is shown with the horny covering (which serves it for a shell) lifted up to show how entirely separate it is from the ribs.

can be bent this way and that. In like manner the ribs are, or should be, " well covered." So it was in times past with the forbears of the Tortoise. But in them the skin was provided with an armature of bony knobs, to serve, perchance, as a coat of mail to withstand the blows of enemies. In course of time this armature grew more seeming perfect. The separate " bosses " of bone became welded to form a shield : hard and inflexible : a very efficient protection, no doubt, yet one for which a heavy penalty was to be exacted from the Tortoise yet unborn ! For as the shield increased in perfection so it limited the movements of the back ; and so, in proportion, the muscles of the back degenerated, growing less and less in volume in each generation. As the muscles shrank, so the shield descended, till at last it rested on the spines which project from the backbone, leaving but a small space between the shield and the ribs. Then the spines became reduced in length from lack of use as supports for muscle, and the shield sank lower still, till it rested on the ribs themselves. Then, last stage of all, the ribs and the shield merged, the one into the other, until to-day the ribs themselves have vanished, leaving only their heads and their lower ends, which may readily be made out in a dried skeleton of any Tortoise or Turtle. While of the backbone little more than a tube remains for the passage of the spinal cord.

Now is this capable of proof ? I hear someone ask. Of reasonable proof—yes. And the evidence may be obtained from certain marine Tortoises commonly known as Turtles. In the species known as the Leathery Turtle, or " Huth," the shield is made up of innumerable little discs of bone closely interlocked by means of toothed edges to form a kind of mosaic, and over this is stretched, as it were, a leathery skin. When this shield is raised there will be found underneath a thin layer of muscle or " flesh," covering the backbone and ribs ; and when the skeleton is examined these ribs, and the backbone itself, will be found to have preserved their integrity. The accompanying photograph should make this much clear. Now when we turn to the examination of any other member of the Tortoise-tribe—be it Tortoise or Turtle—we shall find a very different state of things. The shield can no longer be raised without exposing the whole of the internal organs of the body ; and when we come to inquire as to the reason of this, we find that the shield and the ribs have come to fuse into one common mass of bone, the only means of identifying the ribs being, as we have said, by the traces of their lower ends and of the " heads," whereby in times past they were jointed on to the backbone. That this is so is no mere guess-work, for every Tortoise, in the course of its growth from the egg to the adult, passes through the ancestral stages. So that if we select the right period of growth for our examination we shall find a stage answering to that seen in the adult Leathery Turtle ; that is to say, a stage displaying complete ribs distinct from the outer shell.

But this is not all. In ourselves, for example, the scapula, or blade-bone, lies *outside* the ribs, and just under the skin. But in the Tortoise and Turtle the shoulder-blade, and certain other bones attached thereto, and the hip-bone, lie *inside* the body and *underneath*, instead of outside the ribs! This strange state of things takes place during the growth of each Tortoise. At first they lie outside the body, but as the shell grows it extends forwards and backwards, further and further, at first overhanging and finally enclosing both blade-bones and hip-girdle.

So far, we have described only the upper part of the shell. The lower portion is formed in part by the breast-bone, and in part by bony plates belonging to the skin. These, like those of the back, gradually increased in size, causing a gradual wasting away of the muscles of the belly till nothing was left but bone, as in the case of the great back shield. Thus is formed the box-like chamber which makes the Tortoise's dwelling-house. The two ends are open, but the apertures are very small, so that when the head and legs are drawn in they completely block up the entrances. And to make the barricade the more effectual the skin of the legs is studded with bony nodules, forming a very effectual armour. Some Tortoises, however, seem to have found an even more excellent way of securing themselves from unwelcome visitors, and this—either by forming a hinge across the breast-plate, allowing the front end to be drawn up, or by a hinge across the hinder-part of the upper shell, allowing the tail end to be drawn down like a portcullis.

Finally, a word as to the horny plates which cover the shell of the Tortoise. These are arranged, it will be noticed, in a perfectly symmetrical manner—a single row down the back, and in pairs along each side; while around the edge of the shell runs a series of smaller plates. These horny plates have been formed of the outer layer, or surface, of the once flexible skin. But the bony plates of the shell are also symmetrically arranged, and, curiously enough, the two systems of symmetry bear no relation to one another. To get a proper grip of what this means, take the case of the Crocodiles and Alligators. These have a very serviceable armature of bony plates along the back, but arranged in transverse rows, permitting the body to bend freely; and each plate is horn-covered. That is to say, the bony plates and their horny coverings correspond in size and shape.

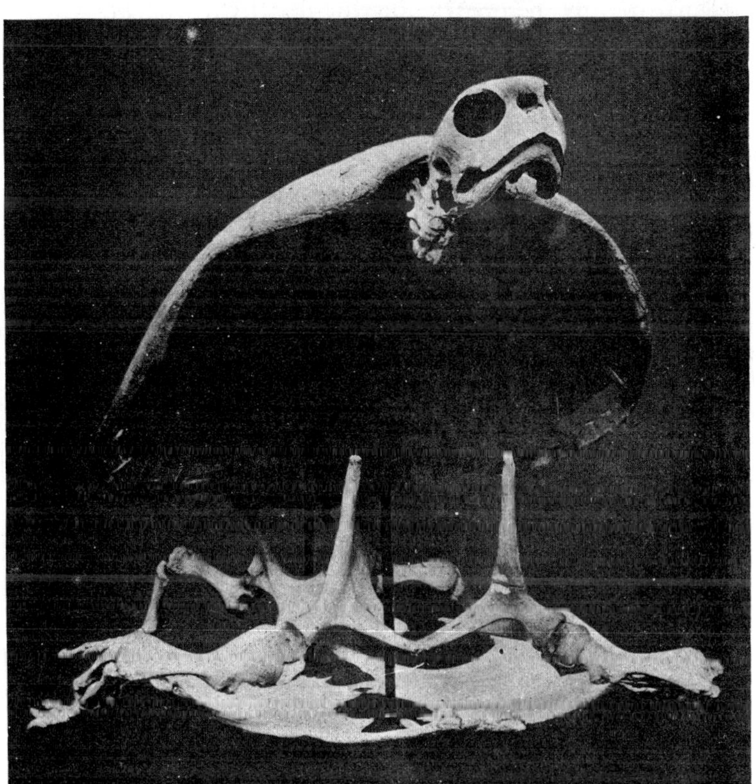

THE SHELL OF A TORTOISE.

This is the skeleton on the Green Turtle. Notice that here the weight of the hide has caused it to settle upon the ribs until the two distinct parts of shell and ribs are merged into one. This supports the idea that the shell is really formed of "nodules" of bone merged together to form a solid "plastron," or shield, into which the ribs have been incorporated.

But while we can readily enough piece together the story of how the Tortoise got his shell, no one has yet satisfactorily interpreted the difference between the shell of the Leathery Turtle and that of all the rest of the

Tortoise-tribe. The first made up of irregular, closely-interlocking discs of bone; the other of no less closely interlocking, symmetrically-disposed plates. A glance at the accompanying photographs should make this point clear, but it will hardly afford a solution. In the case of the Leathery Turtle the bony shield is covered by a leathery plate answering to the original flexible skin; in all the rest of the Tortoises and Turtles this skin has developed with symmetrically-arranged plates of horn. Perhaps some of our readers may be able to read the riddle.

THE SWORD OF THE SWORD-FISH.

This piece of timber has been pierced to a depth of no less than twenty-seven inches by the sword of one of these fishes, a fact which demonstrates the effective use to be made of this weapon in attacks on other fish.

THE SWORD-FISH

BY W. P. PYCRAFT, F.Z.S., Etc.

THE sea, without doubt, holds many stranger fishes than the Sword-fish, if we take as our standard of comparison bizarre colouring, or grotesqueness of shape, for example. But it would be difficult to find anywhere in the wide oceans of the world a finer embodiment of irresistible strength and undoubted courage, though tinged, maybe, with ferocity. The Sword-fish combines the speed of lightning with the driving power of a thunderbolt, as one might imagine from its clipper-built lines.

But under this common term, "Sword-fish," are two distinct types. One, represented by *Xiphias gladius*, which is of almost world-wide distribution, not even our own narrow seas being left unvisited; the other, by the type known also as the "Sail-fish" (*Histiophorus gladius*). This is of a somewhat less wandering disposition, though since it haunts both Indian and Pacific Oceans its quarters can scarcely be called cramped! In size they are about equal, attaining a length of about fifteen feet, which places them among the largest of bony fishes. But it is not on a matter of size that the Sword-fish depends for his fame, but on the irresistible thrust of his sword. This weapon is fashioned out of the fore-part of the creature's head, which is drawn out into a long, pointed rod, dense as ivory. No other armature does he possess, not even teeth in jaws, nor a coat of scale-mail to protect his glistening sides. Nature evidently thought the sword-point protection enough for him, and teeth are unnecessary, since his victims are swallowed whole. Cod and Tunny are all chased and spitted at the sword's point with ease. They constitute the chief food of these

THE SWORD-FISH.

It is not on a matter of size that the Sword-fish depends for his fame, but on the irresistible thrust of his sword. This weapon is fashioned out of the fore-part of the creature's head, which is drawn out into a long pointed rod, dense as ivory. No other armature does he possess, not even teeth, nor a coat of scale-mail; his sword is sufficient protection.

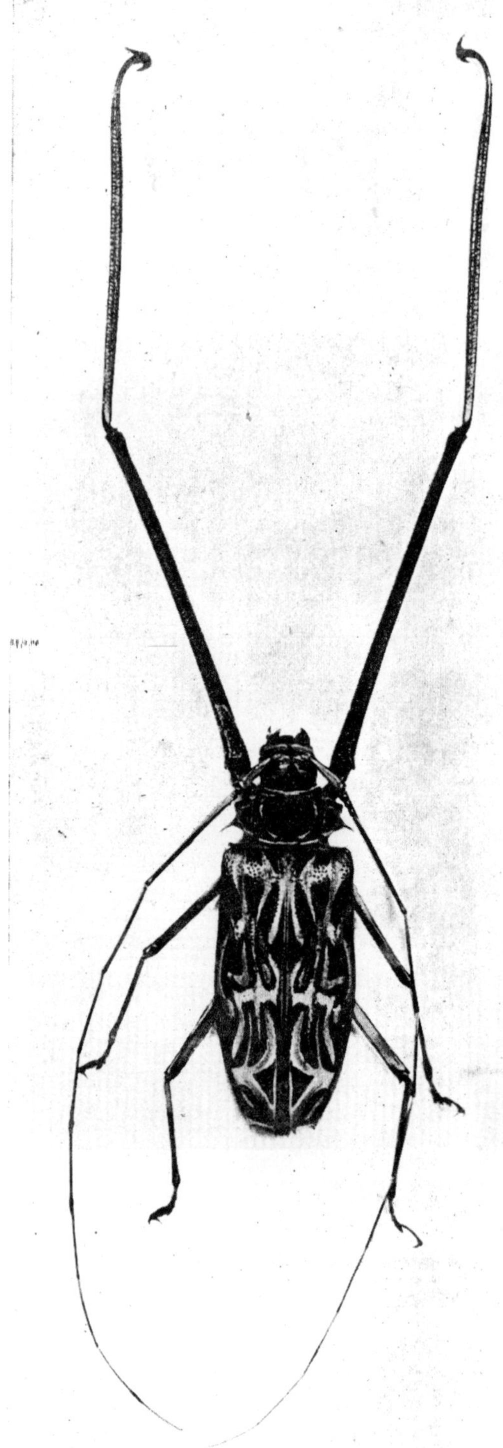

Photo by] [*J. J. Ward, F.E.S.*

THE HARLEQUIN BEETLE.

The development of the forelegs is peculiar to the male insect. Its colouring is red, white and black, and the total length of the Beetle is over a foot.

creatures; though how, when spitted, they are removed from the sword and conveyed to the mouth is a mystery. But besides whales are often attacked, and with savage ferocity. They are stabbed again and again, till at last they die of exhaustion and loss of blood. Yet it is difficult to grasp why such monsters as whales should be slain thus, for the Sword-fish certainly cannot swallow his victim whole, and it is equally hard to see how he could do so even piecemeal, being toothless But the fact remains that whales are slain thus. Occasionally, however, the Sword-fish makes a disastrous mistake, as when, for example, full of fury, he drives at the hull of a ship apparently in mistake for the body of a whale. For such an unfortunate error there is no recovery: the attack was unprovoked, and the penalty incurred is a heavy one—either the sword must be left behind, or the owner must be ignominiously drawn after the ship till he dies! And this because, from the tremendous force of the blow, the weapon is driven in so far as to make withdrawal impossible, for solid timber will not yield like living flesh. In the Museum of the College of Surgeons is a section of the bow of a South Sea whaler transfixed by the end of one of these swords, measuring one foot long and five inches in circumference. In the British Museum of Natural History two similar illustrations of this kind are to be seen. In the one case a sword has pierced no less than twenty-seven inches of timber; in the other are remains of *three* swords all close together, and broken off short in the middle of a piece of timber a foot thick. When so long ago as 1795 H.M.S. *Leopard* was repairing, a sword was found which had gone through one inch of sheathing, a three inch plank, and beyond to a depth of nearly six inches into solid timber. The men engaged on the work of repairing estimated that it would have taken nine strokes of a twenty-five-pound hammer to drive a bolt of similar size and form to the same depth into this hull—yet this was accomplished by a single thrust of the fish!

To withstand the tremendous lateral shock to which the vertebral column must be subjected by such charges a peculiar and remarkable mechanism has been provided. The spine, which in most fishes rises from the top of each segment of the backbone, is in the Sword-fish fashioned into a broad, square plate. And each vertebra grips the base of the plate next in front of it by a pair of flattened blades, so that, in consequence, an almost rigid rod is formed. Thus the body forms a sort of flexible arrow, which is discharged, not from a bow, but by its own volition—by violent side-to-side movements performed by the muscles along the shaft, which are as powerful as bands of steel.

The Sword-fish known also as the "Sail-fish" derives this name from the huge fin which runs along its back, and which is, at times, thrust out of the water and used as a sail. To preserve its balance during such feats the paired fins which answer to the hind-legs of land animals are produced into rods of great length, thus acting somewhat as centre-boards.

Finally a word as to young Sword-fish. These, when newly-hatched, bear no resemblance to the graceful Sword-fish which is to be. A short body, short jaws, armed with tiny teeth, and a huge pair of eyes, are the most striking features of this stage; but a little closer examination will reveal an enormous spine projecting from the gill-plate behind the eyes, its tip projecting above the back, and a second below this runs along the lower border of the belly, with its point as far back as the base of the tail. Gradually, as the form of the adult appears, and the sword begins to grow, the spines behind the eye undergo a gradual decline, the upper disappearing, the lower remaining to form a long spur on the gill-cover.

A RADIOLARIAN.

These minute creatures range in diameter from the tenth to the one-hundred-and-fiftieth of an inch. Though classed amongst the lowest forms of life they are by no means simple in structure. This specimen has had the outer network of fine threads removed to show the internal structure. This network can, however, be distinguished round the edge.

THE HARLEQUIN BEETLE

BY K. G. BLAIR, B.SC., F.E.S.

THE Harlequin Beetle is a native of Tropical America, ranging from Mexico down to Southern Brazil. It has received this name on account of the varied pattern of red, white and black with which it is adorned, though perhaps a more striking feature of the insect, and the one from which it derives its scientific name, is the extraordinary length of its front pair of legs. The beetle itself is a large one, its body attaining a length of about three inches, but its front legs, in the case of a well-developed male specimen, are about double this length, giving the creature a reach from finger to finger, so to speak, of over a foot in extent.

It might be supposed that its size and vivid colouring would make the beetle a very conspicuous insect, but the meagre and somewhat diverse accounts of its habits given by travellers hardly support this idea. It is fairly often found on tree-trunks, and is then, indeed, a very conspicuous object; but it seems to be chiefly nocturnal in habit, remaining concealed during the day within the burrows that the larva has made in the trunks of trees, principally those of a kind of indiarubber tree. From these it emerges in the early morning to feed on the exuded sap of the tree, of which it is very fond.

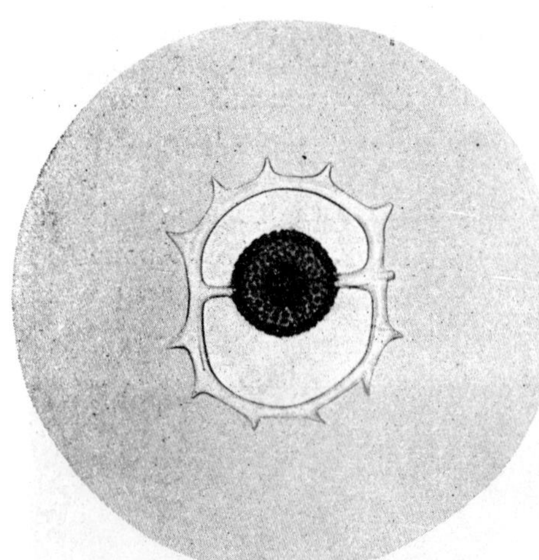

Photo by] [*Flatters & Co., Longsight.*

A RADIOLARIAN.

It has been named Saturn from a resemblance of the creature to the ringed planet. Like many of its kind, its soft body is of many colours.

One writer describes it as living amongst the branches, where the length of its limbs gives it a most graceful swinging motion as it climbs from twig to twig, analogous to that we find in the gait of certain other animals dwelling in similar situations, such as the Stick Insects among insects, or the long-armed Gibbons among vertebrate animals. Another account states that in walking its movements are slow and awkward, and that when disturbed its long forelegs adopt a "rigid attitude of defence." That this immense development of the forelegs is confined to the male sex only would indicate that their chief function is to hold the female in some peculiar way, but no observations on the subject are to hand.

RADIOLARIANS

BY RICHARD KERR, F.G.S., F.R.A.S.

If the Radiolaria were as large as those shown in the accompanying illustrations they would command our attention and admiration. We should provide special glass cases for them, or cabinets in which we could display them to the best advantage. At times we should take them out carefully and show them with as much pride as if they were of equal value with the marvellous concentric ivory balls of Canton, but, unfortunately, they are only known to a very few of our population, although there are teeming millions of them in all the oceans of the world.

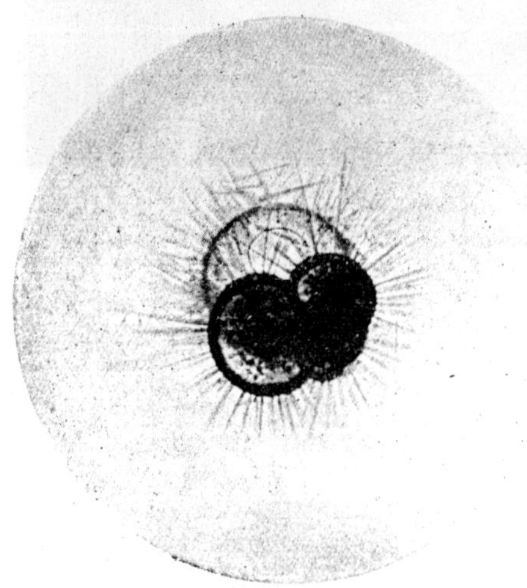

Photo by] [*Flatters & Co., Longsight.*

A RADIOLARIAN.

Showing the extended radiating feelers which serve for catching the minute food particles as well as for protection

The Radiolaria receive their name from the fact that several parts of their structure, and especially their feelers, radiate as from a central point. They are all microscopic. This fact alone is sufficient to make them all the more wonderful, for they range in diameter from the tenth to the one-hundred-and-fiftieth of an inch! The living creatures are of many colours and are by no means of simple structure, notwithstanding that they are classed among the lowest forms of life.

There is a richness of design in the Radiolarian not found in any other form of minute life. The geometric sphere must be reckoned as the chief form of

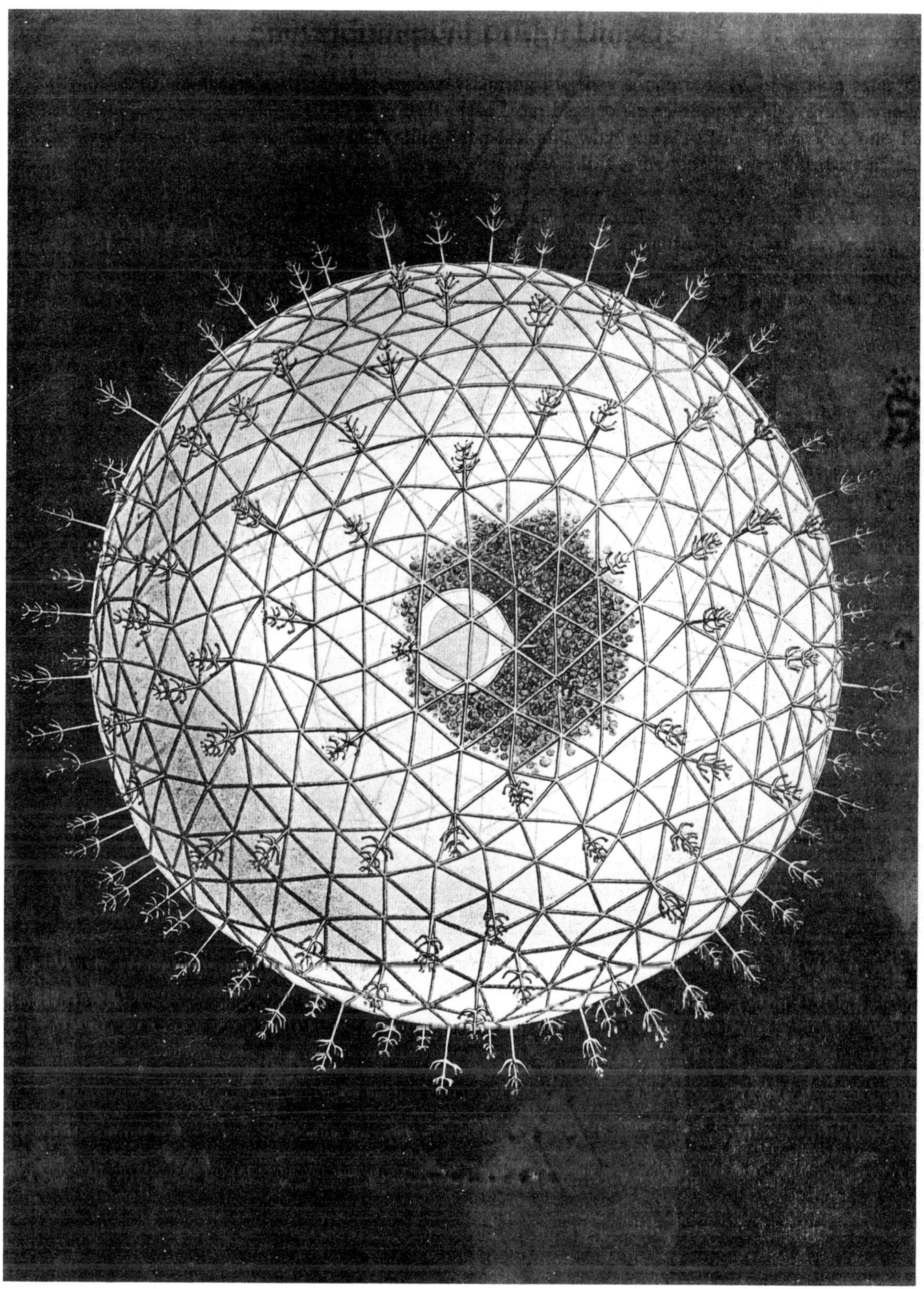

A RADIOLARIAN.

A splendid specimen of elaborate workmanship. No group of animals has skeletons developed in such a complexity of form and it is supposed that the object of such intricacy is the protection of the delicate body within All this surprising detail, we must remember, is comprised "within the dimensions of a point."

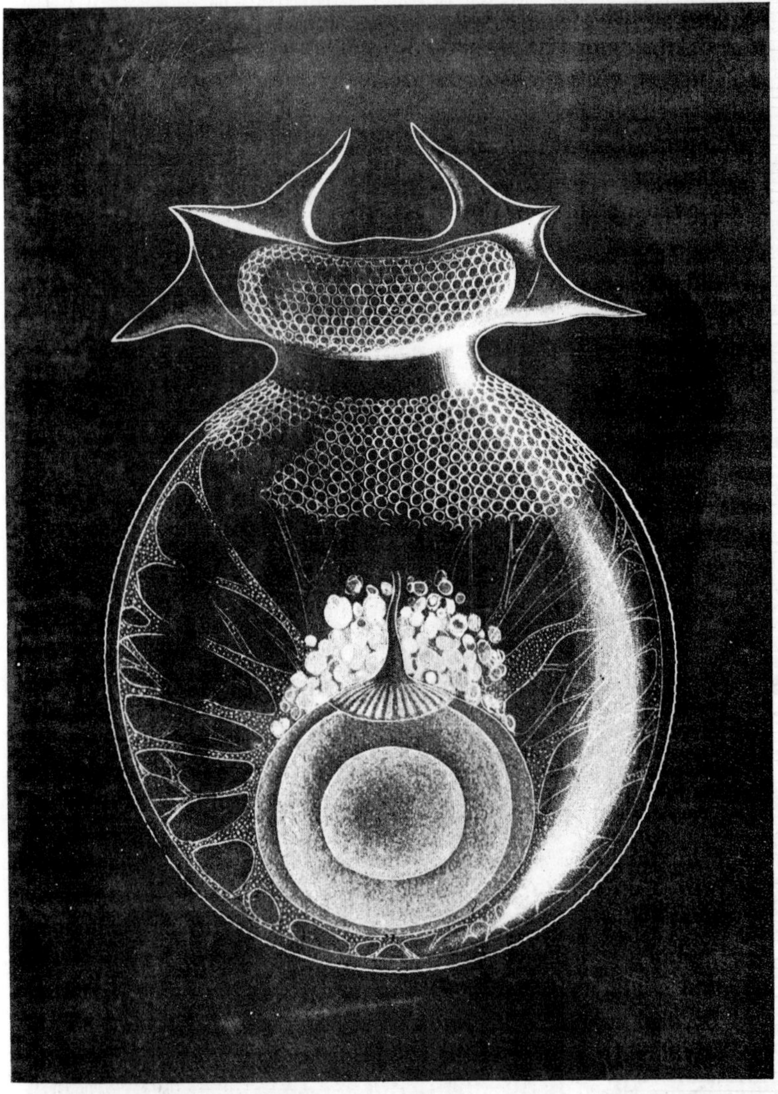

A RADIOLARIAN.

This vase-shaped form was only recently discovered. It was found with other varieties distributed over the floors of the Indian and Pacific Oceans.

the skeleton, perforated and decorated in innumerable ways. But there are helmet-shapes, bird-cage patterns, crowns, cones, pyramids, etc. The skeletons of the Radiolaria are of exquisite construction and are composed of translucent flint (silica), with the exception of one group, at least, the skeletons of which are composed of strontium sulphate. The members of this group—the Acantharia—live in the upper waters of the ocean and seem to have a greater range of life, inasmuch as they are found in all parts of the oceans, from the Arctic to the Antarctic Seas. The skeletons of this group generally consist of twenty rods united in a variety of ways. No group of animals, land or marine, has skeletons developed in such a complexity of form and variety of detail as the Radiolaria. The simple life of these exceedingly small creatures does not, apparently, demand, or need, such magnificent and elaborately constructed skeletons. The usefulness of such intricate and beautiful structures to these humble creatures cannot be fully explained by our greatest marine biologists. Stimuli, stresses and strains in the ocean waters may account for a good deal, by their internal and external action upon the delicate, soft body of the creature, but the wonderful skeleton can hardly be looked upon as a response to all these. All its surprising detail, we must endeavour to realize, is comprised " within the dimensions of a point."

Many groups of Radiolaria are adapted to life on the surface of the waters, with power to descend to certain limited depths and to come up again.

According to Brandt, they descend and ascend by functions that are purely hydrostatic. In some groups the delicate covering of the animal is supported loosely on the radiating skeleton. This covering can be tightened by elastic threads (filaments) and thus raised up. This allows the tiny layers of water which surround and enclose small cavities (vaçuoles) to expand and so raise the creature in the water to a higher level. It can descend again by the contraction of the delicate covering as the elastic threads are relaxed. In other groups there are other organs which perform similar duties and which will be noticed presently.

Nearly all the Radiolaria are capable of emitting a phosphorescent glow. The luminosity presents the same phenomena as that of other marine organisms that are luminous. The light is sometimes greenish, sometimes yellowish, and appears generally to radiate from central fatty spheres. These tiny spheres also unite several functions, inasmuch as they serve, firstly, as reserve stores of nutriment; secondly, as hydrostatic apparatus for lowering and raising the creature; and thirdly, as luminous organs for the protection of the creatures. The radiating feelers are also a protecting arrangement. They also act as sensory organs for apprehending food particles.

By far the most important and most varied means for the protection of the soft body is to be seen in the endless modification of the skeleton. This being so, we have one answer supplied to our query as to the usefulness of the complicated skeleton.

The Radiolaria have power of movement in a limited degree, but, according to Haeckel, we are at present equally ignorant whether all the movements are simply the direct consequences of irritation, or whether they are in part truly spontaneous. This movement, as I understand it, does not include the descending and ascending movements already described.

Radiolarian ooze from the Pacific Ocean was first obtained in 1859, by Lieutenant Brooke, between the Philippine and Marianne Islands, from a depth of three and a half miles. Ehrenberg who first examined and described it, found seventy-nine different species in it and reported "that the quantity and number of different forms increased with the depth." The Pacific Ocean seems to be the ocean which is richest in quantity and quality of Radiolarian

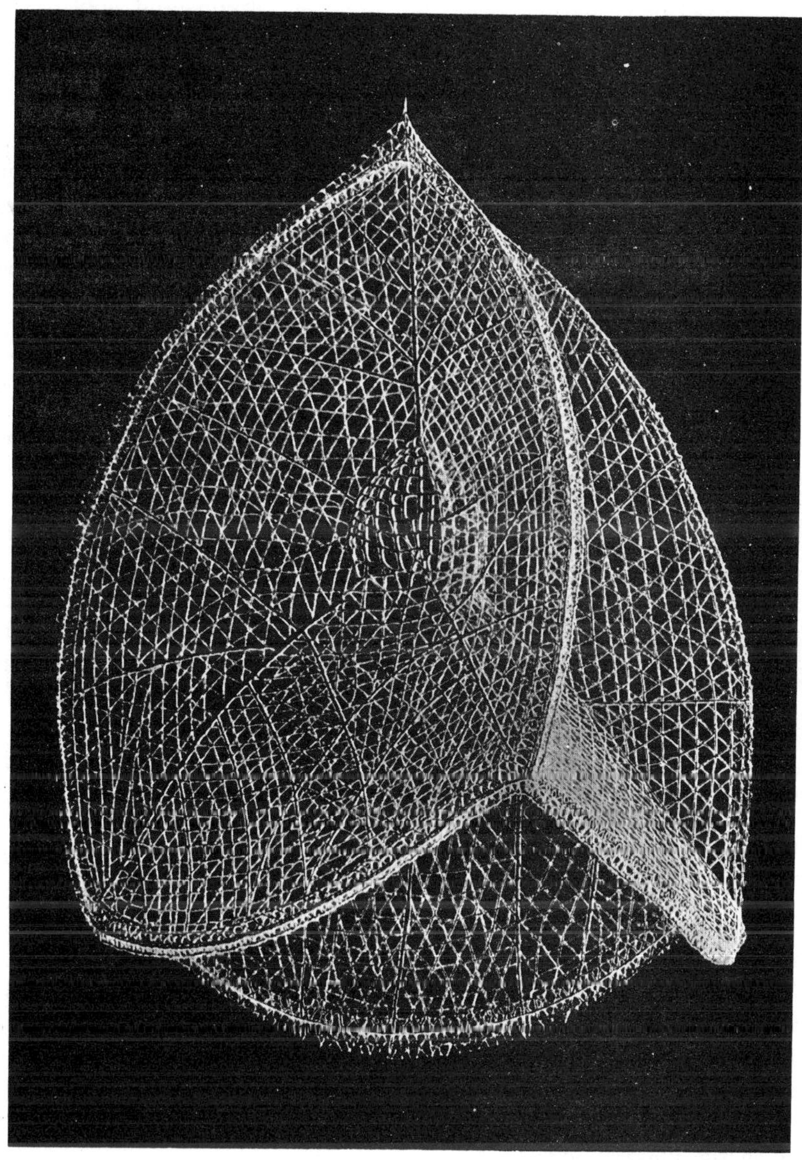

A RADIOLARIAN.

The "Wicker-turban" is a good name for this beautiful specimen. The body has been removed. The diameter of this and the Vase-shaped Radiolarian is about the two-hundredth of an inch.

life. The micro-life of the Indian Ocean is the least known of the three ocean basins. The creatures of this class in this ocean approach more nearly those of the Pacific than those of the Atlantic, both as regards abundance and variety of shapes of the Radiolarians. From many sources of research undertaken at various times, and in several parts, it may be assumed with great probability that the major part of this basin, like that of the Pacific, is covered with Radiolarian ooze.

The Atlantic Ocean, in all parts of which the microscopic life has been examined, has shown the constant presence of Radiolaria, inferior both in quality and quantity, and pure Radiolarian ooze such as has been found on the floors of the Pacific and Indian Oceans has not yet been discovered.

The naturalists of the *Challenger* found Radiolarians universally distributed throughout the waters and spread over the floors of the two great oceans—the Pacific and the Indian—the relative abundance and the species differing, however, with change of locality.

Photo by] [*Tickner Edwardes.*

AN ENGLISH QUEEN BEE.

The mother of the hive. Her long body distinguishes her from the drones and workers.

Dr. John Murray found rich supplies of Radiolaria in the alimentary canal systems of several marine animals, while Jelly-fish and crustacea which live partly at the surface and partly at various depths, swallow large quantities of them.

Within the last few weeks Dr. John Martin found a little patch of beautiful iridescent colour on the surface of the calm water off Whitby, after a heavy storm. It consisted of a number of extremely minute shells of Radiolaria, which, as we have seen, are capable of sending out extensions of their body substance for the purpose of locomotion and for drawing in food. The shells were of magnificent forms and identical with those found in the West Indies and along the coasts of Florida and the Gulf of Mexico. Like the majority of Radiolaria, they were composed of flint. The finder rightly believes they were carried by the agency of the Gulf Stream to the Yorkshire coast.

The coloured illustration which accompanies this article contains a collection of characteristic Radiolaria. They are all possessed of long Latin names which describe the characteristics of each species; for instance, the central figure is called the "Wing-helmet." That on its left, like the letter Y, is the "Star-shell." That above the central specimen, with long twisted spikes, is the "Shell with Six Points.' That to the north-east of the centre is the "Basket-urn," or "Wicker-pitcher." The thorny specimen under this is a "Beautiful Crown." The specimen occupying the central position on the lower line is a "Net-goblet." The average *diameter* of these specimens is about the two-hundredth part of the diameter shown in the illustration!

Detailed descriptions of the Radiolaria will apply almost equally well to the Polycystina, because Müller, Nicholson and several other marine biologists classify the Polycystina as a Family of the Order Radiolaria. Generally speaking, however, fossil Radiolaria are spoken of as Polycystina.

Photos by] [*Tickner Edwardes.*

THE QUEEN-CELLS.

Shown in two positions That on the left is a little above the centre of the comb, and is almost covered by a builder-bee; that to the right is in the more usual position, hanging down from the lower edge of the comb.

At the same time there are, as has been shown earlier in this work, living specimens of Polycystina. There is no definite agreement among biologists that all living specimens should be termed Radiolaria, and that the name Polycystina should be restricted to the fossil forms of Radiolaria. Hence, at the present time, a line of demarcation cannot be drawn between these more than interesting forms of micro-marine life.

THE QUEEN BEE

BY TICKNER EDWARDES

THE life-story of a Queen Bee begins really at the moment when the mysterious governing spirit within the hive decides that a swarm shall issue—the emigration of a large part of the population take place.

As the old Queen is to be sent away with this swarm, the first and most important preparation for the event will be the provision of a successor for her. We now see an extensive work begun in the hive. Queen-cells must be built. The building of these queen-cells is undertaken in a very different way from the method employed with those destined to contain ordinary workers, or ordinary male bees. Instead of being packed together in as little space, and constructed of as little material as possible, the queen-cells are made separately, of generous size, and are placed usually on the edge of the comb, where abundant space and fresh air are obtainable.

At first this cell looks very like an inverted acorn-cup, in the centre of which depends a minute white speck like a microscopic grain of rice, wherein is enshrined the germ destined, in all likelihood, to develop into the future mother of a whole nation. From this egg, after the lapse of three days,

Photo by] [*Tickner Edwardes.*

THE BROOD-COMB.

A very old Brood-comb. The cells, notably those towards the bottom of the comb, have almost lost their hexagonal form owing to the accumulation of pupa-skins within.

emerges a tiny grub ; and immediately on its appearance the Worker Bees deluge it with a thick, glistening, white substance, jelly-like in character, upon which it feeds and grows at an amazing rate. At the end of five days, this grub, which has been kept fairly wallowing in rich nutriment all the time, has grown to the size, and largely exceeds the weight, of a fully developed Queen Bee ; and the cell has grown with it, until its dimensions have reached probably those of five or six ordinary worker-cells. Now the super-feeding of the grub is stopped, the cell is sealed over, and the grub changes to a chrysalis, lying quiet in its cradle for nearly a fortnight longer, when it becomes a fully-grown Queen Bee ready and more than willing to take up the important duties that lie before her.

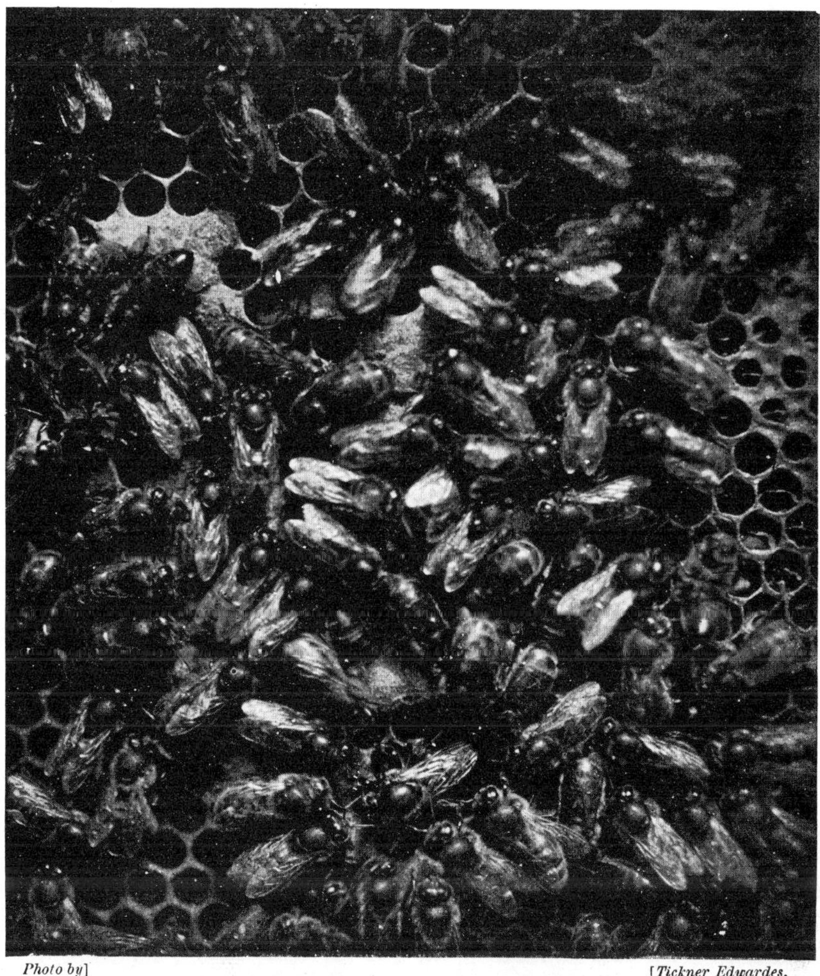

Photo by] [*Tickner Edwardes.*

THE QUEEN BEE AT WORK.

The Queen Bee laying her eggs, surrounded by her attendants. These will carry off the newly-laid egg, deposit it in a cell (right way up), supply it with a store of bee-bread and seal up the opening.

But here very often a check is placed on the ardour of the young princess by the directing powers of the colony. If the swarm has issued and the old Queen has left the hive, no obstacle will be placed in the way of the young Queen's emergence from the cell. But times may not be propitious, cold and windy weather may prevail ; and in that case the issue of the swarm will be delayed, and the appearance of the young Queen with it. From the moment when the preparations for her successor become an accomplished fact, the old Mother Bee has grown daily more restless and suspicious. A fundamental quality in the nature of Queen Bees is their inveterate hatred of one another. Left to her own inclinations, the old Queen would long ago have torn down the queen-cell and destroyed its occupant. But the royal cell has been zealously guarded by the workers, and now, when all preparations for the future are completed, and a crisis has been reached in the affairs of the hive, the vigilance of the guard-bees is redoubled. The hostile approaches of the old Queen are sedulously foiled ; at the same time, the attempts of the young one to break from her prison are frustrated. A hole is bored in the side of the cell, through which she is fed ; but she will be retained a captive until the swarm is gone, and her safety no longer imperilled.

The Bees, however, seldom trust the future welfare of the community to one chance alone. The young Queen may die, or meet with some mishap at any stage in her development. So the almost

Photos by] [*Tickner Edwardes.*

THE QUEEN WALKS ABROAD.

In the upper picture it is the off-season, and she wanders about amongst the bustling workers without any special attention being given her, but in the lower picture she is replenishing the hive, and is seen in the lower part on the right-hand side of the picture parading the comb surrounded with her retinue.

invariable policy is to raise, not one new Queen, but several. As many as a dozen queen-cells have been observed in a hive just before swarming-time, in varying degrees of progress. Only one Mother Bee, however, is allowed to exist in a normal hive. As soon, therefore, as the workers of the colony find themselves possessed of a new Queen capable of attending to all a Queen's duties, they withdraw the guards which are always placed around the surplus queen-cells, and allow the new ruler to work her natural will upon them. This she does with fierce alacrity, breaking down the cell-walls, and putting her royal sisters at once to the sword.

We must now—supposing the observer, though deeply interested, to have none but a very superficial acquaintance with bee-life—look into the brood-chamber of a hive a week or so after the new Queen has been permanently installed, and note what plain eyes can see there of her surroundings and daily activities.

Photo by] [*E. Step, F.L.S.*

PLANT OR ANIMAL?

The only English name that has been given to this group of organisms is Slime Fungus. Some authorities say they are animals, others that they are plants related to the Fungus. The mass shown consists of thousands of individuals that have united prior to breaking up into dust-like spores.

It is early summer, when the honey season is at its height, and the fullest demand is being made on the Mother Bee's resources. All the brood-combs are covered with a busy crowd of Worker Bees, and the combs themselves are full either of young grubs in process of development, of cells sealed over and containing the nymphs, or of cells, newly vacated by their occupants, being cleaned out for further use; some, indeed, already furnished with a second batch of eggs.

Over these combs the Queen incessantly wanders, and whenever she stops at an empty cell, thrusting her long body down into it to deposit an egg at its base, the Bees in the immediate vicinity at once gather round her, forming a complete ring, each Bee having her head ceremoniously turned towards the Queen. As soon as the egg is laid and the Queen moves on, this ring breaks up; but it forms again directly she makes her next halt.

It is generally asserted that the Bees composing the ring are the Queen's attendants or body-guard, and that they keep about her during her egg-laying journeys, alternately driving or leading her, and stopping her at each vacant cell in the path. But we now see that this is not so. The Bees composing the rings are different Bees in every case. The Queen, in fact, is never led over the comb by one set of attendants, but is merely handed on from group to group in the crowd, each making use of her presence as the immediate occasion demands. In this way, day follows day, week after week passes; and whenever we look into the hive, still she is going her unwearied rounds, fed and fondled incessantly, the docile servant of all.

Yet, simple and straightforward as this one duty of the Queen Bee at first sight appears, a little farther study of the matter soon brings us into the midst of a complication of wonders. We see

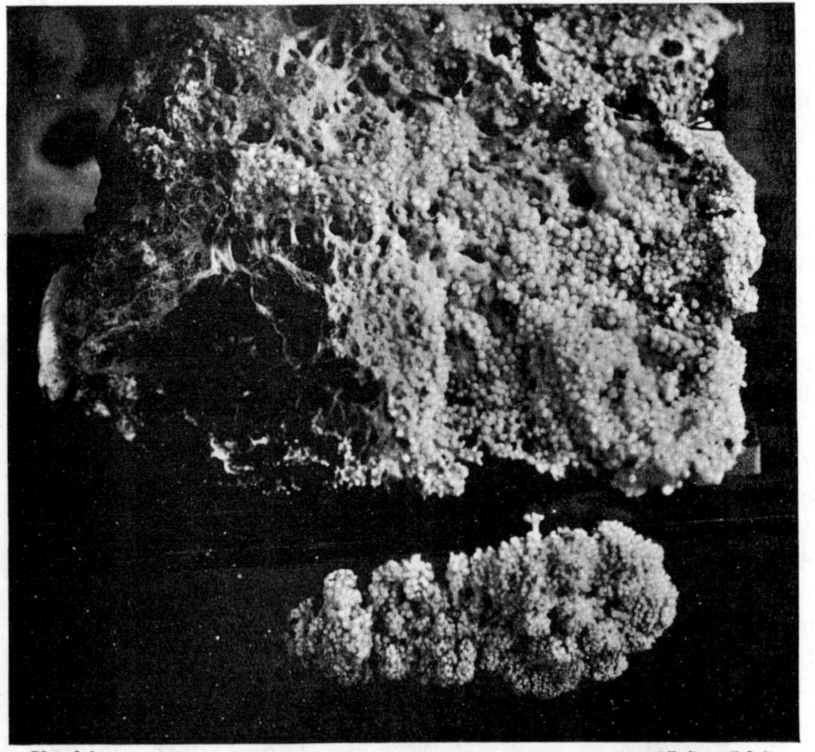

Photo by] [*E. Step, F.L.S.*

PLANT OR ANIMAL?

The upper mass here shown was shut up in a tin box with perforations in the side. The organisms, attracted by light, passed through the perforations to the outside of the box—quite an animal proceeding.

that the comb-cells are of two kinds, a small cell wherein are bred the Worker Bees, and a large cell from which the drones are produced. We watch the Queen dealing with a group of worker-cells, and then see her pass to the neighbouring drone-cells, into which she thrusts her body in exactly the same way. From these presently she may go on to more worker-comb; and again, a little later we may find her on drone-comb once more, as busy as ever.

But we have seen that from the eggs placed in the small cells little Worker Bees are always hatched, and from those laid in the larger cells proceed invariably great, full-bodied drones. And, again, we have watched the egg in the queen-cell develop into a magnificent Queen Bee, a creature differing from the other two kinds of hive-inhabitants in half a dozen ways plainly evident to the naked eye. We cannot now escape the conclusion, therefore, that the Mother Bee—in addition to her power of producing an enormous number of fertile eggs for years in succession, has also this further astonishing faculty:—that she can lay three totally different kinds of eggs at will, changing from one variety to the other just when and where it pleases her.

There is no doubt of the marvel of this, nor of the mystery; and even when we arrive at an understanding of how it all comes about, though the seemingly miraculous side of the matter disappears, we find ourselves face to face with a greater marvel than ever.

PLANT OR ANIMAL—WHICH?

BY EDWARD STEP, F.L.S.

AMONG the many living creatures that are known only to the naturalists, there is a small group of organisms that for long have been somewhat of a worry to those who like to put everything into its proper place—allied plants in appropriate classes and orders, and allied animals in other classes and orders. Some authorities say these are animals; others that they are plants related to the Fungi. The only English name that has been given to them is that of Slime Fungi, which is, of course, objected to by some, as it settles the question of their nature off-hand. Scientifically, they have been variously styled Myxomycetes, Mycetozoa and Myxogastres.

It was the great De Bary who started the trouble. Himself a botanist, he said, in the work that added so greatly to our knowledge of these organisms: "I have placed the Myxomycetes, under the name of Mycetozoa, outside the limits of the vegetable kingdom." Yet, in spite of this, he continued to include them in his later botanical works. In this country Lister supported De Bary, and Saville Kent went so far as to include them in his "Manual of the Infusoria." Cooke and Massee, on the other hand, have contended strongly for their retention as an aberrant group of Fungi. Lister wrote a big "Monograph of the Mycetozoa" for the British Museum Trustees, in which he claimed them as animals; but the queer thing is that Lister's collection is exhibited in the *Botanical* Gallery of the Natural History Museum!

The Myxies—as we may call them shortly—are propagated by invisible spores or germs, much after the manner of fungi, mosses and ferns; and when these spores alight on the moist surface of decaying vegetable matter, whether leaves or wood, they germinate and out creeps a microscopical speck of living jelly—a swarm-cell—very similar to the low form of life known as an Amœba. We cannot say that it has any definite form, for its outline is constantly changing; but under the higher powers of the microscope we can make out a denser portion known as the nucleus, which appears to be the most important part of the creature, and several clear spaces, called vacuoles. There is no mouth, no internal organs, no eyes, no limbs. Soon it develops a long, lash-like extension of one end, and this we know to be the front part from the fact that the lash moves forward and the rest follows. If it comes across bacteria, or any other of the minute organisms that are engaged in breaking up the dead wood, it flows around them, gets them into a vacuole and there digests them. Then the nucleus divides into two, and the division extends across the rest of the jelly, so that we have two Amœbæ instead of one. Scores or hundreds of them may divide in the same manner, and again divide a little later; so that soon we may have thousands of them. This goes on for a few days, then they all withdraw their lashes, and amalgamate into one mass of jelly —the protoplasm of which all living things, animal and vegetable, are formed. The mass is termed a plasmodium.

This amalgamation takes place in the decayed tissues of wood or leaf, but at a certain stage

Photo by] [*E. Step, F.L.S.*

PLANT OR ANIMAL?

The creamy mass from inside the box has passed through the perforated side, and concentrated itself into two cushion-like masses over which a thin brown crust forms, whilst the contents breaks up into dust-like spores.

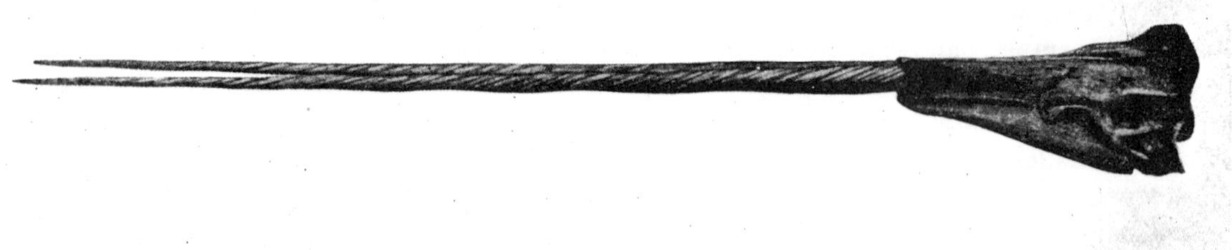

THE SKULL OF A TWO-HORNED NARWHAL.

The Narwhal usually has only one horn, with the second only slightly developed. Here the rudimentary second horn is fully grown

the mass becomes actuated by a common impulse to seek the light, and it moves towards it with a rolling motion.

One day, at the end of October, I came across the stump of a large pine-tree. The interior was decayed, and from it was pouring a sheet of cream that flowed over the bark. Several square feet of the bark were thickly coated with it, and there was evidently more to follow. With a strong knife I cut out a large piece of bark, and placed it carefully in a tin box. On reaching home, I at once took the photograph shown on page 681, which is the natural size. Then I placed it again in the box and closed the lid tightly. The bark was so placed that it nowhere came into contact with the sides of the box, as I hoped by this treatment the plasmodium would be induced to stay on the bark.

The tin-box had contained tobacco, and the manufacturers had perforated its sides with their initials. The light, of course, found access through these perforations, and this low organism enclosed within, that had no eyes or other apparent sense organs, knew thereby the way out. By some means a portion of the plasmodium stretched across the gap between the bark and the box, forming a bridge through or over which more could follow; and, to my surprise, next morning I found that part of the mass had gathered on the exterior of the still tightly closed box. I then opened the box to expose the principal mass as well, and took the second photograph (page 682).

The box was again closed, but ultimately the whole mass made its exit through the small holes, and consolidated itself into two cushions on the outside, and these invested themselves with a thin, purple-brown crust, as shown in the third photograph (page 683). The whole of the contents of these cushions—which are known as æthalia—then broke up into a fine, snuff-like powder of a dark purple-brown colour, among which ran a number of microscopic threads with knots of lime at intervals upon them. The powder was a vast assemblage—probably millions of millions—of spores, such as originally produced the swarm-cells. The crust, in drying, contracted, so that it split across

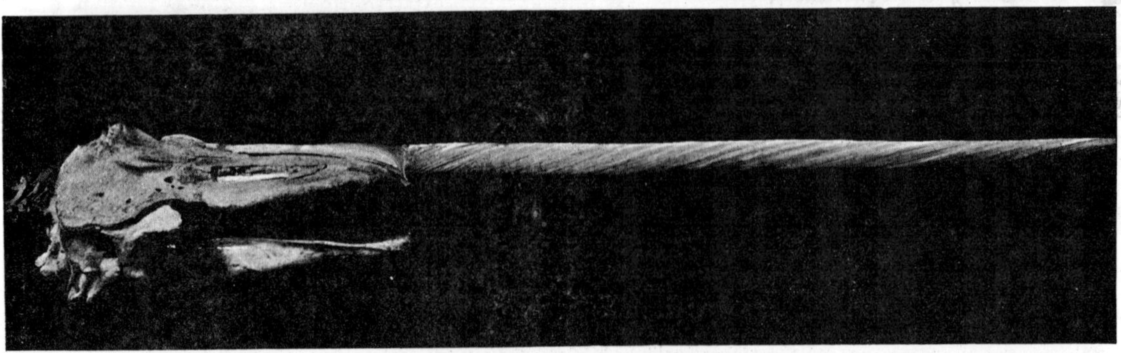

THE SKULL OF THE NARWHAL.

Showing the development of the horn from the frontal bones of the head, and the indications of a second.

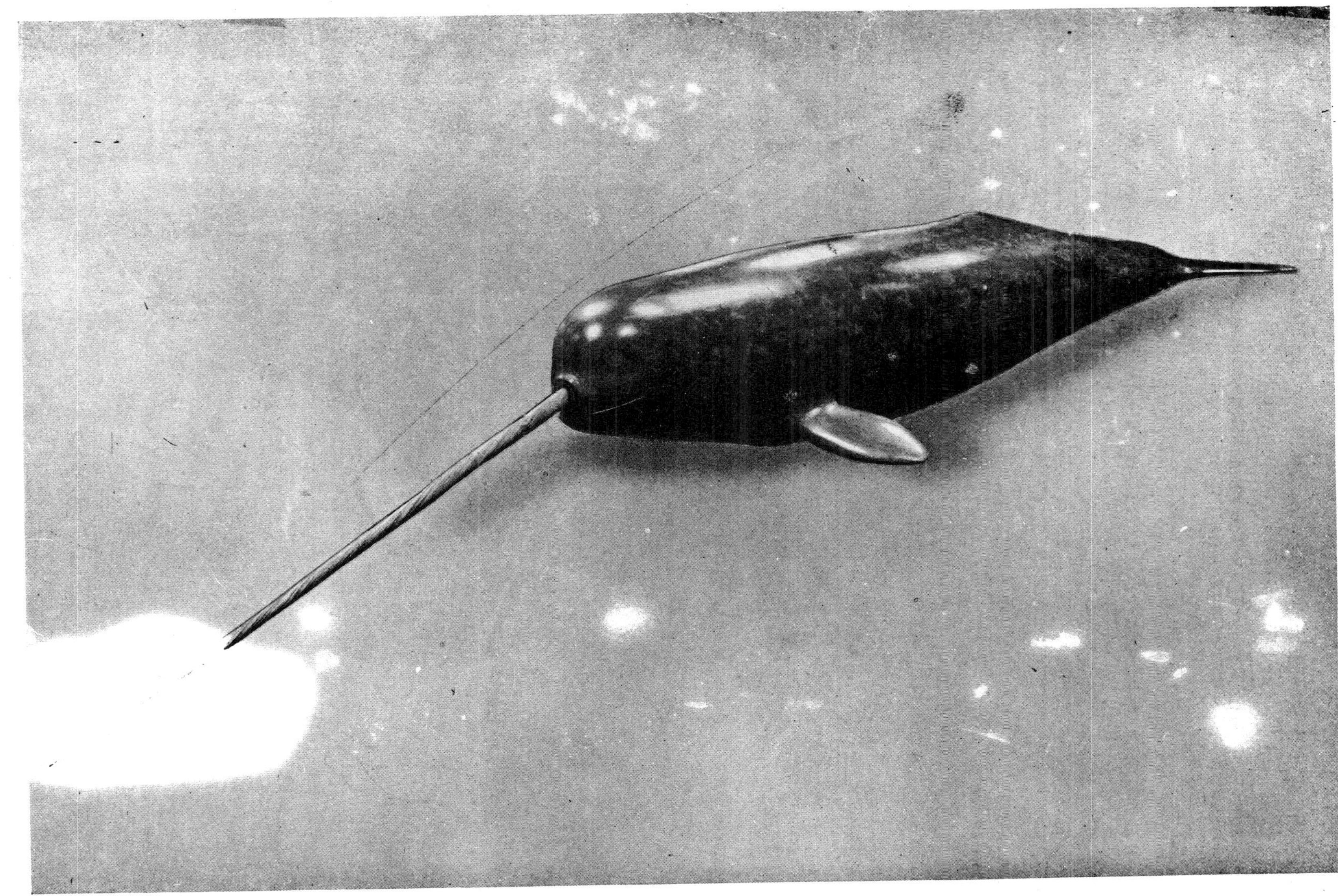

THE NARWHAL.

During its early years the Narwhal might be mistaken for an unusually pale-coloured porpoise, and becomes toothless soon after birth. But in due course the upper lip, of the male at any rate, becomes pierced by a tusk which eventually attains a length of some seven feet, half the length of the entire animal.

and gradually fell away, allowing movements of the air to disperse the light spores and carry them far and wide.

I have selected this Myxy—known as Brefeldia—because of its large size ; but most of the species differ from it in being small and stalked. The reason for considering the " Myxies " as animals is based upon their movements and method of feeding as swarm-cells ; whilst the formation of the sporangia and production of a multitude of spores is more akin to the behaviour of plants.

THE SPIRAL SPEAR OF THE NARWHAL

BY W. P. PYCRAFT, F.Z.S.

ALL the whales are marvellous, but there are some species which impress us as being more so than others, and the Narwhal is one of these. During its early years it might well be mistaken for an

Photo by] [*Doubleday, Page & Co.*

WARNING MARKINGS.

The violent contrast of the black and white rings and the fine markings of the head tell their own tale of danger.

unusually pale-coloured porpoise, and it is further remarkable for its toothless condition—such teeth as it had having disappeared soon after birth. But in due course a curious transformation takes place, at any rate, in the male—the upper lip becoming pierced by a tusk which eventually attains a length of some seven or eight feet—half the length of the entire animal. This tusk, like those of the elephant, answers to one of the incisors, the left, but it differs from all other known teeth in being spirally coiled ; the spiral being left-handed. On rare occasions animals with two tusks of equal size are met with, but such occasions are very rare : usually the right remains within its socket, a mere shrivelled peg which never even cuts the gum. Where two tusks are present the spirals are usually both left-handed ; but specimens are on record where the right tusk has a right-handed spiral.

Naturally one asks of what use is this remarkable tooth ; is it a weapon, or is it merely ornamental ? So far it is impossible to answer this question with anything like certainty, as is shown, indeed, by the several attempts which have been made to explain its use. According to some authorities it is employed to break open breathing-holes in the ice, for it is a dweller in the Arctic Seas. But since the female is tuskless this explanation does not find much favour. Others say it is used by the males in fighting for females, as stags use their antlers ; and males have actually been seen crossing their weapons like rapiers in a fencing match. If anything more than mere fencing takes place the results must be serious for one or both of the combatants, that is certain. The old Arctic explorer, Scoresby, having taken a large skate out of a Narwhal, believed, perhaps from the appearance of wounds in the fish, that the tusk was used as a spear to impale such living prey. But having spitted the fish, how was it removed from the spear ? Perhaps by a violent and sudden " back-water " movement—a reversal of its engines, so to speak.

Photo by] [*H. Dixon.*

A WARNING ATTITUDE.

The Cobra, when on the point of attack, raises its head and puffs out its neck to display its peculiarly terrifying markings.

Another strange feature of this creature is the asymmetry of its skull, the nostrils leading to blowholes being twisted strongly towards the left. It might be supposed that this was due to the great development of the socket for the tusk, which makes the jaw-bone of this side much larger than that of the right, where there is no tusk. But the same twist is found when both tusks are equally developed ; and, further, it occurs again in other species, like the bottle-nosed whale, which is toothless.

WARNING COLOURS

BY E. B. POULTON, F.R.S.

ANIMALS of all kinds, in temperate countries such as our own as well as in the tropics, are, as a rule, coloured so that they harmonize with their surroundings and are concealed. Among these, certain exceptions stand out with startling distinctness, species which seem to be coloured so as to attract attention. Walking along a hedgerow in summer, we shall fail to see numbers of caterpillars resembling the twigs or green foliage, and moths hidden on the bark or among dead leaves ; but the red, black-spotted ladybirds or the cream-coloured caterpillars of the Magpie moth, with their black and orange marks, will at once arrest the attention. The explanation of these conspicuous colours has a very interesting history. Darwin was thinking over the bright colours which adorn so many male animals, and was working out his celebrated theory of Sexual Selection, when he was

Photo by] [*A. S. Rudland.*

THE SKUNK.

Protected by an offensive fluid secreted in certain glands, the Skunk is marked in such a way that it can be seen—and avoided—by even the most unwary.

suddenly puzzled by the tints and patterns of some of the caterpillars. It was, of course, impossible to invoke the aid of Sexual Selection to explain the colours of these immature stages of insects. He asked the help of Alfred Russel Wallace, who, after thinking over the subject, suggested to his friend the theory of Warning Colours, predicting that these brightly-coloured caterpillars would be distasteful to most insect-eating animals, while the well-hidden ones would be extremely palatable. Experiments which were made almost at once verified Wallace's prediction, and since that time immense numbers of further experiments have been made on animals of many kinds, especially insects, and it has been found that species which are conspicuous in nature, both during rest and in movement, invariably possess some special kind of defence, such as an unpleasant taste or smell, irritating hairs, the sting, or the poison fang. It does not follow, however, that bright colours necessarily mean conspicuousness. In our own country we have small caterpillars which feed on brightly-coloured petals, and spiders which lie in wait for their prey in the corolla of buttercups and other flowers ; and both of these are concealed by tints which match their brilliant background. It is also true that, at a considerable distance, patterns made up of strongly-contrasted tints tend to become invisible : indeed, Mr. Abbott H. Thayer is so impressed with this fact, that he believes Warning Colours do not exist, but that all the colours of nature tend to concealment. Probably every species in nature has some foes to fear, and it is a special danger of Warning Patterns that they are so easily seen by those enemies which can put up with the special defence. Hence, the tendency of sharply-contrasted colours to become invisible is doubtless an advantage, especially as they become the very reverse of invisible long before the ordinary enemy is within striking distance.

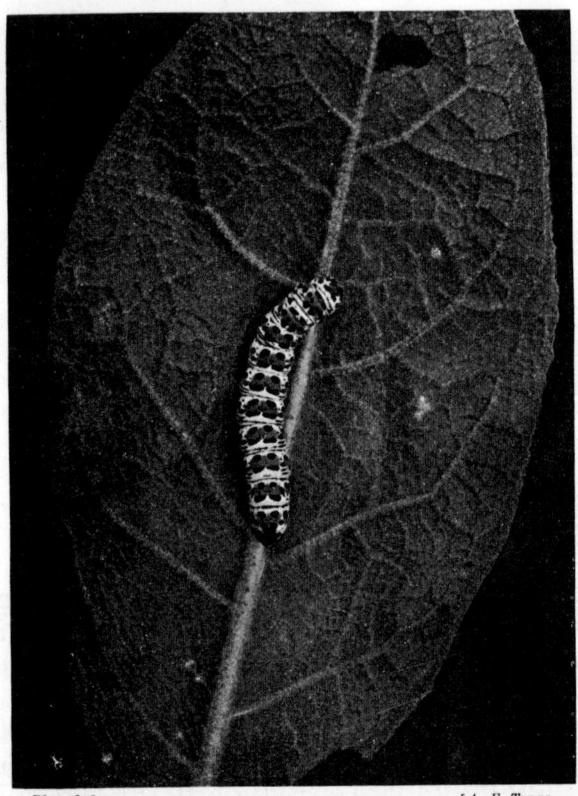

Photo by] [*A. E. Tonge.*

THE CATERPILLAR OF THE SHARK MOTH.

Its striking colouring—greenish-white, bright yellow and black—warns birds of its bitter taste. It is therefore avoided, and having no need to hide itself, feeds with impunity on the broad leaf of the mullein.

On the accompanying plate the artist has represented a great variety of Warning Patterns, selected from many parts of the animal kingdom and from many lands. The caterpillar in the centre at the top of the plate is that of the Cinnabar moth. Here we have an example of a pattern that is by no means wholly conspicuous ; for the orange-and-black-ringed caterpillar feeds on the orange-flowered ragwort. But the caterpillars are gregarious, and many may often be seen a long way off, crowded together on a single plant nearly eaten bare by their united exertions. The meaning and the value of the Warning Colours of this species have been shown in a striking experiment by Professor Lloyd Morgan. He fed some young pheasants on meal " doctored " with various unpleasant substances—quinine, dilute nitric acid, and very dilute strychnine—placing the food on glass slips beneath which oran[...]l-black stripes had been pasted. He soon found that the birds would not touch even undoct[...]meal when it was put upon the patterned glass, and that they regarded with the greatest suspicion and would barely touch the

[By Theo. Carreras.

THE TERMITES' NEST.

A sectional view, not showing the systematic arrangement of corridors and compartments presumed to exist by early writers, but the true state of affairs—a mass of irregular and closely crowded magazine chambers, with a central core of smaller cells, probably nurseries. This type of Ant-mound will reach a height of about twelve feet.

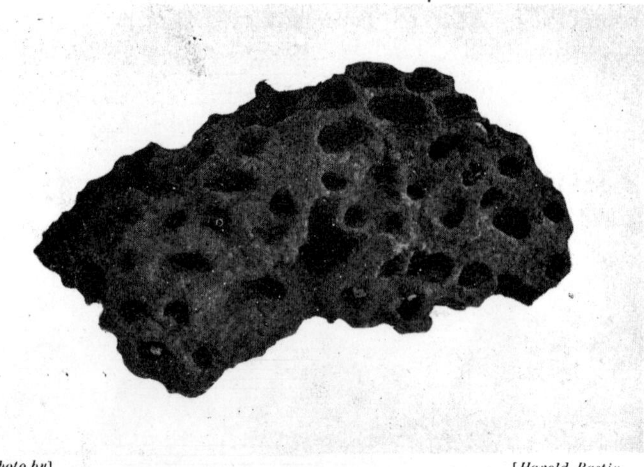

Photo by] [*Harold Bastin.*

"CAKE."

A lump of "chewed" wood, built up by the White Ants for the purpose of growing a fungus which serves for food.

Cinnabar caterpillar when introduced to them for the first time. Pheasants which had not been educated in this way treated the caterpillar far more unceremoniously until they, too, learned from the taste of the caterpillar itself that the black-and-yellow rings meant something decidedly unpleasant.

To the left of the Cinnabar is the female hornet, an insect with a pattern rather like that of the Cinnabar, but warning of a far more terrible defence—the most formidable sting possessed by any European insect. It is interesting to observe that the stingless male hornet bears the same pattern as the female, and moves its body as though about to sting. It may be said to mimic its own female. To the right is a conspicuous Oriental moth, belonging to the Geometridæ, a vast group, of which the great majority are well concealed. The two other moths on the plate, on the contrary, belong to specially-protected day-flying groups, of which the species, as a whole, bear conspicuous Warning Colours. That on the right, belonging to the Burnet group, is the Beautiful Erasmia, from China and North-East India, while the one on the left, belonging to the Agaristidæ, is the Spotted Eusemia, from Assam. A closely-related African species is mimicked by a butterfly, called, in consequence, the Eusemia Mimic. It is far commoner for moths to mimic butterflies, so we have reason to believe that the Eusemia is specially distasteful.

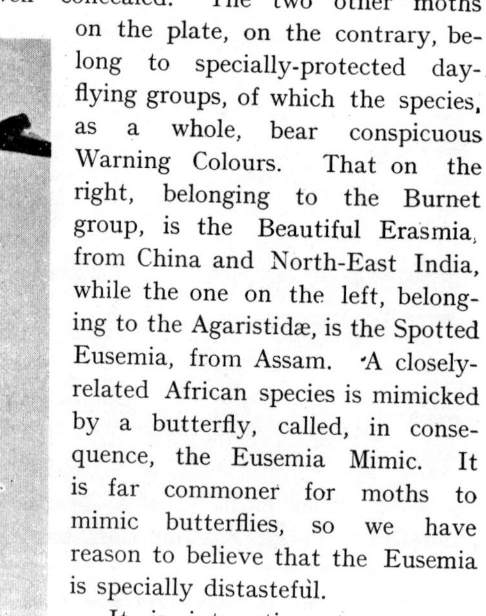

It is interesting to compare the Warning Colours of vertebrate animals with those of insects. The venomous heloderm, shown above the salamander on the right, provides the one justification in fact for the popular dread of lizards. In order to determine how far the pattern is warning and how far concealing, it would be necessary to

Photo by] [*Harold Bastin.*

THE HANGING NEST OF A TERMITE COLONY.

These arboreal nests are composed of finely chewed wood particles, cemented by the juices of the Termites into a remarkably hard substance

study this lizard in its desert home in Mexico or Arizona. The startling pattern of the European salamander gives warning of a defensive secretion poured out by glands in the skin. The toad is defended in the same way, and although ordinarily well concealed, it can adopt the most remarkable warning attitude. I have seen a toad, attacked by a dog, entirely disconcert its adversary by swelling out its body to an enormous size, at the same time rising to its fullest height on all four legs. The venomous tropical American coral snake, on the left at the bottom of the plate, possesses a very remarkable and unusual pattern, although one that is more often seen in caterpillars. Mr. Thayer considers that it promotes concealment, as, indeed, is probably true at a distance. But the pattern is mimicked by harmless snakes in the same locality, and there can be little doubt that it belongs to the class of Warning Colours. Other venomous snakes, which are ordinarily inconspicuous, warn their enemies, on a near approach, by warning attitudes, as with the cobra and hamadryad, or warning sounds, as with the rattlesnake, or the Indian Beaked Echis. It must be remembered that it is of no advantage to a snake to bite an enemy it cannot eat. It will, on the contrary, be weakened for a time by the loss of its last line of defence—the store of venom. Warning Colours and sounds and attitudes enable the snake to preserve this weapon when oftentimes it would be wasted.

WHITE ANTS

BY HAROLD BASTIN

White Ants are "Ants" only in name. The true Ant undergoes the complete metamorphosis of grub and chrysalis, while the newly-hatched White Ant is very like its parents, except in size and the absence of wings, arriving at the adult state through a gradual process of development unmarked by abrupt form-changes. It is, therefore, wise to substitute the name "Termite" for that of White Ant and avoid ambiguity. Termites are remarkably ant-like, however, in their manner of life. Most of the species dwell together in populous communities, the social organization of which is highly com-

Photo by] [*Harold Bastin.*

THE RAVAGES OF THE TERMITE.

A piece of wood riddled with tunnels and completely pulverized by the Termites. No woodwork—and little else—is immune from their ravages. The thing attacked—be it table, or box, or even a living tree—is always hollowed completely out and only a husk left.

TERMITES.

The male Termite is here shown with wings outspread; underneath to the left is one type of soldier Termite, while to the right is the worker. (All are a little larger than nature.)

plex, for there is probably no system of social life throughout the entire animal kingdom so elaborate as that of the Termites. But whereas the communal life of the Ant hinges upon the helplessness of the young—each Ants' nest being virtually a huge co-operative nursing establishment—that of the Termite does not admit of a like explanation, because the newly-hatched young are well able to fend for themselves. The members of a " termitarium "—*i.e.*, a community or state of Termites—vary in number from a few hundreds in some species, to millions—it may be hundreds of millions—in others. They are always divided into several grades, or castes, which can be distinguished readily by their appearance. In some instances the differences are so great that it seems well-nigh incredible that the various castes belong to the same species. The number of the castes varies greatly in different kinds of Termites, but there are always two sharply distinguished general classes—to wit, the propagators and the workers. The latter frequently include a number of individuals, easily known by the great size of their heads and jaws, which are termed " soldiers." It is commonly supposed that the soldiers are charged with the defence of the community, but recent investigations have cast grave suspicion upon their ability to accomplish this task. The worker Termites, indeed, appear to be better fighters than the so-called soldiers.

These soldier Termites, at least in the case of certain species, constitute a kind of scientific puzzle. With their great jaws they are unable to gnaw wood, or to avail themselves of the other sources of food-supply open to their companions. Thus they are condemned to long periods of starvation, broken at intervals by cannibal feasts. Not only do they devour the dead of the community and kill off the sick and ailing, they also, in times of excitement, exhibit a kind of mania, and destroy six or eight of their fellows with a few blind strokes of their huge jaws, incidentally providing themselves with a hearty meal. Yet they decline to eat the dead of alien tribes. In a word, these warriors seem rather to impose upon than to protect their community, and one marvels that they should be tolerated within its walls.

The worker Termites are true to their name. They build up and repair the nest, tend the eggs, collect and store food, and cherish the king and queen. These royal personages call for a special introduction. They are the only perfectly developed individuals in a termitarium—all the other members of the community being wingless and incapable of reproduction. When royal Termites first reach maturity they possess ample wings. At a certain season of the year these young princes and princesses are produced in very large numbers, and on a given day they all leave the nest, after the manner of true Ants, and disperse in all directions. Most of them are snapped up

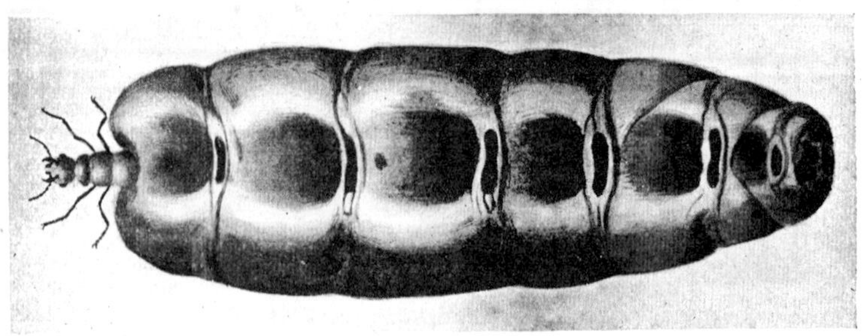

THE QUEEN TERMITE.

She is the mother of the hive, laying eggs at the rate of sixty a minute, that is eighty thousand a day. She is usually about three inches long.

Photo by] [*W. Saville Kent.*

TERMITE MOUNDS, ALBANY PASS, NORTH QUEENSLAND.

The neighbourhood of Albany Pass is noted for its colonies of Ants' nests, which have given the name of Ant-hill Point to one of its promontories. Notice the grass from which the mounds rise and which forms the chief food of these insects, for they are not wood-borers.

by birds, reptiles, and other insectivorous creatures ; but a few members of the swarm escape, and go off in pairs, male and female together, to found new communities. It is a curious fact that these royal Termites tear off their own wings immediately they reach the ground after their virgin flight.

A termitarium, when once established, contains ordinarily a single royal pair. The king, when he has discarded his wings, is not an imposing personage ; but before her reign is far advanced, the queen becomes the most amazing of all insects. She attains an almost incredible size and fertility, and her enormous body is vastly disproportionate to her head and thorax. She may produce as many as sixty eggs per minute.

As the death of a queen Termite completely disorganizes the community of which she is the very source and mainspring, one is not surprised to learn that provision is made against this contingency. The workers rear a certain number of " reserve queens," each being capable, should occasion arise, of assuming the dignity and functions of a reigning queen. Science cannot yet tell us in what manner these reserve royalties are produced. They are indistinguishable from the rank and file when they leave the egg ; but the worker Termites possess a mysterious power of modifying, checking, or accelerating the development of the young at will ; and from eggs which appear to be identical when they are laid by the queen, can produce royalties, soldiers, workers, or any of the intermediate forms which play a part in the economy of the community—just as they wish, or in accordance with the demands of the moment. Probably a dietary, selected both in quantity and quality, will bring about this result ; possibly, also, there is a system of massage and surgery.

The nests of Termites vary greatly in appearance, and many of them are very remarkable. Some of the less populous communities dwell within the decayed stumps and branches of trees,

Photo by] [*W. Saville Kent.*

A CURIOUS TERMITE NEST.

This type of nest belongs to a particular species of White Ant, which also lives on grass and has nothing to do with the wood-boring species. These nests will, when fully developed, attain a height of fourteen feet.

[By Theo. Carreras.

THE ROYAL CELL.

The Royal Cell is a compartment in the innermost part of the Ants' nest. It has only a narrow opening which allows the ingress and egress of workers and soldiers, but not of either king or queen. They are, however, always surrounded with a guard and a court; while a band of workers carries off the eggs which the queen lays to the Ant nurseries.

and merely construct a few barriers across the natural hollows of the wood, or make burrows into the heart of the rottenness. Some form elaborate hanging nests among the branches after the manner of wasps; others rear enormous structures. For instance, the African warlike Termite, already referred to, erects conical hills, which sometimes attain a height of from six to ten feet, and a circumference of from twenty-three to thirty-three feet; while the towers built up in process of years by certain Australian Termites occasionally attain a height of twenty-three feet. Another Australian species, the "Compass Ant," sets up inverted, wedge-shaped masses of dark-grey mud, from four to five feet high. They invariably have the same orientation—the narrow faces of the wedge pointing nearly north and south; possibly this is to avoid the extreme heat of the sun. To appreciate the marvel of their construction, we must remember that these huge piles are built up fragment by fragment, each atom set in place by the surprising effort of a single insect.

Termites commonly employ either earth or wood for building purposes, and the raw materials are always swallowed before being used a habit which ensures thorough grinding and amalgamation, and the extraction of any nourishment which they may contain. For Termites are frugal to a fault. Not only do they subject their very "bricks and mortar" to a preliminary digestive process; they actually eat over and over again the same food until (so we must suppose) its last atom of nourishment has been extracted. Moreover, the insects devour all refuse matter, and so maintain a spotless cleanliness in the nest.

In some of the more advanced of the termitaria mushrooms are grown for food. The workers heap up spongy, yellowish masses of wood pulp in the larger compartments of the nest, and these masses soon become plentifully sprinkled with a small, white fungoid growth of a peculiar kind, apparently induced by Termite cultivation. The fungus extracts from the wood-pulp its nitrogenous matter, and the insects thus secure a supply of concentrated food, which is said to be employed chiefly for feeding the younger members of the community. When a portion of prepared wood-pulp (or "mushroom-cake," as it is called) becomes exhausted, it is removed and replaced by a fresh supply.

Some Termites also feed upon the secretion of their own salivary glands, or that of their fellows. A tiny globule of liquid appears in the mouth. When it has increased to about one millimetre in diameter, it is either swallowed, used for building purposes, or, if the Termite is neither hungry nor in the mood for work, passed on as a dainty to another individual.

Photo by] [*W. Saville Kent.*

A MERIDIAN TERMITES' NEST.

So named from the position always occupied by this kind of nest, which has its narrow sides, one of which appears here, facing directly north and south. East and west the sides are abnormally long, and it is possible that the nest is so constructed as to get as little as possible of the sun's heat. These nests are found chiefly in Australasia.

In conclusion, reference must be made to the destructiveness of Termites. Wherever they pass they leave a trail of devastation. Furniture, clothing, books, flour and grain are quickly devoured, while woodwork is rapidly reduced to a heap of powder and chips. Only iron and the hardest stone appear to daunt them. It is even said that they are able to injure glass by means of their corrosive saliva. Years ago a species of Termite was accidentally introduced into the island of St. Helena. It increased and multiplied to such an extent that Jamestown was soon reduced literally to ruins, and new buildings had to be erected. Physically, the Termites are a feeble folk, much relished by birds and beasts, and can take no undue risks. They, therefore, approach the object of their desire through tunnels and covered ways; so that in tropical countries it is not an uncommon occurrence to find doors and window frames crumbling away without any visible signs of decay. When Termites obtain access to a building, they eat away all the interior of the woodwork, always leaving a thin outer shell intact, and never once exposing their persons to the vulgar gaze. In countries where they abound it is a rare thing to *see* a Termite, although the evidences of their existence are everywhere all too apparent.

ROCK-BORERS

BY BERNARD C. WHITE

THE purpose of a shell is obvious. It is primarily intended to serve as a protection to its soft-bodied owner, who would otherwise be left "naked to his enemies." Perhaps molluscs with clasp shells (known as Bivalves) are the most unprovided against attack, for however hard their shells may be they are not proof against well-directed force, as we have seen in the case of the oyster and the starfish (page 605). The Rock-borers, however, are amongst the wisest of their kind, for they have found a means of securing themselves against all intruders. They seek out a rock not too hard in texture, and with their heavy foot, which they use after the fashion of a centre-bit, they scoop out long, narrow, vertical pits which will in time be deep enough to entirely cover them. As a provision against being buried alive, they are supplied with particularly long siphons which extend beyond the orifices in the rocks. Almost every

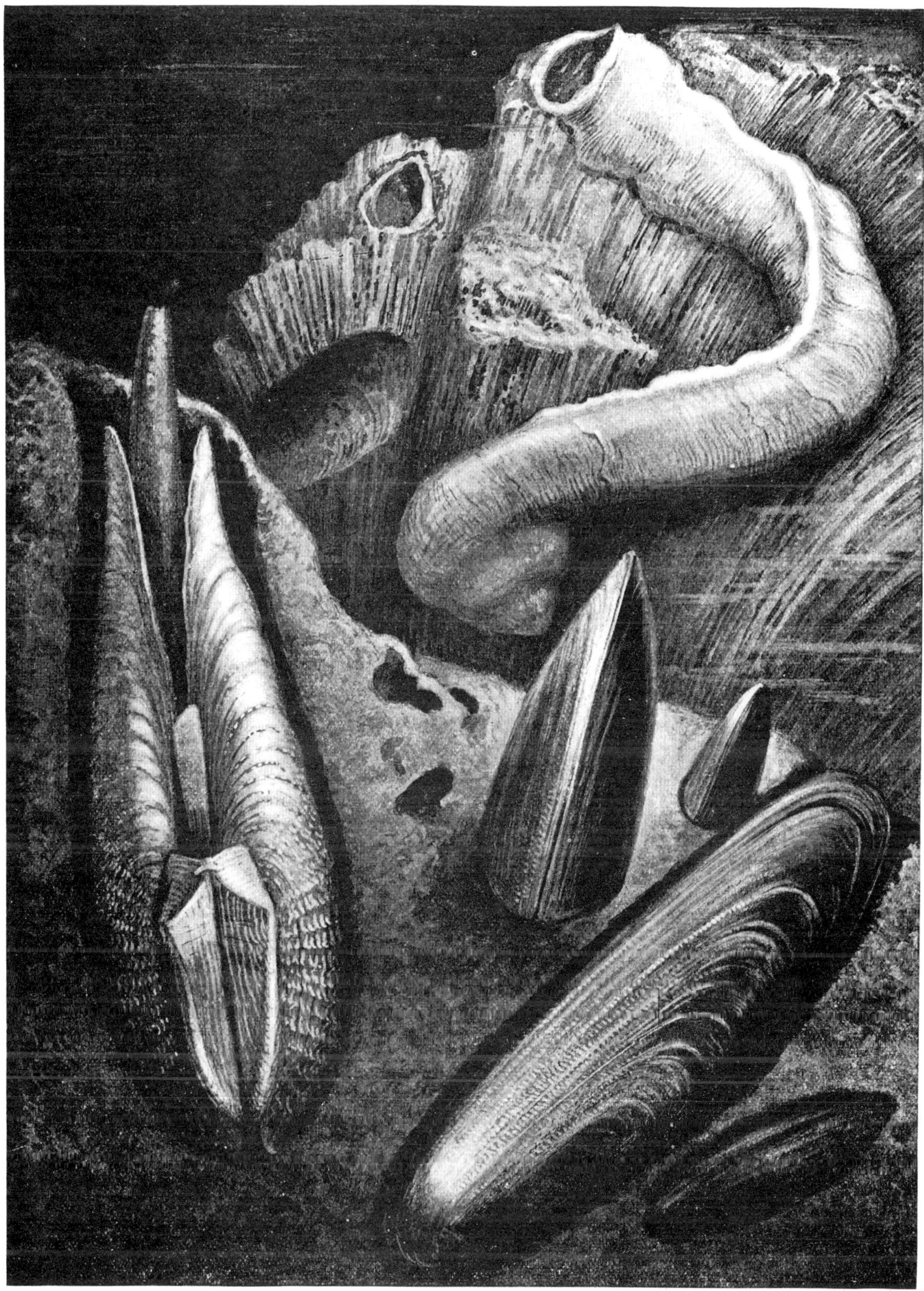

[*By Theo. Carreras.*

THE ROCK-SHELLS *IN SITU.*

Three kinds are here grouped together. In the background is the Magilus, who has extended his shell to prevent interment in the steadily growing coral. To the left is the Common Piddock in his burrow, showing the hinge-plates which strengthen his hard but brittle shell. To the right is the Date Mussel, which, in order to suit itself to its burrowing habits, has departed from the usual mussel shape and developed into a long cylinder with rounded ends.

bivalve mollusc has a pair of these siphons, one admitting fresh supplies of water to the gills, the other carrying it off.

Sometimes these siphons are united to form one single tube, and certain Rock-borers have enclosed them in an outer tube or sheath of a horny substance, such as composes the hinge of the two valves of the shell. The Streaked Martesia, shown on this page, is an instance of such a provision. The horny tube lines the whole length of the passage which it has bored and merges into the hinder end of the shell. In the Paper Piddock, this tubing has only been partially developed and forms a horny cup surrounding the siphons.

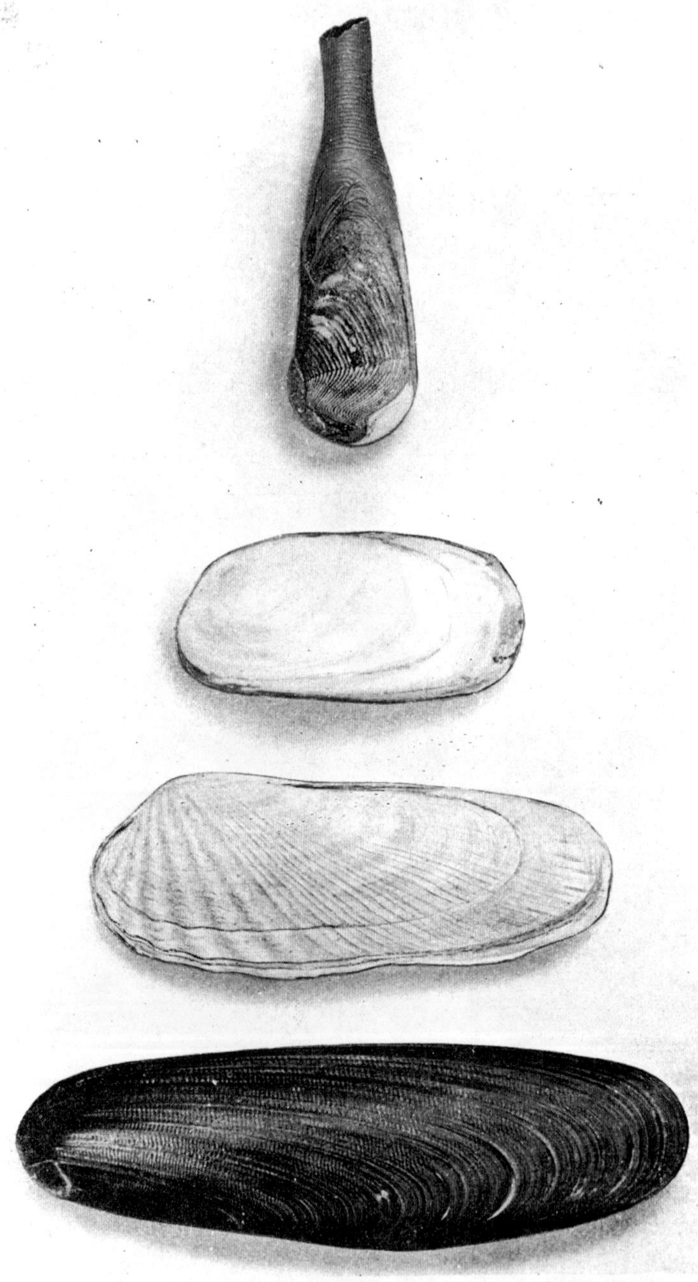

Photo by] [*Bernard C. White.*

BORER SHELLS.

The topmost shell is that of the Streaked Martesia, showing the horny tube which envelops the upper parts of the shell and serves as a protection for the extended siphons of the snail. The next is the Coral-eater; the third one of the Chalk-borers, and the lowest is the Date Mussel. Notice the similarity in shape and the file-like ridges of three of the shells. (Natural size.)

The chief of the Borers is the Common Piddock, which is shown in its rocky burrow in the illustration on page 697. Its shell is a remarkable example of Nature's architecture, for it has been planned, and its fabric has been strengthened, to bear the extra strain of the engineering work practised by its inhabitant. It possesses four shields over and above the two regular shells. Two of these are placed side by side at the hinge, like the buckles of a belt, and a third, which is considerably stouter, locks them together from behind; the fourth, which is shaped something like the pen of a squid, runs upward along the division from the hinge shield to the rounded extremities of the shell and protects their upper margins.

The whole of the pressure of the Piddock as it lies in its hollowed chamber rests upon the beak or foremost margin of the shells. Now, the Piddock uses his shell as a rasp to work away the rock, and a good deal of pressure is necessarily brought to bear on this beak; so we find that it is not rounded as in most clasp shells, but is shaped to something like a point, while a further provision for wear and tear is made by the construction of the hinge plate which turns back over the beak and is united to the shell proper by a series of partitions or ribs. To complete the system of structural support a detached and slightly curved rib, on the principle of a "flying buttress," has been added; it curves inside from the hinge plate and to it the muscles of the mollusc are attached.

It will necessarily be understood that since the foot is the chief instrument used in excavating these tunnels, it becomes an enlarged organ in the Rock-borers, and occupies a proportionately large space within the shell. This causes the two valves to gape at both ends, a characteristic very noticeable in all burrowing shells, both Rock-borers and Mud-burrowers; while it has given a name to a section of the latter group—the Gapers. On the other hand, the Date Shell, shown to the right of the full-page and again on the opposite page, which is also a Rock-borer (although of a distinct species), shows no tendency to gape. The reason is that the Date Shell burrows much less deeply than the Piddocks or the Gapers, so that the foot is not abnormally developed and affects only slightly the structure of the shell. But to the average bivalve such a gape would mean certain destruction. The Rock-borers, however, are differently situated. The gape occurs where little or no strain is felt, and securely lodged as they are in their rocky fortresses, they can defy invasion and can neglect with impunity this flaw in their coat of mail.

By permission of] [*Sir H. H. Johnston, G.C.M.G.*

THE LARGEST WOODPECKER IN THE WORLD.

This bird is from twenty-one to twenty-two inches in length. The male has a curled crimson crest, while that of the female is black. This handsome species is known as the Imperial Woodpecker.

One word as to the shape of the shells. The tendency is for the deepest burrowers amongst the Borers to have the most cylindrical shells, so we find the Common Piddock is much more elongated than the Piddock-shaped Borer or the Coral-eater (shown on page 698), and the Date Mussel, than any of the true Mussels; while the Wood Piddock, which is a near ally of the infamous Shipworm, and only buries itself in floating timber to a depth of about one and a half inches—that is, just sufficiently to cover the shell—is almost globular in shape, each valve being very deeply moulded.

The shells of some of the Rock-borers are further provided with ridges and points, which serve as rasps to help in the excavation of the burrow and to file the sides smooth. The Date Mussel and the Common Piddock have this characteristic to a marked degree.

There are two Borers which shouldbe specially mentioned. One is the Flask Shell which, besides making a burrow in sandstone, is " beautifully finished off as though lined with porcelain," and covers himself with a jacket composed of odd pieces of stone and sand stuck together by some secretion of the mollusc.

In the background of the illustration on page 697 is shown Magilus, an interesting coral dweller of Eastern Seas. This snail is not one of the deeply-boring molluscs. In its younger stages it has the habits and shape of the average snail, but as its age increases it seeks a permanent home in the coral-reef, and grooves out a little hollow sufficiently large to hold the shell with the mouth-piece open at the edge to allow the ingress and egress of the snail. But the coral grows and the snail is compelled to extend the limits of his cavern if he wishes to preserve his communication with the outer world ; he therefore adds whorl after whorl to the spiral, but not symmetrically as do most of his brother snails, with the result that, as he grows older, his shell develops into an irregular tube running through the coral mass with the original small spiral at the end. He is forced to continue adding to his structure or he would soon be completely enveloped in a coral grave, so that if he is able to survive till he attains a good age he may have a tube some seven or more inches long. He cannot occupy all this long corridor, so he leaves the spiral portion and comes out to live in the outer tube and blocks up the disused chambers of his home with a solid deposit of lime.

Photo by] [*J. J. Ward, F.E.S.*

THE NEST OF A MASON-WASP.

The most curious habit of the Mason-Wasp is its provision for the grub. A number of grub-cells are here shown, one of which has been broken open to reveal the caterpillars stored by the mother Wasp for the food of the yet unhatched grub.

THE LARGEST WOODPECKER IN THE WORLD

BY SIR HARRY JOHNSTON, G.C.M.G.

THE Woodpeckers are a sub-order of birds of almost world-wide distribution, the only areas of land surface from which they are absent being Australasia and the Arctic regions. For the most part they are smallish birds, ranging in size from that of a finch to a jay ; but the example here illustrated is almost the size of a crow, the adult being twenty-two inches in length. This form—the Imperial Woodpecker—belongs to the group of Ivory-billed Woodpeckers, the white beak being very hard and like ivory in appearance. It is a handsomely-coloured bird, mainly black

Photos by] [*J. J. Ward, F.E.S.*

THE NEST OF A MASON-WASP.

A sandy bank, between the bricks of a house, or the sunny side of a wall, may be taken as a site for the mud nest. This the mother Wasp builds up of little pellets of moistened sand and mud which she chews into a kind of mortar and moulds into a tube-like cell. There are eight cells shown here, each holding the grub of a Wasp. In the lower photograph the end cell has been opened and shows the grub, who has eaten up the eight or nine caterpillars provided for food.

[By Theo. Carreras.

THE NEST OF A MASON-WASP.

The nest of another species of these Wasps is shown here. The chief difference is that the cells are situated in a group at the bottom of a long burrow which projects from the surface of the bank in a sort of tube with jagged ends and a number of holes along its sides. The grub-cells number three or four in a group.

and white (the black being tinged with dark slate colour and green), and both male and female have quaintly-curled crests, this adornment being a vivid crimson in the male and black in the female. The male is slightly larger than the female.

The Imperial Woodpecker is still a native (in spite of constant persecution for its plumage) of Western Mexico, with a near relation—the Princely Woodpecker—in Texas and Florida. It is, of course, only found in the forest, and frequents mainly the great conifers (cypresses and pines) and the evergreen oaks in its search for the large grubs of beetles that are to be found in tunnels or crevices of the bark. The Ivory-billed Woodpeckers of the Southern United States and of Cuba resemble the Imperial Woodpeckers in general coloration, but are not such big birds.

THE LIFE STORY OF A MASON-WASP

BY JOHN J. WARD, F.E.S.

THE Mason-Wasp is a Solitary Wasp, and differs entirely in its habits from the Social Wasps. Amongst the Solitary Wasps we find only perfect females and males; for in their economy there is no room for the worker, or imperfect female. Both sexes appear together early in the summer, and their nest-building habits constitute some of the most astounding revelations of insect life.

This particular group of insects has been given the name of Mason-Wasps to distinguish it from another group of insects called Sand-Wasps. They are smaller and darker-coloured than the common Ground Wasps, and the abdomen is spindle-shaped, tapering at both ends. They are marked with yellow, usually of a brighter tone than that of the common Wasp, but the black colour predominates. A technical characteristic which readily divides them from the Social Wasps, and which may be verified with a pocket lens, is that their feet have two claws, while those of the Social group are simple.

Having distinguished our quarry, the next thing is to watch its movements. If there is a pump in the garden, that is a likely place to find the female of the species here described gathering materials to build her nest ; for it is built of mud. She is a wonderful worker and quickly gathers and carries away little pellets of moistened sand and mud which she agglutinates with mucus, producing a kind of mortar which she then plasters into a tube-like cell.

The site of the nest may be the crevice of a window-frame, between the bricks and the woodwork ; or along the top of a doorway ; the sunny side of a summer-house is another favourite position. Sometimes, however, a perfectly flat wall—and even that of a newly-built house—may be used, the Wasp using the mortar between the bricks as a base for her work, projecting her masonry outwards, as shown in the accompanying photographs.

The cell may be from three-quarters to one inch in length, and when it is formed, the Wasp changes her labour and becomes a hunter, and presently you see her returning with a live caterpillar, which she places inside the cell. Again and again she flies away and returns with other caterpillar victims until nine or more are closely packed in the cell (see photograph), together with one of her eggs. It should be observed, too, that although the caterpillars she places in the cell are alive, yet they are inactive; for when she effects their capture she stings them in some peculiar manner which is just sufficient to paralyse them, but does not kill them outright. In this manner she provides her progeny with living prey that will last it throughout its feeding period. When

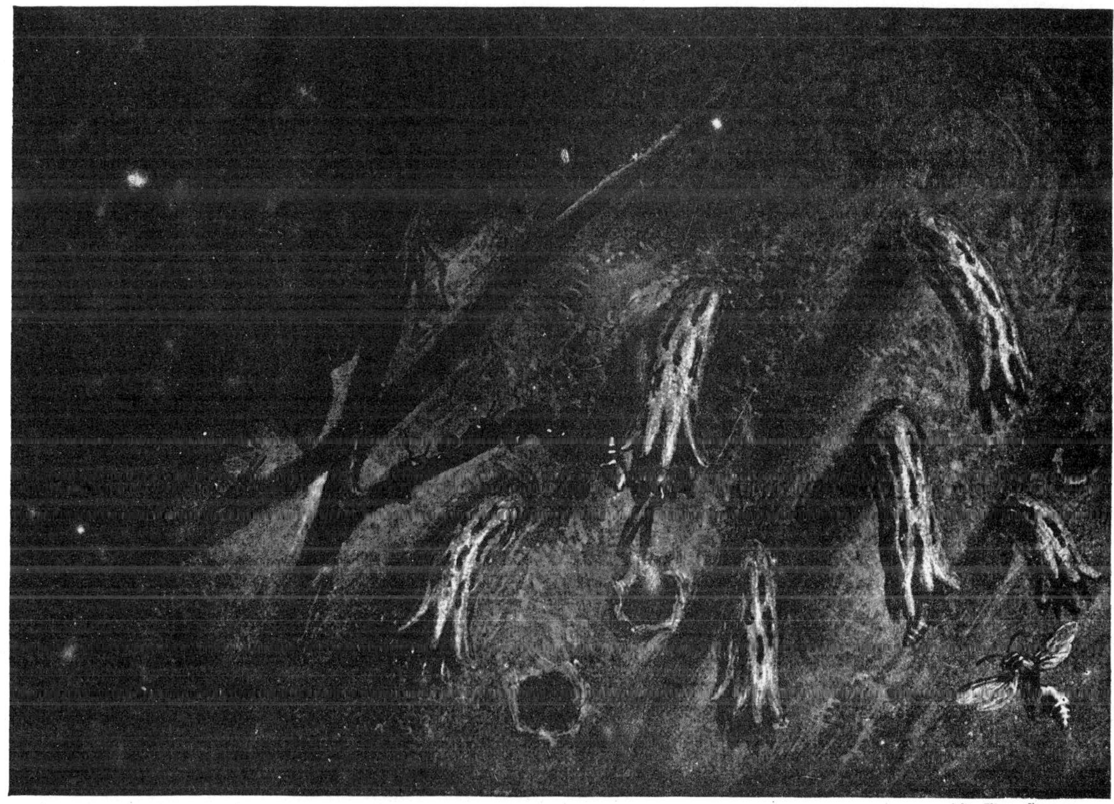

[By Theo. Carreras.

THE NEST OF A MASON-WASP.

The Wasps are here shown entering the nests. They are storing up food for the young grub, which for this species consists of a multitude of little green caterpillars. As many as one hundred and eighteen have been taken from a group of three cells. Ants are always ready to attack these Wasps, but they will never venture within the nesting-tube.

Photo by] [*J. J. Ward, F.E.S.*

THE MASON-WASP.

The insect, which belongs to the class of Solitary Wasps, can readily be distinguished from the better-known Social Wasp by its small size, its spindle-shaped body and the two claws on the feet.

the first cell is complete she seals it up and commences to build another immediately adjoining it ; sometimes the cells are made in a group, but more often they are placed in a line. With the closing of the cell the mother's interest in her offspring ceases ; she has done all that is needful for its welfare. The little wasp-grub emerges from the egg before many hours have elapsed, and quickly devours its provender. In eight or ten days it has become a large, pale-coloured grub which occupies the whole of the cell. It then lies immured in the cell until the early months of the following spring, and in the meantime changes into the pupal stage. About the end of May it bites its way out and emerges as a pretty little black and yellow Wasp, almost exactly like the parent Wasp that laboured so persistently on its behalf ; the male Wasp, when he emerges, may be distinguished by his much smaller size.

Such in brief is the life-story of this little Mason-Wasp when all has gone well ; sometimes, however, before the Wasp seals up her cell, she has a visitor—a most handsome, but, nevertheless, an uninvited guest.

The lady on whose house the Mason-Wasp had built her nest-cells shown in the photographs, told me how she and her children had watched a " lovely green and red fly " which sometimes came to the wasp-cells. From its description it was, doubtless, the Ruby-tailed Fly, which is like a gorgeously-attired house fly, its head and thorax, or middle portion, being of a vivid metallic green, while its abdomen glows with ruby fire. It should be observed, however, that it is a four-winged fly ; indeed, it is related to the wasps on which it preys.

This insect will deposit its eggs in the cells of almost any wasp or bee to which it can gain access, and while the mother Wasp is searching for caterpillars, this parasitic visitor awaits its opportunity to enter her nest. Sometimes it is caught in the act, and attacked by the Wasp, when it immediately rolls itself into a ball and presents its glittering armour to the sting of its attacker, which in this case is useless. If it escapes unobserved, the mother Wasp carefully seals up the cell, enclosing the egg deposited by the fly. Then if, a little later on, the cell is opened, its larva will be found to have become full fed, but no trace will be found of the wasp-grub that should rightfully occupy the cell. It is highly probable that it would be the first and last meal of the parasitic grub, for its method is, I think, to emerge from the egg rather late, when the wasp-grub has finished its food supply, and then devour it.

The same Mason-Wasp may have several " nests," the number of cells largely depending on the capacity of the site selected. Sometimes, too, this same species will build its nest in holes in sand-

CURIOUS HAILSTONES.

Taken from a very early photographic print in the possession of the Royal Meteorological Society.

[By M. Quénisset.

A SHOWER OF ENORMOUS HAILSTONES.

After a drawing made by M Quénisset of a particularly severe Hailstorm which descended upon Juvisy. The individual stones were as large as oranges.

banks; and other species of the same genus with somewhat different habits may also be found in the same situation. Another species selects hollowed bramble and raspberry stems, making its cells in the pith. Holes in old gate-posts may provide still another species. Various species will select all sorts of singular places when they enter houses: a pistol-barrel, folded paper, the edges of books, the mouth-piece of a horn, the drawer of an old-fashioned looking-glass, and a house-keeper's work-box, are all objects in which cells of these interesting insects have been found.

Furthermore, we have much to learn regarding the curious habits of these insects, and likewise those of their parasites, as but little knowledge is available regarding several British species. They are insects which, after one's attention has been called to their habits, may be observed by any amateur student of insect life, and should the reading of this article be the means of any of my

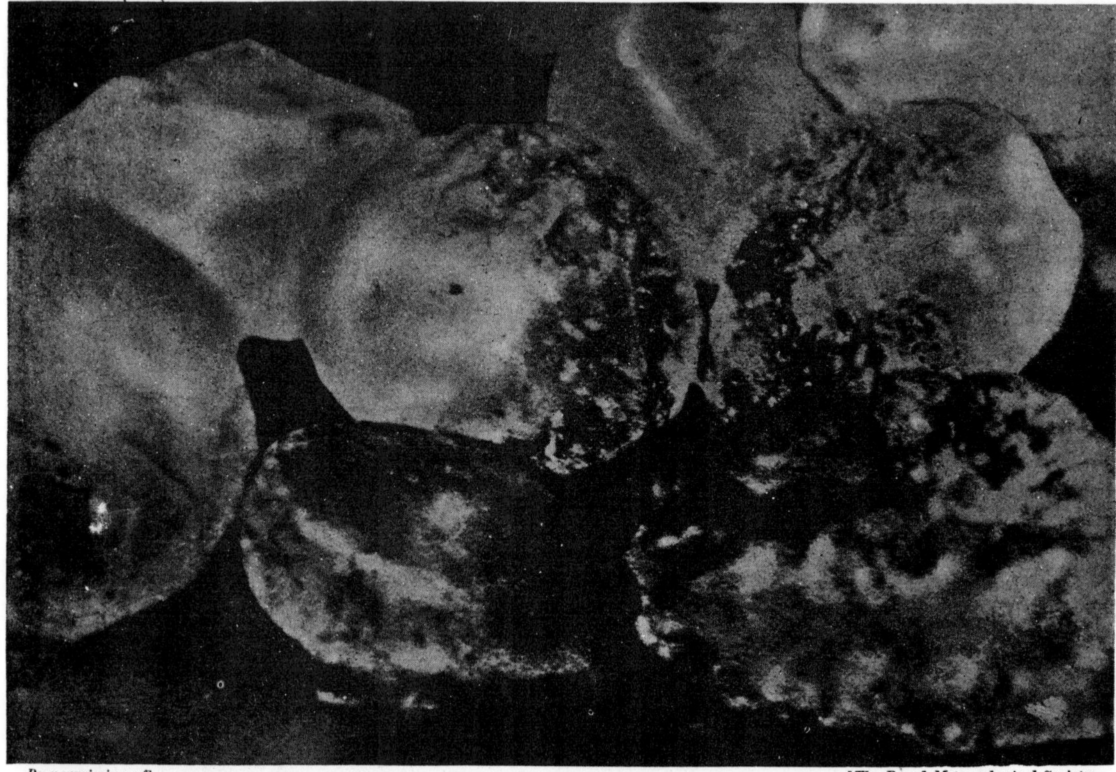

By permission of] [*The Royal Meteorological Society.*

ENORMOUS HAILSTONES.

A photograph of Hailstones which fell at Seaford, May 30th, 1897. They are shown actual size.

readers recognizing one or more species of these insects during their nest-building season, I should feel grateful if they will carefully record the sites selected, material used, and, as far as possible, the size and markings of the insect, together with the place and time of its appearance, and address these particulars to me through the publisher of this work.

HAIL

BY RICHARD KERR, F.R.A.S.

As regards hailstones, we are more fortunate than our friends on the Continent. We very rarely have to experience a bombardment of these frozen missiles the size of a turkey's egg. Yet, our neighbours, according to most reliable accounts, are at times subjected to falls of hailstones larger than a man's fist, and some surprising descriptions of enormous hailstones and tremendous hailstorms are recorded by several French writers, including the famous astronomer Flammarion. So we find insurance offices against damage by hail are to be seen in most Continental countries.

We used to be satisfied with the statement that Hail was nothing more than frozen drops of rain. But this definition does not satisfy the scientific mind of to-day. Every department of science is undergoing the scrutiny of keen specialists. Nothing in Nature is, to them, too small or too insignificant for close examination. This being so, Hail is raised to a position of equal importance with the snow-crystals and other beautiful forms of water.

[By M. Quénisset.

A transverse section of the central Hailstone shown on page 709, showing its formation of a series of hexagons round a central hexagonal core.

Hail is often the accompaniment of a storm at a time when the surface temperature of the earth is extremely high, and the process of cooling is correspondingly rapid; this cooling of the air is continually going on, for all hot air ascends, and, consequently, becoming cooler, is absorbed by the sun's rays in the form of vapour; but it is only when this condensation takes place with peculiar rapidity that Hail is formed. The greater cold necessary for freezing is provided by the rapidity of the evaporation taking place over the earth surface; for evaporation has always a tendency to lower the surrounding atmospheric temperature. Now at the time of a storm the air-currents are often of very varied temperature, so that the substance falling from the storm-cloud, be it either a rain-drop or a snow-flake, is affected by the varying temperatures through which it has had to pass. It is melted and congealed often many times before it reaches the earth, and this rapid series of transitions causes it to take the ultimate form of an ice-mass or Hail. It is not necessary that these variations in temperature should be remarkable; a few degrees above and below zero would be sufficient to affect the transitions from ice to water. Not only does the atmospheric temperature affect the form in which the water vapour returns to earth, but the electricity present in the storm-cloud is in some degree responsible for the phenomenon. Hail-storms are noticeable for the presence of electricity in a high degree.

It may possibly surprise some of our readers to learn that *large* hailstones are not simply frozen quantities of water that have fallen directly from the clouds, but that in the process of growth, or of enlargement in bulk, they have fallen as small particles from clouds of enormous height; have descended, it may be, a thousand or more feet; have *ascended* again to a great height; have fallen

[*The Monthly Weather Review.*

A Hailstone which was taken up at Morgantown, W. Virginia. It measured roughly two inches in height and had a volume of about one cubic inch.

[By M. Quénisset.

Transverse sections of Hailstones, showing their ordinary internal structure.

again; and this upward and downward journeying process may have to be repeated several times before they are sufficiently heavy to enable them to break away from the grasp of the controlling power in which they are held. This is only a small part of Professor Wilson A. Bentley's theory. It is well worth closer examination, inasmuch as he has devoted many years to the study of the phenomena of snow, ice and hail. The results he has arrived at are original, ingenious and reliable.

The central particles, or nuclei, consist of either tiny pyramid-shaped or star-shaped granular snow within the upper portions of very lofty clouds, which are violently expanding vertically. The ascending air-currents may develop into whirlwinds sufficiently rapid to enable them to carry *upward* with them all the raindrops and granular snow that come within their grasp. These are released or expelled from the grip of the powerful upward air-currents only at or near the summit of a spreading cloud. The smaller ones are blown out horizontally, and fall to the earth as ordinary-sized hailstones. The larger ones, caused by the melting and re-freezing, fall first and nearest to the central vortex, and as they fall earthward and become partly melted, they encounter strong horizontal indraught air-currents, and are drawn by them again into the *ascending* vortex and are again carried far upward and converted into ice and re-coated with granular snow, and eventually expelled again, as in the first instance, from the summit and spreading portions of the shower-clouds This theory assumes that the summer hailstones begin as granular snow, and are partly melted by partial descent into warm air, and are congealed through ascent within the higher clouds. Powerful air-currents must blow upward through hail-producing clouds to sustain and buoy up the larger hailstones for the length of time necessary for them to " grow " to so considerable a size. The rapidly rotating air-currents would account for the spiral markings displayed on many of the larger hailstones, and there is no known limit as to the number of times the hailstones may descend to lower strata of air and be forced upward again to colder regions.

The hailstones shown below, which were picked up by M. Quénisset (to whom we are indebted for the illustration) on the lawn of the Observatoire Flammarion at Juvisy, seem to substantiate this theory, for they were made up of *layers* of ice alternately opaque and transparent, and were without a central core—a fact which points to their growth under different atmospheric conditions, such conditions being caused by variation in the air-currents.

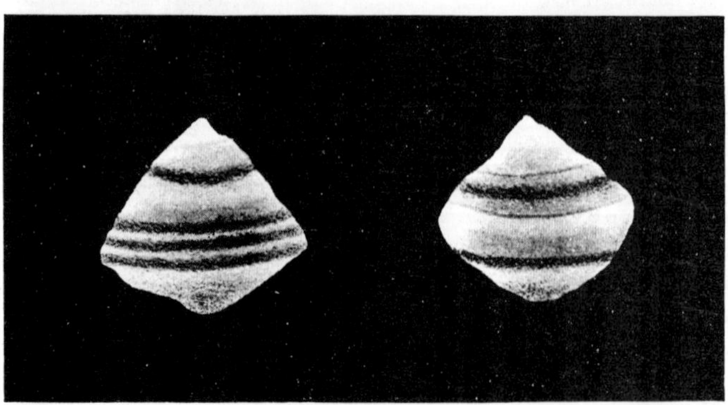

[By M. Quénisset.

Hailstones which fell at Juvisy in 1909, showing their composition of different layers of ice They are natural size.

All this line of reasoning applies chiefly to the formation of summer hail. Unlike winter hail, summer hail occurs over very small areas and is purely a local phenomenon. In general, winter hail is the product of wide-spread general storms. The melted snow, that descends as liquid rain-drops from the lower side of a warm

[By M. Quénisset.

DIFFERENT FORMS OF HAILSTONES.

The four largest of these Hailstones were picked up after a severe storm at Auxerre. The two central ones fell just outside Tiflis on the 12th September, 1863. The first four are remarkable for the combinations of both transparent and opaque ice, and the central two for their division into numerous facets like a badly-cut diamond.

air-current, freezes into hail while the drops are falling through the cold substratum of air lying between the warm stratum and the earth. Winter hail usually occurs simultaneously over large areas, and the hailstones frequently contain tiny air-bubbles. Some possess a few large bubbles and air-tubes which radiate from the centre outward.

Hailstones assume many fantastic shapes. The most usual form is the sphere, though oval and pear-shaped stones are found in considerable numbers. These latter are due to the fact that the larger round ones, because of their greater weight, fall downwards faster than the smaller hailstones and smaller cooled raindrops, and hence overtake and merge into these. Cone-shaped hailstones are likewise no uncommon variety. They usually have a more or less spherical base, and a particularly fine specimen of this type is shown on page 707. This, with a number of similar stones, fell at Morgantown, West Virginia, on April 28th, 1877. Each measured roughly two by one and a half inches, with an average volume of almost one cubic inch. Perhaps some of the most abnormal shapes among hailstones fell at Auxerre in a terrific storm, which swept over the city on the 29th of July, 1871. The stones fell in the full blaze of sunlight, and owing to the extraordinary conditions — for there was no apparent atmospheric disturbance — they preserved their original shape. The most characteristic forms were sketched by M. Naudin, and are shown in the four corners of the full page. In the centre of the same page are two stones which were picked up on the 12th of September, 1863, just outside Tiflis. Their ellipsoidal form is broken up into a number of facets much after the manner of a roughly-cut jewel, and a transverse section of one of these stones is shown in the upper illustration on page 707. Other peculiar formations are shown on page 704. They have been taken from a very early photograph belonging to the Meteorological Society. Here the conical form has been broken up into jagged projections, much as the spherical stone shown in the bottom right-hand corner of page 709 has been altered by the agglomeration of other ice particles.

THE SLUG-LIKE CLION.

One of the Shell-less Sea-snails known as "pteropods" or "wing-footed," from wing-like expansions on each side of the head. During life it is of a beautiful purple colour and swims in vast shoals at the surface of the Arctic ocean, where it forms the main source of the food supply of the Greenland Whale.

In France hailstorms, as a rule, follow the courses of rivers and valleys, and this is especially the case when the storm-clouds are low. This fact is borne out by investigation in the countries particularly subject to such storms. If, however, there is nothing to cause the storm-cloud to deviate from its course, the usual direction is from the south-west to the north-east; that is to say, the direction of their origin—the Atlantic currents. Hailstorms appear to occur in cycles of ten years, and those of the greatest violence have occurred in the evenings.

THE FOOD OF WHALES

BY W. P. PYCRAFT, F.Z.S.

THE achievement of the impossible has been accomplished again and again by Nature! And no more convincing illustration of this fact could be found than that furnished by the bill of fare of certain of the Whales. The Rorquals, or Fin-whales, which include the largest animals which the world has ever seen, feed for the most part on herrings, smelts, and other small fish, engulfing them in shoals. As the huge cavern of a mouth is opened the luckless crowd, striving to flee from the wrath to come, are suddenly and irresistibly carried in with the intake of water. And to increase the capacity of this mouth still further, an enormous pouch of skin, recalling that of the pelican, is suspended from the lower jaws. To enable its walls to resist the tremendous pressure of the volume of water taken in, strengthening bands of tendon, as strong as steel and without its brittleness, run from one end to the other; thus it is that when the mouth is empty the walls of the pouch are thrown into great longitudinal folds or pleats, the pleats answering to the silk pleats of a closed umbrella, the tendons to its steel ribs. As soon as the mouth is full, muscles attached to the tendons draw in the walls, and the great fleshy tongue is raised towards the roof of the mouth. The consequent reduction in the size of the chamber drives out the water between the lips. To prevent the food captured at the expense of so much energy being driven out with the stream, huge plates of "whalebone," tapering to a point and having their inner margins frayed out, are suspended from the sides of the upper jaw; that is to say, along each side of the roof of the mouth. Between the interspaces of these plates the water is forced, the frayed inner edges forming a most effective strainer and preventing the escape of even the smallest particles of solid matter.

The full force of the importance of this mechanism was not realized until the amazing discovery was made that some species, such as the Common Rorqual, subsist, during many weeks, at any rate, on a diet of small shrimp-like crustacea, which it hunts with its still larger relative, the Blue Rorqual—the largest of all the Whales—during August, among other places, off the coast of Ireland. In spring, this giant—the Blue Rorqual—contrives to find

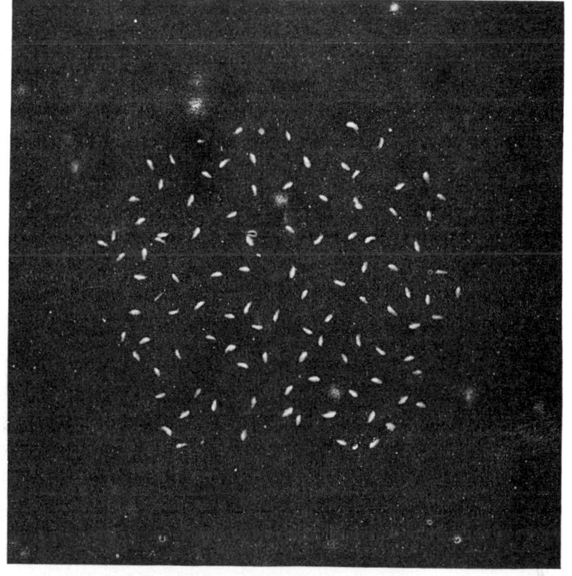

THE LONG-HORNED TEMORA.

These minute creatures are common in the English Channel and other coasted waters. The tribe contains a large number of species, and on them Rudolph's Rorqual is apparently entirely dependent.

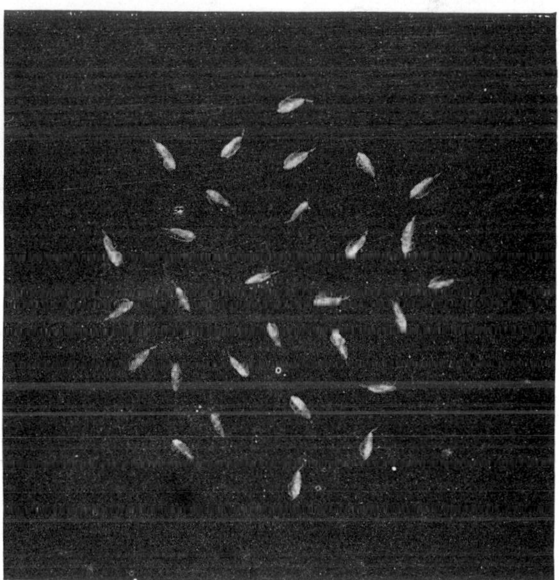

THE FINMARCH CALANUS.

Its home is in the sub-Arctic seas, and it is carried down in spring by the East Icelandic Polar stream to its spawning place south of Iceland. The enormous shoals produced here are carried back, multiplying as they go, along the coasts of Norway during the summer and autumn.

an ample subsistence on the small crustacea known as Euphausia, or Unarmed Rhoda, and a nearly related species which is eagerly hunted during August. But even more remarkable is the case of Rudolph's Rorqual, a species which attains a length of some fifty feet, which is said to subsist entirely on crustacea, not only of the species already named, but on the much smaller Finmarch's Calanus, and the still more minute Long-horned Temora, shown in our illustrations.

No less wonderful is the case of the huge Greenland Whale, which feeds mainly, if not entirely, on minute surface-swimming animals ; and especially the pteropod mollusc known as *Clione limacina*, which, during life, is of bright purple colour. That creatures so huge should be able to subsist on the bodies of animals so minute as those shown in our illustrations seems incredible. And, indeed, it would not be possible were it not for the fact that these creatures swarm in the sea in such vast quantities as to be almost beyond human conception. But almost more wonderful still is the very existence of these creatures in such swarms ; the power of multiplication which they display is unparalleled in the animal kingdom. It seems incredible that they could withstand such appalling inroads on their ranks. Yet, when we come to reflect upon the matter, we cannot resist the conviction that the very existence of these giants of the deep is to be traced to the presence of these feeble folk. In earlier times there were no whale-bone whales, and their becoming is to be traced to Nature's effort to relieve competition among the earlier-made, tooth-bearing whales by transforming some in such a way that they should be able to avail themselves of this superabundant food-supply. The truth of this view is shown by the fact that these whalebone, or " Baleen " bearing Whales were once on a time toothed Whales ; for even to-day the vestiges of teeth are to be found in the jaws, though they not only never cut the gum, but disappear before birth. The history of the origin of whalebone has yet to be traced. The best, that is to say the most elastic, is that obtained from the " Right Whales "—the Greenland, or Arctic Right Whale of the Polar Seas, and the North Atlantic Right Whale, the South Atlantic Right Whale and the North Pacific Right Whale. But according to some authorities these are all but geographical varieties of a single, widespread species. But be this as it may, these Whales differ conspicuously from the Rorquals, and especially in this matter of the whalebone, the longest blades of which may measure as much as twelve feet long.

A NORWEGIAN "SHRIMP."

It swarms in fairly deep water in many parts of our seas, and forms no small part of the food of the Common and Blue Rorqual. If a jar of sea-water containing specimens of this creature be brought into a dark room a tap on the glass will cause a flash of phosphorescent lights like a row of tiny lamps along the sides of the body.

But during the course of ages a change seems to have overtaken all the Whales in the matter of their food ; for among what are known as the " Toothed Whales " many are practically toothless, as the Bottle-nosed Whale for example. The huge Sperm Whale, or Cachalot, has lost all the teeth in the upper jaw, but has retained a row of conical pegs in the lower jaw. These

Photo by] [*H. J. Shepstone.*

THE NAOSAUR.

A photograph of a restoration of one of the wonderful extinct reptiles. Particularly noticeable is the development of the spines which project from the vertebrae of the back. During life they supported a fold of skin after the fashion of the fin of a fish.

Photo by] [H. J. Shepstone.

THE SKELETON OF THE NAOSAUR.

The Naosaur was not much larger than a large dog. Notice that from the abnormal spines were also developed small spines at right angles to the main shaft, for what reason it is impossible to conjecture.

may serve to hold its prey, which consists almost entirely of squids and cuttlefish—some of enormous size. That this is so is rather surprising, for cuttlefish are agile creatures, and have a trick of covering their retreat by squirting out an inky fluid. Somehow the Sperm Whale seems to have broken down this means of defence. We cannot tell what takes place in the mysterious depths which form the hunting-ground of the Sperm Whale, but we must assume that these cuttlefish and squids swim in great shoals, and that somehow the Whale must have devised a means of approaching such shoals without revealing his near presence. From the bodies of his victims the Sperm Whale distils the precious "ambergris," which is expelled from the intestines in the form of a solid, nearly black mass composed largely of the horny beaks of squids and cuttles. This precious matter is largely employed in perfumery, and commands a high price, commonly as much as ninety shillings per ounce.

It seems incredible that bodies so huge should be nourished on food so unsubstantial, for it must contain a large percentage of water; but fact is stranger than fiction.

NAOSAUR—THE RECORD BREAKER

BY W. P. PYCRAFT, F.Z.S.

A MORE extraordinary creature than the Naosaur has never lived. A glance at its skeleton will suffice to make this much certain; it looks more like an ingenious human "fake" than a piece of Nature's handiwork. Nevertheless, there is no getting away from the fact that the Naosaur was once very much alive, and that the remains which have been so skilfully pieced together were made in Nature's workshop. Naosaur was a reptile, not by any means a large one as ancient reptiles go, but of about the size of a large dog; and it throve a few million years or so ago—during the period known as the Permian. But it must have been fairly numerous, and must certainly have existed over an immense period of time, for its remains are found scattered half over the world, since they have been recovered from rocks in Bohemia and in France, as well as in Texas, U.S.A., from whence the finest specimens have been obtained. Its nearest living relation is the strange Tuatera, or Sphenodon, of New Zealand.

The most remarkable feature of this most remarkable animal is the enormous growth of the spines which project from the vertebræ of the back. In some living animals to-day, as in the Bison, these are of great length; but never before or since in the world's history have spines been seen such as Naosaur succeeded in developing. During life they supported a fold of skin after the fashion of the fin of a fish. But this is not all; for from base to tip these spines developed smaller spines at right-angles to the main shaft, like hat-pegs, and what purpose these served no one has been able to divine. After its back-fin—for such we may call it—the most striking feature of this strange animal was perhaps its head, whereof the jaws were armed with most formidable teeth.

The only animal to-day which displays anything at all comparable to this strange excess in the growth of the skeleton is the Flying Lizard, wherein the ribs are of enormous length, projecting for a great distance on each side of the body, and similarly support a fold of skin. But in this case the result is not merely "ornamental," for these skin-folds form a parachute, enabling their owner to take flying leaps through the air.

THE JEW'S EAR FUNGUS

BY EDWARD STEP, F.L.S.

THERE is a certain spot on top of the Surrey Downs where there is a considerable thicket of elder-trees. A person with susceptible nerves might consider the spot uncanny when he noticed that these barkless elders had apparently got swarthy human ears sticking out from their bare wood, reminding him of the days when it was the fashion to demand an ear as a legal fine and to nail it to the pillory. These ears are not all of good form, but some of them are perfect at the distance of a few feet, provided that one sees them on a damp day.

If, dismissing all gruesome ideas, we boldly approach a dead elder and make a close examination of one of these objects, we shall find that it has the stiff but yielding consistence of our own ears, velvety to the touch, but cold and thin. It is really the fruiting stage of a fungus, whose threads have been ramifying through the trunk and breaking up its tissues, so that a good kick would serve to snap one of these thick stems; and now it is manifesting its presence outwardly by developing these ear-like gristly flaps, upon which myriads of reproductive spores are ready to be carried off by the wind. It belongs to a group of fungi that are characterized by gelatinous flesh that hardens into thin horn under the influence of sun and wind, but plumps up again into its old softness after an hour's rain.

Photo by] [*E. Step, F.L.S.*

JEW'S-EAR FUNGUS.

Sticking from the bare wood of a dead elder-tree, this fungus looks like a human ear nailed to the pillory, or whipping-post, of former days.

THE CROSSBILL

Uses its beak for wrenching open pine-cones.

The other day I was showing a friend some "specimens" of fungi, and he came across some dry and shrivelled Jew's Ears. On telling him what they were, he was manifestly disappointed. "Not much in the name! That's nothing like an ear." I placed a specimen in water and told him he should see it later. After about half an hour's soaking I removed it, and dried it on a towel. When I placed it on his hand he let it fall to the ground at once, and showed repugnance to the thing. He admitted it was far too like the real thing to be pleasant.

THE PEEWIT.

Its beak serves as a pair of forceps to pick up worms and insects.

The Jew's Ear is not restricted in its growth to elder-trees. It is said to grow also on the elm, but I do not remember ever to have met with it save on elder, and then only on dead trees that had lost their bark. I have several times attempted to inoculate a healthy elder with the fungus, but have not succeeded in doing so.

Respecting the name, the reader may be tempted to ask why *Jew's* Ear? I believe the name has some connection with the traditionary belief that Judas hanged himself on an elder-tree. Judas' Ear in later days would get shortened into Jew's Ear. I have not consulted the philologists on the point, or made independent research, but on the face of it, that looks like the probable evolution of the name.

SOME MARVELS OF BEAK TRANSFORMATION

BY W. P. PYCRAFT, F.Z.S.

THAT the lower animals, at any rate, live to eat, rather than eat to live, is made manifest by the often elaborate mechanism which has been devised to secure this all-important end. The birds demonstrate this fact better than most animals, and the evidence they present is seen by their beaks. Take that of the Raven, for example. This can be used with equal facility as a dagger, a hammer, a pair of pincers, an

THE MACAW

Has an immensely powerful beak for cracking nuts.

THE NIGHT-JAR

Is provided with a bag under the beak to hold the prey, which is gulped down as it flies.

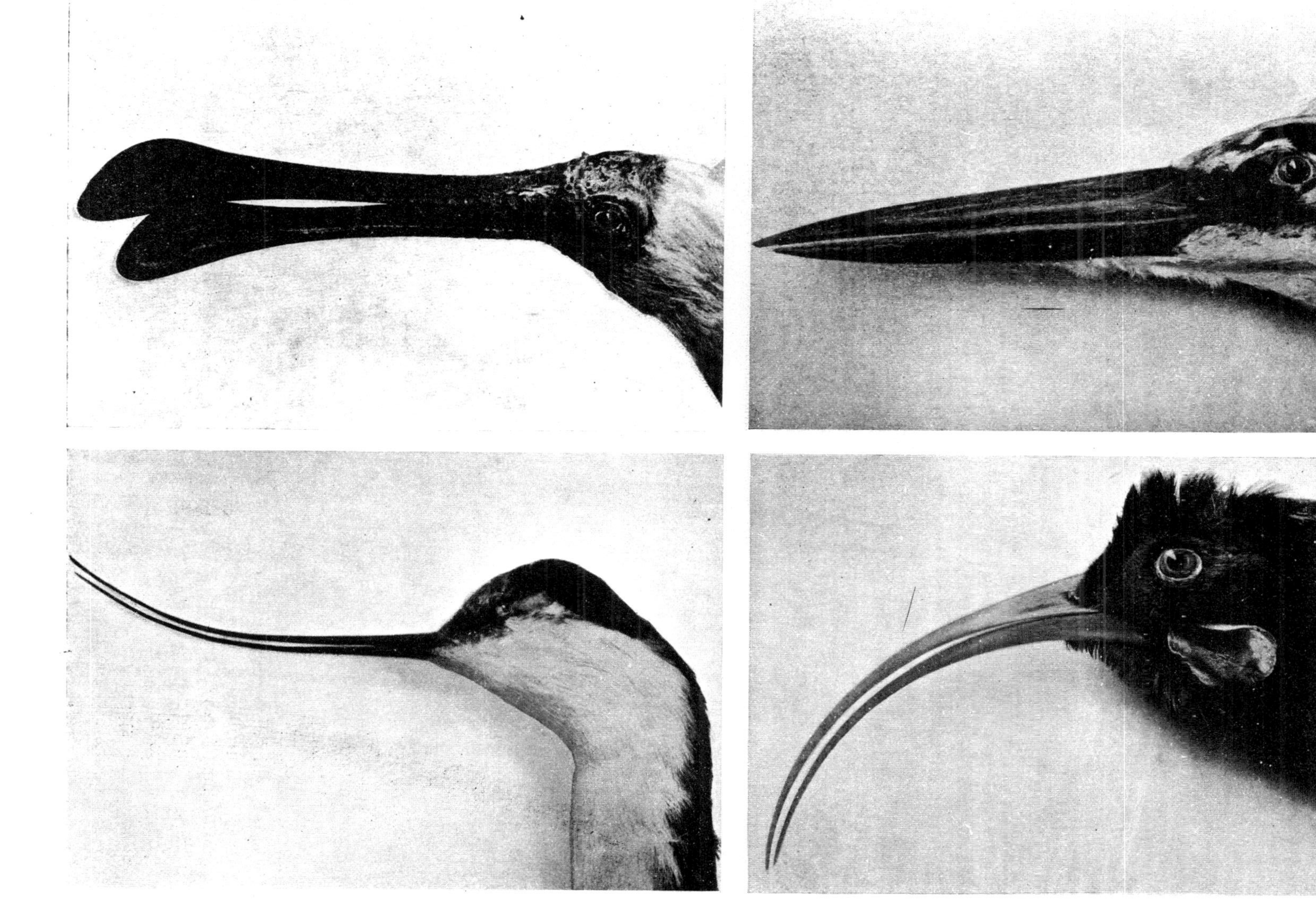

BIRDS' BEAKS.

According to the nature of the food supply and the difficulties of obtaining it, so the beak of the bird is adapted to overcome the difficulties and capture the prey. The beaks here shown are those of the Spoonbill and Heron, and underneath of the Avocet—all these three are fish-catchers. The last is the female Huia bird, who uses her beak as forceps to catch insects.

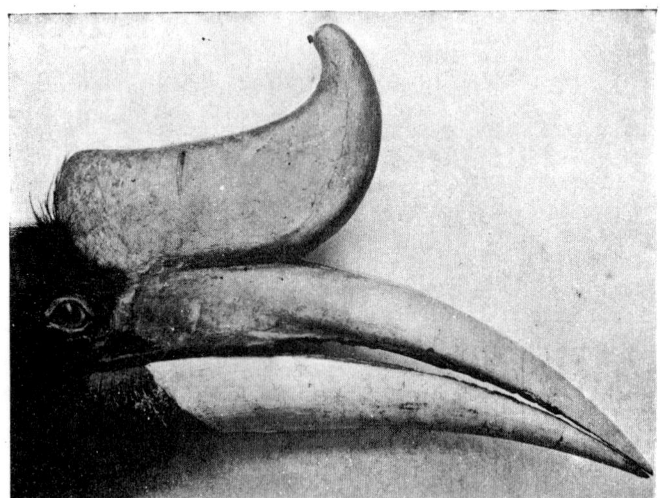

The Rhinoceros Hornbill's ornamental beak.

The Puffin's beak—fashioned for holding fish.

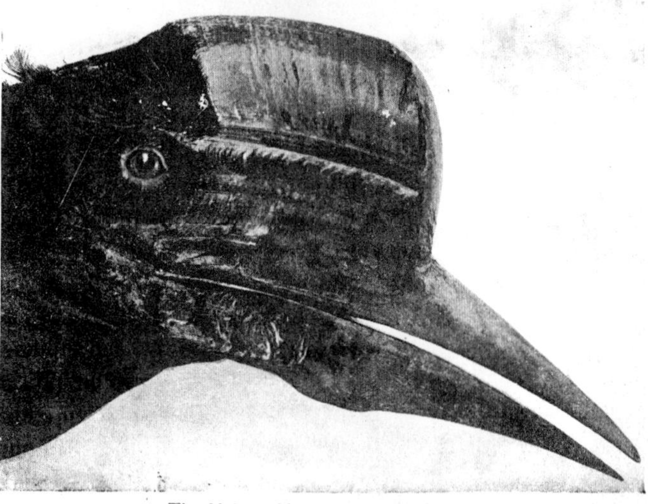

The Helmet Hornbill's useful beak.

instrument for rending and tearing; hence, from its multifarious uses, it forms what we call a "generalized" type of beak, as distinct from a specialized type, such, for example, as that presented by the Scissor-bill. The beak of this bird is unique. The lower jaw, it will be noticed, is longer, considerably longer, than the upper; but more than this, each is flattened from side to side, till it presents the appearance of a couple of paper-knives! Only at the gape do these blades expand to form the cavity of the mouth. The Scissor-bill is a fish-eater, and catches its prey after a quite peculiar fashion, skimming along the surface of lakes and rivers, with the long, lower blade in the water, the upper raised, so that in careering over shoals of small fish one victim after another is caught as in a cleft stick.

The various uses to which the Raven's beak is put can be studied in action separately in certain widely different birds. Thus the beak as a dagger is seen in the Heron, where it is used to spear water-voles and fish; it is also a dagger, and a pair of forceps as well, in the Kingfisher. The Peewit's beak is also used as a pair of forceps, for picking up worms and insects. In the Woodcock, we have a beak which is formed on the model of, and is used like, a pair of surgical forceps: that is to say, the tip only is used to grasp objects out of sight, as when the beak is thrust down into soft, saturated soil for the purpose of securing worms at the end of the thrust. The beak as a hammer attains the maximum of perfection in the remarkable Helmet Hornbill. In all the other Hornbills—and they are all remarkable birds—the beak is extravagantly, not to say grotesquely, developed, as may be seen in the photograph of the Rhinoceros Hornbill; but in the Helmet Hornbill the horny beak-sheath, in all other species of great thinness and delicacy, is of excessive density, as hard as ivory and very thick, as is shown in the adjoining figure.

The beak of the Raven can at need, it was remarked, be used as a pickaxe; but in the Woodpecker it has become a far more efficient weapon, which can splinter even the hardest wood. By intuition, the Woodpecker knows which trees have fallen victims to the insidious trickling of rain down the stumps of broken boughs to the very heart of the tree, and strikes unerringly through the hard outer casing of wood to gain the desired hollow within, wherein he may make a nursery for his family. Another bird skilled in woodcraft is the strange New Zealand Huia bird: strange not so much, indeed, from the form of the beak, as from the fact that the male and female have beaks of different types. In the female it is long, slender and downwardly arched, or "decurved," while in the male it is short, stout, and only slightly arched. Now these birds are very fond of a fat beetle-grub which lives in decaying trees, and the female drags these forth by thrusting in her long, curved forceps. Occasionally she detects one of these morsels in a cavity too narrow for her beak, when, it is said, she calls her mate, who promptly enlarges the hole with his more powerful weapon—and bags the morsel for himself!

Some beaks make most efficient shears. Two very different types are those of the little Crossbill and the great Macaws. In the Crossbill the jaws are strongly arched in opposite directions, and the points cross to opposite sides of the beak. They prove most efficient instruments in wrenching open pine-cones and fruit. The Macaw's beak is immensely

The Scissor-bill—a fish-catcher.

The Raven—a flesh-tearer.

The Woodpecker's beak is adapted for splintering bark.

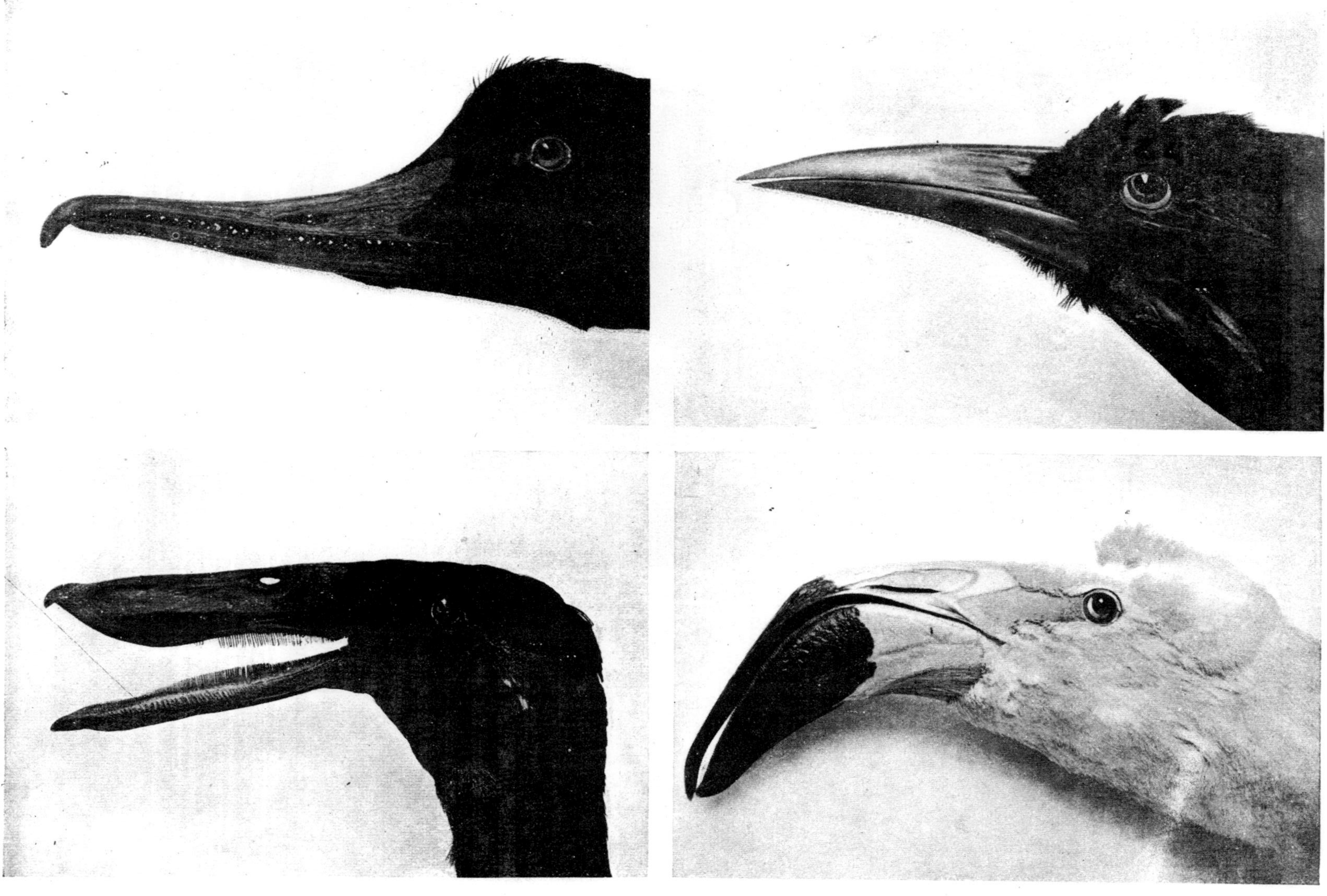

BIRDS' BEAKS.

The two upper beaks are those of the Merganser, a fish-catcher, and the male Huia, a wood-insect eater. Below are the Shoveller Duck and the Flamingo—both feeders on minute vegetable and animal organisms.

powerful, and will shear through even the hardest nuts with but a very slight effort. Beaks for tearing living flesh vary in shape somewhat. But the most efficient forms are such as are seen in the Falcons, where, in addition to the sharp, hooked point, a tooth is developed on the cutting edge.

So far, all the beaks of which we have spoken play an important part in securing the daily bread. But there are some birds which may almost be said to be able to dispense with a beak. In the Night-jar, for example, this is little more than a horny knob at the front edge of the rim of the mouth, which is enormous, being cleft back to beyond the eyes. Now the Night-jar does not capture its prey—beetles and other insects—by means of its beak, but gulps them down as they fly, his open mouth forming a remarkably capacious bag.

The number of birds which have acquired peculiarly-shaped beaks as aids to wringing subsistence from the waters that cover the face of the earth is legion. One of the most remarkable of these, the Scissor-bill, we have already described; but another, surely no less curious, is that of the "Open-bill," of which one species is Indian, one African. Herein, the jaws display a great gap along the middle. Now this most curious organ is apparently used for the capture of shellfish, though in what it proves more serviceable than the more typical beak of, say, the Oyster-catcher, which also lives largely on shellfish, it is difficult to say. The Pelican, the Merganser and the Shoebill are all fish-eating birds, and their beaks display most surprising diversity. But the Pelican loves to thrust his capacious jaws into a shoal of fish, pouching his victims within a most capacious bag suspended from the lower jaw. The Merganser chases his victims under water, holding his captive till he reaches the surface by means of the horny, tooth-like edges of his jaws. In the White Pelican, by the way, the ridge of the beak in the male develops during the courting season strange, horny excrescences of an irregular, oblong shape. They are probably "sexual adornments," but of a kind quite unique. The Avocet and the Spoonbill present us with very diverse types of beak, yet which are used for a similar purpose—skimming small animals from the surface of smooth water. In the Avocet the beak is curved upwards and drawn out to a fine point; in the Spoonbill, on the other hand, it has become transformed into a pair of flattened, spoon-shaped blades. The Shoveller Duck and the Flamingo, again, feed on similar food—that is to say, on minute vegetable and animal organisms, which are caught by passing a large quantity of water rapidly through the mouth and straining off

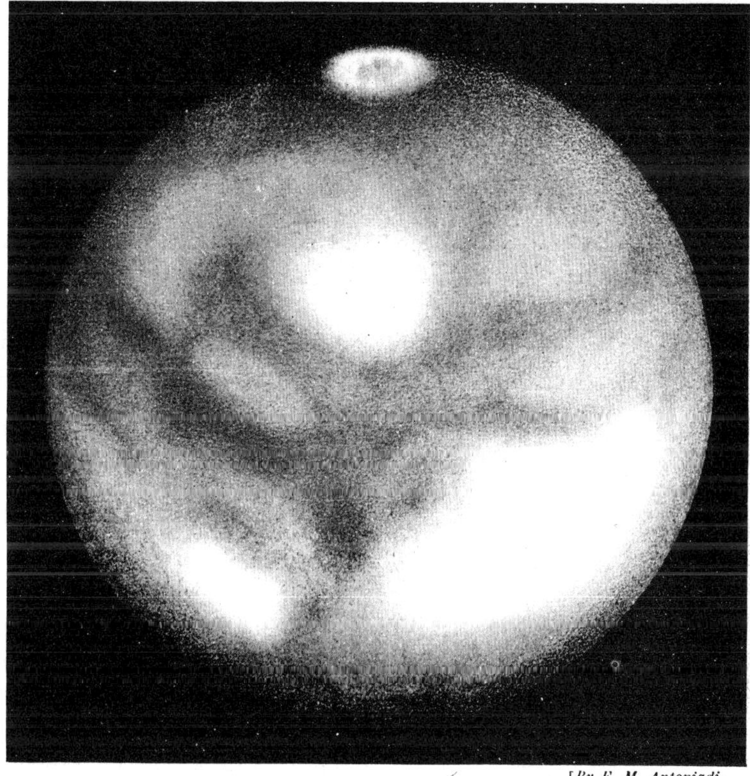

[*By E. M. Antoniadi.*

MARS.

This drawing of Mars by M. Antoniadi shows many variations in light and shade when compared with his beautiful water-colour, which is reproduced as a coloured plate.

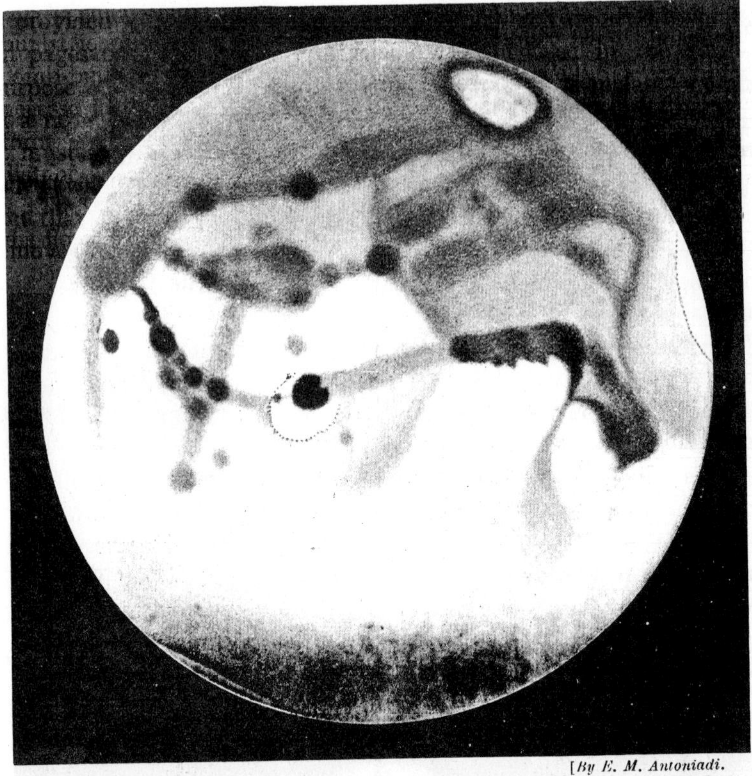

[By E. M. Antoniadi.

MARS.

Still further fluctuations are observable here, and it is this variety which militates against the idea that the dark lines seen on the surface are really artificial and permanent waterways.

the solid matter by means of horny plates which run along the edges of the jaws. And the process in both is aided by a great fleshy tongue. The curious angle at which the beak is bent in the Flamingo is perhaps associated with the great length of the neck and legs. For though the Flamingo can swim well enough, it feeds when wading; and it would seem that, this being so, the peculiar shape of the beak fits best with the position of the head.

Finally, we come to the beak of the Puffin. This, like the beak of the Merganser, is fashioned for holding fish, and widely though it differs in shape, it is still as effective in use. But we are concerned here with another of its aspects, and that a curious one. Briefly, at the autumn moult, the large, gaily-coloured plates at the base of the beak are shed, to be assumed anew in the spring; herein we are reminded of the beak of the White Pelican, which sheds and renews the horny plates which project from the ridge of the beak, as shown in our illustration.

THE "CANALS" OF MARS

BY E. WALTER MAUNDER, F.R.A.S.

FORTY-EIGHT million miles! This is the distance from the earth of the planet Mars, on November 17th, the date of its nearest approach this season. Its nearest approach; but forty-eight million miles is a very distant nearness! How can we form any real conception of such a number, of such a distance? One second of time quickly passes; forty-eight million seconds stretch out to much more than a year and a half. Six thousand worlds as large as our own, placed side by side, would not bridge the gulf between ourselves and Mars, even at its nearest approach. Yet eight thousand miles, the diameter of our earth, is not a short distance; it is the distance traversed by a steamer in going from London to Buenos Ayres; the distance in direct flight of London from Honolulu.

The beautiful painting of the planet Mars by the gifted Greek astronomer Antoniadi, which accompanies this number of "MARVELS OF THE UNIVERSE," if set up three-quarters of a mile from the observer, will appear larger than the actual planet when nearest to the earth. What possibility is there that we should ever learn anything about the handiwork of its inhabitants, if there be any such, upon a world so small and so remote?

True, we have the telescope to help us, but this at its best only means that we can see Mars as well as we could see M. Antoniadi's drawing at distances of from twelve to twenty feet, or as we can see the sun or the full moon with a child's shilling spy-glass. The difficulty of the enormous distance of the planet is indeed reduced: it is not overcome.

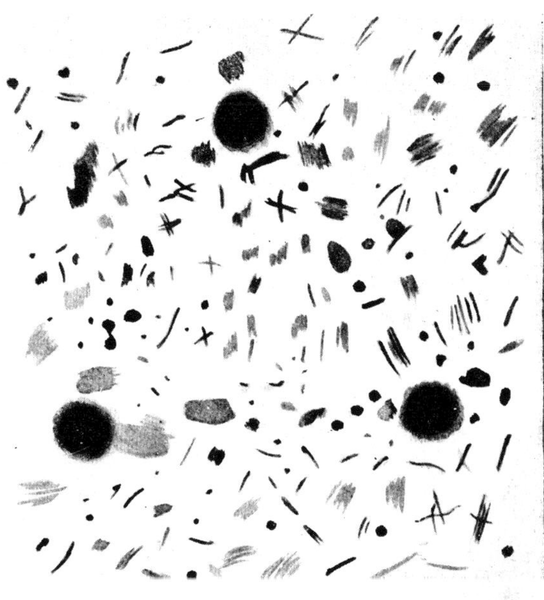

CANALS IN MARS?

These two drawings support the contention that it is only distance which causes the markings on Mars to appear as straight lines. Here are shown a number of irregular patches—

But Mars, when seen under good conditions and in a perfect telescope of eight inches aperture and upwards, shows a remarkable structure, well-brought out in the accompanying sketch (page 724) made by the late Major Molesworth on April 2nd, 1903. The planet is covered by a network of long, narrow lines, arranged in a kind of geometrical pattern, roughly resembling a spider's web; and this net-work appears to Professor Lowell, the astronomer of Flagstaff, Arizona, U.S.A., so manifestly "unnatural," that he considers it to be proof of the agency of highly intelligent beings. He claims that these narrow lines—the notorious "Canals"—must be artificial waterways; or, rather, the belts of vegetation lining the banks of such waterways and irrigated from them.

In a previous article it was shown that from a consideration of two facts relating to Mars that we know with certainty—the feeble action of gravity at its surface, and its distance from the sun—it may reasonably be inferred that the mean temperature of the planet is everywhere below the freezing-point of water, so that its seas are ice to the bottom. Such a world can be no fit home for life; except, perchance, in its lowest, most rudimentary forms; and a world in the grip of a frost so intense could not be one where titanic hydraulic engineering works would ever be dreamed of or carried out with success.

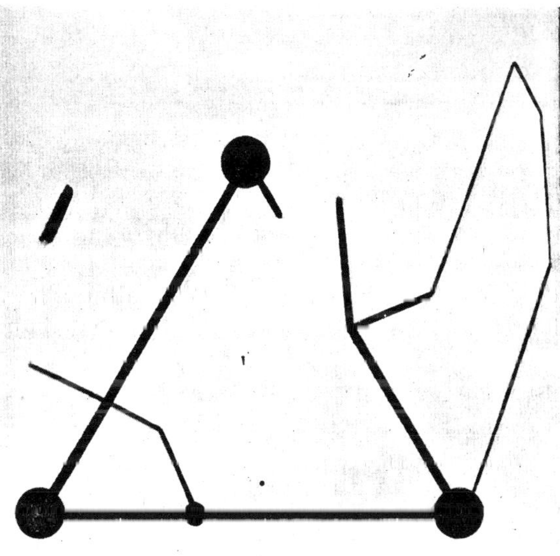

CANALS IN MARS?

—which contract into a continuous series of lines with circular points of junction when viewed at a distance of thirty feet.

But the network of the "Canals" exists; if it is not the outcome of the efforts of Martian engineers and "navvies," how can it be explained?

This explanation is simple. These "Canals" look so regular because Mars is too far away for us to see their irregularities; they look straight, because Mars is too far away for us to see their crookedness. A straight line is the smallest object that we can perceive, and the smallest object that we can perceive necessarily looks like a straight line. The explanation of the narrow, straight "Canals" lies in the forty-eight million miles that extend between them and ourselves.

Everyone knows how easy it is to detect a telegraph line seen against the background of the sky. Not everyone knows that to a person of normally good sight such a wire can be perceived at a distance from it of two hundred thousand times its breadth ; or to use a technical expression of the astronomer, " It subtends an angle of one second of arc." So in September, 1896, a long stream of spots was seen on the sun, that to the naked eye appeared a long, straight, narrow gash on its surface ; but in a good astronomical telescope the long, straight line resolved itself into scores of spots, separate and distinct, and each full of exquisite detail.

A single round, black spot can be perceived, if the observer is six thousand times its diameter from it ; *i.e.*, if " it subtends an angle of thirty-four seconds of arc." But for fully-defined vision, for the object to be truly seen in its actual shape and form, the observer must not be more than one thousand seven hundred times its diameter away ; it must " subtend an angle of two minutes of arc." Within this wide range, from one second to one hundred and twenty seconds of arc, all small objects will tend to show themselves like Martian " Canals " and " oases "—that is, as straight lines and round dots.

CANALS IN MARS?

The above sketch of the planet, which was made on April 2nd, 1903, shows the presence of the supposed canals in a marked degree. It was made by the late Major Molesworth when Mars, under good conditions, was seen through a telescope of eight inches aperture.

The two diagrams on page 723 illustrate this principle : looked at close at hand, they appear utterly different ; but from a distance of thirty feet or more they look almost exactly alike. The irregular dots and scratches adopt the appearance of the straight " Canals."

The nearer we approach an object, the better we see it, and increase in the power and efficiency of our telescopes effects an analogous improvement.

Schiaparelli, with telescopes of eight and eighteen inches aperture—Lowell generally used twelve inches—saw " Canals " by the score. Barnard, with telescopes of thirty-six and forty inches, Antoniadi with one of thirty-three, find the spider's web of straight " Canals " giving way to an infinitude of intricate detail in which there is no hint of " ruled lines," or artificial arrangement. Finally, photography with the great five-foot reflector of the Mt. Wilson Observatory has added its word, and the restless canal-diggers of Mars have gone to join the gnomes and fairies, the trolls and kobolds of our dead mythologies.

WATER BUGS

BY EDWARD STEP, F.L.S.

BUGS, in the entomological sense, are insects that are continuously active throughout life, not changing their form, but developing four wings which are fully formed and usable only in the last phase ; and whose mouth-parts take the form of a fine-pointed rostrum or beak, which they insert into the flesh of animals or plants in order to suck blood or sap. This rostrum has a hinge at its base which allows it, when not in use, to be folded down under the breast and between the basal joints of the legs. This order (the Hemiptera) contains thousands of distinct species, over four hundred of them inhabiting Britain.

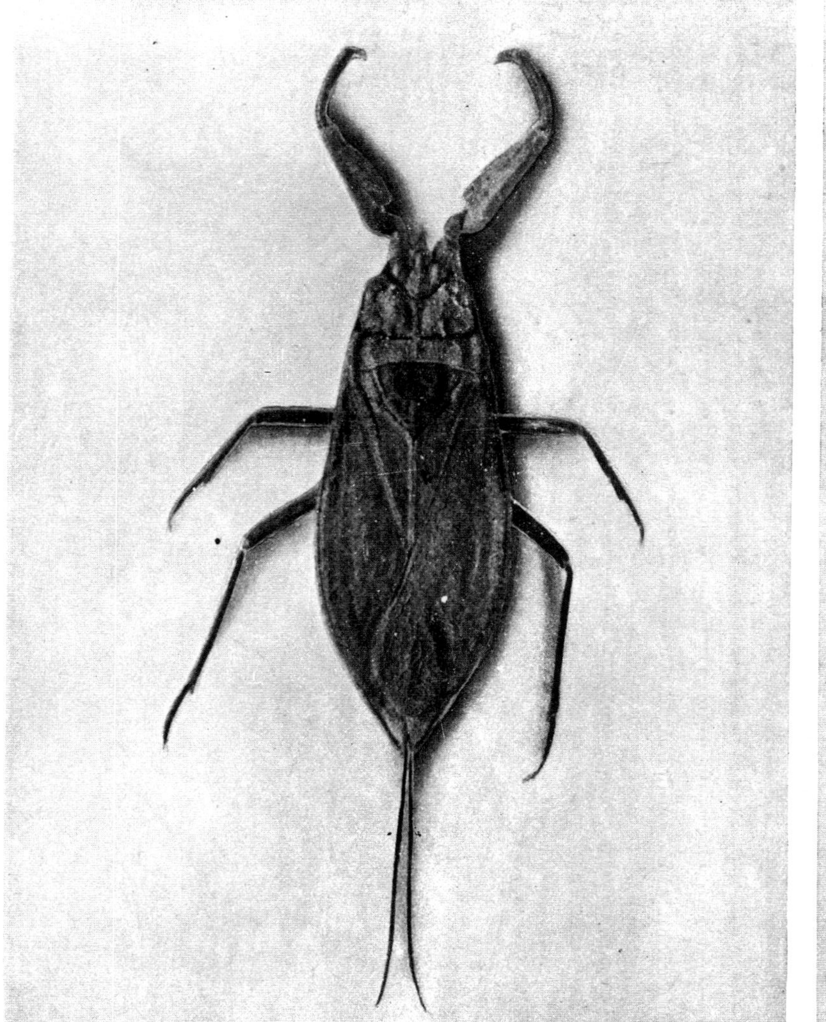

Photos by] [*Hugh Main, B.Sc., F.E.S.*

THE WATER SCORPION.

Both upper and under side aspects of this badly-named Water Bug are given. That of the upper side shows the wings closely packed upon the back and the small size of the head, almost hidden between the basal joints of the forelegs. On the underside are seen the overlapping of the segments of the body, the attachment of the limbs, and the great distance between the first and second pairs of legs, the former occupying almost the position of antennae, which are absent from this species. (Four times the natural size.)

Photo by] [*H. Main, F.E.S.*

WATER SCORPION.

The nymph or pupal condition, in which only rudimentary wings are evident. (Twice the natural size.)

But it is not our purpose in this present writing to attempt to deal with Bugs generally. We propose to take as examples a few that are addicted to an aquatic life and which may be studied in any pond. Several of these are known to country boys, not as Water Bugs, but as Ditch-skaters. They have long, thin, spider-like legs, which are probably the natural prototypes from which man has evolved the idea of ski and snow-shoe. The principle is, at least, the same: that if a sinkable weight is distributed widely over a fluid or other yielding surface, it will not sink. Look at these Ditch-skaters moving nimbly over the surface of the pond, and you would imagine that they are unsinkable. So active are they that if you wish to carry away anything more than a generalized idea of their forms and structure you must catch a few and examine them at close range. I say a few, because a single specimen would be very misleading, as you might therefrom get the notion that they are all alike. There are really a number of distinct species, so distinct in some cases that they represent different genera, though at a glance most of them are alike; but it will be found no easy matter to capture an individual, so quick are they to elude hand or net, doubling on their tracks with astonishing facility.

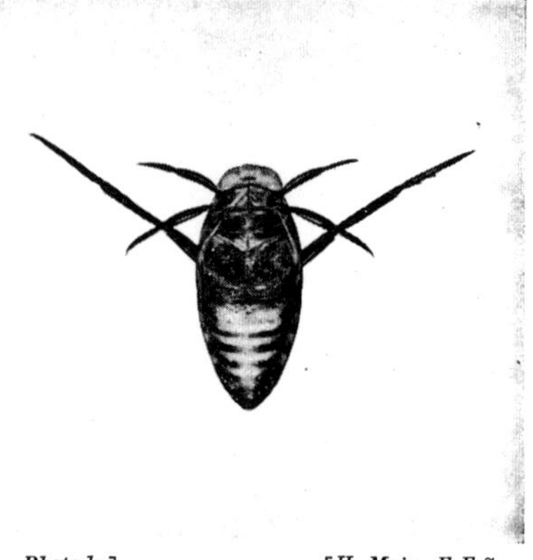

Photo by] [*H. Main, F.E.S.*

THE BOATMAN.

Empty pupal skin from which the fully developed Boatman has emerged. (Twice the natural size.)

One of the most plentiful of these Ditch-skaters is Gerris, whose dull, almost black back is margined with a thin reddish line. Its length is only a third of an inch. One would hardly suspect it of having wings, but there are four of them, neatly packed and lying close to the back. The underside is clothed with short, silvery-grey hairs. There is a peculiarity about the legs. In most insects the bases of these all come close together along the centre of the underside. In Gerris the first pair are only just behind the head, whilst the second and third pairs appear to be attached half-way through the insect's length, but the appearance is really due to the first joint being closely pressed to the body; and instead of having their attachment in the middle line they arise from the sides of the body. This arrangement has its importance in connection with the peculiar habits of the insect, and

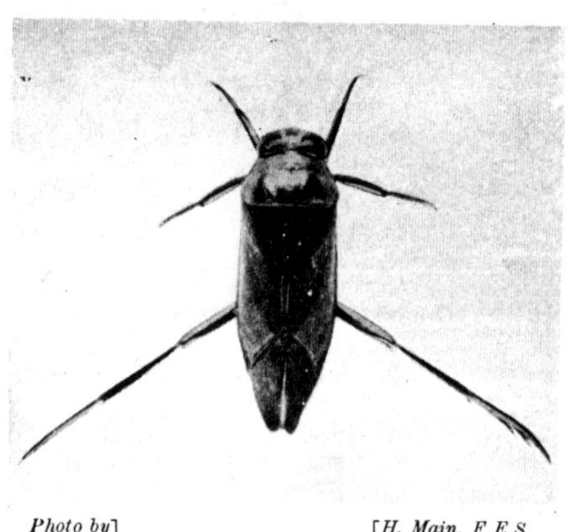

Photo by] [*H. Main, F.E.S.*

THE BOATMAN.

The stout body slopes from the sides to a ridge along the back. As the Boatman swims on his back this ridge becomes a steadying keel. (Twice the natural size.)

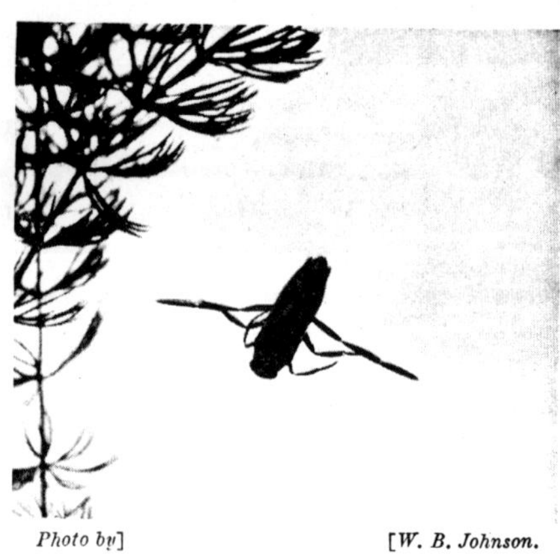

Photo by] [*W. B. Johnson.*

THE BOATMAN.

This Water Bug swims on his back, and uses the long third pair of legs after the manner of sculls. (Slightly reduced.)

suggests comparison with the outriggers on racing skiffs. The rowing of Gerris is effected by the middle pair of legs, which are much longer than the others. The purpose of what looks like their purposeless movements on the water is to keep a sharp look out for the larvæ of other aquatic insects, which are secured by the front pair of legs until the sharp beak can be inserted like a dagger and the victim's blood sucked.

More remarkable than Gerris in appearance is Hydrometra. Only half an inch long, its length is twelve times its greatest breadth. Its head is drawn out to such an extent that it accounts for two-fifths of its entire length, and the eyes are situated half-way along the head. Its general appearance is much like a miniature copy of a Stick Insect, to which, however, it is in no way related. In this species the wings are undeveloped, only the merest rudiments of them being discoverable. The legs, though of considerable length, are reduced to the minimum of thickness, but they are used for walking instead of rowing. These insects are so extremely slender that they require to be looked for very closely.

The Water Cricket of the angler is not a cricket at all, but one of these Water Bugs known to entomologists as Velia. It is built more on the lines of Gerris than of Hydrometra, and it is very like Gerris in its movements. It may be distinguished from the latter by its greater breadth, its broadest part being between the bases of the wings and forming the middle line of a diamond-shaped area including the head and a space between the folded wings. I should say, if the wings are there to be folded, for it is only occasionally that the wings are developed. It appears as though a small percentage of individuals are set apart by Nature to be explorers, and provided with wings that they may leave the piece of water where they were born and seek unoccupied ponds for the purpose of populating them. When the wings are present they are dark-brown with several white spots; in their absence the insect may be known by the presence of two orange ridges along the back. Somewhat similar related forms to Velia are Microvelia and Mesovelia, which must be looked for among the water-weeds and bog-moss; but they are only about a fourth of the size of Velia, and so must be searched for very closely.

No such difficulty, however, attends the detection of Ranatra, if it is present in the pond you are investigating, for it exceeds two inches in length, from the tip of the short beak (which is always kept extended), to the end of the tail bristles—which, however, account for a third of its length. But its apparent length is considerably increased by the

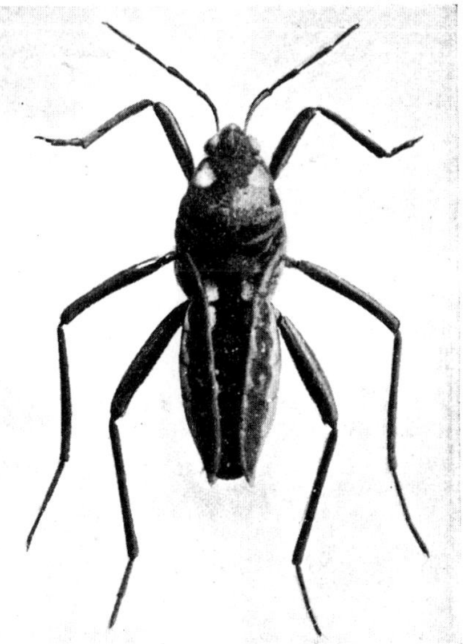

Photo by] [*H. Main, F.E.S.*

"WATER CRICKET."

An adult that has not developed wings.

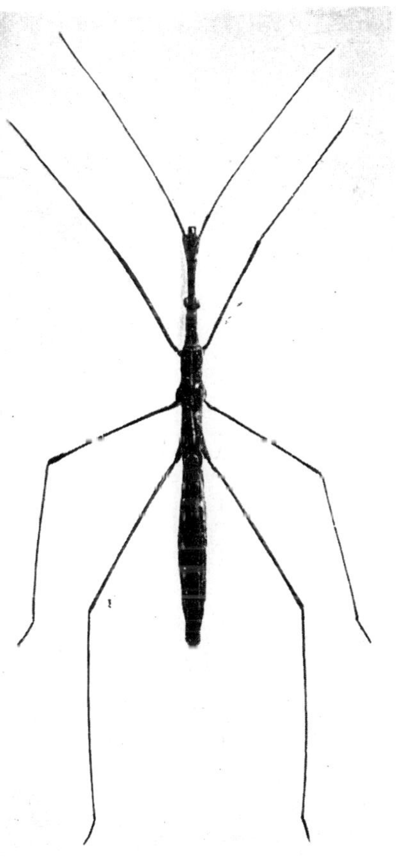

Photo by] [*H. Main, F.E.S.*

HYDROMETRA.

This Water Bug might be taken for a newly hatched Stick Insect.

Photo by] [*J. J. Ward, F.E.S.*
FOOT OF HYDROMETRA.
From one of the walking legs.

Photo by] [*J. J. Ward, F.E.S.*
FORE-FOOT OF HYDROMETRA.
Modified for holding prey.

position of the first pair of legs, which are attached just below the prominent eyes. The entire insect reminds one very forcibly of the Stick Insects, much more so than Hydrometra does because of the superior size of Ranatra. The fore-part of the creature, again, suggests relationship with the Mantis, for the first pair of legs end in sickle-shaped joints which fold down like the blade of a pruning-knife into a slit in the next joint. At one side of this sheath is a strong, sharp spine, serving to hold more securely any unfortunate insect imprisoned by the closed sickle. The hinging of these leg-joints enables such prey to be brought against the point of the beak, where its juices may be sucked.

The two long bristles that constitute Ranatra's tail are hollowed along their inner faces so that when they are brought together they constitute a tube connecting with the respiratory system of the insect. Ranatra can skim along the surface like the Skaters, but it will usually be found clinging to the weeds in an almost inverted position with the tip of its tail tube just out of the water. In this position it can remain for an unlimited period, for the whole of its air-supply can be obtained through this tube. Ranatra has two pairs of long and serviceable wings, but from the very neat way in which they are folded and packed upon the back, they are not at all noticeable. In the ordinary way Ranatra is an inconspicuous, monotonously mud-coloured creature; but if we have the rare fortune to see it with wings expanded for flight, we shall be astonished to find that under the wings the back proper is coloured bright red.

The Water Scorpion is a sluggish creature closely allied to Ranatra, and though as dissimilar in general appearance as possible, the fore-legs are developed in much the same manner and with the same purpose in view. It has also the tail bristles; but there all external resemblance ends. The mud-coloured body is broad, but exceedingly thin. Under the wings the broad back is coloured red. The name Water Scorpion is not a good one, for the fore-limbs of *the* scorpion are pincers like those of the crabs, whilst these are more akin to those of Mantis.

One other of these Water Bugs must be mentioned. This is the Boatman—usually pleonastically styled Water Boatman. When seen at the surface he appears to be a rather lazy boatman, for his oars, though ready for action, are still. It will be noted that he is back downwards with the head below the surface and the tail just above. His great eyes enable him to see above and below at the same time, and if you make a movement he at once brings his oar-like hind legs into play and takes a rapid inclined

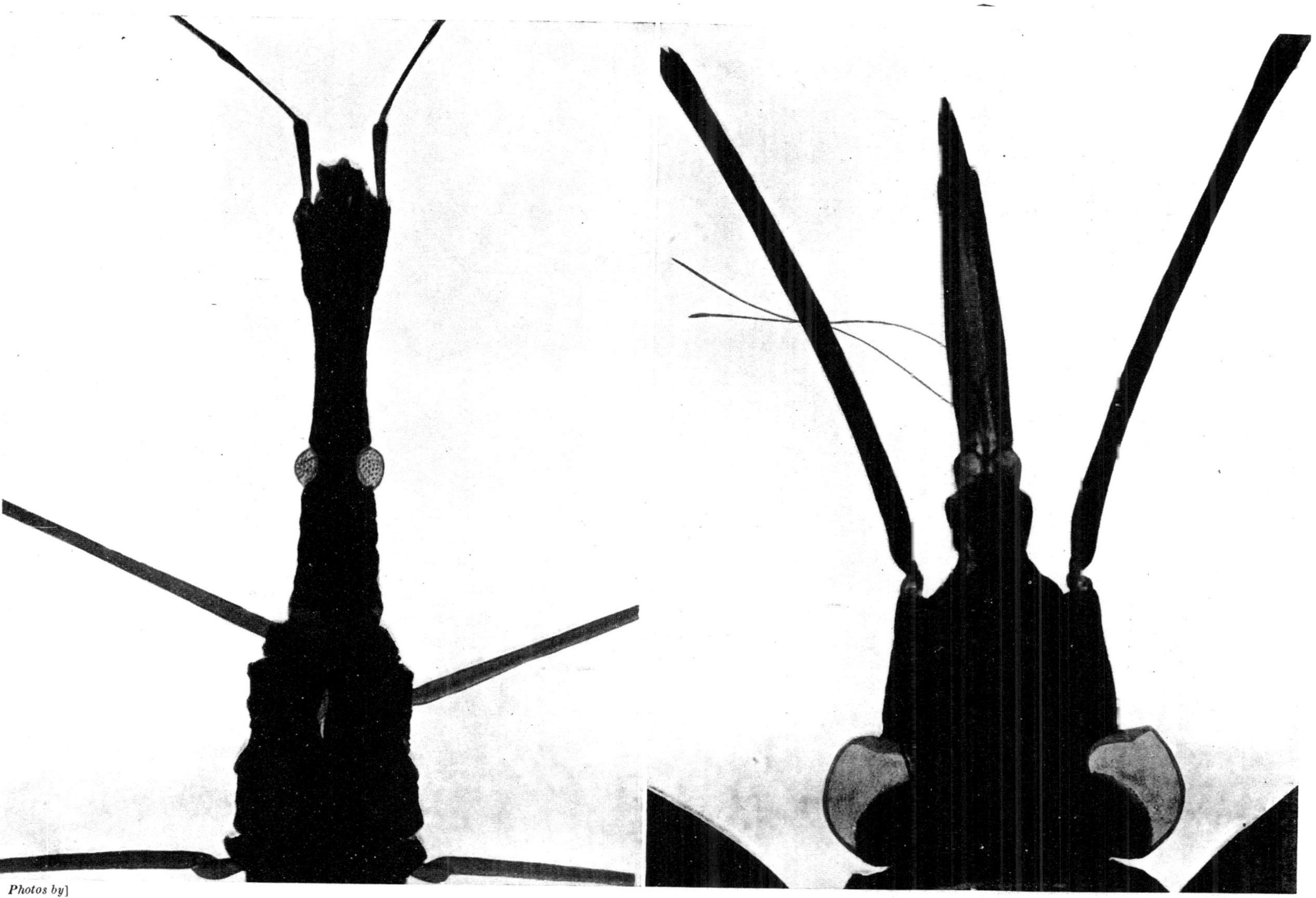

Photos by] [*J. J. Ward, F.E.S.*

HEAD OF HYDROMETRA. HEAD OF GERRIS.

The heads of these two Skaters, enlarged to twenty-five times the natural size, are shown side by side for comparison. In Gerris the dagger-like tongue is shown extended in front of the head, but in Hydrometra it is folded underneath, out of sight. Note the respective lengths of the head and position of the eyes.

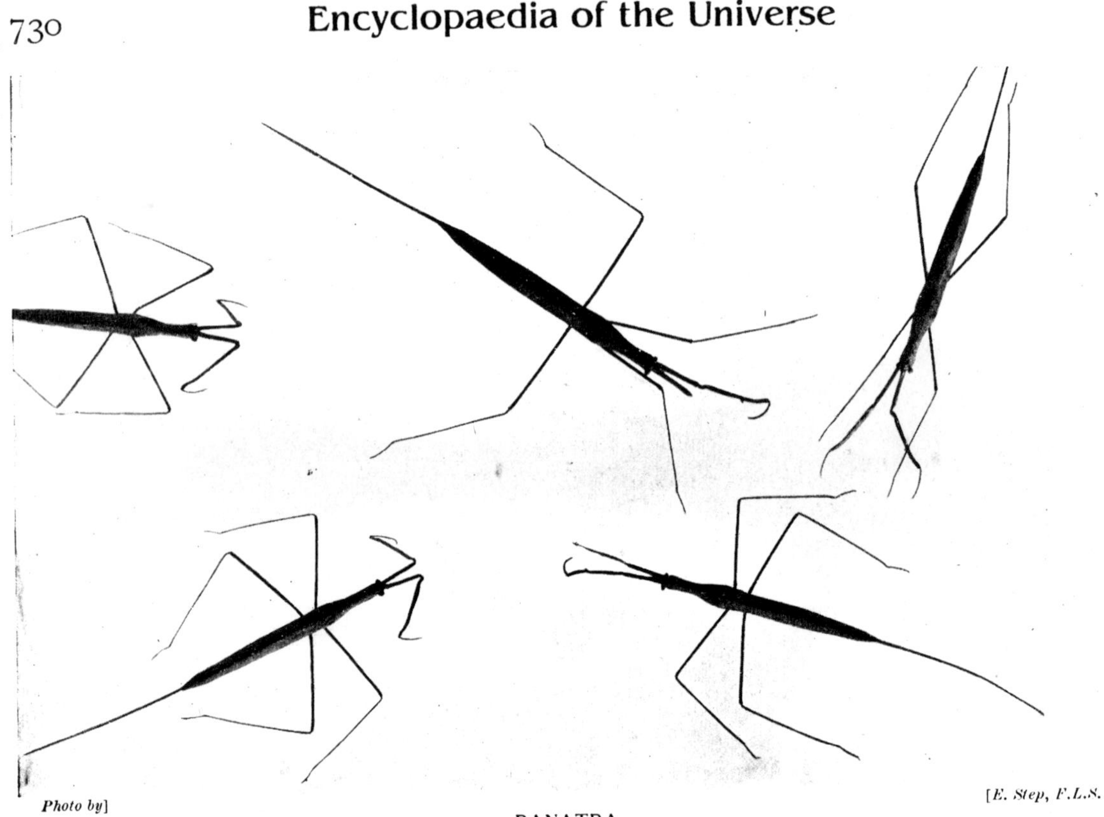

Photo by] [*E. Step, F.L.S.*

RANATRA.

A group of these remarkable Water Bugs in natural attitudes and life size. The curious resemblance to the unrelated Stick Insects is strikingly shown. Note the Mantis-like development of the first pair of legs.

dive into the depths. It is noteworthy that neither Ranatra, Water Scorpion, nor Boatman is provided with antennæ.

SAND-WAVES AND SAND-RIPPLES

BY EDWARD A. MARTIN, F.G.S.

ONE'S thoughts travel instinctively to the waters of the ocean when mention is made of waves, and it seems at first sight to be a contradiction of terms to associate waves with sand. But one quickly realizes that a sea-wave does not necessarily involve a transportation of water, and that the water of a wave may itself be stationary, since only the energy which passes through it is translated. One then sees that waves can take place in other media than water. The water in a water-wave has, however, a movement in a vertical direction, and a wave in a cornfield has an oscillating forward and backward movement within certain limits. But in the latter case no one supposes for a moment that the ears of corn pass from one side of a field to another, although one can see the wave-impression do so. It is when sea-waves approach a coast that there is an actual translation of the water itself, and the forward movement of energy dashes the last breaker which is conveying it with a thunderous roar upon the strand.

As water-waves are waves of the wind, so sand-waves are begotten of the wind also. Take a smooth water-surface or a smooth sand-surface and let a puff of breath impinge upon either. There is immediately a derangement of the surface, even though the current of air be parallel to it, for even then friction will do its work. But let a gust of air beat upon it at an angle, and a depression is immediately formed, and when a depression is formed, a corresponding ridge must be also formed. But the similarity of the two cases ceases here. In water a series of ripples is formed by a single gust. In sand there must be a steady blow to produce ripples. Water, after being rippled, will in time again attain a level. Sand will remain rippled.

As sea-waves are merely big ripples, so sand-waves are but magnified sand-ripples. The big is the magnification of the small in both cases, and the formation of ripple-marked sand is the first stage towards that of sand-waves, and these may rise in the form of sand-hills to a height of two hundred and fifty feet. Some have occasionally reached even five hundred feet.

The whole of the Western Sahara is more or less smothered in sand. Vast areas of sand move with the trade-winds from east to west. The rocky surface is hidden below, and the sand forms wave after wave, shifting gradually with every wind more and more to the west, until the sand becomes blown out to sea to form sand-banks along the shores of Western Africa.

Each particle of sand is on the move. Immediately a depression in the sand has been formed by the blowing-out of the sand-grains, the tendency is for the next supply of sand to pile up on the windward side of the depression until the ripple-crest reaches a height of instability before the wind, when the further supplies will be carried over the crest and drop on the lee side of the ripple. Any slight irregularity in the surface will give rise to a bank, or if the solid rocks are too far below the surface to be of any effect, a bank or wave may be given rise to by intermittent winds acting at various angles or with different degrees of intensity. The sand will thus be arranged in more or less parallel waves running at right angles to the source of the material, and when they reach the coast will give rise to what are known as coast-dunes.

Looking away across a sand-desert where the banks of sand rise and fall with great regularity as far as the eye can reach, the appearance presented resembles nothing so much as an ocean whose waves have suddenly been solidified in a position of rest. The sand of the desert has been sorted

SAND-WAVES.

Both sand-waves and sea-waves are formed by the action of wind, but while a series of water-ripples may be produced by a single gust, in sand there must be a steady current to cause the same effect. The photograph shows a sand-sea on the Algerian Frontier.

out, so that the larger, rocky fragments and particles from which the sand has been derived, have been left behind at the edge of the desert. The farther the sand has travelled, the finer the condition of dust at which it has arrived. The slightest wind-movement raises it again and carries it forward, and nowhere is the sand so finely-divided and rounded as in the heart of the desert. The wind, where it strikes anything of a rocky, solid nature, uses the sand-dust as an emery-powder to wear down all prominences, at the same time adding to itself the material of which it has robbed the solid rock.

Sand-storms are made up of this sand dust. Almost as fine as the impalpable clay-dust which blows along our roads, it has the biting, pitting characteristic which sand has, and travellers in such a storm can but shield their hands and faces against its onset.

It is this fine sand-dust which, acting with a proportion of grains of larger size, gives rise to the beautiful sand-ripples which characterize the slopes of the larger sand-waves. One is of course

Photo by] [*H. J. Shepstone.*

SAND-DUNES.

Heaped up by the prevailing west winds on the fields of the Plain of Sharon. Many a wood or forest has been wrecked, and a fertile country made absolutely desolate by these inroads of moving sands.

accustomed to the rippled sands at the seaside, but those that have been formed directly by the wind are even more regular and more beautiful than those formed beneath the rippling waters on a sandy shore. It has been found that desert-ripples are formed at any time when there is a direct and steady wind, and up to a certain limit they increase in size and regularity so long as the regularity of the wind is maintained. Given a steady wind of great excavating power and the ripples become dunes and sometimes valleys arranged in a similar direction to that of the wind, as in the great desert of North-Western India.

By the continual attrition to which blown sand is subjected, all sharpness in the particles is worn away, and under a microscope they are seen to be perfectly rounded. Great as is the wearing-action to which sand in a river or on the seashore is subjected, it is but slight as compared with that of sand conveyed by the wind. It has been estimated that a hundred years of wind-action will produce in sand what would take three thousand years to produce by water action. Consequently many an ancient rock, when made up of fine, rounded particles, has been judged to have been originally laid down as desert-dust, or have been blown out to sea to form sand-shoals. Such may

Photo by] [*Dr. Vaughan Cornish.*

ÆOLIAN SAND-WAVES.

Two different forces of wind have combined to form this beautiful sand panorama. The stronger and more persistent force has drilled out the numerous hollows and piled up the larger ridges, while a gentler current of air sweeping over the whole plain has produced the delicate ripples over the whole surface.

Photo by] [*H. J. Shepstone.*

SAND-DUNES IN THE COLUMBIA VALLEY.

These waves may sometimes be piled up to a height of five hundred feet. As it is the wind that forms these ridges, so it is the wind that furnishes the sand. Wind-action will produce in one hundred years an amount of sand that would take three thousand years of water-action.

have been the case with some pre-Cambrian rocks of Charnwood Forest and the St. Peter Sandstone of the Upper Mississippi.

Moving sands have been responsible for the wrecking of many a wood or forest. A prevailing wind will constantly tend to carry the sand from the windward side of a dune over its crest, and strew it out on the leeside. This, of course, results in the steady forward movement of the whole mass, and in this case it is a veritable moving wave of sand. In the beginning of the last century a church in the neighbourhood of the Kurisches Haff was separated from the sea by sand-dunes. By 1839 the church was completely covered by the moving dunes, and by 1869 the sand had proceeded still further inland and left the ruins of the church exposed.

Just a word, in conclusion, as to what it is that sand is composed of. Reduced to its simplest form it is just silica, or the oxygen of the air combined with silicon. The most familiar forms of silica are flint and quartz, and these, in their finely-divided state, give rise to almost all of the waves of the desert and the dunes of the seashore. It can be imagined, when this is realized, what a large proportion of the earth's surface is made up of silica. It is, in fact, the most abundant solid constituent of our globe.

SCALE INSECTS

BY HAROLD BASTIN.

If the reader will examine apple or hawthorn twigs during the winter months, he will almost certainly find some that are thickly encrusted with minute objects which, when examined through a lens, resemble tiny brown mussels. These are the shells secreted by a well-known orchard pest called the Mussel Scale insect, which has become practically cosmopolitan in its range, having been carried on nursery stock from one apple-growing region to another.

The Scale Insects, or Coccidæ, constitute a very remarkable group, most of its members being highly injurious from an economic point of view, although a few—such as the Cochineal insect, to which reference has already been made—are directly useful to mankind. The Mussel Scale has been mentioned chiefly because its metamorphosis is well known and probably more or less typical of the life-histories of scale insects in general. If we lift up a large Mussel Scale in winter, and place it beneath the microscope, we shall find that it shelters from fifty to one hundred brilliant white eggs. At one end of the scale we shall also find the dead and shrivelled remains of the adult female insect. In the springtime the eggs hatch, and the minute, six-legged larvæ creep from under the scale, and roam over the twigs and branches. Each eventually settles down at a spot where the cuticle is tender, inserts its " beak," and begins to pump up the life-blood—*i.e.*, the sap—of the tree. Of course, the amount of sap extracted by the individual scale insect is infinitesimal ; but when, as is commonly the case, millions of the pest are working together, the loss to the tree is very serious. Not infrequently the scales are found upon the leaves, and occasionally upon the fruit. One scale insect, closely related to that of the apple, may frequently be found upon the skin of oranges.

The larvæ, when once settled, become sedentary in the fullest sense of the word. Each begins to secrete a waxy substance, which rapidly hardens and forms the shell or scale. This grows by successive layers of secretion as the insect increases in size. The female larva of the Mussel Scale insect eventually becomes an inert, grub-like creature, which never leaves the scale, but perishes therein after depositing its eggs. The final moult of the male larva, however, is preceded by a kind of pupa state, and eventually the adult insect issues from the scale as a minute " fly " with long

Photos by] [*Harold Bastin.*

SCALE INSECTS.

To the left is a branch covered with Ash Scales, nearer to the right are two bearing a number of Willow Scale-insects, while the right-hand photograph is of the Mussel Scale. They are shown a little less than natural size.

Photo by] [*Harold Bastin.*

THE LARGEST SCALE INSECT.

The female insect is often over an inch in diameter, while the male is a mere speck. It is a native of South Africa, and is particularly abundant on the M'sasa tree.

antennæ and well-developed wings. Mr. Robert Newstead, the well-known authority on scale insects, has found male Mussel Scales upon broom and whortleberry plants, but they appear to be unknown upon the apple in this country.

In the case of another common scale insect, a frequenter of the ash, alder and willow, the males actually predominate. The minute male scales of this species may be seen crowded on the ash stem to the left of the preceding photograph, while the larger female scales are scattered over the two remaining stems. As a rule, however, male scale insects are greatly in the minority as individuals, while science has quite failed to discover the males of many species.

Two conspicuous scale insects are found upon currant bushes. The species known as the Cottony Cushion Scale is especially noticeable where it occurs on account of the cushions, or pads, of white, waxy filaments which the females secrete as coverings for their eggs. No " scale " in the ordinary sense of the word is formed, the brown object lying above the cushion being the hardened body of the adult female. This species, or a closely allied form, is sometimes found as a pest on vines and peaches in glasshouses. The Brown Currant Scale is widely distributed, and the reader should have no difficulty in finding specimens upon cultivated currant bushes, or upon the so-called "flowering currant." Here, again, no scale is formed, the eggs being protected by the hard, shell-like body of the dead parent. No cottony cushion is secreted.

Photo by] [*Harold Bastin.*

THE COTTONY CUSHION SCALE.

The peculiar white "cushions" are formed by pads of white waxy filaments used to cover the eggs. The brown object lying above the "cushion" is the dead body of the mother.

The largest known scale insect is not unlike an enormously magnified " Currant Scale." It has been named the Giant Coccid, and is indigenous to South Africa. A shrunken specimen in possession of the writer measures more than one inch in diameter.

A scale insect familiar to most people, by sight if not by name, is the Felted Beech Coccus. Owing to the whiteness of the covering secreted by the females, and its exposed position on the boles and main branches of beech trees, the species is most conspicuous. At a distance, the trunks look as though frozen sleet had drifted against them in a gale of wind. Time was when foresters attached no special significance to this infestation ; but the insect is now recognized as a serious pest, capable, if unmolested, of working wholesale destruction in beech woods ; and several years ago the Board of Agriculture issued a leaflet describing the approved methods of treatment.

Photo by] [*Harold Bastin.*

THE BROWN CURRANT SCALE.

Perhaps the most familiar of all the Scale Insects. As many as a hundred eggs are secreted under this hard brown shell, which is the dried body of the parent. These eggs hatch in spring; six-legged grubs creep out. Each eventually settles down on the "green" bark, inserts its beak and commences to pump up the sap from the branch. The scales are here shown about four times larger than nature.

FISHES OF THE OLD RED SANDSTONE.

At an exceedingly remote period—and no one can say within a few million years how remote it was—these fish-like creatures existed, possessed of neither limbs nor lower jaws. Amongst these was the subject of the above photograph—Coccasteus.

While it is impossible to deny that scale insects are capable of incalculable damage, it is interesting to note that certain species secrete substances valuable to mankind. The "manna," endowed by the wandering Israelites with a miraculous origin, was almost certainly a kind of honey-dew produced by the Manna-bearer, a Coccid found on tamarisks in the Mediterranean district. This substance is still used as food by the Arabs, who call it "man." The white wax of China is derived from *Ericerus pe-la*—another scale insect, found upon an indigenous ash tree. It was formerly greatly prized, but is now falling into disuse owing to the introduction of kerosene. In India, a similar wax is secreted by a scale insect known as the Wax-bearer. The wax secreted by other species of the latter genus has also been utilized by mankind.

Again, a fatty substance containing a peculiar acid is produced by a Mexican Coccid. It is used as a varnish, which dries and hardens on exposure to the air; also as an external medical application in various ailments. The curios known popularly as "ground pearls" are the outer shells of the Pearly Coccids. In South Africa, and the island of St. Vincent, they are collected, strung together as necklaces, and sold to tourists.

The most important products of scale insects, however, are lac and the crimson dye known as cochineal. Lac is secreted in large quantities by the Lac Coccid upon the twigs of its food-plants, fig, buckthorn, and other trees. It is imported to this country from India in its natural state, and after treatment forms the basis of varnish, French polish, and many other important materials.

FISHES OF THE OLD RED SANDSTONE

BY R. I. POCOCK, F.R.S.

ALTHOUGH the term "fishes" is ordinarily used in a vague and elastic sense for most animals that live in the sea, its scientific equivalent, "Pisces," is much more restricted in its application. In the zoological class "Pisces," for example, the lamprey and its allies find no place, being rigorously excluded therefrom because they differ from all species above them in the scale of creation, like sharks, sturgeons, herrings, salmon, and the like, as well as from all other vertebrated animals, in some very remarkable features, the most remarkable, perhaps, being the complete absence of limbs and of a lower jaw. The lampreys, indeed, are supposed to have branched off from the vertebrate line of descent before these important organs had been acquired ; and it may be explained in passing that the limbs of the higher vertebrates are represented in all typical fishes by the two pairs of fins on the lower surface of the body.

These preliminary remarks about the lamprey are a necessary introduction to an account of the fishes of the Old Red Sandstone, because at the exceedingly remote period—and no one can guess within a few million years how remote it was—when the rocks of that geological system were laid down on the bottom of the then existing seas or fresh-water lakes, there lived some marvellous fish-like creatures which seem to have been related to the lampreys, inasmuch as they appear from the available evidence to have possessed neither limbs nor lower jaws. Nevertheless, in one important particular at least these ancient fishes differed widely from the soft-skinned lampreys, namely, in having the fore part of the body and head protected by a durable and sometimes elaborately constructed investment of armour ; and it is to this armour that their preservation in the sedimentary rocks is due, and from this that their scientific name, *Ostracodermi*, or oyster-skinned, has been derived.

Without entering further into the classification of the Oyster-skinned Fishes, it may be said that they belong to, or may be grouped around, three main types, known as the Wing-shield, the Helmet-bearer, and the Winged Fish. The Wing-shield, which has been found in the Upper Silurian or Lower Old Red Sandstone of Herefordshire, was about six inches long, and seems to have been somewhat torpedo-shaped. The hinder part of the body is believed to have been covered with rhomboidal scales, while the forepart, which had a pointed snout, carried a dorsal plate, ending behind, at all events in some species, in a sharp, backwardly-directed spine ; there was a ventral or breast-plate as well, and an eye was set on each side of the head. The shape of this fish, which has been claimed to be the earliest vertebrate yet discovered, suggests that it was a rapid swimmer ; but an allied form, known as the Sickle-shield, was apparently rather a crawler than a swimmer, if we may judge from the great width of the body with its nearly square-cut angles, straightish

FISHES OF THE OLD RED SANDSTONE.

A side view of Coccasteus shown on the opposite page; it belongs to an ancient group of armoured lung-fishes. It is from this group that the higher creatures originated, the stages being Amphibian, Reptile, and, from the Reptiles, Mammals and Birds.

front edge and comparatively narrow scaly tail projecting from the middle of its hinder border. The body armour of the Sickle-shield consisted of dorsal, ventral and lateral shields separated by strips of integument strengthened with a pavement-like coating of close-set scales. The eyes seem to have been set at the sides of the head, as in the Wing-shield, and the mouth probably opened in front of the large ventral shield.

The structure of the Head-shield, too, suggests that this fish probably crept about the bottom and buried itself in sand or muddy sediment. The great head-shield, shaped like a saddler's knife, with its semi-circularly rounded margin and acute angles, was curiously reminiscent of the head-shield of the trilobites, and the eyes were placed close together near its summit, so that when the bulk of the fish was buried, the eyes were free of sediment, and able to keep watch upon enemies or prey passing above in the water. The rest of the body was comparatively narrow and was covered with plate-like scales; and the tail, which had smaller scales, carried a fin beneath at the tip and another above at the root.

FISHES OF THE OLD RED SANDSTONE.

The Cephalaspis, or "Head-shield," was in all probability a "bottom" fish, so that the eyes, placed close together at the top of the head, would be of service while the remainder of the fish was buried in the mud.

This creature was apparently coeval with, and about the same size as, the more fish-like Wing-shield. The later and still stranger form, the Winged Fish, had a scaly, fish-like tail, with a terminal fin below, while the main part of the body was enveloped in a great buckler composed of large overlapping plates, and a composite sub-circular head-shield, carrying a pair of dorsal eyes, was somewhat sharply marked off in front. Very curious, too, was the pair of jointed, limb-like appendages articulated to the sides of the body behind the head, which were certainly used for locomotion, although their true significance is a matter about which there is no unanimity amongst zoologists; some holding that they represent the front fins of fishes, others that they are merely movable outgrowths of the head-shield, while others think they correspond to the paddle-shaped limbs of the scorpion-like Euryptermis that lived in those ancient seas.

This raises another very interesting question connected with these strange fishes of the Old Red Sandstone. The resemblance of the head-shield of the Head-shield to that of the trilobites has been already alluded to. It might equally well have been compared to that of the king crab, an ally of the trilobites. The likeness, indeed, between these ancient fishes and the still more ancient trilobites and contemporaneous king crabs, which were related both to the crustaceans and the arachnids, is too striking to be disputed; but while most authors hold that the likeness is purely superficial and due to adaptive convergence, there are others who consider it to be an outward and visible sign of inward and genuine kinship, and who see in these ambiguous creatures actual links in a chain of affinity between the vertebrated and the articulated animals, pointing to the descent even of man himself from species more nearly related to king crabs and scorpions than to any other living animals.

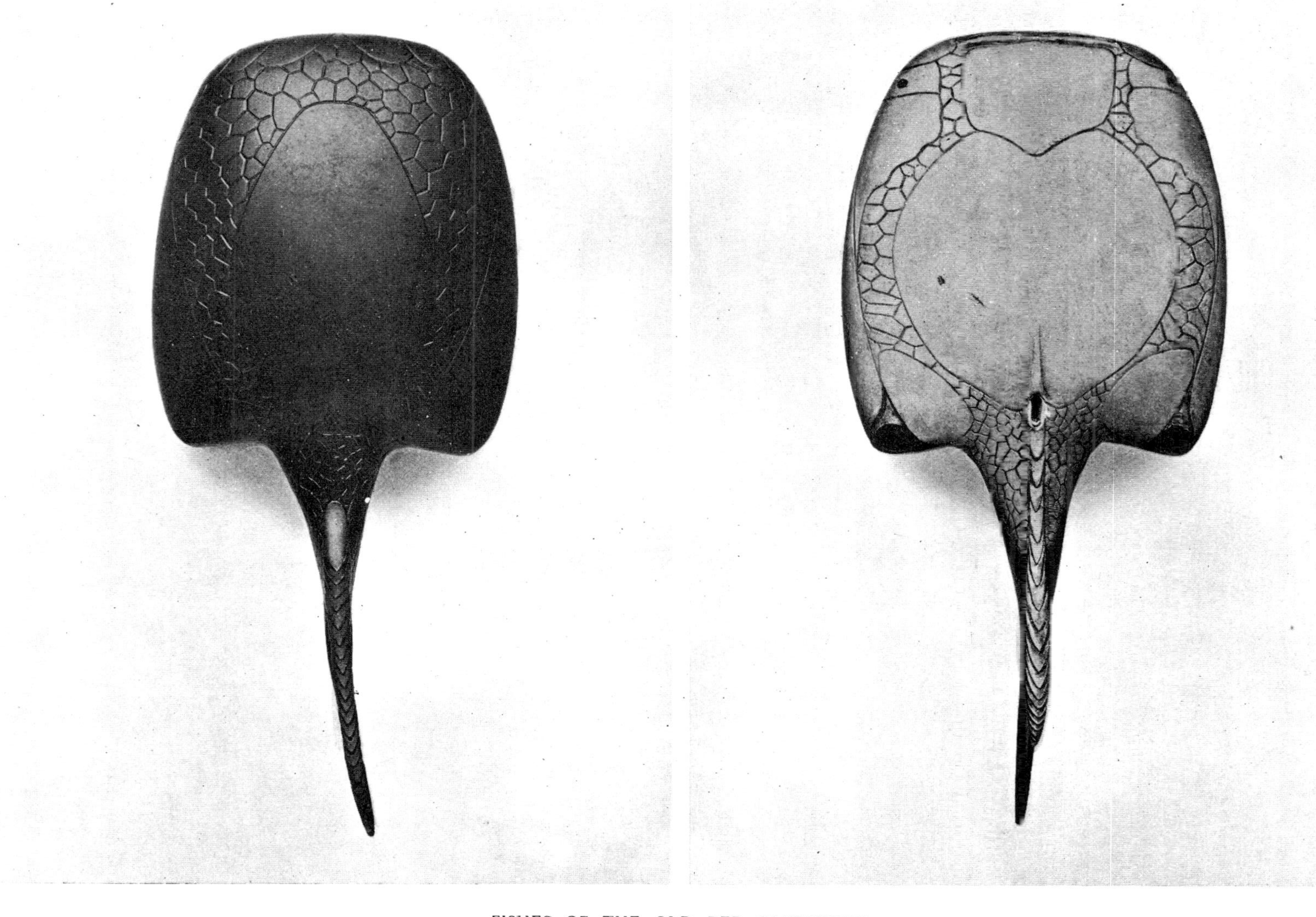

FISHES OF THE OLD RED SANDSTONE.

From the peculiar shape of this Fish—Drepanaspis or Sickle-shield—it is supposed to have been a crawler rather than a swimmer. The larger segments marked on the body show the position of the armour-plates; these were united together by a thick skin, strengthened by a pavement of closely-set scales. These are the smaller divisions between the shields.

It is not here possible to do more than allude briefly to this seductive hypothesis which puts the descent of the vertebrates upon a higher plane than that occupied by the "worms," which have hitherto been accorded the distinction of being their primitive stock; but it is quite clear that the theory in question, which has been most forcibly and ingeniously urged by Dr. Gaskell, invests these fishes of the Old Red Sandstone with quite a new interest.

These, however, are not the only fishes that lived in that primeval period. There were others belonging to the existing orders, containing the sharks, the lung-fishes, and the gill-breathing fishes with bony skeletons, but those of the latter group were all of a lowly type more or less, serving to link the existing bony fishes with the lung-fishes.

Now the lung-fishes are of peculiar interest, because it was from them that the higher vertebrates originated: first the amphibians, then the reptiles, and from the reptiles, the mammals and the birds. They are represented at the present day by only three types, one of which lives in

FISHES OF THE OLD RED SANDSTONE.
One of the armour-plated fishes remarkable for the peculiarly elongated snout.

Australia; another in South America, and the third in tropical Africa. They are thus a good illustration of the wide dispersal on the earth of many ancient but waning types. Some of the Old Red Sandstone lung-fishes, like Dipterus, were not so very different from the living Australian kind; but others were so distinct as to be set aside in a special group by themselves. Coccasteus, shown in our illustration, is one of these. The special resemblance shown by Coccasteus to the latter lies in the investment of the fore part of the body, both above and below, by bony plates. The head, which carried a pair of large lateral eyes, was similarly protected above, and was movably jointed to the back plates. A Scotch species was only a little more than a foot long; but in North America there lived some huge examples akin to it, one of them, known as the Titan Fish, quite rivalling large existing sharks in size, if their computed length of about twenty-five feet be correct, while a smaller type, the Terrible Fish, perhaps ten feet long, must have been a predatory fish equalling the sharks in voracity, to judge from the size and sharpness of the teeth with which its jaws were armed.

Reference has already been made to the existence in the Old Red Sandstone of fishes linking in a measure the lung-fishes with the more highly-organized fishes known as the Teleostomi, of the present day. These are the Crossopterygians, or Thick-finned Fishes, which have many points in common with two existing genera of African river-fish, namely, the Reed Fish, from Senegambia and Calabar, and the Many-finned Fish of the Nile. These two fishes stand quite by themselves amongst living forms; and, like the mud-fishes above referred to, are evidently the sole survivors of a group which in the past was represented by a host of interesting and diversified types; some of the best known having been discovered in the Old Red Sandstone of Scotland.

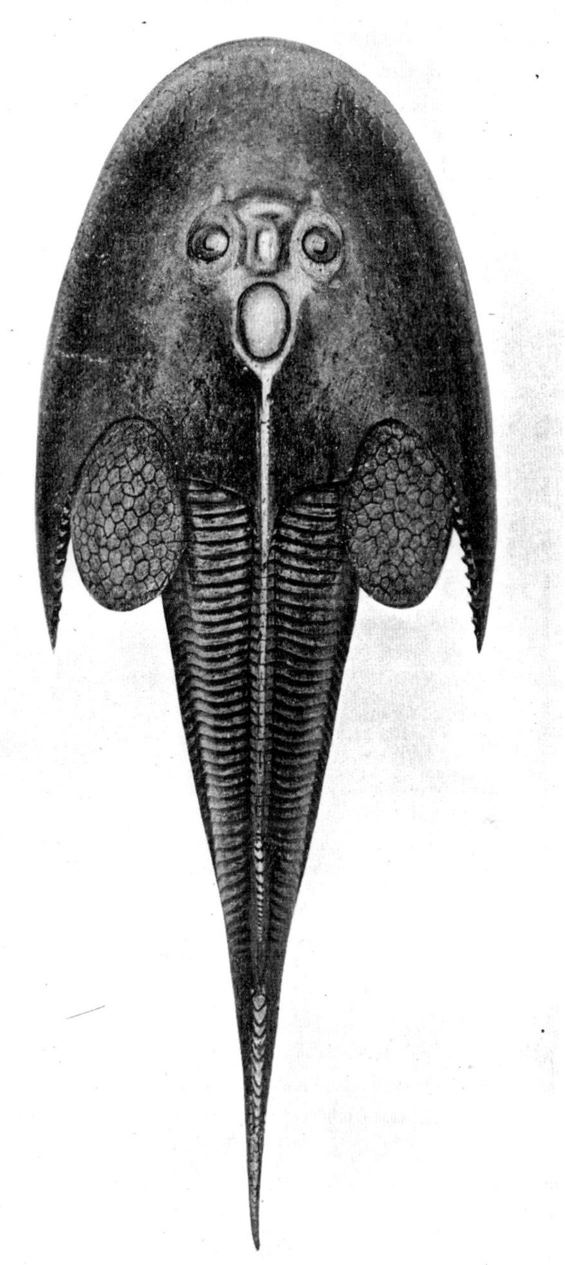

FISHES OF THE OLD RED SANDSTONE.

This fish, the Pterichthys, or Winged Fish, was enveloped in a great buckler of overlapping plates and a composite sub-circular head-shield. The most curious features, however, are the jointed limb-like appendages, which were certainly used for locomotion, but exactly how it is impossible to determine.

Of the shark-like fishes of the Old Red Sandstone, one of special interest claims attention This is Climatius, and the importance of this fish arises from the circumstance that it throws some light upon the origin of the paired fins of other fishes, and, incidentally, of the limbs of the higher vertebrates. It has already been explained that the limbs of the latter correspond to the two pairs of fins, called the pectoral and the pelvic, which are present in all typical fishes. In other words, since the higher vertebrates were derived from fishes, fin-like limbs were the primitive type of the human arms and legs. But whence came the fin-like limbs? Many facts in the structure and development of fishes suggest that the earliest types developed a continuous flap-like fin along each side of the body, which served to balance it in the water and was in all respects similar to the unpaired fins running along the back and on the tail, both above and below. At two points, one in front and one behind, the lateral fins became differentiated and enlarged, giving rise at a very early stage in the evolution of fishes to the paired fins, as the intervening portion of the fin in question disappeared as a separate structure. Now the interest of Climatius is this. It shows unmistakable signs of having possessed a series of at all events five paired fins, or rather of spines representing them, for in this fish, as in all its allies, the fins were supported in front by large spines. And if the theory of the origin of the limbs of vertebrates from lateral fins of fishes be true, the condition of things found in the ancient fish Climatius is exactly what one might expect. Thus the study of the fishes of the Old Red Sandstone helps to solve many difficulties connected with the structure and origin of the higher vertebrated animals.

THE EGG-EATING SNAKE

BY R. LYDEKKER.

MANY kinds of snakes, like the larger kinds of lizards, are exceedingly fond of a breakfast or dinner of birds'-eggs whenever they have the good fortune to come across a nest. But these tasty banquets are, as a rule, few and far between, and the eggs devoured not too large for the capacity of the mouths of the reptiles by which they are eaten. There is, however, a certain South African

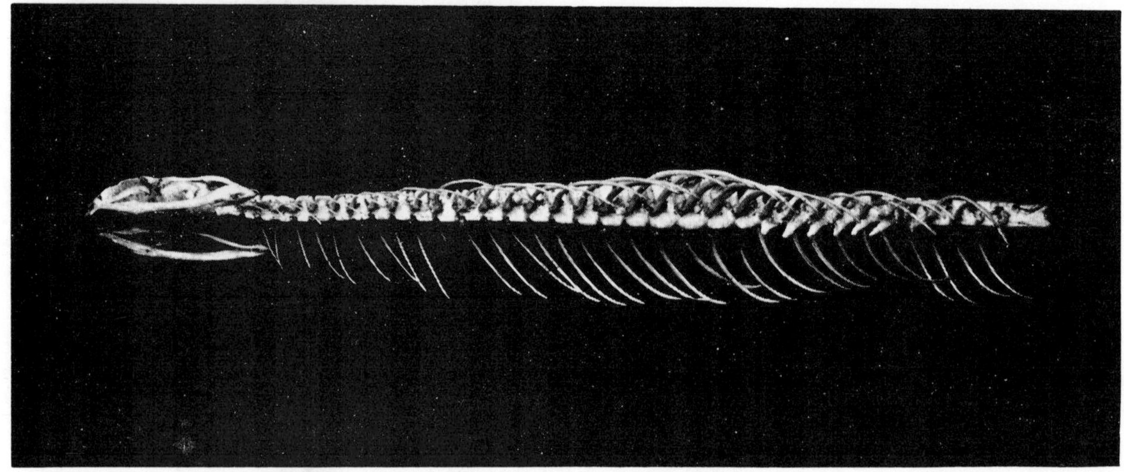

THE EGG-EATING SNAKE.

The upper portion of the skeleton of the neck, showing (towards the right of the photo) the spines projecting from the under surface of the vertebrae. The tips are capped with a hard, enamel-like substance and protrude into the gullet for the purpose of crushing the egg's shell.

snake which may be said to have made a profession of egg-eating, and beats any sword-swallower or other juggler in respect of the size of its boluses.

For although this Egg-eating Snake apparently never exceeds a yard in length, and is generally a good deal smaller, with a body of which the calibre is in no wise in excess of that of other members of the tribe of the same approximate length, yet, nevertheless, an individual considerably short of the maximum dimensions can swallow a hen's egg, while smaller ones will devour the egg of a pigeon.

Unlike most other snakes and lizards, the Egg-eating Snake does not break the eggs in its mouth, but gets them well into its gullet before cracking the shell, in order that not a drop of the savoury contents may be lost. But how, it may be asked, does a snake with a gullet-capacity suited under normal circumstances to swallow substances about the size of an acorn, manage to get a hen's egg into its gullet, and when it has succeeded in this, how does it crack the shell ? Owing to the fact that the bones composing the upper and lower jaws remain separate from one another in front, all snakes are able to extend their mouths to an almost inconceivable degree, as anyone may see for himself who watches an ordinary English grass-snake devour a frog. It is merely an excessive development of their stretching capacity that enables an Egg-eating Snake to get the egg of a pigeon or a fowl safely ensconced in its gullet. But, to judge from the struggles and

THE EGG-EATING SNAKE.

This Snake apparently never exceeds a yard in length and is usually a good deal smaller. The pushing of the egg into the gullet appears to be a difficult process, for the Snake performs many queer contortions as the widely distended throat stretches over the egg like a glove. When once in the gullet, however the egg is easily broken up and dissolved.

contortions made during the process of getting the egg into its mouth and then pushing it down into the gullet, the snake does not have by any means an easy time of it while feeding, and its dinner, one would think, is not an event to be looked for with pleasure. However, after the snake has, so to speak, drawn or pushed itself over the egg like a glove, the latter at length glides slowly into the upper part of the gullet of its devourer.

From a photograph] [*Kindly lent by Herr Carl Hagenbeck.*

THE WHALE-HEADED STORK.

This strange bird, which has affinities with the Storks and Herons, is most nearly allied to the Tufted Umbre of Tropical Africa. It is called by the Arabs "The Shoe-beak," or "Father of a Shoe," from the shape of the broad beak, and is found on the Upper Nile, the Victoria Nyanza, and the Upper Congo.

Most animals, even if they did not choke right away, would be able to do nothing with such a hard and big object as a hen's egg jammed tightly in their throats, the walls of which have no strong compressive power. The snake has, however, a special mechanical arrangement for overcoming the difficulty. This is brought about by the spine-like projections from the under-surface of the vertebræ of the neck, which are ordinarily embedded in the muscles, being elongated to such an extent that their tips project into the gullet. Nor is this all, for the projecting tips are capped with a hard, enamel-like substance, so that they form a kind of saw which rips and crushes the shell of the egg in its passage down the gullet, thus allowing the contents to be swallowed without the loss of a drop. As the egg breaks up and collapses, the throat of the snake suddenly resumes its normal calibre.

It may be added that a closely similar structural peculiarity characterizes a certain Indian snake which belongs to a totally different group of the Ophidian order; and if there is anything in analogy, it may be inferred that this snake is likewise an egg-eater, although this requires to be verified by actual observation.

THE WHALE-HEADED STORK

BY SIR HARRY JOHNSTON, G.C.M.G.

THIS strange-looking monster was first revealed to us in the forties of the nineteenth century, when such bold pioneers as the late John Petherick set themselves to explore the Upper Nile, and, above all, that region of interminable marsh known as the Bahr-al-Ghazal. The Whale-headed Stork may have been known to the Ancient Egyptians, but it is not represented in any of their pictures. As it is so essentially a marsh-frequenting bird, it is possible that its range did not extend northwards into Egypt proper, and therefore, unlike so many other Nile water-birds made famous to us by Egyptian art, it never came much under the observation of this wonderful Asiatic-African people.

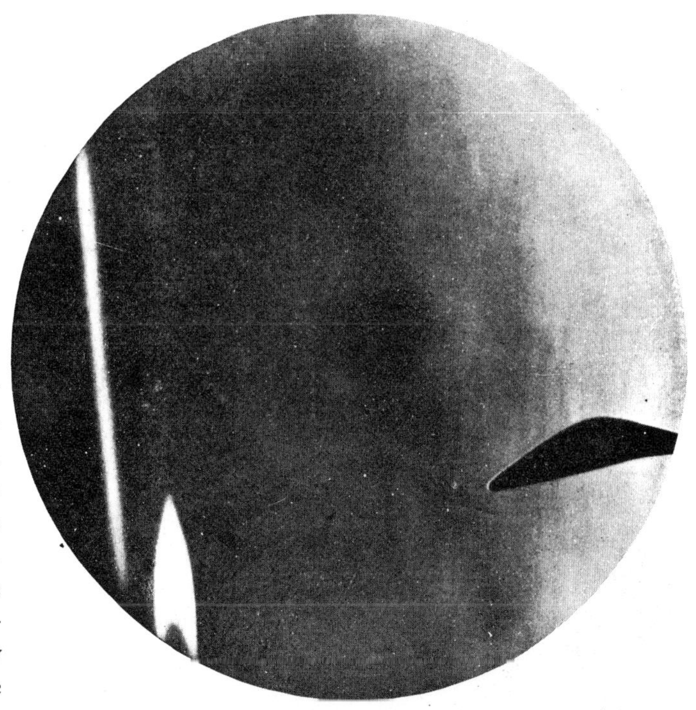

THE SUBSTANCE OF AIR.

A propeller fifteen centimetres in diameter is shown mounted on a vertical shaft in front of a lighted candle and motionless.

For a long while—in fact, nearly sixty years—it was believed that the Whale-headed Stork only inhabited the region of the Bahr-al-Ghazal and the adjoining Mountain Nile. The present writer fancied that he saw it in a great congeries of water-birds in the marshes of the Kunene River in South-West Africa, and the late Sir Henry Stanley declared repeatedly that he had seen this unmistakable bird on the Northern Congo. It would really appear as though both observers may have been correct; for, firstly, I myself discovered the bird to exist in the kingdom of Uganda (from which I sent numerous specimens home), and later travellers have obtained it from the north-eastern bend of the Congo, and even from as far south as the country of Katanga near the borders of Northern Rhodesia and the Upper Zambezi.

In its structure the Whale-headed Stork offers almost equal affinities with both Storks and Herons, and it is difficult to say to which of the two groups it should be more correctly assigned: perhaps the Herons have the best claim. But its nearest relation is that equally unplaced stork-like bird, the Tufted Umbre, likewise an inhabitant of Tropical Africa and Madagascar, but of far more extended range than its monstrous whale-headed relation. The Tufted Umbre, or Hammerhead, is quite a small bird, of a uniform umber-brown, but with a rather broad, hooked beak that might well have been the originator of the monstrous beak of the Whale-headed Stork. This development of jaws has been likened by the Egyptian Arabs to a shoe, and the Whale-headed Stork is often called in Sudan-Arabic "Father of a Shoe"; others, again, compare it to the hull of a ship, and call the bird *Abu markab* for this reason. On its outside this beak is not unlike tortoise-shell in appearance and colouring. The attitudes of the bird and the carriage of neck and body in flight resemble those of the Herons; its habit of clattering the beak more recalls Storks. The huge head is surmounted by a ridiculous little curled crest, and the plumage ranges in general colour (according to age, sex and local type) from brownish-grey to bluish-grey, the centres of the feathers being generally dark, which gives the plumage a longitudinally striated appearance. It has struck me that the Uganda form of the bird is more bluish-grey in plumage than the Whale-headed Stork of the Sudan. In fact, in some Uganda types the colour of the plumage was almost beautiful in being of such a delicate ashy blue, with an inclination to greenish iridescence on the browner feathers of the wing. The food of this bird seems to consist of fish, frogs, snakes, land and water snails and fresh-water bivalves. The nest is made on the ground, though the bird will freely perch on trees.

AIR CURRENTS AND WAVES

BY F. W. HENKEL, B.A., F.R.A.S.

THE air, as a fluid, is almost constantly in motion, the state of rest being a rare exception and one only found existing over limited areas and for a short time. Our earth is everywhere covered with this atmosphere of usually invisible gas, whose height is not less than fifty miles, in which " we live and move and have our being," which presses on every square inch of its surface with a force equal to the weight of 14.7 lbs. The unequal heating by the sun of different regions of our planet on account of their varying presentment towards the central body is the primary cause of the great atmospheric movements or currents. Every disturbance communicated to the air, however small, gives rise to greater or less movements of the latter. The vibrations of a sounding body give rise to the states of alternate compression and rarefaction in the air, whereby the sound waves are propagated, which, falling upon the tympanum of the ear, set the latter in motion, and thus ultimately the brain receives the impression to which the name of "sound" is given.

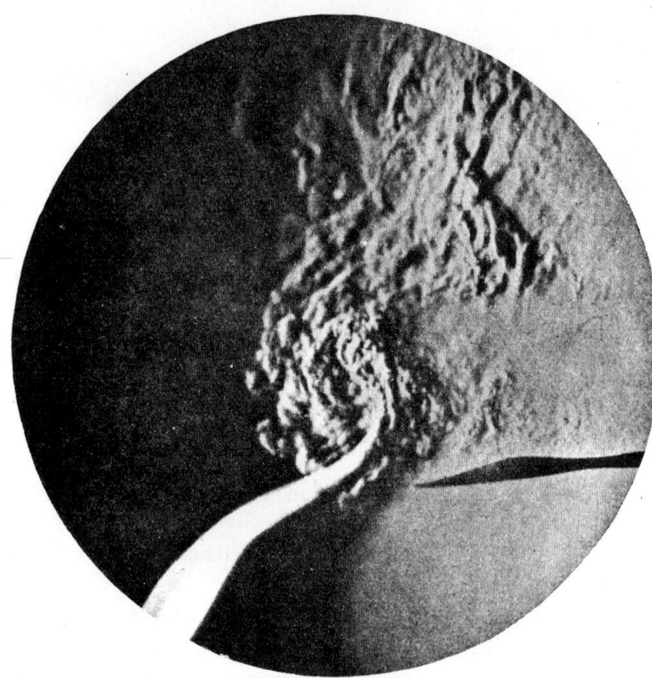

THE SUBSTANCE OF AIR.

Here the propeller is making 1,300 revolutions a minute. Two vortices will be noticed in the air, caused by the action of the two blades; these are spiral and extend into space, only a vertical section of each spiral appearing in the photograph.

The greater heating of the equatorial regions and the deficient supply of the polar ones give rise to the two main currents of our atmosphere—the equatorial warm current northwards and southwards from the equator towards either pole and the polar cold currents from the latter towards the equator. Air being heated becomes lighter and ascends, its place being taken by colder air from surrounding regions. But this is not all. The earth is all the while turning round on its axis from west to east (the diurnal motion of rotation), and this motion is quicker near the equator than in higher latitudes. The air moving polewards, since it is coming from a region of quicker rotation towards one of slower motion, is deviated towards the *east*, and conversely a current coming from the region of the poles will lag behind in the opposite direction. Poisson and Herrel developed this theory, and gave a general explanation of the great atmospheric currents. Air-currents coming from the equator northwards are deflected, as stated above, and thus in our hemisphere the south wind becomes a *south-west* one, and the polar current flowing equatorwards becomes a north-eastern wind. The latter is known as the *north-east trade* wind, and blows with great persistency over the North Atlantic and North Pacific Oceans. Similarly in the southern hemisphere we have the *south-east* trade winds. In England, as is well known, the south-west (rain-bearing) wind is more common than any other, and the late James Glaisher, for a long time in charge of the Meteorological Department at Greenwich Observatory, estimated that on the average this wind blows for over two thousand seven hundred hours per annum.

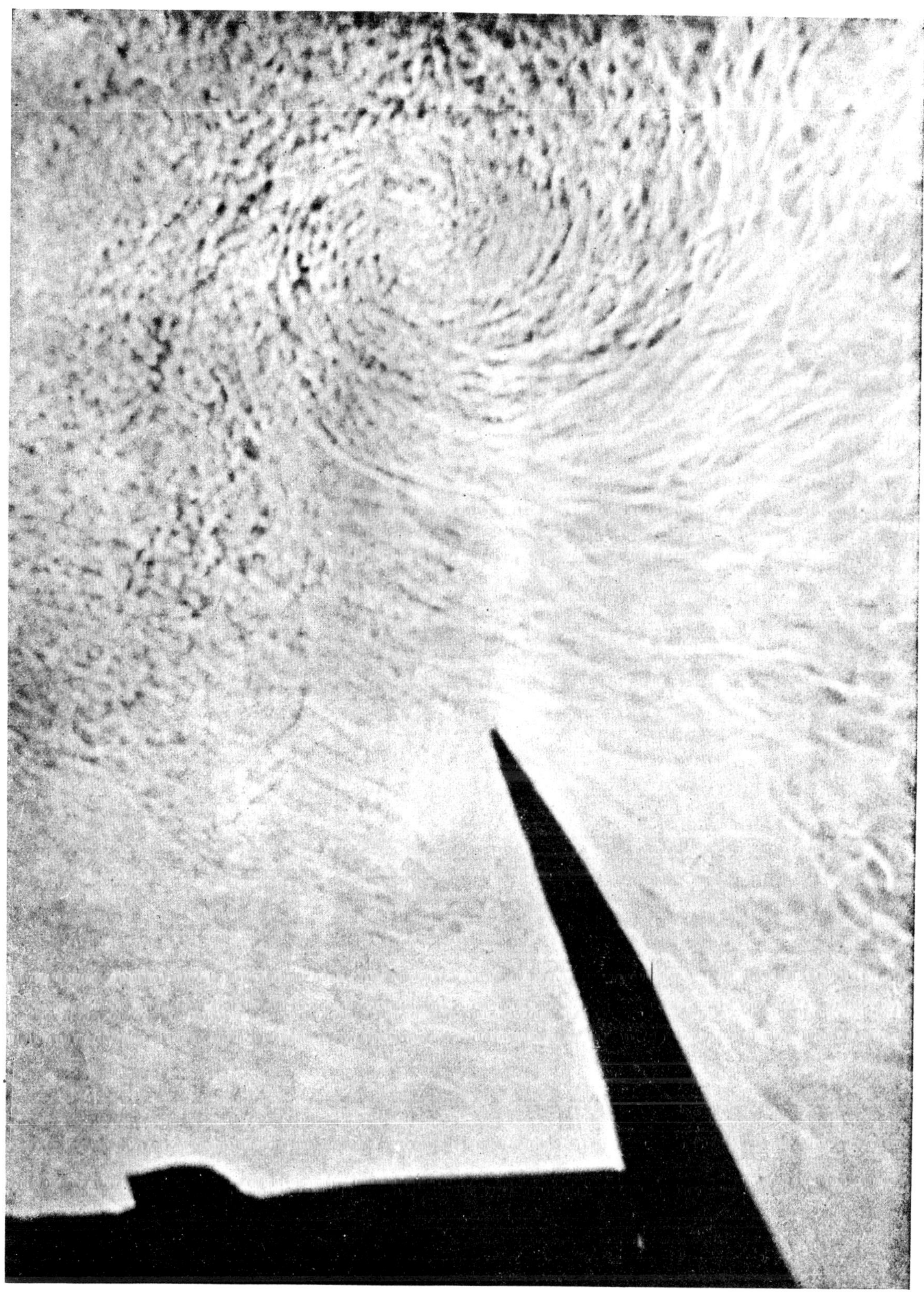

THE SUBSTANCE OF AIR MADE VISIBLE BY PHOTOGRAPHY.

The photograph was taken by an eminent Japanese savant, M. Tanakadate, Professor of Physics in Tokio University. In order to obtain the effect, the air of a dark room has been heated with Bunsen burners and a propeller has been set rotating in front of the hot air. In this way eddies of hot and cold are produced, and these are illuminated with an electric flash and photographed. The propeller was fifty centimetres in diameter and mounted on a horizontal shaft

The " brave westerly winds " of the southern ocean, by their constant blowing, " pile the water in such heaps," that by the help of these winds and waves the clippers in the old sailing-ship days were enabled to attain such speeds as to arouse the wonder of all, though of course such speeds are far inferior to those of Atlantic liners nowadays, gained by means of a reckless expenditure of coal and at heavy risk to human life.

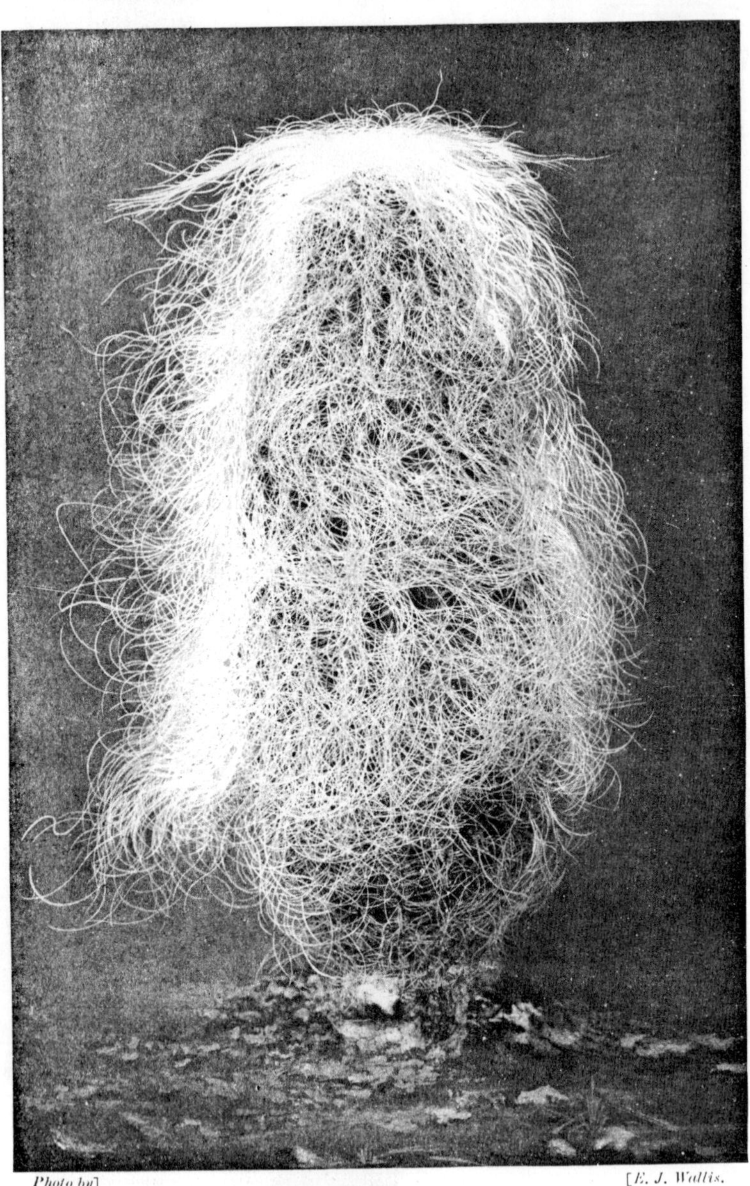

Photo by] [*E. J. Wallis.*

THE OLD MAN CACTUS.

One of the most singular-looking of the tribe. Some of the specimens present a close resemblance to a back view of an aged man's head clothed with long silvery locks.

Summed up in a very few words, the general circulation of the air is as follows. Over the equator there is a zone of medium pressure, having on either side (lats. 30° N. & S.) a belt of higher pressure, within which the trade winds blow throughout the year, except over the north part of the Indian Ocean, where in July we get the " south-west monsoons " and in January the " north-easterly " winds. Further north and south (polewards) of these regions the surface winds in *general* blow polewards, in Europe the south-west wind being the most prevalent one. The trade wind zones shift slightly during the course of the year, owing to the southward and northward motion of the sun. There is an upper current blowing from the equator polewards, known in tropical regions as the " anti-trade," at a height of some ten thousand feet above the surface ; but this gradually descends till in latitude 40° its lower portions are not far from the ground. Recent researches by means of kites with self-recording instruments have revealed the presence of very rapid currents in the upper air, whose directions and speeds differ much from those observed nearer the surface, and it has been suggested that by taking advantage of these swift-moving upper currents (above five thousand feet from the ground) flying machines may be able to attain speeds greater than those of an express train !

Of the smaller air movements or sound waves many methods are given in works on physics whereby these may be rendered visible and measurements as to their length taken.

An apparatus known as Kundt's tube, inside of which the air is set vibrating, and which is strewn with lycopodium, or other light powder, is one of the most familiar for this purpose. The air set vibrating is at one point in rapid motion, at another almost at rest, and the powder collects at these latter, called "nodal" points, leaving the rest of the tube. The number and distance between these points will vary with the nature of the rod which imparts its vibration to the air in the tube, and by filling the tube with other gases instead of air the relative speeds of these latter may also be ascertained.

Photo by] [*S. L. Bastin.*

A MAMMILLARIA.

The tips of the tubercles which bear the spines can be seen covered with white hair-like prickles.

A body of air rotating often forms a vortex ring, and, as such, possesses many of the properties of an elastic solid. Vortex rings in air are sometimes accidentally produced from the bowl of a pipe, but can be more regularly formed in the following manner. A box is taken, covered at one end with a cloth, whilst a round hole is cut at the other end. This box is filled with smoke, or, better, hydrochloric acid gas and ammonia are introduced through separate apertures. These instantly combine and produce dense white fumes of sal ammoniac. A smart tap on the back of the box sends a ring through the opening, and this ring persists for some time; if the experiment is carefully performed it may travel right across a room before dissipating. A second ring encountering the first does not coalesce with it, but rebounds, bouncing away as though an elastic ball.

Air-currents and waves have been rendered visible by photography, by taking advantage of the difference of refraction between hot and cold air, a phenomenon which is said to account for the "fuzziness" of a heated poker. From every aeroplane and other flying machine when in motion there issues a continual series of waves, like those in water from the propeller of a ship, but in ordinary cases they are not

Photo by] [*S. L. Bastin.*

HEDGEHOG CACTUS.

An example of a numerous family protected by close rosettes of fine spines from which longer ones stand out.

visible. The formation of the different types of clouds, each at about an average general level, however, indicates the existence of some of the great air-currents whose extent and speed have been more definitely shown by kite observations. A thorough knowledge of these is evidently a desideratum in these days of aerial navigation, since they will prove of far greater importance to the aeronaut than the winds and ocean currents are to the sailor.

VEGETABLE HEDGEHOGS

BY EDWARD STEP, F.L.S.

PLANTS have such a marvellous power of adapting themselves to their environment that, however adverse the conditions may be in any particular spot, you are almost certain to find that some form or other of plant-life has discovered that it is unoccupied territory and has determined to annex it. There it is flourishing in its own way. The adaptations necessary to existence there have brought out such well-marked characters that if you show an unknown plant to an experienced botanist, he will be able to tell you within a little under what conditions it was growing. The varnishing of a leaf, the turning of its surface into velvet or clothing it more sparsely with long hairs; the spreading of its substance out broad and thin, or contracting it into a round, fleshy lump—all these varying forms and conditions have direct relation to the circumstances surrounding the plant's growth.

Photo by] [*S. L. Bastin.*

A CACTUS IN FLOWER.

Many of this family bear magnificent wax-like blooms often of brilliant colouring. In this case, however, the flower is pure white.

As examples of one particular form of adaptive change, let us take the family of plants of which a few species are shown in the accompanying photographs. They are a very numerous family, known under the general name of Cacti; but the botanist has had to divide them up into smaller groups or genera, each with a distinctive name, such as Cereus, Opuntia, etc. The Cacti are all inhabitants of hot, arid places, where rain falls on very few days in the year, and where animal life is scarce because there is not enough vegetation—the basis of all animal life—to maintain it. It follows that the few plants that have by their ingenuity contrived to exist under such forbidding conditions are much sought after as food, so that they have had to meet a two-fold difficulty—how to conserve sufficient moisture in their tissues to last them from one rainy season to the next, and how to ward off the attacks of animals.

An ordinary plant growing under average conditions, where its roots can suck up as much moisture as it requires—usually more than it requires—is constantly giving off large quantities of water in the form of vapour from innumerable minute pores in its leaves. If the water-supply at the root falls off or fails, the transpiration from the leaves still goes on, with the result that they and all

Photo by] [The Detroit Photographic Co.

A CACTUS GARDEN AT RIVERSIDE, CALIFORNIA.

Such a view gives a good idea of the height to which these vegetable hedgehogs can grow. The tall upright shafts of the Torch-Thistle, which can be distinguished easily in this group, have been known to reach as high as sixty feet.

Photo by] [*S. L. Bastin.*

A MAMMILLARIA.

A second example, in which the longer spines are more evident.

the soft new shoots become limp, and in a long continued drought shrivel up or fall off. Leaves, under the conditions prevailing where the Cactus grows, are a dangerous possession on this account; and so the Cactus has learned to do without them. The vegetative body that ordinary plants spread out in the form of stems, branches and leaves, offers in that form so enormous a surface for evaporation that the Cactus has had to concentrate it all into a globular or cylindrical form, so that the external surface is reduced to a minimum. Then all the cells and tissues of which it is constructed have had to be provided with extra thick walls, so that transfusion of fluids from one to the other and to the exterior shall not be so rapid as in other plants. Having given up the luxury of leaves, the entire surface of this globular or cylindrical body has had to carry out the functions of the leaves. So we find that the skin of the Cactus is always green and soft, and microscopical examination would show that it is provided with stomata—the breathing pores through which it takes in gaseous food, and gives off gases evolved in the process of nutrition. We have said the Cactus has given up leaves. Here we feel we must hedge a little: the statement is too sweeping and must be qualified. Leaves, as such, it is true, are not to be found, but they are present in a changed form. They have all been converted into spines and prickles; some of them stout and long, others fine as finest hair, but all ending in long points of almost microscopic fineness. Often from every cushion of short, fine spines there arise several spines of much greater length and standing out straight. Whilst the long spines are evidently designed to jab the muzzles of the larger mammals, the shorter ones would prevent smaller animals like mice from pushing their slender muzzles underneath the long spines and so getting a bite. It is a common practice for plants of many

Photo by] [*S. L. Bastin.*

THE BROOM CACTUS.

The plants photographed do not show the normal form, the ridges having grown in such a way as to produce curved crests analogous to the familiar garden Cockscombs.

different families, when growing in dry places where nutritive vegetation is scarce, to protect themselves from being eaten by turning leaves or shoots into spines or thorns. Examples that come readily to mind are the Acacias, and our own Furze, Blackthorn and Hawthorn.

Imagine a herbivorous animal parched with thirst, and knowing, perhaps, by its sense of smell that the plump Cactus has its tissues full of cool liquid. He rushes at it to take a bite of the succulent flesh. But before his teeth can reach the rind his tender lips and tongue are pierced by hundreds of the needles, and he retires, howling with pain. It is only by occasional experiments of this sort that animals have learned that it is dangerous to attack such plants, and the experience goes into the stock of memories that passes by inheritance from generation to generation and commonly goes under the name of instinct.

With a dry, loose surface there is often a fair amount of moisture in the subsoil, and the Cacti have long roots to send down on explorations for water. Even where there is very little buried moisture, there is, of course, need for an extensive root system to find what little is available.

Many of the Cacti are boldly sculptured into alternate thick ridges and deep grooves. The ridges bear cushions at regular intervals, and these are surmounted by rosettes of spines. In some the spines are supplemented by long, white hairs, which give the plants so venerable an appearance, that one species in particular (shown on page 750) is known as the Old Man Cactus. These hairs have a protective function different from that of the spines which also are present for they tend to keep the surface cool, and so minimize the loss of moisture through evaporation.

We must not get our notion of the Cacti from the diminutive plants in tiny pots that are sold for conservatory decoration. These are often mere rooted offshoots from plants that grow to a considerable size. A better idea of size may be obtained from the view of a Cactus garden in California (page 753), where a number of different species are seen growing side by side. The tall, upright shafts belong to a species of Torch-Thistle, or Cereus, one of

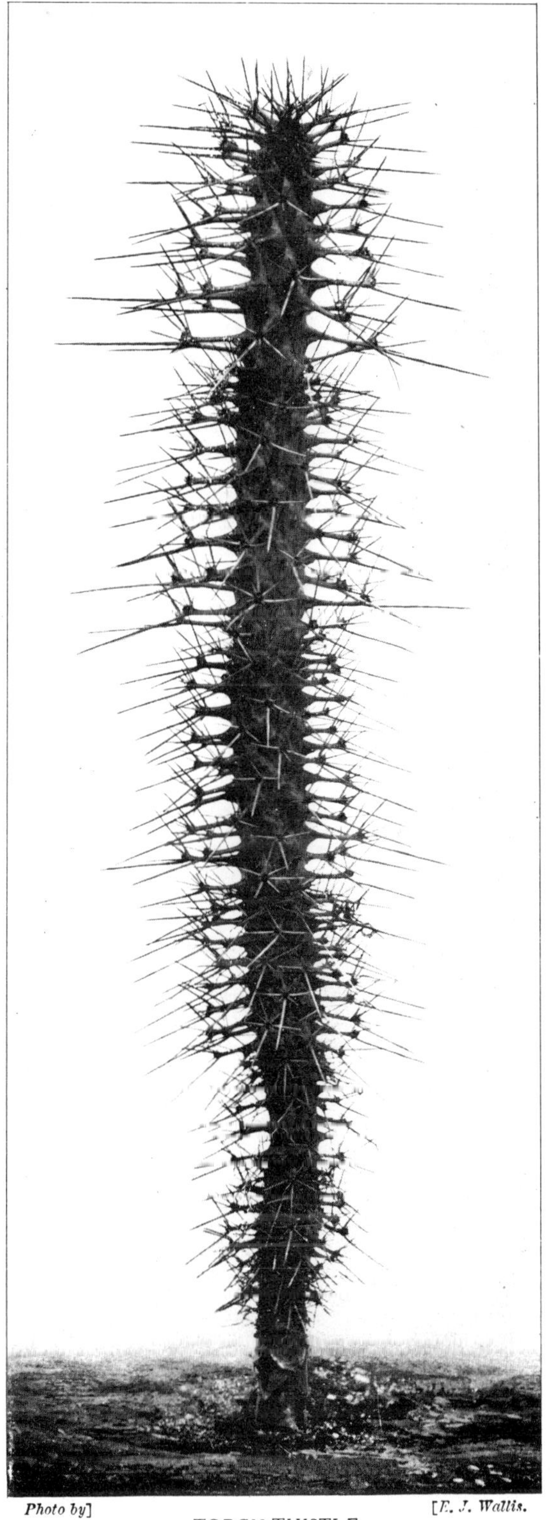

Photo by] [*E. J. Wallis.*

TORCH-THISTLE.

Many species of Torch-Thistle attain a great height. The character of their protecting spines is well shown in this photo.

Photo by] [S. L. Bastin.

TURK'S CAP CACTUS.

The fez-like extension of the plant proper consists of a woolly substance beset with soft spines. From it the flowers are produced.

which, known as the Suwarow, attains a height of fifty or sixty feet. Its habitat is the almost desert mountains of New Mexico, where little else in the way of vegetation is seen, and these look more like telegraph-poles than living plants. As a rule, they are unbranched, and are fluted throughout their length with from twelve to twenty deep channels and a corresponding number of ridges. Along the ridges at intervals of about an inch are thick yellow cushions, bearing each about half a dozen long spines and many smaller ones. It bears cream-coloured flowers, three or four inches across, from near the summit of the stem, and these are succeeded by egg-shaped fruits containing black seeds embedded in crimson pulp, which the Indians gather for food. It is believed that the Suwarow lives for two or three centuries. Some other species of Cereus are noted for the beauty of their flowers, which open at night. We have given photographs of two of these in our first volume.

A quaint form is the West Indian Turk's Cap, or Melon Cactus, shown on this page. The plant proper is globular in form, about a foot across, with from twelve to twenty ridges running from top to bottom, and surmounted by a cylindrical crown or fez of woolly substance thickly set with soft spines. The small rosy flowers spring out of the fez.

Another genus of short, broad and very symmetrical forms is Mammillaria, so called because the plants are ornamented with teat-like tubercles, which bear the clusters of spines at their tips. Two examples of this genus are shown in the photographs on pages 751 and 754.

Then there are the Hedgehog Thistles, which are egg-shaped or round, some of the latter resembling the Turk's Cap, except that they want the fez. They have strong ribs and rounded bosses, which bear the long spines. These are arranged in rosettes with usually a straight central spine. Some of them are truly terrible in their array of fixed bayonets. The Toothpick Cactus is a member of this genus. Between sixty and seventy years ago a large specimen of this Cactus, nine feet high and nine and a half feet in girth, was on view in Kew Gardens for a short time. It weighed a ton, and the difficulty of handling so prickly and heavy a succulent plant, in bringing it from its native Mexico, caused such injuries to the soft flesh that it never got over, and it soon died. Considering how slowly these plants increase in size, this specimen must have been very old. It was estimated to bear over fifty thousand toothpicks or protecting spines. The flowers of all this genus are produced from the top of the plant. As examples of the Hedgehog Thistles, we give photographs of Wislizen's Cactus, which is rather squat in early life, but later assumes a cylindrical form, and becomes very large. Its spines, for hardness and toughness, might be made of steel, so that the plant has to be approached with caution by man or beast.

Photo by] [S. L. Bastin.

WISLIZEN'S CACTUS.

Clad with spines as tough as steel. Radiating series of shorter white spines protect the flesh, and from the centre of each cluster stand out longer and stronger spines, which keep animals at a distance.

[By Arthur Twidle.

THE MOON'S HALO.

An effect produced by innumerable prism-shaped crystals of ice suspended in the atmosphere and distant forty-five times the diameter of the moon's disc from the outer edge of the moon. The *halo*, or glory, has prismatic colourings, the red tint at the edge nearest the moon and the blue on the outer margin. This distinguishes the phenomenon from the more usual *corona*, which is due to aqueous vapour and is coloured blue nearest the luminary and red towards the exterior rim.

THE MOON'S HALO

" Last night the moon had a golden ring," sang the poet, and he was no doubt alluding to the phenomenon which we are about to discuss. It is of sufficiently rare occurrence to require a description, and, though as a spectacle it is more particularly associated with the moon, it can, nevertheless, be observed in the sun. The reason for its greater apparent frequency round the lesser luminary is that its effect is usually overwhelmed by the strength of the solar rays. This "golden ring," in a complete state, is a circular band of light concentric to the moon's disc, and is distant forty-five times the moon's breadth from its edge. It has all the colouring, though in a fainter degree, of the rainbow, with the red tint at the edge nearest to the moon and a blue or purple one at the opposite margin—a particular characteristic which distinguishes the halo from the lunar or solar " corona." In the "corona" the ring comes much nearer the luminous orb—about eleven times nearer—and has the prismatic colours ranged in reverse order to those of the halo, that is to say, the blue is nearest to the moon (or sun), and the red forms the outer margin. It is an effect of light seen through an atmosphere surcharged with aqueous vapour, and is more frequently seen than the true halo.

This halo is due to entirely different causes, of which the chief factor is the refraction of light. At the time of its appearance the higher regions of the air are charged with minute prism-shaped crystals of ice. A prism catches the light, breaks it up into rainbow colours and refracts or bends the ray at a given angle, the angle varying according to the position of the prism. But there is one position at which the prism can sustain a longer phase of turning without causing any perceptible variation in the angle of the refraction of transmitted light, and it is when myriads of the ice-prisms are placed in this position, between the light and the eye of the observer, and throwing the rays directly towards the eye, that the phenomenon is apparent. The position of the crystal

CRYSTALS.

These six examples of crystals are typical of the six systems of Crystallography. Reading from left to right, those in the upper row are: Cubic, tetragonal, hexagonal; and in the lower, rhombic, monoclinic and anorthic. The bases of the crystals will be the same as those of the models.

STIBNITE CRYSTALS.

Crystals most dissimilar in outward appearance may be found to have the adjacent sides forming equal angles, and are therefore considered to belong to the same family. Stibnite is a member of the hexagonal group.

in the atmosphere is determined by its particular shape, that is to say, all prism-shaped crystals will occupy the same position, while all other shaped crystals will occupy different planes, and will therefore be negligible quantities in the effect produced. Again, the light must vibrate a definite number of times before it is in a fit condition to be absorbed and reflected by the ice-prisms, and that means that only those crystals which are distant twenty-three degrees from the luminary's disc are available for "mediums" in the transmission of light. To sum up, it is the presence of innumerable prism-shaped ice-crystals in the atmosphere, situated at a given distance from the sun or moon's disc and held in suspension at a particular angle, and so refracting the light directly to the observer, that produces the solar or the lunar halo.

CRYSTALS

BY RICHARD KERR, F.R.A.S., F.G.S.

CRYSTALS, owing to their peculiarly matchless beauty and durability, are the objects of costly ambition to the rich as well as the admiration of all classes of people.

We are indebted to Pliny the Second for putting together in a work an almost complete and compendious account of all the stores of information regarding the mineral kingdom extant in the first century A.D. From this justly-celebrated work, the *Historia Naturalis*, a true idea may be formed of the high estimation in which ornamental stones were held by the Greeks and Romans.

It is noteworthy that all the stones that are most highly valued are either transparent or translucent.

Great heat on a scale beyond the powers that man can employ, together with a pressure equally unattainable by man, must have been two, at least, of the great agents employed in Nature in order to reduce the elements of which precious stones are constituted to a condition of vapours or liquids. The substances destined, if we may so put it, to become the composition of precious stones or crystals and to render them homogeneous throughout, could not have been mixed together mechanically by the grinding together of solid materials. Nature is an extensive laboratory with ample means for effecting marvellous changes. In conjunction with the two agents already named, electricity and magnetism may have been employed.

AN EMERALD CRYSTAL.
A magnificent specimen from the Natural History Museum, over an inch long, embedded in corundum.

The numerous varieties of crystals are the result of wonderful natural laws which are so constant and invariable that almost all the many materials of which the earth is composed have been classified under six groups or systems as represented on page 758. The most precious of all gems, because the most beautiful and the hardest substance known, is the diamond. It has been proved to consist of pure carbon, but the exact manner of its production cannot be said to be fully ascertained. Mr. Gardner Williams, a great authority on diamonds, when asked as to his theory of their origin, replied: "I have no theory. . . . All that can be said is that in some unknown manner carbon, which existed deep down in the internal regions of the earth, was changed from its black and uninviting appearance to the most beautiful gem which ever saw the light of day." In addition to the diamonds formed at considerable depths in the earth, there are many known instances of diamonds in meteorites, some of which are exhibited in the mineral department of the Natural History Museum.

The largest known diamond, which may be only a portion of a very much larger one, is the famous "Cullinan." Its weight before cutting was 3,025 carats, or, expressed in avoirdupois, 1 lb. 5 oz. 14 drs.! It was found about twenty miles from Pretoria.

Several beautiful gems, such as the white and the blue sapphires, topaz, ruby, the oriental amethyst and the oriental emerald, are derived from alumina—the principal element of clay. In a crystalline state it is remarkably hard and durable, and is known as corundum.

THE CULLINAN DIAMOND.
The largest known Diamond weighed before cutting 3,025 carats, or 1 lb. 5 oz. 14 drs. avoirdupois. It is shown in its untouched state.

Quartz, the commonest of all constituents of the rocks of the globe, is the chief constituent of many beautiful stones—rose quartz, common amethyst, common topaz, chalcedony, avanturine, heliotrope, plasma, onyx, sardonyx, opal, jasper, carnelian, cairngorm, moss agates, agates, chrysoprase, and many others.

Tourmaline consists of alumina and silica. It assumes a great number of colours, and is of more value for its scientific services than for its ornamental purposes. Opticians by its use can tell

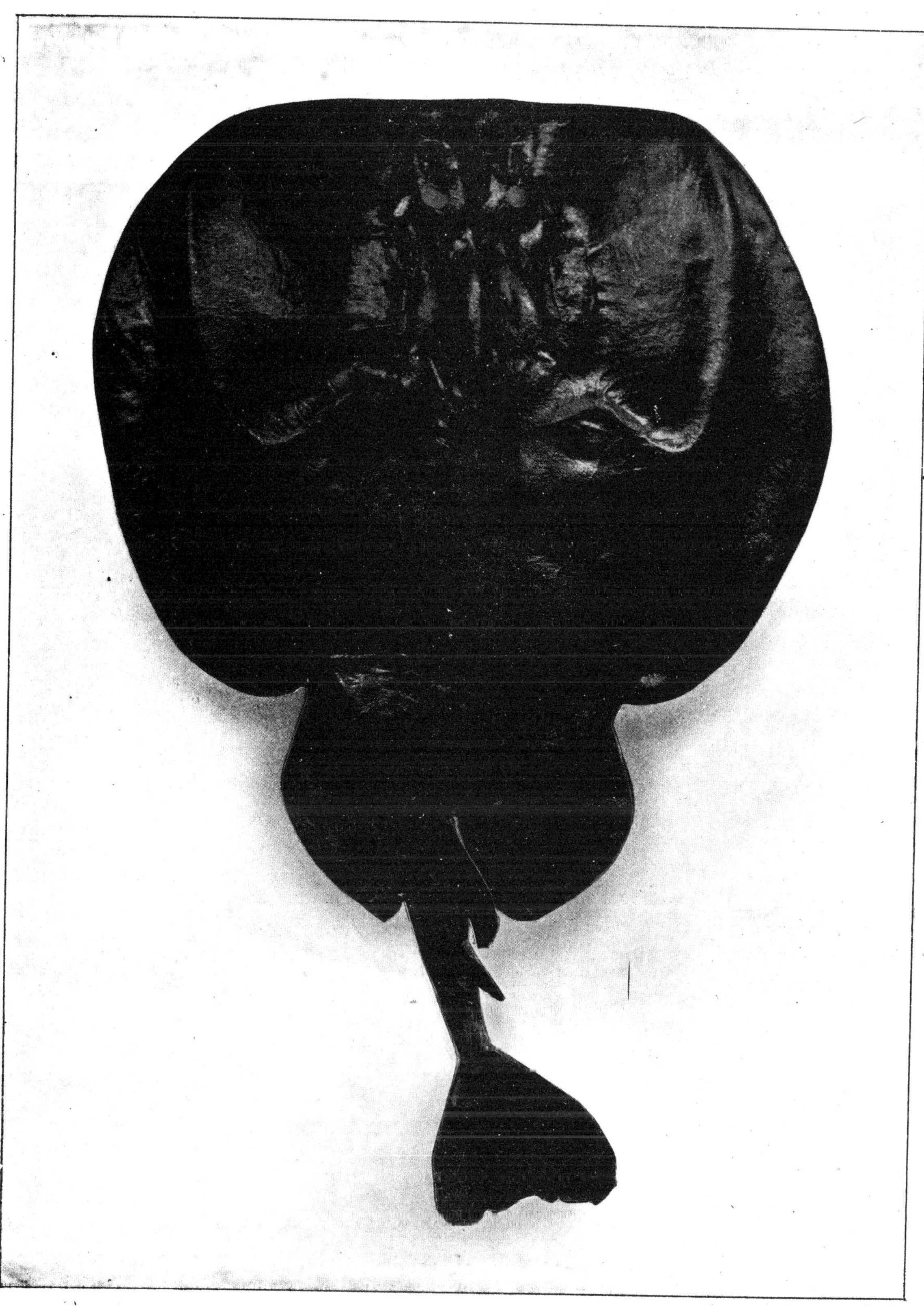

THE ELECTRIC RAY.

Notable for its power of generating electricity. The electrical organ consists of two large masses of hexagonal prisms looking like the comb of a honey-bee, situated between the head and the breast fins. The discharge depends on the size, health and vigour of the fish, but that of a large Ray may be sufficiently strong to temporarily disable a man.

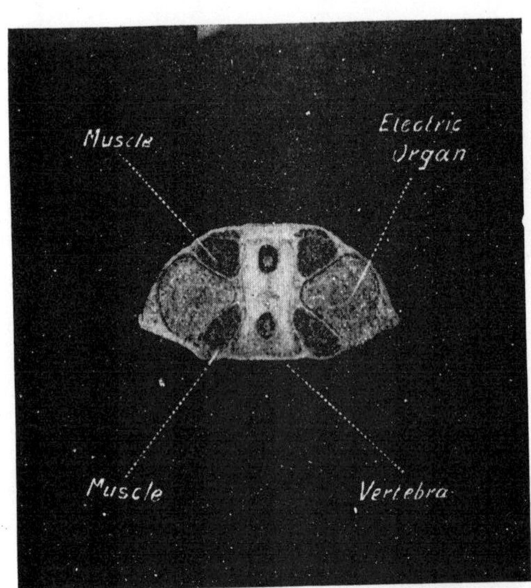

THE SKATE.

A transverse section of the Skate showing the nature and position of the electric organ.

in a moment the difference between glass and pebble lenses. It is also useful on a larger scale for polarizing light.

Sulphur crystals consist of pure sulphur, an element found in a native state in proximity to volcanoes. Copper pyrites consist of a combination of sulphide of copper with bisulphide of iron. It is from this mineral that a great deal of the copper of commerce is obtained. Sulphide of antimony crystals are also known as stibnite. The composition is seventy-four parts of the element antimony and twenty-six of sulphur. The type used by printers is made of an alloy of antimony and lead.

Fluor spar consists of calcium and fluorine in nearly equal parts. It is the source of hydrofluoric acid. It can also be worked in the lathe and polished for ornamental purposes. Felspar consists of silica and alumina with potash. It forms one of the component parts of granite. Decomposed felspar is a clay called kaolin, used for making porcelain. A beautiful variety of felspar, known as labradorite, gives at various angles marvellous varieties of colour. Beryl is closely allied to the aquamarine, and is composed of silica, alumina and glucina. The finest beryl known was in Mr. Hope's collection and cost five hundred pounds.

Chrysoberyl is composed of alumina and glucina. It is cut as a gem. The specimens from Siberia are of an emerald-green colour and form the variety named Alexandrite. It shows remarkable colour effects. The hyacinth, resting in basalt, is a variety of zircon and is composed of silica and zirconia. Turquois is composed of alumina, phosphoric acid and water. Lapis lazuli, also called ultramarine, is composed of alumina, lime, soda, iron and sodium, etc.

I am indebted to Dr. Max Bauer for the following illustrations: the diamond, emerald, topazes, amethyst, hyacinth and chrysolite. And to Dr. G. F. Kunz for those of the two tourmalines, ruby, sapphire and turquois.

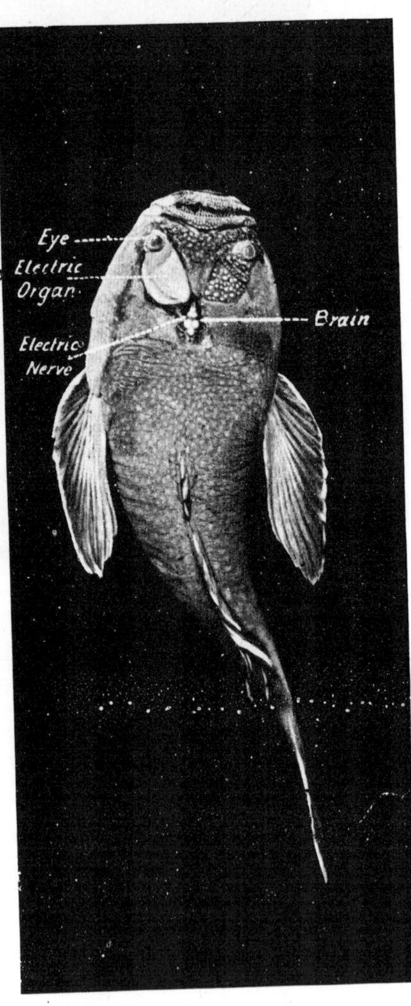

THE ELECTRIC STAR-GAZER.

With the outer skin removed to show position and connection between the various organs of the head producing the electric current.

LIVING ELECTRIC BATTERIES

BY W. P. PYCRAFT, F.Z.S.

It is now just over one hundred and twenty years since Luigi Galvani made his famous discovery of animal electricity. The facts he brought to light were accounted then among the marvels of Nature; and though, in the course of years, an immense advance in our knowledge of the phenomena he revealed has been made, we are as ignorant as ever as to how this electricity is produced. We have yet indeed to discover what electricity is. In the animal kingdom, then, it still remains one of the marvels of Nature, and this in the deepest sense of the word. In pursuing this theme the savants of to-day have the advantage of Galvani, for wonderfully complex electric organs, such as Galvani never dreamed of, have been found to exist in many species of fishes.

In one or two species these electric batteries are capable of giving very powerful shocks. This is especially true of the great scaleless Electric Eel of the Orinoco and the Amazons and swamps draining into them. As a matter of fact, it is in no way related to the eels, but to the carps and cat-fishes. The eel-like shape has been acquired by a process of transformation, wherein the likeness to the eel grew in proportion as it assumed the eel's habits. But this by the way. When full grown, this wonderful fish may attain a length of eight feet, and the thickness of a man's thigh. It is not, however, with its bulk, or its affinities, that we are concerned, but with its electrical powers, which are considerable, exceeding those of all other similarly endowed fishes. The electric battery runs along each side of the tail, and thus extends along about four-fifths of the whole body. It consists of two huge masses of bands, or columns of cells filled with a jelly-like substance, and richly supplied with nerves. All that careful dissection can tell us further of this is that this wonderful tissue has been formed out of ordinary muscle.

The Electric Cat-fish, a native of Africa, is only a little inferior in its powers to the Electric Eel, for the shocks it is capable of giving are severe; hence this fish is called "Raad" by the Arabs—a name meaning thunder. Kept in an aquarium with other fishes, it soon kills its companions. Unlike all other electric fishes, its battery is formed by a transformation of the skin, and not of the muscles. It envelops the whole body like a mantle between the skin and the muscles.

Third in order of these living batteries comes the *Torpedo*, or Electric Ray. Here the electricity is formed in two large, oval masses of tissue lying on either side of the head. There are several species of these Electric Rays, but all agree in their electrical properties. They give, Dr. Günther

THE ELECTRIC CAT-FISH.

Its capacity for producing shocks is only slightly inferior to that of the Electric Eel. Its battery envelops the whole body between the skin and the muscles, and is itself a transformation of the skin.

tells us, a shock voluntarily, either in self-defence, or to kill prey, and that shock is painful. Of two specimens captured some years ago in the estuary of the Tees, the stomach of one was found to contain an eel weighing two pounds, and a flounder weighing one pound ; and of the other, a salmon weighing nearly five pounds, and in no case was there mark or blemish of any kind found in the victim's body.

The curious and ugly fish known as the Electric Star-gazer, like the Torpedo, has the electric organs one on either side of the head behind the eye, and it seems that they have come into being by the transformation of some of the muscles of the eyeball. Each contains no less than one hundred and fifty layers of electric plates, or "electro-plaxes," piled up horizontally, in series. The controlling nerves lie in the upper part of each layer, the blood-vessels in the lower part.

But there are fishes living within our gates, so to speak, which also possess electric organs; they are the Skates. These organs can be found in the Common Skate, for example, if a section be cut through the body just in front of the back fin. In such a section the electric organs will

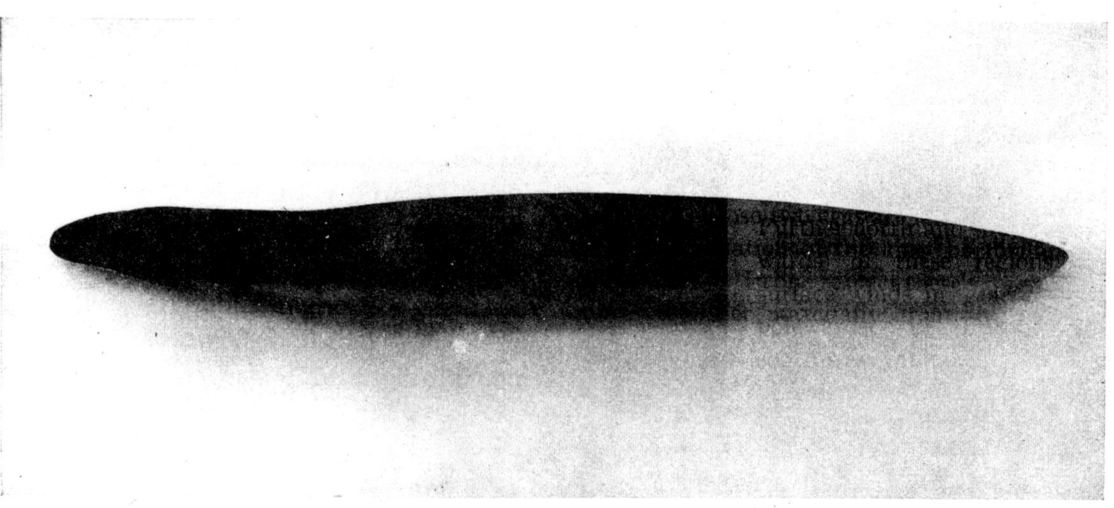

THE ELECTRIC EEL.

The discharges given off by this creature are sufficiently powerful to seriously inconvenience so large an animal as a horse. The battery extends along four-fifths of the whole body, and consists of two huge masses of cells filled with a jelly-like substance and supplied with an infinite number of nerves.

be found on either side of the backbone, as shown in our illustration. Unfortunately, however, for experimenters, the shocks which these fish give are very feeble. However, with more delicate instruments of research responses may perchance be obtained, which may materially enlarge our knowledge of this mysterious subject.

There are one or two further points to which reference must be made. Firstly, the shock imparted by an electric discharge increases with the number of electric plates included in the circuit. Thus in the Electric Eel the maximum shock is given when the body of the fish is so curved that the head and the tail are in contact with different points on the surface of some other fish. Secondly, the discharge may be reflex or voluntary. Repeated discharges induce fatigue and weaken the shocks.

A FLY THAT IS BORN IN ANTS' NESTS

BY HORACE DONISTHORPE, F.Z.S., F.E.S.

THIS beautiful Fly, of which we give a photograph, is superficially very like a "bumble" bee in appearance (though, of course, it only possesses two wings, instead of four, as in the latter), being clothed in a sort of pile of golden and metallic hairs. The female is of a more reddish colour, and

Photos by] [*H. Donisthorpe, F.Z.S., F.E.S.*

A FLY THAT IS BORN IN ANTS' NESTS.

The upper photographs show, to the left, the upper side, and to the right the under side of the strange flat grub of this Fly. Below is the Fly itself, very like a "bumble" bee except that it has two wings instead of four; and on the right is the "ash-coloured" ant in whose nest it is usually born.

Photo by] [*H. J. Shepstone.*

SKELETONS OF THE DUCK-BILLED DINOSAUR.

These vegetable-eating Dinosaurs belonged to the closing period of the Age of Reptiles. They were fifteen to sixteen feet or more in height and measured about thirty feet in length. They had no protective armament, and would probably escape the Tyrannosaur, the king of the flesh-eating Dinosaurs, by taking to the water.

the male bluish. In the perfect state it is generally to be found lurking at the roots among long grass and plants, and it flies reluctantly. There are four European species, all of which occur in Great Britain. The most interesting facts concerning this insect are to be found in its life-history, and especially its earlier stages.

The grub of the Fly is a curious-looking object, white, oval and convex, with a flat creeping-sole and a neatly reticulated upper surface. It has a little tail, and the head is hidden in front. In appearance it is much like a slug, and, indeed, it was originally described as such. In 1823 von Heyden described and figured this grub, but he did not give it a name. He concluded, however, that it was a *mollusk* of an extraordinary new genus. In 1824 Spix named it as a slug. Although about this time its real state as the grub of this Fly was discovered, it was described in 1839 as a scale-insect, and even as late as 1907 the grub of an allied species abroad was described as a new genus of Mollusca! The grub passes its whole life in the nests of ants. We give a figure of the "ash-coloured" ant, in the nests of which species we have always found these grubs. They are to be found under the bark on trunks of trees, in boughs, or under stones, wherever their hosts, the ants, may have made their home. They are always to be found among the ants. They move slowly, with a wave-like motion on their creeping-soles. The ants pay very little attention to them; if, however, they can get a grub on its back, they will bite at it and kill and devour it. They seem unable to hurt it as long as it is right-side up. The writer has noticed in his artificial nests where he has reared these flies the ants gently bite at the grub and sit on it. When the ants have moved from one chamber into another for more warmth or moisture, or any other cause, the grub has slowly followed them. More than one grub is generally to be found in a nest which harbours them, and the writer has found as many as twelve in one nest of this ant at Porlock, in Somersetshire. He has now discovered what the grub feeds on, which was unknown hitherto. It was suggested that it consisted of the soft parts of wood, but this could only be the case where grubs occurred in nests in wood, and not when in the bare galleries in the earth where they are frequently found. Ants possess a special pocket in the head in which they retain parts of substances which they have eaten and not digested, and these they spit out. It is these minute pellets (which, though hard, must still contain a considerable amount of nourishment), and also on the droppings of the ants, on which these grubs feed. The young grubs shrivel and die if kept separate from the ants. They change to a chrysalis at the end of April. The integument hardens and becomes more brittle, and the whole creature turns brown and remains stationary. Two little horns are pushed out in front in two or three days, which are connected with the breathing of the fly. In twenty to thirty days the perfect insect emerges. It breaks off the front part of the chrysalis case and crawls out. It is then without wings, but these grow to their full size in a very short time. The Flies generally hatch at night, when the ants are in a more dormant state. When an ant approaches the Fly before its wings are developed it kicks out with its back legs to send the ant away. They leave the nest as soon as possible and fly away. The female Fly returns after courtship in May and June to lay her eggs in the same or some other nest. The ants have been noticed to drive a Fly away, which has returned each time to lay her eggs.

This species is considered rare, but is widely distributed in Great Britain.

A DUCK-BILLED DRAGON

BY W. P. PYCRAFT, F.Z.S.

THAT the Land Dragons or Dinosaurs have had their day and ceased to be is, perhaps, something to be thankful for; since even the comparatively sluggish vegetarian species would have

Photo by] [*H. J. Shepstone.*

THE SKIN OF THE DUCK-BILLED DINOSAUR.

The outer covering of this Dinosaur was thin, a fact which tends to confirm the opinion that it was an aquatic creature. It was, further, completely studded with larger and smaller tubercles, much after the style of the lizards of to-day. These tubercles varied in size and arrangement according to the different species

occasioned something more than a mere diversion if encountered when bathing or boating, for they were, most of them, more or less aquatic in their habits.

Measured by the Dinosaur standard, the Duck-billed Dragon, Trachodon, or Thespesius, as it is variously called, was a beast of medium size, attaining a length of about thirty feet and a height of twelve or thirteen feet. All that we know of this creature has been gleaned from specimens collected during the last forty years or so from the Laramie Sandstone of Montana and the Upper Cretaceous of Wyoming, U.S.A., and to-day that knowledge is almost as complete as we can ever hope to make it.

Briefly, the Duck-billed Dragon flourished towards the end of the Great Age of Reptiles; it was indeed one of the last surviving species of its race. In its skeleton it presents several very remarkable characters, and a high degree of specialization, a feature which has proved the undoing of so many types in the past. From its appearance during life it must well have deserved to rank

Photo by] [*H. J. Shepstone.*

THE DUCK-BILLED DINOSAUR.

These mummified remains, from which the enveloping cast of the Dinosaur's skin was obtained, must have been rapidly embedded in fine river sand and clay: not, however, before the decomposition of the dead Dinosaur had set in, for the skin had already shrunk away from the bones.

as a Dragon, albeit one of the harmless type. Its body seems to have been covered with a tough hide studded with irregular bony plates capped by horn, much as in crocodiles to-day, but of a more irregular shape. Its method of locomotion was bipedal, though, like the kangaroos, the forelegs were doubtless used to support the body as occasion demanded, as when feeding on shore. The upright carriage, as might be supposed, made a great demand on the muscles, especially those of the loins and the base of the tail; and these were supplemented by tendons like great steel bands, but flexible as no steel can ever be. This we know because the central portion of these bands became converted into bony rods, and these have been found in plenty associated with the skeleton. We find precisely similar bony rods in the tendons of a fowl's feet and the lower end of the "drum-stick." The forelegs could have been but little used; they served, perhaps, to convey food to the mouth, and, as we have suggested, on occasion to support the fore-part of the body. As a consequence, they were degenerate, not only in size, but also in the fact that the number of fingers was reduced to three, and a little more than a vestige of a fourth. When swimming, the body must

Photo by] [*H. J. Shepstone.*

THE DUCK-BILLED DINOSAUR.

This species of Dinosaur was provided with webbed feet, a long tail with much-compressed sides, and a huge duck-like bill. Al these characteristics indicate a more or less aquatic existence. The jaws were provided with a marvellous grinding apparatus, and more than two thousand separate teeth have been counted in the upper jaw alone.

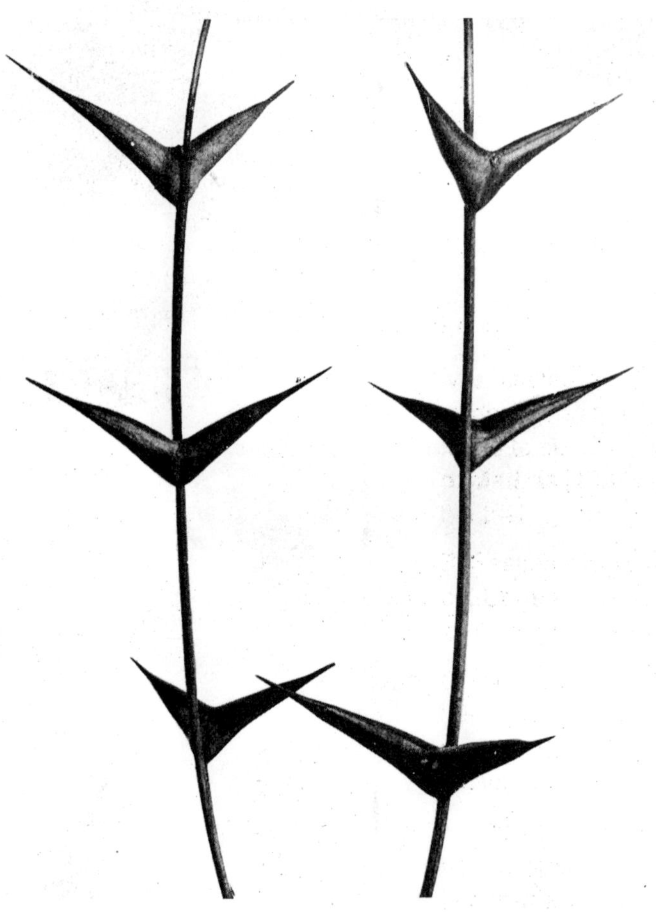

BULL'S-HORN ACACIA.

The interior of these spines is used as food by the ants that adopt them as nests. One kind of ant makes an entrance near the tip of one spine and bores through the base to the other spine. Another ant makes an entrance at the base leading to both spines.

have been propelled by undulating movements of the long and immensely powerful tail. The fore-limbs, on such occasions, would serve rather as balances. The great hind limbs, like those of all the bipedal reptiles, were bird-like, and bore but three toes.

But the most remarkable feature of this strange beast was its head. All the later Dinosaurs displayed a tendency to lose their teeth, replacing them with horny plates forming a beak. In the Duck-billed Dragon this new-fangled notion of suppressing the teeth was associated with a successful attempt to break the record in the matter of teeth borne on the jaws at the back part of the mouth, answering to the "grinders" or "cheek" teeth. For here they existed in row upon row one above the other, small, spear-shaped, and tightly wedged together like a pavement; over two thousand belonged to the upper jaw alone. But, it must be remembered, they were not all in use at once; only those along the cutting edge of the jaws were functional. As these got worn, or broken, they were shed, and replaced by those in the row immediately below. Thus an almost endless succession of teeth was possible. These teeth were admirably adapted to cutting up grass and rushes and submerged weeds, which were evidently seized by the curious, toothless, shovel-like "beak" which formed the fore-part of the jaws.

ANTS AND THE BULL'S-HORN ACACIA

BY EDWARD STEP, F.L.S.

To the average person of intelligence the animal world is separated from the plant world by a great and unbridgable gulf. Animals of many kinds, including man, feed upon plants. Man admires the flowers of plants, and occasionally shows an interest in the structure and development of plants; but though he may have read of the remarkable behaviour of insectivorous plants and the inter-relations between flowers and pollinating insects, he is too apt to regard these things as mere fairy tales, invented by the naturalist to amuse him. On the whole, his view of the nature of plants remains pretty much what it was a hundred years ago; and to those who have no actual intimate acquaintance with the plant as a living thing, it is probably easy to hold such a view. But it is difficult for those who have studied the plants, not as dead and dry herbarium specimens, but as they carry out their natural functions in the woods and fields. There are so many points at which certain plants come into contact with certain animals, and have been seriously modified by the relationship, that it is no longer possible to regard them as mere automata, or to deny them the possession of some glimmerings of sense. The evidence for this unorthodox view is far too voluminous and varied to be even hinted at here; but one sample may be given which is only one of many such that have been carefully investigated.

The case we have selected is that of the Bull's-horn Thorn or Acacia, a plant of the Central American savannahs. As in most of the Acacias, the long leaf is broken up into a number of leaflets, arranged in pairs along the midrib, and the branches and twigs are protected from the depredations of browsing animals by the development of thorns and spines. In the Bull's-horn Acacia these take the form of long, stout curved thorns set in pairs in such relation to each other that they present much the appearance of the horns of a bull, to which circumstance the plant owes its name. But like most other vegetation of that region, the Acacia has a worse foe than the herbivorous mammal to contend with, and one against which the sharpest spines are no protection. This is the Sauba or Leaf-cutting Ant, concerning which we gave some particulars in the first volume of this work. The Sauba comes in hundreds, and in a few hours completely defoliates the tree or bush it has fixed upon for attack, carrying the leaves home to convert into leaf-mould upon which to grow fungi for food, being a vegetarian ant.

By some means the Acacia has learnt, through the experience of its ancestors, that there are carnivorous and stinging ants as well as herbivorous ants, and that the former is the best antidote to the latter. If we were to judge by results in this matter we should have to confess that the "insensible" plant has more intelligence than man. Man the cultivator can only see in insects of all kinds active enemies, and impartially kills all, friends and foes alike. The Acacia is more discerning—encourages the friend, pays him for his protection and so holds the foe at bay.

Photo by] [*S. L. Bastin.*

AN AFRICAN ACACIA'S SPINES.

In this species the spines are much dilated at the base, and used as nests by protecting ants.

Out in the savannah there lives a little ant called Pseudomyrma, which, in addition to the business-like pair of jaws usual in ants, possesses a sting like a bee or wasp, whose effects on the human victim last for twenty-four hours. During the dry season this ant appears to live in the ground; but when the wet season arrives and the Acacia puts forth new shoots and unfolds new leaves, Pseudomyrma will be found on the Acacia, and you cannot touch the plant without having a number of the ants swarming over your hands, biting and stinging at the same time. At the base of each pair of leaflets on the new leaves there is a little hollow, and at this time the plant pours into the hollow a sweet fluid much like the nectar with which most plants attract bees to their blossoms. The purpose is the same here, only it is ants instead of bees whose visits are sought. The ants love sweets and largely live upon them; so numbers climb up the stems and twigs and run over the leaves in the search for these little saucers of sweet drink. Now, also, comes the great Sauba, intent on carrying home a load of leaves. But the alarm is spread among the much smaller stinging ants, and they immediately do battle with the Saubas and drive them off. There is no enemy the Sauba appears to dread so much as these little stingers. So does the Acacia benefit by its outlay in nectar.

Photo by] [*S. L. Bastin.*

THE ACACIA ANT.

The little stinging ant that lives in the hollowed spines of the Acacia and keeps off leaf-cutting ants. (Much enlarged.)

But it might be only a matter of chance that the stinging ants were in the way when the Saubas called; how much better if the Acacia could induce its stinging friends to take up residence on its branches, and so be always on duty, patrolling the ways that lead to the leaves. The Acacia puts forth a new set of thorns, and fills them with a substance that will serve as ant food. By some mysterious means—possibly it is inherited knowledge—the ants know this, and set to work to bore a hole in the upper part of one only of each pair of thorns. This aperture is large enough to serve as an entrance for the ants, and they soon clear out the interior. Strange to say, this treatment does not, as might be anticipated, stop the development of the thorn; instead, it appears to stimulate it. Mr. Belt, to whom we are indebted for our knowledge of the relations between the plant and the ants, reared the Bull's-horn Acacia from seed in the Nicaraguan forest, where there are none of this species of ant; and he found that the unattended thorns did not attain their full size, but turned yellow and dried into small persistent prickles. Where the ants have access, however, they become larger at the base after they have been emptied. There is a natural division between the bases of a pair of thorns, but this the ants bore through, so that they have the run of both thorns with only one means of entrance and exit. In these thorns they construct nests and rear their young, and the moment any noisome leaf-fretting insect sets foot upon the plant the ants pour out of these receptacles and make an onslaught upon the intruder, who is stung and driven off if large, and, if small, is killed and eaten.

Now the sweet fluid is only secreted by the leaves whilst they remain in a young and tender condition. Probably they are incapable of doing so when the leaves are fully developed; but it is clear that this flow having ceased and the stuffing of the thorns having been consumed, the ants—supposing they stayed on the same Acacia—would have to make food-hunting forays, leaving the Acacia unprotected whilst they were away. The plant appears to have realized this difficulty, for

Photo by] [*S. L. Bastin.*

SPINES OF THE BULL'S-HORN ACACIA.

The Bull's-horn Acacia is a Central American shrub which protects itself from browsing animals by the production of these spines. Certain stinging ants bore into the spines and convert them into nests, from which they pour in order to drive off leaf-cutting ants.

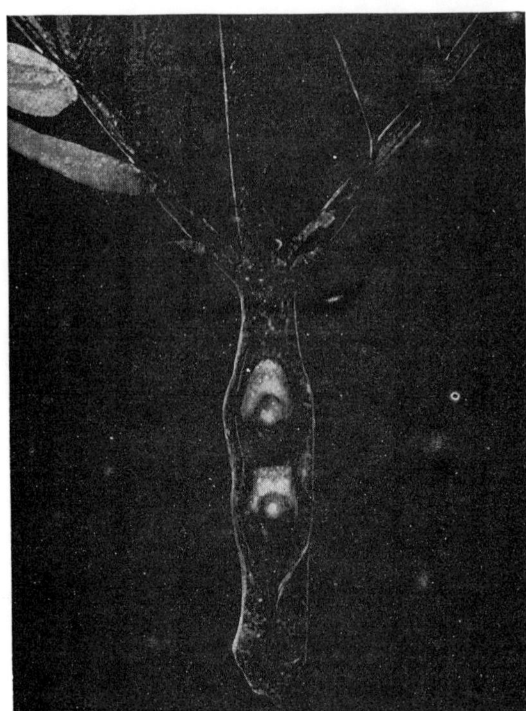

HONEY-GLANDS OF ACACIA.
The base of the compound leaf is here enlarged to show the glands.

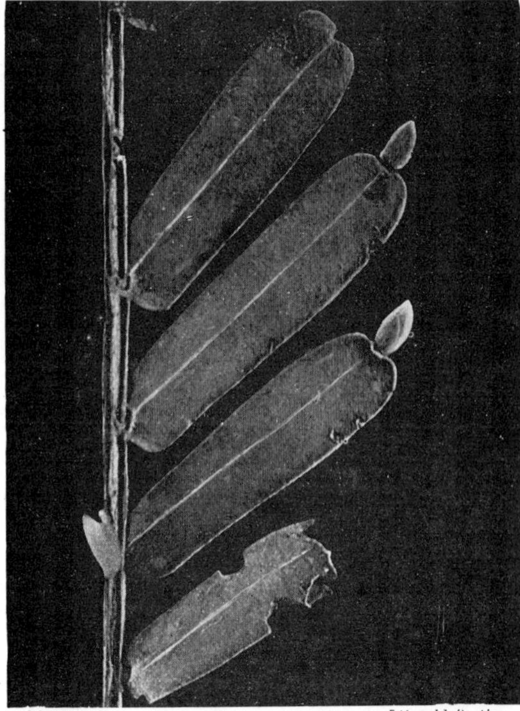

Photos by] [Harold Bastin.

"BELT'S BODIES."
Leaflets of an Acacia, much enlarged, to show the pear-shaped bodies first made known by Mr. Belt, which now bear his name.

it has, in addition, developed special food-bodies for the ants which have no analogy in the growth of other plants, and apart from their value to the ants are of no use whatever to the plant. Every small division of the compound leaf has at its extremity a little yellow, pear-shaped body growing out from it. When the leaf unfolds these "pears" are not ripe, and they do not ripen all at once, but in succession. The ants may be seen going from one fruit to another to ascertain their fitness for gathering, and on finding one that is ripe, an ant nips through the stalk and bears the "pear" off to the nest.

Belt found that his seedling Acacias, grown in his forest garden, though they were visited by various local species of ants, were not really tenanted by them. These ants knew nothing of the needs of the Acacia, nor the price it was willing to pay for their protection. They left both thorns and "pears" alone. The Sauba ant came among others, and stripped the plants of their foliage, for there were no ant-police to keep them off. Out on the savannah, Belt says, he never found one of these Acacias that was inhabited by Pseudomyrma effectively attacked by the Saubas. One other species of ant—Crematogaster—he found inhabiting the Acacia and nesting in the thorns. Their methods differed somewhat from those of Pseudomyrma, for they made their entrance to the thorns from the base instead of near the point.

The photograph on page 771 shows an African species of Acacia which has the base of its thorns dilated into a spherical chamber which is similarly tenanted by a protecting ant. At first the thorns are white and soft, then green, and later become brown or black and hard. They are bored by the ants, who are almost invariably found in possession of them. This species is never found to be eaten by cattle, and this immunity is believed to be due solely to the presence of the ants.

From all this it is clear that there is so definite an understanding between the Acacia and the ants, established in the far past and perfected in the course of many generations of both plants and ants, for their mutual benefit, that this commensalism has now become a fixed habit. To many people such an understanding between an animal and a plant will be unbelievable; but strange as it may appear, it is true, and it does not stand alone.

THE EGGS OF SNAILS AND THEIR EGG-BASKETS

BY BERNARD C. WHITE

INSIGNIFICANT and unworthy of consideration as the snail undoubtedly is to the casual observer, there are hidden away in the coiled shell and mean-looking exterior wonders of coloration, of construction, of mechanism, of expedients for the maintenance of life which amply repay patient and careful inquiry. One of these wonders is the methods employed by snails for the careful deposition of their eggs. These are not scattered broadcast to take their chances with the multitude of dangers which beset them, as might be expected from the lowly position of the mollusc in the scale of nature, but are, as we shall see, often provided with special means of protection and given fairy cradles to dwell in.

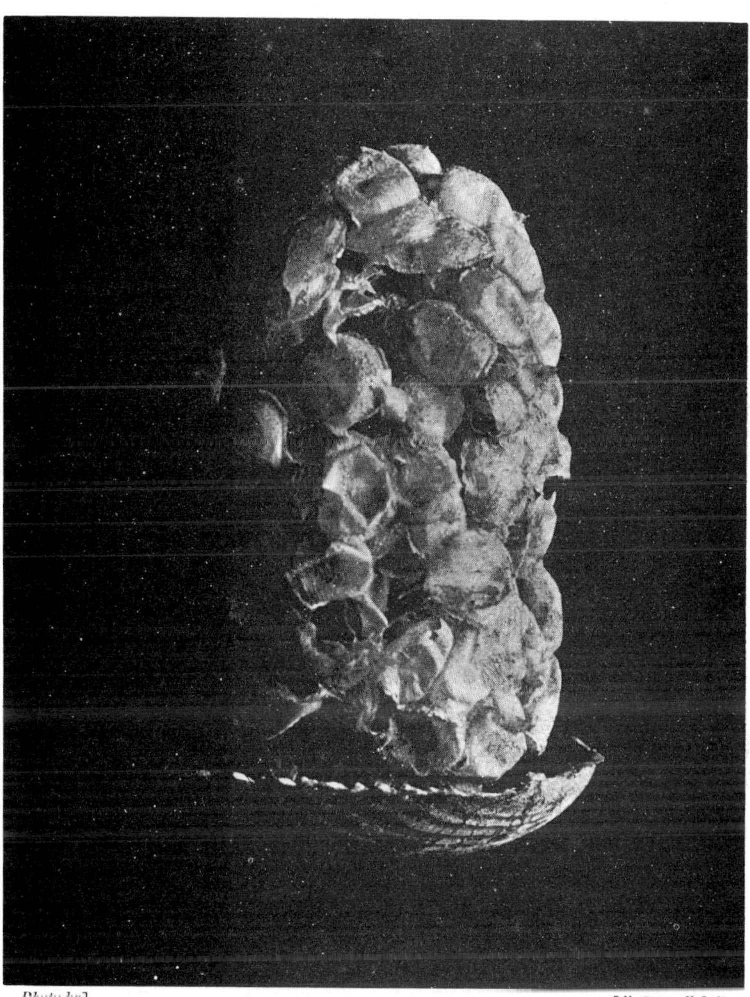

Photo by] [*E. Step, F.L.S.*

THE EGG-CAPSULES OF THE WHELK

As they appear after the young have emerged. They are here attached to a scallop-shell. Each capsule contained a number of eggs; but few would reach maturity, as the first-developed young devour any unhatched eggs they may find within the capsule

But the first remarkable point with regard to the eggs of snails is their number. It is no uncommon thing to find a batch of eggs from one mother numbering several thousands. The Swan Mussel of English ponds and rivers will have as many as twenty thousand; while the Painters' Mussel often bears ten times that number. Yet it is the oyster who is without her compeer in this matter; for she is the mother, not of thousands, but of hundreds of thousands, and a single individual has been known to produce as many as six million eggs. There is a reason for this prolific brood. Many are the enemies of the oyster, throughout its career, for nearly all marine life has learned to appreciate its esculent qualities, and as a result the race of oysters is continually suffering a heavy depletion which can only be rectified by an abnormally high birth-rate.

Some snails, on the other hand, produce comparatively few eggs, and such are usually of considerable size. The Giant Snail of South America, which has been mentioned earlier in these pages, has an egg which may measure as much as one and three-quarter inches in length, while that of the Left-handed Agate Snail is as large as a pigeon's. The shells of nearly all such eggs are brittle and white, an exception to this colouring being found in the Black-mouthed

Bulimus, which is tinted a beautiful pale blue-green ; while the large egg of the Worm-eating Slug (see page 261) is flexible and will rebound if lightly dropped.

But the most curious method of egg-protection is that adopted by the marine flesh-eating molluscs. These enclose their eggs in horny capsules or purses, each purse containing a varying number of ova. These purses are often united in a single mass, or if not joined are deposited together on any available rock or stone.

The most familiar of such capsules are those of the Common Whelk, which are shown on page 775 attached to an empty scallop-shell. Large bunches of these parchment-like capsules—mostly after they have been evacuated by the young whelk—are to be found on every coast. The Dog Winkle forms capsules for all the world like yellow grains of rice set very close together on tiny stalks.

Somewhat similar of arrangement are the " egg-baskets " of the Murex family, which are shaped like little cones standing close together on their apices, the base of each being bordered with an elaborate horny frill which suggests the corolla of minute auricula, for the largest capsule is hardly half an inch in expanse.

A SAND-SAUCER.

It is the egg-ribbon of the Necklace Shell, and is composed of sand and mucus; in this the eggs are laid, arranged in quincunx order. In time the sand hardens and effectually protects the eggs.

The Fresh-water Snails known as Ampullaria or Bottle-shells produce very beautiful capsules, which are united together round any suitable reed-stalk or blade of water-weed, and form a mass of tiny, frail, concave chalk " bubbles." Another curious egg-cluster is that of the Necklace Shell. It is not an uncommon object of the South and West coasts of England, where it is known as a Sand-saucer. The mucus of the mollusc is massed into a flat or slightly thickened ribbon and curved into a rough circle, bearing some resemblance to a girdle, while the glutinous nature of the material causes the sand to adhere thickly. The eggs are deposited quincunx fashion in the girdle while the composition is still pliable. Soon it begins to harden, and forms an impenetrable fortress for the harbourage of the young Necklace Shells.

The Violet Shell has adopted the principle of a float for the safe deposit of her eggs. She excretes a quantity of transparent slime which, as it hardens, imprisons bubbles of air. To the under-surface of this float she attaches the egg-capsules. Of these capsules it has been observed that usually " those nearest to the animal contain eggs ; those in the middle, young with fully-formed shells ; while those at the further end are broken and empty."

But the most elaborate of all constructions are the egg-baskets of the Spindle-shell family. " Basket " is hardly the term for these huge ruffles, which rise coil upon coil from the rock where they usually are affixed. Two such coils are shown on the opposite page. They are composed of a series of very flat, round capsules like diminutive talc watch-cases with frilled or spiked edges, and graduated in size, running on a thread which may attain a length of three or more feet. The girth

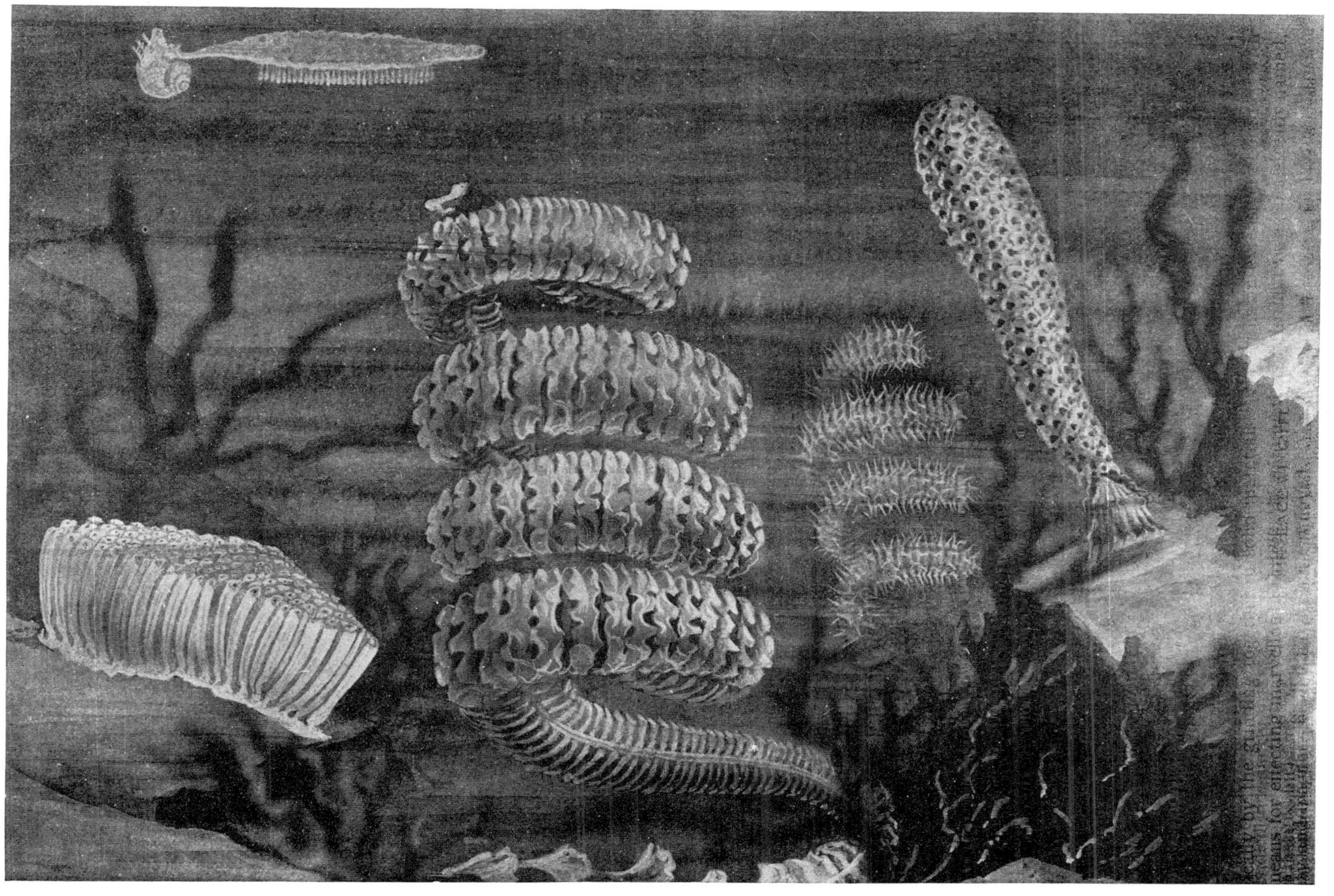

THE EGG-CAPSULES OF SNAILS.

In the centre is the long coiled stalk of the Fusus Snails, each ring containing a number of young. Behind, to the right, is the thorny coil of a species of Sycotypus; further to the right is the "egg-basket" of an unknown species. It is hard, pink, honeycombed, and after the nature of a coralline, and is attached to a top-shell. The snow-white mass of capsules to the right is from Mauritius. Above is the egg-float of the Violet Shell.

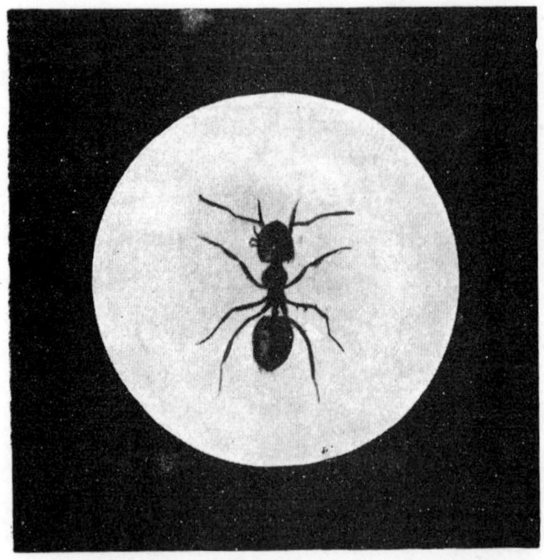

An ant with a Mite attached beneath the chin.

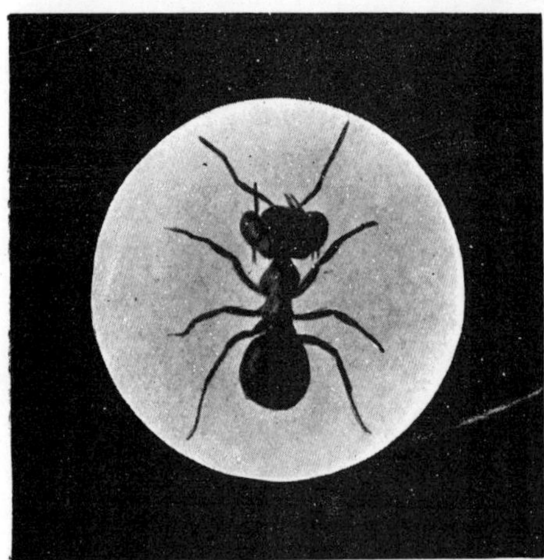

Two Mites, one on each side of the ant's head.

Photos by] [*H. St. J. Donisthorpe, F.Z.S.*

A scarce yellow ant with a large red Mite.

of some of these is, considering the size of the snail, immense, especially in the case of those of the Proboscid Spindle, where they may be as much as two inches in diameter. Each contains a number of eggs. Possibly not only the horny integument but the similarity of the coil to sea-weed may be a protection for the young Spindle-shells against their enemies.

MITES THAT LIVE IN ANTS' NESTS

BY HORACE DONISTHORPE, F.Z.S., F.E.S.

To most people Mites do not exist at all. If looked for, however, they are to be found everywhere, and their habits are both interesting and very variable.

They are related to spiders and scorpions, and generally possess eight legs. They have commonly only three pairs of legs at birth, and obtain four pairs after a moult and metamorphosis. Probably the most generally known is the "Cheese Mite," which is microscopic. Some are aquatic, and live in both fresh and salt water. The ticks, of which the "Sheep-tick" is a serious pest, occur on mammals, birds, and even on snakes and turtles. The "Itch Mites," one of them causing the disgusting complaint called the "Itch," burrow within the skin of man and other mammals. A few occur in the tracheal passages of seals, and one has been found living within the lungs of a monkey. There are several species attached to the nests of moles; others are sometimes found on beetles; the "shard-borne" beetle of Shakespeare being called the "Lousy Watchman," on account of the numbers of Mites with which it is often infested. Many species feed on living plants, and produce galls on the leaves and twigs. It is not, however, generally known that a large number of species are only to be found in the nests of ants. Many only occur with one particular kind of ant, but others frequent the nests of several different species. There are some twenty-seven different kinds of Mites now known to occur in ants' nests in Britain, of which nineteen were first discovered here by the writer. Many of them live as scavengers in the nests; they are not fed, or even noticed, by the ants. They devour the dead bodies of their hosts and other refuse in the nests. They are generally to be found in the galleries and underground chambers of the ants. We give a photograph of a large red species which was observed by the writer in some numbers in Cornwall in the nest of a scarce yellow ant, its proper host. This Mite is usually rare, both here and on the Continent. There are many smaller species of the same family which may be noticed as minute red

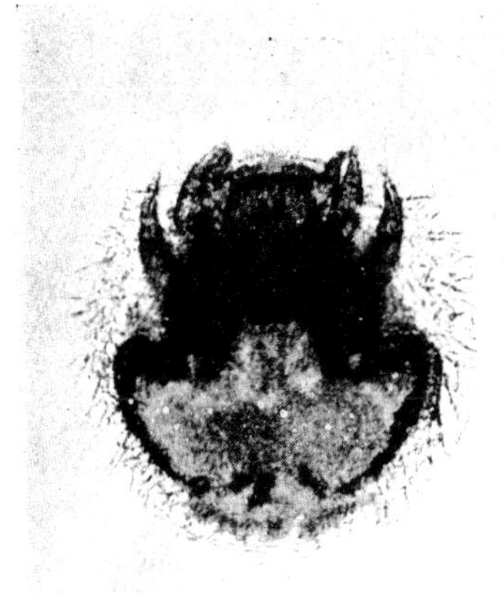

A Mite that is often found fixed on the grub of the ants.

specks in the nests as they move very slowly about. Some of the most interesting species belong to a family which are called "Antennæ bearers," because the first pair of the Mite's legs are long and are waved about in front and look like antennæ. We possess all the four known European species, each of which has a different kind of ant for its host. These Mites are always attached to the bodies of their hosts. In the species which is found on the jet black ant, the Mite is always on the head of its host. It is nearly as large as the ant's head, and its favourite position is under the chin. When it wishes to be fed it scrapes the ant's mouth with its front feet and the ant lets out a drop of liquid which the Mite sucks up. They can hold on very tight, having suckers to their feet, and the ant is unable to get rid of them. When an ant rubs the chin on the ground, the Mite runs round to the back of its head. It is curious that the other ants never try to rid each other of these parasites. When an ant is feeding at some honey, and has a Mite on its chin, the latter gets to one side of the ant's head to allow it to feed. When two ants feed each other, if one has a Mite on it, the Mite leans forward and shares the meal. In the species which is found on one of the larger yellow ants, the Mites are frequently to be found on the body of the ant. Sometimes one ant will have as many as three or more Mites on it at once.

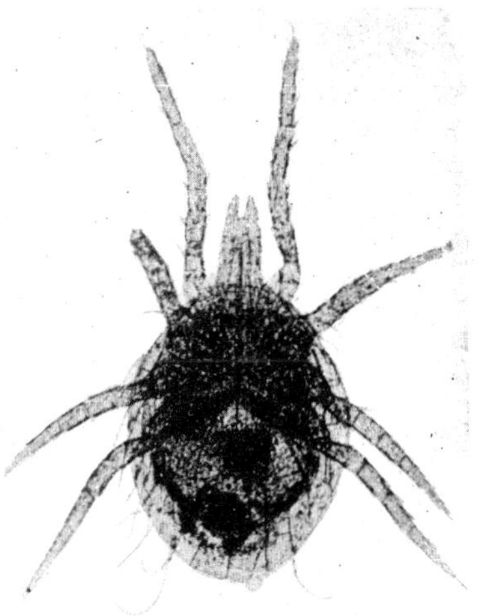

The "rider" Mite, which jumps on and off its host as it runs along.

The ant looks a very curious object with two Mites, one on each side of the body, and one or more on the head, either on each side, as shown in one of our photographs, or on the top of the head and under the chin. The Mite is often fed by other ants besides the one on which it may be. They solicit another ant for food with the front feet, and are then fed by it from its mouth.

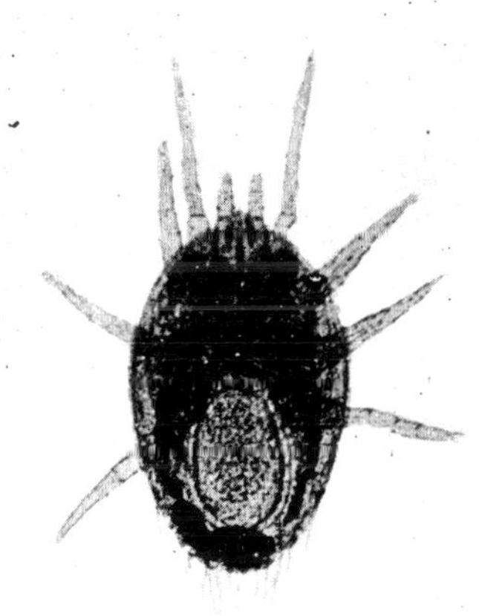

Photos by]

A Mite that lives among the egg-masses of the ants.

Several other kinds of Mites occur on the bodies of their hosts. One is called the "rider," because it jumps quickly on and off the ant as it is running along. Another very small allied species lives amongst the egg-masses of the ants. It does not feed on them, but is fed by the ants licking their eggs, which they do frequently. Another species fastens itself by its teeth to the comb-like spur on the ant's front legs. A round, scale-like Mite is often found

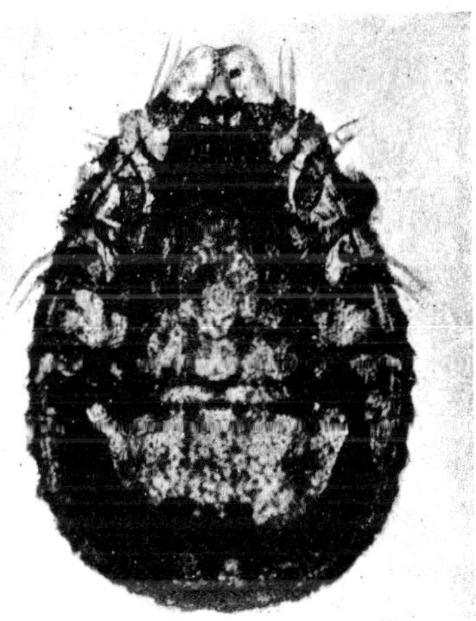

[*H. St. J. Donisthorpe, F.Z.S., F.E.S.*

A Mite which acts as scavenger in the ants' nest.

attached to the grubs of the ants, but it also fastens itself to the sides of the ant's hind body, where it causes a scar. Another yellow species, with long legs, is of exactly the same colour as its host, and it much resembles the hind body of an ant as it runs about the nest.

It will thus be seen that Mites are not uncommon in ants' nests, and that they play many different rôles.

FOUR-FOOTED NEST BUILDERS

BY R. LYDEKKER, F.R.S.

We are so accustomed to associate the habit of building nests for the rearing of the young (and in their case for the incubation of the eggs) with birds, that we are very apt to overlook the fact that the practice is likewise common to a large number of the smaller mammals. Indeed, the greater proportion of those species which live in holes or burrows and produce their young in a helpless condition, such as foxes, marmots, rabbits, moles, and rats and mice, construct nests at the bottom of these chambers, where their offspring are safe from harm. The same is the case with the duck-billed platypus of Australia, whose two eggs are laid in a chamber at the termination of the long burrow in the river's bank. In one respect these subterranean nurseries differ materially from the nests of birds, since they are in most cases used as permanent places of residence for the parents, while in not a few instances, as among marmots and hamsters, they are employed as granaries for storing grain and other food.

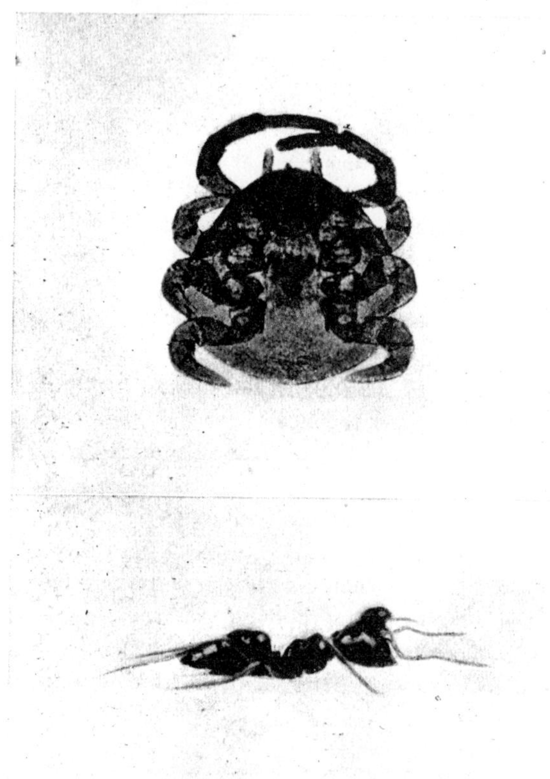

A MITE AND ITS HOST.

The upper photograph shows one of the "Antennae-bearer" Mites, much enlarged; beneath, it occupies a very favourite position on the head of the ant.

From these subterranean nests there is a transition to nests of dry grass and leaves constructed in "runs" or amid tussocks of coarse grass, such as those made by field-mice, hedgehogs, and shrew-mice; these being for the most part, at any rate, employed only as nurseries, the winter quarters of the parents being constructed in situations affording better protection against the winter cold. From these, again, we may pass to breeding-nests, like those of the harvest-mouse, constructed amid the growing stems in a corn-field, and then on to nests built in bushes or trees, like those of dormice and squirrels. It is to nests built in situations raised above the ground that attention is restricted in the present article, subterranean nests forming much too large a subject.

Before proceeding further, it will, however, be well to mention that one group of mammals, namely, the kangaroos and many of their relatives, which produce their young in an almost embryonic condition, have solved the nursing problem by developing pouches in the bodies of the female parents, in which the young are nurtured and carried until able to shift for themselves. A similar expedient is resorted to by the Australasian spiny ant-eaters, or echidnas, in which the female develops a temporary pouch for the reception of her eggs and the subsequent helpless young.

Photo by] [*Chas. Hose.*

THE NEST OF THE ORANG-UTAN.

Two nests are shown in the photograph, one almost in the centre, the other somewhat to the right; their height from the ground was between fifty and sixty feet. The photograph was taken in the Rejang district, Sarawak, and at the time of snapshotting an Orang-utan was on one of the nests

The most familiar, and at the same time one of the most beautiful, of the nests made by quadrupeds is that of the graceful little harvest-mouse, a species ranging from Great Britain to Japan. This structure was described years ago by Gilbert White of Selborne, who, after mentioning that these mice never enter houses, goes on to state that they " are carried into ricks and barns with the sheaves ; abound in harvest, and build their nests amid the straws of the corn above the ground, and sometimes in thistles. They breed as many as eight at a litter, in a little round nest composed of the blades of grass or wheat.

" One of these nests I procured this autumn, most artificially platted, and composed of the blades of wheat ; perfectly round, and about the size of a cricket-ball ; with the aperture so ingeniously closed, that there was no discovering to what part it belonged. It was so compact and well filled, that it would roll across the table without being discomposed, though it contained eight little mice that were naked and blind."

The writer then expresses his wonder with regard to the manner in which the mother mouse suckles her young—a problem which, so far as I know, is still as undecided as it was in White's day.

Nests which would appear to be very similar to these of the harvest-mouse are built by the little mouse-lemurs of Madagascar—curious, big-eyed, nocturnal creatures, looking very like dormice in general appearance, but cousins of the ordinary lemurs of their native isle, and thus distant relatives of monkeys. One of these mouse-lemurs, inhabiting the forests of the Betsileo district, is

THE ORANG-UTAN ON ITS NEST.

This is one of the two nests photographed on the previous page. It is roughly constructed to form a large platform where the builders can remain dry and comfortable till the nocturnal dews and vapours rising from the ground have been dispersed by the sun.

reported to frequent the topmost boughs of the tallest trees, in which it constructs amid the terminal twigs a structure closely resembling from below the nest of a bird, and requiring some discrimination to recognize as the abode of a mammal. This nest, which is beautifully constructed of twigs and leaves and lined with hair, is used both as a place of repose for the parents and as a cradle for the young. Some of the other species, such as Coquerel's mouse-lemur, make nests of as much as a foot and a half in diameter, which appear to be chiefly used as places of repose for the adults during the heat of the day, or throughout the season of heat and drought, when several of these lemurs become completely torpid, although there can be little doubt that in most cases, at any rate, these or similar nests are likewise utilized for nursing purposes.

THE NEST OF THE WATER-VOLE.

Like many animals whose new-born young are perfectly helpless, the Water-Voles make a chamber at the end of a long burrow, usually lined with down from the mother's fur. But, unlike the nest of a bird, it serves as a permanent abiding-place for the parents.

Much better known in this country is the nest or "drey" of the squirrel, which is a well-built structure of leaves, moss, and fibres, carefully interlaced, and placed either in the fork of a tree or in a hole in the stem or branches. It is ready for the reception of the three or four blind and naked "squirrelets" by June; and, when constructed in a tree-fork, is made in such close accordance with its surroundings as to be as difficult to detect from below as is the nest of a chaffinch when similarly situated.

The fact that the dormouse constructs two different nests—the one for the use of the parents themselves during the long winter sleep, and the other for the reception and nurture of the young in summer—suggests that the same thing may occur in the case of the mouse-lemurs; and it is, at any rate, interesting to note that both dormice and mouse-lemurs accumulate a large store of fat at the root of their tails to be used as a food-supply during the long period of torpor. The winter nest of the dormouse is a globular structure of twigs, leaves, moss, and grass, generally placed in low bushes, but sometimes in gorse, and furnished with a small supply of food. At times small colonies of these nests, to the number of a dozen or so, are constructed within a very limited area. The summer nest for the three or four helpless young is practically similar in character to the winter one, but appears to be always built at a height of a yard or so above the ground, whereas the latter may be situated either on or in the ground.

THE NEST OF A HARVEST MOUSE.
The most beautiful of all the nests of quadrupeds, built amid the straws of corn.

Of late years, owing to the installation of actual specimens of these curious structures in several of our museums, the public has been made familiar with the fact that the orang-utan, or mias, the great red ape of Sumatra and Borneo, makes a practice of building large platforms of boughs and twigs high in the forest trees which form its abode. As a rule, the smaller trees are selected as the situation of the nests, in consequence of being protected from the wind and storms by their taller neighbours, and they are, in consequence, generally placed at a height of between twenty and fifty feet from the ground. Not infrequently several of these nests, or, rather, platforms, are built in a single tree; and it appears to be a very general belief among the Dyaks that each pair of apes constructs a fresh nest every night, although it has been remarked that if this were really the case more nests would be found than can actually be seen. The main use of these platforms is as sleeping places, where the owners can remain dry and comfortable till the morning sun has dispersed the nocturnal dews and vapours at lower levels; although it may be regarded as certain that the maternal mias avails herself of the structure for the accommodation of her offspring.

That various other mammals, and especially rodents, make nests will of course be readily understood; but what has been stated with regard to the species forming the main subject of this article will apply with but little modification to the others.

THE LIVELY FLEA

BY JOHN J. WARD, F.E.S.

THERE are fleas and fleas. The dog flea is not the same species as that which patronizes the domestic cat; and the common flea which inflicts its surprise attacks upon mankind without the slightest respect of person, is still another distinct species; indeed, each kind more or less restricts its attention to its respective host—although an occasional misadventure may prompt a species to make the best of a bad job.

The entomologist is able to produce

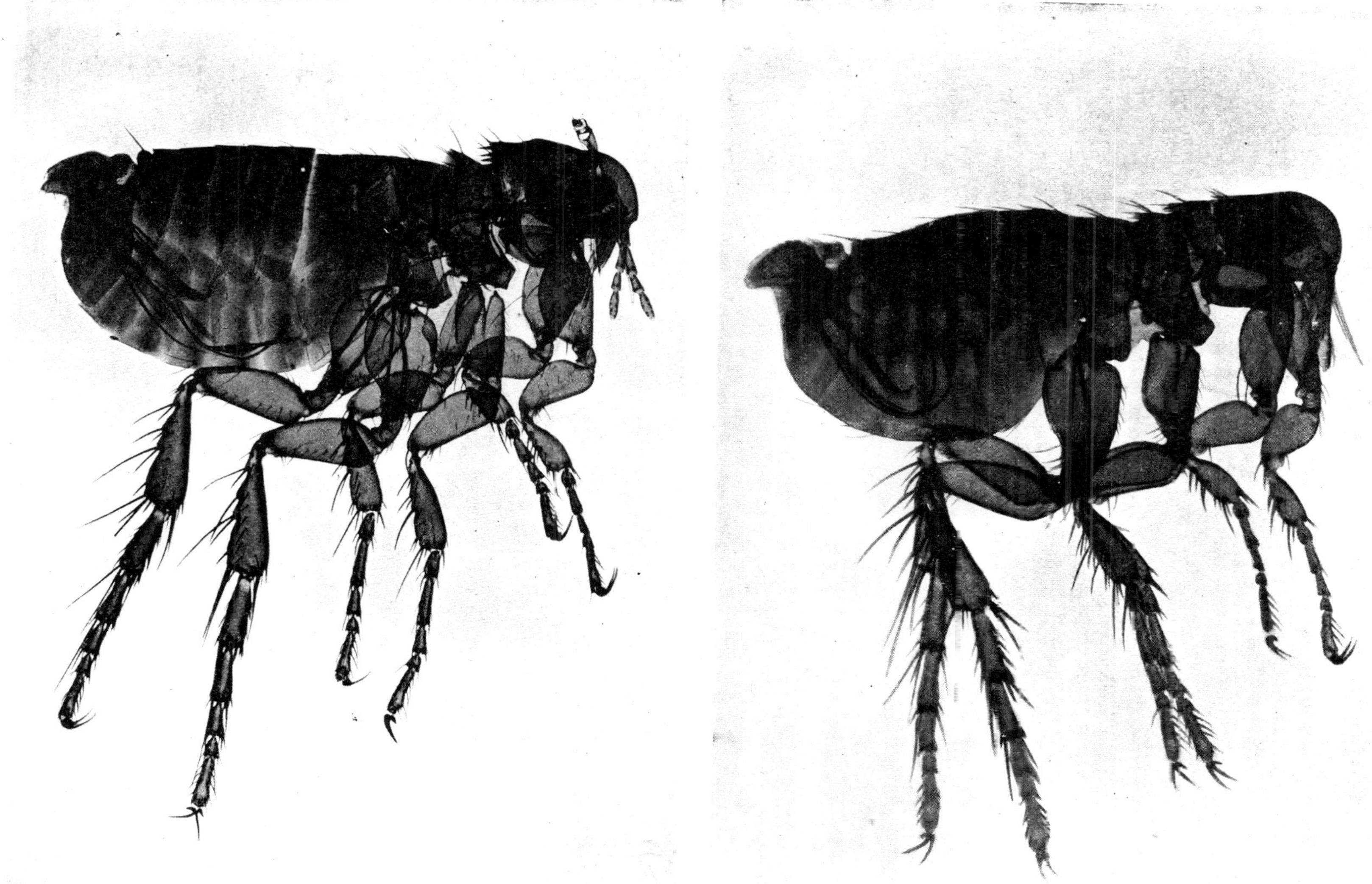

Photo by] [*A. Banfield.* *Photo by*] [*J. J. Ward, F.E.S.*

FLEAS.

Fleas were once upon a time two-winged flies; but ages of parasitic habits have caused the loss of limbs rendered unnecessary while the legs, on the other hand, have become marvellously developed, so that the insect is enabled to perform prodigious jumping feats. The species shown here are: on the left hand, the male Flea of the dog, and, on the right, the male Flea of the hedgehog, both magnified about fifty times

Photo by] [*J. J. Ward, F.E.S.*

THE JUMPING LEG OF A FLEA.

Notice the strongly-developed muscles of the thigh.

evidence which tends to show that fleas were once two-winged flies, which, owing to the development of the parasitic habit, have so completely lost their functionless wings that even the entomologist himself can now scarcely trace their rudiments. That explanation probably throws some light on the manner in which the various flea species became associated with special and widely-removed hosts, for coming of a family of winged flies they would have ample opportunity of seeking and selecting those animals best adapted to their requirements.

With the parasitic habit degeneration necessarily commences. It does not follow, however, that the subsequent development of an organism which falls from the standard of its race, is without complexity. As a matter of fact, a glance at the anatomy of a flea very conclusively shows how intricate and complex its evolution has been while on its downward path. With the loss of its wings a marvellous development of the legs has taken place; indeed, fleas are the greatest jumpers known in nature—although the locusts run them closely in this direction. The human flea has developed the jumping habit to a greater perfection than some of the species which are found on the lower animals. If a man of about six feet in height was as expert in jumping as the human flea, he would cover a mile in four springs and would reach an altitude of nearly two hundred feet each leap.

Photo by] [*J. J. Ward, F.E.S.*

THE FEMALE HUMAN FLEA.

She can be distinguished from her mate by the downward curve of the hinder segments of her body.

The hedgehog is an animal which is very abundantly supplied with parasites, and amongst these it possesses its own particular flea. I was somewhat surprised a short time ago to discover that the hedgehog flea when placed on the ground rarely attempts to jump, although it is provided with powerful jumping legs. The explanation of this curious characteristic is, I think, associated with the habits of the flea's host. The hedgehog usually has a regular track to his hiding-place, and ground freshly dug will show the track even the next morning, for the feet of the hedgehog quickly flatten down the soil where it runs. Consequently, when the hedgehog flea loses its host, its safest policy is to remain on the track until the hedgehog returns (for it runs over its track many times throughout the night), when, I have no doubt, it can jump quite well. That, I think, is the explanation of the hedgehog flea's peculiar habit. I may also add that I do not think the hedgehog flea will migrate to man.

The flea of the fowl presents another example of a flea which does not leap; it, however, runs swiftly.

The body of these degenerate insects has, likewise, become marvellously adapted to their requirements. It is clothed in a suit of mail, perfect in fit and rhythmic in action. Its horny segments are provided with bristles all pointing backwards, and it is these bristles, combined with its smooth surface, which account for the flea's extraordinary power of slipping through closely-pressed fingers; for, if it can only wriggle forward, these bristly segments prevent it from slipping back again. Also the body is laterally compressed, being so flat that it can bear enormous pressure on the soft body of its victim without ill effects. This compression of the body is common to many forms of animal parasites, but the fleas are remarkable for their lateral flattening, for, as shown in the illustrations of my previous article on "Bird Parasites," the bodies of most other forms of parasites are compressed from above and below.

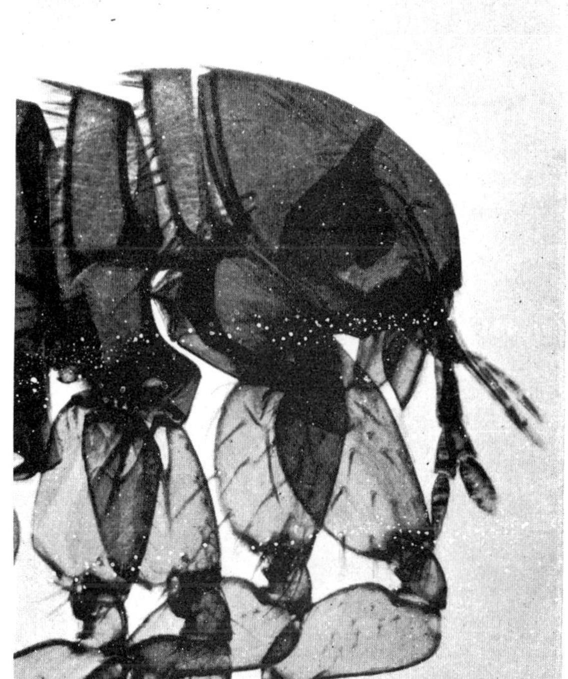

Photo by] [*J. J. Ward, F.E.S.*

THE HEAD AND SHOULDERS OF THE FLEA.

Notice the bristles which fringe the hinder rim of each segment. They point backwards and prevent the insect from slipping.

On the terminal segment of a flea's body is seen the curious organ known as the "pygidium." This is a most extraordinary piece of mechanism when seen highly magnified, being provided with disc-like orifices from which arise long sensitive bristles. It is very doubtful what sensory function this organ serves, but it has been shown with some probability that it represents the flea's auditory apparatus.

It now remains to consider the head of the flea, for in this part of its anatomy are contained the weapons with which it inflicts its so-called "bite," which is really a minute puncture opened by a

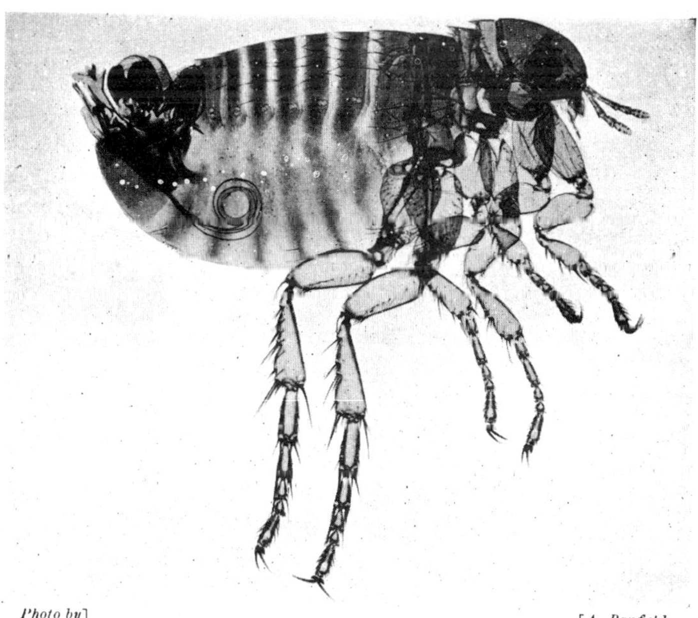

Photo by] [*A. Banfield.*

THE MALE HUMAN FLEA.

The body of the Flea is laterally compressed, and this flatness enables it to bear enormous pressure without hurt.

pair of strong and flexible knives toothed along both their edges. These blades vary somewhat in form in the fleas of different animals, and between them is a suctorial tongue, also coarsely toothed, by means of which the flea increases and absorbs the flow of blood from the wound made by the saw-like lancets.

The flea, like other insects, deposits eggs, which are of a comparatively large size, and in the case of the human flea twelve is a maximum batch. The larvæ which hatch from the eggs are bristly worm-like animals, and they will feed on almost any organic dust and dirt that happens to be near, such as that rubbed from clothes, fur, feathers, etc. ; they will also thrive on powdered bread crumbs and similar food materials. As a consequence, they are often reared between the boards of houses where floors are left unswept. After ten or twelve days' feeding they change into the pupa stage, and a week later emerge as perfect fleas. In cold weather their development takes longer. It is a curious fact that the human flea proper, although common in Europe, yet does not thrive in America, and is there very rare. A common species which infests both the cat and dog there, also becomes a pest in houses. There are a hundred or more kinds of fleas known to infest mammals and birds.

Photo by] [*J. J. Ward, F.E.S.*

THE HEAD OF A FLEA.

Showing the complex drill which probes into the flesh of its victim. An outer sheath of two strong knives with toothed edges encloses a suctorial tongue, also coarsely toothed.

THE MIRAGE

BY F. W. HENKEL, B.A., F.R.A.S.

THE name of Mirage is commonly given to certain illusory optical phenomena whereby objects are apparently seen in positions far removed from their true places, sometimes clear and distinct, at others distorted, the images being erect or inverted (*i.e.*, seen upside-down). Special local varieties of Mirage, such as the Fata Morgana of Italy, the "serab" of the desert, are well known, and the phenomena known to the sailor by the name of *looming* are of kindred nature.

In ordinary cases an object is seen by light transmitted directly from itself to the eye, and when the intervening medium (usually air) is of approximately uniform density and the object is not very far above or below the level of the eye, it is seen in its true direction, since light then travels in straight lines. If, however, the medium is *not* of uniform density, or if a change of medium takes place (*e.g.*, if, for instance, the light passing from an object to the eye travels first through air and then through water), the path of the rays is no longer straight but broken or *refracted.* In a similar manner a straight stick held obliquely partly in water and partly outside will *appear* bent at the surface of the water, the immersed part being apparently bent upwards. If a coin be placed at the bottom of a vessel with opaque

[By Theo. Carreras.

MIRAGE.

Most kinds of Mirage are examples of total reflection from layers of air. Rays of light are reflected with *changed* direction to the observer's eye, while it often happens that other rays not passed down into the reflecting layer will be present also, so that the phenomenon will be observed of what looks like an object with its image reflected in water; sometimes the positions are reversed and the reflection appears inverted in the air. Such a case is shown here.

[*By Theo. Carreras.*

MIRAGE.

In Arctic regions Scoresby and others have related that images of ships have sometimes been seen *in inverted position* in the air at a time when the ships themselves were too far off to be seen.

sides, in such a position as to be *just* hidden from the eye of an observer a short way off, it will be found that by pouring water into the vessel the coin will become visible. All rays of light passing from a rarer transparent medium to a denser one, as from air to water, will pass into the latter, but will suffer this change of direction, each being bent nearer towards the perpendicular to the common surface of separation. Conversely, rays passing from the optically denser medium into the rarer one will be bent away from this perpendicular, and make a greater angle on emergence than they did before entry. When, however, the angle of incidence reaches a certain value called the critical angle, the emergent ray just grazes the surface in passing out, and no ray making a greater angle than this can emerge, but must remain within the denser medium. For air and water this critical angle is about $48\frac{1}{2}°$, and the phenomenon of *total reflection* occurs when a ray of light travelling in the denser of two media (or in a denser part of one and the same medium) arrives at the surface of separation, making an angle greater than the critical value with the perpendicular. The reflection is called total, because, in all ordinary cases of refraction, only a *part* of the light is refracted and passes out, a part being also reflected and another part scattered ; but in *total* reflection the incident light is almost completely reflected (none of it being refracted) as though from the surface of a very perfect mirror.

If a glass vessel containing water be held above the level of the eye, and the *upper* surface of the water be looked at from below, it will appear, when seen by total reflection, as a brilliant silvery reflecting surface.

Most kinds of Mirage are examples of total reflection from layers of air. The sand of the desert is intensely heated by the tropical sun, and, in consequence, the layers of air nearest the ground are much hotter and lighter than the air somewhat higher up. Rays of light from an object entering the rarefied layers of air obliquely downwards are bent up more and more by refraction the further they penetrate, and at last fall on a stratum at an angle greater than the critical angle. This they cannot penetrate, but total reflection takes place and the rays reach the observer's eye with *changed* direction, seeming as if they came from a point as far below the reflecting stratum as the object itself is above it, whilst it often happens that the latter is itself seen directly by rays which have not passed down into the reflecting layer. Thus an appearance is presented like that of an object and image reflected in a lake of calm water.

Such a Mirage is sometimes seen across lakes on calm autumn mornings. Here it is the water that heats the lower layers of air. In Arctic regions, Scoresby and others have related that images of ships have sometimes been seen *in inverted positions* in the air at a time when the ships themselves were too far off to be directly seen. Scoresby once recognized his father's ship by seeing its inverted image through a telescope when the object itself was over thirty miles off. In such cases the ice and sea cool the lower atmospheric strata so that their density rapidly increases downwards. Rays passing obliquely upwards into the rarer layers are bent more and more downwards (owing to differences of density), till they suffer total reflection, and their images appear in the air as far above as the objects themselves are below the horizon.

Allied to these appearances is the phenomenon known as *looming*. Distant objects are seen abnormally elevated above their true position, and appear especially near, sometimes in an erect position, sometimes inverted. It is recorded that on one occasion the top of Etna "appeared by refraction like an island in the sea." Several ships sailed from Malta to take possession of this supposed new island, but soon the image vanished, and the quest was seen to be in vain (Archibald).

A necessary condition for the production of such phenomena is that the air should be at rest, and then, if for any reason, as by undue heating or undue cooling, the regular rate of decrease of temperature upwards is disturbed, total reflection at the surface of a rarer stratum will occur. The rays of light passing through the air will be gradually deflected in direction and then bent *down* or *up*, towards the side on which the denser strata lie, thus often affording a view of objects which would not otherwise be seen. What is seen over *heated land* is a portion of the sky apparently on

[*By Theo. Carreras.*

MIRAGE OF BREAKERS ON THE BARRIER REEF.

After a phenomenon instanced by the late W. Saville Kent, when the invisible breakers beating on the further shore of the lagoon were clearly visible reflected in the sky above.

the ground; the denser strata are above and the rays bent *upwards*. In the Polar phenomena the denser air is below, the rays are bent downwards and objects below the horizon are thrown up (Everett). An appearance similar to Mirage may be obtained by gently pouring alcohol or methylated spirit upon water (without mixing) in a vessel with glass sides. The spirit, though specifically lighter, is *optically denser* than the water and rays traversing the layer of transition are bent upwards, so that the surface appears like a mirror when looked at very obliquely by an eye above it.

SPIDERS' WEBS

We recently dealt in these pages with the wonderful spinning apparatus of the Spider, so that there is no need on the present occasion to explain how the substance of which the web is composed is elaborated. We desire rather to call the attention to a few of the many diverse types of architecture adopted by the different groups or genera of Spiders. There are two of these types that are known to all of us, though to some of us, it may be, only in a vague sort of way, because these objects are too commonplace to have attracted our serious attention. Some spiders employ their silk-producing faculty for the purpose of making underground tubes and tunnels which they conceal by artfully contrived trap-doors in which the silken threads are also used for holding the earthen door together and for forming its hinges. Others, which are hunters rather than trappers, construct no snares and reserve their silk for the elaboration of their egg-cocoon only.

Photo by] [*J. Holmes.*

WEB OF GARDEN SPIDER.

This photograph shows the beaded character of the spiral threads, which converts the web into a trap to which flies adhere, apart from any entangling of their limbs.

One can get some idea of the evolution of the web-weaving skill of the Spider by watching the weeds along a hedgerow, the branches of the hedge itself, young trees on the edge of the wood, the furze-bushes on the heath, which will show us webs of different degrees of complexity ranging from the simple single line to the large wheel-like structure of the Garden Spider, with its radiating "spokes," its short connecting lines which turn it into a net, and the deadly effectiveness of the additional cross-lines with their minute spherules of gum that make it very difficult for any trapped insect to make its escape. This elaborate structure, which we unthinkingly destroy with a slash of our walking-stick, has been constructed with considerable labour, and on a definite plan. Similar works schemed and carried out by man are said to be the result of much serious thought, the application of scientific principles, and so forth; but when done by the lower animals, the philosophers tell us it is purely the result of certain reflex actions of an automatic nature, and with that lucid explanation we have to be content.

If we stand before a web of the Garden Spider, such as that shown in the photograph on page 795, we shall note that the whole structure is suspended from a more or less horizontal line that extends from one bush or branch to another; and the throwing this line across the gap is the initial difficulty that this little engineer has to overcome. Relative to the size of the Spider the space to be bridged is enormous. But the Spider effects her purpose in a very simple way. Climbing out

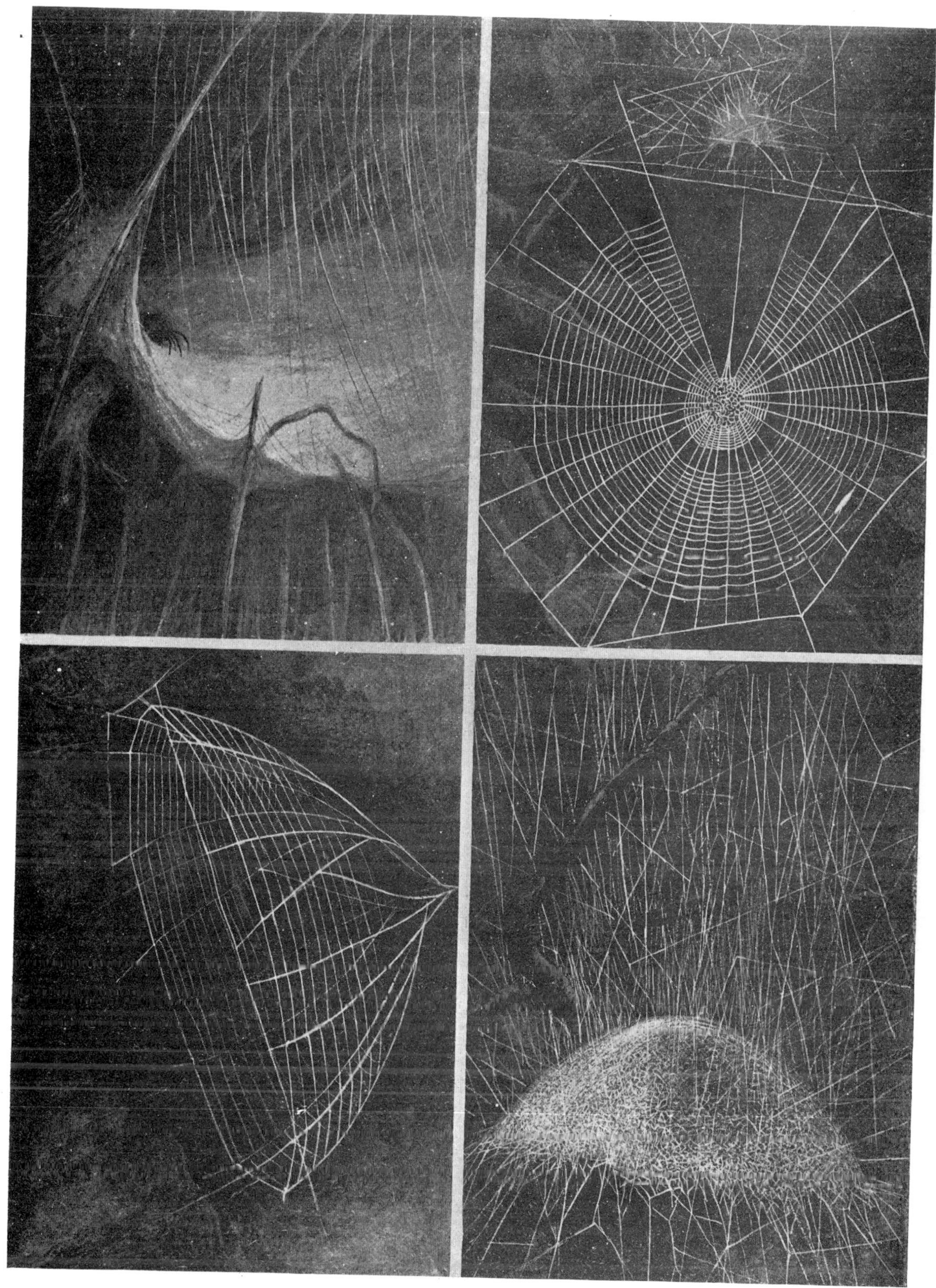

[*By Theo. Carreras.*

SOME WONDERFUL SPIDERS' WEBS.

The web in the top left-hand corner is constructed with a silken tunnel, which is the Spider's retreat. In the second web this retreat is bell-shaped and is attached to the web by a single thread. Below, on the left, is a snare that "letting in, lets out no more," for to right and left the curved strands hem the quarry in, while the disordered arrangement in the threads of the last web makes an almost invisible and effective trap.

Photo by] [H. J. Shepstone.

A TENT-SHAPED SNARE.

Webs of this character may frequently be found on furze-bushes and other shrubs. A sheet of silken tissue forms a sort of floor below the tent.

to the end of a twig that is on the windward side of the gap she wishes to bridge, she lets out a length of silk which the breeze carries across, and its own adhesiveness attaches it to a twig on the opposite side. By experimental pulls the Spider assures herself that the attachment is good, and then runs across it, paying out a second thread as she goes to strengthen the first. One or two similar lines are thrown across below the first, and other more or less vertical lines are run across to connect them together and make them taut. Guy lines are run out from near the ends to be attached to adjacent twigs and leaves to steady this framework, and from a point near the middle the top and bottom lines are again connected. From half-way down this vertical line she commences to construct the "spokes" or radii of her wheel, attached at pretty equal distances (which depends upon the size and the length of stride of the engineer), until she has them radiating in all directions. When this elaborate wheel is complete she begins to convert it into a net by walking round and round in an ever-increasing spiral, paying out as she goes a line of a special character. This special line is dotted at very short intervals with minute globules of a gummy liquid, as is beautifully shown in the photograph on page 792. In this photograph the Spider will be seen sitting in the centre, with her feet on the principal lines that she may know by her exquisite sense of touch when any insect has blundered into her net.

Other species of the same genus construct their webs on similar lines, but not all so elaborately. Smaller webs, intermediate between simple lines and these symmetrical or geometrical webs, as they have been termed, may be found along the hedgerows. One such is shown in the illustration on page 793. Much more imposing, and really beautiful if seen early on an autumn morning when the dew makes it much more evident, is the tent-like structure shown above. Here an upright shoot of the furze has been adopted as the tent-pole, and from its terminal spine long lines have been carried to the tips of more horizontal branches all around the bush, the intervening spaces filled in with an irregular network of branching and cross-lines. It suggests an advance upon the last-mentioned structure in the direction of the geometrical web, but quite deficient in the orderly neatness of the latter. It will be seen in the photograph that beneath the tent is spread a sort of floor of closely-woven tissue, as though to intercept any victims that, struggling to free themselves from the upper meshes, fall through into the tent.

This sheet of tissue reminds us of the tough fabrics the House Spider is so fond of weaving in all sorts of corners of the house, and which appear as though specially designed to catch all the dust that may be floating around, and so to give an aspect of neglect and untidiness wherever the housemaid's brush has failed to reach. The tidy housewife, when she gives expression to her annoyance, does not always realize that the friendly spider is trying to do her a good turn by keeping down the fly nuisance. It is upon such fare that she nourishes her great body and the long, bristly legs which give you such a " creepy " feeling when they run over you. Look carefully at that nest before you ruin it with the feather-broom, and you will see the artfully-contrived retreat in which the Spider waits concealed ready to pounce out and tie up her latest victim in such a thorough fashion that escape is impossible. That retreat is the " parlour " immortalized in the nursery song.

Somewhat similar arrangements will be seen in two of the webs figured in the full-page illustration. In one we have a wheel-shaped web with the spiral line cut out of a segment, in the centre of which a strengthened line runs from the middle of the web to a snug little retreat to one side, in which the Spider sits hidden, watching for the arrival of the first blundering insect. In the adjoining figure a tough web of fairly close texture is spread over grass, and at one side it is finished off as a

Photo by] [H. J. Shepstone.

THE WEB OF THE GARDEN SPIDER.

In this photograph the primary cables upon which the web is constructed and suspended are clearly shown: and by reference to it the method of construction is made plain.

funnel running downwards, in which the Spider waits and watches. Such webs may be found in abundance on heaths and along hedgebanks in autumn. In some other species, easily to be seen if we are resting on a bank in uncultivated places and casting our glances over the ground, there is a silken tube only, constructed between small clods, which, of course, serves no purpose in catching prey, but only as a retreat from which the Spider lying hidden can pounce out upon any insect so luckless as to pass by its open end. Tunnels of similar character will be found in the mortar of old walls in which a rough hole may have been found by the Spider, which she carefully lines with a silken tube to make more comfortable as a retreat.

Photo by] [*W. Saville Kent.*

A WAYSIDE WEB.

Webs of not very symmetrical form may be found spun on hedgerows, shrubs, and other plants. Although they suggest a stage in the evolution of the geometrical web, the rays are loose and irregular, and lack the spiral line.

A MYSTERY OF FISH LIFE

BY W. P. PYCRAFT, F.Z.S.

THE Sail-bearer is a most remarkable fish ; it is also a most exasperating fish. The merest glance at its outlines challenges curiosity as to what its mode of life may be, what the need of its enormous fins—larger in proportion to the body than those of any other living fishes. Yet so far this creature has succeeded in evading all our efforts to discover its life-history. It is without a history, and without a name in common speech. So far all that can be said of it is that it is a " pelagic " fish ; that is to say, it haunts that great wilderness of waters known as the open ocean, and lives near its surface, swimming, it is said, not vertically as one would suppose, but obliquely ; that is to say, with the body inclining to one side with a " heel," so to speak. But whether this " heel " is always on the same side, and always the same in all individuals, there are no records to show. But even so, *why* the " heel " ? Has the development of these fins overstepped the mark, so to speak ? Do they play the part of the " Tail that Wagged the Dog " ?

This excessive development of the dorsal and anal fins, as these great sails of the back and

THE SAIL-BEARER.

So named from the enormous fins which the fish bears. It inhabits the mid-waters of the ocean, and is said to swim obliquely near the surface. Whether this position is advantageous to the fish or whether it is caused by the over-development of the fins is a question yet to be decided. From tip to tip of the fins measures about two and a half feet.

A FOSSIL SEA-BAT.

That this development of the fins is no new thing these fossil bones of a species of Sea-bat clearly show; for here the peculiarity is carried to an even greater pitch than it is in the Sail-bearer.

belly are called, is no new thing; for we know from the remains of fossil fishes that ages ago, far back in the Cretaceous epoch, there lived fishes which in this respect possibly exceeded what now obtains in the Sail-bearer. A finely-preserved specimen of one of these, which has many living representatives, known as Sea-bats and common in the Indian Ocean and Western Pacific, is to be seen in the Fossil-fish Gallery in the British Museum; and a photograph thereof is given here. This, too, we must assume was a "pelagic" fish. That existence in the ocean wilds is not directly responsible for this remarkable shape is amply demonstrated by other fishes living in the same areas which are the exact opposite in shape, being drawn out to an inordinate length, so as to form mere ribbons! And here, again, we find the same thing in the records of the rocks, for in the Fossil-fish Gallery just referred to may be seen a specimen with a long, tubular, or spindle-shaped head, and a ridiculously attenuated body.

Finally, mention must be made of a fish from the West Coast of Africa which in outline agrees with the Sail-bearer; but here the body itself and not the fins has been changed, so that it is twice as deep as it is long, a distinction shared by no other fish living or extinct.

WATER MITES

BY CHARLES D. SOAR, F.L.S., F.R.M.S.

Amongst other inhabitants of fresh water, ponds, ditches and streams, are a number of small spider-looking creatures known as Water Mites. Although closely related to spiders, they can be readily recognized from spiders in having the body in one segment. They require a microscope to see them properly, and when properly lighted and viewed under, say, an inch and a half objective, they form a very fascinating object. They vary much in size according to the species, from a pea to a very small pin's-head. Red is the most common colour, but some exhibit every conceivable hue. The coloured plate with this article will show this. They have been found in fresh water in all parts of the world, and there are upwards of a thousand species known to science. In the British Isles we have recorded over two hundred distinct species, and it is more than probable there are more to come. As far as we know, they all lay eggs; none bring forth the young alive. Different species have different methods of depositing the ova, some in streams or swift-running water deposit the eggs on stones, others in ponds on the stems and leaves of water plants; some bore holes in the outer skin of reeds in the water and deposit the eggs underneath. Then one or two species deposit their eggs on the mantle of fresh-water mussels, and another of the same genus always deposits its eggs in the fresh-water sponge. The time from the deposition of the ova to the appearance of the larvæ varies with different species from a few days to over a month; and when the larvæ is examined, it will be seen they are totally unlike their parents, both in structure and colour. Unless we had bred the larvæ at home, like the late Saville Kent succeeded in doing in so many cases, we should be quite unable to recognize them, or give them their proper name. In this stage, unlike the adult, they only have six legs. Some species are quite transparent, but all the larvæ have very delicate colouring. It is during this stage they become parasitic, attaching themselves to all kinds of hosts; different species of larvæ selecting their own particular favourite. Certain species are found in great numbers on water scorpions and water beetles. The Large Water-Beetle is hardly ever found without one or more of these parasites attached to its under-side. When once the larvæ have firmly attached themselves with their mouth organs to their particular host, the six legs they started existence with are of no further use to them, so they usually fall away. The larvæ gradually get larger and larger, until in the following spring they are ready to burst the envelope in which they are confined and come away again as independent free swimming creatures. The empty envelope and mouth organs of the Mite remain on the host for some little time after the lodger has left. The Mite has now entered on its intermediate stage known as the nymph. It has now much the appearance of the adults and it has eight legs. Its colouring is still weak, and we cannot judge to what sex it belongs, but we can make a good guess at the species, and, if not the species, we are able to place it in its right genus. It grows a great deal in this stage and it is always free, never parasitic. When about to turn to the adult it usually gets into some out-of-the-way place on a water plant, the fork where the leaves grow from the stem being a favourite place; it attaches itself firmly and remains for a few days in an inert state; its legs very often again falling away as useless. At the proper time the skin splits and the Mite again takes to swimming, this time as an adult Mite. At first it is not fully developed, the skin being soft and the body small in proportion to the legs and other

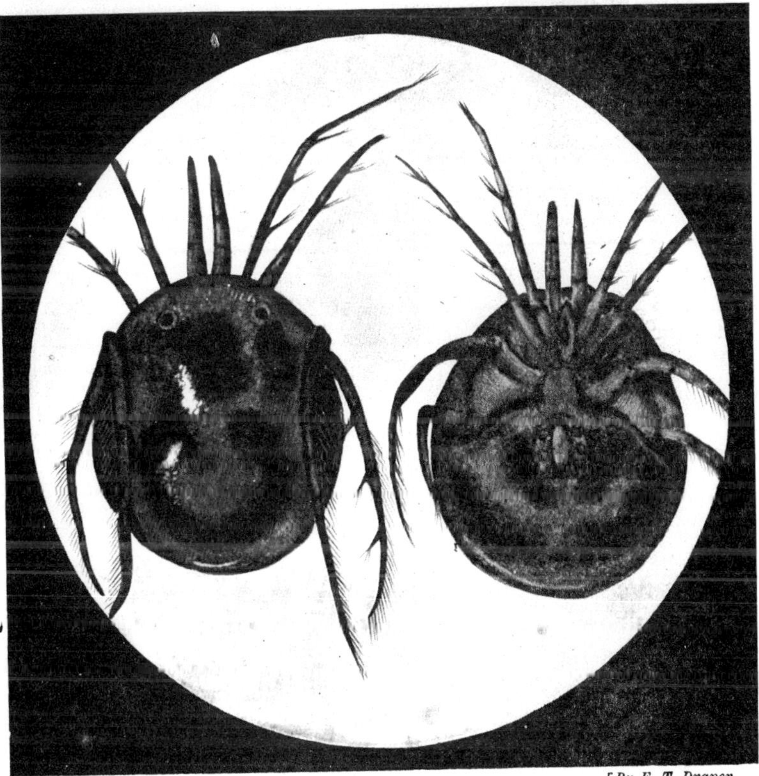

[*By E. T. Draper.*

ONE OF THE WATER MITES.

The upper and under sides of one of the commonest of the British Water Mites. It is a large Mite of bright red colour with black markings and dark blue legs and "feelers."

chitinous parts of the Mite. But a few days puts all this right, and we now have the adult Mite in all its beautiful colouring and perfect form.

Water Mites feed on decaying vegetable matter and on infusoria (other minute forms of animal life), and if kept in a small glass tank or aquarium at home, they can always be detected browsing about on the bottom and on the stems of the water plants. One genus of Water Mites are also very fond of sucking the juices from members of another genus of large, soft-bodied mites, very common in the British Isles. Anyone attempting to keep these two together in the same tank will soon find this out for themselves.

Water Mites are not all swimming mites. The swimmers can generally be recognized by the long so-called swimming hairs near the joints of the legs; but there are exceptions to this rule, as there are several powerful swimmers found in fast-running water that only have one or two spines near the joints on the legs. Besides the Mites which swim, there are crawlers. These are nearly always red, and slow and sluggish in their movements. The males and females are also much alike. But in the swimmers in some genera the males are so different in shape to the females, that unless they are taken together it is almost impossible to pair them at the first capture. Then there are hard-skinned and soft-skinned Mites.

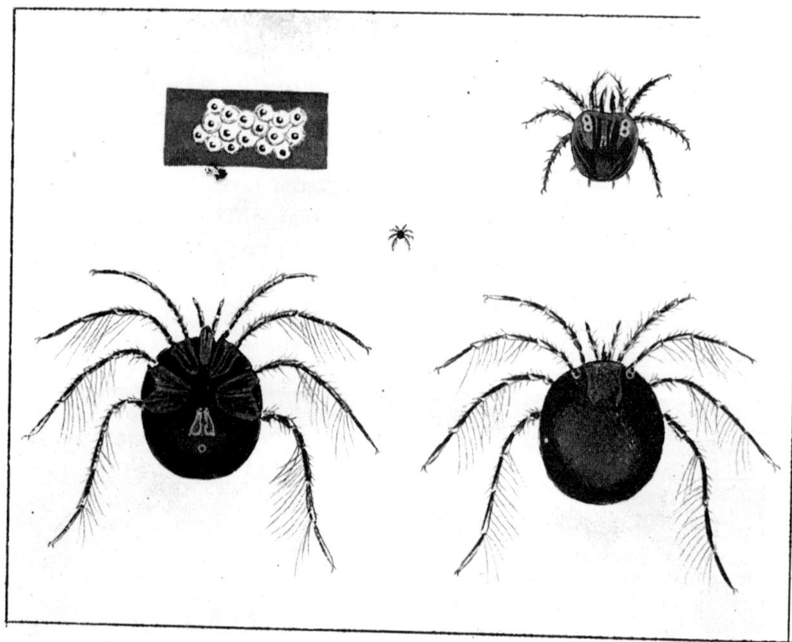

[*By W. Saville Kent.*

A WATER MITE.

Mr. Saville Kent was one of the first to make a systematic study of these Mites and to rear them through their various stages, and it is from his sketches that we are enabled to illustrate the life-history of a Mite. This brilliant scarlet species is shown natural size in the centre of the sketch.

TYLOSAURUS

BY REV. H. N. HUTCHINSON, F.G.S., F.Z.S.

AMONG the many wonderful things revealed to geologists as they search the stratified rocks for "lost creations," few, perhaps, are more strange and unlike modern types than certain huge aquatic reptiles who flourished during what is known as the Cretaceous period. That period is represented in England by the well-known chalk and certain other formations. These great marine reptiles propelled themselves by means of their tails and elongated bodies rather than by their limbs. If they were living at the present day, they would doubtless be called "sea-serpents," for their skulls are very much like those of the modern lizard known as Varanus. Their fossilized remains have been found in Europe, North and South America and New Zealand. The late Professor E. D. Cope, of Pennsylvania University, bestowed on this group the name Pythonomorphs, on account chiefly of a remarkable arrangement of the jaws which enabled them to swallow their prey whole. One of the best known of these huge marine reptiles was the Mosasaurus, already described in this book.

[*By J. Green.*

THE TYLOSAUR.

The Tylosaur possessed small scales like those of a snake. It flourished in the sea which covered the western part of North America, relying on the abundance of small fish for food.

Photo by] [*W. S. Berridge, F.Z.S.*

THE MIDWIFE TOAD.

The Toad is an inhabitant of France and Southern Europe. It is the male Toad which cares for the eggs, carrying them about with him in the manner shown above.

It must have attained a length of seventy-five feet or more. Another was Clidastes, which was forty feet long; Platycarpus was another. The Tylosaurus, here depicted, was of about the same size as Clidastes, only not so slender and snake-like. Dr. S. W. Williston, who has of late years discovered and described most valuable specimens of these old sea-serpents, was fortunate enough to discover one specimen of a skeleton complete in almost every detail. Such good fortune is a rare thing, as all fossil-hunters know only too well. In the case of Tylosaurus now under consideration, Dr. Williston found an impression of the skin near one of the paddles, showing small scales, rather like those of a snake. In Western Kansas a great many specimens have been found, showing that these reptiles flourished exceedingly in the sea which covered that part of North America in the Cretaceous period. They doubtless played the part of whales in those days, during the "Age of Reptiles," before the mammals had attained to the grandeur of their present exalted position. There were plenty of fishes to serve them as food, so doubtless they had a good time and multiplied over the earth. But their days could not be prolonged, for in the great march of upward progress, or evolution, higher types appeared on the scene, and so they were doomed to extinction. Our knowledge of this order of reptiles is greatly due to Professor Cope's arduous labours. The accounts which he has given of his explorations and the great difficulties he encountered, are most interesting, and such as every geologist can appreciate. We in England, who visit clay-pits, quarries, railway cuttings, etc., during an afternoon walk, and return home at our leisure with a few small specimens in our pocket, or a bag on our back, can hardly realize how arduous must be the work of digging out the skeletons of such beasts as Tylosaurus, packing up the separate bones and transporting them for long distances.

THE MIDWIFE TOAD

BY R. LYDEKKER, F.R.S.

It is a remarkable circumstance that while in some representatives of the great tribe of frogs and toads the care of the eggs, as in the case of the Surinam Toad (to which a special article is devoted), is confided to the female, who may develop special structural modification for their accommodation;

in other species this office is undertaken by the male alone, who may likewise possess special appliances to aid him in the task. Chili, for example, is the home of a species named Darwin's Frog, in which the male has the singular habit of taking charge of the eggs laid by the female and transferring them —in what precise manner is unknown—to a special breeding pouch in his own gullet, which communicates with the mouth by means of a pair of slits in the floor of the mouth, one on each side of the tongue. Here the eggs remain till they are hatched into miniature frogs, which appear to skip the usual tadpole stage, and eventually hop out of their father's mouth to try their own luck in the world.

The so-called Midwife Toad of France and Southern Europe was for more than a century the only species other than the Surinam Toad known to carry about its eggs on its own body.

Unlike the great majority of frogs and toads, which deposit their spawn in water, the female Midwife Toad lays a long string of from about twenty to fifty eggs on land in early spring. As she lays them, she is closely attended by her partner, who, when the whole string is completed, proceeds to attach the mass to his hind legs by thrusting his feet into the midst. When the packing is completed, the eggs form a grape-like mass covering the upper part of the hind legs and loins.

In proportion to the size of the parent toads, the eggs, which are bright yellow in colour, are remarkably large. When first laid by the mother they are as round and regular as grains of sago, but by mutual pressure they soon assume an ovate form. When fully laden with his burden, with which, however, he gets about actively enough, the male Midwife retires to his usual abode, which may be a hole in the ground, or a chink between the stones in a heap or a wall, where he has to get through a period of about three weeks before the tadpoles make their appearance.

In order to prevent the precious eggs from drying up, the careful parent makes most of his peregrinations abroad in search of food by night; and if he finds that even this care does not keep them sufficiently moist, he refreshes them by an occasional bath in the nearest pool or stream. When the three weeks of probation are over, some instinct impels the burdened father to take to the water; and when once there, the tadpoles, which are by this time fully formed, bite their way through the tough envelope in which the eggs are wrapped, and make their appearance in the outer world.

Photo by] [*W. S. Berridge, F.Z.S.*

THE MIDWIFE TOAD.

There is little in the appearance of the female Toad to distinguish it from other members of the group. As is the case with another species (Darwin's Frog), the mother leaves the care of the eggs to her mate.

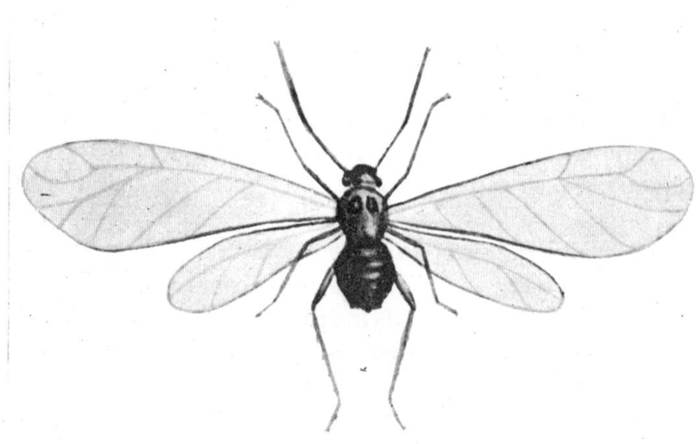

Photo by] [*H. St. J. K. Donisthorpe.*

A WINGED ROOT-APHIS.

Found in a Wood-ant's nest at Oxshott.

When newly hatched, they resemble ordinary tadpoles in being devoid of limbs; but whereas the latter are at first provided with branching external gills, in the tadpoles of the Midwife Toad these gills, of which there is only one on each side, are shed before hatching and replaced by internal ones, which again give place in due course to lungs.

DAIRYING ANTS

BY H. ST. J. K. DONISTHORPE.

ONE of the most important items in the menu of many kinds of ants is obtained from Plant Lice. The latter are known as Aphides, "Green Fly," "Blight," etc. They are insects, having six legs and a long delicate rostrum, with which they pierce the integument of plants and suck up the juices. They are well called "ant cows," as ants both milk and breed them, keeping them in herds and building sheds for them and walls to protect them. The liquid which the ants obtain from these "cows" is voided in colourless drops, and contains a fair proportion of sugar. When voided on to the leaves of plants it is called "honey-dew." Some species of ants lick it off the leaves, but others actually milk the Plant Lice. They stroke them with their antennæ, when a droplet is exuded, which the ants suck up. Many Aphides possess two tubes, which are situated on the top of the back near the end of the body. It has been incorrectly stated that it is from these tubes that the droplet is obtained, and, indeed, many text-books on entomology repeat this error. It is, however, from the extremity of the body that the sweet secretion is supplied. The tubes contain a thicker and more sticky substance, and this is used by the "cows" to protect themselves from the "Aphis-lion," the grubs of lady-birds, and other insects which prey on them. The "cows" also obtain protection from the ants, which drive away their enemies.

Photo by] [*H. St. J. K. Donisthorpe.*

THE "BUG-LIKE" APHIS.

The most truly ant-loving of all Aphides. It is never found anywhere but in ants' nests, and there it seems to have the "run of the place." When the nest is disturbed it seeks the safer retreats of its own accord

Some ants collect the eggs of Plant Lice when laid, and keep

[*By Theo. Carreras.*

THE APHIDES OF THE DAIRYING ANTS.

"Green Fly," or "Blight," as the Aphides are often called, are a race that find much favour with ants, for by them they are supplied with a colourless liquid containing sugar, which is much appreciated. For this purpose "ant cows" are bred, kept in herds, and milked by the ants, who will also build sheds and walls to protect them and will drive away their enemies. They are here seen milking the "cows" by stroking them with their antennæ.

them in their nests during the winter ; and when they hatch, they carry the young Plant Lice out of the nest and place them on their proper plants. The writer has often found these eggs, which are covered with a thick black membrane, in large numbers in ants' nests.

Of the underground species which live on roots, many are only found in ants'-nests, and some of the yellow subterranean ants live exclusively on the excrement of Root Aphides. Some of these have not even been described, and the writer has collected many species which have not yet been identified. When winged forms are hatched in the nests of ants, the latter clear a way for them, to allow them to escape into the open. We give a figure of a winged Aphis which was found by the writer in a Wood-ant's nest at Oxshott. It was new to science, and was described by the late Mr. Buckton, who wrote a monograph on the British Plant Lice. When an ants'-nest is disturbed

[*By Theo. Carreras.*

ROOT APHIDES.

Many Aphides live on underground roots, and these are similarly cherished by the ants. Some ants will collect the eggs of Aphides and keep them in their nest during the winter. When the young Aphides are hatched out they will then carry them out of the nest and place them on their proper food-plants.

which contains subterranean Aphides, their hosts may be seen to pick up the "cows" and carry them down the galleries into safety. The most truly ant-loving Aphis is a species which is called the "bug-like" one, as it is superficially very like a small bed-bug. This insect is never found anywhere except in the nests of ants. It chiefly inhabits the nests of a little brown ant, often found under stones on our coasts. It has occurred to the writer in numbers with its hosts in Cornwall. It is more active than the rest of the underground species ; it walks about freely in the chambers of the nest, and is not always fastened to underground roots. When the nest is disturbed, it hurries down the galleries of its own accord into safety. It also salutes its host with the back legs when approached by an ant. A large white species, found with the little garden ant, has a similar habit.

These insects produce vast quantities of their "honey-dew," and it is not surprising that ants should have found out their value, and utilized them as cows.

SEA MATS

BY EDWARD STEP, F.L.S.

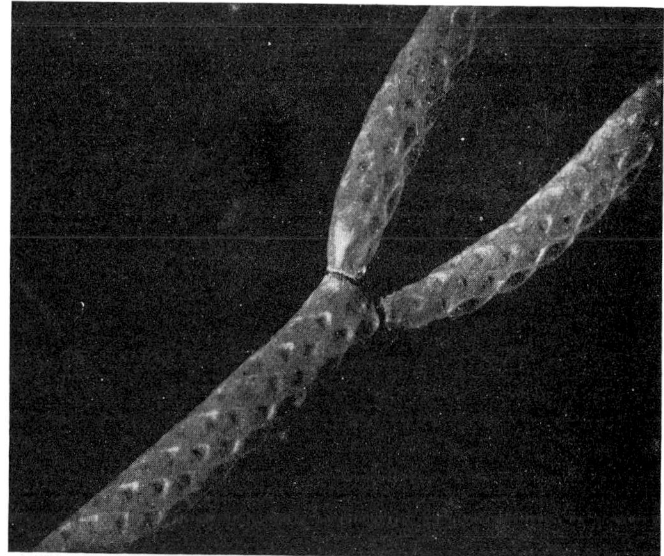

Photo by] [*W. West.*

BUGLE CORALLINE.

One of the finest of our native species. It is three inches in length, and its chalky stems have elastic joints.

IF one were content with knowledge derived from a cursory superficial examination, we should certainly class the Sea Mats with the Sea Firs; but surface appearances are often deceptive; and though the makers of the Sea Mats dwell in little cells or pockets much after the manner of the Sea Firs, a closer study of structure and organization shows that they are in no wise related or really alike. The Sea Mats occupy a position in the scale of organization on a much higher plane than that of the Sea Firs. Scientifically, they are known as Polyzoa, from their being found associated in fixed colonies. In this, of course, they resemble the Sea Firs, but whereas the latter are true socialists, each member of the community benefiting from any bit of good fortune that falls to one of them, the Sea Mat community is made up of strict individualists, each keeping to himself what he can secure. The community is so closely compacted because it owes its origin to a single individual, who emerged from an egg and passed through a free-swimming larval stage in which locomotion was made possible by its possession of a fringe of short gossamer threads, and by lashing these in succession, it rowed itself about. After a time it settled on a seaweed, a shell or a stone, glued itself down, dropped its gossamer fringe, and developed a thin, horny crust or cell, with an opening near the top. Around its mouth it developed a crown-like circle of tentacles with which it might catch any minute creature that passed by that would serve as food.

If we were to cut through one of these horny cells and the little polyzoon within, we should find its structure to be of a very simple character, but still showing a great advance upon that of the Sea Fir zoophyte. First of all, we should find that the horny cell is lined with a soft living bag, and in the centre a tube, which hangs down from the mouth to about two-thirds of the depth of the cell, then suddenly turns upon itself upwards again and makes exit through the cell-wall a little below the mouth and crown of tentacles. This U-shaped tube is the creature's digestive system, the lower and most roomy portion forming a sort of stomach. To one side of this tube, between the

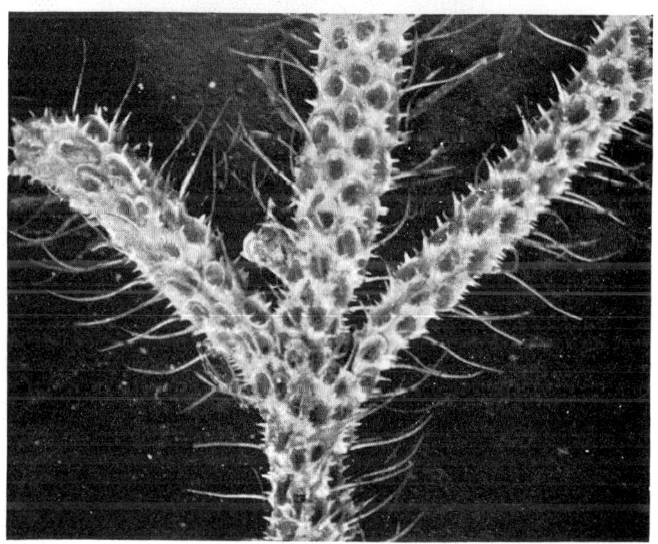

Photo by] [*W. West.*

FOLIACEOUS CORALLINE.

A silvery phosphorescent thread that creeps over red seaweeds. It is provided with numerous lashing bristles.

two extremities, there is a single knot of nerves—its entire nervous system—and lower down there are muscles which attach the tube to the base of the body wall. These muscles are the machinery by which the mouth and tentacles are at will withdrawn into the cell. There is little else in the creature's anatomy; reproductive organs, of course, but no heart, no lungs, no liver, no eyes.

This is a generalized account of the creature—which is spoken of as a polypide; in the numerous groups into which the Polyzoa are broken up there are many variations of the details. The horny cell is really the polypide's own outer skin, hardened for its protection. In some groups it is gelatinous and soft; in others it is impregnated with lime and becomes stony. From the coral-like appearance of the latter forms they were formerly called corallines, a name shared by stony-coated seaweeds and some of the Sea Firs. The Tertiary beds known as the Coralline Crag owe their name

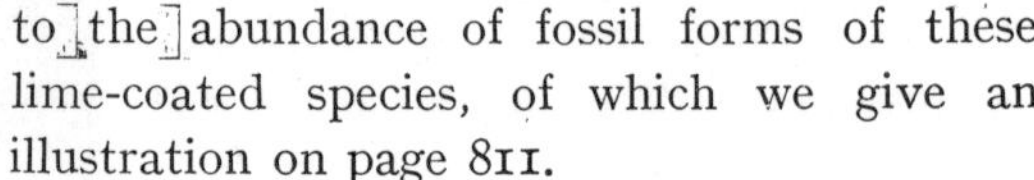

to the abundance of fossil forms of these lime-coated species, of which we give an illustration on page 811.

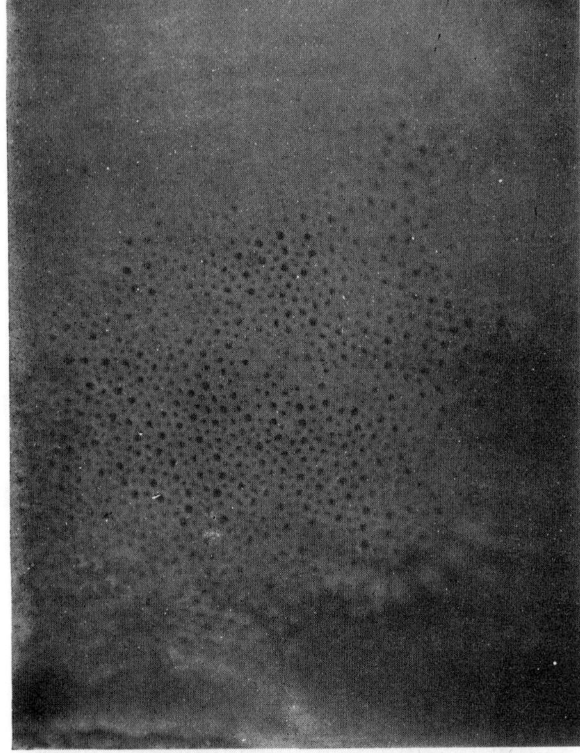

Photo by] [*W. West.*

THE BRISTLY DISH CORALLINE.

The magnified central portion of a disc whose entire width is about an inch. The occupied cells come on the margins outside the field of the photograph.

The Sea Mat proper, whose English name has been applied as a popular designation of the whole group, is the pale brown, spongy-looking, though thin structure that branches like a seaweed, and is frequently thrown up on the sea-shore with the seaweeds that have been torn from the rocks by a rough sea. In this case it will be found to be made up of two layers of cells—or zoœcia—cemented back to back. Other species form long, bristly threads that creep over the stems and fronds of seaweeds, and often when they are attached to the smaller weeds it is difficult to distinguish them from their host-plants.

In the upper part of many of the zoœcia there will be found a little chamber containing eggs, and by the embryos hatching from these new colonies are formed. The single zoœcium resulting from one of these grows into a "mat" by the polypide throwing out buds on each side, and these becoming fully formed throw out other buds, and so *ad infinitum*, each bud remaining attached to its parent until the colony gets large enough to offer such resistance to the waves that it gets swept off by superior force and destroyed.

The tentacles are hollow, and when they are extended the fluids of the body-cavity flow into them, and are thus brought into close contact with the water of the sea, from which oxygen is absorbed. They thus serve as gills. The openings of the zoœcia can be closed by hinged doors or lids.

In many of these Sea Mats there are certain cells of a remarkable character which must be mentioned. They are smaller than the others and project beyond the general level; they have thicker lids, and in some species they present a remarkable resemblance to the head of a minute bird, on which account they are known as avicularia. On exploring the cell we find there is little inside beyond the muscles that effect the opening and closing of the lids. These are particularly prominent

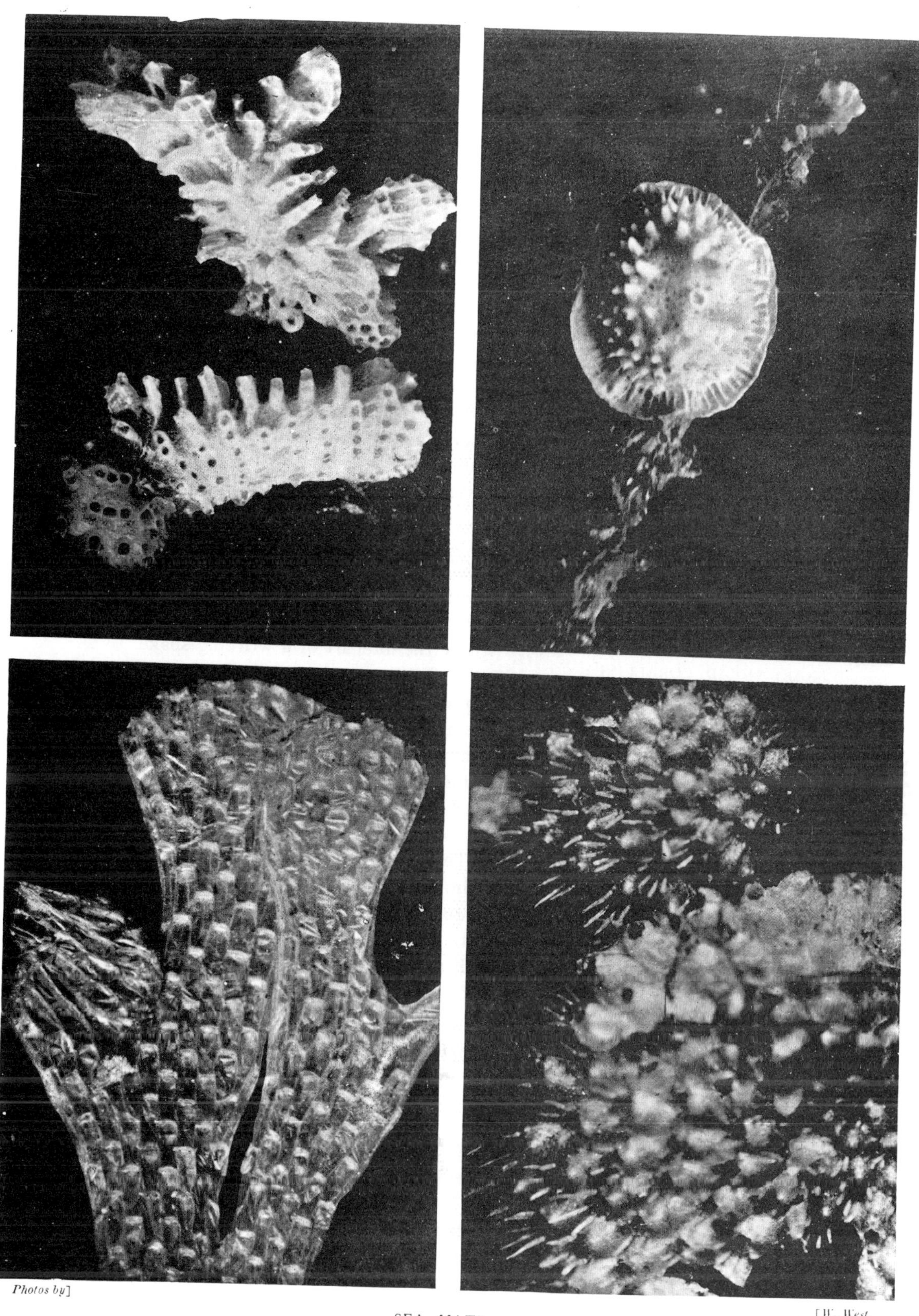

Photos by] [*W. West.*

SEA MATS.

The top photographs show, to the left, the Purple Idmonea, with coral-like colonies about one inch in length of a chalky substance with a purple-tinted and glassy surface: to the right, the Snowy Dish, whose colony measures about a quarter of an inch across and forms a stony disc. Below, to the right, is the Fringed Sea-Scurf, which lives in tufts half an inch high on seaweed, and to the left the Paper Sea Mat, often mistaken for a seaweed.

Photo by] [*W. West.*

SNAKE'S-HEAD CORALLINE.

The creeping stem supports curved branches, each ending in a single cell, which appear like the foreparts of writhing snakes.

Photo by] [*Flatters, Longsight.*

FREDERICELLA.

An example of the fresh-water relatives of the Sea Mats.

in the species known as the Bird's-head Coralline. The colony forms flat branches arranged in spiral tufts an inch or two high, growing upon rocks and shells. The "bird's-heads" are attached by short stalks to the sides of ordinary cells. They sway backwards and forwards on these stalks, snapping their "jaws." Experiments have shown that these jaws can take firm hold of hairs, worms, small crustaceans and other small animals held near them, but whether they are engaged in catching prey remains to be proved.

In other species these modified zoœcia have undergone further degradation, the snapping jaw having been reduced in thickness and lengthened until it resembles a bristle, when it is known as a vibraculum. This lashes the water, and sets up a current which is supposed to direct food particles to the tentacles. In the so-called Snake's-head Coralline each zoœcium terminates a long branch from a creeping main stem, and when the tentacles are withdrawn the curving branch looks like the forepart of a writhing snake.

Our illustrations are all greatly enlarged from portions of Sea Mats under the microscope, and serve to show some of the many types of form and structure to be found in this interesting family of animals. The Paper Sea Mat is one of those that stand up independent of support and are most frequently mistaken for seaweeds. The Foliaceous Coralline is a long, slender, branching form that runs over shells and red seaweeds, and is one of those that possess numerous bristles or vibracula, and is phosphorescent. The Fringed Sea-Scurf forms little tufts, about half an inch high, and of a pearly white colour, on seaweeds and hydroids. The Snowy Dish is an example of the section in which the cells are composed of lime. They are united to form a saucer-shaped disc, about a quarter of an inch across. Also of a chalky substance, but with a glassy polish, is the beautiful Purple Idmonea, whose colonies are about an inch in length. The Bugle Coralline is one of the finest of the British species. The joints of the forking stems are elastic, which enables the three-inch length to sway with the movements of the water without risk of breaking. The photograph adjoining is given as an example of the species that occur in fresh water.

THE COCK OF THE ROCK

REMARKABLE both for plumage and habits is this gorgeous inhabitant of the interior parts of South America. On the head is displayed a peculiarly compressed and rounded crest, and the whole body of the bird is coloured a beautiful yellow orange, which at times approaches a brilliant crimson, for the hue appears to vary according to the geographical distribution. The hinder parts of the wing are grey, while their margins and the tail-feathers are black. It is a solitary and shy bird, feeding before sunrise and after sunset, and usually remaining in hiding throughout the day. Salmon has given us the following description of its breeding haunts: "I followed up a mountain stream where the gorge became gradually more enclosed till I arrived at the mouth of a cave with high rocks on each side and overshadowed by high trees, into which the sun never penetrated. All was wet and dark and silent. We had hardly become accustomed to the gloom when a nest was found, a dark bird stealing away from what appeared to be a lump of mud upon the face of the rock. This was the nest of the Cock of the Rock, containing two eggs; it was built upon a projecting piece, the body being made of mud or clay, then a few sticks, and the top lined with green moss. It was about five feet from the water. I did not see the male bird, and, indeed, I have rarely seen the male and female birds together, though I have seen both sexes in separate flocks." The colour of the female is a dull olive-brown, and the eggs are a pale buff, marked with blotches of various sizes shading from red-brown to lilac. The Cock of the Rock has always attracted the attention of travellers from the peculiar dances which take place during the mating season. Only the male birds are performers, for the hens supply the interested audience. Here is Robert Schomburgk's description of a dance he witnessed. "The stage, if we may so call it, measured four to five feet in diameter; every blade of grass had been removed and the ground was as smooth as if levelled by human hands.

Photo by] [*A. E. Smith.*

FOSSIL SEA MATS.

Formerly considered as small corals, these fossils are so abundant in certain strata of the Tertiary period that the beds are known to geologists as the Coralline Crag.

On this space we saw one of the birds dance and jump about, while the others evidently played the part of admiring spectators. At one moment it expanded its wings, threw its head in the air, or spread out its tail like a peacock, scratching the ground with its foot; all this took place with a sort of hopping gait, until tired, when, on emitting a peculiar note, its place was taken by another performer."

THE ALLOSAUR

BY W. P. PYCRAFT, F.Z.S., Etc.

If the reptiles of to-day possessed a written record, or the ability to explain the records of the past, they might well swell with pride, for no group of the vertebrate or backboned animals has produced so many bizarre and gigantic types. In any comprehensive survey of the giant types the Dinosaurs must always rank first, and this because so many were land dwellers, and, consequently, their remains display a marvellous wealth of structural variation, determined in part by the mere mechanical requirements of locomotion on land, and in part by the all-important need of providing provender for their daily needs. This last factor, the hunger-stimulus, furnishes us with some curious and interesting data. Thus, in the first place, it is clear mere bulk, in those far-off days, no less than to-day, had its disadvantages, and limits were placed upon size in the case of the carnivorous types sooner than with the vegetarians. The Brontosaurus, Atlantosaurus and Diplodocus, ranging as they did up to, and over, eighty feet in length, could never have achieved this distinction had they not been eaters of grass. They were slow, unwieldy, contemplative creatures: dull of wit, and defenceless. For long generations, no doubt, this was a matter of no consequence. But it was not to be expected that Nature would go to the trouble of producing such mountains of flesh without, sooner or later, producing some other creature which could and would appreciate its juiciness when newly killed. This, then, determined the evolution of the eater, in the form of creatures like the Allosaurus, Ceratosaurus, Megalosaurus and Tyrannosaurus. These were the lions and tigers of those far-off days, but immensely more powerful, and of huge bulk. The Allosaurus was probably about forty feet long.

THE COCK OF THE ROCK.

Most beautiful is the colouring of this strange bird, which is a native of the interior of South America. Save for the wings and tail, the bird is completely clothed in orange, scarlet, or blood-red, the colour varying according to geographical distribution. During the mating season the male bird performs a curious and elaborate dance.

In their structure they present some curiously interesting features. Allosaurus, for example, had comparatively slender, hollow bones, indicating lightness with strength; further, it was evidently bipedal, and probably displayed remarkable agility. The form of the hind leg, and the great thickness of the tail, show that its method of progression was not kangaroo-like, by leaps, but it was

Photo by] [*H. J. Shepstone.*

A RESTORATION OF THE ALLOSAUR.

This was one of the most formidable of the flesh-eating Dinosaurs. As the restoration shows, the creature was built for speed and strength. When attacking he probably advanced on the powerful hind limbs, using the short fore-limbs with their terrible claws to grapple with and probably tear the throat of his victim.

Photo by] [*H. J. Shepstone.*

THE ALLOSAUR.

The original skeleton from which the restoration of the Allosaur was drawn. This Dinosaur lived in the same age as did the Brontosaur. The latter was herbivorous and probably often fell a victim to its blood-thirsty foe, in fact joints of the backbone of the Brontosaur have been found furrowed with deep grooves made by the teeth of (probably) the Allosaur.

capable probably of running with some speed, on occasion. The forelegs could certainly have borne no share in the support of the body, for they are, in proportion to the rest of the body, ridiculously small. Nevertheless, it is believed that they served as grappling-hooks of no mean power. When commencing the attack on, say, an eighty-foot Brontosaur, it probably opened hostilities by a tremendous leap, then clutched at its victim's fore-quarters with the long-clawed fingers, and followed this up by swift and vicious bites at the base of the neck. Or it may have adopted the tactics of the Cape hunting-dog to-day and have bitten at the flanks till a hole was torn in the side, releasing yards and yards of entrails. So unwieldy a creature as the Brontosaurus could have no prospect of success against such an opponent, though doubtless it would make a show of resistance, if only by biting ; and to provide against a piece being bitten out of the middle of his back on the occasion of such an attack, Allosaurus and his kindred had an armature of body plates protecting the spine and upper surface generally. This armature may, however, have been developed to secure a measure of protection against infuriated members of its own species, squabbling, perchance, over the right of ownership of the same " kill."

In both hind- and fore-feet, it will be noticed, the number of the toes has been reduced to three on each foot ; the fore-foot, however, retained a vestige of a fourth toe, as did the hind-foot of the smaller, nearly-related Ceratosaurus. We may surmise that the missing digits were removed to afford increased development of the survivors. This reduction accords with what is found among birds, and many mammals, but it is significant to note that no predatory mammal has evolved a similar type of foot, nor is any predatory mammal bipedal.

Professor Henry Osborn, an American naturalist of renown, inclines to the view that Allosaurus and his kin were rather hyæna-like in their habits. That is to say, they were carrion-eaters. This may well have been the case, though the evidence, on the whole, seems to favour the view that they generally, at any rate, killed for themselves. But a glance at the teeth seems to show that they did not greatly concern themselves with their meat, but swallowed it whole, in great mouthfuls. This is probable, because all the teeth have the form of great pointed spikes, and none are modified for chewing, as are the great scissor-like cheek-teeth of lions and tigers. We may further safely assume that these cold-blooded hunters had no movable tongue ; and we know they had no lips. Thus they were not blood-drinkers.

Looking at the photograph of the skeleton it will be noticed that the neck-vertebræ, like those of the back, bore long ribs, which were probably freely movable. This is a common reptilian feature, but unusual among these giant reptiles.

TSETSE-FLIES

BY ERNEST E. AUSTEN.

IT has long been known that the bites of Tsetse flies are, or may be, fatal to horses, cattle, dogs and other domesticated animals. Scarce ten years have elapsed, however, since it was proved by the brilliant investigations of Surgeon-General Sir David Bruce and other naturalists, British and foreign, that the insects with which we are dealing are associated with and solely concerned in spreading the terrible human disease called *Sleeping Sickness*, which at the present time is the scourge of many parts of Tropical Africa, and, by restricting the supply of native labour, is even threatening the continued prosperity of commercial enterprises in the south of the continent.

Photo by] [*H. J. Shepstone.*

THE SKULL OF THE ALLOSAUR.

It is more than three feet in length. Notice how the framework of bone is as light as possible. Such a provision would be very necessary if the hind limbs of the Dinosaur (however powerful) were to support this huge bulk in an erect position, when the head would be about twenty feet from the ground

We must now, with the help of the accompanying illustrations, all of which have been photographed from actual specimens, endeavour to give the reader some idea of the outward appearance, habits, and life-history of Tsetse-flies, and also indicate where they are found. Before doing so, however, it is necessary to point out that it is altogether wrong and misleading to say " the Tsetse-fly " (just as we speak of " the " lion or " the " tiger), as though there were but a single species or kind in existence. Naturalists to-day are acquainted with no fewer than *fifteen* existing species or kinds of Tsetse-flies, all agreeing with each other in their essential characteristics, but differing in minor details of structure, markings, and habits, and, what is more important, also it may be differing in their disease-carrying properties.

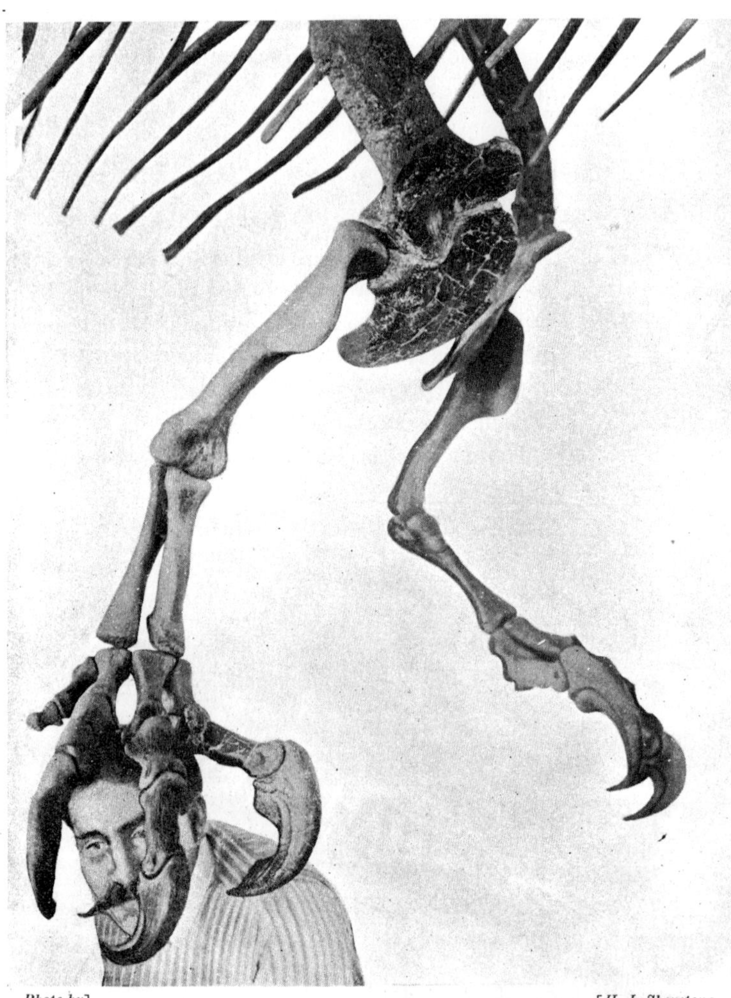

Photo by] [*H. J. Shepstone.*

THE CLAWS OF THE ALLOSAUR.

That such a claw could do enormous damage goes without saying. The longest digit measures over twelve inches. The fore-limb, however, was short, and could not possibly have been used for walking.

Although at an earlier stage of the earth's history at least two species of Tsetse-flies are known to have existed in North America, at the present day these insects—with the exception of one kind, the range of which extends into the south-west corner of Arabia—are found only in Africa, where they are practically restricted to the Tropical portion of the continent. It must not be supposed, however, that Tsetse-flies occur everywhere within the area mentioned; on the contrary they usually show a peculiar limitation to particular tracts, which are often of very limited extent, and are known as " fly-areas," " fly-belts," or, in Africa, south of the great lakes, " the fly." Some kinds of Tsetse are met with only close to water; others may be encountered at a considerable distance from any river, stream, or lake, even on low hills and up to an altitude of some five thousand five hundred feet. Open grass plains, however, are always free from Tsetse-flies, since cover in the shape of more or less thick bush or forest, shady trees, or reeds, is essential to their existence.

A good idea of the general appearance of Tsetse-flies, as seen when at rest upon a twig or with the wings extended in the attitude of flight, may be gained from our illustrations, which represent the two kinds of Tsetse known to convey Sleeping Sickness from man to man. While differing, as already mentioned, in certain respects according to the species, the insects with which we are

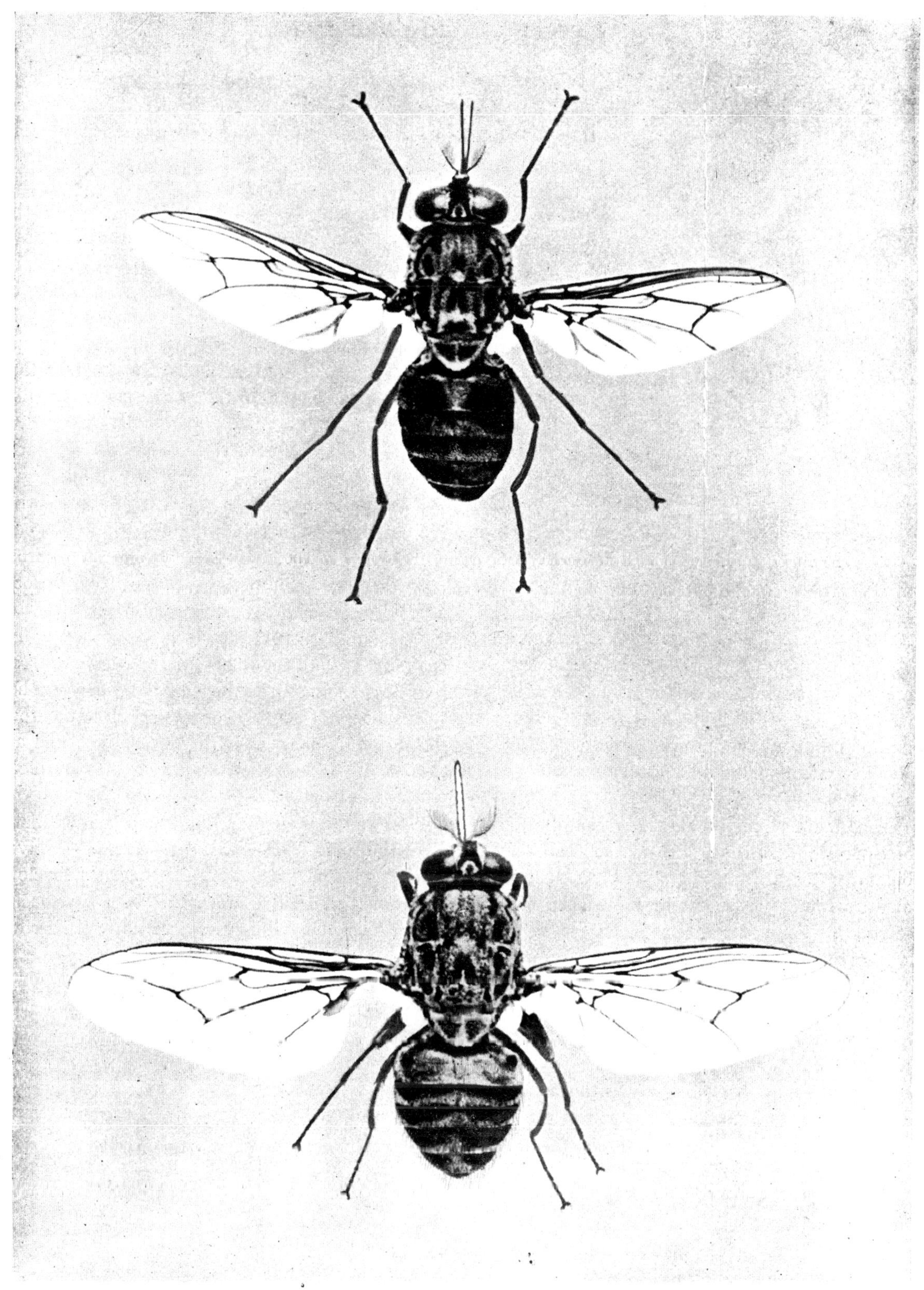

TSETSE-FLIES.

The upper photograph shows the Tsetse-fly which carries Sleeping Sickness in West Africa, the Congo Free State and Uganda. It is the darkest of all Tsetse-flies. The kind shown in the lower photograph is to be found in Nyasaland and North-Eastern Rhodesia. It infects domestic animals with Tsetse-fly disease; is much lighter in colour than the previous fly, and has the hinder half of the body conspicuously banded. (Both are shown five times natural size.)

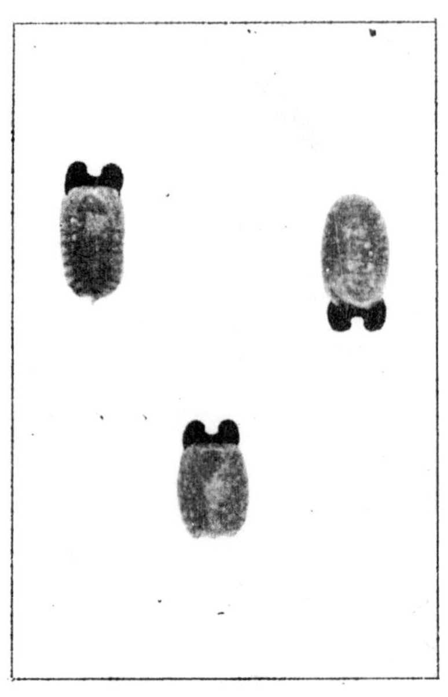

GRUBS OF THE TSETSE-FLY.

The maggot is not deposited by the mother until fully grown, and only one is produced at a time. Note the notched black hinder end characteristic of Tsetse-fly maggots. (Four times natural size.)

dealing may be described as ordinary-looking dark-brown, blackish or black, yellowish-brown, or yellowish flies, varying in length from about a quarter of an inch or rather more in the case of the smallest kind, to about half an inch in that of the largest, with in all species a prominent proboscis, or beak, which, when not in use, projects horizontally in front of the head. The position of the wings in the resting insect is a point of the highest importance, since, coupled with the prominent horizontal proboscis, it affords a ready means of distinguishing a Tsetse from any other blood-sucking fly for which it might otherwise be mistaken ; instead of diverging at the tips like those of a house-fly, the wings of a resting Tsetse-fly lie closed flat over one another down the back, like the blades of a pair of scissors.

All Tsetse-flies feed on blood, and the blood-sucking habit is common to both sexes. Human beings, wild and domesticated animals, birds, and even reptiles, are attacked with apparent impartiality, and with such avidity that, if permitted to feed undisturbed, a Tsetse, in the short space of from twenty to forty seconds, may become " swollen out like a balloon with bright red blood." Tsetse-flies are sometimes met with literally in swarms ; on other occasions, even within a fly-belt, few or none may be seen ; they are most active on bright, sunny days, and may be equally blood-thirsty throughout the day or (in very hot weather) be specially aggressive at certain periods—before eight a.m., and again after four p.m., and about sunset. To some extent, no doubt, these details vary with the species, and are also dependent upon the locality and season ; certain kinds of Tsetse are known to bite by night as well as by day.

The manner in which Tsetse-flies reproduce their kind is not the least remarkable characteristic of these wonderful insects. Instead of laying eggs like the majority of flies, a female Tsetse gives birth at intervals to a full-grown maggot, each maggot being produced singly, and, until its development is complete, being retained within the body of the mother, and there nourished by a special secretion. As a rule the maggot is deposited on loose, sandy soil, into which it promptly burrows and forthwith changes into a hard, dark-brown or black chrysalis, from which the perfect fly makes its escape some five weeks later.

For nearly fifty years it was believed that the death of domestic animals after being bitten by Tsetse-flies was due to the action of a peculiarly powerful poison, which was supposed to be elaborated by the insect itself and injected into its victim while the fly was sucking its blood. In 1895–96, however, the popular theory was completely exploded by Bruce, who showed that Tsetse-fly disease in animals is caused by a microscopic living organism, which exists in the blood of many different kinds of wild animals in Africa without apparently doing these creatures any harm. When a

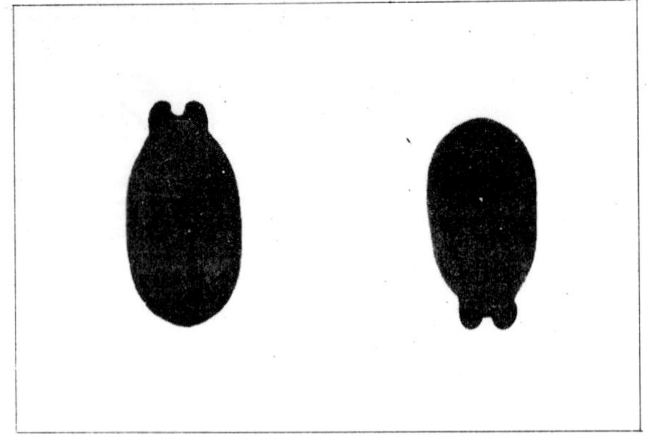

PUPÆ OF THE TSETSE-FLY.

The second stage of a Tsetse-fly—a hard barrel-shaped case, from which the Fly eventually makes its escape by pushing off a cap from the rounded end. (Five times natural size.)

Tsetse-fly bites such animals it sucks up some of the minute organisms with the blood, and if, after the lapse of a certain time, the fly then bites a horse, ox, or dog, the organisms, which have in the interval undergone certain changes within the body of the fly, escape into the blood of the new victim, multiply rapidly and eventually cause death. Sleeping Sickness in man, which within the last ten or twelve years has swept away many hundreds of thousands of people in Uganda and the Congo Free State, besides being prevalent in other parts of Tropical Africa, is caused and conveyed in a precisely similar way.

THE FRILLED LIZARD

BY R. LYDEKKER, F.R.S.

THE Frilled Lizard is to be found throughout the greater part of Northern and North-Western Australia, and in the course of an expedition under the command of Captain P. P. King, R.N.,

GORGED TSETSE FLIES.

A Sleeping Sickness-carrying Tsetse-fly after feeding, shown in two positions and six times natural size. Note how the hinder half of the body, containing the stomach, is swollen out with blood.

during the years 1818–22, a botanical collector named Allan Cunningham captured on the branch of a tree at Careening Bay, Perth, what he regarded as a very curious and remarkable lizard. When it came to England this unique specimen was given to Sir Everard Home, by whom it was transferred to the Museum of the Royal College of Surgeons. Here it was seen by Dr. J. E. Gray, at that time the leading naturalist at the British Museum, who described it as a new genus and species, under the name of *Chlamydosaurus kingi.*

Now as this wonderful lizard is specially characterized by possessing a frill-like cloak round its neck, which can be opened and closed like an umbrella, the name *Chlamydosaurus*, meaning the "lizard with a cloak," was most appropriate and suitable. In naming the species after the leader of the expedition, instead of after its discoverer, Mr. Cunningham, Dr. Gray was, however, guilty of a piece of petty and inexcusable snobbishness.

The neck-frill, which is supported by cartilaginous ribs analogous to the metal ones of an umbrella, renders the Frilled Lizard quite unmistakable, and distinguishable at a glance from all the other members of the scaly four-footed tribe. When at rest, the frill or cape is not particularly conspicuous, lying folded back along the reptile's neck and shoulders. It needs, however, only a shock of surprise or rage to cause the whole frill to be suddenly erected and expanded, so as to form a large disc-like structure, some ten inches across, standing out at right angles to the neck, and having the open-mouthed head in the centre. When thus expanding its frill, the reptile usually sits up, and turns boldly towards the intruder by whom it has been disturbed, with the evident intention of intimidating him by its formidable appearance ; the effect being intensified by the formidable series of teeth with which the jaws are armed, the bright yellow colour of the interior of the mouth, and the loud hissing sound which proceeds from the throat. That all this has a terrifying effect on other animals can scarcely be doubted, and in many instances it probably serves to protect the reptile from attack. In the case of captive specimens the frill is often expanded on very slight provocation ; and in some instances the opening is only of momentary duration, so that the whole action is almost flash-like.

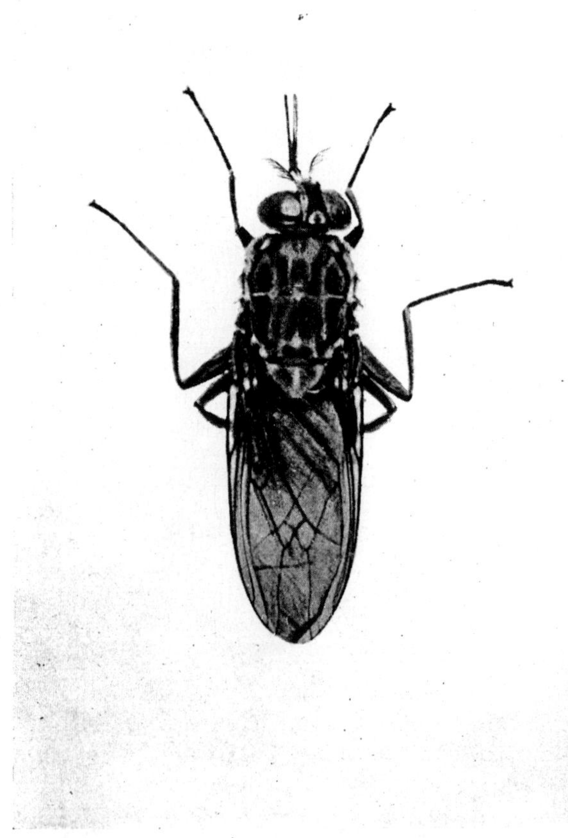

A TSETSE-FLY WITH FOLDED WINGS.

A Sleeping Sickness-carrying Tsetse-fly as seen when at rest. (Five and a half times natural size.) Note the wings closed one over the other like the blades of a pair of scissors.

In human beings the Frilled Lizard is calculated to inspire curiosity rather than terror, as its total length is only about a yard, of which eight or ten inches is occupied by the head and body, and the remainder by the long whip-like tail. The general colour of the rough granular skin is a mixture of olive, brown, and dull slaty grey, with tinges of yellow.

As already mentioned, the original specimen captured by Mr. Cunningham was found on the branch of a tree, and according to an American naturalist, Mr. Ditmars, who has kept specimens in captivity, this is the general habit of the species ; his examples persistently scampering over the boughs of an old cedar-tree placed in their cage, and seldom descending to the gravel with which the floor was strewn.

If removed from the branches and placed on the floor of the cage, and then suddenly startled, they would, however, rear themselves on their hind-limbs, with the frill tightly folded and the fore-limbs pressed close to the chest, and start running in this posture. And it appears from the graphic accounts given by an Australian naturalist—the late Mr. Saville Kent—that in the more desert districts of Queensland the Frilled Lizards are habitually terrestrial, and may frequently be seen running at a great pace on their hind-legs, with long strides, in a manner which gives them an irresistibly comic appearance. It has, however, to be confessed that Mr. Kent made rather too much of this peculiar action, being apparently unaware that it is common to several species of lizards belonging to totally distinct groups ; and when he went on to argue that the assumption of

Photo by] [*W. Saville Kent.*

THE FRILLED LIZARD.

The Lizard, which is a native of Northern and North-Western Australia, is here in a characteristic attitude. It generally frequents the branches of trees, and when startled it faces the cause of its annoyance, erects the large frill and opens the mouth, showing its formidable teeth and the bright yellow colour of the interior of the mouth, producing also a loud hissing sound in the throat—all for the purpose of terrorizing its foe.

the upright posture by the Frilled Lizard indicates kinship with the iguanodon and other extinct giant reptiles, or dinosaurs, of a long-past epoch, he started an idea which every naturalist, possessing even a slight acquaintance with anatomy, will recognize as preposterous. He is, however, entitled to the credit of having written a very interesting account of these strange lizards.

The Frilled Lizard is one of those structures which it is so difficult, or rather so impossible, to explain on any hypothesis of fortuitous evolution by means of slight and gradual development. For what use would an incipient frill be in terrifying enemies ; and if of no use in its early stages, why should it have continued to develop until it attained the effective dimensions it now possesses ? It appears, indeed, to be a case of " the whole frill or none."

VEGETABLE BREAD

In those survivals of Paradise known as the South Sea Islands, there grows a tree whose fruit has almost the consistency of a newly baked loaf, and it serves the natives as a good substitute for the baker's bread.

The tree is of only moderate size, and its leaves are large and rough, of a dark green colour, with their edges cut into bold, teeth-like divisions. The flowers are simple and individually inconspicuous, but owing to the manner in which they are clustered they are obvious enough. Those that bear stamens are crowded on a club-shaped spike ; but the pistillate flowers form round buttons much like those of the familiar plane-tree of our city parks and streets. These button-like clusters ultimately enlarge until they are as big as melons, the exterior rough, with six-sided knobs, which are the remains of the flowers. The interior is filled with a dough-like pulp, produced by the swelling up of the receptacle which supported the flowers. This constitutes the bread, but, like the

Photo by] [*W. Saville Kent.*

THE FRILLED LIZARD.

The frill is here admirably displayed. The skin which forms it is supported by cartilaginous ribs very similar to the whalebone ribs of a primitive umbrella. When fully expanded this frill will measure as much as ten inches across

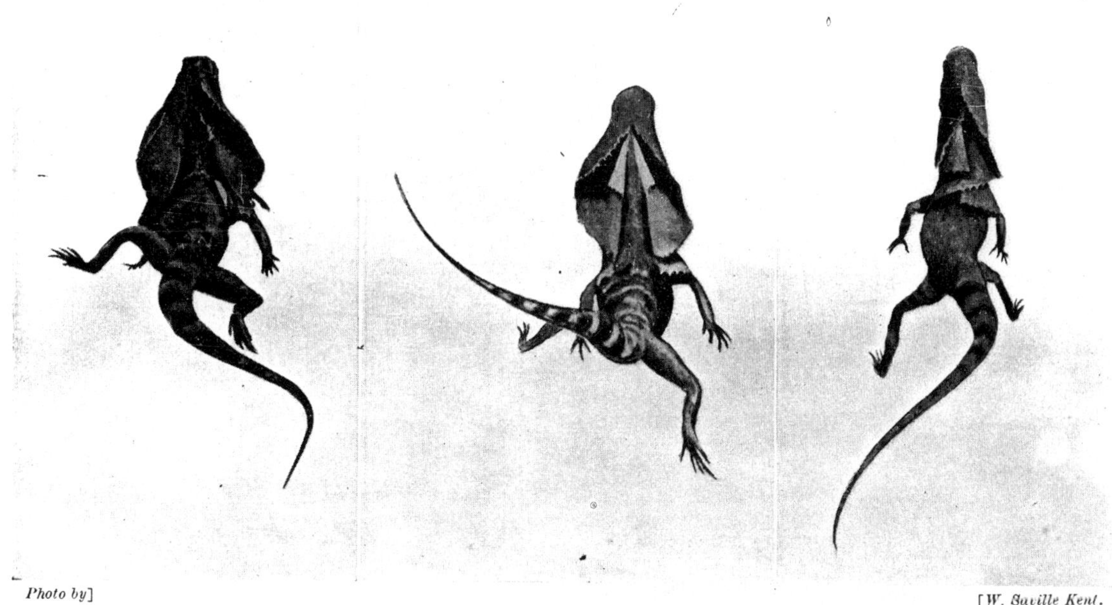

Photo by] [*W. Saville Kent.*

THE FRILLED LIZARD.

According to Mr. Saville Kent these Lizards are, in the more desert parts of Queensland, habitually terrestrial, and may frequently be seen running in the curious fashion illustrated in the above photographs. The long strides and rapid motion of the short hind legs give them a very comic appearance.

dough prepared from wheaten flour, it requires to be cooked before it can be eaten. This is effected by heating stones and putting them with the Bread-fruit into an oven.

Another species, the Jack-fruit tree, can be distinguished from the Bread-fruit tree by its leaves having uncut edges, and by the fruit being less spherical in shape ; but it is equally useful as a food Besides timber, the Bread-fruit tree yields caoutchouc, which the natives use for caulking the seams of their canoes, whilst a coarse kind of cloth is elaborated from the bark of young trees.

THE DRY ROT FUNGUS

BY EDWARD STEP, F.L.S.

THERE are many of the smaller fungi that cause annoyance and loss to man by spoiling his garden and field crops, or ruining his food-stores indoors. But probably a more-to-be-feared fungus foe is the Dry Rot, which most people have heard of, but few have seen. It is more insidious than most of our fungus foes, because it works in the dark, bringing our house down about our ears, and perhaps giving no sign of its presence until its evil work is fairly complete. Its work may be compared with that of the termites in the tropics, who eat the solid core of supporting beams, leaving a mere outer skin to simulate strength until the crash comes, and shows that what was thought to be, and looked like, sound timber is merely a thin tissue full of dust.

In the case of Dry Rot the crash comes before the beams and rafters have been thoroughly demoralized, for the attack has usually been made where they rest upon the wall. The Dry Rot fungus, in its beginnings, is a weak and innocent-looking thread that it seems absurd to put in competition with the strength of solid oak ; but give it the conditions it loves—a still, close atmosphere—and it will reduce the strength of the oak to the weakness of tinder. It appears to proceed by applying an exudation to the microscopical cells of which the timber is built up, and this renders the firm cell-walls soft and brittle, whilst the contents are dissolved out to supply the fungus with food. So fed, it increases the length and number of its threads, until they form a thick, soft, white felt-work, two inches thick, and spreading in circular fashion from the original centre. So long as it has a base on woodwork the felted sheet can spread over brickwork or masonry, even penetrating mortar, in order to reach other woodwork many feet away. When once it has attained a footing in a building, it can only be got rid of with much difficulty and at great expense. When it becomes mature, it produces fruiting organs analogous to the mushrooms and toadstools of some of the better-known fungi. These sporophores, as they are termed, are usually fan-shaped or half-circular, prettily marked with zones of brown and silvery grey, and having a yellowish margin. Sometimes these sporophores measure six or seven feet across at their broadest part. The wedge-shaped sheet in the photograph on page 826 is a segment of one such large sporophore, which had unfortunately been broken up by workmen before I had the opportunity for photographing it. On the under surface there are numerous pits sunk from which are produced millions of rust-coloured spores, by which this pest is disseminated.

Sometimes the Dry Rot may be found attacking standing trees—but rarely, if ever, in this country—before they are converted into timber, and any wood so used is likely to produce trouble if used for constructional purposes where the atmospheric conditions are at all favourable. Wood that is suspected of contamination in this way should be treated before use with either creosote or sulphate of copper, which are absolute enemies of all forms of fungi.

Photo by] [*W. Saville Kent.*

THE FRILLED LIZARD.

Showing it in its ordinary posture with the umbrella-like frill folded up over the neck.

"SEA-GOOSEBERRIES"

BY J. SINEL.

Of the myriads of living things that paddle and pulsate, gyrate, glow and scintillate, in the great world of waters, few can rival, and none can surpass for wondrous, delicate and modest beauty, the little organisms popularly known as "Sea-Gooseberries."

These radiant things are tentatively placed by naturalists in the class of animals with a simple stomach sac—the class to which belong the sea-anemones, jelly-fishes, and coral polyps—but they present so many deviations from the type that there is now a tendency to place them in a class by themselves. Several species occur around the British coasts, the most plentiful of which are the little globose forms that have suggested the popular name. The most abundant one, on, at least, the northern side of the English Channel, is named Cydippe, while on the southern side this species is to a large extent replaced by the Plumed Hormophora.

If the visitor to the coast will, on some calm summer day, sail or row a little distance from the shore, and proceeding very slowly, lean over the side of the boat and closely scan the surface of the water, he will scarcely fail to see numbers of these beautiful things floating past. Let one or two

Photo by] [*F. Skeen & Co.*

GATHERING JACK-FRUIT.

The Jack is a species of Bread-fruit which grows naturally in the Indian Archipelago. It is a principal article of food among the natives.

Photo by] [*E. Step, F.L.S.*

DRY ROT FUNGUS.

Showing the sheets of felt-like mycelium spreading over the back of oak panelling. The dark triangular portion is a segment of the spore-bearing organ.

be dipped up and transferred to a glass of clear sea-water and viewed, if possible, with a pocket magnifier, and then some insight into their wondrous structure and beauty will be obtained. Take for example the one termed Hormophora. This is about the size of a hazel-nut, clear as if made of glass, and of about the consistency of a ripe cherry, globose in form, but slightly conical at the upper part. From the centre of the base, and extending about two-thirds of the way up the sides, are eight muscular bands which are closely set with rows of fringed paddles arranged transversely, like the rungs of a ladder. These paddles look like miniature combs, and hence their family name of "Comb-bearers." From the sides, near the base, proceed two long threadlike tentacles fringed with little feathery appendages of two patterns—one a simple little feather, another long and curled. At the upper end, termed the "Oral pole," is a depression, which leads to the mouth, and at the base another depression which is a sense-organ and which contains one or more calcareous bodies termed *Otoliths*. Around this are four other bodies termed "balancers," the function of which is to enable the little creature to keep upright in the water.

Although these organisms consist of about ninety-eight per cent. of water, yet their faint and delicate tissues comprise muscle and nerve, and they seem to evince volition.

When viewed as I have suggested, it will be seen that from the stomach sac four tubes radiate to the surface, but these are not excretory, for, as in the sea-anemones, unassimilated or undigested food material is ejected by the mouth. Lower down are two pouches into which the feathery tentacles can be instantly withdrawn and stowed away when their owner suffers the least alarm. Then there is a series of tubes and vessels that in some way represents a circulatory system ; finally, on either side of the vertical bands that support the comb-like paddles are the reproductive glands—both male and female in the same individual. One would imagine that creatures so ethereal would have no gross appetites, yet the reverse is the case ; like things made of coarser clay, they are *Carnivores*, and prey greedily on the little crustaceans and thousands of other things that live at the surface of the sea. Anything small enough to be swallowed is considered by them lawful prey.

Unlike their cousins, the jelly-fishes and sea-anemones, that have "stinging cells" wherewith to kill or paralyze their prey, the Comb-bearers (except one foreign species) are without these weapons; but instead of such they have "glue-cells," that, like the sticky hairs on some insect-feeding plants, cause the adhesion of any little living things that come in contact with them. In the tentacle-bearing species, like Hormophora, the long streamers are abundantly set with these glue-cells, and, acting like animated "fly tapes," they reap a rich harvest in the teeming microscopic life of the sea. How the captures are transferred to the mouth has not, I believe, been observed, but it is quite evident that these tentacles are the chief purveyors.

Locomotion, in the globose forms, is performed chiefly by the waving of the comb-like paddles that fringe the vertical bands, and the motion of these, proceeding with rhythmic precision from base to summit, gives an appearance of waves of light flowing continuously along them. Their colouring is borrowed direct from the sun, for they reflect the rays of light, which are split up and wrought into new combinations, so that a radiance of silver and gold, and delicate shades of crimson and blue, violet and emerald, ripple along the rows of paddles, and are reflected from the smooth spaces between. Then, as a crowning glory, the majority of these creatures shine at night with a soft, effulgent light.

A species frequently met with in the English Channel is Beroë. In shape this is like a Phrygian cap, and, like Ocyroe and Venus' Girdle, it is without tentacles; its average size is a little below that shown in the figure, but it sometimes reaches a length of five or six inches. It is pigmented, and its iridescence is reflected from a surface of pink and buff, whilst a net-work—the course of its circulating canals—is spread like an ornamental design over the whole surface.

A very beautiful one—a native of warm seas—is the Many-horned Eacharis. This is nearly of

Photo by] [*E. Step, F.L.S.*

DRY ROT FUNGUS.

Back of oak panelling, showing the fungus *in situ*, spreading upwards in a semicircular manner. The soft margin is about two inches in thickness The two portions above have been removed from other parts of the woodwork.

the form of a typical jelly-fish, and, like a jelly-fish, it progresses through the water by the alternate contraction and expansion of its cup-like body.

Two other species—again natives of tropical or sub-tropical seas and closely similar in form—do not swim, but creep, travelling, as a rule, slug-like over rocks and stones or over the mud of the sea-bottom, but often quitting the shore, they creep, back downwards, on the under-side of what is known as the " tension-film " that forms all water surface. The one shown on this page is not floating, but thus " creeping " ; and if it were to let go it would sink. They are about an inch in diameter and are brilliantly pigmented with scarlet, crimson, or yellow, and their long, retractile, feathery tentacles are snowy-white.

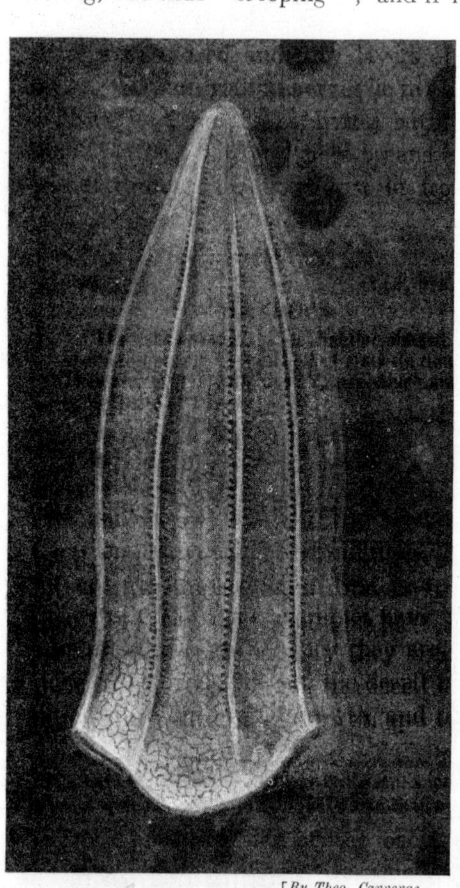

[By Theo. Carreras.

BEROË.

One of the Sea-Gooseberries, though its shape is rather that of a Phrygian cap. It will attain some five or six inches in length, and in colour it is iridescent with a ground colour of pink and buff, overspread with a delicate tracery.

In reproduction the Comb-bearers are peculiar. Eggs are set free in due season, and develop in the open sea, the young being very like the parent ; but the first batches of eggs in the season are from undeveloped and otherwise immature individuals. Then there is a cessation of reproduction ; after this first effort at plenishing the earth—or rather the sea—they sink to the bottom and assume the fully adult form ; then rising to the surface once more, they repeat the process.

These little Comb-bearers are indeed worth the trouble of study. They are not the least of the Marvels of the Universe.

THE STRUCTURE AND ORNAMENT OF SHELLS

BY BERNARD C. WHITE.

INFINITE variety is the first impression received in the study of shells. This variety falls into the three groups of structure, ornament and colour.

Definite causes would most certainly produce these results ; but while the variation from the original type can easily be detected, it is seldom possible to point out the causes, so varied and so subtle are their influences and so indefinite the period over which these influences have extended. Let us investigate these structural changes a little. Probably the most primitive form of shell was shaped like a tent, after the style of a limpet. Such a covering made for security, and all would go well until the creature within the shell desired to supplement its primitive organization. More room would then be wanted, and another story would be added, so that the tent-shape would become an elongated cone. Then the pinnacle, filled with the creature's body, would possibly become too heavy to support its own weight, and would develop into a coil. In process of time this coil would be repeated as characteristic of the family, and the spiral shell would be evolved.

At the top of page 831 are shown varieties of the tent-like construction. First is the Tooth Shell, which, however, must not be taken for anything *allied* to a limpet, for it is of an entirely different class. The snail has taken on entirely different habits, is a sand burrower, and has a different internal organization ; but it is quite possibly a development of a primitive form of tent-shell, narrow, elongated, as the homes of all the burrowers are, and inclining to a curve. Next is the

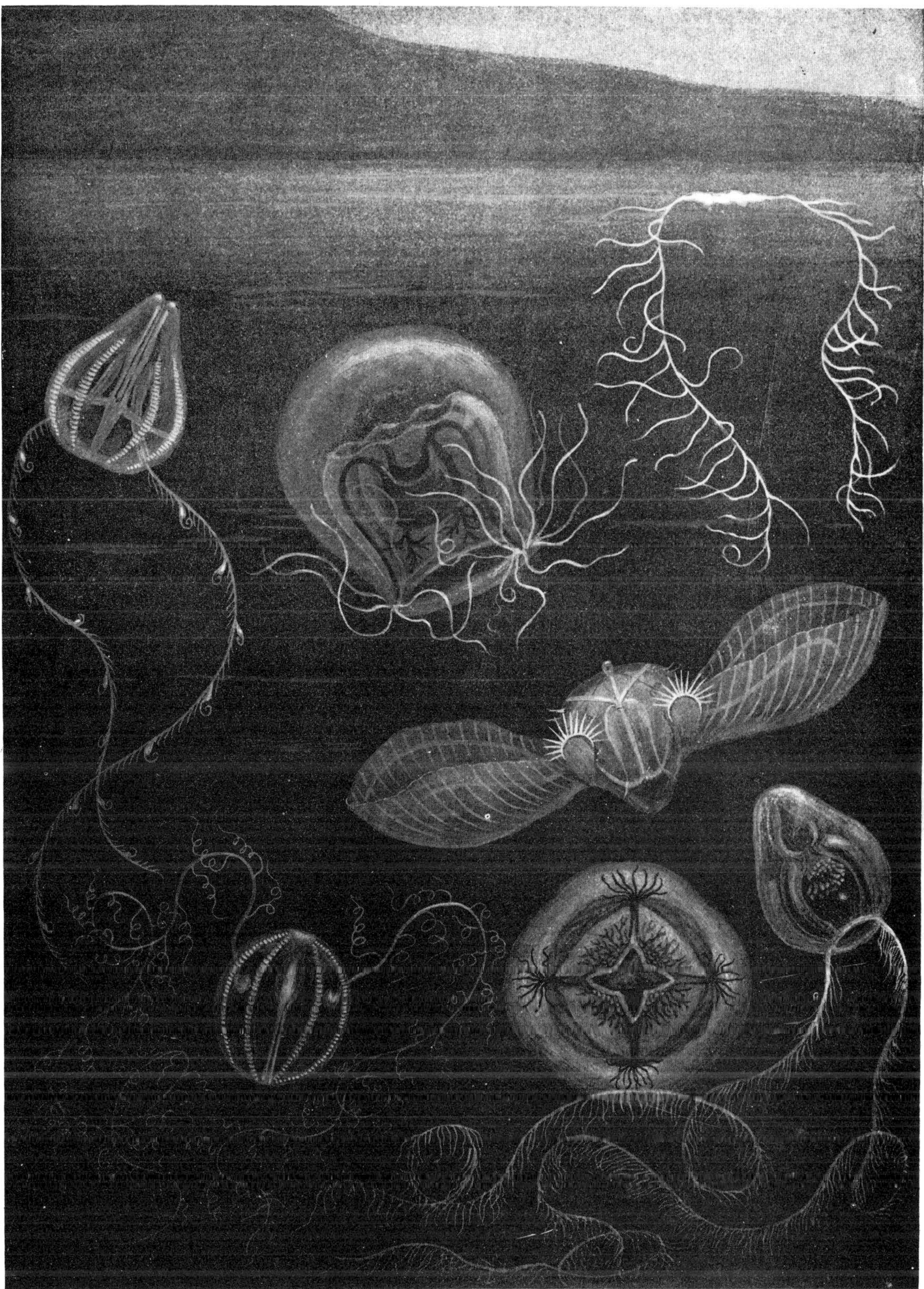

[*By Theo. Carreras.*

SEA-GOOSEBERRIES.

A peculiar class of Jelly-fish very abundant in the surface waters of the open sea. Although ninety-eight per cent. of their substance consists of water, the delicate tissues in which this water is contained is made up of muscle and nerve. These ethereal creatures prey greedily on the little crustaceans and any other marine life which comes within their reach. These specimens are about natural size.

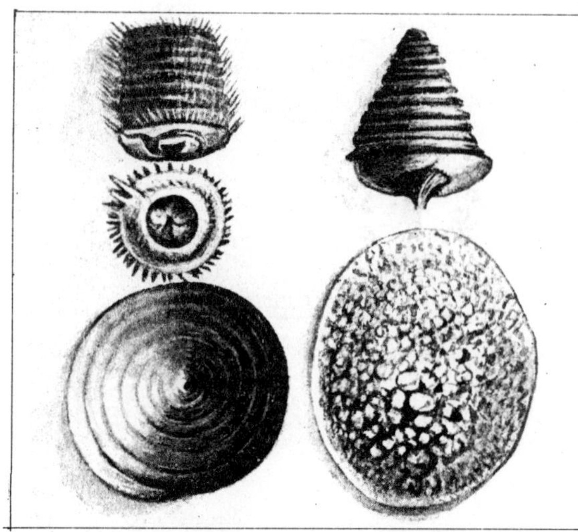

[By C. E. Swan.

The doors of shells. The largest is of a stony substance, the others are of horn.

[By C. E. Swan.

The snails have the doorways of their shells inverted and drag the apex of the shell along the ground.

[By C. E. Swan.

The first two of these shells show breathing channels, which are useful when the door of the shell is closed down; the others are notable for the teeth which guard the entrance to the shell

Umbrella Shell, heavy and very flat; then come instances of more elaborate structures (the Slipper and the Cup-and-Saucer Limpets), where an interior platform has been introduced for the attachment of the muscles. The habits of the Slipper Limpets are curious. They live piled up one on the top of the other, in the same way as bowler hats may be seen stacked up by a hatter.

With regard to the coiling of the spiral, some of its many variations are shown on page 833. First is the Hungarian Bonnet, which is almost limpet-shape, save that its apex has begun to coil backwards. Now most spirals are formed round a central column or axis of the shell, which runs through the whole length, just as a winding staircase is constructed round an iron shaft. Typical examples of the spiral are the Top Shells, where the coil narrows somewhat suddenly to a point, or the Augur Shells, where the narrowing takes place very slowly, and so produces a slender and sharply-pointed shell. These are the usual forms. Now let us turn back to our illustration on page 833. Here we have a Worm Shell (7), for all the world like a coil of tubing pulled out to its fullest extent. This shape is possibly an adaptation of the snail to life amongst seaweeds and corallines, where the creature could coil itself into the stem and branches of the weed and remain securely fixed. It should be noticed that the central column has been dispensed with; so has it in the Staircase Shell (6), only here the coils are more compact and built neatly one on the top of the other, forming a graduated spire. To demonstrate this, the shell has been photographed from below. If a spiral shell be held in the hand with the opening uppermost, this opening will generally turn to the right of the spire. But here are shown two shells (3) of the same species of mollusc, each of which coils a different way; such cases as this are not uncommon, but are restricted to certain species, while certain other species persistently coil towards the left.

It is no uncommon thing to find a shell in which the last whorl of the spire is developed out of all proportion to the remainder. Of such a type is the beautiful Harp Shell (4), and this

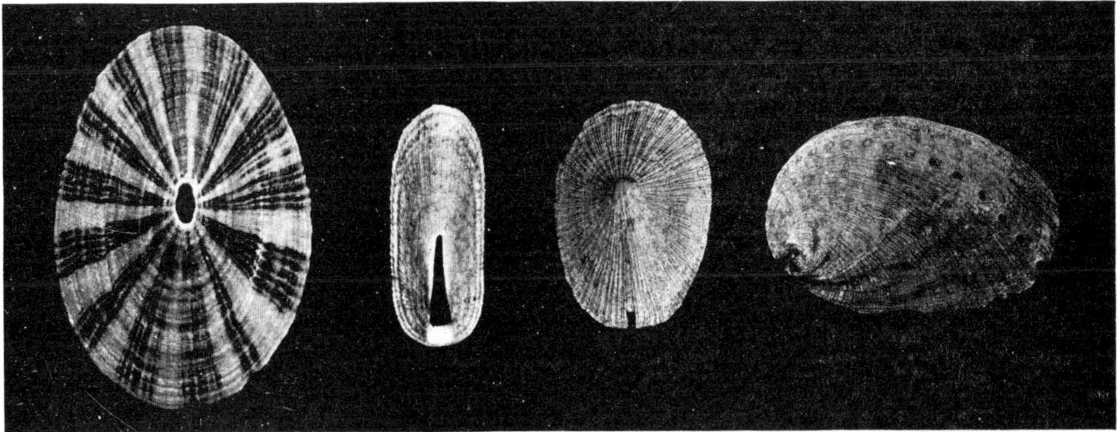

THE KEYHOLE LIMPETS AND THE ORMER SHELL.

Noteworthy for the "slit" in the shell which allows for the passage of fresh sea-water and waste material to and from the shell. In the Ormer the slit develops into a series of perforations.

peculiarity is still further developed in the neighbouring Egg Shell (5), where the outward semblance to a spire is completely lost, and the aperture is enlarged till it occupies the whole length of the shell. This shell is one of the Cowry family, and it is noteworthy that the race, in their early stages, develop normal, though small, spirals, which are swallowed up in the aggregation of the later whorls.

Now let us examine the clasp shells on the lower part of the page. The variety is not so great as in the univalve shells, and is chiefly concerned with the position of the umbones, or beaks, situated near the hinge. The normal position for these is shown in the Scallop Shell (1), though its tendency to lean to one side, as in the Venus Shell (3), is very common. Sometimes a shell is so formed that one valve is out of all proportion to the other; and when this happens, the beak of the larger valve is always very prominent, and curves towards the margin, while that of the smaller valve is merely rudimentary. This style of shell architecture reaches its highest fulfilment in the Lamp Shell (2), of which a fossil is shown on page 833, and becomes a perversion of all recognized ideas of bivalves in *Hippurites cornu-vaccinum*. What the laws are which have governed these developments it is at

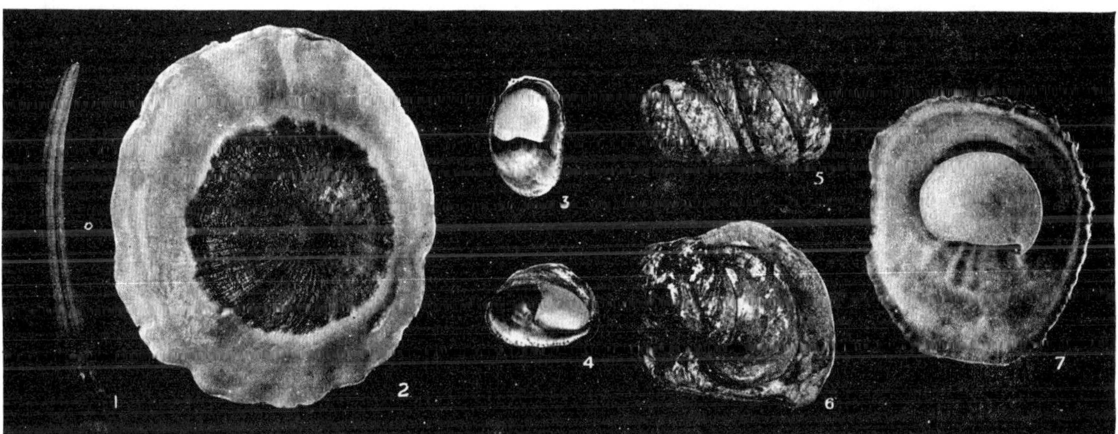

TENT SHELLS.

No. 1 is a Tooth Shell, which is probably a highly developed form of a primitive "Tent," such as the Umbrella (2). The Slipper Limpets (3 and 4) and the Cup-and-Saucer Limpet (7) show a more elaborate internal structure. The packing habit of the Slipper Limpets is shown in 5 and 6.

present impossible to determine ; but as the study of shell-life increases, an answer to the riddle may not unreasonably be hoped for. We do know, however, the causes which have influenced the form of the Watering-pot (4). At a glance, this curious shell can hardly be recognized as a bivalve. The rudimentary valves are marked by two scars on the under-side of the tube. The creature that inhabited these valves took to burying itself in sand or mud ; this caused the valves to curve more and more, and as they were never opened (the creature finding ingress and egress from the *ends* of the shell), the edges united and formed a single hollow tube. In this tube it lived, developing all the characteristics of the borer shells, such as were described on page 696, and working its way by means of the enlarged foot into the mud and sand. But some means was required for preventing the sand from falling in and filling up the further opening, and so the perforated doorway, like the rose of a watering-can, was added. Possibly the frills with which the shell is ornamented are the survivals of the corrugations which are the familiar ornaments of many clasp shells, such, for instance, as the Scallop.

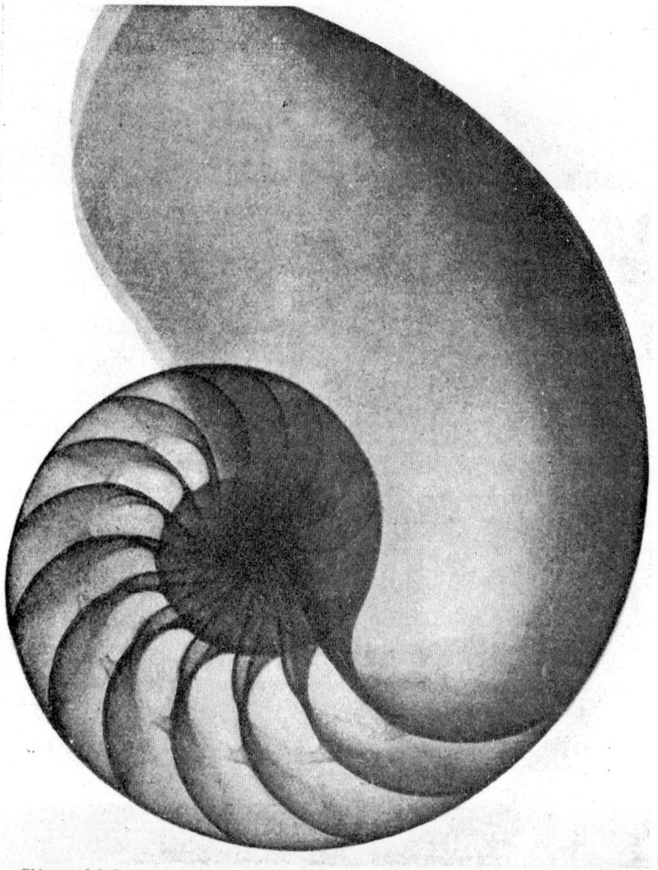

Skiagraph by] [*J. Green.*

SHELL STRUCTURE.

This beautiful skiagraph shows the intricate plan upon which the Nautilus is built. Notice how the whole of the structure coils round a central core. This core in shells of graduated whorls, such as cone shells, develops into columns; variations from this principle, however, are the Staircase Shell and Worm Shell shown on the opposite page. These have *no* central column.

In certain molluscs perforation in the shell will be noticed ; some examples of such shells are shown in the upper illustration on page 831. It is supposed that these apertures allow the escape of waste material, and in some cases the entry of fresh supplies of sea-water for the gills of the snail. The position of this slit may vary, as in the case of the limpets here shown. In the Ormer Shell, however, the slit is replaced by a series of perforations, which run down the whole length of the whorls ; as the animal extends its shell and makes further perforations, the earlier holes are closed up. The most singular example of the slit is that of the very rare Pleurotomaria, where it forms an open channel through the centre of the later whorls, and can be traced throughout the entire coil of the spire.

The entrance to the snail's home demands a little attention, for unless it be guarded, it is a source of danger to the shell's owner. Many snails, therefore, adopt the device of the Winkle and make horny doors for closing their doorways. This door, or *operculum*, is attached to the foot, so that when it is withdrawn it completely fits into the aperture. All doors are not of the same construction ; some are heavy and solid, like polished stone ; others are curved and horny, and some are shaped just like a screw stopper. But some snails cannot breathe when the door is shut, and so these have made a little tubular side-entrance which allows the passage of the necessary

SHELL ARCHITECTURE.

In these photographs variations in the forms of the two great classes of shells, the univalve (in the upper) and the bivalve (in the lower) are shown. The names of the univalve specimens are: (1) the Bonnet Shell, (2) the Coil Shell, (3) the Twisted Shell, (4) the Harp Shell, (5) the Egg Shell, (6) the Staircase Shell, (7) the Worm Shell. Those of the bivalves: (1) the Scallop Shell, (2) the Lamp Shell, (3) the Venus Shell, and (4) the Watering-pot Shell.

water, but is too narrow to admit enemies. Again, just as barbed wire and broken glass are used by humans to prevent intrusion on private property, so the shells of some molluscs are provided with jagged and twisted teeth at the mouth, which serve the same purpose. Finally, there are the curious shells at the centre of the illustrations on page 830, which have these doorways completely inverted, so that the snails walk with the points of their spirals touching the ground.

Only a brief mention can be made of the ornament of shells (see below). Often this ornament serves a protective purpose, as, for instance, the spines, which render their owners disagreeable to predatory marine life, for, if swallowed, they would cause the serious laceration of an unwary gullet. Such molluscs, therefore, as the Thorny Oyster and Thorny Woodcock are left undisturbed, to live out their lease of life. All ornament was probably developed from ridges. These ridges first took the form of corrugations, then became more pronounced and of the nature of frills; while the frills in time became elaborately goffered, until the corrugations closed together and settled definitely into spines. It is usually the margin of a shell which is frilled, and when the creature adds a later margin, the frill remains as a crest or a ridge of thorns on the exterior of the shell.

Of coloration we have here little space to write—and who can describe the indescribable? When it is remembered that most of this colour is hidden by masses of weeds and living matter, which becomes attached to a shell at the bottom of the sea, it is difficult to determine the purpose of such elaborate beauty; and it is equally difficult to conceive Nature to be so prodigal that she lavishes the beauty on a mollusc for a mere whim. Sometimes, it is true, coloration serves the purpose of protective resemblance and mimicry, but such cases are rare, and do not include the dazzling reds and purples and golds, or the pastel shades and peacock colourings which distinguish so many varieties of shell-life.

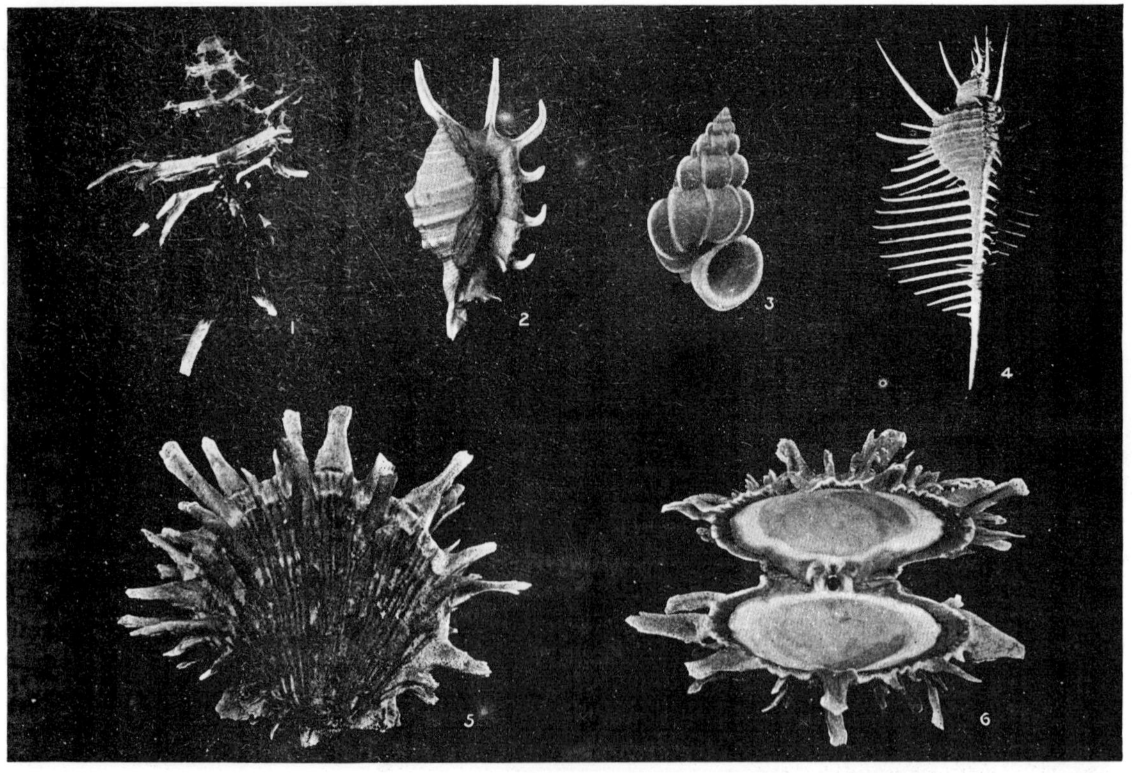

THE ORNAMENTS OF SHELLS.

They can be roughly classed as "frills" and thorns. Colour, which is the chiefest characteristic, cannot be shown in photographs; but the coloured plate gives a few examples of this most beautiful form of decoration. The shells on this plate are: (1) the Endive Shell, (2) the Scorpion Shell, (3) the Wentletrap, (4) the Thorny Woodcock, (5 and 6) the Thorny Oyster, closed and open.

Photo by] [*W. S. Berridge, F.Z.S.*

THE WHITE-HEADED SAKI.

The morose and grotesque appearance of this South American monkey belies its natural character, for it is shy and inoffensive, and does not long survive captivity. Except for the head, the monkey is entirely black, with coarse matted hair and a very bushy tail. The large white beard which surrounds the face adds to the effect of human caricature.

THE WHITE-HEADED SAKI

THE kindred tribes of monkeys and apes are notable for the unexampled ugliness of many of their members. Their position in the scale of creation is, no doubt, responsible for this notoriety, for while they are sufficiently allied to man to invite comparison with the higher animal, man himself has attained so much higher a level of mental development and physical beauty that the want of such qualities in the ape is made all the more apparent. When, therefore, we come across monkeys which look for all the world like some human caricature, our curiosity and astonishment cannot fail to be aroused, and so it is that certain species, as, for example, the Chimpanzee, the Orang-utan and the Gorilla have become notorious for their grotesque human characteristics.

A more uncommon case of peculiarly human appearance is that of the White-headed Saki, which, with its squat nose, its bushy white beard and shaggy eyebrows, bears a striking resemblance to a fat, heavy-jowled and ill-favoured misanthrope. Appearances, however, belie the gentle and inoffensive nature of the creature, for the Saki is shy and silent and too delicate to long survive captivity.

The Saki race inhabit Guiana and the valleys of the Amazon. They are distinguished from their near allies, the Douroucolis, and from all other American monkeys, by the position of the teeth in the lower jaw, which are thrown forward out of the vertical. The jaw itself curves out considerably as it nears the socket. Like the Douroucolis, the tails of the Sakis are not adapted for climbing purposes. The colouring of the White-headed Saki is, as the name implies, white or yellowish on the forehead, with a black streak in the centre, the rest of the body is completely black. The hair,

Photo by] [Fred Enock.

A FAIRY FLY

Caught in the web of a spider and held fast by the globules of viscid fluid with which each thread is supplied.

which is thick and coarse, sticks out all round the head, and the beard is heavy and undivided. The tail is particularly large and bushy.

"FAIRY FLIES"

BY FRED ENOCK.

FAIRY FLIES do not occupy much space, for they are the smallest four-winged insects known to science—the largest species measuring just a millimetre from head to tail, and the smallest, *one ninety-fifth of an inch long!!* Just try to realize what this animated atom of an insect is made up of. The organs of flight, two upper and two lower wings geared or hooked together, as perfectly as those of the largest bee or wasp—of exquisite form and graceful curve, the outer margins surrounded by a fringe of long hairs, giving it the appearance, under certain light, of a peacock's tail spread open; each of these tiny organs of flight kept in motion by a powerful set of muscles, attached to the inside covering of the thorax, which is divided into plates, artfully joined together by the most delicate membrane, so that the Fairy Fly can twist its body in any direction.

The head, too, is full of marvels. Simple and compound eyes, double-toothed jaws and tiny tongue and those wonderful organs aptly termed "feelers," for they are the most sensitive organs in the whole fly; those of the male being thread-like, while the female has shorter and broader-jointed ones, each terminated by an oval club-shaped joint. The under side is covered with sensory hairs—with these it identifies its host-egg, though it may be embedded beneath the covering skin of a plant-stem. If ever there was a fairy-like waist, surely these Fairy Flies have such, supporting the abdomen, terminated by its wonderful egg-laying apparatus. The latter is composed of a number of delicately-barbed lancets forming a tube *so* small and delicate that microscopic measurement shows the internal diameter to be but one eight-thousandth of an inch!! and through this tube the germ or egg is conducted into the host-egg—for greatest of all insect marvels, these Fairy Flies lay *their* eggs *within* the eggs of injurious insects!!

Photo by] [Fred Enock.

THE EGGS OF THE FAIRY FLY

Shown, as they are laid, in the eggs of the Frog-hopper. One to four may be deposited in each host-egg, which is itself buried in the bark of a reed. (Magnified thirty times.)

Now the question presents itself: *How* do these Flies find the right host-egg? Only one answer can be given to this, viz., by their unceasing perseverance. We have only to follow *their* example and, in the course of a few years, we may possibly find out the life-histories of one or two from start to finish.

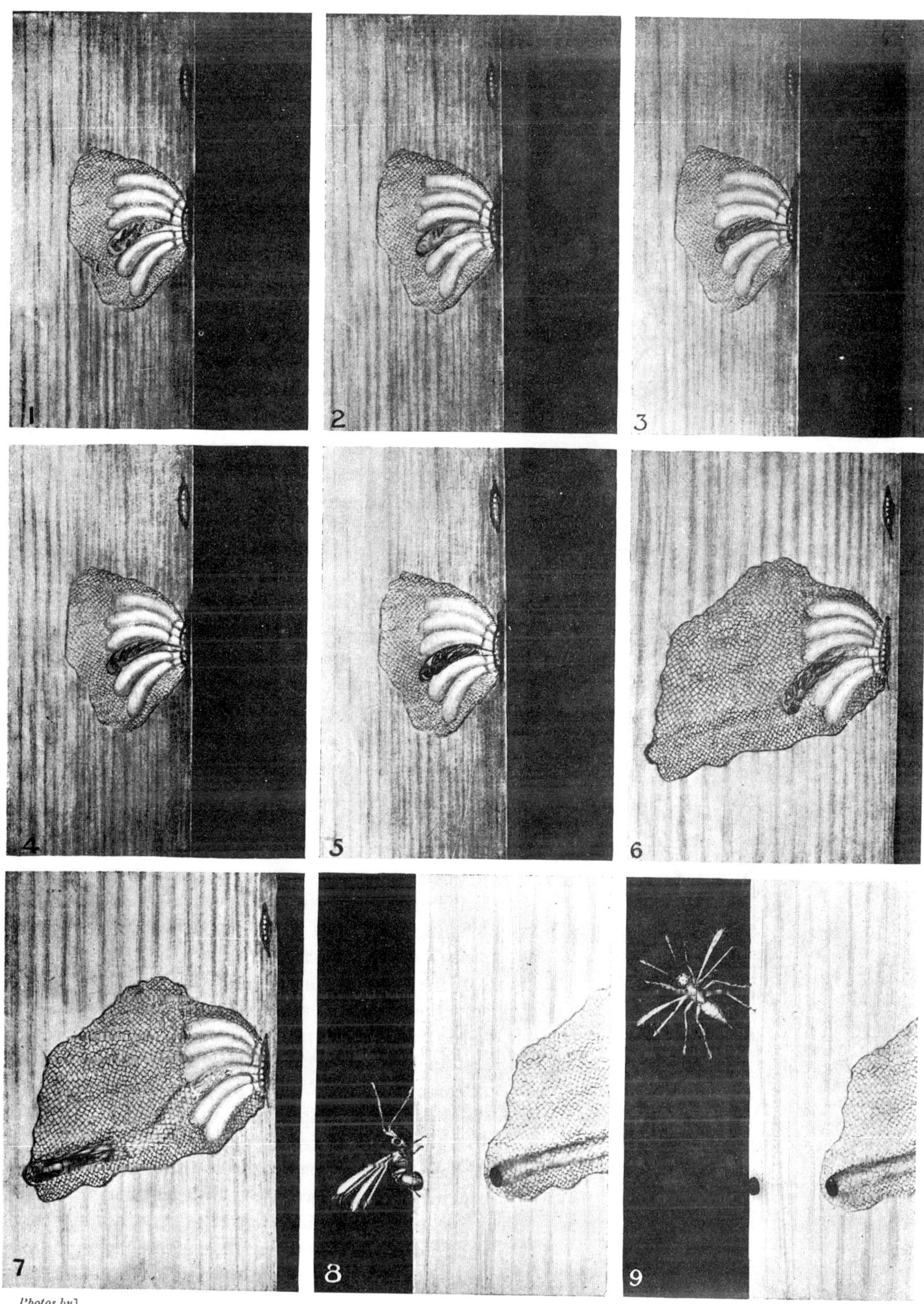

Photos by] [*Fred Enock.*

THE LIFE-HISTORY OF THE GOLDEN FAIRY FLY.

In the first illustration it is to be seen passing its earlier stages in the Frog-hopper's egg within a reed-stem. At No. 5 it is fully developed, and at 6 it has eaten away the skin of its host's egg and is commencing to tunnel through the pithy stem. Finally (No. 8) it bores through the thick bark of the reed and emerges (No. 9) to take its virgin flight. Throughout the reed is shown white against a black background, with part of the outer covering torn away to show the eggs and fly.

Photo by] [*F. Enock.*

One of the most beautiful of the flies—the Fairy-winged Fly. (Magnified thirty times.)

Let us take "The Golden Fairy Fly"—whose haunt is among grass and rushes.

An examination of the lower part of one of these rushes will show that the green skin is peppered with minute dark-coloured specks. These are the punctures made by a near relative of the "Cuckoo-spit" insect, and underneath we find a number of white, flask-shaped eggs radiating from the dark puncture. Our next move is to lie flat on our stomach, with head low down among the rushes and to *quietly* watch the stems—and possibly we shall be rewarded by seeing a Fairy Fly running up and down, her feelers in rapid motion, rapping the stem. Suddenly her activity ceases, for her feelers have touched one of the dark specks, and with a tremulous motion, she gently "sounds" with her clubbed feelers to ascertain the species and condition of the hidden eggs. If these prove satisfactory, she quickly brings the tip of her egg-laying lancets to the head of an egg, and pressing with all her might, she forces it between the feelers, and her whole body remains motionless for some minutes, while she is laying one of her eggs in one of those of the Cuckoo-spit or Frog-hopper. Withdrawing the weapon, she attacks another egg, and "stings" that, and sometimes the whole cluster, or, maybe, one here and there.

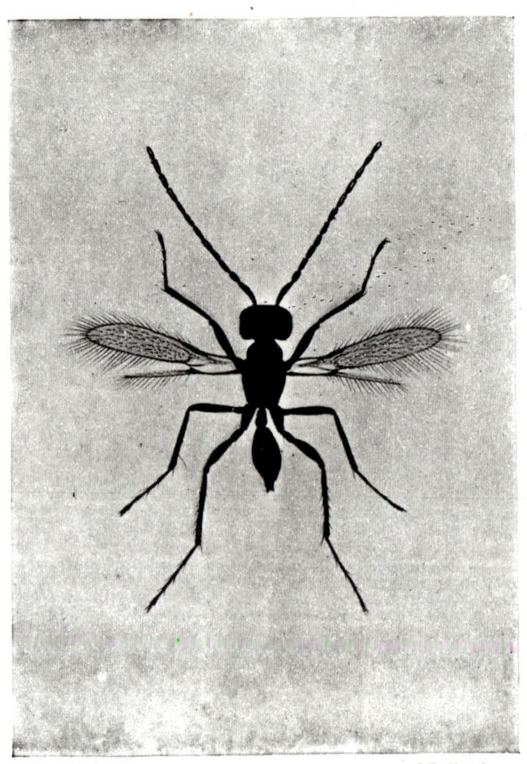

Photo by] [*F. Enock.*

The Aquatic Fairy-fly, which receives its name from its ability to *fly* under water. (Magnified thirty times.)

Of course, the act of laying its egg in that of the injurious Frog-hopper's has doomed the host-egg to destruction. In a few weeks, according to season, the germ hatches to a legless grub, which, feeding upon the fluid, becomes full grown, when it can, by most careful illumination, be seen moving about. Next commence those mysterious changes from pupa to fly only to be seen by hours of watching; but to *see* the actual building up of the various organs by the coming together of the cells is a sight never to be forgotten. From out of a mass of pink circular "cells," or rings, with darker centres, appear others, which "come together" somewhat after the fashion of bubbles in a cup of tea! And at first we see before our eyes certain indistinct tracings, which increase in compactness, and as we strain our aching eyes, we recognize that these masses of cells have assumed distinct form, and the head, thorax, limbs and abdomen are all marked out—in *pink*, until all the cell-material is used up and we see distinct organs, all assuming their special colours; the head changing to dark brown, the eyes deep crimson, the tiny double-toothed jaws almost black, the limbs yellowish, with hairy joints, the reddish abdomen, and the wonderful wings with the fringe of long hairs all turned up and lying upon the surface—until the final exit from the host-egg, when they turn over on their hinges, each one taking its allotted place. The final change is now taking place, the delicate jaws are moving, until the tips press against the inner side of the host-egg, through which a hole is bitten, and gradually the Fairy Fly forces its way through the soft pith, until it reaches the opposite side; then the thick covering of the stem has to be bitten through before the Fairy Fly can escape and take its virgin flight.

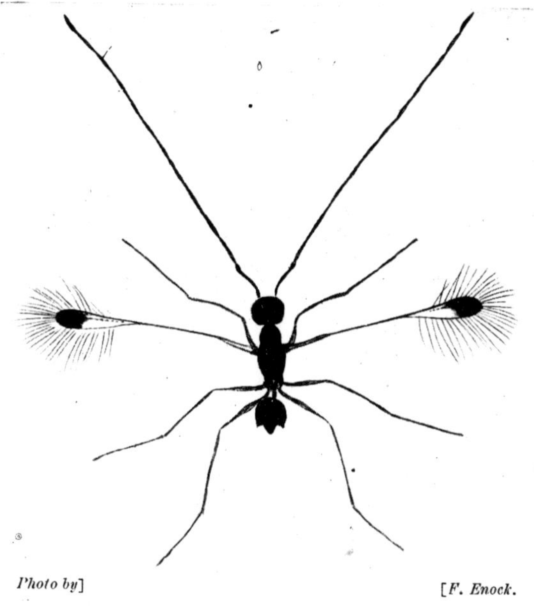

Photo by] [*F. Enock.*

The Battledore-wing Fly is, like most of the Fairy Flies, very small: the body measuring from head to tail one-fortieth of an inch.

THE "ELEPHANT-FISH"

BY W. P. PYCRAFT, F.Z.S., Etc.

THE grotesque-looking Chimæras are represented only by a few species, but all amply justify the family name, for in appearance they seem indeed to embody all the attributes of the fabulous, fire-spitting monster after which they have been named by the savants. But of all the Chimæras, perhaps the species commonly known as the "Elephant-fish" is the most bizarre; and this because of the curious flap of skin which projects from a short beam in front of the mouth. The purpose of this strange appendage is not certainly known, but it seems to serve as an organ of touch. This curious fish is found only in the Antarctic basin and the South Pacific, and among other things it is remarkable for the singular form of its egg—though, as a matter of fact, all the Chimæras lay peculiar eggs. That of the Elephant-fish measures between nine and ten inches in length, and about five inches in breadth; while its upper surface is covered with a hair-like growth, which gives to the whole a close likeness to a piece of seaweed. The centre of this egg is occupied by a pear-shaped cavity, containing first the yolk of the egg, and, later, the embryo. One end of this cavity is guarded by a valve, which forms the doorway to the outer world, from which the young in due course makes its escape. While still within the chamber the young fish develops long, filament-like gills, which project like tufts of crimson threads from each side of the head, and carry on the work of breathing.

The Elephant-fish has a near relation in the American "Spook-fish," which, unlike the rest of its relatives, dwells near the surface of the

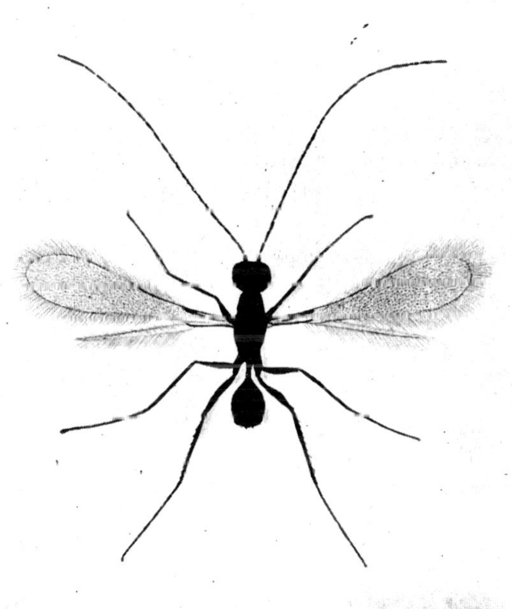

Photo by] [*F. Enock.*

The Yellow-legged Fly (male). Notice the long feelers and slender "waist." (Magnified thirty times.)

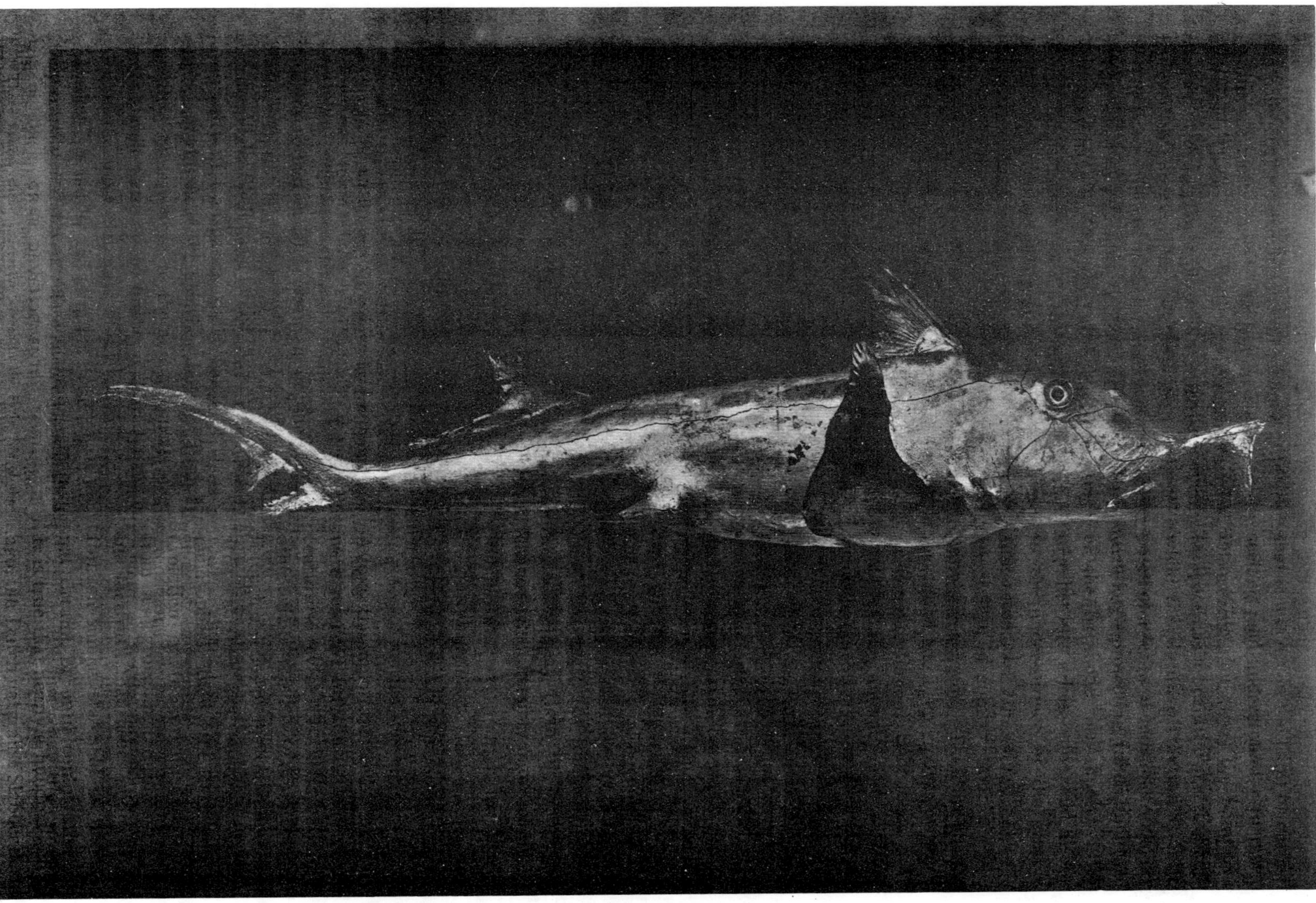

THE ELEPHANT-FISH.

A fish remarkable for the curious flap of skin projecting from a short beam in front of the mouth. It probably serves as an organ of touch. This largest of the Chimaeras occurs occasionally in the deep waters of our own seas

sea. But even so, it goes down deep to lay its eggs, which measure about six and a half inches long, and have the lower end produced into a long "tail," which, it is believed, is thrust down into the mud, so as to anchor the egg in a vertical position. But this is not all. When first laid its walls are impervious; but soon they become pierced by a number of slits through which currents of water pass to the growing embryo.

BEAUTIFUL PHEASANTS

BY SIR HARRY JOHNSTON, G.C.M.G.

IF the different groups and families of birds competed for a beauty prize it would be difficult to award the gold medal to any one section, and declare this or that family to be more beautiful than any other. Possibly, we should decide that in regard to graceful shape, vivid beauty of coloration, and exquisite elaboration of pattern the Pheasants held the chief position. Humming Birds and Sunbirds are exquisitely beautiful, but they are very small. Parrots are gorgeous, but they are often clumsy or grotesque in shape. The Birds of Paradise perhaps reach a climax in wonderful structure and exquisite tints of plumage, but the Pheasants are more numerous, are almost world-wide in their range, and play quite a spectacular part in human civilization; for do they not include the Peacock, the Turkey, the Guinea-fowl, the Gold, Silver and Amherst Pheasants, the nearly equally beautiful Pheasant of our woods, and the many handsome breeds of Domestic Fowl, all of which are very much mixed up with the surroundings of humanity?

[*From a specimen in the Royal Collection at Stuttgart.*

BULWER'S PHEASANT.

This rare and beautiful form is found in Western Borneo, both in Sarawak and in the Dutch territory along the Baram river. The male has a heavily plumed tail of snowy white and the rest of the plumage is metallic-blue and sepia-brown.

These domesticated "Pheasants," however, though in the forms of Peacock and Turkey they are among the marvels of the bird class, are nevertheless so familiar to us that they have ceased to be marvellous, and to excite fresh wonderment we must present much rarer forms of the Pheasant family from the jungles of easternmost Asia.

A Pheasant which is a marvel for the length of its tail is Rheinhardt's Pheasant, which is found in the mountains of Tonquin and Annam. The total length of the specimen here illustrated, which is to be seen in the Stuttgart Museum in Southern Germany and which is probably the finest in any collection, is nearly nine feet, of which length the tail measures nearly six feet. It is really the middle pair of the tail feathers which are so broadened and enormously prolonged. They are boldly marked with ocellations of chestnut-brown on a pale-grey ground, some of the smaller markings being black rings with white centres. The plumage of the body and wings is mainly

[From a specimen in the Royal Collection at Stuttga t

RHEINHARDT'S PHEASANT.

This rare and truly remarkable bird, with a tail nearly seven feet long in some examples, inhabits the dense forests of Tonquin and Annam.

dark-brown, here and there tinged with russet and marked with white spots or lines, while there are touches of black about the head and crest.

This bird, like its wonderful relation the Argus Pheasant, is much given to dancing and parading before the females in the breeding season, and for this purpose clears with its feet a circular space of ground under the trees. Around this it races at full speed, stopping every now and then to lower its wings and elevate its wonderful tail. The Argus Pheasant is equally remarkable in its way and is more beautifully marked with bolder ocelli or "eyes" than Rheinhardt's Pheasant. In fact, it derives its name from the Greek myth of Argus with the hundred eyes. These wonderful markings, as in the case of all Pheasants, are confined to the male bird. In size the adult male of the Argus Pheasant is about six feet long. It is not the tail which is the chief feature in this bird, but the enormous development of the secondary quills of the wing, though the two middle tail feathers are very long.

Closely allied to the Argus Pheasants and Rheinhardt's Pheasant, though not nearly so large, are the Peacock Pheasants, with two or three spurs on each leg. Of these, two are illustrated in this article, one of them—the largest of the group—being the Chinquis Peacock Pheasant, and the other, almost the smallest, Nehrkorn's Peacock Pheasant. The Chinquis is a bird which has frequently been shown in our own Zoological Gardens and is fairly common in the great aviaries of European pheasant fanciers. Nehrkorn's Pheasant, however, is a very rare bird, even as a museum specimen. The Chinquis has a comparatively wide geographical range from Sikkim in British Tibet through Assam to Lower Burma and Indo-China. Nehrkorn's Peacock Pheasant, however, is confined in its distribution to the small island of Palawan between the Philippines and North Borneo. This last-named bird is of great beauty in plumage, the adult male having the greater part of the body purplish-blue changing into dark-green, many of the feathers being edged by margins of steel-blue or golden-green. The tail has a ground colour of buff or reddish-brown, but each tail feather is ornamented with a pair of large bluish-green-violet ocelli surrounded by rings of black and grey. The face is marked with white and there is a vivid patch of naked crimson skin round the eye, so that it is a bird of very striking coloration. The Chinquis Pheasant, though it has the splendid metallic blue-green ocelli on its back and tail and the coverts of the wing, has the rest of the plumage whitish-grey and brown.

The Chinquis is a bold bird of vicious temper (in the male), so that it will often attack human beings entering the aviary in which it lives. The male, moreover, is very spiteful towards the female, so much so that most of the Peacock Pheasants of this type that one sees in European aviaries are living as widowers, having killed a succession of females which had been provided for them as mates. It is a great pity that such a handsome bird should have such a disposition, as it is quite well able to stand the cold of a European climate and might otherwise have been as easily domesticated as the gorgeous peacock, its near relation. The Peacocks, Peacock Pheasants, Argus and Rheinhardt's Pheasants form a small group by themselves, with distant relationships both to the Guinea-fowl of Africa and the Turkeys of America, though, of course, they are all "Pheasants."

Another group nearly allied is that which includes the Domestic Fowl and its wild relations, the Silver Pheasants, the Fire-back Pheasants, the Crossoptilons, and Bulwer's Blue-wattled Pheasant, a very

NEHRKORN'S PEACOCK PHEASANT.

The handsomest and the smallest of the group. It is confined in its range to the island of Palawan between Borneo and the Philippines.

rare and very beautiful form confined to Borneo. In the adult male of Bulwer's Pheasant the naked skin of the face and the wattles is bright azure-blue, the legs are red, and the neck and chest dark crimson ; the rest of the plumage of the body and wings is more or less black and brown, tinged with purple, and with beautiful steel-blue edgings to the body feathers. An abrupt contrast with this dark-tinted body is the tail, which is an absolute white. The tail feathers are also crimped and curled in a fantastic manner. This bird only comes from the western parts of Borneo, where it inhabits the dense forests.

Amongst other Pheasants are the Gold and the Amherst, and the many species of the genus to which the Pheasant of our woods belongs, especially the splendid Reeves' Pheasant, with markings of black, white and gold, and a tail sometimes four feet in length. But these birds are so well known that we have become used to their marvellous beauty and it is not necessary to illustrate them in a special manner.

THE CHINQUIS PEACOCK PHEASANT.

This is the largest of the Polyplectron group and is the size of a large domestic fowl. It is also the commonest species and the most widely spread.

Although Pheasants may have existed anciently in Europe before man came into existence, they are at the present day (with the exception of the Common Pheasant) exclusively confined to Asia, and mainly to the easternmost parts of Asia, from Manchuria on the north to Borneo and Java on the south. Unfortunately, their extinction is proceeding at a rapid rate, and it is quite possible that only a few species of these marvellous birds may ultimately survive, those which, like the Gold and Silver, Amherst, and Reeves' Pheasants, will breed in captivity, and which now exist in numbers in aviaries all over the world. All the shy and rare kinds, like the Argus and Rheinhardt's Pheasants, are diminishing rapidly in numbers through the destruction of forests which is going on in South-Eastern Asia, in order to plant rubber, coffee, tea, etc. In China they are being slaughtered on a tremendous scale, partly for their plumage, still more for their flesh, as they are sent in a frozen state for consumption as food in Europe. The construction of the Siberian Railway across Central Asia has assisted much to further this trade in frozen game from China.

SEA-SCORPIONS

BY REV. H. N. HUTCHINSON, F.R.G.S., F.G.S, F.Z.S.

FOSSILS are often puzzling things to those who find them, especially when the specimens which the geologist brings home in his bag are only fragments of some now extinct creature of long ago. Fanciful names have frequently been given to fossilized organic remains found in quarries, or brickyards, and other places. The Belemnites, so familiar to all fossil-hunters, were at first called "Thunderbolts," from their long pointed shapes ; but, in reality, they are the internal bones of cuttle-fishes.

From "Extinct Monsters."] [*By kind permission of Messrs. Chapman & Hall.*

SEA-SCORPIONS.

These are the giant crustaceans of a day gone by, whose nearest living descendants are by some naturalists **considered** to be the King-crabs shown on page 293. They were possessed of a coat of armour and limbs for crawling, swimming, or seizing their prey, and some reached a length of **nearly six feet**. Three types of these Scorpions are shown here. Note especially the one to the left with the feelers each armed with a pair of large pincers.

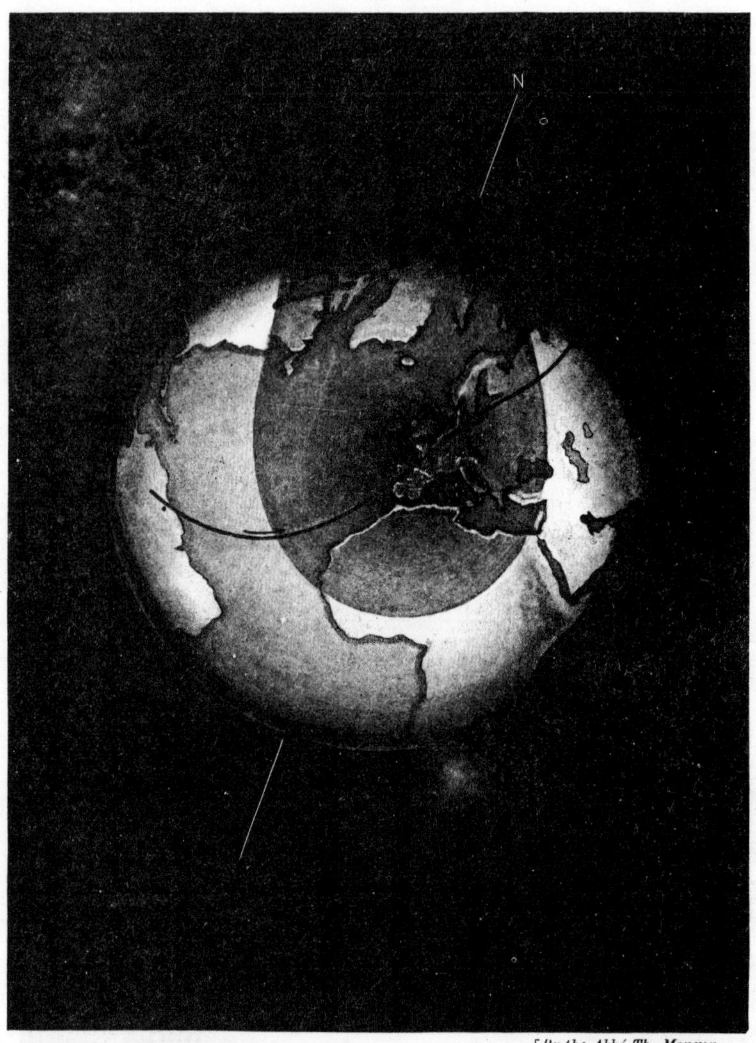

[*By the Abbé Th. Moreux.*

THE ECLIPSE OF THE SUN, APRIL 17th, 1912.

Showing the area of the Earth covered by the shadow-cone at noon as it would appear to a spectator in space, who would be looking straight towards Les Sables d'Olonne in the Vendée department of France

So with the strange creatures from the Old Red Sandstone strata, represented in our plate by a restoration by Mr. J. Smit. The men who worked in the sandstone quarries in Forfarshire along with the illustrious Hugh Miller, used to discover fragments of this very strange crustacean. Its hard coat of jointed armour bore on its surface curious wavy markings, that suggested to their minds the sculptured feathers on the wings of cherubs—of all subjects of the chisel the most common. Hence they christened these remains "Seraphim." The part to which this name was given turned out to be part of the under-side below the mouth. They greatly puzzled Hugh Miller, who was inclined to ascribe them to some unknown fish of the period; for there were, as every geologist knows, certain very "queer fish" in those days. Some of these have already been described in this work by Mr. Pocock. However, Hugh Miller submitted some of his fragments to the celebrated naturalist, Agassiz, a pupil of the famous Cuvier, of Paris, who had already written a famous work on fossil fishes. He, with his great knowledge of living types of animals, was able to solve the riddle by declaring that the remains belonged to a huge lobster! It was not, strictly speaking, a lobster; but we shall not be far off if we say it was first cousin to a lobster. At that distant period (the Old Red Sandstone) there were no lobsters, but this Pterygotus, with its long "jaw-feet" and powerful tail, was something between a king-crab on the one hand and a scorpion on the other. In front was a great shield, or carapace (as we see in crabs and lobsters), bearing two large eyes, made up of numerous small lenses. The five pairs of appendages were all attached under or near the head-shield. Behind this follow twelve rings, or segments, the last of which forms the tail; two at least of these carried gills for breathing. Of these appendages (or limbs), the lowest and largest evidently were used for propulsion through the water. In all other crustaceans the antennæ are used for feeling about, but in this case they were used for catching the prey, being provided with claws. Dr. Henry Woodward, F.R.S., who has written a learned monograph on these creatures, concluded that our Sea-Scorpion was a very active animal. The whole family to which it belongs seem to have been fitted for rapid movement ; and in this case the tail-flap also served as a propeller. One of the chief points of interest in Pterygotus, which was six feet in length, is that naturalists see in it a resemblance to the crab in its young, or larval, state. This is only in accordance with a law, known as the law of recapitulation, that living animals in their younger stages show resemblances to extinct animals of long ages ago in their adult state. Naturalists have also noticed a close resemblance to land-scorpions and spiders. We may, therefore, not be far wrong in venturing to call these creatures Sea-Scorpions.

ECLIPSES

BY G. F. CHAMBERS, F.R.A.S.

ECLIPSES of the Sun and of the Moon are totally different in principle A total eclipse of the Sun implies that the Sun totally disappears from sight : a total eclipse of the Moon implies that

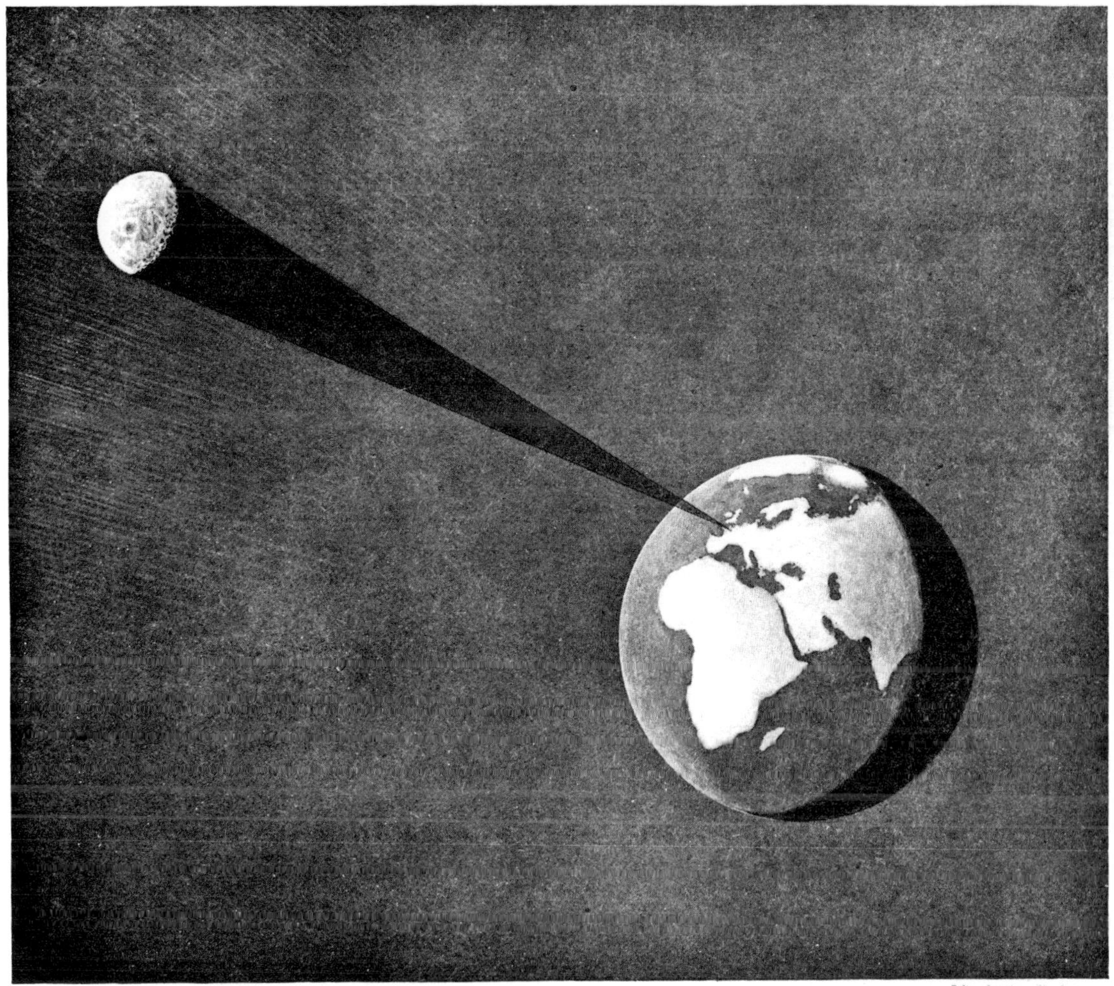

[By Lucien Rudaux.

THE ECLIPSE OF THE SUN, APRIL 17th, 1912.

If lines be drawn touching the surfaces of both Sun and Moon and extended till they touch the surface of our globe these lines will be found to form a cone. Now that part of the cone which is between the Moon and the Earth will be in deep shadow, for the Moon will have eclipsed the Sun's rays, and it is at the point on the Earth that forms the apex of this shadow-cone that a total eclipse will be observed. On April 17th, 1912, this point of the cone swept the neighbourhood of Paris at ten minutes past noon.

the Moon also totally disappears from sight (though sometimes there are exceptions to this rule). Here the resemblance ends, all the other features being altogether dissimilar. The Sun undergoes total eclipse because the Moon passes in front of it, and absolutely blocks it from our view. On the other hand, when the Moon undergoes total eclipse nothing comes in front of it, but it disappears (when it does disappear wholly) because it passes into the dark shadow cast by the Earth into space. The philosophy of a lunar eclipse may be realized on any evening when a man walking down a street sees another man turning a corner not reached by a lamp erected at a short distance round the corner but short of the corner. In such a case, though the man who has turned the corner could be struck by a bullet fired by the first man, yet the first man might not be visible to the second man, because the lamp did not shine on him. Moreover, the total phase of a solar eclipse seldom lasts more than three or four minutes, and can never last more than seven minutes, whilst the Moon may be absent from view for more than an hour.

Total eclipses of the Sun as astronomical events used to stand in a certain sense on the same footing as Comets: that is to say, to the ancients and during many centuries they were not known to be subject to any definite laws, but were deemed celestial portents of terrestrial misfortunes of

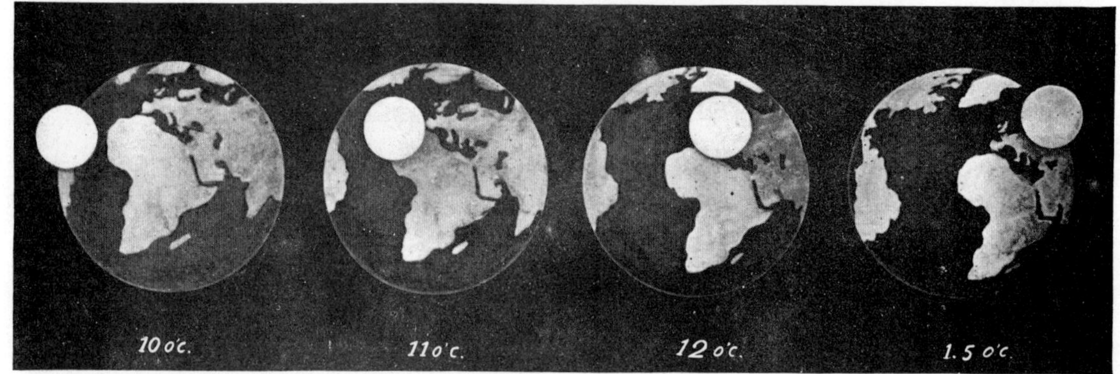

[*By Lucien Rudaux.*

THE ECLIPSE OF THE SUN, APRIL 17th, 1912.

What a spectator in the Sun would see of our Solar Eclipse between the hours of 10 a.m. and 1 p.m. The Moon as a brilliant disc would be brought into relief against the Earth as it passed along a course determined by its own rotation and that of the Earth.

various kinds. An eclipse which happened on May 28, B.C. 585, though not the oldest on record, is one of the most interesting in ancient history, for it occurred in the middle of a battle between the Lydians and the Medes, and resulted in the stoppage of the fighting, followed by a treaty of peace. It should, however, be stated that there was one man to whom this eclipse did not come altogether as a surprise. Thales, of Miletus, a distinguished Greek astronomer, predicted the eclipse, though the exact date was long disputed by moderns, who suggested other dates than 585.

A solar eclipse commences by a black notch becoming visible in the Sun's disc. This notch increases until it becomes a large and growing slice, showing much black and little bright in the Sun's surface. If this condition subsists until the eclipse comes to an end and the Sun reappears in all his glory the phenomenon goes by the name of a "partial eclipse of the Sun." But the grand climax is reached where the Sun totally disappears; the effects at that stage are very grand and awe-inspiring. Day suddenly becomes night, and to see what o'clock it is by your watch you must strike a match or have a lamp at hand. If the darkness is qualified, that is to say, not absolutely intense, people and things, and especially the distant horizon, acquire a very lurid, or, as some would say, a very "creepy" appearance.

The special sights to be looked for in connection with a total eclipse of the Sun by an ordinary observer using his eyes or a small telescope may now be mentioned. When the Moon passing across

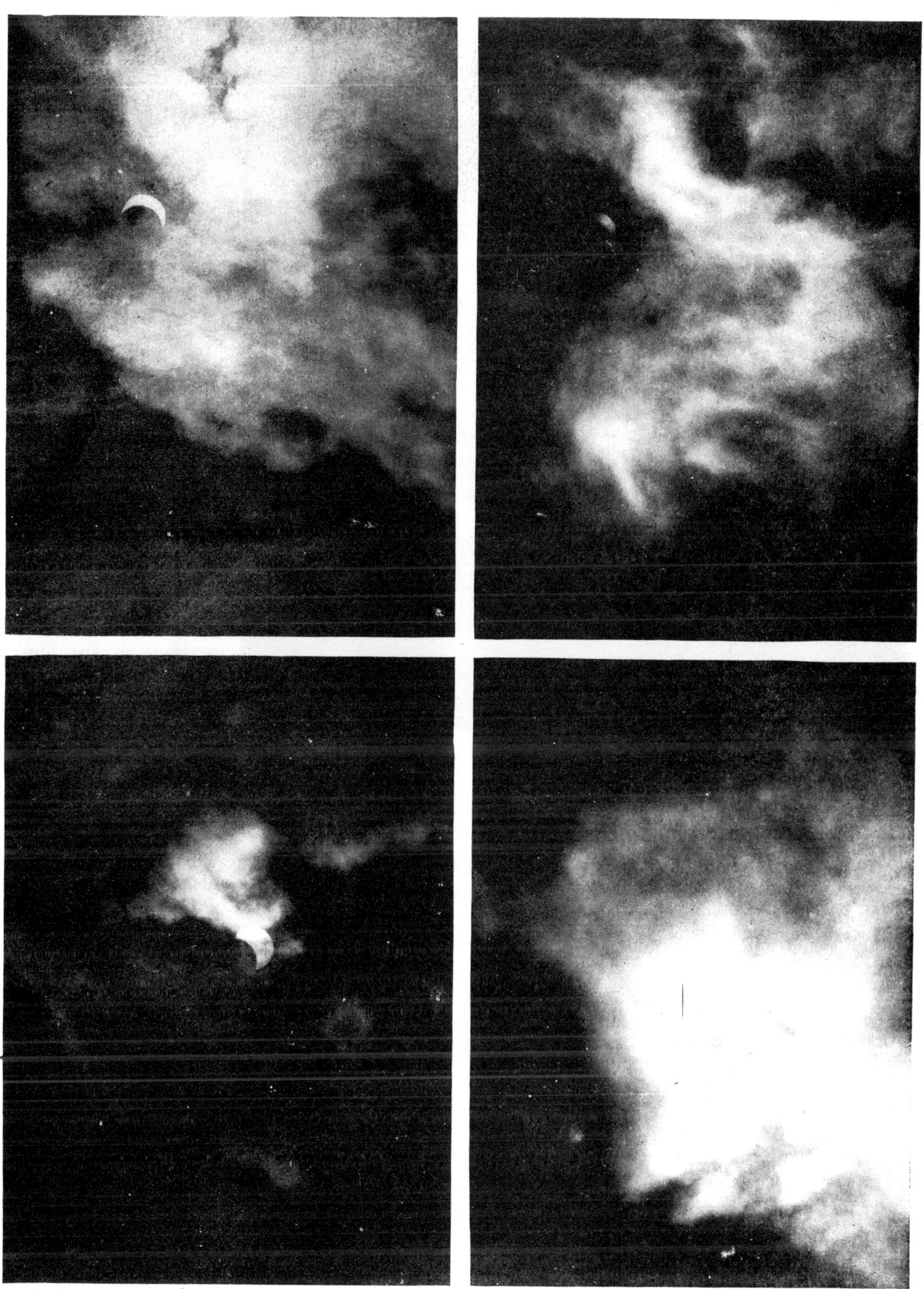

Photos by] [*Priestley & Sons.*

THE ECLIPSE OF THE SUN, AUGUST 30th, 1905.

The thin veil of clouds which was drifting across the sky made the outline of the Sun during the partial eclipse very visible. The times at which these photographs were taken are (commencing with the upper and reading from left to right)—1.1 p.m., 1.8 p.m., 1.30 p.m. and 2 p.m.

THE CORONA OF MAY 28th, 1900.

The Corona, or halo of light, usually appears a few seconds before the total extinction of the Sun's light and continues visible until a few seconds after the luminary first reappears. It is in the nature of an atmosphere surrounding the Sun and extends outwards for several millions of miles.

the Sun has almost, but not quite, blocked out the last trace of the Sun itself, just as it does so the forward edges of the two bodies exhibit alternate points of light and darkness which are known by the name of "Baily's Beads," so-called from the English astronomer who first particularized them, though he was not actually the first to record them. It is generally considered, though the explanation is not perhaps wholly satisfactory, that they are due to the mountains of the Moon blocking out absolutely portions of the Sun's surface, whilst other portions of that surface not so obstructed peep through the valleys between the different lunar mountains. Almost immediately after the disappearance of the Beads, which of course is very sudden, red flames flash out at various points round the circumference of the Moon. These "prominences," as they are called, have been dealt with in an earlier article (see page 1). But the most interesting phenomenon connected with these eclipses is the "Corona," or halo of light which can be observed surrounding the black disc of the Moon. It usually appears a few seconds before the total extinction of the Sun's light, continuing visible during the whole period of the Sun's absence and for a few seconds after its reappearance. Its nature is probably that of an atmosphere surrounding the Sun and stretching outwards to a distance of several diameters of the Sun; in other words, to several millions of miles.

There is one peculiarity connected with the Corona too important to be ignored—it varies in extent and outline from time to time, seemingly dependent for some occult reason upon the prevalence or absence of spots on the Sun. These, it is well known, vary in number during a clearly recognized period of about eleven years. If a total eclipse occurs when the spots are at or near a minimum the Corona extends in a marked manner right and left of the equatorial diameter of the Sun. If, on the other hand, an eclipse is coincident with a period of maximum spots, the Corona is contracted on both sides opposite to the equatorial diameter of the Sun, and is much more compact taken as a whole all round the Sun's circumference. These differences seem well established, but a satisfactory explanation is not yet forthcoming.

One other concomitant of a total solar eclipse may be mentioned. The Sun being on the farther side of the Moon, the Moon casts a shadow into space : that shadow strikes the Earth as the eclipse progresses. Observers who are very adroit with their eyes may often catch sight of the shadow approaching the place where they are standing, passing over them, and then passing away on the opposite side. The sensation is very weird, but so momentary that it often escapes notice. It is momentary because the speed with which the shadow traverses the Earth's surface is one thousand eight hundred miles an hour, or about thirty miles a minute.

In an eclipse of the Moon the total phase comes on very slowly during a period of, it may be, an hour or more, and when totality has arrived, though the Moon usually becomes wholly invisible, such is not always the case, for sometimes our Satellite's disc can be traced, exhibiting a ruddy hue. This is a matter which seems to depend upon the meteorological condition of the Earth's atmosphere.

PORCUPINES AND THEIR IMITATORS

BY R. LYDEKKER, F.R.S.

THE partial or entire conversion of ordinary hairs into bristly spines, and these again into stout horny spines of great strength, is such an obvious and easy method of obtaining immunity from the attacks of a large class of foes, that it can scarcely be a matter of surprise to find that it has been independently resorted to by several totally distinct groups of warm-blooded quadrupeds. And, as if to show us how the process has been accomplished, there is an almost complete gradation from

ANOTHER TYPE OF CORONA.

Notice the difference between this and the Corona on the opposite page. The variation in the shape of the Corona is due presumably to the prevalence or absence of spots on the Sun. If a total eclipse occurs when the spots are at or near a minimum the Corona extends to a much greater distance left and right of the equatorial diameter, as is shown here. If, however, the spots are at a maximum the Corona extends more or less equally all round the Sun's circumference

ordinary soft-haired mammals to the complete *cheveux de frise* with which the " fretful porcupine " is equipped.

In the deserts of Syria and North Africa there are, for instance, certain " spiny mice," whose bristly coats, doubtless, render them an exceedingly unpleasant morsel to owls and other birds; while there are likewise spiny rats in South America. Again, the Tree-Porcupines of the latter country, in which the spines partake largely of the nature of bristly hairs, serve to lead on to their more spiny Canadian relative; while much the same thing may be observed when the relatively small Brush-tailed Porcupines of the Malay countries and West Africa are compared with the larger true Porcupines of the Old World.

But Porcupines are by no means the only mammals which have developed a complete spiny armour; for we have a familiar instance of another such development in our own hedgehog and

Photo by] [*W. S. Berridge, F.Z.S.*

THE CANADIAN PORCUPINE.

Although a tree-climbing species, where one would expect to find a long tail that could be used for climbing, the Canadian Porcupine is possessed of a short, thick tail, and is besides very heavily built. It is a voracious creature, and will entirely strip a tree when the foliage is to its liking.

its Old World relatives, which have nothing to do with Porcupines properly so-called, but are more nearly akin to moles and shrewmice. Madagascar, again, has a spiny animal in the shape of the so-called hedgehog-tenrec, which is likewise a member of the shrewmouse order, although belonging to a peculiar group restricted to the great African island. Neither have Australia and New Guinea been left out in the cold in this matter, since they are the home of the so-called Porcupine Ant-eaters, of which more anon.

Judging from one's own feelings, immunity from attack would seem to be somewhat dearly bought by the acquisition of a spiny coat like that of the Porcupines; but as the creatures have been brought up to such a garb, they are obviously used to it, and, therefore, do not pine for the velvety dress of the mole.

" Porcupine," or " Porpentine," as it was once spelt, is a corruption of *porcepin*, or " spiny pig," the South European name of the typical representative of the whole group. That the creature has nothing to do with swine will be obvious to anyone who takes the trouble to look at a Porcupine

Photos by] [*W. S. Berridge, F.Z.S.*

CURIOUS PORCUPINES.

The upper photograph is of a Brazilian Tree-Porcupine, the lower of the Brush-tailed Porcupine. The first, like most of the Tree-Porcupines, has a long tail, which the owner can curl round a branch and thereby hang suspended. The Brush-tail is a West African species, and is much smaller and more rat-like than the Porcupine proper.

Photo by] [W. Saville Kent.

THE ECHIDNA, OR PORCUPINE ANT-EATER.

Dissimilar in every other particular, the Echidna, like the Porcupine, possesses parti-coloured spines. Besides the platypus it it is the only other egg-laying mammal. The single egg and the newly-hatched young is carried by the mother in a pouch developed on the underside of her body

at the Zoo, and observes the pair of chisel-like teeth with which the front of each jaw is armed. As a matter of fact, Porcupines are akin to rats, beavers, and other members of the rodent order. But there seems a popular tendency, due, perhaps, to their mobile snouts, to call rodents pigs, as witness the familiar " Guinea Pig."

The Porcupine is one of the largest of rodents, weighing something like twenty pounds ; and since practically everyone is familiar with the marvellous armature of black- and white-banded spines, it will be unnecessary to describe them, although it is important to mention that certain wholly white quills in the tail are hollow and open at the ends.

As a Porcupine walks, the quills in its tail make a loud rattling, and there has been considerable discussion as to whether this noise is of any advantage to the animal. It is, however, now the opinion that its function is to serve the purpose of a rattle, and thus to warn other creatures of the presence of a dangerous animal.

To refer to all the different groups of mammals which develop a panoply of spines in their hide would spin this article out to a greater length than permissible ; and my few remaining remarks must therefore be restricted to the Echidnas, or Porcupine Ant-eaters, of Australia and New Guinea.

Except in possessing parti-coloured spines, which are comparatively short, Echidnas are utterly unlike Porcupines, having long, beak-like, toothless snouts, short, squat bodies, rudimentary tails and very stout limbs, armed with powerful claws for digging. In the large Bruijni's Echidna of New Guinea the beak is much longer than in the typical species, while it is also nearly as much arched as that of a curlew. In this species, at any rate, when standing the body is elevated on the columnar legs in much the same fashion as that of an elephant.

Echidnas, as mentioned in another article, are the sole living relatives of the platypus, and thus the only other mammals which lay eggs. In place of depositing her single egg at the end of a long burrow, like the platypus, the female Echidna carries it in a pouch developed on the under-side of her own body during the breeding season ; this pouch likewise forming a warm and snug nursery for the young.

As mentioned in the article on Ant-eaters, where additional particulars in regard to Echidnas are recorded, the latter appear to feed exclusively on ants—either black, red, or white.

I have headed this article " Porcupines and their Imitators," but it might perhaps have been better had I written Echidnas in place of Porcupines, as it is perfectly evident that the former represent a much more ancient group than the latter. The ancestral Echidnas were, however, almost certainly provided with teeth, while it is quite likely that they were devoid of spines. Consequently, it is possible that Porcupines developed spines earlier than Echidnas, although this is one of the many problems in nature which we cannot solve—at any rate, in this world.

THE PEA SEED PEST

BY HAROLD BASTIN.

THERE is a small beetle—a kind of weevil to be precise—which does much mischief to peas. Its life-story presents some points of unique interest. The adult female insects are alert at the season when peas are blossoming. They lay their amber-coloured eggs on the pods when the latter are very young—in fact, as soon as the flower has withered. The minute grub which hatches from the egg has six short legs, while its thorax is armed with a series of cutting spurs. These spurs assist it in making its way through the substance of the pod to one of the swelling peas. Sometimes it burrows straight through the pod, from the egg to a pea. But if it hatches before the peas are sufficiently developed for its purpose, it lives for a time as a miner in the pod. Sooner or later, however, it penetrates a pea. The wounds made in the pod and the pea are so slight that they soon heal over, while the scars can only be detected by diligent search. Up to this time the tiny larva

Photo by] [*W. S. Berridge, F.Z.S.*

THE EUROPEAN PORCUPINE.

The finest quills and the most formidable array of these sharp weapons are to be found on this species. A remarkable point about the Porcupine is that as it walks the quills in the tail make a loud rattling, which possibly serves as a warning signal to other animals.

has been capable of rapid movement. But when once it reaches its goal within a pea, it changes its skin, and with its skin go its legs. It is transformed, in fact, into an inert maggot. Meantime the pod and its peas continue to grow as if nothing out of the common had happened. Nevertheless, the thief has forced an entry and does not remain idle. It feeds within the growing pea—cautiously, we must suppose, for its gnawings seem in no way to affect the development of the seed.

Only one grub is found in a pea. Moreover, the pea provides more than sufficient food for the creature's needs ; so that when the insect at length changes to a pupa, the interior of the pea is only partly demolished. Before transforming to the dormant pupa state, the grub eats its way to the rind of the pea, leaving only the thin outer skin stretched across the end of its burrow. Thus, when the beetle is mature, it has only to push through this diaphragm in order to make good its escape. It has been shown experimentally that the beetle would be unable to liberate itself but for the instinctive preparations made by the grub.

Photo by] [*Harold Bastin.*

THE PEA WEEVIL.

The adult insect is about one-fifth of an inch long. Its ground colour is black, but this is more or less hidden under a brownish or greyish down. The female lays her amber-coloured eggs on the tender pods of the pea as soon, in fact, as the flower has withered.

Some of the beetles emerge in the autumn. But a large number of them, probably the majority, lie up in the peas throughout the winter. Many do not appear until after the peas have actually been sown. This means much to the practical gardener, because many beetle-ridden peas, being only partially destroyed, still retain the power to germinate. But such seedlings are naturally weak, and few, if any, of them develop into good plants. It follows that all peas which contain beetles should be discarded. How are they to be known ? In the case of peas from which the insect has already emerged, the tell-tale exit hole is sufficient evidence. But when the beetle still lies perdu within the pea, detection is not so simple. Such peas, however, may be known by a round spot on the surface, slightly different in colour from the rest of the rind—that is, by the patch of thin skin beneath which the beetle lies. This spot must not be confounded with the circular depressions, or dimples, natural to many peas. A little practice enables the gardener to recognize the sign-manual of the beetle pest.

Many gardeners aver that by throwing seed peas into a vessel of water, those which are infested may be at once discovered because they will float on the surface, the sound peas sinking at once to the bottom. Experiment has shown, however, that this statement is not wholly trustworthy. Probably the only safe plan, when valuable seed is concerned, is to examine the peas separately. In this way one may assure oneself that each pea sown has at least a fair chance of developing into a strong plant.

The adult pea-beetle is black in colour, more or less covered with brownish or greyish down. It has a very near relation, which attacks broad beans—two or more larvæ being sometimes found in

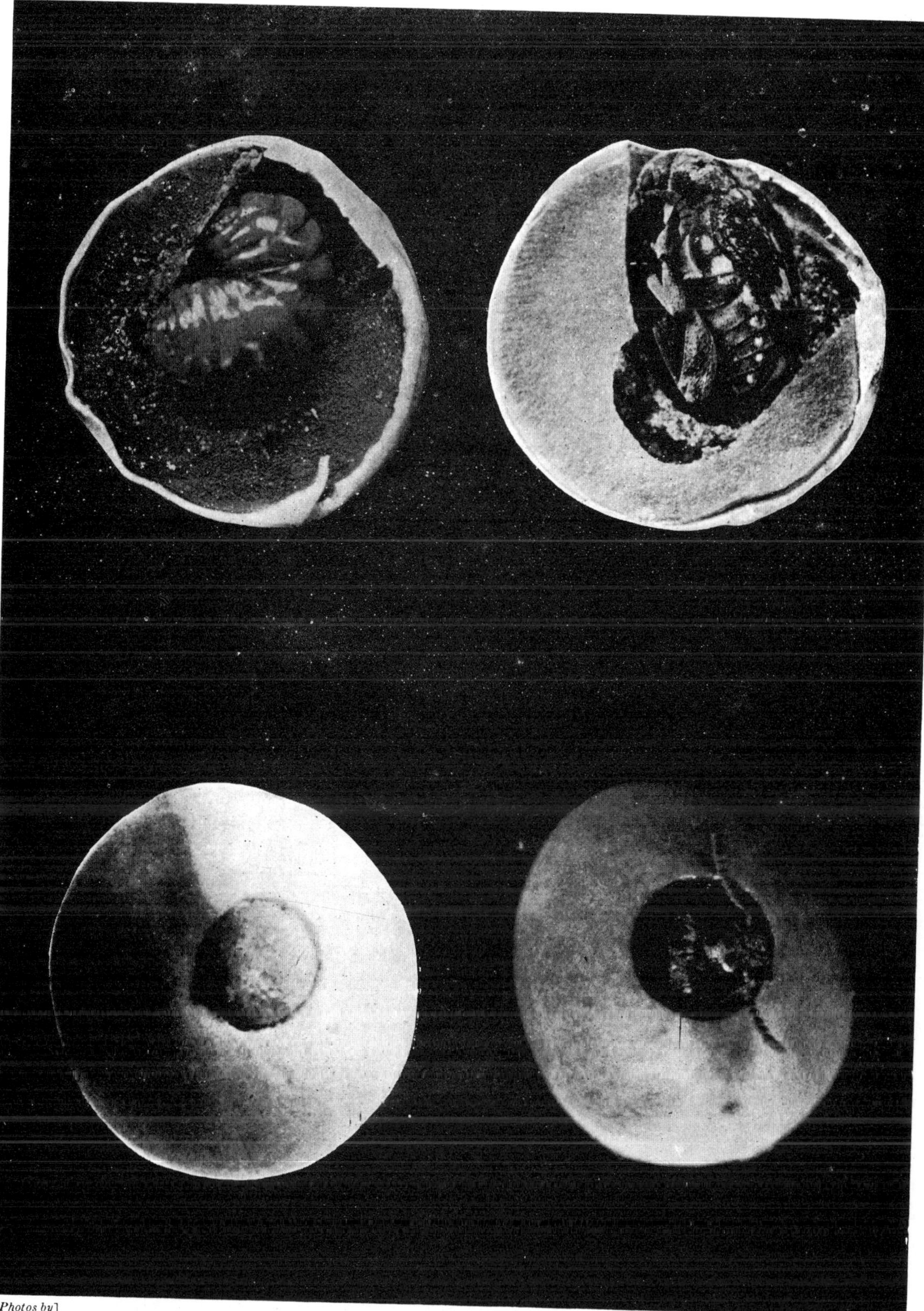

Photos by] [*Hugh Main, F.E.S. and Harold Bastin.*

THE PEA WEEVIL.

Taking the upper photographs first and reading from left to right, the first shows the grub of the Weevil in the pea. Notice how it has bored right up to, but has left intact, the rind of the pea. In the second photograph is the perfect beetle almost ready to emerge, and in the third the exterior of the pea is shown, with the round, discoloured and slightly raised patch which indicates that the beetle will soon make its appearance, as it does in the fourth photograph

one bean. In tropical countries allied species attack seeds of various kinds. The larvæ of some, it is said, are found in the cocoa-nut. It is a remarkable fact that many larvæ of this family are content with a very small amount of nourishment. For example, the portion of the pea eaten by the pea-beetle grub is only slightly larger than the full-fed grub itself. On the other hand, certain species require two seeds to enable them to complete their growth. One such species is said to migrate, when half grown, from one seed-pod to another—laboriously dragging itself along the ground by means of its mandibles in order to effect its change of residence.

BUTTERFLY DISGUISES

The theory that animals are capable of imitating for themselves the nature of their surroundings in order to conceal their identity is now almost universally accepted, and examples of protective coloration, as this characteristic is called, are to be found on every hand. Insects possess this faculty in a marked degree, and much careful study has been given to the many methods of protective coloration they have adopted.

Now, to a large extent, the world of an insect is a world of flower and foliage, of blue sky and shadowy boskage, and it naturally results that the prevailing forms and colourings of such a world are assumed by the insects themselves.

Mention of protective coloration has already been made in these pages, and reference to the article on "How Nature protects Insects" (see page 351) will show that this method of concealment is, to all appearances, more general among moths than among butterflies. It is well to add the proviso "to all appearances." The closer observation of the ways of moths in this respect is due in a large measure to the greater facilities of watching them. During the day they remain in a state of quiescence, with folded wings, and clinging to the bark of trees, old wood fences, or other hiding-places, where the dark tones and markings of their upper wings against the uneven and

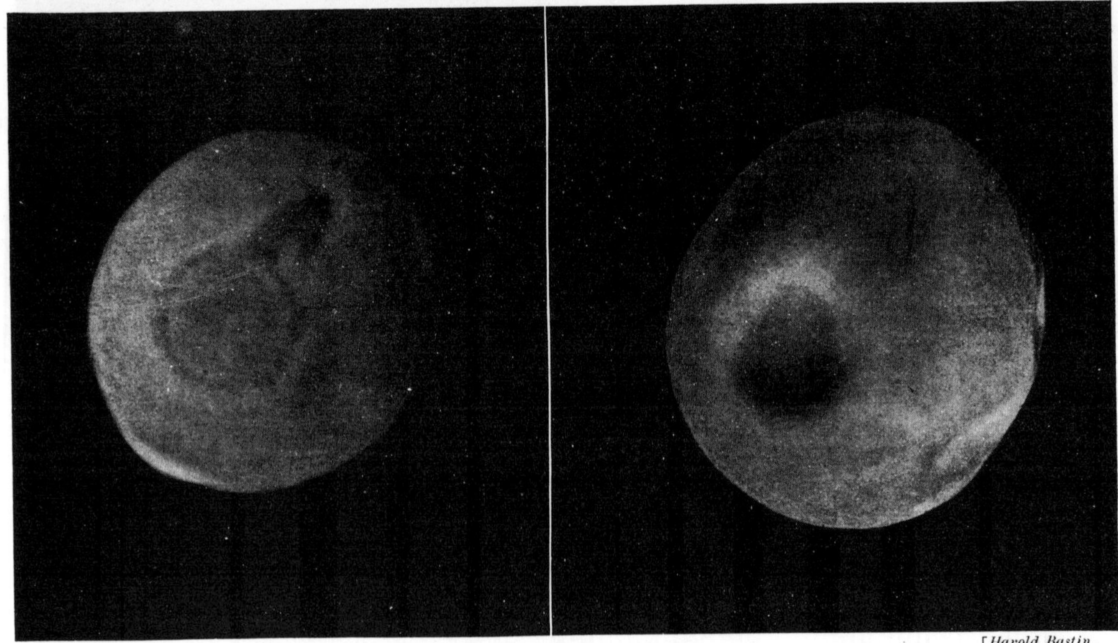

Photos by] [*Harold Bastin.*

THE WORK OF THE PEA WEEVIL.

The first pea has been attacked; but the small discoloration is the only sign of the enemy that lurks within. It must be carefully distinguished from the second pea, which is entirely unaffected, for dimples such as are shown here are frequently to be found in stored peas and will disappear as soon as the pea begins to germinate

sombre background render them invisible to other than accustomed eyes. All this time the Nature-lover can be searching them out in their accustomed haunts and taking notes undisturbed.

But butterflies are abroad in the sunlight, and an entirely different scheme of colouring is necessary to render them invisible. Of course, there are some which appear to scorn danger, and flash their colours as they fly; but there are many others which take advantage of Nature's secret alchemy to render themselves as immune as may be from attack. The little Blues, which frequent open meadow-land and are such children of the sun, have wings that might very well be fragments of the summer skies themselves, for in them is the same deep, clear colour, while even the vague shimmer of a summer haze is imitated by the long silken plumage which covers them. Watch these little summer sprites as they flutter in the air, watch them as they dance giddily over the meadow. How soon they are lost to sight! They seem to melt into the sky and in a twinkling become invisible as the atoms of the atmosphere. There is the secret of their colouring: their blueness renders them invisible in their flight. A similar case of invisibility is afforded by the Apollo, only here the effect is produced by the transparency of the wings, which are as fine as gauze, slightly marked with a few grey pencil-lines, and one or two well-defined red eyes bordered with black on both upper- and under-wings. These eyes, however, are inconspicuous in flight. Another butterfly with strong protective coloration is the Orange Tip, with wings that on the under-sides are strongly mottled with green and white. This at first glance would appear to be anything rather than of a "protective" tendency; but wait till the butterfly is resting on a flower, such as the wild parsley, where innumerable white petals and little green sepals are clustered together; then it is impossible to distinguish which is flower and which is butterfly.

THE COMMA BUTTERFLY.

The ragged appearance of the wings and the dull grey-brown of the under-sides admirably imitate the dry and withered appearance of a fallen leaf.

The Green Hairstreak, again, is a butterfly which in flight has the appearance of burnished bronze; but at rest the upper-side of the wings is hidden and only the underneath is visible. The surface is thickly powdered with a metallic green bloom, which so nearly approaches the colour of foliage that the Green Hairstreak is then very difficult to discover. An instance was brought to the writer's notice of a beech hedge where a great company of Hairstreaks was congregated. The butterflies arrested the attention as they danced hither and thither in the sun; but when, one by one, they settled down among the green beechen leaves, they disappeared as completely as though they had been phantoms.

Some of these clever mimics—and this is especially the case with such butterflies as come out in the autumn—adopt the appearance of withered leaves. Two good examples are here illustrated.

One is that of the Comma, whose under-side is a speckled grey-brown, and the other is the pale Brimstone. In the first, the uneven edge of the wing accentuates the ragged appearance of a leaf. This is a not unusual form of mimicry which many species have adopted, including the Tortoiseshell and the Peacock. The Brimstone was photographed when hibernating in a holly bush; very cleverly it was concealed, for the dull yellow colour and the pointed edge of the wings produced a perfect resemblance to a sapless holly leaf which was ready to fall from the bough.

But the most striking example of protective resemblance is that of the Indian Leaf Butterfly. The expanded wings are a beautiful combination of dull orange and dark golden browns, with a strong violet iridescence; but when the wings are closed, as in the illustration on the opposite page, the similarity to a dried leaf is remarkable. The dull brown colour, the strong veining, even the little discolorations such as are to be found on a decaying leaf—all have been carefully imitated; while the butterfly's under-wings are provided with a short tail, which when the wings are closed, does duty for the short stalk attaching the leaf to the stem, and completes the deception.

Photo by] [*A. E. Tonge.*

THE PALE BRIMSTONE BUTTERFLY.

The Butterfly is here shown hibernating in a holly bush, where the sharply-pointed wings and the pallid hue of yellow-green resemble a faded holly leaf; the drooping position of the wings also assists in the deception.

THE CRUST OF THE EARTH

BY EDWARD A. MARTIN, F.G.S.

IMMEDIATELY the earth was sufficiently far from, and disconnected with, the sun to radiate away from itself more heat than it was generating or receiving, there commenced that slow and tremendously gradual process known as the formation of a crust. You can watch the formation of a crust in a very homely way, by observing the cooling at the surface of a pot of cobbler's wax, or on the top of a pannikin of porridge, when the heat beneath is slowly withdrawn, the crust forming as soon as it fails to receive as much heat as it is sending off into the air above. But Nature was more irregular in her actions. The seething mass of molten rock, as it was whirling round in space, and whirling round the sun at the same time, whilst it already began to bulge at the equator, was subjected to great waves of a tidal nature; and as soon as a crust was formed in one place, the pressure beneath of the shifting contents would here and there break through the tender crust again and again, and only after the lapse of an enormous period of time would the loss of heat cause such a crust to form as to enable it to make a secure stand against the tremendous assaults made upon its internal concavity. Even now, as we know, the crust is still liable to be broken through, and the weak parts of the earth's surface follow well-recognized lines which contain the world's volcanoes.

When once the crust-forming influences obtained the upper hand, nothing was possible that could interfere with the process going forward. Whatever rival schools may say, the earth did cool, and a crust did eventually form. Sometimes, no doubt, portions of the crust would disappear into

THE INDIAN LEAF BUTTERFLY.

The best-known and most striking example of leaf-mimicry. The Butterfly with wings expanded is both gorgeous and striking. The tips are a velvety-brown that merges into a broad band of dull orange; this in turn is shaded into a blue-brown—the principal colouring—which often appears as a rich glowing purple owing to vivid iridescence. When the Butterfly is at rest, however, it can scarcely be distinguished from a dried leaf, so exactly do the under-sides of the wings simulate the shape and the dull, withered appearance, while even the delicate nervures imitate a leaf's veining.

Photo by] [*F. F. & Co.*

The once-horizontal strata at Stair Cove, Dorsetshire, have been folded so that portions have assumed a vertical position.

the boiling cauldron below. The results of millions of years might be undone in a short time, as we measure it. Thousands of square miles might be swallowed up almost in a moment. But the process, begotten of the radiation of heat, could not cease. Time was ever on its side. What was lost here was gained there, and in the end a recognizable crust appeared. Watch again your pannikin of porridge. Turn on the heat supply again but slightly, shift its position, and your miniature volcanoes will appear. With the application of a moderate supply of heat these volcanoes will disgorge their contents, and lava-flows will cover the whole, only again to form a thickening of the original crust when the heat-supply is diminished. So in the far-off early days of the earth's crust fearful eruptions of molten rock material took place, and covered up the early-formed crust, until, as we look at them now, metamorphosed as they are, it is really impossible to say, in the case of the earliest layers of the crust, which are igneous rocks and which are altered sedimentary rocks.

We may be sure, however, that as soon as there was an appreciable crust at all, then the effects of running water were seen. Boiling water it may have been. The complex atmosphere above the crust would be densely charged with steam, which would be discharged on the crust as water—laden, of course, with mineral contents; and the water, as soon as it commenced to flow, would carve its channels in the crust, and carry away the sediments it had collected to a happier spot, where the crust was sufficiently thick to receive them. So, fire-formed rocks and water-formed rocks would together be in process of formation, and the two would be almost inextricably intermingled where we examine them to-day. Can anything

Photo by] [*R. Welch.*

Ancient rocks of carboniferous limestone on the banks of the Boyne, which have been squeezed by tremendous lateral pressure, and have yielded as though composed of quite soft material.

be seen anywhere nowadays of this earliest-formed crust? Probably not. The process we have described lasted for so long a time, and the earliest crust may have so frequently become involved repeatedly in the general melting-pot, that we cannot safely say of any definite rock that this is a part of the fundamental crust. Most geologists say that the fundamental rock can nowhere be seen. Some American geologists point to a great series of rocks, known as the Archæan rocks, in America, and suggest that here we have what represent portions of the original crust. But only portions, for there can be no doubt that the original surface has long since been dismantled, only the lower portions appearing after being laid bare by erosion, whilst the eroded materials have gone to form the earliest sedimentary rocks (Keewatin) resting upon them. These Archæan rocks are, anyway, the most ancient that have anywhere been discovered. They are contorted, doubled up, and foliated into extraordinary shapes, and the material of which they were composed has been exposed to such intense heat that their structure has been changed, and their crystalline condition has obliterated all that might give us any clue as to whether life of any kind existed in these very early days of the earth's history. Gneisses and schists of all kinds, which may have been sediments, have been cut into by other rocks of a volcanic nature, and the whole have been so crumpled and crushed that the interlocking masses are not separable from one another. There is a thickness of these unfossiliferous rocks in the Grand Cañon of the Colorado of no less than a third of a mile, and this is probably a good deal less than there was originally. This gives us a vivid idea of the length of time during which this formation was being laid down, and during this time, so far as we know, no life had appeared upon our earth. But in the

Photo by] [*C. J. Watson.*

Crystalline vein filling up an ancient crack in igneous rock at Oidfjord Vaud.

Photo by] [*R. Welch.*

Great folds of crushed and contorted quartzite much penetrated by volcanic dykes and diorite "sills."

Keewatin rocks there is so much carbon that it is probable that at least seaweeds had begun to grow.

Above these rocks, in the same Great Cañon, there rests another great formation, the Algonkian, and this in some places is itself scores of thousands of feet thick. Nor was it deposited uniformly on the Archæan. The latter masses had in the interim become raised out of the sea or out of reach of great volcanic influences, and had formed possibly a part of a continent. Denudation of its surface had planed away much of its irregularities, and the crumplings and contortions were scarcely visible. We cannot tell how long this old land-surface lasted. We only know that in the course of time it underwent the rigours of an Ice Age, before it went down beneath the sea again, and on its planed-down edges the great Algonkian series was laid down unconformably. In the Great Canyon there are no less than twelve thousand feet of these rocks, which are themselves divided about half-way by a great unconformity, and contain several sheets of lava interbedded with them. The two great formations which we have mentioned were formerly known as the Laurentian system, and they cover an area which is estimated at two million square miles. In Britain the rocks we have been dealing with are represented by the Lewisian gneisses of the Hebrides and the Torridonian of Scotland.

Photo by] [*R. Welch.*

Old Red Sandstone boulder supported on a pillar of rock at Kenmare, Kerry. The boulder has protected the pillar from denudation, whereas the surrounding rock has disappeared.

In passing on to the well-recognized Cambrian rocks, we reach a time when Mother Earth had to some extent settled down from the early and impetuous days of her hot youth. If the rocks we have dealt with were now piled upon one another, and a further series were piled on the top of these, representing a length of time about equal to all that had gone before, we should have a towering cliff about fifty miles high. In imagination we might see a picture something like that shown in our coloured illustration. But in real fact this exists nowhere. At no place are all the formations to be seen together. But the sequence is one which is never departed from. The order is always the same. The base of the Archæan has never been seen, and it probably stretches at varying distances below every formation on earth. It is the bottom of what has come to be known as the Crust of the Earth, and this has been put by varying authorities of recent years at between twenty-five and

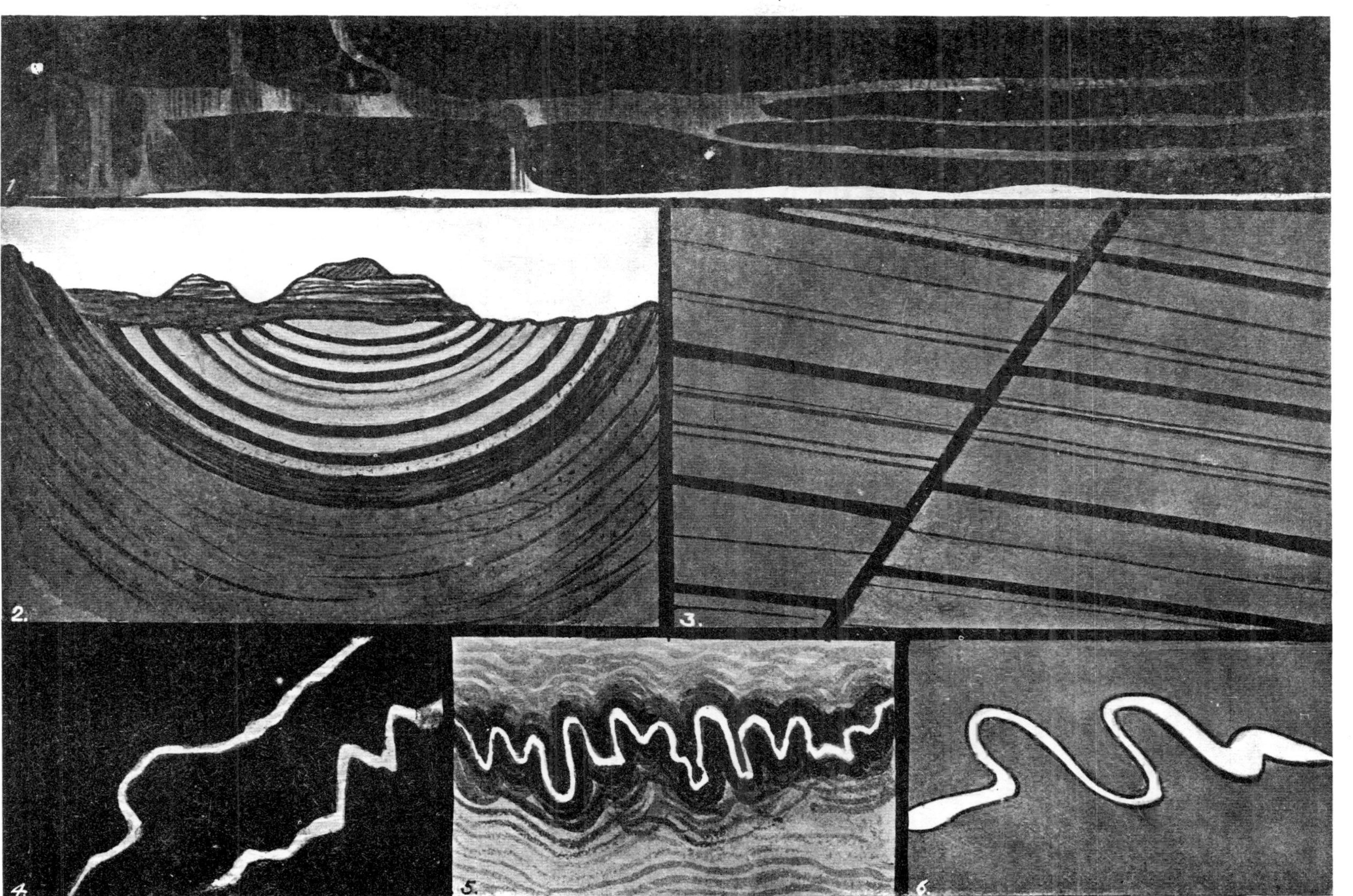

THE EARTH'S CRUST.

Some of the phenomena which are to be observed: (1) Sandstone rock in the Isle of Skye pierced by three intrusive veins of igneous material, all having their origin in one dyke. (2) Bristol coalfield showing coal-measures in a trough, bounded by ancient rocks on each side and overlain unconformably by New Red Strata. (3) This fault or crack in the coal-measures has resulted in the coal-seams being thrown down out of their normal position. They give much trouble to miners. (4) Granite veins in Micaceous Sandstone; in 5 the vein conforms to the layers of the Mica-schist, while in 6 the vein is contorted and cuts across the cleavage of the slates, which cleavage is brought about after the intrusion of the vein.

Photo by] [*H. J. Shepstone.*

An old line of sea-cliffs at Wady Firan, from which the sea has long since retired.

forty-five miles. Roughly speaking, we may say that we live on a thin crust which is about one two-hundredth of the earth's radius. As to what supports that crust, a great change has come over the views of the scientific world in recent years. Considerable light has been shed upon this subject by observation of the paths of waves set about by earthquake shocks. It is undoubtedly a fact that the centre of the earth contains a solid core of heavy metal, possibly to a great extent nickel-iron, such as is found in meteorites, but from the manner in which shocks lose their velocity therein, or even are completely lost, it is possible that the real centre is of some less-conducting material, and this centre may be as much as two-fifths of the whole radius of the globe. But above the metal sphere, the " barysphere," there is room for that mass of highly-heated, molten material which finds its way at times to the surface through the weak places in the crust which give rise to volcanoes.

So, then, at all events the crust is not so very thick after all. There is one central fact which seems to be retarding the thickening of the crust. Whatever may have been the extent to which heat may have been radiated into space in the ages that are gone, it is an undoubted fact that radium and its allies are now making up for any loss which the earth may be now suffering. We have it on good authority that below forty-five miles there is no radium; and as nickel-iron meteorites contain no radium, it is possible that here we begin to come within the range of accumulated masses of melted meteoritic matter. But the radium which is found in our thin crust is, in the opinion of some of the best authorities, sufficient to make up for the loss of heat which goes on by radiation. In fact, a state of equilibrium has been

Photo by] [*R. Welch.*

Tops of five-sided and six-sided columns, forming a portion of the Giant's Causeway, Antrim. The rock is igneous dolerite. The concave tops of some of the columns act as salt-evaporating pans.

reached. We need in that case no longer fear that we may soon reach a Mars-like condition, when the greatest care is necessary to conserve the water-supply. We need not feel apprehensive lest we are fast racing towards a condition such as we see in our own satellite—the moon. We shall not die of refrigeration. One authority goes so far as to say that our radio-active substances will not only prevent that, but may actually bring about the danger of such heat as will possibly result in catastrophic action of the earth's surface. It is strange, indeed, that all the activities of human life are carried on at the surface of so thin a crust, and stranger that we still know so little of what is going on more than a mile from the surface.

Photo by] [*Mrs. Aubrey le Blond.*

In this—the Aletsch—Glacier there is a central moraine of rock-debris which has flowed down from the mountains, and is deposited with boulder-clay on the lower grounds, where the glacier melts. The valley is scored with striations caused by blocks imprisoned in the ice.

We have already described what an unconformity is. Such an one as occurred between the Archæan and the Algonkian again appears now at the bottom of the Cambrian rocks. An immense length of time must have elapsed before these came to be laid down upon the upturned edges of the Algonkians. With the Cambrians we can turn to those more familiar British areas where the rocks were first noticed and studied. But what we cannot see in Britain are the clear and distinct evidences which appear in Norway and in far-off China of an Ice Age having actually occurred before they came to be laid down. The Cambrians give us clear evidence that the forces of nature which we see going on around us now were then fully in evidence also. Rains and frosts, winds and water-action carved out the land into hill and dale. Above the Cambrian rocks we find next the Ordovician, and above that the well-known Silurian

Photo by] [*C. A. Byers.*

This view of the Great Cañon illustrates the gradual lowering of the upland plain by the carving of the river through the rocks, which were subsequently degraded by atmospheric agencies.

Photo by] [*E. A. Martin, F.G.S.*

The raised beach at Brighton now stands twenty feet above the present beach. It rests on chalk, and together with the rubble cliff above it has been raised in human times out of reach of the sea.

rocks. The mountains of Wales are probably some of the oldest mountains in existence. Right away on the very top of Snowdon one may collect sea-shells. Think of the time that has elapsed to enable the rocks containing them to be thrust up into such a position ! And to judge how this came about, we have to consider the great elevatory forces which are hidden from our view. The cooling of the crust has caused in the past whole masses, almost whole continents, to slide as though they were in a liquid condition. Nothing is so solid that it cannot under sufficient pressure be made to flow as a liquid. In Scotland these very early formations have been sometimes doubled into such folds that older rocks have been pushed, for miles, over younger rocks. In the coalfields of France and Belgium older Silurian rocks have been folded over later-formed coal-measures. Such great deformations cannot take place with impunity. Fractures take place beneath such strains, and great thicknesses of rock slide away from those with which they were formerly in contact. Faults result, and in the case of the great Northumberland fault, there has been a sliding of no less than five hundred and forty feet. But it is conceivable that many of the long-existing faults may extend right down through the whole crust, and when these give evidence of their existence by a further sliding, we have earthquakes as a result. The further movement of the well-recognized fault which runs through San Francisco caused the great quaking which led to such disastrous results to that city.

Our coloured illustration will give a better general idea of all the sedimentary formations than could possibly be given in a few words. Above the Silurian we pass to the Devonian, with its beautiful limestone corals, and its Old Red Sandstone equivalent elsewhere, with its wonderful development of armour-plated fishes. Then we see the next phase, in which over wide areas of the earth's surface thick forests of lowly-organized tree life flourished, whose decay has furnished us with our coal-supplies. Above these we have the two New Red Sandstone formations, the Permians and the Trias, in which frequently desert conditions are portrayed, and from which our rock-salt supply is principally derived. In the next age we are introduced to the world-wide dominion of the reptiles, of which the Lias Clay has furnished abundant proof. We have the dominion of the air conquered by the flying reptiles, whose wing-membranes gave a spread, in some cases, of several feet, whilst at the height of their power, the earliest true bird appeared, a menace to the power of the reptiles. Up through the oolites, we have constant types of life appearing

Photo by] [*E. A. Martin, F.G.S*

One side of a mountain one thousand feet high has been gouged out to supply slates of Cambrian age, at the Penrhyn quarries, North Wales. The worker is here seen shaping the slates.

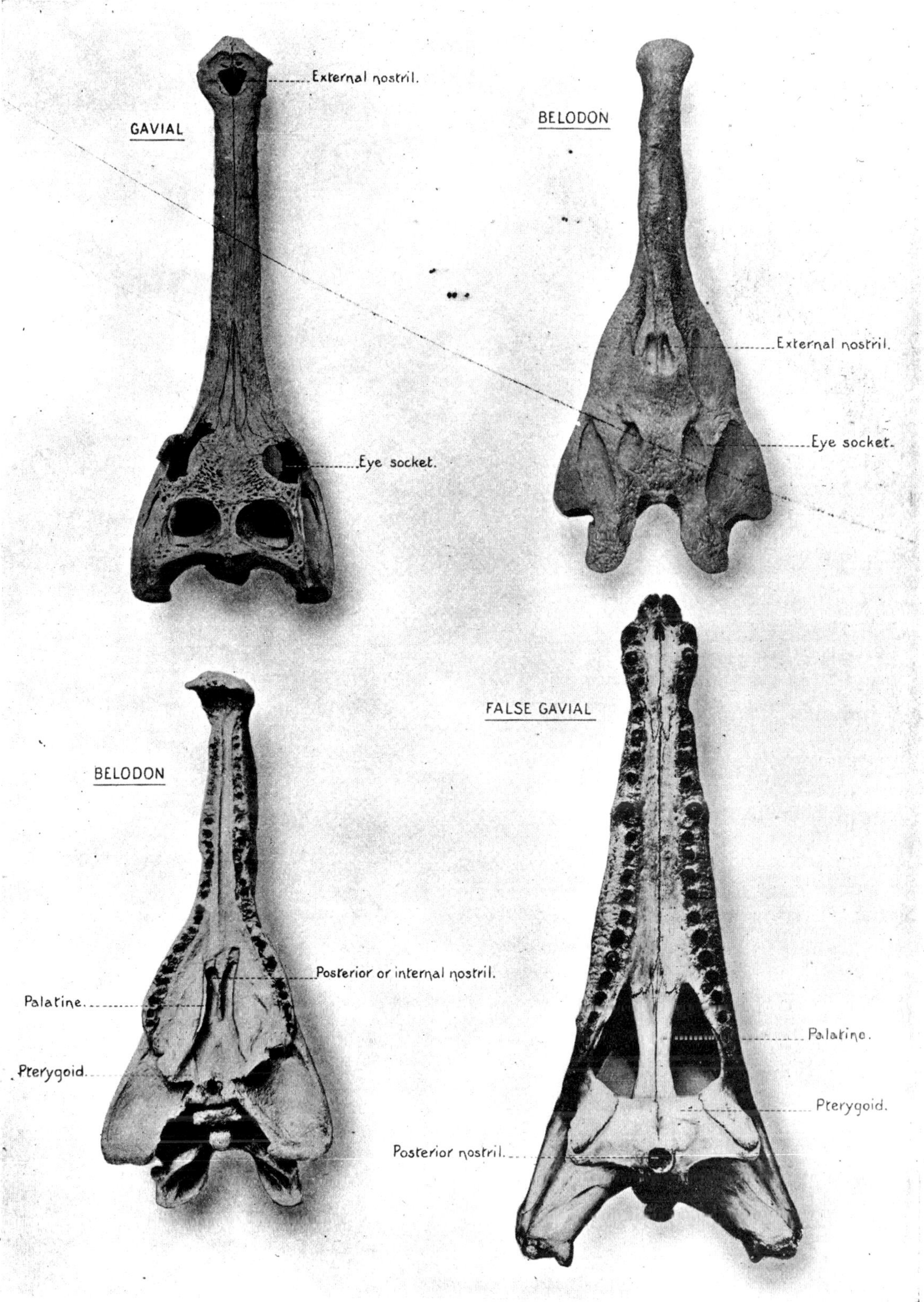

NEW NOSES FOR OLD.

The above illustrations show how the modern Crocodiles and their kin acquired the mechanism which enables them to breathe under water while drowning their prey. In the extinct Belodon the external nostrils, it will be seen, open at the base of the snout, in the Crocodiles and Gavials they open at its tip. In Belodon the passage of the nostrils opened into the mouth by passing straight downwards from the upper surface. Here they were guarded on either side by plates of bone forming ledges: these were the Palatines. In course of time these ledges met in the middle line, involving a similar meeting on the part of the bones next behind them—the Pterygoids. Finally, the Palatines and Pterygoids from forming an open channel formed a closed tube opening at the extreme base of the skull, as in the False Gavial and all other kinds of Crocodiles to-day.

THE GEOSAUR.

A restoration of a marine Crocodile of the Jurassic period. It shows a transition stage in the evolution of the race, for here the limbs are converted into paddles and there are no bony plates on the back.

and disappearing. In these we view the Portland beds, so valuable economically, and interesting by reason of their trees of the cycad type. Then, as we reach the Chalk Age, we see the inrush of trees of the modern type of dicotyledons, living, indeed, within the Arctic Circle, where for half the year all Nature is wrapped in darkness. Very evidently it was not so in those days. Then the Chalk, which for very many centuries had been forming under the sea by the agency of minute foraminifers, was uplifted into dry land, and was also folded into wave-like folds, only again, after great denudation, to be let down into the sea once more, to receive upon it the London Clay and other tertiaries which are now found in the London, the Hampshire, and the Paris Basins.

But a great change once again came upon the scene. The Ice Age had arrived. First covered with glaciers, and finally, by their union, by one great ice-sheet, the land sank beneath its burden, until Britain was lost to view, save a few islands where the peaks of her former mountains showed above the sea. Then after repeated oscillations, she finally rose " from out the azure main," and many of her beaches have since been raised high and dry above the present sea.

The history of the earth's crust is one of frequent movement. Uprisings and subsidences are still in evidence in many parts of the world now. Volcanic action is far from quenched. What has been formerly may be again. Hills and mountains are born but to disappear. But one constant factor alone remains, and that is the shifting, ever-restless sea.

THE EVOLUTION OF THE CROCODILE

BY W. P. PYCRAFT, F.Z.S., ETC.

CROCODILES seem remote enough from the daily lives of most of us who are dwellers in the British Islands, but time was when these creatures swarmed in the rivers which then traversed this land of ours. This we know, because their remains have been found in the London Clay near Sheerness, in Hampshire in the Upper Eocene sands of Hordle Cliff, in the Middle Eocene of Bracklesham Bay, Sussex, the Purbeck stone of Swanage, the Stonesfield slate near Oxford, and the Upper Lias of Whitby. These remains tell us, among other things, that in those far-off days, stretching back millions of years, the climate must have been sub-tropical. They tell us, too, another story no less interesting; and this concerns the evolution of the modern Crocodiles, Alligators and Gavials, which live, happily for us, in regions far remote.

The tale of the coming of the Crocodile would take long to tell if we attempted to begin at the beginning ; we must be content with a fragment or two bearing on the diabolical ingenuity with which Nature has fashioned the skull of the modern Crocodile, so as to enable it to drown its prey without discomfort, to say nothing of the risk of drowning itself. If the under-surface of the skull of a modern Crocodile be examined, it will be seen that the hinder passage of the nostrils, or "posterior nares," has been carried back to the extreme hinder end of the skull, while they open forwards at the extreme tip of the snout ; so that the air necessary for breathing is brought right back to the very mouth of the windpipe in a closed tube. To further perfect the mechanism of breathing, the fleshy palate which forms the roof of the mouth sends down a tightly-fitting flap or fold, which fits on the tongue. Thus it is that the scaly monster is enabled to drown his victim, even though his own mouth is full of water ; for he can breathe readily enough if once in a while he can manage to shove his snout just above the surface of the stream. We must suppose that Crocodiles throughout all the ages adopted a similar method ; but in their case the mechanism was far less perfect, since the bony tube leading to the back of the throat was conspicuous by its absence ; and this passage could have been formed only by fleshy membranes liable at any time to yield to pressure, and either admit water or close up the air-passage. This much the older fossils, such as we have already alluded to, show us. In all these, save the oldest, to be presently described, the anterior nostrils opened at the tip of the snout, as in their descendants to-day ; but the posterior

[*By Theo. Carreras.*

A SPIDER-HUNTING WASP.

It is the mother Wasp who is the hunter. When she has found a suitable spider she stings it, and either paralyses or kills it ; then she hangs it up out of harm's way in the forked stem of some plant while she digs a burrow, into which she finally drags her victim. Then, depositing an egg, she leaves it as a food-supply for the yet unhatched Wasp-grub. A section of one of the burrows is shown in the foreground. The species is known as the Tornado Wasp.

nostrils, so far as the bony skull is concerned, opened into the very middle of the mouth. To get to the very bottom of the story, we have to turn to a very primitive Crocodile known as Belodon, from the Keuper of Würtemberg, a creature nearly allied to a species which has been found in the Elgin Sandstone of Scotland. In Belodon, as is shown in the photograph on page 869, the external nostrils opened near the eyes, instead of at the top of the snout, while the internal, or posterior, nostrils opened directly underneath. In this case it is a matter for debate whether victims could be dispatched after the method adopted by later Crocodiles; perchance, they were killed and eaten ashore.

Photo by] [*J. J. Ward, F.E.S.*

THE SPIDER-HUNTER'S VICTIM.

A paralysed spider has here been photographed hanging on the fork of some grasses, where it has been deposited by the Spider-hunting Wasp out of the way of possible marauders. The huntress has gone off to find a suitable place in which to dig her burrow; she will then carry off her prey.

But how did the bony tube conveying the air to the back of the throat come into being? It will be noticed that the inner, or posterior, nostrils of Belodon lie in a shallow trough, and it is easy to see how, by the gradual approach of the edges thereof, the trough became converted into a tube, which, gradually extending backwards, by the approximation of more and more of the edges of the trough, eventually extended to the very end of the skull, as in the Crocodiles and Alligators of to-day.

All the ancient Crocodiles so far referred to were amphibious creatures, and lived in rivers, deep pools and estuaries, just as do their living relatives. But two extinct species are known which had forsaken these inland waters for the open sea. As a consequence, like the old Ichthyosaurs, and Plesiosaurs, or "fish-lizards," the Turtles, Whales and Porpoises lost the use of their limbs as supports on land by their conversion into paddles. One of these, the Geosaur, is shown in the accompanying restoration; another, very like it, *Meliorhynchus*, has been found in the Oxford Clay of Peterborough.

SPIDER-HUNTING WASPS

BY JOHN J. WARD, F.E.S.

A PRETTY little Wasp, barely half an inch in length, with shining black head and thorax, and an abdomen two-thirds red and the remaining part black, was resting on a thistle leaf while arranging her toilet. She had been so occupied for fully ten minutes, and was now smoothing out her fore-wings, which were smoky at their tips, excepting a conspicuous pale-coloured spot. So far as I know, this particular Wasp has no common name, although it, or some nearly-related insect, is, I think, known as the Path-Wasp. It, however, belongs to a family of the Solitary Wasps, and is recognized by the entomologist as the Spider-Hunter.

[*By Theo. Carreras.*

SPIDER-HUNTING WASPS.

Two species of Spider-Hunters are shown here. The name of the upper species is Trypoxylon, and the female is shown carrying her booty home to the burrow in a tree, where her mate awaits her return. Below is the Robber Wasp, who burrows in the earth; she is dragging her spider to the burrow. Notice how the unfortunate victim is held so that the egs shall not hinder its transit across the rough ground. Neither species is more than half an inch in length.

Photo by] [E. J. Wallis.

PLANT MIMICRY.

A South African Mesembryanthemum that grows where vegetation is scarce and succulent plants unprotected by spines are in danger of being exterminated. This species exactly resembles a pebble to escape notice. It has been photographed among real pebbles to show how closely it mimics them.

Suddenly the sun blazed forth from behind a cloud, when instantly the little Wasp struck up a shrill, piping note and darted into the air. She performed a peculiar jerky flight for a moment or two, and then returned to the thistle leaf again, still singing her high-pitched note for all she was worth. She did not rest, but travelled rapidly along the leaf down the stem to the ground, where she appeared to become very excited, although at first it was not obvious what the excitement was all about.

She moved in and out amongst the leaves near the ground with her feelers curved at the tip and quivering and bristling with activity. Into any little holes in the soil that appeared this lively insect promptly popped her head, withdrawing it almost as quickly; indeed, the whole of her movements seemed to indicate that she was in a most terrible hurry.

In the course of five or six minutes the search had covered several yards of ground, every leaf-fold and crevice being closely examined. Then she met with a stone embedded against the root of a tree, and between the root and the stone she disappeared. While wondering what she was doing beneath the stone, some object appeared at the mouth of the hole by which the Wasp had entered—it was the black and red body of the Wasp herself slowly backing as she pulled out the rightful owner of this retreat, a Wolf Spider.

The Wolf Spider had itself on this occasion been surprised; for the plucky little Wasp had boldly entered its retreat and there attacked it. The spider, finding an enemy within its home, would probably act as if it were dead, and then the Wasp, who is thoroughly acquainted with this manœuvre, would carefully examine it, and being satisfied that it was not so dead as it appeared to be, would seize it and sting it in its nerve-cord, and so paralyze it.

When it was clear of the hole she appeared to become very nervous, continually looking behind her as she hurriedly walked backwards dragging the spider, which was as large as herself. It was obvious that she suspected an enemy was near, for her attention seemed divided between her treasure and her surroundings. After she had travelled with her spider about nine feet, piping excitedly the whole of the time, she suddenly left it and flew into the air, as if taking a bird's-eye view to see if she was clear of all danger. Apparently she was satisfied, for she suddenly swooped down upon her victim and hastily dragged it to one side amongst some grass blades, and then, seizing the base of one, slowly climbed up it, also dragging up her spider. At last she reached a node from which a blade forked, and in the angle she carefully placed her treasure, touching and pulling it here and there to test if it was safely balanced. Then with a parting glance, she flew into the air and disappeared from view.

About fifteen minutes later she returned, but only to look at her prize, for she flew off again almost instantly. Twenty-five minutes later she again returned and rushed hurriedly to her spider, but did not touch it. Instead, she arranged her toilet for some eight or ten minutes. Evidently she had been digging her nest in the sand-bank somewhere near, and was now having a brush-up.

Suddenly she set up her shrill piping and literally tumbled to the ground with her spider, and then she seemed possessed of extraordinary strength, and the spider was dragged at a reckless pace for more than twenty feet, and then laid down while she fluttered her wings and quivered her feelers—perhaps she was taking her bearings. Another sharp pull sideways for a few inches then followed, when she suddenly let go, turned and plunged her head into a neat round hole at the foot of the bank.

A moment later the Wasp and spider disappeared into the hole, the last view of the latter being its vanishing legs. Two or three minutes later the head of the Wasp appeared, and perhaps at that moment she was depositing her egg on the spider's body. Slowly she emerged, and then quickly raked up the sand with legs and mandibles and neatly filled up the hole—so neatly, indeed, that I found it quite impossible to locate the spot afterwards—and then hurried from the place (as if afraid of being seen there) to brush up her wings before taking flight.

So the Spider-hunting Wasp labours for her offspring, which she will never see. Nevertheless, it is provided with an ample store of living food material to supply its wants when it hatches from the egg ; and, if no parasite has surreptitiously placed its egg on the spider, or within the nest, there will some day next season emerge from that hole a red and black active little Wasp like the one whose labours we have witnessed.

PLANT MIMICRY

THERE are many persons who believe that the flowers of certain of our native Orchids offer remarkable cases of the mimicry of insects. Such are the Bee Orchis, the Fly Orchis, the Spider Orchis and the Butterfly Orchis. But just why such mimicry should exist has never been explained satisfactorily. Until such explanation is forthcoming, we tear that these " counterfeit presentments " must be classed as merely fortuitous and having no definite relation to the life of the plant. A much better case is that of the

Photo by] [*E. J. Wallis.*

PLANT MIMICRY.

Another species of Mesembryanthemum that escapes notice by mimicking the angular fragments of rock among which it grows.

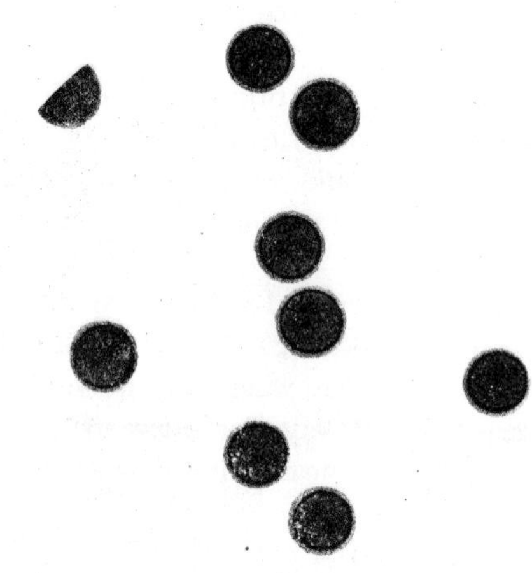

Photo by] [*W. West.*

HOLLYHOCK.

The pollen grains shown slightly enlarged in another photo are here magnified sixty-six times. They are covered with short spines, which cling to the hairs of insects.

Stinging Nettle being mimicked by the White Dead-nettle and the Nettle-leaved Bellflower. Here the advantage to the mimicking species is very evident. The Stinging Nettle, by reason of its stinging hairs, is avoided by all browsing animals, and, therefore, it is a distinct advantage to other plants to be mistaken for it and left alone. The Stinging Nettle has inconspicuous green flowers that hang in strings from the upper parts of its stem. The Dead-nettle is entirely unrelated to it, and has large conspicuous white flowers; but until these appear anybody but a botanist would accept the plant as a Stinging Nettle, and avoid it accordingly. To make the resemblance more effective, the Dead Nettle selects as the place of its growth similar situations to those affected by the Stinging Nettle—field borders and waysides chiefly. Not merely so, but it will often be found actually mixed up with a colony of Stinging Nettles.

In the case of certain succulent plants the mimicry is of inanimate things—stones to wit. The photographs show two species of Mesembryanthemum—allies of the little Ice-plant of our conservatories. These plants, like other succulents, grow in hot, dry situations, where vegetation is rare and eagerly sought after by herbivorous animals. Instead of protecting themselves as the Cacti do under similar conditions, by developing a hedgehog-like panoply of spines, they have hit upon a simpler plan: that is, to get themselves up in the likeness of pebbles and rough fragments of rock. Our examples have been photographed in the midst of such material to show what admirable cases of mimicry they are. Of course, when the rains come and these plants put forth their conspicuous flowers, the deceit is at an end; but at this season a good deal of ephemeral vegetation starts into rapid growth, and tender leaves and shoots would be more in demand than pretty flowers apparently springing from among rocky débris and pebbles.

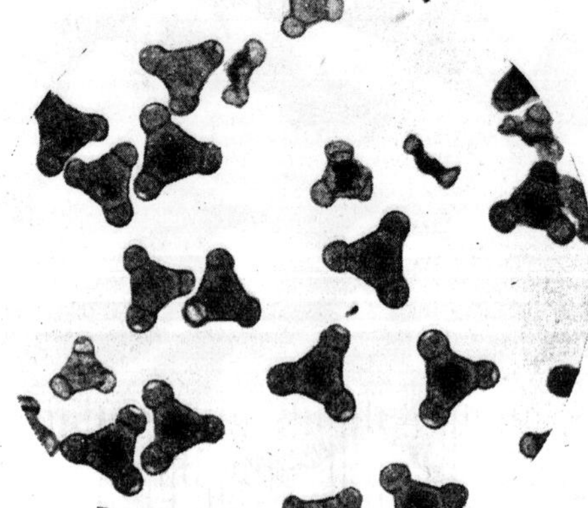

Photo by] [*W. West.*

EVENING PRIMROSE.

The pollen grains are here magnified sixty-six times. They are triangular, with a protuberance at each corner. Viscid threads attach them more securely to the heads of moths.

POLLEN

BY EDWARD STEP, F.L.S.

EVERYONE has noticed the yellow, meal-like powder which clusters around certain of the central organs of a flower, which the bee kneads up and packs into the hollowed part of her leg to carry home for feeding the grubs. Most of us know also that this is not the principal object the flower has in view when producing this *pollen*, and that its real purpose is the fertilization of the incipient seeds contained in the swollen base of the most central of all the flower organs. If a few grains of this powder are examined under the microscope every particle of it will be found to have a definite form, agreeing with all its fellows from that particular flower—though different species of flowers differ remarkably in the character of the pollen they produce.

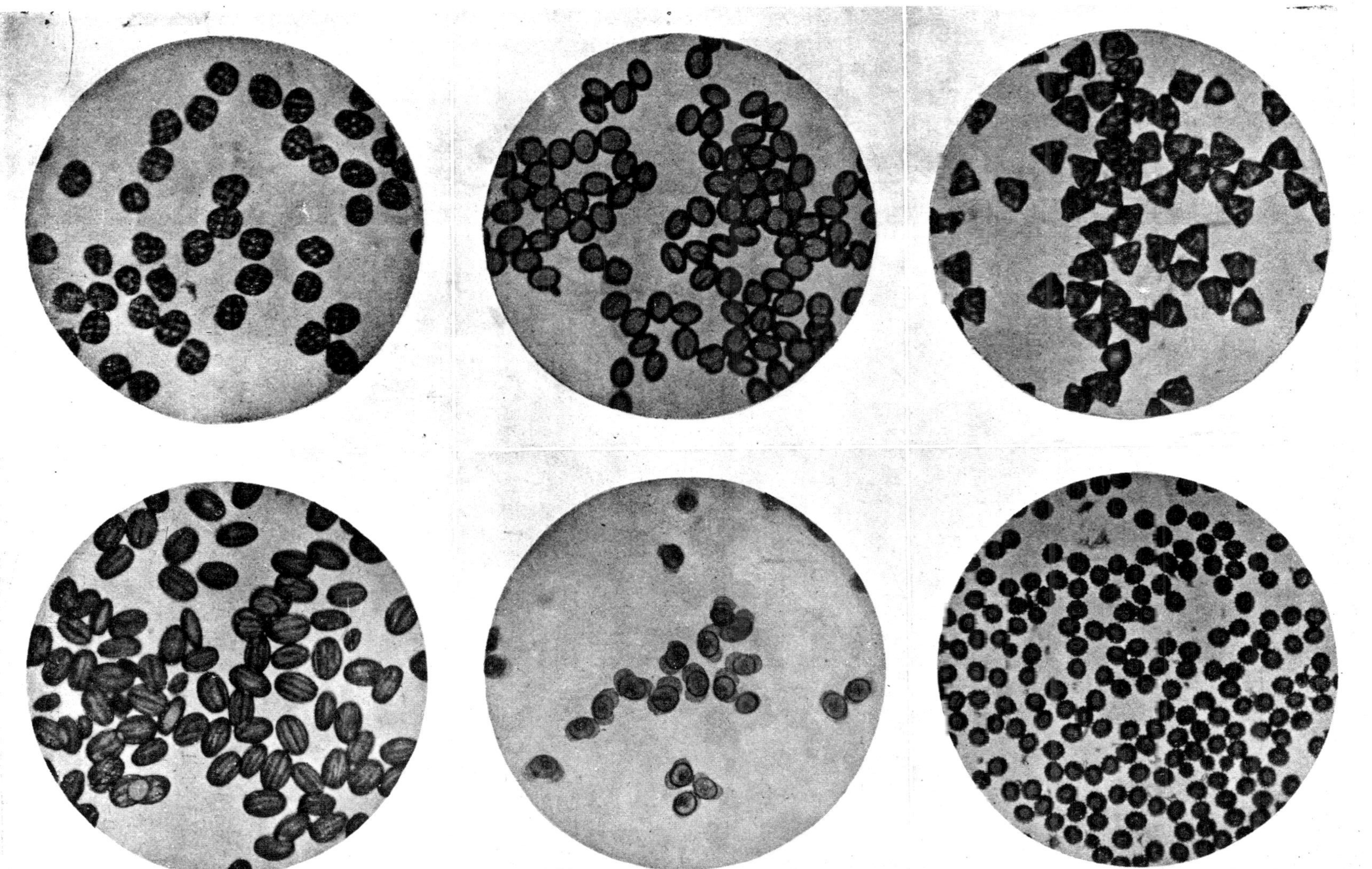

Photos by] [*W. West, L.D.S.*

POLLEN

Here are shown, through the microscope, pollen taken from six separate plants. The upper row includes that of Mimosa, Lesser Bindweed and Rosebay Willow-herb. The lower row shows pollen of Sage, Cedar of Lebanon and a species of Thistle. The two photographs to the left show the pollen magnified one hundred and twenty-five times; the other four are eighty-three times larger than the natural size.

Each minute grain of pollen will be found to consist of a bag filled with a speck of the living matter called protoplasm, which is the basis of all life, animal and vegetable.

The central organ of the typical flower is the pistil consisting of three parts—the ovary at its base, containing ovules, above it the style, whose upper extremity is the sticky or hairy stigma. The ovules are intended to develop into seeds, but they cannot do so unless the contents of the pollen-grains reaches them and mingles with their own substance. In many flowers it is easy for the pollen as it falls from the anthers to become attached to the stigma ; but in many other flowers there are various ingenious arrangements to prevent this happening—a mixture of the protoplasm of two distinct flowers or two distinct plants of the same species being desired. The contrivances for insuring this cross-pollination, as it is called, excite the admiration of all who have paid attention to the subject, but it is not our purpose at present to enter upon any details concerning these. We merely mention them here to explain why pollen taken from many different species of flowers should give us considerable variety of form and ornamentation.

To get this pollen carried from the anthers of one flower to the stigma of another, the plant has by subtle enticements enlisted the services of many kinds of insects ; but in the cases of our forest-trees, our grasses and some other plants with inconspicuous green flowers, reliance is placed upon

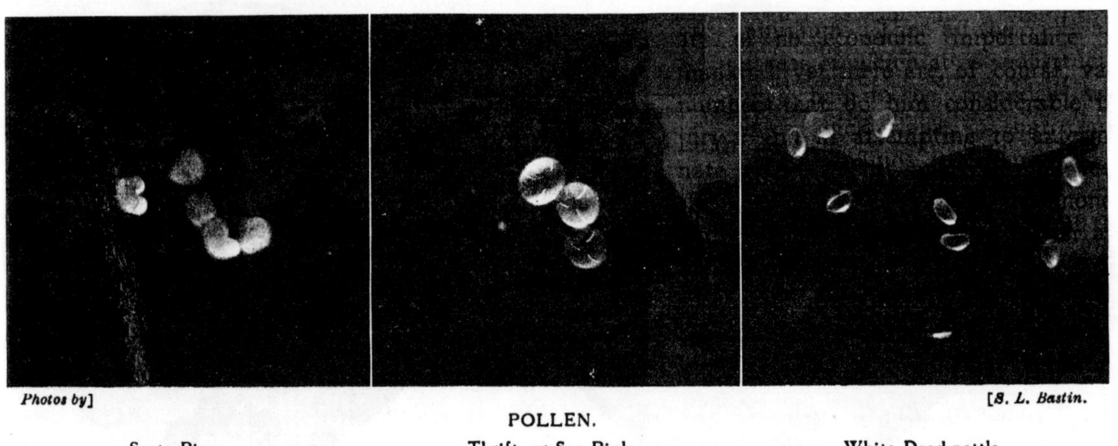

Photos by] [*S. L. Bastin.*

POLLEN.

Scots Pine. Thrift, or Sea Pink. White Dead-nettle.

the wind as a carrier. It will easily be understood that pollen of quite different characters is required for these two chief classes of carriers ; in one case, a light, buoyant pollen grain that will not cling to its fellows ; in the other, a sticky grain that will easily cling to the hairs of an insect's face and legs. Wind-pollination is a speculative and wasteful method, and, therefore, we only find this method adopted by flowers of a more or less primitive type, such as the pines, oak, beech, birch, hazel, etc. The quantity of pollen these trees have to produce in excess of that actually effective in the fertilizing of ovules is prodigious. If set out in numbers, the figures would be almost unbelievable. Walk in spring along the edge of a pine wood, and look into the hollows and pools in the direction in which the wind has been blowing. You will find the yellow pollen lying in heaps in the hollows and drifted to one side of the pool as a floating scum. Nearly every year there are letters in the newspapers describing these heaps as due to a rain of brimstone ! so little do most of us know of the common and ever-recurring phenomena of our own little planet. The more highly-organized flowers, however, that have adapted their structure to the visits of insects—and in many cases to special species of insects—are more economical in their production of pollen. Their carriers are more reliable, and so their pollen is relatively scanty, due provision being made for the considerable percentage the insects will appropriate to their own use.

In the photographs reproduced as illustrations to this paper the examples have been drawn mainly from insect-pollinated flowers, and the names are set under them. It will be seen that most of these are spherical, their surface bristling with points that enable them to cling to their carriers ; but if any of them are brought into contact with the viscid stigma of the next flower visited the stickiness causes them to adhere there. The pollen-grain so detained sends out a tubular shoot which pierces the stigma, penetrates the whole length of the style, enters the ovary and unites with an ovule there. Its mission has been accomplished successfully.

Some, instead of being spherical, are rather flattened and three-cornered, as is the case with that of the evening primrose ; others have the appearance of a cylinder spirally coiled upon itself ; some have chimney-like openings with lids ; in some the surface is ornamented with depressions like the cells of honeycomb. Viscid threads are attached to those of the evening primrose, which ensures their sticking more readily to the moths that carry them ; and in the orchids similar threads from each grain unite to form a stalk ending in a sticky knob which adheres to the head of the insect. The buoyant, wind-borne type is shown in the pollen of Scots pine and cedar. It will be seen that the power of remaining suspended in the air for as long as possible is given by equipping the pollen-grain with two little air-bladders.

Photo by] [*A. E. Smith.*

POLLEN.

The pollen of the Mallow here shown is very similar to that of the related Hollyhock.

THE CHEESE-MITE

THE mite belongs to the lowest order of that group of animals which includes the spiders. Already numerous members of the family have been described in these pages, chief amongst them being the brilliant little Water-mites shown in the coloured plate opposite page 763, and a large number of the ticks which infest birds and are first cousins to the mites.

A mite, as we have seen, agrees with the spider in having eight legs ; but there is a considerable difference between the two in their earlier stages, for while the spider is born with all the characteristics of the parent, the mite, at its first appearance, has never more than six legs, and sometimes only four. It is hatched from an egg, and changes its appearance so remarkably during its career that the different forms of the same mite were at first classed as different species. It is the modern representative of an ancient race, for some of its progenitors are to be found royally entombed with the far-famed fly, in amber.

The Cheese-mite belongs to the most important group of this family, all the members of which

Photo by] [*W. West.*

HOLLYHOCK.

The clustered stamens are shown with the anthers discharging the pollen. The round dots are the pollen-grains (shown on a larger scale in another photo).

have soft skins and oval bodies, with the feet frequently terminating in adhesive vesicles. Looked at with the naked eye a number of these minute creatures have only the appearance of a little brown dust in the mould of a cheese, but when examined under a microscope they can be seen scrambling about on stout legs which terminate below the "foot pads" in sharp hooks. From the body radiate a number of long bristles, while the rostrum, or beak, is provided with two pairs of very serviceable pincers. A dry cheese and one which has a rind broken into crevices and cracks is most likely to be attacked by the mite, and as a fully-ripe Stilton possesses these qualities in a high degree, it is seldom to be found without a numerous company of mite devotees.

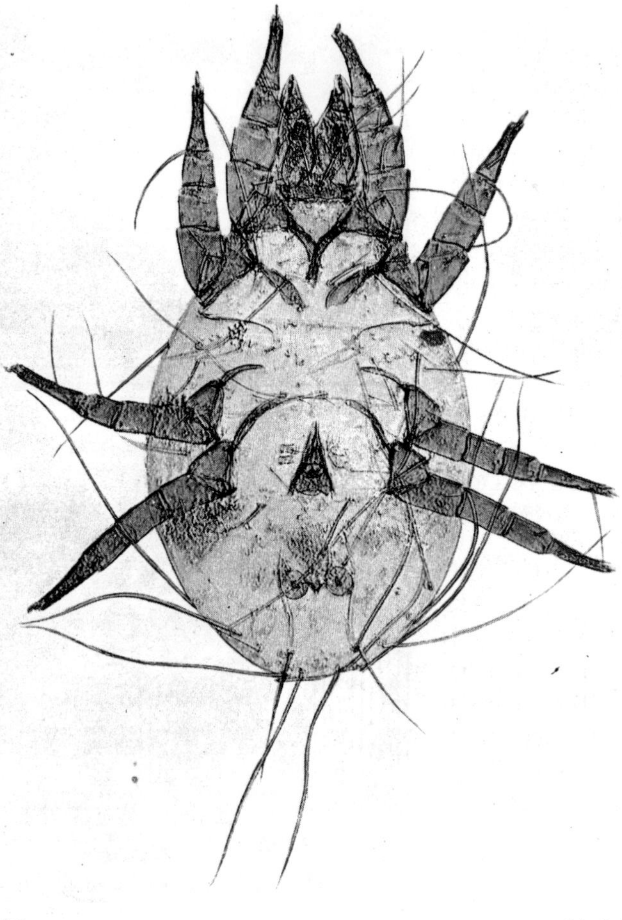

Photo by] [*A. E. Smith.*

A CHEESE-MITE.

Looked at with the naked eye a number of Cheese-mites appear only as a little brown dust in the mould of a cheese, but seen under a microscope each Mite becomes the repulsive-looking creature, with a soft oval body sparsely furnished with bristles, shown in this photograph.

THE GLYPTODON

BY REV. H. N. HUTCHINSON, F.R.G.S., F.G.S., F.Z.S.

THE continent of North America has yielded many remains of extinct animals and greatly added to our knowledge of the past development of life on the earth. To the palæontologist these discoveries of the last sixty years or more have, as it were, opened out a new world. Countless new mammals, new reptiles and new birds have been dug out from American strata by Leidy, Marsh, Cope, Hatcher, Osborne and others, and the results of all their labours have been of the highest value to students of Evolution. But South America has also contributed its share. For example, the Megatherium, that great extinct sloth, already described in this book, came from the Pampas region of South America, that strange region of which Mr. W. H. Hudson has given such a wonderful account in his famous book, "The Naturalist in La Plata." The Glyptodon, of which we give a restoration by Mr. J. Smit, the well-known animal artist, received its name from Sir Richard Owen, in reference to the sculptured aspect of the grinding surface of its teeth. To the eye it resembles more or less an armadillo. The picture shows a wonderful and highly ornamental cuirass, or large plate of armour, covering the whole body, but allowing the head to show in front, while the legs come out beneath. Both head and tail were also protected with armour. This creature, possessing no movable bands, as living armadillos have, could not have rolled itself up into a ball; and we may also infer from its thick, short toes

From "*Extinct Monsters.*"] [*By kind permission of Messrs. Chapman & Hall.*

THE GLYPTODON.

The forefather of the Armadillo. Its modern representatives, however, have their armour divided into movable bands, which enable them to roll themselves into a ball. The shell of the Glyptodon, on the contrary, was composed of bony plates, inflexibly united to one another at the edges. A curious bony cap protected the head, and the tail was likewise encased in armour.

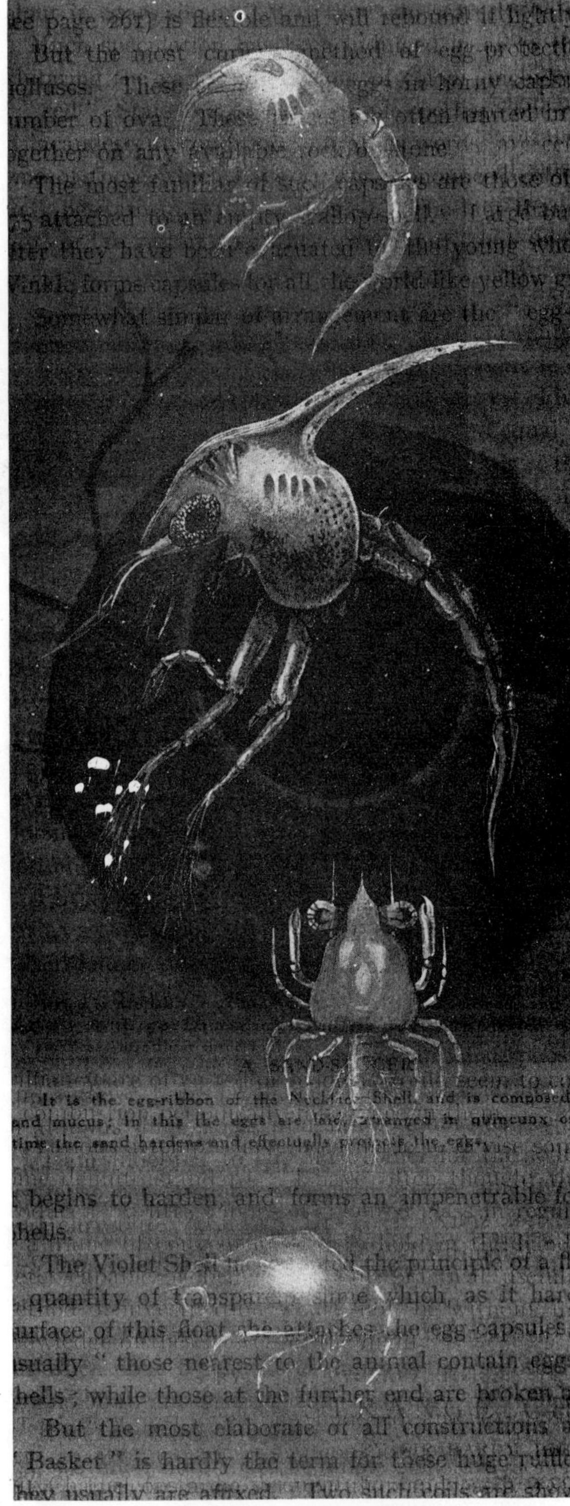

CRAB DEVELOPMENT.

These figures show successive stages in the early life of the common Shore-crab. The third and fourth show the same infant Crab from above and from the side respectively.

that it was not in the habit of burrowing, or of seeking food underground. In this particular form (*Glyptodon asper*) the bony scutis of the carapace has a beautiful rosette-like sculpture, while the sheath of the tail is entirely composed of a series of movable rings, ornamented with large projecting tubercles; the vertebræ of the backbone are almost entirely fused together into a long tube, and also are joined to the under-surface of the great shield, to which the ribs are united. The cheek-teeth are sixteen in number, four above and four below on each side. These are channelled with two broad and deep grooves, which divide the surface into three distinct lobes. Hence the name of the animal. The anatomy of this armadillo greatly resembles that of a giant sloth. In fact, sloths, ant-eaters, and armadillos all belong to the same family of so-called toothless animals with no front teeth, though one or two forms are really toothless. Their diet was vegetarian, but we do not know much about their habits. One would think they had but few enemies, being quite safe from snakes. Monkeys also could not hurt them. But why they died out is a mystery. Were they too big? or did the little modern armadillo beat them in brain-power? We cannot tell. Possibly man was the exterminating agent.

THE TRANSFORMATIONS OF CRABS

AMONG the important secrets of Nature which were reserved for the naturalists of the last century to reveal was the fact that all animals, other than the most primitive forms, pass through a series of metamorphoses before they arrive at their final adult state. The ancient philosophers knew the fact concerning the evolution of the bright and airy butterfly from a "grovelling worm," the caterpillar; but the truth that fishes and crustaceans pass through similar states of incompleteness was not dreamt of until comparatively recently. It is true that nearly a hundred and fifty years ago, the Dutch naturalist Slabber declared that he saw a minute marine creature known as a

Zoëa turn into another creature known as Megalopa. But nobody took Slabber's statement seriously until fifty-four years later, when Vaughan Thompson showed that Slabber was in the main correct, and that both Zoëa and Megalopa are stages in the evolution of the Crab.

Even then, controversy raged warmly around the question whether this metamorphosis was general among the Crab kind, or restricted to a few special species.

The fires of that controversy have died out long since, and to-day we know that Crabs and Lobsters and their kind—with a few exceptions where the changes are undergone in the egg—pass through several larval stages of existence, corresponding to what were formerly known as distinct species under the names of Zoëa and Megalopa. The accompanying illustration (page 882) will make this plain. It shows the successive stages in the infancy of the common Shore-crab.

When one remembers that the eggs are about the size of a mustard-seed it will be realized that the larval Crabs on hatching are very minute and perfectly transparent—as, indeed, they remain until they develop true crab-shaped shells. In these earliest stages they are remarkable for having an enormous "head," and a long, thin tail. The "head" a little later produces a very long spine from its upper side.

In the Megalopa stage there is a much nearer approach to normal crab-form, with large, stalked eyes, and nippers to the front pair of legs; but the long, slender "tail" sticks out behind.

All this time the larval Crab swims through the upper waters, where its transparency gives it a chance of escaping the numerous enemies that would feed upon it. When the last larval skin is thrown off, and an opaque "shell" takes its place, the miniature Crab sinks to the bottom and tucks its flattened tail tightly against the under-side of its body. As the Crab grows the shell becomes too small, and from time to time is thrown off, a new and at first soft shell having formed beneath the old one. Our illustration on this page shows the whole series of such shells that were thrown off successively by the same Shore-crab in its progress from infancy to adult life.

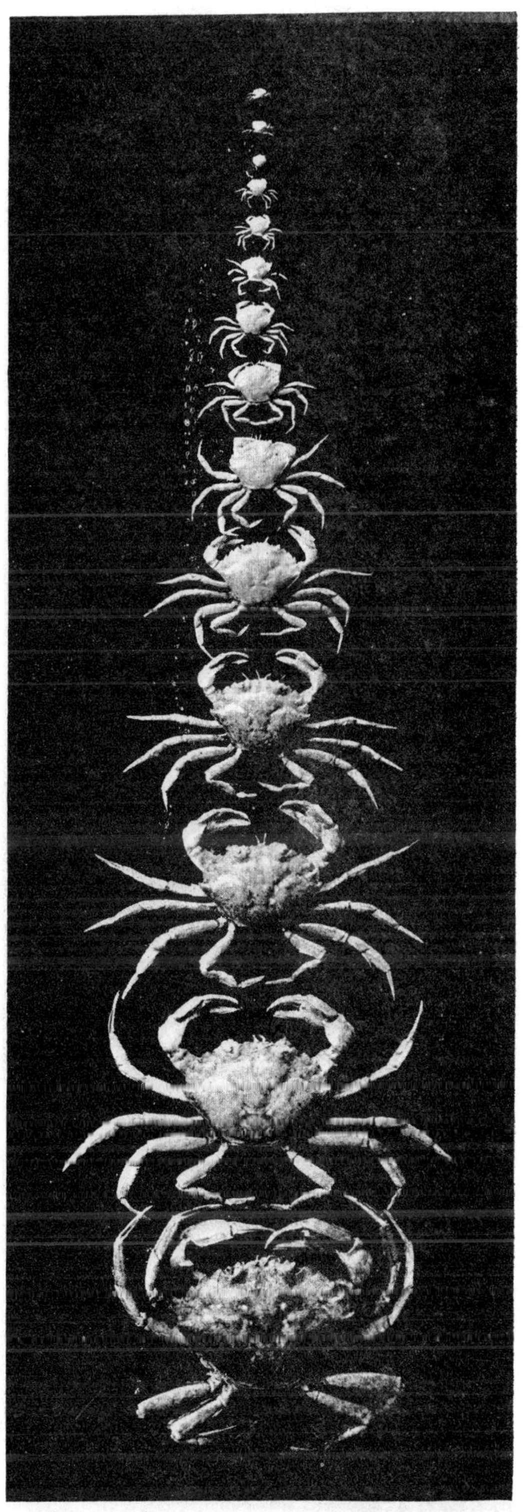

TRANSFORMATIONS OF CRABS.

The photograph represents fourteen empty shells, all cast off successively by the same individual Shore-crab to accommodate its increasing bulk.

ENCYCLOPAEDIA OF THE UNIVERSE

ENCYCLOPAEDIA OF THE UNIVERSE

A POPULAR WORK ON THE MARVELS OF

THE HEAVENS :: THE EARTH :: PLANT LIFE :: ANIMAL LIFE :: THE MIGHTY DEEP

WITH 1000 FINE ILLUSTRATIONS

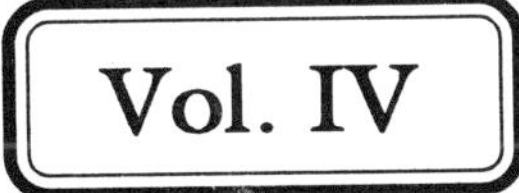

Editor

WALTER HUTCHINSON

B.A., F.R.G.S., F.R.A.I., Etc.

WITH AN INTRODUCTION BY

LORD AVEBURY, P.C.

AND WITH CONTRIBUTIONS BY LEADING SPECIALISTS, INCLUDING :

Sir Harry Johnston, G.C.M.G., K.C.B.
Richard Lydekker, F.R.S.
Frank Finn, F.Z.S.
Edward Step, F.L.S.
E.W. Maunder, F.R.A.S.
W.P. Pycraft, A.L.S., F.Z.S.
Russell F. Gwinnell, F.C.S.
Frank T. Bullen, F.R.G.S.
Tickner Edwardes
J J Ward, F.E.S.
C. Flammarion
Lionel E. Adams, B.A.
Harold Bastin, F.E.S.
S. Leonard Bastin
K. G. Blair, F.E.S.
Sir Herbert Maxwell, F.R.S.
E. B. Poulton, F.R.S.
G. F. Chambers, F.R.A.S.
H. St. J. K. Donisthorpe, F.E.S.
Richard Kerr, F.G.S.
W.F. Kirby, F.L.S.
Hugh Main, F.E.S.
Ernest Marriage
E.A. Martin, F.G.S.
T. E. R. Phillips, F.R.A.S.
R. I. Pocock, F.R.S.
Dr. E. J. Spitta, F.R.M.S.
E. W. Swanton
J. Sinel
W. Mark Webb, F.L.S.

DPH

DISCOVERY PUBLISHING HOUSE PVT. LTD.

NEW DELHI-110 002

Published by:
Namit Wasan
DISCOVERY PUBLISHING HOUSE PVT. LTD.
4383/4B, Ansari Road, Darya Ganj
New Delhi-110 002 (India)
Phone : +91-11-23279245; 23253475; 43596065
E-mail : discoverybooksindia@gmail.com
discoverypublishinghouse@gmail.com
namitwasan9@gmail.com
web : www.discoverypublishinggroup.com

Edition: 2020

ISBN: 978-81-8356-873-9 (Set)

Encyclopaedia of the Universe

Printed at:
Infinity Imaging Systems
Delhi

CONTENTS OF VOLUME IV.

Vol. IV

DINOTHERIUM

BY REV. H. N. HUTCHINSON, F.R.G.S., F.G.S., F.Z.S.

THE great extinct creature of which a restoration is here given was a relation of the elephants. The reader may perhaps have already guessed that much from the large proboscis and elephantine limbs. It lived in the Miocene and Pliocene geological periods, and seems to have ranged over Europe, East Africa, and perhaps India, but is unknown in America. Its history, considered as a fossil, is an interesting one. First of all (as has often been the case), isolated teeth were found and described in 1715, 1775 and 1785. Kaup, in the year 1829, gave the name Dinotherium to a lower jaw; but not understanding it, put it upside down. Then in 1835 a more or less complete skull was found

Photo by] [E. Step, F.L.S.

THE SHORE-CRAB.

It has here attained its full development. The Shore-crab must be distinguished from the Edible Crab, which reaches a much greater size, and for this reason its alternative folk-name of the Green Crab is perhaps more suitable.

at Eppelsheim, Hesse-Darmstadt. This head, when exhibited in Paris, excited great interest among the naturalists in that city. M. de Blainville pointed out its most remarkable points, and concluded that it was related to the manatees and dugongs. Recent researches tend rather to confirm that view. Kaup placed it between the tapir and rhinoceros. But it was evidently a proboscidian—that is to say, elephant-like. The skull, lower jaws, teeth, and some limb-bones of this great and strange beast may be seen in the Natural History Museum at South Kensington. The skull and lower jaws are in a separate glass case, and it is evident that the roof of the skull has been considerably crushed in. It is important to bear this in mind; otherwise one may get quite an erroneous idea of the shape of the head. But the most conspicuous feature of this specimen is the downward curve of the lower jaws—so different from those of any other big quadruped. The downward curve of the tusks reminds one of the walrus. That shrewd naturalist and geologist, Dean Buckland, concluded that it was almost mechanically impossible for such a jaw, nearly four feet long, to have been supported in air by its possessor; he therefore concluded that it was destined to

From "Extinct Monsters."] [By kind permission of Messrs. Chapman & Hall.

THE DINOTHERIUM.

This huge creature was probably a progenitor of the modern elephant. The huge lower jaws, which were nearly four feet long, were provided with tusks which curved downwards. It is difficult to determine the reason for this exception to the general rule for the tendency of tusks to be upward. Possibly Nature was only folding them *down* because they were no longer wanted.

live in the water. The aquatic habits of the tapirs (more or less allied to elephants) seem to point to the conclusion that the Dinotherium inhabited fresh-water lakes and rivers. Some writers have pointed out that the tusks might have been useful for grubbing up by the roots large aquatic plants, and possibly also to hook on to the bank and assist the creature to drag its body out of the water. The present writer is inclined to think that there must have been tusks in the upper jaws. Time alone can settle that point ; but should that prophecy be confirmed, he would like to suggest that Nature was only folding down the lower tusks because they were no longer wanted ! Unfortunately the only known skull is in a fragile condition, and partly mixed up with plaster, which prevents such a thorough examination as might otherwise be made by the geologists in charge of it.

Photo by] [*H. J. Shepstone.*

THE MIDNIGHT SUN.

It is here approaching the horizon, but it will not disappear beneath the sea, for within the Arctic Circle the sun, during a certain period, will describe complete circles in the sky *above* the horizon.

With regard to the size of Dinotherium, there can be but little doubt that it was a big creature, like most of the modern elephants. The biggest yet known is from the Balkan peninsula. The remains found there indicate an animal quite as big as the largest elephant, possibly even bigger.

THE MIDNIGHT SUN

BY F. W. HENKEL, B.A., F.R.A.S.

It is well known that on account of the diurnal rotation of the earth on an axis whose extremities are the north and south poles respectively, the heavenly bodies (the stars) apparently describe circles of greater or less angular magnitude around the north and south celestial poles, those nearer these points moving in smaller circles than those further off, but all in the same time, one sidereal day.

The sun, moon and planets share in these motions, but have, in addition, others, partly due to the earth's annual motion and partly to their own independent movements. The sun, changing his place amongst the stars during the course of the year, owing to the earth's annual motion, is visible for a greater or lesser time during the various months. For London he rises at shortly after eight a.m. on December 21, attains his greatest altitude (of 15°) when due south at noon, and sets a little before four p.m., being then at his greatest (angular) distance south of the equator. By March 21 he has reached the celestial equator, rises due east at six a.m., attains the altitude of 38½° at noon, and sets at six p.m. when due west, the day being thus twelve hours long, or equal to the night (equinox). After this, coming further northwards, on June 21, he rises before four a.m., reaches a noon-day altitude of 62° (more than two-thirds the way from horizon to zenith—*i.e.*, the point overhead), and sets after eight p.m., this being the longest day. After this, the days diminish and the phenomena of the daily motion are reversed. Further northwards, however, the days in summer are yet longer than for London; for Edinburgh, seventeen and a half hours long on June 21, and over eighteen hours in the Shetland Islands (latitude 60° N.).

Further north still the sun will be above the horizon for a yet longer time, and at latitude 66½° N. it will not set at all on the longest day, but will merely touch the horizon at "midnight" and the phenomenon of the Midnight Sun will occur. For places yet further north, within the Arctic Circle (latitude 66½° N. to the pole), there will be periods during which the sun will not set at all, but like a circumpolar star, will describe complete circles in the sky above the horizon, having its greatest altitude at noon, and being lowest down at "midnight."

Thus at Hammerfest, latitude 70½° N., the sun will be *above the horizon continuously* from the end of May to nearly the end of July, having an altitude of 43° (its maximum amount) at noon on June 21 and of 4° only at midnight, not far above the

Photo by] [*E. Step, F.L.S.*

THE SMALLER HORN-TAIL WASP.

This species differs from the other in colouring as well as size. It is of a uniform dark steely-blue colour, and its "horn-tail" is less prominent.

A PARASITE OF THE HORN-TAILED WASP.

By some strange instinct this Ichneumon can discover the borings of the Horn-tails and therein lays its eggs, from which the young hatch out and devour the Horn-tail grubs. On the left is a piece of wood in which an ovipositor of a Horn-tail was found fixed.

vapours of the horizon. Thus, notwithstanding that the sun is then continuously visible, on account of its low altitude the amount of heat is not excessive; whilst, on the other hand, its complete absence during several weeks of winter partly accounts for the severity of the climate within the Arctic (and Antarctic) zones; for *perpetual day* in summer has its correlative of *perpetual night* in winter. At either pole the sun is above the horizon for six months at a time, being below the circle for the rest of the year, circling round the sky almost parallel to the horizon ("parallel sphere"), its altitude varying with extreme slowness, but never (on midsummer day for each hemisphere respectively) exceeding 23½°.

All places from latitude 66½° N. to the pole have perpetual day for a shorter or longer period, between March 21 and September 23, those not far within the Arctic Circle for a few days in June only; whilst perpetual night reigns for a corresponding period in the winter. In the southern hemisphere (Antarctic zone) the phenomena are exactly reversed; perpetual day occurs sometime between September 23 and March 21, perpetual night sometime between March 21 and September 23.

THE TUNNELS OF THE HORN-TAILED WASP'S GRUB.

The grub of the Horn-tail will complete the task commenced by the mother and will bore a tunnel of considerable length. When fully grown it spins a cocoon, weaving into it the chips of the gnawed wood. These grubs work havoc amongst the trees they favour.

THE HORN-TAIL WASP

BY EDWARD STEP, F.L.S.

THE Horn-tail, though called a wasp for convenience, is more nearly akin to the saw-flies whose grubs reduce the foliage of gooseberry-bushes and other plants to hideous skeletons. But the Horn-tail, though in size and appearance much mor formidable, is a solitary insect that is rarely seen, and its grub carries on its depredations out of sight in the timber of growing pine-trees. There are cases on record in which whole plantations of coniferous trees have been rendered worthless as timber owing to their being so riddled with the burrows of this grub.

The size and appearance of the perfect insect may be gathered from our photographs, which represent two species. The larger of the two has its body banded with orange and black, and its body ends behind in the stout horn-tail.

The "horn-tail" is really a beautifully-contrived auger within its sheath, by means

Photos by] [*E. Step, F.L.S.*

THE HORN-TAIL WASP.

These photos show the female insect one and a half times the natural size. In the left-hand picture she is just alighting on the trunk of a pine-tree; in the other she has folded her wings and is searching for the best spot on which to bore and deposit an egg.

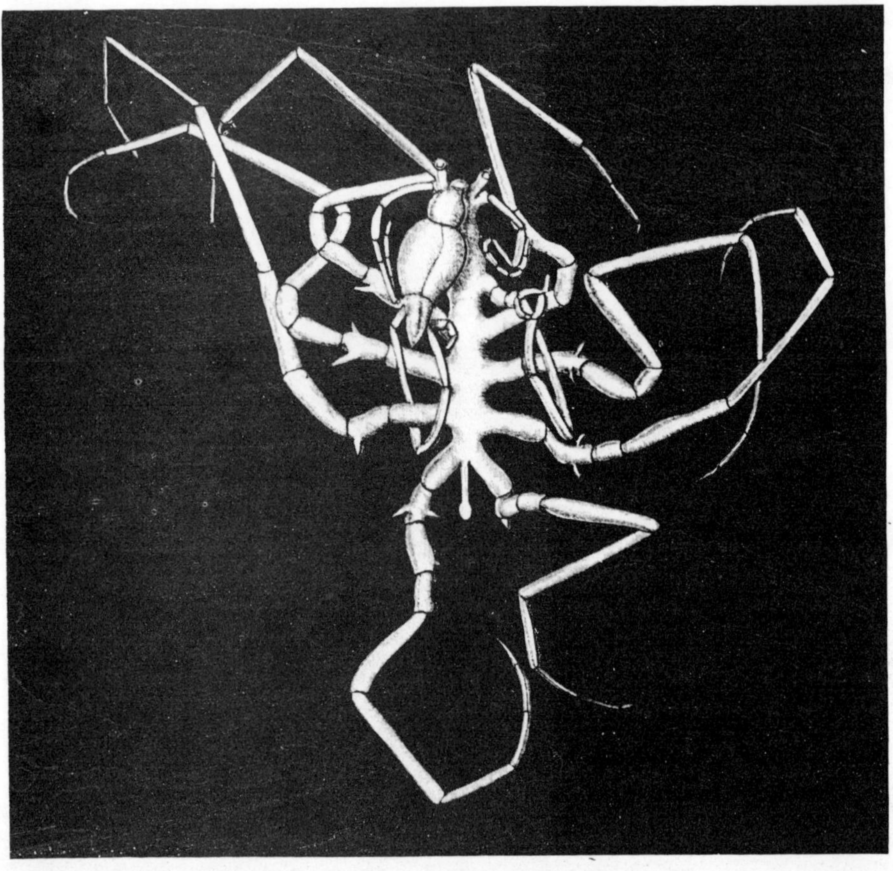

A SEA-SPIDER.

Only a few species of these mysterious animals are found near the shore, but many appear to flourish in the deep sea. The specimens on these two pages were discovered during the *Challenger* Expedition. This is a small species, and has been magnified about fourteen times. It is without eyes.

of which the female insect bores holes in trees, in each of which she deposits an egg. That is the worst that can be said of it, and to the timber-grower it is bad enough; but so far as attacks upon our tender bodies are concerned, the Horn-tail is a perfectly harmless creature. The male, which is more rarely seen than the female, might be mistaken for a different species, as it lacks the black band across its body, and ends in quite a little point instead of the auger and its guard. The prominent orange spine which terminates the last ring of the female's body is not really connected with the boring apparatus, which is some distance below it and attached to the underside much further forwards. The borer is hinged so that it can be used at right angles with the insect's body. It is protected by a couple of yellow sheath-blades, which, when at rest, are sunk in a groove of the body. The borer itself will be seen between them, a hard, polished, black needle, three-quarters of an inch long. With this instrument, then, a hole is bored through the bark, and an egg deposited.

In due course the egg hatches and a tiny white grub issues, who finds himself in a land of plenty, for his proper food is all around him. He sets his infant jaws to work at once, and soon has the beginning of a boring that he will ultimately extend far into the solid wood. How long it takes him to become full grown is not accurately known, owing to the secluded nature of his existence, but it is known to extend over several years, from the fact that after a house has been built three years, the perfect insects have emerged from timber used in its construction. It is presumed that, in these cases, the grub was in the wood when the tree was cut down, but so small that its burrows did not come in the way of the saw that cut the tree into planks or rafters, and that its presence, therefore, was not suspected. Knowing how much agitation can be caused by the presence of an ordinary wasp at a tea-party, one can realize the feelings aroused in a family when a number of these big Horn-tails keep on appearing from holes in the nursery floor!

The grub feeds on, it may be presumed, leisurely and continuously. It has not got to hurry, like leaf-feeding larvæ, to get its full size before autumn comes with its falling leaves. Shut up in the tree, winter and summer are much the same to it ; but at length it arrives at the full-fed stage. Before becoming a chrysalis it makes all the preparation it can for its emergence in the winged form, and to that end carries its burrow close up to or partly through, the bark, leaving little thickness to be cut through by the jaws of the perfect insect. When the latter splits up the chrysalis skin it finds itself restricted to creeping through a tunnel, whereas it wishes to exercise its newly-acquired wings. Finding a certain thickness of bark bars its way, it sets to work with its jaws and cuts through the impediment. But if the tree has been cut up and made into roof-beams, floor-boards, shelves or boxes, the poor Horn-tail may find all kinds of obstacles substituted for, or added to, the natural ones.

Thus, infected beams have been covered with sheet-lead, but the Horn-tail has worked through it. Their burrowed wood has been worked up into a box, the box filled with ball-cartridges, and the Horn-tail has been so unfortunately placed that it has eaten through the wood only to find that the bullet end of the cartridges lay in the way ; but the indomitable perseverance of the Horn-tail has carried it through the bullets even.

We believe that a package of cartridges so treated in the arsenal at Grenoble is still to be seen in one of the French museums. In a similar establishment—also in France—military uniforms were stored upon a shelf containing one of these insects, and on emergence the poor creature had to make its way through a pile of woollen trousers.

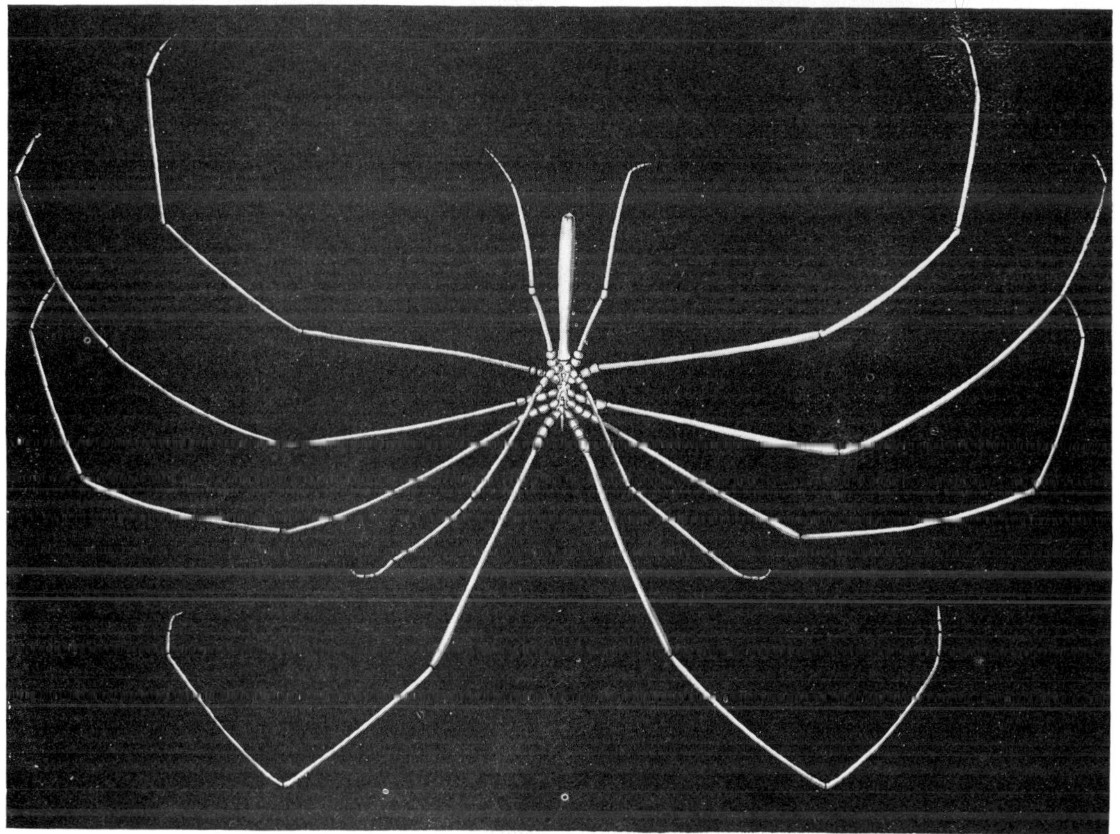

THE MIGHTY SEA-SPIDER.

This specimen with its long slender limbs can cover a space of nearly two feet. It is found at depths varying from one thousand four hundred to one thousand six hundred fathoms. These deep-sea forms of Sea-spiders are usually of an orange or bright red colour, and are generally larger than the in-shore species.

SEA-SPIDERS

BY JOHN J. WARD, F.E.S.

THE Water-spiders of the pond are true spiders which have developed aquatic habits, but such cannot be said of the Sea-spiders ; indeed, it is a question whether they have any claim at all to the title of spider. Their anatomy proves them to be a complex zoological puzzle, and perhaps their greatest resemblance to spiders lies in the fact that they possess four pairs of conspicuous jointed legs and mouth appendages bearing a superficial likeness.

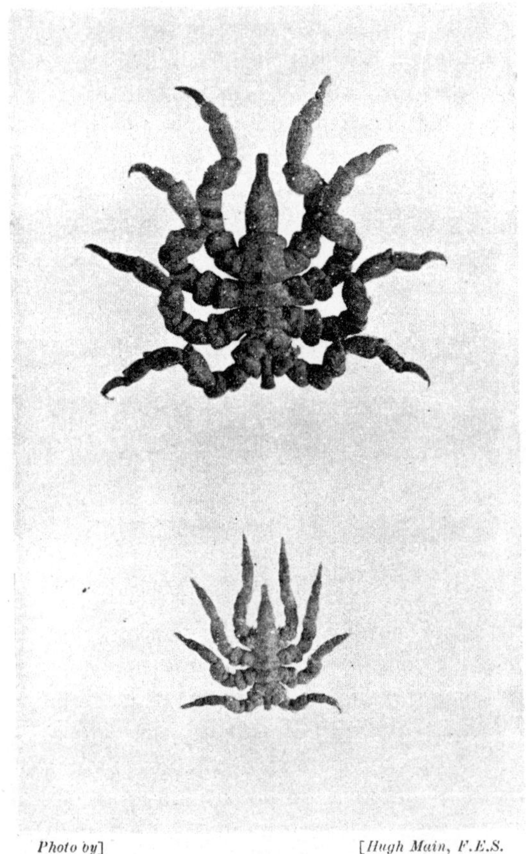

Photo by] [*Hugh Main, F.E.S.*

SEA-SPIDERS.

Two fine specimens magnified about three times. Notice the large head, the huge body-joints to which the limbs are attached, and the minute body proper.

When we come to examine their anatomy in detail, however, then at once difficulties begin to arise. The most striking feature is that the large abdomen, so characteristic of spiders, is reduced to a tiny structure so small as to be scarcely noticeable. The apparent body-trunk is not a body at all, but consists of four rings to which the legs are attached. If we think of a spider without an abdomen, and with that part to which its eight legs are attached divided more or less into four segments, each of which carries a pair of eight-jointed legs bearing a single claw, and which legs may vary in size from about the length of the segmented-trunk to seven or eight times its length (according to the species), we begin to get a rough idea of the general anatomy of a Sea-spider ; all excepting its head and its egg-carrying legs ; although it should be mentioned that two Antarctic genera are made still more extra-ordinary (and unspider-like) by possessing five pairs of legs.

The head is a short column terminating in a proboscis, which may be either much shorter or much longer than the whole body, including the leg segments and the abdomen proper. Attached to the sides of the proboscis are a pair of pincer-like organs, often armed with teeth, and a pair of spinous, jointed palps, or feelers ; the latter organs, however, are absent in some adult forms. Also there are four simple eyes arranged in two pairs, these being the only special sense-organs known in these creatures.

Finally, just below the head, and attached to the under-side, are a pair of egg-carrying legs. It is in the males that the latter organs are most developed, the care of the eggs being their special occupation. The females of some species are provided with these legs, but only in one species are they known to carry the eggs ; in several genera they are quite absent in the females.

It is obvious, then, that Sea-spiders are extremely queer animals, and consequently the zoologist, in classifying them, is compelled to provide them with a special place in a group which he designates the Pycnogons. Their anatomy seems to reveal characters of some of the highest members of the spider family combined with some very degenerate features, and both made still more complex by comparatively recent specialization of many parts.

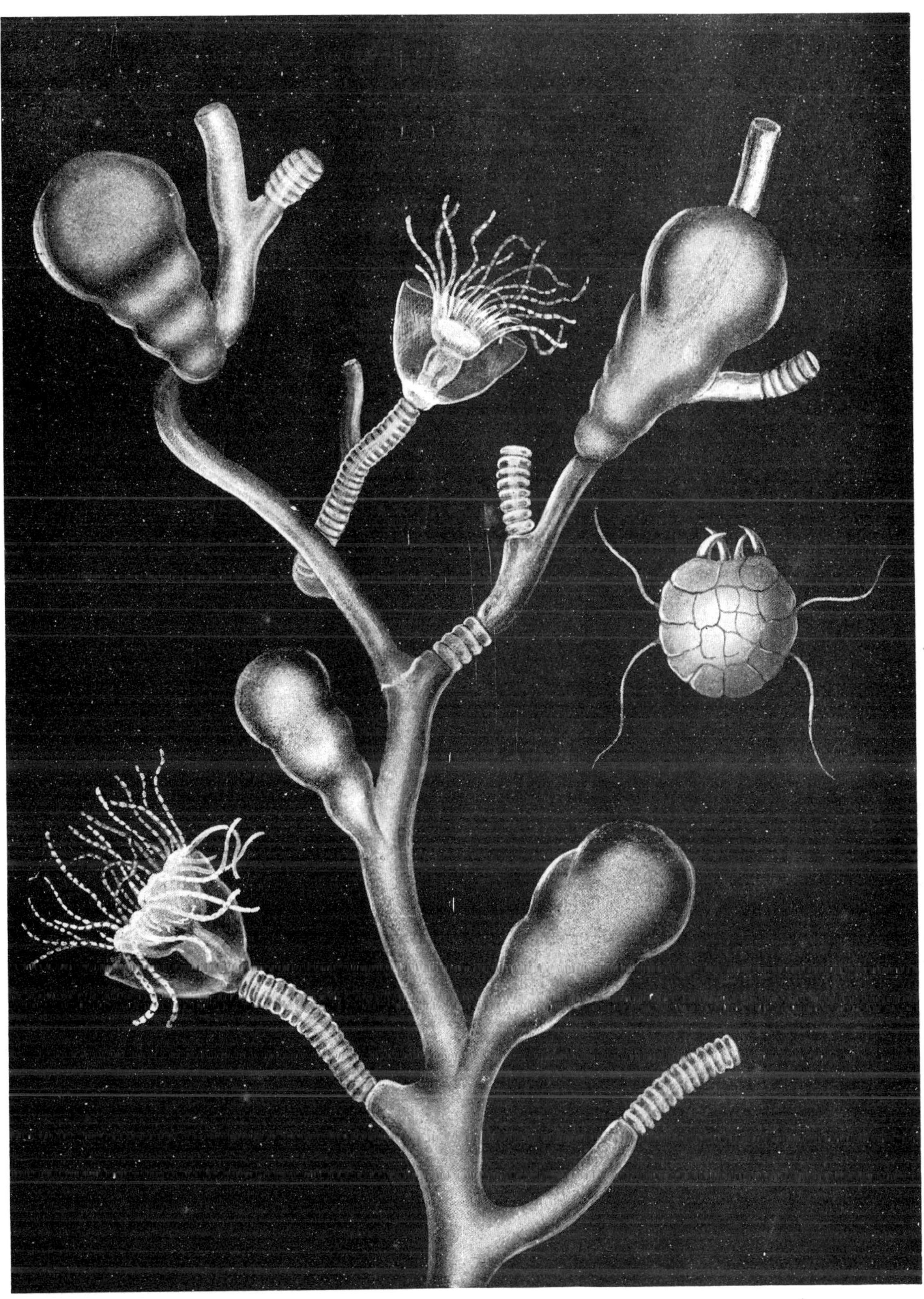

THE GALLS AND YOUNG OF ONE OF THE SEA-SPIDERS.

The young of this family of Sea-spiders (Phoxichilidium) live a semi-parasitic life within the digestive organs of hydroid colonies, such as the Bell Coralline here shown, and form a growth after the nature of a gall. The young are hatched from an egg, and are provided with a proboscis, feelers, and from one to four pairs of legs. It is here greatly magnified.

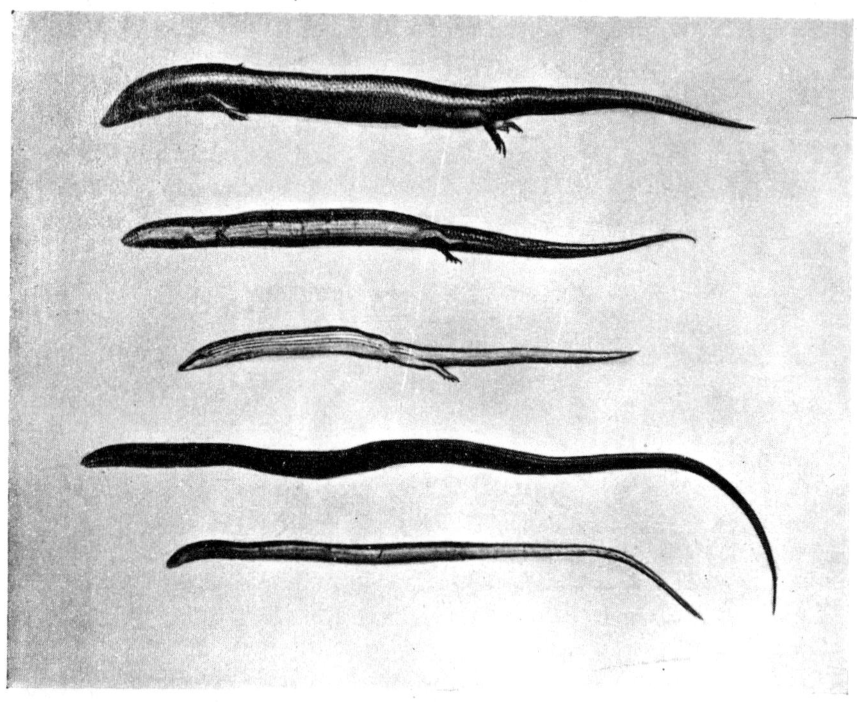

THE EVOLUTION OF THE LIZARD.

These five species of Skinks illustrate the gradual reduction and disappearance of the limbs and the transition from a lizard-like to a snake-like bodily form. From the topmost specimen, which has both pairs of limbs present, each with five toes, and with the tail distinct from the body, it passes through an evolution by gradual stages till in the last the limbs are reduced to minute undivided vestiges and the bodily form is snake-like.

On the foreshore only a few species are found, but, as revealed by the voyage of the *Challenger*, these mysterious animals abound in the great abysses of the ocean, often to the extent of providing a recognized character of the fauna over vast areas. Even at a depth of two miles or more they make their home, and what is more remarkable, the deep-sea forms are by no means dwarfed by their environment; indeed, some of them attain a comparatively large size, as, for example, the Mighty Sea-spider of the Southern Ocean, which is illustrated on page 891, can extend its legs to cover nearly two feet of space. These deep-sea forms are usually of an orange or bright red colour; the example Nymphon, which appears amongst the illustrations (page 890), is also a denizen of the ocean depths. Another feature worthy of note regarding these deep-water species is that their eyes are often imperfect, being destitute of both lens and pigment. In one species which was dredged by the *Challenger* at a depth of only thirty-eight fathoms, eyes were found to be lacking. On the other hand, many species dredged from a depth of over two miles have their eyes normally developed.

The male takes the ball of eggs from the female and may carry several masses, each one representing the whole brood of one female. The eggs of the various species differ considerably in size, some masses containing more than a hundred, while those of the small members of the genus *Pallene* are so large that there are usually not more than two eggs in each mass. When the larvæ hatch out, they cling to the body of their parent, who carries them about until they can take care of themselves. The young, unlike spiders, undergo a true metamorphosis before completing their development.

THE EVOLUTION OF LEGLESS LIZARDS

BY W. P. PYCRAFT, A.L.S., F.Z.S., ETC.

THAT animals may undergo profound transformations of shape in response to the demands of a new environment is an undoubted fact, though the process of change is so slight as to be imperceptible, even in a hundred generations. But there are some curious exceptions to the rule. Thus aquatic animals reveal their habitat by their feet, which are commonly webbed. Often the whole leg undergoes transformation. In the seals, for example, the limbs are little more than flippers; but those answering to the fore-limbs have changed less than the hind-legs, which have become excessively shortened, and incapable of being flexed, as are the hind-legs of terrestrial animals. In the whales this transformation has proceeded yet further. Herein the fore-limbs have become so changed as to be totally unlike those of land-dwellers. No fingers are visible, and when the hand is dissected these fingers are found tightly bound together and enormously lengthened. The hind-legs have vanished completely. In the old fish-lizards, and in the penguins, among birds, we find the fore-limbs changed in response to this aquatic life. But the change has not been effected to attain quite the same ends; for in the fish-lizards and whales the paddles serve simply as balancers; in the penguins they are propellers, for the penguin swims by means of its paddles, which are modified wings, which its ancestors used to fly with. This being so, it is curious that the hippopotamus, one of the most aquatic of mammals, displays not the slightest evidence of this fact in the shape of its legs.

Now what is true of aquatic is true also of land animals, though the causes, or stimuli, to the changes are often difficult of discernment. What, for example, determined the evolution of the snakes? These creatures, we know, once possessed limbs like those of the lizard; and evidence of this fact is to be found in vestiges of the hind-limbs, which, in the pythons, for instance, are to be found at the base of the tail. And this loss of the limbs, it is to be noted, has gone hand in hand with a lengthening of the body, the number of the vertebræ which make up the backbone having been enormously increased. More on this head of the evolution of the snake we cannot say; the theme has been introduced because of its important bearing on what follows.

Briefly, there are certain lizards which, superficially, are so like snakes that only an expert can tell the one from the other. Our common Slow-worm is one of these. Every trace of

Photo by] [*W. B. Johnson.*

THE SLOW-WORM.

Another degenerate from the Lizard type. In some specimens of the Slow-worm signs of rudimentary legs may be distinguished, but in others external marks have completely disappeared. It should be noted that the folk-name of this animal is entirely a misnomer, for it is no worm, but a true Lizard.

limbs has disappeared long since, so that a careful examination would be necessary, for those who did not know the Slow-worm at sight, to decide whether it were a lizard or a snake. The distinguishing marks apart from coloration are not numerous. In the first place, the Slow-worm has a movable eyelid ; the true snake has no power of closing its eyes, which are protected from injury by a transparent horny plate, like a watch-glass. Hence the stony stare of the snake that is so disconcerting. In the second, the Slow-worm has a slit marking the aperture of the ear, which is not found in the snake. Thirdly, the belly of the snake is provided with a series of horizontal plates which play an important part in locomotion. These are wanting in the Slow-worm. The Glass-snake, or Scheltopusik, of Southern Europe, is another of the lizards, which, in like manner, has lost its legs. In this snake the body bears along each side a deep groove, the purpose of which is not precisely known. Those who might desire further evidence as to former existence of legs in the Slow-worm would have to make a careful dissection, exposing the shoulder-girdle and hip-girdle, of which, at any rate, vestiges remain. In the Glass-snake small flaps of skin mark the last traces of the hind-legs ; the fore-legs have vanished utterly.

But even after the examination I can imagine some being still rather doubtful as to the inter-

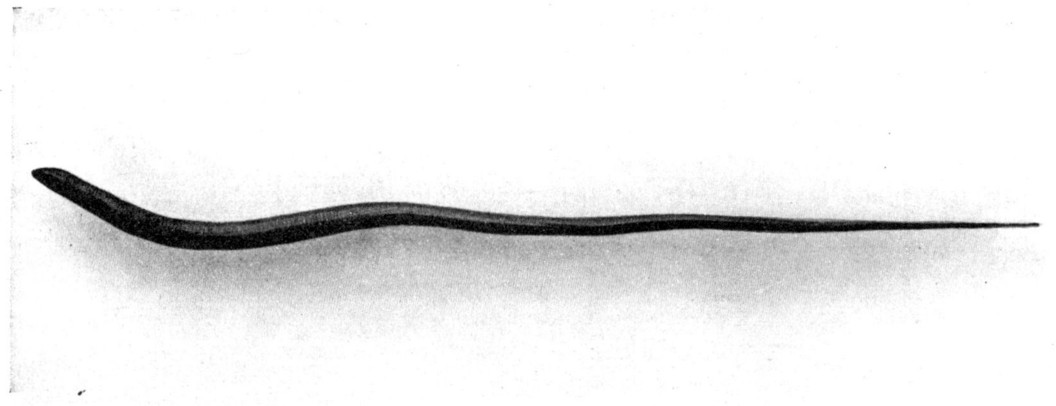

THE GLASS-SNAKE.

This is one of the most interesting of the American Lizards. It owes its name to the fragile jointing of the backbone, for the bones of the tail are so slightly linked together that it will break off at the slightest blow. This footless Lizard is therefore seldom captured without an injury.

pretation of the evidence. Such will find all hesitation vanish, I imagine, after an examination of certain lizards known as skinks, wherein every gradation from the condition of four-leggedness to complete limblessness will be found. In the Biskran Skink both pairs of limbs are present, and each foot has five toes. It is, as may be seen in our illustration, a typical lizard. In the Atlas Mountain Skink the legs have all dwindled amazingly, the front limbs having become reduced to useless remnants ; yet all have four toes. In the Suakin Skink this process of degeneration has proceeded a stage further ; and the fore-leg, now difficult to see, has still further reduced the number of its digits, only three remaining. But we have two stages further yet to go. The Three-fingered Skink has passed so far towards the limbless condition, that its legs are represented only by short, cylindrical stumps, cleft into three at the free end ; digits these clefts can hardly be called, though they answer thereto. Finally we have Günther's Skink, wherein minute stumps, barely traceable in the living animal, and not at all in photographs, are all that remain of once functional five-fingered hands and feet.

Only one other lizard need be mentioned in this connection, and this is the Scale-footed Lizard, wherein the fore-limbs have vanished completely, and all that remains of the hind-limbs is a pair of scale-like outgrowths. In all else, like the legless skink just referred to, the creature is, to the uninitiated, a snake—having, like the snakes, an excessively elongated body and no legs.

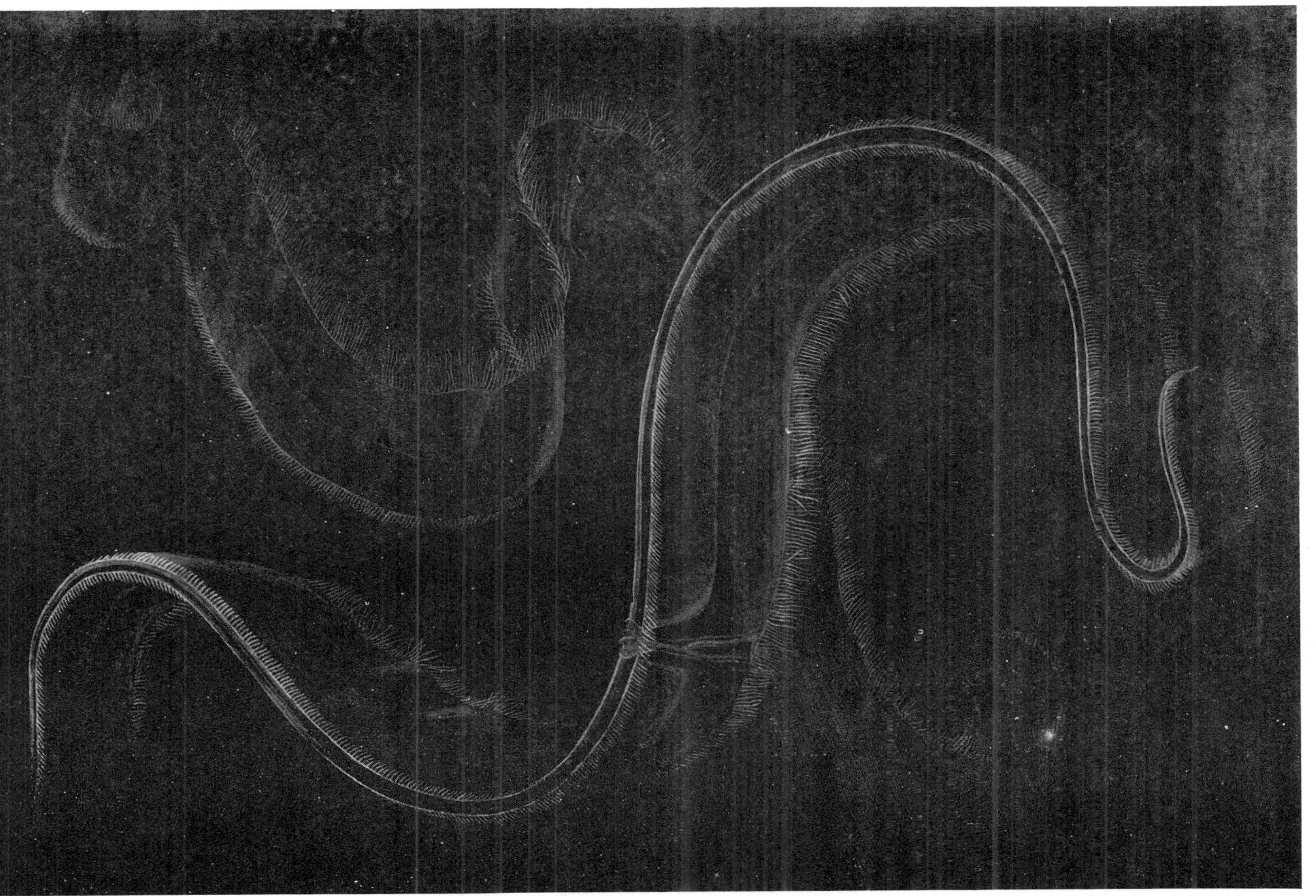

[By Theo. Carreras.

THE GIRDLE OF VENUS.

This is the most beautiful of all the comb-bearing Jelly-fish. It differs much in shape from its near allies, the Sea-Gooseberries, but its habits are the same. It floats at the surface of the water, or moves along by means of an undulating movement and the vibration of the combs. In the light it glows with iridescent colourings, but of itself it is transparent and almost colourless. It will reach a length of from two to three feet.

Photo by] [*Robert Whitbread.*

A SANDSTORM.

A photograph taken near Kartum in the Egyptian Soudan of an approaching sandstorm. The chief factor in such a storm is the simoon, or whirlwind, which accumulates in itself vast clouds of sand and dust which it cannot support and which it releases on its onward course.

THE GIRDLE OF VENUS

THE Girdle of Aphrodite would perhaps be a more fitting title to apply to this jewelled riband of the sea; for the Greek goddess was more definitely connected with the sea than her later rival, the Roman Venus. Aphrodite, as her name implies, was born of the foam; to her were credited the dual divinities of Beauty and the Sea, and, therefore, to her were dedicated all the fairest treasures of her realm. But to-day Venus has usurped her rank and her attributes, so that we frequently meet with such names as Venus' Flower-basket, Venus' Ear, Venus' Looking-glass and Venus' Girdle. But what may be this Girdle? Imagine a transparent riband some five feet in length, coiling and uncoiling as it curves its way through the water, the long edges of the riband fringed with small thread-like tentacles or *cilia*. Imagine it, too, with the sunlight playing on it, when it will reflect a thousand wonderful tints of silver and azure, pink and amethyst, or, on a dark night, when its rainbow hues will be changed for brilliant phosphorescence and it will appear as a glowing riband of lambent flame. Surely so beautiful an object might not unworthily be ascribed to the goddess of the sea. But the nature of this wonderful riband has not been described. It is a jelly-fish and belongs to the group of Sea-Gooseberries, which have already been described (see page 824). Although in shape it differs so peculiarly from these little sea-balloons with their long tentacles, it is a difference which is more apparent than real; for, as a well-known writer has aptly remarked, if one of these little balloons could be rolled out in the same way as a cook rolls out pastry with a rolling-pin, the result would be just such a riband as we have here. The mouth of this queer creature is situated in the middle of the lower edge and opens out into a large stomach. To right and left extend long canals which carry the nutriment into all parts of the body. Since the body proper extends across the shortest width, Venus' Girdle must be considered to be short and very wide rather than long and narrow. It is to be found in the Mediterranean and is particularly plentiful in the Bay of Naples, where the fishermen have likened it to a gleaming sword and given it the name of *sabre de mer*.

THE WIND

BY F. W. HENKEL, B.A., F.R.A.S.

THE Wind, that potent, invisible agent whose variability is in common parlance a synonym for all that is most changeable, is of importance in more ways than one. Though even yet our knowledge of the movements of the vast atmospheric ocean at the bottom of which we dwell is very imperfect, by the observations of meteorologists all over the world a considerable body of important information has been obtained as to the main features of its circulation both in the lower and in the upper regions. The action of the Wind as a determining, perhaps the main, factor in the important movements of the ocean known as currents, its influence as a geological agent in promoting and assisting the *weathering* of the rocks and in removing abraded material from one place to another, piling it up as loess, sand dunes, etc., will also come in for our consideration in this brief sketch.

Though in our own country, as we have already mentioned, the very name of Wind is usually regarded as a symbol for all that is variable, yet there is nevertheless a certain amount of regularity in it unsuspected by the chance observer. In England and over Western Europe generally, the south-

Photo by] [*G. R. Ballance.*

WIND MARKINGS ON SNOW.

Very beautiful are the ridges and patterns made by the Wind on the soft surface of the Alpine snow-fields. Sometimes the snow is swept into long ridges, an effect that results from a strong and persistent Wind. Here the uneven surface of the snow is due to a strong and choppy wind.

west Wind is far more prevalent than any other ; the late Mr. Glaisher, of Greenwich Observatory, from the records of ten years' observations, giving it an average prevalence (annually) of two thousand seven hundred and thirty-seven hours—more than twice that of any other. In Asia and Eastern North America the north-west Wind is perhaps more frequent. The Trade Winds (north-east or south-east, according to the hemisphere), blow regularly for many months at a time over wide areas ; the monsoons of the Indian Ocean blow for six months in one direction, and then again for the other six months of the year their direction is exactly reversed. Two main directions of air currents, due to the difference of temperature between the equatorial and polar regions, are especially to be noticed, the equatorial warm currents northwards and southwards towards either pole, and the polar cold currents from each pole towards the equator. It may here be noted that the *Wind* is named with reference to the points of the compass *from* which it comes, *e.g.*, a south Wind is a Wind *coming from* the south and thus *blowing towards* the north. The intensity of the Wind is estimated in practice by experience of sensation, though various instruments, known as anemometers, have been devised from time to time, but their indications are even yet not very reliable. According to its intensity, the Wind is variously estimated as light, moderate, fresh, strong, gale, storm or hurricane. A scale proposed by the late Admiral Beaufort, and consequently known by his name, is still in very general use amongst sailors and practical men generally. On this a series of numbers from 0 to 12, for "calm" to "hurricane," is taken, the former for cases when there is no Wind, the latter when its velocity is one hundred miles per hour or upwards, speeds not uncommon in tropical latitudes, but, fortunately, almost unknown in our too-much-maligned British climate.

Photo by permission of] [*The One and All Association.*

A TORNADO IN OKLAHOMA, U.S.A.

A tornado is a mass of air in rapid rotation round a nearly vertical axis. The cylinder of air, or "spout," has a violent upward motion. The central column of rarefied and cooled air becomes condensed into vapour, which often takes the form of a water-spout.

The name of cyclone, popularly applied to some violent but comparatively rarer storms of the tropics, is given by meteorologists to regions of atmospheric circulation, usually more or less circular or oval in form (whence the name—κύκλος, Greek=circle), in which the air moves round and inwards towards a central region of comparatively lower pressure, and the main difference between the various kinds of disturbances of this nature is a difference in size and sometimes in shape.

Photo by] [*George Wharton James.*

THE EAGLE'S HEAD, ARIZONA.

One of the landmarks in the world-famous Petrified Forest. It is a splendid example of erosion, the process by which rocks are crumbled away and which is chiefly due to the action of the Wind.

[By Arthur Twidle.

A CYCLONE IN AUSTRALIA

When these cyclones, or "willi-willis," take place the damage caused is enormous. Ships are often thrown inland by huge tidal waves, while in the swirl of the ebb thousands of cattle perish.

The pressure at the centre of a cyclone is always lower than at its borders. When this fall is gradual we have the more or less unsettled weather conditions with which we are all familiar; when it is very rapid over a short space we get the storms of greater or less violence to which various local names are also given. The tropical cyclones, besides being more intense, are usually much smaller than those of temperate regions. Of a somewhat different character is the *tornado*, which may be described as a mass of air in rapid rotation round a nearly vertical axis; in height it may be as much as one hundred and fifty to two hundred feet, but its breadth is usually very much less. The cylinder of air, or funnel "spout," as it is called, has a violent upward motion; it sways to and fro, and the whole system also has a forward motion. The speed of the rotary motion may sometimes run up to five hundred miles per hour; the upward motion, though slower, often exceeds eighty miles per hour. The forward movement is much slower than this, and usually in a north-easterly direction. The central column of rarefied air being cooled by expansion, any vapour within it is condensed, and so we sometimes have a "water-spout" accompanying it. Great damage is often done by tornadoes—houses, trees, fences, churches, etc., being blown down and carried up into the air for a considerable distance.

The "dust-whirl" of the roadside is a miniature tornado, whilst the "simoon" of the desert is a whirlwind carrying large masses of sand and dust, under which caravans are sometimes buried. The "pampero" of the Argentines is accompanied by much moisture, whence arises thunder, rain, and a typical form of cloud, the "cloud wreath." In northern latitudes a similar kind of disturbance is known as a line squall, and such a line squall it was which caused the capsizing of H.M.S. *Eurydice* off the Isle of Wight in 1878.

Of various other local varieties of Wind distinguished by special names, we may mention the Föhn Wind of Switzerland, the Sirocco of Italy, the Mistral of Southern France, the Chinook Wind of Canada, the Blizzards and "Barbers" of the States, the Nortes of Mexico, the Harmattan of Northern Africa, the Nor'-Wester of India, and the "Arched" and Bull's-eye squalls of the tropical seas. The Föhn Wind is a hot, moist current of air which, after crossing the Swiss mountains, has some of its moisture condensed, and becomes compressed and hotter.

The "Sirocco" of Italy and Sicily is commonly described as a dry, hot wind, and the "Harmattan" of Western Africa, is an east wind blowing from the desert, carrying clouds of red dust far out into the Atlantic. The cold, often snow-bearing Winds of North America known as "Blizzards," or sometimes as "Barbers," are familiar by name to the newspaper reader, though the name has come to be popularly applied to *any* cold wind during the winter which is accompanied by driving rain, snow or sleet. The "Nortes" of Mexico, the "Bora" of the Adriatic, and the "Mistral" of Provence, are cold, northerly Winds, coming at the rear of cyclonic disturbances. Another series of Wind phenomena, the land and sea breezes, common on tropical islands, but not altogether unknown elsewhere, may also be briefly mentioned. The common explanation of these phenomena as due to the varied heating of the land and adjacent sea only partially accounts for the observed details of the circulation; but we must refer to the author's "Weather Science" (pp. 158–9), for further particulars. By day the air over the land is heated and *rises*, its place being taken by cooler air from the sea; we thus get the *sea* breeze; by night, the action being reversed, the air near the ground flows outwards as the *land* breeze. On a low, flat coast such breezes are usually of slight force, but when there is a range of mountains not far from the sea the resulting winds are much stronger. Thus the sea breeze of Jamaica, owing to the proximity of the Blue Mountains, is often very powerful.

The action of Wind as a geological factor now calls for our consideration. The surfaces of rocks are exposed to various agencies, and undergo the changes known by the general name of *weathering*. The rocks themselves, when not too hard, are worn down into fine particles by the action of frost, rain and running water, and this material is blown by the wind to lower levels or carried away

A SHELL-BORING SPONGE.

It is no uncommon thing to find oyster-shells pitted with holes made by this boring Sponge. Sometimes, as in this instance, the Sponge bursts the confines of the shell and develops externally into large cork-like masses.

Photo by] [*A. Leal.*

SPONGE SPICULES.

The variety of these flint needles which support the structure of the Sponge is limitless. Sponges are grouped according to their spicules.

altogether Rocks consisting of harder and softer particles undergo curious changes, the former parts persisting, the latter being removed by the denuding agencies. The *tors* of Cornwall afford striking examples of this weathering effect, a number of blocks piled one upon another often having a very top-heavy appearance, their softer materials having been removed, leaving the harder granitic masses remaining. On shores where stretches of sand are left dry the action of wind in modifying the surface is well shown. Sand *dunes*, or hillocks, are thus produced by the accumulation of wind-borne particles, sometimes deposited in a manner resembling stratification, in roughly horizontal layers piled upon one another. In Western Europe blown sand occupies a large part of the coast from Spain to the Baltic, advancing at the rate of three to twenty-four feet per year. Sand dunes of a height of over one hundred feet are met with on the shores of Lake Michigan, and forests and even low hills have been buried by the forward movement of the wind-driven particles (Geikie's " Geology," p. 443). The deposit of yellowish clay or loam, first noticed in the valley of the Rhine, but met with in many other places also, known as *loess*, consisting mainly of hydrated silicate of alumina (clay), with grains of quartz, flakes of mica, and sometimes lime, was considered by Richthofen to be due to the long-continued drifting and deposit of fine dust by wind over areas more or less covered by grassy vegetation, but Dr. Davison has since suggested that loess is best explained on the supposition that it has resulted from snowdrift.

The great ocean currents are due, in part at least, to the action of Wind. The Winds not only agitate the sea and produce irregular waves, but, where they are continually blowing in one direction, they cause the surface water of the ocean to take a similar course. Of these resulting (ocean) *currents* the best known, and to us the most beneficial, is the Gulf Stream, a body of warm water coming from the Gulf of Mexico. Passing northwards, it runs parallel to the eastern coast of the United States almost as far as Newfoundland, then turns north-eastwards towards the Azores, where it divides, one portion going past these islands and to the north of Norway; the other part bends more to the right, passes the coast of Portugal, and then runs back in a westerly direction to the West Indies. Partly as a result of this warm current, but still more as a consequence of the influx of moisture-laden air from these warmer regions, which, condensing, sets free heat, our climate is so much milder than that corresponding to similar latitudes in other parts of the globe.

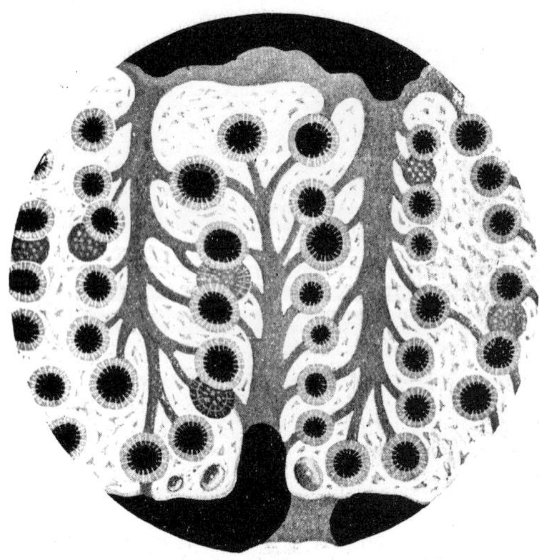

SECTION OF A SPONGE.

Showing the channels for the in and out currents of water and the whip chambers where the absorption of nutrition takes place. This section is of a complex Sponge.

The polar current of the North Atlantic, on the other hand, is a cold-carrier, whose course southwards from Baffin's Bay influences the temperature of the Atlantic coast of America in the opposite sense to that in which the Gulf Stream

NEPTUNE'S CUP SPONGE.

These huge Cup Sponges from the East Indies are the giants of their kind. They will sometimes reach a height of over five feet, and their girth is large in proportion. The skeleton is composed of a dense network of bundles of pin-shaped spicules.

A "CHALLENGER" SPONGE.

The "fronds" are bowl-shaped. On the convex sides are the oscules, on the concave the incurrent pores.

affects the corresponding latitudes of Europe. In the Southern Hemisphere, the Peruvian or Humboldt current of the South Pacific, and the Agulhas current of the Indian Ocean, the former a cold current, the latter a warm one, are best known. The currents of the Indian Ocean in general are "wind products," governed in their direction by the prevailing winds, flowing westwards under the influence of the north-east monsoon, and eastwards when under the influence of the south-western. Thus, by these currents flowing in various directions, some carrying water hotter than the surrounding sea, some carrying cold water and sometimes icebergs, there is produced a constant circulation of the waters of the ocean.

Variations in atmospheric pressure also give rise to fluctuations of the level of lake surfaces, to the discharge of springs, and modify considerably all tidal phenomena. The impulse of the Wind upon a water surface throws the latter into pulsations which vary from mere ripples to huge billows. The latter often effect much destruction upon shores or banks exposed to their action.

The transport of volcanic dust by Wind may be here referred to as another example of the geological work done by the atmosphere. After the great eruption of Krakatoa, it is estimated that clouds of fine dust were hurled from that volcano to a height of seventeen miles; and by the agency of the Wind this material was transported for thousands of miles. It is even thought that the presence of this matter in the air was the cause of the remarkable sunsets witnessed in Europe during the autumn and winter of 1883 and the spring of 1884.

SPONGES

BY BERNARD C. WHITE.

It would be as well to start this article with two cautions: firstly, don't look upon sponges as plants; and, secondly, don't think that the living creature is at all similar in appearance to the prepared sponge of domestic use.

Sponges are animals, or rather colonies of animals, assembled in one indivisible body for their mutual well-being. Certainly they are a very low form of animal-life, for their methods of reproduction are very plant-like. Here is an easy test by which their animal nature can be proved: subject a small piece of sponge to the flame. It will not burn brightly, but will char, and when entirely burnt up, will leave a small charcoal-like residue. All the time that it is smouldering it will give off a smell very similar to the burning of *hair*, a result which is seldom to be observed in the burning of vegetable matter. Again, chemical analysis has shown that the composition of sponge is very little different from that of silk, which is definitely an animal substance.

As regards the appearance of sponges, there are so many different varieties and forms that it would be impossible to enumerate them all; in fact, it has been well said that by reason of the wide diversity of forms sponges might with equal justice be considered formless. Some, like the "Crumb of Bread" Sponge, will cover the surface of flat rocks with ranges of miniature volcanoes; others, such as the Tree Sponge, also of our own shores, divide up into branches with strong suckers, by which they are attached to the rock. Others are spherical and entirely unattached, and are rolled about on the sea-bed by the motion of the waters, the toughness and elasticity of their fibre protecting them from damage.

There is a species of Glass Sponge which exactly resembles a mushroom, while yet another sponge is merely a floating mass of fine white and semi-transparent tubes united together by a loose network, and other strange varieties are illustrated in these pages. But the best-known forms of sponge are the "Cup" Sponge and the Bath Sponge of commerce. The first, as its name implies, is cup-shaped, of very fine texture and innumerable small apertures. It is sometimes called the Common Bath Sponge, is altogether more loosely knit, has larger apertures and no very definite shape. Every sponge in its natural state is covered with a thick gelatinous membrane, entirely concealing the fibrous material which is usually recognized as "a sponge" (but which is really only the skeleton of a sponge). In this flesh is stored the protoplasm that goes to build up the wonderful structure; it is a thick, milky and glutinous substance, and has to be extracted before the sponge can be placed on the market, for it quickly putrefies as soon as the living organisms composing the fabric are destroyed; the process of extraction is called "taking the milk out" of a sponge.

In spite of its lowly position in the scale of animal life the structure of a sponge is very complex. The material which composes the cup may be divided into two sections, the outer or fleshy part, and the inner or fibrous part. The body tissues are composed of various kinds of cells, some of which assist in the process of nutrition, while others are employed in

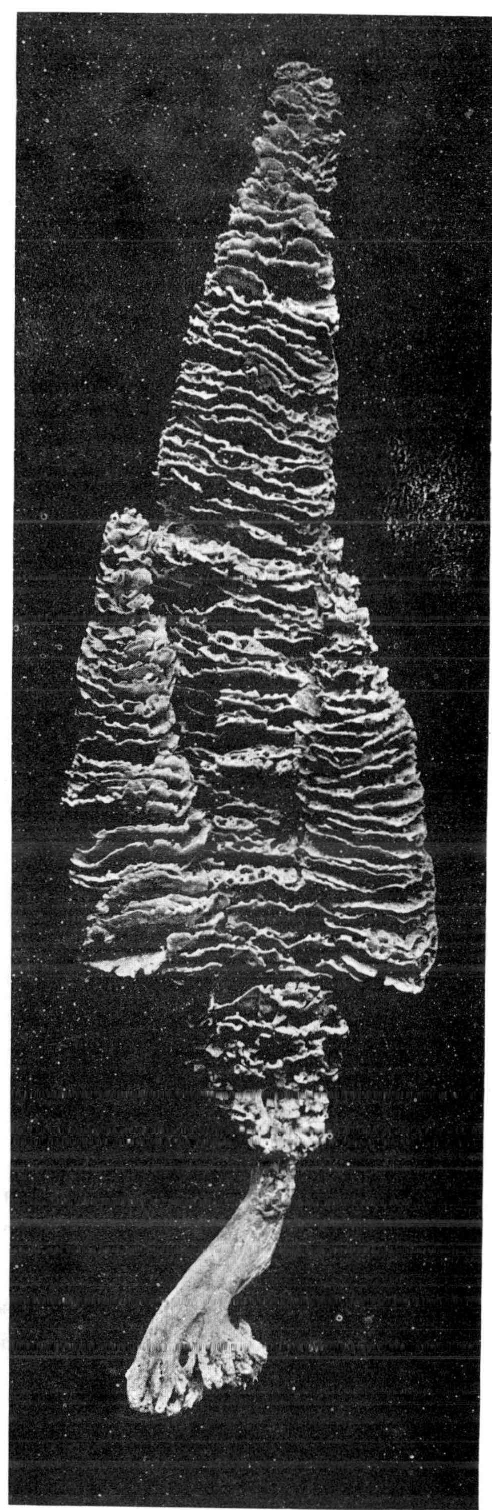

A STALKED SPONGE.

This beautiful Sponge has a thick main stem with three branches. From these branches rise spiral blades of sponge-tissues diminishing in size as they ascend.

developing the fibrous skeleton, and yet others form a kind of lining to the body substance. At certain seasons this " body" produces egg-cells, which ultimately develop into little oval embryos, with a convex base covered with tiny threads, or " cilia," which by their lashing movement drive the larval sponge freely about the water.

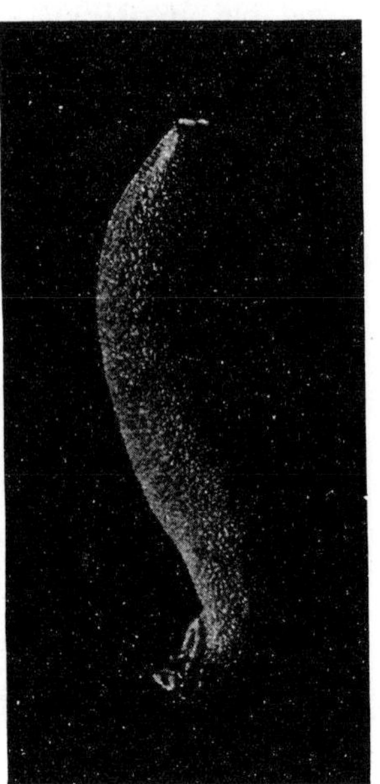

Photo by] [*W. Plomer Young.*

One of the simpler forms of Sponges and of limited growth.

The inner part of a sponge consists of yellow solid and interlacing filaments or threads, which have already been in part described. In some species these threads are extremely fine. For instance, in the best Turkey Sponge they are about the nine-hundredth part of an inch in diameter. They are very tough and will resist the action of practically any acid; but they, nevertheless, are not sufficiently firm to balance and support many sponge growths, and are, therefore, assisted in their work by numberless minute flint needles, which are known as spicules. Very marvellous is the diversity of form to be found in the spicules. They are shown as they appear under the microscope on page 904, but though the examples are numerous, they form by no means an exhaustive collection of these strange architectural pins. They are to be found scattered through the bulk of the sponge, sometimes as single particles; in other cases, hooked one into the other to form a complete flinty framework. As a general rule, sponges are classified according to the composition and structure of the skeleton, and are again subdivided according to the various types (single-rayed—four-rayed, etc.) of their spicules.

A TREE SPONGE.

This wonderful Sponge is to be found on the banks of the Amazon. It attaches itself to branches of trees which are submerged during the floods, and is so constructed that it can survive the period of drought when it is quite out of reach of the water.

But no mention has yet been made of the internal economy of the sponges, for this is not the least marvellous part of their structure. Let us take a simple sponge, such as the Sycon shown in the upper illustration on this page. You will notice that at the upper end is an open mouth much like the mouth of a well. This is known as the oscule, and if you were to examine the sides of this sponge you would find a great number of little pores. It is by reason of these characteristic pores that the whole group of sponges has been given a collective Latin name, which signifies pore-bearers. Careful examination will show the use of these pores; they will be seen to allow the passage of minute streams of water, which pass through diverse channels and passages until they meet in the central hollow and pass out in one continuous stream through the oscule. But the water that passes away has been changed; it has been deprived of the minute particles of matter which were held in it, and it has also exchanged its oxygen for carbonic acid gas. How, then, has this change taken place. Let us cut a horizontal section of the sponge, and we will find neatly-

THE WALLACHIAN SHEEP.

By a process of natural selection, in which man has recognized the tendency and by careful breeding assisted Nature, many remarkable types of domestic animals have been evolved. It is not all animals that are capable of domestication, but the Sheep is eminently fitted for the service of man and has undergone many strange transformations. The Wallachian Sheep is remarkable for its long twisted horns.

arranged rows of cells, which are internally covered with still smaller flask-shaped cells, which are attached to the wall of the " whip chamber," as the cavity is called, by their convex bases, while from the open mouth of each flask protrudes a single cilia or thread, which is about one three-thousandth part of an inch long. All this elaborate arrangement is, as can be imagined, on the minutest scale possible, yet every single flask is a separate entity—a complete individual. By the motion of the cilia it causes a constant flow of fresh water. From it, it sucks out all available nutriment, and again by a contrary motion of the cilia it passes away the impoverished stream.

A very much magnified section of a complex sponge is shown in the lower illustration on page 908. Here can be seen the circular (or rather polygonal) whip chambers and the branched channels which lead from the pores and to the oscules of the sponge. The difference between a simple and a complex sponge is not so great as it at first appears. To put it briefly : The whip chambers are ranged neatly round the oscule, or central mouth, of the simple sponge, and the water channels lie directly between the external and internal surfaces. In the complex sponge the sponge wall has become elaborately folded, so that the water channels are also folded, and branch out into further auxiliary canals, which, in their turn, connect with auxiliary oscules.

THE FOUR-HORNED PIEBALD RAM

In this species the horns, which in the domestic Sheep are non-evident, have attained a maximum of growth

The late Professor Huxley summed up the nature and organization of a living sponge in a paragraph which is worth remembering. He described it as " a kind of sub-aqueous city, where the people are arranged about the streets and roads in such a manner that each can easily appropriate his food from the water as it passes along."

SOME WONDERFUL SHEEP

BY W. P. PYCRAFT, F.Z.S.

All the various breeds of animals which man, in every quarter of the globe, has gathered around him, can be traced to different species of wild animals. And we commonly lose sight of one very significant fact in this connection ; and that is, that only certain species of wild animals have proved capable of domestication, and these have given rise to races which are often characterized by the extravagant development of one or more parts of the body.

These singular forms most of us seem to regard as out of court when considering the marvels of Nature. They are held to be " unnatural " products. In so far as they are not found in a state of nature, that is to say, in a wild state, this is true; but all the same, they are not artificial products, as the Chinese dragon, for instance, is artificial. On the contrary, they are as "natural" as lions and tigers. Their several peculiarities are perfectly natural "expression points" of innate or inborn tendencies of growth, which have been fostered by man's supervision. In some cases he has seized upon tendencies, which, greatly magnified, would increase his profit; in others, upon features whose only merit is in the greater extent of their freakishness and grotesqueness. But in no case has he made the character in question, he has only fostered its development by persistently selecting for his breeding-stock those individuals which possessed the coveted character in greatest degree.

THE FAT-TAILED SHEEP.

This South African species requires to store sufficient nourishment to withstand long periods of scanty pasturage. The tail has therefore been abnormally developed for the storage of reserve supplies of fat. It will weigh as much as seventy pounds.

The history of our domesticated sheep well illustrates this. The sheep, to most of us, is, before all things, a wool-producing animal. Now this wool is entirely the product of domestication: of the process of selective breeding. Wild Sheep, and most breeds of sheep in hot countries, are hairy, like goats, with a more or less well-developed "under-fur." In the domesticated sheep man has gradually bred out the hair, and enormously increased the under-fur, the Merino Sheep representing the maximum attained in this direction. This was originally a Spanish breed. But in 1783 Louis XVI. of France started an extensive Merino farm on an estate of his at Rambouillet. In the course of a century, on this estate, by careful selection, flocks remarkable for their great size were raised, and these were known as "Elephant Sheep," and famed for the excellence of their wool and their superior hardiness. These Rambouillet sheep have since been spread over nearly the whole of the sheep-growing world.

More remarkable, in some respects, are the Fat-tailed and Fat-rumped Sheep. The former is a South African breed, and is one of the most remarkable breeds in the world. The tail is of great length and excessive width at the base, and at its maximum may weigh as much as seventy pounds. In such cases the poor sheep is relieved, as much as possible, of its

THE FAT-RUMPED SHEEP.

Here the rump serves the same purpose as the tai[illegible] Fat-tailed Sheep. This Sheep is found in many parts of C[illegible]a, Persia Arabia, and North-East Africa—all countries, be it ob[illegible] subject to long periods of drought.

burden by fastening the tail to a board, often mounted on wheels. Here, again, neither freakishness nor ornament has been aimed at, but use, for in times of drought the fat on this tail is slowly absorbed, so that the tail, like that of certain wild animals, is a reserve store of food.

The Fat-rumped Sheep is found in many parts of Central Asia, Arabia, Persia, and North-East Africa. Here the tail is short, and the fat is accumulated on the buttocks. In Central Asia both a black and white strain are kept, the lambs of the former yielding the famous Astrachan " fur." Usually these sheep are piebald.

The Wild Sheep is a horned animal. Under domestication these horns have been bred out in some breeds, and multiplied in others, as in the case of sheep found in such remote regions of the world as Africa and the Hebrides. In the strange Wallachian Sheep they have become twisted like a cork-screw, so as to recall the horns of the Kudu antelope. This duplication of the horns is due to the duplication of the original pair, and a skull in the British Museum shows this process of splitting in its initial stages.

THE MANED SHEEP OF THE CAMEROONS.

The smallest Sheep in existence. An adult ram stands only nineteen inches high at the withers. Its black colour is remarkable, and rather points to the theory that domestication has rendered protective coloration unnecessary, and so the more conspicuous colour has been allowed to spread.

The smallest breed of sheep in existence is the tiny Maned Sheep of the Cameroons, and it is significant that limb-bones agreeing in size and conformation with those of this diminutive race have been discovered in Wiltshire. An adult ram of this dwarf breed stands only nineteen inches high at the withers. Most of the body is covered with hair about an inch long, but in the neck it is lengthened to form a kind of mane, and there is also a ruff on the throat. In the general coloration it is red, but the ears, face, throat ruff, buttocks, the under-parts, and the greater part of the legs are black. This extensive area of black is somewhat remarkable; and seems to have been derived from the extension downwards of the black band found on the flanks of the wild Moufflon, whose under-parts, for protective purposes, are white. Domestication has rendered protective coloration unnecessary, and thus the black has been allowed to spread without check.

BLACK BOYS AND GRASS-TREES

THE Black Boy here alluded to is a vegetable production, a native of Australia, like the human black boy. It is probable that the early British settlers in the colonies, ever on the look-out for the natives, constantly imagined that they saw them lurking in the bush, and found, when they were better acquainted with their surroundings, that what they had often taken for a " black fellow " with a long spear raised ready for use, was nothing more dangerous to them than a Grass Gum-tree. Apart from this remarkable resemblance, which has struck many modern travellers, the plant is

Photo by] [*W. Saville Kent.*

THE DRUMSTICK GRASS-TREE.

The Grass-trees are the last survivals of an age when rushes and sedges grew to enormous size. They are only to be found in the temperate districts of Australia. The Drumstick will sometimes reach a height of thirty feet, though the average growth is little more than twelve feet. The peculiar shape of the flower-stalks gives the name to the tree.

Photo by] [*W. Saville Kent.*

THE UNDERGROUND GRASS-TREE.

So called because the long black trunk which is apparent in the Drumstick is here hidden underground. The flower-spikes are shown fully developed; they are completely studded with tiny white star-like florets.

sufficiently interesting to merit a few lines of description.

There are several different species of these Grass Gum-trees, some having quite short stems, others tall, thick trunks. Seen from a little distance, the short-stemmed kind when in flower might easily be mistaken for a near relation of our common reed-mace or cat's-tail—often erroneously called bulrush. As a matter of fact the Grass Gum-trees are lilies, but that fact is only to be discovered by dissecting out one of the flowers from the densely packed assemblage of them that makes up the tall cylindrical flower-spike, often fifteen or twenty feet high, formerly mistaken for the black boy's spear. The leaves are very long and grass-like, and form a dense tuft at the top of the stem, whence they curve gracefully outwards and downwards. The white heart of the top of the stem, consisting mainly of tender, undeveloped leaves, is esteemed as food by the natives, whilst the colonists make use of the "grass" as fodder for their cattle. The taller species have rough-coated trunks resembling those of palms, the rough exterior being due to the bases of former leaves which are cemented together by a red or yellow resin which they pour out freely, and from which they derive part of their name. These resins are fragrant, and the yellow kind is known commercially as Botany Bay Resin, the red as Black-boy Gum.

Another group of Australian plants are known as Grass-trees—without any mention of gum—and they have a close superficial resemblance to the Black Boys, though the flower-spike is globose instead of cylindrical. Strange to say these Grass-trees are not related to the others, being great rushes instead of lilies. But then some botanists regard the rushes as degenerate lilies.

THE LARGEST OF THE BATS

BY SIR HARRY JOHNSTON, G.C.M.G.

Our [illegible] in Great Britain of a Bat is a small animal, not much bigger than a mouse; in fact [illegible]use" is an early name for Bat in English, and, in scarcely different pronunciation and sp[illegible] German. The Insect-eating Bats are for the most part small, though some of them have bodies as large as a rat. But the Fruit-eating Bats, which form a sub-order very distinct from the Insect-eating section, attain to a much greater size. A photograph of what is perhaps the largest known Bat is here given, the specimens in question being exhibited in the Natural History Museum at Munich. The species to which this Bat belongs is said from time to time to produce examples measuring as much as five feet right across the expanded wings. The examples which I have photographed at Munich are not so big as this, perhaps only three feet or a little over, but they are of the same kind, the well-known Kalong, or Fruit Bat, of Java, Sumatra and the Malay Peninsula. These huge Fruit Bats are a russet-brown in the colour of their fur, and the leathery wings are blackish-brown. They emit a strong musky odour. Their chief food consists of bananas, mangoes, and other fruits sufficiently soft in their outer covering for the Bat with its strong teeth to penetrate through the rind to the pulp. They are noisy beasts, especially at night-time, when they quarrel a great deal over their feasting. There is something very dislikable about Fruit-eating Bats. None of them are good to eat (except to hungry savages), and most of them are haunted by a repellent musky odour, alive and dead. Some genera in Africa and elsewhere develop heads of such monstrous ugliness that no sprite or goblin more frightful in appearance has ever been conceived by the artist of a fairy-tale book. When seen flying against an evening sky (for they are mainly nocturnal in their habits, though more wide-awake in the day-time than Insect-eating Bats), they answer well to the description applied to them by Arabs of "The Devil's Birds." Like most other Bats, their fur is tenanted by large, disgusting-looking lice; they are, as I have said, so noisy as to be a perfect nuisance (I well remember sleepless nights and disturbed days passed

Photo by] [*W. Saville Kent.*

THE ARBORESCENT GRASS-TREE.

The finest specimen in Western Australia. To the right hand is shown a single flower-spike. Another name given to the Grass-tree is the "Black Boy," on account of their supposed resemblance (at a distance) to a bent and wizened native with a shock head of hair

at Zanzibar through my tenanting a room near which grew a fig-tree harbouring immense numbers of Fruit Bats), and, lastly, they are most destructive in the gardens and orchards of natives.

Some Fruit Bats in Malaysia and West Africa develop tongues of great length covered with horny teeth. With these they rasp the rind of fruits so as to suck out their pulp.

Except for the usual absence of a tail (though this organ is developed in one or two types), the Fruit-eating Bats are somewhat more generalized in their structure than the Insect-eating sub-order, and perhaps approximate a little more to the Lemurine type, which seems to have originated the Bats at a far-off period in the earth's history when the primitive Lemurs and the Insectivores scarcely formed distinct orders. There is a good deal about the head of the Fruit Bats which suggests affinities with the Lemurs and also with that strange creature of the Malay Peninsula so difficult to classify, the Colugo, which does not precisely fly, but glides down, and even up, through the air on its greatly expanded, skinny membrane between the limbs.

Photo by] [*Sir Harry Johnston, G.C.M.G.*

THE FRUIT BAT.

This is the largest member of the Bat tribe. The fur is russet-brown in colour, and the leathery wings are blackish-brown. The Fruit Bat emits a strong musky smell and feeds chiefly on bananas, mangoes and other soft-skinned fruit.

THE TUATERA

BY R. LYDEKKER, F.R.S.

THE claim of a comparatively insignificant-looking species of lizard-like reptile, now restricted to two or three small rocky islands off the coast of New Zealand, to a place among marvellous animals is based on the fact that it is the sole living representative of an otherwise long-extinct group. A casual observer, on being shown one of these Tuateras, as they are called by the Maori, would probably remark that it looked a very ordinary sort of lizard; and, as a matter of fact, it presents a considerable superficial resemblance to one of the larger Tropical American Iguanas, measuring, as it does, from about eighteen inches to something over a couple of feet in length, and having a flexible crest of spines running along the back and tail. Its internal anatomy, as well as its teeth, differs, however, *in toto* from that of any true lizard; and to such a degree that the creature has no right to be included under that name, although it may be so designated in popular language.

For, despite its Iguana-like form, the Tuatera comes in many respects nearer to a crocodile or a tortoise than it does to a lizard; and it is a member of a very primitive and ancient group of reptiles, of which some of the other representatives have left their fossilized remains in the New Red

Photos by] [*Sir H. H. Johnston, G.C.M.G., and W. Saville Kent.*

FRUIT BATS

The Fruit Bat is a repellent creature. It has a disagreeable odour and is a noisy and quarrelsome creature, while some of the African genera have heads of such monstrous ugliness that nothing more frightful in appearance could well be conceived. The Arabs have well named them "The Devil's Birds"

Sandstone of Shropshire and Cheshire, as well as in the corresponding formations of Central India and elsewhere.

The Tuatera is, therefore, eminently entitled to the designation of "living fossil," which I have elsewhere proposed for such survivors from earlier periods of the earth's history.

In addition to its stout bodily form and spinal crest, the Tuatera is characterized externally by the generally dark olive colour, sprinkled with paler dots along the flanks, of its granular skin, as well as by the cat-like shape of the pupil of its brilliant eye.

But the creature has yet another claim on our attention—and this of far greater import to the evolutionist than any hitherto noticed—for deep down in the tissues of the brain is buried the rudiment of a cyclopean eye, which was, doubtless, functional in the Tuatera's ancestors, the position of this so-called pineal eye being still indicated by a hole in the bones of the forehead of the modern Tuatera. Probably, indeed, all primitive vertebrate animals possessed such an eye, of which the vestiges are better preserved in the Tuatera than in any other living animal.

Photo by] [*W. S. Berridge, F.Z.S.*

THE TUATERA.

The Tuatera might aptly be termed "a living fossil," for it is the sole representative of an ancient group of reptiles which flourished in the age of the New Red Sandstone. It is, besides, notable for a rudimentary third or cyclopean eye situated deep down in the tissues of the brain.

Formerly the Tuatera, whose claim to a place among "marvellous animals" my readers will, I think, no longer doubt, was found on the New Zealand mainland, where it appears to have been mainly exterminated by pigs. On its present island home it stands in sore jeopardy at the hands of rapacious collectors, and it is only too probable that we shall ere long be shown the last of the Tuateras in a bottle in some museum!

CORAL-FISH

BY W. P. PYCRAFT, F.Z.S.

AMONG those of us who have not had the good fortune to explore tropical and sub-tropical seas there is a general impression that fishes, in the matter of coloration, suffer considerably by comparison with, say, the birds. This impression is due partly to the fact that, with a few exceptions, our native fishes are lacking in striking colours, and partly because in our museums stuffed fishes, from whatever part of the world they may have come, are of the same uniform dull-brown hue. Yet, as a matter of fact, no group of animals contains more striking illustrations of the lengths which coloration may run to. The fishes which swarm amid coral-reefs, such, for example, as that which forms the Great Barrier Reef of Australia, afford some of the most remarkable illustrations of this fact. And it is evident that the vivid and bizarre combinations of colour which they present have been developed to harmonize, as it were, with their surroundings, for they live in a very riot of colour. The lowly creatures which have formed this enormous reef—itself one of the world's marvels—are themselves gorgeous in hue, and the formidable bulwark they have raised affords a home for myriads of anemones, sponges, molluscs, star-fishes, and so on, which have become, so to speak, compelled to live up to the tone of their neighbours, if only to avoid being conspicuous! For, striking as any of these gaily-coloured animals may be when seen apart from their surroundings, they attract, individually, by no means the same amount of attention when seen amid their native surroundings. But not the least remarkable fact about this wonderful coloration is the extraordinary speed with which it can be changed, at least in many species; for rapidity of change in this respect, indeed, the chameleon is a slow-coach. As a case in point, take that of one of the File-fishes, which, when taken out of the water was of a dark greenish-black, scribbled all over with fine yellow lines. But when turned into a white bucket of sea-water the yellow lines gradually expanded till the whole body was of a pale yellow minutely speckled with black. But changes of coloration, more or less rapid, are almost universal among fishes, though less noticeable in those of more subdued colours.

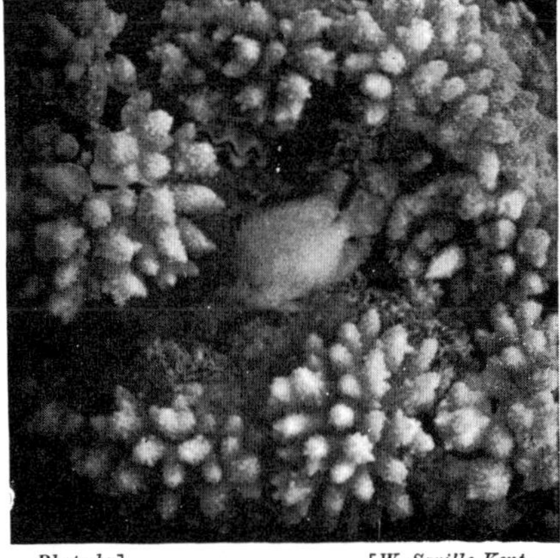

Photo by] [*W. Saville Kent.*

A Coral-fish in natural surroundings, showing its similarity to the coral amongst which it lives.

A CORAL-FISH.

One of the most beautiful of the inhabitants of the coral-reef, resplendent with blues and gold and green.

A reference to the accompanying coloured plate will do more to convince the reader of the barbaric splendours of these fishes than any verbal description. Take the central figure, which is one of the Scorpion-fishes. Even though it cannot be called beautiful, judged by human standards, it must be admitted that its coloration is handsome, while it is further remarkable in the excessive development of the breast-fins, which, from their great size, induced many to believe that this fish was a species of flying-fish. Such, however, is not the case. Like so many of its relatives, it is a dangerous creature to handle, for the spines of the back-fin are provided with poison glands and inflict most painful wounds.

The upper left-hand figure and the

lowermost are Wrasses, a family represented by about four hundred species, all of which are brilliantly coloured. Julis is often caught on coral-reefs off Cape Flattery by carefully lifting out of the water, and breaking open, growing clumps of a branching species of coral known as Porites. Such specimens are citadels of refuge when the tide is down, and yield a rich harvest of small fish of all sorts, as well as other forms of animal life. But the Wrasses, which live amid coral-reefs, like Pterois, are mostly carnivorous, and feed upon other fish and free-swimming animals. The coral-forming animals are apparently not eaten. But their near relatives, the Parrot Wrasses, live largely on coral polyps; and this because they have developed a special masticatory apparatus enabling them to deal with such stony morsels.

Balistes and Tetrodon exhibit a particular fondness for coral-reef, feeding largely on the polyps which rear these stony bulwarks against the fury of the ocean. Balistes, sometimes known as File-fish and Trigger-fish, feeds also on mollusca, and bores holes through the shells of some of the giant species by the aid of powerful teeth. Tetrodon is one of the Globe-fishes, which are almost

[By W. Saville Kent.

CORAL-FISHES.

Many of these fishes are very small, and are often caught amongst the reefs by breaking open growing clumps, where they are to be found lying in the crevices waiting for the return of the tide.

invariably represented in natural history books in the inflated condition which they assume when threatened by enemies. At such times the creature rushes to the surface, and filling the belly with air, turns turtle. The abdomen then projects like a bladder above the surface of the water, where, being blown by the wind, the frightened creature is wafted to a place of safety. In most species the body is armed with powerful erectile spines, but in some the spines are immovable. The flesh, by the way, of most of the File-fishes and Globe-fishes is highly poisonous, at any rate so far as human beings are concerned.

One could easily write a volume or two on the fishes which haunt coral-reef. But we can find space here only for one or two of the more remarkable types. Among the most celebrated and most remarkable of these are the Butterfly-fishes, represented by many different species. Those of the genus Chœtodon, Heniochus and Holœanthus are the best known. Some, like the Big-scaled Heniochus, are remarkable for the possession of a long whip-like appendage to the front of the back-fin, and, so far, of unknown function.

Though tinted with all the colours of the rainbow, and some to spare, and living amid a never

CORAL-FISHES.

Their colouring is the chief feature of Coral-fishes, and many striking examples are shown in the coloured plate. A most remarkable fact about this coloration is that in many species it can be quickly changed. Here is an instance: One of the File-fishes when taken out of the water was dark greenish-black, scribbled with faint yellow lines, but when it was turned into a white bucket of sea-water the yellow lines gradually expanded till the whole body was a pale yellow colour minutely speckled with black.

ending pageant of brilliancy, these fishes have their moments of anxiety and bursts of wild excitement. They spend their lives, indeed, in eating their smaller and endeavouring to avoid being eaten by their larger neighbours. Some, like the Wrasses and the Giant Serranus, are formidable foes indeed. One of the latter is the so-called Rifle-fish. Not only is it extremely ferocious and gorgeously coloured, but it is further said to be able to bring down prey from the upper world by jets of water expelled from the mouth. We repeat the story, but do not believe it has any foundation in fact!

MINUTE DETAILS OF BUTTERFLIES AND MOTHS

BY JOHN J. WARD, F.E.S.

THERE are more wonders in a butterfly or a moth than at first meet the eye. The charming forms and bright colours of these insects gratify our æsthetic tastes so completely that we are apt to forget those wonderful details of anatomy by which such pleasing effects are produced. Let us, therefore, by means of the microscope, look a little more deeply into the mechanism of these familiar insects.

The head of a butterfly or moth is largely composed of four pairs of organs: The two large eye-masses, the two feathery palps in front of the "face," a pair of antennæ or feelers, and a long coiled sucking tongue, or proboscis, consisting of two marvellously-constructed tubes, which can be separated and united at the insect's pleasure.

Considering the eye-masses first, we find that each of these is composed of an infinite number of six-sided facets, or lenses; indeed, each facet represents a perfect eye in itself, being the base of a lens-like pyramid, whose apex is directed towards the interior, terminating at the extremity of the optic nerve. It is, therefore, able to receive an image of an external object and focus it on the retina, just as a camera lens produces a picture on a photographic plate. Also, each pyramid-lens is coated on the sides with a dark pigment; it therefore acts quite independently of the others. From the human standpoint, the question then arises why an insect should need so many lenses in its "eye"? Does it see thousands of objects with each eye-mass, or only one? There is doubtless an answer to the question, but probably man will never know it; for his nervous organization is so different from

Photo by] [*E. J. Spitta, F.R.M.S.*

THE TONGUE OF A MOTH.

Very delicate is this beautiful apparatus which enables the Moth to gather the honey hidden in the deep recesses of tubular flowers.

that of an insect that any comparison is almost useless. He, however, at least knows that while his own two eyes each take up an image, yet only one picture is conveyed to his brain. Probably the Swallow-tail Butterfly, with eye-facets estimated in round figures at seventeen thousand in each eye-mass, and the Death's-head Moth, whose lenses likewise number about twelve thousand, both view the objects around them much as we ourselves do, or it may be that their different visual organs give them an understanding of form and colour that to us would be incomprehensible. The matter is still more complicated from the fact that moths, in addition to their compound eyes, often possess on the summit of the head ocelli, or single eyes. When they are present, there are two, but sometimes they are absent. Furthermore, in butterflies these single eyes are entirely wanting.

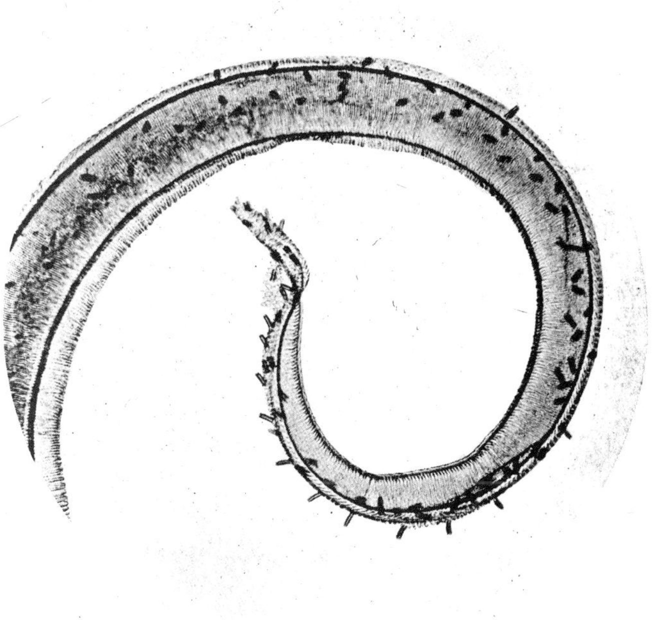

Photo by] [*E. J. Spitta, F.R.M.S.*

THE TONGUE OF A MOTH.

The scale of magnification is much larger than in the preceding photograph, and the little tubes which project from the tissues of the tongue can be easily seen. These tubes are probably taste organs, as are also the leaf-shaped excrescences clustered together at the tip of the tongue.

As most of the higher forms of insects possess ocelli, as well as compound eyes, it would seem that these single eyes have entirely lost their function and disappeared in the butterflies, and are now following the same course in the moths. Probably the feathery scales which clothe the heads of these insects may have played some part in their suppression by obscuring the light from them.

In spite of their extraordinary visual organs, there is no striking evidence to show that either butterflies or moths possess very keen sight. I am inclined to think that their power of distinguishing colour is much greater than that of recognizing form. The other day I was much interested in watching the familiar small white butterflies amongst some sweet-peas in my garden. Only a few flowers were blooming, and their colours were white, pink and purple. While watching for over an hour, I found that nearly every small white butterfly which flew near was attracted towards the few white flowers; and, as they do not attempt to sip nectar from the blooms, but immediately fly off again on reaching the flower, I could only assume that they mistook the flowers for others of their species, and had to approach closely before discovering their mistake. It was a significant fact that not a single one visited either the pink or purple flowers.

Then there is the behaviour of moths and other insects before a bright light, especially in the case of some of the hawk-moths, which sometimes become so reckless as to plunge at the globes of electric arc-lights with such force as to cause them to be hurled stunned and helpless to the ground —a fact which seems to show that their sight-organs differ from our own. It is remarkable, too, that some moths are never attracted by light. In explanation of this, I would suggest that a bright light probably presents to the eye of an insect a succession of rapid waves, or perhaps vortex movements, which exert on it a fascination very similar to that which a large, revolving flywheel of an engine, or even the continually rippling waves on the surface of water, does on many sensitive people. Just as there are people who can coolly face these things without losing control of their

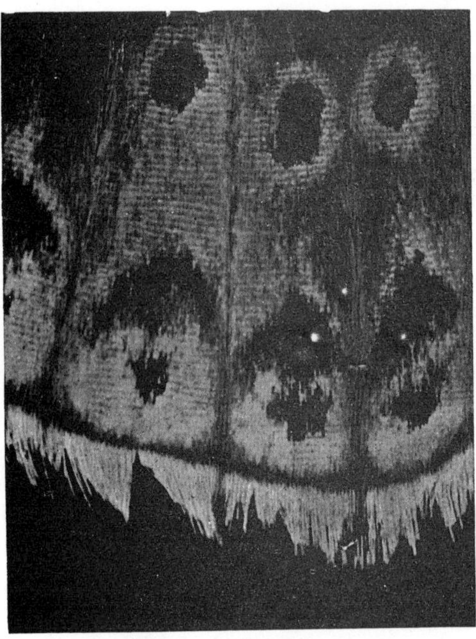

THE WING OF THE PEACOCK BUTTERFLY.

To show the marvellous arrangement of the countless scales which go to make up the beauty of a butterfly's wings.

nerves, so it may be that some moths are not attracted towards the lights, while to others every bright light becomes an irresistible vortex.

On the other hand, it may appear that butterflies and some moths possess very keen powers of vision, judging by their rapid manner of taking flight when we approach and try to capture them. We must not, however, necessarily conclude that these insects have seen or even heard us coming towards them, because in that case we are judging their movements by our own senses. Slight currents of air may have affected some of their problematic organs of sense, and they may have *felt* our approach, or even detected it by senses with which we are not familiar. As a somewhat analogous instance, it is well known that bats, when their eyes are covered, or completely destroyed, can avoid intricate obstacles during their flight, and find their way back to their nesting-holes. The whole subject of the senses of insects is a very fascinating one, and I shall again refer to it when considering the feelers of these animals.

Immediately in front of the face are the three-jointed palpi, which vary considerably in form with different genera, and are often beautifully clothed with scales and hairs. These organs again serve sensory functions, but it is extremely difficult to locate their particular economy.

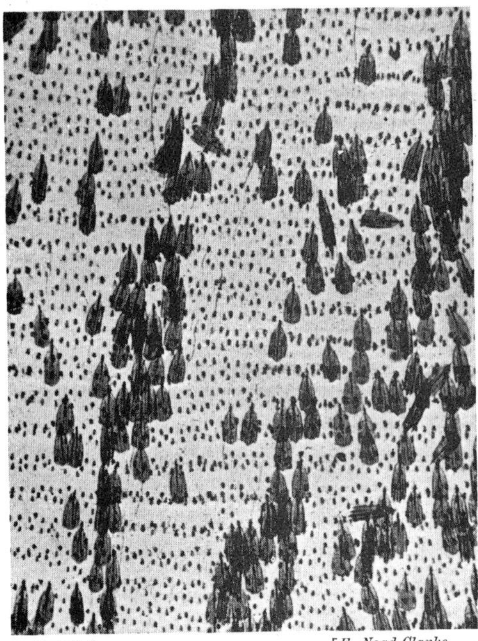

Photo by] [*F. Noad Clarke.*

THE WING OF THE CURRANT MOTH.

It has been partly denuded of scales to show the method by which each is set into the membrane of the wing.

From between the palps the long, flexible proboscis appears, which is uncoiled to sip the sweet nectar from the depths of flowers, and in some cases to feast at a filthy puddle, or upon the juices of some putrefying carcase of a dead animal. The majestic Purple Emperor Butterfly, which soars at lofty heights above the tops of tall oak trees, often descends to the earth in search of such carrion, and then becomes so engrossed in its feast that it may be captured without difficulty.

Although often no thicker than a fine hair, the proboscis consists of two gristly tubes which connect on their inner side by means of innumerable minute fibrils, much after the manner of the barbs of a feather. The inner sides of the tubes are grooved, and, when attached, they form a canal through which the fluid food is absorbed, the tubes themselves not being used for absorption. When a butterfly or moth is cleaning up its mouth-parts after feeding, it will sometimes be seen to separate these tubes and extend each one individually. The tip of the proboscis is provided with a large number of raised papillæ which may serve as organs of taste, and they possibly assist the insect in securing a grip when probing into the various shaped nectaries of flowers.

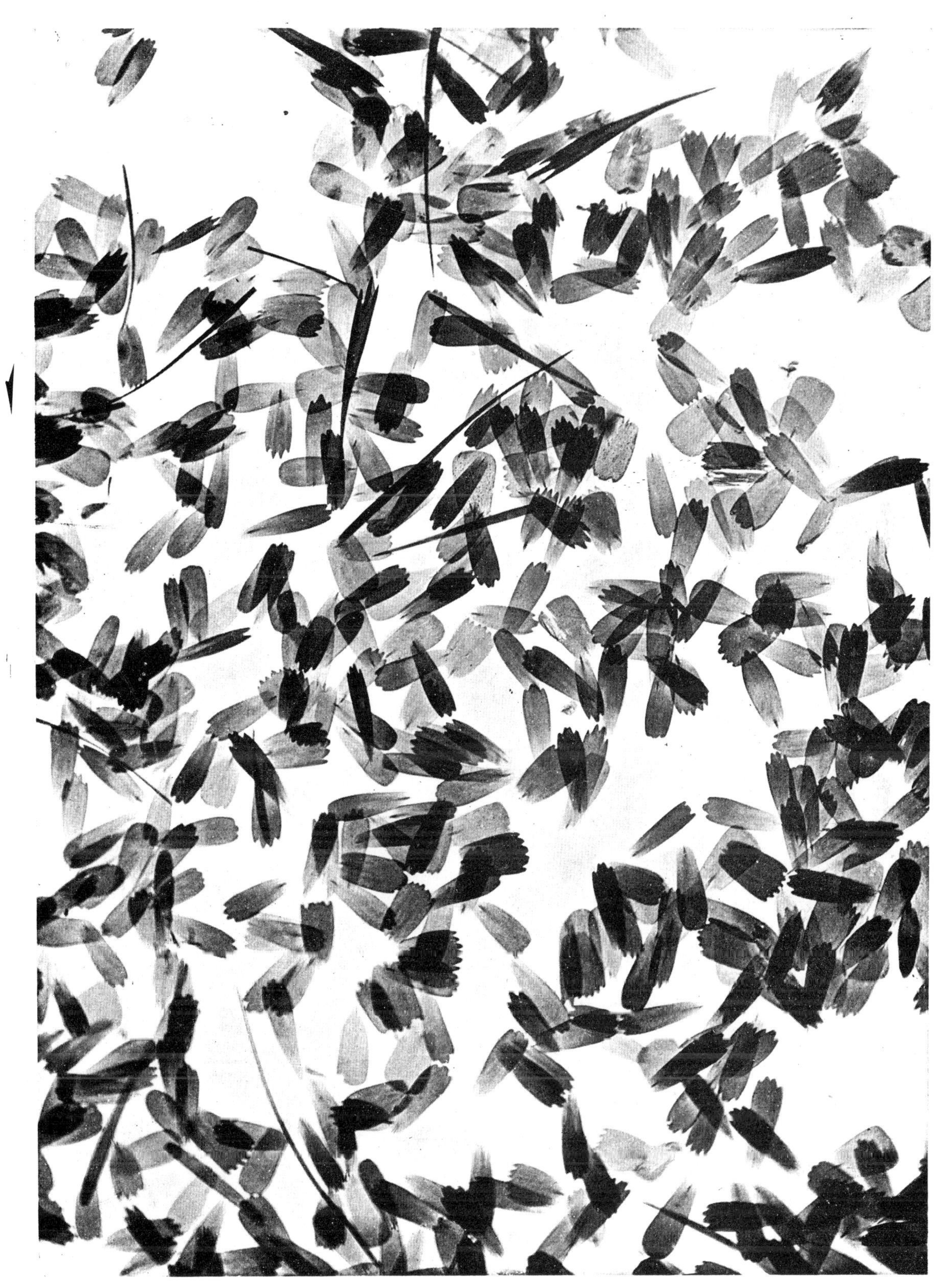

SCALES FROM A BUTTERFLY'S WING.

Many varieties of scales are to be found on Butterflies, and only a few are shown in the accompanying illustration. Each scale is delicately wrought and of definite shape. There are besides the wonderful blending of colours and the elaborate variety of patterns which make of a butterfly's wing a "miracle of design"

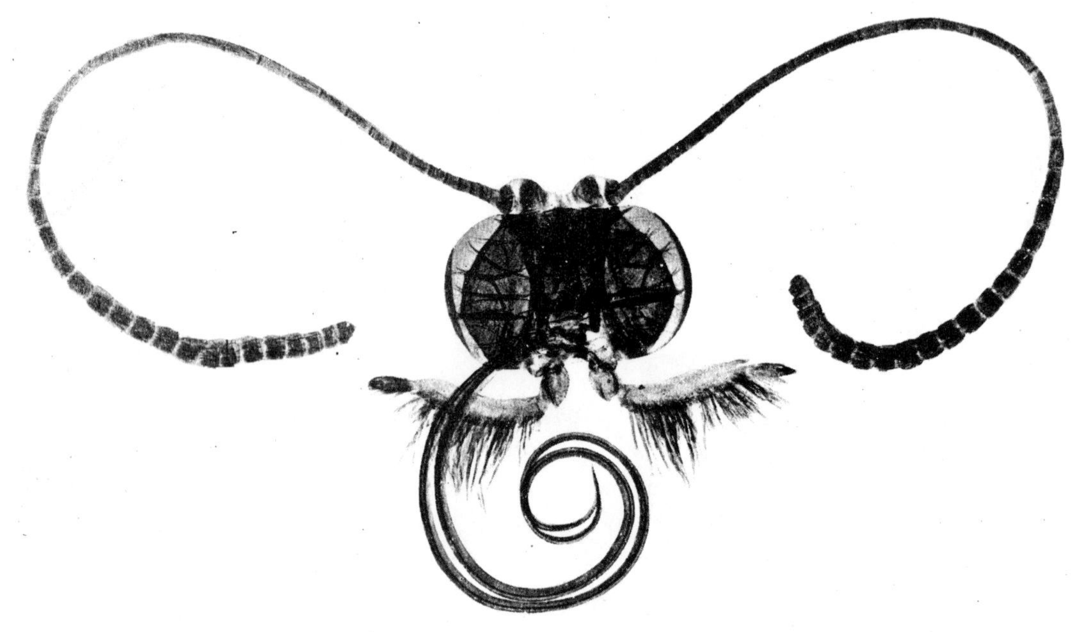

THE HEAD OF A BUTTERFLY.

It is dissected out to show the many parts of which it is composed. First are the long jointed antennæ; beneath are situate the eyes, which are very large, prominent, and composed of innumerable facets; lower down are the palpi, or feelers, which guide the action of the proboscis, or tongue. When at rest the proboscis is curled up between the two palpi.

From the structure of the mouth-organs it is, of course, obvious that a butterfly or a moth cannot eat green leaves as food; all its nutriment must necessarily be of a liquid character; the sweet nectar of flowers, the juices of ripe fruit, and in some cases the unsavoury material previously referred to, constitute its entire diet. All the green food which it needs is consumed while in its caterpillar stage, for then it is provided with a pair of biting mandibles; the change in the mouth-parts from those of the caterpillar to those of the butterfly or moth is a very remarkable one.

In some genera of moths the proboscis is developed to an enormous length. For instance, the Convolvulus Hawk-Moth probes the depths of the flowers of the tobacco-plant, petunias, etc., with a tongue nearly five inches in length—more than double the length of its body. With these long tongues there is no need to alight upon the flower; the moth simply hovers before it on its rapidly-vibrating wings and thrusts in its proboscis. In contrast to such examples, some moths take their last meal in their caterpillar stage, being entirely destitute of mouth-parts afterwards; and, consequently, their life in the winged state is very brief—simply a period to exercise sexual functions.

The final organs of the head are the antennæ, or feelers. In moths they usually terminate in a point, while in butterflies they are generally distinctly clubbed at the tip. These again are sensory organs, probably serving several senses, and possibly senses which we cannot interpret from the human standpoint; for undoubtedly insects are subject to sensations of which we are not cognizant.

To take a concrete example. It is well known to entomologists that male moths can be attracted from long distances by a newly-emerged female moth enclosed in a box with a gauze covering, and it is generally explained that they are guided to the female by the acute sense of smell possessed by their antennæ. We, however, cannot detect any smell being given off by the moth even when close to it, yet the male moths are attracted from all directions, and often from considerable distances.

Nevertheless, these same moths are lured to flowers by means of sweet smells which we can fully appreciate, apparently showing that they have a sense of smell similar to our own.

In such cases (and often when the species have wingless females) the male moths are found to have large and feathery antennæ bearing very delicate and complex structure, while those of the female are much more simple. Possibly, then, these organs may be natural transmitters and receivers of some form of wireless telegraphy, by means of which the sexes propagate waves of communication one with another. Whether my suggestion be correct or otherwise, these organs when viewed through the microscope present to the eye some marvellous mechanism, as witnessed in the accompanying photograph of the tip of the antenna of the Pale Tussock Moth.

The feelers of butterflies at first glance do not reveal such complex structure as those of the moths; but it does not follow that they are any the less complex on that account. If we closely examine one of these organs—say, of the familiar Small Tortoiseshell Butterfly—we find that even the club at the end contains twelve joints, while the slender stem which connects it with the head is also jointed throughout its length. Its upper surface is beautifully clothed with brown and white scales, and the under-side is bare, but coloured bright orange at the tip, shading to brown towards its base. It is a marvellous object when seen under the microscope, well illuminated against a dark ground; but we cannot fathom the meaning and function of its numerous joints, and hollow club, with its thousands of minute depressions, and their artistic clothing and colouring.

Leaving the head, we may now turn to the body, which consists of thorax (the portion next the head) and abdomen. The thorax is made up of three closely-united rings, or segments, each of which on its lower side bears a pair of legs. The second and third segments also carry a pair of

Photos by] [*J. J. Ward, F.E.S.*

THE ANTENNÆ OF TWO BUTTERFLIES AND A MOTH.

In the centre are the beautiful fringed fronds of a male moth; to the right is the club-shaped antenna usual amongst butterflies and to the left is that of the Small Tortoiseshell Butterfly, which is distinguished by the dark and light scales on the upper surface.

wings on their upper surfaces. The abdomen really consists of nine rings, although only seven or eight (according to whether it is a male or female insect respectively) are usually visible, the hidden segments being more or less withdrawn into the body.

Both thorax and abdomen are thickly clothed with hairs, some moths being distinctly tufted along the back ; and in butterflies, the males sometimes possess clasping organs at the tail. In those moths which deposit their eggs in the crevices between the bark of trees, and in similar situations, the females are provided with a protruding ovipositor. The spiracles, or breathing-holes, through which they take in air for respiratory purposes, are arranged along the side of the body, but are generally hidden from view amongst their fur-like covering.

The legs need no special reference, as they are of the usual five-jointed type characteristic of insects generally. In some families of butterflies the front pair of legs are more or less rudimentary, and particularly so in the males. They may be naked, or covered with short or long hairs, and the

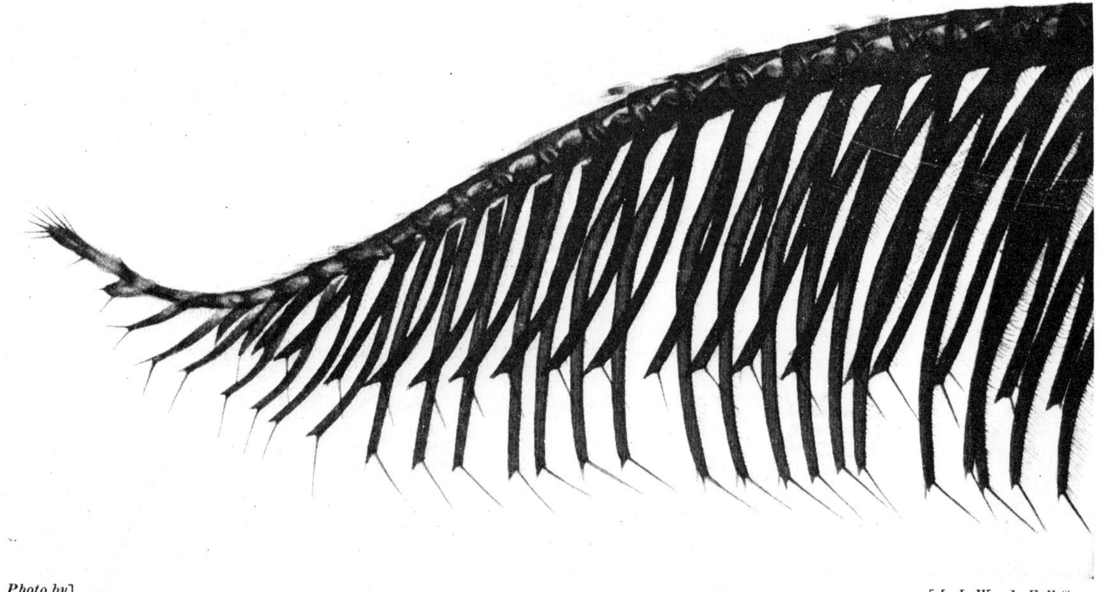

Photo by] [*J. J. Ward, F.E.S.*

THE ANTENNA OF THE PALE TUSSOCK MOTH.

A part of it has been magnified twenty-five times to show the elaborate jointing of the main stem and the delicate fringe of hairs which supplements each minute branch.

fully-developed leg terminates with a foot bearing a pair of claws, which sometimes are forked at their tips.

Finally, we have the wings to consider, organs which do not offer any very unusual structural details considered from an insect point of view, excepting that they are beautifully clothed with hairs and scales. They are really transparent, but are covered on both sides with innumerable opaque-coloured scales, and which, when handling a moth or butterfly, we rub off as " dust " upon the fingers. Even these tiny scales often possess a most complex structure, although morphologically considered they have to be regarded as nothing more than minute flattened hairs.

These minute scales are arranged on the wings in regular order, like the tiles on the roof of a house, being attached to the wing-membrane by little stalks, or pedicels, the latter being implanted in tiny cups. Although they are placed in lines, yet scales of different colours are so situated as to form circles (sometimes concentric circles), spots, streaks, irregular blotches and bands, etc., and in this manner a kind of mosaic design is built up, producing the markings which often prove so fascinating on the wings of these insects.

THE HEAD OF A MOTH.

Smaller eyes than ours must view the approach of such monstrous ugliness with terrified astonishment. The fluffy covering of the moth greatly enhances the effect of size, while the bulging eyes and formidable feelers create an impression of rapacity which is entirely at variance with the gentle habits of a moth.

THE LEG OF A MOTH.

Notice the long silky hairs fringing the upper part of the leg and the thick woolly scales covering the leg itself. (Magnified sixteen times.)

The colours are produced by two methods, the brilliant metallic effects are due to iridescence, and the sombre hues largely to the presence of pigment contained in the scale. As previously stated, the scales, although so minute, have a complex structure; indeed, they were formerly much used as test-objects for microscope lenses. Not only are they very variable in form, but they are also delicately marked with fine striæ, or lines, rows of bead-like dots, and other markings which to resolve test the powers of the best lenses.

Sometimes they are slightly undulated or irregular, and this structural detail, together with their line-like markings, produces the iridescent colour-changes seen on the wings of many butterflies. The lines and curved parts of the scale cause a shadow to fall in a greater or less degree on the opposite side to that of the source of light; and so with every movement of the insect the rays of light are caught at different angles, causing refraction, and consequently giving origin to a variety of colours—just as a glistening diamond does.

Some of the Morpho Butterflies of tropical America are the most dazzling and beautiful objects in the insect world. Their wings, when expanded, extend from six to seven inches, presenting lovely iridescent sheets of the palest blues to the sunlight, which at the slightest movement suddenly change into deep blues and purples. It is said that on the Amazons the flashing of the wings of these insects can be seen half a mile away, giving the effect of small mirrors catching and reflecting the sunlight.

It has been calculated that on the wings of a butterfly of the genus Morpho there are arranged no less than a million and a half of these light-refracting scales; some butterflies have no fewer than one hundred thousand to the square inch.

The males of many species of butterflies have, in addition to the ordinary scales, some of special form, which are known as plumules. These scales are not infrequently of a battledore shape, although they vary in different species, in some cases being quite hair-like. They are situated on various parts of the wings, sometimes along the veins, but more often in bands and patches in particular areas, according to the species. They are believed to be associated with scent-glands contained in the wing structure, and on that account become very interesting objects.

We have already observed that butterflies and moths carry on sensory functions which are difficult to understand when measured from the human standpoint. It is, therefore, extremely interesting to find that the male butterfly, when love-making, resorts to human methods, so far as to make himself attractive with perfumes that are quite familiar to our own sense of smell.

It is still more interesting to find that each species of butterfly has its characteristic odour. For example, the familiar Small White Butterfly of the garden favours sweetbriar or ripe-apple perfume; while its near relative, the Large White Butterfly, delicately puffs the air with a scent somewhat like violet-powder. Another near relation, the Orange-tip Butterfly, fascinates his spouse with a fairly strong smell of lemon; indeed, the scents of the different species are as varied as those of the flowers which supply them with liquid food.

The female insects also appear to emit similar odours, but it is in the males that they are most perceptible. If the wings of a freshly-captured male butterfly are gently rubbed, so as to remove some of the surface scales, its characteristic perfume may sometimes be readily detected. They appear, however, to be able to control the emission of these odours, and a specimen which has been exercising its wings may not prove so strong-smelling as one which has just commenced flight in the morning sunlight. Possibly the fluttering and circling movements around each other seen when these insects are flying are owing to the male insect's endeavours to get on the wind-side of his mate, or that he may get into a favourable position to fan his perfumes towards her.

Photo by] [*E. Step, F.L.S.*

THE PEA CRAB.

This is one of the smallest of Crabs, and is an intimate associate of the mussels, in whose shells it lives. Here are shown: topmost, the male; below the female, and lastly the female with the large tail-plates disclosed, for these are usually folded up and packed away under the shell.

It is possible that some butterflies may favour scents similar to the flowers which they most frequent when in search of food. In that case, it would be of material advantage to them both in attracting and in finding their mates, as their sense of smell for their choice food-plants would necessarily be highly developed. That, however, is a point which, so far as I know, has not been investigated.

In any case, these curious scent-glands only present another instance of what marvels of nature may be hidden away, even amongst the extremely minute scales which clothe a butterfly's wing.

The scales themselves are marvellous, both in structure and in the charming colour effects which they produce ; and, as we have now seen, they cover still other hidden marvels within the wing-structure.

The wings themselves, when denuded of their scales, appear as nothing more than frameworks of hollow tubes and ribs, covered with an upper and lower transparent membrane ; but when we see these apparently simple pieces of mechanism in use, as the butterfly soars over the flowers in the sunlight, or flashes its colours along the woodland glades, we have to recognize that even they, too, are structures none the less marvellous, although their marvels appear in the finest adjustments of combined strength, lightness and utility.

Finally, everyone who has seen a butterfly or moth emerge from its chrysalis and quickly unfold and fill out its wings, with all their myriads of coloured scales, every one of which is placed in perfect order, will readily agree that there are still some extraordinary details of these insects which need explanation.

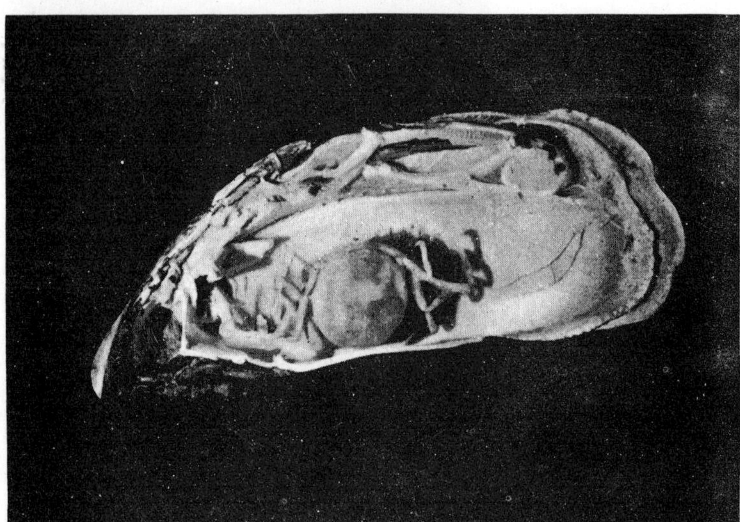

Photo by] [*W. B. Randles.*

THE PEA CRAB

Shown as it is usually to be found, that is in the shell of the mussel. Both mussel and Crab benefit by this partnership, probably because the Crab feeds on the larger organisms which enter the shell of the mollusc and, defying the mussel's digestion, would yet decay in the shell and set up a diseased condition.

PEA CRAB AND MUSSEL

BY EDWARD STEP, F.L.S.

Among the most anciently known of the strange partnerships that exist between creatures unrelated to each other, there is none more remarkable than that subsisting generation after generation between the little Pea Crab and various species of bivalves. The ancient philosophers dwelt upon the fact, and gave their own interpretation of it ; and the poets of antiquity enshrined it in immortal verse. Naturally, they differed in their explanations, and some of them disagreed about the facts ; but they did agree that this strangely assorted couple lived together constantly in an association that was only severed by the death of one of the two. In recent times the story has gained additional interest, because it has been found to be only one of a large number of partnerships of similar character, to which modern science has given the name of Commensalism—which signifies table companionship ; in another word, messmates.

The Pea Crab is so-called on account of his size, for he is, when mature, one of the smallest of the crabs. He is not one of those that may be found upon the seashore by turning over stones or hunting among the seaweeds. If you wish to make acquaintance with the Pea Crab as a living entity you must dredge for large-sized mussels, especially the Horse Mussel and the Fan Mussel. It was the latter that was well known to the ancients in this connection, the Fan Mussel being a common bivalve in the Mediterranean. The Pea Crabs—of which there are several species—differ from crabs that live the ordinary free life of their class, in not having their shelly coats armed or decorated by spines, hooks, or bristles. Instead, all their spines have been suppressed, their sharp edges rounded off, and the rough exterior smoothed and polished. There are two reasons for this uncrablike condition : first, such roughness of exterior and defensive spines are quite unnecessary to a crab that lives wholly under the protection of the mussel's much harder roof ; and second, they

[By C. E. Swan.

THE ERYOPS.

Naturalists are not yet agreed as to the exact position the Eryops occupied in the scale of creation; but in the shape of the upper bone of the foreleg it is closely associated with an extinct beaver and the still extant Duck-billed Platypus. It must therefore be supposed to have been a burrower and of aquatic habits.

THE ERYOPS.

One hint of the reptile is to be found in the presence of blunt spurs on the ribs of Eryops, such as are met with to-day only in crocodiles, alligators, the Tuatera, and the birds. The position of the ribs denotes also that this strange creature breathed like a frog.

would be distasteful and positively harmful to the delicate mollusc. So natural selection and the survival of the fittest has resulted in the evolution of a smooth and unarmed crab that can jostle against the soft tissues of the mussel without doing them any harm. Some may contend that this is not so ; that it is more reasonable to suppose that the Pea Crab took up residence in the mussel-shell because it was unfitted by its unarmed condition for a life in the open. Against this view, it should be pointed out that when it leaves the egg as a zoea, the young Pea Crab is well armed with three long and stout spines, and that the margins of his hind-body run into spiny projections, which are thrown off when he takes up residence in the shell of a mollusc.

Another small point, though not without importance in this connection, is the fact that the male Pea Crab has a harder shell than the female, whose covering is really more like parchment than shell. The reason for the difference is this : the female, we believe, has never been found except in a bivalve shell ; the male, who is more rarely seen than the female, has been taken at large. To seek the female he has to wander from mussel to mussel, and run the risks attending such adventures ; therefore, it is advisable that he should be clad in a stouter dress than the stay-at-home female. So his shell has not deteriorated to the same extent as that of his mate.

It may be presumed that when the mussel opens the valves of its shell, and sets up the current that brings in the microscopic organisms upon which it feeds, there are larger particles flow in of which the mussel cannot make use, and which might by decaying in the shell set up a diseased condition. It is probable that the little Pea Crab takes all such as its share, and that by consuming it a benefit is conferred upon the mussel, which justifies it in giving house-room to its partner. In cases of true commensalism there is always some such reciprocal arrangement as the basis of partnership.

The ancients variously regarded the Pea Crab as a burglar, who forces his way in when the valves are slightly open and eats up the mussel at leisure ; and as a friendly monitor, who by a timely nip indicated to the mussel when to close his doors with a snap, and so capture some curious visitor upon whom the confederates could banquet.

There are many species of Pea Crab spread over the oceans of the world, and they have each some similar working arrangement with divers kinds of sea beasts, not necessarily molluscs.

ANCIENT UGLINESS

BY W. P. PYCRAFT, F.Z.S.

By no stretch of the imagination could Eryops be called beautiful—an interesting beast, however, it certainly is. A creature of immense antiquity, and, so far, known only from the Permian formations of Texas, it stands at the parting of the ways between the Amphibia—frogs and toads and newts, and their kin—and the Reptiles By some authorities, indeed, it is regarded as a primitive reptile; on the whole, however, the evidence seems to show that it must be regarded as a side branch of the group which gave rise to the higher group, which, in turn, gave rise to the mammals at the head of which stands Man himself, at once the wisest, and most brutal of all living things. When doctors disagree, who shall decide? Let us leave the problem of the relationship of Eryops to other and higher groups to those whom it may concern, and turn our attention to the creature itself.

There is something of the crocodile in its appearance, and this likeness is due to the armature of bony, horn-ensheathed plates which covered its body, as well as to the shape of the head. From the short, massive, twisted character of the upper bone of the fore-limbs we may infer that Eryops was an aquatic beast, which lived in burrows on the banks of the streams in which, doubtless, it swarmed. At any rate, this upper bone agrees very closely with that of creatures so unlike as the extinct beaver, Trogonotherium, and the living Duck-billed Platypus. Both these are aquatic types, and since we know the Duck-billed Platypus, or Ornithorhynchus, to be a burrower, and the modern beavers are burrowers, we may fairly infer that Eryops was no less skilled as a digger. About the coat of bony mail we have no doubt, for samples thereof have been found with the bones of the skeleton. From the teeth, and the fashion of the skull, we gather that Eryops was a near relation of those strange

Photo by] [*S. L. Bastin.*

THE BROOM.

The Broom explodes its seeds from the pod so that they may fall on ground which is not overshadowed by the parent stem and the young plants have a chance of growth. On a warm afternoon in early autumn a continuous crackle caused by explosions of the ripened pods will be heard in the vicinity of a Broom.

creatures, the "Labyrinthodonts"—the ancestors of the modern Amphibia. These great beasts exceeded Eryops in size, and Eryops is about half as big as a big crocodile.

A curiously reptilian feature about the skeleton of Eryops, as may be seen by a reference to the accompanying photograph, was the presence of blunt spurs upon the ribs. These are met with to-day only in the crocodiles and alligators, the lizard-like Tuatera, and the birds.

Eryops seems to have played the part of crocodiles in the rivers and deep pools of those far-off days; just as did the Labyrinthodonts in similar scenes in what are now the British Islands. Some fine specimens of these queer-looking creatures have been found in the coal measures both of England and Scotland. They have no name in common speech, and their rather formidable scientific name is derived from the fact that the teeth, when seen in section, present a complete and very beautiful pattern, the harder substance of the tooth being arranged in serpentine coils radiating outwards from the centre of the tooth. Hence the name "Labyrinth-toothed" or "Labyrinthodont." Eryops shared this peculiarity. From the fact, too, that the ribs of Eryops, and its kindred, do not reach down to the middle line of the chest we may be pretty sure breathing took place after the fashion of that of the frog. That is to say, air is drawn in through the nostrils by depressing the great fleshy tongue, thereby creating a vacuum which the air rushed in to fill. When the tongue was raised a pair of valves closed the passage from the mouth through the nostrils and forced the air down the windpipe to the lungs.

Photo by] [*Edward Step, F.L.S.*

THE MONKEY'S DINNER BELL.

Such is the name given to this curious fruit of a South American plant. The seeds are contained in separate capsules, which when ripe explode with a loud report and are shot far away from the seed-cluster.

VEGETABLE GRAPNELS AND SHOOTERS

BY SIR HERBERT MAXWELL, F.R.S.

THE sheep-killing grasses already described are not the only plants which inflict suffering, and even death, upon the animals which they employ to transport their seeds. Dr. Livingstone described how the fruit of the South African "Devil's Horns" afflicts grazing animals. The seed is enclosed in a hard, spiny envelope, ending in two long curved beaks like goats' horns. When these get into the nostrils of an ox or antelope the animal is helpless to remove them, and must suffer intensely until relieved by death, or, in the case of domestic animals, by the herdsman.

Still more horrible is another South African herb, the Grapnel-plant, the fruit of which is set with claws an inch long., Even the lordly lion is sometimes destroyed by its agency. When he rolls or basks in the sand the spines may pierce his skin, and when he tries to tear out the fruits with his teeth, his mouth gets filled with them, and he must die of starvation.

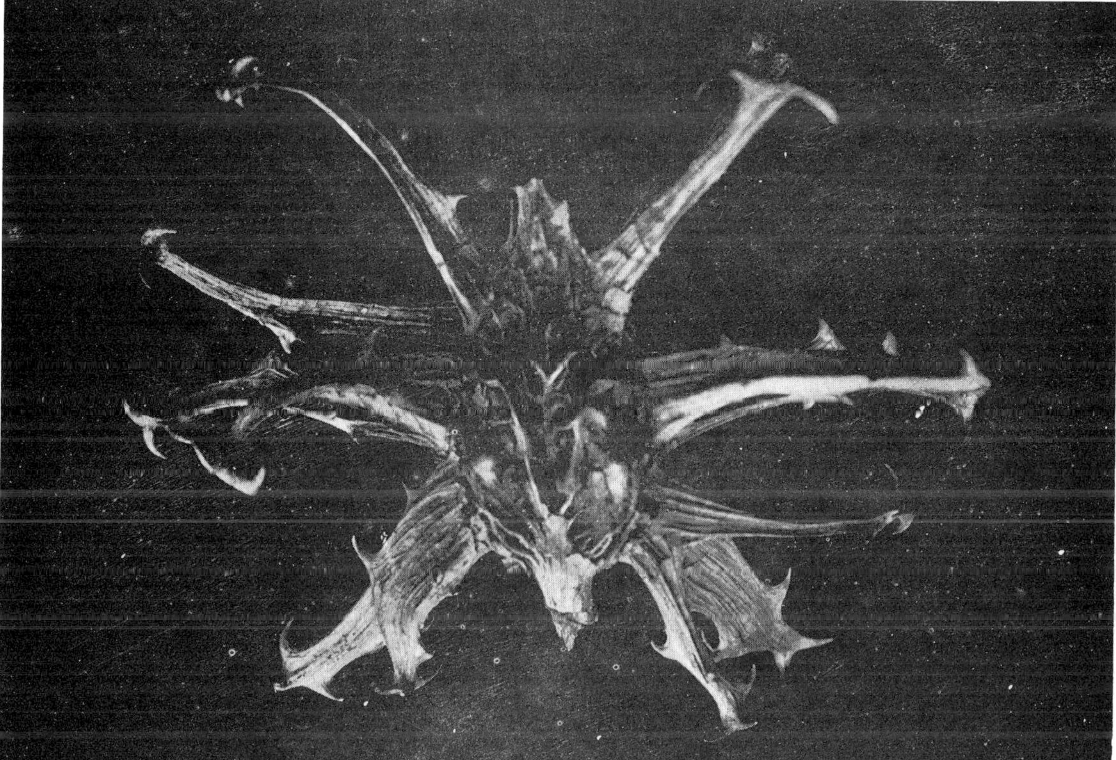

Photos by] [*S. L. Bastin.*

SEED GRAPNELS.

The lower photograph shows the Grapple Plant Fruit armed with long claw-pointed spines which become entangled in the hide of animals so that the fruit is carried further afield. In the upper photograph are the Burweed—to the left—and the Burdock. The seeds of these plants, besides being armed with hooks, are slightly covered with a viscid fluid which gives them a firmer hold on any moving substance they may light upon.

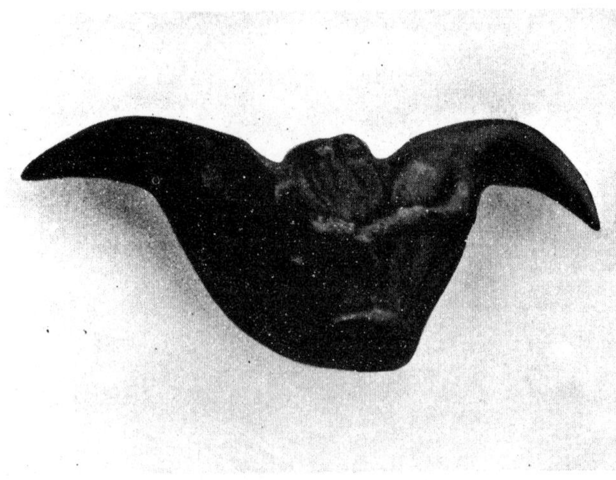

THE BULLOCK-HEADED NUT.

It is the fruit of a species of Trapa which grows on the river-bank. The seed usually falls into the river among rushes and water-plants, to which it is safely anchored by the strong horns. The seed would probably perish if it were to fall on the ground, as it needs the hard case to be soaked in the water before germination can commence.

Many of our common British weeds, though less formidable in their armament than these African terrors, are equally persistent in their demand for a free passage and equally ingenious in the means by which they effect it. It has been remarked already that most of the vast Composite family—thistles, groundsel, dandelion, etc.—rely upon the wind to waft their plumed seeds hither and thither. But in the common burdock the plumes are too short and stiff to act as sails for the heavy seeds, so the plant arms each of the numerous scales surrounding the flower head with a strong hooked point which readily catches upon the hair of animals or the clothes of man, and so the whole seed-head is carried away and gradually drops its seeds to colonize fresh ground.

Another member of this family which has obtained a footing (not, let us hope, a permanent one) in this country as a weed of cultivation is the burweed, a coarse annual about two feet high. It has discarded aviation altogether as a means of seed dispersal, providing no wind-hairs, but relying instead upon the hooked prickles with which the female flowers arm their burs. The burs are small, but get into sheep's wool in such numbers as to prove a sore trial to merchants, and this has been the means of disseminating the plant into the whole of the temperate world. Originally indigenous to the Mediterranean region, the loss it has caused to Australian sheep-farmers has been reckoned in some cases as high as fifty per cent. The traveller Frauenfeldt saw horses in Chile in 1860 with their manes and tails so heavily tangled with these burs that they could hardly walk. In 1871 it was first noticed in this country growing near Edinburgh.

Photo by] [*E. Step, F.L.S.*

THE DEVIL'S HORNS.

A good name for this terrible grapnel. The seed-pod lies in the grass with its formidable horns curved upwards. These hook themselves into the nostrils of grazing animals, and irritation is set up which causes the poor creatures to rush about wildly shaking their heads, and so dispersing the contents of the half-opened pod.

The common carrot, forget-me-not, goose-grass or Robin-run-the-hedge, and enchanter's nightshade are among other herbs which have recourse to hooks to secure the means of colonization; and although familiarity may blunt our admiration for the ingenuity of the mechanism, it is not any easier to account for its development than it is for that of less common or more elaborate armature.

We come now to plants that shoot their ripe seeds away to a considerable distance. The scentless dog-violet is a good example of this. The capsule containing the smooth seeds hangs down at first, but when ripe stands erect above the leaves, and opens in three equal valves, exposing the seeds. Then the valves begin to shrivel, thrusting out the seeds which continue to cling to their foothold till it gives way with a jerk and the seeds are flung out. Lord Avebury records having seen them thrown to a distance of nearly ten feet, which is twenty or thirty times greater than the height of the plant itself. This is the more remarkable because the scented violet is content to lay its packet of seed on the ground beside it, and so allow the seeds to fall out when ripe.

The pretty cyclamen seems unwilling to part with its offspring, for it hugs the seed-vessels to itself by rolling the flower-stems into a tight coil, so that the seeds ripen in the bosom of the parent, lying there throughout the winter. But the flower stems rot away, so that when the dry winds blow in March the capsules are ready to be rolled along the ground, discharging the seeds as they go.

Mechanical catapults, on the principle of the dog-violet's, are common enough in British botany. Herb-robert and other wild geraniums are accomplished shooters ; the delicate little woodsorrel

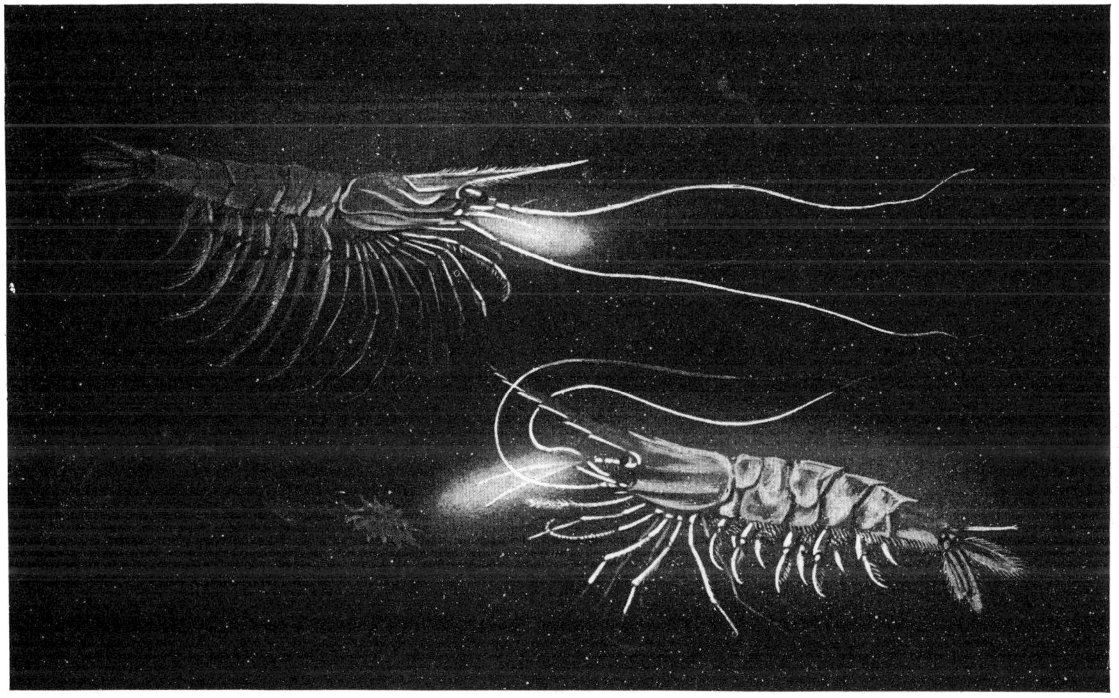

LUMINOUS SHRIMPS.

These two species inhabit the deep waters of the Indian seas. Of the two the lower has the greater luminosity, but in both the phosphorescence appears to proceed from the base of the antennae. In some species the light is a brilliant white, in others inclined to green, while the parts of the body which are luminous differ according to the species.

covers its seeds with a neat cup, which, by suddenly turning inside out when dry, scatters the ten shining black seeds to a considerable distance. If you stroll quietly over the common on a hot day in May or June, you may hear a series of detonations like fairy sharp-shooting. They come from the pods of the gorse and broom, bursting with an audible report and shooting out their seeds.

The squirting cucumber, which furnishes the drug elaterium, discharges its projectiles, not by mechanical means, but through chemical action affecting the juices so as to cause internal pressure within the fruit, detaching it suddenly and violently from its footstalk and ejecting the contents.

LUMINOUS SHRIMPS

In the depths of the ocean where the light is poor or non-existent, we find a remarkable difference in the effect this condition has upon the animals living there. Many of them have their eyes reduced to mere rudiments, and the ordinary function of eyes lost altogether. On the other hand, some creatures under the same conditions have their eyes enormously enlarged, apparently that they may receive as large a share as possible of the weak light available. It is characteristic of many of the blind creatures of the deep sea that they are provided with very long feelers and hair-like appendages of a highly sensitive nature, by means of which they appear to feel their way about, as Professor Moseley puts it, " Just as a blind man does with the aid of his stick."

The light that does exist in these depths below five hundred fathoms is not the light of day, but a phosphorescence or luminosity such as we are accustomed to see at the surface occasionally. We have already shown in this work a number of the deep-sea fishes that possess luminous organs ; but some of the crustaceans are equally luminous, size for size. Among them are a number of shrimps and prawns which, unfortunately, have no English names by which one can refer to them separately. Although living in darkness or semi-darkness, where colours cannot be distinguished, many of these are of beautiful rosy or green tints, in which they differ greatly from the deep-sea fishes, which are mostly black.

[*By Theo. Carreras.*

HONEY-GALLS.

A twig of the Oak showing the galls, which exude droplets of a sweet and viscid fluid. This is collected by the workers of the Honey Ants and stored in the living honey-pots which form a part of the ant community.

One of these prawns has prominences on one pair of the foot-jaws which are luminous ; others have a special development of the base of the antennæ, rendering it phosphorescent. These creatures, therefore, go through the deep waters much like a motor-car on land, shedding a light before them on their path. One family contains so many luminous species that it has been given a Greek name signifying " a bright light."

Of this family several genera are represented in British waters, the species having luminous globules present on various parts of their bodies. The two species of prawn figured on page 939 are from deep water in the Indian seas, and although they represent different families, they agree in producing their luminosity from the base of the antennæ.

"ANTS AS HONEY-POTS"

BY H. ST. J. K. DONISTHORPE, F.Z.S.

THE true Honey Ants live in Mexico and the south-western of the United States and have long been esteemed as an article of food by the Indians. The best known species is the Honey Ant of the Garden of the Gods, a place near Manitou in Colorado.

[*By Theo. Carreras.*

ANTS AS HONEY-POTS.

Set aside from the time they are "callows" to act as storage-cells for honey, these ants never leave the hive, but are fed with the collected sugar by the general workers of the nest. A special sac is used for keeping this honey, which by a process of culture in the early stages of the ant's existence becomes distended out of all proportion to the rest of the ant's body. When called upon by an inmate the ant will disgorge a drop of the precious fluid and feed its fellow-worker.

Before we deal with its dwelling-place and habits, let us consider what is a Honey Ant, and why it is so called. As is well known, many ants are in the habit of collecting honey-dew and storing it in their crops, which become distended, and when they reach home they feed their brood and the rest of the colony by disgorging the sweet liquid. Ordinary ants may often be seen returning home with the abdomen much distended. Now, in the Honey Ants this habit has been developed until a class of workers have been produced which are called " repletes." Their bodies are enormously swollen ; in fact, they are literally " honey-pots." The size and rotundity of the abdomen is due to the crop, and not the stomach. The former being filled with honey, forces back all the other organs against the walls of the abdomen, which are stretched to their utmost capacity. The ordinary worker is a quick and graceful creature, but the repletes, or " honey-pots," are only able to waddle along. The replete form is confined to certain workers, and these are only formed when they are callows. That is to say, when they are newly hatched from the chrysalid into which an ant grub changes before it becomes a perfect insect. The integument of the body is then soft, but when they are older it becomes hard, and they are unable to be formed into " honey-pots." This has been proved in observation nests in captivity, the callows being fed with maple sugar and other sweet liquids. These they greedily devoured, and some of them became semi-repletes, whilst others acquired the complete rotundity of a genuine " honey-pot." It is chiefly the larger workers that act as " honey-pots," but the smaller ones sometimes play this part. The " honey-pots " never leave the nest, but store the sweet liquid which is fed to them by the other workers which go out to forage for food. In return they disgorge some of their store when the other workers wish to be fed. They are thus the living casks, or barrels, of sweet stuff, kept by the ant colony as a

[*By Theo. Carreras.*

THE HOME OF THE HONEY-POT ANTS.

The nests are found on the top of the stony ridges in the Garden of the Gods, Colorado, and other suitable localities in Mexico and the south-western of the United States. The nest is composed of a mound of small pebbles, which have been collected by the industry of the inmates. In the centre of the mound is a large circular entrance, which is always guarded by a ring of ant-sentries.

store in time of want. They live a quiescent life, hanging to the roofs of the subterranean chambers in the nest. In the words of Professor Wheeler, the celebrated American myrmecologist: "Those who, in anthropomorphic mood, are wont to extol the fervid industry and extraordinary feats of muscular endurance in ants, should not overlook the beatific patience and self-sacrifice displayed by the replete Honey Ant as it hangs from the rafters of its nest, month in, month out—for years, perhaps—a reservoir of temperamental as well as liquid sweetness." They can move about a little, but should they fall from the ceiling, they are unable to return to their former position without assistance. There are often as many as three hundred "honey-pots" in the different chambers in a large colony.

THE JAWS OF THE PORT JACKSON SHARK.

The Port Jackson Shark is a species inhabiting the Australian seas and is the last representative of a very early race. Its food consists, amongst other things, of shell-fish, and the beautiful teeth are especially adapted for crushing hard shells.

The nests are found on the top of the stony ridges in the Garden of the Gods and other suitable localities. The large circular entrance to the nest is in the centre of a cone-shaped crater composed of small pebbles. A gallery goes down into the ground and is divided into smaller ones. These lead to underground chambers which have smooth, flattened floors and rough, vaulted ceilings. The honey-pots hang from these ceilings side by side, and the inequality of the roof enables them to hold on by their claws.

The workers are nocturnal in their habits and do not go out during the day. A guard, however, is set at the doors of the nest to prevent other ants, or spiders, etc., from forcing an entrance. About half past seven in the evening the workers leave the nest and visit the thickets of slim oaks which are in the neighbourhood. The twigs of these oaks are covered by roundish galls caused by a small four-winged insect.

At night these galls exude small drops of a sweet, watery fluid. These drops are eagerly sucked up by the foraging ants, and when they return home they feed the repletes with the contents of their crops.

The galls provide probably only a portion of the ants' food, and they no doubt obtain a large quantity of their honey from the scale insects and plant lice which occur on the oaks and other plants in the neighbourhood.

Other species of Honey Ants occur in Australia and South Africa, but the economy of the American species has been most fully worked out by McCook and other observers.

SOME REMARKABLE SHARKS

BY W. P. PYCRAFT, F.Z.S., ETC.

THE Shark tribe have an evil name; nevertheless, they are a most wonderfully interesting group of fishes. They have a history which can be traced back to the most remote periods of time, and from the naturalist's point of view are second to no other fishes in the light they throw on the evolution theory.

That there are dozens of different species of Sharks in existence to-day few people probably suspect, and fewer still realize how extraordinarily interesting some of these are. Undoubtedly the two most remarkable to-day are the Hammer-head and the Elfin Shark. The first-named has already been described in these pages. The last is an extremely rare and little-known species, discovered in 1898 in Japanese waters. Its most striking feature is a long snout, which, curiously enough, is separated from the jaws by a deep cleft. What purpose this serves no one has been able to fathom, and it is absolutely without parallel among fishes.

THE GREAT GREENLAND SHARK.

Its average length is about twenty-six feet; but it is chiefly noteworthy for the small size of its eggs, which, furthermore, lack the horny case characterizing the eggs of most Sharks.

No less wonderful, in its way, is the strange Frilled Shark (*Chlamydoselache*), a species which, on its discovery in the deep sea off Japan some thirty years ago, created tremendous interest. And this because of its primæval character. The mouth, it will be noticed, unlike that of typical sharks, is at the end of the snout, while the gill-openings are bordered by curious flaps of skin. The teeth are remarkable, and so also is the tail, since it does not turn upwards as in other Sharks. It more nearly resembles some fossil sharks, in short, than any living species. A dweller in the deep seas, it has, since its discovery, been met with in many parts of the world; but, so far, little has been discovered as to its mode of life.

Up to the time of the discovery of the Frilled Shark the pride of place, in point of interest, was held by the Port Jackson Shark, a relatively small species, living, as its name implies, in Australian seas. It is the last representative of a race which can be traced back to remote geological times; and which, moreover, as is shown by its teeth, differs in no way from its forbears of the coal period. These teeth, as will be seen in our illustration, are remarkable in their shape and arrangement. They cover the jaws in a series of nodules of varying size, which, in shape, recall orange-pips! As might be supposed, such teeth are used for crushing purposes, these sharks feeding on mollusca, whose shells are readily smashed between their grip. Of its habits nothing is known.

Somehow the very name "shark" conjures up visions of sudden and awful death, but species

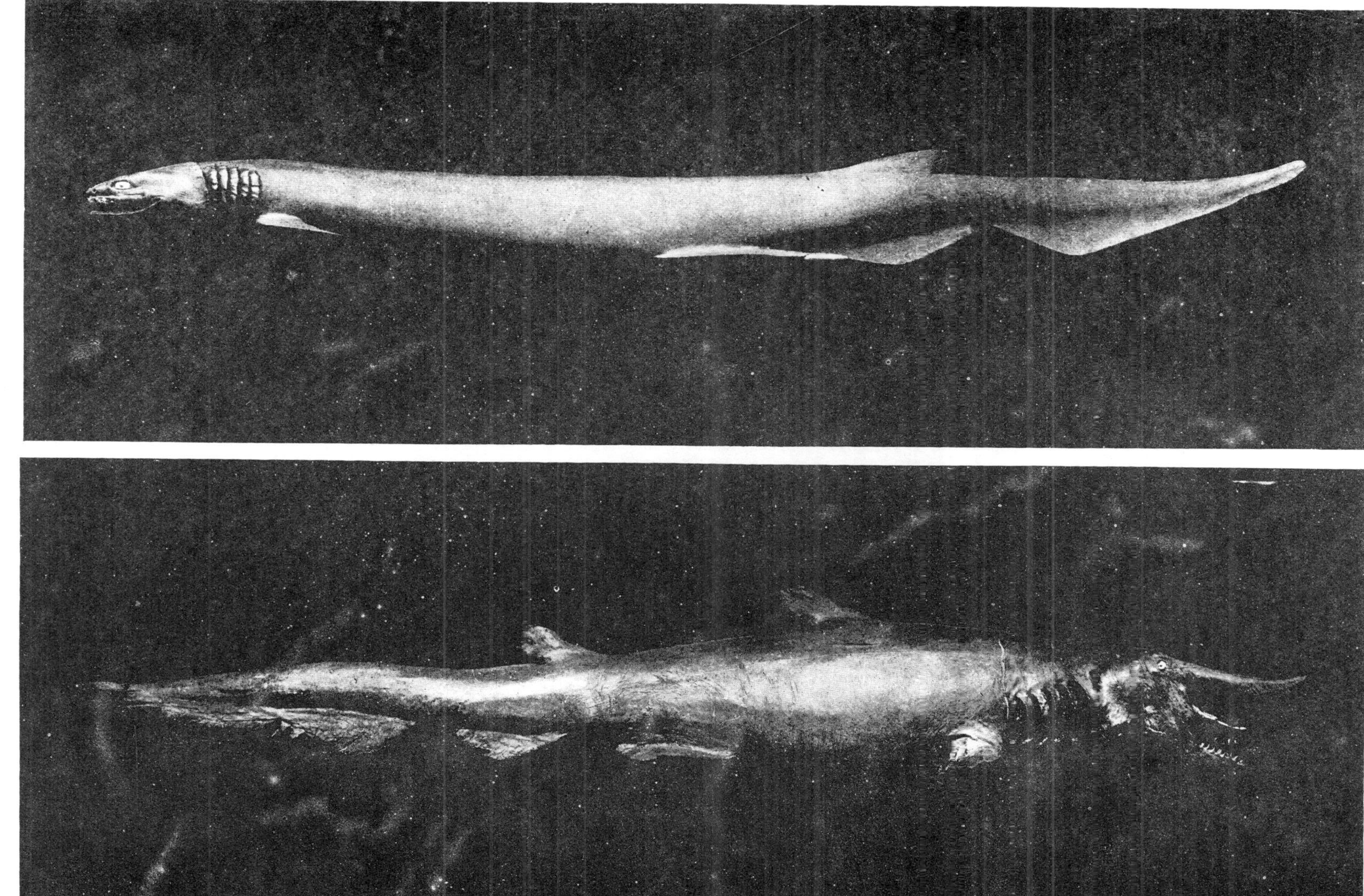

TWO CURIOUS SHARKS.

The upper is the Frilled Shark and the lower the Elfin Shark. Both species have lately been discovered in Japanese waters. In the first the position of the mouth at the end of the snout and the frilled borders to the gills are curious, while the second is remarkable for the long nose, which is separated from the jaws by a deep cleft, a feature that is without parallel among fishes.

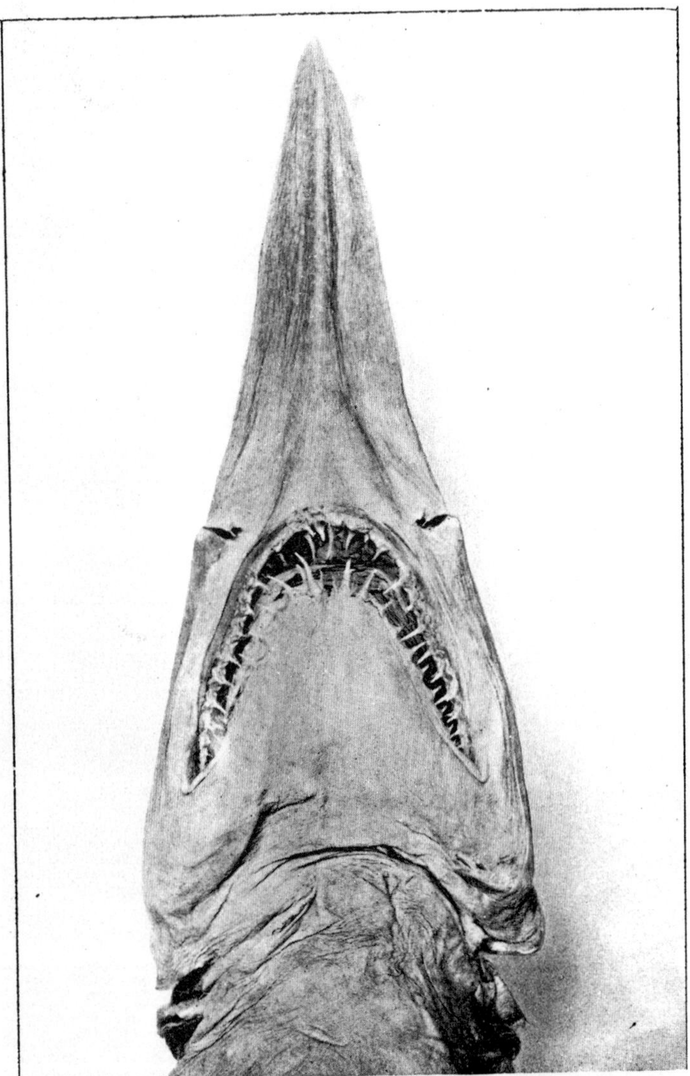

Photo by] [H. J. Shepstone.

THE HEAD OF THE ELFIN SHARK.

The under-surface has been photographed to show the position of the mouth, which is that of the average Shark. How this came about is not known; possibly the body of the Shark has in the course of time become more elongated, while the mouth has remained in its primitive position.

which inspire that nameless dread are known only to the shipwrecked in tropical seas, or those who, in these regions, find themselves suddenly overboard. The most dreaded of all these hounds of the sea is the Great Blue Shark, or "Man-eater" Shark, as it is often called. Specimens of forty feet long are on record, but these are dwarfs compared with an extinct species, which, judging from the size of teeth in the British Museum, must have attained more than twice this length. But no less dangerous is the smaller Common Blue Shark of the English Channel, which attains a length of about twenty-five feet. Young sharks of this species, seven or eight feet long, are not unfrequently taken off our coasts. The fins of this shark, and allied species, are much used for making soup in parts of India and China, though less now than formerly. Half a century ago as many as four hundred tons were exported from Bombay to China. It is a somewhat surprising fact that in places like Aden, where sharks of various kinds abound, the natives will swim and dive fearlessly in the open sea, while a European following their example would be instantly devoured. Does the white skin of the European serve as a lure?

The largest of living sharks is the Rhinodon, or Whale Shark, of the Indian and Pacific Oceans, which attains a length of as much as sixty feet. It is, however, a quite harmless species. It is curious that the Great Greenland Shark of the Arctic seas, a clumsy-looking creature of some six-and-twenty feet long, should never have been known to attack shipwrecked men, since it is a most determined foe of the huge Right Whales, which formerly abounded in these seas. The species stands unique among sharks for the small size of its eggs, which, furthermore, lack the horny case so characteristic of the eggs of the Shark tribe.

One of the most characteristic features of the Shark is the position of its mouth, which, as everybody knows, is placed on the under-surface of its head. Why is this? By way of answer, we may hazard a guess that it has come about in the transformation of the body towards gaining increased speed. To effect this end the snout has been thrust forward beyond the jaws, giving the body a pointed, torpedo shape, greatly facilitating the flight through the water during the chase of prey.

In the Frilled Shark, which represents the primæval shape of these fishes, it will be noticed the mouth forms a cleft at the extreme apex of the head, which forms a flattened cone. In the Whale Shark the mouth is similarly at the extreme end of the snout, and this species is of sluggish habits, and feeds only on small fishes. According to some, indeed, it lives largely on seaweed. This, however, is probably a mistake. The small size of its teeth, anyway, show that it is harmless enough. Thus, then, it is more than probable that our interpretation as to the significance of the shape of the head in the typical Sharks is correct.

SAW-FLIES

BY FRED ENOCK, F.L.S.

THE family of Saw-flies is a very large one, and, unfortunately, they all come under the head of insects that are more or less injurious to vegetation.

The caterpillars are gregarious in their habits, and it is no uncommon sight to the observant naturalist to see on various trees, or in the hedgerows, broods of these peculiar grubs, sometimes fifty or sixty feeding within quite a small compass of leaves, which they soon reduce to bare ribs —passing on to another spray, to repeat the process of defoliation until the shrub is sometimes quite disfigured by their depredations.

The form of the caterpillars is quite distinct from that of ordinary butterflies and moths. The first thing that strikes one is the uncomfortable way in which they cling to the edge of a leaf when

[By Fred Enock, F.L.S.

THE WILLOW SAW-FLY.

Many of the British Saw-flies are more or less gaily coloured. Some are deep violet, others of a metallic green, and others are banded with red and yellow, while all have beautifully iridescent wings. They vary in expanse of wing from three-eighths of an inch to two inches and a quarter.

feeding. Taking hold with the six true legs and the few central claspers, the hinder third of the body is without any apparent claspers, and is generally held aloof over the back or sticking straight out; at other times holding on the edge by its front legs it will curl up the other part into a perfect coil, and then much resembles a flat-coiled snail shell.

These characteristic poses of the Saw-fly grubs will always distinguish them from those of butterflies and moths.

Another common character is the number of black spots on each division of the body, and two very prominent ones on the head.

It is amusing and quite startling at times, when passing a large brood of Saw-fly caterpillars, to note how, directly one's shadow falls upon them, they all start wagging their tails in the most vigorous manner.

[*By Fred Enock, F.L.S.*

THE WILLOW SAW-FLY.

Here the Fly has laid an egg in the lower Willow-leaf, where the ruddy-coloured gall has already formed. The Fly cuts a minute slit in the leaf, therein lays an egg, then with a special saliva hermetically seals up the egg. Soon a rust-coloured ring appears round the point of incision, which in a month or so develops into a blister or gall.

When full grown, and some attain to this condition in a fortnight or three weeks, the caterpillars leave their food-plant and solemnly walk down the twigs and branches and trunk, until they reach the ground, into which they laboriously burrow a short depth, and by twisting and twirling their bodies about, form a small oval chamber. They then line this with coarse silk, and resting for two or three days upon their back, change to pupæ — the perfect Saw-fly emerging sometimes the same year, while others remain till the spring.

The greater number of Saw-fly grubs form these subterranean cocoons; but others, and among this series we find the largest species—the giants of the family—are internal feeders.

On page 887 is a photograph, natural size, of one which lays its eggs on the pine-tree *trunks*, into which the grub eats a hole, feeding upon the solid wood, until it reaches full growth in a year or two—during that time it is boring holes, larger and larger, until they are from a quarter to nearly three-eighths of an inch in diameter. They form their cocoon of the small pieces bitten from the burrows and close up to the bark. In July or August the matured insect bites a hole through the hard cocoon and bark and emerges, resting on the trunk until its beautiful wings and steel-blue body are fully developed, and then taking its flight. The noise of its powerful wings in motion is considerable, and it is often an object of terror to those who see it and imagine it to be a hornet.

Another large species affects the willow, while a still larger one is often found on the hawthorn. The caterpillar feeds at night, and is a conspicuous object when the light of a lantern is thrown on it—the whole body is almost pure white, on a delicate green ground. This one appears in April, and, when caught, exhibits tremendous power in its jaws, which it uses in self-defence on the fingers of its captor. The hard, oval cocoons may frequently be seen in the winter on the bare twigs

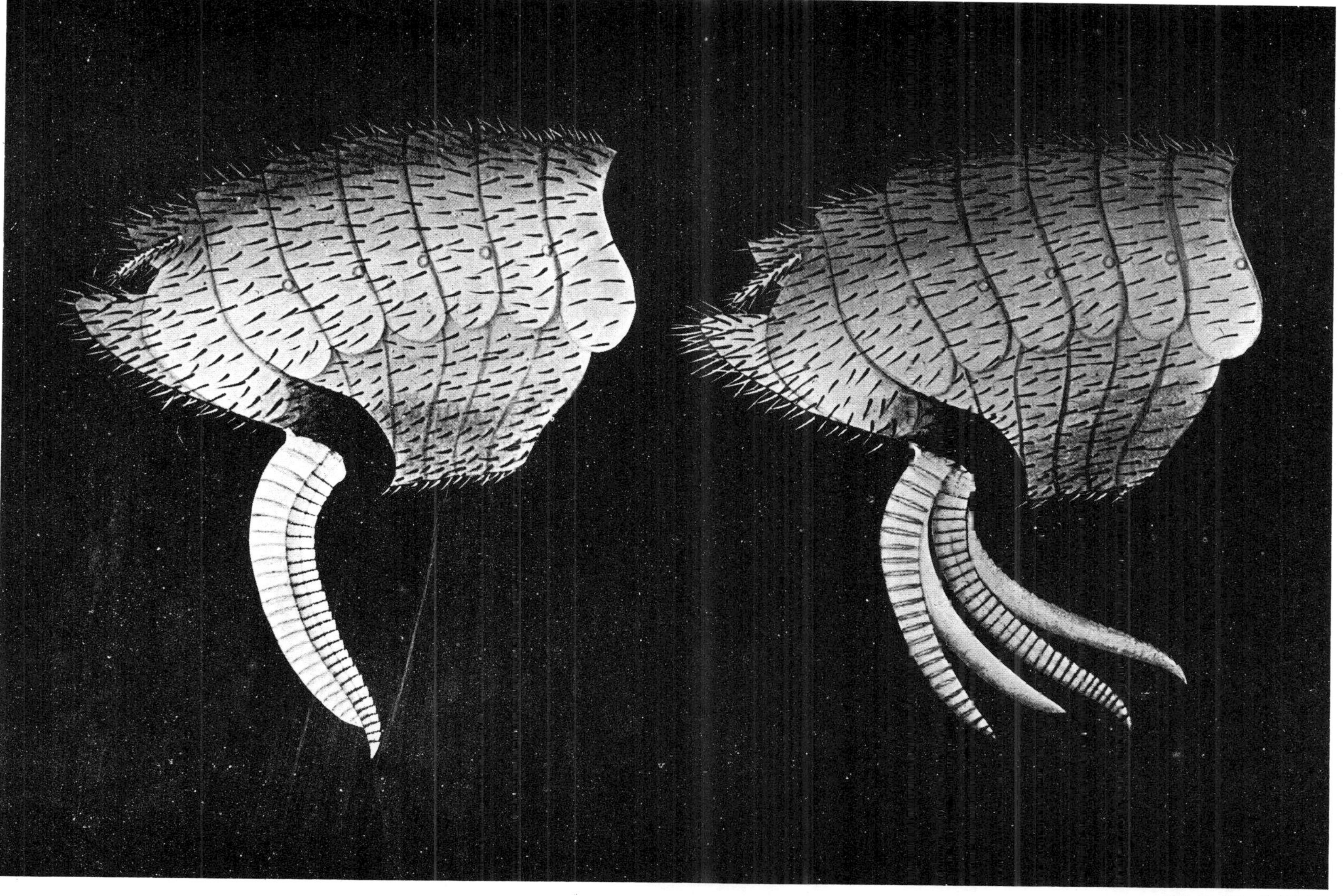

[By Fred Enock, F.L.S.

THE "SAW" OF THE SAW-FLY.

Very marvellous is the construction of this delicate instrument, which is used not only for drilling holes but for the laying of the eggs. In the left-hand photograph the saw is shown enlarged in its natural state; in the right-hand it has been dissected to show the two saws and two tenons, which after the cutting of the leaf is effected, combine to form a tube, or egg-laying apparatus.

of the hawthorn, and sometimes, instead of the Saw-fly emerging, a dozen or more Ichneumon flies pierce the cocoon—the parent of them having laid *its* eggs in the body of the caterpillar of the Saw-fly.

In the case of those caterpillars which are gregarious, the female lays her eggs in the most curious positions : sometimes twenty or thirty in the leaf-stem of a poplar ; in another, in the ribs at the back of a leaf ; and again in the stem of a soft rush ; and very often the Saw-fly makes a tiny incision between the upper and under cuticles of a leaf and therein lays an egg.

Perhaps the most familiar objects on the willow-leaves growing in the hedgerows from May to September are the crimson swellings, or galls, sometimes a dozen or more on one leaf. These galls are caused by the Willow Saw-fly (page 947), which cuts a minute slit in the upper skin, or cuticle, of the leaf (page 948) with its saws and lays an egg, the two saws and two tenons then forming a tube, or egg-laying apparatus. Immediately the egg is laid in its place, the Saw-fly turns round, and with saliva from its mouth, hermetically seals up the egg. In from eight to ten hours, around each incision appears a small rust-coloured ring, which, as time goes on, increases, and in a month or so the two cuticles separate, and a slight blister appears on both sides, which slowly forms into a red spongy gall about a quarter of an inch in diameter. In the centre of this the egg hatches to a minute white caterpillar, feeding upon the inside material of the gall. A hole is bitten on the under-side, so that all waste material can fall out. In the photograph below the gall is shown in section, with the full-grown caterpillar of the Willow Saw-fly. In a month or more the grub becomes full-grown and escapes from the gall, and walking down the twigs and branches, it reaches the trunk ; and if it can find a loose piece of bark, it crawls to the back of it and spins a brown silken cocoon. Sometimes we have found a cluster of twenty or more all spun up together. In these cocoons the grub changes to a chrysalis and emerges in

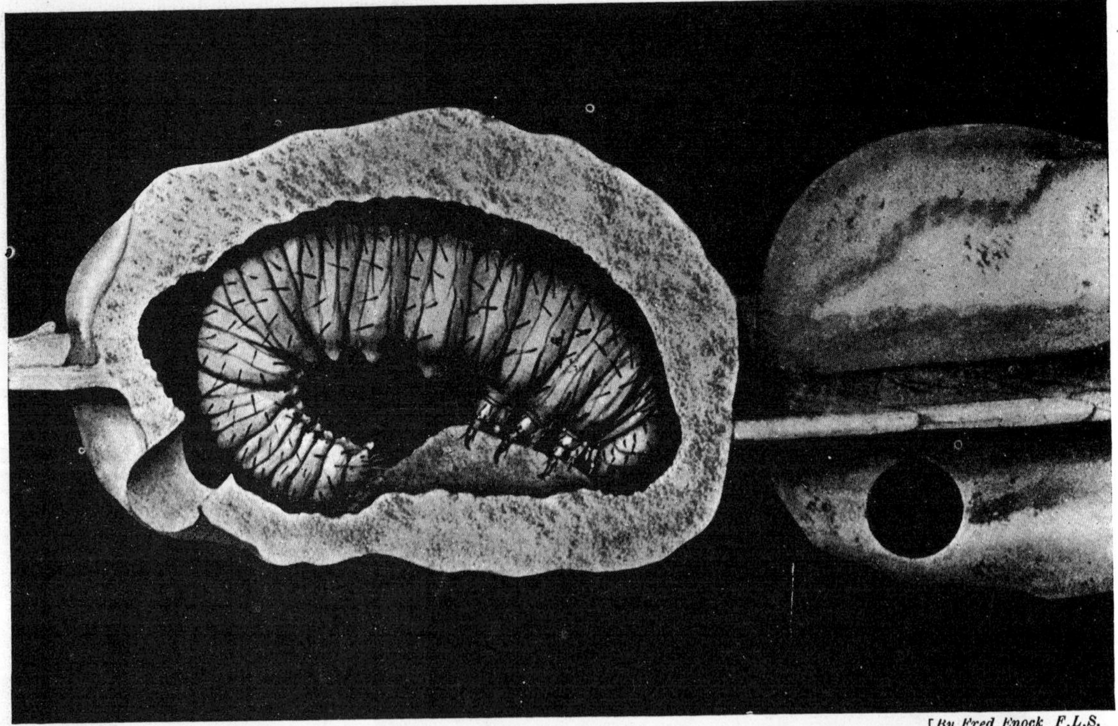

[*By Fred Enock, F.L.S.*

SECTION OF A WILLOW-GALL.

The grub is shown inside the Willow-gall. To the right is seen the exterior of part of the gall, with the tiny hole made by the grub from inside. The passage indicated by the hole is shown in the section as leading to the cavity in which lies the grub curled up in a characteristic attitude.

summer, and quickly commences to lay eggs, starting a second brood, which, after making their cocoons, hibernate during the winter, emerging early in May. Many parasitic flies attack the caterpillars during their feeding-time, piercing through the gall-grub, into whose body they first inject a dose of formic acid, to paralyze it. The parasite then lays an egg in the hole on the under-side of the gall, and this hatches on the fourth day into a white grub, which attaches itself by its powerful jaws to the body of the paralyzed Saw-fly grub, and there it remains for a fortnight, slowly but surely devouring the body. When full fed, it just rests by the side of its late host and changes to a naked chrysalis; and when mature, the fly emerges, bites its way through the gall, and is soon at its work among the second brood of Saw-flies.

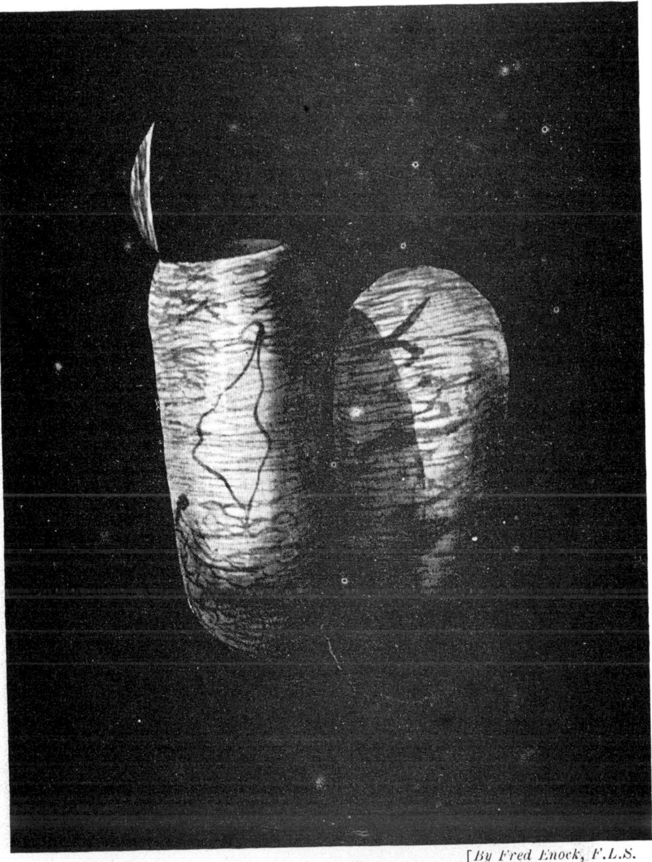

[*By Fred Enock, F.L.S.*

COCOONS OF THE WILLOW SAW-FLY.

Crawling down from the Willow to the wood-strewn ground beneath, the grub crawls behind a suitable loose piece of bark and spins a brown silken cocoon. These cocoons are sometimes to be found in clusters as many as twenty together. From the left-hand one in the illustration the Fly has already emerged.

Many of our British Saw-flies are more or less gaily-coloured, some being of a deep violet or green colour; others with red or bright yellow bands across their bodies and the wings beautifully iridescent. In size they vary from two inches and a quarter to three-eighths of an inch in expanse of wings.

A number are very injurious to both our root crops, such as turnips, on the leaves of which the caterpillars of the "Black Jack" Saw-fly sometimes feed in immense numbers, doing a great deal of damage. Then the gooseberry bushes are attacked by another species, which completely denudes the bushes of every leaf and small gooseberry. Another attacks the pear-tree, and much resembles a small black slug.

Perhaps the worst Saw-fly pest is the one which, as soon as hatched from the egg, eats its way through the half-grown plums, right into the soft stone, and is not content until it has completely destroyed the kernel; and then it bites its way through, falls to the ground, where it spins its cocoons, and waits for the next spring crop.

THE DELICIOUS MONSTER

THE tribe of plants to which belongs our common Cuckoo-pint or Lords and Ladies, includes some remarkable tropical representatives. For that matter our own native species is remarkable, but it is one of the weaknesses of human nature to despise what is common, and to value what is not easily accessible. If the Cuckoo-pint could not be obtained nearer than the Andes or the Malay Archipelago it would be grown in our greenhouses as a curiosity; but as it may be found in every copse and hedgerow only the botanist knows that it is both curious and interesting.

In Tropical America there grows a climbing plant of this family, to which Adanson gave the name of Monstera deliciosa, without explaining the first word; the second has reference to the flavour of its fruit. The plant has large, heart-shaped, leathery leaves, in which respect it makes no great departure from others of its tribe; but if you look at the upper leaves you will find it rather difficult to say what shape they are. The more central area of the leaf is perforated with round, oval and oblong holes of various sizes, and the margin is cut into irregular lobes. Now, such a leaf has been as well and truly fashioned as the still heart-shaped ones below it; and these perforations of the centre and cuttings of the margin are due neither to the attacks of insects nor any other extraneous agency. The upper leaves alone, after attaining their normal form, have developed these openings in a perfectly healthy manner. The form and arrangement of the leaves on the stems of all plants appear to have relation to the necessity for every leaf to receive its due share of light and air, in order that it may carry out its proper functions for the nourishment and growth of the plant as a whole. The upper leaves of Monstera, if they retained their original entirety, would so shut out the light from the lower leaves as to prevent their efficient working; so they begin to develop perforations of the centre and margin, and these continue to widen until the leaf would come under the schoolboy's definition of a net—"a lot of holes sewn together."

Photo by] [*Fred Enock, F.L.S.*

THE SAW OF THE PINE SAW-FLY.

Showing the beautiful and complicated structure of the apparatus by which the Fly can bore into the trunks of Pines.

The minute flowers of this plant, like those of our Cuckoo-pint, are arranged round a central spike called a spadix, which is protected by a loose outer sheath called a spathe. The flowers of our Cuckoo-pint develop into separate red berries; those of Monstera grow so crowded that the fruits become compressed into six-sided forms and the whole assemblage becomes a fleshy, cylindrical body, the interior of which, when ripe, is pulpy and edible, with a flavour of pine-apple. Occasionally a few of these may be seen in the best class of fruit-shops, but their price prevents them from becoming well known.

THE RAT-TAILED MAGGOT

It is called the Rat-tailed Maggot because the maggot stage is the most interesting part of its life-history, but really It, when perfectly developed, is a fly, and the maggot-stage, therefore, forms a quite subsidiary part of its existence. This fly, which is known to the naturalist as Eristalis, belongs

Photo by] [*C. B. Waite.*

THE DELICIOUS MONSTER.

This mighty plant of Tropical America is a near relative of the common Cuckoo-pint of our woods and lanes. It is a climber, as the strong aerial roots in the foreground of the photograph testify. But the most remarkable characteristic is that of the perforation of the upper leaves, a method devised by Nature to allow light and air to reach the lower foliage, which would otherwise be doomed to decay.

to a group of these insects that bear a marked resemblance to a bee, so much so that even an expert may be for a moment deluded, and, as a result, the fly obtains a certain amount of cautious respect which is not entirely its due. But a bee has four wings, while the fly has only two, so that with a little care this imitation can be distinguished from the bonâ-fide bee. But to come back to the Maggot. It can best be described as a fat, white grub, about an inch long, with a tail that may reach four inches or a trifle more, and the best place to find it is in a rather stagnant pond or stream where the water is not very deep and where there is plenty of mud. The more the mud is composed of decayed vegetation and putrefying material, the more abundant and well-nourished the Maggots will be. They will be seen standing on their heads with their long tails stretched up to the surface of the water, not necessarily perfectly perpendicular, for some will curve slightly to the surface, and perhaps one or two could be seen crawling about on the bottom with the "tail" at right angles to the body and still reaching up to the water level. Suppose a rainy season augmented the water in the pond; the Maggots would still be found with their tails reaching to the surface, and the depth would have to be increased to something over six feet before the Maggots would be inconvenienced, and then they would crawl up any available object until they could once again protrude their tails from the water. And why is this? The reason is that the tail is a breathing-tube; in its centre it contains two narrow channels by which oxygen is conveyed to the owner. These channels are wrapped up in external tubes; the nearest is known as the respiratory tube. At its extremity are five feathered hairs, which are spread out, and which cling to the surface film of the stagnant water, and in this way help to support the long swaying length of tail. This tube folds into the channelled sheath, which is only visible when the tail is fully extended. Like the respiratory tube, it is channelled or furrowed, as is also the last section of the tail, the wrinkled sheath, which is further distinguished by tubercles bearing little tufts of minute hairs arranged in pairs along its length. This last section gradually merges into the body of the grub. So the Rat-tail, which to all appearance is so insignificant and repulsive a Maggot, is possessed of a wonderfully elaborate breathing-

Photo by] [*E. Step, F.L.S.*

THE FRUIT OF THE DELICIOUS MONSTER.

Arranged like the berries of our Cuckoo-pint, but with each berry compressed by over-crowding into a six-sided form.

organ, which can be folded up into a small compass like a telescope, or be stretched out till it is nearly six inches long and as thin as a horsehair.

When the grub changes into the pupal or intermediate stage, similar to that of the chrysalis in the butterfly, its skin first of all becomes hard and opaque, then appear two pairs of horns at the head end, the fore pair about half the length of the other. These are to serve as breathing-organs, supplying the place of the tail, which is now retracted, and, like the body-skin, has become quite hard. They are not long wanted, for, if the season be favourable, the fly will be ready to emerge after eight days.

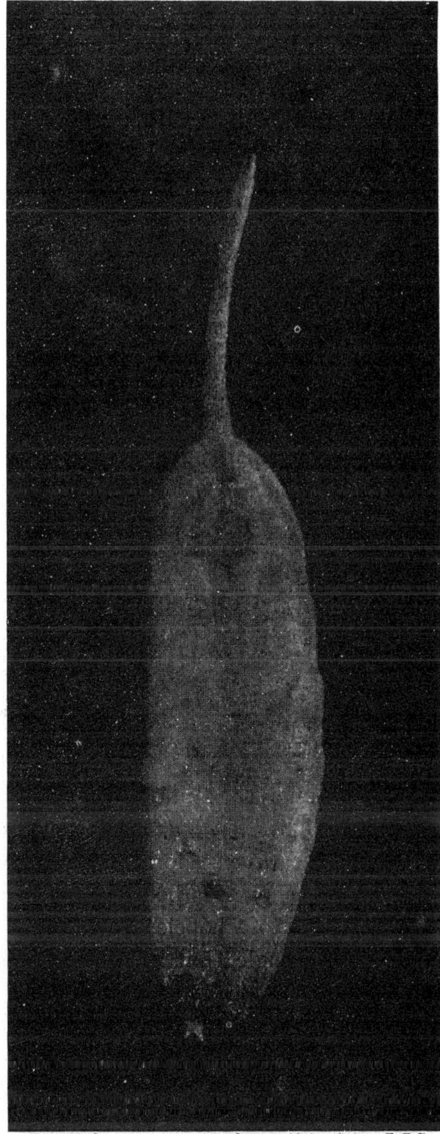

Photo by] [*Hugh Main, B.Sc., F.E.S.*

THE RAT-TAILED MAGGOT.

The Maggot is remarkable for its telescopic "tail," or breathing-tube, which within certain limits can be extended according to the depth of water in which the creature lives so that the aperture remains above the water and allows a free passage of oxygen to the Maggot.

SEA-WORMS

THE word "worm" is generally held in abhorrence as indicating something ugly, slimy and wriggly. But if we would learn something of the wonders of Nature we must not be deterred by a word; and since Darwin showed, in his volume on the Earthworm, how important a work that despised creature does for the human race, that word should have for ever lost its terrors for the most sensitive of us. But the word "worm" covers a multitude of lowly creatures of the most diverse form and habit, and the worms that dwell in the sea include some of the most exquisitely-coloured as well as beautifully-formed of the smaller animals. They are of several distinct types, and their habits vary as much as their forms. Some, which we hope to deal with on a future occasion, are sedentary, building fixed tubular homes of sand or porcelain. Others are free to rove the vast domain of the sea at will. It is of a few of these that we desire to speak at present.

Several of these Sea-worms are distinctly unwormlike in form. Such, for example, is the beautiful Sea-mouse, so-called because its rounded back is completely covered with a kind of long hair. But no terrestrial beast was ever clothed in fur of so glorious a character. It is only when you turn it on its back that you get an idea of the Sea-mouse's real nature. Then you see that it is built up of rings with side feet; above it is really covered with broad, overlapping armour-plates, but these are almost hidden by a thick brown "underfur" from which emerge long iridescent hairs and bristles that reflect almost every conceivable brilliant tint. The despiser of worms should see this Sea-mouse; but probably then he will not admit that it is a worm. Several other of the Sea-worms are covered with similar plates, but without the addition of the coat of fur.

What we have spoken of as feet in the case of the Sea-mouse are shared by many other Sea-worms, though they are not jointed limbs, but fleshy protuberances from the sides, which, in turn, bear bunches of bristles—often iridescent—which are used as organs of locomotion by pressing them against the surface they are gliding over. Some of them have these feet protected

Photo by] [*Hugh Main, B.Sc., F.E.S.*

The intermediate stage of the grub. The outer skin hardens into a tough envelope, and the breathing-tube, which is no longer wanted, becomes contracted. As this stage progresses two horns appear. These are breathing organs.

from above by a leaf-like scale, such as may be seen on the big, many-hued worm that occupies the centre of the coloured plate. This magnificent worm is a representative of the Leaf-worms, which are so-called because of these organs, of which they have one on each side of every ring of their long, centipede-like bodies. Some of them also have gills upon these lateral protuberances.

So various are the Sea-worms that, whilst some have heads ornamented with few, or many, antennæ, two or four eyes, jaws of varying number; others have no head, or a head minus antennæ, or jaws, or eyes. They swim in the sea, twine round sea-weeds, explore empty shells, or live in crevices of the rocks—some even bore holes in the rock for a habitation. One of those that live in crevices is shown to the left of the coloured plate. He belongs to the more worm-like section, which includes the fisherman's "Lug." Its feet are branched, and from the upper branch he can protrude a long thread which appears to be of the nature of a feeler. These threads are especially abundant near the head, and they are drawn out to an enormous length, proportionately to the size of the worm. At night they are luminous, and we have often surprised our friends by gently prodding the worm and so stimulating it to glow through all these hair-like threads with a phosphorescent light.

One other diverse form must be mentioned—the black Long-worm shown on page 958. He looks like an enormous boot-lace that by long soaking in the sea has become very soft and flabby, and has got into a hopeless tangle. One hesitates to mention its length. Many have tried to unravel the tangle, but when a few feet have been carefully straightened out this part begins at once to tie and twist itself up again, and the task is usually given up for lack of time, and perhaps of patience. Professor Mackintosh essayed to unravel one that "half filled a dissecting jar eight inches wide and five inches deep. Thirty yards were measured without rupture, and yet the mass was not half uncoiled."

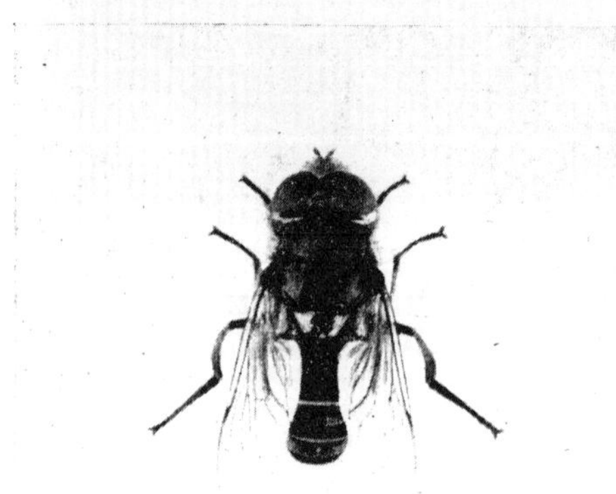

Photo by] [*Hugh Main, B.Sc., F.E.S.*

The perfect insect is often mistaken for a bee, though it can readily be distinguished by its single pair of wings. It feeds on the nectar of sweet-smelling flowers.

THE DIMORPHODON

BY W. P. PYCRAFT, F.Z.S., ETC.

THOSE who wander through the galleries of fossil animals at, say, the British Museum must often be tempted to ask: "Can these dry bones live?" And every now and again the artist, seeking new themes, will make an attempt at the difficult task of resuscitation. In the accompanying plate the accomplished animal painter, Mr. C. Whymper, has conjured up for us a scene such as may well have taken place a few

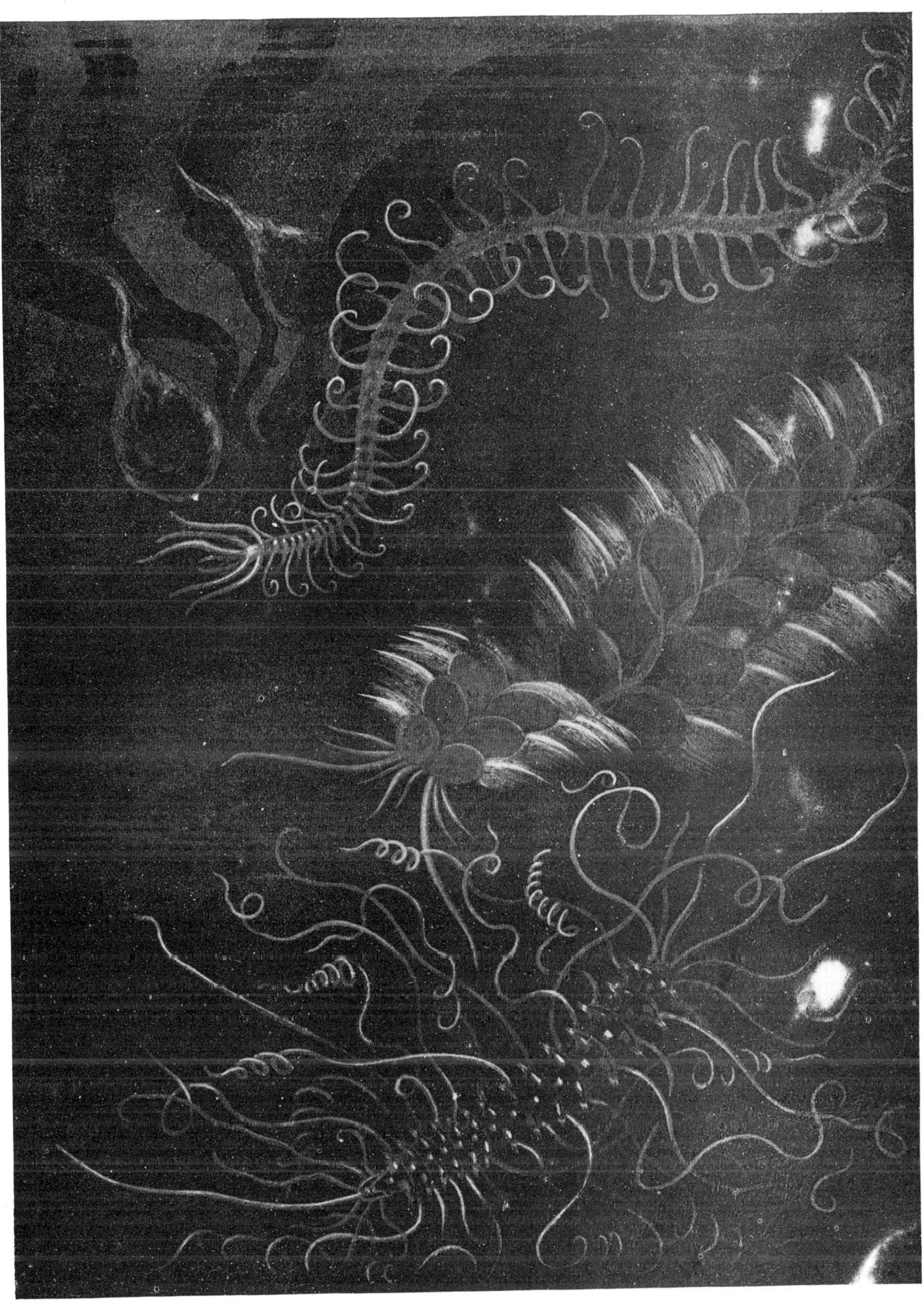

SOME SEA-WORMS.

The worms of the sea are remarkably diversified in form and habit. Those illustrated above are made up of many joints, each joint provided with long tentacles or bundles of gleaming, glassy bristles, which serve as organs of locomotion. The central figure has its upper side protected by a double series of armour-plates

A LIVING GORDIAN KNOT.

A soft black Worm without external organs of any kind, and of extraordinary length. As commonly seen, coiled up in some empty shell its principal occupation appears to be tying itself into a hopeless tangle—from which, however, it glides with absolute ease.

million years ago. He has chosen the period known as "the Age of Reptiles," when these creatures were the dominant types of life on the earth, and when the Mammals, which trace their descent from the Reptiles, were just appearing.

The most striking figure in his picture is naturally the flying dragon, or Pterodactyl, Dimorphodon, which was undoubtedly a carnivorous creature. It has been suggested that it was also warm-blooded, which is quite possible, though, this being the case, it differed from all other reptiles, which are a cold-blooded race. The Dimorphodon is depicted as about to wrest a meal from the jaws of one of the ancient crocodiles, evidently a near ally of the marine armoured crocodile Teleosaurus, a type bearing a strong resemblance to the long-snouted Gavials of to-day.

And what about the "meal," the subject of the dispute. Here was a difficulty. It is certain that the day of the mammals had begun at the time these reptiles were living, for the records of the rocks proclaim as much, though only by fragments too imperfect to enable us to reconstruct a convincing picture. Accordingly the most primitive of living mammals to-day, the Duck-billed Platypus, or Ornithorhynchus, has been chosen to represent these ancient prototypes. The Duck-bill is well-named *paradoxus*, for while it is undoubtedly primitive, retaining many characters derived from its reptilian forbears, in others, as in the duck-like beak, it is highly specialized. Among its primitive characters the most striking, of course, is the fact that it lays eggs, a peculiarity shared only by the Echidna, among mammals, and there can be no doubt but that the early mammals were oviparous, as the reptiles remain to this day.

In his handling of the Pterodactyl Mr. Whymper has shown great restraint—at any rate, in so far as coloration is concerned; for it is quite probable that these agile creatures displayed vivid and rapidly changing hues, as is the case with many reptiles to-day, even among quite sluggish species.

Dimorphodon, it must be remarked, represents the oldest, most ancient of all the known flying dragons, and remains thereof have been obtained from the Lower Lias of Dorsetshire. Remains of Teleosaurus, with its sprawling, interlocking teeth and short fore-limbs, have been found in our Stonesfield slate, which was deposited in a period remote from the Lower Lias; but there is good evidence to show that, nevertheless, Dimorphodon stretched the span of the ages, and loved to squabble with the upstart crocodile, as has been so vividly portrayed by Mr. Whymper in our coloured plate.

LANTERN-FLIES

NATURE seldom indulges in freaks merely for the purpose of display; and yet it is difficult otherwise to account for the strange formations of the Lantern-flies. An explanation would not be difficult if the tales of the luminosity of these insects were to be believed; but although the insects are by no means rare, so little confirmatory evidence is forthcoming on this subject that there is every reason to doubt the assertion that the Lantern-fly shown on page 961 "has four luminous spots, two on the upper surface of the abdomen and two on the under surface, the former giving a purple and the latter a green light." On equal grounds, the power of the Torch-fly on page 960 to carry a light "like a star" at the end of its long, tubular head is open to discussion. Certainly it is from these supposed light-giving qualities that the popular name has been derived; for once given the attribute of luminosity, the large empty and semi-translucent head-cavities of the insect would soon in the popular imagination become a lantern in which would glow a subtle fire during the night hours.

But the cautious naturalist will find a possible solution to the riddle of such abnormal growths in

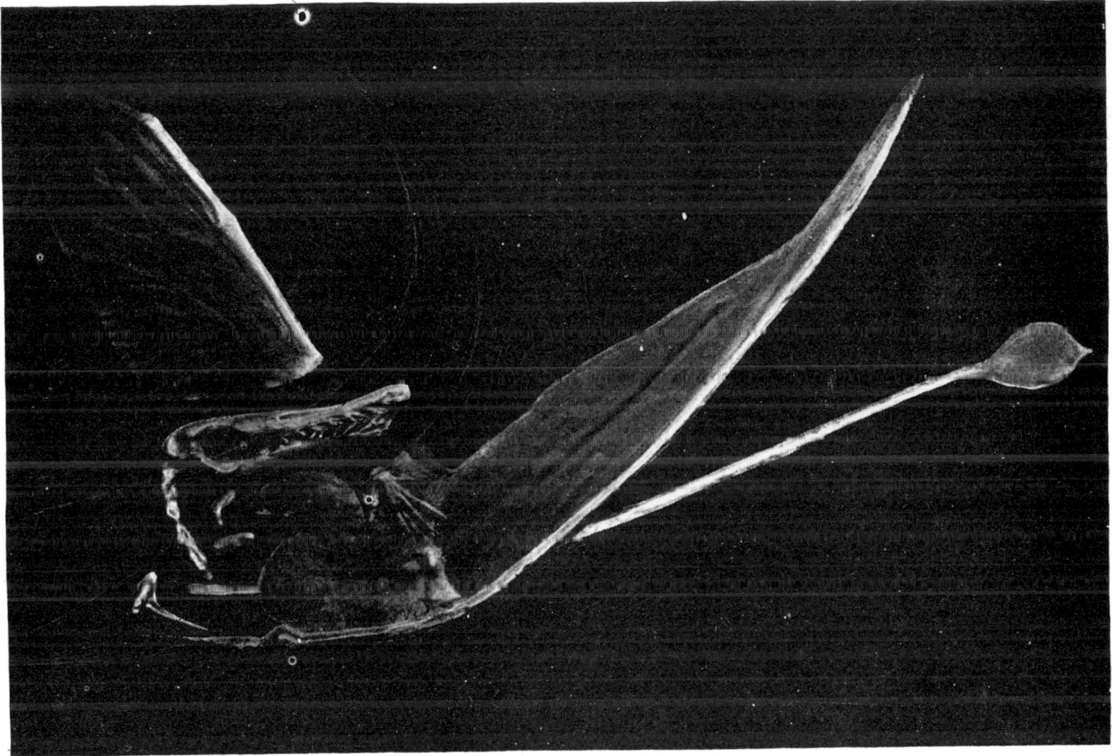

A PTERODACTYL.

One variety of the great family of winged lizards which flourished at a remote period in the world's history. Notice the peculiarly racquet-shaped tail-wing distinguishing it from all others of its tribe. The giant of the Pterodactyls, the Pteranodon, is shown on page 341, while the Dimorphodon of the coloured plate is yet another species.

a less fantastic field of research. The development of warning sounds, markings and colours, as well for the purpose of defence as for danger-signals, is a well-recognized means used by Nature for the protection of her species. By this means the porcupine preserves itself from aggressive enemies and by it the caterpillar of the Cinnabar Moth is made immune from the attacks of birds. But if colour can be adopted for protective purpose, why not form? and possibly herein lies the reason for the assumption by the Lantern-fly of such a grotesque head-dress; for not only are they curious of shape and disproportionate of size, but in the case of the Lantern-fly the mask is ornamented with vivid markings of red and yellow and blue—essentially warning colours. These colours, however, soon fade after death.

Photo by] [*Bernard C. White.*

THE TORCH-FLY.

The head of this curious insect is developed into a hollow and semi-transparent horn. The upper wings are transparent and green, with yellow ringed dots; the under wings are orange, with a broad black band. It is a native of China, and is shown a little smaller than natural size.

There are about twenty known species of Lantern-flies and all are inhabitants of tropical countries. The two best known species are those shown in the illustrations. Of these, the Torch-fly comes from China. A good specimen will be about two inches long. The colourings vary from pale apple-green of the wings to the deep, ruddy amber colour of the head, and the brilliant orange of the under-wings. The Lantern-fly is a much larger species and will measure as much as five inches across the expanded wings. Except for the head, of which mention has already been made, the colouring is subdued; olive greens, yellows and browns being the chief constituents. It is a native of tropical South America.

THE EVOLUTION OF THE HORSE

BY REV. H. N. HUTCHINSON, F.G.S., F.Z.S.

THE theory of evolution in the world of life, now accepted by nearly all naturalists and geologists, has been greatly strengthened since the days of Buffon, Lamarck, Erasmus Darwin and Charles Darwin, by the great and remarkable increase in our knowledge of extinct animal types, brought about by geological discoveries. Out of all this mass of evidence, no more complete case can be found than that afforded by the series of fossil horses discovered in Europe and America, but chiefly in the latter country. Through these fortunate discoveries, we can now read the history of this noble and useful animal more fully than that of any other quadruped, reptile, or bird. It is perhaps hardly necessary to point out that the modern horse walks on one toe only; but our readers may be surprised to learn that his earliest ancestor had five toes, corresponding to the five toes seen in many other mammals and the five fingers of the human hand. Two, however, of the lost toes are represented by the two "splint-bones," which for a long time were a great puzzle to all who in any way made themselves familiar with the horse's skeleton. Now they are no longer a

Photo by] [*Bernard C. White.*

THE LANTERN-FLY.

What exactly may be the reason for the abnormal development of the head is uncertain; possibly the creature has adopted such a head-dress in order that its enemies may be terrorized. The insect is supposed to have luminous patches on the body, but this peculiarity has not been definitely proved. It is a native of Brazil, and is here shown enlarged about twice the natural size

mystery, for we recognize in them "rudimentary" digits (fingers or toes) inherited from a three-toed ancestor, the other two having entirely disappeared in the course of ages. The development of the horse seems to have taken place along two separate lines, one in Europe and the other in America. The record in the latter continent is far more complete; for it so happens that the physical conditions prevailing there throughout nearly the whole of the Tertiary period were singularly favourable to the preservation of those creatures which lived on land. In what is now the Rocky Mountain region the members of the horse tribe were specially numerous, and their fossilized remains are sealed up in the strata of those regions.

Only in the Pliocene deposits does the true horse appear, and then the genealogy is more or less complete. The present race of wild horses which roam over the pampas are not the descendants of the fossil horse of South America, but have sprung from those introduced by the Spaniards more than three hundred years ago. It has often been stated that the direct descent of the horse has been satisfactorily traced by Marsh and others, but that is not true; for there are some still unsolved problems in this field of research. Professor H. F. Osborn, of the American Museum of Natural History, has been for years working at this subject, and through his labours important results have been obtained. It was necessary once more to explore the geological hunting-grounds of the Rocky Mountain region, and for this purpose expeditions were sent out between 1890 and 1904. Mr. J. W. Gidley, who was placed in charge of this work, was successful in finding a large number of fossil skeletons. The result is that five complete skeletons are now to be seen *mounted*, in the museum (not merely lying on slabs of rock). These are of very great value, constituting an unique collection. The museum has been enriched by thousands of specimens, representing the more or less complete remains of seven hundred and seventy-one horses. After

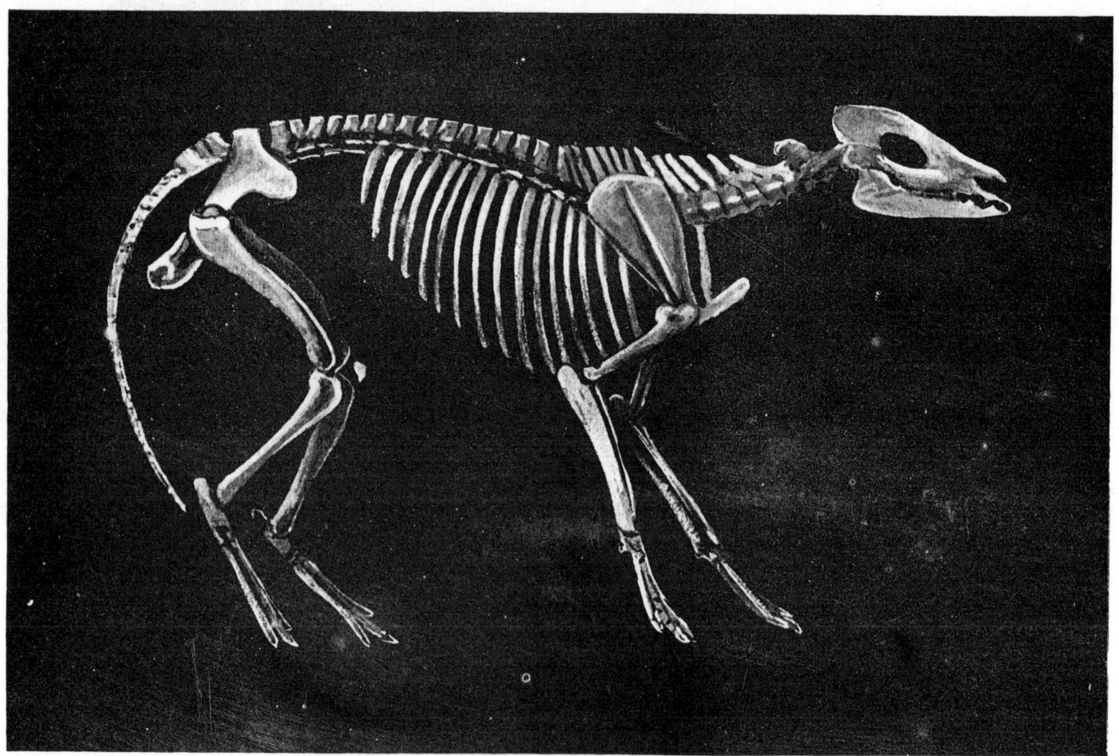

THE SKELETON OF A PRIMITIVE HORSE.

Protorohippus and Hyracotherium are the names by which it is usually known. It was a little animal about fourteen inches high at the withers. Notice the four toes which preceded the more modern hoof.

working on this vast amount of material, Professor Osborn came to the conclusion that there is evidence of at least three (possibly four) collateral lines of development in the evolution of the horse, some large, some small, all more or less intermingled. His conclusions may be summed up briefly as follows: First, we have the little Eohippus (or "Dawn Horse"), only eleven inches (or two and three-quarter hands) at the withers, with its wrist, or knee, near the ground. The hand was still short, with four little hoofs, and part of a fifth. It had a wide range in America and Europe.

THE EVOLUTION OF THE HORSE.

Successive stages in the believed ancestry of the Horse. The first of the series (top) is the Dawn Horse, the second is the Hyracotherium, the third Mesohippus; then come in order Protorohippus, Hipparion and the modern Horse.

The next stage is seen in Hyracotherium, also known as Protorohippus. Its skull and fore-foot are shown in our illustration. It was a graceful little creature, about fourteen inches (three and a half hands) high at the withers. The next stage is seen in the Mesohippus. One finger has been dropped, and the skull is decidedly bigger, as our illustration will show. Here we see the middle, or third, digit getting larger. Unfortunately the skull of Protorohippus, which is the next stage, is unknown; but we have the feet, of which a fore-foot is shown in our illustration on the next page. Here the centre digit is much bigger, and this increase becomes still more striking in the Hipparion; until at last, in the modern horse, this one digit is supreme, and the others are only represented by the two feeble "splint-bones" already referred to. The outline restorations (2, 3, 4, 5, and 6 in our illustration on this page) may be studied in connection with the bones of the skeleton on page 962. The Hipparion was of the size of a modern donkey.

At the same time that these changes were taking place certain other important changes, or adaptations, were going on in other parts of the body (according to the Law of Correlation), especially in the number and structure, etc., of the teeth. These may be briefly summed up by saying that as horses grew bigger and swifter, their dental apparatus at the same time became improved, so that the food was more perfectly ground down, and thus more vital energy was provided. This fascinating subject is well illustrated by models and bones in the Natural History Museum, South Kensington.

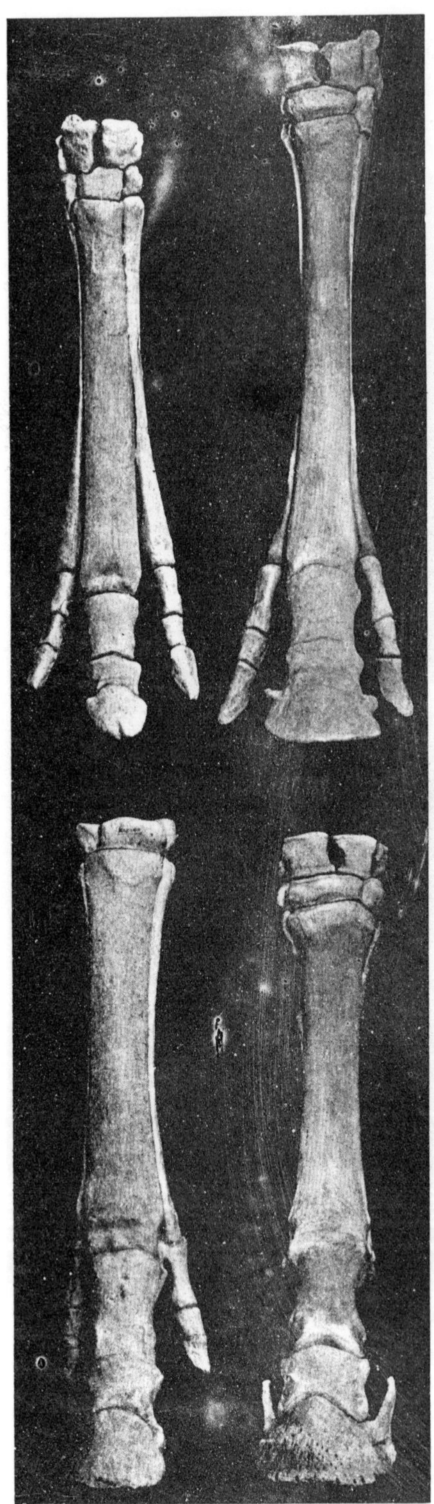

EVOLUTION OF THE HORSE'S HOOF.

The earliest known ancestor of the Horse had five toes. These were gradually reduced to three, later to one—the present day hoof.

MONITORS, OR GIANT LIZARDS

BY R. LYDEKKER.

THE claim of the Old World lizards, commonly known as "Monitors," to a place among marvellous animals is based on the fact that they are by far the largest members of their order, one Indo-Malay species attaining a length of about eight feet, and weighing so much as sixty pounds. Even this monster is, however, a pigmy compared to an extinct Queensland species, which probably measured some forty or fifty feet in length, and was therefore about double the size of the biggest living crocodile.

The origin of the name "Monitor" is not a little curious. The Egyptian representative of the group is known, in common with other lizards, by the Arabic name of *waran*, or *ouran;* and by some unusually clever person this has been identified—although, of course, incorrectly—with the German *Warner*, signifying a warner. From this it was but a step to substitute the equivalent Latin term *monitor* —a warner! In spite of the absurdity of its origin, the name is, however, a convenient and distinctive designation for these great lizards.

Monitors are represented by rather less than thirty species, ranging over Africa, Arabia, India, the Malay countries, and Australia; the two species represented in the accompanying illustrations—namely, Gould's Monitor and the Variable Monitor—being natives of the last-named country.

In general appearance Monitors are very like overgrown ordinary lizards; but they are specially characterized by the unusual elongation of the neck, and the great length of the deeply-cleft tongue, which, when these lizards are active, is kept in constant motion, being alternately protruded and withdrawn with lightning-like rapidity in the same manner as that of a snake. In the matter of diet Monitors are carnivorous, consuming lizards and snakes smaller than themselves, as well as small mammals and birds. A particularly favourite *bonne bouche* is a clutch of birds' eggs or callow young. In devouring eggs, each one is swallowed whole; and in captivity the large Indo-Malay Monitor will think nothing of devouring eight or ten hens' eggs at a meal, which are swallowed in such rapid succession that they may be heard clicking against one another as they glide down the creature's capacious gullet.

In all cases the tail is of great length and flexibility, but whereas, in the species inhabiting dry and sandy districts it is whip-like in shape, in those frequenting

Photo by] [*A. A. White.*

THE VARIABLE MONITOR

The Monitors, or Giant Lizards, are the largest of existing Lizards. The species represented is a native of Australia.

swampy districts, like the Sandarbans of Bengal, it is laterally compressed, and thus serves as an efficient swimming instrument. The great Indo-Malay species—generally known in Bengal, in common with other lizards, as *Go-samp*—will, indeed, take readily to the water, and has even been known when pursued on the seashore in the Sandarbans to plunge headlong into the surf and disappear. In the Bengal rice-fields these lizards abound, and when disturbed start off with a violent rush—much to the alarm of inexperienced snipe-shooters.

Photo by] [*A. A. White.*

GOULD'S MONITOR AND THE VARIABLE MONITOR.

The whip-like tails of both these Australian Lizards indicate that they inhabit dry, sandy districts; those that frequent swampy neighbourhoods have the tail flattened from the sides to aid in swimming.

GNATS AND MOSQUITOES

BY K. G. BLAIR, B.SC., F.E.S.

EVERY year, just about the holiday season, reports appear in the newspapers of the plague of Mosquitoes from which some well-known seaside resort is suffering, and visitors from the colonies or the tropics are said to assert that the insects in question are just the same as those which they know at home. Instead, however, of such cases being, as they are usually considered, abnormal, the Mosquito, like the poor, is always with us, though in private life he is called a Gnat. The Gnat is not a single species of fly, but one of a little group or family, the separate species of which are distinguishable only to the expert; and the same family in other parts of the world is known as the Mosquito, a Spanish word, literally meaning a " little fly " (Latin—*musca*, a fly).

Gnats or Mosquitoes, whichever we like to call them, are always more abundant in swampy, marshy districts than in dry, well-drained countries, a circumstance which follows necessarily from the fact that they breed in stagnant pools; for the same reason, also, a period of wet weather during the summer gives them an excellent opportunity for rapid and prolific reproduction, which, in due course, leads to their appearance in such numbers as to constitute a veritable plague. Indeed, it is not unlikely that the flies with which the Egyptians were plagued were Mosquitoes that had bred in the pools left by the overflowing of the Nile. That the land of Goshen should have been exempt lends support to this theory, for that country is a rather higher, dry, grazing district, not subject to be overflowed, and, consequently, would not serve as a breeding-ground for the "flies."

Gnats are particularly suitable subjects for the study of life-history and development, for they are common and easily procured in all stages; they require very little care while under observation; and, finally, their life-cycle is fairly quickly run through. The perfect insects are found from early summer onwards; even during the winter the hibernating females may be seen in numbers on the roofs and walls of cellars and outhouses. To begin at the beginning, the eggs are readily found as little brown boat-shaped masses floating on the surface of any open water-butt; with a little patience, the Gnat may be observed preparing the raft. Resting usually on the edge of some floating object, or sometimes on the surface of the water, with her hind legs she arranges the eggs side by side in such a manner that each egg is upright, with one end exposed to the air and one end in the water. As each egg tapers slightly towards the upper end, the resulting egg-mass forms a slightly concave raft, which in case of accident is self-righting, so that it always floats with the hollow side up. In a few days the eggs hatch, the young larvæ escaping from their lower ends straight into the water.

The larvæ wriggle actively about in the water by means of lashing their tails violently from side to side. They are almost black in colour, and when full-grown, about half an inch long. The head is attached by means of a distinct neck, and is furnished in front with a pair of feathery organs, with which it sweeps up the microscopic organisms upon which it feeds into the mouth. Behind, the neck is a conspicuous bulge, in which are situated the parts that give rise to the wings and legs of the fly. Near the tail is a moderately long, straight tube pointing directly upwards, the significance of which we shall see directly, and behind it is the fan of fine hairs which serves as a powerful oar.

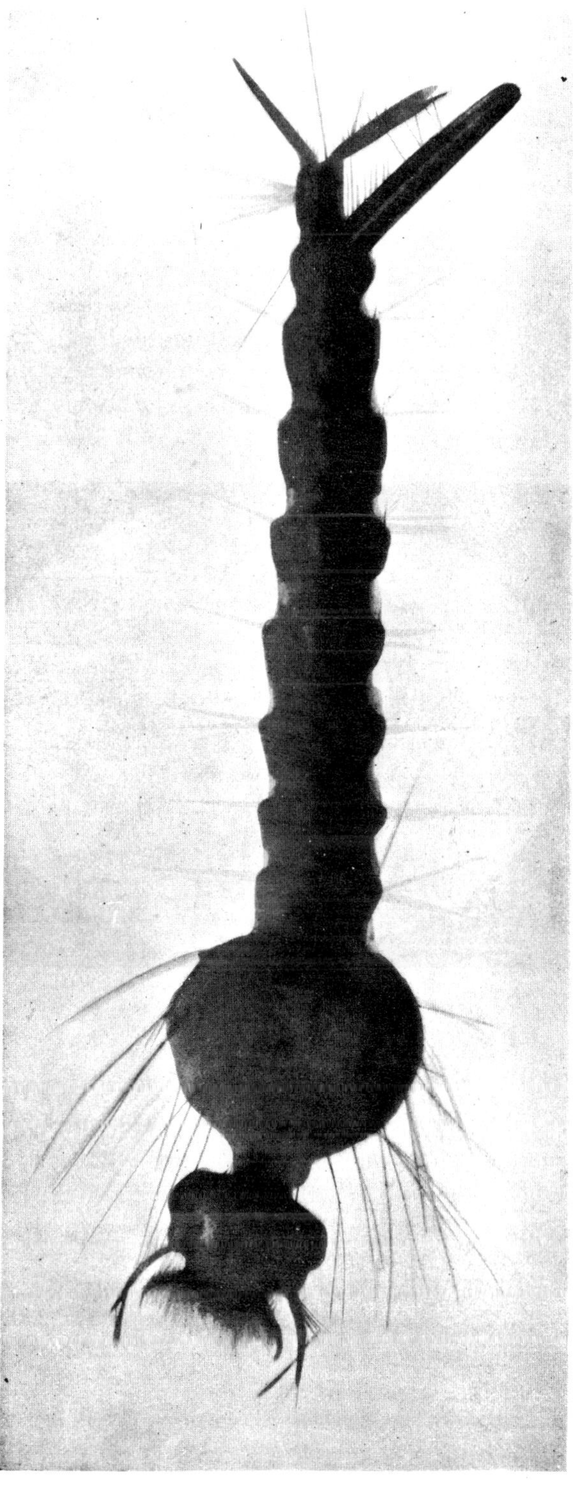

GNAT GRUB.

The grub or larva of the Gnat floats at the surface of the water head downwards. It is shown here enlarged about eighteen times the natural size

When undisturbed the larva usually hangs head downwards from the surface of the water, with only the tip of the tail-tube breaking the surface. In this position the openings of the two main air trunks of the body are exposed to the atmosphere and the larva is breathing; at the same time it is also feeding, for the brush-like organs at the mouth maintain a constant stream of water which brings with it the organisms which form the food of the creature. On the approach of danger the larva closes five little flaps at the end of the breathing-tube, by this action losing the support of the surface-film, when it sinks by its own weight. As soon as the danger is past and the larva wishes to return to its former position, it has to swim violently upwards. This condition, it will be noticed, is just the reverse of that which obtains with the larvæ of water-beetles; these are lighter than the water, and in order to remain below have to hold on to submerged bodies; when they want to come up to breathe they merely let go their hold and automatically rise tail first to the surface.

When full-grown the larva moults and becomes a pupa. In addition to the rudiments of the limbs of the perfect Gnat being now visible, other great changes have taken place. The creature no longer breathes through its tail, but through two small tubes, or trumpets, that have now appeared behind the head. Consequently the pupa has to adopt a different position in the water, and floats back upwards, with its tail curled underneath it and only the two breathing trumpets breaking the surface. In a few days the fly is ready to emerge; the pupa skin splits down the middle of the back and the body of the fly bulges out through the slit. Slowly and gradually it pushes itself further and further out; legs and wings can be observed to be pulled out of their several sheaths, until the creature seems to be standing on its tail, with its extremities still caught in the floating pupal skin. It now rests for a few moments, then pulls its legs clear; the wings and body follow, and after a very short rest either on the floating skin or on the surface of the water, it flies away. The whole process only occupies a few minutes from first to last, the wings being expanded as they leave their pupal envelopes.

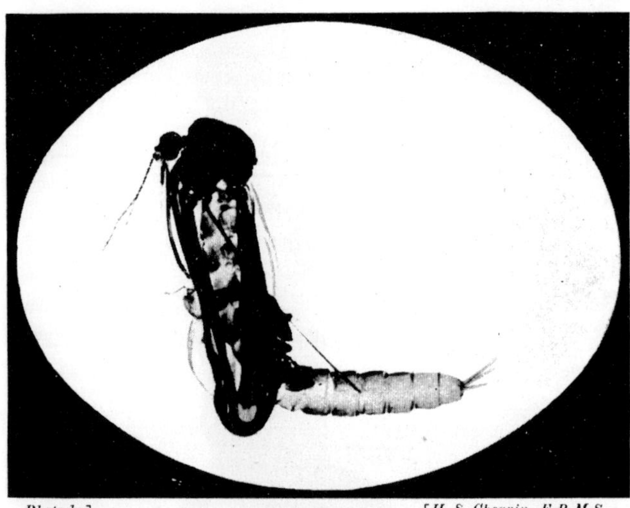

Photo by] [*H. S. Chearin, F.R.M.S.*

THE GNAT.

The new-born Gnat is emerging from the floating skin of the chrysalis, maintaining its balance whilst withdrawing its body, wings and long legs. (Enlarged about ten times.)

The Gnat is a beautiful object under the microscope, or even for observation with a good lens. The delicate scales with which the body and wings are somewhat sparsely clothed, the plumose antennæ, especially those of the male, and the complex proboscis, consisting of a strong supporting sheath enclosing a whole armoury of exquisitely delicate lancets, are all objects whose beauty cannot fail to arouse our admiration, whatever may be the effect produced by the operations of the latter upon our persons. As a matter of fact, the male Gnat is perfectly harmless; it is the female alone that does all the mischief.

Irritating though Gnats may be, we in this country should be thankful that Gnat-bites are nothing worse than Gnat-bites. In warmer climates there is always the possibility that they may mean infection with malaria, yellow fever, or other diseases due to a microscopic parasitic organism living in the blood. The parasite is sucked up by the Mosquito from some person already infected with the disease, and in its body it undergoes a certain development until it in turn is communicated to some healthy person bitten later. Now we see how important it is that open

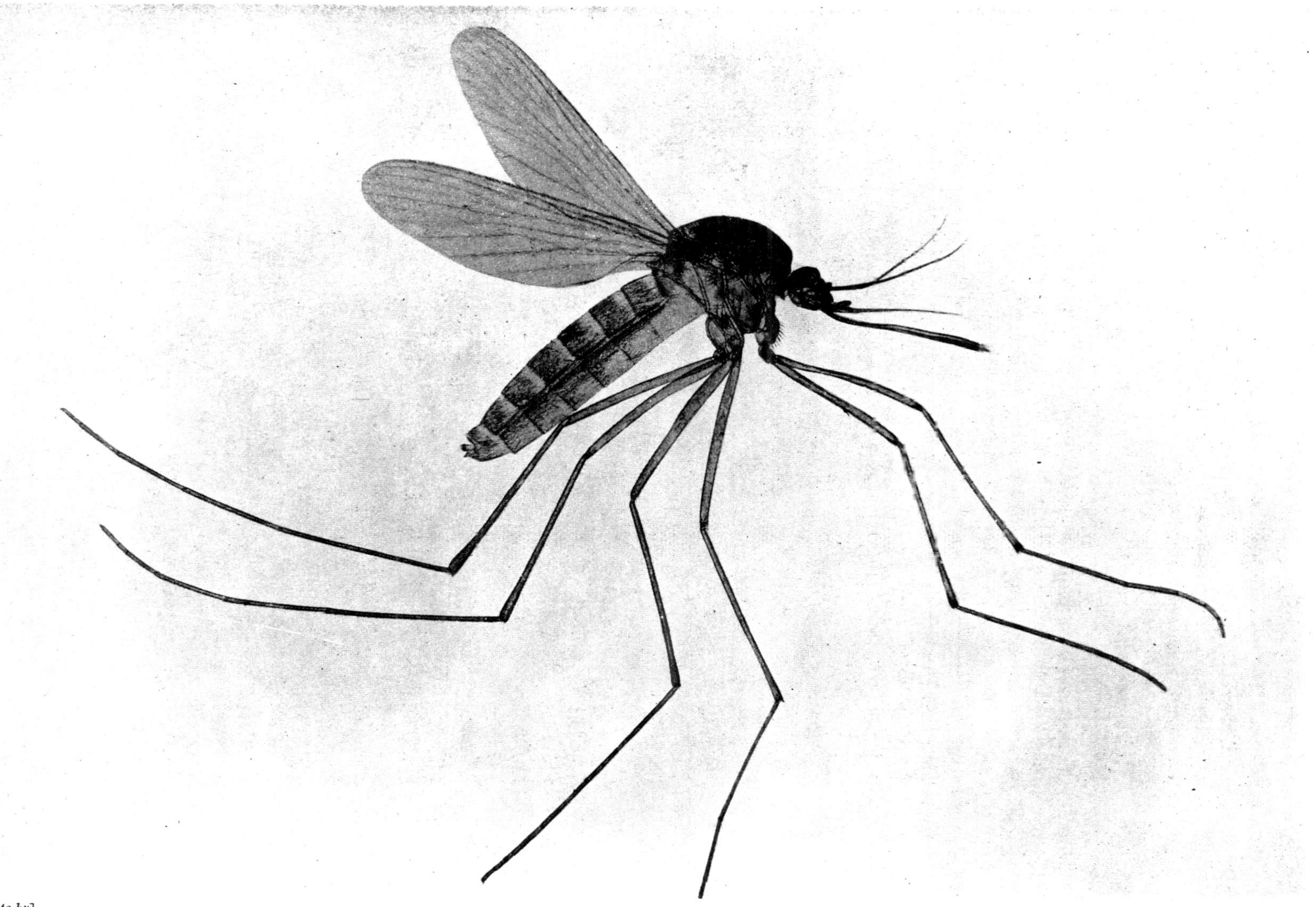

Photo by] [*A. E. Smith.*

THE COMMON GNAT.

The female Gnat (here shown) is known by the simplicity of her antennae (compare with photo of male's antenna on next page). She is also to be identified by her power of inflicting painful stings.

Photo by] [*H. S. Chenrin, F.R.M.S.*

GNAT'S ANTENNA.

An enlarged photograph of part of the feeler of a male Gnat. The corresponding organ in the female is of much simpler structure.

pools or tanks of water should not be allowed near human dwellings in districts where these diseases are rife. Fish in the tanks are fairly efficacious in destroying the Gnat larvæ, and a film of oil on the surface of the water wets and clogs the openings of their breathing-tubes and is absolutely fatal to them. It is attention to points such as these that has done so much of recent years in improving the conditions of life for the white man in the tropics.

AN ANCIENT RELATION OF THE RHINOCEROS

BY W. P. PYCRAFT, F.Z.S., ETC.

In South Dakota, where the temperature in the shade registers 106°, American savants, for years past, have endured torments unflinchingly, that they might hunt in one of Nature's ancient graveyards for the creatures entombed there millions of years ago. Their privations and devotion have been richly rewarded, for there are few spots on the earth's surface which have yielded beings so strange.

The giant beast which forms the subject of our illustrations (pages 972 and 973) is one of these. This is the Great Titanothere, or Brontops, a creature nearly related to the rhinoceros of to-day, but considerably larger, and differing, among other things, in having a pair of huge horns on the snout, set side by side instead of one behind the other, as in the African rhinoceros to-day. These fearsome weapons were largest in the bulls, and that they were used to exemplify the old adage, "Might is right," is attested by the signs of broken ribs which these fossil remains sometimes display.

Big as he was, Brontops was, compared with others of his kind which succeeded him, a slender, agile creature. His forbears, however, were smaller and yet more agile, and, as the skull of one of these, known as Megacerops, shows, the horns on the snout were smaller. In course of time, Brontops, in turn, disappeared from the face of the earth, and was succeeded by the yet larger Symborodon, with larger horns; and Symborodon, in turn, gave place to the most gigantic of all, the ponderous Brontotherium, wherein the horns on the snout attained a huge size.

The legs of Brontops were intermediate in size between those of the rhinoceros and the elephant; and the beast stood nearly ten feet high at the withers, and measured nearly fifteen feet in length.

All kinds of speculations have been made to account for the sudden extinction of the Titanotheres, such as vast floods, excessive droughts, cold waves and epidemics ; but, as likely as not, a change in the nature of the vegetation had much to do with the matter, for the teeth of this creature were not of a kind that could readjust themselves to new demands in the matter of grinders.

One can gather some idea of the vast period of time and the enormous changes which have taken place on the earth's surface since Brontops roamed amid the reed-beds of the ancient lake which formed his haunts when one reflects that the Rocky Mountains were then only beginning to rise, while the lake-bed, then at the sea-level, is now raised up some four or five thousand feet !

In those far-off days the shores of this ancient lake harboured crowds of alligators like those of to-day. In spring the Judas-tree and the acacia bloomed ; in the autumn the persimmon sweetened. Canes and palms gave grateful shade along the alkali levels, while on the hills birch, chestnut, ash, hickory and sumac throve. The huge extinct otter, Patriofelis, preyed upon the alligators, and the last of a series of giants even, if possible, more remarkable than Brontops, probably occasionally put in an appearance. This was the strange Uintatherium. But, so far as the evidence goes, these disappeared just as the forbears of Brontops were putting in their claim to a place in the sun.

The search for these monsters of other days is not, as we have already indicated, one for those who like to do their exploring in comfort ; and it would be rendered still more arduous but for the fact that most of such remains found in this region have become partially exposed by wind and rain. The skull of the specimen forming the subject of our illustration, for example, was found by Professor Osborn protruding from a hard sandstone cliff, and the chest and limbs were chiselled out by men under his direction. The position of the bones showed that the beast had wandered into bog and there been swallowed up !

Photo by] [*Walter Bagshaw.*

GNAT'S WING.

An enlarged view of part of the Gnat's wing, showing the fringes of the nervures and of the border

THE FROGBIT

BY JOHN J. WARD, F.E.S.

THE Frogbit is one of the prettiest and most interesting plants in the British flora. Let us, on this autumn afternoon, hie to the pond where it lives, and peep into the marvels of its economy. In so doing, we shall discover that this curious little plant prepares for winter just as surely as do the water-beetles, the large newts, the frogs, some of the snails, and other aquatic creatures now hiding amongst its floating leaves ; indeed, we shall find that it hibernates just like its animal compeers, coming out with them again in the spring-time.

In the first place, we observe that it is a floating plant, its roots, thickly clothed with root-hairs, merely being suspended like streamers in the water below. Its whole weight is carried by its flat

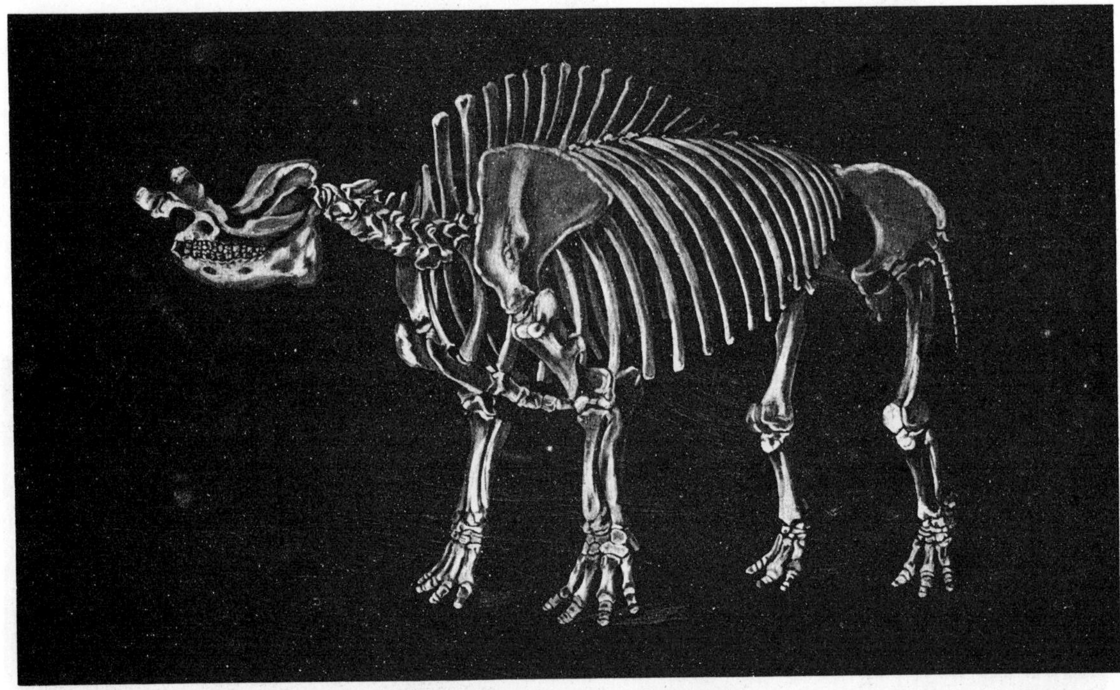

BRONTOPS

This ancient forerunner of the Rhinoceros stood ten feet high at the withers, and was about fifteen feet in length. Our illustration is after the restoration by Marsh.

leaves, which rest upon the surface. Like the water-lilies, and most plants which grow in the open pool or river, its leaves approach a circular form, for that shape is the best adapted for buoying up the plant ; and, as space in such situations is of no consequence, each leaf can spread out its tissues to the sunlight undisturbed ; for the plants of the open waters are few, and there is little or no competition for air and sunlight amongst them, as in the case of their crowded neighbours on the land.

So we find the Frogbit's leaves spreading out often larger than a five-shilling piece in diameter, which is a comparatively large size for so small a plant. Above, the leaves are of a bright green, but early in the year, and in shady and cool places, their submerged sides are reddish purple, a feature which at once indicates that the plant needs warmth. Such purple colouring in plants is generally associated with tender tissues, for it possesses the special property of absorbing light rays and converting their energy into that of heat. If we think for a moment of the large number of plants which in their buds, catkins, and early stages show red and purple colouring, we can then

[By J. Smit.

BRONTOPS.

Otherwise known as the Great Titanothere, a remote ancestor of the Rhinoceros, distinguished by a pair of horns, side by side, on his snout. (See page 970.)

Photos by] [*J. J. Ward.*

THE FROGBIT.

In these photographs and the two on the opposite page—

understand the Frogbit's purple absorbing layer beneath its leaves ; for, like the early and similarly coloured shoots of the colt's-foot and the asparagus, it is capturing all the available heat from the sun's rays, and preventing its escape, thus correcting the chill which the water gives to its leaves.

Indeed, it needs all the heat-energy it can possibly secure, for it is a most active little plant, continually throwing out runners, bearing new tufts of leaves, throughout the summer months, sometimes producing pretty white flowers, but rarely ripening its seeds ; and then, the moment the temperature begins to cool down, and business gets slack, as it were, suddenly shutting up shop and reserving its gains just as the seaside caterer does at the end of the busy season. How skilfully it economizes we will now proceed to see.

Even on this afternoon of mid-autumn, we can see it preparing for those early-morning frosts which so quickly ruin its floating leaves. Some long and strong shoots have grown downwards into the water, each one bearing a large bud on its apex. (See photo on page 976.) These buds are encased with strong protective scales, and are stored with rich starches and living material ; indeed, the plant is steadily conveying all its valuable food substances from its leaves and stems into these large buds ; and when the surface of the water suddenly becomes frozen over, it will not be the Frogbit that becomes frozen, but, instead, its more or less empty skin—its skeleton leaves and stems.

We have yet to witness, however, how marvellously this matter is arranged. The long stems on which the buds ripen extend well down into the water ; for while the surface of the pond is warmest during summer, yet it is coldest during winter. The Frogbit seems fully aware of this fact, and lest an early frost might take it by surprise, it pushes its richly-stored buds well below, out of harm's way. Then at last real frosts begin to come, and we immediately see the large buds, one by one, quickly drop off and steadily sink to the bottom. Their density is just a little greater than that of the water, so that they slowly sink with their apex pointing upwards, and so alight upon the mud at the bottom standing upright. In the accompanying photograph (page 976) an example is shown with a bud attached, and two others detached and resting on the bottom.

At the bottom of the pond it is several degrees warmer than at the surface, and in this cool, but unfrozen, area the Frogbit lies by throughout the winter months for warmer days to come, well wrapped up in its suit of protective scales; like many of its animal neighbours in the pool, it hibernates until the dull and cold days have passed.

Winter floods, and other factors, may carry the buds about the bottom of the pool, lake, or ditch on which they rest, moving them into new surroundings, which explains why the Frogbit does not necessarily appear the following year in the same area where it grew in abundance the previous season; but that is of little consequence, for the buds do not leave the bottom.

About the end of March, though, they become extremely buoyant, the least movement of an awakening animal, which happens to poke its nose out of the mud, causing them to behave like a bouncing ball. Some days, during warm sunlight, a bud will get a start and rise slowly and steadily to the surface, and remain there, apparently sunning itself. If the sunlight fails, or the cool of night approaches, this particularly buoyant bud slowly sinks to the bottom again.

That is the beginning of the spring awakening—something analogous to the movement of the hedgehog, who has remained immovable from November until a mild day in March, and who then suddenly awakens for a moment, only to rearrange its bed and make itself more comfortable to sleep again

Photos by] [*J. J. Ward.*

THE FROGBIT.

—is shown the development of the bud from March 25th to April 4th.

for several cold days which follow ; nevertheless, it is a sign that the forces of life are becoming impatient.

A few consecutive mild days work wonders on these extra buoyant buds, whose tissues have developed some special cells, which during sunlight, in some curious manner, become filled with air, causing the bud to become lighter than the surrounding water. Just what happens to them is shown in the four photographs on page 977. There they are seen quietly rising to the surface, with a little bubble of oxygen at their base.

On reaching the surface, they float about and rapidly increase in size ; and it becomes apparent that the scale-leaves, which remained so tightly wrapped around the buds while at the bottom of

Photo by] [*J. J. Ward, F.E.S.*

THE FROGBIT IN AUTUMN.

The Frogbit prepares for winter by the formation of buds at the end of long shoots. Into these buds all the materials from the leaves is packed, and ultimately they are buried in the mud at the bottom. There they lie snug and safe when their pool is thickly frozen, and await the coming of spring.

the pool, are now beginning to expand, revealing the tender green leaves within, which have been so thoroughly protected.

It is then most interesting to watch the daily development of one of these floating buds. In the accompanying set of photographs (pages 974 and 975) I have endeavoured to show how in the course of eleven days (from March 25th to April 4th) a bud has changed into a little Frogbit plant, which now only has to expand its leaves more fully, and to throw runners into the water bearing tufts of leaves, to resemble almost exactly the autumn plant which gave it its origin. We cannot say the parent plant, for it is not a new plant produced from the seed of two parents ; it is simply a detached portion of the original plant, which has emerged from hibernation with all the vigour of a young Frogbit, and will soon display its characteristic white flowers upon the surface of the pool.

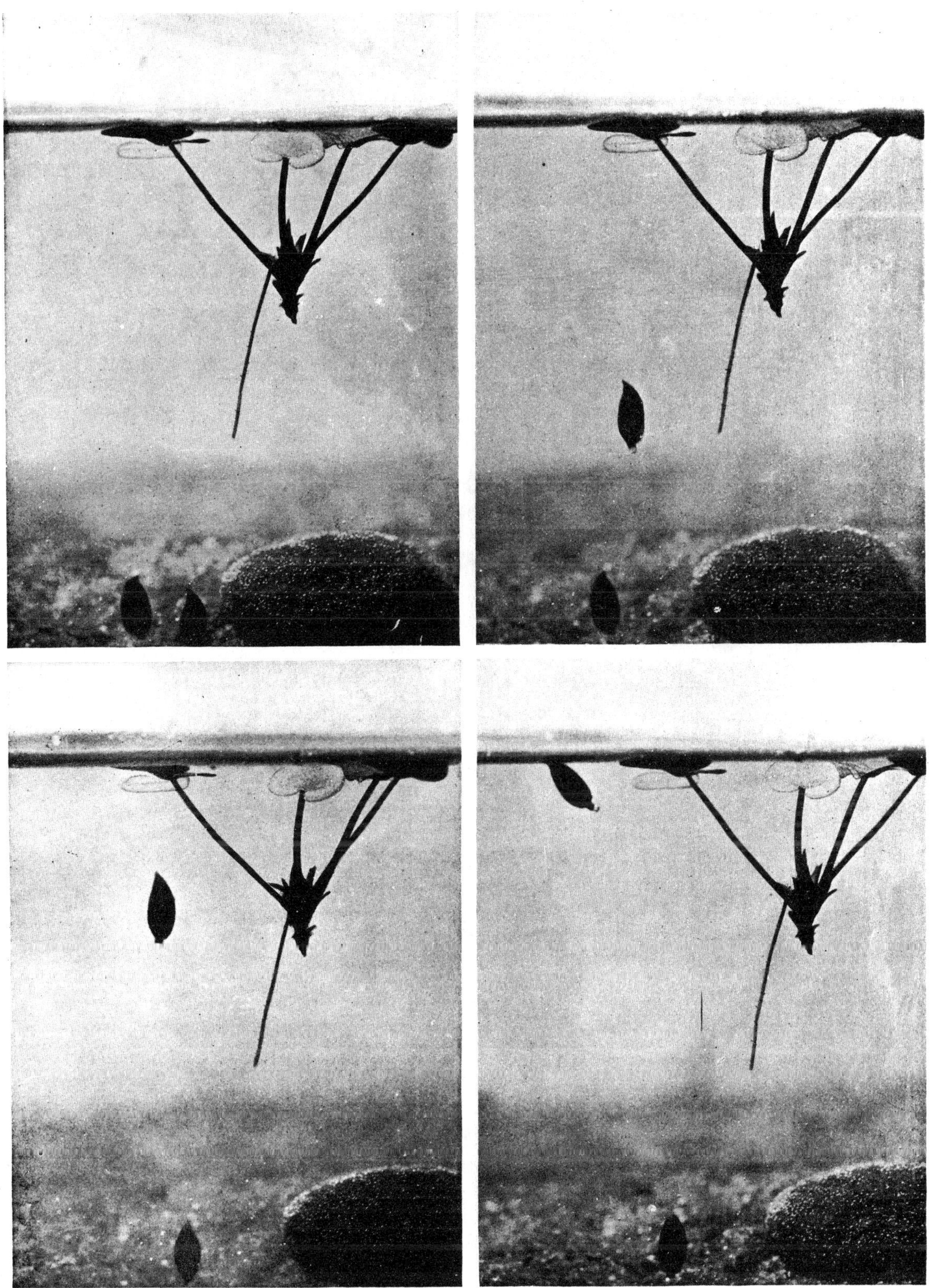

Photos by] [*J. J. Ward, F.E.S.*

THE FROGBIT.

In the first mild days of spring these buried buds become buoyant and one after the other they rise up to the surface, there to develop into Frogbit plants as shown in the photograph on page 976.

Photo by] [*E. Step.*

LONG-LEGGED SPIDER CRAB.

The limbs are drawn out to great length and fineness, so that the Crab can walk and climb with ease over the most delicate seaweeds.

CRAB FORM AND HABIT

BY EDWARD STEP, F.L.S.

ONE of the most engrossing studies in natural history is the consideration of the way in which Nature, without altering the general structure, adapts the various members of a group to different habits and habitats. It does not matter whether you take beasts, birds, fishes, or any other of the principal classes of animal life, you will find similar methods adopted to produce the vast variety of forms with which we are acquainted. All the backboned creatures are built on essentially the same plan; yet by the extra development or the partial suppression of certain bones, or sets of bones, the utmost variety of external form is brought about, and, with it, adaptation to a special mode of life.

The same rule holds good with other classes. Here is the great Crab family—the Crustacea, so-called because their outer skin is so impregnated with lime, that it forms a hard, unyielding box, which does away with the necessity for a complex internal skeleton. Owing to the shell-like character of the integument, the Crabs have been lumped with oysters and whelks, etc., under the name of Shell-fish—one of the most unfortunate and misleading terms ever invented, because none of the diverse forms included under it is a fish. In spite of this hard, unyielding box, which has to be thrown off entirely from time to time to allow of the creature's increase of size, the type is seen to be perfectly plastic. Because it is the best-known crab-form, let us take the common Jack Crab of the fishmongers' shops as the type, and point out a few of the directions in which it has been modified to adapt it to habits differing, more or less, from those of the Jack Crab. This familiar crab has the principal part—the trunk—broad in front and narrowed behind, and to its underside are attached five pairs of jointed legs. There is also a thin hinder part, popularly known as the "tail," which attains a considerable size in the Lobsters, but in the Crabs is usually folded underneath to be out of the way. The foremost pair of legs are so hugely developed, that they are of no use for locomotive purposes, but of great value for defence and offence, the last segment ending in a powerful pair of nippers, with tooth-like knobs and points on their opposed edges. The other four pairs of legs are used for running over the sea-bottom, and their last joints are almost solid, sharp points, useful for digging into sandy or gravelly floors and retaining a firm hold. The wedge-like form of the body, viewed from the side, is useful for thrusting backwards under a rock, where, like a British soldier with his back to the wall, he can defy a dozen adversaries. If there is no rock handy, he can bury himself quickly in sand or gravel by the same means; then by packing his bristly limbs close to his body, they will act as a rough filter for the water that is being drawn under the edges of his upper shell to keep his gills supplied.

In the common Shore Crab (illustrated on page 884), which is more given, when at rest, to burying itself in the sandy shore, with its eyes and antennæ exposed, the front edge of the shell on each side of the eyes is cut into five tooth-like lobes to assist in this same object—the filtering off of overlying sand whilst allowing water to pass between the teeth, and beside the folded limbs to the gill-opening below. The hinder pair of legs have the last joint flattened out, and in the Shore Crab's very near relations will be found illustrations of the way in which this flattening has been increased to fit the limb for use as a paddle. In the Velvet Fiddler Swimming Crab, shown on page 985, the last two joints have been expanded to a far greater extent, in order to fit the crab for swimming. This process reaches a climax in Henslow's Swimming Crab, which has had the entire shell made thin and light, to enable the crab to mount to the surface-water and pursue schools of fish upon which it preys.

These locomotive limbs of the crabs are variously manipulated by Nature to serve different purposes. In some of the Spider Crabs they are drawn out to a remarkable length and fineness, to enable them to climb among the more delicate seaweeds and hydroids. An example of this treatment is given on page 978, where the Long-legged Spider Crab is shown with limbs five or six times the length of the body. Just as the long, slender legs of the Garden Spider enable it to climb over the rigging of its own snare and to run lightly over the grass and other small leaves and shoots, so these Spider Crabs are able to climb and crawl over weeds so fine that when they are taken from the water they collapse into a shapeless mass. These Spider Crabs are more or less coloured to

Photo by] [*E. Step.*

THE YELLOW CRAB.

A heavily built Crab found among stones on rocky shores and easily mistaken for a stone when at rest with its limbs contracted.

harmonize with the red weeds among which they live, and they attach fragments of the weed to their limbs and bodies by means of natural hooks and spines, so that these minute ribbons and laces, waving in the water, take off all resemblance to a crab.

Another genus of Spider Crabs is represented on page 985, in which the limbs, though slender, are more proportionate to the size of the body. This is known as Gibbs' Spider Crab, and his legs have swollen joints, which make him look like a sufferer from rheumatic gout. But it is little of either legs or body that is seen by his submarine associates, for Nature has provided him with a fur of club-shaped hairs, from which spring rows of hooked bristles, which enable him to attach tawdry finery in the shape of bits of weed, corallines and living sponges. The individual photographed (a female) has induced a mass of sponge to grow upon her back, under which she goes abroad perfectly disguised. This sponge is not stuck on—it probably was when very small—but grown to the natural rough coat. Most of the Spider Crabs utilize extraneous matters for their disguise, and Nature has made the deception easy for them.

Photo by] [*E. Step.*

THE POLISHED CRAB.

The trunk is thin, narrowed behind and highly polished, so that it easily slips down into the sand.

Somewhat similar in the extent of two pairs of its legs is the Crab whose portrait is given on page 981, and which is known to the Japanese as the Demon's-face Crab. The resemblance to a grotesque mask of the human face is solely due to the front and back elongation of the upper shell. If you compare it with the broader back of the Shore Crab, you will find that all the elevations and depressed lines of one are repeated on the other; but their lateral compression in the Demon's-face Crab produces the effect of lips, broad nose and slanting eyes. An allied species is found both at Hong Kong and in the Mediterranean, where the Italians have named it the Knave. The shortening of the last two pairs of limbs and bringing them practically on the back is no doubt due to a habit of holding a sponge or other foreign object over its back. This has not been recorded as a fact in the Knave's economy, but the similar arrangement of the corresponding limbs of the Gnome Crab (see page 485) leads one to suspect it.

On our own coasts we have a Crab that is similar in respect of the grotesque human features on its back, and they are produced in the same way. This is known as the Masked, or Long-armed Crab. In the male, whose photograph is seen on page 983, the first pair of limbs are very long (much shorter in the female), and it uses these as levers to push its wedge-shaped body into the soft sands in which it takes its rest. It will be noted that the antennæ, which in the majority of Crabs are very short, are here as long as the body. In ordinary these are used by Crabs as sense organs, much as the cat's whiskers are said to be used, to obtain impressions of their surroundings. But in the case of the Masked Crab they have undergone a remarkable modification to fit them for an entirely different purpose, necessitated by the crab's peculiar habits. Looked at closely, each of these feelers will be seen to have along its inner face two rows of stiff hairs, and when the antennæ are brought close together, side by side, the hairs of both interlock and so constitute a tube. When the Crab has buried itself deeply in the sand, the tips of these antennæ stick out into the water above, and down the tube the Crab draws water to supply its gills, channels under the

Photo by] [*C. J. Mogridge.*

THE DEMON'S-FACE CRAB.

Owing to the contraction of the trunk from the sides, the ordinary elevations and depressions of the surface form a resemblance to a grotesque mask. It is a native of the coasts of Japan. A similar Crab, but of another species, is found in the Mediterranean.

Photo by] [*H. Main.*

BROAD-CLAW CRAB.

The trunk and pincer-claws are broad and thin, fringed with long coarse hair that collects mud and disguises the Crab.

front part of the lower shell conducting it to the gill-chambers. This is a great and important change from what is usual in the majority of Crabs, and yet it is effected by very slight modifications of the prevailing structure.

In the so-called Squat Lobsters—which are not lobsters, but crabs—we have a similar lengthening of the trunk; but in this case it is much broader behind than in front; and the tail, instead of being reduced and folded beneath the body, is here kept expanded behind, ready to be suddenly flexed in true lobster fashion, to send the animal rapidly backwards through the water to some convenient hole in the rocks. The shell of the back is crossed by hairy lines, which give it the appearance of being composed of overlapping plates, and the first four pairs of legs are covered with half-circular, spine-fringed scales. The fifth pair of legs are very slender, with a brush-like development at the extremity. In ordinary these weak limbs are not seen, for they are kept folded up and tucked under the edge of the shell in the gill-chambers.

A nearly-related crab, known as the Broad-Claw, is of such exceeding flatness, that he looks as though he had been mechanically flattened. This is especially noticeable in respect of the pincer-claws, which are remarkably spread out, and their apparent size increased by fringes of long, shaggy hair. This gathers particles of mud, and as the crab is fond of clinging to the rocks to which his flat under-side closely adheres, he looks like a mere daub of mud. If you take him off—at the cost to yourself of a very sharp nip from the pincers—you will find that he is not so dirty as he looks, for the under-surface is clean and white, and polished like porcelain.

A very near relative is known as the Porcelain Crab for a similar reason, though he differs from Broad-claw in having a very circular, flat little body, and long, rather eccentrically-shaped pincers. His habit is to pack himself away in all sorts of crevices where flatness is desirable, but the very broad claw might be in the way.

Photo by] [*H. Main.*

BROAD-CLAW CRAB.

Here is shown the white and clean polished under-side. The upper side always looks muddy and dirty.

But the smooth finishing of a crab is carried to much greater lengths in the Polished Crab, whose portrait appears on page 980. Flat and broad it is, and well fringed with hairs, its white surface as smooth and shiny as a piece of glass. It lives under the sand, and its smooth body, narrowed behind, is well calculated to make descent easy. Its antennæ are fringed like those of the Masked Crab; but as they are comparatively short, we can see at a glance that it does not bury itself far below the surface.

In the opposite direction we have crabs that are as rugged as possible. A little species found in deep water in our own seas is of this description. It lives among corallines, and is got up to resemble them so exactly that when its limbs are folded up at rest it is to all appearance just a boss of coral or coralline. Then the Yellow Crab whose portrait appears on page 979, though not exactly rugged, is very stony-looking. Like the Jack Crab, he is fond of getting under real stones of a larger size, and if he cannot manage to get below them, settles down close to one, and packing up his limbs close to his body, passes for a water-worn stone. Somewhat similar in the purpose of their design are the two larger Crabs in the lower portion of the coloured plate. The one on the right, known as Calappa, is said to have caused grave concern as well as astonishment to hardy mariners on the South American coast, who have picked up what they considered a pretty mottled pebble on the shore and determined to include it among their hoard of "curios" to be taken home in due course. The astonishment comes when the supposed pebble puts forth limbs and scuttles away. A glance at its figure in the plate will show that the big pincer-claws are so fashioned that when they are folded in front of the mouth they increase the resemblance to a stone, but their roughly-toothed upper edge allows water to pass between them and the body-crust to reach the gill-opening. Note also how the eyes peep out just above these closed heavy door-like claws and inform the Crab when it is safe to cease being a stone and to show some activity.

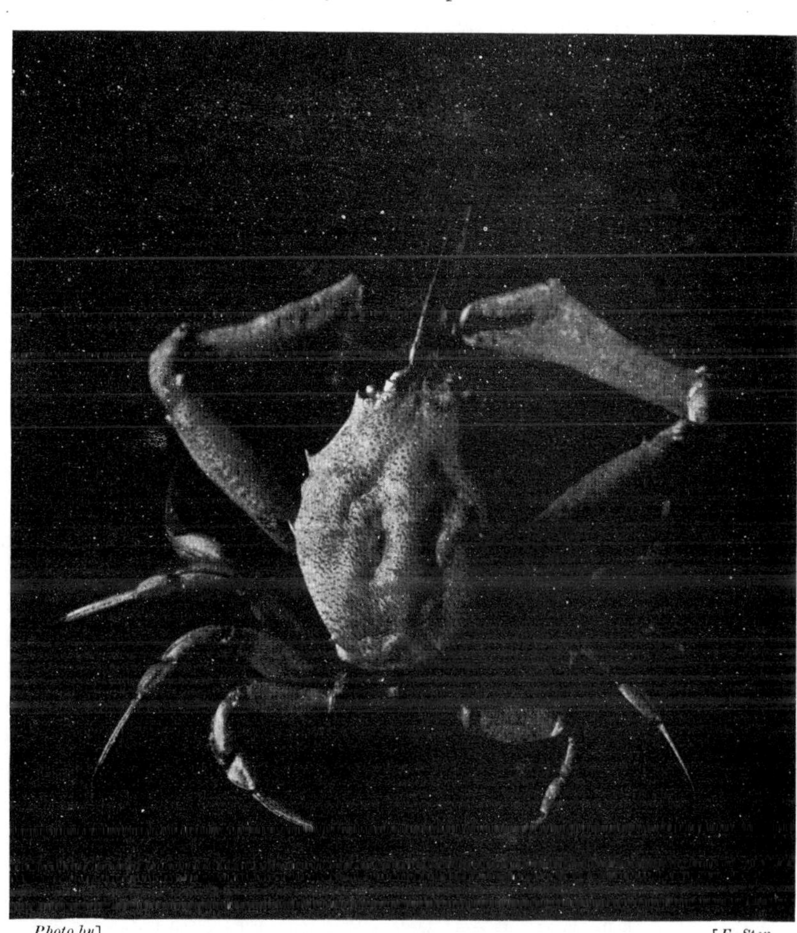

Photo by] [*E. Step.*

THE MASKED CRAB.

The contraction of the trunk has brought about a resemblance to a mask. This Crab spends much of its life buried in the sand. Its long antennae applied together constitute a tube through which fresh supplies of water reach the gills.

The pincer-claws of the Swimming Crab at the top of our coloured plate are also designed for use as strainers when buried beneath the sand, as is usual with Swimming Crabs at rest. A glance at the coloured plate will show other forms of Crab exteriors from various parts of the world, which considerations of space will not allow us to describe separately; but in the light of what has been said, the reader may be interested in making conjectures as to their mode of life based on a study of their forms.

The point we wish to emphasize is that this almost endless variety of form is obtained without violating the general plan, which is adhered to, it will be seen, in all the illustrations we have given. If one keeps this point in view when examining any creature, whether Crab or any other form of animal life, it will be found to add enormously to the interest of our study.

LIGHTNING

BY F. W. HENKEL, B.A., F.R.A.S., *Author of "Weather Science."*

THE developments of atmospheric electricity on a large scale resulting in the phenomena of a thunderstorm are amongst the most impressive of natural phenomena, and in many countries the wielding of the thunderbolt has been considered an especial attribute of the gods (Jupiter Tonans, etc.). The atmosphere is always in a condition of electric instability; in fine weather generally positively charged, in stormy and rainy weather, often negative. It is probable that an electrically charged cloud is made up of a large number of charged drops of water which have derived their charges from evaporation. These drops condensing and uniting, the strength of their charges gradually increases, whilst at the same time the surface of the ground below becomes oppositely charged by the process known as induction, the layer of air between keeping the two opposite charges separate, serving, as it were, as a condenser. By degrees the difference of potential will become so great, and the intervening air will be so highly strained, that it will collapse, and a disruptive discharge take place, a spark or series of sparks will partly relieve the electric tension. Flash after flash then passes, till equilibrium is restored; the *light* being due to the intense heating causing the air to become luminous. This produces a sudden expansion (partial vacuum), followed by a rush of air from surrounding regions, whence comes the sound, thunder. If the path of the spark be fairly straight and not very long, one sharp, loud clap is heard; if it be long and broken, a succession of rolling and rattling sounds follows, with echoes from the clouds around. Since the speed of light is enormously great (one hundred and eighty-six thousand miles per second), the flash is at once seen, but the speed of sound being only about eleven hundred feet per second, or about one mile in five seconds, the reports from the various clouds at different distances will be heard in succession for some time after the lightning flash is seen.

Photo by] [*E. Step.*

THE SQUAT LOBSTER.

A Crab whose shelly covering is rough with spiny scales. Like a real Lobster, it leaps rapidly backwards by the sudden flexure of the broad "tail."

By counting the interval of time between the flash and thunder following, from the longer or shorter interval between these, the observer may ascertain whether the storm clouds are receding from or approaching towards his position.

Three different kinds of lightning are commonly distinguished: (1) *forked*, or zigzag lightning;

Photos by] [*E. Step.*

SWIMMING CRAB AND SPIDER CRAB.

The upper photograph shows the Velvet Fiddler Crab, in which the hindmost pair of limbs are flattened out and fringed with hairs to adapt them for use as paddles. The lower photo is of Gibbs' Spider Crab, which has a living sponge growing over and concealing the body.

Photo by] [*W. Plomer Young.*

EFFECTS OF LIGHTNING.

An oak-tree of large size whose trunk has been split to the base by a lightning stroke.

(2) *sheet* lightning; (3) *ball*, or globular lightning. The first, or zig-zag lightning, is the most common form, but as has been shown from photographs, the discharge should more properly be described as branching or ramified in most cases. The line of least resistance is probably followed by the "current," and the presence of obstacles, solid particles, or local electrification, etc., sometimes makes a crooked path, though longer than the direct one, easier for it to follow. Of recent years remarkable instantaneous photographs of lightning have been taken by Dr. Lockyer and other observers. The flash itself never lasts longer than $\frac{1}{100{\cdot}000}$ of a second (Thompson), but is sometimes of an oscillatory character, being in more respects than one similar in nature to the discharge of a Leyden jar. The second form, sheet lightning, commonly seen at night times on the horizon and without thunder following, is probably only the reflection on the clouds of flashes taking place at some other region. When seen reflected from a storm too far away to produce audible thunder it is often known as "summer lightning." The third form, globular or *ball* lightning, is very rare. Here the discharge appears to take the form of balls of fire which move slowly along and then explode. It is probable that some cases of the appearance of such balls of fire are not really electrical phenomena, but meteoric masses from outer space rendered incandescent by passing through the atmosphere and finally bursting into fragments on reaching the ground or in the lower air, but since similar phenomena have been occasionally observed by Planté, Cavallo, and others in the discharge of electrical machines, there seems no reason to doubt that this kind of discharge occasionally occurs in Nature, though the accounts have perhaps been exaggerated.

The physical effects of lightning, as is well known, are of a most potent character, often causing death to men and other animals, blasting trees, and more or less completely destroying lofty and unprotected buildings. M. Flammarion, in his interesting popular work, gives examples of some very remarkable and often erratic performances of this intense form of atmospheric electricity. Instances are on record of persons being struck by lightning but receiving no other injury than the destruction of part of their clothing, their bodies being uninjured; whilst in other cases its effects have been positively beneficial. An innkeeper at Remaines (France), who suffered severely from rheumatism, and was quite unable to walk without the help of a stick, was struck by lightning and rendered unconscious for a time. For some hours afterwards his sight was affected, but he ultimately recovered, and, strange to say, was completely cured, being able to do without his stick and to pursue his avocations quite comfortably. In 1872 a Kentucky negro, condemned to death for having murdered a fellow workman, was about to be executed. At the moment of his setting his foot upon the scaffold there was a terrible clap of thunder and the condemned man was struck dead by the lightning (Flammarion). The sheriff was so much moved by the occurrence that he resigned office.

The appearance of the sky previous to a thunderstorm is usually very characteristic, the " thunder-clouds " being well-known forms. The heavy, cumulus clouds with admixture of cirro-stratus above and " cloud curtain of loose texture " beneath produce intense darkness. The cirro-stratus incoming clouds are sometimes twenty to fifty miles in advance of the storm. Next come the cumulus clouds, " thunder-heads," under this the rain curtain and squall clouds below. The wind now changes in direction, the air becomes cooler and the barometer rises slightly. Then the rain or hail begins to fall, the lightning flashes and the thunder rolls, till the centre of the storm has passed.

Mr. Shaw, in his recent work on " Forecasting Weather," states the conditions necessary for the causing of atmospheric instability. The undercutting of the warm surface layer of southerly air by a more westerly or northerly one creates the instability necessary for violent showers, hail, and possibly thunder and lightning, the flow downwards of the air behind the squall front and the ascent of the displaced air.

In general, two classes of storms are fairly well defined, (1) heat thunderstorms, and (2) cyclonic thunderstorms. The former are mainly confined to the Tropics, but are also experienced during the summer time in more temperate climates ; the latter may occur also in winter, are characteristic of our own Atlantic coasts, and are so named from their connection with cyclonic disturbances. They

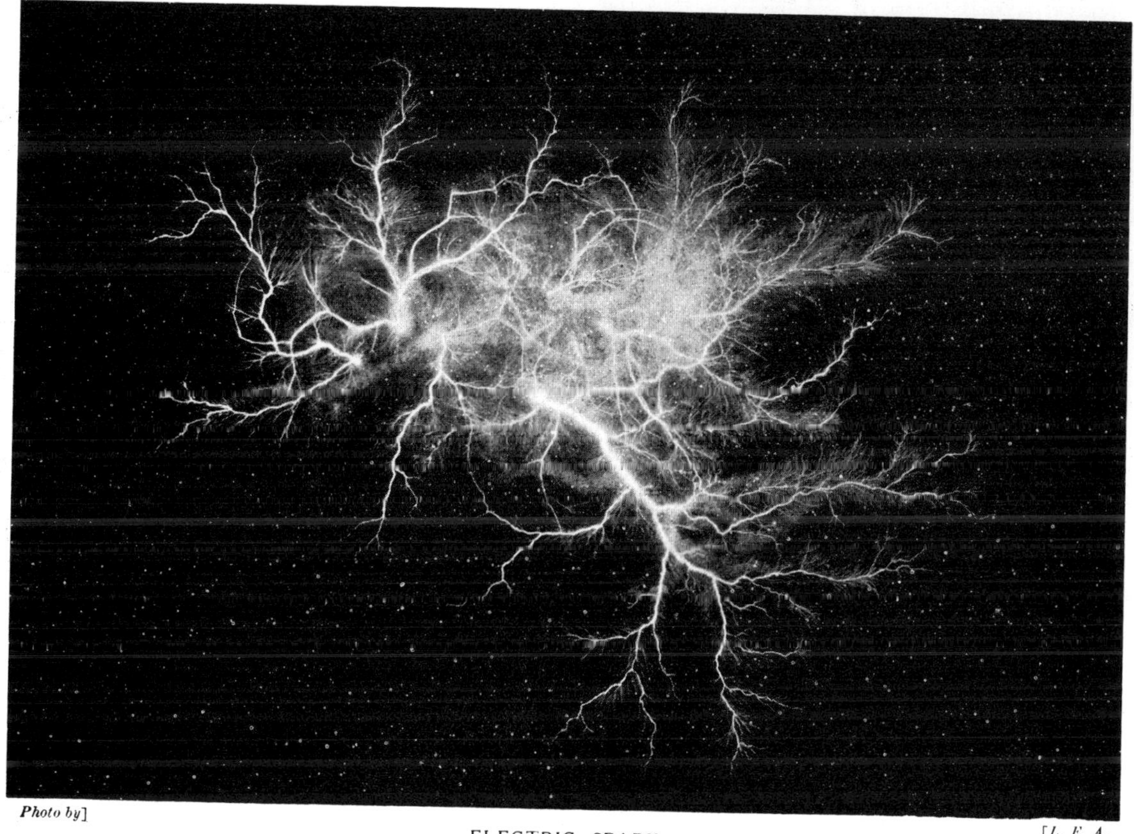

Photo by] ELECTRIC SPARK. [*L. E. A.*

This photograph of an electric spark from an induction coil shows on a small scale the character of the electric discharge in Nature—the flashes of lightning.

are not so dependent on solar heat as the other kind, frequently take place at night and come along with gales of wind. They are perhaps more dangerous than the heat thunderstorms, since though they are not so violent, yet, as the clouds from which they come are at a lower level, the lightning discharges more frequently strike the ground.

A theory as to their occurrence attributes them to the ascent of warm, moist air by convection. The cloud is often isolated and continuous in "summer storms," its height varying from one thousand feet to the level of the cirrus clouds (say, thirty thousand feet at times). Spreading out in all directions in a sheet, it often presents the appearance known by the name of "anvil shape." Cyclonic thunderstorms are considered to be due to the cooling of the upper air, producing similar effects to the heating of the warmer air, in disturbing electrical equilibrium.

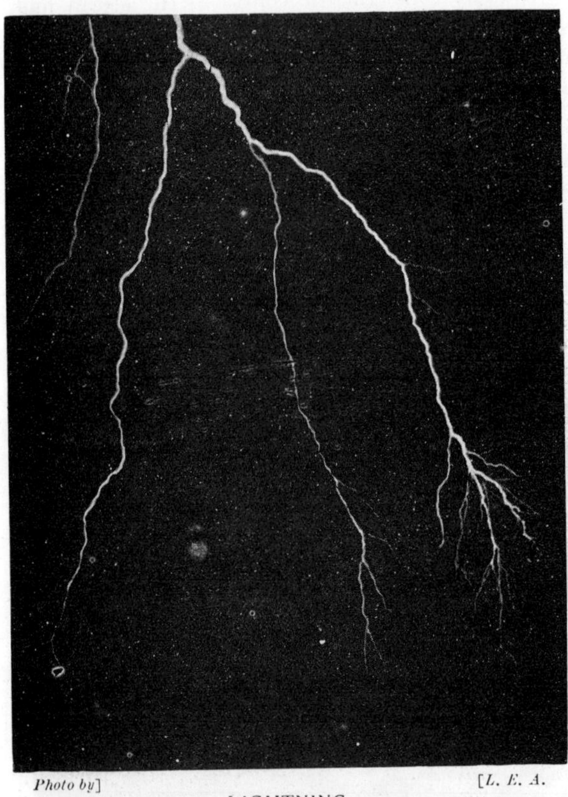

Photo by] [*L. E. A.*

LIGHTNING.

What is commonly known as forked lightning is here shown. It would be described more correctly as branched lightning.

There is both an annual and a diurnal periodicity in the frequency of thunderstorms. More storms occur in summer than in winter, the summer storms most frequently in the early afternoon at the time of the day when the temperature is the highest. The winter "cyclonic" storms occur at all hours, frequently during the night, more common on the western coast (Atlantic side) than on the other, because, owing to the proximity of the sea, great contrasts of temperature at different heights in the atmosphere less frequently arise. That storms are, however, associated with sudden changes of temperature is evident from the fact that hail so often falls during their occurrence, and that the thunderclouds are mainly of the cumulus type, whose presence indicates the existence of a colder and drier layer of air above that in which the clouds form, and, lastly, changes of wind occasioning a rapid fall of the thermometer also take place.

Dr. Scott (*Meteorology*) remarks that changes of temperature form an essential condition for the generation of a thunderstorm. "To a certain extent we may regard fogs and thunderstorms as inverse phenomena. The surface fog is characteristic of conditions of stability in the stratification of the atmosphere; the thunderstorm of marked instability" (Shaw). The peculiarity which the weather seems to display in taking on "moods" for special types of occurrence is also remarked by this author.

If a thunderstorm occurs in any part of the country on one day it is quite common for other parts of the country to be visited by similar phenomena on the next day or the day after, though the other weather conditions do not specially point to this. Thus, though thunderstorms are peculiarly local phenomena, yet, after a spell, few places seem to have been exempt, the storm bursting out here to-day and to-morrow elsewhere, the next day further off, and so on. The thunderstorms of Iceland all occur in the winter, though it was at one time thought that they were quite unknown in Arctic regions. In Abyssinia, over a period of observations during which two thousand storms were registered, nearly all these happened in the afternoon hours, only

Photo by] [*H. J. Shepstone.*

LIGHTNING.

A wonderful photograph of an electrical storm and the long branched and undulating flashes of lightning. Each flash las s no longer than the one hundred-thousandth part of a second of time.

twenty-two between midnight and eleven a.m. Thunderstorms are of daily, almost hourly, occurrence in the Tropics on a scale which we in more temperate zones can scarce form any conception of.

Yet for all their frequency, our knowledge as to the development of electricity in these outbursts is by no means satisfactory; though from time to time more or less plausible hypotheses have been put forth, the subject is still a mysterious one, and we must be content to wait for further light. Dr. Simpson, of the Indian Meteorological Department, has outlined one theory. He supposes that upward currents of air prevent rain that would otherwise be deposited from falling. The raindrops increase in size "through cycles of growth," then break up with separation of electricity, until their charge is so great as to produce a gradient of more than thirty thousand volts per centimetre. Then the lightning flashes, the thunder rolls and the accumulation is neutralized over a limited area; this process continues, another flash, and then another, and so on till equilibrium is attained and the storm ceases.

THE HORNED CURASSOW

BY SIR HARRY JOHNSTON, G.C.M.G.

THE Curassows and Guans belong to a family (entirely restricted in its distribution to Central and South America) of the same order as the Pheasants and the Australian Mound-builders. Some of the Curassows, indeed, will hybridize with certain old-world pheasants; but, of course, the forms derived from this cross are infertile mules. The differences between pheasants and turkeys, on the one hand, and Curassows, on the other, lie chiefly in the position of the hind-toe and other structural points, and in the life-habits, which in the Curassows are arboreal. These handsome South American birds usually (but not always) build their nests in trees.

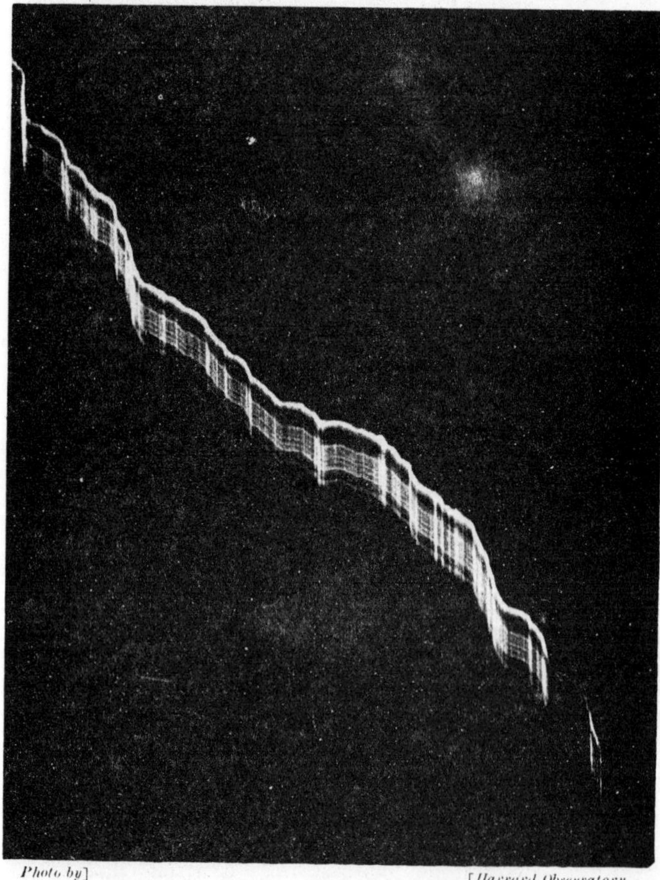

Photo by] [*Harvard Observatory.*

LIGHTNING.

This photograph shows the spectrum of a lightning flash, by which means the electric light is broken up into its component colours as indicated by the lines of differing intensities.

Of the three sub-families of Curassows, the one which most nearly approaches the turkeys, pheasants and guinea-fowl in affinities is the rare and beautiful Lord Derby's Mountain Pheasant of Guatemala, which I have illustrated in colour from two remarkably fine specimens—male and female—to be seen in the Museum of Natural History at Leiden, Holland. Probably no living specimen of this species has ever been exhibited in the London Zoological Gardens, though it seems that the celebrated Lord Derby of the middle of the nineteenth century (after whom the bird is named) had examples in his wonderful menagerie at Knowsley.

The Mountain Pheasant—sole representative of a peculiar sub-family—is apparently restricted in its range to one small patch of Southern Guatemala, the forests on the slopes of a mighty volcano, the Volcan de Fuego, nearly thirteen thousand feet high. This was an active volcano three hundred years ago, but has since been quiescent, and in the rich volcanic soil a forest of great denseness and variety harbours a wonderful fauna of birds. The Mountain Pheasant inhabits mainly the upper regions, the woods of evergreen oaks, cypress and pine, magnolia and bamboo, where the vegetation is a good deal like that of Florida. It nests in the boles or forks of trees, and is very difficult to see and shoot, in spite of its very striking plumage of glossy burnished green-black and ivory-white, its coral-red legs and scarlet horn. This strange excrescence in the male often attains a length of three inches. In the female it is shorter. The two sexes differ in size, but not in plumage, though the tints of the male are more vivid.

HORNED CURASSOW.

Otherwise known as the Mountain Pheasant, and a native of Guatemala. The head of the male bird is here shown with its distinctive scarlet horn.

THE BECOMING OF A CHICKEN

BY W. P. PYCRAFT, F.Z.S., ETC.

Few of us probably regard an egg in the light of one of Nature's marvels; nevertheless that distinction might well be claimed for it. For it is surely nothing short of marvellous that the fluid contents enclosed within the white porous shell should, under certain conditions of temperature, become transformed in the space of about nineteen days into the pulsating body of a bird.

To understand the course of this transformation it is necessary to bear in mind the fact that the fluid contents to which we have referred, the "white" of the egg—which in its raw state is not white, but transparent—and the "yolk," exist only for the sake of that little disc which everyone must have noticed lying on the top of the yolk of an egg when broken into a cup. This disc is the "germ," the matter which is to give rise to the chick. All the rest is destined to serve as food for this germ, for being imprisoned within the shell no nourishment from the parent is possible. Very well. Now let us take a brief survey of yet other contrivances for the welfare and successful development of this germ. We have broken an egg into a cup, and have seen the germ; now notice the little hailstone-like bodies lying within a twisted, transparent cord, which is itself hardly distinguishable from the rest of the "white" of the egg, of which it forms a part. These cords, one at each pole, are the chalazæ, and serve to keep the yolk in position. The yolk itself is not a uniform liquid, but is composed of alternate, concentrically-arranged layers of white, and yellow yolk, though this fact is not easily demonstrated. Then, as to the shell!

This is lined with a delicate, very tough, double-layered skin at the broad end of the egg: as incubation proceeds, the inner layer comes away from the outer, and the space formed becomes filled with air to enable the young chick to breathe. The air finds its way through the shell, which is porous, and composed of two layers, a very thin outer layer covering a thicker inner layer, which is pierced by numerous vertical canals. This shell is formed by the walls of the oviduct as the mass of yolk and its enveloping "white" passes down, and, in some birds, as in hawks and guillemots, for instance, rich pigments and curious patterns are added to its outer surface.

Now as to the coming of the chick. After about sixteen hours' incubation the circular disc or "germ" which we have already seen, becomes pear-shaped, and grooved down its middle; and the foundations of its backbone and nervous system are laid. By the first half of the second day the heart and principal blood-vessels have put in an appearance, though as yet, of course, they are but imperfectly developed, and towards the end of the day two structures peculiar to embryonic life give their first signs of growth. These are the amnion and the allantois. The first-named eventually grows into a delicate transparent bag filled with a clear fluid, and surrounding and enclosing the embryo, protecting it from injury. The allantois is a bag which grows out from the hinder end of the gut and receives the waste given off by the developing kidneys. But more than this, its walls growing upwards, spread under the shell, and become richly supplied with blood-vessels. These, absorbing the air which passes through the shell, serve to oxygenate the blood, so that the allantois really does the work of the lungs which have not yet come into being.

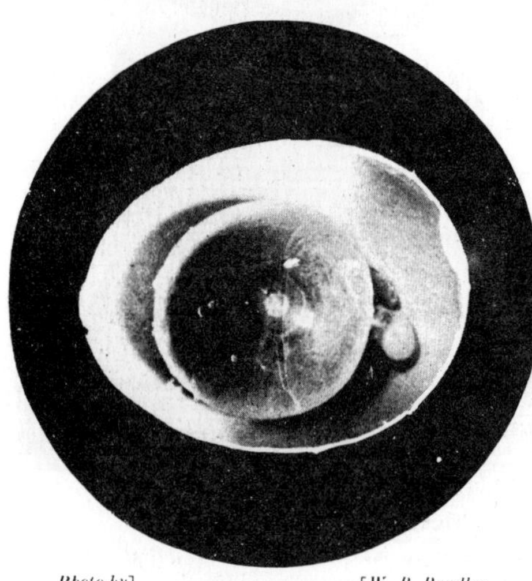

Photo by] [*W. B. Randles.*

HEN'S EGG.

After eighteen hours' incubation. The germ is seen to be changing to a pear shape.

During the third day the brain, the eyes, the nerves, the lungs, liver, and other glands begin to appear. And by now, be it noted, the "white" of the egg has largely disappeared, having been absorbed by the blood-vessels.

During the fourth day the nose, ears and jaws begin to take shape; the wings and legs appear in the form of tiny buds, like large pins' heads. Twenty-four hours later these limbs have lengthened, elbow and knee can now be recognized, and faint traces of digits are visible enclosed between a web.

The sixth day marks a new epoch. Till now the growing body had not developed sufficiently far to enable one to trace any likeness to any particular animal, much less to a bird, but now the bird-like features are unmistakable. Yet the head is still huge in proportion to the rest of the body and there is no beak, which does not take shape till the twelfth day. The day following nails appear on the toes and scales on the legs.

As early as the ninth day feathers begin to make their appearance, in the form of tiny button-like, or stud-like projections over the body, and by the thirteenth day these have assumed the form of long rods, which soon will burst, and release the nestling down.

By the nineteenth day our developing chick has become a "ripe embryo." The white of the egg has entirely disappeared, and the remains of the yolk are very fluid and much diminished. Just before hatching, which is now imminent, the rest is drawn up into the body, and stowed away under the intestines. The long stalk, or neck of the yolk bag, is in open communication with the gut, and through this passage the last of the yolk is taken up. Hence it is that for some twenty-four hours after hatching the young chick requires no food. He still has the remains of "his bottle" to finish!

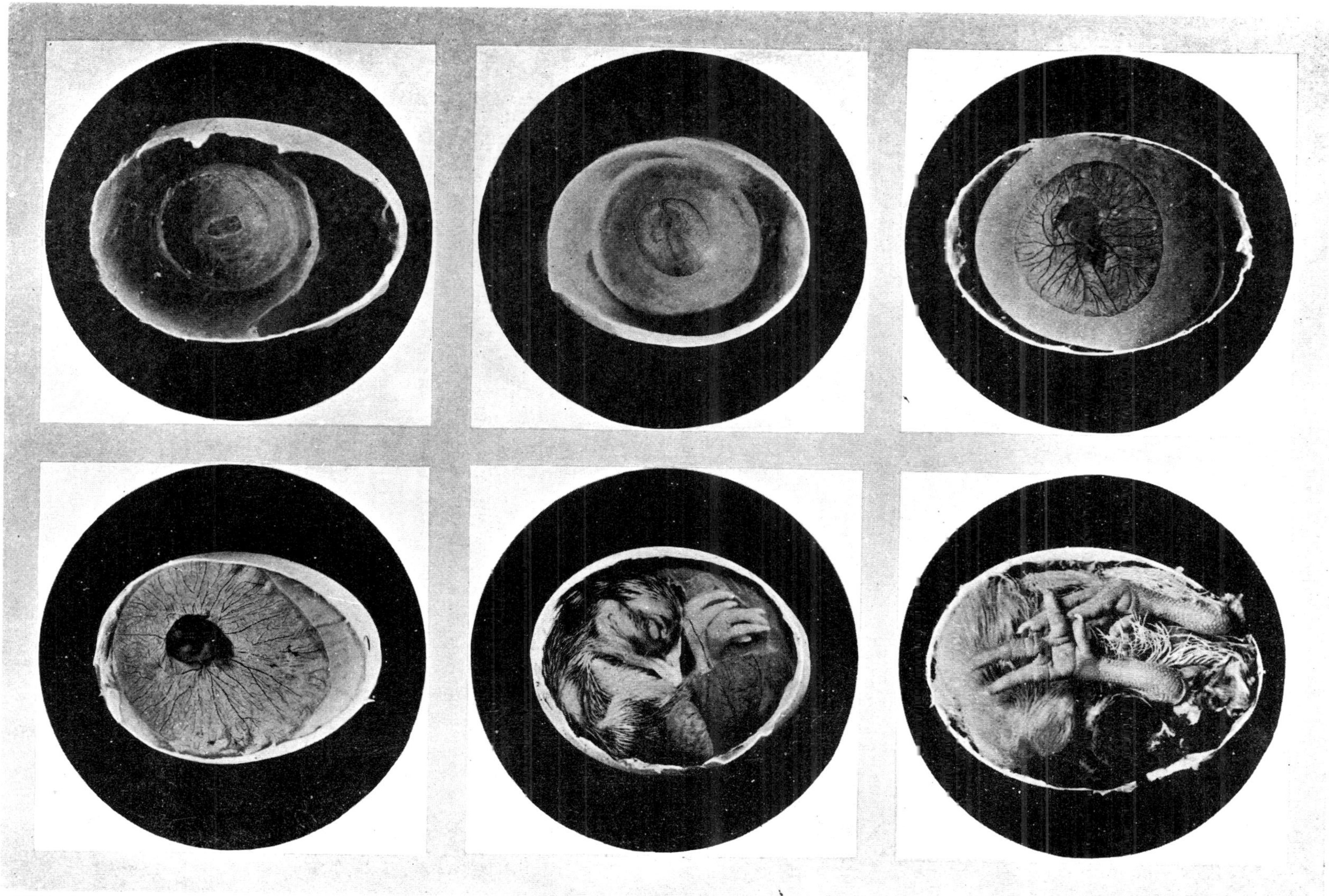

Photos by] [*W. B. Randles.*

STAGES IN THE EVOLUTION OF THE CHICK.

The germ becomes much longer than broad, then blood-vessels extend from it in all directions, the backbone appears, the chief internal organs, brain and eyes are formed, then the nose, ears and jaws, the wings and legs sprout out, and on the nineteenth day the chick is complete and well clothed.

The chick is now lying curled up lengthwise in the egg, with its beak touching the shell-membrane where this forms the wall of the air-chamber already referred to, at the broad end of the egg. By the twentieth day a great and momentous change takes place. A little conical stony knob has been growing at the top of the beak. It is now ready for use, and is applied, first of all, to break through the inner shell-membrane, and so enable the lungs to get their first draught of air, which has been stored at the large end of the shell, and in the chamber between the shell and this membrane, to which we have already referred. Thereupon a great change takes place. The pulmonary circulation starts into being—to cease only with life itself—and the blood at once ceases to flow through the allantoic vessels which we have already remarked, spreading out in a membrane lying just under the shell, and drawing air through its pores. The lungs having begun to discharge their proper function, this temporary lung is no longer wanted. The egg-tooth is now used for the last time, and this to file through the walls of the shell. Once a hole is made the breach is speedily enlarged and the bird—is hatched.

It has now only to get dry, and as this process goes on the down gradually opens out and the fluffy little mite, which we know as the "chick," has begun a career which the demand for "spring-chickens" may speedily end, or he may have entered upon a long and joyous existence in some delightful farmyard.

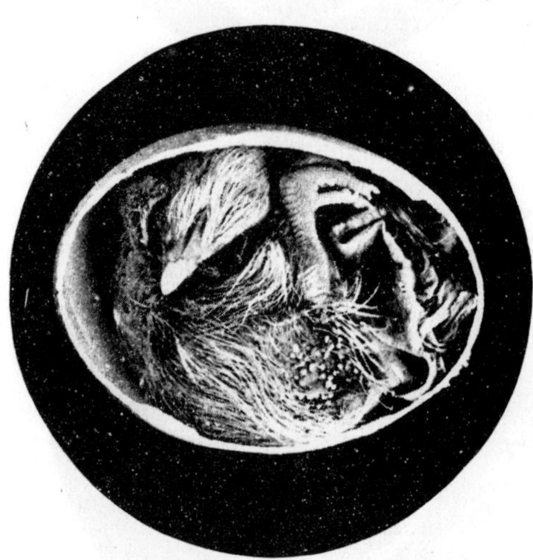

Photo by] [*W. B. Randles.*

THE CHICK IN THE EGG.

The nineteenth day of incubation. The chick is now ready to break the shell and obtain air for the inflation of its lungs.

THE SHIP-WORM

BY BERNARD C. WHITE.

Now that England has ironclads to guard her shores, and now that piers, built on iron girders, and jetties composed of great boulders or masses of granite have taken the place of the old wooden quays, the fear of the Ship-worm has become a thing of the past. But it was not so long ago that the Ship-worm was one of the most dreaded scourges of maritime enterprise. The cause of all the terrors was, as far as appearance is concerned, a very harmless-looking creature—just a worm: certainly of considerable length, but slender and pale and very soft to the touch, with a curiously-shaped "head" and what looked like a double tail. The trouble lay, not in any dangerous quality of the creature itself; but in its social habits, in the multitude of its brethren, and, chiefest of all, in its fatal propensity for tunnelling into wood. The Teredo (for such is the title by which the naturalist knows the Ship-worm) is doomed to die unless it can find a home for itself in some bit of water-logged timber; and as a result, no fragment of wave-washed timber was secure from its attack, and when once a Teredo had found a home in the core of the wood, others would immediately claim a share in the newly-found domain, and what before might have been a sturdy baulk of timber would rapidly be turned into a worthless piece of sea-drift as rotten as tinder. The worst was that the havoc wrought would be indiscernible until the damage would be too far gone to be remedied and the apparently sound wood ready to crumble to the touch. In fact, the action of the Teredo on sea-weathered timber is very similar to that of the termite in its ravages of dry woodwork on land.

If the ravages of the Teredo be on a very large scale the resulting damage may be very serious; for instance, the carefully-built-up coast-line of Holland has on several occasions been threatened with destruction on this account, as the dykes have been found to be so badly riddled with the tunnels of the Ship-worm that their collapse and the consequent flooding of the country was feared. No really satisfactory methods of preventing these attacks is known, though a certain amount of success was obtained by steeping timbers in a solution of ferric oxide before their immersion in the sea.

It would be unfair, however, to dwell on the evil work of the Ship-worm without mentioning that a great engineering feat was once successfully accomplished by their aid. In the adjoining photo you will see that the bore of the Teredo is lined with a thick coating of porcelain; some of this has been broken away and shows the hollow channelling, but in certain places the covering still remains, and it is quite thick, and, for the size of the creature that made it, exceedingly strong. The method by which the Teredo obtains this result was once carefully watched by Sir Isambard Brunel, and it is said that the same plan of working was adopted by him in the construction of the Thames Tunnel. He found that the Teredo bored out only a small piece of tunnelling at a time and then set to work to plaster the sides well with a kind of cement. Brunel therefore arranged that only a few feet of tunnel should be excavated at a time, and that it should be firmly bricked round before the boring was carried further. It is to this method that the great engineer owed the success of the enterprise, and it was due in a great measure to the application of the hints derived from watching the tactics of the Ship-worm.

THE SHIP-WORM.
Burrows of the Ship-worm in wood. Note how they are lined with cement—the prototype of man-made tunnels.

We must now turn our attention to the creature itself, and find out all we can of its family and structure. First of all, we must rid ourselves of the idea that this worm is a worm in very truth. It is a mollusc and belongs to the great family of bivalves and is intimately connected with the Piddocks, which have been described in the article on Rock-borers (see page 696). In fact, it might almost be considered to be an elongated variety of the Wood Piddock, so akin is it both in formation and characteristics ; and this is especially true of it in its early stages. Then it possesses a well-formed but very round clasp shell, from the hinder margin of which protrude the long siphons enclosed in a stout leathery tube. Near the junction of the siphons with the body proper the mantle is developed into a frill with a heavy fringe of cilia.

This stage in the life-history of the Teredo is shown on the lower right-hand side of the full-page illustration. The foot is protruded, for the creature is ready to settle down and take up the final stage of its existence ; that is, as an inmate of a burrow such as is shown perforating the wood. When it has found a suitable place it commences to bore, always with the grain of the wood, and then to line its burrow with a kind of cement which is exuded from the mantle and as it hardens becomes like porcelain. But soon it will undergo a very decided transformation ; for once again if we look to our illustration (page 997) we shall see that the globular shell and fringed mantle have entirely disappeared and, instead, we have the long thin " worm " shaped like a nail with a rounded head, and all that is left of the shell are two thin, chalky shields. These have been drawn on a black background in order to show their peculiar shape more distinctly. It should be noted that two distinct types of Teredo—the Common Ship-worm and the Norwegian Ship-worm—shells are shown, for the species of this strange family are numerous. The larger of these two is the Norwegian Ship-worm, whose siphons may be developed to more than a couple of feet in length.

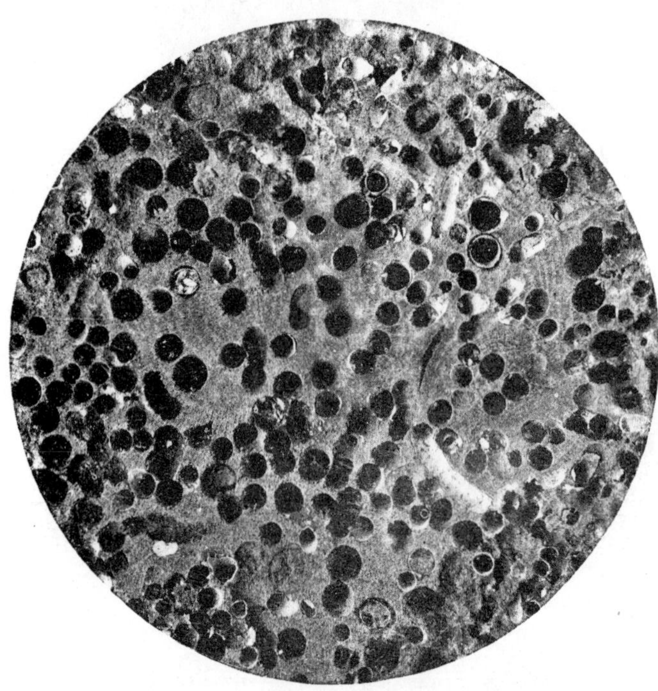

THE SHIP-WORM'S WORK.
Cross-section of a piece of timber tunnelled by the Ship-worm.

In likening the Ship-worm to a nail which works its way into a plank, the simile of a well-known writer has been used ; but if we adopt this description we must remember that the nail bores into the wood head foremost. The exact means employed to effect ingress into the hard material is not known, but as far as we can tell all the boring is done by the action of the foot. Possibly, also, the double " tail " already mentioned, of which each branch terminates in a paddle-shape valve or pallet, is used to enlarge and smooth the sides of the burrow. The Teredo does not, as is popularly supposed, feed on the wood which it excavates. All this is passed out as débris, while it nourishes itself on the microscopic life of the water.

Although the wooden ship has almost gone, and of those that remain patent compositions and copper sheathing have rendered the majority useless to the Ship-worm, there is little likelihood that he will disappear from the high seas. He may have to change his name as being no longer appropriate, but it is probable that he will always find floating logs and planks to riddle ; and there are always temporary timber structures about our docks and quays that will serve to keep a remnant of the race, from which in due course—who knows ?—may be evolved a mollusc that can bore steel plate !

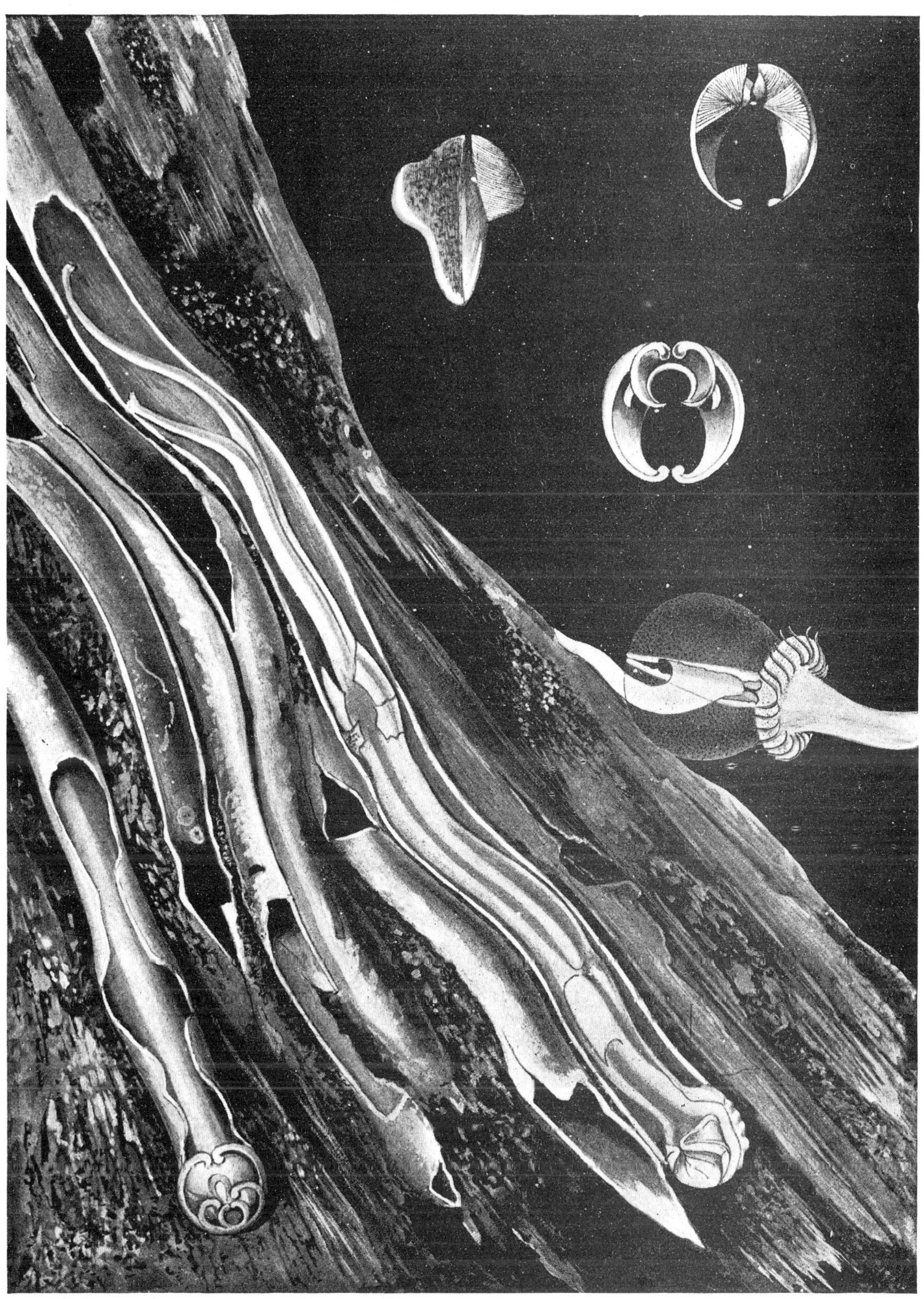

THE SHIP-WORM, OR TEREDO.

Section showing Ship-worms in their burrows, with the boring-shield in front and the respiratory siphons behind. Details of the shell are shown above; also a greatly enlarged figure of the larva attaching itself to the wood.

The Maggot taking its rest between meals. (Three times natural size.)

Photos by] [*H. Main.*

THE CATERPILLAR-HUNTING MAGGOT.

The Maggot hunting between the leaves of Asphodel, which show the ravages of the caterpillar. Some of the silk-threads spun to draw the leaves together are noticeable. (Three times the natural size.)

A CATERPILLAR-HUNTING MAGGOT

BY HUGH MAIN, B.SC., F.E.S.

THE life-history of a Hover-fly was told and illustrated in a preceding article. There we saw how the grubs hunted around and devoured large numbers of the aphis, or green-fly, that infests sweet-peas and other plants.

Although the food of the larvæ of many different species of Hover-flies consists of green-fly, this is not an invariable rule in this group of insects.

There is found in various parts of the South of England, but by no means commonly, a rather large and handsome Hover-fly. It has no common English name, so we must give its scientific name, *Xanthandrus comtus*, which may be translated as "The fly whose male is adorned with yellow markings." As in so many of the lower animals, the female is less ornamented than her mate.

Nothing seems to have been observed in this country of the life-history of this particular Hover-fly. In the South of France, however, where it is more commonly met with, it may be found and followed through its various stages.

In April the asphodel is throwing up its bright green leaves and beautiful spikes of flowers. It is abundant in the neighbourhood of Hyères. Like other plants, it has a host of insect life to support, larvæ of moths, beetles, and other small fry, finding in it a temporary home and a ready supply of succulent food.

It can be noticed by anyone whose eye is not entirely attracted by the flowers that many of the triangular green leaves are spun together by threads of silk. On gently pulling these apart a greenish, wriggling caterpillar may be discovered, which lives as in a tube, with its food all round it, and protected, one would imagine, from the observation of any enemies that would like to make a meal of it. If we wish to find this caterpillar, to rear from it the beautiful little moth, *Hastula hyerana*, we have to hunt for it where we notice the spun-together leaves. Perhaps, however, we shall find that we are not the only hunters, and that a long, greenish, semi-transparent maggot, with the blackish central area of its back surrounded by lighter parts, is crawling about and pushing its narrow head in between the drawn-together leaves, evidently also on the hunt for something. A careful watch may reveal one of these hunters in the act of finding its prey—namely, the caterpillar of which we also are in search. When *we* find one of these caterpillars it commences to wriggle away immediately; but when one is discovered by this other hunter, it does not seem to make any effort to escape, although a much more terrible fate is in store for it than at our hands. Our object would be to keep it supplied with food, so that it might complete the changes which would produce the perfect moth. The object of the tiny hunter, however, is a dinner, and that with little or no ceremony. It at once attacks the caterpillar with its little black, hard jaws, and soon eats a hole through the soft skin. It then continues to eat the internal parts, pushing its sinuous body in one direction and then in the other, till the whole of the poor caterpillar, except its now shrivelling skin, is consumed.

The caterpillars of some moths and butterflies are horrible cannibals, and the entomologist wishing to rear these must keep them in solitary confinement. Like the caterpillar above described, those attacked seem to make no objections to being eaten alive, and remain quite still so that the other may enjoy his dinner at his leisure.

The illustrations show the "hunter" at rest during an interval between meals (page 998). Then we see it on a leaf evidently much eaten by a caterpillar, and with its head protruded out of sight. On separating the leaves, a sight is obtained of the caterpillar being devoured by the maggot, whose head is inserted through the hole in the side of its passive victim (page 999).

It would be difficult to say how many caterpillars are consumed by one of these hunters, but from its persistency and voracity the numbers must be considerable.

When full growth is reached and feeding is over, the maggot settles down on some chosen spot. It becomes fastened down at its tail-end, which broadens out. The long body contracts and becomes rounded off, and the skin hardens (page 1000). Inside the skin the wonderful changes go on which produce from the materials of the larva the altogether different form of the pupa, and then the fly.

Photo by] [*H. Main, F.E.S.*

THE CATERPILLAR-HUNTING MAGGOT.

The Maggot has found a caterpillar, and here has its head inside the skin of its passive victim, which is being rapidly consumed. (Four times the natural size.)

When the fly is fully formed, it pushes off the front part of the old skin, and crawls out to unfold its wings and spend the rest of its life hovering in the sunshine, sipping nectar from the flowers and taking its share in the duty of carrying on the race (page 1002).

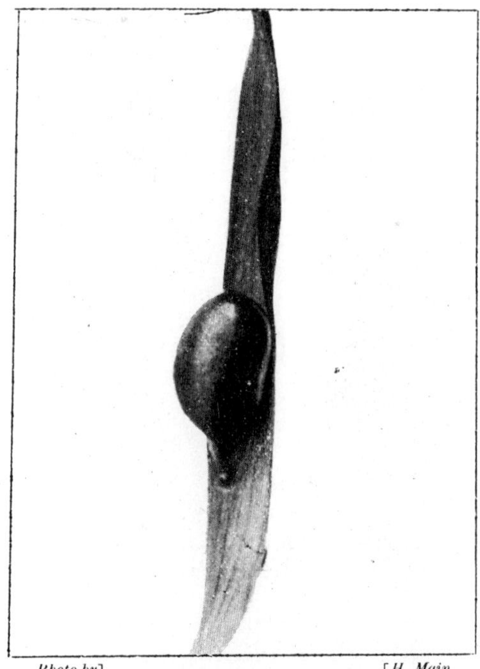

Photo by] [*H. Main.*

THE CHRYSALIS.

The Maggot contracts and its skin hardens. Within the pupa or chrysalis has formed.

PARIASAURUS, AN ANCIENT REPTILE

BY R. I. POCOCK, F.R.S.

TIME was, and that not so very long ago, when zoologists thought they had sufficient evidence to justify the theory that the mammalia were descended directly from the class of vertebrata to which the toads and salamanders belong, and that the reptiles, which were guessed to be the progenitors of the birds, occupied a side branch of the tree of life. It is now, however, believed that the characters mammals have in common with salamanders have been transmitted to the former through a group of reptiles, now wholly extinct, which, from its many mammalian features, is commonly called the Theromorpha.

One of the best-known of these is Pariasaurus, a nearly complete skeleton of which is mounted in the Natural History Museum at South Kensington. Although only a few feet high and some eight or nine feet long, Pariasaurus was evidently a beast of great strength and weight; and the general massiveness of its build and its short, but powerful limbs suggest anything but agility of movement, while the shortness of the tail coupled with the structure of the feet leaves little doubt that it was purely terrestrial, and neither aquatic nor arboreal in habits. That its food was vegetable is indicated by the shape and setting of its teeth; and that the plants it fed on were tough and fibrous may be inferred from the stoutness of its jaws.

Pariasaurus lived during Triassic times, and is one of the most ancient reptiles yet discovered. The interest of this circumstance is increased by the many points of resemblance it shows to certain great creatures, the Labyrinthodons, which are still older, and which are not true reptiles at all, but belong to the class Amphibia, which comprises the frogs, toads, salamanders and newts of the present day. Pariasaurus, therefore, is, in a measure, an annectant type, serving to bridge the passage between these two great classes of cold-blooded vertebrates. On the other hand, as has already been said, it indicates the line of descent of the mammals. It must not be supposed, however, that it was the ancestral stock of our own class. This distinction must be assigned rather to another group of reptiles, related, but not very closely related, to Pariasaurus, but of the same age and horizon.

These are the Cynodontia, or dog-toothed reptiles, which had tearing teeth of a highly-perfected, predatory type; and the reason for mentioning them in this connection is the certainty that they were the only living enemies that Pariasaurus and its allies had to fear. In those ancient days, indeed, these two groups of reptiles fulfilled the rôle that is now played by carnivorous and herbivorous mammals, thus attesting the fierceness of the struggle for existence in the dim dawn of vertebrate evolution.

Another interesting point about this great reptile is the wonderful extent of its longitudinal—that is to say, of its north and south—distribution in the world. It was first discovered in Cape Colony; but since then its remains have been unearthed in abundance in the valley of the Dwina, close to the shores of the White Sea. Now if Pariasaurus were a mammal, we should have no right to assume that its occurrence in such widely-sundered localities proved identity, or even close

[*By C. E. Swan.*

THE PARIASAUR.

An ancient reptile eight or nine feet in length. It roamed the land and fed upon vegetation of a tough and fibrous character, as proved by the structure of its teeth and jaws.

similarity, in their physical features and climatic conditions, because we know that closely-allied species of that class are adapted to extraordinarily diverse surroundings. But since reptiles, unlike mammals, have very limited powers of temperature adjustment, the heat of the blood rising and falling with that of the medium in which they live, it may be assumed that in Triassic times, when Pariasaurus flourished, the climates of Northern Europe and Southern Africa were, to all intents and purposes, the same. Moreover, although it is possible these monsters hibernated like European reptiles of the present day, the apparent necessity nowadays for tropical or sub-tropical heat in developing reptilian bulk perhaps justifies us in thinking that the conditions under which Pariasaurus lived, whether in the northern or the southern hemisphere, were such as to produce and maintain a perennially luxuriant vegetation. Thus a study of one isolated type of extinct animals may throw a flood of light upon the conditions of existence in two remote parts of our planet, perhaps a score or more of millions of years ago.

THE SCALLOP

BY BERNARD C. WHITE.

THERE is no class of shell which has such variety and brilliancy of colour as the Scallop. Some species are glittering orange. Another species, which is particularly lovely, glows with red and purple; another is covered with crimson and white ribbon patterns, while others are noticeable for their delicacy—pale pink, with dainty white frills; pale saffron, with a pink flush; lemon-coloured Scallops, and Scallops of daintiest green. But description fails to paint their glories, and if the reader desires to realize their beauty, he cannot do better than visit the Shell Gallery of the Natural History Museum, where he will find them under their family name of Pectinidæ, a group of dazzling colour.

Photo by] [*H. Main.*

THE FLY.

The pupa skin has been thrown off and the insect appears as a beautiful fly.

It should be noticed that the shell of a Scallop covers a large superficial area, but is comparatively thin and light. Such a combination of qualities would not, under ordinary circumstances, go for strength, and the inmate of the shell would have little chance of withstanding the stress of the ocean.

But Nature has provided for such a contingency. It will be noted that except in abnormal cases the shell of the Scallop is marked with hollow ribs, which vary in number according to the species, and run from the beak to the margin of the shell. It is these ribs which give additional strength to the shell's light structure. But why is it necessary that the shell should be light? Well, the Scallop, differing from all other bivalves, takes swift journeys through the water. It does this by rapidly opening and closing the valves of the shell, and if it is to pursue this habit, its shell must necessarily be less solidly built than that of a sedentary mollusc.

This power of locomotion in the Scallops is a fact not as generally known as it should be, even among those whose daily business is in the shallow waters of our coast. Mr. Step has told of the astonishment exhibited by a couple of fishermen to whom he had shown some Scallops in active flight. The men included oyster- and scallop-dredging among their avocations, and thought they were being imposed upon when he spoke of Scallops flitting through the water. When some small species of Pecten were disturbed and caused to pursue their zigzag flight before the eyes of the fishermen they wondered how they could have lived so long without discovering the fact for themselves.

There is yet another purpose served by the broad and hollow ridges of the scallop-shell—that is security. The grooves, it should be noted, of the under-valve fit into the ridges of the upper, and vice versa, so that when the inhabitant has determined to close up the two valves their edges are inexorably locked one in the other, and there is no chance of them slipping apart, whatever force may be brought to bear on them. It is hopeless, therefore, to try to wrench the two valves asunder, and the fortress can only be entered either by leverage or boring. The star-fish employs the former method: attacking the unfortunate victim in the same way as he does the oyster (see page 605), but the majority of the Scallops' enemies favour the boring process, which does not call for the same amount of strength.

Most Scallops are perfectly symmetrical and have equal valves, but exceptions to this rule are of some importance, as they form a large group in the family. These, like the Great Scallop, or the

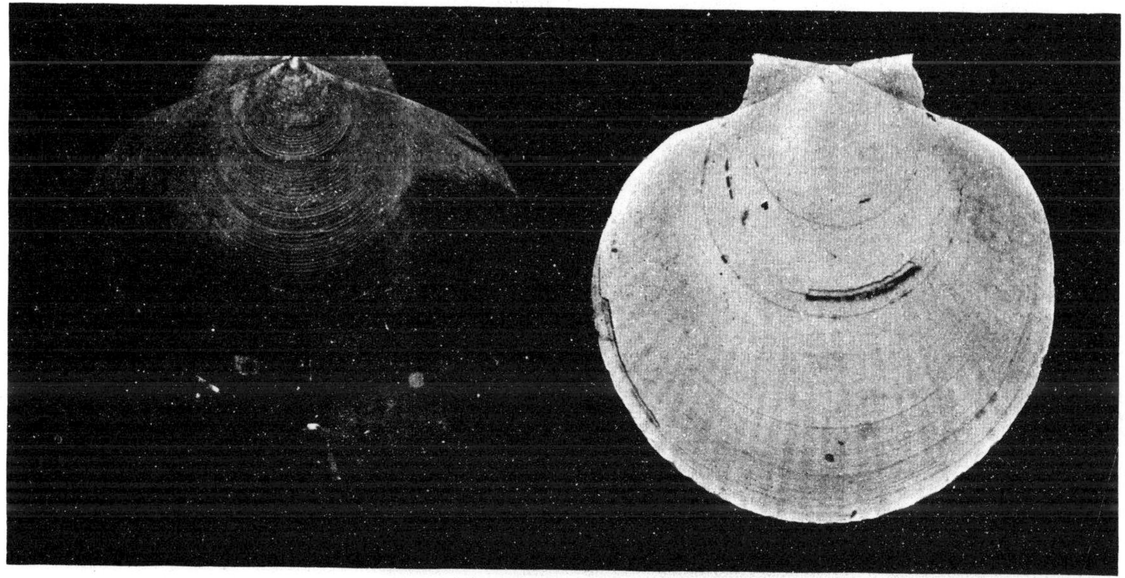

SCALLOPS.

The lower valves of two species in which the sculpturing takes the form of fine concentric grooves and ridges.

variety shown in the illustration on page 1004, have the upper valve flat and the lower deeply curved, so deeply, indeed, that the thrifty housewife has long looked upon them as excellent large spoons which Nature has provided for domestic use.

The Scallop family can also provide some excellent examples of a remarkable feature in shell coloration; that is, the unequal distribution of tint on the upper and lower shells of a bivalve. Take, for example, the specimen here given. The outer side of the lower valve of the shell is white, the upper, red; and among other instances of this characteristic is that of the Great Scallop, or Clam, where the upper shell has red ribs with brown interstices, and the lower is white suffused with pink and yellow. No doubt the reason for such coloration is that the upper shell is alone conspicuous, and therefore assumes the tints of its surroundings in order to escape notice, while the lower is allowed to follow the natural coloration of the species; but it is wonderful that the mollusc should be able to vary, as it undoubtedly can, the colouring of the various parts of its shell.

There is another point which ought to be mentioned before we leave our subject, and that is the Scallop's possession of ocelli, or little eyes, which lie beneath the outer fringe of its mantle. This mantle is double-fringed and borders the whole lip of the shell. In the upper mantle-border the fringe is very delicate, but in the lower it develops into long tentacles, and between these gleam the eyes, which are usually deeply coloured, either indigo-blue or green, and appear to be endowed with optical powers of a comparatively high order.

THE INTELLIGENCE OF ROOTS

BY S. LEONARD BASTIN.

WHEN studying the question of root-movement the great Darwin came to the conclusion that the underground organs of plants often behaved in a manner which it was not easy to explain. Indeed, he actually went so far as to commit himself to the opinion that the root-tips of some seedlings evidenced as much intelligence as was to be observed in the brains of the lower animals. The student of the living plant is continually coming across instances in connection with the movements of roots which are really very difficult to explain on the lines of mere cause and effect.

SCALLOPS.
Side views of two species to show the difference in the upper and lower valves, and the natural position when at rest on the sea-bottom.

Many of the old-fashioned mysteries in connection with root-behaviour have, of course, been entirely explained. One instance by way of illustration may be taken. It has often been observed that the roots of plants will travel a long distance towards a moist portion of the soil. There is really nothing very strange about this, seeing that, generally speaking, roots follow the line of least resistance. For a long way round the moist part in the soil the ground will have been loosened by the filtering through of minute particles of water. To our eyes the condition of this soil might hardly appear to be damp, yet to the sensitive root-tip the ground is a suitable medium through which to grow. As opposed to this, the hard, resisting character of dry soil discourages the extension of the roots in that particular direction.

A great many cases of root-movement are not so easily explained. In the present article a few of the most striking of these which have come under the notice of the writer have been collected. Whether we are right in using the word " intelligence " in connection with cases such as the following or not, it must be admitted that these instances are quite unexplainable by ordinary means.

[*By Theo. Carreras.*

SCALLOPS IN FLIGHT.

Scallops flit through the water in a zigzag course, by the rapid alternate opening and closing of their valves.

Photos by] [*S. L. Bastin.*

HAZEL GROWING UPON WILLOW.

A Hazel bush growing upon a pollard Willow has to send its roots ten feet in order to reach the ground. In the lower photo, part of the bark has been removed to show the roots.

About ten years ago a seedling Hazel bush started life in the decayed vegetable matter right at the summit of a pollard Willow. The bush grew quite happily for a time and increased considerably in size. After a while, however, it seems that the limited root-accommodation was not sufficient for the plant's needs. Accordingly it started to send down roots on the inside of the trunk of its host to the ground. The distance travelled was no less than ten feet. Finally, the roots of the Hazel embedded themselves in the soft mould collected in the inside of the hollow trunk. Another very singular case of roots travelling a long way, and with unerring instinct, to reach the soil occurred in South Africa. In this instance a species of climbing Cactus was rambling over the roof of a shed. In a certain part of the roof there was a hole, and when the stem of the Cactus reached this spot an enormous quantity of roots were sent out. These travelled down (through the air, of course) a distance of eight feet to the ground below. Actually, they were never destined to reach their goal on account of the fact that the shed was infested with rats, and these creatures gnawed off the roots and in this way prevented them from touching the ground. In both the cases mentioned above it is not easy to explain how it was that the roots of the plant knew that the ground was within reach, whilst they were still many feet away. In the case of old hollow Beech-trees, it is quite a common experience to find that these send down roots on the inside of their trunks through the air to the ground (see page 1008).

In greenhouses it is often easy to find cases of root-intelligence. A little fern in the writer's possession was growing in a pot with some other plant. The pot was always kept standing in a saucer of water, and it would seem that the fern desired to draw upon this supply. Accordingly it sent down a root on the outside of the pot to the water beneath, travelling a distance of about four inches. Some plants of

Monstera deliciosa (the great tropical Aroid, which is often trained about warm greenhouses) suffered from an insufficiency of water. In order to obtain a further supply they sent down their roots with a remarkable instinct to a tank of water. This meant a journey of several feet through the air. How they could possibly have discovered the right quarter to which to direct their growth it is very difficult to suggest.

It is not always very easy to study roots when they are growing in the soil, but opportunities do sometimes occur. Gravel- and sand-pits are always well worth visiting in this respect. A curious case of a Fir-root came under notice recently; this had been brought to light owing to the removal of a quantity of soil. In its earliest days the root had grown along horizontally through the top layer of mould. Eventually this course brought it into a stratum of pure sand; this material seems to have been too dry for the well-being of the root, and, as a consequence, a very strange thing happened.

The growth of the root in the direction in which it was going ceased abruptly, and the tip evidently turned off at right angles and grew downwards. The root persevered through two feet of pure sand, extending downwards to a layer of moist soil beneath. It is highly improbable that the sand would be affected upwards to any extent by the moist layer underneath, and the root must have had a bad time when passing through the dry region. Yet in some way, of which we can have little knowledge, it knew that finally a moister region would be reached.

It may be straining a point to talk about the intelligence of roots, but such cases as the foregoing seem to show that these organs possess a sense of being able to feel things before they actually come into contact.

Photo by] [*S. L. Bastin.*

CACTUS ROOTS.

A trailing Cactus in South Africa, growing up over a shed, sent down roots through a hole in the roof. The ground was eight feet beneath.

CORALS

BY EDWARD A. MARTIN, F.G.S.

THAT corals are both beautiful and marvellous objects no one will gainsay. They have been familiar to all from childhood's days. One's earliest recollections of corals have generally been drawn from coral necklaces, made up of pieces of what has been called "precious coral." Then one remembers that a friend returning from "India's coral strand," or elsewhere, has brought a lump of massive coral as a memento, and perhaps some interest has been felt as to the origin of the many objects known as corals. The forms they take are very numerous, but, in general, there are three or four chief types of structure: the branching corals, such as the madrepores; the massive corals, which form coral reefs; the single corals, as the cup coral and the mushroom coral; and the red coral, which is used so much for necklaces and bracelets. There are many varieties of each of these, some made up of parallel pipes like the reeds of an organ, whilst others look like stags' horns, or like thin, crinkled paper. The word "coral" is not a strictly scientific term, for many of the shapes formed by the work of the coral-polyps, or, as they have been called, the coral-anemones, are not made of hard limestone, such as is the case with those mentioned, but are made up of horny material, such as in the case of the sea-fans, which we have already dealt with in MARVELS OF THE UNIVERSE. We are more particularly concerned now with the hard, stony corals, and some of these are shown in our illustrations. All of these are made up of carbonate of lime.

Photo by] [*S. L. Bastin.*

INTELLIGENT ROOTS.

A Beech whose trunk was hollowed by decay sent down new roots into the cavity on their way to the earth.

It must be remembered that in looking at these beautiful objects we only see the marvellous skeleton which the polyps have formed. We human beings have the power of taking from the food we eat the limy material which is used to form our internal skeletons. It is quite an involuntary action on our part. We form our skeletons without knowing it. The lowly-organized polyp extracts just the same limy material from the sea, and perhaps from its food, and makes its skeleton. These skeletons are the "corals" which we handle. The polyps have gone. The thin rind of flesh in which they were seated has dried up, or has been decomposed, and nothing remains of the hundreds of wonderful creatures whose work survives them for so long.

Nothing is more remarkable than the fact that the coral-polyps form colonies containing thousands of individuals, all of which are joined together into one united kingdom of corals. The jelly-like flesh which surrounds the hard part is perforated by canals through which circulates what corresponds to the blood of higher animals. Seated in clefts in this flesh, the polyps are just where these canals open out on to the surface, and there these radiant flowers of the sea expand their petals in the shape of tentacles. That they are not flowers, of course every one knows; but to an observer seated in a flat-bottomed boat whose bottom is not wood but plate-glass, a coral reef is a perfect garden of lovely tints, whilst amongst the coloured polyps themselves there dart here and there those brilliantly-painted fish, which have taken upon themselves the very colours of the coral bank. "Zoophytes," or animal-plants, was the name given to many of these lowly creatures

[*By Theo. Carreras.*

LIVING CORALS.

Several of the photographs that follow show the various forms of Coral when cleaned and divested of their builders. Here a group of diverse forms are shown as they appear in the process of building, with the polyps expanded.

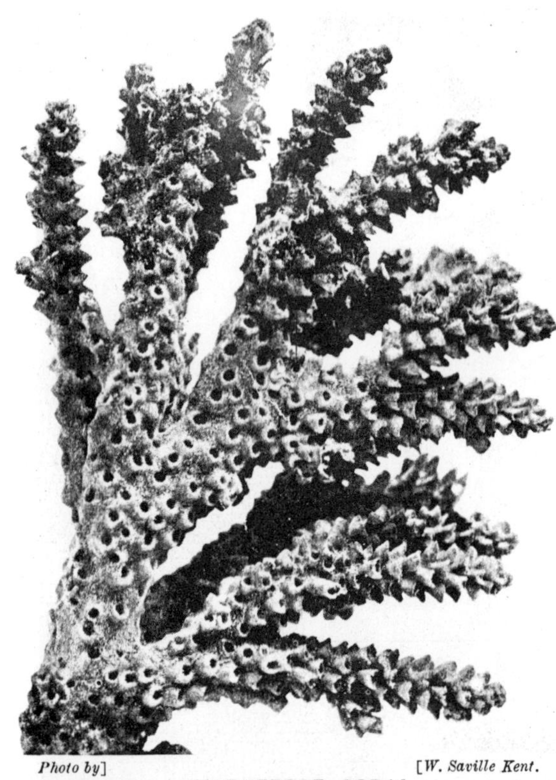

Photo by] [*W. Saville Kent.*

A MADREPORE CORAL.

In this species the polyp-tubes are large and very plainly seen.

at a time when the boundary between animal and vegetable was not clearly defined; whilst to the creatures which we are now discussing the poetical name of "daughters of the sea" was given, for that is what the Greek words mean from which we derive our word "coral." The hard skeleton gives us but a poor idea of the wonderful appearance of these creatures when living, for in some of them, as in the case of the red coral, not even the clefts remain in which the polyps were seated; these made no impression on the solid skeleton. In many others the clefts were in the "corallum" itself, and the hard part grew in the shape of thin plates right up inside the outer portion of the creatures themselves. Thus in most of the corals we see this arrangement of thin, leaf-like plates, all seemingly radiating from the common centre.

The coral animal is nothing but a miniature sea-anemone. Its tentacles surround its mouth, and this leads down into a stomach and a body-cavity. Its young, strangely enough, are altogether unlike the parent. When an ovum is discharged and has been fecundated, it becomes a tube-like object, and swims about rapidly by means of a vibrating hair-like process. It finally settles down on a rock, and it begins to deposit a limy base. It develops a mouth and tentacles, and will, in course of time, become an adult coral. It then gives rise to a colony by sending off buds and developing many thousands of individuals. On the other hand, there are some corals which, although they grow and grow, yet remain simple and single corals all their lives. Such an one is the cup-coral, which in times long past grew in profusion; in fact, owing to the numbers in which this genus existed in former days, they must have formed what are practically great reefs of fossil coral. It is extremely interesting to remember that amongst the one or two true British corals, one is the unique little Devonshire cup-coral, which is found off Ilfracombe and elsewhere. The metamorphoses through which a coral passes are just the reverse of the familiar changes through which an insect passes. The coral is free in its larval state and fixed in adult life. The chrysalis of an insect is almost immovable, but the butterfly is as gay as the wind.

Photo by] [*H. J. Shepstone.*

CUP CORALS.

Though differing in form, these are constructed like those shown on the opposite page and are nearly related to them

Little causes beget big results, and how true this is of the corals we can see in the coral islets which bedeck many a wide waste of ocean. Building upon the submerged mountain-tops of the Pacific Ocean, they have reared through untold generations reef upon reef, until they have reached the surface of the water. Bank upon bank of their ancestors they have entombed, the work of myriads who have laboured before them, piling up cell upon cell into one vast structure of calcareous rock. Then dying when they reached the surface, their banks have proved the resting-place of flotsam and jetsam of the ocean. Seeds of trees and plants have been wafted across the seas; drift-wood has been cast upon its shores; seaweed, with various forms of littoral life in its larval stage, has been cast up on it. The coral is ground down to a fine sand, and birds make their appearance. Life of all kinds begins to inhabit the new island, and perhaps

Photo by] [*H. J. Shepstone.*

LEAF CORALS

Similar in the type of structure to the Mushroom Coral, the ridges are here more open and convoluted. The centra figure is an example of the Noble Leaf Coral

finally man himself comes to occupy it. And for all this the little coral-polyp is in the first place responsible.

Amongst the more remarkable forms of single corals is the Mushroom Coral. The resemblance of this coral to the upturned top of the common mushroom is most exact. But when living, it would have been noticed that the diverging plates, to which we have before referred, were on the top. These plates look like the gills of a mushroom, but, of course, in the latter they form the under part of the vegetable organism. The mouth in the centre is surrounded, when living, by very short tentacles, and these are contracted within the ridges of the gills. In the young state the Mushroom Coral has a stalk, and one can often see in the older forms the hard knob where the stalk formerly joined it. Then finally it became severed, and fell to the bed of the sea. Sometimes the mushroom has become elongated, until it is twice or thrice as long as it is broad. I have a "slug-coral," as it is then called, which is nearly a foot long.

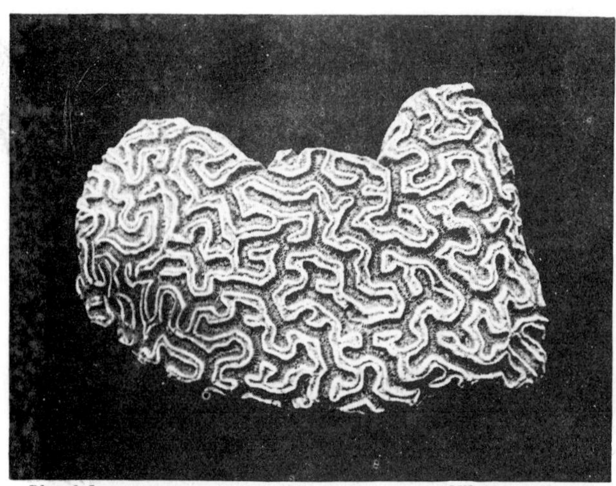

Photo by] [*W. Plomer Young.*

BRAINSTONE CORAL.

The labyrinth-like form of this beautiful Coral is due to the way in which the polyps have branched and extended by budding.

Although corals continue their species often by ova, they as frequently extend their colonies by means of fission, or even by budding. The method known as fission is responsible for many of the varied shapes which their skeletons exhibit. In the *Mussa*, we find that by the splitting-up of the individuals, others have been formed side by side with the founders of the colony, and the hard skeletons have consequently remained attached to one another. In a further development a number of tortuous skeletons have been formed, all still united to the parent stalk, but leaving between them channels as tortuous as the coralla. The result is seen in the coral known as the *Euphyllia*, which we illustrate (pages 1010 and 1011. Somewhat similar in build, but with the process carried still further, we have the *Trachyphyllia*, shown on each side of the photograph on page 1011. But in others the rows of corallites tend to become united at their bases more rapidly than in others, and when they grow outward at their sides as quickly as they grow upward, they become massive, and we have the form known as the Brainstone Coral (here shown). In such forms the combined corallites wind in a complex manner over the whole mass, and at once suggest the convolutions of the human brain. In other corals we find that the greater growth is in a vertical direction, and then we have the *Dendrophyllia*, in appearance like a living tree; or the *Isis*, whose skeleton is made up of alternate pieces of horny and calcareous material; or the madrepores, and the millepores, popularly known as "Stag's Horn Coral." We have called any of these anemone-like creatures by the name of coral if they formed a hard limy skeleton; but it must be remembered that in the millepores, the pink *Stylaster*, and the brick-red *Distichophora*, there are certain differences of organization which separate them somewhat from the true corals, and ally them with the little green Hydra of our ponds and ditches. They are, therefore, sometimes called "hydrocorallines."

Photo by] [*W. Plomer Young.*

MADREPORE.

A section cut lengthwise through a small branch of Coral, much enlarged to show the tubes into which the polyps withdraw.

We have mentioned the madrepores, and in this family we have a very large number of species Their structure is perhaps the most familiar of all the corals, save the precious red coral.

Many of the names of corals terminate in the syllable "pore," and it is interesting to consider why this is so. We know

Photo by] [*W. Saville Kent.*

ON THE GREAT BARRIER REEF.

Many different species of Corals have contributed to the building of this vast reef, and a careful examination of the photo will reveal some of their varied forms. Others from the same reef will be found photographed separately on neighbouring pages.

Photo by] [H. J. Shepstone.

MUSHROOM CORAL.

In this photo the polyps are alive and active. It will be seen that they effectually hide the stony base of the Coral and look like large Anemones. The upper figure with the polyps partially retracted gives better indication that it is a Coral.

what mother-of-pearl is, and it seems probable that the "Madrepore" coral was really called so because it was the "mother-stone." But it soon came to be noticed that different corals had different forms and arrangements of the little holes, or cups, in which the polyps lived. So the Greek word which meant "stone" came to be confused with a similar word, which meant a "little hole," which differed from the other in one letter only. So we had *millepora*, or thousand-pore; *heliopora*, or sun-pore; *pocillopora*, or little-cup pore, and others.

We have already referred to the wonderful Barrier Reef off the north-eastern Australian coast. This is a coral reef on a colossal scale. The very mention of it brings to mind Saville Kent's monumental book, and one of his photographs of a portion of the reef is shown in illustration of this article. The great reef extends for over twelve hundred miles, at a distance from the shore of twenty or thirty miles, or even more. Here, in the greatest profusion and variety, are found all the chief forms of corals, and, indeed, our knowledge of them has been inconceivably increased since the Barrier Reef has been so thoroughly explored. More recently coral-reef exploration of another kind has taken place, in order to prove the truth or otherwise of Darwin's famous theory of coral-islands. At Funafuti, one of the Ellice Islands, several borings have been made, one of which penetrated to a depth of no less than one thousand one hundred and fourteen feet. The result of the close examination of the cores of coral brought up was to re-establish the truth of Darwin's theory. The island of Funafuti was shown to have been gradually formed during the subsidence of the foundations on which it rests, and this subsidence amounted to at least eight hundred and seventy-seven feet. Nothing but coral-reef rock was met with throughout the whole of the boring. It would be interesting to know exactly the rate at which various corals grow, but with the exception of one or two records, we have little information on the subject. Agassiz, from his observations on the

Florida coast, thought that it would take over a thousand years for coral to grow from a depth of seven fathoms (forty-two feet) to the surface of the sea; but, of course, the rate would vary with the species, by the prevalence of currents, and possibly would be regulated by the supply of food. Darwin records that a ship which was sunk in the Persian Gulf had her copper bottom covered in the course of twenty months with a thickness of two feet of coral. The case of another sunken ship which was encrusted with coral showed only a growth of three inches in a year. On the other hand, some corals which were planted on the Madagascar coast were seen to have reached a height of nearly three feet in the short period of six months.

VON DER DECKEN'S LOBELIA

BY SIR HARRY JOHNSTON, G.C.M.G.

IF the classification of plants were based on purely external appearances no botanist would be daring enough to place in the same genus the little blue Lobelia of our gardens and the extraordinary developments of Lobelias to be found on the loftiest mountains of Tropical Africa and on lesser heights in South Africa; for these last, instead of being little herbs only a few inches high, assume the appearance of tree-lilies, of aloes, or of cacti, and may grow to a height above

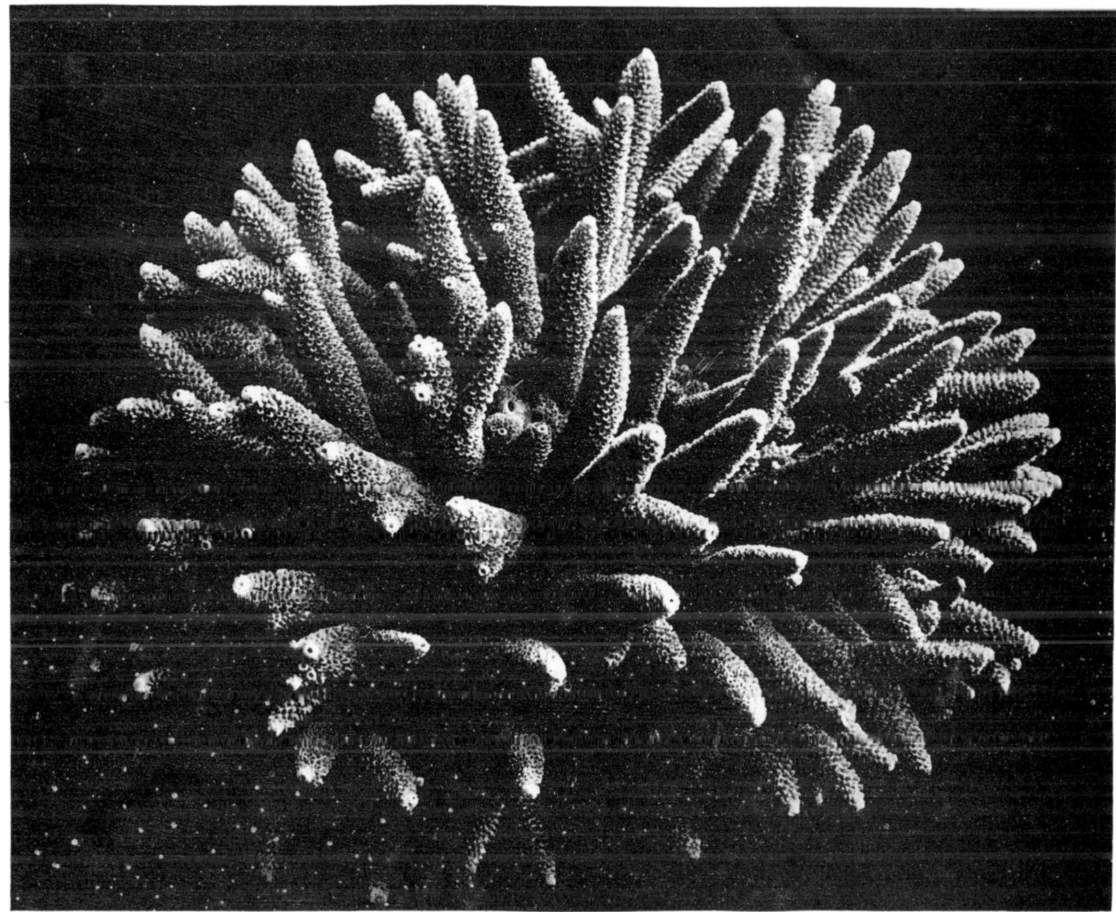

Photo by] [*W. Saville Kent.*

STAG'S-HORN CORAL.

The popular name covers many distinct species of Madrepores and other Corals. The rasp-like character of the surface is due to the mouths of the numerous tubes or cells in which the polyps dwelt.

the ground of nearly thirty feet. Yet the structure of their flowers shows them to be true Lobelias and nearly related to the dwarf species, which is not only one of the most common plants in cultivated gardens, but as a wild flower produces gorgeous effects on South and Central African mountains, where it grows in veritable carpets of ultramarine or azure blue.

When the great German explorer, Baron von der Decken, made a minute examination of the slopes of Kilimanjaro in Eastern Equatorial Africa, though he did not succeed in reaching the snow-line, he attained to altitudes where an Alpine vegetation grew, and here he observed the giant Lobelias which were subsequently named after him. Later, the present writer, visiting the same regions, likewise collected the flowers and leaves of these monstrous Lobelias. But this plant was—to his thinking—a far more striking and noteworthy object on the Ruwenzori Mountains, almost in the heart of Africa. Here they grew at an altitude of between ten thousand and thirteen thousand feet, with the appearance shown in the accompanying photographs—a tall column, the upper part of which consisted mainly of shining, scaly bracts, directed downwards, while the lower part was of stiff dark-green leaves directed upwards. Under each of these scaly bracts of light-green was a large flower of intense ultramarine blue. When the sun grew strong in the daytime and the flowers were ripe, the bracts rose till they were horizontal in position and thus revealed beautiful blue flowers below, which were at once visited by numerous bees.

By permission of] [*Sir H. H. Johnston.*

ON RUWENZORI.

A natural garden at twelve thousand feet above the sea, in which Von der Decken's Lobelia is a conspicuous feature.

There are other giant Lobelias on these same Snow Mountains of Eastern Equatorial Africa, as, for example, Stuhlmann's Lobelia, the flowers of which are a pale reddish-green, and also grow in tall columns concealed by green bracts. This, or an allied species, would develop a bare trunk devoid of leaves but crowned by a great tuft of spear-shaped foliage, finally surmounted by the slender column of bracts and flowers, so that they looked very much like a dracæna or tree-lily.

Von der Decken's Lobelia is such a marvel of vegetation, and at certain times so beautiful, that it is surprising efforts have not already been made to introduce it into European gardens, where it ought to stand the climate well, as it must be accustomed to occasional frosts and cold, wet weather.

WINTER WHITENING

BY W. P. PYCRAFT, F.Z.S., ETC.

WHY is it that beasts and birds which have to face a prolonged and severe winter turn white as the snow which surrounds them? This is a question more easily asked than answered. But all the evidence seems to favour the Darwinian interpretation, which is summed up in the two words "natural selection." And herein we have a perfectly wonderful illustration of the way in which

Photo by] [*Underwood & Underwood.*

VON DER DECKEN'S LOBELIA.

This giant Lobelia, which sometimes attains a height of thirty feet, is a near relation of the dwarf bedding-plant so well known in our gardens. The broader, darker portion of the column shown in the photo is the leafy part of the plant; the narrower part above is the flower-spike.

animals act and react upon one another. Briefly, it is assumed that time was when this whitening did not take place, and that, in consequence, when the snows of winter fell, the garments of fur and feather of every inhabitant of the region showed up as conspicuous dark blotches against their white setting, and thus became an easy mark for every raptorial bird and beast in the neighbourhood; and similarly, of course, the prowling enemy was descried by his intended victim. Now if this interpretation be correct, then, some of my readers will say, a very different result should have followed. In a word, a season or two should have rid the preyed-upon of their enemies, and permanently. They must have died of starvation, for being dark, they would have been detected before they could get within range of the victim, and thus only by the merest accident would they

NORTHERN ANIMALS IN SUMMER DRESS.

The Mountain Hare, Grouse and Ptarmigan are here shown in summer fur and plumage whose colours harmonize with those of their surroundings.

ever get a meal—that is to say, when they caught a victim by surprise. Thus, there would be no need for the whitening process.

But suppose these imaginary circumstances be reversed: that is to say, that the carnivorous types were the first to assume a white garb at the approach of winter. Then, unless their victims could respond in like manner they would speedily be wiped out. That natural selection alone could not have originated these white vestments goes without saying; and we must assume, therefore, that this factor seized, as it were, upon those individuals which, during some ancient past, tended to suffer some physiological check at the fall of the temperature ushered in by the advent of autumn, whereby the formation of the normal pigment for the coloration of the fur or feathers, as the case may be, was checked, more or less completely. Such individuals would, at once secure an advantage over those which were not thus affected. The latter, from their conspicuousness, would be killed off; the whitening individuals would survive and hand on their inherent peculiarities to their offspring.

In England we see very little of this curious seasonal change, for our winters rarely bring much snow. In Scotland, however, we may watch the whole process of change, for the Mountain Hare, the Stoat or Ermine, and Ptarmigan all assume white garments for the winter months, and farther north the Arctic Fox behaves in like manner. And in all these cases it will be noted the summer liveries are no less protectively coloured, and for the same reasons. That this assumption of a white livery is not entirely to be explained by temperature is amply demonstrated in the case of the Red Grouse, which, even in Scotland, does not turn white in winter, though for weeks it may have to burrow deep in snow to find its daily bread. Visitations of prolonged snow, however, in the haunts of the Red Grouse are rare ; hence, no change is made. With the continental Willow Grouse, however, which is almost identical in plumage with our Red Grouse—save only in the wings—a white dress is annually assumed, as in the case of the Ptarmigan ; and for the same reason—to wit, because a long winter and deep snow have to be faced.

FIRE-FLIES

THE Fire-fly is not a fly at all in the sense in which naturalists understand the word. Like our Glow-worm, that is not a worm, the Fire-fly is a beetle. Our forefathers went largely by casual appearance when bestowing names, and seldom troubled their heads about strict accuracy. The luminous insect of our lanes was always seen crawling on or near the earth, therefore it was a glow-*worm*. The luminous beetles of the tropics were mostly seen on the wing in the evening, therefore they were flies.

Fire-flies are of several species, all inhabitants of tropical America, but the inhabitants do not bother themselves with the fine distinctions of the entomologist. To the native mind they are all *cucuyos*. The Fire-fly proper is a beetle about an inch and three-quarters long, in general form, as will be seen from our illustration, very much like our common click-beetles or skip-jacks, whose grub is the notorious wire-worm that tunnels our potatoes and renders them worthless.

WINTER WHITENING.

In these photographs the Mountain Hare the Grouse, Ptarmigan and Stoat or Ermine are all shown in winter dress, which again harmonizes with the now prevailing snowy conditions.

Its colour is a rusty-brown shading to black, and on the shield-shaped section of its body, popularly known as the head, but more analogous to shoulders, there are two oval, yellowish-white spots. These are two of the Fire-fly's lanterns, for it is these areas that give out a light like that of our glow-worm, but much more brilliant. There is another lantern on the under surface, but situated at the junction of the two larger divisions of the body, and it is this organ which emits the principal light when the beetle is in flight. The emission of light appears to be quite under the control of the beetle, for when the latter is either feeding or at rest the light is suppressed ; but when in flight for courtship the light is produced strongly. The Indians of Vera Cruz make a business of catching and selling them. They are caught by guile. The Indian attaches a glowing piece of charcoal to a stick, and whirls this around in the air to make it glow, when, being mistaken for a Fire-fly of extra brilliance, the real beetles are attracted to it and caught. They are stored in wire cages, fed with crushed sugar-cane, and kept clean until purchased by ladies for personal adornment. These fasten them to their hair and attach them as jewels to their dresses.

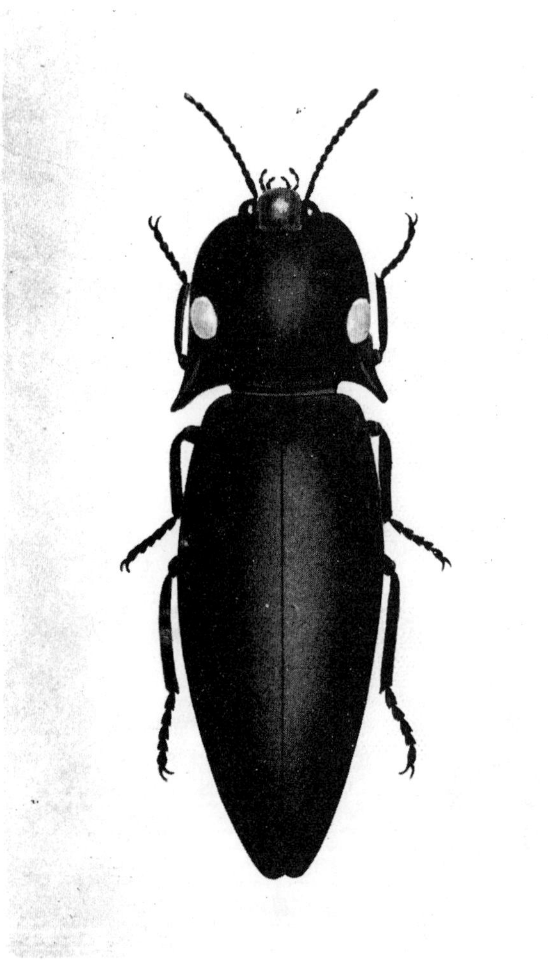

FIRE-FLY.

A beetle of tropical America with luminous patches on both upper and under sides. It is here shown twice the actual length.

THE ARTIFICER IN THE HIVE

BY TICKNER EDWARDES.

THROUGHOUT all ages, as far back as records exist, mankind has been extolling the ingenious works and ordered communal life of the honey-bee ; but perhaps the most wonderful thing in a bee-hive, to-day as ever, is the bee itself.

At the very threshold of investigation we are brought face to face with a seeming impossibility concerning her—an impossibility, however, which is nevertheless an uncontrovertible fact. Of the tens of thousands of worker-bees in a hive not one exists to-day as nature fashioned her. But for the deliberate artifice of her nurses when she lay in her cradle-cell, each worker-bee could and would have become a fully-developed queen, a creature differing from her present self in a dozen distinct and fundamental ways.

Of this fact, that she originates from an egg which is identical with that producing the widely dissimilar queen, there is no doubt. Over and over again the experiment has been made of transposing the eggs, or even the young larvæ, deposited in worker- and queen-cells ; and the queen-egg put into a worker-cell develops into a common worker-bee, just as surely as the egg taken from the worker-cell and placed in a queen-cell will hatch out eventually as a fully-equipped queen.

And the more we compare the two creatures, the less credible becomes this truth of their identical origin. The worker is by nature a devotee of the open air and sunshine ; after the first fortnight of her life, nothing but the most intemperate weather, or fall of night, will keep her in the hive. But all the queen's instincts seem to be directly contrary to this. Except for the one hour when she goes on her mating-flight, and again when she accompanies the swarm, nothing will tempt her abroad. Again, the worker-bees are sociable, peace-loving, their every act infused

WINTER WHITENING.

In the upper photograph the Arctic Fox is shown in his summer pelage on a northern moor. Below he is seen in his white winter coat, which renders him inconspicuous among his snowy surroundings.

with the true communal spirit ; but if two queens meet, there is nothing for it but a duel to the death.

In physical attributes there is an even more marked divergence between mother-bee and worker. The queen is much larger in body, though smaller in head and considerably shorter of tongue. The queen's mandibles, or sidelong jaws, are curiously notched, while those of the worker-bee are smooth and spoon-like in form. The sting of the queen is much longer than that of the worker ; it is also strongly decurved. The worker's sting is straight.

On the worker's legs exist several ingenious implements—a pair of neatly-contrived pincers, a sort of curry-comb for removing pollen-grains from the hairy surface of her body, a basket-like device wherein the pollen is collected and borne to the hive—of all of which the queen's legs are wholly destitute. The wonderful wax-generating organs, under the plates of the worker-bee's abdomen, do not exist in the queen ; while the queen has two enormously-developed ovaries with their accompanying parts, which in the worker are so atrophied and rudimentary as to be practically non-existent.

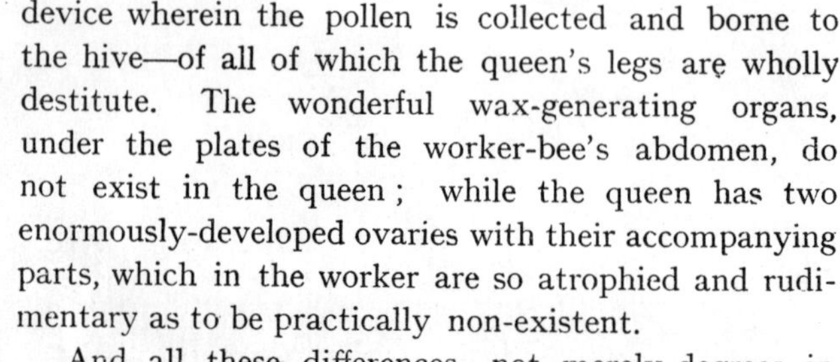

Photos by] [*T. Edwardes.*

BEE'S STING.

The Bee's sting is here contrasted with the point of the finest needle, both being magnified to the same extent. The upper figure is the sting, the lower the needle point

And all these differences—not merely degrees in development, but veritable organic changes—are due, we have no alternative from believing, to treatment and food-modification in the larvæ hatched from one and the same kind of egg.

It is easy to set the words down on paper, but to grasp their full significance is a wholly different thing. Biologists tell us that, with all forms of life, in the germ of the ovum lies potentially each completed being. Yet how to reconcile this axiom with the known facts in the growth of queen- and worker-bee, seems a problem not only of difficult, but of inconceivable, solution. And to admit an exception to an universally accepted law, is merely to increase the dilemma ; for in science an exception does not prove, but wholly discredits and destroys, a rule.

Thus a mystery invests the very beginnings of life in the honey-bee, and never entirely leaves the mind throughout all our subsequent study of her intricate anatomy. We need, however, no miraculous tinge to stir up a spirit of wonder when we come more nearly to examine her extraordinary equipment.

Perhaps the thing we notice first about a worker-bee is the way in which she employs her antennæ. These are two flail-shaped appendages attached to the forehead, and she incessantly waves them about, making use of them apparently in all she undertakes. To the naked eye they are no more than two slender, flexible, thread-like attachments, but under the microscope they exhibit innumerable complications. They bear upon their surface at least half a

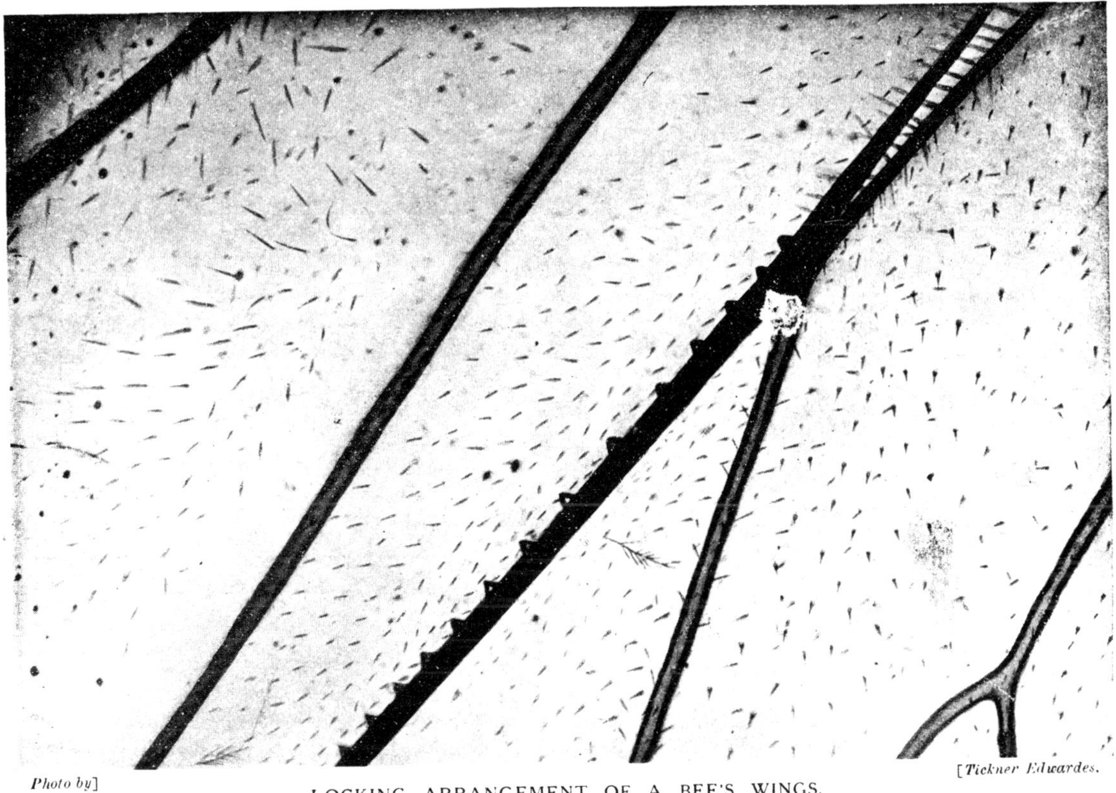

Photo by] [*Tickner Edwardes.*

LOCKING ARRANGEMENT OF A BEE'S WINGS.

The Bee has two wings on each side, but in flight they lock together, forming one continuous plane. This junction is effected by a series of small hooks on the under wing which engage with a flange on the upper wing. The photograph shows part of the wings on the left side with their edges so joined.

dozen distinct kinds of delicate instruments, each of which is obviously the organ of a different sense.

There are simple hairs in abundance, evidently organs of touch. Other hairs are short and stunted, ringed at their base, and connected with a nerve-fibre. Others again are hollow, their shafts enclosing the end of a nerve. There are curiously-shaped pits with deeper conical depressions beneath, and other pits closed with delicate membranes like drum-heads, within each of which a nerve-end stands up free.

Some of these intricate organs on the bee's antennæ are manifestly feeling implements, but as to the uses of the others it is possible only to hazard a guess. The open pits are conceivably organs of smell, and those covered in by membranes are probably auditory in function. The hollow bristles, like Pierrot's hats, enclosing extended nerve-fibres, have never yet been satisfactorily accounted for. But to assume that a bee has only five senses like ourselves, is to assume altogether too much. In all probability she receives impressions by means of these subtly-made antennal organs, of the nature of which we can form no idea.

The compound eyes of the honey-bee, each consisting of thousands of separate lenses, have been often described. They are for distant vision, and are employed by the bee in her long journeys to and from the hive. But it is not generally known that, for use at close quarters, and especially in the gloom of the hive-interior, she has three other and simple eyes set in a triangle upon her forehead.

Under a strong glass these eyes show up big, dark, and lustrous, with all the solemn wisdom of an owl's. They must have extraordinary powers of appreciating very faint rays of light. Even in the daytime the interior of a hive must be plunged in what would be to us almost total darkness ;

but it is principally at night that the honey-bee carries on the minute and complicated work of comb-building, just the kind of work, one would think, in which the faculty of sight would be most necessary. Yet the combs grow steadily downward cell by cell, each cell keeping its hexagonal form, right dimensions, and extreme attenuity, and each comb being built exactly the required distance from its neighbours.

The popular conception of a bee's tongue as a hollow instrument by which she sucks nectar from the flowers in no way prepares the observer for what is seen when a near examination of the part is made. It is at once apparent that the bright brown, glistening appendage of the mouth is no tongue at all, but a continuation of the under-lip. It is one of the most complicated organs found in insect-anatomy. It consists of half a dozen distinct parts separated longitudinally, which are capable of being brought together in various combinations, so as to form not only a single, large, hollow organ when a bulk of liquid has to be dealt with, but smaller tubes necessary for absorbing minute quantities of fluid.

Photo by] [*Tickner Edwardes.*

PART OF A BEE'S TONGUE.

Small quantities of liquid are sucked or mopped up by the central process, tipped with fine hairs; but when a greater quantity is to be dealt with the semi-circular side-pieces are brought together, forming a larger tube.

To the naked eye the six legs of a bee are very much like each other except in size, but under the microscope they appear as three pairs of elaborate implements specially designed for use in the various tasks in which the worker-bee engages.

Each leg consists of nine separately-jointed parts. On each of the forelegs is situated a very ingenious device by which the antennæ are cleaned. This is a semi-circular recess in the thigh, lined with a crescent of stiff bristles. The antenna is placed in this, and the knee-joint closed, when the brush engages with a curiously-shaped projection on the next upper leg-part, the whole forming a complete ring. The antenna is then drawn through this, and is thus scraped free of all adhering matter.

The second legs have their special equipment of appliances useful in the minute commerce of the hive; but it is the worker-bee's hind-legs which carry the principal tools employed in her foraging labours.

All the pollen—which, in some form or other, is an indispensable part of the food of every member of the colony, young and old—is transported to the hive on these hind-legs. The thigh is broadened

Photo by] [*Tickner Edwardes.*

THE COMB APPROACHING COMPLETION.

On the next page the busy workers will be seen in a much earlier stage of their work. Here their task is nearing completion. All the material used in comb construction is generated in the bodies of the worker-bees.

Photo by] [*T. Edwardes.*

BUILDING A HONEY-COMB.

The Bees are here shown forming themselves into a living plummet while making the comb, an ingenious device which ensures the verticality of the structure, though it is built in almost total darkness.

out and hollowed, and the hollow is surrounded by a rim of strong, thick-set bristles curving inward, so that the whole device forms a kind of basket. The part of the leg below this basket is also broad, and its surface is covered with ten parallel rows of short stiff spines—a perfect curry-comb.

But in addition to the pollen-baskets and the curry-combs, the worker-bee carries on her hind-legs two implements more ingenious still. These occur at the fourth articulation of the leg, and are nothing more nor less than pairs of pincers as artfully contrived as any tool fashioned by man. The one joint has a cup-like process placed in the angle of the elbow, as it were; and the sharp bristled edge of the next joint above it opens and closes upon this cup. It is obviously meant to grasp some delicate, slippery, and fragile substance; and though the fact is perhaps not entirely substantiated, it is most probable that the worker-bee uses these pincers to draw out the newly-formed wax-scales from their generating places under the abdominal plates.

The wings of the bee, and her peculiar system of flight, form in themselves an engrossing study. The honey-bee has four separate wings, but she makes use of them for locomotion as if they were only a single pair, the two wings on each side locking together at their inner edges when they are extended for flight. This locking arrangement is very curious. The lower edge of each front wing is turned sharply up, forming a flange; and the upper edge of the rear wing is furnished with a row of tiny hooks which automatically engage with the flange, as the bee elevates her wings for flight.

Perhaps, however, the sting of the honey-bee is the most cunning, most intricate thing in the whole anatomy of this extraordinary creature.

Man fashions engines of war that can hurl destruction over half a country-side. Rapiers he has made so fine that they will slay and scarce reveal a wound. But never in his most destructively diabolical moments did he conceive a weapon like the sting of a bee. It is a blade so slender that, beside it, the finest needle ever made is but a roughly-welded bar. And yet not one blade, but three together: not only that, but each blade furnished with an edge of jagged barbs, and drenched with a venom doubly compounded. Nor these alone; for the three keen-toothed, poisoned darts are so contrived that, after the initial common thrust is made by the wielder, they take to themselves an individual energy; and go on thrusting, each in turn, deeper and deeper, though the whole contrivance may have been torn from its possessor's body, and she herself may be dead before the death-stroke she launched has worked her will.

THE BRITISH TRAP-DOOR SPIDER

BY FRED ENOCK, F.L.S.

In some of the oldest books on Natural History we find fairly good woodcuts of the nests of the Trap-door Spider, and by the tourists to the South of France these " funny little things " are brought home as souvenirs of their visit; but it is an extraordinary fact that though these nests have been known to naturalists and tourists for over a century, no one has yet told how the Spider makes and fixes its wonderful trap-door. Some years ago the late Sir Sydney Saunders presented to me a nest of the Ionian species, together with the living occupant, which did not long survive. The nest was four inches deep by an inch in diameter, fitted at the top with a

Photo by] [*T. Edwardes.*

AN OLD-FASHIONED BEE GARDEN.

The straw skeps shown in the photograph are probably almost identical with those used by bee-masters for a thousand years back. With hives of this pattern it is impossible to learn anything of bee-life. Not until the movable-comb hive was invented in the middle of the last century was the wonderful communal life of the honey-bee revealed to investigators.

most wonderful lid nearly a quarter of an inch thick with bevelled edge fitting exactly to the mouth of the nest.

Though the nests of the Trap-door Spiders differ so much in appearance, the Spiders are very much alike. On page 1031 is shown our indigenous species. The first point to strike the observer is the enormous muscular development of its jaws and limbs, and if further proof be desired the Spider will soon give it should the finger be placed too near. A slight touch rouses the apparently sleepy Spider, and in a moment the huge jaws are raised and the long fangs driven deep into the flesh, causing blood to flow freely.

The male is of much lighter build than the female, with long limbs fitted for climbing over rough

Photo by] [Tickner Edwardes.

COTTAGE RUINS WITH RECESSES FOR BEE-HIVES.

Down to the seventeenth century sugar was practically unknown as an article of diet except to the very wealthiest classes. The mass of the people depended on the honey-bee for the whole of their sweet-food. Care was taken of the hives, and the skeps were frequently placed in recesses in the walls of the homestead for warmth and shelter and protection from robbers. The ruins shown probably date back four hundred years.

ground, among the roots of heather, etc., where the nests are generally found. The nests of our British representative are from nine to fifteen inches in length, by about three-quarters of an inch in diameter. On escaping from the maternal nest the young Spider excavates with her jaws a hole in the sand one inch deep by one-sixteenth in diameter. This she lines with a close fabric of woven silk, carrying up the tube a short distance above ground, and fastening the aerial part to the grass-stems or other herbage—as shown on page 1030 by a section of a mature Trap-door Spider's tube. Unlike the common Garden Spider, whose life is begun and ended in less than six months, our Trap-door Spider lives to quite an uncertain age—the writer having kept a number from infancy to *nine years of age.*

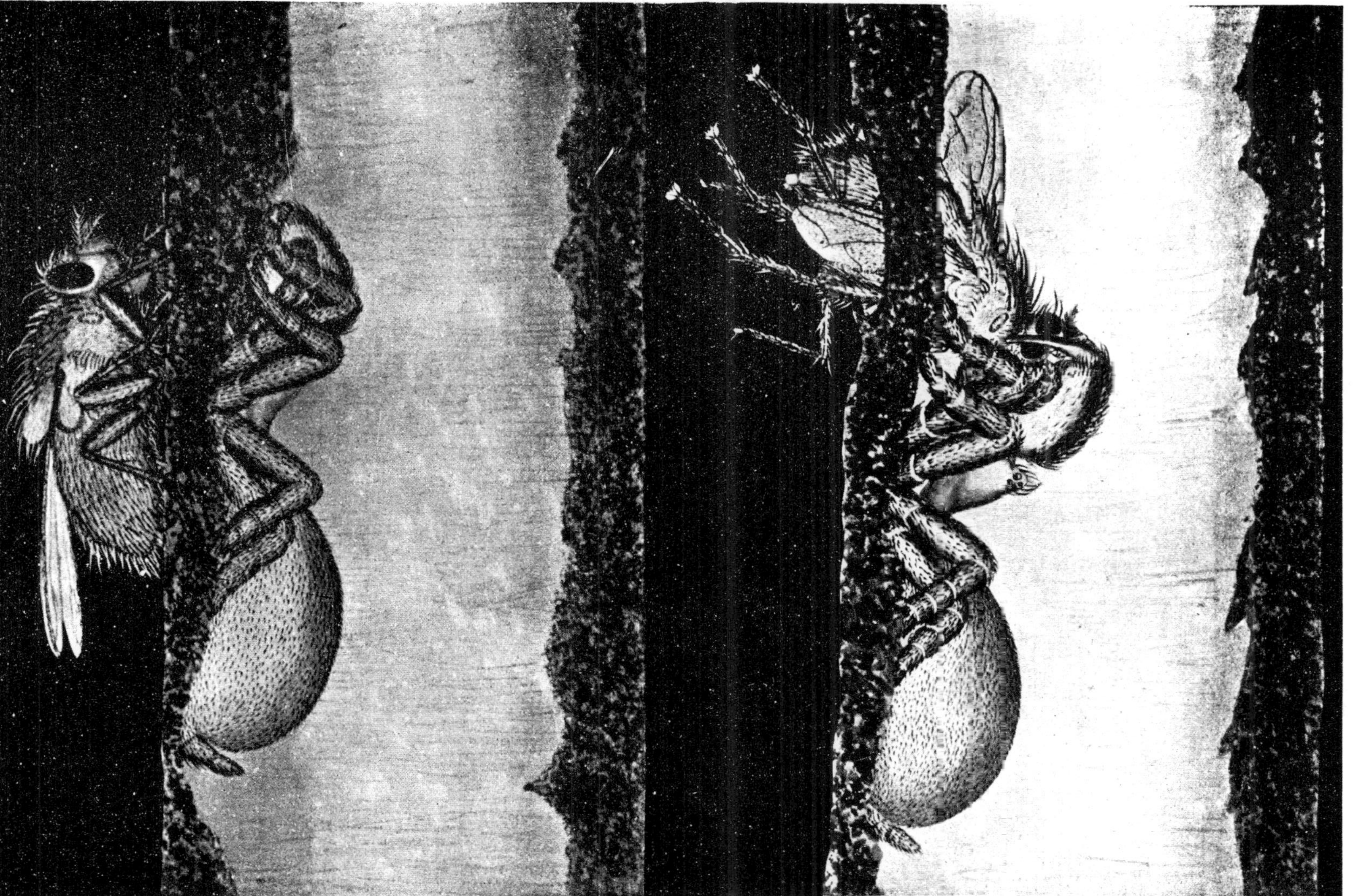

[By F. Enock, F.L.S.

TRAP-DOOR SPIDER AND BLUEBOTTLE.

A Bluebottle alighting on the outside of the Spider's silken tube, the Spider rushes to the spot, and thrusting her fangs through the silk, secures the fly and draws it through a rent made for the occasion, which is afterwards repaired.

The nest is enlarged each year, the aerial part representing the snare, and when the tube above ground is finished it bears a considerable resemblance to its environment.

Supposing we dig up a mature nest and forcibly dislodge the occupant—for unless this be done the Spider never leaves her nest. We must be prepared to "stand by" for perhaps some hours before the evicted one will deign to move, for with limbs drawn close up to her body she "sits" sullenly peering out from the eight small eyes; then suddenly awakes and makes as quick a dash for some rough cavity in the sand, in which she immediately commences to work, using her immense jaws for excavating. Opening the jaws as wide as possible, she takes a big jawful of sand and drops it a short distance on one side; this process being repeated until a decided hole is made—when the Spider quietly climbs up, pushing the sand out before her.

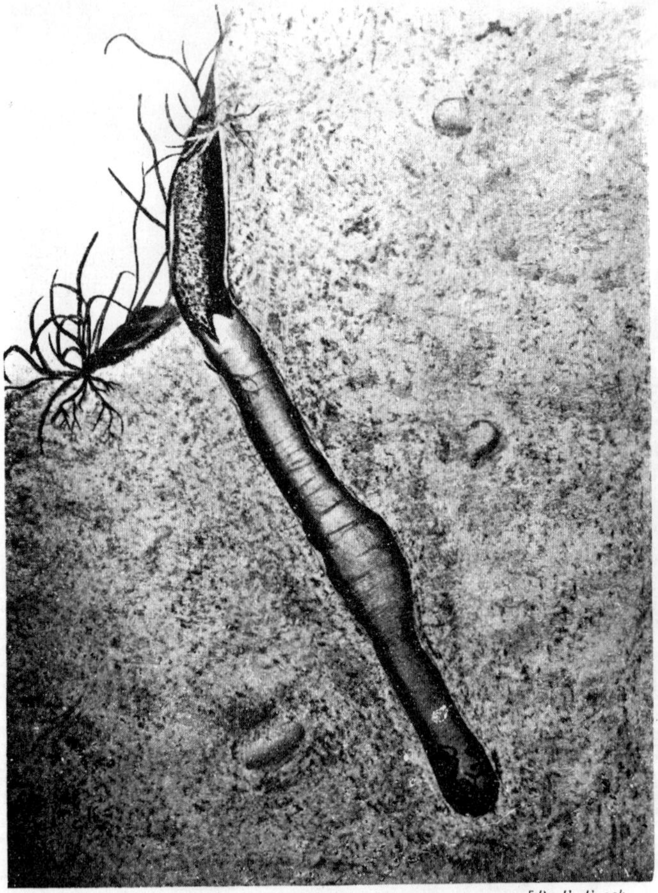

[By F. Enock.

NEST OF THE BRITISH TRAP-DOOR SPIDER.

This species, though one of the Trap-door Spiders, constructs no definite door, but opens one in the silk tube where required, and afterwards spins the rent together again.

This task occupies some hours and is frequently not commenced until night-time. As soon as a depth of an inch has been sunk, the Spider lines the inside with silk, carrying up the structure an inch or two above the ground-line. When the ends have been properly fastened to the side of the bank or herbage the Spider descends, and taking a small mouthful of sand from the bottom, ascends, and driving her long fangs through the aerial part, she bends them down on the outside and then pushes the sand through the silk and pit-a-pats it against her jaws, causing it to adhere to the somewhat sticky silk. So she goes on until the whole of the aerial part is "sanded" outside to make it the same colour as the ground. Next she reverses her position and fastens strong silken threads to the inside and right up to the top, and quietly walks down to her pantry. So she goes up and down until these telephonic communications are all properly constructed. Now she is ready to take a rest, but first she attaches her spinnerets to the threads and takes hold with her claws and falls asleep, when suddenly her rest is broken—there is someone or something "ringing up" above (page 1029), and as the lines are not engaged, she quickly, but oh! so stealthily, climbs up until she reaches the exact spot, where *outside* the fly is resting in blissful ignorance—for neither can see through the opaque tube—whilst the Spider quickly opens wide those terrible jaws, the fangs are raised until at right angles to the vertical tube, and like a lightning-flash are driven through right into the head of the fly!

A few moments are spent in sampling the fly, before the Spider begins to pull it through the rough trap-door opened beneath. Sometimes this is a tough job, but the strength of the Spider overcomes; a large rent is torn, and head first the fly is dragged down until Spider and fly reach the parlour or pantry (page 1033), where the victim is quickly hung up; for the Spider shows her wonderful forethought by running up to mend the hole torn in the snare. This she repairs in the most scientific manner, drawing the edges together with her broad spinnerets and then darning them together in so neat a way that it excites feelings of wonder. After this she runs down with light heart and steps to her prey, which, as soon as sucked dry, is once more seized between the Spider's jaws, carried straight up to the top, forced through the tube and dropped outside.

The male Trap-door Spider is, as might be supposed, more handsomely coloured—in deep purple—and when he arrives at maturity, and has cast his last suit of juvenile clothing, he escapes for ever from his short silken tube, and goes a-courting! For Spider, like we bipeds, has a soft heart, and is thereby led into trouble; for having found the nest of one of the "gentle sex," he runs up the outside of the aerial part and drums a tattoo and knocks for admittance; but the lady will not be disturbed, until the gay cavalier, wearied by his long waiting, bursts through all ceremony and silken nests and enters (page 1032), introduces himself, and they dwell together for three or four months, until the day arrives when they have had enough of each other, and the female, to put an end to all differences, kills her mate, eats what she can of him, finally picking up his remains, which are carried up to the top and thrown out!

A short time after the shocking tragedy, the Spider climbs half-way up the nest, and in the upper side forms a hollow, across which she spins a beautiful hammock of floss silk, filling it with about one hundred spherical eggs. These hatch, and the large family of spiderlings remain with their mother for some months, until April, when a small hole is made in the aerial part of the nest, and the young are all driven up close together into this part, until the hole is reached—whence they escape; the greater part to "balloon away" and start new colonies, while others remain near the maternal home.

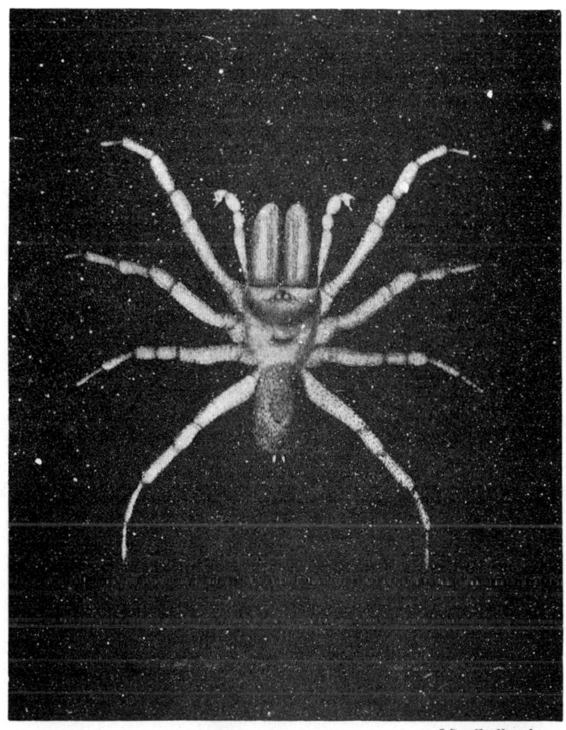

[By F. Enock.

BRITISH TRAP-DOOR SPIDER—MALE.

Three and a half times the natural size.

[By F. Enock.

BRITISH TRAP-DOOR SPIDER—FEMALE.

Three and a half times the natural size.

EROS

BY G. F. CHAMBERS, F.R.A.S.

THE discovery of the planet Eros by Witt, of the Urania Observatory, Berlin, on August 13th, 1898, caused a great sensation in astronomical circles, though, perhaps, it would be more correct to say that the sensation was due not to the mere fact of the discovery of the planet, but to the discovery of the wholly unforeseen orbit in which it moved. Minor planets have been so plentiful during recent years that all pleasure and romance attaching to them have long since ceased, and we must go back to 1801 and 1845 to find a flutter at all reminding one of the excitement caused by Eros. It is just worth while to recall what happened in 1801, and to do that properly we must go back another twenty years to the discovery of Uranus, by William Herschel, in 1781.

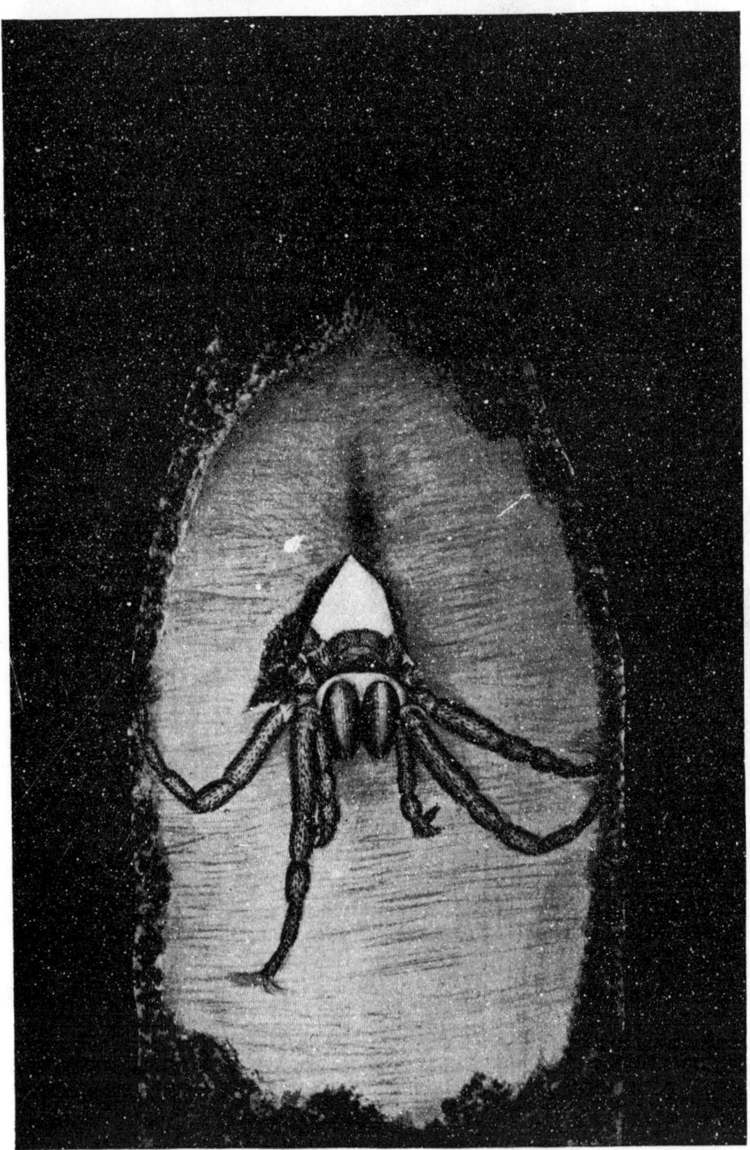

[By F. Enock.

SEEKING A WIFE.

The male Trap-door Spider unceremoniously breaks through the silk wall when he goes a-courting.

At about that period a German professor at Wittenberg noticed certain coincidences in the distances of the then known planets which seemed to indicate a very regular law of progression, but which left an unexplained blank between Mars and Jupiter in the series of planets revolving round the sun. This blank seemed to suggest that an undiscovered planet existed, and was worth looking for; and a little society was formed of practical observers to carry out a systematic search. Even before it got into full working order, a new planet was found on the first day of the new century—that is, on January 1st, 1801—by Piazzi at Palermo, which received the name of "Ceres." This was followed in 1802 by the discovery of "Pallas"; in 1804 by "Juno"; and in 1807 by "Vesta." No more unknown strangers seeming to be forthcoming, the search was discontinued. But it was resumed in 1830 by a German amateur named Hencke, at Driessen. No success, however, attended his labours until 1845, when he at last

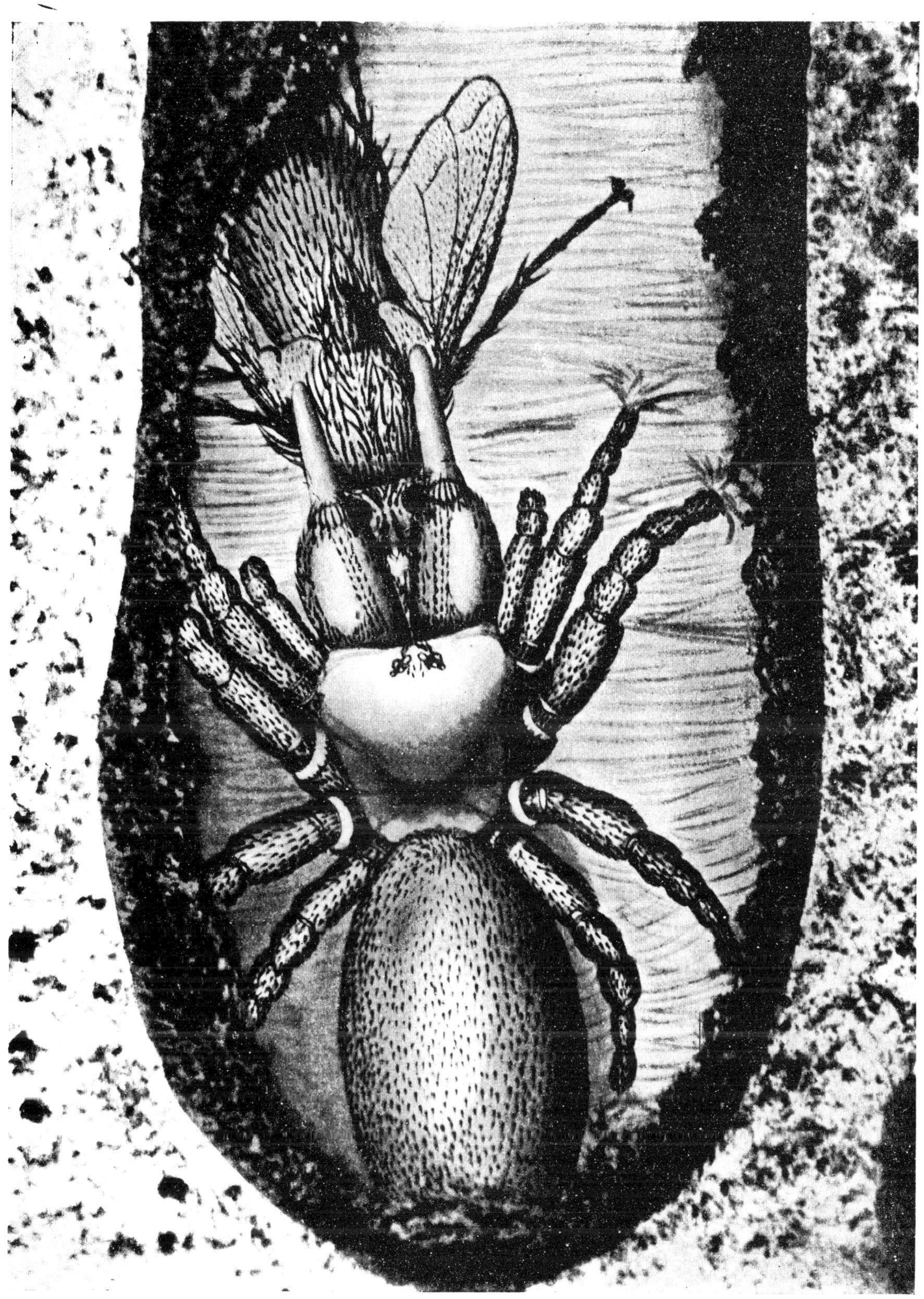

[*By F. Enock, F.L.S.*

THE SPIDER AND THE FLY.

Having caught her Bluebottle in the manner shown on page 1029, the Spider backs down to the bottom of her tube, which serves as her pantry, and there deposits her prey until she has time to consume it.

discovered a new planet, which received the name of "Astræa." This discovery clearly suggested that there were other small planets in the solar system which might be found if they were looked for, and Eros was one of the fruits of this expectation, though very long down the stream, both in date and succession.

It was very quickly seen after the discovery of Eros that there was something altogether unprecedented in its movements, for it was passing along the heavens at a most unusual pace; and so soon as it was possible to determine its orbit, astronomers discovered that they had in tow a minor planet the orbit of which was wholly unlike that of any yet known. Within a short time of its discovery it was retrograding in the heavens at the rate of $\frac{1}{2}°$ a day, which proved at once that its orbit must differ in a marked way from the rest of the family of minor planets, and was of a very sensational character, calculated quite to revolutionize some of our ideas of the solar system. We had always looked on Mars as our nearest outside neighbour, but the mean distance of the new planet from the sun being only 1.46 as compared with Mars' 1.52 it was evident that the new planet (which, by the way, for some time was only known as "D Q") would be under certain circumstances nearer to us than Mars, and would have, at any rate, some portion of its orbit within that of Mars. The novelty and sensationalism attaching to the matter were intensified by its being found that, as the eccentricity of the orbit was as great as 0.23, the perihelion distance would be only 1.12 (in radii of the earth's orbit), and the aphelion distance as much as 1.79; whilst the least distance from the earth's orbit is only 0.15, as compared with 0.27, the least distance of Venus when in transit, and with 0.38, that of Mars when in perihelion. The perihelion and aphelion distances of Mars being 1.38 and 1.67 respectively, it will be seen at once (especially by a study of the annexed diagram) how curiously the orbit of Eros compares with that of Mars. It follows from all this that Eros will thus on certain occasions be far closer to the earth than any other member of the solar system; and thus it becomes available as an excellent means for determining the sun's parallax, which means, in other words, the distance of the earth from the sun, for the parallax of Eros when nearest to us will be more than 1′ of arc. It has been estimated that these circumstances connected with Eros enable it to be utilized for determining the distance of the earth from the sun with an accuracy three times greater than that available by any previous method.

So soon as Eros became a recognized member of the solar system, the interesting question arose: Had it been seen before? This question was the more interesting because, owing to the great eccentricity of its orbit, it would vary very much in its visible brightness estimated in star magnitudes. Its magnitude when found in 1898 was no more than the 11th, whilst, in January, 1894, when it had been at its minimum possible distance from the earth, its Opposition brilliancy was estimated to have been as great as that of a star of the 7th magnitude. The strange thing was that it should have escaped notice under such circumstances. However, an examination of a large number of photographic plates taken in various years beginning with 1889, and at various observatories, showed that it had left its mark on a considerable number of plates. This has greatly facilitated the accurate determination of its orbit, and its period of revolution round the sun is now known to be 643 days at a mean distance of 135,442,000 miles. It has also, unfortunately, been ascertained that the planet will not be again in its most favourable position for observation until January, 1924.

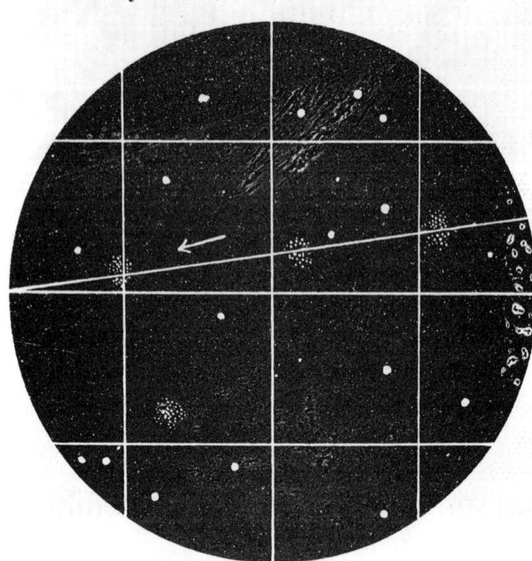

Diagram illustrating the method of distinguishing a moving star-like point in a field of fixed stars.

André, of the Lyons Observatory, has suggested that Eros is *double*, comprising two bodies revolving round each other, and that this would explain certain regular changes in its brilliancy which are otherwise inexplicable.

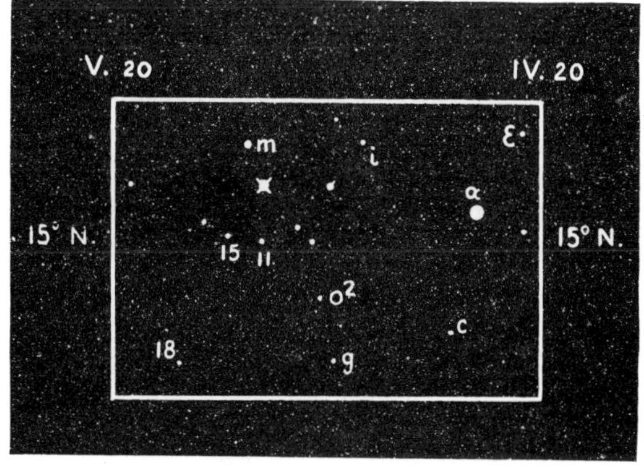

VESTA AT OPPOSITION, DEC. 8th, 1914.

The bright star is Aldebaran. The upper half of the diagram contains stars in Taurus; the lower half stars in Orion.

THE CERATOSAUR

THE extinct huge reptiles to which the general name of Dinosaurs, or Terrible Lizards, has been given, were a numerous family, and although the prevailing form is similar to that of the kangaroo among mammals, there is also considerable variety among them which has suggested to their discoverers distinctive names. The feature that at once distinguishes Ceratosaurus from other Dinosaurs is the possession of a stout horn behind and above the nostrils. This, though perhaps the most obvious, is not the only point in which it differs from its nearest relations. There are peculiarities of the bones forming the spinal column, and the hip bones are united much as we find in birds.

This Horned Dragon, as we may call him, was about twenty-two feet in length, judging from the almost complete skeleton which it was the good fortune of the late Professor Marsh to disinter from the Jurassic strata of North America. Its upper surface was protected in great part—and probably sufficiently—by a series of hard plates embedded in the skin, much as we find in the skin of the crocodile to-day, and extending from behind the head all down the neck and back. The skull has a large brain cavity, and the orbits of the eyes have thick, bony ridges over them, as shown in the illustration. The nostrils have separate openings far forward on the snout, and just behind them in the middle line rises the bony core of the great horn from which the Ceratosaur gets its name (*keras*, horn; *sauros*, lizard). The teeth, it will be seen, were tolerably formidable weapons, and indicate that the Horned Dragon was no vegetarian. The contrast in size between the hind- and fore-limbs was very pronounced, and it is probable that the fore-limbs were scarcely ever used for locomotion.

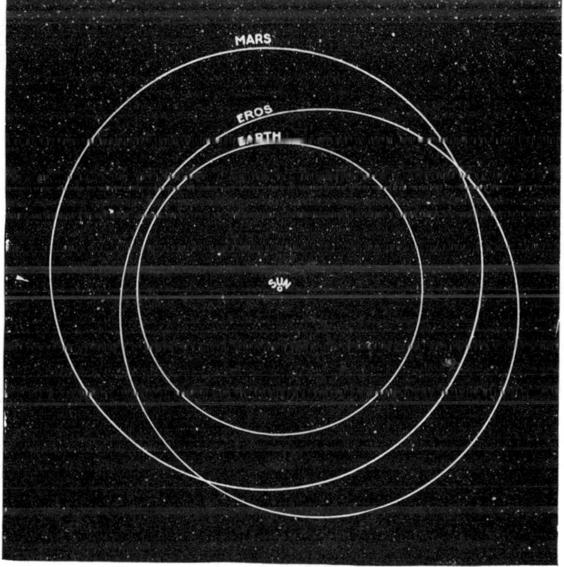

THE ORBIT OF EROS.

Showing that under certain circumstances Eros would be nearer to the earth than Mars.

THE RIVAL BARRAMUNDAS

BY W. P. PYCRAFT, F.Z.S

THE name Barramunda has been productive of not a little confusion to the student of fishes; for it is commonly applied to two totally unlike and unrelated species, both natives of Australia. As a rule, perhaps, this name denotes the strange Lung-fish (*Ceratodus*), of which a living specimen is to be seen in the gardens of the Zoological Society of London. But it is quite commonly used for the fish illustrated on page 1038, which is also known as the Dawson River Salmon. Inasmuch as there is a general superficial likeness between the two, the mistake can be understood. But the Dawson Salmon belongs to the typical "bony" fishes, and is known to the expert under the name *Osteoglossum leichardti*.

As a food-fish it is excellent, and it has the further advantage of attaining a considerable size. But it has a still further interest, and this because of the fact that another species of the genus also occurs in Australia, one in Sumatra and Borneo, and one in South America.

Since all are river-fish, this distribution is somewhat remarkable. Another near relation

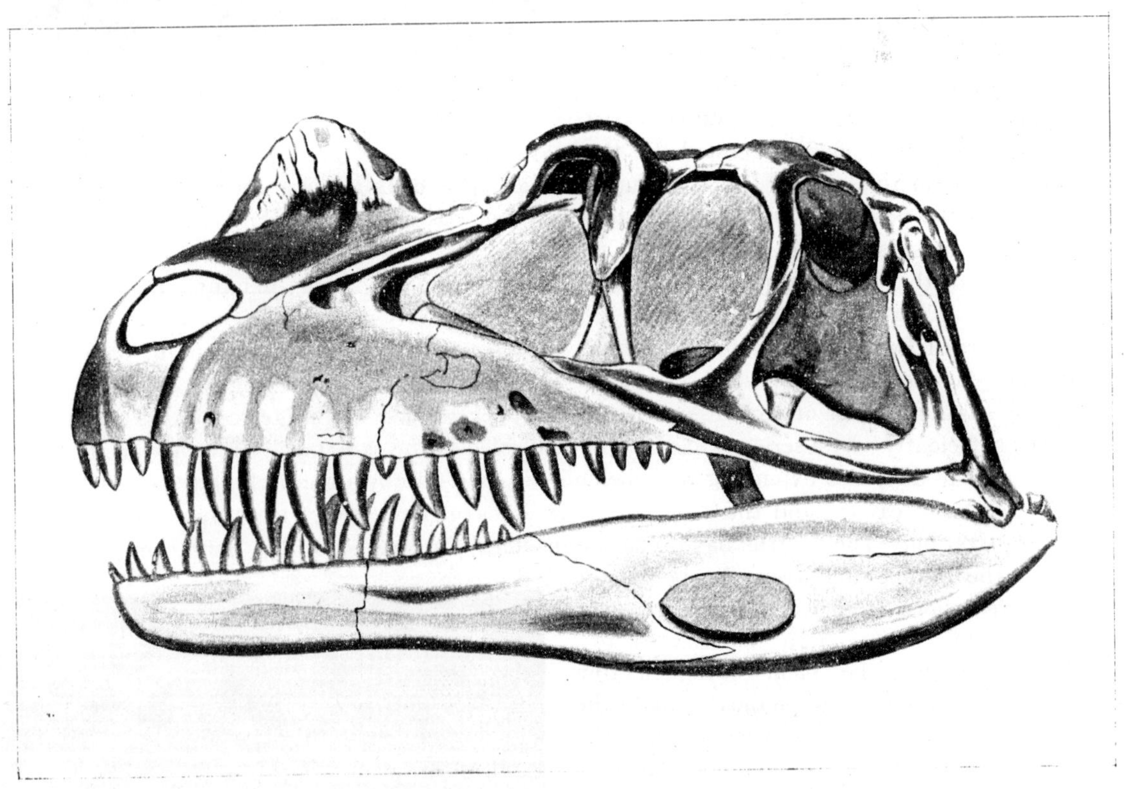

THE SKULL OF THE CERATOSAUR.

Note the thick bony ridges over the eye-sockets and the bony core of the horn standing in the middle line behind the nostrils.

of the Australian Osteoglossum is the African Heterotis, a fish remarkable for its curious nest-building abilities, and the fact that its air-bladder is cellular and probably used as a lung.

During the breeding season the Heterotis constructs a large nest in a swamp, about two feet under the surface of the water. The young are as singular as their parents, since, for a time, they breathe by long, slender, filament-like gills which project in tufts from each side of the head.

The Ceratodus, the true Barramunda, is, as we have remarked, one of the lung-fishes, with relations in Africa—the Protopterus, and in South America—the Lepidosiren. A glance, however, at the form of the fins and the shape of the tail would at once reveal the difference between the true Barramunda and the fish which has been erroneously so-called—that is, the one whose portrait we have inserted as illustration to this description.

By permission of] [*Messrs. Chapman & Hall, Limited.*

THE CERATOSAUR

A huge extinct reptile of the Jurassic period, with enormously developed hind-legs and feeble fore-legs. He was about twenty-two feet in length. An almost complete skeleton was discovered in North America by Professor Marsh.

Photo by] [*W. Saville Kent.*

THE QUEENSLAND BARRAMUNDA.

Also known as Dawson River Salmon. It measures about two feet in length.

THE MAPLE-LEAF CUTTER

BY EDWARD STEP, F.L.S.

FOR many years we have all been familiar with the so-called Jumping Bean from America, which has at intervals aroused sufficient public curiosity to make it a marketable commodity in the streets of London. All the time we have had in our own fields and gardens quite as remarkable a natural curiosity of very similar character ; but the phenomenon has been—and up to the time of writing remains—unknown to all but a few naturalists.

In the case of the " Jumping Bean," a seed with two flat sides and a convex side has been entered by a caterpillar, which feeds upon its substance and renders it useless as a seed. By some obscure action it is enabled to cause this seed to turn from one flat side to the other, this movement being the so-called jump. In the case under consideration, instead of a seed we have a circular disc cut out of the leaf of a Maple or Sycamore, and this disc moves about with short jerks in a startling fashion. The story of its fabrication we consider so interesting that we believe a brief recital of it will set numerous readers searching for the disc and its fabricator in their own neighbourhoods.

We have already in this work given accounts of certain Saw-flies, but they are a numerous family and have not all the same habits. The propulsive power in what we will call the Jerking Disc is the grub of a Saw-fly. As usual with Saw-flies, its mother cut a slit with her saw in the leaf of a Sycamore or a Field Maple and deposited an egg in it. The egg hatches and the minute maggot-like grub that issues from it begins to excavate the juicy flesh between the upper and lower cuticles of the leaf.

It does not share the devastating character of many of the Saw-fly tribe, who reduce many leaves to grim skeletons. In this case, at the end of his term of residence the Saw-fly grub has not consumed a fourth of the leaf-material, although he has succeeded in disfiguring the leaf with a yellow or brown blotch. The accompanying photographs of Sycamore leaves will make clear the extent of this disfigurement. Sometimes there is only one Saw-fly grub at work in a leaf, sometimes there are two or three, and occasionally four. The Sycamore leaf has five sharp-pointed lobes, and the grub always commences its operations at one of these points, and cuts away on all sides from that point, leaving an area behind him from which the green material has been completely cleared.

At last he appears to feel that he has eaten all that is necessary to enable him to work out the remainder of his destiny fasting. He now sets to work without compasses to describe a true circle in the upper cuticle of the leaf. With his jaws he makes a series of cuts, each one forming a segment of the circle, but separated from the next cut by a minute interval of unbroken skin. He finishes within the breadth of a pin at the point where he began, and the circle has taken nearly an hour to cut.

The circle is complete, but it remains attached to the rest of the leaf much as a postage-stamp is attached to its fellows in a sheet of stamps. One can be removed from the centre of the sheet by a little pressure. The perforations in the leaf are short gashes instead of the dots of the stamp-sheet.

Having all but separated his disc, our grub proceeds to weave a tissue of fine silk from the edges of the disc and beneath himself. This appears to us to be a remarkable piece of work, much more so than the weaving of the oval cocoon cleverly performed by the caterpillars of many moths. There the spinning is in all directions and the caterpillar can move as he pleases to effect it. In the beginning the caterpillar attaches threads to all available points to form an outer scaffold inside which he can fashion his egg-shaped cocoon with comparative ease. In the case of our grub, not one of his threads must extend beyond the cut edges of his disc, or his purpose would be defeated. The fabric to be woven is a flat tissue, and this part of the performance may be likened to a weaver lying in bed and weaving a complete bottom-sheet under his body. The grub is equal to the task ; and the next we know is that the disc has separated from the rest of the cuticle, and is now the upper side of a circular bag, the lower side being of closely-woven silk, and the grub lies snug between. How the separation from the leaf is effected is by no means clear. It has been suggested that it is due to the pull of the silk threads beneath ; but this can scarcely be so, as the silk floor sags slightly in the middle, which indicates that it is not taut, whilst the vegetable roof is flat. Our own theory is that the almost complete severance of the disc from the rest of the leaf cuts off so much of the supply of moisture that the disc contracts and so effects complete severance.

Photo by] [*H. Main, F.E.S.*

SYCAMORE LEAF.

Showing the discoloured patch where the Leaf-cutter is still at work. The long dark mark in the upper part of the light patch is the larva under the leaf cuticle.

Photos by] [*Hugh Main, B.Sc., F.E.S.*

THE MAPLE-LEAF CUTTER.

In these two photographs the work of the larva has progressed. In the first of these he has marked out the disc that is to be his roof during winter In the second photo he has made a silken floor to the disc and has separated it from the leaf.

You may be looking at one of these discs, still in position on the leaf, when suddenly it separates and drops to the ground. This is the act of the contained grub, but how accomplished has not yet been discovered. The disc falls to the ground, and if it has alighted upon an unsatisfactory spot it begins to set up those jerking movements already mentioned. Judging from the fact that the way to induce the movements is to expose the disc to strong sunlight, they appear to be a provision against the grub getting dried up. It seeks a shady position, where it can lie safely all through the autumn and winter. In spring it changes to a chrysalis in its cell, and, later, when the Maple and Sycamore are in leaf, a four-winged Saw-fly emerges and sets about providing for the continuance of the race.

Photo by] [*H. Main, F.E.S.*

THE SEPARATED DISC.

Both upper and lower sides of the disc are here shown, enlarged to six times the natural size. The first figure shows the upper side, the second the lower or silken side. The larva shows through.

SEA SQUIRTS

BY J. SINEL.

DOWN at the lowest limit to which the tide recedes, on every rocky shore, wherever blocks of stone are supported on others at some little distance from the ground, so as to form little grottos, or where portions of rock project as overhanging ledges, the observant visitor will not fail to have noticed, close packed on the under side of the roofs thus formed, masses of globular leathery bodies which, in size, colour and shape somewhat resemble good-sized cranberries and which, when touched, emit little jets of water.

These, in popular nomenclature are the Red Currant Sea Squirts (page 1042). If not actually the most abundant, they are at least the best known of that strange class of animals which forms the subject of this article. They are classed as Tunicata, because they are invested with a

Photo by] [*H. Main, F.E.S.*

THE MAPLE-LEAF CUTTER.

The upper and lower sides of the larva are here shown about six times the natural size.

leathery coat or tunic, and they are also known as Ascidians, because, in outward form, they bear some resemblance to the old-time skin wine-bottle, or *Askos.*

Two thousand two hundred years ago, in his great work, " History of Animals," the naturalist Aristotle wrote about the Ascidians, and naturalists have been writing about them almost ever since, but it is only in these latter days that their place in nature has been discovered. Until a few years ago they were thought to be somehow related to the molluscs, and were classed as " Mollus-coidea." But modern science has unravelled their life-history and through embryology has discovered their place in nature.

These lowly-looking animals that, to the ordinary observer, are just leathery bags filled with something soft, are now shown to hold a kind of cousinhood with the highest of nature's children, the vertebrates. When this fact was first discovered it was supposed that they represented an ancestral vertebrate form; but it is now held that such is not the case, but that they are just unsuccessful competitors in the evolutionary march. Their ancestors in the far past had started on the vertebrate road and had made good progress when they took a wrong turn. This much embryology, which is an epitome of the history of the race, clearly shows.

Immensely varied in shape, in size, and in habit as are these Sea Squirts, they are all built on one plan, which is as follows: In adult life they consist of a bounding membrane more or less strengthened with a leathery or even horny tunic. Within this is a branchial sac which acts as the gills do in a fish. At one side of this is a grooved, rod-like organ, termed the endostyle, which secretes an adhesive mucus and the function of which we shall see in a moment. At the upper part of the body is an opening termed the inhalent aperture, and at one side is another, termed the exhalent aperture. The linings of these are clothed with fine vibratile microscopic threads or cilia, by the

Photo by] [*J. Sinel.*

RED CURRANT SEA SQUIRTS.

A natural group upon the rock is here shown one and a half times the actual size. It is one of the commonest forms of these interesting animals.

waving of which the sea-water is drawn into the inhalent aperture to bathe the branchial sac and there yield its free oxygen; thus the Ascidian breathes. Whatever particles of organic matter the water contained are then secured by the mucus of the endostyle and passed on to the mouth, which is lower down in the body and on the opposite side. The food thus secured—chiefly minute floating plants such as diatoms, spores of seaweeds and tiny animal forms—is then passed into the stomach, the waste products passing through the intestine into a space between the branchial sac and the body wall, known as the atrial cavity; thence they are carried out, together with the water which has served for respiratory purposes, by the exhalent aperture. Close by the stomach are the reproductive glands, where eggs are formed, which, after fertilization, pass through a duct into the atrial cavity, there to hatch and develop.

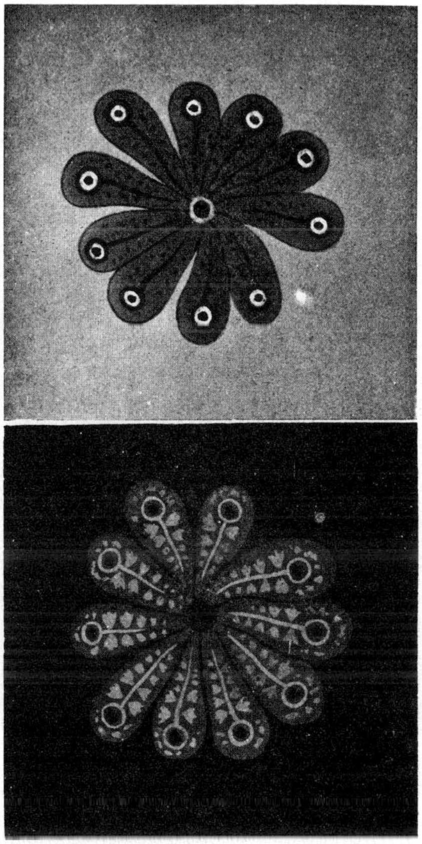

Near the bottom of the sac is the heart, an organ the function of which is very remarkable, and to which there is nothing analogous in any other living thing. Instead of sending the blood coursing constantly in one direction, as do the hearts of all other animals, the Ascidian's heart throws from twenty to forty consecutive currents in one direction, then stops for a few seconds and starts again, sending the same number of currents in the opposite direction, and so on constantly. In some of the small transparent forms, such as Clavellina (page 1044), this curious operation can be watched by means of an ordinary pocket-magnifier.

Thus much for the structure and functions of the adult, which a glance at the diagram on page 1047 will make clear and which will be seen to be simplicity itself. But now comes another phase in the life-history of these animals. The eggs which have passed into the atrial cavity develop into young. But these young are not like the parent. Instead of being simple bag-like organisms, each little Ascidian is to all intents and purposes a young vertebrate. They vary a little in outward form, according to species, but they are all on the same general plan, which is approximately that of a tadpole. It is in this larval or tadpole stage that the structural details which give high rank to these animals are seen.

[By J. Stuel.

COMPOUND SEA SQUIRTS.

The numerous species of Botryllus are always found associated in rosette-like "systems" immersed in a leathery layer on the rocks or seaweeds. There are a number of "systems" in each colony. Two species are shown above.

Referring to the diagram on page 1046, we see that in the tail there is a jointed rod or notochord, and a *spinal nerve*, foreshadowings of the backbone and spinal cord of the vertebrates. Also a rudimentary brain, an optic nerve and an eye.

Now these little tadpoles, after being launched into the outer world through the exhalent aperture of the parent, lead an independent, active, roving life, and seem full of promise. But this is only for a few days; lassitude seems to overtake them, and soon they settle down upon a rock, a stone or a seaweed. They alight upon their heads and by means of the three little suckers that project from this, they make fast to whatever object they have alighted on and settle down for life. Little by little the tail, the incipient backbone and spinal cord, the brain and the eye, are absorbed and disappear. Then, of every grace and every virtue shorn, the active little tadpole becomes a mere Ascidian or Sea Squirt.

There is, however, one exception to this rule; in a genus termed Appendicularia, which consists of one or two minute species, the tadpole form is retained through life. These are shown of the natural size in the upper right-hand corner of page 1045.

Now, whilst all the Ascidians have this method of reproduction, viz., the production of eggs from which free-swimming larvæ are developed, many species have a second method, viz., reproduction by budding, *e.g.*, in Clavellina. Hollow, root-like threads are thrown out from the base of one individual, and from these, buds arise and develop into new individuals—like young strawberry-plants on their "runners," until a colony is formed as shown below, or as those shown on page 1042.

In what are termed the Compound Ascidians, such as those shown in the coloured plate, this reproduction by budding results in the formation of groups so symmetrical that each group or colony can be taken as one organism, especially as the leathery tunic in these runs continuous from one to the other, so as to present an unbroken leathery surface in which the little individuals appear to be simply embedded (page 1043).

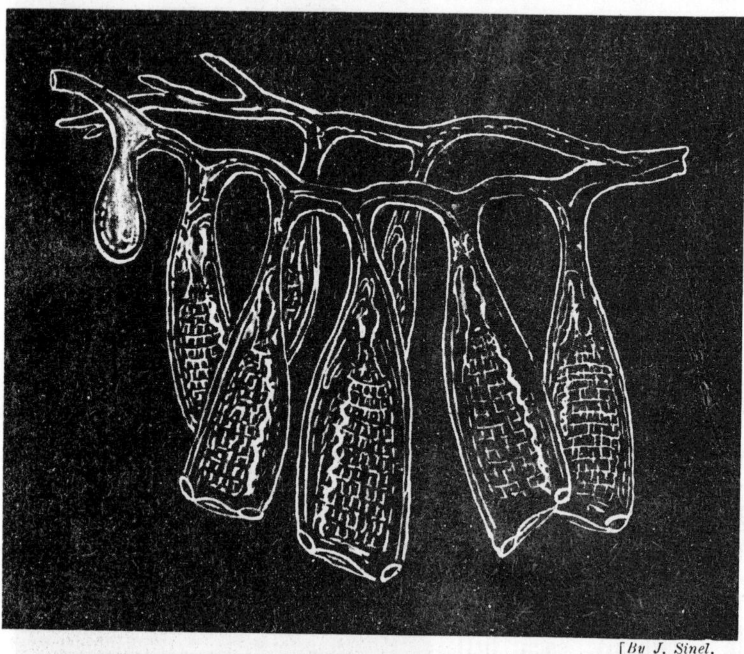

[*By J. Sinel.*

CLAVELLINA.

A group of this Sea Squirt, the members connected at the base. A developing bud is seen on the left. Three times the natural size.

In some species, such as Morchellium (shown in the coloured plate), they are set side by side so as to form pear-shaped masses, which are usually grouped side by side and suspended by their small ends from some overhanging portion of rock. In the beautiful Strawberry Ascidian (also in the plate), they are arranged in double rows in zigzag method, the little bright-red individuals, each with a tiny collar of white, forming a pretty design on a flattened strawberry-shaped mass which is affixed to the rock by a short stout stalk.

In some species the separate individuals are so small as to be invisible to the unaided eye, and they appear to be set without order in a leathery skin which encrusts either rock or seaweed. One remarkable species named Leptoclinum lacazii forms encrusting masses sometimes a foot square and of a brilliant crimson colour (see plate). This species is found only at the very lowest range of the tide in rock grottos or deep oozy recesses.

Those who have read Victor Hugo's "Toilers of the Sea" will perhaps remember that in describing the marine cave in the Channel Island in which one of his heroes had been eaten by an octopus, he says: "The walls were splashed with crimson stains, as if giants had been fighting there." It was Leptoclinum lacazii that the renowned novelist had seen, but he did not know what these red splashes were. In some of the thin, stone-encrusting ones, such as Didemnum, the leathery skin or communal tunic is closely packed with myriads of microscopic many-rayed stars of carbonate of lime or of silica, and thin slices of this, seen under the microscope, present a spectacle like the starry sky at its best.

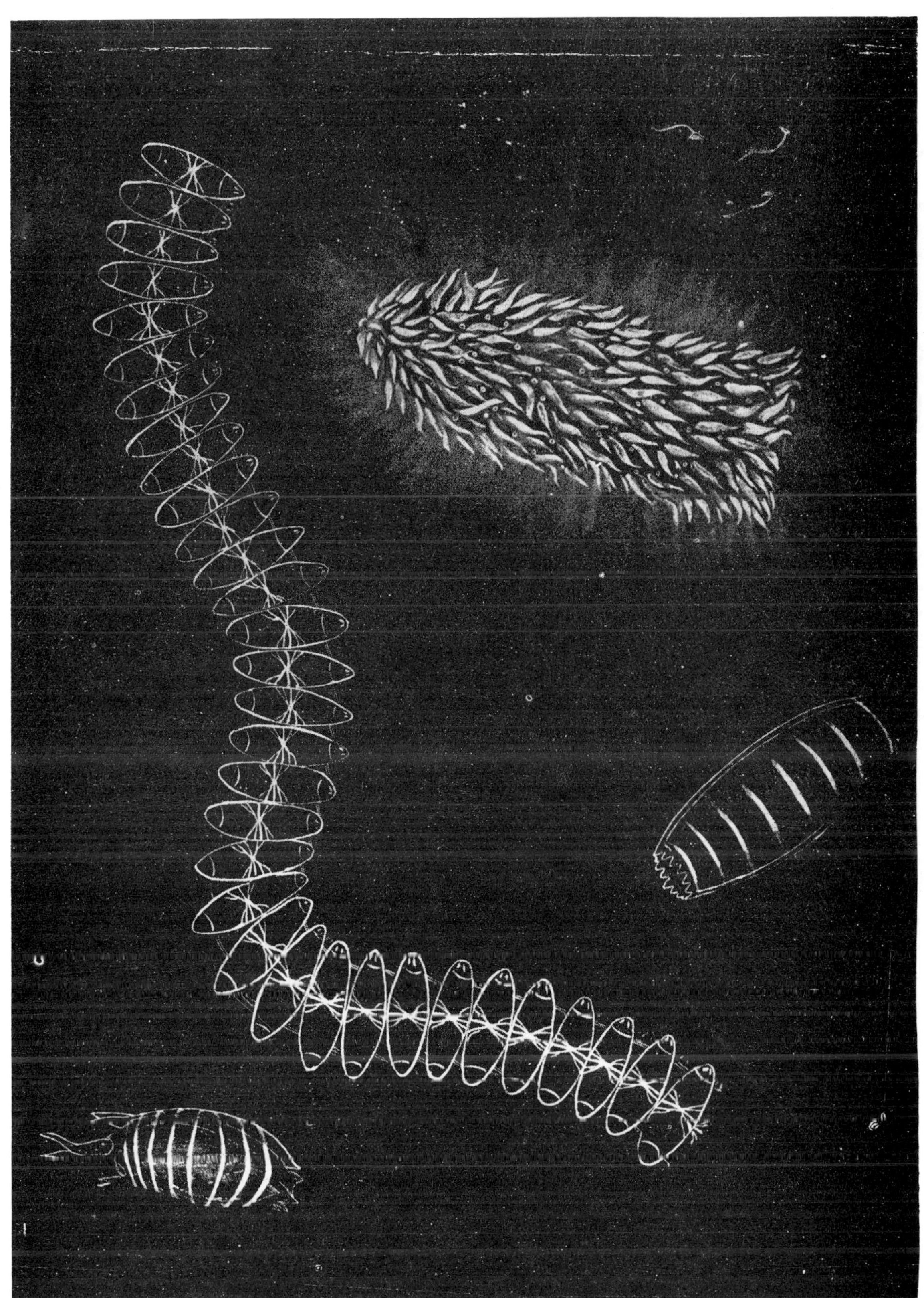

[By J. Sinel.

SWIMMING SEA SQUIRTS.

The figure in the lower left-hand corner is the Solitary Salps. Above it is the long chain of another species, a great number of individuals being connected. The barrel-shaped form to the right of it is Doliolum, represented life-size for British waters, but in tropical seas it is four or five times larger. The bullet-shaped form above it is Pyrosoma, in which again many individuals are united to form a hollow cylinder. It is remarkably luminous. In warmer seas it has been found measuring four feet in length.

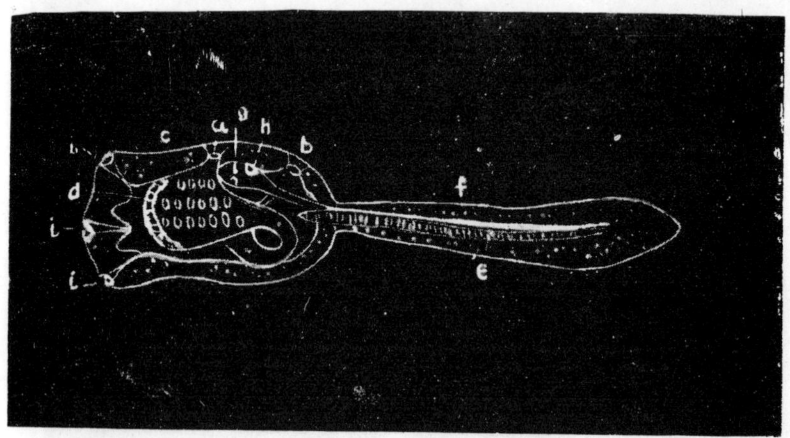

LARVA OF SEA SQUIRT.

A consideration of this diagram will make clear the peculiar structure of these animals. Currents of sea-water enter at *a*, and exit at *b*, after passing through the gill chamber *c*; *d* is the endostyle; *e* the notochord; *f* the spinal nerve. The position of the brain is indicated by *g*, the eye by *h*, and *i*, *i* are the suckers.

But best known of all the compound forms are those popularly termed Starry Botryllus. In this division of the Ascidian family, the individuals are not loosely scattered in the investing communal tunic, but are disposed in little sets with a star-like arrangement. Each little group or set thus arranged is termed a "system" (page 1043), and usually consists of from five to ten, but sometimes as many as twelve or fifteen individuals. They are always so arranged that their inhalent apertures are outward—that is, at the points of the rays—and their exhalent apertures are united into what is termed the "cloaca," in the centre of the star. Dozens, or even hundreds, of stars are sometimes grouped into one large colony, which, in favourable situations—as in sheltered rocky caverns—may form patches eight or ten inches square. All these Starry Botryllus are of beautiful appearance. The communal test or tunic—which has a texture very much like glossy patent leather—is in some species crimson, in others green, in others purple, and in one very common form yellow; whilst the star-like systems present an endless series of colorations. A few—but only a very few—of these are shown in their natural size and colours in the coloured plate.

Allied to Botryllus is a form termed Botrylloides. In this the individuals, instead of being arranged in star pattern, are set in double rows, and in these there is not much variety in coloration, their tints ranging simply from orange to brick-red. A group of these also is shown in the plate.

But whilst all the Ascidians so far mentioned are dwellers in the quiet shelter of rocky grottos and the still waters of shady rock pools, there are others whose home is the bounding wave. Many species are never attached to rock or weed, but live, either idly floating, or slowly travelling, in the open ocean. The best known of these are the Salps. Some of these live singly, such as the Solitary Salps (page 1045). Others are in a way analogous to the compound Ascidians, and are affixed together in long double chains. In the same group as the Salps is a peculiar barrel-like form termed Doliolum, which, in British waters, is usually found of about the size there shown, but in tropical seas it is sometimes eight or nine inches in length.

A wonderful form, which is in the "compound" series, is named Pyrosoma; many individuals are here united into a hollow cylindrical mass, pointed at one end and flat at the other—very like a modern cannon-shot.

The inhalent apertures of all the individuals are outwards as in Starry Botryllus—as may be seen in the plate—and the exhalent apertures open inwards into a central cavity which opens out to sea at the base, so that the water which is being constantly expelled, propels the Pyrosoma colony slowly along. This species is highly phosphorescent, in fact it is probably the most brightly luminous of all living things, hence its name Pyrosoma or Fire-body. The individual or "colony," as we may choose to term it, is shown in the plate of about the size it occurs in these waters, but it grows much larger in tropical seas, and there is record of one being netted which was four feet in length.

From this brief outline, it will be seen that the Sea Squirts form a vast and varied group; they range from arctic to tropical seas, and from sea-surface to almost abyssal depths. The brightly coloured ones present a perplexing problem to the student of colours and coloration in relation to environment.

In all higher forms, colours and their arrangement play a definite and important part, either as protective and mimetic coloration, when they tend to render the individual inconspicuous in its surroundings; warning colours, when the individual is either armed or is not good to eat; sexual coloration and recognition markings, when colours or their arrangement have some relation to sex during the courting season, and so forth. But in these Ascidians, none of these conditions seem to be implied—bright reds and bright greens vie in splendour side by side, as conspicuous as can be, and yet they are good to eat, and masses of those that affix themselves to weeds are frequent in stomachs of dog-fishes and other predacious things of the sea. Thus protective coloration and warning coloration do not apply. The Ascidians do not court, therefore sexual selection or recognition markings do not come in. Nor are their colours due to the food they eat, for the reds, the greens, the purples and the yellows all feed on the same material, viz., whatever comes to their net. This problem therefore offers an interesting field for investigation, for up to now, the reason for these striking features we do not know.

THE OAK-APPLE AND ITS MAKER

BY HAROLD BASTIN.

THE Oak-apple, or "King Charles' apple," as it is often called in commemoration of an oft-told incident in this monarch's life, is among the most familiar of all those natural objects which reappear with unfailing regularity as one season succeeds another. Yet comparatively few people know the true history of an Oak-apple, or how it is produced. The writer has heard it described as a kind of fruit; and again as a species of fungus; while some of the old naturalists invented far more fantastic theories to account for its origin. Even when it became known that an Oak-apple contains numerous white grubs which in course of time give rise to a brood of "flies," the mystery was not lessened, for nobody seems to have thought that the insects could be in any way responsible for the substance upon which they feed. Indeed, the grubs were supposed to occur spontaneously in the Oak-apple—exactly as maggots were at one time believed to be mysteriously generated in meat as

PYROSOMA.

A single individual detached from the colony on page 1045. The letters *a* to *d* have the same significance as in the previous diagram; *e* is the mouth, *f* the stomach, *g* intestine, *h* ovary, *i* oviduct, *j* atrial cavity, *k* luminous organs.

a concomitant of decay. The experimenter Redi exposed the latter fallacy by covering meat with gauze, thus proving that without flesh-flies to lay their eggs there could be no maggots. In like manner the careful investigations of naturalists more anxious to establish facts than to formulate theories have revealed the secrets of the Oak-apple ; and the sequel demonstrates in a striking manner the correctness of the adage which avers that truth is stranger than fiction.

The story of the Oak-apple begins in the dark days of winter when a small, wingless insect, not unlike an ant in appearance, creeps up the trunk of an oak-tree. This usually occurs during the months of December and January. Whence she comes we will inquire subsequently. For the moment we will limit our attention to her behaviour. She makes her way along the branches, thence to the small twigs, and begins to examine the fat, brown buds with the air of an expert, tapping them knowingly with her antennæ in much the same way that an ichneumon wasp "sounds" a promising caterpillar before implanting an egg in its plump carcase. The resemblance is more than passing, for the Oak-apple insect, as we may now call her, is actually preparing to deposit her eggs in the heart of a bud. Often she chooses a terminal bud for her purpose, but she selects auxiliary buds as well. She deliberately mounts the bud, cleverly inserts her long ovipositor, or egg-laying instrument, between the closely-wrapped scales, and plants her eggs right down among the tightly-packed baby leaves which lie snugly waiting for the call of springtime to rouse them from their slumber. The eggs are pear-shaped, translucent and excessively minute. Sometimes a great many are placed in a bud, sometimes only a few, but the average number appears to be about thirty ; thus, one is inclined to think that the insect, guided by instinct, lays in each case as many eggs as the bud is capable of supporting. If the insect does not find room for all her eggs in one bud, she goes to another, and so on until her task is completed. For to inoculate oak-buds is the sole object of her being, and when her store of eggs is exhausted, she dies.

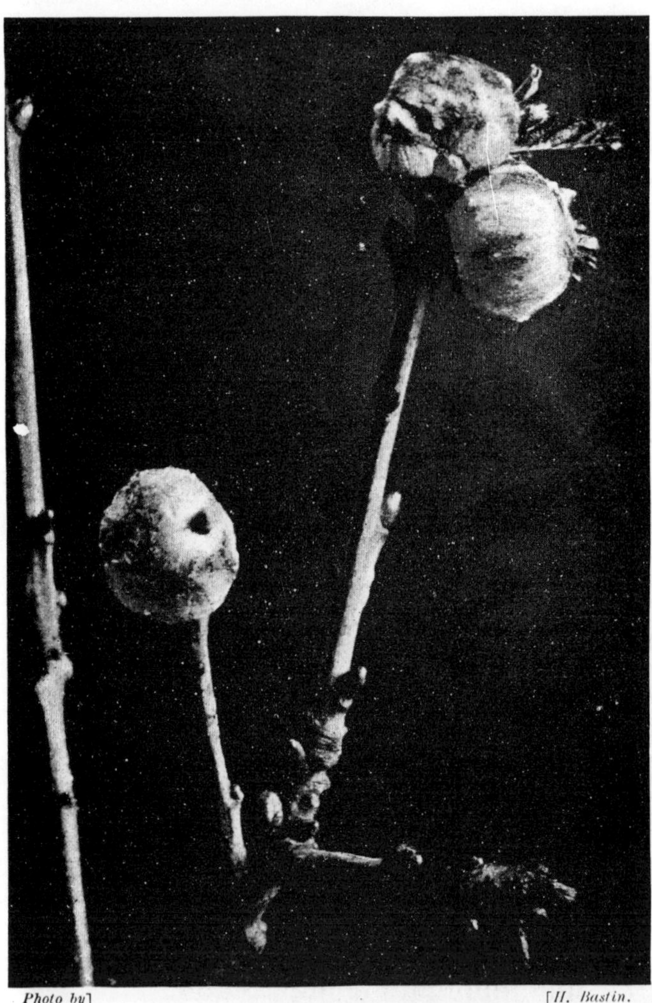

Photo by] [*H. Bastin.*

OAK-APPLES

Three newly-formed Oak-apples are here shown. It will be noted that they are all at the extremity of twigs.

The next act in this wonderful drama is not easy to interpret. All the oak-buds that contain eggs develop into Oak-apples when the tree begins to put forth its leaves in May. How are we to account for this ? Many theories have been suggested, including the plausible one that the insect injects a minute drop of poison into the bud, and that this stimulates the vegetable cells, so that they undergo rapid multiplication, and produce this abnormal outgrowth which is known to naturalists

Photos by] [*Harold Bastin.*

OAK-APPLES.

In the upper picture the wingless "winter female" is crawling up a twig in order to reach the terminal bud, in which her eggs will be deposited (see page 1051). The lower photo shows a fully-developed "apple."

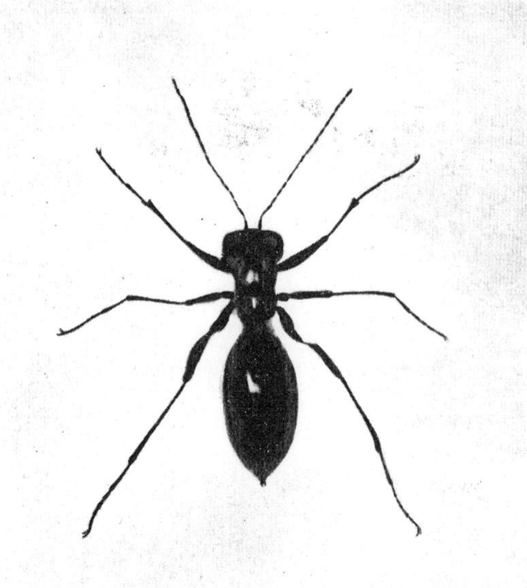

Photo by] [*H. Bastin.*

FEMALE GALL-WASP

This is the "winter female" without wings, which emerges from the root-galls, and by her egg-laying causes the production of the Oak-apple.

as a gall. But scientific observers have found that although the gall-insect actually injects fluid into the wounds which she makes in the bud, she does this either to provide a lubricant for the working of her ovipositor, or as a varnish to seal up the injury. The fluid, in fact, has been proved to be innocuous. We are left to assume, therefore, that the presence of the eggs, and subsequently the gnawing of the tiny grubs, are the factors which determine the growth of the gall. It must be remembered that the eggs are invariably laid in actual contact with what botanists call cambium or meristem tissue—*i.e.*, the layer of delicate cells which, by rapid multiplication, bring about the phenomenon which we call " the growth of the plant." These cells are known to be highly sensitive, and it is not, therefore, unreasonable to suppose that their behaviour is influenced by the proximity of foreign bodies—viz., the eggs and young grubs. The gall, in fact, is due to an irregular and excessive growth of the plant tissue, and we may regard it as a kind of non-malignant tumour. The harm done to the oak is not serious, being limited to the destruction of one bud; but the gain to the insect is very great indeed, for the Oak-apple is at once the nursery and the feeding-ground of the little family of grubs. Each grub lies in a chamber surrounded by the spongy tissue of the gall, bathed in the sap upon which it feeds. It grows rapidly, changes to a pupa, and in June becomes a perfect " fly," which bites its way to the surface of the gall, and so escapes into the air and sunlight.

Photo by] [*H. Bastin.*

FEMALE GALL-WASP.

The winged female that emerges from the Oak-apple, and lays her eggs in the roots of the Oak, thus causing the production of root-galls. See page 1052.

Strange to say, the insects which emerge from the Oak-apple are totally unlike their parent—the ant-like creature which laid the eggs in the oak bud. Apart from minor differences, they have wings, while their ranks are made up of males and females, whereas the wingless winter generation consists entirely of females which are endowed with the mysterious power of parthenogenesis, or virgin reproduction. In a word, the life-cycle of the Oak-apple insect exhibits that remarkable phenomenon which naturalists call " alternation of generation "—a phenomenon that has nullified the old-time theory that like invariably produces like. Moreover, the habits of the winged insects which come from the Oak-apple are quite different from those of their progenitors. After pairing, the females resort, not to foliage buds, but to the roots of the oak. They work their way beneath the soil, and oviposit usually in the more tender rootlets. As a result, another kind of gall is formed—this time an outgrowth from the root. At first it is soft, while it may be pink or reddish in colour; but it soon becomes hard and nut-like. These root-galls are typically unicellular—*i.e.*, each contains only one larval chamber; but they are often crowded closely together in masses which look like one large gall. As in the case of the Oak-apple, the grubs change to pupæ in the galls, and during the winter and early spring the perfect insects emerge—these, of course, being the wingless females with which our life-history commenced. To recapitulate: Oak-apples are abnormal productions of the oak-tree, due to the activities of a small insect which lays its eggs in the buds. They give rise to a summer brood of "flies," comprising males and females, which differ in many ways from the parent form. The females lay their eggs in the root of the oak, and thus induce the formation of another kind of gall, from which in due season the winter generation of the species emerges—this generation consisting of wingless females, capable of virgin reproduction, which complete the cycle by laying their eggs in oak-buds.

Thus this wonderful alternation of generation—this interchange of personality, as it were—goes on year after year; and in this way we get our Oak-apples. These facts were not known to the old naturalists, who gave the insect which emerged from each kind of gall a different name. Thus, while the generation which is bred in the root-galls was appropriately dubbed *aptera* (*i.e.*, without wings), their offspring were termed *terminalis*—because the Oak-apple commonly arises from a terminal bud. In order to avoid confusion, these names are still retained by writers; they do not, however, stand for distinct species, but for two recurrent forms of one and the same insect.

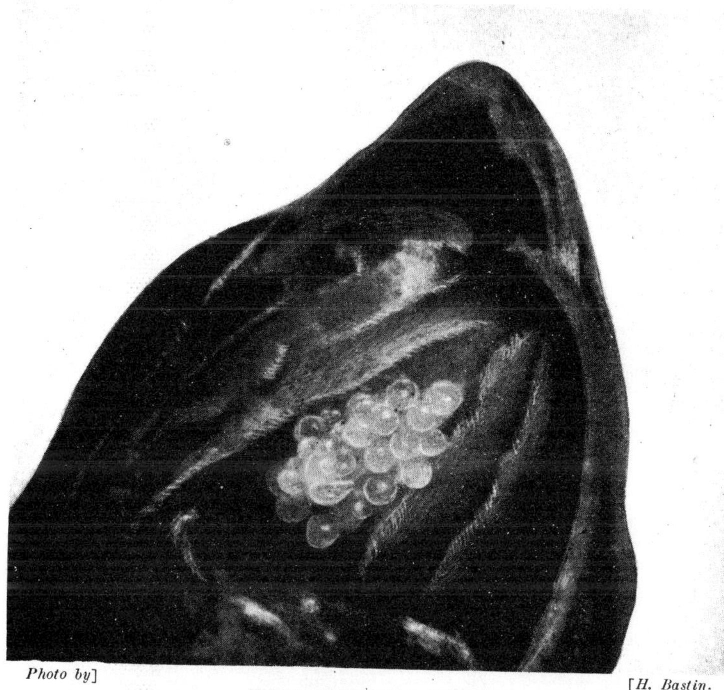

Photo by] [*H. Bastin.*

EGGS OF GALL-WASP.

A section through a terminal bud of an oak-twig, showing the eggs laid by the "winter female." Their presence so acts upon the bud that instead of developing into a shoot it swells into the familiar Oak-apple.

In the foregoing paragraphs the writer has employed the word "fly" to describe these wonderful gall-insects because the term is in common use. We must remember, however, that it is really a misnomer. Strictly speaking, a fly—*e.g.*, the blue-bottle or the daddy-longlegs—has only *one* pair of wings; whereas these gall-insects have *two* pairs of wings, like wasps and bees, to which, in fact, they are not distantly related. Thus they are more fittingly referred to as *gall-wasps*.

A very large number of insects live at the expense of the Oak-apple, or of the gall-wasps to which it rightly belongs. Some of these are parasites which devour the larvæ or pupæ. Others are so-called inquilines, or guest insects. The latter are nearly related to the true gall-wasps, but instead of forming galls on their own account the parent insects insert their eggs into existing galls. Thus, the mere fact that an insect emerges from an Oak-apple is no proof that its parent was the originator of the gall. Oak-apples are also burrowed into and eaten by the grubs of beetles, the caterpillars of moths, and the young and adult forms of numerous other insects. A mass of Oak-apples kept under observation produced no less than seventy-five different species of insects, the total number of specimens obtained being upwards of fifty-five thousand. We perceive, therefore, that an Oak-apple is not merely a beautiful natural object, but a populous centre of insect activity.

HOW ROCKS ARE BUILT

NATURE has fashioned the rocks which form the crust of the earth in a great variety of ways, and has drawn upon a great number of materials for their composition. The old geologists considered that the first crust which formed on the surface of our molten earth was granite, but this has been

Photo by] [*Harold Bastin.*

ROOT-GALLS.

These underground galls do not appear to have connection with Oak-apples, but it is from these that the "winter females" emerge whose egg-laying produces the "apples" on the oak-twigs.

proved a mere speculation, and the study of petrology brings it home to us that the crust has been built up in many stages by the aid of fire, water and wind. Chemical forces have also been at work, since most of the elements present in rocks are combined with one another to form comparatively few common minerals, which may be recognized without artificial aid when large enough. The microscope can, however, often tell more than a complete chemical analysis, because it shows us the internal method of arrangement of the minerals present, and this is of the utmost service in estimating the value of the rock, and also in forming a fairly accurate surmise of the life-history of the formation.

The fascinating study of the microscopic structure of crystals as indicating the origin of minerals and rocks was first put on a practical basis in this country by Dr. Sorby, of Sheffield, fully fifty years ago, and he employed thin sections which had been polished down until they were transparent. Although the portion under the microscope is very small, yet such investigation affords most valuable help in understanding how the rock itself came into being. It can indicate the order in which the

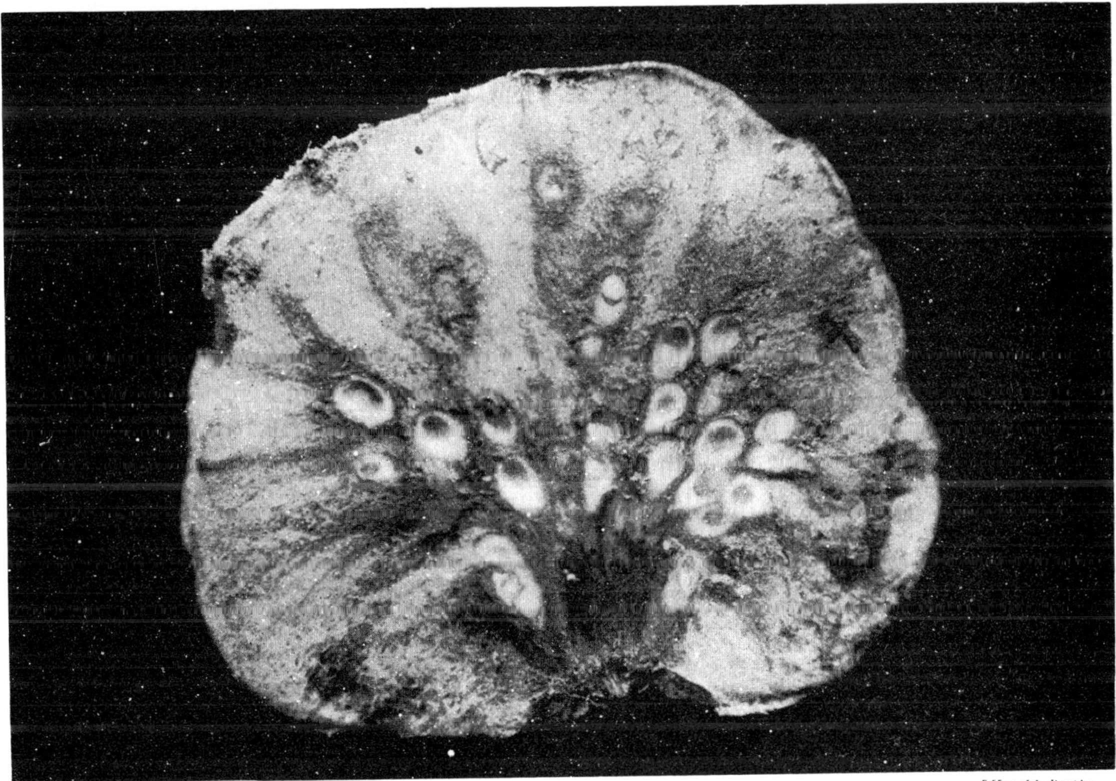

Photo by] [*Harold Bastin.*

OAK-APPLES.

The upper photo shows a little heap of Oak-apples from which the perfect insects have escaped, the perforations marking the points of their exit. The lower photo is an enlarged representation of an "apple" that has been cut through to show the numerous cells in which as many Gall-wasp grubs are feeding.

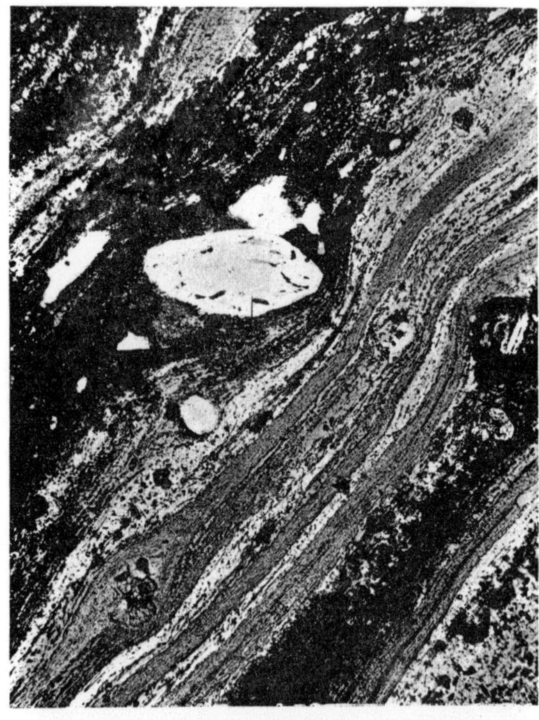

RHYOLITE.

Showing fluidal structure of the differently-coloured glassy bands. Magnified 15 times.

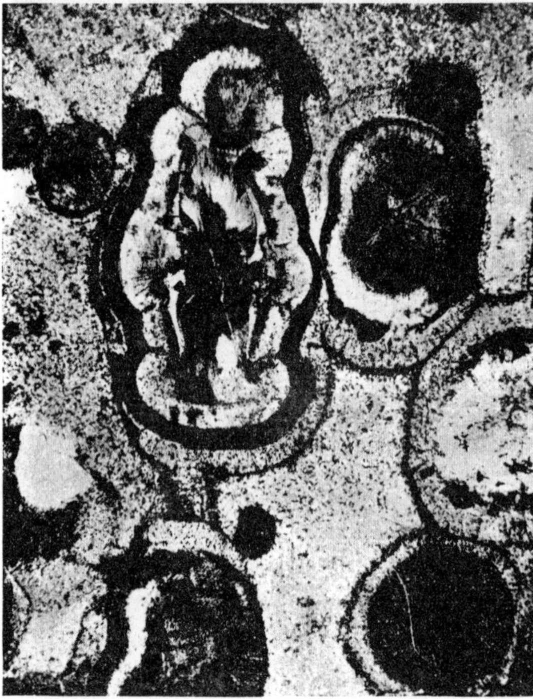

Photos by] [*P. Primrose.*

SPHERULITIC RHYOLITE.

Showing the radial structure of the glassy matter. Magnified 15 times.

crystals grew out of the originally fused mass or magma; from the size of the individual crystals we may judge the rate of cooling of the mass, and even form an idea of the depth at which it has solidified; and subsequent chemical and physical changes leave their unmistakable mark upon the crystalline arrangement.

It is difficult to classify rocks precisely, but two broad divisions commonly accepted are: (1) Igneous rocks, in the formation of which fire has had an important part; and (2) Sedimentary rocks, which are made up of broken materials and organisms deposited by water and wind. The first set are often subdivided into groups according to their chemical composition—acid or basic; their physical character—vitreous or crystalline; their mode of occurrence, as deep-seated (plutonic) or surface (volcanic) rocks, and their mineral constitution and order of sequence in time may also have to be taken into consideration. The second set are often called clastic or fragmental rocks, and they are grouped according as their formation has been brought about by mechanical, chemical or organic means.

The study of the Igneous rocks is perhaps the most interesting, and they can be likened to slags, since they are chiefly silicates which have solidified from the molten condition with varying proportions of silica (the acid portion) and oxides of various metals (the basic portion), and the ratio of these two determines whether the rock is said to be of an acid or a basic nature, or it may be intermediate, when it is called neutral.

Granite is the most common of the acid rocks, and its structure indicates that it has been cooled from fusion very slowly and under great pressure, so that the whole mass is composed of crystals, of which the first to form is felspar, then mica, and lastly the quartz or excess silica which fills in the interspaces left by the others. Sometimes the cooling has been so slow, extending to several thousand years, that some of the crystals attain a length of several inches; as in Dartmoor granite. They can then be examined without the aid of the microscope, but occasionally the quartz has crystallized simultaneously with the felspar in a very fine-grained, graphic intergrowth to which the name of Granophyre is given. Such granite is found in the Island of Skye, and the fracture has facets which resemble characters in old Hebrew writing.

When a portion of the molten magma, from which the granite is eventually formed, manages to escape by a vent and cause an eruption of acid lava, the composition of the rocks thus formed may be identical with the granite, and yet their structures are completely different. In the volcanic neck itself, the lava will take the longest time to solidify, but although cooled slowly, it is not now under the great pressure under which granite is formed, and the structure is called micro-crystalline or porphyritic. This is a ground-mass of such small crystals that only the higher powers of the microscope reveal it as different from a glass, and in this there is only a small proportion of the felspar and quartz separated as complete crystals. This quartz and felspar Porphyry occurs as long veins or dykes branching out from the Cheviot Hills, and similar dykes are called Elvans by the Cornish miners.

If the acid volcanic lava has been poured out in sufficient quantity on the older surrounding rocks, it will solidify less gradually than in the vent, and before the flow has been completely stopped the first-formed crystals render the mass pasty. The still fluid portions drag past these crystals until they become solidified as a glass with a characteristic fluidal structure. This formation is called Rhyolite, and is well marked in the lava flows of Iceland, which exhibit the differently-coloured glassy bands flowing round the numerous minute crystals of felspar and quartz, and carrying along in themselves crystallites of other minerals.

As the speed of the lava flow becomes less and less, this flow structure gradually disappears, and the crystallization begins at a nucleus, which may be a small particle of magnetite, and extends radially in all directions. This forms a number of Spherulites, which are simply segregations into spots of partially-devitrified glass possessing a different colour from the surrounding matrix. This structure, known as Spherulitic Rhyolite, is so highly developed in some of the Jersey rocks that the

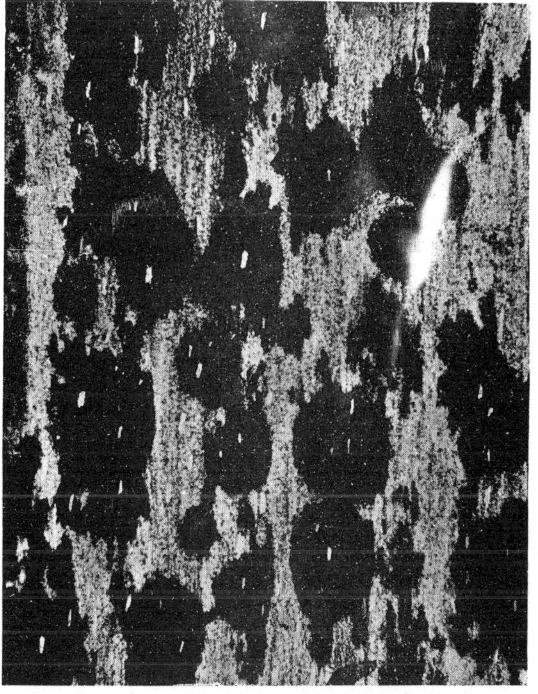

OBSIDIAN.

From Iceland, showing microlites of felspar (white), and elongated spherules of brown glass. Magnified 20 times.

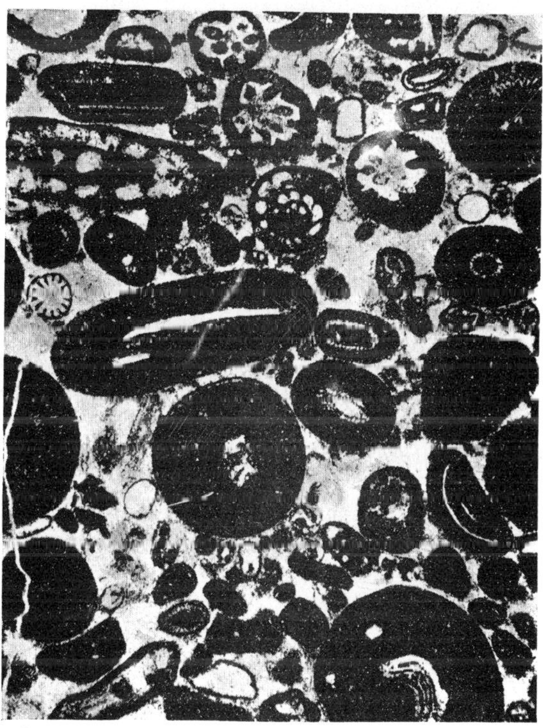

Photos by] [*P. Primrose.*

OOLITIC LIMESTONE.

From Shropshire, showing shells and rounded grains of calcite. Also called Roe-stone. Magnified 40 times.

deep yellowish-brown spherulites are closely packed and some of them have been compressed into ovoid shapes.

Where the lava in its flow traverses any colder solid material, it is subjected to a fairly rapid cooling which leaves it almost completely void of crystalline structure. In this vitreous condition it is called Obsidian, and even when black it possesses quite a glassy appearance. It frequently contains small incipient crystals or microlites of felspar arranged in the direction of the lava flow and these may serve as nuclei around which spherulites of dark reddish-brown glass have formed. This structure is shown in a remarkable degree by some of the Obsidian found in Iceland, and when this rock is rendered frothy by steam it is called pumice.

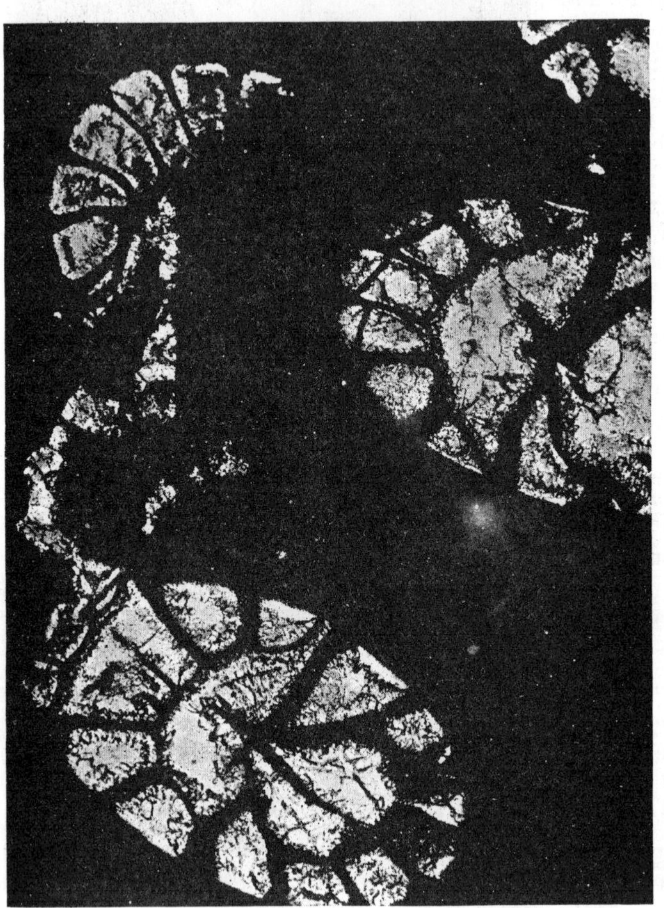

Photo by] [P. Primrose.

CORAL LIMESTONE.

From St. Monans, Perthshire, showing sections of coral skeletons, and the septa of the polyps filled with crystalline calcite. Magnified 45 times.

Should the lava encounter water, it is not only very rapidly cooled, but it is slightly altered in structure so that it loses its vitreous lustre and becomes wax-like, when it is called Pitchstone. This peculiar rock is found in greatest abundance along the west coast of Scotland, and the variety from Corriegills on the Island of Arran is world-famous because of the very beautiful arrangement of the embryo crystals throughout the ground-mass of greenish-coloured glass. These microlites of hornblende assume the shape of plumes and fronds of ferns, so delicate in tracery that it requires a high magnification to reveal their full beauty. The separation of these microlites indicates that the rock containing them has the glass slightly devitrified, and it therefore occurs at a slight distance from the full chilling effect of the water.

The Sedimentary rocks are generally of more use for industrial purposes than the Igneous rocks, but their structure is much less regular. Thus the sandstones are more or less compacted aggregates of sand grains, and the slates have been formed from clay particles. Of greater interest microscopically are the limestones, which may be of chemical or organic origin. In the latter case the rocks are composed of innumerable shells and skeletons of minute water animals, usually marine, and of a very low order in the scale of life. Although little more than shapeless masses of jelly, they have the power of secreting calcium salts from the water in which they live gregariously, and of converting these into carbonate of lime to form their shells, etc., which in time accumulate as masses of limestone. In some of our limestone rocks and their soft equivalent, the chalk beds, shells of mollusca are found which are large enough to handle, but the more delicate protozoa called Foraminifera require the microscope to identify them. These shells constitute the Globigerinan ooze, and the animals themselves are considered to be the earliest form of life which existed on the

Photo by] [*P. Primrose.*

SECTION OF PITCHSTONE.

This section shows the glassy matrix of quickly cooled lava, and fern-like embryo crystals of hornblende beginning to devitrify the greenish glass. From Corriegills, in Arran. Magnified 200 times.

Photo by] [*E. Step, F.L.S.*

LAND CRAB.

This Crab, a native of the West Indies, spends the greater part of the year on land, visiting the sea only during spring for the purpose of laying its eggs

earth. They have numerous representatives alive in the waters of the Mediterranean at the present time.

Another type of limestone is called Oolitic because of the occurrence in its structure of numerous round and egg-shaped particles resembling cod-fish roe, which give it the alternative name of Roe-stone.

The concentric structure of these small pin-heads of crystalline calcite is often arranged round a grain of sand or a shell fragment, but it is likely to be of organic origin like the pea-grits of the lower Oolites. The remains of coral polyps sometimes form the basis of limestone rocks, and although consolidated, they still show in many cases the shape of the skeleton built up of carbonate of lime, and the septa is usually filled in with crystalline calcite.

Only a few of the most interesting examples of rock-structure under the microscope have been here touched upon, but even these show how marvellous are some of the ways in which Nature has built, in her wonderful laboratory, the rocks which we so frequently regard with only the very slightest interest.

LAND CRABS

To anyone who has observed the habits of the common Shore Crab which spends most of its adult life above tide-mark on our shores, there should be nothing positively surprising in the statement that there are crabs which live habitually out of the water, though they may not have considered what serious modification of structure is implied in this departure from the usual habits of the race. There are several species, however, that have so far marked their renunciation of a marine life that they have ceased to develop the gills that are proper to that habitat, and have converted the gill chamber into a kind of lung, enabling them to breathe air instead of water.

Photo by] [*E. Step, F.L.S.*

LAND CRAB.

A front view of the same species at rest. These Crabs are only active at night, retiring into holes for the day, or burying themselves in the soil. They are very destructive to plantations.

The best-known of these is the Land Crab of the West Indies, which visits the sea only once a year—in the spring—for the purpose of laying its eggs. They march down in myriads, allowing no obstacles to impede their direct course. The rest of the year they spend inland, even ascending the mountains. They burrow in the ground and remain in their holes during the day; but by night come out to feed, when they inflict great damage on sugar plantations and other cultivated ground. The noise made by the rattling of their shells against stones and each other is such that, at least on one occasion, it has caused alarm to a portion of the English army.

The curious may read about this affair in Whistler's "Account of Admiral Penn's Expedition to the West Indies, 1654–5," and further particulars concerning the habits of the Land Crab may be gathered from Patrick Browne's "History of Jamaica."

Seeing that they are crabs, it is necessary that, like all other crabs, they should purify their blood by means of gills. These are contained in chambers on each side under the upper crust, and in the marine species a constant flow of water sets in under the edge of this crust, caused by the incessant movement of the gills. The oxygen is extracted from this current, which then passes out near the jaws. The Land Crab has been so far modified in structure that the gills extract oxygen from the air, but they must still be kept moist to ensure their efficient working. To secure the

Photo by] [*E. Step, F.L.S.*

PAINTED GRAPSUS.

This Crab prefers to climb perpendicular cliffs to running along the shore. It lives in cliff holes. At St. Paul's Rocks, in the Atlantic, Darwin observed it raiding the nests of sea-birds.

necessary moisture the gill chamber is lined with a spongy membrane which holds a supply of water.

Another Land Crab is the notorious Robber or Coco-nut Crab, described in early pages of this work.

There are other species that are, so to speak, Land Crabs in the making—crabs that, though they live almost entirely above high-water mark, do not venture far from the sea. Such is the Beckoning Crab, of the mangrove swamps; such is the Racer Crab, which cannot live in water for a single day and collapses after an immersion lasting only for a few hours. Then there is the beautifully-marked Grapsus of the Azores, which prefers to climb and run about the perpendicular sides of the cliffs, and lives in holes and clefts. Darwin has described how this species, at St. Paul's Rocks in the Atlantic, would raid the nests of the terns, stealing the fish which the parent birds had brought for feeding their young. They have even been accused of stealing the young birds. Prof. Moseley, during the voyage of H.M.S. *Challenger*, made many observations on these crabs, at several places. He says they are more wide-awake than the birds and keep well out of reach, so that they are difficult to capture—which the birds are not.

HUMMING-BIRDS

BY SIR HARRY JOHNSTON, G.C.M.G.

THE Humming-birds are certainly marvels of avian development, in some respects the most elaborately evolved of all birds. Their mode of life has made smallness an ideal to be aimed at rather than size, as they have devoted themselves more especially to the search for tiny insects in the cups of flowers as well as, no doubt, the sweet liquids which flowers develop in their innermost recesses for the attraction of insects. Consequently, it is in this group that we find the smallest known bird (*Mellisuga minima*),* only two inches and a half long, which is peculiar

By permission of] [*Sir H. H. Johnston, G.C.M.G.*

HUMMING-BIRDS.

The largest and the smallest of the Humming-birds are shown in this photograph.

to the large West Indian island of Hispaniola (Haiti-Santo Domingo), and Jamaica. But not all Humming-birds are so minute. The largest of them, which comes from the Andes of South America, from Ecuador to Chili, is eight and a half inches long, has large, strong feet, and flaps its wings instead of hovering about the flowers with a whirring flight like almost all other Humming-birds, a form of flight peculiar to the group. The largest of the Humming-birds, however, is the plainest coloured, being only greenish-grey brown and a dull white. Humming-birds specially noteworthy for gorgeousness are members of the genus *Eulampis* in the Southern West Indies and South America, one of which is golden-green, with dazzling blue chest and lower back. *Eulampis jugularis*, from the Windward Islands, has a crimson throat and *green* wings, and this peculiarity (found in one other genus of Humming-birds) is an exception to the general rule which provides that, however gorgeous may be the bodies of these gem-like birds, the wings are a sepia-brown or

* It is possible, however, that there is an even smaller bird, whose total length is under two and a half inches, in either Peru or Cuba.

By permission of] [*Sir H. H. Johnston, G.C.M.G.*

HUMMING-BIRDS.

A group of the Fan-tailed Humming-bird and its nest. A native of China.

By permission of] [*Sir H. H. Johnston, G.C.M.G.*

HUMMING-BIRDS.

Humming-birds with long tails and gorgets, of the genus Sparganum.

a dusky black. But in a large species of *Pterophanes* (which is remarkable for its musky scent and inhabits the Central Andes Mountains) the whole of the wing is a vivid blue, except for the tip, which is black, the rest of the body being dark green. The Rainbow Humming-bird has a golden-green forehead and orange-scarlet crown, with a violet-blue stripe in the centre, intensely black feathers in the nape of the neck, a vivid spot of lilac in the centre of the throat, and the rest of its plumage chestnut and green, with dusky-coloured wings. The Sappho Comet Humming-birds, ranging from Peru to Argentina, are bronzy-green, with crimson and fiery orange. Almost every tint of gems or of flowers is reproduced in Humming-birds, but this gorgeousness is almost invariably confined to the adult male, the females being often merely greenish-brown and white, with spotted or streaked under-parts, though sometimes they exhibit a duller copy of the incredibly gorgeous tints of the male, tints which are described under such terms as "shining cobalt," "bright scarlet," "vivid orange," "deep beryl-green," "blazing emerald," "sapphire-blue," "amethyst," "deep purple," "topaz," and so forth.

Yet this coloration is seldom due to actual pigment, but rather to the prismatic surfaces of the feathers, so that in some lights the gorgeousness is not visible at all, and the bird would seem to be black, grey, or greenish-brown. Perhaps, therefore, as regards actual effulgence of colour, Humming-birds must take a second place in comparison to the quite unrelated Sun-birds of the Old World, in whom many of the gorgeous tints are due to actually pigmented feathers. In fact, colour for colour, the Sun-birds are more gorgeous than the Humming-birds and than any other members of the avian class. But although they resemble Humming-birds in their long, curved beaks, they are more commonplace in outline and structure; for the Humming-bird is not merely vivid in tint (putting aside the few examples of the family that are plain coloured), but develops other adornments or eccentricities of shape, which render him further remarkable to the observer. The wings run mainly to primary quills, and are long, narrow and pointed, like the wings of the Swifts (with which family the Humming-birds are closely allied; in fact, they are little more than gorgeous, marvellously developed, flower-haunting Swifts). Unlike most other birds, the secondary quills are reduced to only six or seven in number as against ten primaries.

The breast-bone is enormously developed in its keel, in order to furnish a strong base for the wing muscles which are intended to keep these narrow wings in a continual vibratory motion, so that when the biid hovers before a flower the whirring wings are quite invisible to the spectator, or, at any rate, only show a vague blur of grey. By this vibrating motion the bird is able to poise itself for quite a length of time before the object it is examining, and the whole of the flight recalls that of insects rather than of birds with its whirring and rapid swoops. In young birds the beak is quite short and with a wide gape, but in the adult is always long and sometimes is developed to a length more extravagant and disproportionate than is met with anywhere else in the Bird class. For instance, in the Swordbill Humming-bird the nearly straight beak (curved slightly upwards) is over four inches long; and in the female more than four and a half inches, or half an inch longer than in the male. The total length of the bird's body is eight and a half inches. Other Humming-birds develop one or more pairs of feathers on their tails to an extravagant degree. Remarkable amongst these is the Racquet-tailed Humming-bird, in which the second pair of tail feathers are long and wiry and their tips expanded into large racquet-shaped terminations. The beautiful "Doctor" Humming-bird of Jamaica also develops a second pair of its tail feathers to an extravagant length, and these are scalloped along the outer edge.

Others, again, have their tail feathers arranged in two long "ladders," or in other quaint shapes, illustrated in the accompanying photographs. Some species here illustrated develop extraordinary ruffs and frills about the neck, crests on the top of the head, tufts behind the ear, pointed beards of feathers, or other exaggerated gorgets, or puffs of white down round the

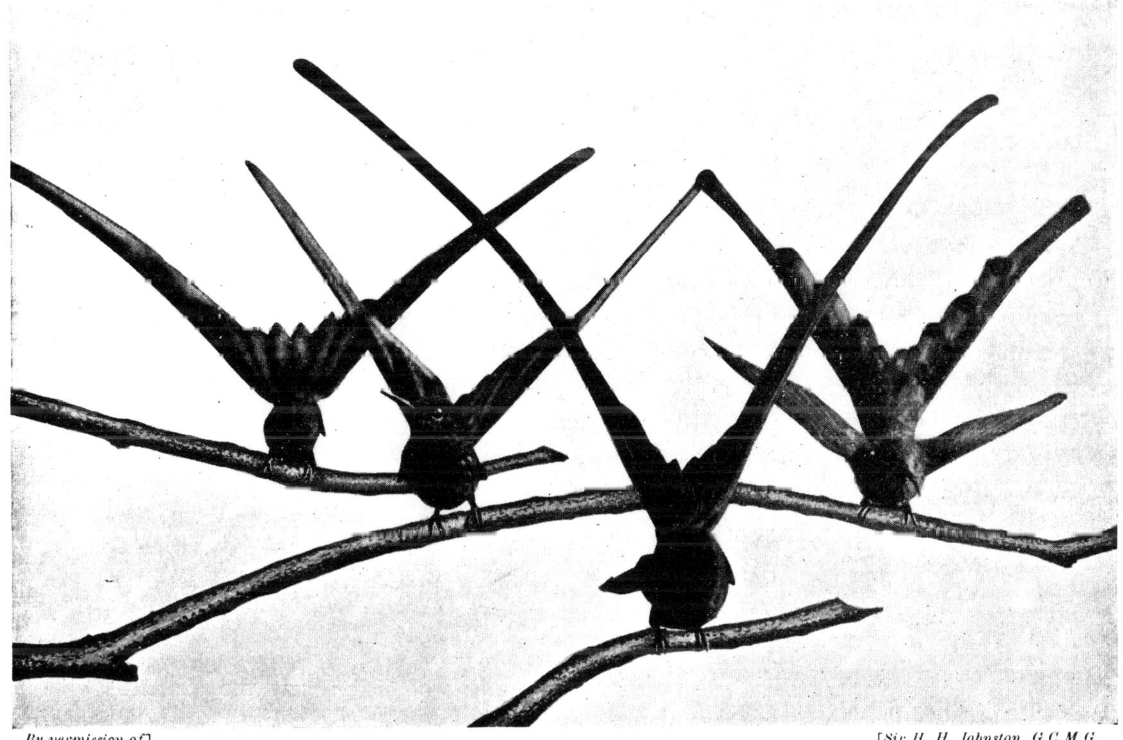

By permission of] [*Sir H. H. Johnston, G.C.M.G.*

HUMMING-BIRDS.

Long-tailed Humming-birds of several species: Cynanthus, Cometes and Lesbia.

knees of the tiny little legs. The tongue of Humming-birds is a characteristic feature something like that found in Woodpeckers. It can be suddenly protruded for a considerable distance and as suddenly drawn in again, carrying with it on its sticky saliva the many small insects for which the bird probes the flower. Many Humming-birds have serrated edges to their beaks, with which, no doubt, they crush some of the insects before swallowing them.

Photo by] [*A. W. Dennis.*

MISTLETOE.

The Mistletoe is here seen in its natural position growing from the branch of an apple-tree.

MISTLETOE

BY EDWARD STEP, F.L.S.

We are too apt to regard familiar things as commonplace and lacking the quality of marvellousness. We buy boughs of Mistletoe at Christmas to decorate our houses because it is the custom to do so ; and we revive the old Norse fairy-tales in which it figures, and do not forget its supposed connection with the mystic rites of the Druids. Yet some of the later antiquarians would tell, as Lord Avebury does, that its association with the Druids is very doubtful. In spite of this blow to our cherished beliefs, there are still some things connected with the " mystic plant " that are worthy of consideration.

For one thing, the Mistletoe is almost singular among native flowering plants in being an epiphyte—a plant that grows upon a plant and has no direct connection with the soil. It is quite singular in being an epiphytal parasite that possesses green leaves. These leaves, too, have a character all their own—their shape is peculiar, their consistence leathery, their surface dull, and their particular yellow-green unusual. Their presence is a certificate of character: they testify that the Mistletoe is not altogether abandoned to piratical courses. On the strength of its possession of green colouring matter there are not wanting those who claim that the Mistletoe gives as well as takes: that in winter, when its host is leafless and therefore unable to elaborate new material, the evergreen Mistletoe is active and supplies some of its surplus produce to its host.

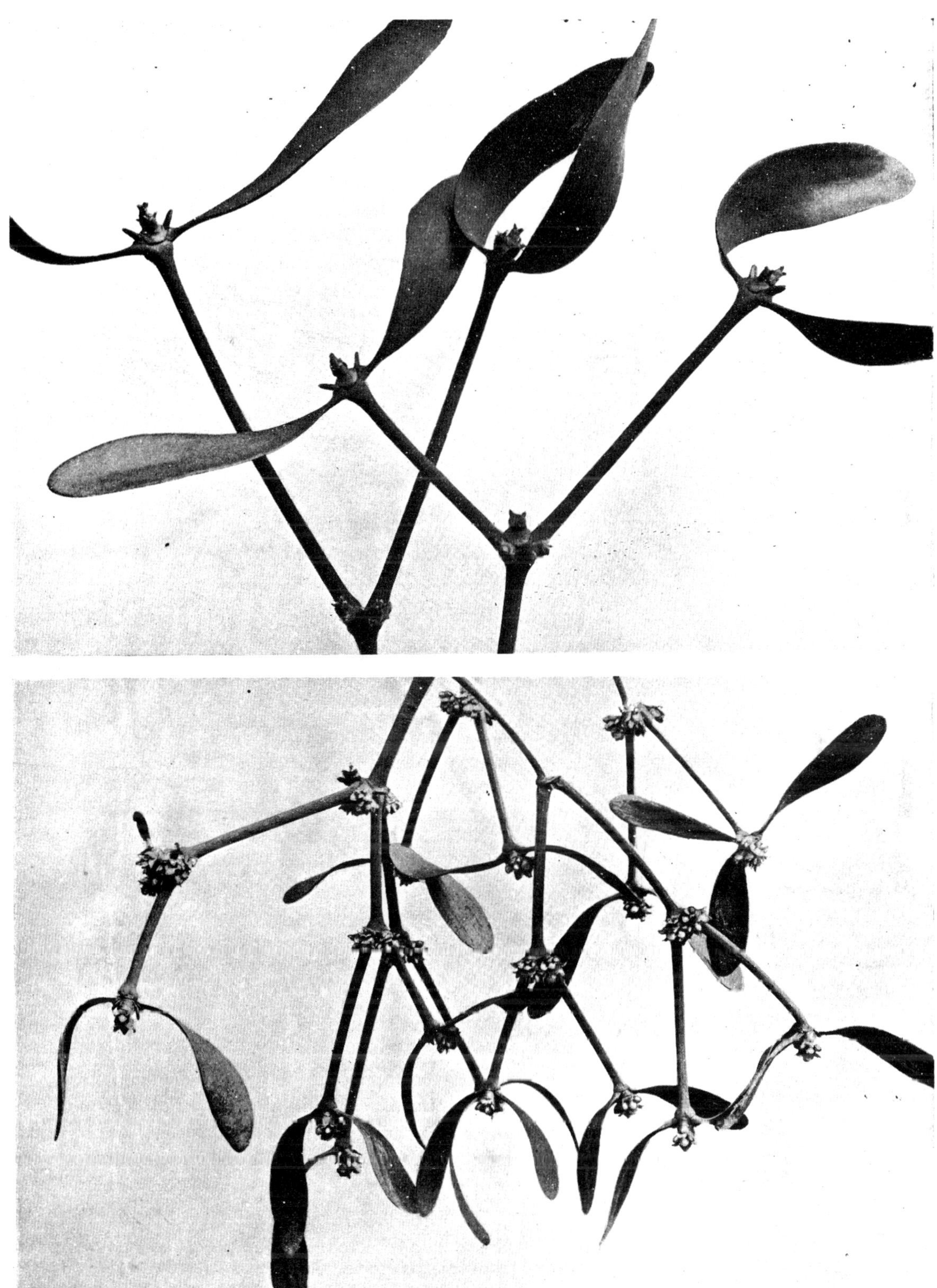

Photos by] [*A. W. Dennis.*

MISTLETOE FLOWERS.

The flowers of the Mistletoe are not very conspicuous, being quite without the petals that make other flowers attractive. The upper photo shows the female flowers, which are succeeded by the white berries. The lower photo is of male flowers, which only produce pollen for the fertilization of the female flowers.

The beginning of the story is with the flowers. Some persons who have taken great care to cultivate the Mistletoe upon an apple or poplar have been disappointed in getting a fine bush of the parasite, but never a berry. This is due to the fact that the flowers are male or female, and that the bush which produces the male flowers very rarely bears a female flower. The female flowers alone develop into berries.

As flowers are popularly considered, those of the Mistletoe are very poor affairs. They are small and inconspicuous, being quite without the petals that make other flowers attractive. Their envelope consists solely of four little triangular green sepals surrounding an equal number of stamens, or a simple little ovary with its stalkless stigma. In spite of this un-

Photo by] [*A. W. Dennis.*

MISTLETOE FRUIT.

Female branches in winter when the inconspicuous flowers have developed into pearl-white berries. These are shown about one-third less than the natural size. Each berry contains a single seed.

attractiveness, the flowers (which make their appearance in spring) are visited by flies and bees, who carry the pollen from the male bush to the female bush, and so aid in the fertilization which results in the pearly-white berries. Each berry measures a third of an inch across, and contains one seed with a thick coat of slimy pulp.

The pulp is beloved of the missel-thrush, which has got its distinctive name from the fact ; but the seeds resist the action of its digestive apparatus, and are sown by the bird on branches. Getting into a crevice of the bark, the seed germinates and sends a shoot through the bark and into the wood. This internal shoot develops into a thick woody peg, and an external part becomes a tough stem with leaves in pairs. At first it appears to be content with the thin watery sap that rises from the roots of the tree through the heart wood, but, later, it sends out branches into the bast, that lies between wood and bark, through which descends the rich, elaborated sap from the leaves. It probably thrives mainly upon this descending sap, but perhaps, as a sort of recreation, works up the crude ascending sap in its leaves.

The favourite host-plant of the Mistletoe in the wild state is the black poplar, but it has been more widely cultivated on the apple in Herefordshire and Normandy. It occurs naturally on the wild crab and on various conifers, but is very rarely found upon the oak, and its seeds are more reluctant to germinate upon that tree than upon any other. When artificially propagated its entire history can be recorded, and in this way it has been ascertained to attain the respectable age of forty years. This in itself is evidence that its host does not seriously feel the drain of a Mistletoe-bush, although it was recently stated that the Herefordshire apple-growers (who derive large amounts from its sale at Christmas) have decided that the Mistletoe must go.

Photo by] [*E. Step, F.L.S.*

AN INSECT IMPOSTOR.

The Lunar Hornet Clearwing Moth in form and colour presents a very close general resemblance to a Wasp.

AN INSECT IMPOSTOR

THE family of moths known as Clearwings may be regarded as living under false pretences. The man in the street who happened to see one of them in flight or at rest on a flower would never dream that he was looking at a moth, but would decide that the creature was some kind of fly, or one of the number of stinging insects known as Hymenoptera. The entomologist, accustomed as he is to nice points of distinction that are invisible to all but naturalists, has been struck by these counterfeit resemblances, and has enshrined his view of the matter in the popular or scientific names bestowed upon the insects.

In this country we have fourteen representatives of the family, and they all agree in having wings you can see through as easily as you can through those of a fly or a wasp. Their scientific names associate them with likeness to hornets, wasps, bees, ants, ichneumon flies, gad-flies, craneflies, gnats, etc. It must be confessed that in most of these species the resemblance to the insect named is rather remote; but they *have* all a general likeness to some insect that stings—though it is hard to say honestly which insect. There are at

Photo by] [*E. Step, F.L.S.*

QUEEN WASP.

A Wasp photographed at the same time as and with similar surroundings to the above, for comparison.

least two of them, however, in which the likeness is strong, though not exactly to the particular insect whose name they bear.

These two are the Hornet Clearwing and the Lunar Hornet Clearwing. The latter we show in our photograph, which is from a female newly emerged from the chrysalis. Frankly, there is no resemblance to a hornet, for the hornet is marked with black on a brown ground, whilst the Hornet Clearwings are banded with black on a clear yellow ground, which gives them a strong superficial resemblance to a wasp. When the wasp and the Hornet Clearwing are placed side by side the veriest tyro could probably point out the Simon Pure and the Pretender ; but on a cursory glance almost anybody would say, " There goes a wasp," when a Hornet Clearwing passes by.

When these Clearwings are still in the chrysalis skin their wings are clothed with scales like those

By permission of] [*Doubleday, Page & Co.*

BLACK ANGEL-FISH.

A food-fish which attains the length of two feet. It is common in the West Indies and on some parts of the American coast.

of other moths. Many of them are rubbed off in the act of emergence and the others go almost in the first flutter of their wings. These insects have found that it pays to keep up this resemblance to a stinging insect, for their enemies—chiefly birds—who know the consequence of stings, have learned to avoid them. Those that most nearly resemble stinging insects are the ones that are most likely to survive the perils of insect life long enough to transmit their characters to another generation ; and so through the history of the race the tendency has long been in the direction of perfecting the imposture.

ANGEL-FISH

BY W. P. PYCRAFT, F.Z.S.

BEAUTY and ugliness are relative terms, and the very thought of the one suggests the other. This association is possibly answerable for the use of the term " Angel " fish, on the one hand, for a

By permission of] [*Doubleday, Page & Co.*

YELLOW ANGEL-FISH.

The photo shows a young specimen of a fish that grows to a foot and a half long. It gets part of its name from the broad yellow margin to each of the scales, and the other part from its beautiful appearance. Like the Black Angel, it is a food-fish.

member of the shark tribe which might almost be regarded as ugliness personified, and, on the other, for the use of the same name for certain fishes of tropical seas remarkable for the brilliancy of their coloration. These are members of a family sometimes called Butterfly-fishes; while their scientific appellation Chætodontidæ has been bestowed in allusion to the peculiar form of the teeth, which are minute, closely set and extremely slender, suggesting rather bunches of bristles than teeth.

But while, as we have remarked, several species of these "bristle-toothed" fishes (Chætodontidæ) are described indifferently as "Angel-fish," the name seems originally to have been bestowed on the species known as the Yellow Angel-fish, and to ichthyologists as *Holacanthus ciliaris*, which, however, in the matter of coloration is less striking than its near relation, the Japanese Paradise Fish (*Holacanthus imperator*). The black and yellow bands, which give these fish the appearance of wearing "blazers" of an unusually striking pattern, as a matter of fact produce a quite opposite effect, causing the wearer to blend so completely with its surroundings as to seem a part thereof, for they live in an environment where strange and striking contrasts of colour obtain on every hand.

Photo by] [*E. J. Spitta, F.R.M.S.*

THE MIDGE.

In this greatly magnified portrait of the Midge, the lancets are seen hanging down from the head. These are the instruments with which it tortures its human victims.

Near relations of the Angel-fish are found in American waters. And these, in their adult stages, afford a most interesting contrast in the matter of coloration, as may be seen in the case of the Black Angel-fish (*Pomacanthus arcuatus*) shown in the accompanying photograph. But while the adults are invested in sombre shades of black and brown, the young wear a livery in which broad bands of yellow play a conspicuous part, wherein they recall their adult relations of tropical seas.

THE MIDGE AND ITS CURIOUS EYE

BY E. J. SPITTA, F.R.M.S.

Of all the abominations to be found in insect life that attack the human subject, one of the most trying is certainly the little Midge. For so small an insect, its bite seems peculiarly irritating, and whether this is caused by the insertion of a lancet poisoned from without, or by the injection of an irritating fluid at the time of biting from within, seems yet doubtful.

The insect is so extremely minute and excessively tender, that it is no easy matter to mount one whole without doing it some sort of injury; but such, however, has been successfully accomplished, and is shown highly magnified in the first illustration. The lancets can just be seen hanging down from the head, but they are better shown in the still more magnified photograph given on this page. In this the head is shown to be in a great measure formed by a peculiarly-shaped overlap of eyes—a mass of them on each side of the head being like a flap-eared cap such as one sees used in windy weather. The portion intervening between the eyes is prolonged downwards into the biting apparatus, the lancets and tongue being plainly visible, and shown to be supported by a very strong and thickened base. On either side are a pair of sensory organs, whose properties are at present unknown. Probably only one on each side is a true antenna. They are all supplied with beautiful details, and two of them consist of curiously-shaped pieces jointed together, the bulbous ends of each section being marked by the presence of several sensory hairs and several "pits," which in insects are usually thought to be organs of sense.

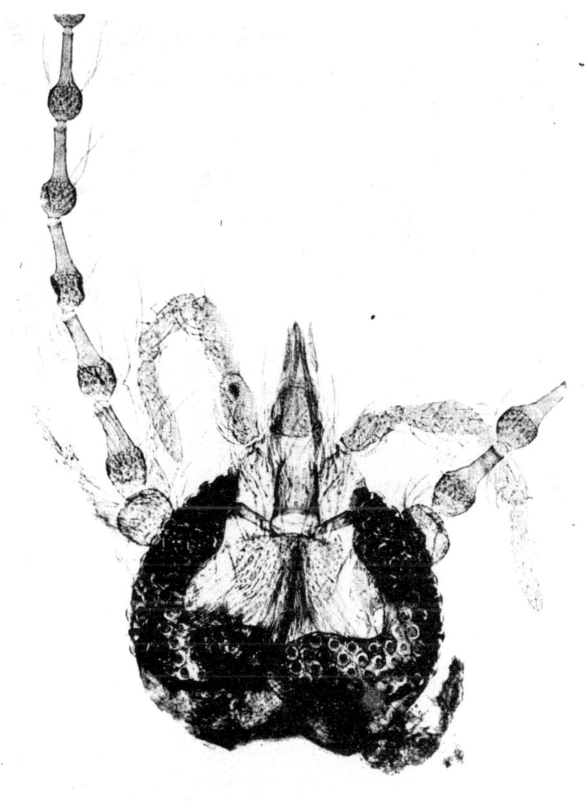

Photo by] [*E. J. Spitta.*

HEAD OF MIDGE.

The head is here viewed from below, and parts of the remarkable antennae as well as the mouth parts are shown.

Looking at the head from the side, as in the photo below, the curious eyes—over a hundred are to be counted—are well seen. These are of remarkable design. Most insects that have compound eyes show them to be a group of facets—more like the section of a honeycomb—being many thousand (usually about four thousand) in number, each one of which is a complete eye in itself, as witnessed by the fact that each will furnish a distinct image of an object placed beneath it.

In the case of the Midge, however, each eye is visible as a distinct *protrusion* rather than a flat facet, and this is shown to be the case very plainly in the photo overleaf, which is much more magnified. In the ordinary compound eye of the insect, if it be looked at sideways, the facets seem all on one level—quite even, in fact, and uniformly flat, having

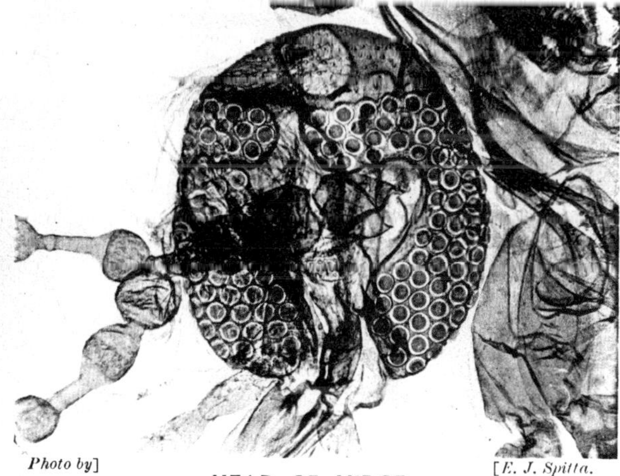

Photo by] [*E. J. Spitta.*

HEAD OF MIDGE.

Here the head is seen from the side, with the curious eyes and the base of the lancets and antennae.

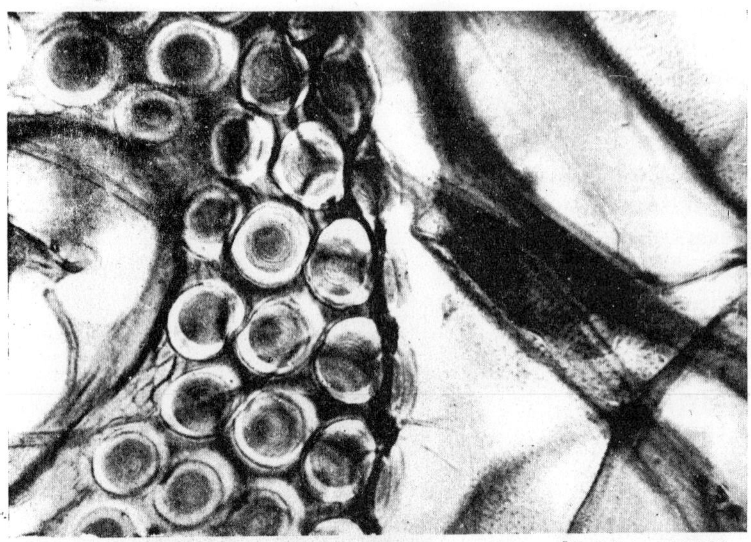

Photo by] [*E. J. Spitta, F.R.M.S.*

MIDGE'S EYES.

A small number of the eyes are here seen greatly magnified, showing that each is separate from its fellows, unlike the usual compound eye of insects.

no projecting portion whatever; but in the case of the Midge it is quite different, for if the centre of this illustration be looked at, three eyes are seen edgeways, and it is quite evident they stand out—like a watch-glass from its case—from the surrounding tissues in a manner at once very striking. Nothing is known as to the use of this arrangement, and some have thought that it is an effect produced by the shrinking of the fluids of the eye after death, which makes the actual lens, so well seen very highly magnified in the illustration below, appear to push forward the membrane enclosing it. This, however, would not account for the amount of protrusion, and may therefore be regarded as insufficient to explain so extraordinary a construction. In the same illustration attention is called to the tessellated membrane supporting these eyes, which is very beautifully executed; it seems very curious such beautiful workmanship should be for ever hidden from our gaze save with the use of the microscope.

AN EOCENE MONSTER

A RHINOCEROS with the legs of an elephant is about as near as one could go in a popular brief definition of what the Tinoceras was. In the British (Natural History) Museum may be seen a cast of the skeleton which Professor Marsh discovered in Wyoming. This skeleton we reproduce, together with an attempt to clothe it with flesh and hide, and give a reasonable idea of its living appearance, based upon the clues which are furnished by the bony structure. The general resemblance is to a rhinoceros, but the legs are quite of the elephantine type, and the shape and positions of the toes suggest that the feet were provided with thick pads. It was about twelve feet in length, without including the tail, and it has been estimated that its weight when alive was more than two and a half tons.

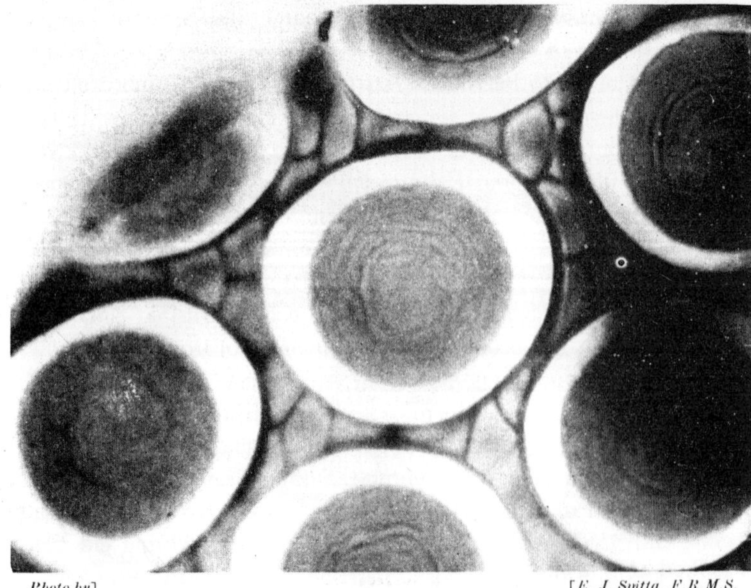

Photo by] [*E. J. Spitta, F.R.M.S.*

MIDGE'S EYES.

A few of the eyes are here subjected to still higher powers of magnification.

By permission of] [*Messrs. Chapman & Hall.*

THE TINOCERAS.

An animal that flourished in the Eocene period. Though his head, with its six incipient horns, suggests relationship with the Rhinoceros, his limbs are nearer to those of the Elephant, and his tusks have relation to neither.

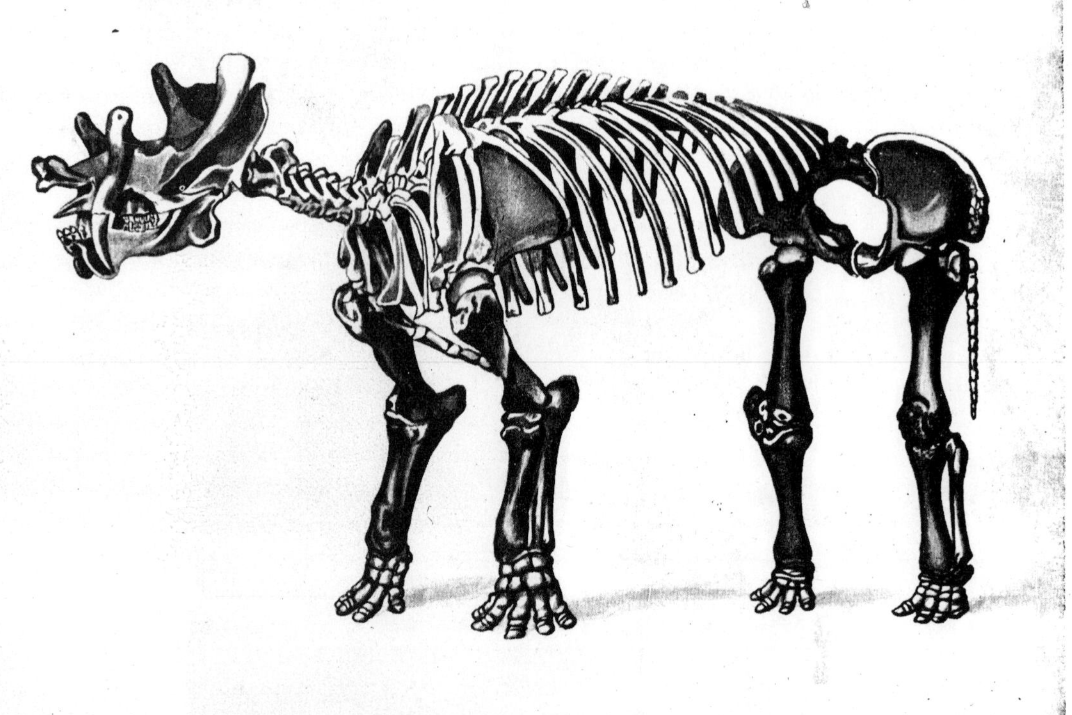

SKELETON OF TINOCERAS.

As set up by Professor Marsh, who discovered its fossil bones in Wyoming. It is twelve feet long

An examination of the remarkable skull shows that it had six stout prominences on the head, which may have been horns of some sort, of which only the cores, or bony supports, are left. Professor Marsh thinks that it is possible that natural casts of these horns may have been left in some of the beds to reward further research. On the other hand, the bosses on the skull may only represent that stage in mammalian evolution when horns were beginning to develop. There is a suggestion of Brontops in the head of our restoration, though it will be noticed that, quite apart from the profusion of horns or bosses, there is a striking difference in Tinoceras possessing a pair of tusks that are dagger-like.

There was not much accommodation for brain in that skull, but the mammals of the Eocene period are remarkable for their small brains. The five toes show that Tinoceras had no close connection with the modern rhinoceros, but probably represents an early stock from which both rhinoceros and elephant branched off.

THE GUACHARO, OR OIL BIRD

BY SIR HARRY JOHNSTON, G.C.M.G.

THIS is one of the strange " stranded " types of bird for which South America is so remarkable. South America as a continent is only less isolated and primitive in its fauna than Australia. During a long period of the earth's history it must have been completely cut off from North America and intermittently united across the Atlantic with Africa, and by Antarctica even with Australia and New Zealand. Consequently it received intermediate generalized half-and-half forms of beast, bird, and reptile, which, when it became an island-continent, continued to exist there, whereas elsewhere they were killed out by competition. Amongst these indeterminate forms of birds is the Guacharo here illustrated and shown with its nest made of clay and its eggs. (The specimens have been photographed from the collection in the Natural History Museum of Munich.) This strange bird was first described to the scientific world by the great German traveller, Humboldt, who met with it in some extraordinary caverns in Venezuela, and for a long time it was thought that this Oil Bird (so called because it subsists largely on oily nuts and develops, especially in its nestling young, much oil or fat in its skin glands) was confined in its distribution to this one particular part of Venezuela; but it has since been met with elsewhere in caves in Northern South America and the Island of Trinidad. In structure the Oil Bird is related to the owls, the goatsuckers, the trogons, and other picarian birds, and is the survivor, no doubt, of a generalized group from which most of these forms originated. Its coloration is very like that of goatsuckers—a mixture of chocolate-brown, cream-yellow, grey and white; brown predominating. The bird forms colonies of many individuals, which congregate and build their nests of plastered clay in dark caverns. They are mainly of nocturnal habits, and fly out in the dusk to seek for the fruits and nuts on which they feed. Their nestlings are much sought after by the Amerindian savages on account of their oily flesh, but to Europeans they seem to have a disgusting taste, like the smell of cockroaches. The adult birds are also killed for the oil their bodies contain. This oil keeps well and burns brightly in lamps.

THE ANIMATED OAT

BY EDWARD STEP, F.L.S.

OCCASIONALLY, one may still see in country cottages the quaint little toy-house with two doors, from one of which a man emerges when there is a prospect of rain, and from the other a woman

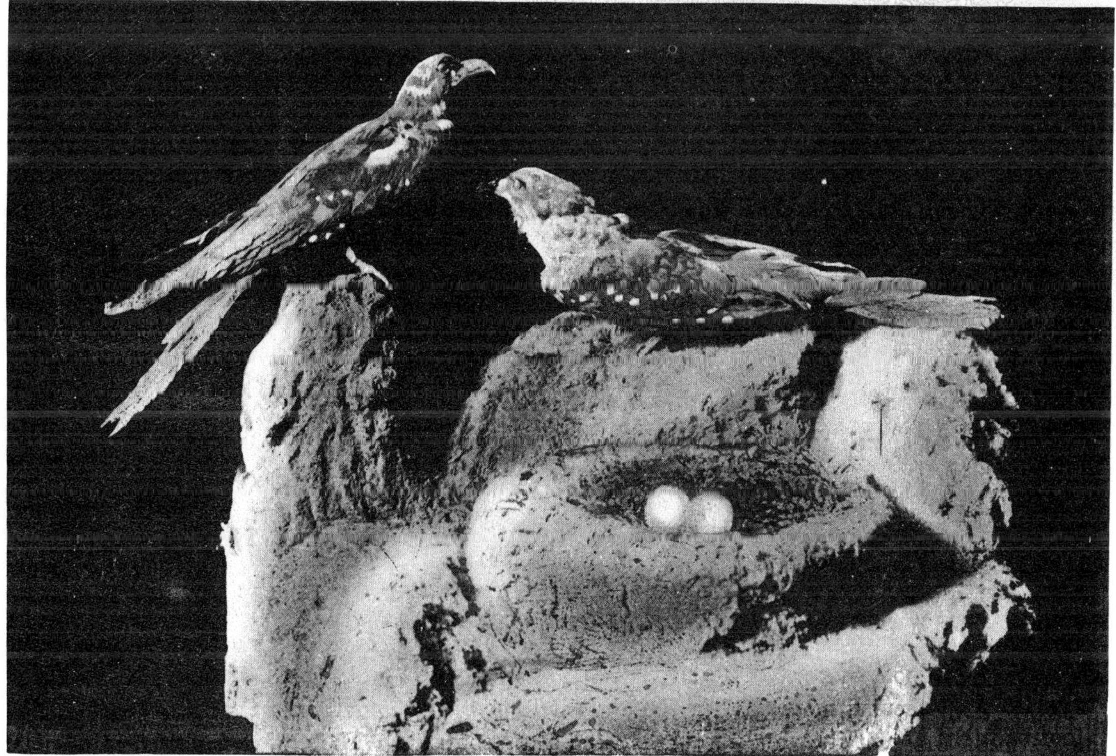

By permission of] [*Sir H. H. Johnston, G.C.M.G.*

GUACHARO, OR OIL BIRD

A kind of Goatsucker peculiar to South America; of night-flying habits, and making its clay-plastered nest in dark caverns.

on the approach of fine weather. This is really a hygrometer worked by pieces of twisted catgut, which expand when the air is dry and contract in the presence of moisture. More rarely we may come across a simpler and more natural hygrometer in the shape of a large oat-fruit mounted on a piece of card, which the cottager will assure you is an infallible weather indicator. This, they will tell you, is not the common oat, but one of a rare kind hard to come by and, therefore, to be treasured accordingly. We heard recently of the possessor of one of these who declared she had been offered five shillings for her "barometer," but she knew better than to part with it!

Photo by] [*E. Step, F.L.S.*

ANIMATED OATS.

The photo shows three different positions of the long awns which are spirally twisted below the "knee." The spiral contracts and expands according to the amount of moisture in the atmosphere.

Everyone knows that in the growing of the common oat the fruit contains two or three seeds or oats, and that certain of the enveloping scales or chaff end in long, slender bristles or awns.

There is a species introduced from North Africa in which these awns are a couple of inches long with a sort of knee half-way. Below the knee the awn is thick, unbending, and spirally twisted; above it tapers from the knee to a very fine, hair-like point.

The cottager will tell you that when rain is coming both awns will bend over to the left. When they bend to the right fine, dry weather is sure; and when a change is in progress the awns cross each other in the centre. There is a certain amount of truth in this, for the twisted awn is highly hygrometric. But every oat-fruit gathered from the same plant and similarly mounted will not behave in the same way. One will be pointing to the right and another to the left under the same meteorological conditions. Still, by mounting one specimen only and noting its behaviour under different conditions, it will serve as a guide to the humidity of the air.

But the real marvel of this Animated Oat does not consist in its weather indications, but in its natural purpose.

Its hygroscopic property is really an ingenious plan for burying the seeds. The bent-kneed awns are levers for pushing the seed into the soil.

As the ripe fruit falls from the oat-plant to the ground the tip of one awn and the point of the fruit are in contact with the soil. The base of the fruit is furnished with brushes of golden-brown stiff hairs, arranged in semi-spirals, and these are also acted upon by moisture and dryness, closing up or standing out at an angle. The effect of the lengthening movement of the awns is to screw this point into the soil, and the hairs act as barbs to prevent its withdrawal when the awns contract. The action is much like that of the Stork's-bill fruit already described in this work (page 193). Its power may be demonstrated by holding one in the closed hand, when it will be felt making its way between the fingers.

Photo by] [*E. Step, F.L.S.*

THE ANIMATED OAT.

A large kind of Oat, a native of North Africa, with long awns to the seeds which bend and twist with varying degrees of humidity in the air. Such movements assist in the natural sowing of the seeds, and have been used as weather indicators.

SOME REMARKABLE INFANT FISHES

BY W. P. PYCRAFT, F.Z.S.

SOME of the most striking, and at the same time the most puzzling, facts which the study of animal life presents are furnished by young animals; and by way of illustrating this point, it would be difficult to find two more remarkable instances than those furnished by the infant Ribbon-fish and Stalk-eye, whose portraits accompany these lines.

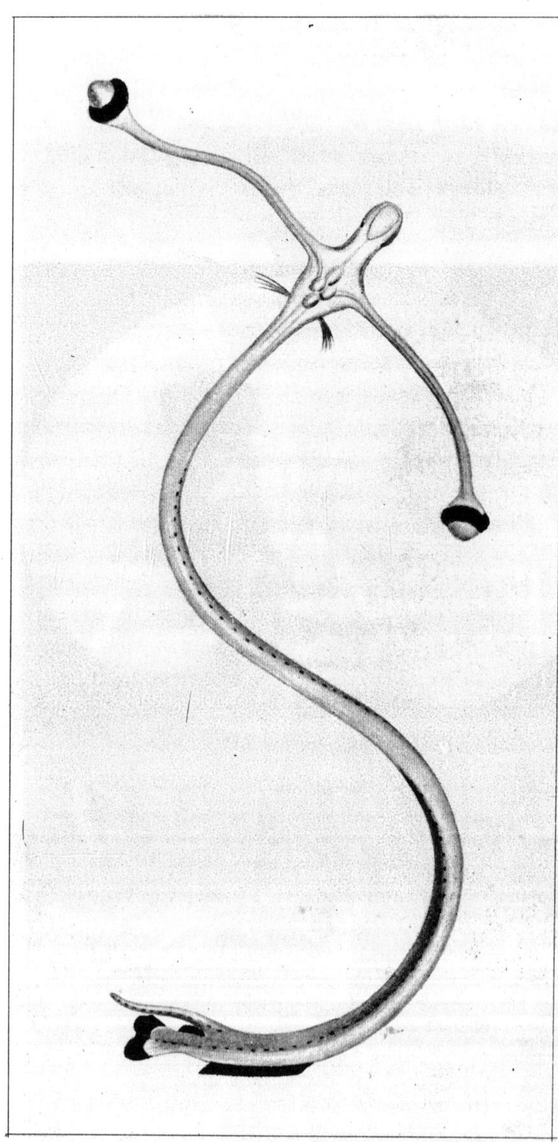

YOUNG STALK-EYE.
Its eyes are mounted on long, flexible and retractile stalks, much like the eyes of a snail.

The adult Ribbon-fish, or Deal-fish, which is sometimes cast up on the coasts of Scotland, it should be remembered, is excessively elongated, and flattened from side to side, while the fins are its least conspicuous feature. Between this and the larval stage it would be difficult to find a stronger contrast, either in the matter of the form of the body, or in the development of the fins! Nothing is certainly known of the life-history of the young Ribbon-fish, but we may safely assume that it is spent in the ocean abysses, where the only semblance of light is that of the pale, thin rays emitted by the bodies of the inhabitants of these regions of eternal night.

In these awful depths there are no disturbing currents, and hence the possibility of long, thread-like fin rays such as distinguish the young Ribbon-fish. But what purpose do they serve? I hear my readers ask. Judging from what obtains among other, and adult, fishes of these regions, they serve as organs of touch, and the curious arrow-head tabs of skin which run up these rays probably give increased sensitiveness, either to touch or temperature. In the deep sea Bathypterois some of the rays, or supporting rods, of the breast-fins have been produced into long, delicate, bony threads covered with a very sensitive skin, and these are moved about—up, down, forwards, sideways, backwards—as the fish moves slowly about, seeking at once both food and direction. They remind us of the way the blind man guides himself by the tapping of his stick. But we have yet to discover whether the young Ribbon-fish is luminous and what are its haunts in later life. There are several species of Ribbon-fish, it may be remarked, and all alike have so far evaded our investigations, but all seem to haunt the deep sea. In the typical Ribbon-fish (*Regalecus gladius*), which may attain a length of twelve feet or more, the foremost rays of the back-fin, and those of the belly-fin—answering to the hind-legs—are excessively lengthened, though not so much so as in the larval form of the Deal-fish; and their likeness in this respect may indicate relationship, or merely adaptation to like requirements.

In its way the young Stalk-eye is more remarkable than the young Ribbon-fish. Nothing is known of its life-history save that it is a dweller in the abysses of the ocean: we do not even know what shape it assumes when adult. But this we do know, that among the vertebrates it is unique in the matter of its eyes, which are borne on long, flexible, and retractile stalks recalling those of the snail. This is really a most extraordinary fact, for these eyes must be of tremendous importance, though they may achieve no more than distinguishing the pale phosphorescent light of their surroundings from the darkness which may be felt. This being so, it seems strange that they should be constantly exposed to such a risk of being bitten off. Possibly they are extremely sensitive, so that the vibrations in the water set up by an approaching enemy may serve to give warning in time to permit of their being instantly withdrawn to the level of the head, or, at least, far enough in that direction to be out of harm's way.

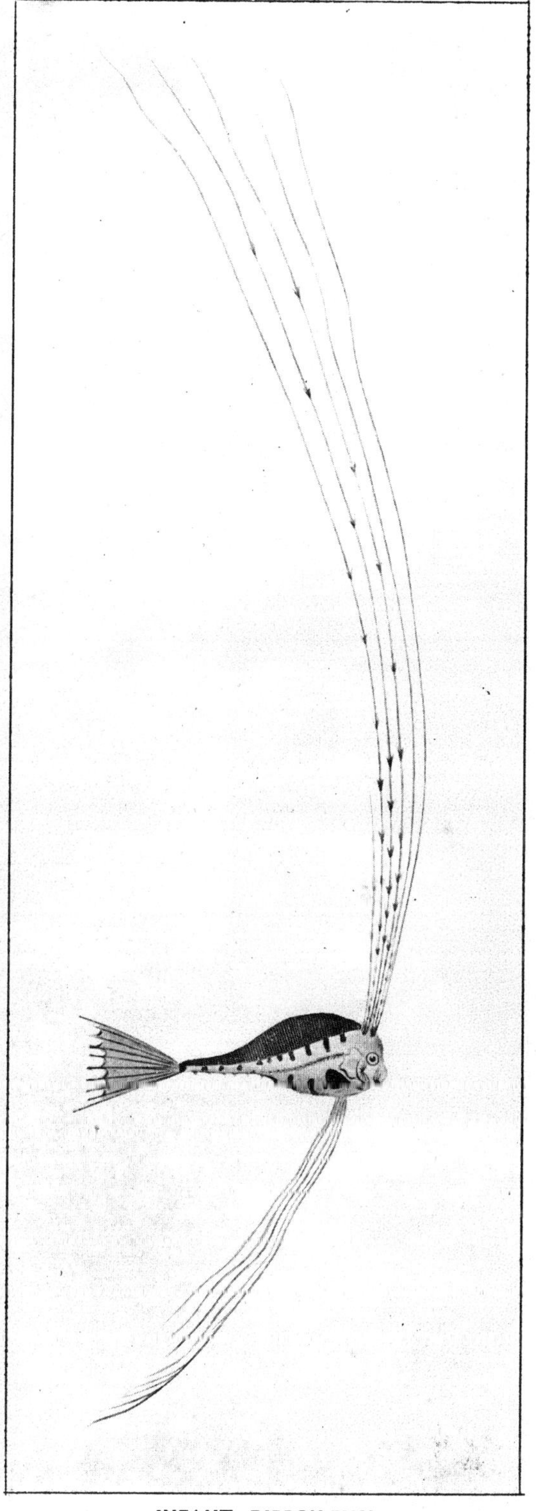

INFANT RIBBON-FISH.

Remarkable for the enormous length of certain of its fin-rays. The arrow-head appearance is given by little tabs of skin.

THE LIFE HISTORY OF THE FROG

BY WALFORD B. JOHNSON.

When the night frosts of autumn set in Frogs seek out suitable places in which to pass the winter. Moss-lined crannies and hollows in the stumps of trees are the places most favoured, and there they remain till the spring recalls them to activity.

During this period of hibernation these creatures are in a state of torpor, the mouth and nostrils are closed and respiration is all but absent, being carried on then entirely by means of the skin. It is only healthy Frogs which can successfully withstand the rigours of the winter, for the weaklings die during their hibernation; giving another instance of Nature's method of weeding out the unfit.

With the fitful sunshine of March, these Frogs issue from their hiding-places and congregate in considerable numbers in the ponds, and there they commence to spawn. Curiously enough, the

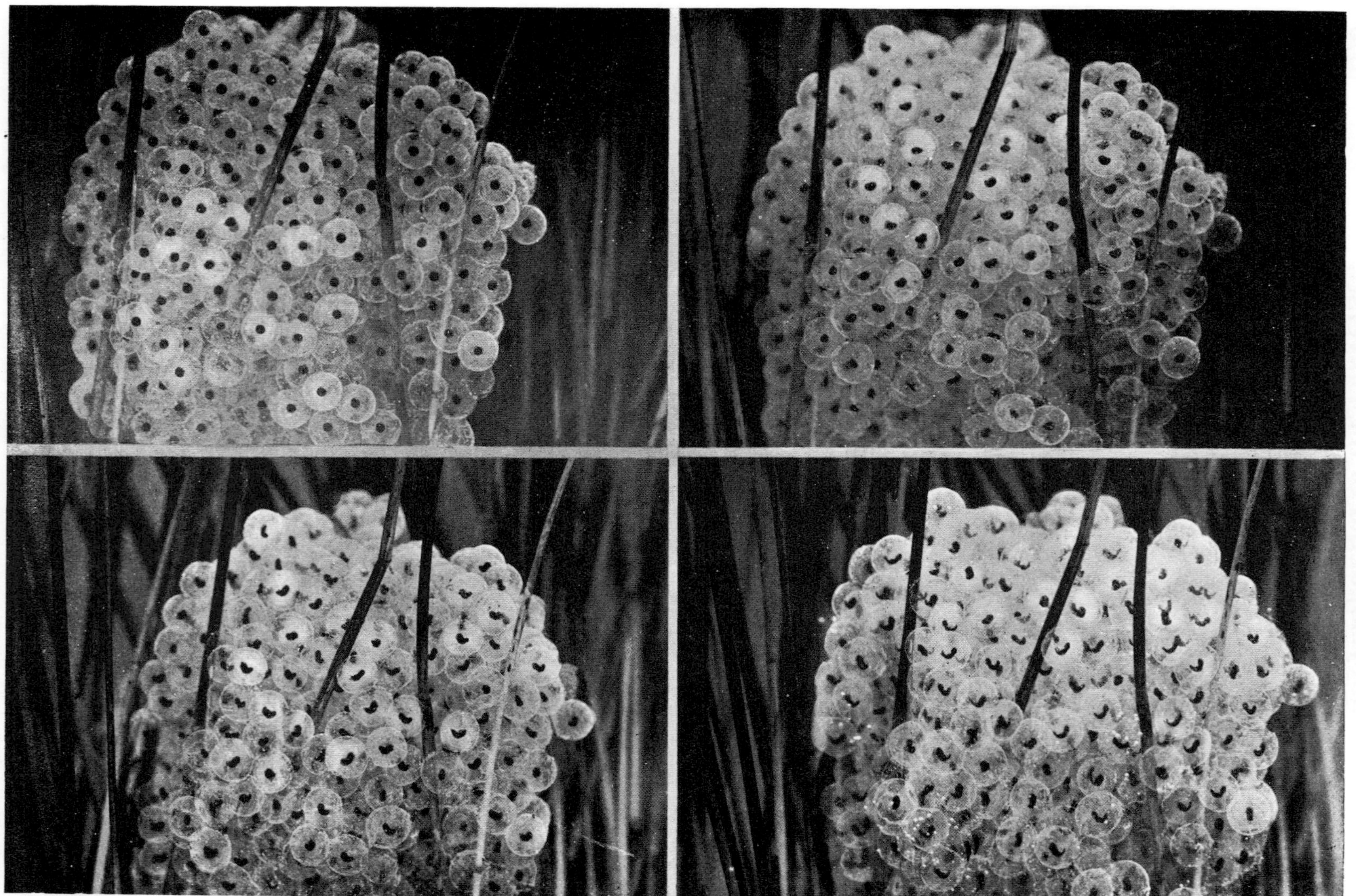

Photos by] [*Walford B. Johnson.*

THE DEVELOPMENT OF THE FROG'S EGGS.

In the first of these four photos the mass of spawn is seen when the eggs have absorbed sufficient water to become distinct from their fellows. In the second the black central embryo is beginning to elongate The third is more advanced, and in the fourth there is some approach to the future tadpole form.

same water is chosen year after year, and, too, the same part of the pond ; while other pieces of water, seemingly just as suitable, are invariably ignored.

The female Frog deposits her eggs in the water in the form of a small black mass, not more than an inch and a half in its longest dimension. If this were laid out of water it would feel to the touch much the same as tacky rubber solution ; and the mass on close inspection would be seen to consist of a quantity of tiny jet-black spheres, all tightly pressed together. The tackiness is due to a minute coating of a sort of gelatine on each sphere. Now, when laid in water, as is the case in nature, this gelatine-like substance, which entirely envelops each egg, rapidly absorbs water and becomes

Photo by] [*W. B. Johnson.*

FROG'S EGGS HATCHING.

Many of the spherical envelopes have burst and released the small tadpoles.

a slippery jelly. In a few hours this original black mass swells up to a lump of jelly, in which are embedded at regular intervals the tiny black bodies already mentioned. These are the actual eggs, the jelly being merely a protective covering.

Our first illustration is a mass of frog-spawn at this stage and the black eggs will be seen, each in its own covering of jelly. This gelatine covering plays a very important part in the subsequent welfare of the eggs. They require a constant supply of the oxygen dissolved in the water, and as the jelly is very porous it keeps the eggs apart and allows the water free access. However, even in nature, the innermost eggs seldom hatch out of a big mass, on account of their being blanketed by the more fortunate ones on the outside.

Anyone who attempts to pick up a quantity of frog-spawn will readily see how ably the slippery jelly foils the attacks of birds and fishes, which would gladly prey on the black eggs were they not thus prevented. The spawn is always laid in some shallow where it cannot sink more than half an inch or so below the surface; for the eggs require light and heat from the sun to hatch them; and their black colour largely aids in this absorption.

In some ten days or more after they were deposited the eggs show signs of changing shape. No longer are they practically spheres, for these black bodies appear to unroll. Reference here to the illustrations will make this perfectly clear. As the embryo tadpoles develop, the jelly gradually loses its toughness, as it were, and becomes slimy and mucilaginous, so that when the young tadpole is ready to leave the spawn it is able to slowly wriggle its way through the jelly out into the water. It is a curious little creature now, jet black and just a head, body and stumpy tail, but without a suggestion of limbs. Under its head is a gland which secretes an adhesive substance that causes the tadpole to stick to anything the gland touches. In this way the creature fastens itself to the outside of the mass of jelly or to the weeds amongst which the spawn was deposited. Gradually it works its way upwards until the spawn is covered on top with these young tadpoles.

They have no mouths yet, and are incapable of any movement beyond an occasional wriggle They breathe by means of tiny gills which, at this stage, somewhat resemble ears at the side of the head. In a few days the tail lengthens and the gills grow to be feathery-looking appendages, and then the creatures begin to swim freely. Gradually these new gills disappear, for a skin grows over them until they are no longer visible and a horny, beak-like mouth is formed.

Photo by] [*W. B. Johnson.*

FROG'S EGGS HATCHING.

In this photo, though some of the eggs in the lower part of the mass have not yet hatched, the free tadpoles are wriggling their way to the top to enjoy the greater light and warmth of the surface waters.

Respiration is carried on now by taking in water at the mouth, passing it over the concealed gills and expelling it at the single gill-hole at the left side of the head. Lungs, too, are beginning to form, and the tadpole frequently visits the surface to exercise its new power. As development proceeds the gills tend to fall into disuse and the lungs begin to predominate; so that when the creature leaves the water as a Frog the gills will have disappeared altogether.

This we may regard as the first true tadpole stage, for the creature now remains much the same shape for most of its aquatic life. By this time the part of the pond in which the spawn was laid is a seething mass of wriggling black; and it will be obvious to all that it was never Nature's intention that all these tadpoles should reach maturity. They are the scavengers of the pond, feeding on anything vegetable or animal, bruising and sucking the food with their horny mouths, but preferring those things that have already been softened by decay.

These tadpoles also constitute the staple food of the carnivorous creatures which share their haunts, such as fishes, newts and the numerous tribe of predaceous water insects. Consequently their ranks are sadly thinned by the time they assume the proper form of the Frog, and leave the water. Until now our tadpole was quite black, but with the loss of the external gills its colour begins to change. Its body becomes closely dotted with minute golden spots, which, as time goes on,

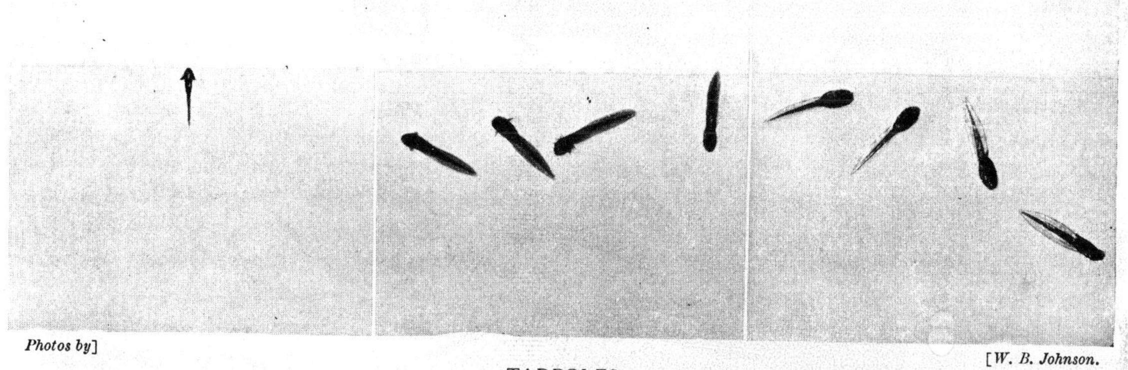

Photos by] [*W. B. Johnson.*

TADPOLES

The early stages of Tadpole-life are here shown of the natural size. The single Tadpole to the left has recently hatched, and the gills are just noticeable behind the head. The central four are a little older, and the four to the right still older. The tail has lengthened, the gills become feathery-looking, and the Tadpoles swim freely.

gradually expand, and give the creature the familiar brown colour. This admirably protects it from unwelcome observation, while lying at the bottom of a muddy pond. Tadpoles are always active and feeding. Their mission is to feed, and on the refuse which would decay and poison the water were it not removed by them. Naturally our tadpole grows rapidly, and after a few weeks from the date of hatching the hind toes become visible at the root of the tail. At the same time within the gill chambers the front legs begin to develop, although they are as yet invisible from the outside.

Steadily these legs lengthen and grow stouter, and by the middle of June the hind pair become folded in the position so characteristic of a Frog.

Then the left fore-leg suddenly pushes its way out through the gill-hole, but the right one has to forcibly make an exit in the skin and, in consequence, appears some hours later. The little creature now is a veritable Frog, but with a tadpole's tail. This, however, is not for long, for as soon as it is the proud possessor of four legs the tail begins to wither. A sort of inflammation sets in at the tip and gradually the tail shortens to a stump, and in three or four days is absorbed and so vanishes altogether.

The little Frog now wanders into the grass at the waterside and in time reaches the meadows, feeding on the tiny insects it meets with. A summer shower will cause the simultaneous exit of

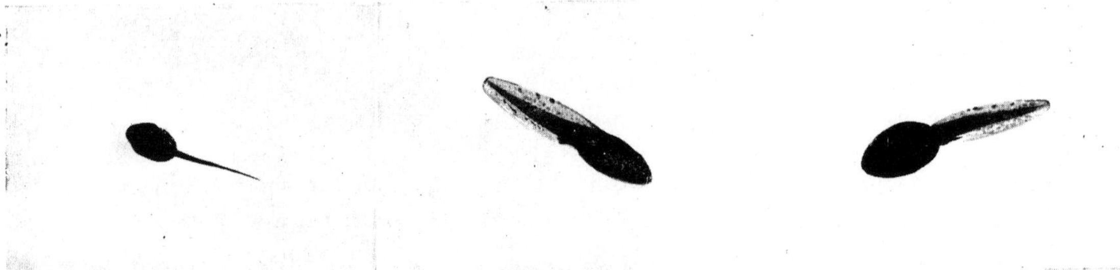

Photos by] [*W. B. Johnson.*

FROG-TADPOLES.

In this series further advances are evident. The gills have disappeared, and in the middle figure the hind legs have put in an appearance, the colour is changing from black to brown, and the tail-fin has become high and spotted. (Natural size.)

thousands of these young Frogs from the ponds, and the sight of so many actively hopping little creatures leads the uninformed to declare that they have descended from the clouds with the rain.

Henceforth it is a terrestrial creature, not an aquatic. A certain number of Frogs will always be found in the neighbourhood of the pond, because of the moist condition of the lush waterside vegetation and the abundance of insect life about it. For the Frog restricts itself to an animal diet, and its food must be taken alive—flies, caterpillars, worms, slugs, beetles. All small things that show activity will arouse the interest of the Frog. He sits with great stolidity until the moving creature comes within range. Then the tip of the long sticky tongue is shot out, and returned with its quarry adhering.

In about three seasons, if it escapes the jaws of its many enemies, our Frog will be full-grown and able to breed ; and then, if the fates are still kind, the creature may live on till it reaches the age of seven or eight years, or even more.

A FOSSIL SEA-LILY

BY FREDERICK CHAPMAN, A.L.S.

THE accompanying photograph is not that of a palm, nor a " fossil fern " as the discoverer called it ! It belongs to a group of animals allied to sea-urchins and star-fishes, and is nearly related to the living feather-stars, which were described on page 643. We return to the subject only for the purpose of showing a complete fossil of a distinct type. The rock in which this interesting and

Photos by] [*W. B. Johnson.*

TADPOLES—ALMOST FROGS.

The hind-limbs now show that they are jointed and have fingers at the extremities. In the photo to the right the fore-limbs are just perceptible. The tail is beginning to dwindle (Natural size.)

Photos by] [*Walford B. Johnson.*

THE COMMON FROG

The two upper figures are slightly enlarged. They show the young Frog, four-footed, with a tail and without. The latter differs in little more than size from the adult whose portrait occupies the lower half of the plate.

beautifully perfect fossil was found is the brown sandy mudstone of Brunswick, a suburb of Melbourne. Many æons ago, in the older stage of the Silurian period, the locality was occupied by a wide shallow sea, in which these animals flourished, together with numerous extinct types of shell-fish. Along the mud-flats and sandy shores of that epoch various strange forms, like large wood-lice (trilobites) crawled, and numerous pod-shrimps leaped about amid the shallows, chasing other smaller fry. There were no true fishes in those days, for even the sharks had not then made their appearance.

The Helicocrinus, as the present fossil is named, consists of a cup-like body surmounted by five branched arms, beset with tiny jointed pinnules or filaments, by the waving of which currents were caused and food brought into the mouth which they surrounded. The body is mounted on a long, jointed and once flexible stalk, the lower end of which terminates in a coil, which was probably used for the purpose of twining itself round the stalk of a sea-weed or other rooted object. This unique specimen is in the collection of the National Museum at Melbourne.

Photo by] [*F. Chapman, A.L.S.*

"STONE-LILY."

A beautiful example of a type of Feather-star that flourished in the Silurian period. This very perfect fossil was found in the neighbourhood of Melbourne, Australia.

LUMINOUS METEORS

BY G. F. CHAMBERS, F.R.A.S.

FROM time to time, usually at night, but not invariably so, certain luminous objects suddenly appear in the sky, which are popularly known as "Shooting Stars" (or "Falling Stars"), or they may be "Fire-balls," or else "Thunderbolts." These objects bear astronomically the generic title of "Luminous Meteors." They differ very much in size, in frequency and in appearance; but for astronomical purposes all come under the family name of "Luminous Meteors."

The most familiar of them is the "Shooting Star," and I will start with that as a point of departure to lead up to the others. Though often recorded in ancient times, Shooting Stars were only regarded as casual manifestations of some unknown forces of Nature, and not subject to any

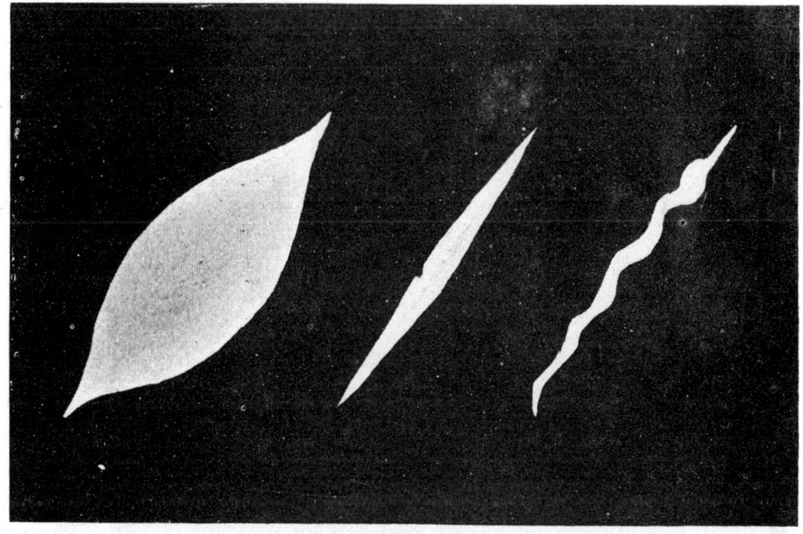

METEORS.
Three views at intervals of a Luminous Meteor of the Fire-ball type.

recognizable laws, either of time or place in the heavens. It is only within the last hundred years or so that these vague ideas have been replaced, and we are now in a position to know for certain that these luminous visitors are, indeed, for the most part, subject to distinct laws, both of time and place. If a would-be observer of Shooting Stars is provided with a properly-prepared calendar for the twelve months of the year, he will learn from that, that year after year Shooting Stars may be observed on particular dates and in particular regions of the sky, these dates being now fairly well defined, and the particular places likewise. Modern observations spread over more especially the last fifty years enable us to say that showers of Shooting Stars of known regularity amount in number to upwards of seventy; that is to say, if an observer, provided with such a calendar, carefully scrutinizes between January and December of any year, seventy particular points amongst the constellations on seventy specified dates, he will be more or less sure to notice a few, or it may be dozens, or, under certain particular circumstances, hundreds of Shooting Stars.

Attentive observation will disclose the fact that in such cases every star of a particular group seems to start from one and the same (or nearly the same) point. Hence the phrase "Radiant Point," which is now in common use in this department of astronomy. Though on any particular occasion no more than perhaps half a dozen Shooting Stars may be seen on a given night, and a superficial observer may think they pass haphazard across the sky; yet such is not always the case, though the paucity of their number may make it difficult for him to believe there is any method or regularity in their movements. I have mentioned the fact that some seventy Radiant Points have been determined, but it must not be supposed therefrom that they are all equally recognizable with

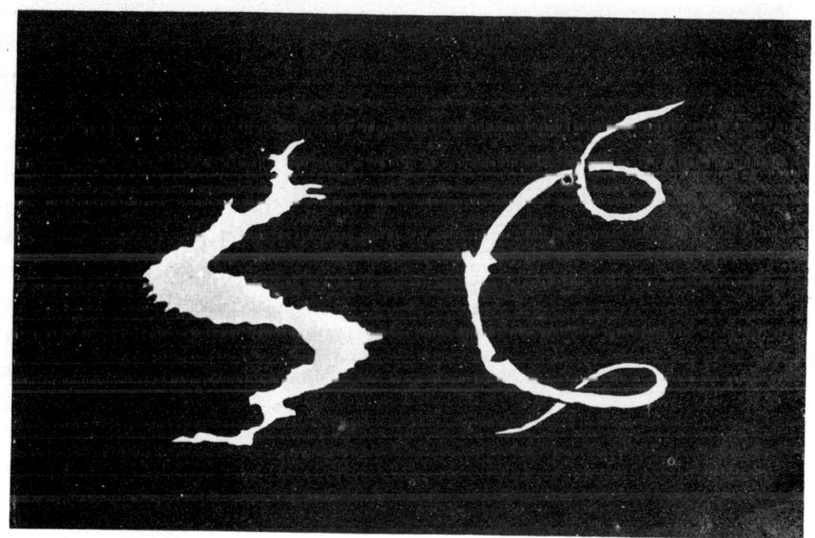

METEORS.
Observed developments of a Luminous Meteor of the Fire-ball type October 19th, 1877. The second figure shows effect ten minutes later than the first.

Photo by] [*E. E. Barnard.*

A METEOR IN FLIGHT.

The luminous object gradually increased in brilliancy and as gradually faded out, ending in two short flashes of light. The above is the trace left on a photographic plate.

ease. There are, however, several which are especially prominent, and the dates and positions of these are given below.*

All these showers appear annually except the Andromeda shower, which only shows itself about every six and a half years, and is supposed to be associated with the lost Comet of Biela, or, indeed, to be the lost Comet reduced to fragments!

The showers of August and November are often on such a large scale, that they have frequently been recorded by non-scientific historians through many centuries, and the special attention given to them by astronomers during recent years enables us to speak somewhat confidently about them.

In addition to the Shooting Stars, which are referred to in the foregoing paragraphs as being subject to laws with respect to time and place, it must be mentioned that we may often see single stray Shooting Stars, of which it cannot yet be said that they belong to any of the known groups. They may seem to have no links with any particular group, and may appear to be quite erratic, yet it is probable that in real truth they are nothing of the kind, though we may be unable, for lack of sufficient observations, to particularize their point of origin and their associations.

Independently of their attachment to particular groups, it must be stated that there is a distinct variation in the visible number of these objects, depending both on the season of the year and the hour of the night. For instance, during the last six months of the year Shooting Stars are much more numerous than in the first six months, almost to the extent of being double in number. This is a perfectly established fact, and applies not only to Shooting Stars, commonly so-called, but to the larger meteors to which I shall have to allude directly. As regards the hourly variation between sunset and sunrise, it is quite certain that these objects are more numerous between 2.0 a.m. and

* Date.	Constellation.	Position in R. A. H.	M.	Position in Declination. °
January 2	Quadrans	15	20	53 North
April 20	Lyra	18	0	33 North
July 28	Aquarius	22	36	12 South
August 10	Perseus	3	4	57 North
October 18	Orion	6	8	15 North
November 13	Leo	10	0	22 North
November 20	Taurus	4	8	23 North
November 27	Andromeda	1	40	44 North
December 10	Gemini	7	22	33 North

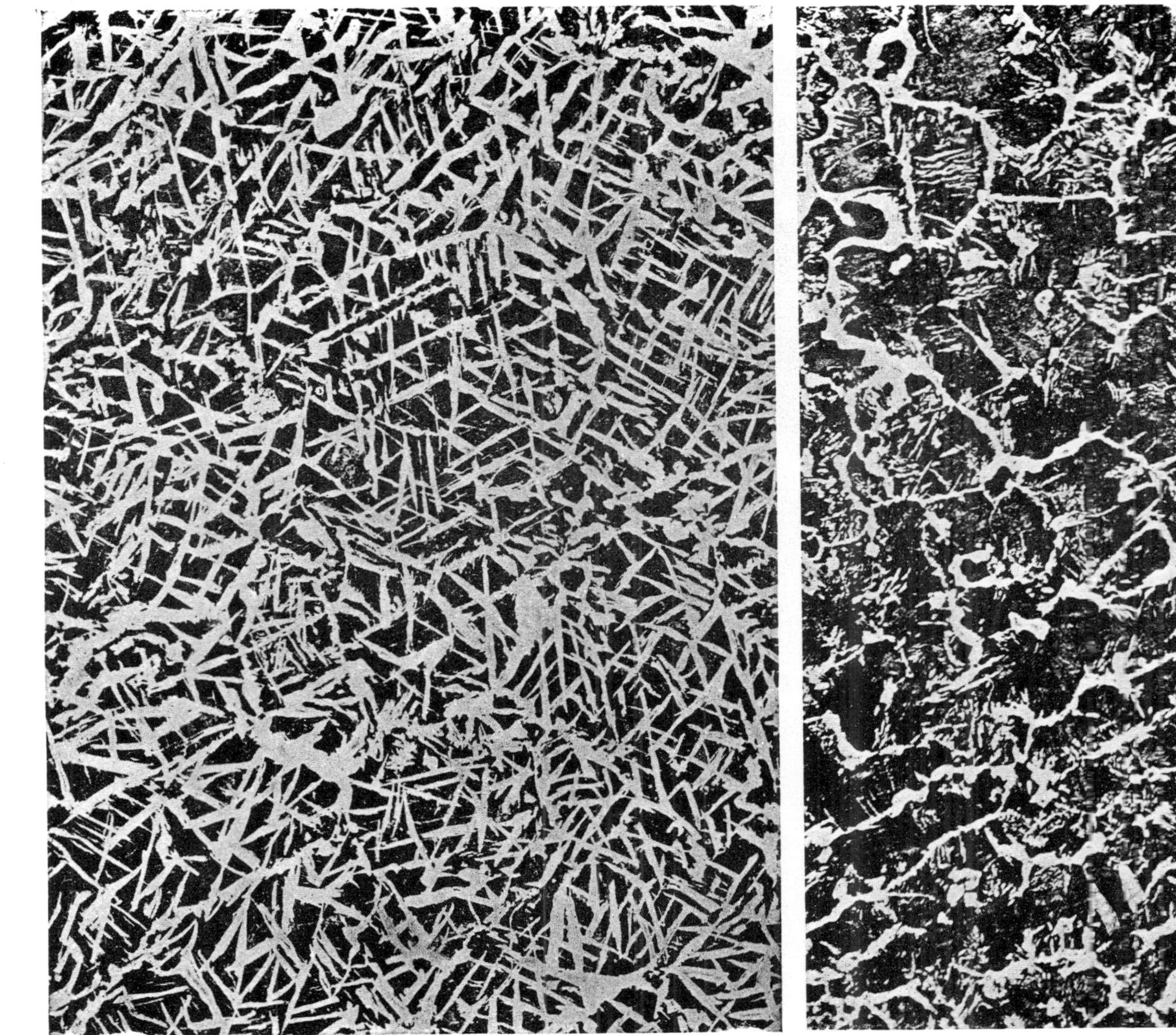

Photos by] [*P. Primrose.*

METEORIC METAL.

These photos show sections cut through portions of Meteorites to show their structure. That to the left is a polished section of Meteoric Iron, enlarged about three and a half times, showing Widmannstätten figures indicative of very slow cooling of the nickel-iron minerals present. The photo to the right is nickel-iron metabolite, magnified a hundred and thirty times, showing the result of heating in passing through the air

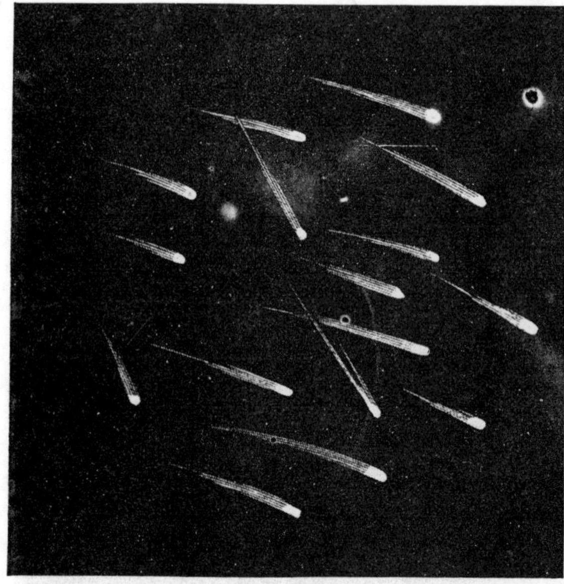

METEORS.
A flight of telescopic Meteors.

3.0 a.m. than at any other time. It is not easy to explain the meaning of these definite variations, and the possibility of explanation is rendered the more difficult by reason of the fact that the Earth is nearer the Sun in the winter than in the summer. Moreover, the distribution of the Radiant Points is very unequal over the visible heavens. They are far more numerous at and around the IInd Hour of Right Ascension than they are at and around the XIIth Hour of Right Ascension. These discrepancies are no doubt partly real, but are also in part due to the difference in the duration of night, when those two portions of the heavens are above the horizon in the Northern hemisphere, where meteor observations have hitherto been chiefly carried on.

It is time to pass on to those other Luminous Meteors which are probably identical with Shooting Stars in their nature, but which manifest themselves in a much more striking and pretentious form. These we call "Fire-balls." They appear suddenly and are usually noiseless, though sometimes an explosion is heard, which may or may not result in the descent to the Earth of an Aërolite or the broken-up fragments of an Aërolite. Their form is generally pear-shaped, the thin end of the pear giving forth sparks seemingly of incandescent matter proceeding from an incandescent nucleus.

These sparks, whatever may be their nature, often remain visible for many minutes, and drift slowly away from the point of apparition, doubtless under the influence of wind-currents.

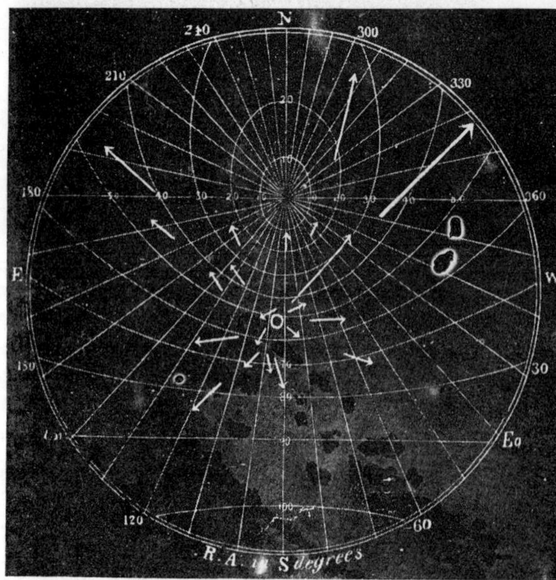

METEORS.
The Radiant Point of the Geminids, December 12th.

The question will naturally be asked, What are these objects which we have been considering? Have they any physical existence when we see them, or are they mere optical manifestations? It cannot be doubted that at some moment they have a physical, and probably solid, existence; that Shooting Stars are small masses of matter which first become incandescent and burn away as they enter our atmosphere; whilst Fire-balls are very much larger masses of more solid material, which also become incandescent and burn and burst into fragments of sufficient size to hold together and fall to the Earth. This last statement seems sufficiently proved by the fact that not only many Aërolites (*i.e.*, stones from the air—Gr.) have been found upon the surface of the Earth, but have been found while still warm to the touch. As to where they originally came from and why they should visit the Earth we have no information—at any rate, as to the former point. It has been suggested, but the idea is too fanciful for acceptance, that they are fragments of some extinct and broken-up Planet wandering about space. Less fanciful, and with some basis of fact to support it, they have been regarded as in certain particular cases to be fragments of some broken-up Comet. As regards their coming to the Earth, that would be simply the result of the attraction of Gravitation, akin to the reason why Sir Isaac Newton's apple fell to the Earth and other apples fall to the Earth whilst philosophers look at trees.

A CURIOUS FUNGUS

At a little distance the photograph on page 1093 has always been taken for a representation of a group of Japanese ladies. It is really a row of a peculiar fungus, very little known except to experienced toadstool-hunters, and many of these have never had the pleasure of seeing it in its native haunts. The White Helvella is not one of those species that you meet with everywhere. It is what naturalists describe as extremely local; but, like many other local species, when you have once discovered its haunt, you may visit the place year after year at the proper season with every prospect of finding it.

A METEOR

This large Meteor was photographed November 14th, 1901, with the telescope pointing to ϵ Leonis. The Meteor is seen half-way up the photo, a little to the right of centre.

The Helvellas are near relations to the Morels, which are favourably known to epicures. They differ from the ordinary types of toadstools, in the fact that their microscopic reproductive spores are not produced on "gills," or in tubes, but from the naked surface of the strangely lobed and twisted head. These spores are so minute that they cannot be seen as such without the aid of the microscope. But if a Helvella is carefully gathered and stood in a quiet room where the air is still, you may observe on approaching that it gives off from its wrinkled upper surface tiny puffs of almost imperceptible smoke. That smoke consists of thousands—perhaps millions—of the microscopic spores, which are puffed out of elastic cells in the apparently smooth surface. The plant is so sensitive that the mere approach of a person, without actual contact, is sufficient to cause this discharge of spores.

It must be confessed that these Helvellas are not so substantial as they appear to be. The solid-looking stem is hollow, and the epicure has to have it stuffed with force-meat. Even the strengthening ribs of the stem are hollow. The texture is thin and brittle. At first it is pure white, but, later, it becomes cream-coloured. There is another species that offers a strong contrast to it, in being coloured a sooty black. It does not present the inviting appearance to the toadstool-eater that attaches to the White Helvella, but it is considered equally good by those who are not wholly guided by appearance in the choice of their food.

PARENTAL CARE AMONG THE FISHES

BY W. P. PYCRAFT, F.Z.S., ETC.

ONE can hardly suppress surprise at the striking demonstrations of affection which one finds among the cold-blooded fishes in regard to the care which they display for their offspring. For in this respect they are in advance of the Reptiles, though their distant relations, the Amphibia—the frogs and toads and their kin—include within their ranks no less striking instances of parental care.

Two illustrations of this aspect of fish-life must suffice for our present purpose. The first of these shall be furnished by the American Bow-fin, which, in the preparation of the nursery for its young, displays a remarkable solicitude. In the spring the male and female come up from the deep water to the reedy shallows to spawn. But, unlike most river-dwellers, they take elaborate precautions to secure the safety of their future offspring. This they do by clearing away the reeds over a large circular area, biting through the stems of all plants that cannot be broken down. Then a shallow depression is formed at the bottom of the pool thus created, and in this the eggs are deposited. These develop with remarkable rapidity, not more than eight days dividing the interval of laying and hatching.

Photo by] [*E. Step, F.L.S.*

WHITE HELVELLA.

A singular fungus of delicate appearance and structure. The ribbed and furrowed stem is hollow. (Natural size.)

The young are most jealously guarded, but by the male only. He keeps the swarm together by circling around and above them continually, and his task can be no light one, since he may have to secure the safety of as many as a thousand individuals.

The part played by the male is the more striking, since he does not exceed more than one and a half feet in length, while his mate attains a length of two feet, a disparity in size recalling what obtains among the birds of prey. The likeness is further accentuated by the fact that the male Bow-fin is the more brightly coloured of the two, the dark mottled-green colour common to both sexes being relieved in the case of the male by a round black spot margined with yellow at the base of the tail.

The Bow-fin is a most voracious fish, feeding largely upon crustacea and insects; but besides, whenever opportunity affords, on other fish. It is also one of those remarkable fish wherein the air-bladder has developed a rich supply of blood-vessels, thereby enabling it to perform the work of a

Photo by] [*Edward Step, F.L.S.*

WHITE HELVELLAS.

A group showing various forms assumed by the "head" of this beautiful fungus. The example to the extreme left has not developed a head at all. The upper surface of these folded lobes bears minute embedded cells from which the reproductive spores are cast out.

BUTTER-FISH

The male of the Butter-fish, also known as the Gunnel, when the eggs have been deposited, carefully gathers them into a heap, and coils his body around them until they hatch.

lung. This change in function has come about to enable the fish to live long out of the water, which it is often called upon to do.

In the Butter-fish of our coasts the care of the family is similarly thrust upon the male. In this species, however, no nest is made. As soon as the eggs are laid, the male gathers them into a heap and coiling his body around them, patiently awaits the time when they shall hatch. In this he resembles that strange and little-known member of the frog-tribe, the ichthyophis, and the python among the snakes. But here the female plays the rôle of incubator. There is, however, this difference between them. In the ichthyophis there is no rise in temperature of the body embracing the eggs, while in the case of the python a marked rise has been observed. More rarely, as in the case of the sea-snakes, the mother coils her body round her young ones till long after hatching ; probably she guards her eggs in a similar fashion, but of this we have no certain knowledge.

The development of the parental instinct among the fishes is by no means easy to understand, especially since this instinct occurs in such a sporadic fashion, and is, having regard to the total number of species, so rarely met with. What is true, however, in the case of the fishes is true also of the frog-tribe. In the majority of cases the eggs are just shed into the water and left to their fate. But, as these pages have already shown, there are a number of singular instances when the young are most jealously guarded. One could understand like habits in nearly-related species, but precisely similar behaviour in animals often not even remotely related demands more of our powers of interpretation than we can yet respond to.

THE BEARDED LIZARD

BY R. LYDEKKER.

Reference has been made in a previous article to the apparently threatening gestures assumed, under the influence of irritation by the Australian Frilled Lizard ; the particular structure in that instance being the expansible frill or collar from which the species takes its name. There is, however, another Australian lizard, generally known in this country as the Bearded Lizard, but in its native land as the Jew Lizard, which is also in the habit of assuming what appears to be an attitude calculated to terrify predacious birds and other enemies. In this case, however, the mechanism by which the terrifying action is produced is of a somewhat different, although kindred, type. For the Bearded Lizard is one of a group of about fourteen exclusively Australian and Tasmanian species, all of which are furnished with a strongly-developed transverse fold on the neck. Under the influence of excitement, this fold, at any rate in the case of the Jew Lizard, is capable of being inflated in such a manner as to produce the semblance of a frill comparable to that surrounding the whole neck of the Frilled Lizard. Indeed, in parts of the country where the latter species is unknown, the colonists are in the habit of applying its name to the Bearded Lizard.

Both the Frilled and the Bearded Lizard belong to a family (technically known as the Agamidæ) which takes the place occupied in Europe by the typical lizards (Lacertidæ) ; the most distinctive feature of the former group being that the teeth are attached to the summits of the jaws instead of growing up on one side of an elevated parapet.

The bearded species, which appears to range over the greater part of Australia, has the depressed form of body characteristic of ground-dwelling lizards in general, the base of the long tail being likewise flattened in a similar manner, although it is elsewhere cylindrical. In length a full-grown specimen grows to about twenty-one inches. As regards colour, the Bearded Lizard is either uniformly brown above or brown with symmetrically arranged darker markings ; as a rule, there is a black spot on each side of the neck, the lower surface is brown or brownish, either of a uniform colour, or marked with darker or lighter spots, and in adult males, in which alone the fold attains full development, the throat is blackish.

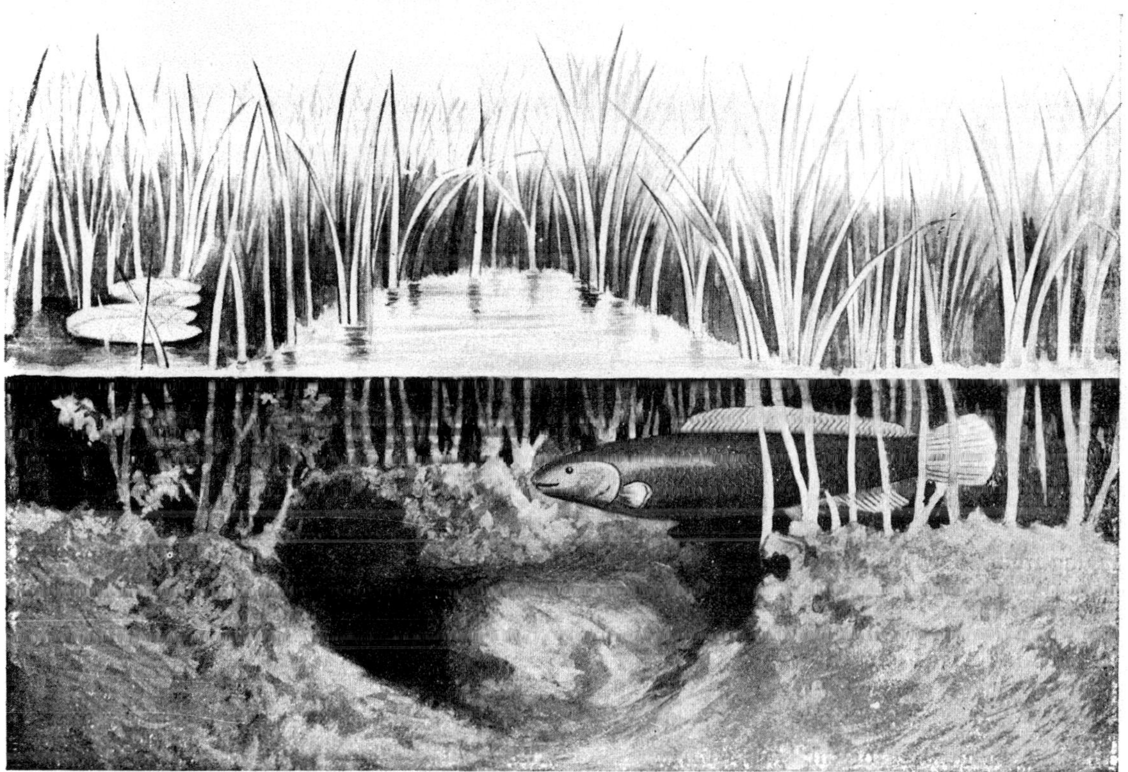

THE BOW-FIN.

The Bow-fin is a large freshwater fish of American rivers. Both sexes unite to clear a considerable space of weeds, etc. in order to make a nursery-ground ; but when the eggs have hatched the care of the young devolves upon the male entirely. He is about one and a half feet in length ; the female about two feet.

As details of the habits of these lizards appear to be lacking, it must suffice to add that the inflatable throat-fold of this species, like the frill of the Frilled Lizard, is strictly comparable to the so-called hood of the cobra.

GROWING GOLD

In the early days of gold-washing, the miners always kept a sharp look-out for "nuggets" among the sand and gravel in which they got the pay-dust. The fine, flattened particles were quite evidently deposited by the river, which had carried this gold from the exposed and weathered surface of the original vein; but the only explanation possible to account for the heavy masses of gold,

Photo by] [*W. Saville Kent.*

THE BEARDED LIZARD.

A native of Australia which has a strongly-developed transverse fold of the skin on the neck, which it is in the habit of inflating in such a manner that it terrifies predacious birds and other enemies.

ranging from a few ounces to several pounds in weight, was that they had "grown" in the pockets where they were found. Although comparatively rough and apparently water-worn on the outside, the nuggets were certainly not carried down by the stream which laid down the gravels; but the fact that the nugget-metal was always purer than the reef-gold and, unlike it, highly crystalline in structure seemed to prove the truth of the old miner's saying. Many organic salts found naturally in the earth are capable of dissolving gold, and this solution would be robbed of its gold by contact with quite a number of minerals, such as iron pyrite. The examination of a large number of Australian nuggets has led to the conclusion that they have not been built up of concentric coatings round a nucleus, and although they possess a crystalline structure like fused gold, they have not been so formed; but rather deposited from solution in pocket-like cavities in the rocks where the chemical constituents favoured this precipitation.

Photo by] [*W. Saville Kent.*

THE BEARDED LIZARD.

The Lizard is here seen from above. It will be noted that it has the depressed form of body characteristic of ground Lizards in general. It is about twenty-one inches in length, and of a brown colour.

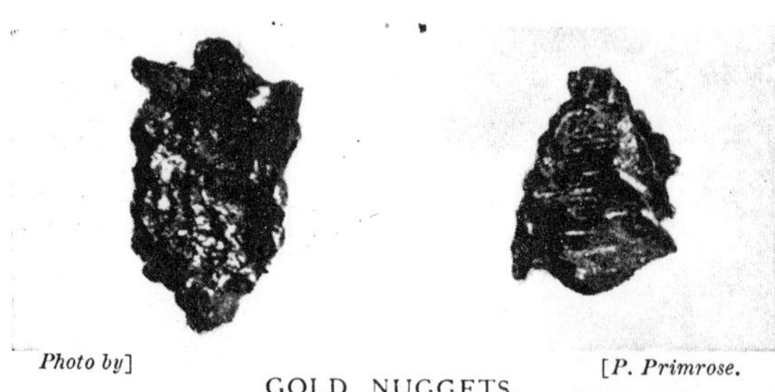

Photo by] [P. Primrose.

GOLD NUGGETS.

In this photo of nuggets from the United States of Colombia, the leaf-like appearance of the surface is well shown.

The wonderful crystalline structure of nugget-gold is not very apparent from an inspection of its knobby surface; but on cutting it through and polishing the precious metal, the mode of arrangement or orientation of the crystals is well developed by a slight etching with aqua-regia. The photograph reproduced in the illustration shows a section of a small gold nugget, taken from a pocket in the banks of a stream in Colombia, South America, and this has been magnified about twenty times. The surface of this nugget presented a peculiar foliated, or "graphic" appearance, which the local miners attribute to the presence of a small quantity of the element Tellurium alloyed with the gold. Two similar nuggets from the same locality are shown slightly reduced from their natural size, and these also exhibit the characteristic laminated external structure, and in some of the interstices small particles of pure quartz have become embedded.

JELLY-FISHES

In one of our coloured plates we present the portraits of a few of the remarkable animals popularly known as Jelly-fishes. The name is "good in part," like a certain famous egg. The good part is the first word, for it very aptly describes the material of which these fairy-like creatures are built. Often the summer visitor to the seaside will find all that remains of one of the larger forms that has been left by the receding tide high and dry on the beach. The water of which it was largely composed has evaporated and left a mere film of hardened gelatine. Out of such slight material Nature fashions some of her most exquisite living forms. We have already dealt with several distinct types of the Jelly-fishes in this work, and it would be idle to go over the ground again, while so many other things await attention. We will content ourselves with mentioning the strange friendships which exist between certain species of Jelly-fishes and the fry of real fishes.

Photo by] [P. Primrose.

NUGGET-GOLD

A section through nugget-gold, magnified twenty times. This gold has been deposited from the decomposition of auriferous solutions. Some authorities claim that this must have been fused, but this is not possible under the conditions of the surrounding country.

The illustration on page 1099 shows one of our larger Jelly-fishes, known as Rhizostoma, sheltering among its extensive ribbons a number of young horse-mackerel. Many of the Jelly-fishes possess long stinging threads, with which they paralyze the small creatures upon which they feed. Rhizostoma does not possess this power in so high a degree as many other species, but doubtless it benefits from the known deadly character of these others. Among these is the Common Stinger

(*Cyanea*), which in the same manner shelters the fry of the whiting.

An American species is made similar use of by a kind of herring; and in Australian waters there is a fish called Nomeus, which is never found except under the protection of a fleet of the Portuguese Man-of-War.

These are strange things, but they are well-authenticated by reliable observers, though these have not yet succeeded in showing *why* the Jelly-fish tolerates companions that it could so easily destroy.

Although on our coasts Rhizostoma is usually something under a foot across, it grows to a much larger size in suitable waters. It is more solidly built and not so graceful as some of its companion species shown in our coloured plate.

Chrysaora, for instance, that is at the surface of the water in the plate, has delicately-frilled streamers nearly a foot in length depending from the margin of its mouth, and from the edge of its umbrella there are a great number of long thread-like tentacles for catching food.

These threads in many of the species are furnished with stinging hairs, which paralyze small creatures by contact, and affect the tender human skin much like the stings of the nettle. Some writers have given rather terrible accounts of the after-effects of these stings, but much depends on individual sensitiveness.

[By T. Carreras.

JELLY-FISH AND HORSE-MACKEREL.

The fry of the Horse-mackerel commonly follow and swim around this Jelly-fish, known as Rhizostoma, keeping under the shadow of its big umbrella and when danger threatens sheltering among its trailing ribbons.

THE WOOD ANT

BY H. ST. J. K. DONISTHORPE, F.E.S.

The Wood Ant is widely distributed in Britain, living chiefly in pine-woods, where it builds up its well-known hillocks. These are chiefly composed of pine-needles, but the ants also use small bits of stick, leaves, and any vegetable or other material handy. We have seen a nest chiefly composed of small pebbles, and the ants are very fond of carrying in lumps of yellow resin from the pine trees, called "ant amber." Some nests in the north are often full of it. In Leicestershire, where this ant occurs in oak woods, the nests are built chiefly of oak twigs, and in some places nests built of grass may be found. A near relation always uses grass and ling.

In the course of time the Wood Ants' hillocks, if undisturbed, grow to a very large size, and as a nest can exist for many years (the writer has known one at Weybridge for over twenty), the number of its inhabitants often reaches a total of many hundreds of thousands.

The habit of building hillocks in this ant and its allies shows their descent from a more ancient Arctic life, as the heaping up of these mounds better attracts the rays of the sun and also concentrates the warmth in the nest.

These ants are very strong, and can lift enormous weights compared with their size; and when one of their ant-hills has been overthrown, they very quickly build it up again. Under the hillock is a large central chamber, which is connected with many other chambers by galleries into the ground, and other parts of the hill.

A colony of the Wood Ant often extends over a very large area, and consists of many nests bound together by roads and tracks, and all the ants are on friendly terms with each other. If, however, some from a strange nest and a different locality are introduced, they are at once attacked and killed.

Photo by] [*H. Bastin.*

NEST OF THE WOOD ANT.

These nests, which are constructed of twigs, pine-needles and other woodland debris, are often conspicuous objects in the pine-woods, as they reach a height of two or three feet, and are connected by roads and tunnels with other nests at a distance.

These ants do no harm—in fact, they are of much use, as they clear all the trees and bushes of caterpillars and grubs, besides capturing any insects they come across. These are dragged back to the nest, to be cut up and given to the ants' brood. They partly paralyze their prey with acid which they eject. When a nest is disturbed, the smell of the acid squirted out by hundreds of ants will be found to be exceedingly powerful. This acid is very strong, some I had tested containing sixty to seventy per cent. of pure formic acid.

Unlike some ants, these do not possess a sting, but rely upon their jaws and acid for their defence. They are exceedingly courageous, and will attack anything. They travel long distances in search of food, and climb all the trees in the neighbourhood, using regular tracks and runs.

On a fine day hundreds of workers may be observed leaving the nest, whilst many more are returning laden with food of all kinds, three or four ants often helping to carry home the more heavy burdens. They also seek out and milk any plant-lice that may occur on the neighbouring trees and herbage. Others drag up fresh pine-needles and other objects to enlarge the nest. At night the different doors in the hillock are closed by arranging bits of the nest materials over them. Sentinels are also always at the various entrances to the nest.

The females of this species have lost the power to found colonies in the normal way; that is to say, a young female is unable to start a colony alone. In many ants, after the marriage flight, the young female removes her wings, seeks a suitable spot under a stone, or in a hole in the ground, and digs a cell. There she lays her eggs and brings up her first brood. In the Wood Ant the

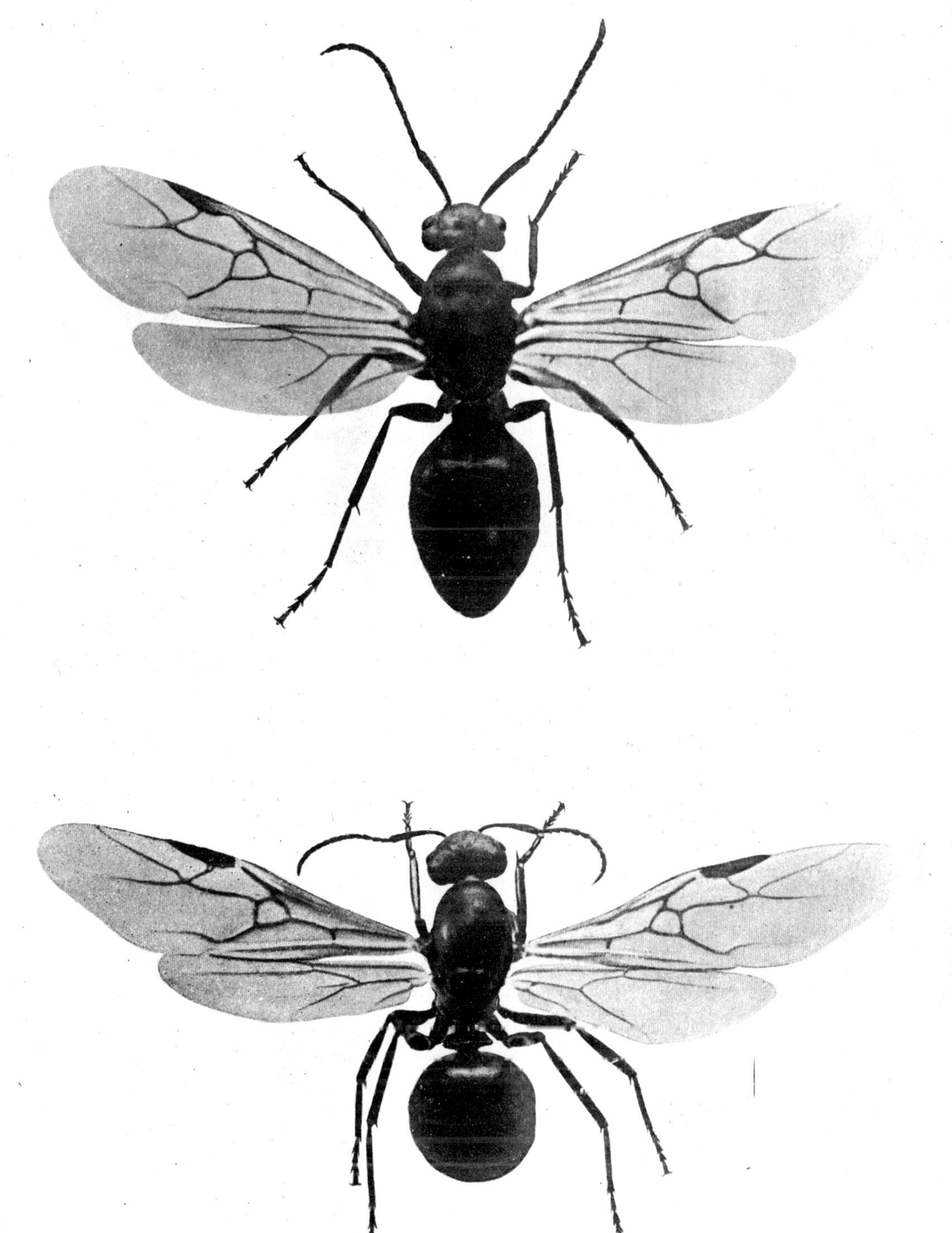

Photos by] [*Harold Bastin.*

THE WOOD ANT.

Here are shown photographs of the male and female Wood Ants in their winged state. The wings serve the purpose of the nuptial flight and are then abandoned by the female. The workers are never winged.

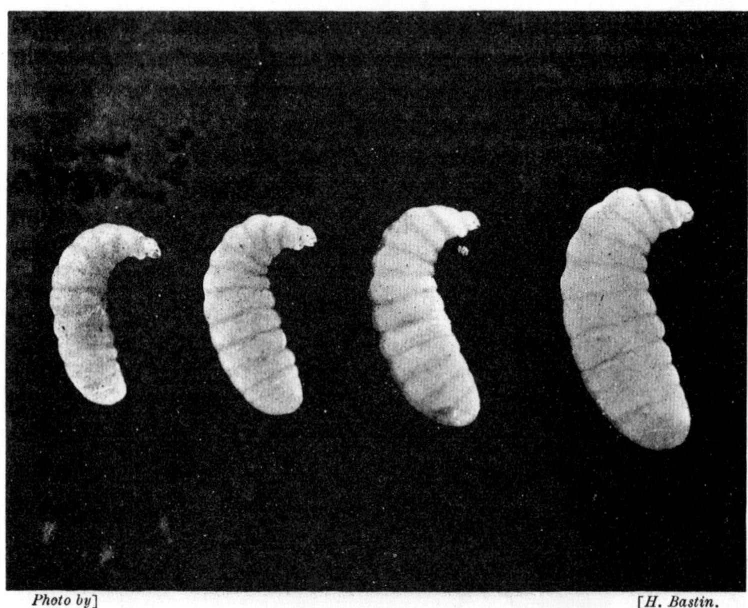

Photo by] [*H. Bastin.*

THE WOOD ANT.

The white grub of the Wood Ant is here shown in four stages of growth. It will be seen that it is legless and helpless, and it depends upon the workers for food and movement.

females are unable to do this, but are received back, after the marriage flight, into their own or some other of the many nests in a large colony.

The fact that numbers of branch nests often extend over a large area is probably the cause why these females have lost the power of colony founding alone. Should she have flown to a place where there are no nests of her own species, she seeks a small incipient or depauperated colony of a Black Ant and endeavours to enter its nest. She is sometimes killed, but often succeeds in inducing the Black Ants to accept her, when they bring up her brood. Should a black queen be present, she will be killed by the Wood Ant. The writer has seen young females entering the nests of the black species, and has found small, incipient mixed colonies of the two species in nature. He has also proved by experiments with observation nests, that the black species will accept a young female Wood Ant as their queen. In the course of time the Black Ants die off, and a pure Wood Ant colony is the result. Under these circumstances the female is a temporary social parasite, and by such humble means the vast and wide-reaching nests of a populous colony may be brought about.

The males of the Wood Ant are black, and possess four wings, as is the case with most male ants. The females and workers are red and black, the former having wings until after the marriage flight, but the workers never have any. The marriage flight takes place in the spring and early summer, after which the males die, and the females start to lay eggs.

The eggs, which are licked and carried about by the workers, grow a little in size before the young grubs hatch. The latter are fed by the workers, both from their crops and by providing them with cut-up animal food. When full grown, the workers cover them with small particles of earth, and they spin cocoons in which they change to chrysalids. These are the so-called " ants' eggs," which are collected and given to partridges and pheasants. The true eggs are, of course, very much smaller. Sometimes the grubs turn to chrysalids without first spinning a cocoon, and are then naked, as is always the case with a great number of other kinds of ants.

The workers carry the brood about to the different chambers in the nest, for more moisture or heat, as the case may be. They help the newly-hatched ants, which are called " callows," out of the cocoons, and free them from the chrysalis skin.

A large Wood Ants' nest is also an asylum for many other creatures besides the ants themselves. A small Yellow Ant, which has a wingless male, only lives in these nests. It is also the home of a tiny spider, numbers of different beetles, some flies, and various other small insects and mites. Some of these prey on the ants; many are scavengers in the nest; others, again, are true parasites, laying their eggs and living in the bodies of their hosts. Some beetles only pass the earlier stages of their existence in the nest. One of these is a large green " chafer," the female of which bores her way into the nest and lays her eggs there. The young grubs feed on the nest materials, etc., and when full-fed, make a hard cocoon of earth and vegetable refuse in which to pupate. When the perfect beetle emerges, it flies away.

Another beetle, Clythra, which is red spotted with black, somewhat like a " lady-bird " in appearance, has a most interesting life-history, which has been elucidated by the writer. The beetle itself feeds on the young leaves and shoots of birch-trees, etc. Its female seeks a tree or shrub, above or near a nest of the Wood Ant, and drops her eggs on the ground. She covers the eggs, which she holds with her back feet, with excrementitious matter, which she arranges in layers. This makes them resemble a small bract, or part of a plant ; in fact, they look exactly like the end of a birch catkin. The ants pick this up and carry it into the nest, as they do with bits of vegetable refuse. When the young grub hatches, it builds a small black case on the covering of the egg. This case has V-shaped ridges on one side, and is constructed of earth mixed with excrementitious matter. As the grub grows it enlarges its case, by scraping off portions of the inside with its jaws, and plastering more on the outside. It feeds on the vegetable refuse of the nest, and also on the pellets of the ants. When full-grown, the grub fastens the mouth of the case to a bit of wood, or other object in the nest, and turning round inside the case, changes to a chrysalis. When hatched, the perfect beetle cuts a rim round the end of the case with its jaws, forming a cap which it forces off. It then crawls out of the nest and flies away.

A SEAWEED THAT MAKES LIMESTONE

BY F. CHAPMAN, A.L.S., F.R.M.S.

THERE are some seaweeds, like the jointed Halimeda and the nullipores of the coral reefs, as well as the corallines of the British shores and the Mediterranean, which have the remarkable faculty of encrusting their fronds with carbonate of lime.

Halimeda is a green seaweed, which flourishes luxuriantly on the outer reef slopes of coral islands, and also in the lagoon. At Funafuti, in the South Pacific, it was found from a foot or so below sea-level to forty-five fathoms. It forms great tufty masses, and where the conditions are favourable, the separate triangular joints of the seaweed accumulate to form a thick bed of limestone.

Limestone rocks due to the agency of this little plant have been found in the Solomon Islands, Christmas Island, at Funafuti, and in the New Hebrides. (Collected by Dr. Douglas Mawson, of the present Antarctic Expedition.) In the deep boring at Funafuti the detached joints of this seaweed

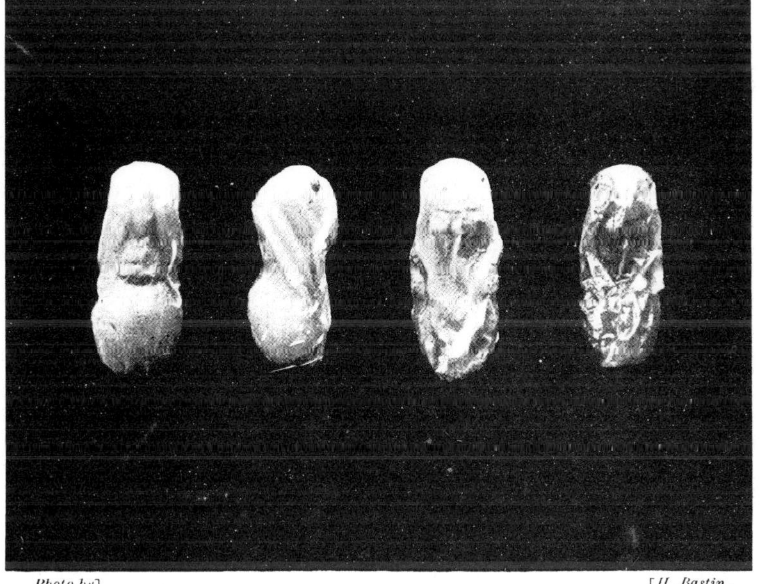

Photo by] [*H. Bastin.*

THE WOOD ANT.

Here are shown the cocoons containing the chrysalids, whose forms partly show through. It is these cocoons that are sold as ant-eggs for feeding pheasants and other birds.

were easily recognized at six hundred and fifty-two to six hundred and sixty feet, showing the characteristic cell-structure beautifully preserved, and similar to the magnified figure given on page 1107.

EARWIGS : HOW THEY OPEN AND CLOSE THEIR WINGS

BY FRED ENOCK, F.L.S.

CAN anyone be so far gone in his love for insects as to sing the praises of Earwigs? Those "horrid Earwigs," that, according to those wonderful repositories of facts the oldest inhabitants, walk into our bedroom in broad daylight, and when we are wrapt in sweet slumber, creep over our beds and into our ears, and endeavour to find a way to our brain! Speaking from personal experience, I have, when beating foliage for larvæ, frequently had all kinds of insects, including Earwigs, fall on my neck and face, and as hurriedly endeavour to find cover. Those that are active flyers, soon spread their wings, but spiders are more deliberate in their actions, and attempt to make snares around one's head. It is astonishing how soon we can recognize the tread of a spider, or an insect, and how deliberately a looper caterpillar seems to walk round one's neck, measuring its paces until a sudden movement of the head causes it to drop down somewhere between our shirt-collar and collar-bone.

Photo by] [*H. Bastin.*

ANT-NEST BEETLE.

Clythra, a beetle with wing-covers spotted much like those of a Lady-bird, lays its eggs in the nest of the Wood Ant, and the young pass their grub and chrysalis stages therein.

At such a time there is no mistaking the subject of our notes, the Earwig, for it drops down rather heavily, and, being of a somewhat retiring nature, it quickly endeavours to hide itself, either in the deep recesses of our flannels—up our sleeves, and not infrequently it falls upon our ears, and—creeps in—and out again. It wants to hide, so takes the first chance that offers, ears, collar or shirt-sleeves.

As usual, there has been considerable discussion among philologists as to which should be its right name, Ear*wig* or Ear*wing*. To many it will be news that these insects have *wings*; for the organs of flight are not seen except by those who study and search out the marvels of insect life; but when fully expanded, the Earwig's wings resemble in form the lobe of our own ears.

Sometimes there is no accounting for an insect's name. Some years ago, when in the North of England, I learned that the local name for an Earwig was a "Worzlebug"! It is a well-known fact that in the United States of America all insects are called "Bugs."

It is always a pleasure to speak well of much-maligned insects, but few will be inclined to give any credit to an Earwig—which every gardener knows eats out his chrysanthemum buds, nibbles the petals of his favourite phloxes, and takes up apartments in every petal of his dahlias. We are bound to admit that Earwigs do all the damage put down to their account, but they also help to keep down green fly, as I can testify, for I have often, when preparing an Earwig for microscopic examination, found its viscera full of the remains of hundreds of green flies.

Photos by] [*Harold Bastin.*

THE WOOD ANT.

The Ant to the left is an egg-laying female; that to the right is a typical worker.

This beneficial work is always carried on at night. It is no uncommon sight to the entomologist to find a goodly company of Earwigs feeding upon the " sugar " spread on the tree-trunks to attract the moths. On several occasions I have found them with their wings spread wide open, and stuck fast to the sticky sugar. On warm, windy nights, a trap consisting of a tarred board set up in a garden may arrest the flight of these Earwigs. We have an Earwig much smaller than our familiar common species, that flies actively in the hot sunshine, and whose wings are very similar, but not so delicately formed.

Now that we have given our evidence to the fact that Earwigs have wings, let us spend a short time in examining these wonderful organs, and I venture to think that we shall find that we can admire the marvellous way in which they are fashioned, spread out and folded up.

Looking at an Earwig as it walks along, it appears to be divided into four parts—the head, showing the eyes, antennæ or feelers and the mouth ; next, the shield-like thorax, to which are attached the six legs, the two horny wing-cases, and the two wonderful wings, of which the tips only are seen, resting on the shiny brown body, which is terminated by the two powerful pincers. The latter are used as organs of defence, and also to assist in unfolding and folding up the wings.

[By F. Chapman.

HALIMEDA.

A jointed seaweed whose fronds are encrusted with lime, which helps in decay to build up thick beds of limestone.

It is a delicate operation to unfold and spread out these wings, and requires the greatest care lest the membrane be torn ; but if the student desires to understand how they *are* folded, he must be prepared to spend considerable time and patience before he can unfold one out to its fullest extent, and then fold it up again !

We will suppose that, after sacrificing a large number of Earwigs in our endeavours, we have at last succeeded in opening out one to its fullest extent, as shown on page 1108, and our astonishment at the extent of each wing is very real, and more so when we realize that this beautiful piece of mechanism adorns the much-despised Earwig !

Now, let us take in all the detail of the wing. First, we find it is attached to the body by a leathery-looking piece forming part of the " armour " of the thorax, the internal ends of this piece being connected with the flexor and extensor muscles which control the opening and closing of the wings, together with all the delicate connections to each nerve, in a truly marvellous manner.

The two horny wing-cases are held open in front, to be out of the way, so that the true wings are, when expanded, flat and flimsy. Immediately in front and joined to the thorax, is a horny rib or stiffener, which acts the same as a spring in an umbrella, which, as we all know, must reach a certain point before the umbrella is fully expanded or before it will remain open. At the dip of this stiffener, we see a number of ribs radiating to the outer margin of the wing. These ribs or creases do not *all* reach the centre point, and at about one-third of the width of the wing a crease runs all round parallel to the outer margin.

Now, so long as the Earwig desires the stiffener keeps the wing taut—just as does the spring in an open umbrella. If we desire to close the umbrella we must press the spring, and the ribs and material immediately fall into folds.

So it is with the expanded wing of the Earwig, which, directly the muscle is relaxed, begins to fall into a number of pleats, exactly like a lady's fan ; in fact, there are two fan-like actions taking place as the wing closes. The tip of the wing drops, bringing the radiating creases closer together. As the movement continues, they close like a fan, and in so doing start another central point on the transverse crease. As the first fan is closing, the ribs are turned to the back of the wing, and draw the second series of ribs and creases underneath the stiffener until the main part of the wing is folded up so tightly that only a small part of the wing beyond the transverse crease is flat. At this stage the Earwig requires a little help, and turning up her body, she opens wide the pincers as shown on page 1110, the points just press upon the flat part of the wing, and the act of closing the pincers gives the wing the necessary fold, and the tail straightens out again, the folded wings are dropped and the wing cases assume their natural closed position (page 1112).

Photo by] [*F. Chapman.*

SEAWEED LIMESTONE.

A section of limestone-rock composed of joints of Halimeda, from Malekula, New Hebrides. (Natural size.)

Apart from its wonderful wings, the Earwig shows that it has maternal instincts, in its remarkable care for its eggs, which, in spring, are often turned up and scattered when digging over one's garden. It is most interesting to watch how carefully the mother Earwig will collect these eggs, taking them one by one to some place of safety underground—where she broods over them until hatched. The young Earwigs remain with the mother until large enough to take care of themselves.

It should be added that all our figures are taken from the female Earwig. The sexes may be distinguished at a glance by the shape of the forceps, those of the female being almost straight, their inner edges meeting when closed. Those of the male are of more pleasing shape, taking the form of a pair of calipers, their edges not being capable of meeting except at the tips. In both sexes they are habitually carried open, with their tips higher than their base—the most impressive attitude, because it enables them to be seen at once by an approaching possible enemy. But the Earwig is by no means an aggressive insect, one might say that appearances are in favour of regarding it as a coward, for in daylight it is always seeking cover, and appears to share the belief attributed to the ostrich, that if it can only hide its head in some crevice its body will not be seen. It may be, however, that this trick is due to a sensitiveness of its eyes to daylight, it being essentially a nocturnal insect. It is then that it gnaws at the edges of the

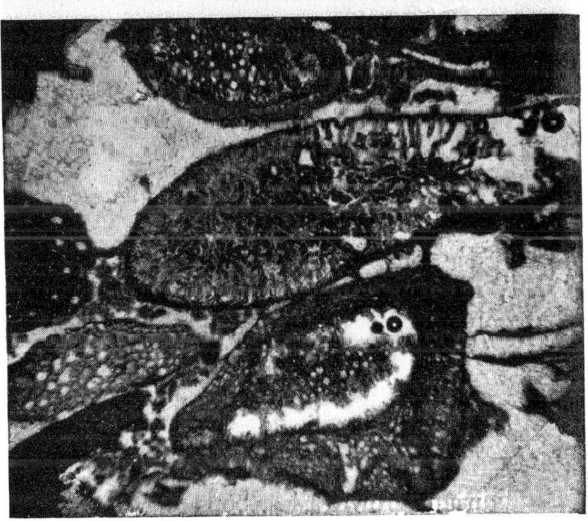

Photo by] [*F. Chapman.*

SEAWEED LIMESTONE.

A section of the rock shown above, but here seen under the microscope, highly magnified.

gardener's most promising flowers, and only with the strong morning light does it seek to hide in the traps that he has cunningly set nice and handy for it.

GUEREZA MONKEYS

BY R. LYDEKKER, F.R.S.

THE forests of Abyssinia are the home of a rather large black-and-white, long-tailed, and long-haired monkey known to the natives of the country as the Guereza. This monkey, which belongs to a rather large and exclusively Tropical African group of species characterized by the rudimentary condition or practical absence of the thumb, presents a very striking type of coloration, a band across the forehead, long tufts on the cheeks, and a mantle of still longer hair extending from the shoulders to the hind region of the flanks, as well as a long tuft clothing the terminal third of the tail, being glistening white, and thus forming a striking contrast to the deep black of the rest of the coat. This monkey ranges some distance southwards of Abyssinia, through the forests of East Africa; but still further south, more especially in the neighbourhood of Mount Kilimanjaro, its place is taken by a closely-allied species, characterized by the still greater development of the white tufts and fringes, and accordingly known as the White-tailed Guereza. In this beautiful monkey the pendent white mantle is much longer and more extensive than in its Abyssinian relative, while the tail, which, with the exception of a very small section near the root, is wholly white, is clothed with very long drooping hairs, recalling the "flag" of a setter's tail. On the other hand, the long tufts on the cheeks of the Abyssinian species have altogether disappeared, so that the head is covered only with short hair, which is white on the chest and throat.

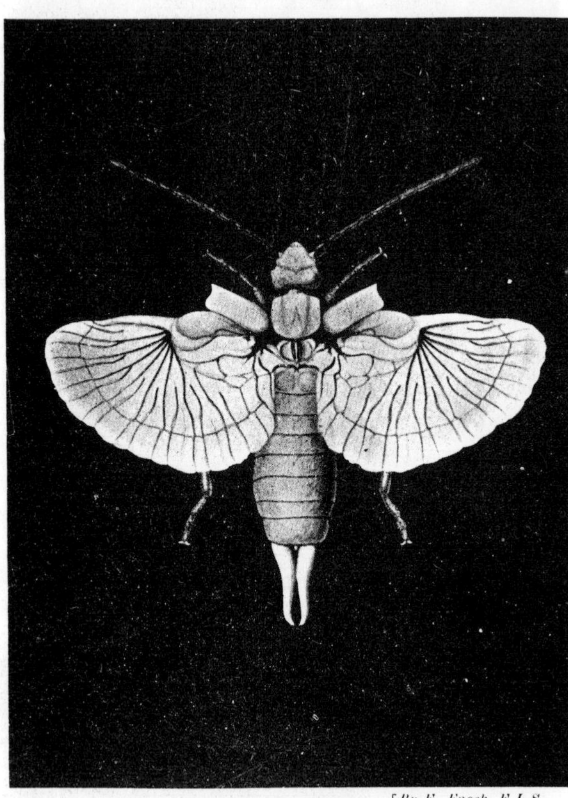

[*By F. Enock, F.L.S.*

COMMON EARWIG.

The Earwig is here shown with its wings fully extended, and the wing-covers raised out of the way.

Returning to the Abyssinian species, it is interesting to note that there is a gradual gradation from this species, by means of various intermediate forms, to the Black Guereza of the forests of the West Coast.

That such a striking type of coat and colouring as is presented by the White-tailed Guereza has some meaning and object can scarcely be doubted; and explorers of the tropical forests of East Africa tell us that the long white mantle on the flanks and the heavily "feathered" tail have a protective value. For we are also told that the branches of the lofty forest-trees of the Kilimanjaro district are clothed with great masses of a long pendent, greyish-white lichen, and that amid this luxuriant growth the troops of White-tailed Guerezas which make these forests their home are practically invisible. It is, in fact, a striking case of the protective resemblance of an animal to its inanimate surrounding; and it is this which justifies the inclusion of these White-tailed Guerezas among "Marvellous Animals."

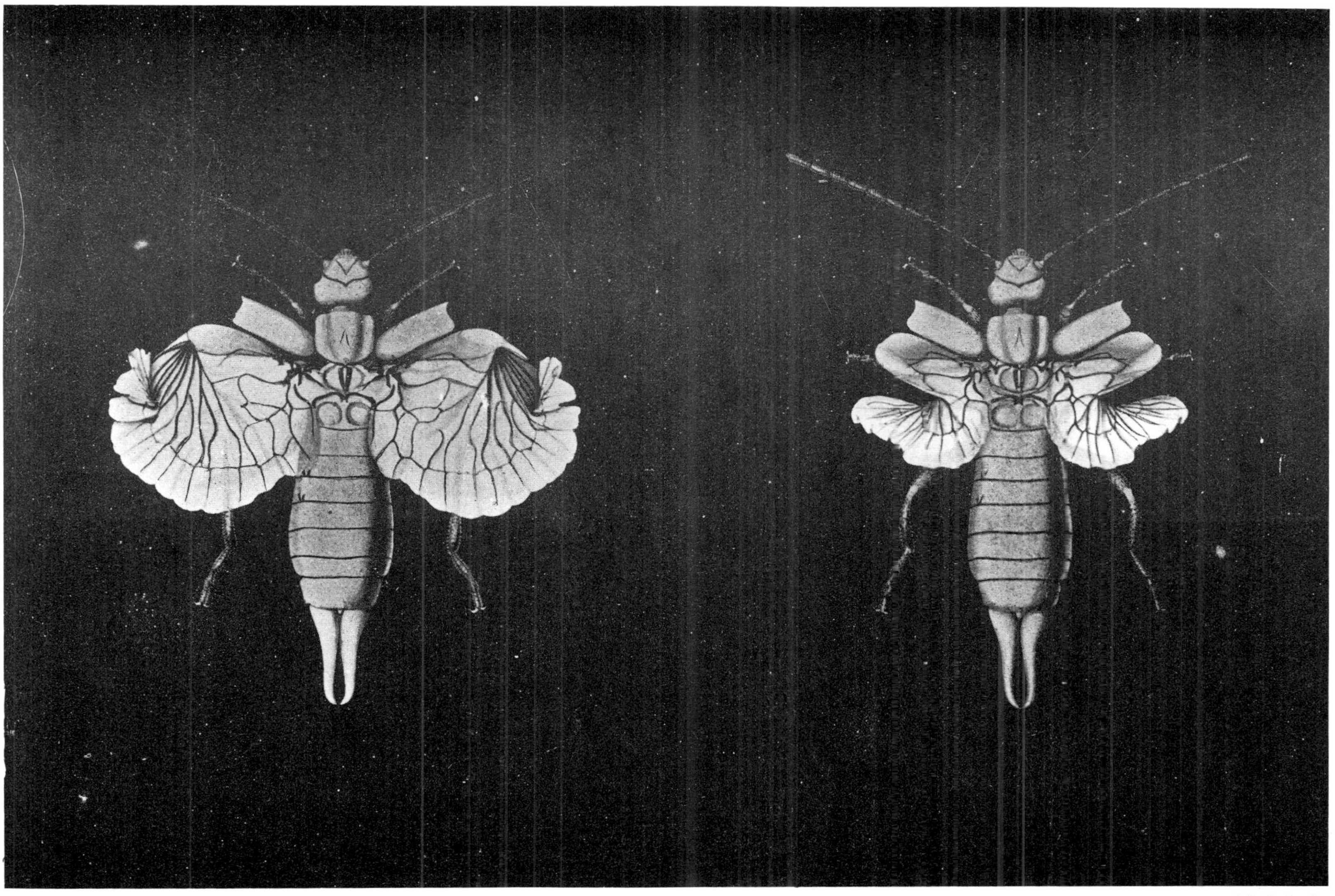

[By F. Enock.

THE EARWIG BEGINS TO FOLD UP ITS WINGS.

The wings fold up like the pleats of a lady's fan. There are really two portions of the wing where this fan-like folding takes place: the tip of the wing drops, bringing the radiating creases closer together, and as the first fan closes the ribs are turned to the back of the wing and the second folding begins.

That these monkeys should thus harmonize with their surroundings cannot be regarded as a very surprising, although undoubtedly a very interesting, fact. The real difficulty, as in so many analogous cases, occurs in connection with the intermediate colour-grades between this species and its sable relative of the West Coast. So far as we know, their colouring is not specially protective, and it is impossible to imagine that they represent the intermediate stages in a kind of blind natural selection culminating in the white-tailed species.

SPIDERS' EGG-BAGS

In addition to the net-like snares and household tubes spun by spiders, there is another form of weaving adopted by some of them for the protection of their batch of eggs. These are in many cases closely-woven tissues; and each of the groups into which the Spider family is divided has its own special pattern of cocoon. Some of these are quite familiar objects in the angles of board fences; but others, though tolerably common, are only evident to the eyes of a naturalist. Such, for example, is the egg-cocoon of the Wolf Spider, a spherical affair about the size of a pea. This spider, being a nomad, has no home, and apparently cannot trust herself to leave her precious eggs where she may not be able to find them again. So she carries them constantly about with her, either holding the cocoon in front of her or attached to her under-side. It then looks like a part of her body, and is generally accepted as such, except by those who know her habits. If she is deprived of it, she is manifestly much distressed, and appears to show great relief when it is restored to her.

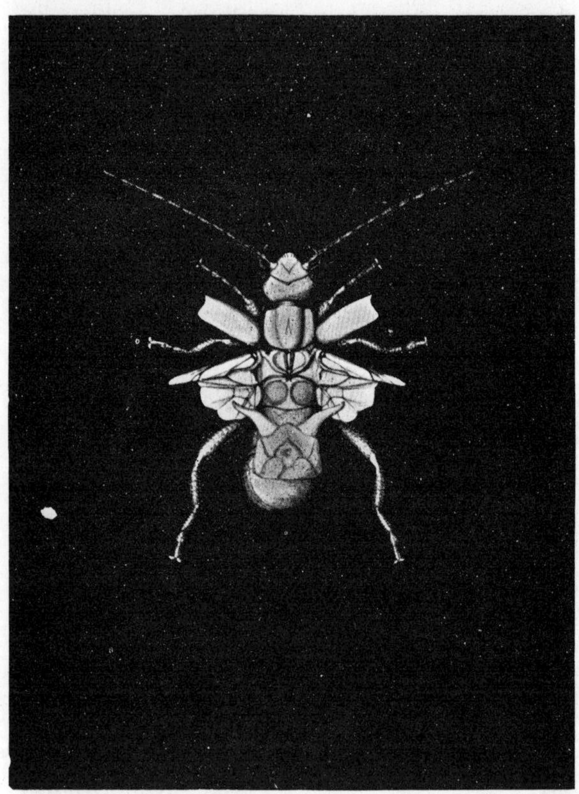

[*By F. Enock, F.L.S.*

THE WINGS HALF-FOLDED.

The front portion of the wing is now packed up so tightly that only a small part remains flat. The wide-open pincers are brought into use to fold this portion.

When the eggs hatch and the young spiders emerge from the cocoon, they mount to the mother's back and there crowd together for about a week, when they disperse and each sets out upon a separate hunting career. The illustration of Pardosa on page 1115 shows the young spiders crowded on the parental back. Until their dispersal they take no food. They do not merely walk off in different directions; this would probably result in many of them remaining so near their parent that in hunting they would compete with her. Instead, each diminutive spider scrambles up the nearest plant to the highest point, and then, standing on tip-toe, with the hind-body raised in the air, it pays out from its spinnerets a long filament of gossamer which floats out on the slightest air-current; and when a sufficient length of it is afloat the little spider lets go his foothold, and, hauling in a little of the gossamer, sails off at the end of it. In this way they are often carried to great heights and to long distances horizontally, so that the chances are that they never again meet with the parent who was so solicitous about them whilst they were as yet only eggs.

The common Garden Spider, or Cross Spider as it is often called, makes a more spreading cocoon, which is quite a familiar object where there are close wooden fences. The egg-nest is made in an angle where one board overlaps another, or in a crevice of bark. She first spins a pad of silk upon which she deposits her yellow eggs, then another sheet of silk is laid over them, and over this a great deal of silk of looser texture, many short threads running off on all sides to secure it to the fence. When this cradle is finished, the spider sits with her legs spread across it to guard the precious contents, as shown in our photo. There is good reason for this watchfulness, for these nests are visited by ichneumon-wasps, who lay their eggs in the spider-eggs, if the parent is not on guard. But supposing all goes well and the eggs hatch, there will be seen a vast concourse of tiny black atoms—hundreds of them—who congregate in a compact ball, and take no food for a time. If alarmed, they run off in all directions, but return when the danger is past and again form up into a ball. Such a disperson, due to the attack of a small black ant, is shown in our photo on page 1114.

Some spiders, after elaborating the egg-cocoon, hang it up in a corner of their snare-web, a procedure which enables them to attend to ordinary business whilst keeping guard over the eggs. One of the species of spiders that haunt the insides of our houses is known as Pholcus, distinguished from the thick-bodied, hairy-legged common House Spider by its narrow body and very long, slender legs, that remind one of the Harvestman. Pholcus makes no egg-cocoon, but she glues her eggs into a compact ball and carries them about wherever she goes. She is not a nomad like the Wolf Spider, and therefore there seems, at first sight, no reason why she should secure her eggs at home. But she has experience of the housemaid's brush, and though a snare may be easily renewed, that is not the case with a ball of eggs. Another spider, named Ero, fashions her egg-bag into a ball and suspends it by long threads, apparently to prevent creeping enemies from getting at it (see page

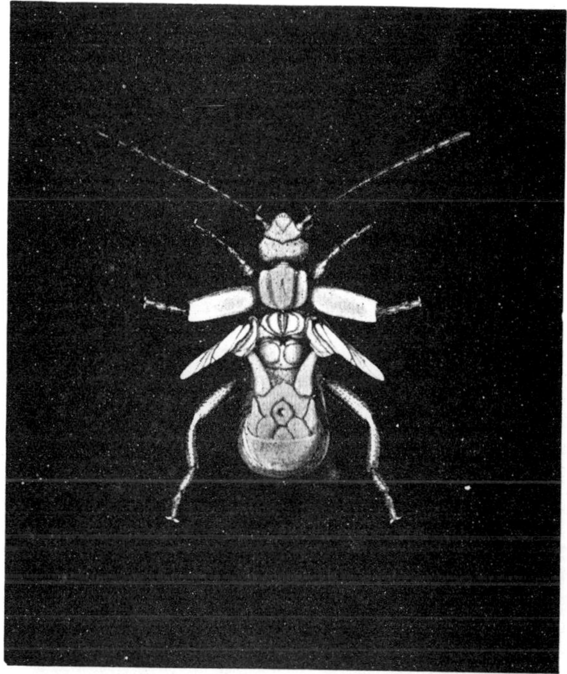

[*By F. Enock, F.L.S.*

FINAL TOUCHES.

By the aid of the pincers the folding and packing is completed, then—

[*By F. Enock, F.L.S.*

BACK IN POSITION.

—the wings are neatly placed in their resting position, the body is extended, and all that remains is the closing of the wing-covers.

1116). A number of loose loops hang about the ball and are probably intended to make it difficult for ichneumon-wasps to attack it. This spider only lays about six eggs.

A remarkable egg-cocoon is that made by Agroeca, a spider that is plentiful in pine-woods. With the stalk by which it is suspended, it is the shape of a wine-glass, with the mouth closed by a flat sheet of tissue. When new, it is perfectly white, and in contrast to the dark-red trunks of the pines to which it is attached is a very conspicuous object. But the spider has no intention of its attracting the notice of passers-by. She hunts about on the bark and finds all sorts of minute odds and ends, which she industriously sticks all over her nice clean-looking work, until it appears to be merely some bits of dirt that have become caught in stray wisps of an old web. We show photos of the two stages (page 1117), and it will be noticed that in the later condition it is difficult to see the outline of the disguised cocoon against the dark background supplied by the pine-trunk. This has always appeared to us a remarkable case of spider-wisdom—the spiders are all well endowed with wisdom, even though it be only inherited wisdom. First of all, there is the clever manipulation of the silk filaments to form this well-designed cocoon, and then the wisdom of disguising it that it may not attract the attention of birds, ichneumon-wasps, etc. It is a psychological heresy to attribute to such creatures any knowledge of the reason why they do these things ; but the observant naturalist is bound at times to be a psychological heretic. There are two species of ichneumon-wasp that endeavour to lay their eggs in the eggs of this spider in spite of all her care.

[*By F. Enock, F.L.S.*

THE WINGS COVERED.

The wing-covers are closed down securely over the wings, and the Earwig is now prepared for activity as a running insect.

Somewhat similar in texture to the cocoon of Agroeca is that of Theridion. The latter spider constructs a papery cocoon, of which the ground plan may be said to be spherical with the addition of several spikes sticking out from it. It is attached to the under-sides of leaves, where it is easily overlooked. Now this is wonderfully like a gall that is formed on the under-side of the rose-leaf, and one wonders whether the egg-cocoon of Theridion is intended to resemble that gall (see page 1140), and so to be passed over by those birds that hunt for spiders. An argument against this is that some galls are systematically explored by certain birds for the grubs they contain, and in their case the resemblance would tell against the spiders. Still, the resemblance is remarkable and worthy of note.

Some spiders construct their egg-cocoon within a tent formed by spinning two or three wayside leaves together; and several species that live under stones or in tunnels in the earth or old wall keep their egg-cocoons near them in their dens. Some North American representatives of our Garden Spider string their egg-bags to the ropes of their snares, so that they may have them under close observation whilst they are attending to their business of trapping insects.

The sub-aqueous, bell-shaped home of the Water Spider may also be considered as an egg-cocoon, for it is there that her eggs are kept, and it serves for a time as a nursery for her young ones. A

Photo by] [*W. S. Berridge, F.Z.S.*

GUEREZA MONKEY.

The most beautiful member of the Monkey tribes is found chiefly in the forests of Abyssinia. Its coloration is black and white, and the long white fringes of its mantle and tail harmonize with the long pendent grey-white lichens with which the trees are covered.

species of Theridion, which makes an irregular snare, is more careful in the planning of her nursery. Adjoining the snare, she constructs a conical tent open at the bottom only. Then she fabricates globular cocoons of yellowish-white, in each of which she encloses from twenty to sixty small eggs of similar colour to the silk, and hangs these up in the roof of her tent. When the eggs hatch the young spiders remain in the tent for some time, and the mother-spider brings them food. Another member of the same genus, distinguished by having her upper side marked by delicate black lines on a pale greenish ground, makes a greenish or bluish egg-bag, and secures it to a leaf, and by means of properly placed threads, draws the edges of the leaf towards the middle, and so hides her treasure. A somewhat similar procedure is adopted by Sparassus, a green hunting spider. Her egg-cocoon is, like herself, green, and she hides it by curling two or three leaves around it, and securing them in position by uniting them with a few threads. Clubiona, which has an olive-green hind body, with a red-brown shield marked on it, and makes her snares on the branches of furze, constructs a nursery tube, in which she places her egg-cocoon.

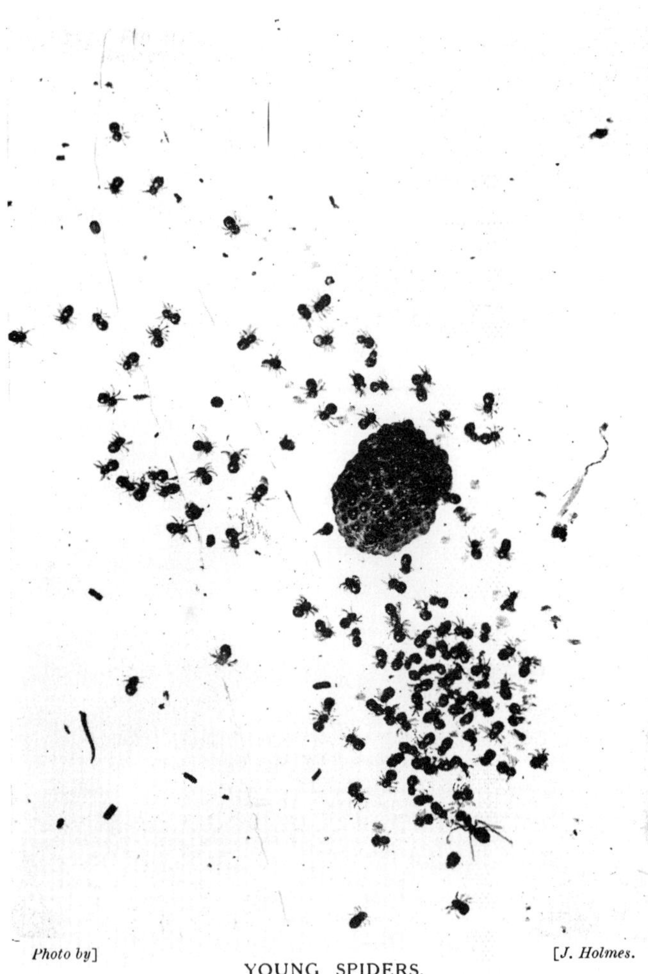

Photo by] [*J. Holmes.*

YOUNG SPIDERS.

The newly-hatched young of the Garden Spider crowd together into a ball. They have here begun to disperse in alarm, probably caused by the presence of an ant (lower right-hand corner). The danger past they will again crowd together.

CRABS AND ANEMONES

In early pages of this work we gave an account of the Hermit Crabs, and their universal habit of making friends with anemones, sponges, and other marine creatures for their mutual benefit. Of necessity, we did not then exhaust the subject, and among the species we omitted to mention was Anderson's Hermit Crab, a deep-water (one hundred fathoms) species of the Indian Ocean, which, true to the family traditions, has established an *entente cordiale* with an anemone, or rather, with a whole colony of anemones. The anemone in question (Mammillifera) is a social one, and by continuous budding forms a sheet of connected anemones large enough to cover the crab. This particular Hermit does not utilize an empty shell for the protection of its soft hinder parts, mainly, we suspect, because at the depth it affects mollusc-shells are not a plentiful commodity. So this cloak-like colony of anemones serves instead, and when danger threatens the crab seizes the front edge of its coverlet and draws it over its head.

Alcock tells us that in the deep waters of these Indian seas the Hermit Crabs differ from most other deep-sea animals in having normally developed eyes, and that the anemones they carry about with them are frequently luminous, so that the light they give out enables the crab to see what he is doing. As in all these arrangements, the anemone benefits by being carried to fresh hunting-grounds.

One of the most astonishing of these arrangements is the case of a shore crab (Melia), also from the Indian Ocean. It may or may not be a reciprocal affair—there is at present not sufficient evidence to decide—but it is at least a remarkable example of crustacean intelligence. When Mobius some years ago told how the crab seeks about for certain anemones, and goes hunting with them in his pincer claws, naturalists at home received his account with some reserve; but more recently Borrodaile has fully substantiated the story. The crab detaches the anemones from the rock and goes about under the water with one in each pair of pincers. The anemone catches small creatures that they come across, and the crab hooks his share out of the anemone's mouth with the claw of his first walking limb. Mobius says that if the crab be robbed of its anemones, it appears to be greatly distressed, and hunts about for them.

[By T. Carreras.

SPIDER AND YOUNG.

The female Pardosa carries her egg-cocoon until the eggs are hatched; then accommodates the young spiders upon her back until they are fit to look after themselves.

Our own Thornback Crab, the largest of our native species of Spider Crab, has its upper shell thrown into valleys and mountain peaks, and closely studded with hook tipped spines and bristles, which serve to keep on the heterogeneous coating of rubbish with which it delights to disguise itself Not only does it use dead matter, such as severed branches of seaweed, for this purpose, but it also deliberately plants living things, such as anemones and sponges, in the valleys. We have frequently

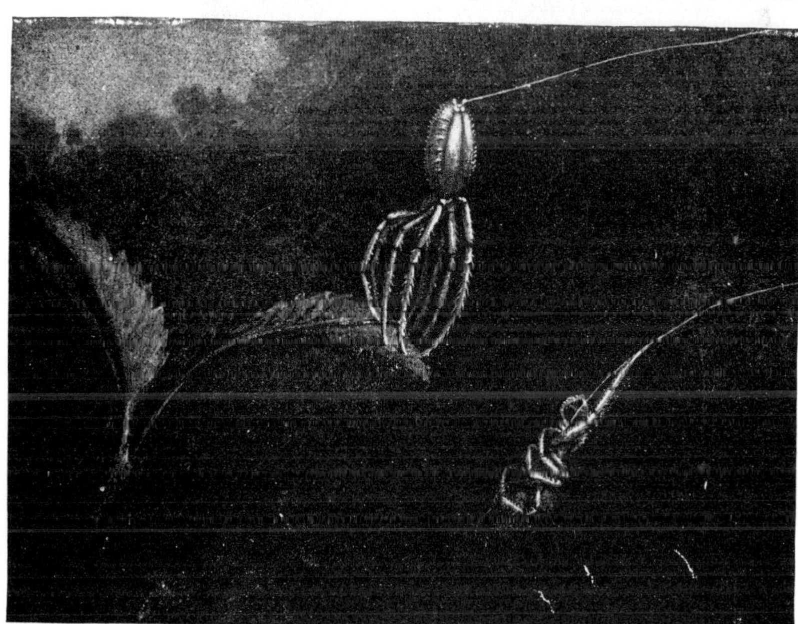

[By T. Carreras.

SPIDERS AS AVIATORS.

The young spider standing on tiptoe emits a considerable length of gossamer from its spinnerets, and when there is sufficient pull from this buoyant line, lets go and seizes the line with all its legs. In this way the minute spider travels great distances through the air.

availed ourselves of this knowledge when hunting for marine life, and done part of our collecting from the Thornbacks thrown out of the crab-pots by the fishermen, who set no store upon this crab, as it is not marketable. On one occasion we took no fewer than half a dozen specimens of the Snake-locked Anemone from the shell of one half-grown Thornback. It is true that in this case there is no evidence of friendship between the two animals. The crab has simply placed the anemone on its shell for its own private ends, and the anemone appears to regard its perambulatory base as being quite as good, probably better, than a fixed rock.

[*By T. Carreras.*

EGG-COCOONS OF SPIDERS.

The spiky egg-cocoons attached to the underside of a leaf are those of a spider named Theridion. The hanging cocoon on the right is that of Ero.

NEWTS OF THE PAST

THE Newts and Salamanders of the present day, with their near relations, the Frogs and Toads, form a comparatively small class of animal life; but there was a time in the history of the globe when they were far more numerous than they are. Only a few pages back we have shown the stages by which the modern frog progresses from a fish-like, water-breathing creature (the tadpole) to a reptile-like, air-breathing one. What we may all witness every spring in almost every pond is a repetition in brief of the evolutionary changes that went on through the Carboniferous and Permian periods by which a higher type was produced from the fishes. That was the day of the Newt family, but, later on, the true reptiles appeared, and the Newts began to lose importance in competition with the better equipped, scale-armoured reptiles. Sir Richard Owen bestowed the name of Labyrinthodonts upon some of these ancient Newts, because of the minute structure of the teeth as seen under the microscope—the so-called cement of the teeth being folded in a complex manner suggesting the turns of a labyrinth.

Our illustration includes five different species about a pool in the coal forest. The largest and central member of the group is Archegosaurus, of which we also give a view of the forepart of a fossil skeleton found in the Permian at Saarbrücken. In the hinder-part of the skull will be seen the aperture or socket of the middle eye, referred to in our account of the Tuatera Lizard (page 916).

Ceraterpeton, on the rock in the foreground, looks much like a modern newt with a very long tail—longer than its body. Dolichosoma resembles a snake far more than a newt, but, nevertheless, its organization places it among the amphibious scaleless Salamander tribe. Loxomma is shown with its head protruding from the shelter of a fern-like plant of the Coal period. The restoring artist has been discreet in this matter, for the very good reason that only its head is known in a fossil state at present. Someday no doubt the trunk and limbs will be discovered by some getter of coal. Actinodon there can be no doubt about, for a fairly complete skeleton—lacking only the right fore-limb—is in existence. It is rather short and thick, and the short, wide skull has a broad muzzle.

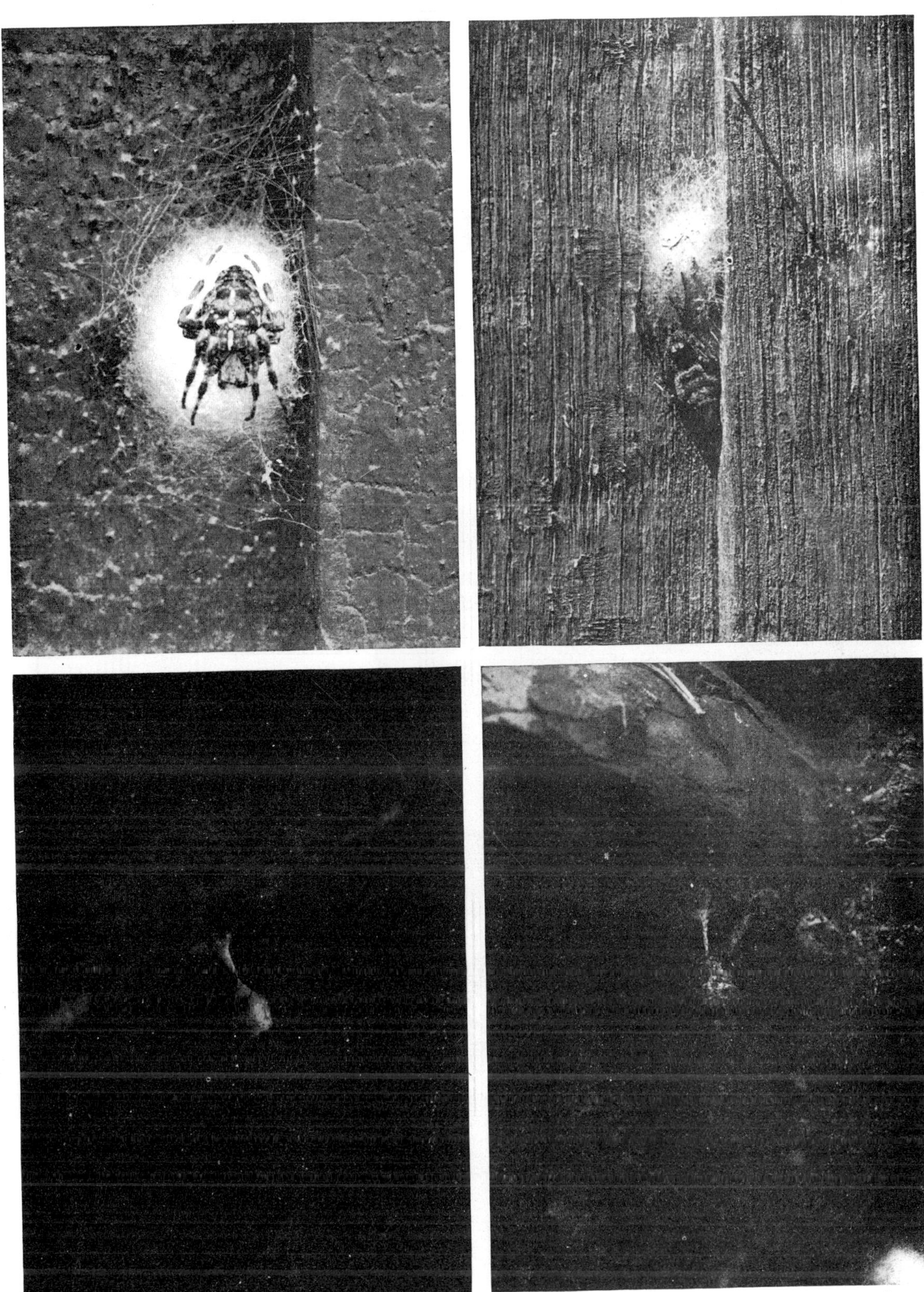

Photos by] [E. Step, F.L.S.

SPIDERS' EGG-COCOONS.

In the upper row the left-hand photo shows the common Garden Spider guarding her egg-cocoon, spun up in the angle of a fence. The right-hand picture is of a nearly related species whose cocoon is of similar construction. Below is the singular wine-glass-shaped egg-cocoon of a woodland species of Agroeca immediately after completion. The second photo shows a cocoon after the spider has disguised it by covering it with bits of rubbish.

Although, as stated, some of these ancient members of the Newt tribe were legless and had more likeness to snakes than to Newts, the majority had four legs, of which the hinder pair was commonly the larger. We have called them Newts instead of using the family name Amphibia, which includes the Frogs and Toads as a separate group from the reptiles with which they are popularly classed. Some of the ancient Amphibians were, of course, like toads of enormous size. Their footprints have been well preserved in the Triassic rocks of Europe, and bear striking testimony to the great inequality between the fore and hind feet. The impression of the hind foot is very like that of the human hand. One of these toad-like creatures, whose footprints have come down to us, is named Cheirotherium on that account (*Cheiros*, hand, Gr.). The size of some of the largest of these may be gathered from the fact that the skull of a Mastodonsaurus is upwards of three feet in length. A peculiarity in which they differ from all modern representatives of their class (which have soft, scaleless skin) is that their bodies were protected by an armour formed of oval plates of bone.

[*By T. Carreras.*

ANEMONE-CARRYING CRAB.

Melia, a crab of the Indian Ocean, detaches anemones from the rocks and goes hunting with one in each of its pincer-claws The anemone catches food and the crab shares the spoils.

SOME CURIOUS SHRIMPS

BY EDWARD STEP, F.L.S.

ALL Shrimps are curious and worthy of consideration, even the Shrimps of commerce—Blackwall Shrimps, Gravesend Shrimps, and the lordly Shrimp that is distinguished by the name of Prawn and sold by the dozen instead of by the pint. Having reached this point we have practically exhausted the list of species that have got English names, which means that these are the only species that are known to the public.

If the common Brown or Blackwall Shrimp be examined before being peeled and eaten, one can get some idea of its structure, though zoological specimens that have been boiled are not the best for the purpose of study. Still, we can discover several facts from the examination of such specimens. First, we learn that a Shrimp is built on much the same lines as a Lobster, though its armour-plating is not fortified with such thick deposits of lime as in that gigantic Shrimp. The Prawn and the Red or Gravesend Shrimp come much nearer to the form of the Lobster, the "head" being furnished with the long, toothed sword that stands out from between the eyes. The value of that embellishment, as an element of personal beauty, can be estimated by placing a Brown Shrimp and a Red Shrimp side by side. The Brown Shrimp by comparison appears to be deficient, to have lost his most prominent feature by having the front of his head cut off. The so-called "head" of the Shrimp is head and trunk combined, and bears the beautiful breathing machinery under the sides of the shell, whilst the long, thick hinder-part corresponds with the thin, flat "tail" that the crab tucks under his body. The complicated mouth-parts of the Shrimp are well worth examination; also the two pairs of antennæ and the stalked eyes. The Brown Shrimp when alive is grey, dotted with brown, and when it sinks down upon the fine sands it chooses for its living-ground, it can only be seen by very sharp eyes. The moment it touches the sand the swimming feet and the broad scale at the base of each outer antenna throw up the sand around it, and this settling completely hides its outline.

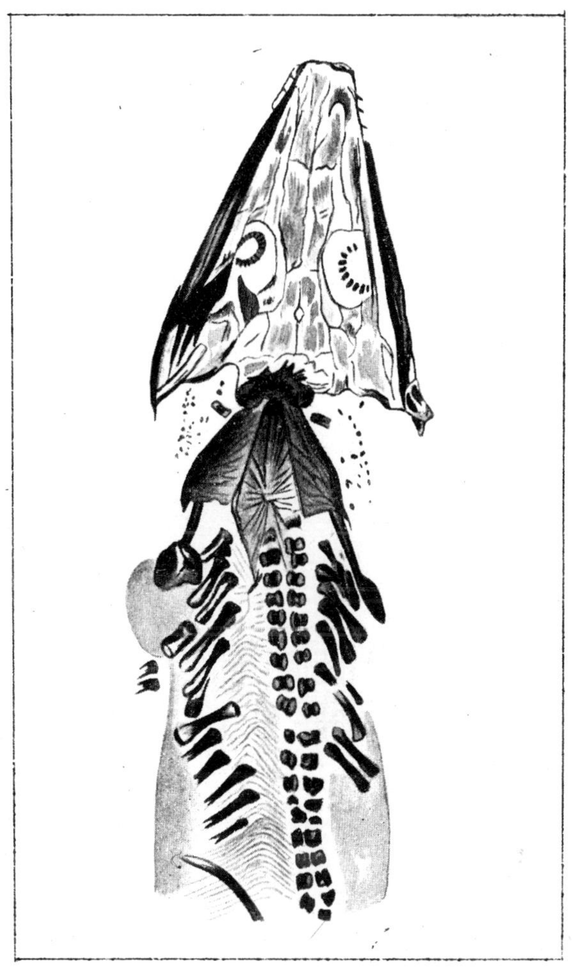

A FOSSIL NEWT.

The forepart of the fossil skeleton of Archegosaurus, found in the Permian rocks at Saarbrücken. In the hinder-part of the skull will be seen the aperture of the middle eye formerly possessed by newts and lizards.

But this is only the common Shrimp—the Shrimp that is sold wholesale by the hundred-weight and retailed by the pint—and, therefore, too common to have interest for some people. Let us look at a few of the less common species that are known only to the crab and shrimp specialists, otherwise carcinologists.

Here is one called Nematocarcinus (see page 1122), whose body conforms pretty much to the Prawn type, though its "rostral spine" between the eyes has not the length it attains in the Prawn. It has long, lash-like antennæ three or four times the length of the body, and most of the walking legs are drawn out to an enormous extent. The hinge connecting the third and fourth joints of these limbs is peculiar, adapted to support the muscular strain consequent upon the great length of the joints.

As this Shrimp is only obtained from great depths in the ocean the purpose of these long limbs is not known.

More remarkable is the burrowing Shrimp named Thaumastocheles, which the trawl of the *Challenger* brought up from a depth of four hundred and fifty fathoms in West Indian waters. It is, like many of these abyssal Shrimps, a sightless creature, with possible vestiges of eye-stalks and sockets, but no eyes; indicating that once upon a time it had eyes, but taking to deep water where there is no light to see with, the eyes were given up as a useless encumbrance. The most striking point in its appearance is the disparity in size between the right and left "hands"—the pincers of the first pair of legs. They are both slender and elongated, and their inner edges armed with spine-

By permission of] [*Messrs. Chapman & Hall, Ltd.*

NEWTS OF THE COAL FOREST.

A restoration of five species of Amphibians or Newt-like creatures whose fossil remains have been found in the Coal Measures and the deposits of the Permian period The central figure is that of Archegosaurus, whose skull is figured on the previous page

like teeth of varying lengths; but the "finger and thumb" of the right are as long as the animal's body, with a corresponding development of the teeth. It appears probable that when the Shrimp lies buried in the sand this formidable implement, partly closed, sweeps the surface like a rake and draws in all sorts of good things to be turned over and picked up by the shorter pincers of the other limbs. This species will be found figured on page 1124.

Then we have the assortment of strange creatures figured on page 1125. It is taking a liberty with names to label all these as Shrimps, but seeing that there is no popular term applicable to them, what else is one to do? One cannot expect to be allowed to refer to them as Flabellifera and Asellota, which are their scientific family names; so we take the liberty of including them all under the name of Shrimps in a broad sense implying small crustaceans that are not crabs or lobsters.

The one that carries his eyes at the extremity of long stalks is known by a scientific name that signifies Oar-Shrimp. It is another of the *Challenger* finds, and it has the distinction of being the creature whose eyes are farther away from its head than in any other animal. Our portrait of it is much larger than life; the actual length of the Shrimp without its eye-stalks is only a quarter of an inch. The eye-stalks by themselves exceed that length.

The Slender-bodied Shrimp, to translate its name Ischnosoma, is a quaint form that would remind one somewhat of the Stick-insects were it not for the too abundant legs. Apart from these organs, it might serve as model for a very deadly form of truncheon. Some of the joints of the hind-body or "tail" have become fused together, which gives an air of stiffness and solidity to it, whilst the "head" portion, which, in proper Shrimps shows no joints, consists of four rings. One point that will be noticed is the absence of eyes, in which respect the Slender Shrimp offers a strong contrast to the Oar-Shrimp. It is, of course, a deep-water species, and was found in the neighbourhood of the Azores.

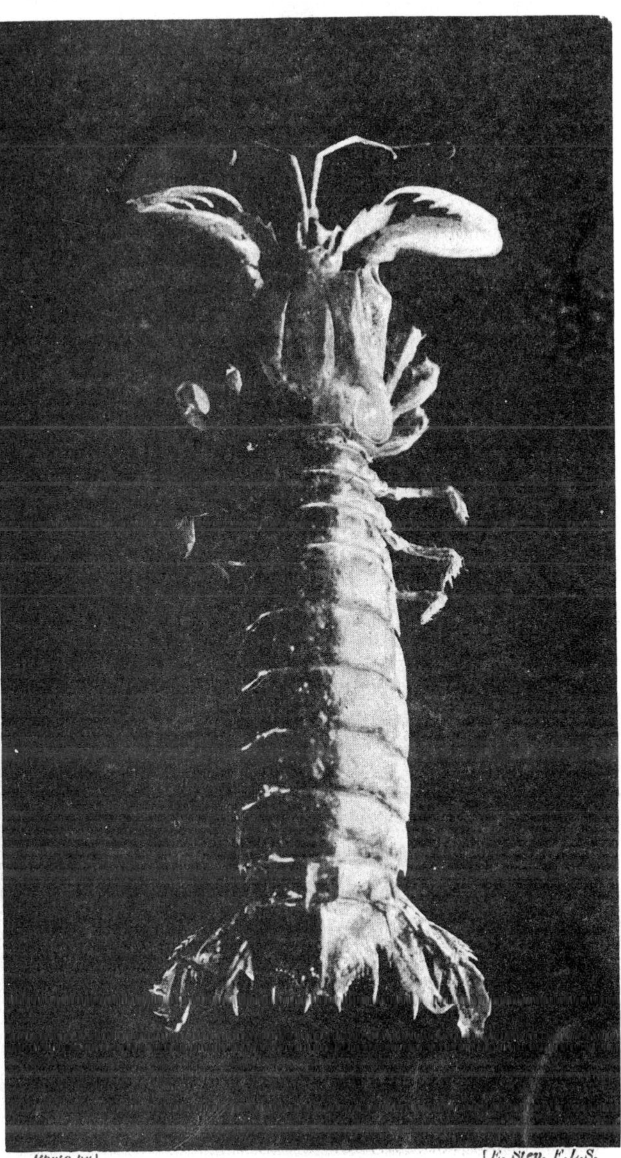

Photo by] [*E. Step, F.L.S.*

MANTIS-SHRIMP.

Part of the jaws of this remarkable shrimp is so developed as to resemble the front limbs of the Mantis, and is used for similar purposes. The Mantis Shrimp is about three inches long, and it burrows in sand and mud

The Hatchet-jawed Shrimp looks like a creature suffering from swelled head, this division of the body being equal to the other two parts in length, and much more in bulk. The two very prominent appendages of the jaws are much like hatchets in shape. We have a closely-related species on our own shores, whose shape is very like that of a big ant with very powerful nipper jaws. It has not such enormous eyes as its hatchet-jawed relation.

Neasellus, the remaining species on the same page, looks like a ragged woodlouse. Little is known of its habits. It came from the island of Kerguelen.

The Mantis Shrimp, whose portrait is shown on page 1121, is well-named, for its foremost legs —really part of the masticatory apparatus—have a very close resemblance to those of *the* Mantis, and are used in the same manner and to the same purpose. The end joint has its inner edge furnished with long teeth, which close down on a row of shorter teeth along the edge of the next joint, so that any creature imprisoned between the two joints is in a bad way. The succeeding three pairs of legs, though much smaller and differing in shape, are also furnished with a hook which closes down on the next joint and serves to hold smaller prey. Compared with the Shrimps we have been considering, this is a gigantic creature, its length being more than three inches. That is the length of specimens taken on our own southern shores; in the Mediterranean it attains to much larger dimensions. It burrows in sand, and lies with its head and big arms out of the burrow, into which it shoots backward with great rapidity on the slightest alarm. There are a number of closely-related species, similar in form and habit (with differences), and some of them are regularly hunted as articles of food, though owing to their alertness they are not caught easily.

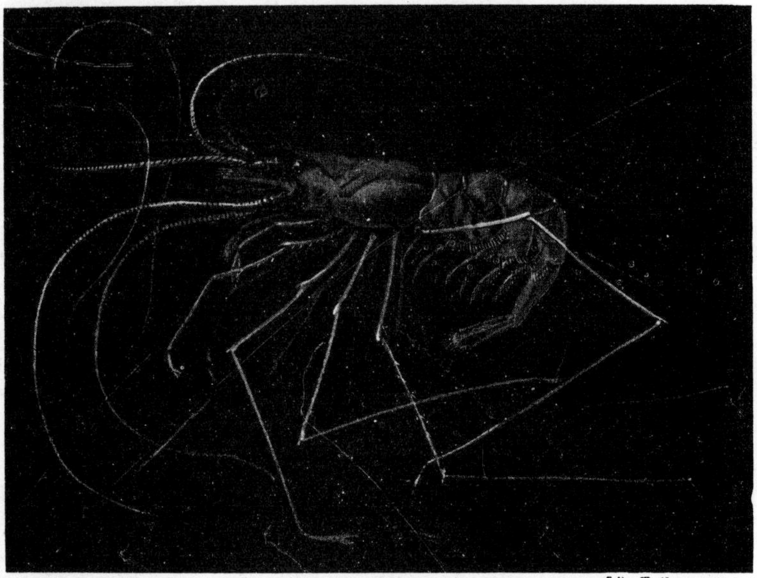

[*By T. Carreras.*

A LONG-LEGGED SHRIMP.

What with the abundance and length of its antennae and the length of its ten walking-legs this shrimp appears to have enough to do to avoid a tangle. But in the deep-sea it has room to spread them all out and keep them apart.

Another singular Shrimp of comparatively large size is the Mud Burrower, though if the scientific name be translated we should call it the Beautiful Queen. It is rather lobster-like in form, though its " shell " is thin and parchment-like. The most striking thing about it is the disparity in the size of the two " hands," one of them being developed to an enormous size. Sometimes it is the left hand that is enlarged, sometimes the right. As the Mud Burrower, alias Beautiful Queen, lies in her burrow this big hand serves as a street-door, barring entrance to any undesirable callers and ever ready to be thrust out to seize anything palatable that may be passing. One can appreciate at a glance the narrowness of the " head " in both these species when the burrowing habit is borne in mind. For a quick retreat to the rear into the burrow any enlargement of the foreparts would cause wedging that might be disastrous. This species, whose photograph is shown on page 1123, is coloured a beautiful bright pink, which fades after death, so that museum specimens fail to give a proper idea of its beauty. From what we have said about its burrowing habits, it must not be inferred that the species is incapable of anything else. As a matter of fact, when occasion demands or tastes incline that way, the Beautiful Queen can show herself as an active and graceful swimmer; but as her shell is thin and not much protection against the powerful jaws of some of the big fishes, she wisely spends the greater part of her life in the security of her burrow.

This burrow is excavated chiefly by means of the third pair of legs. Mr. Say, describing an American relative of this species, says that he found the burrows proper about eighteen inches below the surface of the sand, communicating with the water by a more or less vertical tube which finished off above the sand level with a sort of chimney. These tubes and burrows do not collapse, as might be expected, when the excavator is not at home or is dead. The constant pressure of use, and possibly some exudation from the Shrimp's body, causes the sand to become indurated, so that when, after the decease of the architect, they fill up with sand from the entrance, they can still be traced by these indurated walls and the little chimney at the top.

Gebia, the Mud-borer, is much like the Mud-burrower, but has not such finely-developed "hands." It is a yellowish-white Shrimp, ornamented with star-shaped orange spots. Sometimes it saves itself the trouble of doing its burrowing for itself by occupying the vacated burrows of the Razor-shell.

Leach's Shrimp, or the Variable Shrimp as it is sometimes called, is more shrimp-like than some of the species we have mentioned. A rather abrupt bending of the hind-body gives it something of a hump-backed appearance. It is chiefly worthy of note because of the quick-change power it possesses. It is common in rock-pools on our own coasts, but is not easily seen by the uninitiated by reason of this trick of changing colour to suit its environment. If a red Shrimp is swimming between your eye and some red seaweeds, it is not easily seen, and if in the course of its swimming it comes among green weeds, though for a moment it is well contrasted with its background, it at once begins to disappear by turning green. It may, in fact, be almost any colour according to circumstances red, pink, salmon, emerald green, cobalt-blue, grey, chocolate or white. These colour-changes are most marked in the younger specimens. When it grows old and possibly settles down to some particular station, it appears to have a colour-pattern which has become fixed in relation to its environment, for mature individuals from different situations, though they will be found to differ remarkably one from another, have lost the power of changing when artificially introduced to new surroundings.

Photo by] [*F. Step, F.L.S.*

THE MUD-BURROWER.

A lobster-like shrimp of a bright pink colour that lives in a burrow and uses its larger pincer-claw to block the entrance.

An interesting group are the Skeleton Screw-shrimps, quite small creatures, whose bodies are slender and thread-like and give one the idea that he is looking merely at the backbone of some small creature to which a few of the limbs have remained attached. Gather a bunch of the small lime-encrusted weeds known as corallines from any pool and examine it in a basin of sea-water, and after a time you will see some of these Skeleton Shrimps sprawling in an ungainly fashion over the branches, which they closely resemble in form and colour. Their climbing is accomplished by three pairs of hooked feet at the hinder extremity of the long, pointed body. A little behind the head is a pair of longer limbs ending in an enormously swollen " hand," which is larger than the head, and reminds you strongly of a boxer with the gloves on. Their movements, as they sprawl over the coralline branches, and throw these big hands about, are suggestive of comic business, and though you know better, you feel that they are performing solely for your amusement. If you consult a carcinologist about this strange creature, he will tell you that almost the entire length of its drawn-out body consists of head and chest (thorax), and that the hind-body or abdomen is represented solely by the almost imperceptible last joint.

THE CRESTED IBIS

BY SIR HARRY JOHNSTON, G.C.M.G.

THE Ibises are a very interesting group of stork-like birds which, no doubt, represent pretty nearly the original stock from which were evolved in course of time the storks and herons, as well as the closely-allied spoonbills, and there may even be some affinity between the Ibises and that vast assemblage of wading birds, the plovers, curlews, snipe, and stilts. The Ibises are literally of world-wide distribution, except that they are absent from the cold regions of Europe, Asia, and North America. Otherwise, they inhabit the Americas from California and Florida to the very extremity of Tierra del Fuego. At one time they even frequented England, and were known in early English as " Livers," having thus given their name to the great city and port of Liverpool, to say nothing of the numerous Livermeres and Liversedges scattered about England. The same species —the Glossy Ibis—still frequents Holland and is a common bird throughout Southern Europe. A large and handsome Ibis is found in China, and is very like in appearance a large grey Ibis found on the coasts of Mexico. The Sacred Ibis of Tropical Africa is famous for its association with the ancient Egyptians, and from its very effective colouring of inky black and snowy white. Ibises are found nearly all over Australia; they are not even absent from New Zealand, or from many oceanic islands in tropical seas.

[*By T. Carreras.*

A SHRIMP WITH A RAKE.

A blind deep-sea shrimp from the West Indies The pincers of the right "hand" are as long as the shrimp's body and fitted with long rake-like teeth. It is believed that when almost buried in sand or mud the shrimp sweeps the surface with this rake and so obtains its food.

One of the handsomest of this tribe of beautiful birds (which, it must be remembered, includes the vivid Scarlet Ibis of South America) is the Crested Ibis, peculiar to Madagascar. It may be seen by my illustration (which is drawn from a very fine specimen of this bird at Stuttgart) that the coloration is rich and peculiar, the beak being a shining beryl-green, the crest glossy black-green and creamy white, and the silky plumage of the body and wings a charming contrast of cinnamon-brown and pure white, whilst there is an effect of vivid coral-red brought into the scheme

[*By Theo. Carreras.*

CURIOUS SHRIMPS.

The shrimp whose eyes are in the upper right-hand corner is the Oar-Shrimp, one of the *Challenger* discoveries. It has the distinction of having its eyes farther from its head than any other animal. Below it to the right is the Slender-bodied Shrimp, an eyeless species found in deep water at the Azores. On the left is the Hatchet-jawed Shrimp with head as bulky as the rest of its body. The small specimen at the top of the page is Neasellus from Kerguelen Island.

Photo by] [*W. S. Berridge, F.Z.S.*

WHITE IBIS.

Structurally identical with the Scarlet Ibis and only distinguished by the different colour of the plumage.

of colouring by the bare space round the eye, and by the legs and feet.

THE DADDY-LONG-LEGS, OR CRANE FLY

BY E. J. SPITTA, F.R.M.S.

It is more than probable that everyone has seen a Daddy-long-legs, but although so commonly met with, comparatively few may be aware of the terrible devastation brought about in certain years by its grubs, often called " leather-jackets " from the toughness of their skins. Whole pastures have been literally eaten up by them, and almost permanently destroyed in some instances, for, be it understood, the larvæ not only ruin the crops of grass during years of bad infestation, but what is so much more serious, they materially injure and sometimes utterly destroy the roots. Well for the farmer whose pastures are affected if he allows the rook and the starling to patrol his fields without being driven off by gun or crow-clapper. These birds can at least alleviate his trouble by destroying large numbers of the " leather-jackets."

Photo by] [*W. S. Berridge, F.Z.S.*

GLOSSY IBIS.

A common bird in Southern Europe, and formerly in Britain, where it was known as the Liver. It is supposed to have given its name to Liverpool and several other places where it had its haunts.

Space forbids further reference to *antecedents*, so the remarks that follow are limited for the most part to the peculiar habits and construction of the little insect itself.

The Daddy-long-legs is a member of that order of insects known as Diptera, from their possessing only a single pair of wings, whereas the members of all other orders have two pairs. In explanation of this departure from the rule in insect structure, it should be said that it is really due to the non-development of the hinder pair, of which the vestiges may be seen in a pair of minute organs called balancers, just behind the base of the developed wings. When the insect is flying these balancers, as well as the wings, are kept in rapid vibration, but what purpose they actually serve is not very clear. It has been suggested that, having lost their function as wings, they have been developed into sense organs.

There are at least two kinds of Crane Fly to be found in England (some describe a third): the Tipula oleracea and the Tipula paludera, but their life-history and habits seem identical, save that the former appears for the most part from May to August, whilst the latter is mostly met with from July to September. From this it seems that in August both varieties are fairly common, and this accounts for the greater number of these insects that are met with in that month.

Photo by] [*W. B. Johnson.*

CRANE FLY.

Also known popularly as "Daddy-long-legs." The grubs, known as "leather-jackets," are very destructive to grass-lands. Note the "balancers" behind the wings, which are supposed to be vestiges of a second pair of wings.

It should be mentioned in passing that the male is smaller than the female, and has a blunt-ended body, whilst that of the female terminates in a point which is, in fact, the egg-laying apparatus.

Passing over the silly and apparently aimless behaviour of the Daddy, one must take note of the curious way it has of casting off its legs or a portion of one upon the slightest touch. The accident seems to cause it no inconvenience whatever, for it flies away as merrily as ever. This matter will be referred to later.

Further, mention should be made of the curious way this insect has of depositing its eggs, which are often several hundreds in number, in shape always elliptical and black in colour. Instead of depositing them while resting on all its feet, in the ordinary horizontal position, the Crane Fly moves over the place she selects to lay her eggs upon, *in an upright position*, maintaining her body vertical by the hind-legs, or as much as remains of them, and raising a pair of fore-legs up in the air wherewith to balance herself. In point of fact, the ovipositor and the two hind-legs resting on the ground remind one of a three-legged stool!

Turning to the structure of the Daddy-long-legs, we see above an entire male specimen laid out for study with the microscope. The head apparatus is shown highly magnified (page 1128), and it will there be noticed that beside the proboscis lie four other bodies that look like antennæ. Three of these are magnified more in the lower photo, and their terminations show great beauty in design. Which are true antennæ is difficult to say.

Near by them, in the upper photo, we see the large compound eyes, which are apparently grossly out of proportion to the rest of the body.

It is to the curious arrangement of the joints of the legs, however, that our special attention should be directed. Two are shown on page 1129. If casually looked at the joint resembles what

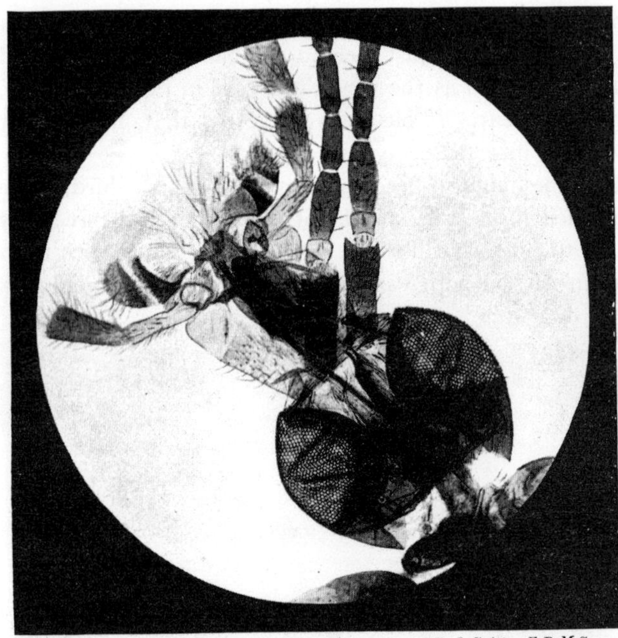

Photo by] [E. J. Spitta, F.R.M.S.

HEAD OF CRANE FLY.

Showing the eyes, proboscis, antennae and other appendages. (Highly magnified.)

is known as a ball-and-socket arrangement, but such a joint, if properly constructed, would not be so easily parted as we know to be the case in that of the Daddy-long-legs, for the slightest touch is sufficient. The reason is not far to seek if we carefully dissect a limb.

In the second photo on that page this has been done, and the mystery is at once solved, for the lower limb will be seen to have no ball at all to articulate with the socket of the adjacent member; it is, in fact, a very poorly-defined and tapering *spike*. This moves with great freedom, no doubt, in the roomy socket into which it articulates, but, it is almost needless to point out, is very easily disarticulated. What happens, then, when a leg comes off, is not a fracture of a limb, as commonly supposed, but a coming-off of a portion at one of these joints. No doubt the arrangement is for self-preservation.

One more point of interest must not be overlooked. The Daddy when at rest seems to compose itself with legs or portions of legs at curious angles, which one would think a very exhausting and fatiguing position to maintain. But if we examine and compare the two lower photos on page 1129, we shall see (by their special arrangement for the purpose) that the position is maintained by resting one part of the leg upon two stout bristles attached to an adjacent limb; hence the resting act is truly one of resting, and not one obtained by a protracted muscular effort.

In this portion of the leg the joint is of a more incomplete nature still, and is no doubt so arranged for still easier dislocation, doubtless to prevent the insect being caught, if the legs become adherent to the object upon which it is resting.

Photo by] [E. J. Spitta, F.R.M.S.

CRANE FLY.

These are three out of four appendages of the head, but it is not clear which pair are the true antennae. (Highly magnified.)

GIANT FIGS

BY EDWARD STEP, F.L.S.

The Figs are a numerous family which includes several species worthy of special description. Two of them are probably well known to every reader. These are *the* Fig, whose fruits, green and dried,

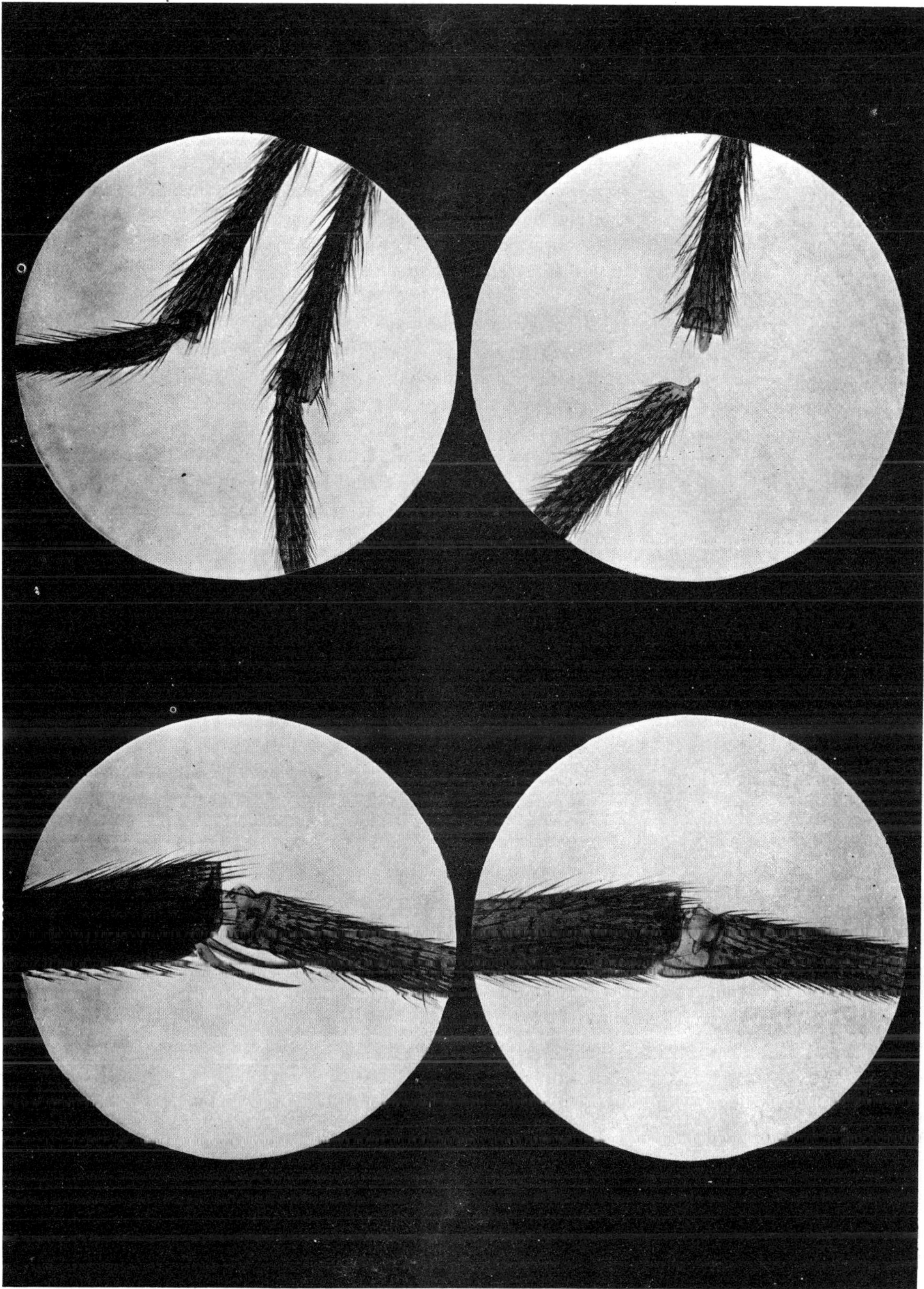

Photos by] [*E. J. Spitta, F.R.M.S.*

LEG-JOINTS OF THE CRANE FLY.

The Crane Fly readily parts with its legs or parts of legs. The two upper photographs explain this—one joint being connected by a mere peg that fits into a socket of the joint above. In the two lower circles will be seen the pair of stout bristles at the end of a joint which are brought into use as rests in certain positions of the legs.

are familiar on our tables, and the India-rubber Plant so frequently used as a room decoration in towns on account of the hard-wearing qualities of its varnished, leathery leaves, which are able to survive dust and the dry atmosphere of our living-rooms. The Fig has been successfully naturalized in our gardens for nearly four centuries, and in the South of England even ripens its fruit out of doors. Hundreds of tons of the dried fruit are imported into this country every year. Few persons who know the foliage of the Fig-tree would at sight associate it with the India-rubber Plant, the characters of the leaves being so widely different in shape and texture. The Fig is worthy of note, because what we eat as fruit is really a swollen and hollowed flower-stalk. The flowers were all inside the hollow and the real fruits are there later, so small as to be unrecognized. The desirable portion as a food is the swollen stalk.

In tropical countries there are Figs that become enormous trees, and chief of these is the Banyan of India, old specimens of which will cover an area of two thousand square feet, with hundreds of "trunks"—really natural supporting props to its extensive branches. The old idea—embalmed in Milton's famous lines—was that "the bending twigs take root" to nourish the ever-extending circumference of its aerial growth; but this is not strictly accurate. What really happens is that the horizontal branches send down thin, succulent roots towards the earth, and where there are no cattle to browse these off, as they like to do, the tips enter the soil and ramify as ordinary roots, sucking up supplies of raw material for the foliage to elaborate above. The above-ground portion

Photo by] [*F. Skeen & Co.*

TRIMEN'S BANYAN.

This tree is very similar to the common Banyan, and has the same habit of sending down roots from the horizontal branches. The position of the main trunk may be seen, but it is almost hidden by roots which have become part of it. All the other upright portions are roots in various stages of enlargement.

thickens into a woody support for the long, heavy branch, and looks like an ordinary tree-trunk.

It is a remarkable fact that few of these giant Banyans have grown from a seed germinated in the soil, as is usual with trees and most other plants. In its origin the Banyan is much like the mistletoe. A bird, having fed to repletion upon the Banyan's fruit, flies to a palm or other high tree to rest and digest. As a preliminary the beak is cleaned and the plumage preened, and some seeds left on the palm; or they may pass uninjured through the digestive organs. Anyway, they get left, and in the rainy season germinate. The seedling cannot get sufficient nutriment in the crown of the palm, so it sends roots down the palm-trunk to the ground. These roots on all sides of the palm-trunk, as they grow, squeeze and strangle the poor palm, and the growth of the Banyan's leafy boughs above it shuts out the light, so that the palm is soon killed. The Banyan's branches are stretched forth far and wide on every side, and as they advance they send down the roots that will harden into props, enabling the branch to extend farther and farther from its base. In time the central starting-point of the Banyan will decay, but although it has no real trunk, the tree continues to flourish.

Photo by] [*Max Ferrars.*

THE EMBRACE OF THE FIG.

A masonry image, after being partly wrecked by the roots of the Peepul, or Sacred Fig, is finally held together by the roots forming a close network around it.

The Banyan has an oval leaf with blunt tip, and so in respect of form it comes between the leaf of the India-rubber Plant and the Peepul, or Sacred Fig-tree of the Buddhists, which is another Indian tree, and has heart-shaped leaves, with a long, slender point. These leaves, which hang downwards, are always on the quiver, like those of our aspen, and, therefore, the Peepul is reputed to give grateful shade on the hottest days.

The Peepul, although the most sacred tree in India, is not above strangling other trees, much as the Banyan does. It has similar habits, too, in often beginning life high in the air and sending its first roots from its elevated birthplace to seek the earth. On page 1132 is a photograph of a factory chimney that was disused. A bird had probably planted a Peepul seed between the topmost course of bricks, which on germination had sent down its roots to earth. Our British trees have been known to take possession of disused chimneys in the same way; but in their case they grew simply as trees on the summit. The Peepul, by virtue of its roots in the ground far below, grew apace and spread on all sides, so that after the lapse of only fifteen years from the time the chimney had been in use it presented the appearance shown in the photograph.

These roots do not always go straight down the outside of the structure upon which they have begun life. Sometimes they penetrate the joints of the masonry, and then by their expansion they lift stone from stone. On page 1131 is a photograph of a figure of Buddha which had been built up of masonry; but a seedling Peepul got possession, and by the expansion of its roots was in a fair way to wreck it. Then the horrid character of its action seems to have struck it. Should the image of Buddha be destroyed by Buddha's own tree? So it set to work to repair the evil it had done, and now the once toppling Buddha is held firmly together by interlacing roots of the Peepul that bind its stones securely together.

Photo by] [*Max Ferrars.*

A FACTORY CHIMNEY.

The photo shows a factory chimney after fifteen years of disuse. In the meantime a Peepul seed has germinated at the top and sent down roots which hold the masonry together.

There is also a big Fig in Australia — the Large-leaved Fig — which has somewhat similar habits. Mr. Backhouse tells us of one whose trunk was forty feet round at the highest point he could reach. He says it often springs from a seed deposited by a bird in the cavity of another tree at an elevation of fifty feet or more. Its roots also spread out like walls along the ground, so that the area occupied by these buttresses is about thirty feet across.

Mr. Rodway, too, has told us of the deadly work of certain Figs in Guiana, and Professor Moseley relates with satisfaction that in the Aru Islands he saw a species of Fig that had been too clever

[By C. E. Swan.

AN EXTINCT GIANT TORTOISE.

Miolania was a giant tortoise that flourished in the Pleistocene period. In addition to the shields and scales common to all tortoises, Miolania had great flanges and bosses of bone on the skull and spiked rings of bone on the tail.

by half, and had choked itself, or rather the original part of itself, by winding its pendent roots spirally around the parent stem. " It seemed just that a Fig, so accustomed to choking other trees, should thus once in a while choke itself."

A GIANT ARMOURED TORTOISE

BY W. P. PYCRAFT, F.Z.S.

TORTOISES, as we have already remarked in these pages, are justly entitled to a place among the world's marvels. And there are some species which stand out, as it were, from the rest of their kind on account of one or more conspicuous departures from what we may call the traditions of the tribe. The extinct Giant Tortoise, Miolania, is one of these. In the first place, it was an armoured Tortoise ; we might, indeed, without violence, describe it as a super-dreadnought among Tortoises, since all Tortoises are armoured. But in Miolania the accessory armature took the form of great flanges and bosses of bone on the skull, and of formidable spiked rings of bone on the tail, recalling those of the ancient Giant Armadillo Glyptodon. Was this additional armour demanded by reason of the creature's environment, or are we to regard it in the light of ornamentation ? To such questions no answer seems possible.

By almost common consent Miolania is regarded as one of the " Side-necked " Tortoises. So-called because the head is drawn under cover of the shell, not by the backward pull of the neck as

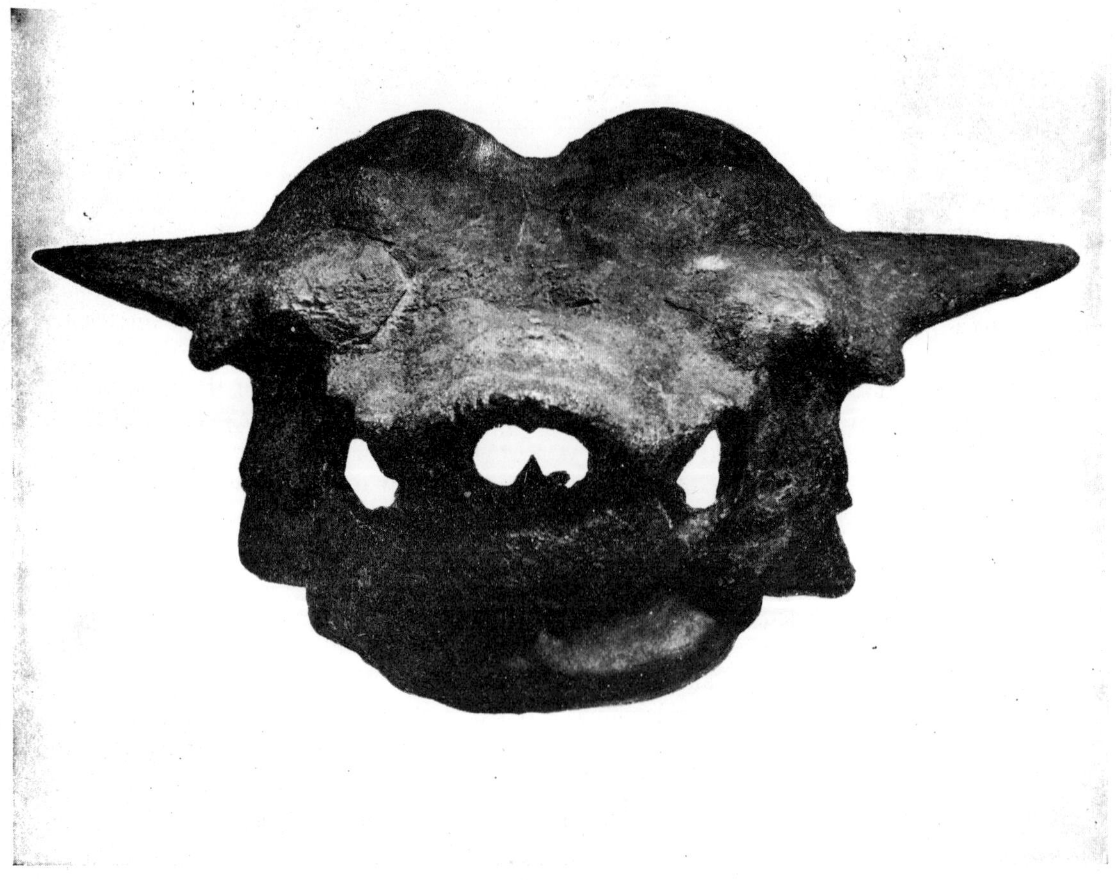

SKULL OF OWEN'S MIOLANIA.

This fossil skull was found in Queensland, and is now in the Natural History Museum. In this photograph the pair of stout horns are seen to greater advantage.

Photo by] [*W. Saville Kent.*

A STINGING ANEMONE.

These Anemones—the first partially closed, the second fully expanded—are coloured to harmonize with tufts of seaweed, in which tints of yellow, green and light orange-red are variously blended. They sting as fiercely as nettles. The species is found on the coast of Queensland, the Lacépède Islands, etc. (Here shown about half the actual size.)

in the typical Tortoise, but by turning the head sideways. Now though there may be many anatomical features which favour this view, there is one objection to it which seems insuperable. This is the fact that the great frill and the spikes which encircle the back of the head in Miolania would effectually prevent any such lateral movement, at least to a degree sufficient to be effective. Similarly, the view that Miolania was an aquatic Tortoise must be regarded as un warranted by facts. At any rate, it is highly improbable that a creature with so formidable an armature, and so heavily weighted a tail, would be aquatic.

Not the least remarkable thing about this wonderful creature is the fact that its fossil remains have been found in the Pleistocene formations of countries so remote as Australia and Patagonia. Remains of the largest species, *Miolania oweni*, have been found both on the mainland of Australia and on Lord Howe's Island, four hundred miles distant. *Miolania argentina* is the species found in Patagonia. One can only explain this curious distribution on the supposition that these now remote continents were once part of a common land area. But that is another story!

ANEMONES OF SOUTHERN SEAS

BY EDWARD STEP, F.L.S.

On page 425 will be found an account of Sea Anemones in general, accompanied by a coloured plate depicting some of the more striking species that occur in Northern Seas. As that account gave the principal facts relating to the structure and development of the Sea Anemones as a class, there is no need to go over the ground again. Most of the species represented in the present coloured plate are members of genera not found in northern waters; but though in form, size, and colour they may differ from those well-known on European coasts, their habits are much the same. The central Anemone at the top of the coloured plate is the Noble Cerianthus, and this is closely related to one of our rarer northern forms—Lloyd's Cerianthus—figured at the foot of our former plate.

In the Southern Seas, Anemones are largely associated with their relatives the Corals, and the coral-reefs shelter a great variety of species. Others, however, like the two illustrated in these pages, are to be sought on sand. Condylactis (see below) has a broad column six or seven inches long, but buries it to the full extent in sand, its base clinging to a rock or stone below. When alarmed it can fold up its tentacles and retire wholly below the surface. These tentacles have a remarkable knotted appearance, reminding one of the seed-pods of the radish, where every seed causes a bulge in the narrow pods. In the Anemone these alternating swellings and constrictions are coloured with grey, brown, and olive-green, while the hidden column is usually bright scarlet. The tentacles are gathered into about six sets, an arrangement due to the puckering of the central disc, which includes the mouth, into as many lobes. Mr. Saville Kent found this species at Port Darwin and on the coast of Queensland.

Photo by] [*W. Saville Kent.*

SEED-POD ANEMONE.

An Anemone that buries its column in the sand and spreads its tentacles on the surface. The column is bright scarlet, and the tentacles are mottled with grey-brown and olive-green. Their knotted form gives them a resemblance to the seed-pods of the radish. (The photo is half the natural size.)

One of the most striking facts in connection with these Anemones of Southern Seas is the well-attested friendship between certain of them and some fishes. The Anemone is named Stoichactis, and three species or varieties, discovered by Mr. Saville Kent, are shown at the foot of the coloured plate. The upper of the three was named after its discoverer, so we may refer to it as Kent's Anemone. These Anemones are monsters, measuring eighteen inches or more across their expanded tentacle-fringed discs. But size is not the most remarkable thing about Kent's Anemone; there is the friendship to which we have alluded. Here is a huge Anemone, armed with hundreds of pliant tentacles which will bend over and adhere to any living prey sufficiently small to be accommodated by the greatly distensible mouth; and yet there are certain fishes which it will allow to shelter, not merely under the lobes of its disc, but even in its mouth. When the temporary danger which caused its retreat has passed the fish reappears perfectly unharmed. Each species of Stoichactis has a distinct species of fish it takes under its protection. What service the fish renders in return

Photo by] [*E. Step, F.L.S.*

ROBIN'S PINCUSHION ON WILD ROSE.

Owing to the irritation or stimulation caused by a Gall-wasp depositing her eggs in the Wild Rose, the tissues grow out in a form much like moss. Beneath this "moss" is a more solid core in which the Gall-wasp's grubs feed.

Photo by] [*A. H. Bastin.*

BEDEGUAR GALL ON WILD ROSE.

At first like a ball of green moss, the Bedeguar later turns crimson and is then known as Robin's Pincushion.

does not appear from the evidence so far ; but it is almost certain that it is useful to the Anemone in some way, for the essence of such arrangements is reciprocity, as the present writer has attempted to show in his book, " Messmates in Nature."

THE GALLS OF WILD ROSES

BY E. W. SWANTON.

THE leaves of the majority of wild roses are attacked by gall-wasps (small insects popularly known as " flies ") of the same family as those which are associated with the oak-apple and other well-known galls which occur on oak-trees. In these wasps there is no alternation of generations, and parthenogenesis is not uncommon. The most familiar of all rose-galls is the " Robin's Pincushion," " Moss Gall," or " Bedeguar " (see above). The last name is probably derived from the Persian *bād*, and Arabic *ward*, meaning " Wind-rose." The name " Moss Gall " is applied to it because of the mossy outgrowths which completely conceal the fleshy part. The moss rose, so well known everywhere, received its popular name from the moss-like fringe into which the edges of the calyx is divided. This tendency to produce " moss " seems to be inherent in our wild roses, and is stimulated into activity when certain tissues are irritated by the presence of the gall-wasp Rhodites rosæ. It would appear that the tendency may be communicated to the branch bearing galls, though very rarely. In 1886 Professor Gautier made the highly interesting observation that a moss-rose branch had appeared upon a smooth sepal-rose bush, which had been growing for fifty years in the Garden of the Luxembourg at Paris. Upon examining this abnormal branch, he found that it bore a certain number of Bedeguar Galls, having the characteristic mossy surface, but no galls occurred on the normal branches.

As in the case of all gall-insects, young and growing tissues only are attacked. The female pricks a bud in spring; with the hatching of the larva abnormal growth begins, continues throughout the summer, and ceases only with the decline of the sap-flow in autumn, by which time the gall is fully developed. The gall is at its best in early September. It is then frequently very conspicuous, for the "moss," which was at first green, has taken on a bright red tint. A large "pincushion" is the result of the coalescence of several galls. If one of them is cut open, white larvæ will be found inhabiting circular cavities (see below); but as the gall is usually infested with numerous parasites, the determination of its rightful occupants is a matter of considerable difficulty. If the gall is removed with sufficient length of branch below it, the latter placed in a tumbler of water, and some gauze is tied carefully around the former, the inhabitants may be easily secured as they emerge. But as the Rev. J. G. Wood observes: "Even should the experimenter find the right species of insect in the gauze bag, he has to go through the wearisome task of searching through the family of Cynipidæ and identifying the species—a process which every entomologist is rather apt to postpone until the visionary period when he shall have leisure." Bedeguar Galls are of frequent occurrence on the Dog Rose and Sweet Briar. They were held in high repute by our forefathers and used medicinally. As late as the early part of the past century, the farmers in certain parts of Yorkshire made an infusion of them to cure diarrhœa in their cows.

The leaflets of various wild roses sometimes bear a little greenish-white gall about the size of a pea. It may be found occasionally on a stem or a sepal, and is always attached by a small stalk.

Photo by] [*A. H. Bastin.*

BEDEGUAR GALL INTERIORS.

The mossy galls have been cut through to show the grubs of the Gall-wasp at work inside, excavating the fleshy substance upon which they feed (Magnified about five times.)

Photo by] [*A. H. Bastin.*

PEA-GALLS OF ROSE.

The greenish-white galls are produced on the leaves of the Wild Rose by the punctures of a species of Gall-wasp different from that which causes the Bedeguar. (Natural size.)

It exhibits no tendency to produce "moss," and is usually quite smooth. It is caused by the presence of the larvæ of Rhodites eglanteriæ.

Another spherical gall occurs on the leaflets of the Dog Rose and Sweet Briar. It is very distinctive in having spiny outgrowths of variable size. It results from the attack of Rhodites rosarum. It is generally greenish, but is sometimes beautifully tinted with pink or red. It occurs chiefly on the midrib, is usually gregarious, and is often deformed through the presence of parasites.

The three galls described above may sometimes be found on one bush. That three closely-allied insects attacking the same plant, and the same part of that plant, cause three dissimilar galls to arise, each of which is constant in form, whatever may be the species of rose visited, clearly indicates that the larva of each causer has its own particular irritant, probably a fluid excretion, which causes the cells to deviate from their normal mode of growth. It must be remembered that the insect does not make the gall; the term "gall-maker," though frequently applied to it, is a most unfortunate and, indeed, misleading one.

Photo by] [*A. H. Bastin.*

SPIKED PEA-GALL.

The galls here shown in different stages of growth are very noticeable on account of their remarkable form. Note their similarity to the Spiders' egg-cocoons shown on page 1116. (Natural size.)

Brief allusion must be made to some rose-galls not illustrated in this work.

The Burnet-leaved Rose is often attacked by Rhodites spinosissimæ. This wasp lays its eggs in all parts of the plant, excepting the root. A single gall is an ovoid swelling, about the size of a small pea, soft and greenish at first, becoming very hard and bright red at maturity. These galls are frequently coalescent, forming large masses.

The Leaf-rolling Sawfly, Blennocampa pusilla, oviposits in the leaves of the Dog Rose. The margins of the leaf roll inwards, ultimately providing two pouches in which the larvæ live. Cultivated roses are frequently attacked by this insect, which is becoming a serious pest in rose-gardens.

A minute fly, Perrisia rosarum, attacks various wild roses, ovipositing in the leaflets, the edges of which fold upwards until their margins meet. The

Photo by] [J. J. Ward.

THE SEA-MOUSE AT HOME.

The remarkable marine worm known as the Sea-mouse is clothed on the upper side with long hairs of iridescent hues, which reflect every ray of light and completely hide the rings of which its body is built up. (Slightly enlarged.)

leaflet is also somewhat thickened and blistered. The pod or pouch thus formed is often tinted red or brown at maturity. The larvæ are yellowish red.

A mite, to which the name Eriophyes rosæ has been given provisionally, appears to be the influencing factor in the production of curious nodular excrescences attaining the size of a small walnut occasionally seen on stems of the Trailing Rose. The swellings arise at the nodes, are persistent, and grow very slowly.

A minute parasitic fungus, Coniothyrium fückelii, attacks the stems of various wild roses, causing them to become swollen and rugged at the point of infection, producing a condition akin to that popularly known as " canker." It also occurs on cultivated roses.

I am not aware of any European records of galls on the roots of wild and other roses. A gall-wasp, Rhodites radicum, oviposits in the roots of wild roses in North America ; the gall which results is globular, about two inches in diameter, smooth and reddish-brown. To those in search of a hobby which will demand inquiry in many of Nature's by-paths, and at the same time allow the student to get together an interesting collection which will probably escape condemnation by Nature protectionists, I can cordially recommend the study of galls.

THE SEA-MOUSE

BY JOHN J. WARD, F.E.S.

THERE is no mouse which inhabits the sea-depths ; indeed, the Sea-mouse is really a sea-worm. Probably the word " mouse " has been incorporated in its popular name on account of its furry

Photo by] [*J. J. Ward.*

THE SEA-MOUSE.

A view of the Sea-mouse on edge in the act of turning over. Here the bright silky hairs along the sides are better displayed. Among them are dark-coloured bristles which contain a harpoon for defensive purposes.

Photo by] [*J. J. Ward.*

THE SEA-MOUSE.

Here is seen the underside, showing the rings of which the body is composed. There are about forty of them, and each bears a pair of foot-stumps at the sides, by means of which the Sea-mouse crawls. (Natural size.)

coat and its habit of creeping on the sea-floor ; for this curious animal is not a swimmer, as at first glance its anatomy might seem to imply.

Its home is in fairly deep water, where it hides amongst the stones, often half-buried in mud. Consequently, it is only after a storm, when one of these creatures has been thrown ashore, that the seaside visitor is likely to meet with it. Then, if it is discovered in a clear pool on which the sunlight is playing, it proves a wonderful surprise for those who view it for the first time, for its hairy covering converts it into a veritable rainbow of colour.

Each of the numerous silky hairs, which clothe the edges of its body, acts as a prism, gleaming with rich gold, fire-like red, and vivid green. Indeed, while writing I have before me a wide-mouthed bottle of methylated spirit in which I have preserved two of these Sea-worms, and their colours are just as beautiful and iridescent as when in the living state. The specimen only needs well washing in fresh water and then placing in the spirit, and securely corking ; after which one has an object of beauty for the museum, which may be viewed by day, or by artificial light. A slender spray of some zoophyte or sea-moss, enclosed with the object, helps to give a natural effect.

Why such glorious colours should be bestowed upon a creature of such sluggish habits, and which for the greater part of its life lies hidden amongst stones and mud at the sea-bottom, is not easy to understand. Probably they have some connection with its natural enemies, for, in addition to these gorgeous silky hairs, it is provided with strong purple or blackish bristles of a most complex character, each consisting of a barbed harpoon enclosed in a horny sheath. The bristles are arranged in groups, and are retractile, being more or less withdrawn into the body when the creature is peaceful, but when annoyed are erected, becoming then somewhat like a miniature porcupine, and anything but a choice mouthful for a hungry foe ; although such mouths as those of the cod, or the dog-fish, can relish the Sea-mouse even with its full armature of spines.

On the back is a felt-like covering composed of interwoven hairs, under which a series of overlapping plates are arranged in pairs, these covering the gills. The felt covering acts as a filter for the sand and mud, allowing only the pure water to reach the breathing organs.

The underside reveals the body divided into about forty rings, each bearing a pair of bristly foot-stumps, which the animal can elongate or contract at its will. The head bears eyes and a pair of feelers, and its mouth is a kind of tube which, to capture the small crustacea, sponges, etc., on which it feeds, it protrudes from the inside, somewhat like the inverted finger of a glove, enveloping the prey and then reversing the action.

SPIDER COURTSHIP

THE display of their charms of form, colour or voice by male animals in their efforts to secure the approval of the females, is quite an ordinary phenomenon. Birds in the mating season delight us with their exuberant vocal efforts, though the concert is not intended for us, but for the allurement of the hen-birds. Certain male birds that lack the power of song, but have fine plumage, strut and dance around the hens, spreading their wings and tail-feathers, and behaving generally in an extravagant manner. Something of the same sort occurs with certain fishes, who get up specially bright liveries for the courting season, and even resort to battles royal for the possession of a desired female, in this latter respect copying the procedure of many mammals. Some insects do the like, and the coquetting of butterflies in the air is of much the same purport. The male butterfly is, as a rule, much more brightly coloured than the female ; just as we find in the case of the birds, and the male butterfly likes to sit on a flower, or in some cases on the ground, and spread its wings in the sun for the enjoyment of passing females.

It is, therefore, not surprising to find that the same rule applies among spiders, where the female is by far the superior of the two sexes. The males, as a matter of fact, are rarely seen. Nearly all the spiders that we see on webs are females, and the builders thereof. The males are not domesticated, are not proper web-weavers, and appear to be regarded in general by the other sex as necessary evils to be quietly disposed of at the end of the honeymoon. But the preliminary question is, shall there be a honeymoon, and to settle this point the males of different species of spider set to work in different ways. Some minute males hang around the circumference

[By T. Carreras.

SPIDER COURTSHIP.

The male of Astia dancing and displaying his fine form and ornamentation before the less brilliant female.

of a big web, trying to attract the attention of the gigantic lady who has built it, the lady in question being in some species a thousand times greater in bulk than her wooer. She may look favourably upon his advances or she may—eat him! In cases of this kind the very smallness of the male may stand to his advantage by enabling him the better to elude the desired one, and it is thought that this may be the reason why male spiders are, as a rule, so much smaller than the females. To illustrate this disparity in size and weight, take the case of a species of Nephila. The female measures two inches in the length of body, leaving the long legs entirely out of account. Her mate is about one-tenth of an inch. If put into the scales it would take thirteen hundred males to balance one female. This being so, it is not surprising to learn that the male is somewhat hesitating in his advances and hangs about the borders of her web for hours, apparently trying to screw up his courage. If repulsed, he does his best to get out of the way as speedily as possible, having some instinctive knowledge of the fate that awaits him if caught. Emerton says: "In these encounters the males are often injured; they frequently lose some of their legs; and I have seen one, that had only four out of his eight left, still standing up to his work."

[By T. Carreras.

SPIDER COURTSHIP

The male of Icius posturing and dancing before the coy female, who is slyly regarding him over the edge of her retreat.

Against this there is not wanting evidence that some species show at least an approach to kindly relations between the sexes—for a time, what we may call the honeymoon period.

But to come to the display by male spiders of their beauties as a means of softening the hard heart of the female: Mr. and Mrs. Peckham, who have made a close study of American spiders especially, have given us an account of the behaviour of one of the Jumping Spiders, closely related to the spider shown on page 1144—whose conduct under similar circumstances is much the same. They say: "When some four inches from her he stood still, and then began the most remarkable performances that an amorous male could offer to an admiring female. She eyed him eagerly, changing her position from time to time, so that he might be always in view. He, raising his whole body on one side by straightening out the legs, and lowering it on the other by folding the first two pairs of legs up and under, leaned so far over as to be in danger of losing his balance, which he only maintained by sidling rapidly towards the lowered side. . . . Again and again he circles from side to side, she gazing towards him in a softer mood, evidently admiring the grace of his antics. This is repeated until we have counted a hundred and eleven circles made by the ardent little male. Now he approaches nearer and nearer, and when almost within reach whirls madly around and around her, she joining with him in a giddy maze. Again he falls back and resumes his semi-circular motions, with his body tilted over; she, all excitement, lowers her head and raises her body so that it is almost vertical; both draw nearer." He has won!

A similar process is gone through in other species, always in those where the male is more brightly coloured or ornamented than the female, and the attitudes adopted are always such as are calculated to display his beauties to the greatest advantage, whatever part of his organization may be the most finely decorated or developed. Sometimes it is a crested head, a painted thorax or an iridescent hind body; sometimes a beautifully-feathered pair of mouth-parts known as palps, or similarly decorated forelegs, as in the dancing male shown on page 1145. Sad to relate, even these exhibitions sometimes fail to achieve the object aimed at, and the female expresses her displeasure by making a savage onslaught on the rejected suitor, an attack which may preclude him from trying his fortune in another direction. This, however, is not always the unhappy ending to the romance. Some males meet with such encouragement that they straightway set to work on the construction of a nuptial tent in which the honeymoon is passed.

RED SNOW

"As white as snow!" Such a phrase is in everyday use to express unexampled purity, and points to the general recognition of the dazzling whiteness of a snowfield. What, then, must have been the surprise of the unprepared traveller who first saw stretched out before him an expanse of blood-stained snow! He would stoop and examine the stains more closely, and would find little enough to satisfy his curiosity—no ghastly corpse beneath the surface, only a few handfuls of reddened snow. To him such a mystery could only be accounted for by black magic, and he would see in the phenomenon the ominous forewarning of the shedding of human blood.

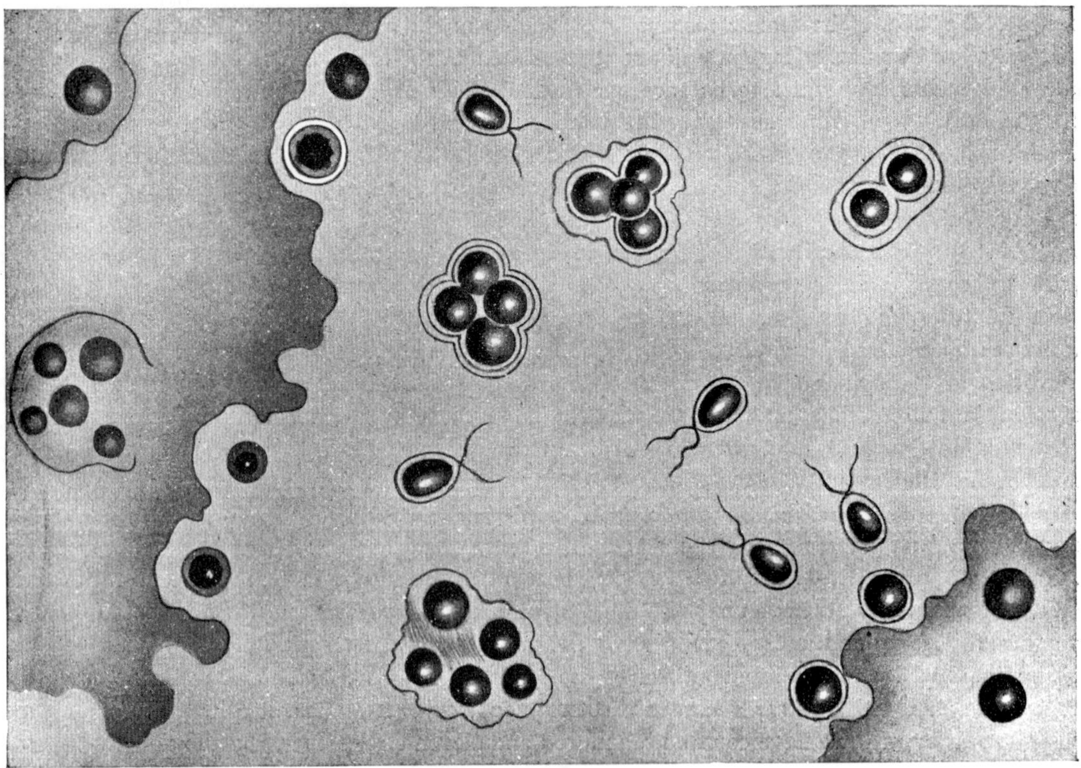

RED SNOW PLANT.

Sphaerella, which gives its red colour to snow, is one of the simplest of plants, consisting of a single cell of microscopic smallness. At one stage it has the power of locomotion, propelling itself through water by the lashing of the gossamer-like organs shown in the illustration.

But the first attempt at describing Red Snow in a scientific manner was made by De Saussure, who came upon it during a tour in the Savoy Mountains in 1760, and started a series of inquiries which were carried on more or less intermittently for the best part of a century. At length it has been satisfactorily proved that the phenomenon is due to the presence in the snow of innumerable hosts of little algæ or primitive plants, which take on a red hue, or, as is more usual, a green. The condition which produces the growth seems to be the temporary melting of the snow by the sun's rays, for the plant is only to be found on the topmost layers of the snowfield. Its name is Sphærella, so called from the spherical shape of each member of the ruddy community. It is very minute and can only be seen distinctly under a microscope; then it will be found that each individual consists of a single cell, but this may be round or oval. The round are the maturer members of the company.

By permission of] [*Sir H. H. Johnston, G.C.M.G.*

HARPY EAGLE.

The largest existing bird of prey. It is a native of Tropical America, and preys mainly upon four-footed animals of fair size. Its claws are far more terrible weapons than its beak.

Their substance is completely made up of protoplasm; and careful observation will detect a slight tint of green, due to the presence of chlorophyll, but this is almost entirely overcome by a rich red pigment contained in the cell which is sufficiently brilliant to make noticeable the presence of the minute Sphærella on the snow.

The oval, or young, cells are supplied with a couple of moving threads or cilia, which are constantly rotating and so give the plant the power of motion, a quality that the round cell does not possess. The reproduction of the species is provided for by the subdivision of the protoplasm within the cell. Such subdivision is shown in the illustration on page 1146. After a time the parent cell ruptures and the young cells escape into the outer world, where, by means of the cilia mentioned above, they can enjoy a brief period of excitement and motion. Soon they lose the oval and assume the rounded shape, and pass through the same stages of maturity as their progenitor.

Red Snow has been found in many parts of the world—the Swiss Alps, the Tyrol, the Carpathians, Northern Scandinavia, and on the Sierra Nevada, but it is more definitely connected with the Crimson Cliffs of Greenland, which were so named by the explorer Ross because of their markedly rich coloration due to the presence of the Red Snow plant.

THE HARPY EAGLE

BY SIR HARRY JOHNSTON, G.C.M.G.

BIGGEST in body of all existing birds of prey (and perhaps as large as the extinct Harpagornis of New Zealand) is the magnificent Harpy Eagle of Tropical America. The Condor Vulture may measure more across its wings, but its body is not so big as that of the Harpy. The largest and finest specimen I have ever seen is in the collection of the Natural History Museum at Stuttgart, and a photograph of this bird is given here. The Harpy Eagles are mainly represented in South America, but they have a relation in New Guinea. They are nearly all of them very big birds of prey, with a development of claws larger even than that seen in the true Eagles. An outlying relation is the extraordinary Monkey-eating Eagle of the Philippine Islands, and another member of the group may have been that monstrous bird of prey, Harpagornis, once inhabiting New Zealand. The food of the Harpy Eagles in Tropical America consists mainly of beasts—monkeys, squirrels, large rodents, the fawns of deer, wild dogs, raccoons, and the cubs of pumas and jaguars (when they can be seized in the absence of the parent). I have seen an African ally of the Harpy Eagle kill a strong, full-grown cat in an instant by a lightning dart of one powerful leg and the instantaneous driving of the long, curved, sharp claws into the cat's heart. It has been several times reported by travellers that the Harpy Eagles will snatch up, kill, and eat the babies of the Amerindian natives, if these are accidentally exposed by the mother.

Photo by] [*W. S. Berridge, F.Z.S.*

MONKEY-EATING EAGLE.

This extraordinary bird, whose food largely consists of monkeys, is considered by some to have had its facial expression modified by its association with its prey, until it has come to have a very monkeyish appearance.

Yet these magnificent birds are not entirely recalcitrant under kindly treatment, but will get to know their keepers and allow themselves to be handled without striking out with beak or talon. As a matter of fact, their claws are far more terrible weapons than their beaks. When a Harpy Eagle is wounded or captive and feels itself to be at bay, it turns over on its back, with its clawed feet in the air, and with these it fences.

Photos by] [*A. Harold Eastin.*

A LEAF-ROLLING BEETLE AND ITS WORK.

The Beetle begins by making a peculiar cut in a Birch-leaf on each side of the midrib, as shown in the first photo. Then it rolls the larger portion of one half as in the second photo, until it reaches the stage seen in the third photo, when it starts rolling the other half around the first half.

A MARVEL OF INSTINCT

BY HAROLD BASTIN.

MANY caterpillars roll up the leaves of their food-plant, thereby forming retreats in which they take refuge when danger threatens. But although these insects well repay attention, they are completely outclassed in skill by certain small beetles—weevils, to be precise—which are grouped by naturalists in two families known as Rhynchitidæ and Attelabidæ. These weevils also roll up leaves, each species adopting a characteristic method; but, unlike the caterpillars, they do not spin silk threads to help them in their labours. Moreover, their habits are not inspired by any personal need for food or shelter, but by what is called—for want of a better term—parental instinct. In a word, the rolled leaves are to be regarded as elaborately contrived nurseries; for the female weevils lay their eggs among the folds, while the grubs, when they hatch, feed upon the partially-decayed leaf-tissue.

Photo by] [*A. H. Bastin.*

A LEAF-ROLLING BEETLE.

The second half of the leaf is now loosely wrapped around the first half.

One of these insects—a pretty, coral-red species—is associated with the oak, while another, scarcely less showy in appearance, is to be sought among the foliage of hazel-bushes. For our present purpose, however, we will trace the habits of the small blue-black species known as Rhynchites betulæ—a double-barrelled name which refers to the insect's characteristic rostrum, or snout, and to the fact that it is commonly found on the birch. Young birches, growing at the outskirts of plantations, are especially liable to harbour this little leaf-roller, although it is sometimes found upon other trees, including the alder.

Photo by] [*A. H. Bastin.*

A LEAF-ROLLING BEETLE.

Both coils are now tightened up

According to the common custom of insects, the male weevil takes little or no interest in the well-being of his offspring. The female does the leaf-rolling quite unaided. She sets about the work in a very systematic manner, and there can be no doubt that in her punctilious attention to detail lies the secret of her success. For it is no light task that she sets herself to accomplish. A suitable leaf having been chosen, the insect makes her way to a point at the edge, not far from the leaf-stalk, and, working upon the *upper* surface, deliberately makes a curved cut from the edge to the midrib.

Photo by] [*A. H. Bastin.*

THE BEETLE.

A magnified view of the Weevil whose clever leaf-rolling work is shown in the other photos.

She then crawls a little way up the leaf, and from the other side of the midrib makes another curving cut to the opposite edge. In one of the accompanying photographs this stage of the work has just been reached, and the weevil is seen taking her customary spell of rest. Probably she remains inactive because, by waiting, the lower part of the leaf wilts somewhat, and thus becomes more manageable. Yet the success of her subsequent operation depends mainly upon the accuracy of the cuts which she has made in the leaf. These cuts have been examined by mathematicians, who have pronounced the curves to be exactly those which are necessary to render the rolling of the leaf most easy for the insect to accomplish. Nevertheless, there is reason for thinking that the weevil's instinct is not infallible, for one may often see half-rolled leaves which have been abandoned, apparently because the insect's strength proved inadequate for the completion of the work. Probably in such cases, the original curved cuts were at fault.

When the leaf has become sufficiently pliable, the weevil goes round to the back—*i.e.*, the under-surface of the leaf—and, beginning at the point where she first cut the tissue, gradually rolls one side of the leaf inwards to the midrib. This part of the work can quite easily be watched by an observer who has the necessary patience at command. Subsequently, however, the weevil changes her tactics, and draws the second half of the leaf round that which has already been rolled, incidentally wrapping herself over, so that we can only judge from the slow tightening of the folds that the work is still progressing. It is at this stage—*i.e.*, when the weevil is hidden from view—that the eggs are laid. Thereafter, the insect creeps out, tightens up the roll, and fastens down the overlap with her snout.

The appearance of the leaf below the cuts is now very much like the "screw" of paper which a grocer makes to hold a pound of moist sugar; and the insect completes the resemblance by making the tip of the leaf into a smaller funnel, which is eventually tucked in with due care. Then comes a curious little ritual, the significance of which is not quite clear. The weevil climbs up the leaf, and deliberately wounds the midrib at the point where her first curved cut meets it. It has been suggested that she does this in order to promote the particular state of decay which will render the leaf suitable as food for her grubs. Be this as it may, the fact remains that she often completely severs

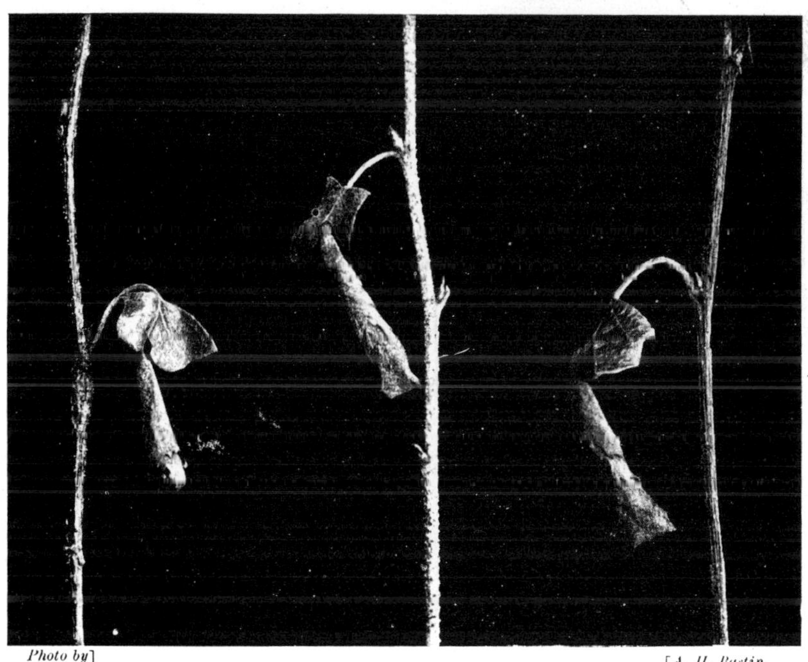

Photo by] [*A. H. Bastin.*

FINISHED WORK OF THE LEAF-ROLLER.

The three examples here shown should be considered from right to left. They show the final stages of tightening up, and tucking up the previously open lower end.

the midrib, with the result that the rolled part of the leaf falls to the ground—the precious eggs, of course, going with it.

Now the marvellous thing about the birch leaf-roller is this : The grubs feed upon the decaying leaf-tissue, and eventually turn to pupæ in the soil. Here they lie all through the winter, appearing as perfect weevils in the early summer. They at once creep up the birch stems, and the females, after pairing, immediately—without education or opportunity to profit by observation—begin to cut and roll leaves after the manner of their ancestors. Whence comes their skill ? How do they know the way in which the curved cuts should be made ? We call this inborn knowledge " instinct," but the word is really a cloak for our ignorance. For what instinct is, and how it originates, are unfathomed mysteries.

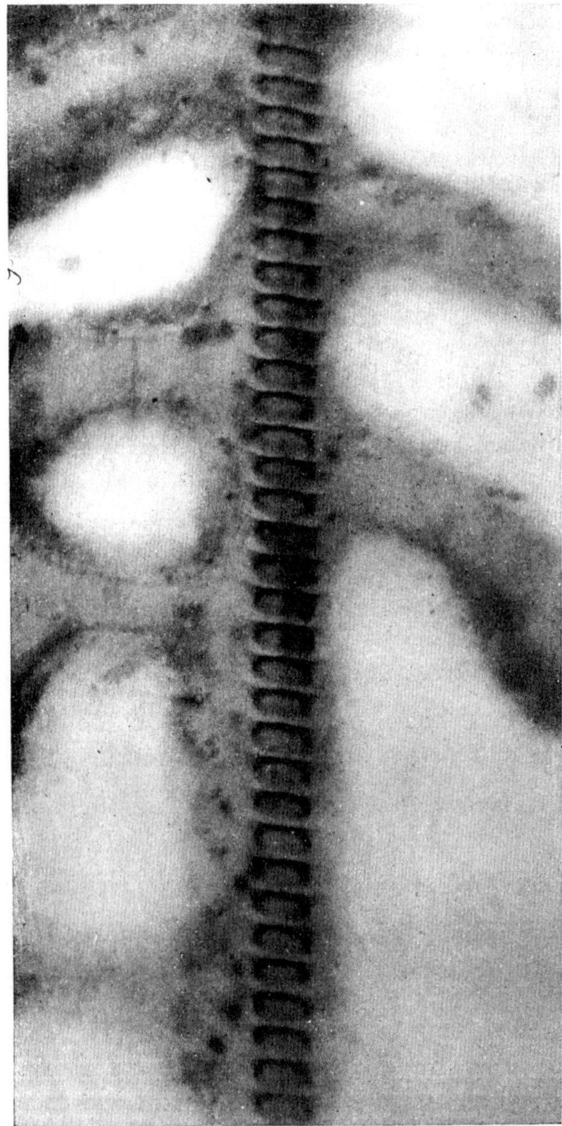

Photo by] [*E. J. Spitta.*

THE CRICKET'S FIDDLE-BOW.

On looking at the Cricket's wing-case a strong nervure will be seen running across part of it, marked by transverse notches. This bow when drawn across the drum of the opposite wing produces the high-pitched "chirp."

THE CRICKET

BY DR. E. J. SPITTA, F.R.M.S.

THE little Cricket that chirps on the hearth is an interesting insect, and seems always to have attracted attention, for its curious habit of hiding away in holes during the daytime and of coming out at night to chirp has even claimed the notice of the poet :

> " On a lone winter's evening, when the frost
> Has wrought a silence, from the stove there shrills
> The cricket's song."

Putting aside the general appearance of the insect itself, the points of interest to the worker with the microscope centre around the consideration of the chirping apparatus; the gizzard and the tongue.

Commencing at the first, which is shown on page 1154, there cannot fail to be noticed a clear space something like a drum in the elytron, or wing-case. Looking nearer the body or thick end of this case, there will also be seen a strongly-marked line—or nervure, as it is called—running across part of the wing-case. This can be recognized, even in the photograph, which is not very highly magnified, by its having a series of notches or transverse file-marks upon it, running its entire length. It is shown much magnified on this page, and is usually called " the bow." It is with this arrangement, used in conjunction with the wing-case of the other side, that the chirp is made. Messrs. Kirby and Spence thus describe it : " The friction of these nervures of the upper or convex surface of the base of the left-hand wing-case—which is the undermost—against those of the lower or concave surface of the base of the right hand—which

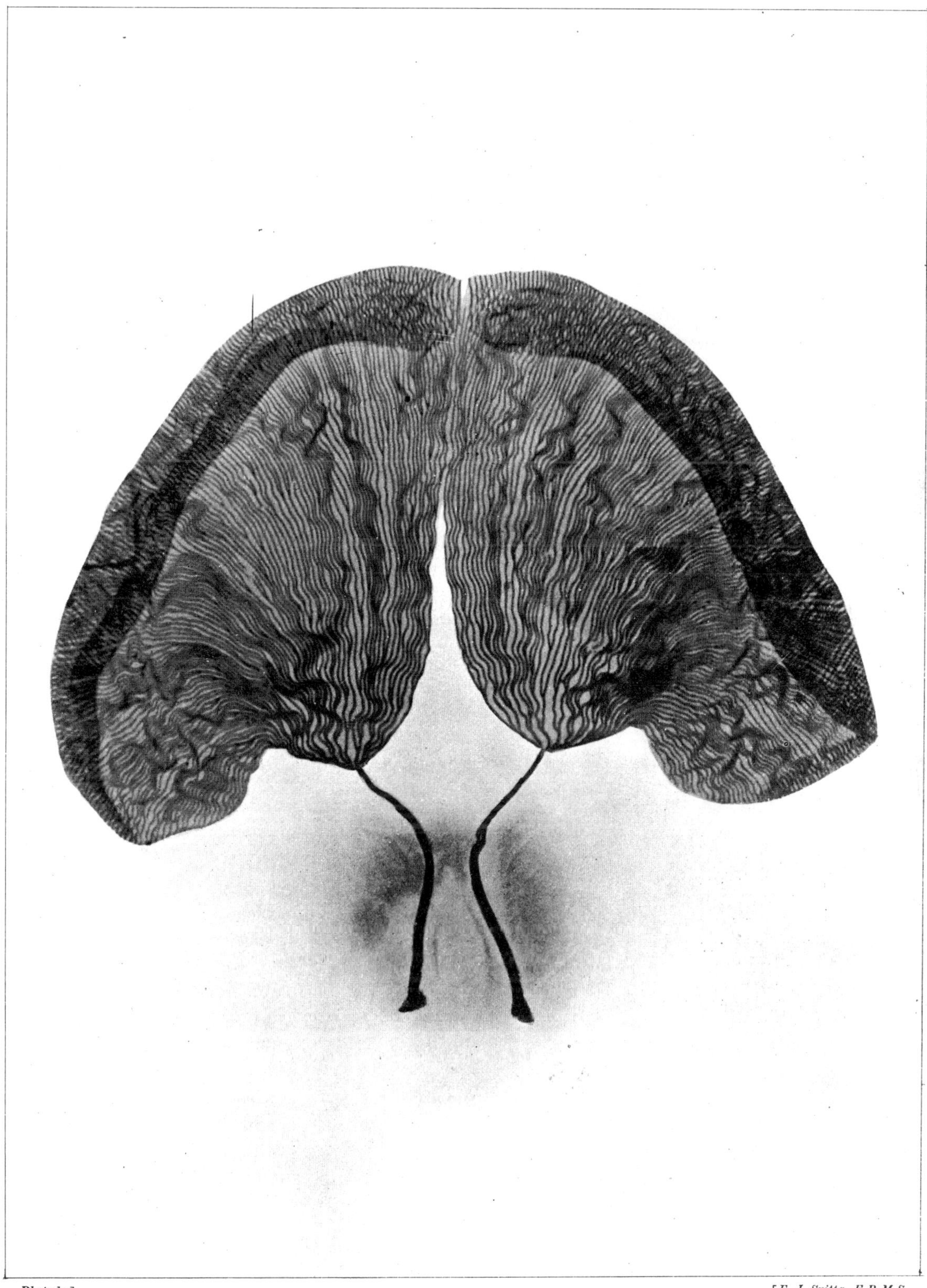

Photo by] [*E. J. Spitta, F.R.M.S.*

THE CRICKET'S TONGUE.

The tongue, here seen highly magnified, has no traces of suctorial tubes such as are seen in the proboscis of the Blow-fly. It is not clear in what manner it is used by the Cricket.

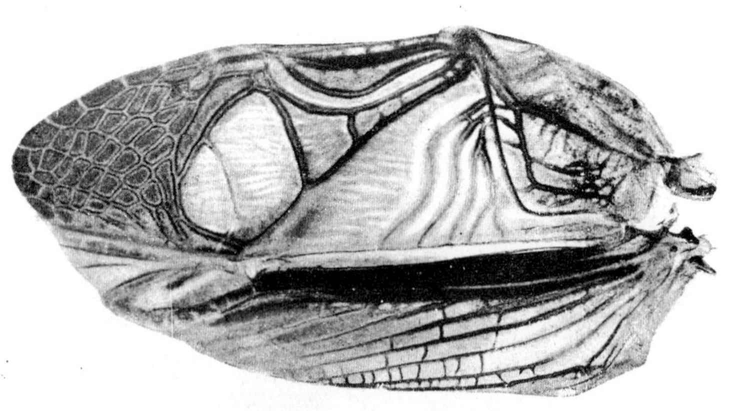

Photo by] [*E. J. Spitta.*

THE CRICKET'S DRUM.

The clear space a little to the left on the wing-case is an area of tense membrane like the head of a drum. Upon this the friction of the "bow" produces the Cricket's chirping note.

is the uppermost—will communicate vibration to the areas of membrane (the drums) more or less intense in proportion to the rapidity of the friction, and thus produces the sound for which the creature is noted."

But M. Goureau, who has paid much attention to all classes of noise made by insects, goes still further into detail, for he believes that the "bow" already mentioned really plays on the other nervures which he even names "the treble strings," and that by this peculiar arrangement the sound is produced.

It does not seem to be known what object the little creature has in making this noise, for if disturbed it ceases it at once.

The gizzard, which is a bag for grinding up the food—taking the place in the internal economy of the insect of the functions performed by our teeth and mastication—is a muscular organ placed in the Cricket at the end of the food-pipe, and is peculiarly powerful for so small a creature. In the photo it is seen to consist of horny teeth arranged in this bag more like the spokes of a wheel than anything else. The teeth are decidedly powerful, and this has been explained to be necessary because of the hard substances which are eaten by this little insect.

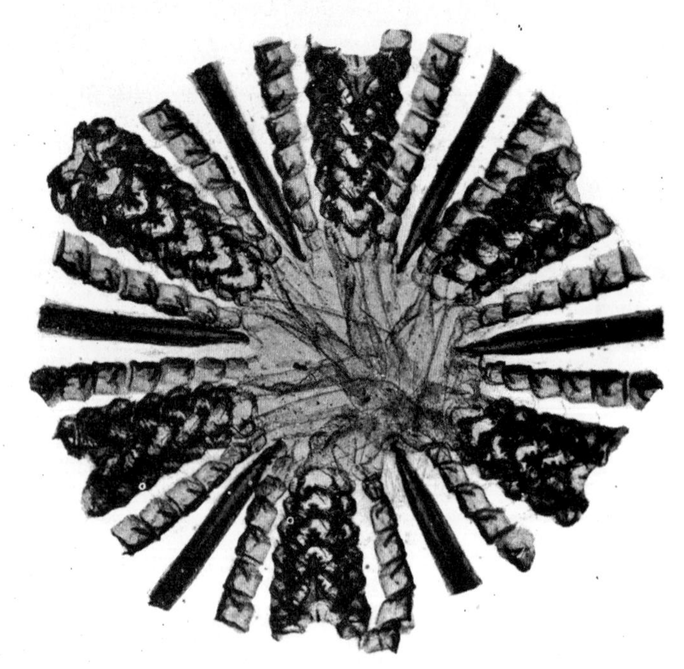

Photo by] [*E. J. Spitta.*

CRICKET'S GIZZARD.

A bag-like organ fitted with rows of powerful horny teeth which tear and break down the food before it is passed to the stomach.

The tongue, as seen highly magnified on page 1153, is extremely elegant and beautiful, but differs very materially from the proboscis of the blow-fly and other suctorial insects, inasmuch as it has no trace of suctorial tubes. What its exact use can be is not certain, unless simply to lick up food much in the same way as a dog uses his tongue, for it seems impossible it can absolutely absorb nourishment in the way that the proboscis of the blow-fly does, for example, seeing that there is an entire absence of any arrangement for so doing.

A LIMESTONE COMPOSED OF WATER-FLEAS

BY F. CHAPMAN, A.L.S., F.R.M.S.

In our ponds and runnels swarms of tiny bivalved crustaceans are always to be found. These are known as Water-fleas (see page 498). They are sometimes so numerous as to form thick beds of sediment as they die and fall to the bottom with the mud. The organic matter of their bodies not infrequently gives rise to quantities of petroleum, which remains locked up in the shales of certain formations like the Carboniferous of Scotland. Some of the limestones of the Isle of Purbeck are mainly composed of the little valves of these lowly crustaceans. The genus to which they belong is now extinct, but we know from the accompanying mussel-shells, turtle-bones and the remains of crocodiles that the prevalent conditions were of the nature of fresh-water lakes and brackish or estuarine river-mouths and swamps bordering the sea of the Jurassic period. The accompanying photographs show a magnified view of the weathered surface of a piece of limestone, and a thin section of the same, from Swanage Bay, Dorsetshire.

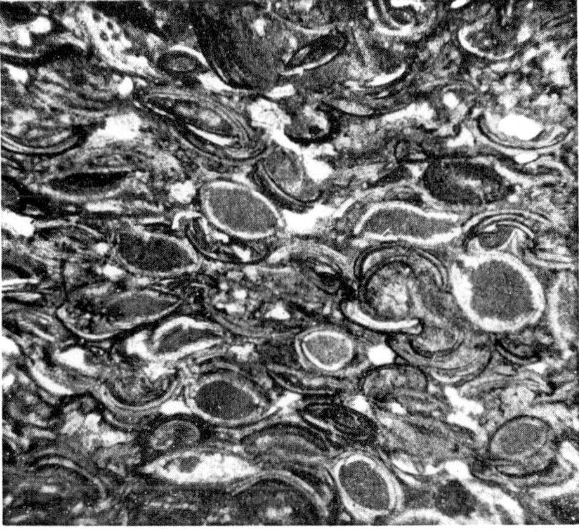

Photo by] [*F. Chapman.*

CYPRID LIMESTONE.

A thin section of this Limestone from Dorset, as seen under the microscope, showing the shells of ancient Water-fleas of which it is made up.

CADDIS-CASES

In clear-water ponds and shallow streams, one may see from the bank a number of more or less cylindrical objects moving over the bottom. These cylinders are variously composed according to the situation where they are found. Some are constructed of grains of sand cemented together, others of small fragments of gravel, cut leaves, short lengths of twigs, small bivalve shells, snail shells, and so forth. They are well-known to every rural schoolboy as Caddis-cases, and to have been constructed as a protection by a tender-bodied grub. In the course of time this grub becomes a chrysalis, and later emerges as a four-winged Caddis-fly.

There are many species of Caddis-fly, and a naturalist who has studied the group can tell by the Caddis-case to which species the contained grub belongs. Each kind has its own special taste in the matter of building

Photo by] [*F. Chapman.*

CYPRID LIMESTONE.

The weathered surface of the Limestone is seen more highly magnified, and exhibits the general form of the fossil Water-fleas more clearly.

84

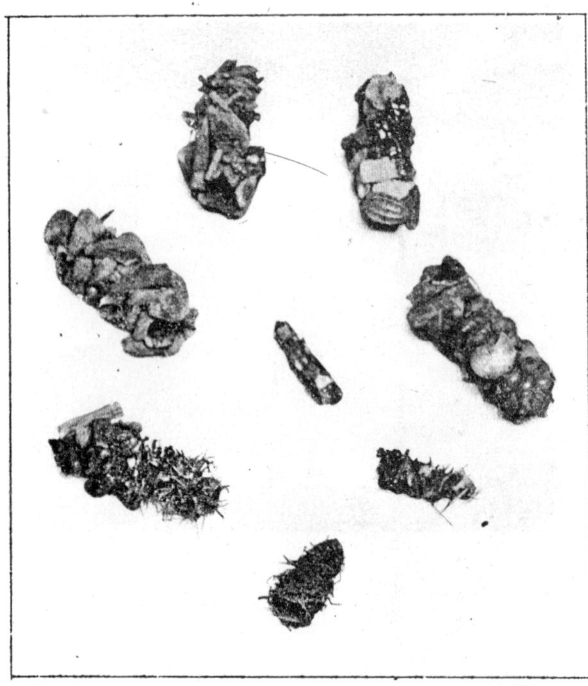

Photo by] [*J. Holmes.*

CADDIS-CASES.

Some of the miscellaneous materials with which Caddis-worms fortify their silken tubes.

Photo by] [*E. Step, F.L.S.*

CADDIS-FLY.

The Caddis-worm ultimately develops into a fly with four brownish wings clothed with hairs.

materials, and also its own ideas architectural. But whatever the material employed or the ultimate appearance of the finished case, the basis of all of them is a silken tube, spun under water by the infant Caddis-worm, to which the exterior ornamentation of whatever kind is attached.

Some of the Caddis-cases of heavy material have a counterpoise in the shape of a little stick of wood, longer than the case, and extending beyond each end. The Caddis-worm protrudes only its head and six legs out of the front opening, and so drags its case about, but can instantly withdraw its head on sighting danger. In due course it changes to a pupa, but first makes ready by attaching the case to some fixed object, and then closing the open ends by silken nets or by small stones, according to species. At the proper time, the pupa escapes from its case and crawls up to the surface of the water, or beyond, and emerges from the pupa-skin as a fly with four brown wings, which are sparsely covered with hairs, and a pair of antennæ longer than the body.

A FOUR-EYED FISH

BY W. P. PYCRAFT, F.Z.S., ETC.

In the matter of their eyes fishes display some very extraordinary modifications, some of which, at any rate, seem to be directly related to the mode of life of their possessors. But of all the instances of this kind so far collected the strange South American "Four-eye," or Anableps, is perhaps the most remarkable. Briefly, each eye is divided horizontally into an upper and a lower portion, the one adapted for vision in the upper world, the other for vision in the water. So far as our knowledge of these strange fishes goes, they seem to find the bulk of their food by a survey of the surface of the water, and not, as is usual, by sighting it more or less distant from the surface. In seeking food, at any rate, they appear always to swim at the surface, working their way upstream, and in small "schools."

[By Theo. Carreras.

THE FOUR-EYED FISH.

The two sexes of this remarkable fish are here shown. The upper and distant figure is that of the female, showing the natural swimming attitude in which the divided eyes come into use. Note how the fore-parts are flattened from above, whilst the hind-parts are compressed from the sides

[By T. Carreras.

FOUR-EYED FISH.

The fore-parts of the fish are here shown on a larger scale than in the preceding picture, to make clear the peculiar division of the eyes which practically doubles them

Now and then one may be seen to dart forward, seize a floating particle and resume its place in the school, but they never seem to leap out of the water to snatch at insects, and rarely, if ever, to dive in search of small crustacea or other minute organisms. But a more careful investigation into the food and manner of life of Anableps is certainly necessary before we can hope to gain any real insight into the purpose and use of this extraordinary eye. Since it constantly swims at the surface, with the upper half of the eye above water, we must assume that this ability to survey the upper world is of vital importance to its well-being. It would seem that the greater part of its food must consist of small animals which run along the top of the water, rather than of organisms beneath.

Many attempts have been made to explain the origin of this wonderful eye, but there seems to be no need to seek for any better explanation than that afforded by natural selection and variation. How the variation possibly arose is a theme which cannot be discussed in the space at our disposal. But it is significant to note that one of the Blennies, Dialommus, a marine species from Panama, displays an exactly similar eye and a similar mode of life.

But Anableps, the double-eyed, has yet other features of interest. The sexes differ conspicuously in size, the males being the smaller. The females bring forth living young, instead of eggs. After birth the young are said to be carried about by the mother in a thin-skinned bag divided by a partition, and thus enclosed, they remain until able to fend for themselves.

By permission of] [*Pearson's Oil-fields.*

PART OF A MEXICAN OIL-FIELD.

In parts of America the deposits of petroleum are so vast and under such great pressure that any well sunk into them is converted into a fountain whose outflow forms lakes of oil, as here shown.

MINERAL OIL

Petroleum, which in its numerous forms has become an indispensable item of domestic and industrial use, is a mixture of hydrocarbons, whose origin has been variously explained by several opposing theories. One view is that it has resulted from the action of water on iron carbide ; another has it that carbon dioxide in water has acted on alkali metals in a molten state in an early condition of the earth. Others insist that the natural petroleum has resulted from the decomposition of organic remains—the animals and plants of the earlier periods represented by the respective geological strata from which the oil is now obtained. These strata differ in various parts of the world. In parts of America, where the petroleum deposits are so vast that any well sunk into them becomes converted into a flowing fountain, the rocks are of Carboniferous, Devonian and Silurian age. The outflow from these wells forms lakes of petroleum, as shown in our view of part of the Pearson oil-field in Mexico. The deposits appear to be under great pressure from water or gas to cause them to rise in such volume wherever a vent is made in the overlying rocks. The oil-fields of the Caucasus owe their supply to rocks ranging from the Cretaceous to the Miocene periods. The Scotch oils are obtained by distillation of carboniferous shales from the Lothians. In this case there is no underground store of the oil to be tapped ; only by the breaking of the shale into small pieces and their subjection to heat can the petroleum be obtained.

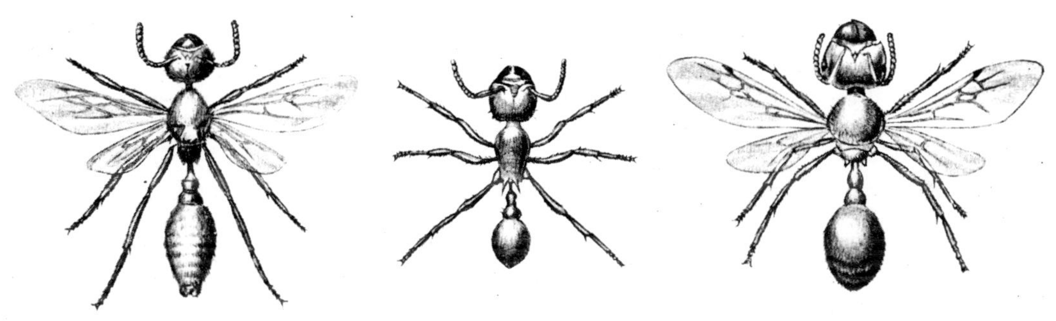

THE AGRICULTURAL ANT.
The insect to the left is the winged male Ant; in the centre is the large worker; and to the right is a young female.

THE AGRICULTURAL ANT OF TEXAS

BY H. ST. J. K. DONISTHORPE, F.Z.S., F.E.S.

It was originally recorded by the ancients that ants stored seeds in granaries to serve as food in time of scarcity. This was doubted in the early part of the last century, being regarded as a fable or myth. Later, however, the fact was thoroughly re-established, and it is now well known that a variety of ants of many different and distantly related genera do harvest seeds for food.

It can be readily understood that in hot, arid countries, where insect food may be lacking, or very scarce for many months in the year, and competition with other insect-eaters may also be very keen, that it would be of great advantage to ants to become, in part, vegetarians. Moreover, carnivorous ants possessing powerful mandibles for crushing the hard integuments of their prey, are well fitted to deal with seeds. In many of the harvesting ants, the larger workers, or soldiers, have enormously developed heads to support the powerful muscles by means of which their large jaws are worked. These workers are literally living " nut-crackers " for the rest of the community, and they crush up the hard seeds when required for food. At the time when these large workers are of no further

THE AGRICULTURAL ANT.

One of the clearings made and maintained by this Ant amidst grass and high weeds. The nest is immediately below the cleared space, and the hole in the centre is the entrance thereto. Roads radiate from the clearing in various directions.

use to the colony, the other ants cut off their heads and throw them on the refuse-heap. A very drastic, but effective, method of getting rid of a superfluous working class.

As has been said, it is now established without doubt that many ants do collect grain and store it. They bring up the seeds to dry when damp, to prevent them from germinating, and if any seeds sprout they gnaw that part away. Moreover, ants play no small part in the distribution of plants, as even in England several ants collect seeds which they carry home. Many seeds get dropped by the way and thus give rise to plants at some distance from their original spot.

The Agricultural Ant of Texas, with which this article deals, has been stated, however, not only to collect seeds of a plant called "Ant-rice," but actually to plant and cultivate them, and gather the seed next year. The seed resembles the grain of oats, and tastes like rice.

The ant itself is of a ferruginous red colour, and possesses, in common with many species that live in deserts and dry ground, a beard of long, curved hairs, and is able to endure prolonged droughts.

Their nests are variable in construction, according to the nature of the ground on which they occur, but are always fully exposed to the sun. They generally occur among the herbage, and all plants growing on or around the nest are cut down, and a flat area, or disc, is cleared for ten or twelve feet round. Should a growing tree be near, they move the nest, or else strip all the leaves off the tree. Roads five inches broad near the nest, but narrower further from it, are made through the herbage for hundreds of feet in different directions, and these roads are weeded and kept as bare as the disc. If the soil is of a gravelly nature, the ants use the gravel to build a cone or crater in the centre of the disc, about three feet high, which is sometimes surrounded by a low mound, two or three feet in diameter. There may be from one to five entrances to the nest. The nests are perforated beneath with flat chambers, some of which form the granaries, connected by galleries. The granaries are said never to occur deeper than two and a half feet, but the galleries have been stated to reach a depth of fifteen feet. Some of the bits of gravel which form the cone are of immense size and weight as compared to the ants.

THE AGRICULTURAL ANT.

The entrance to the nest here takes the form of a perforated cone. In most cases where the cone form is adopted the grass grows right up to its base.

Although they also feed on insects, and other seeds, the "ant-rice" is the only plant stated to be allowed to grow in a circle round the nest, and all other plants are cut down as soon as they appear. This rice is supposed to be regularly sown before the wet season, and about November 1st the young green shoots spring up, and the seeds ripen in June of the next year. The ants carefully weed and attend to these plantations and gather the seeds. After malting and harvesting the seeds, the stubble is cut down and removed. They throw away the chaff and refuse, and the seeds are brought up to dry if damp.

The founding of a colony is as follows: The marriage flight of the bright red females and darker males takes place at the end of June or the beginning of July. After which the female, having removed her wings, digs down into the earth, and closes up the opening. She gradually brings up her first brood of some twelve very small, timid workers. In the spring the workers open up the nest, but are careful to conceal the entrance with small pebbles and bits of stick, which look as if they might have been brought together by the wind. It is not until the second year, when the large workers are produced, that the ants begin to cut down the vegetation around the nest and establish the circular discs. These discs increase in size as more workers are produced. The discs are evidently for the purpose of ensuring as much dryness as possible, to help to prevent the germination of the seeds. The grubs of the ants are fed with portions of crushed seeds, which the workers first coat with saliva to ensure the converting of the starch contained in the seeds into sugar. The sting of this ant is exceedingly painful.

The most recent study of these ants in nature, seems to point out that the intentional planting of the seeds by the ants has been much exaggerated, even if it is not entirely erroneous. Wheeler writes in 1910: "The Texas Harvester has attracted no little attention on account of Lincecum's

THE AGRICULTURAL ANT.

A section of the underground nest showing the galleries and rooms arranged in stories. In the centre an Ant is seen working at the extension of one of the chambers.

statement that it actually sows the seeds of the 'ant-rice' around the periphery of its discs or mounds, and cultivates the crop in addition to harvesting and storing it in its granaries.

"If the nests of this ant can be studied during the cool winter months—and this is the only time to study them leisurely, as the cold subdues the fiery stings of their inhabitants—the seeds which the ants have garnered in many of their chambers will often be found to have sprouted. Sometimes, in fact, the chambers are literally stuffed with dense wads of seedling grasses and other plants. On sunny days, the ants may often be seen removing these seeds when they have sprouted too far to be fit for food, and carrying them to the refuse-heap, which is always at the periphery of the crater, or cleared earthen disc. Here the seeds, thus rejected as inedible, often take root, and in the spring form an arc or a complete circle of growing plants around the nest. Since the ant feeds largely, though by no means exclusively, on grass seeds, and since, moreover, the seeds of Aristida are a very common and favourite article of food, it is easy to see why this grass should predominate in the circle."

THE END

INDEX

Index

Index

Index

Index

Index

Index

Index